Space Sciences Series of ISSI

Volume 42

For other titles published in this series, go to
www.springer.com/series/6592

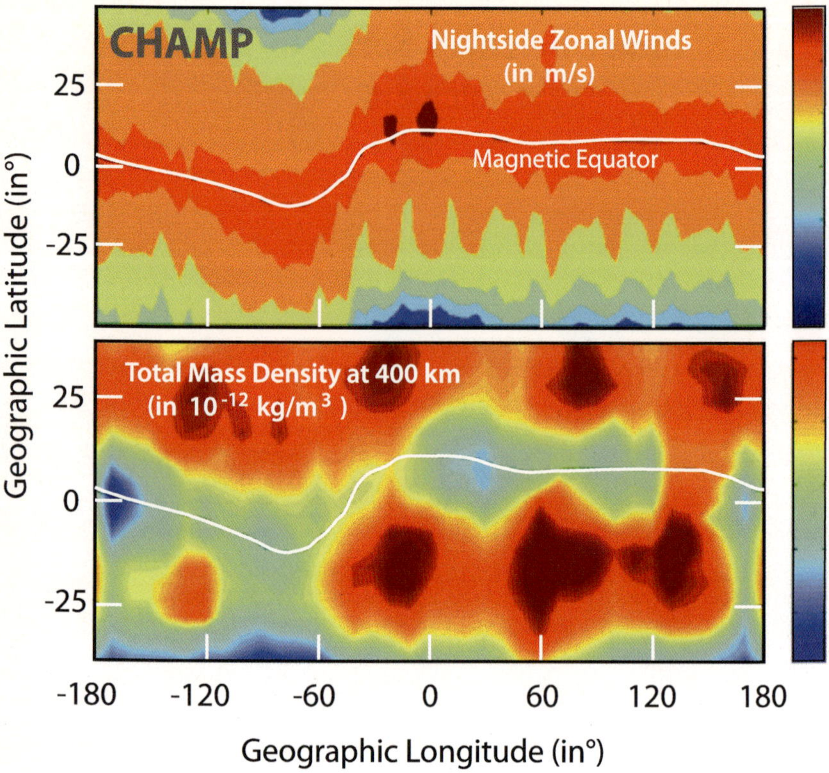

Measurements of upper atmospheric mass density N (*upper part*) and zonal wind velocity V (*lower part*) by the CHAMP spacecraft as function of geographic longitude (*abscissa*) and latitude (*ordinate*) in the latitudinal interval ($-50°$ to $50°$). Velocities have been obtained during the December Solstice and for geomagnetically quiet conditions ($Kp = 0 \ldots 2$). The data are colour coded with *blue* lowest and *red* highest values. The *wavy dark* line in centre around 0° geographic latitude is the geomagnetic dip equator. The intriguing finding is that both density of the neutral atmosphere and zonal winds are organised with respect to the stationary geomagnetic field. Highest zonal speeds concentrate around the equator, while mass densities organise into two bands north and south to the equator (see the paper of Lühr et al., pp. 241 and 243)

Tilmann Bösinger • James LaBelle •
Hermann J. Opgenoorth • Jean-Pierre Pommereau •
Kazuo Shiokawa • Stan C. Solomon •
Rudolf A. Treumann
Editors

Dynamic Coupling Between Earth's Atmospheric and Plasma Environments

Previously published in *Space Science Reviews* Volume 168, Issues 1–4, 2012

Springer

Editors
Tilmann Bösinger
University of Oulu
Oulu, Finland

James LaBelle
Dartmouth College
Hanover, NH, USA

Hermann J. Opgenoorth
Swedish Institute of Space Physics
Uppsala, Sweden

Jean-Pierre Pommereau
National Center for Scientific Research
Paris, France

Kazuo Shiokawa
Nagoya University
Nagoya, Japan

Stan C. Solomon
University Corporation
for Atmospheric Research
Boulder, CO, USA

Rudolf A. Treumann
International Space Science Institute (ISSI)
Bern, Switzerland

ISSN 1385-7525 Space Sciences Series of ISSI
ISBN 978-1-4614-5676-6
Springer New York Heidelberg Dordrecht London

Library of Congress Control Number: 2012947529

Cover illustration: A rare optical photograph of a single sprite event over France showing the multiple elements of a sprite including the topside halo. This picture was taken with a video camera during the Eurosprite 2007 campaign (courtesy Michal Ganot, The Open University, Israel; Figure taken from Pasko et al., this issue, p. 479).

Printed on acid-free paper

Springer is part of Springer Science+Business Media (www.springer.com)

Contents

Earth's Interacting Atmospheric and Plasma Envelopes: Conceptional Remarks

T. Bösinger · J. LaBelle · R.A. Treumann

Abstract We briefly describe the relation between the atmosphere and the external plasma of Earth. After a few words on history we refer to the content of the present issue. The main *mechanical* energy content is located in the dense atmosphere. However, the presence of the magnetosphere hosting Earth's plasma envelope causes a large number of most interesting effects which were completely absent if the Earth lacked its geomagnetic field. A list of such effects includes magnetic substorms, magnetic storms, the optical aurora, the wealth of auroral plasma phenomena and many dynamical effects on the upper atmospheric layers like heating, thermal expansion in the vertical and horizontal winds. Particle precipitation produces excess ionisation, affects recombination, causes electric currents, electric fields, changes in chemical conditions in the atmosphere which modify chemical reactions and may become important in the climate. Observations identified magnetic effects on the neutral atmospheric mass content which so far have not been properly understood. The coupling from downward up leads to mixing in the ionosphere and turbulence, modifies the ionospheric currents and contributes to transfer of electromagnetic energy down to the Earth's surface.

Keywords Atmospheric-plasma coupling · Ionospheric plasma—thermosphere relations · Substorm effects on atmosphere · Auroral region · Neutral gas-plasma coupling

1 Introduction

Since the advent of the space age, i.e. for now about half a century, it is known that our planet together with its atmosphere is embedded into an extended fully ionised plasma envelope, the magnetosphere, reaching far out into Earth's spatial environment. The vertical transition from the neutral atmosphere through the ionosphere into Earth's magnetosphere is schematically shown in Fig. 1.

The behaviour of the plasma in the magnetosphere is rather different from that of the atmosphere. Its electrically charged components—electrons, various types of positively charged ions and, in a narrow lower-ionospheric layer also an admixture of negatively charged ions—sensitively react to the presence of electromagnetic fields. The Coulomb and Lorentz forces exerted on the free electric charges are much stronger than Earth's gravitation

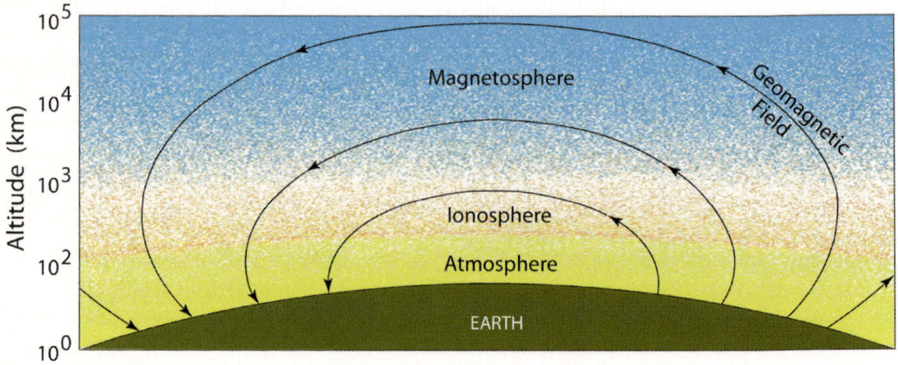

Fig. 1 An extremely simplified height compressed schematic of the gaseous envelope regions of Earth: the neutral atmosphere below ~500 km altitude, the transition region between ~80 km and ~1000 km altitude, named ionosphere, and the plasma envelope known as the magnetosphere at >1000 km altitude

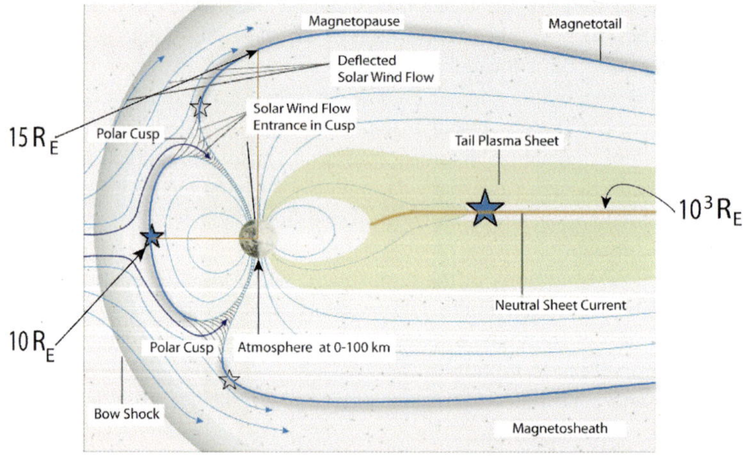

Fig. 2 A somewhat ugly view of the entire magnetosphere-atmosphere system embedded into the solar wind, showing the huge dimensions of the magnetosphere compared to the atmosphere. The atmosphere and also the ionosphere on this picture are just of the size of the circle around the terrestrial sphere. The *stars* indicate the main source sites of magnetospheric activity: reconnection at the nose of the magnetosphere and its equivalent in the tail neutral current sheet, the main responsible for magnetospheric substorms, aurorae and magnetic storms, and two open stars north of the polar cusps which are activated during northward interplanetary magnetic field

rendering their dynamics independent of Earth's gravitational attraction. Electrically the atmosphere itself is not completely neutral; it is penetrated by the geomagnetic field; and it is weakly ionised, mostly by the impact of energetic cosmic rays which disintegrate a few of the neutral constituents and cause atmospheric electricity (cf. Leblanc et al. 2008, for a review). But lower atmospheric processes are dominated by the neutral atmospheric gas.

The magnetospheric plasma, on the other hand, is governed by the magnetospheric magnetic field (Fig. 2). When considering the various atmospheric effects related to weather

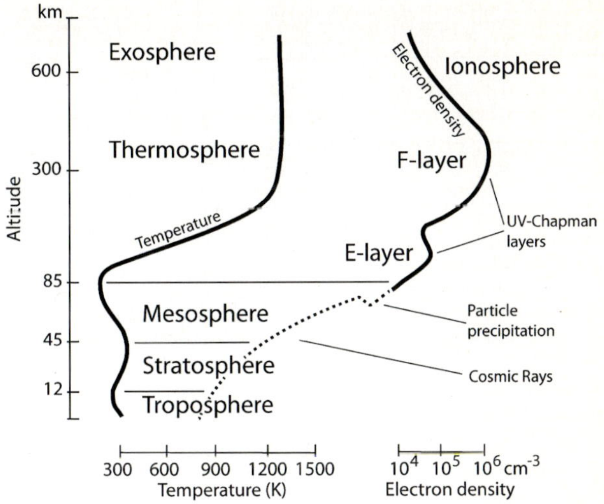

Fig. 3 Atmospheric temperature and electron density profiles. The temperature shows the horizontally layered structure of the atmosphere. The electron density profile shows the ionospheric structure above 80 km altitude and its extension to lower altitudes produced by particle precipitation and at lower altitudes by Cosmic Rays

and climate and the by now well known dynamics of the magnetosphere one is led to the impression that atmosphere and magnetosphere were completely independent, even though, quite naturally, they overlap in the ionosphere, at altitudes above $h \sim 80$ km. Taking the viewpoint of the external plasma, the ionosphere just constitutes the bottom of the magnetosphere. Viewed at from the atmospheric, it does not belong to the atmosphere even though they overlap in the thermosphere. The ionosphere is ionised, with a pronounced altitudinal structure of ionisation shown in Fig. 3. It got its own about independent life, the subject of ionospheric physics. By it atmosphere and magnetosphere are taken as confining semi-transparent lower and upper boundaries. Clearly the ionosphere is the region where both atmosphere and plasma couple most intensely. It is a region of strong current flow perpendicular as well as parallel to the geomagnetic field. Here, substantial frictional drag among its neutral and ionised constituents couples the neutral atmospheric gas to the ionised plasma and leads to mass, momentum and energy exchange.

Ironically, the ionosphere is known for much longer than the magnetosphere; it had been suggested to exist by Carl Friedrich Gauss (see Fig. 4) who, in the fourth decade of the 19th century when stumbling across Laplace's equation $\nabla^2\Phi(\mathbf{r}) = 0$ for a scalar potential Φ outside its sources, developed potential theory including Gauss' theorem as the theory of its solution—the first field theory at all. Searching for applications apart from Earth's gravitational field he, with his colleague Wilhelm E. Weber in Göttingen, campaigned around Göttingen, mapping the geomagnetic field.[1] Being aware that the magnetic measurements were performed on an area, Gauss split the geomagnetic potential into external and internal parts, to his surprise finding that the external part was significant, even when referring to the measurement error with his own least-square method. Gauss speculated that electrical

[1] For this purpose, they invented the portable bifilar magnetometer and the first electromagnetic telegraph between his observatory and the institute. For the data analysis Gauss introduced spherical harmonics and the technique of expanding a function on a curved surface, which led him to the proposal of the general theory of curved surfaces and speculation about the curvature of space which he unsuccessfully tried to detect experimentally by field measurements of gravitation. Not expecting experimental confirmation, he left the extension of surface theory to space as a mathematical exercise to his student Bernhard Riemann who later gave it the final mathematical form widely used since Albert Einstein 1916 in General Relativity.

Fig. 4 The three pioneers in the discovery of the ionosphere and the construction of its physics

currents should flow at some altitude in the atmosphere high above the highest mountains, tentatively putting them somewhere between 10 km and 100 km. It took another century, the formulation of Maxwell's equations and discovery of radio waves until Edward Appleton (1924, 1928a, 1928b) experimentally confirmed the existence of the ionosphere from radio wave reflection, winning the 1947 physics Nobel prize. It also needed atomic physics for Sidney Chapman (1930, 1931a, 1931b) to formulate his theory of generation of an ionised layer by solar ultra-violet radiation ionisation of the uppermost atmosphere, the *Chapman layer*, capable of the current flow whose magnetic fields were anticipated by Gauss.

The present volume illuminates aspects which more rarely come into focus. From the viewpoint of plasma physics, for instance, the effects caused in the atmosphere are customarily ignored when talking about precipitation. The atmosphere around 100 km appears as the surface of no-return. Otherwise one considers the ionosphere as a source of outflow of sufficiently hot ionised material at polar latitudes, the polar wind. Similarly, in atmospheric physics the precipitating charges are taken as a tiny input ionisation, and the small amount of escaping ions is ignored. The questions tackled in the present volume instead ask for mutual effects when plasma and atmosphere interact.

The material to be covered is rather dispersed. No overwhelming effort has so far been put into the investigation of coupling processes. Few of them have been identified. The present volume should, therefore, be considered as the attempt of collecting some of the facts and ideas. It is the result of a workshop organised in autumn 2010 at the International Space Science Institute in Bern, Switzerland, and it is intended to ignite interest in the importance of the interrelation between atmosphere and plasma.

2 Relevance of Plasma Effects in the Upper Atmosphere

Table 1 shows that solar irradiation is the dominant stationary energy source in the atmosphere. There is an overlap region from 100–130 km where the energy inputs by auroral electrons and EUV become comparable. Direct Joule heating affects the atmosphere only above 100 km altitude. The vertical profile of the electron density in Fig. 3 is a measure of the ionisation state of the atmosphere. The ionosphere above 80 km altitude consists of E- and F-layers with Chapman layer structure around maximum densities caused by EUV.

Table 1 Energy Inputs into Earth's Atmosphere (courtesy Gang Lu)

Source	Energy input (W/m^2)	Deposition altitude (km)
Solar radiation		
Total irradiance	1368	Surface
UV 200–300 nm	16	0–50
UV 120–200 nm	0.1	50–120
EUV	0.003	100–500
Particles		
Solar energetic protons	0.002	30–90
Auroral protons	0.001–0.006	100–130
Auroral electrons	0.003–0.03	70–130
Galactic cosmic rays	0.000007	0–90
Joule heating		
$\mathcal{E} = 10 \sim 100$ mV/m	$0.0015 \sim 0.15$	100–500
$\Sigma_P = 15$ mho		

Barth et al. (2009); Cosgrove et al. (2009)

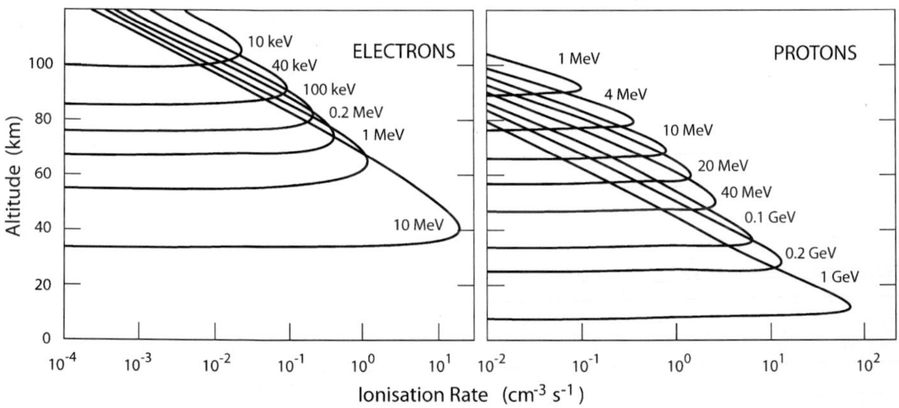

Fig. 5 Ionospheric ionisation rates of magnetospheric precipitating particles (electrons and protons) as function of particle energy and altitude

At lower altitudes ionisation and energy deposition (cf. Fig. 5) is contributed by energetic particle precipitation and, at the lowest altitudes, by the weak Cosmic Ray ionisation.

The dynamics of the lower atmospheric layers is dominated by the atmospheric mass $M_a \approx 5.2 \times 10^{18}$ kg and mechanical energy $\mathcal{E}_a \approx 6.2 \times 10^{20}$ J contents. Decay of atmospheric density with altitude and increase in ionisation let the plasma gain importance in the thermosphere. Its energy sources are solar EUV and the energy input from the magnetosphere the ultimate origin of which is the solar wind which hits the cross section $\sigma_{\text{MSP}} \approx 3 \times 10^{16}$ m^2 of the magnetosphere (Fig. 6) with energy flux $F_{\text{SW}} \approx 2 \times 10^{-3}$ J/ m^2 s, imposing a power $P_{\text{SW}} = \sigma_{\text{MSP}} F_{\text{SW}} \approx 6 \times 10^{13}$ W. The fraction $P_{\text{in}} = \alpha P_{\text{SW}} \sim 3 \times 10^{12}$ W entering the magnetosphere depends on the reconnection efficiency $\alpha \sim 0.05$ between solar wind and magnetospheric magnetic fields at the magnetopause.

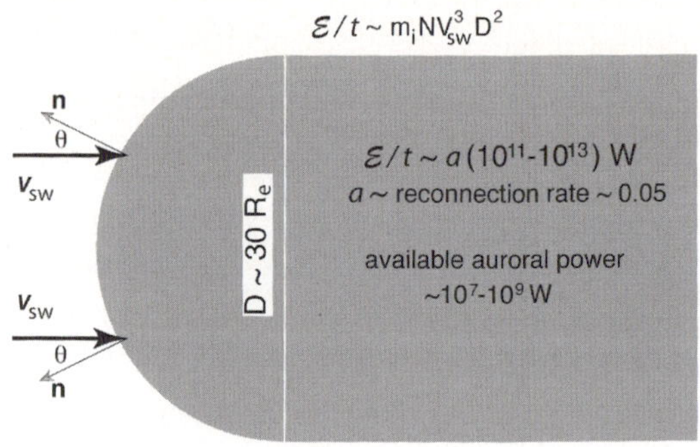

Fig. 6 Estimated global energy flux across the magnetosphere. Given is the power which is nominally added to the magnetosphere. Most of it is not used in processes which affect the aurora but flows across the magnetosphere to escape back into the solar wind. The available power feeds aurorae and substorms

The geomagnetic field connects the magnetospheric plasma to the atmosphere. It maps the convection electric fields into ionospheric altitudes, permits field-aligned electric current flow between the magneto- and ionospheres and serves as a guide for particle beams and plasma waves. Such coupling becomes strongly amplified in the violent magnetospheric disturbances like substorms and magnetic storms. During substorms, a broad latitudinal and longitudinal interval, the auroral oval, is affected. The fraction $\mathcal{E}_m \sim 10^{12}$ J of energy is deposited here within $\lesssim 10^3$ s, giving rise to the rich variety of auroral phenomena. The down-mapped electric field couples the plasma convection via ion drag to the thermospheric gas. Electric fields induce horizontal plasma flow and, via ion drag, horizontal winds. Dissipation of electric currents causes Joule heating in the atmosphere as the source of atmospheric expansion, which drives zonal and vertical winds, generates turbulence and leads to gas-plasma mixing. Low frequency plasma waves are involved into field, energy, and momentum transport; and various types of plasma waves contribute to energy re-distribution, generation of anomalous transport and nonlinear localised and temporally variable processes. These processes culminate and cumulate during the violent magnetospheric disturbances related to substorms and aurorae.

Many of the coupling processes in the upper atmosphere are not sufficiently well understood. Understanding requires identification and analysis of the relevant interaction modes between plasma and atmospheric gas under the conditions of the upper atmosphere. It also requires an intensification of experimental and observational efforts, the use of low-altitude upper-atmospheric spacecraft like CHAMP, DEMETER, FORMOSAT and C/NOFS for monitoring the transition region from the magnetosphere to the atmosphere, the development of new in-situ instrumentation suited for these regions and capable of measuring the variations in the neutral upper atmospheric gas as well as the plasma component. These should be added by focussed ground based and rocket observations.

3 Structure of this Volume

The 21 review papers in the present volume are about half of the presentations that were given at the ISSI-Workshop on "Coupling between the Earth's Atmosphere and Its Plasma Environment". The first two papers by Anne Smith (2011) and Robert Pfaff (2011) are intended to provide brief introductory reviews of the state of interactions in the atmosphere and magnetosphere/ionosphere, respectively. The former is addressed to the space plasma scientist, the latter to the atmospheric physicist.

These introductions into the two main fields, whose interaction and cross-fertilisation is the item of this issue, are followed by a review of the long-term temporal trends in the parameters of the neutral atmosphere and its upper ionised layers (Laštovička et al. 2011), also including extrapolations of those trends into the future. Clearly, since anthropogenic effects have become important during the last few decades, these trends cannot be made independent of them. Rather they also include their effects in several directions: warming and addition of aerosols to the atmospheric constituents which are shown not to restrict to the lower atmosphere but to propagate to higher altitudes where they contribute to the ionic component and modification of atmospheric chemistry. The latter questions are reviewed in detail in two papers (Grygalashvyly et al. 2011; Arnone and Hauchecorne 2011) later in this issue, putting emphasis on the important effects of the minor atmospheric chemical constituents; one may also consult Barth et al. (2009).

Still staying with the focus on the neutral atmosphere, the next paper (Qian and Solomon 2011) considers the temporal and spatial variations, in particular during magnetospheric disturbance, in thermospheric density, i.e. the density in the broad overlap region of the atmo- and ionospheres. Here the neutral density is gradually replaced with that of the ionised component and the collisional coupling gradually fades away with height. Pancheva and Mukhtarov (2011) review in the next contribution the role of atmospheric tides with altitude in this coupling based on experimental results finding that the effect of the tides on the ionosphere has always been underestimated by a factor of $\sim$1.5.

These investigations are supported by a review of the effects of solar atmospheric tides on the ionosphere in low latitudes (England 2011), planetary-scale waves with periods of harmonics of one day and caused by solar heating which clearly is strongest at low geographic latitudes. The paper focusses on non-migrating tides which create longitudinal ionospheric variations. In situ observations by the CHAMP (Lühr et al. 2011), and FORMOSAT-3 (Wang et al. 2011) spacecraft add important experimental results to these considerations which were based on ground-based measurements including theory.

CHAMP discovered pronounced variations of the thermospheric mass content with magnetospheric substorm activity which are still unexplained. It also discovered a strict and as well badly understood control of the thermospheric density and mass content by the static geomagnetic field. These observations were taken at 400 km altitude. At much lower altitudes, radio occultation in combination with GPS was used on the FORMOSAT-3 spacecraft to detect warming in the tropo-stratosphere system in relation to variations in the higher atmosphere-ionosphere and effects of solar eclipse. Residual circulation driven by waves in the strato- and mesospheres, i.e. meridional circulation caused by atmospheric waves, is proposed as an explanation for circulation in the middle atmosphere (Becker 2011). Such gravity waves could be produced in several scenarios one of them solar proton events which are supposed to heat the middle atmosphere.

The final ten papers focus on the ionospheric and magnetospheric plasmas, i.e. on the plasma aspect of the coupling, starting with a review of the global electric circuit (Rycroft and Harrison 2011) which includes the atmosphere and atmospheric electric field into the electrodynamic relations, presents an electric-current picture of the coupling and focusses on effects in the atmosphere, like variations in atmospheric electricity and lightning, and in the ionosphere, like the ionospheric current system which during lightning is shown to couple to the atmosphere. One may note that, recently, two related publications appeared, a concise review on the electrodynamic coupling between the ionosphere and lower atmosphere (Singh et al. 2011), mainly dealing with lightning related effects, and a special issue on atmospheric and solar-terrestrial coupling processes (Pallamraju et al. 2012).

Such coupling is also given during precipitation events in the course of magnetic storms and substorms being local, varying temporarily and not being a persistent stationary process. This kind of coupling is rather different between high and low latitudes as it is governed by the direction of the geomagnetic field, solar radiation and magnetospheric activity. The latter is generally large at high magnetic latitudes where the magnetic field is about vertical and solar radiation intensity is less than at low latitudes. In low latitudes the magnetic field is about horizontal, magnetospheric (plasmaspheric) plasma density is high, and magnetic activity is low while solar irradiance is strong. Various low- to high-latitudinal effects which are related to coupling the lower magnetospheric plasma to the ionosphere and the thermosphere are reviewed in the four papers by Chau et al. (2011), Makela and Otsuka (2011), Haldoupis (2011), and Oyama and Watkins (2011).

The first two of these reviews consider the forcing of the equatorial ionosphere from below, i.e. the effects of atmospheric wind perturbations on the ionosphere during sudden stratospheric warming events under solar minimum conditions when these events are not covered by solar irradiance. These effects, being caused by vertical drifts in the course of the atmospheric expansion which follows the warming, are observed mainly in the evolution of zonal ionospheric electric fields, changes in total electron content and in neutral densities. The other effect is the generation of travelling ionospheric disturbances in the night-time ionosphere in mid-latitudes which are seeded from below.

The focus changes in the next two reviews where effects from above are estimated. One such effect is the occurrence of sporadic E-layers which is reviewed here (Haldoupis 2011) and is being identified as a paradigmatic process where coupling by outside forcing occurs between the ionosphere and atmosphere. Similarly, forcing of the atmosphere in the auroral region causing atmospheric warming, expansion and related effects is reviewed by Oyama and Watkins (2011).

The last five reviews in this book deal with the electromagnetic part of the coupling between atmosphere, ionosphere and magnetosphere. This aspect includes phenomena like lightning in its various modes, and electromagnetic wave propagation. One, of course, expects that lightning is an important factor in the coupling while wave phenomena should play a secondary role mostly because very little energy is transported by waves while lightning is indeed a violent phenomenon. The discovery of sprites and their related effects has thrown new light on lightning. These latter phenomena are the content of the reviews by Pasko et al. (2011) and Shalimov and Bösinger (2011). The former reviews the entire phenomenology of all the luminous transients which may occur during, in the wake and the aftermath of lightning, sprites, blue jets and elms. The latter considers the generation of low-frequency wave phenomena during lightning and sprites thereby building the bridge to the vast variety of wave phenomena.

Finally, the last three reviews of this group by Simões et al. (2011), Demekhov (2011), and Pilipenko (2011) deal with wave phenomena proper: the impulsive coupling effects which could be attributed to the ionospheric Alfvén resonator, waves in the ultra-low frequency range which are related to geomagnetic pulsations, and the propagation of very weak signals like Schumann resonances across the atmosphere and entering into the upper ionosphere where they seem to have been detected recently by the C/NOFS satellite. If these observations can be confirmed in future experiments and will be theoretically understood, they may serve as an important diagnostic tool of the atmosphere-ionosphere relation.

Fig. 7 The workshop participants at the entrance stairs of the ISSI building in Bern. From left to right: *Front line*: Anne K. Smith, H. Lühr, F.A. Simões, Eliana Nossa, A.G. Demekhov, Michi Nishioka, Gang Lu. *Second line*: Y. Yair, M.J. Rycroft, E. Turunen, S.C. Solomon, Dora Pancheva, A. Hauchecorne, J. Kletzing, K. Shiokawa, J.L. Chau, V. Pilipenko. *Third line*: E. Arnone, V.P. Pasko, J.J. Makela, D. Fritts, K.-Y. Wang, H. Opgenoorth, E. Becker, Y. Otsuka, S.-I. Oyama, M. Grygalashvyly, A. Yahnin. *Last line*: J.-P. Pommereau, C.-L. Kuo, S.I. England, J. Laštovička, W. Ward, H. Fujiwara, T. Bösinger, J. LaBelle, R.F. Pfaff

4 Conclusions

The existence of the ionosphere is sufficient proof for the factual reality of a coupling between the atmosphere and the plasma in the magnetosphere, if counting the ionosphere as belonging to it as its atmospheric near-Earth boundary. The transition region processes are complicated. they are governed by vertical and zonal winds, collisions, ion drag, electric currents, resistances, Joule heating, atmospheric expansion and chemistry. Dealing with some of them caused disengagement of ionospheric physics from the atmosphere and magnetosphere, respectively. However, viewed as an intermediate semitransparent transition layer, the ionosphere transition plays the role of a connector between the atmospheric neutral gas and the fully ionised extraterrestrial plasma.

It allows transmission of matter, energy and momentum from the spatially narrow inert atmospheric reservoir of mechanical energy to the magnetosphere *in the form of planetary and gravity waves, tides and winds* It, on the other hand, allows for the passage of energetic particles, currents, electric fields, magnetic disturbances and various kinds of waves from the hugely extended magnetospheric energy reservoir into the atmosphere. Part of this energy is absorbed in the ionospheric transition region, including the thermosphere, and is transforms into thermal energy of the neutrals. It causes local expansion of the thermosphere, winds and turbulent mixing of atmospheric layers and plasma. In this way part of the energy can pass to the deeper atmospheric layers. This downward coupling and the effects it may cause on the middle atmosphere are still not well understood. The various forms of such a transfer need further detailed investigation. The present volume hopes to stimulate intensified scientific endeavour into this important field of research.

Acknowledgement We thank all participants of the Workshop for their constructive contributions to the discussion, in particular, the contributors to this volume for their efforts of identifying the subtle signals of any possible coupling between the two gaseous terrestrial envelopes in their work, data and theory. We also thank the collective of Editors for the organisation of the Workshop. We are deeply indebted to the large number of referees for their voluntary and not further recognised work which was substantial during the evaluation process leading to this volume; among them were the following external reviewers: Alan D. Aylward, Gufran Beig, Edgar A. Bering, James Burch, Ivana Cerovecki, Greg Earle, Yigit Erdal, Jingbo Li, Yasunobu Miyoshi, Michi Nishioka, Jens Oberheide, Eugene Rozanov, and Alexander Yahnin. Finally we thank the International Space Science Institute Bern, its staff, and its directors, for hosting the participants of the Workshop and for facilitating the intense discussion of the various different views on the possible coupling between the atmosphere and extraterrestrial plasmas.

References

E.V. Appleton, Geophysical influences on the transmission of wireless waves. Proc. Phys. Soc. Lond. **37**, 16–22 (1924)

E.V. Appleton, Some notes on wireless methods of investigating the electrical structure of the upper atmosphere. I. Proc. Phys. Soc. Lond. **41**, 43–59 (1928a). doi:10.1088/0959-5309/41/1/305

E.V. Appleton, Some notes on wireless methods of investigating the electrical structure of the upper atmosphere. II. Proc. Phys. Soc. Lond. **42**, 321–339 (1928b). doi:10.1088/0959-5309/42/4/304

E. Arnone, A. Hauchecorne, Stratosphere NO_y measured by MIPAS and GOMOS onboard ENVISAT during 2002–2010: influence of plasma processes onto the observed mean distribution and variability. Space Sci. Rev. (2011, this issue). doi:10.1007/s11214-011-9861-1

C.A. Barth, G. Lu, R.G. Roble, Joule heating and nitric oxide in the thermosphere. J. Geophys. Res. **114**, A05301 (2009). doi:10.1029/2008JA013765

E. Becker, Dynamical control of the middle atmosphere. Space Sci. Rev. (2011, this issue). doi:10.1007/s11214-011-9841-5

S. Chapman, On solar ultra-violet radiation as the cause of aurora and magnetic storms. Geophys. J. Int. **2**, 296–300 (1930). doi:10.1111/j.1365-246X.1930.tb05414.x

S. Chapman, The absorption and dissociative or ionizing effect of monochromatic radiation in an atmosphere on a rotating Earth. Proc. Phys. Soc. Lond. **43**, 26–45 (1931a). doi:10.1088/0959-5309/43/1/305

S. Chapman, Bakerian lecture. Some phenomena of the upper atmosphere. Proc. R. Soc. Lond. Ser. A **132**, 353–374 (1931b). http://www.jstor.org/stable/95664

J.L. Chau, L.P. Goncharenko, B.G. Fejer, H.L. Lui, Equatorial and low latitude ionospheric effects during sudden stratospheric warming events: ionospheric effects during SSW events. Space Sci. Rev. (2011, this issue). doi:10.1007/s11214-011-9797-5

R.B. Cosgrove, G. Lu, H. Bahcvian, T. Matsuo, C.J. Heinselman, M.A. McCready, Comparison of AMIE-modeled and Sondrestrom-measured joule heating: a study in model resolution and electric field-conductivity correlation. J. Geophys. Res. **114**, A04316 (2009). doi:10.1029/2008JA013508

A.G. Demekhov, Coupling at the atmosphere-ionosphere-magnetosphere interface and resonant phenomena in the ULF range. Space Sci. Rev. (2011, this issue). doi:10.1007/s11214-011-9832-6

S.I. England, A review of the effects of non-migrating atmospheric tides on the Earth's low-latitude ionosphere. Space Sci. Rev. (2011, this issue). doi:10.1007/s11214-011-9842-4

M. Grygalashvyly, E. Becker, G.R. Sonnemann, The gravity wave mixing and effective diffusivity for minor chemical constituents in the mesosphere/lower thermosphere. Space Sci. Rev. (2011, this issue). doi:10.1007/s11214-011-9857-x

C. Haldoupis, Midlatitude sporadic E: A typical paradigm of atmosphere-ionosphere coupling. Space Sci. Rev. (2011, this issue). doi:10.1007/s11214-011-9786-8

J. Laštovička, S.C. Solomon, L. Qian, Trends in the neutral and ionised upper atmosphere. Space Sci. Rev. (2011, this issue). doi:10.1007/s11214-011-9799-3

F. Leblanc, K. Aplin, Y. Yair, G. Harrison, J.P. Lebreton, M. Blanc (eds.), *Planetary Atmospheric Electricity*. ISSI Space Science Series, SSSI, vol. 30 (Springer, Heidelberg, 2008)

H. Lühr, J. Park, P. Ritter, H. Liu, In-situ CHAMP observations of ionosphere-thermosphere coupling. Space Sci. Rev. (2011, this issue). doi:10.1007/s11214-011-9798-4

J.J. Makela, Y. Otsuka, Overview of nighttime ionospheric instabililties at low- and mid-latitudes: coupling aspects resulting in structuring at the mesoscale. Space Sci. Rev. (2011, this issue). doi:10.1007/s11214-011-9816-6

S.I. Oyama, B.J. Watkins, Generation of atmospheric gravity waves in the polar thermosphere in response to auroral activity. Space Sci. Rev. (2011, this issue). doi:10.1007/s11214-011-9847-z

D. Pallamraju, G. Lu, C. Lin (eds.), Atmospheric Coupling Processes in the Sun-Earth System. J. Atmosph. Solar-Terrestr. Phys., Special Issue, **75–76**, 1–160 (2012). http://www.sciencedirect.com/science/issue/271871-1-s2.0-S1364682612X00026

D. Pancheva, P. Mukhtarov, Global response of the ionosphere to tidal and planetary wave forcing: recent progress based on satellite measurements. Space Sci. Rev. (2011, this issue). doi:10.1007/s11214-011-9837-1

V.P. Pasko, Y. Yair, C.L. Kuo, Lightning related transient luminous events at high altitude in the Earth's atmosphere: phenomenology, mechanisms and effects. Space Sci. Rev. (2011, this issue). doi:10.1007/s11214-011-9813-9

R.E. Pfaff, The near-Earth plasma environment. Space Sci. Rev. (2011, this issue). doi:10.1007/s11214-012-9872-6

V.A. Pilipenko, Impulsive coupling between the atmosphere and ionosphere/magnetosphere. Space Sci. Rev. (2011, this issue). doi:10.1007/s11214-011-9859-8

L. Qian, S.C. Solomon, Thermospheric density: an overview of temporal and spatial variations. Space Sci. Rev. (2011, this issue). doi:10.1007/s11214-011-9810-z

M.J. Rycroft, R.G. Harrison, Electromagnetic atmosphere-plasma coupling: the global electric circuit. Space Sci. Rev. (2011, this issue). doi:10.1007/s11214-011-9830-8

S.L. Shalimov, T. Bösinger, Sprite-producing lightning-ionosphere coupling and associated low-frequency phenomena. Space Sci. Rev. (2011, this issue). doi:10.1007/s11214-011-9812-x

F.A. Simões, R. Pfaff, J.J. Berthelier, J. Kletzing, A review of low frequency electromagnetic wave phenomena and tropospheric-ionospheric coupling mechanisms. Space Sci. Rev. (2011, this issue). doi:10.1007/s11214-011-9854-0

A.K. Singh, D. Siingh, R.P. Singh, S. Mishra, Electrodynamical coupling of Earth's atmosphere and ionosphere: an overview. Int. J. Geophys. **2011**, 971302 (2011). doi:10.1155/2011/971302

A.K. Smith, Interactions between the lower, middle and upper atmosphere. Space Sci. Rev. (2011, this issue). doi:10.1007/s11214-011-9791-y

K.Y. Wang, C.H. Liu, L.C. Lee, P. Braesicke, Observations of stratosphere-troposphere coupling during major solar eclipses from FORMOSAT-3/COSMIC constellation. Space Sci. Rev. (2011, this issue). doi:10.1007/s11214-011-9829-1

Space Sci Rev (2012) 168:1–21
DOI 10.1007/s11214-011-9791-y

Interactions Between the Lower, Middle and Upper Atmosphere

Anne K. Smith

Received: 2 February 2011 / Accepted: 30 May 2011 / Published online: 19 June 2011

Abstract This survey of atmospheric regions is based on a presentation given at the Workshop on Coupling between Earth's Atmosphere and its Plasma Environment. The workshop was sponsored by the International Space Science Institute (ISSI) during September 2010. This survey begins with descriptions of aspects of the troposphere, stratosphere, and mesosphere that affect vertical coupling. Then the tropopause, stratopause, and mesopause are considered, with a discussion of how they inhibit or allow the transmission of atmospheric mass or perturbations. Two transition interfaces are emphasized: the tropopause, which separates the troposphere from the middle atmosphere, and the mesopause, which separates the middle atmosphere from the thermosphere. The survey of coupling processes looks at radiative processes, wave propagation, and mass exchange. Each topic includes one or more specific examples drawn from recent literature.

Keywords Coupling · Solar variability

1 Introduction

This article introduces the vertical structure of the atmosphere and discusses the interfaces between regions or layers. The focus is on the interactions across regions and the impediments to interaction. Brief descriptions of the lower and middle atmosphere are given and the regions are contrasted. Several examples of coupling across regions are presented. The additional articles in this volume include detailed discussions of aspects of these interactions and other processes not discussed here.

2 Definition of the Atmospheric Regions

Atmospheric scientists classify the neutral atmosphere in regions. The regions and the transitions between them are identified by their mean thermal structure. They have traditionally

A.K. Smith (✉)
NCAR, Boulder, CO, USA
e-mail: aksmith@ucar.edu

been considered to be layers but, since the atmosphere is not layered globally, the term regions will be used here. The regions are the troposphere (Earth's surface to approximately 10–18 km), the stratosphere (10–18 km to about 50 km), the mesosphere (about 50 to 100 km) and the thermosphere (above 100 km). The transitions between these regions are the tropopause, the stratopause, and the mesopause. Because of common dynamical and chemical environments, the region that encompasses the stratosphere, mesosphere and sometimes the lowermost part of the thermosphere is also called the middle atmosphere.

2.1 The Troposphere

The troposphere is strongly affected by the Earth's surface properties.

- The orography of the surface affects the generation of atmospheric waves as winds flowing near the surface are displaced horizontally or vertically.
- The landforms, vegetation, and snow cover of land surfaces affect absorption versus reflection of incoming solar radiation and the emission of outgoing infrared radiation.
- The landforms, vegetation, and snow cover of land surfaces affect exchange of water and trace chemical species with the atmosphere.
- The surface properties (for example, surface wave amplitudes) and ice cover of ocean and lake surfaces affect absorption versus reflection of solar radiation and the exchange of water and trace chemical species with the atmosphere.

The troposphere is highly variable on short time scales; the details of this variability are often difficult to predict. Basic physical processes act to create unstable conditions. The responses to the unstable conditions are a variety of dynamical phenomena that act to bring the atmosphere back toward stability.

An important mode of instability is convective instability. The bulk of the solar flux that enters the troposphere passes through the atmosphere and is absorbed at the Earth's surface (IPCC 2007). A substantial portion of the energy is converted to heat. An additional portion of the energy evaporates water that then contributes to the humidity of the atmosphere. The transfer of the heat from the land or ocean surface to the atmosphere occurs by radiative (infrared), sensible, or latent heating. Reflection of solar radiation at the surface or by particles within the atmosphere also affects the energy exchange between the surface and the atmosphere.

The heat input to the lower levels of the atmosphere can create convective instability. In dry convective instability (i.e., under conditions of low humidity), the air will be unstable wherever the temperature structure satisfies

$$\frac{\partial T}{\partial z} < -\frac{g}{c_p} \equiv \Gamma_d \tag{1}$$

where T is temperature, g is the acceleration of gravity and c_p in the heat capacity of air at constant pressure. The so-called "dry adiabatic lapse rate" Γ_d described by (1) is -9.8 K/km.

If dry air is displaced upward, it will cool at the rate of 9.8 K/km. By definition, if dry air is convectively unstable, its temperature falls off more rapidly than that. As a result, air displaced upward will be warmer, and therefore lighter, than the surrounding air. The displaced air feels an upward force rather than a restoring force; it can continue to rise until it reaches an altitude where the temperature profile is no longer convectively unstable. This rising motion is convection. The opposite condition is an atmospheric profile that is convectively stable. In that case, air displaced upwards will cool and become colder than the surrounding air. Gravitational force will then be a restoring force to accelerate the displaced air back towards its original altitude.

When the humidity is high enough, the exchange of heat due to the condensation of moisture has an impact on the convection process. Air that ascends and cools can hold less moisture by virtue of its cooler temperature. As the relative humidity becomes high enough for the moisture to condense into cloud droplets, latent heat will be released. The latent heat release slows the rate at which the air cools. Under saturated conditions the moist adiabatic lapse rate Γ_s is

$$\Gamma_s = \Gamma_d - \frac{L}{c_p}\frac{dw_s}{dz} \tag{2}$$

where w_s is the mixing ratio at which water becomes saturated and L is the latent heat of condensation. The moist adiabatic lapse rate depends on both humidity and temperature. The lapse rate normally becomes steeper in the upper troposphere where both humidity and temperature are lower. The lapse rate is also expected to become less steep as the overall temperature increases due to global warming (Held and Soden 2006).

Due to conservation of mass, convective upwelling must be balanced by downwelling of air in surrounding areas and the upwelling and downwelling air masses must be linked by horizontal motion. The combined effect of the upward and downward mass flow keep the mean atmosphere temperature structure over large parts of the Earth near the moist adiabatic lapse rate. The horizontal scale of convection is small (less than a few km) even though the heating that generates the instability can be widespread. The location and timing of the appearance of individual convective cells is difficult to predict. The atmospheric variables that best predict where convection will occur are discussed e.g. by Sherwood (1999).

Convective instability is one of several important modes of instability. It is most important in the tropics where the heating is normally strong and the horizontal variations in angular momentum are the weakest. However, convection also occurs in midlatitudes, particularly over land surfaces during the warmer times of the year.

The rotation of the Earth is associated with a change in the angular momentum from the equator (largest) to the poles (zero). The change in the angular momentum affects the horizontal path of air displaced either vertically or in latitude. Instabilities in which the conservation of angular momentum play a role are barotropic and baroclinic instability. Baroclinic instability is a key component of the generation of midlatitude storm systems.

Baroclinic instability is an inevitable situation for the Earth's atmosphere. The unstable situation develops because the atmosphere is heated more strongly in the tropics and therefore has an equator to pole temperature gradient. It also depends on the rotating Earth and the importance of the Coriolis effect on horizontal motion. Through the thermal wind relation, the meridional temperature gradient is associated with a vertical shear in the zonal wind.

$$\frac{\partial u}{\partial z} \approx -\frac{R}{Hf}\frac{\partial T}{\partial y} \tag{3}$$

where u is the zonal wind, R is the gas constant for air, f is the Coriolis term, and H is the atmospheric scale height. Small north-south perturbations in the air flow advect cooler air in the equatorward direction and warmer air poleward. Depending on the spatial scale and frequency of the perturbation and on the conditions of the background atmosphere, some disturbances will be unstable and their amplitudes can grow exponentially. During winter, baroclinic instability leads to the steady generation of midlatitude storm systems.

With vigorous moist convection in the tropics, transport of air parcels from the bottom to the top of the troposphere can occur on timescales of minutes. The timescale for vertical transport is also fairly short (a few days) in midlatitudes. As a result, all but water and short-lived chemical species in the troposphere are well-mixed in the vertical.

Due to instabilities, variations in external heating, and other processes, the winds, temperature, and humidity in the troposphere vary with space and time on multiple scales. Some of these variations damp out with time and have primarily a local impact. However, many perturbations can propagate in the horizontal and/or vertical direction. Propagating disturbances are atmospheric waves. Their behavior in space and time can be represented by a wave equation in one or more dimensions. The generation and propagation of waves is a dominant aspect of the atmosphere at all levels and is a key component linking different regions.

Waves generated in the troposphere that have important impacts on the middle atmosphere include gravity waves, thermal tides, and planetary Rossby waves. Examples of the role of these processes in vertical coupling are presented in Sect. 4.2. The propagation conditions for all of these waves depend on the background winds, the thermal structure of the atmosphere, and characteristics of the wave such as wavelength and frequency. The linear wave equations, based on the assumption that the wave amplitudes are small, are often good predictors of wave propagation even when the conditions of linearity are violated.

Although most of the waves that are important in the atmospheric system are generated in the troposphere, waves can also be forced in the stratosphere and mesosphere. Two examples are the generation of the quasi-two-day wave by baroclinic instability (Plumb 1983; Pfister 1985) and the forcing of the diurnal and semidiurnal tide by ozone heating (Hagan 1996).

2.2 The Stratosphere

2.2.1 Heating

The heating in the stratosphere occurs internally, unlike in the troposphere where much of the heating occurs at the interface with the surface. Most of the stratospheric heat comes from absorption of solar radiation by ozone. Although molecular oxygen is less important than ozone in the thermal budget, absorption of radiation in the Herzberg bands by O_2 is the source of O which in turn is the source of ozone (ozone is formed from the reaction of O with O_2). The solar ultraviolet radiation also photolyzes many other molecules and drives the active chemical system in the stratosphere.

Throughout the middle atmosphere, CO_2 radiative transfer is the dominant cooling process. The increase in CO_2 since the industrial revolution is changing the radiative balance and is expected to lead to a continued cooling of the stratosphere. Water vapor and ozone also make contributions to the cooling. Ozone changes affect both the heating and cooling rates. Changes in these three gases over recent decades have led to temperature trends in the stratosphere (Ramaswamy et al. 2001).

The stratospheric temperature structure is not in radiative balance. Heating occurs in the sunlit atmosphere whereas cooling occurs at all latitudes and all local times. There is no solar heating in the darkness of polar night. However, heating due to dynamically driven adiabatic descent occurs and keeps the polar temperature higher than that predicted from thermal equilibrium.

2.2.2 Dynamics

The adiabatic heating that is so important in the middle atmosphere is the result of a wave-driven circulation. Waves propagate from the troposphere. According to theory, waves that are steady and not dissipating will not interact with the background atmosphere (Andrews and McIntyre 1976). The waves that are most important for the stratosphere are planetary Rossby waves, gravity waves, and equatorially trapped waves such as Kelvin waves.

Planetary Rossby waves, which can normally propagate into the stratosphere only during winter (Charney and Drazin 1961), cause much of the variability in the wintertime middle atmosphere, including sudden stratospheric warmings (McIntyre 1982). Their propagation is affected by the background winds and temperature. They are forced in the troposphere and can be dissipated by long-wave radiative transfer or by Rossby wave breaking (McIntyre and Palmer 1983). Quasi-stationary planetary Rossby waves are always present during the winter months. They are readily observed by satellite and are well-simulated in global climate models. Their amplitudes tend to be larger and more variable in the NH.

Gravity waves generated in the troposphere propagate into the stratosphere. Because they are more difficult to observe and cannot be represented in most global climate models, the global impact of gravity waves on the stratosphere is less well-established. Dissipating gravity waves are responsible for the structure of the stratopause and polar night jet in the high latitude winter (Hitchman et al. 1989; Siskind et al. 2007).

Although equatorially trapped waves have a limited geographical domain, their impact in the tropics is important. In the classic simulation of the quasi-biennial oscillation (QBO) of the zonal wind over the equator, Lindzen and Holton (1968) and Holton and Lindzen (1972) were able to generate the oscillation from interactions of equatorial Kelvin waves, mixed-Rossby-gravity waves, and other gravity waves with the background wind. More recent simulations by Kawatani et al. (2010a, 2010b) showed that the spectrum of waves that contributed momentum to the driving of a simulated QBO was quite broad. Waves from planetary scale, such as Kelvin waves, to short horizontal wavelength gravity waves all contributed to the generation of the QBO in the model.

The 2-dimensional mean circulation in the stratosphere is best viewed using the transformed Eulerian framework introduced by Andrews and McIntyre (1976). The transformed Eulerian mean is a good approximation of the zonal mean Lagrangian mass flux. The transformed Eulerian mean meridional and vertical mass fluxes are determined by adding the Stokes drift due to waves to the Eulerian mean meridional and vertical winds. For steady non-dissipative waves, the Eulerian mean winds and the Stokes drift cancel and there is no net mass flux. A residual remains for dissipating waves. Thus a wave-driven circulation is one that owes its existence to dissipating waves. The mass circulation is known as the transformed Eulerian mean or residual circulation. In the stratosphere, it is referred to as the Brewer-Dobson (B-D) circulation after pioneering inferences from observations of water by Brewer (1949) and of ozone column by Dobson (1956). In the lower stratosphere, the circulation goes from tropics to poles (Dunkerton 1978). Air enters the stratosphere through the tropopause in the tropics and slowly moves toward both poles. It can then reenter the troposphere in middle or high latitudes of both hemispheres. In the upper stratosphere, the B-D circulation has a global cell; it moves air from the summer pole toward the winter pole.

One useful variable for assessing the B-D circulation is the age of air (Hall and Plumb 1994). The age of air measures the time that has elapsed since an air parcel entered the stratosphere. In the high latitude lower stratosphere the age can be several years. To determine the age of air from observations, studies have looked at the concentrations of long-lived anthropogenic gases that have a monotonic trend at the surface, such as CO_2 and SF_6. Although these tracers work well for determining the age of air, they are not ideal for investigating trends that might have occurred as a result of climate change. Variations in the production rate over the years make the analysis highly complicated and sometimes ambiguous (Garcia et al. 2011).

The tropopause is highest and coldest in the tropical region. Because cold air can hold little water, air entering the stratosphere is extremely dry. The seasonal cycle in the tropical tropopause temperature (coldest during NH winter) gives a seasonal cycle in the humidity of the air entering the stratosphere. The vertical mixing is sufficiently low that patterns of higher and lower water mixing ratio can be seen ascending in the lower stratosphere. This phenomenon, called the "atmospheric tape recorder", was first reported by Mote et al. (1996).

2.2.3 Composition

The distribution of the dominant molecules in the stratosphere is the same as in the troposphere: molecular nitrogen and oxygen are the dominant gases and are well-mixed (Brasseur and Solomon 2005). The CO_2 concentrations are similar to those in the troposphere but with a lag in the growth by up to 4–5 years. As mentioned in Sect. 2.2.2, the air entering the stratosphere is very dry. Currently, roughly half of the water in the stratosphere enters through the tropopause and the other half is produced within the stratosphere from hydrogen released by the destruction of methane (CH_4). The contribution of methane is believed to have been lower in preindustrial times.

The most important trace gas in the stratosphere is ozone. Ozone protects the Earth's surface from ultraviolet radiation, provides most of the heating of the stratosphere by solar insolation, and is highly active chemically. It has been extensively studied from satellite using infrared or microwave emissions, and ultraviolet absorption or scattering.

The production of ozone depends on the solar ultraviolet flux (which photolyzes molecular oxygen) and the temperature (which affects the kinetic rate of the reaction that makes ozone from molecular oxygen and atomic oxygen). The production rate is therefore not strongly influenced by global changes in the atmospheric composition. An exception to this generalization is the impact of lower temperature due to greenhouse gas increases. During daylit hours, ozone is lost by photolysis but is quickly regenerated when the O combines with O_2 to form another ozone. However, there are a series of catalytic cycles that remove both O_3 and O, and hence cause a net loss of ozone. These involve the molecules and atoms in the hydrogen, nitrogen, chlorine, and bromine families. Anthropogenic increases in chlorine and bromine are responsible for the severe springtime ozone loss in the lower stratosphere over Antarctica (the ozone hole), for ozone loss in the NH winter/spring lower stratosphere, and for ozone loss in the global upper stratosphere (WMO 2010).

Many of the reactions involved in the catalytic ozone-loss cycles have reaction rates that depend on temperature. The overall effect of a decreasing temperature, with no changes to the background composition, would be to increase the ozone production rate and decrease the ozone loss rate. The combination leads to a marked anti-correlation of perturbations in temperature and ozone. This applies not only for perturbations of temperature and ozone due to waves on timescales of days to weeks but also for predictions of the response of the global ozone layer to greenhouse gas increases.

2.3 The Mesosphere

2.3.1 Heating

Like the stratosphere, the heating in the mesosphere occurs internally from solar ultraviolet radiation. The dominant absorber is molecular oxygen. Another molecule that contributes is water, which absorbs Lyman-α and Schumann-Runge radiation. The absorbed energy can photolyze the absorbing molecule, excite it to a higher state of energy, and/or contribute to increasing its kinetic temperature. The energy is not necessarily available as heat at the time of absorption.

Consider the absorption of ultraviolet radiation (<240 nm) by molecular oxygen. The photolysis gives two products: $O(^3P)$ and $O(^1D)$, i.e. atomic oxygen in the ground and excited states. The photolysis uses much of the energy but the remainder is available as heat. The excited $O(^1D)$ atom can participate in one of many chemical reactions ($O(^1D)$ is short-lived and is much more reactive than $O(^3P)$); it can collisionally transfer its energy to excite another atom or molecule; it can be collisionally quenched and convert the energy to heat; or it can spontaneously emit a photon and drop to the ground state.

The $O(^3P)$ atom has a lifetime that ranges from very short in the lower mesosphere to months in the upper mesosphere (Smith et al. 2010b). The reactions that either directly (reaction with O or O_3) or indirectly (catalytic cycles with hydrogen-containing species) convert $O(^3P)$ to O_2 are exothermic. This means, in simple terms, that the energy from the photon that photolyzed O_2 is not available as heat until the products recombine into another O_2 molecule. Substantial transport can occur in the interim. For example, O generated in the lower thermosphere can be transported downward into the upper mesosphere and provide a source of heating even though the energy from the solar radiation did not penetrate to this part of the atmosphere. The heating can be substantial during night due to the impact of exothermic reactions. Some of the energy from O_2 or O_3 photolysis contributes to spontaneous emission by $O(^1D)$ and other species (notably $O_2(^1\Sigma)$ and vibrationally excited OH) as airglow. Much of the energy that is emitted as airglow is lost to space or to the ground and does not contribute to heating. The loss of energy through airglow is substantial (Mlynczak and Solomon 1993 estimate 20–40% for some processes) but is not precisely known.

The delay between the absorption of energy and the heating effect of the absorbed energy has an additional impact on the dynamics. The net heating rate can shift away from having a pronounced daytime maximum. The two exothermic reactions that have the largest impact near the mesopause are $O + O$ (i.e., the self-reaction of atomic oxygen) and $O_3 + H$. The former is similar during day and night since O does not have a photochemically driven diurnal cycle. The latter is actually larger during night because of the very large diurnal cycle in the concentration of O_3 (Smith and Marsh 2005). One net impact of the exothermic reactions is to damp the magnitude of the diurnal cycle in heating in the upper mesosphere.

2.3.2 Dynamics

A defining characteristic of the mesosphere is the cold summer mesopause. The mesopause at the summer pole is the coldest place in the Earth's atmosphere. The mesopause is interesting for its cold temperature in summer and also because it is very far from radiative balance (the summer is much colder than the winter). The cold summer temperature provides an environment for polar mesospheric clouds (also called noctilucent clouds) even though the humidity is very low. The cold conditions are maintained by strong upwelling. The upwelling is part of a global circulation cell that moves air up at the summer pole, horizontally from the summer to the winter, and down at the winter pole (Smith et al. 2011).

The mesospheric circulation is driven by dissipating and/or breaking gravity waves (Lindzen 1981; Holton 1982). The waves are generated within the troposphere and then filtered by winds in the stratosphere. A gravity wave cannot propagate when the wave phase speed is equal to the background wind. The point where the phase speed equals the background wind is know as a critical layer. Because the stratospheric zonal winds are stronger than the meridional winds, the critical layer filtering has the largest impact on waves propagating in the eastward and westward directions. The strong stratospheric jets in the solstice periods filter out westward waves during the summer and eastward waves during the winter. The waves that are not filtered continue to propagate vertically but eventually break or dissipate in the upper mesosphere. The momentum deposition from the resulting asymmetric spectrum of gravity waves causes the stratospheric jets to reverse direction in the mesosphere. The momentum deposition acts together with the Coriolis force to drive the summer to winter flow in both hemispheres.

The strong dependence of the mesospheric temperature distribution on waves from below also means it is susceptible to differences or changes in the regions below. An example of the impact of changes in the lower stratosphere is given in Sect. 4.2.

2.3.3 *Composition*

The composition in the lower mesosphere is similar to that in the stratosphere. One difference is a shift in the catalytic cycles that destroy ozone. In the stratosphere, reactions involving chlorine species and nitrogen species are very important in determining the ozone balance. Beginning in the lower mesosphere, the catalytic cycles that destroy ozone are strongly dominated by those involving hydrogen species. The reactive hydrogen species H, OH and HO_2 are products of the destruction of methane and water.

Reactions involving species in the hydrogen and oxygen families continue to be important through the entire mesosphere. As noted above, many of the reactions are exothermic and they contribute to the distribution of heating. The proportion of heat from these reactions becomes larger in the upper mesosphere lower thermosphere (MLT) region. The composition in the MLT differs from that in the stratosphere and lower mesosphere in several important ways.

- The gases are no longer well-mixed but are beginning to separate based on their molecular masses.
- Ions begin to make a contribution to the chemical system.
- Some free radicals, for example hydroxyl (OH), become longer-lived and play a more central role in the chemical system.
- Atomic oxygen transitions from a very short lifetime below about 80 km to a long-lived and abundant gas in the upper mesosphere.
- Excited states of atoms and molecules affect the chemistry and energetics.
- Ozone has a large diurnal cycle (smallest during daytime) due to the very rapid photolysis rate.
- Several important molecules, most notably CO_2, are not in local thermodynamic equilibrium (LTE) in the mesosphere; see below.

Non-LTE conditions occur when low atmospheric density leads to a low rate of collisions between molecules (Andrews et al. 1987). In general, strong cooling of the atmosphere occurs through emission by vibrationally excited CO_2. However, under non-LTE conditions, the low collision rate limits the exchange of energy from kinetic to vibrational levels. As a result, the population of vibrational levels is not consistent with a Boltzmann distribution and the infrared CO_2 cooling rates are substantially reduced over the LTE situation.

Several of the characteristics of the mesosphere (extremely cold temperatures, importance of radical species and excited states) contribute to uncertainties in the chemical system. Laboratory measurement of the key reaction rates under realistic conditions is extremely difficult. Satellite observations using atmospheric emissions are constrained by the very low densities and therefore weak signals of natural emissions. Retrieval of temperature or constituent densities using measured emissions from a non-LTE atmospheric state requires knowledge of the energy structure of the molecule. For all of these reasons, the detailed knowledge of mesospheric composition lags that of the stratosphere.

For example, observational estimates of one of the most important gases, atomic oxygen, differ substantially. As pointed out in Smith et al. (2010b), the O concentrations from different analyses are not compatible. O is derived during day and night using two different techniques and different radiance channels from the SABER instrument. When tidal variations are taken into account, the concentrations are consistent between day and night, supporting the reliability of the concentrations retrieved from SABER. However, O derived from SABER is larger by about a factor of four than the concentrations derived from WINDII by Russell et al. (2004). The SABER values are also larger than the concentrations in the MSIS empirical model (Picone et al. 2002).

3 Interfaces Between the Atmospheric Regions

3.1 The Tropopause

The tropopause has traditionally been identified as the altitude of coldest temperature or, equivalently, the altitude where the decline of temperature with altitude ceases. However, for certain purposes, it has been more useful to fine-tune the definition. The tropopause can also be identified by isentropic surfaces or by potential vorticity (Holton et al. 1995). Chemically, the tropopause marks the location of sharp decreases in water and increases in ozone. It is the point where vigorous vertical mixing ceases. The altitude of the tropopause is higher in the tropics (equatorward of 30°) than at high latitudes. The altitude ranges from about 18 km near the equator to about 10 km at the poles and is discontinuous at ±30°.

In the tropics, rather than a distinct tropopause boundary, the transition region is known as the tropical tropopause layer, or TTL (e.g., Fueglistaler et al. 2009). This takes into account the finite depth over which the transitions from troposphere to stratosphere occur. The TTL region separates two regions which have quite different dynamics. Since the air that enters the middle atmosphere passes through this region, its impact on transport of trace species is important.

The processes responsible for the tropopause were discussed by Held (1982). The tropopause marks the transition between the vigorous dynamical mixing in the troposphere and the radiative balance in the stratosphere. A general idea of the tropospheric thermal structure and tropopause location can be deduced using the concept of radiative-convective balance. Solar visible radiation passes through the atmosphere and is absorbed at the Earth's surface (land or ocean). Sensible and radiant heat from the surface warm the atmosphere from below. A purely radiative balance is not in general possible in the troposphere because the heating from below, particularly over land surfaces, leads to convective instability near the Earth's surface. The temperature profile of an atmosphere in radiative convective balance is constrained to maintain stability, i.e., to have a lapse rate that is less than Γ_d for dry air or Γ_s for moist air. This concept has been used in simple radiative balance climate models for many decades. In these models, excess heat in the lower troposphere is redistributed in the vertical direction (as if by convection) until the profile achieves neutral stability.

As discussed in Sect. 2.2, the heating in the stratosphere comes from absorption of solar ultraviolet radiation by ozone. The maximum temperature at the stratopause occurs near where the optical depth for the ozone heating (primarily the Hartley band) is unity. The tropopause then marks the transition between the negative temperature gradient of the tropospheric radiative-convective system and the positive gradient of the radiative balance system in the stratosphere.

Recent work has contributed to the understanding of the maintenance of the tropopause. The altitude and temperature of the tropopause are sensitive to the radiative balance in the upper troposphere and stratosphere and the stratospheric circulation (Thuburn and Craig 2000; Shepherd 2002; Schneider 2004; Birner 2010).

Water in the middle atmosphere depends on the transport of both water and methane across the tropical tropopause. As air passes through the very cold temperature of the tropical tropopause, much of the water condenses into ice particles and settles out. This "freeze-dried" air has water concentrations of only a few ppm. The water concentration increases above the tropopause due to formation from byproducts of methane destruction. The methane transport into the stratosphere depends on the tropospheric concentration and the net mass flux across the tropopause; it is not directly affected by the tropopause temperature.

Observations indicate that the concentration of water in the stratosphere has varied over decadal time scales. Part of the variation is a result of the increase of methane, which leads to more water production by photochemistry within the stratosphere. However, the observed variations in water are larger than can be explained by the methane increase and variations in the tropical tropopause temperature. Water increased for several decades until about 2000 and has decreased since then (Randel et al. 2006; Fujiwara et al. 2010) although the reasons are still not understood. Stratospheric water is important because of its impact on the global climate. The infrared effects of stratospheric water increases during 1980–2000 can account for a acceleration of climate warming at the Earth's surface (Solomon et al. 2010). Likewise, the water decrease since 2000 has slowed down the surface warming compared to what it would otherwise be with unvarying stratospheric water.

Changes in the flux of water from the troposphere to the middle atmosphere also have an impact on the chemistry in the mesosphere (Marsh et al. 2003) and on the formation of polar mesospheric clouds (Thomas 1991).

3.1.1 The Stratopause

The transition between the stratosphere and mesosphere is defined by a change in the vertical gradient of the mean temperature from positive to negative. Although there are some differences between the stratosphere and mesosphere (see Sects. 2.2 and 2.3), the stratopause region is not normally a place of major transition in either dynamics or composition. This is the reason that the terminology now often uses middle atmosphere rather than stratosphere and mesosphere.

The stratopause owes its altitude to two processes: the mean radiative balance and the thermal structure due to gravity waves. The radiative balance is due to ozone, which is produced in the stratosphere. Ozone absorbs solar ultraviolet radiation in the Hartley, Huggins and Chappuis bands. The largest heating occurs around 40–50 km; this is the mean altitude of the stratopause. Away from the high latitude winter, the stratopause altitude is well predicted by the radiative balance. This is approximately the altitude where the optical depth of ozone is unity. The maximum heating rate leads to a maximum in the temperature.

In the high latitude winter, there is no sunshine so the heating is small. In this region, the stratopause is strongly influenced by the adiabatic warming due to sinking motion (Hitchman et al. 1989). This is particularly the case in the NH stratosphere, where the winter dynamics is more active. Gravity wave driving of the circulation, which becomes dominant in the mesosphere, can also affect the stratospause region. Gravity wave driving of the polar winter circulation appears to be stronger in the SH.

Another quality that distinguishes the winter stratopause from the tropopause and mesopause is the very large interannual variability in its location. Three well documented events during recent NH winters (2003–2004, 2005–2006, and 2008–2009) show a displacement of the stratopause to much higher altitudes. All three cases were characterized by the sudden displacement upwards and slow return (over a period of weeks) for a region poleward of 70°N (Manney et al. 2005, 2008, 2009). These events highlight the importance of dynamics in controlling the stratospheric structure in winter high latitudes.

The poleward flow within the stratosphere is driven primarily by quasi-stationary Rossby waves. The poleward flux is directly associated with downwelling motion in the polar winter. The stratospheric temperature is also affected by the larger-scale mean circulation that is characterized by downwelling through the entire winter polar middle atmosphere. Gravity waves are a prime driver of this circulation in the upper levels (above 40–50 km). The vertical propagation and eventual breakdown of gravity waves are strongly affected by the winds through which they propagate.

3.1.2 The Mesopause

At the mesopause, the vertical gradient of temperature switches sign again, from negative back to positive. The mesopause region is the coldest part of the Earth system. The overall temperature is low because of a lack of heating. Molecular oxygen is the leading absorber of solar radiation.

In a climatological sense, there are two distinct mesopause positions. In the summer high latitudes, the altitude is about 83–89 km (von Zahn et al. 1996) while at all other latitudes and times, its position is close to 100 km. Model simulations indicate that the lower and colder position of the summer mesopause is due to very strong adiabatic cooling associated with strong upwelling.

On top of this mean description are several details. First, there is a difference in the summer mesopause altitudes between the Northern (NH) and Southern Hemisphere (SH). The SH position is higher and not quite as cold (Xu et al. 2007). Polar mesospheric cloud abundances are lower in the SH (Bailey et al. 2007). Processes that have been proposed to account for this difference include the annual cycle in Earth-Sun distance (Chu 2003) and differences in gravity wave driving of the circulation due either to tropospheric sources or stratospheric filtering (for example, Dowdy et al. 2001).

Another feature of the mesopause is the existence of temperature profiles that contain more than one minimum. The "mesospheric inversion" or "double mesopause" has been documented in observations and models (Meriwether and Gardner 2000). Analysis by Sassi et al. (2001) and Salby et al. (2002) show that, in winter midlatitudes, the double mesopause phenomenon is associated with planetary-scale variations in dynamics. The existence of planetary waves in the stratosphere and mesosphere affects the filtering, propagation, and breaking of gravity waves. The combined impact of planetary and gravity waves can give sharp changes in temperature and other dynamical fields in some phases of the planetary wave. From a ground-based observation, where only a single longitude can be observed, multiple "mesopauses" can appear in the vertical. Although single site observations of a double mesopause are common, this appears to be solely a local phenomenon.

The mesopause is normally considered to be in the vicinity of the turbopause or homopause (Chabrillat et al. 2002). The turbopause is the altitude where molecular dissipation of waves is more important than turbulent dissipation. Molecular diffusion processes suppresse wave breaking and turbulence, hence the term turbopause. The definition of the homopause is related: it is the altitude at which diffusive separation begins. Both the turbopause and homopause represent a transition to conditions where collisions are not frequent enough to keep the atmospheric constituents well mixed. Observations of wave properties (Offermann et al. 2007) and composition (such as CO_2; see review by López Puertas et al. 2000) suggest that there is a gradual change toward diffusive separation rather than an abrupt onset for either the turbopause or the homopause.

4 Coupling Between Regions

As discussed in Sects. 2 and 3, the basic thermal structure and the interfaces between atmospheric regions are affected by dynamical and radiative processes through the depth of the atmosphere. In this sense, vertical coupling is the basis of the mean atmospheric structure. However, here the term "coupling" is used to refer not to the maintenance of the basic structure but to the response of the atmosphere to perturbations.

This section gives brief summaries of some examples of vertical coupling between atmospheric regions.

4.1 Radiative Coupling

The penetration of incoming solar radiation through the Earth's atmosphere varies strongly with wavelength. The absorption of the radiation depends on the concentrations of gases. Above the stratopause, the conversion of the absorbed energy to heat also depends on the concentrations of gases that are involved in exothermic reactions or in quenching excited states. The interfaces between atmospheric regions (the tropopause, the stratopause, etc.) do not in themselves pose any barrier to the transmission of radiation.

The infrared radiation emitted by the Earth's surface and by certain gases in the atmosphere can be reabsorbed and reemitted multiple times, particularly in the troposphere where the density of absorbing gases and aerosols is higher. Absorption can also be a heat source in the lower stratosphere. In the middle atmosphere, emission of infrared radiation by CO_2 is the leading cooling process.

Perturbations to the radiative balance can occur through

- changes in the solar flux (such as the 11-year solar cycle);
- changes to the absorption of solar radiation (such as reduced absorption in the stratosphere due to ozone loss);
- changes in the reflection or absorption of radiation at the Earth's surface (for example due to changes in snow-cover);
- changes in the reflection or absorption of radiation in the Earth's atmosphere (for example due to changes in aerosol or cloud amounts);
- changes in the absorption of infrared radiation in the atmosphere (due to changes in water, CO_2, or other radiatively active gases);
- changes in the composition in the upper mesosphere that affect the energy loss through airglow.

The radiative changes can lead directly to changes in the thermal structure. Changes in the thermal structure, particularly if the magnitude of the change varies with latitude, can lead directly to dynamical changes.

Examples are given in Sects. 4.1.1 and 4.1.2.

4.1.1 Impact of Antarctic Ozone Loss in the Stratosphere on the Troposphere and Mesosphere

The primary heating in the stratosphere is absorption of solar ultraviolet radiation by ozone. Since 1980, the ozone over Antarctica during the spring has been substantially reduced from levels prior to that year. The cause of this ozone hole is well-understood; it is due to ozone destruction by chlorine and bromine containing compounds that become particularly reactive during the special conditions of SH polar winter and spring. One effect of the ozone loss is a reduction of absorbed solar radiation, which leads to reduced heating. The negative trend in temperature in the lower SH stratosphere during winter is clearly evident in observations (Thompson and Solomon 2002; Randel et al. 2009). Due to the thermal wind relation in the atmosphere (3), this polar temperature trend is associated with a trend in the mean jet strength in the SH spring and early summer.

Much attention has been focused on the climate effects of the SH ozone loss (for example, Gillett and Thompson 2003; Baldwin et al. 2007; Son et al. 2010). A useful analysis tool for this is comparisons of past climate simulations (hind-casts) and future climate predictions between models with realistic ozone loss (and recovery in the future) and those without. These show that the effect of the ozone loss in the SH and of the future predicted ozone recovery are felt all the way to the Earth's surface. Models with realistic ozone loss (either through interactive stratospheric chemistry or through specification from observations) show substantially less warming over Antarctica during the 20th century than those that have non-varying ozone or that do not resolve the stratosphere.

Changes in the SH stratosphere due to the ozone hole can also affect the mesosphere. The primary mechanism is changes in the propagation of gravity waves. The ozone hole led to trends in the lower stratosphere temperature and winds during late spring and early summer. The trend in the zonal mean wind, as simulated in the Whole Atmosphere Community Climate Model, have affected the propagation of gravity waves into the mesosphere (Smith et al. 2010a). As a result, the model simulations show that the strength of the summertime mean circulation in the SH decreased at the onset of the ozone hole around 1980. The model also simulated warmer summer mesopause temperatures due to the change in gravity wave interactions.

The previous example illustrates the interlinked nature of some of the coupling processes. In this case, a chemical change in the stratosphere (ozone loss due to increased chlorine and bromine) leads to a radiative change (less heating due to lower ozone). The radiative change affects the temperature which, through the thermal wind relation, affects the zonal wind. The stratospheric wind change affects the propagation of gravity waves from the troposphere to the mesosphere. The breaking or dissipation of these waves then affects the strength of the circulation in the mesosphere and, through changes in the adiabatic heating, the polar mesopause temperature (Smith et al. 2010a).

4.1.2 Impact of Stratospheric Aerosols on Tropospheric Climate

Volcanic eruptions can inject trace chemical species directly into the atmosphere. One of these gases is SO_2. The sulfur that is injected into the troposphere washes out in a matter of days to weeks, so its direct climate impact has limited duration. Over a period of weeks, the sulfur injected into the stratosphere by explosive volcanoes becomes incorporated into sulfate (H_2SO_4) aerosol particles. The eruption of Mt. Pinatubo, the largest in the satellite era, injected about 20 Mt of sulfur directly into the tropical lower stratosphere. The circulation in the lower stratosphere (the Brewer-Dobson circulation discussed in Sect. 2.2.2) tends to move air upwards in the tropics, poleward and then downward at the middle and high latitudes with a timescale of several years. Therefore, the sulfur injected by tropical volcanoes can persist for longer, become globally distributed, and can therefore have an impact over a wider area and longer period (Robock 2000; Stenchikov et al. 2006).

The best known climate impact of volcanic eruptions is cooling at the Earth's surface, especially during summer. The cooling is caused by the radiative impact of aerosols. The particles absorb and reflect some of the solar radiation and therefore reduce the radiation at the Earth's surface. The net effect of eruptions on the surface temperature is difficult to determine because of few eruptions and large natural variability but is estimated to be 0.1–0.2 K (Robock 2000). The stratosphere is a crucial piece in the climate response. The timescale of several years for air to move through the stratosphere keeps the volcanic aerosols present in the system.

The impact of volcanic aerosols on the heating also affects the stratospheric circulation. The enhanced heating in the lower tropical stratosphere can increase the B-D circulation (Tie et al. 1994). The altered circulation due to Mt. Pinatubo increased the strength of the tropical upwelling motion. This branch of the circulation brings up air that is low in ozone; therefore the net effect of a stronger upwelling was to decrease ozone in the tropical lower stratosphere. There is also a photochemical response of ozone to volcanoes; the aerosol particles enhance the rate of ozone loss due to chlorine and bromine.

The distribution of aerosols in the stratosphere varies irregularly over multi-year timescales, due to intermittency in major volcanic eruptions and to variations in their latitude, time of year, and sulfur content. The impact that a given eruption can have depends on the transport that distributes the volcanic products within the stratosphere. If the stratospheric circulation accelerates due to climate change, as predicted by climate models (WMO 2010), this will affect the climate response to eruptions.

4.2 Coupling by Wave Processes

Vertical propagation of atmospheric waves can carry momentum and energy from one region of the atmosphere to another. To maintain constant kinetic energy, wave amplitude grows exponentially with decreasing density. Under favorable conditions for propagation, a small perturbation can grow to a large wave at a higher altitude. However, the converse implies that a perturbation at one altitude will have rapidly diminishing impact at lower altitudes.

As discussed above, waves will not interact with the background atmosphere unless they are transient or dissipating (Andrews and McIntyre 1976). Sections 2.2.2 and 2.3.2 mentioned the importance of planetary waves and Kelvin waves in the stratosphere and of gravity waves throughout the middle atmosphere. While the waves themselves are generated in the troposphere, the propagation and dissipation depend on the atmosphere through which they propagate. The example of the impact of lower stratospheric changes on the wave processes in the mesosphere (Sect. 4.1.1) illustrates this. The message from this is that perturbations due to waves depend on the wave generation, the wave propagation, and the wave dissipation.

Gravity waves cannot propagate beyond a critical line, where the background wind speed is equal to the phase speed of the waves. As discussed in Sect. 2.3, the critical layer effect leads to filtering of gravity waves by the mean stratospheric winds. This filtering then leads to an asymmetric spectrum of gravity waves. When those waves break in the mesosphere, the momentum forcing drives the mesospheric winds and circulation (Holton 1982).

The winds in the winter stratosphere are never zonally symmetric because of the presence of large amplitude planetary waves. An individual small scale gravity wave will propagate vertically through the local "background" wind, not the zonally averaged wind. This means that the filtering of gravity waves changes with longitude. When those waves break, the momentum forcing will also change with longitude. The variations in momentum forcing will then lead to wavelike perturbations in the mesospheric winds. This mirror effect has been seen in observations (Smith 1996, 1997) and in numerical simulations (Smith 2003). In observations, it is manifest as an anti-correlation between waves in stratospheric winds and those in mesospheric winds.

Analysis has shown connections between variations in the middle atmosphere at northern and southern high latitudes. One of these phenomena is a relationship between the amplitude of the semidiurnal tide in the summer hemisphere and the planetary waves in the stratosphere of the other hemisphere (Portnyagin et al. 1998; Smith et al. 2007; Murphy et al. 2009). Simulations by Yamashita et al. (2002) suggest that this is due to a propagation of the tide from the winter to the summer hemisphere. Although the tides are global modes, the details of their generation and vertical or horizontal propagation in an atmosphere with spatial and temporal variations in temperature, winds, and heating are still not well understood.

Another connection between the hemispheres is a correlation between planetary waves in the winter stratosphere and the polar mesospheric clouds at the summer mesopause (Karlsson et al. 2009). The processes that generate this link are not well understood. Analysis indicates that the signal probably goes from the stratosphere to the mesosphere of the winter hemisphere. Transfer of the signal to the opposite hemisphere may involve the mean circulation and/or impacts of wind changes on the propagation or breaking of the gravity waves that drive the circulation.

4.3 Coupling by Mass Exchange

Coupling among atmospheric regions can also occur by exchange of mass. One relevant process is the transport of chemically active gases from the mesosphere to the stratosphere. Another is the transport of water from the troposphere to the stratosphere.

As discussed in Sect. 3.1, the tropopause marks the top of the region of rapid vertical mixing. Movement of mass across the tropopause occurs much more slowly than vertical movement of air within the troposphere. Flow across the tropopause is important for the composition of both the troposphere and the stratosphere. The impact on stratospheric water was discussed in Sect. 3.1. The advection of stratospheric air into the troposphere increases the concentrations of ozone and other gases (Holton 1990; Olsen et al. 2002).

Although the production of chlorofluorocarbons (CFCs) has been phased out by the Montreal Protocol, these molecules still persist in the atmosphere. The most abundant CFCs have very long lifetimes in the troposphere but can be destroyed by photolysis and chemical reactions in the stratosphere (WMO 2010). The slow removal reflects the slow rate at which air from the troposphere is moved into the stratosphere and then returns to the troposphere.

Transport of mesospheric air into the stratosphere has received much attention in the last decade. A motivating factor for this is to assess the impact of reactive nitrogen (NOx, which consists of NO and NO_2) on stratospheric ozone. Downward motion by the mean circulation during winter (see Sects. 2.2 and 2.3) can transport trace species with positive vertical gradients. Perturbations in the composition can occur if the timescale for transport is short compared with the times for mixing with the surrounding air or significant photochemical production or loss.

Perturbations to NO (Randall et al. 2006, 2009), NO_2 (Hauchecorne et al. 2007), CO (Jin et al. 2005; Funke et al. 2009), ozone (Smith et al. 2009), and OH (Damiani et al. 2010) have been identified during periods of strong dynamical perturbation. It turns out that the assessment of NOx is complicated because NO can also be produced *in situ* in the polar mesosphere due to particle precipitation events (Randall et al. 2005). For this reason, NOx is not always a reliable indicator of enhanced transport between the mesosphere and stratosphere.

4.4 Observed Links with Unknown Mechanisms

For a number of decades, there have been studies reporting unexplained but statistically significant links in the atmosphere. Many of these have dropped out of view as additional data contradicted the previous relationship. The most persistent have been associated with solar variability, particularly the 11-year solar cycle.

The first-order atmospheric response to solar variability is fairly straightforward: an increase in the solar flux will increase the heating (assuming absorbing gases or surface properties do not change) and will increase the photolysis rate of some compounds and can therefore affect the composition. The direct heating can be estimated based on satellite measurements of the solar irradiance and knowledge of the absorption as a function of altitude in the atmosphere. The predicted changes in the solar flux penetrating the troposphere due to the solar cycle observed or estimated over the previous century are quite small (IPCC 2007). Likewise, the variability in the solar flux reaching the lower stratosphere is small so little direct response is expected there.

Nevertheless, observational analyses show that various climate indicators are well-correlated with the solar flux. Gray et al. (2010) give a comprehensive overview of the current status of the observations and the search for understanding.

One correlation that has received much attention is a relationship between the 11-year solar cycle, the QBO in tropical winds, and the middle and high latitude winter stratosphere. Analysis presented by Labitzke (1987) showed that the NH winter 30 hPa temperature was anticorrelated with the solar flux if one looked only at winters during which the QBO was in its westerly phase. Since then, many years of observations have been added and the correlation still persists (Labitzke 2005). To date, there has been no explanation of the physical mechanism for the observed relationship. The explanation is confounded by several factors. 1) The Holton-Tan relationship (Holton and Tan 1980) establishes a link between the tropical QBO and the high latitude stratospheric temperature. Separating this mechanism from the solar cycle is not straightforward. 2) The two important volcanic eruptions during this period, El Chichon in 1982 and Mt. Pinatubo in 1991, both came near solar maximum. These had an impact on the stratosphere that, with the short duration of available observations, is not possible to separate statistically from the solar cycle. 3) The QBO is asynchronous with the annual cycle. The beating of these two cycles (the QBO at 24–30 months and the annual cycle at 12 months) gives a longer period variation that is difficult to separate statistically from the solar cycle (Salby and Shea 1991; Smith and Matthes 2008). 4) Numerical models have not succeeded in reproducing the observed relationship.

The scientific community is divided about these correlation studies. Many researchers dismiss them as being spurious correlations. The thinking behind this is that, if one examines enough time series (there are hundreds of possibilities) over what is, for an 11-year cycle, a fairly short data record, there will be some that show correlations. The possibility of spurious but significant correlations is well-established statistically. The atmosphere has several modes of longer term variability (ENSO, decadal oscillations, etc.) that in some circumstances can be mistaken for a response to the solar cycle if extreme care is not taken.

Despite the reservations by many in the community, much effort continues to be expended in looking for additional correlations and trying to explain the correlations that turn up.

5 Conclusions

This document gives a brief description of atmospheric regions and some of the processes involved in their interactions. The coupling between regions has received much attention in recent years as both observations have improved and numerical models have provided better tools for analysis. The coupling described in this paper is categorized into three types of processes: radiative, wave propagation, and mass exchange.

Vertical coupling occurs readily through radiative processes. Radiative processes can work both ways; the less dense parts of the atmosphere can have an impact on the underlying layers. Several examples were presented for the effect of stratospheric changes (aerosol distribution; ozone loss) on the troposphere. Note however that both of these involve local changes to the stratosphere itself (such as its temperature) and therefore include dynamical as well as radiative processes.

Vertical coupling through wave processes also occurs readily. Most of the waves that are important to the middle atmosphere are generated in the troposphere. The conditions for generation of the waves and their propagation through the intervening atmosphere both lead to remote effects.

The downward impact of waves was not discussed here. Because of the exponential decrease in wave amplitude with increasing atmospheric density, downward propagation of waves is not expected to be significant. One proposal for downward wave coupling is the impact of the middle atmosphere on the structure of planetary waves in the winter troposphere. For example, recent studies (Perlwitz and Harnik 2003, 2004; Kodera et al. 2006) have shown that, under some circumstances, upward propagating planetary waves can be reflected in the stratosphere. This reflection has an impact not only on the wave structure in the stratosphere but also on the wave within the troposphere.

Coupling of atmospheric regions through mass exchange is also possible. In this case, we focus on the concentrations of trace chemical species that are perturbed due to changes in the transport. The examples are water transport into the stratosphere and mesospheric air transported into the NH polar winter stratosphere.

This overview gives a description of the regions of the atmosphere and a taste of some of the important coupling processes between them. This survey is not comprehensive but instead reflects the experience and interests of the author. The knowledge of the atmosphere and its interactions is ever increasing due to increased measurements and improved numerical modeling capabilities. These will lead to better understanding of the interactions and better predictions of how they will affect the future of the atmosphere.

Acknowledgements The National Center for Atmospheric Research is sponsored by the National Science Foundation. Support for this work was also provided by the NASA Heliospheric Guest Investigator Program.

References

D.G. Andrews, M.E. McIntyre, Planetary waves in horizontal and vertical shear: the generalized Eliassen-Palm relation and mean zonal acceleration. J. Atmos. Sci. **33**, 2031–2048 (1976)

D.G. Andrews, J.R. Holton, C.B. Leovy, *Middle Atmosphere Dynamics* (Academic Press, San Diego, 1987)

S.M. Bailey, A.W. Merkel, G.E. Thomas, D.W. Rusch, Hemispheric differences in polar mesospheric cloud morphology observed by the Student Nitric Oxide Explorer. J. Atmos. Sol.-Terr. Phys. **69**, 1407–1418 (2007)

M.P. Baldwin, M. Dameris, T.G. Shepherd, How will the stratosphere affect climate change? Science **316**(5831), 1576–1577 (2007). doi:10.1126/science.1144303

T. Birner, Residual circulation and tropopause structure. J. Atmos. Sci. **67**, 2582–2600 (2010)

G. Brasseur, S. Solomon, *Aeronomy of the Middle Atmosphere: Chemistry and Physics of the Stratosphere and Mesosphere*, 3rd edn. (Springer, Heidelberg, 2005), p. 646

A.W. Brewer, Evidence for a world circulation provided by the measurements of helium and water vapour distribution in the stratosphere. Q. J. R. Meteorol. Soc. **75**, 351–363 (1949)

S. Chabrillat, G. Kockarts, D. Fonteyn, G. Brasseur., Impact of molecular diffusion on the CO_2 distribution and the temperature in the mesosphere. Geophys. Res. Lett. **29** (2002). doi:10.1029/2002GL015309

J.G. Charney, P.G. Drazin, Propagation of planetary-scale disturbances from the lower into the upper atmosphere. J. Geophys. Res. **66**, 83–109 (1961)

X. Chu, Lidar studies of interannual, seasonal, and diurnal variations of polar mesospheric clouds at the South Pole. J. Geophys. Res. **108** (2003). doi:10.1029/2002JD002524

A. Damiani, M. Storini, M.L. Santee, S. Wang, Variability of the nighttime OH layer and mesospheric ozone at high latitudes during northern winter: influence of meteorology. Atmos. Chem. Phys. **10**, 10291–10303 (2010)

G.M.B. Dobson, Origin and distribution of polyatomic molecules in the atmosphere. Proc. R. Soc. Lond. Ser. A, Math. Phys. Sci. **236**, 187–193 (1956)

A. Dowdy, R.A. Vincent, K. Igarashi, Y. Murayama, D.J. Murphy, A comparison of mean winds and gravity wave activity in the northern and southern polar MLT. Geophys. Res. Lett. **28**, 1475–1478 (2001)

T.J. Dunkerton, On the mean meridional mass motions of the stratosphere and mesosphere. J. Atmos. Sci. **35**, 2325–2333 (1978)

S. Fueglistaler, A.E. Dessler, T.J. Dunkerton, I. Folkins, Q. Fu, P.W. Mote, Tropical tropopause layer. Rev. Geophys. **47**, 1–31 (2009). doi:10.1029/2008RG000267

M. Fujiwara, H. Vömel, F. Hasebe, M. Ogino, S.-Y. Ogino, S. Iwasaki, N. Nishi, T. Shibata, K. Shimizu, E. Nishimoto, J.M. Valverde Canossa, H.B. Selkirk, S.J. Oltmans, Seasonal to decadal variations of water vapor in the tropical lower stratosphere observed with balloon-borne cryogenic frost point hygrometers. J. Geophys. Res. **115** (2010). doi:10.1029/2010JD014179

B. Funke, M. López-Puertas, M. Garcia-Comas, G. Stiller, T. von Clarmann, M. Höpfner, N. Glatthor, U. Grabowski, S. Kellmann, A. Linden, Carbon monoxide distributions from the upper troposphere to the mesosphere inferred from 4.7 μm non-local thermal equilibrium emissions measured by MIPAS on Envisat. Atmos. Chem. Phys. **9**, 2387–2411 (2009)

R.R. Garcia, W.J. Randel, D.E. Kinnison, On the determination of age of air trends from atmospheric trace species. J. Atmos. Sci. **68**, 139–154 (2011)

N.P. Gillett, D.W.J. Thompson, Simulation of recent southern hemisphere climate change. Science **302**(5643), 273–275 (2003). doi:10.1126/science.1087440

L.J. Gray, J. Beer, M. Geller, J.D. Haigh, M. Lockwood, K. Matthes, U. Cubasch, D. Fleitmann, G. Harrison, L. Hood, J. Luterbacher, G.A. Meehl, D. Shindell, B. van Geel, W. White, Solar influences on climate. Rev. Geophys. **48** (2010). doi:10.1029/2009RG000282

M.E. Hagan, Comparative effects of migrating solar sources on tidal signatures in the middle and upper atmosphere. J. Geophys. Res. **101**, 21213–2122 (1996)

T.M. Hall, R.A. Plumb, Age as a diagnostic of stratospheric transport. J. Geophys. Res. **99**(D1), 1059–1070 (1994)

A. Hauchecorne, J.-L. Bertaux, F. Dalaudier, J.M. Russell, M.G. Mlynczak, E. Kyrölä, D. Fussen, Large increase of NO_2 in the north polar mesosphere in January–February 2004: evidence of a dynamical origin from GOMOS/ENVISAT and SABER/TIMED data. Geophys. Res. Lett. **34** (2007). doi:10.1029/2006GL027628

I.M. Held, On the height of the tropopause and the static stability of the troposphere. J. Atmos. Sci. **39**, 412–417 (1982)

I.M. Held, B.J. Soden, Robust responses of the hydrological cycle to global warming. J. Climate **19**, 5686–5699 (2006)

M.H. Hitchman, J.C. Gille, C.D. Rodgers, G. Brasseur, The separated winter stratopause: a gravity wave driven climatological feature. J. Atmos. Sci. **46**, 410–422 (1989)

J.R. Holton, The role of gravity wave induced drag and diffusion in the momentum budget of the mesosphere. J. Atmos. Sci. **39**, 791–799 (1982)

J.R. Holton, On the global exchange of mass between the stratosphere and troposphere. J. Atmos. Sci. **47**, 392–395 (1990)

J.R. Holton, R.S. Lindzen, An updated theory of the quasi-biennial cycle of the tropical stratosphere. J. Atmos. Sci. **29**, 1095–1107 (1972)

J.R. Holton, H. Tan, The influence of the equatorial quasi-biennial oscillation on the global circulation at 50 mb. J. Atmos. Sci. **37**, 2200–2208 (1980)

J.R. Holton, P. Haynes, M.E. McIntyre, A.R. Douglass, R.B. Rood, L. Pfister, Stratosphere-troposphere exchange. Rev. Geophys. **33**, 403–439 (1995)

IPCC, *Climate Change 2007: The Physical Science Basis. Contribution of Working Group I to the Fourth Assessment Report of the Intergovernmental Panel on Climate Change* (Cambridge University Press, Cambridge, 2007), p. 996
J.J. Jin, K. Semeniuk, A.I. Jonsson, S.R. Beagley, J.C. McConnell, C.D. Boone, K.A. Walker, P.F. Bernath, C.P. Rinsland, E. Dupuy, P. Ricaud, J.D.L. Noë, J. Urban, D. Murtagh, Co-located ACE-FTS and Odin/SMR stratospheric-mesospheric CO 2004 measurements and comparison with a GCM. Geophys. Res. Lett. **32** (2005). doi:10.1029/2005GL022433
B. Karlsson, C. McLandress, T.G. Shepherd, Inter-hemispheric mesospheric coupling in a comprehensive middle atmosphere model. J. Atmos. Sol.-Terr. Phys. **71**, 518–530 (2009)
Y. Kawatani, K. Sato, T.J. Dunkerton, S. Watanabe, S. Miyahara, M. Takahashi, The roles of equatorial trapped waves and internal inertia–gravity waves in driving the quasi-biennial oscillation. Part I: Zonal mean wave forcing. J. Atmos. Sci. **67**, 963–980 (2010a)
Y. Kawatani, K. Sato, T.J. Dunkerton, S. Watanabe, S. Miyahara, M. Takahashi, The roles of equatorial trapped waves and internal inertia–gravity waves in driving the quasi-biennial oscillation. Part II: Three-dimensional distribution of wave forcing. J. Atmos. Sci. **67**, 981–997 (2010b)
K. Kodera, H. Mukougawa, S. Itoh, Tropospheric impact of reflected planetary waves from the stratosphere. Geophys. Res. Lett. **35** (2006). doi:10.1029/2008GL034575
K. Labitzke, Sunspots, the QBO and the stratospheric temperature in the north polar region. Geophys. Res. Lett. **14**, 535–537 (1987)
K. Labitzke, On the solar cycle–QBO relationship: a summary. J. Atmos. Sol.-Terr. Phys. **67**, 45–54 (2005)
R.S. Lindzen, Turbulence and stress owing to gravity wave and tidal breakdown. J. Geophys. Res. **86**, 9707–9714 (1981)
R.S. Lindzen, J.R. Holton, A theory of the quasi-biennial oscillation. J. Atmos. Sci. **25**, 1095–1107 (1968)
M. López Puertas, M.A. López Valverde, R.R. Garcia, R.G. Roble, A review of CO_2 and CO abundances in the middle atmosphere, in *Atmospheric Science Across the Stratopause*, ed. by D.E. Siskind, S.D. Eckermann, M.E. Summers (AGU, Washington, 2000), pp. 83–100
G.L. Manney, K. Krüger, J.L. Sabutis, S. Amina Sena, S. Pawson, The remarkable 2003–2004 winter and other recent warm winters in the Arctic stratosphere in the late 1990s. J. Geophys. Res. **110** (2005). doi:10.1029/2004JD005367
G.L. Manney, K. Krüger, S. Pawson, K. Minschwaner, M.J. Schwartz, W.H. Daffer, N.J. Livesey, M.G. Mlynczak, E.E. Remsberg, J.M. Russell III, J.W. Waters, The evolution of the stratopause during the 2006 major warming: Satellite data and assimilated meteorological analyses. J. Geophys. Res. **113** (2008). doi:10.1029/2007JD009097
G.L. Manney, M.J. Schwartz, K. Krüger, M.L. Santee, S. Pawson, J.N. Lee, W.H. Daffer, R.A. Fuller, N.J. Livesey, Aura microwave limb sounder observations of dynamics and transport during the record-breaking 2009 Arctic stratospheric major warming. Geophys. Res. Lett. **36** (2009). doi:10.1029/2009GL038586
D. Marsh, A. Smith, E. Noble, Mesospheric ozone response to changes in water vapor. J. Geophys. Res. **108**(D3), 4109 (2003). doi:10.1029/2002002705
M.E. McIntyre, How well do we understand the dynamics of stratospheric warmings? J. Meteorol. Soc. Jpn. **60**, 37–65 (1982)
M.E. McIntyre, T.N. Palmer, Breaking planetary waves in the stratosphere. Nature **305**, 593–600 (1983)
J.W. Meriwether, C.S. Gardner, A review of the mesospheric inversion layer phenomenon. J. Geophys. Res. **105**, 12405–12416 (2000)
M.G. Mlynczak, S. Solomon, A detail evaluation of the heating efficiency in the middle atmosphere. J. Geophys. Res. **98**, 10517–10541 (1993)
P.W. Mote, K.H. Rosenlof, M.E. McIntyre, E.S. Carr, J.G. Gille, J.R. Holton, J.S. Kinnersley, H.C. Pumphrey, J.M. Russell III, J.W. Waters, An atmospheric tape recorder: the imprint of tropical tropopause temperatures on stratospheric water vapor. J. Geophys. Res. **101**, 3989–4006 (1996)
D.J. Murphy, T. Aso, D.C. Fritts, R.E. Hibbins, A.J. McDonald, D.M. Riggin, M. Tsutsumi, R.A. Vincent, Source regions for Antarctic MLT non-migrating semidiurnal tides. Geophys. Res. Lett. **36** (2009). doi:10.1029/2008GL037064
D. Offermann, M. Jarisch, H. Schmidt, J. Oberheide, K.U. Grossmann, O. Gusev, J.M. Russell III, M.G. Mlynczak, The "wave turbopause". J. Atmos. Sol.-Terr. Phys. **69**, 2139–2158 (2007)
S.C. Olsen, A.R. Douglass, M.R. Schoeberl, Estimating downward cross-tropopause ozone flux using column ozone and potential vorticity. J. Geophys. Res. **107** (2002). doi:10.1029/2001JD002041
J. Perlwitz, N. Harnik, Observational evidence of a stratospheric influence on the troposphere by planetary wave reflection. J. Climate **16**, 3011–3026 (2003)
J. Perlwitz, N. Harnik, Downward coupling between the stratosphere and troposphere: the relative roles of wave and zonal mean processes. J. Climate **17**, 4902–4909 (2004)

L. Pfister, Baroclinic instability of easterly jets with application to the summer mesosphere. J. Atmos. Sci. **42**, 313–330 (1985)

J.M. Picone, A.E. Hedin, D.P. Drob, A.C. Aiken, NRLMSIS-00 empirical model of the atmosphere: statistical comparisons and scientific issues. J. Geophys. Res. **107** (2002). doi:10.1029/2002JA009430

R.A. Plumb, Ibaroclinic instability of the summer mesosphere: a mechanism for the quasi-two-day wave? J. Atmos. Sci. **40**, 262–270 (1983)

Y.I. Portnyagin, J.M. Forbes, N.A. Makarov, E.G. Merzlyakov, S. Palo, The summertime 12-hour wind oscillation with zonal wavenumber $s = 1$ in the lower thermosphere over the South Pole. Ann. Geophys. **16**, 828–837 (1998)

V. Ramaswamy, M.L. Chanin, J. Angell, J. Barnett, D. Gaffen, M. Gelman, P. Keckhut, Y. Koshelkov, K. Labitzke, J.J.R. Lin, A. O'Neill, J. Nash, W. Randel, R. Rood, K. Shine, M. Shiotani, R. Swinbank, Stratospheric temperature trends: observations and model simulations. Rev. Geophys. **39**(1), 71–122 (2001)

C.E. Randall, V.L. Harvey, G.L. Manney, Y. Orsolini, M. Codrescu, C. Sioris, S. Brohede, C.S. Haley, L.L. Gordley, J.M. Zawodny, J.M. Russell, Stratospheric effects of energetic particle precipitation in 2003–2004. Geophys. Res. Lett. **32**(5) (2005). doi:10.1029/2004GL022003

C.E. Randall, V.L. Harvey, C.S. Singleton, P.F. Bernath, C.D. Boone, J.U. Kozyra, Enhanced NO_x in 2006 linked to strong upper stratospheric Arctic vortex. Geophys. Res. Lett. (2006). doi:10.1029/2006GL027160

C.E. Randall, V.L. Harvey, C.S. Singleton, P.F. Bernath, C.D. Boone, K.A. Walker, NO_x descent in the Arctic middle atmosphere in early 2009. Geophys. Res. Lett. **36** (2009). doi:10.1029/2009GL039706

W.J. Randel, et al., An update of observed stratospheric temperature trend. J. Geophys. Res. **114** (2009). doi:10.1029/2008JD010421

W.J. Randel, F. Wu, H. Vömel, G.E. Nedoluha, P. Forster, Decreases in stratospheric water vapor after 2001: links to changes in the tropical tropopause and the Brewer-Dobson circulation. J. Geophys. Res. **111** (2006). doi:10.1029/2005JD006744

A. Robock, Volcanic eruptions and climate. Rev. Geophys. **38**, 191–219 (2000)

J.P. Russell, R.P. Lowe, W.E. Ward, Atomic oxygen annual and semi-annual variations in the mesopause region for mid and equatorial latitudes. J. Atmos. Sol.-Terr. Phys. **66**, 451–461 (2004)

M.L. Salby, D.J. Shea, Correlations between solar activity and the atmosphere: an unphysical explanation. J. Geophys. Res. **96**, 22579–22595 (1991)

M. Salby, F. Sassi, P. Callaghan, D. Wu, P. Keckhut, A. Hauchecorne, Mesospheric inversions and their relationship to planetary wave structure. J. Geophys. Res. **107** (2002). doi:10.1029/2001JD001525

F. Sassi, R.R. Garcia, B.A. Boville, H. Liu, On temperature inversions and the mesospheric surf zone. J. Geophys. Res. (2001). doi:10.1029/2001JD000756

T. Schneider, The tropopause and the thermal stratification in the extratropics of a dry atmosphere. J. Atmos. Sci. **61**, 1317–1340 (2004)

T.G. Shepherd, Issues in stratosphere-troposphere coupling. J. Meteorol. Soc. Jpn. **80**, 769–792 (2002)

S.C. Sherwood, Convective precursors and predictability in the tropical western pacific. Mon. Weather Rev. **127**, 2977–2991 (1999)

D.E. Siskind, S.D. Eckermann, L. Coy, J.P. McCormack, C.E. Randall, On recent interannual variability of the arctic winter mesosphere: implications for tracer descent. Geophys. Res. Lett. **34** (2007). doi:10.1029/2007GL029293

A.K. Smith, The origin of stationary planetary waves in the upper mesosphere. J. Atmos. Sci. **60**, 3033–3041 (2003)

A.K. Smith, D.R. Marsh, Processes that account for the ozone maximum at the mesopause. J. Geophys. Res. **110** (2005). doi:10.1029/2005JD006298

A.K. Smith, K. Matthes, Decadal-scale periodicities in the stratosphere associated with the solar cycle and qbo. J. Geophys. Res. **113** (2008). doi:10.1029/2007JD009051

A.K. Smith, D.V. Pancheva, N.J. Mitchell, D.R. Marsh, J.M. Russell III, M.G. Mlynczak, A link between variability of the semidiurnal tide and planetary waves in the opposite hemisphere. Geophys. Res. Lett. **34** (2007). doi:10.1029/2006GL028929

A.K. Smith, M. Lopéz-Puertas, M. García-Comas, S. Tukiainen, SABER observations of mesospheric ozone during NH late winter 2002–2009. Geophys. Res. Lett. **36** (2009). doi:10.1029/2009GL040942

A.K. Smith, R.R. Garcia, D.R. Marsh, D.E. Kinnison, J.H. Richter, Simulations of the response of mesospheric circulation and temperature to the Antarctic ozone hole. Geophys. Res. Lett. **37** (2010a). doi:10.1029/2010GL045255

A.K. Smith, D.R. Marsh, M.G. Mlynczak, J.C. Mast, Temporal variations of atomic oxygen in the upper mesosphere from SABER. J. Geophys. Res. **115** (2010b). doi:10.1029/2009JD013434

A.K. Smith, R.R. Garcia, D.R. Marsh, J.H. Richter, WACCM simulations of the mean circulation and trace species transport in the winter mesosphere. J. Geophys. Res. (2011). doi:10.1029/2011JD016083

A.K. Smith, Longitudinal variations in mesospheric winds: evidence for gravity wave filtering by planetary waves. J. Atmos. Sci. **53**, 1156–1173 (1996)
A.K. Smith, Stationary planetary waves in upper mesospheric winds. J. Atmos. Sci. **54**, 2129–2145 (1997)
S. Solomon, K.K. Rosenlof, R.W. Portmann, J.S. Daniel, S.M. Davis, T.J. Sanford, G. Plattner, Contributions of stratospheric water vapor to decadal changes in the rate of global warming. Science **327** (2010). doi:10.1126/science.1182488
S.-W. Son, E.P. Gerber, J. Perlwitz, L.M. Polvani, N.P. Gillett, K.-H. Seo, V. Eyring, T.G. Shepherd, D. Waugh, H. Akiyoshi, J. Austin, A. Baumgaertner, S. Bekki, P. Braesicke, C. Bruhl, N. Butchart, M.P. Chipperfield, D. Cugnet, M. Dameris, S. Dhomse, S. Frith, H. Garny, R. Garcia, S.C. Hardiman, P. Jockel, J.F. Lamarque, E. Mancini, M. Marchand, M. Michou, T. Nakamura, O. Morgenstern, G. Pitari, D.A. Plummer, J. Pyle, E. Rozanov, J.F. Scinocca, K. Shibata, D. Smale, H. Teyssedre, W. Tian, Y. Yamashita, Impact of stratospheric ozone on southern hemisphere circulation change: a multimodel assessment. J. Geophys. Res. **115** (2010). doi:10.1029/2010JD014271
G.L. Stenchikov, K. Hamilton, R.J. Stouffer, A. Robock, V. Ramaswamy, B. Santer, H.-F. Graf, Oscillation response to volcanic eruptions in the IPCC AR4 climate models. J. Geophys. Res. **111** (2006). doi:10.1029/2005JD006286
G.E. Thomas, Mesospheric clouds and the physics of the mesopause region. Rev. Geophys. **29**, 553–575 (1991)
D.W.J. Thompson, S. Solomon, Interpretation of recent Southern Hemisphere climate change. Science **296**, 895–899 (2002). doi:10.1126/science.1069270
J. Thuburn, G.C. Craig, Stratospheric influence on tropopause height: the radiative constraint. J. Atmos. Sci. **54**, 17–28 (2000)
X. Tie, G.P. Brasseur, B. Briegleb, C. Granieri, Two-dimensional simulation of pinatubo aerosol and its effect on stratospheric ozone. J. Geophys. Res. **99**, 20545–20562 (1994)
U. von Zahn, J. Hoffner, V. Eska, M. Alpers, The mesopause altitude: only two distinctive levels worldwide? Geophys. Res. Lett. **23**, 3231–3234 (1996)
WMO, *Scientific Assessment of Ozone Depletion: 2010* (Global Ozone Research and Monitoring Project-Report No. 52, Geneva, Switzerland, 2010)
J. Xu, H. Liu, A.K. Smith, R.G. Roble, C.J. Mertens, J.M. Russell III, M.G. Mlynczak, Mesopause structure from thermosphere, ionosphere, mesosphere, energetics and dynamics (TIMED)/sounding of the atmosphere using broadband emission radiometry (SABER) observations. J. Geophys. Res. **112** (2007). doi:10.1029/2006JD007711
K. Yamashita, S. Miyahara, Y. Miyoshi, K. Kawano, J. Ninomiya, Seasonal variation of non-migrating semidiurnal tide in the polar MLT region in a general circulation model. J. Atmos. Sol.-Terr. Phys. **64**, 1083–1094 (2002)

Space Sci Rev (2012) 168:23–112
DOI 10.1007/s11214-012-9872-6

The Near-Earth Plasma Environment

Robert F. Pfaff Jr.

Received: 1 March 2012 / Accepted: 3 March 2012 / Published online: 20 June 2012

Abstract An overview of the plasma environment near the earth is provided. We describe how the near-earth plasma is formed, including photo-ionization from solar photons and impact ionization at high latitudes from energetic particles. We review the fundamental characteristics of the earth's plasma environment, with emphasis on the ionosphere and its interactions with the extended neutral atmosphere. Important processes that control ionospheric physics at low, middle, and high latitudes are discussed. The general dynamics and morphology of the ionized gas at mid- and low-latitudes are described including electrodynamic contributions from wind-driven dynamos, tides, and planetary-scale waves. The unique properties of the near-earth plasma and its associated currents at high latitudes are shown to depend on precipitating auroral charged particles and strong electric fields which map earthward from the magnetosphere. The upper atmosphere is shown to have profound effects on the transfer of energy and momentum between the high-latitude plasma and the neutral constituents. The article concludes with a discussion of how the near-earth plasma responds to magnetic storms associated with solar disturbances.

Keywords Ionosphere · Upper atmosphere · Electric fields

1 Introduction

In this article we discuss the near-earth plasma environment, for which the ionosphere is the natural focus. Because this ionized gas co-exists with the neutral atmospheric gas to which it is coupled, the ionized/neutral gas system is more properly considered a partially ionized gas rather than a medium consisting of two distinct, independent fluids. We discuss the origins of the ionized component and describe its general morphology and dynamics, both in the framework of the earth's magnetic field as well as with respect to its coupling

Contributed to the ISSI Monograph entitled: *"Coupling of the Earth's Atmosphere and Its Plasma Environment"*.

R.F. Pfaff Jr. (✉)
NASA/Goddard Space Flight Center, Greenbelt, MD 20771, USA
e-mail: Robert.F.Pfaff@nasa.gov

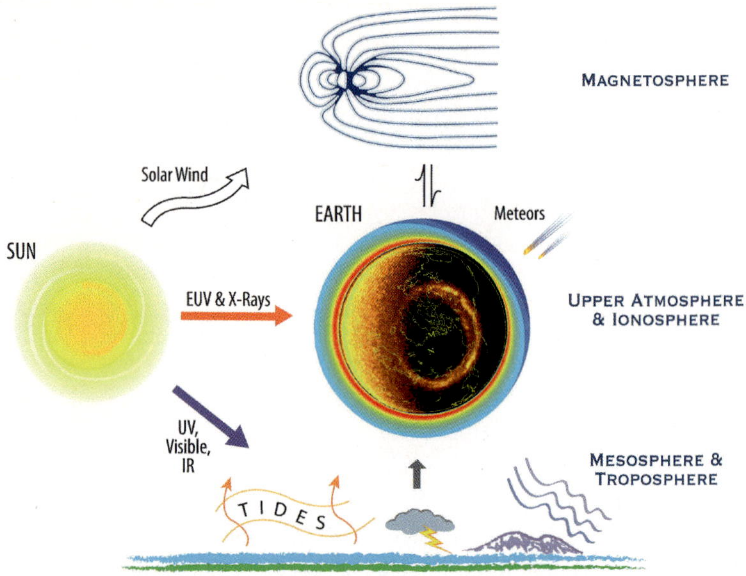

Fig. 1 The various paths of solar energy to the earth's upper atmosphere and ionosphere as well as its coupling to the magnetosphere above and the mesosphere and troposphere below

to the neutral atmosphere and its motions. Because of the emphasis of this monograph on atmosphere-plasma coupling, and in order to keep the article manageable, we do not include a detailed discussion of the earth's magnetosphere plasma, except in its role of exchanging energy and momentum with the partially ionized atmosphere at its base.

We discuss two main mechanisms by which ambient or "thermal" plasma is created in the near-earth environment, each of which depends on an interaction of an external energy source with the earth's extended atmosphere: (1) photo-ionization of the neutral gas via impinging extreme ultraviolet (EUV) solar radiation and (2) impact ionization of the neutral gas via energetic particle precipitation. A second form of impact ionization, due to meteor ablation, is also briefly discussed. Whereas photo-ionization is responsible for the creation of the primary part of earth's ionosphere, particularly at mid- and low-latitudes, at high latitudes impact ionization by precipitating, energetic particles creates thermal plasma with different properties and features. Although their ionizing mechanisms are driven by different energy sources, the resulting plasma in each case is a consequence of energy or momentum impact with the upper atmosphere itself. Indeed, as we shall see, it is precisely *because* the earth has an extended atmosphere that it also has an ionosphere. In fact, any planet with a sensible, gravitationally-bound atmosphere and an ionizing source of radiation of sufficient energy and intensity, will consequently form an ionosphere. For planets which also have an intrinsic magnetic field and a substantial rotation rate, such as the earth, the resulting plasma environment may be expected to be particularly dynamic and rich in its complexity and physical processes.

In order to illustrate the context of the near-earth plasma within the larger sun-earth system, we depict the main pathways by which energy and momentum are deposited and exchanged within the earth's upper atmosphere in Fig. 1. Here, one path shows how the sun directly influences the upper atmosphere via EUV and X-ray radiation, both via heating of the neutral atmosphere, which results in its expansion, and also via photo-ionization of a small fraction of this same neutral gas which thereby re-generates the ionosphere throughout the day. A second path shows how solar wind processes, including plasma flow with embedded interplanetary magnetic fields, interact with the magnetosphere to drive large-scale electric fields, currents, and energetic particles that subsequently interact with the underlying upper atmosphere/ionosphere. Such solar wind/magnetospheric input impacts the near-earth plasma primarily at high latitudes, although it may dramatically influence all latitudes during periods of magnetic storms. Although the globally averaged energy input by the solar wind/magnetosphere is not as high as the total solar photon irradiance, its impact in the upper atmosphere is nonetheless quite significant, and locally it can be the dominant energy source. The diagram in Fig. 1 also includes a sketch showing that the upper atmosphere is also influenced by energy and momentum sources from below, namely from electrical discharges in the troposphere (lightning) and mesosphere (e.g., sprites), as well as via upward-propagating tides, planetary waves, and gravity waves. Such sources have profound effects on the characteristics of the mid- and low-latitude ionosphere in particular. The energy source that heats and sets in motion the lower and middle atmosphere is, again, solar radiation, but in this case, its ultraviolet, visible, and infrared wavelengths.

An outline of this article is as follows: We begin with a description of how the earth's ionosphere is formed via photo-ionization, and its daily variation. We then discuss how these ionized particles are influenced by the earth's magnetic field and how they couple to the neutral atmospheric gas via collisions, as represented by the mobility tensor for electrons and ions. The dynamics and morphology of the ionosphere at low and mid-latitudes are then discussed, followed by a separate section on the earth's high latitude ionosphere including the aurora and the exchange of energy and momentum with the magnetosphere. Finally, we discuss the effects of magnetic storms on the ionosphere and conclude with a brief summary.

There are several excellent references that describe the earth's ionosphere and upper atmosphere. These include: Hargreaves (1992), Kelley (2009), Prölss (2004), Rishbeth and Garriott (1969), and Schunk and Nagy (2009), among others.

2 Creation of the Earth's Ionosphere via Photo-ionization and Its Daily Variation

In this section, we describe how the ionosphere is created by photo-ionization associated with solar EUV and X-ray radiation that impacts the earth's tenuous upper atmosphere. We also describe how some positive ions thus produced may transfer their charge to different neutral species and how others subsequently recombine with free electrons to form neutral gas particles. We conclude with a general description of how the mid- and low-latitude ionospheric density and temperature vary with local time and altitude.

2.1 The Earth's Extended Neutral Atmosphere

In the Introduction, we discussed how the external sources of energy and momentum impinge on the earth's atmosphere to create thermal plasma via two main mechanisms: solar radiation induced photo-ionization and impact ionization from energetic particles and meteors. Before discussing these ionization processes, we first review some of the fundamental features of the Earth's upper neutral atmosphere in order to illustrate how they influence the resulting plasma parameters formed by these ionizing mechanisms.

Whereas the earth's atmosphere near the surface consists primarily of nitrogen molecules (N_2) and oxygen molecules (O_2), the atmospheric constituents change markedly at higher

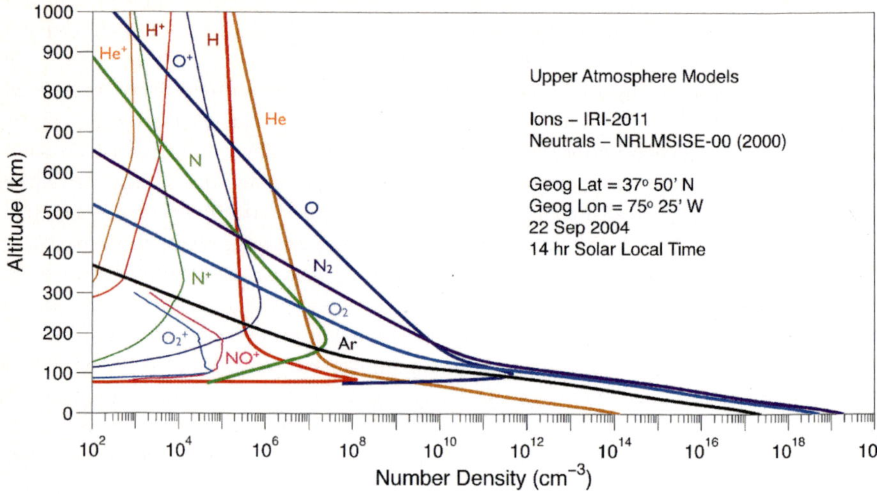

Fig. 2 Model distributions of neutral and ionic constituents as a function of altitude for a mid-latitude location (Wallops Island, Va., USA) at 14 hr Solar Local Time. Ion distributions below 80 km are not provided by the model

altitudes. At about 100 km and above, the atmosphere ceases to be homogeneous and converts to free-molecular flow with a large mean-free path of the particles. Additionally, the atmospheric constituents above about 100 km directly absorb solar EUV radiation. Because of this absorption and the fact that these gases are very ineffective infrared radiators, the neutral temperatures increase significantly with altitude above 100 km. Consequently, this region of the upper atmosphere is often referred to as the thermosphere.

The main process by which the upper atmosphere absorbs EUV energy from the sun is photo-dissociation. Diatomic oxygen is the primary molecule in the upper atmosphere that dissociates. This occurs when it is struck by photons with wavelengths of 242 nm or less, as expressed by the reaction:

$$O_2 + h\nu(\lambda < 242\ \text{nm}) \rightarrow O + O + \text{energy}$$

The creation of atomic oxygen is important because it is readily photo-ionized, as discussed later.

The various species of atoms and molecules that make up the atmosphere are shown in Fig. 2, for mid-latitude, daytime conditions with moderate solar activity, as computed from the NRLMSISE-2000 model (Picone et al. 2002). The concentrations per unit volume of the atmospheric constituents and their variation with altitude affect the ionosphere in two ways: First, they determine the amount of neutral gas available for photo-ionization at different altitudes, and second, they determine to what depth the solar EUV and X-ray radiation might penetrate, as discussed in the next section. For example, a consequence of a decreased neutral atmospheric density is that photons can penetrate to lower altitudes.

Both the neutral and plasma densities in the earth's upper atmosphere vary considerably because of EUV solar radiation variations associated with solar activity. Unlike the main portion of the sun's emission spectrum, for example where the visible light is emitted, the EUV portion of the spectrum originates at the base of the highly dynamic corona, where it undergoes large variations in accordance with solar storms and solar activity. In fact, the EUV radiation typically changes by factors of 2–3 (sometimes larger) within the solar cycle.

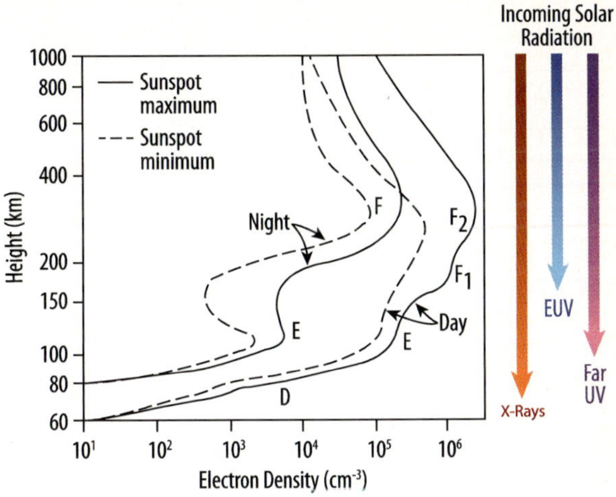

Fig. 3 Representative plasma density profiles for day and night at solar minimum and maximum conditions at a mid-latitude location (Hargreaves 1992). The colored arrows show the approximate penetration depths of the X-ray, EUV, and UV radiation from the sun

The EUV radiation is doubly important for determining the characteristics of the ionosphere as it: (1) is absorbed by the upper atmosphere and thereby heats the atmosphere, causing it expand, as discussed above, and (2) photo-ionizes a small fraction of the gas, creating the ionosphere, as discussed below.

2.2 Photo-Ionization, Charge Exchange, and Recombination

The main mechanism that creates the ionosphere is the photo-ionization of a tiny fraction of the earth's extended, tenuous neutral upper atmosphere during the daytime. During the night, there is a very small amount of radiation input resulting from the high altitude scattering of the solar radiation into the shadow of the planet and other minor sources. Incoming solar EUV and X-ray radiation penetrate the upper atmosphere with sufficient energy to detach an electron from a neutral atom to create an ion and electron pair, as depicted here for oxygen, the main atmospheric gas ionized in this manner:

$$\mathrm{O} + h\nu \rightarrow \mathrm{O}^{+} + \mathrm{e}^{-}$$

This radiation penetrates to different altitudes based on wavelength. Enhanced plasma layers are formed where the intensity of the incoming radiation, which falls off with decreasing altitude, and the amount of a particular gas constituent available to be ionized, which falls off with increasing altitude, are optimal, as discussed by Chapman (1931). For the earth, two main layers of ionization are created in the daytime ionosphere due to the photo-ionization of oxygen atoms: one formed by soft X-rays around 110 km, denoted as the E region and one formed by EUV radiation near 150 km denoted as the F1 region.

The resulting ionization plotted as a function of altitude is shown in the left hand side of Fig. 3, for a mid-latitude location (Hargreaves 1992). Note the strong dependence of the resulting plasma density on solar activity, which directly affects the strength of available EUV and X-ray radiation as discussed above, as well as how the profiles differ between day and night, particularly at the lower altitudes (discussed below). The incoming ionizing radiation is represented schematically on the right hand side of the figure. The actual penetration altitudes at a given latitude vary during the day with the solar zenith angle. For large solar zenith angles, grazing incidence corrections must be taken into account. (See discussion in Rishbeth and Garriott 1969 and references therein.)

The number density distributions of the different ion species which constitute the ionosphere are shown in Fig. 2, taken from the International Reference Ionosphere (IRI) empirical model (Bilitza and Reinisch 2008; Bilitza et al. 2011), for the same mid-latitude, daytime conditions used for the neutral density profiles shown in this figure. The curves show how the various ion constituents change with altitude. Indeed, the distribution of ions within the ionosphere is controlled to a large extent by the distributions of the neutral gas constituents. When comparing the ionized density profile with that of the neutral density profile shown in Fig. 2, notice that, for altitudes below about 300 km, the plasma number densities are 10^4 to 10^6 times *smaller* than those of the neutral parameters.

It is important to emphasize that photo-ionization is sometimes only the first step in a chain of processes by which the ambient ion gases are ultimately formed. For example, oxygen atoms are readily photo-ionized while atmospheric molecules, such as O_2 and NO, do not photo-ionize nearly as efficiently. However, charge exchange readily occurs between the oxygen ions and the ambient nitrogen and oxygen molecules (N_2 and O_2), creating NO^+ and O_2^+:

$$O^+ + N_2 \rightarrow NO^+ + N$$
$$O^+ + O_2 \rightarrow O_2^+ + O$$

At the lower altitudes, the resulting molecular ions are the dominant ones, even though the initial ionization is produced by photo-ionization of oxygen atoms. This is an important aspect of the near-earth plasma because the molecular ionization is lost quickly through recombination when their ionization source is cut off.

As soon as the positive ions and electrons are freed, they are subject to chemical recombination processes as well as charge exchange and transport processes. The most common of these processes is dissociative recombination, which is important for the E-region molecular ions:

$$O_2^+ + e \rightarrow O + O$$
$$NO^+ + e \rightarrow N + O$$

This is a rather fast process in which newly formed O_2^+ and NO^+ typically recombine in seconds. In fact, the loss rate for the molecular species, L, is given by: $L = C_r N_i N_e$, where C_r is the recombination coefficient, N_i is the ion density, and N_e is the electron density. Since the loss rate depends on the ion density, if the molecular species is the dominant ion, the loss rate is very rapid. On the other hand, O^+ recombines via a radiative process:

$$O^+ + e \rightarrow O + h\nu$$

which is very slow. In fact, O^+ is far more likely to be lost via charge exchange or other processes rather than through recombination. Since the molecular ions dominate the lower altitudes (i.e., less than 200 km), when the direct illumination of the sun vanishes at dusk, the lower ionosphere, which rotates with the planet, becomes largely depleted of any significant plasma until local sunrise at which time the main photo-ionization process starts over again. In this fashion, we see that at the lower altitudes, the ionosphere is continuously being reborn every day.

The physical processes and chemistry associated with the ion and neutral gas constituents, with particular emphasis on ionization and recombination, constitutes the general domain of *aeronomy*, the science of the upper atmosphere. A detailed discussion of these processes may be found in Rishbeth and Garriott (1969), Banks and Kockarts (1973), Prölss (2004), and numerous other texts.

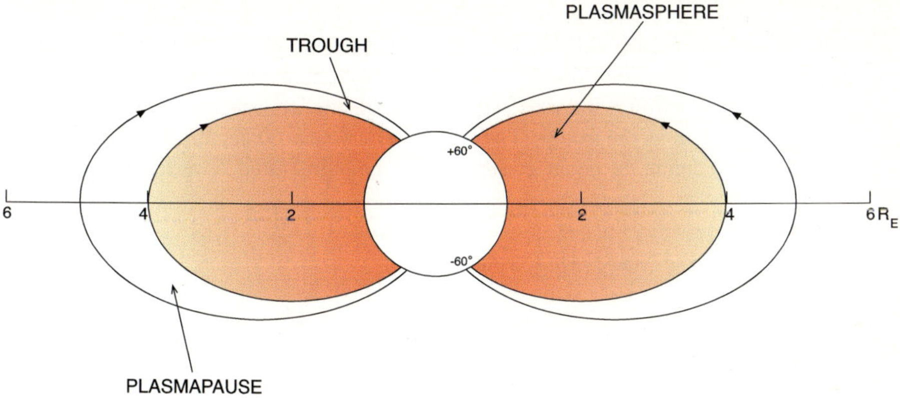

Fig. 4 The plasmasphere, an extension of the ionosphere at mid and low latitudes. The plasmasphere is confined to closed magnetic field lines and extends to approximately 4 earth radii at the equator

2.3 The F2 Peak, Protons, and the Plasmasphere

Returning to Fig. 3, we see that the majority of the ionospheric plasma peaks at altitudes above the F1 layer, known as the F2 peak. Although this layer dominates the ionosphere, it is not formed primarily by direct photo-ionization occurring at this altitude, but instead is the result of other, subsequent processes, including for example, the upwards diffusion of O^+ ions to higher altitudes where they do not encounter significant numbers of molecules with which to charge exchange. In addition, neutral wind drag-driven ion flows are believed to be another significant process that contributes to the F2 ionization peak properties at mid- and low-latitudes, as discussed by Rishbeth (1967). Finally, upward plasma transport due to electric fields, particularly during the daytime, also produces changes in the F2 peak, as discussed in Sect. 4.

Note further in Fig. 2 that the lighter hydrogen ion, H^+, does not become dominant until much higher altitudes. Within the "topside" ionosphere (i.e., above the F2 peak), charge exchange between oxygen ions and hydrogen atoms is common, accounting for all of the protons of ionospheric origin at mid and low latitudes:

$$O^+ + H \rightarrow H^+ + O$$

For moderate solar activity, at altitudes above about 1000 km, the H^+ ion dominates, and this region is often referred to as the protonosphere. The amount of H^+ present varies considerably between day and night and is also highly dependent on solar activity. For example, Heelis et al. (2009a, 2009b) showed that during extreme solar minimum conditions, the crossover altitudes of O^+ and H^+ ions could be as low as 450 km at night. He^+ also increases in importance with altitude, although, except possibly during geomagnetic storms, it is typically minor compared to H^+.

The large concentrations of light ions extend to altitudes of about 4 earth radii at the equator, as shown in Fig. 4. This high-altitude region is called the plasmasphere and is basically a torus that surrounds the mid- and low-latitude portions of the earth and is considered an upward extension of the ionosphere at magnetic latitudes below about 60 degrees. The plasmaspheric plasma is trapped on closed magnetic field lines that rotate with the Earth and thus the plasmasphere co-rotates with them. (The earth's magnetic field is discussed in the next major section.) The outer plasmasphere is composed mostly of hydrogen ions with singly ionized helium accounting for approximately 10–20 % of the plasma (e.g., Craven et al. 1997) and has temperatures of roughly 1 eV (or 11,600 K). The plasmasphere terminates at a sharp decrease in density known as the plasmapause. At lower altitudes, it is sometimes associated with a sharp decrease in the topside ionosphere with increasing latitude that is referred to as the trough. The trough occurs near 60 degrees magnetic latitude, and is also used to define the equatorward edge of the high-latitude ionosphere (see review by Rodger 2008).

2.4 Temperature of the Near-Earth Plasma

In addition to its number density, the near-earth plasma is also characterized by its temperature. For the photo-ionization process described above, when an atom absorbs a photon and emits an electron, the resulting photoelectron typically has an energy of several eV. This energy then dissipates via inelastic collisions with the neutral gas. The electron has a higher velocity than that of the ion at the same temperature due to its much smaller mass and the conservation of momentum. The resulting temperature distribution of each species depends on their collisions with other neutral and ionized particles and thus varies with altitude and geographic position.

Typical average values of electron and ion temperatures vary between 350–3000 K, or 0.03–0.3 eV, as shown in Fig. 5 for conditions at the magnetic equator at dawn, noon, sunset, and midnight as computed from the IRI model. In the lower ionosphere (<120 km), the average values of the electron, ion, and neutral gas temperatures (T_e, T_i and T_n) are similar due to the increased number of collisions with the over-abundant neutrals. The values of T_e, T_i, and T_n then diverge with increasing altitude. The electron temperature is highest in the early morning. The neutrals, rotating into sunlight, have yet to warm and the neutral atmosphere is still contracted from the colder, nightside temperature. Hence, the electron-neutral collisions are less frequent. During the day, the neutral atmosphere expands, and the electron-neutral collision frequency thus increases at the higher altitudes, reducing the

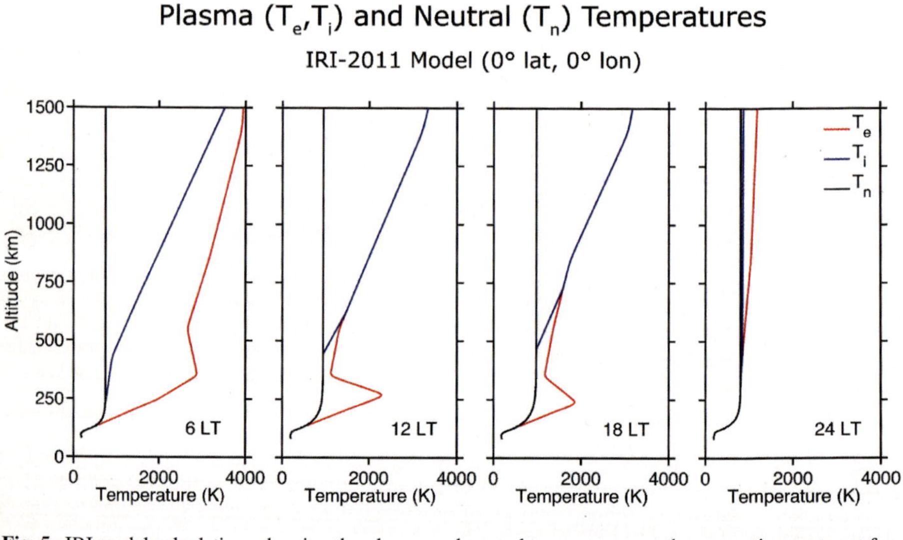

Fig. 5 IRI model calculations showing the plasma and neutral temperatures at the magnetic equator at four different local times

electron temperature. The electron temperature further decreases during the day as it equilibrates with the ion temperature due to electron-ion collisions. The temperatures are largest at the higher altitudes where the neutral and plasma densities are low. The enhanced electron temperature near 300 km during the day results from photoelectrons produced within the upper altitude portion of the layer where the neutral atmosphere is less dense. At night, there are no newly-formed photoelectrons and the electron, ion and neutral temperatures quickly converge to similar values via collision processes. See Schunk and Nagy (2009) for a detailed discussion of the neutral and plasma temperature variations.

At higher latitudes, precipitating particles, magnetospheric electric fields, and electric currents also play significant roles in heating the ionosphere. Frictional heating (Joule heating) between the ions and neutrals in particular contributes significantly to ion and neutral heating at high latitudes, and plasma instabilities at low altitudes (e.g., two-stream instabilities, particularly in the auroral zone) may produce elevated electron temperatures locally.

2.5 Metallic Ions

Another source of ions to the earth's near-earth plasma results from the influx of meteoroids. The earth is laden by $\sim 10^8$ kg/year of meteoritic material (e.g., Ceplecha 1992), depositing dust, grains, and atoms released by ablation. As these particles settle due to gravity, metallic atoms may undergo charge exchange to form metallic ions such as Fe^+, Mg^+, and Na^+, and photo-ionization may also occur (Kopp 1997).

The main source of metallic ions in the ionosphere, however, is impact ionization by meteors striking the atmosphere. This mechanism directly produces ionized gases at the meteor head, which occurs when the kinetic energy of the incoming meteoroid is sufficient to remove electrons from the neutrals (e.g., Vondrak et al. 2008). Whether a meteoroid will undergo impact ionization depends on its mass, velocity, and angle of attack with respect to the atmosphere. Such ionization is observed through radar echoes of meteor trails, as they interact with the atmospheric gas primarily between 80–110 km, forming cylinders of ionizing gas often orders of magnitude denser than that of the surrounding plasma (e.g., Dyrud et al. 2001; Close et al. 2002; Janches et al. 2009). An example of such radar "head echoes" associated with the ionized gas at the head of an ablating meteor is shown in Fig. 6 (Chapin and Kudeki 1994). Notice in particular how the head echo extends for over 15 km in altitude. The head echoes are followed by a radar trail of plasma irregularities with organized scatter that, in this case, lasted for several minutes as they were formed near the magnetic equator. Such head echoes are commonly detected by ionospheric radars at all latitudes and illustrate the prevalence of meteor-produced ionization at low altitudes (e.g., Mathews et al. 2008).

The glowing gas associated with ablation of larger meteoroids produces the familiar "shooting stars" that are observed in the night sky at all latitudes and which are most prevalent in the pre-dawn hours when the earth's proper motion directly sweeps through the interplanetary medium. For such emissions to be visible to the unaided eye, the mass of the meteoroid is generally on the order of milligrams, whereas "radar" meteors have considerably smaller masses, on the order of 1–100 micrograms, and their ablation optical signature is generally not visible to the unaided eye. Ablation of these smaller meteoroids are much more frequent and are believed to be a more significant source of metallic ions, at least compared to that produced by the larger meteoroids (D. Janches, personal communication, 2012).

Metallic ions in the earth's ionosphere have recombination lifetimes for neutralization that are very long. Thus, the long lasting metallic ions generally undergo subsequent motions after they are created. For example, as the metallic ions settle in the lower portion of

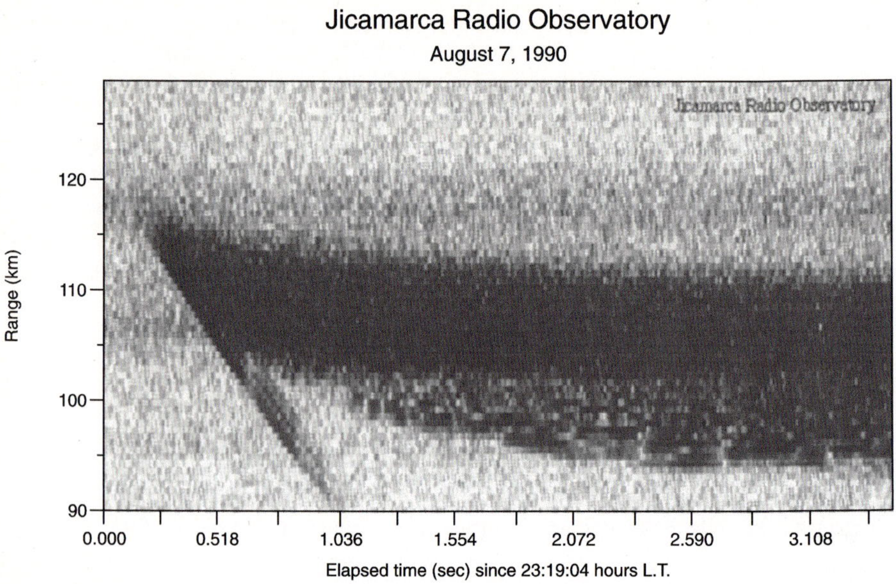

Fig. 6 Example of radar backscatter measurements of a meteor ablating in the upper atmosphere. The *diagonal line* on the left hand side corresponds to the head echo, which creates a plasma density envelope as the meteoroid penetrates the atmosphere. The broad scattering region to the right of the diagonal line corresponds to plasma irregularities in the meteor trail (Chapin and Kudeki 1994)

the upper atmosphere, their distributions are transformed by neutral wind shears to form "sporadic-E" layers, as discussed in Sect. 4, while others contribute to charged aerosols that are responsible for strong radar echoes at lower altitudes in the mesosphere. The long-lived metallic ions may also be transported over long distances via upper atmospheric winds and electric fields and have been measured by satellites at altitudes as high as 1000 km (Grebowsky and Aiken 2002), far from the altitudes where ablation is greatest—near and below 100 km.

2.6 Daily Variation of the Mid- and Low-Latitude Ionosphere

Prior to discussing the coupling of the ionosphere to the neutral gas and its dynamics, model computations are provided to illustrate how the ionosphere varies, at least in general sense, with both local time and altitude. Such a representation is provided in Fig. 7 for both solar minimum and solar maximum conditions. These computations are from the 2011 International Reference Ionosphere (IRI) as discussed in Bilitza et al. (2011), and show conditions for all local times along the geomagnetic equator at the equinox.

The model ionosphere in Fig. 7 shows how the plasma densities are enhanced during solar maximum conditions for all local times, with the peak of ionization (the F2 peak) moving to significantly higher altitudes compared to solar minimum conditions. For both solar minimum and solar maximum, the ionosphere is considerably more dense at higher altitudes (>250 km) for most of the day and well into the night, exhibiting its lowest plasma number densities before dawn. The plasma density at the lower altitudes (<200 km) is very low during the night due to recombination losses of the dominant molecular ions. The apparent decrease in density of the base of the F-region after sunset, which is most pronounced during solar maximum, is due to vertical transport and is discussed in Sect. 4.

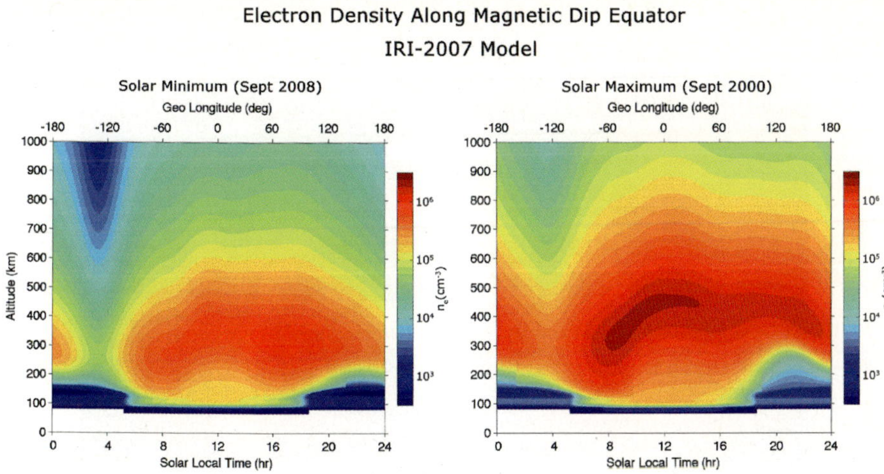

Fig. 7 IRI model calculations of electron density along the magnetic dip equator for 24 hours of local time at solar minimum (*left*) and solar maximum (*right*) conditions at the equinox

Another representation of this same equatorial plasma density distribution is shown in Fig. 8, in which the density is shown in a global distribution around the planet along the magnetic equator, again computed from the IRI. The panel depicts the ionosphere to 2000 km and is to scale with the earth's diameter. Seen as a narrow black circle, a very thin ionosphere-earth cavity (i.e., between the lower edge of the ionosphere and the earth's surface) is present, whose dimensions increase at night as the lower ionosphere goes away due to recombination effects, as discussed above. This cavity is particularly important for the excitation of Schumann resonances (low frequency electromagnetic waves that propagate around the globe) and other phenomena, as discussed by Simoes et al. (this issue).

To illustrate the diurnal variations of the ionosphere at a single location, the *measured* plasma density from 50–500 km is plotted for four consecutive days in Fig. 9 as observed by the Arecibo observatory in Puerto Rico (Figure courtesy N. Aponte). The day numbers at the bottom of the plot correspond to local noon. Clearly, the data show that the lower altitude (<250 km) ionosphere turns on very quickly after sunrise and then it recombines abruptly at sunset. During the night, the ionospheric density is hence substantially reduced at altitudes up to about 250 km, where the lower edge of the F-region can be very "ragged", influenced perhaps by transport effects. During the day, the F-region plasma density is fairly uniform, extending into nighttime where it begins to decrease at the higher altitudes after midnight. The very faint, thin lines visible at night near 110 km are due to sporadic-E layers and will be discussed further on below.

Finally, we show IRI model computations illustrating how the plasma temperature varies with local time and altitude in Fig. 10. Note how the electron temperature increases dramatically in the early morning, as discussed above.

3 Coupling of the Near-Earth Plasma and the Neutral Atmosphere Gases

In this section, we first describe how the ionized gases are organized by the ambient magnetic field and then discuss how the ionospheric charged constituents are coupled to the neutral gas via collisions, which vary markedly with altitude. We describe how this coupling is

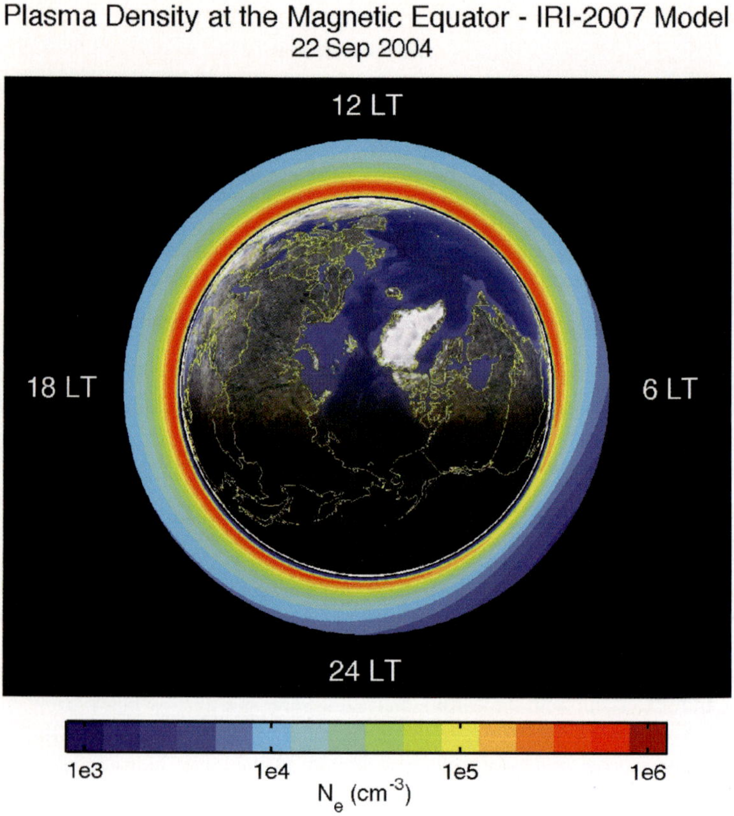

Fig. 8 IRI model calculations of the plasma density at the magnetic equator, presented in the altitude region scaled to the earth's radius

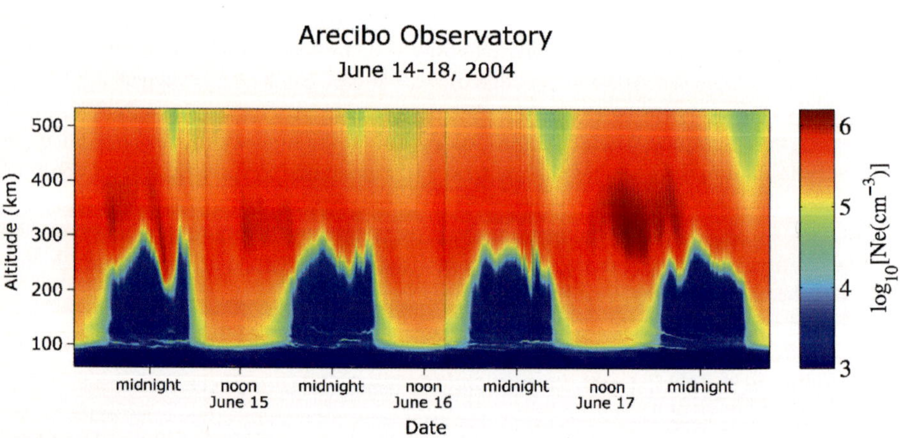

Fig. 9 Incoherent scatter measurements of the plasma density for four consecutive days at the Arecibo Observatory in Puerto Rico (Figure courtesy of N. Aponte)

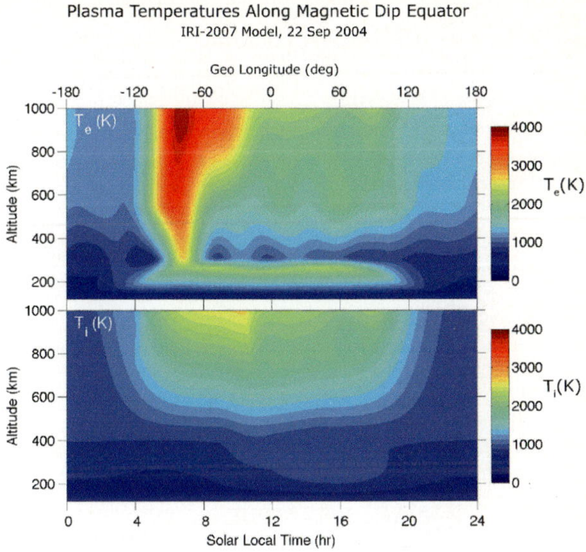

Fig. 10 IRI model calculations of the electron and ion temperatures along the magnetic equator for a 24 hour period

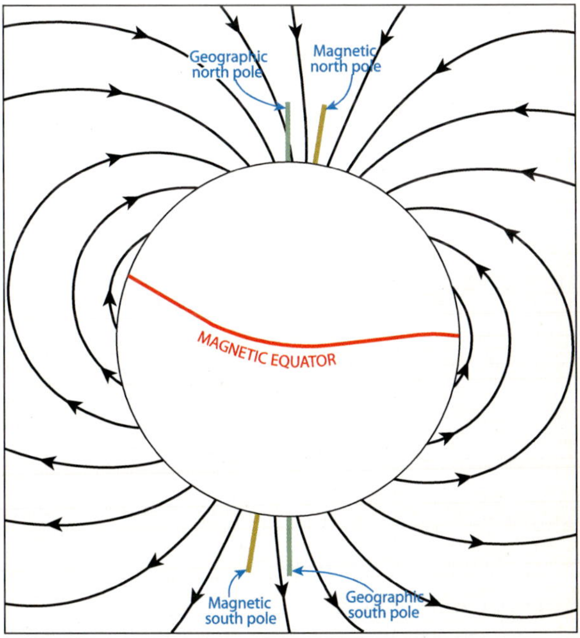

Fig. 11 The Earth's dipole magnetic field. The field lines are nearly vertical at very high latitudes yet nearly horizontal at low latitudes. The magnetic field line geometry has profound effects on the near-earth plasma. The magnetic poles are offset from the geographic poles

represented by the mobility and conductivity tensors, which form the basis of a large number of physical processes that govern the behavior of the partially ionized gas environment.

3.1 The Earth's Magnetic Field

Unlike the neutral atmospheric gases, the ions and electrons that comprise the near-earth plasma are compelled by Lorentz forces to gyrate about the earth's magnetic field lines.

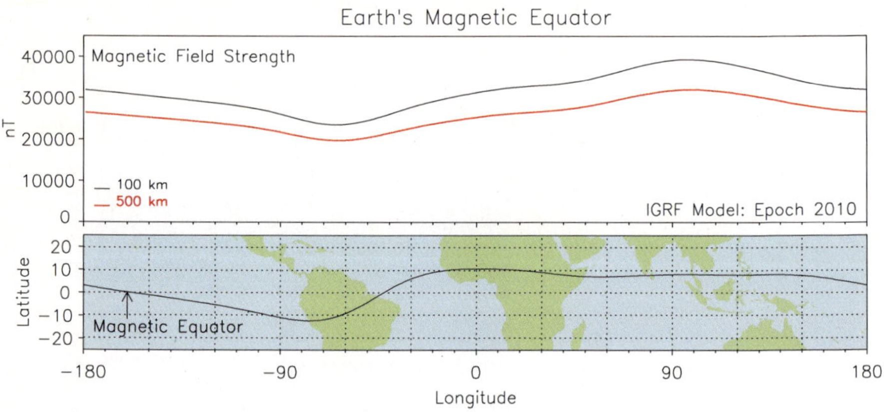

Fig. 12 The *lower panel* shows the location of the earth's magnetic equator. The *upper panel* shows the model magnetic field strength along the magnetic equator at 100 km and 500 km altitude

Thus, an appreciation of the earth's magnetic field is essential to understand the behavior of the near-earth plasma.

The Earth's magnetic field is essentially that of a dipole whose axis is tilted with respect to the earth's axis of rotation by approximately 10 degrees, as shown in Fig. 11. Consequently, the magnetic equator is also offset from the geographic equator. This figure shows how the main field lines are nearly vertical at high latitudes, stretching into the deep magnetosphere, yet remain closer to the earth, subtending much smaller angles with respect to the horizontal below about 45 degrees latitude. Indeed, at latitudes of roughly 60 degrees and below, the field lines "close", meaning that they extend to the conjugate hemisphere at roughly the same magnetic latitude. This condition strongly influences the near-earth plasma, as we shall show.

The magnetic field strength at 100 km altitude near the poles is $\sim$0.65 G, and decreases to about half that value at the magnetic equator, which is defined by the location on the earth where the magnetic field vector is strictly horizontal. In addition to being tilted, the dipole is slightly distorted and hence the magnetic field strength near 100 km altitude may be as low as 0.25 G in the Peruvian sector but near 0.4 G in the Indian sector at the magnetic equator. These differences strongly influence currents and longitudinal effects of plasma flow. Figure 12 shows the geographic location of the model magnetic equator, as well as the strength of the magnetic field at the equator. The declination of the magnetic field at the equator is generally perpendicular to the magnetic equatorial plane. This declination introduces longitudinal influences on plasma drifts and can also be important for magnetic flux tubes whose "footpoints" in the northern and southern hemispheres may be significantly displaced in longitude.

As we shall show, the earth's intrinsic magnetic field has a profound effect on the plasma with respect to directing $\mathbf{E} \times \mathbf{B}$ drifts and currents, as well as connecting the near-earth plasma with energy and momentum sources in the distant magnetosphere.

3.2 Coupling of the Ionized and Atmospheric Gases: Collisions and the Mobility and Conductivity Tensors

Overview. We demonstrated in the previous section how the ionized and atmospheric gases co-exist in the same volume particularly in the altitude range of 90–1000 km. We now discuss how the ionized and neutral gases are coupled, which sets the stage for later sections

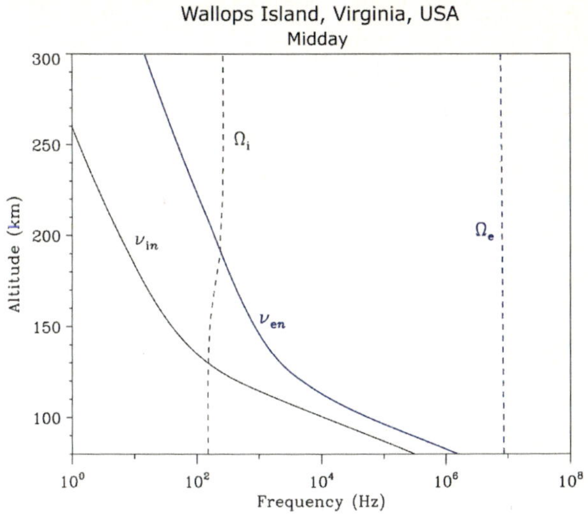

Fig. 13 Altitude profiles of the electron-neutral collision frequency and the ion-neutral collision frequency. Profiles of the mean ion gyrofrequency and electron gyrofrequency are also shown

where we show how the partially ionized fluid responds to different forcing functions, primarily those due to electric fields and winds. In particular, we demonstrate that, although the ionized component is controlled by the earth's magnetic field, both the ionized and neutral gas motions are subject to collisional interactions between the ions and neutrals that vary with altitude. These interactions are specified by the mobility and conductivity tensors, as shown below.

Gyro and Collision Frequencies. Charged particles in the ionosphere gyrate about the ambient magnetic field lines at their cyclotron frequencies, Ω_j, given by

$$\Omega_j = e|\mathbf{B}|/m_j \tag{1}$$

where e is the electric charge, $|\mathbf{B}|$ is the magnetic field strength, and m_j is the mass of the species which is designated by j. For reference, the gyro frequency for O^+ ions in a magnetic field of 0.5 Gauss is 48 Hz, whereas the gyro frequency for electrons is 1.4 MHz for the same magnetic field strength.

Since the upper atmospheric gases are dense (compared to the plasma) and only weakly ionized, collisions between ion and neutrals significantly influence the individual and bulk motions of the plasma. Because of these collisional effects, the statistical collective behavior of the particles may be defined using the equations of hydrodynamics. Various expressions have been put forward for the ion-neutral collision frequency, ν_{in}, and the electron-neutral collision frequency, ν_{en}, that depend on the neutral density, neutral and ion masses, and the temperatures (e.g., Banks and Kockarts 1973).

For the lower altitude regions of the ionosphere, model ion-neutral and electron-neutral collision frequency profiles are shown in Fig. 13 for a mid-latitude location at mid-day. It is clear that below 130 km, the ions are collision dominated by the atmosphere whereas the electrons are able to execute their gyro motions about the magnetic fields and consequently undergo $\mathbf{E} \times \mathbf{B}$ drifts across magnetic field lines. The lower ionosphere, where $\nu_{in}/\Omega_i > 1$ and $\nu_{en}/\Omega_e \ll 1$, constitutes the dynamo region where strong currents flow, as discussed in the next section.

Mobility and Conductivity Tensors. Because the plasma distribution function in the presence of a magnetic field is highly anisotropic, we define a mobility tensor, $\overline{\overline{\boldsymbol{\mu}}}$, which relates the velocity (**V**) and force (**F**) vectors:

$$\mathbf{V} = \overline{\overline{\boldsymbol{\mu}}} \cdot \mathbf{F} \tag{2}$$

where

$$\overline{\overline{\boldsymbol{\mu}}} = \begin{bmatrix} \mu_1 & \mu_2 & 0 \\ -\mu_2 & \mu_1 & 0 \\ 0 & 0 & \mu_0 \end{bmatrix} \tag{3}$$

This tensor has been defined in a geometry such that the magnetic field, **B**, is parallel to the z-axis. For single particle motion subject to an applied force, we have the following mobilities:

$$\mu_{oj} = 1/m\nu_{jn} \quad \text{(parallel mobility)}$$
$$\mu_{1j} = (1/m\nu_{jn})(\nu_{jn})^2/\left(\Omega_j^2 + \nu_{jn}^2\right) \quad \text{(Pedersen mobility)}$$
$$\mu_{2j} = (1/m\nu_{jn})(\nu_{jn}\Omega_i)/\left(\Omega_j^2 + \nu_{jn}^2\right) \quad \text{(Hall mobility)}$$

The profiles of these three mobility components are plotted in Fig. 14 for three different locations, one at the magnetic equator (Jicamarca, Peru), one at high latitudes (Poker Flat, USA), and one at mid-latitudes (Wallops Island, USA). The mobilities shown here are combined for ions and electrons, and use the mean molecular weight for the ions. Note that the Pedersen mobility peaks at around 125 km and the Hall mobility peaks near 105 km. The mobilities are higher at the lower latitudes due to the weaker magnetic field strength, since they are roughly proportional to $\sim 1/\Omega_i^2$.

We now define a conductivity tensor, $\overline{\overline{\boldsymbol{\sigma}}}$, that relates the electric field and current density:

$$\mathbf{J} = \overline{\overline{\boldsymbol{\sigma}}} \cdot \mathbf{E} \tag{4}$$

which can also be expressed in terms of the difference of the ion and electron motions:

$$\mathbf{J} = Ne(\mathbf{V}_i - \mathbf{V}_e), \tag{5}$$

where N is the plasma density and e is the electric charge.

The conductivity may be expressed in terms of the mobilities such that $\boldsymbol{\sigma} = Ne^2(\boldsymbol{\mu}_i + \boldsymbol{\mu}_e)$. The components of the conductivity are thus:

$$\sigma_o = Ne^2(\mu_{oe} + \mu_{oi}) \quad \text{(parallel conductivity)}$$
$$\sigma_1 = Ne^2(\mu_{1e} + \mu_{1i}) \quad \text{(Pedersen conductivity)}$$
$$\sigma_2 = Ne^2(\mu_{2e} - \mu_{2i}) \quad \text{(Hall conductivity)}$$

The ionospheric conductivity depends linearly on the ambient plasma density and the coupled neutral and ionized gases via the anisotropic mobilities as organized by the magnetic field. Equation (4) may now be expressed more fully:

$$\mathbf{J} = \sigma_1 \mathbf{E}'_{\text{perp}} + \sigma_2 \mathbf{b}' \times \mathbf{E}'_{\text{perp}} + \sigma_o \mathbf{E}_{\text{par}} \tag{6}$$

where $\mathbf{b}'$ is the unit vector along the magnetic field direction and $\mathbf{E}'_{\text{perp}}$ is the perpendicular electric field in the earth-fixed frame, such that $\mathbf{E}'_{\text{perp}} = \mathbf{E}_{\text{perp}} + \mathbf{U} \times \mathbf{B}$, where **U** is the neutral wind. This expression shows that if there is an electric field perpendicular to the magnetic

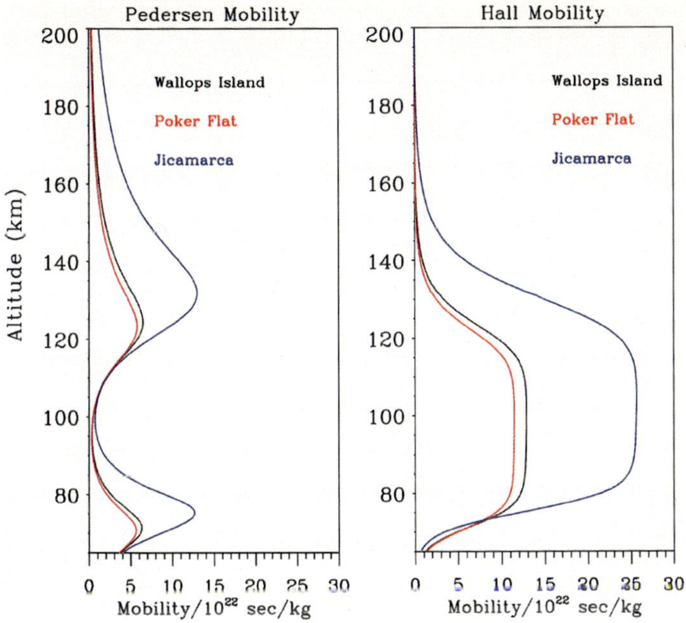

Fig. 14 Altitude profiles of the Pedersen and Hall mobilities at Wallops Island, Virginia; Poker Flat, Alaska; and Jicamarca, Peru

field, **B**, the current will have two components—one along the electric field but perpendicular to **B** (Pedersen) and one perpendicular to both the electric field and to **B** (Hall). Their magnitudes differ by ν_i/Ω_i and the angle, α, between them is given by $\alpha = \tan^{-1}(\Omega_i/\nu_i)$.

Model calculations of the Hall and Pedersen conductivity in the E-region at different latitudes are presented in Fig. 15. (The parallel conductivity for all locations is orders of magnitude higher than that of the corresponding Hall and Pedersen conductivity and is not shown in the figure.) The top two panels show the plasma density at noon and the Hall and Pedersen conductivities calculated using the mobilities as shown in Fig. 14. The lower panels show the plasma density and conductivity at 2100 LT for Poker Flat, Alaska for both quiet (left) and active (right) conditions. The active auroral conditions were selected to provide plasma densities of $10^5/\text{cm}^3$ at 110 km, corresponding to a moderate auroral arc. Clearly, for conditions where there is no aurora, the high latitude conductivity at night is very low. The thermal plasma created by the auroral particles provides a very strong conductivity which is important for current closure, Joule heating, and other important high latitude effects, as we will discuss later on.

4 The Dynamics and Morphology of the Ionosphere at Low and Middle Latitudes

In Sect. 2, we showed how the earth's ionosphere at low and middle latitudes is formed via solar photo-ionization and subsequently varies with altitude and local time, revealing important differences between the daytime and nighttime plasma number density and temperature. In this section, we show how this ionization is significantly altered by dynamics. The dynamics are driven primarily by neutral forcing (winds) of the ions that subsequently set up internal polarization electric fields. In the next section, we will show how the opposite

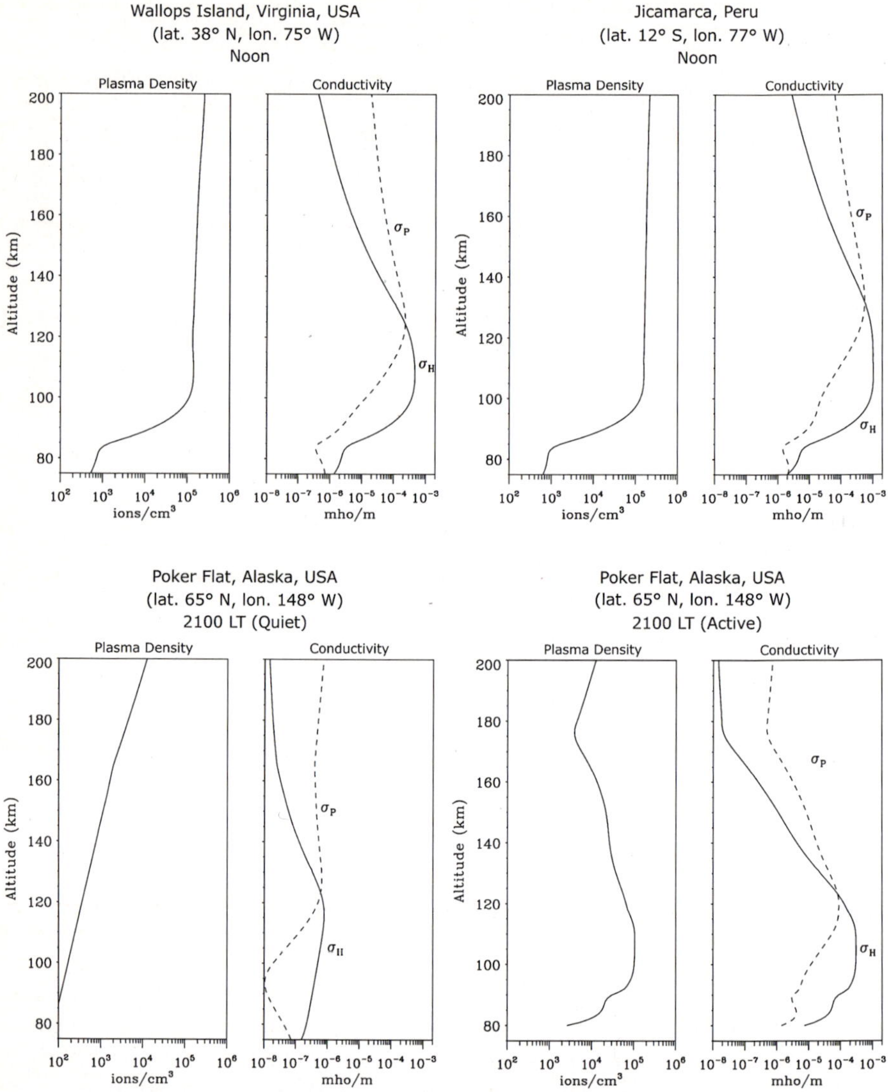

Fig. 15 Altitude profiles of the Hall and Pedersen conductivities for representative plasma density profiles at Wallops Island, Virginia, USA (*upper left*), and Jicamarca, Peru (*upper right*). The lower panels show the Hall and Pedersen conductivities at Poker Flat, Alaska, USA, for quiet conditions (*left*) and moderate auroral conditions (*right*)

occurs at high latitudes, i.e., electric fields drive $\mathbf{E} \times \mathbf{B}$ plasma motion via large amplitude, externally imposed, magnetospheric fields that, through ion-neutral collisions, subsequently drive neutral winds.

4.1 Dynamo Currents and the Equatorial Electrojet

As the solar EUV heats the neutral upper atmosphere, the neutral gases generally expand away from the sub-solar point, creating a system of upper atmosphere tidal motions or winds.

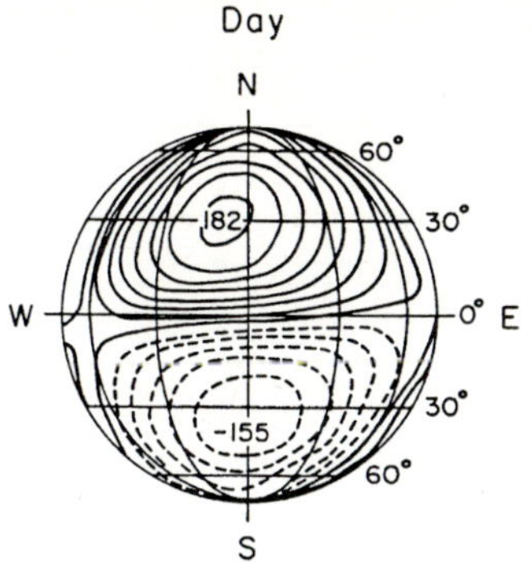

Fig. 16 Current patterns set up by the atmospheric dynamo for daytime (*left*) and nighttime (*right*) hemispheres (after Matsushita 1965)

The neutral motions drag ions across magnetic field lines via ion-neutral collisions, creating a very small charge displacement with the electrons that subsequently sets up relatively weak, ionospheric electric fields. At the lower altitudes ($\lesssim$130 km), the ion-neutral collision frequencies become so large that the ions are unable to complete their gyro orbits whereas the electron cyclotron orbits, due to their much smaller mass, are not restricted and hence, remain "magnetized". Due to the differential drifts of the ions and electrons at these low altitudes, a system of global scale, horizontal current patterns develops, in which internal, polarization electric fields are generated to restrict the current flow to be non-divergent. This worldwide system of currents is called the "solar quiet" (or Sq) dynamo which form current loops at mid-latitudes which close at the magnetic equator, as shown in Fig. 16 (Matsushita 1965).

The strongest current driven by the atmospheric dynamo is the equatorial electrojet. This current system exists as a result of both the dynamo action at all latitudes and the enhanced conductivity set up by the horizontal magnetic field geometry at the magnetic dip equator. At this location, in the altitude range of roughly 95–115 km, significant zonal (east-west) electron drifts are set up in the following way. Vertical $\mathbf{E} \times \mathbf{B}$ drifts due to the ambient eastward electric fields (driven by the solar forcing discussed above) combined with the north-south magnetic field are inhibited for ions because the ion-neutral collisions are sufficiently large such that $V_i = 0$. Hence, only the electrons are magnetized and the resulting vertical charge separation between the electrons and ions establishes a vertical polarization field, $\mathbf{E}_p$. This polarization electric field, in turn, creates a horizontal $\mathbf{E}_p \times \mathbf{B}$ drift which reinforces the small, existing dynamo current and thus creates the strong electrojet current. This process is depicted in Fig. 17. The equatorial electrojet is strongest within a degree of the magnetic equator, where the magnetic field vector is precisely horizontal, thus enabling the vertical, polarization electric field described above to be set up. This current can also be explained in terms of the Cowling conductivity, which is an enhancement, by a factor of σ_2^2/σ_1, of the Pedersen conductivity at the magnetic equator (e.g., Kelley 2009).

At night, the zonal electric field reverses direction and thus the polarization electric field is downward. The resulting $\mathbf{E} \times \mathbf{B}$ drift again enhances the electrojet current since the current is also of opposite direction at night. The magnitude of the nighttime current densities is much lower than during the daytime due to the greatly reduced ambient electron density.

An example of measurements of the polarization electric field, current, and plasma density measured on a sounding rocket launched in the daytime electrojet from Brazil is shown in Fig. 18a. In this figure, the resulting electrojet plasma drifts are calculated independently

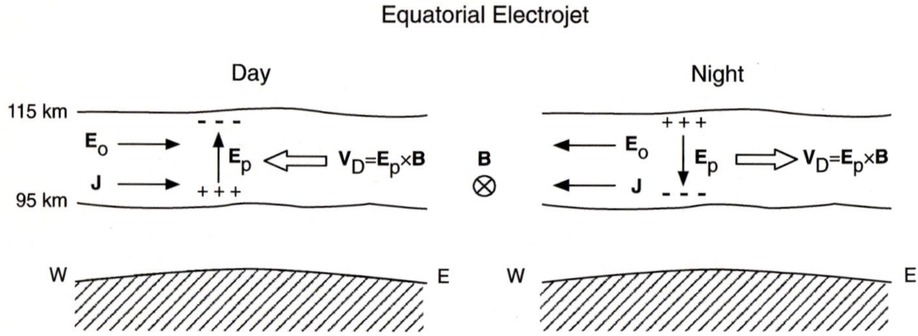

Fig. 17 The zonal electric fields create the vertical polarization fields ($\mathbf{E}_p$) at the magnetic equator in daytime (*left*) and nighttime (*right*). The $\mathbf{E}_p \times \mathbf{B}$ drifts drive the daytime and nighttime electrojets in opposite directions in accordance with the sign of $\mathbf{E}_p$

using both the measured electric fields (i.e., the $\mathbf{E} \times \mathbf{B}$ drift) and the measured current and plasma density (i.e., $\mathbf{V}_e = \mathbf{J}/qN_e$, assuming $V_i = 0$), which show comparable results (Pfaff et al. 1997a). The discrepancy between these two drift profiles may be due to the neutral wind contribution, which was not measured in this experiment and hence not taken into account.

When the relative electron-ion drift speed exceeds approximately the acoustic velocity, collisional two-stream waves are excited (Farley 1963; Buneman 1963). These intense, meter-scale waves are observed in electric field wave measurements recorded by sounding rockets, for example in the equatorial electrojet off the coasts of Peru and Brazil (Pfaff et al. 1987, 1997b). The altitude-spectrogram in Fig. 18b shows an example of such two-stream waves, represented by the broadband emissions extending to higher frequencies between 102 and 110 km, where the instability threshold conditions are met. The intense, lower frequency waves observed in particular at the lower altitudes are driven by the gradient-drift instability and are set up by the vertical (positive upwards) DC electric field in concert with the upward ambient plasma gradient. The lower edge of the daytime ionosphere is unstable at long wavelengths in the electrojet region, as evidenced by structure in the plasma density profile at 105 km altitude and below, shown in the figure and discussed in Pfaff et al. (1987). Indeed, the lower ionosphere in both the daytime and nighttime electrojet regions is typically highly unstable to these plasma instabilities (e.g., Fejer et al. 1975). Distinct, km-scale waves dominate the spectrum when the gradient-drift instability is active, propagating zonally along the electrojet direction. These large scale waves have been shown to change directions between daytime and nighttime conditions as the ambient polarization electric field changes sign (Kudeki et al. 1982, 1987).

4.2 The Fountain Effect: The F-region Dynamo, and Plasma Drift Patterns at Low Latitudes

Given the formation of the ionosphere by photo-ionization from solar photons and the controlling role of the solar zenith angle in the resulting plasma production, it would be logical to expect the ionosphere to have maximum number density at the sub-solar point and that the variations in this density would be fairly predictable with local time, latitude, and season. In fact, this is not the case. The daytime ionospheric density does not peak at the sub-solar point, but actually has two distinct maxima about ±15 degrees from the magnetic equator which can be quite pronounced.

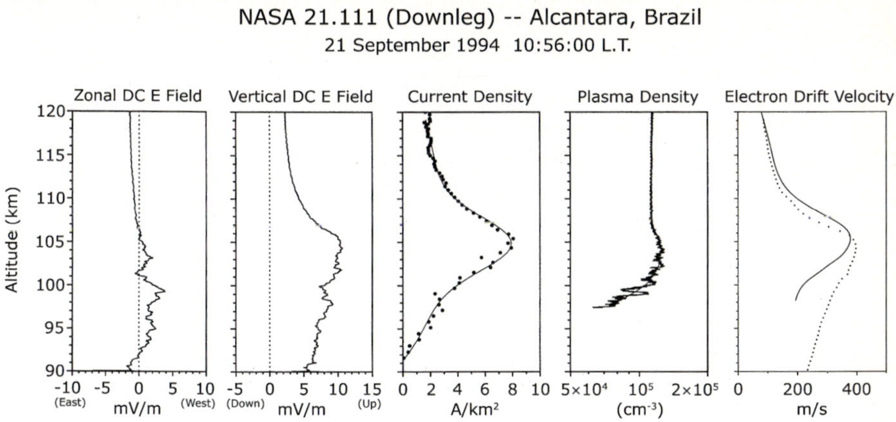

(a)

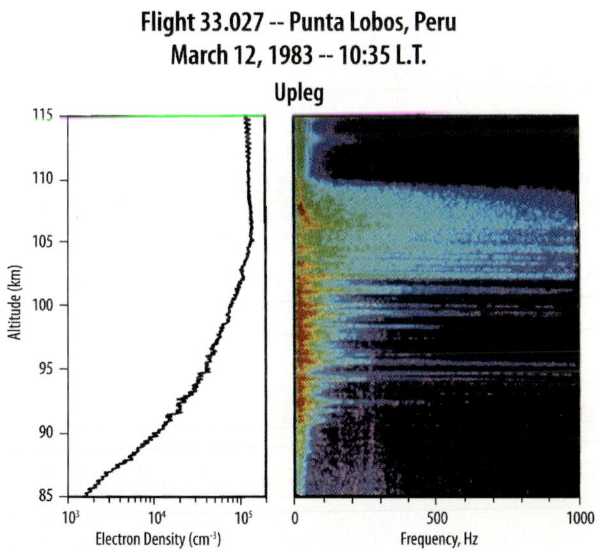

(b)

Fig. 18 (**a**) Sounding rocket observations of the altitude profile of the DC electric field, current density and plasma density in the daytime equatorial electrojet at Alcântara, Brazil. The panel on the right shows independent calculations of the electron drift velocity using (*solid*) the current density and plasma density and (*dotted*) the polarization electric field (Pfaff et al. 1997b). (**b**) Sounding rocket observations of electric field waves displayed in an altitude sonogram (*right panel*) in the daytime equatorial electrojet at Punta Lobos, Peru. The color levels correspond to over 40 dB in wave power, with *red* denoting stronger amplitudes than *blue*. Simultaneous measurements of the unstable plasma density are shown in the left hand panel (Pfaff et al. 1987)

These ionization peaks and the equatorial valley in between them were initially observed in ionosonde data and eventually explained as a consequence of the eastward electric field in the daytime, low latitude ionosphere (Appleton 1946; Martyn 1947). As a result of this electric field, the $\mathbf{E} \times \mathbf{B}$ drift forces plasma upwards at the magnetic equator which then, via diffusion and gravity, "slides" down the magnetic field lines to form the two maxima away from the magnetic equator, peaking at $\sim$300–400 km altitude, as illustrated in Fig. 19.

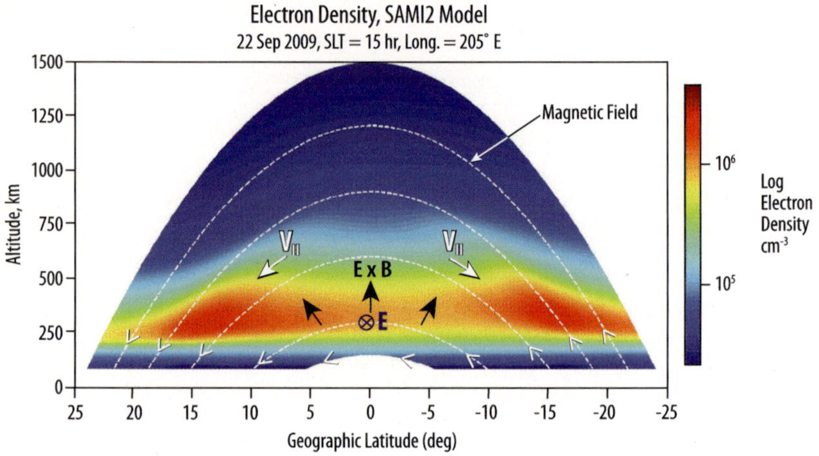

Fig. 19 SAMI2 model calculations versus latitude and altitude of the plasma density for 1500 SLT at 205° East. The upward **E** × **B** drift at the magnetic equator is driven by the eastward electric field, and there is subsequent flow downward along the magnetic field lines

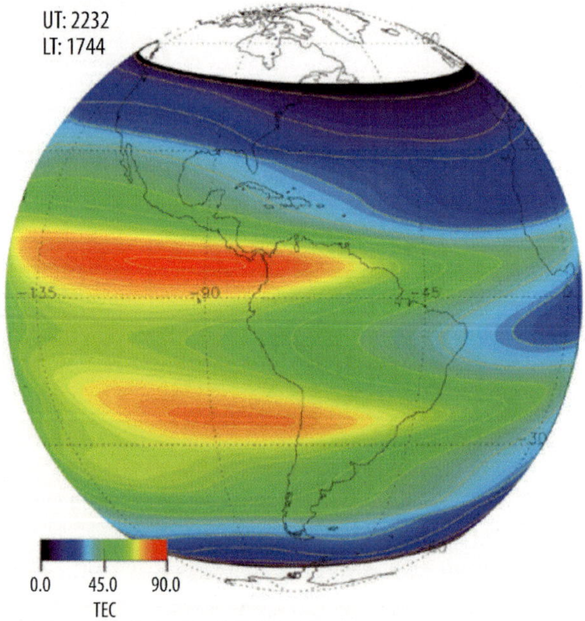

Fig. 20 SAMI2 model calculations of the global total electron content (TEC) at 22:32 UT, showing the equatorial anomaly enhancement in the late afternoon. The center of the globe corresponds to 17:44 LT (Figure courtesy J. Huba)

Referred to as the so-called "fountain effect" or the equatorial ionization anomaly (EIA), this phenomenon represents a significant, fundamental mechanism that characterizes the near-earth plasma environment. Model total electron content calculations for moderate solar conditions are shown in Fig. 20 that were calculated with the SAMI2 model (Huba et al. 2000) and show how the anomaly "wings" dominate the late afternoon ionosphere at low latitudes.

Another process at work at low latitudes that drives the plasma is the so-called "F-region dynamo". At high altitudes in the low latitude F-region during the daytime, wind-driven field aligned currents flow along the magnetic field lines between the F and E-regions. At night, however, the E-region ionosphere is devoid of plasma due to recombination effects discussed previously, and hence, the "feet" of the magnetic field lines, whose apex altitudes are in the F-region, are no longer in high conductivity E-region. As a result, large-scale electric fields are set up in the F-region which drive a low-latitude dynamo in the F-region at night in order to maintain current continuity (e.g., Rishbeth 1971, 1997; Heelis 2004).

An enhanced vertical drift in the F-region after sunset is a characteristic of the low latitude ionosphere, whose intensity exhibits seasonal, longitudinal, and solar cycle dependencies, tending to be largest near the equinoxes at solar maximum (Scherliess and Fejer 1999). This "pre-reversal enhancement" has been attributed to processes also related to differences in the conductivity between the E- and F-regions near the terminator and is generally attributed to a polarization electric field set up to maintain current continuity near the dusk meridian (e.g., Heelis et al. 1974; Farley et al. 1986; Haerendel and Eccles 1992; Eccles 1998).

Although the winds and polarization electric fields can be quite variable day-to-day, an average statistical pattern emerges from *in situ* and radar measurements of meridional (radially outward) drifts at low latitudes which is used in a number of models to represent the low latitude electrodynamics. Measurements of the meridional drifts near the magnetic equator from the ROCSAT satellite (Fejer et al. 2008) averaged for moderate solar activity conditions are shown for all longitudes and local times in Fig. 21. Note how the diurnal variations have a marked dependence on longitude, with the strongest enhancements after sunset occurring in the longitude sector of -120 to $+60$ degrees East at the equinox.

Because the mid- and low-latitude ionosphere is a "closed" system (i.e., the magnetic field lines that encompass this region are all closed), it may appear straightforward to develop a self-consistent electrodynamic model based on the wind driven E- and F-region dynamos, even when electric field mapping and variations of magnetic field declination with longitude are taken into account. However, the dynamo-driven drifts are not the whole story. Tidal effects (discussed next) propagating from below also have profound effects in the ionosphere, and account for the 4-cell pattern in longitude in the morning vertical drifts shown in Fig. 21. In addition, neutral forcing from high latitudes as well as penetration electric fields of magnetospheric origin at all latitudes due to strong magnetic activity will also affect the global drift patterns. Finally, upwards propagating electric field contributions from tropospheric storms are also believed to contribute to the drifts at any given time, particularly at night when the plasma density is greatly reduced and the thunderstorm-generated electric fields couple more efficiently to the ionosphere.

4.3 Effects of Tides on the Plasma Density and Electric Fields

A very important contribution to the low and mid-latitude upper atmosphere is neutral forcing from below. The strongest and most regular forcing is due to solar thermal tides which exist over a wide range of harmonics related to the earth's 24 hour rotation period. Given the strong effect of neutral winds on the low and mid-latitude electrodynamics, it is not surprising that tidal forcing might have profound effects on the near-earth plasma. The significance of these tidal effects and the fact that they are well-organized, for example with longitude, have received renewed interest in the space physics community as new data sets have recently become available.

Solar tidal effects have been observed in the ionosphere in both the electrodynamics and the plasma density. The actual mechanisms by which the tidal signatures are manifest are still being debated. Apart from migrating tides which are responsible for creating day-night variations in the ionosphere, the most distinct tidal signature is attributed to the diurnal eastward wavenumber-3 (DE3) non-migrating tide which produces very clear longitude effects

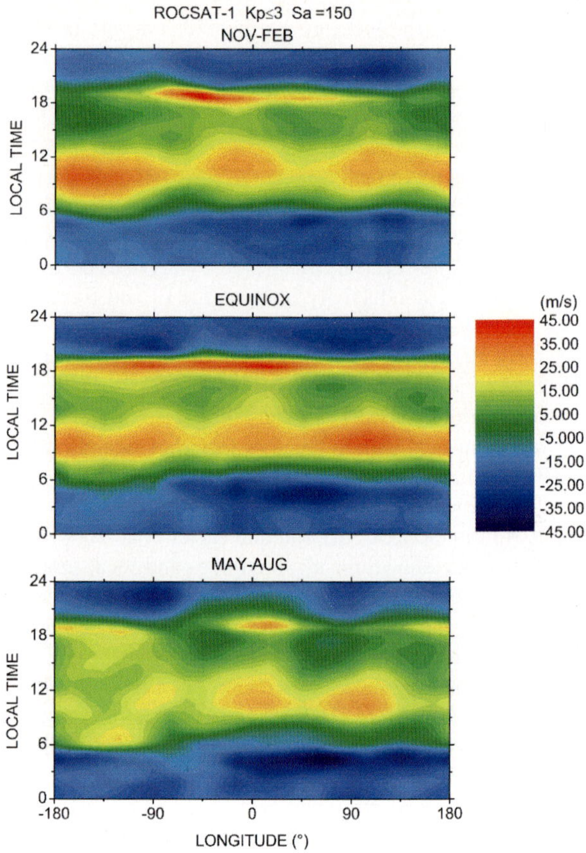

Fig. 21 Average vertical drift velocities recorded by the ion drift meter on the ROCSAT satellite, as a function of local time, longitude, and season, for quiet times during moderate solar flux conditions (Fejer et al. 2008)

in the upper atmosphere. An example of how this tidal structure effects the low latitude, E-region electric fields is shown by England et al. (2006), who compared the equatorial electrojet strength as measured by the CHAMP, SAC-C, and Oersted satellite magnetometers and demonstrated that they each display the same 4-peak longitudinal structures. An example of enhancements in the total electron content at evenly-spaced longitudes at low latitudes for the 12–14 LT sector is shown in Fig. 22 using observations from the COSMIC satellites. (Figure provided courtesy of C. Lin. See Lin et al. 2007, for details.) These authors argued that the tides influence the F-region plasma by affecting the dynamo electric fields which subsequently modulate the vertical plasma transport. To demonstrate this, these authors used the same COSMIC data to create cross section plots of the equatorial ionization anomaly for a peak and a valley period to show that the EIA extended to higher altitudes and higher latitudes in periods of peak tidal forcing, a behavior which can be clearly linked to larger driving electric fields. That the equatorial electric fields show this same wave-4 tidal signature is also evident in the 4-peak longitude variations of the vertical plasma drifts evident in the morning (9–11 LT) period in Fig. 21. A review of the effects of the non-migrating tides on the near-earth plasma is provided by England (2011, this issue).

Airglow enhancements in the crests of the equatorial anomaly peaks at 20 LT provides another representation of this non-migrating D3 tide phenomenon, as shown in Fig. 23 (Immel et al. 2006). This figure shows a reconstruction of 30 days of observations gathered with the FUV imager on the IMAGE satellite. The southern "band" is introduced as a mirror

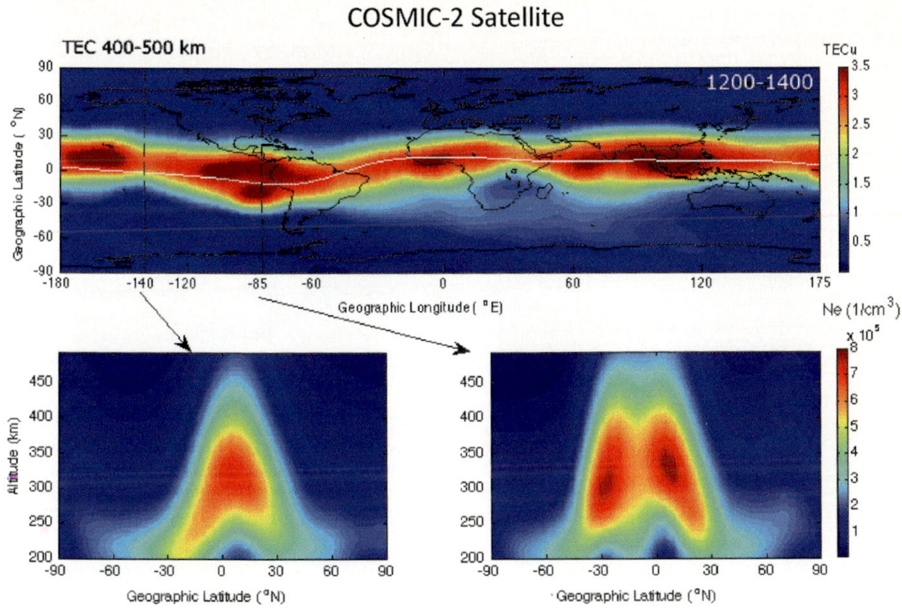

Fig. 22 Average TEC observations recorded by sensors on the COSMIC-2 satellites for the 1200–1400 LT sector as a function of latitude and longitude (*upper panel*). The lower panels show cross sections versus altitude and latitude corresponding to a minimum (*left*) and peak (*right*) TEC region, illustrating how the equatorial anomaly responded to different electric fields in each case (figure courtesy C. Lin. See also Lin et al. 2007)

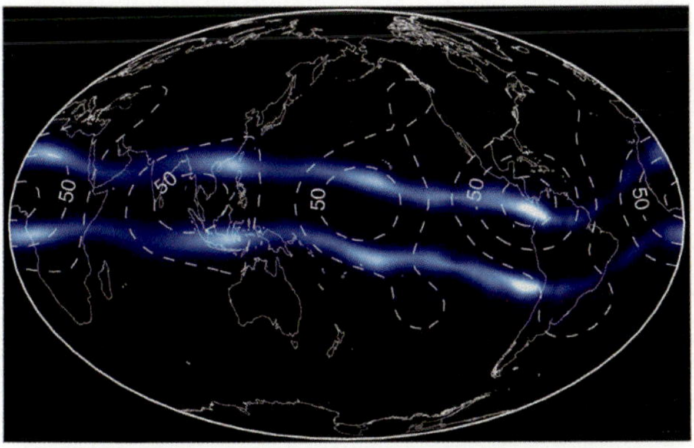

Fig. 23 Reconstruction of nighttime ionospheric emissions for March 20–April 20, 2002, as measured with the IMAGE satellite (Immel et al. 2006), which shows the average location and brightness of the equatorial anomaly. (Because of poor sampling of the southern hemisphere, its emissions are represented by a mirror image of the northern anomaly.) The image represents data at 2000 LT and includes an overlay of white dashed contours which represent model temperatures at 115 km

image of the upper, northern band. The white dashed contours shown in this presentation represent model temperature variations due to the upward propagating tidal oscillations, which correlate nicely with the observed peaks in the equatorial anomaly.

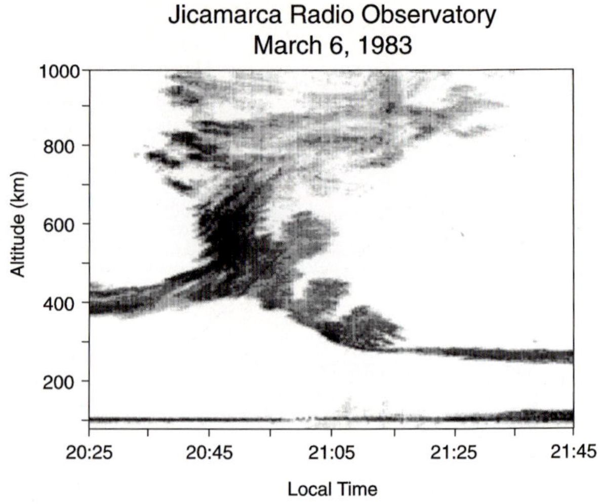

Fig. 24 Backscatter radar power at 50 MHz as recorded on the night of March 6, 1983. *Darker shades* correspond to more power, which is depicted on a logarithmic scale (LaBelle 1985; Kelley et al. 1986)

4.4 Large Scale Plasma Depletions at Low Latitudes ("Spread-F")

At low latitudes at night, major distortions of the near-earth, ionospheric plasma may occur, often of planetary-scale proportions. This phenomenon may be characterized by a series of depleted magnetic flux tubes or wedges that stretch between magnetic conjugate locations in the northern and southern hemispheres. Sometime referred to as "bubbles", these elongated plasma depletions typically have internal, eastward electric fields of 5–20 mV/m (although sometimes much greater) and hence undergo significant upward $\mathbf{E} \times \mathbf{B}$ motions at speeds much greater than those of the background, ambient plasma (e.g., Aggson et al. 1992). The seat of the depleted flux tubes is well established to be the base or lower ledge of the nighttime F-region plasma (e.g., near 250 or 300 km). When fully developed, however, these depletions can rise to altitudes of over 1000 km and extend significantly in latitude. In some cases, primarily off the magnetic equator, narrow plasma enhancements, instead of depletions, are observed (e.g., Le et al. 2003; Klenzing et al. 2011).

A key feature of the equatorial plasma depletions is that they are themselves unstable, and readily display strong electrostatic plasma irregularities with scale sizes that range from over 100 km to less than one meter. Certain scales contribute to Fresnel and Bragg scatter at high frequencies and cause radio scintillations as well as radar and communication outages (e.g., see de La Beaujardière 2004, and references therein). In fact, this general phenomenon was first identified by its associated irregularities, which create a spread in the F-region scatter on ionosonde traces and is generally referred to as "spread-F" in the literature. Recent reviews of large-scale plasma depletions/spread-F may be found in Woodman (2009), Kelley et al. (2011), and Makela and Otsuka (2011, this issue).

An example of equatorial spread-F is shown in the Jicamarca backscatter radar map for 3-m irregularities in Fig. 24 (Kelley et al. 1986; LaBelle 1985). Such radar "range-time-intensity" plots are similar to observations gathered with a slit camera: the radar beam has a fixed, one-dimensional (vertical) orientation through which the two-dimensional, dynamical ionospheric structures are observed. Thus, the reader must use caution in interpreting the image as a two-dimensional map, particularly because the features are not stationary, but evolve in space and time. With that precaution, we can nevertheless draw some important

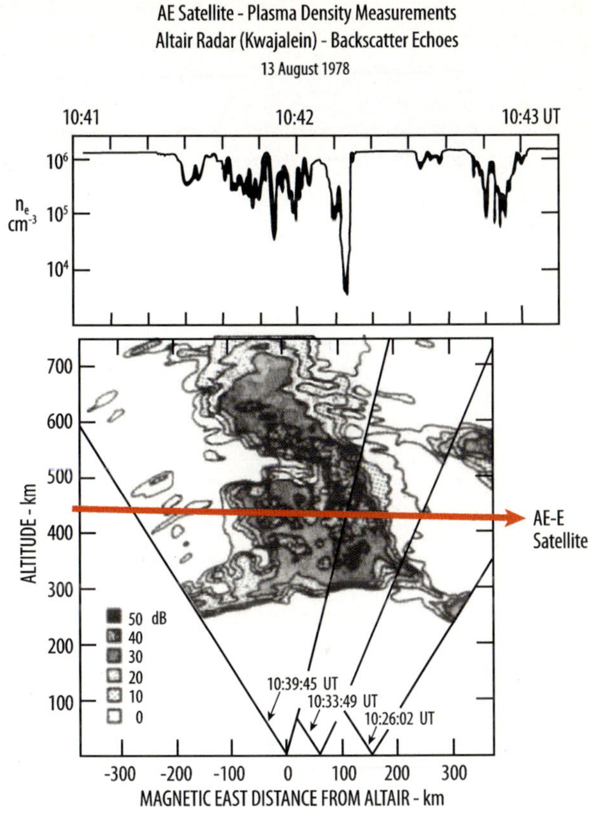

Fig. 25 Spatial relationship of the backscatter power measured during equatorial spread-F by the Altair radar at Kwajalein Atoll (lower panel) with the density measurement by the AE-E satellite on August 13, 1978 (after Tsunoda et al. 1982). The satellite trajectory is shown in *red*

conclusions regarding large-scale ionospheric features and dynamics from this presentation. For example, the strong radar backscatter near 400 km at 20:25 UT almost certainly corresponds to the F-region ledge which then rises near 20:45 UT, presumably due to the pre-reversal enhancement of the eastward electric field, which causes the F-region to rise after sunset at low latitudes. The altitude of the ledge then decreases and settles at a range of about 300 km during the latter portion of the event shown (i.e., near 21:35 UT). This strong scatter that traces out the F-region ledge is known as "bottomside" spread-F.

The salient feature of Fig. 24 is the spread-F "eruption" shown in the center of the panel near 20:45 UT and extends to much higher altitudes, forming so-called "plumes". In fact, in this example, there appear to be 3 or perhaps 4 evenly spaced plumes that emerge from the center and backside of the F-region ledge (between 20:45 UT and 21:15 UT) whose irregularities constitute "topside spread-F". Furthermore, note that above about 700 km, there are very high altitude structures that display a more diffuse character. Such scatter is believed related to smaller scale instabilities that are driven at high altitudes where the collision frequencies are negligible. Finally, we note that the narrow, darkened traces near 100 km correspond to nighttime electrojet irregularities, for which the decrease in intensity near 21:00 likely corresponds to the change in the direction of the ambient zonal DC electric field, as shown for other examples in Fejer et al. (1975).

We now consider another radar backscatter example, in this case from the steerable Altair radar at Kwajalein, which the AE-E satellite traversed with a plasma density probe, as shown in Fig. 25 (Tsunoda et al. 1982). In this case, as shown in the upper panel, the probe clearly

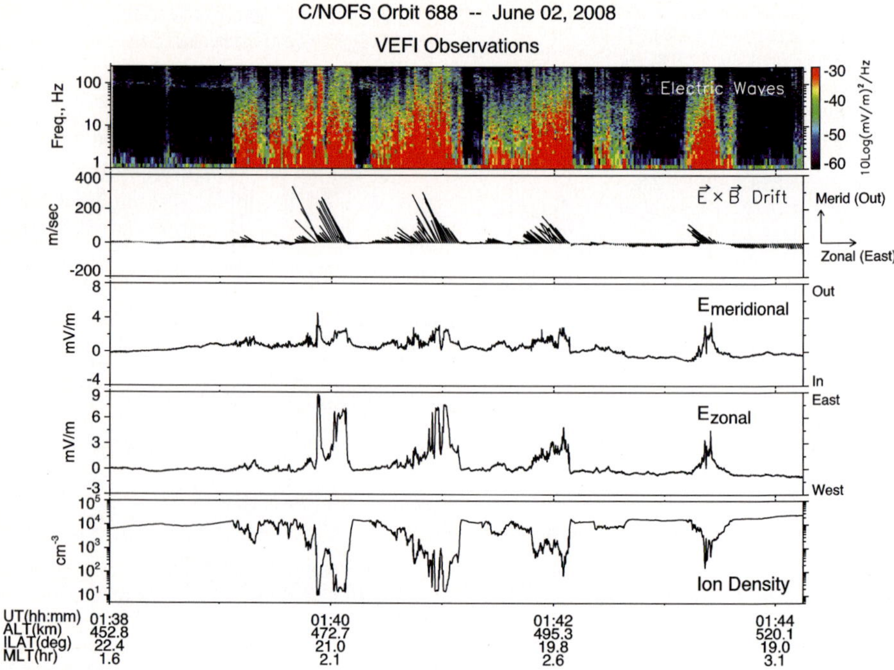

Fig. 26 Electric field and plasma density measurements recorded by the C/NOFS satellite during strong spread-F density depletions, as shown in the *bottom panel*. The two panels above the density show the DC electric field components and the panel above that shows the corresponding **E** × **B** velocity. Notice how the zonal component of the electric field, which corresponds to the outward plasma drift component, is anti-correlated with the log of the plasma density. The *uppermost panel* shows ELF electric field spectra that reveal strong irregularities coincident with the density depletions

observes profound plasma density depletions, decreasing by over two orders of magnitude, which correspond precisely to the regions of strong radar backscatter, as shown in the lower panel. This example and many others have demonstrated conclusively that strong backscatter echoes originate in regions of well-defined, abrupt plasma depletions.

An example of C/NOFS satellite probe observations of plasma density depletions and associated electric fields within the depletions is shown Fig. 26. The data show a series of depletions that extend to very low plasma densities and which have associated upwards and westward **E** × **B** drifts, as measured by the vector electric field detector on that satellite. Note how the zonal electric field and the log plasma density are anti-correlated, even at small scales, indicative of a quasi-linear relation between these quantities, as shown by Aggson et al. (1992) for similar observations using data from the San Marco satellite. The C/NOFS data in the uppermost panel of Fig. 26 show electric field irregularities with the associated depletions. These are the irregularities that disrupt radiowaves, as discussed above.

Although the definitive source of the onset and triggering of such low latitude plasma depletions is still debated, the rising, depleted flux tubes are believed a consequence of an interchange or Rayleigh-Taylor instability (e.g., Woodman and La Hoz 1976). In this case, the plasma density at low altitudes (<250 km) is greatly reduced at night so that there exists a "heavy" fluid (i.e., the F-region ionosphere) suspended over a "lighter" fluid. The instability is particularly susceptible near the magnetic equator because the magnetic field lines in the F-region are nearly horizontal and essentially sustain the higher density plasma at higher

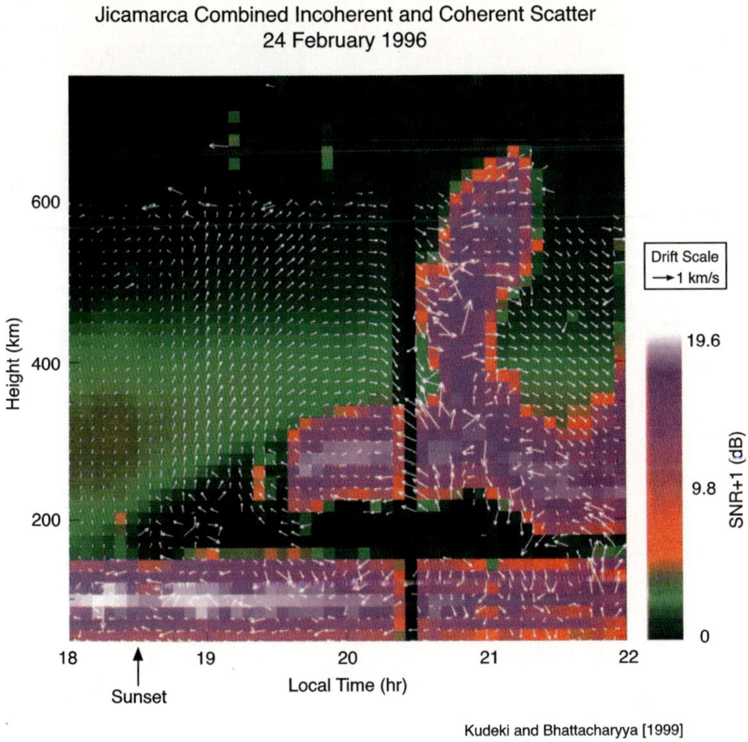

Fig. 27 **E** × **B** drifts (*white arrows*) and Jicamarca backscattered power map for the early evening hours of February 24, 1996 (Kudeki and Bhattacharyya 1999). The scattered power corresponds to incoherent scatter (*brown* and *green shades*), and shows how the lower edge of the ionosphere rises after sunset, which occurred at 18:32 LT. The stronger scatter (*red* and *purple shades*) correspond to coherent scatter of the spread-F and E-region irregularities. Notice the vortex pattern in the **E** × **B** *arrows* prior to the onset of the high altitude plume that begins near 20:30 LT

altitudes. Small perturbations on the F-region ledge would produce an eastward electric field which grows in concert with an associated density depletion, starting the instability process. Another factor believed important for irregularity development is that at sunset, large pre-reversal, zonal electric fields cause the F-region to rise suddenly to altitudes more susceptible to shorter scale instability growth (e.g., Kelley 2009).

Despite the apparent success of the Rayleigh-Taylor instability explanation for the evolution of low latitude plasma depletions, it is widely acknowledged that this instability is too slow to produce the observed prompt depletions after sunset and an additional triggering mechanism is needed to explain their rapid appearance after sunset. Furthermore, the presence of a heavy fluid over a light fluid at the magnetic equator occurs every night, so some additional instigator is required beyond statistical fluctuations. Recent theoretical analysis has shown that spread-F depletions may be triggered whenever eastward neutral winds propagate at different speeds than that of the ambient plasma, as discussed by Kudeki et al. (2007). These authors show how a vortex in the ambient plasma at sunset (in the plane perpendicular to the magnetic field) contributes to the fact that the plasma and neutral gases subsequently move at different velocities. An example of the vortex motion in the plasma drift detected by the Jicamarca radar is shown in Fig. 27 (Kudeki and Bhattacharyya 1999). Shear instabilities also have been suggested to create the large-scale undulations that ulti-

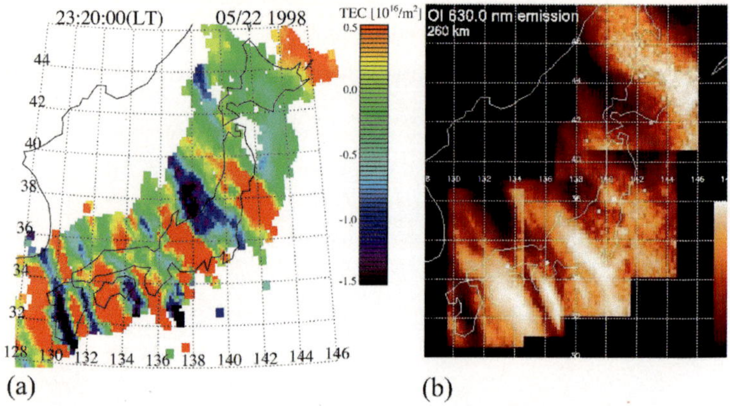

Fig. 28 Simultaneous distributions of total electron content showing gravity waves/TIDS at 23:20 LT on May 22, 1998 using GPS data (**a**) and 630 nm airglow data (**b**) (Saito et al. 2001)

mately seed the Rayleigh-Taylor growth of spread-F (e.g., Hysell and Kudeki 2004). This shear in the zonal plasma drift at the base of the F-region was observed in altitude profiles of the vertical electric field measured on sounding rockets flown from Kwajalein Atoll, as reported by Hysell et al. (2006).

4.5 Acoustic Gravity Waves and Traveling Ionospheric Disturbances

Acoustic gravity waves have been known for some time to affect the ionosphere and, at some level, such oscillations are expected to be ubiquitous in the ionosphere. In general, apart from singular events driven by localized phenomena such as earthquakes, two sources of gravity waves in the mid and low-latitude ionosphere are customarily studied: (1) upwards-propagating gravity waves that originate in the lower atmosphere, such as those generated by thunderstorms and other tropospheric energy sources (e.g., Walterscheid 2001), and (2) horizontally-propagating gravity waves generated at high latitudes, for example by auroral processes, that subsequently propagate equatorward (e.g., Crowley 1991). The neutral motions associated with the gravity waves set the ionosphere in motion where they are observed as traveling ionospheric disturbances, or TIDs. In a situation similar to that of tidal forcing, recent advances in instrumentation and observations have enabled a more in-depth analysis of the role on gravity waves/TIDs in near-earth region of geospace.

Traveling ionospheric disturbances are frequently observed in airglow, radar, and GPS measurements. Recently, arrays of GPS receivers have shown the organization of gravity waves/TIDs traveling equatorward from the poles. Figure 28 shows TIDs evident in simultaneous GPS and airglow data over Japan, clearly revealing the large scale density variations associated with such events (Saito et al. 2001). TIDs are generally grouped into large scales (>1000 km) and mesoscales, (scalelengths between 300 and 1000 km), the latter being termed MSTIDs in the literature. The fact that such waves include electric fields which can easily map along the magnetic field lines explain conjugate MSTID measurements observed by Shiokawa et al. (2004) and Otsuka et al. (2004). The coupling processes between the E and F region ionosphere are now being modeled (e.g., Yokoyama et al. 2009) and underscore the importance of gravity wave sources in understanding the mid-latitude ionospheric structure.

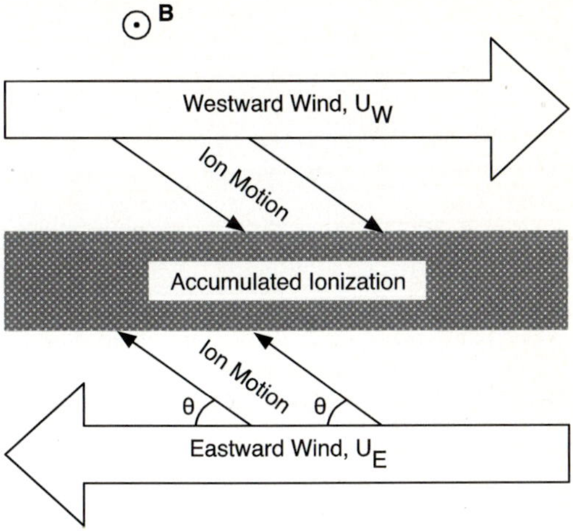

Fig. 29 The shear in the neutral wind creates a layer of enhanced ionization (after Hines 1964)

Using measurements from the Dynamics Explorer-2 satellite gathered near the end of its mission between 250–300 km, Earle et al. (2008) reported gravity wave measurements at mid-latitudes, comparing simultaneous waveform measurements of the neutral density, plasma density, ion velocity, and neutral velocity perturbations. They showed plasma density variations ($\Delta N/N$) of approximately 30 % with scale lengths along the satellite orbit of over 100 km, and argued that because the neutral density fluctuations are observed to lag their neutral vertical velocity by 90 degrees (phase quadrature), the perturbations are gravity waves as predicted by Hines (1964). They also showed that the observations agree with gravity wave dissipation models of Vadas and Fritts (2005) and Vadas (2007). These authors argued that such measurements are more common at the lower altitudes, below where most ionospheric satellite probes generally gather their measurements.

4.6 Distinct Metallic Ion Layers in the Lower Ionosphere ("Sporadic-E" Layers)

In Sect. 2, we showed that the main ions in the lower ionosphere (<150 km) were NO^+ and O_2^+. Missing from this description of the near-earth plasma environment is the fact that this region is often characterized by distinct, heavy ion layers that are common in both the daytime and nighttime ionosphere and may be somewhat more prevalent at mid-latitudes. Such layers are often termed "sporadic-E" layers, since they occur in the lower E-region and may abruptly influence radiowave applications as they drift within the field-of-view of beams of HF radiowave communication antenna. In fact, the layers are actually much more ubiquitous than this name implies. We nevertheless retain this nomenclature for the discussion here.

Sporadic-E layers are dramatic examples of neutral-plasma coupling that routinely occur in the earth's lower ionosphere/thermosphere system. Typically forming in the altitude region of 90–130 km, it is fairly well established that such layers are set up through a combination of neutral wind shear and the effects of long-lasting metallic ions resulting from meteoric input (e.g., Whitehead 1970; Mathews 1998; Haldoupis 2011, this issue). That sharp, distinct sporadic-E layers result from the occurrence of shears in the zonal neutral wind, as advanced by Hines (1964), Whitehead (1970), and others, is shown conceptually in Fig. 29 (after Hines 1964). Such layers may then be modulated by upward propagating gravity waves and shear instabilities (see discussion below), possibly creating polarization electric fields within the crests and troughs of the modulated layer.

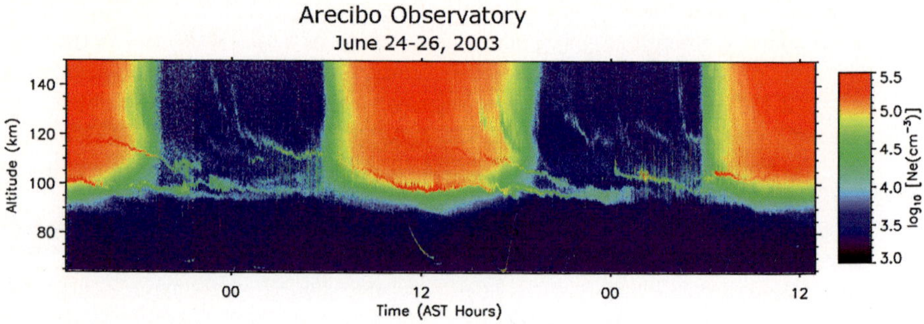

Fig. 30 Incoherent scatter measurements of the plasma density in the lower ionosphere for two consecutive days at the Arecibo Observatory in Puerto Rico. Note the sporadic-E layer continues unbroken between day and night

To illustrate the character of sporadic-E layers, altitude profiles of continuous plasma density observations from the Arecibo incoherent scatter radar are shown in Fig. 30 for two full days, in a format similar to that of Fig. 9, except the focus is now on the E-region. The narrow layers of enhanced ionization that are more distinct at night but are also present during the day are the sporadic-E layers. Of particular importance is the fact that these layers transition from day to night without any apparent break, and they descend, in general, with local time, although occasionally the layers "jump" to slightly higher altitudes.

In addition to research carried out with radars and ionosondes, sounding rocket experiments have provided considerable insight to the formation and character of sporadic-E layers. In particular, large amplitude neutral winds and neutral wind shears have been observed in vapor trails released along near vertical rocket trajectories and shown to exist at the same altitudes of simultaneously observed sporadic-E layers (see, for example, Larsen et al. 1998, and references therein). Those authors also showed calculations of the Richardson number based on the measured wind profile, demonstrating that the altitude range near the layer was highly unstable.

In another experiment, a sounding rocket was launched in conjunction with observations of sporadic-E layers recorded by the incoherent scatter radar at Arecibo, Puerto Rico. The top panel of Fig. 31 presents sporadic-E layers between 100 and 110 km in continuous plasma density profiles obtained with the Arecibo radar during the evening when a sounding rocket was launched from the north coast of Puerto Rico at 21:45 LT, as indicated by the arrow. The rocket payload included an ion mass spectrometer from which data in the lower panel of Fig. 31 clearly shows that the lower density layer was composed primarily of Fe^{+} and Mg^{+} ions (Earle et al. 2000). These results are in agreement with previous mass spectrometer measurements of metallic ions associated with sporadic-E layers (see Grebowsky and Aikin 2002, and references therein). Earle et al. (2000) used the observed separation with altitude of the different metallic layers to estimate the zonal winds necessary to sustain the layer against diffusive effects.

The combined action of neutral wind shears, electric fields, and the sharp plasma density gradients intrinsic to the sporadic-E layer subsequently drive instabilities that include features that are detectable by radars on the ground. In fact, an important consequence of the wind shears is the fact that they can be unstable to the Kelvin-Helmholtz instability. Such a process has been shown by Larsen (2000) to account for the quasi-periodic (QP) radar echoes that have been regularly reported in backscatter radar data associated with sporadic-E layers (e.g., Yamamoto et al. 1991; Hysell et al. 2004). The effects of the Kelvin-Helmholtz

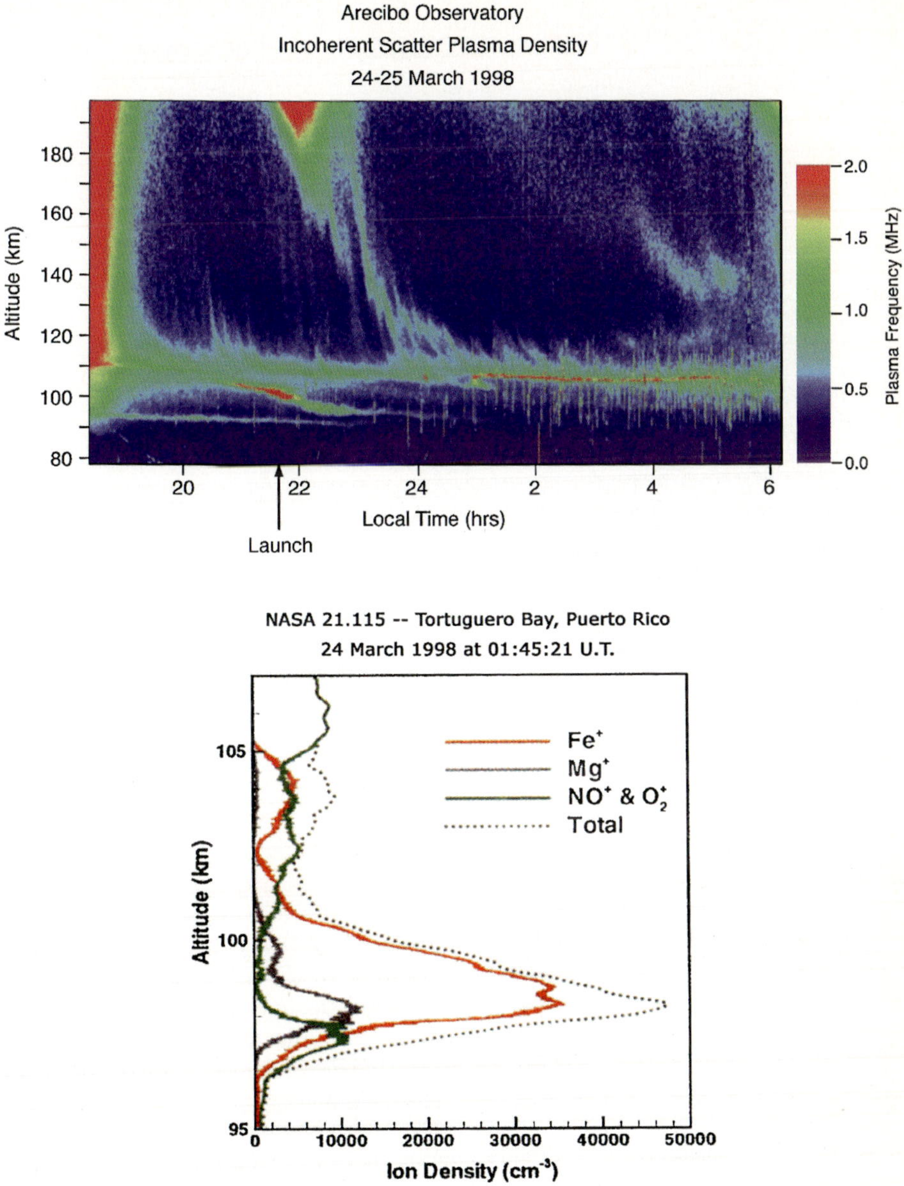

Fig. 31 Incoherent scatter measurements of the plasma density on the night of March 24–25, 1998 at the Arecibo Observatory in Puerto Rico (*upper panel*). A sounding rocket was launched at 01:45:21 UT (see *arrow* in *top panel*) which recorded the ion composition data shown in the *lower panel* (after Earle et al. 2000)

instability on a low altitude, nearly horizontal vapor trail released from a sounding rocket in the presence of sporadic-E and QP-echoes have been analyzed by Larsen et al. (2005). In this experiment, periodic whirls are clearly visible in the vapor emissions emitted along a horizontal trajectory in the E-region, suggesting that the Kelvin-Helmholtz instability was at work during this experiment.

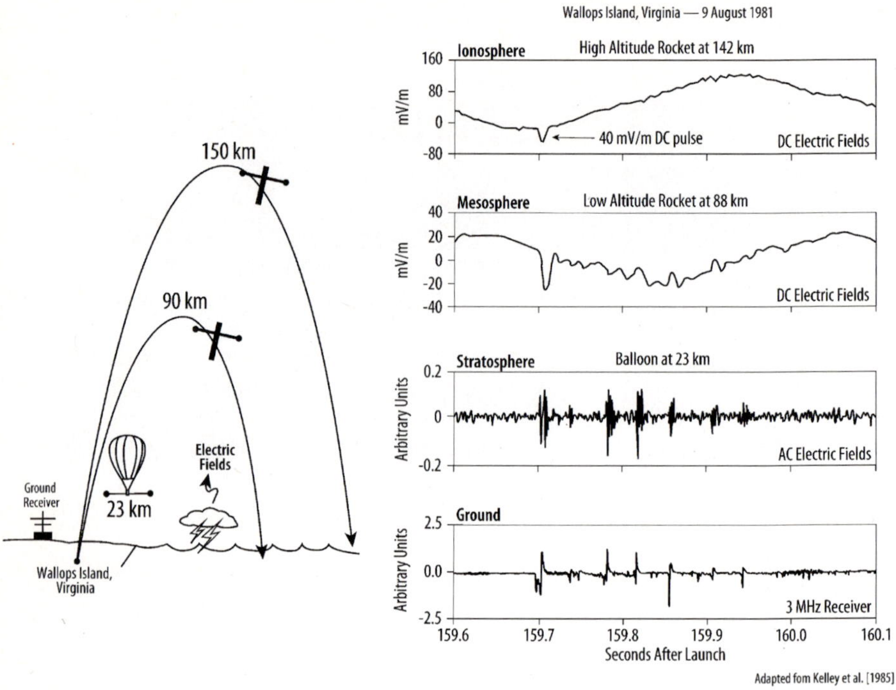

Fig. 32 Electric field measurements recorded over an active thunderstorm by two rockets, a balloon, and a ground receiver (after Kelley et al. 1985). The sketch on the left shows the experiment geometry. The *panels on the right* show data from each platform for one half-second period when one rocket was in the ionosphere and one in the mesosphere. Note how the DC electric field pulse is observed simultaneously in both rocket data sets and appears to penetrate upwards, into the ionosphere. The large amplitude, very low frequency sinusoidal variations in the upper two panels are due to the spin of the rockets

Finally, we mention another layer phenomenon commonly observed in the mid-latitude ionosphere below roughly 200 km—the so-called "intermediate" or "descending" layer (e.g., Earle et al. 1998; Mathews 1998). An example is shown in the Arecibo density profiles in the top panel of Fig. 31. A descending layer appears to evolve from a density bulge associated with the lower edge of the F-region near 22 LT, from which a low amplitude, but distinct, density enhancement transitions to lower altitudes during the next hour or so. The creation and dynamics of such layers are not fully understood.

4.7 Electrical Contributions to the Upper Atmosphere Due to Tropospheric Storms

The lower edge of the ionosphere has been recognized for some time as the conductive upper layer of the earth-ionosphere waveguide. As such, it provides an essential part of the global circuit driven by atmospheric electricity (e.g., Wilson 1920). The ability of low frequency (DC-ELF) electric fields associated with tropospheric storms to penetrate into the ionospheric plasma remains a topic of considerable discussion (see Simoes et al. 2011, this issue).

To investigate the extent of electrical energy deposition in the ionosphere, Kelley et al. (1985) reported observations gathered by a pair of sounding rockets launched simultaneously to apogees in the mesosphere and ionosphere over an active thunderstorm at a mid-latitude location (Wallops Island, Virginia, USA). As shown in Fig. 32, the results clearly demonstrated that the DC electric field "pulse" associated with lightning spherics penetrated into the ionosphere. Besides requiring a revision of models of the transfer of electromagnetic energy across the atmosphere/plasma interface, these results demonstrated that significant energy is deposited in the ionosphere by lightning-generated electric fields. These researchers also reported a significant component of the DC electric field pulse was oriented parallel to the magnetic field direction in the lower ionosphere. Such an electric field would be expected to contribute to localized plasma heating, as discussed by those authors.

The effects of tropospheric storms on the ionosphere may include other consequences. Burke et al. (1992) reported Dynamics Explorer-2 observations of Alfvén waves near 350 km altitude over a hurricane that show a DC electric field pulse of more than 40 mV/m with an associated plasma density depletion and strong magnetic signature. These authors argued that this signature was due to Alfvén "wings" associated with an intense upwards-propagating DC electric field generated by the hurricane forces. Similarly, using data from the San Marco satellite, Farrell et al. (1994) reported large, localized DC electric field "structures" in the ionosphere over tropical storms, providing further evidence that DC-coupled electric fields associated with tropospheric storms propagate upwards and penetrate into the ionosphere. These and numerous other observations underscore how the ionosphere is significantly influenced by energy and momentum sources originating in the troposphere below. We now turn to the high latitude ionosphere and consider energy and momentum sources that originate from above, in the magnetosphere.

5 The High Latitude Ionosphere

The ionosphere at high latitudes differs significantly from that at low and mid latitudes. At high latitudes, the ionosphere exchanges energy and momentum with the solar wind/magnetosphere primarily via the highly inclined magnetic field lines that characterize the earth's polar regions. The magnetic field thus serves as a conduit for both charged particles and their associated field-aligned currents and DC electric fields, thus directly coupling the high latitude ionosphere and upper atmosphere with the magnetosphere. Although photoionization still contributes to the ambient plasma environment, an alternative, often dominant, source of ionization in the high latitude ionosphere is impact ionization of energetic particles with the neutral atmosphere, and the primary drivers of high latitude ionosphere dynamics are electric fields that map down from the magnetosphere. In contrast, the primary source of ionization for the low and mid-latitude ionosphere is photo-ionization from the sun and the primary drivers of plasma transport are neutral winds and small amplitude electric fields created by the neutral winds.

In this section, we introduce the visual aurora and the different populations of precipitating electrons that create the aurora. We discuss the creation of thermal plasma by auroral precipitation and compare this with photoionization. We then review the dynamics and currents of the high latitude ionosphere associated with magnetospheric processes. This is followed by a discussion of the many fundamental processes in which the high latitude plasma environment interacts with the upper atmosphere. We conclude with some brief remarks regarding auroral arc electrodynamics.

5.1 The Visual Aurora and Energetic, Precipitating Auroral Particles

Overview of the Visual Aurora. Optical emissions associated with the aurora result from interactions of energetic, precipitating particles with the upper neutral atmosphere. Typically

Fig. 33 Photograph of the aurora taken in January, 2005 at Bear Lake, Alaska. Credit: Joshua Strang, USAF

Fig. 34 View of the aurora photographed from the International Space Station on March 4, 2012, 17:20:33 UT on a pass over the South Indian Ocean. Figure courtesy NASA

produced within the altitude region of 80–250 km, displays of discrete auroral arcs are perhaps the most striking, visible manifestations of the interactions of space plasma—in this case, precipitating energetic electrons—with atmospheric, neutral gases. Figure 33 is an auroral display photographed from the ground that shows sheets of auroral emissions, which are modulated by large amplitude spatial undulations that have typical durations of seconds to minutes. Another view of such auroral "curtains" is shown in Fig. 34. In this case, the side of the aurora has been photographed from the NASA International Space Station, at an altitude near 400 km. Note that the auroral emissions are characterized by nearly vertical striations that correspond to the ambient magnetic field lines along which the precipitating particles are guided. The green glow at the base of the aurora is the most common auroral hue ($\lambda = 557.7$ nm) and is due to interactions of the more energetic electrons with oxygen at the lower altitudes (100–130 km) whereas the pink-red glow ($\lambda = 630.0$ nm) in the upper portion of the auroral display is from interactions of less energetic electrons with oxygen at higher altitudes (130–250 km). Other, less common, colors of auroral displays include bright red and blue, which correspond to emission lines associated with the interaction of precipitating energetic particles with other gases such as N_2.

An image of the aurora gathered by the DMSP satellite above the southern polar region is shown in Fig. 35 and depicts discrete arcs, diffuse aurora, and auroral forms filled with turbulent-like structure. This image demonstrates that, in the aggregate, auroral emissions and the processes that create them are essentially global in their extent, expanding across the polar region with a variety of patterns in which shorter scale structures are embedded. The ensemble often resembles an auroral oval (sometimes referred to as an auroral belt), as is illustrated in the images with the wider field of view shown in Fig. 36. This time sequence of images, gathered every 12 minutes by the imager on the Dynamics Explorer-1 satellite, illustrates how dynamic the large scale aurora may be, in this case brightening and expanding poleward on the nightside during a magnetospheric substorm. (See Frank and Craven 1988 for a discussion of the auroral images from this instrument.)

The source of the auroral optical emissions shown in the preceding figures is energetic electrons with typical energies of 100's of eV to 10's of keV that precipitate into the high latitude neutral gases where they impact and are "braked". Here, such energetic electrons interact with the neutral atmosphere in two basic ways; excitation and ionization. Excitation occurs when an energetic electron strikes a neutral atom or molecule and excites the neutral to a higher energy state. When the neutral relaxes to a lower energy state, a photon is released, creating the visible aurora. Ionization of the ambient neutral gas occurs when the incoming energetic electrons impact the neutral atoms or molecules and liberate electrons to create ions, thermal electrons, and more energetic secondary electrons. In the remainder of this sub-section, we discuss the different types of aurora and their relative distributions in the high latitude ionosphere. We then discuss the ionization associated with the precipitating particles in the following sub-section.

The aurora may be categorized into three primary types that depend on their respective acceleration or precipitation processes and on the characteristics of their associated precipitating energetic particles: (1) discrete, (2) oscillating or Alfvénic, and (3) diffuse. Each are introduced below, followed by comparisons of their relative number flux and energy flux contributions to the high latitude ionosphere. Although we mention other auroral features as they pertain to the near-earth plasma environment, we do not attempt to review all of the aurora-related processes here. Readers are directed to many reviews of auroral phenomena and related physical processes, including those by Rees (1989), McFadden et al. (1999), Lummerzheim (2007), and Newell et al. (2009). In particular, the ISSI monograph entitled, *Auroral Plasma Physics*, edited by Paschmann et al. (2003), contains a thorough compendium of current understanding of the physical processes associated with the aurora.

Discrete Aurora. The visual aurora shown in Figs. 33, 34, 35, 36 are, in general, associated with discrete auroral arcs. *In situ* measurements by probes on sounding rockets and satellites have shown that the precipitating energetic electrons associated with discrete auroral arcs have narrow, nearly monoenergetic spectra (e.g., McIlwain 1960; Evans 1968). These electron spectra are typically organized in "inverted-V"-shaped energy functions that correspond to spatial variations across the discrete auroral arc (Frank and Ackerson 1971). (The inverted-V's tend to resemble inverted-U's where the energy is displayed as a logarithm.) The peak energies of such energetic electrons typically range from 100's of eV to a few tens of keV. In Fig. 37, energetic electrons detected near 4000 km altitude by electrostatic analyzers on the FAST satellite are shown alongside an image of

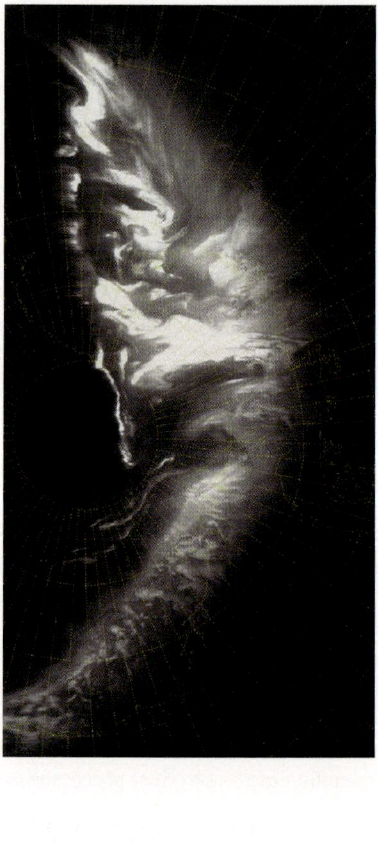

Fig. 35 Image of the aurora taken by the visible-light sensor on the Defense Meteorological Satellite Platform (DMSP), at approximately 880 km above Antarctica on July

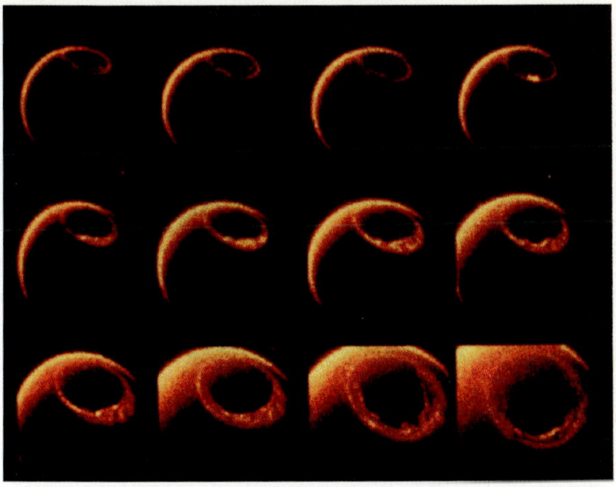

Fig. 36 Consecutive false-color images of the northern auroral oval recorded at the onset and evolution of a substorm as the *Dynamics Explorer 1* spacecraft descends from near its apogee during the period 0529 through 0755 UT on 2 April 1982. Each image is telemetered in a twelve minute interval. The poleward bulge of an auroral "surge" is first seen in the fourth image, which begins at 0605 UT. A westward-traveling surge then propagates along the poleward edge of the oval as highly structured, eastward-moving auroral forms develop in the post-midnight sector. The sunlit atmosphere is visible at the *upper left* in each image. The passband of the optical filter is 123–155 nm, in which the dominant responses are from emissions of atomic oxygen at about 130.4 and 135.6 nm and the LBH bands of molecular nitrogen (figure courtesy NASA. See also Frank and Craven 1988)

multiple auroral arcs simultaneously photographed by an all-sky camera (Stenbaek-Nielsen et al. 1998). Although the optical emissions are from much lower altitudes (near 100 km), the correspondence of the visual arcs with the *in situ* measurements of the higher altitude electron structures is evident.

Significant progress has been made in the last 15 years establishing the relation between accelerated, energetic electrons and the discrete aurora, primarily using observations gathered from satellites that included probes with high time resolution designed specifically to address these processes, such as Viking, Freja, and FAST. As discussed by numerous authors (e.g., Carlson et al. 1998; McFadden et al. 1999; Paschmann et al. 2003; and Marklund 2009), data from these satellites have helped establish that electrostatic "U-shaped" potentials are set up in what is often referred to as the auroral acceleration region, existing at altitudes of roughly 2,000–20,000 km, high above where the electrons interact with the upper atmosphere and the auroral light is emitted. Such potential structures contain electric field components oriented along the magnetic field direction that account for the nearly monoenergetic electrons responsible for the discrete aurora. The same potential structures also accelerate ion beams along the magnetic field in the opposite direction, as observed by these same satellite probes (e.g., Shelley et al. 1976; McFadden et al. 1999).

An important feature of the data in Fig. 37 is that on the equatorward side of the event near 9:18:15 UT, there is an inverted-V region of *upwards* accelerated electrons. These electrons have no atmosphere in which to interact and thus no visual analog in the all-sky camera display. Indeed, the lack of optical emissions in such circumstances has been referred to as the "black aurora" by Marklund et al. (1994). High resolution satellite probes on the Viking, Freja, and FAST satellites (e.g., Marklund et al. 1994; Carlson et al. 1998; McFadden et al. 1999) have identified such upwards accelerated electrons as a regular

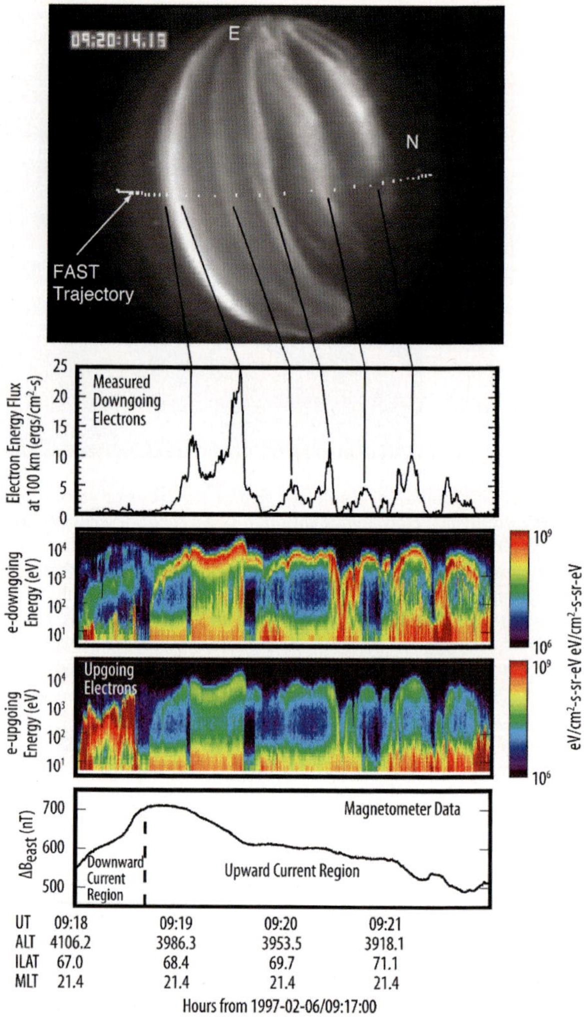

Fig. 37 Multiple arcs in a low-light "all sky" image recorded on an aircraft (*top*) and by instruments on the FAST satellite (*lower panels*) near 21:20 MLT. In the *top panel*, the FAST trajectory is mapped along the magnetic field line to the 100 km conjugate point at 10 s intervals as the satellite passes across from left to right. The *second panel* is the precipitated energy flux on a linear scale. The narrow auroral arcs are clearly displayed here, as are their correspondence to the visible aurora in the *top panel*. The auroral optical emissions, in particular the two arcs to the far right (north), changed over the 4 minutes of the pass. The two color panels show the measured electron energy spectra for downgoing (third panel from bottom) and upgoing (second panel from the bottom) and show a number of inverted-V structures. Note the strong upgoing electrons present in the FAST data at the start of the event near 09:18:15 UT that have no visual analog in the all-sky image. The *lower panel* shows the FAST zonal magnetometer data corresponding to the downgoing current region, where the upgoing electrons are observed, and the upgoing current region, where the downgoing electrons are observed (after Stenbaek-Nielsen et al. 1998)

feature of the aurora. The satellite probes have also measured the perpendicular electric fields in the V-shapes and have shown how they reverse polarity in accordance with the oppositely-directed accelerated particles (e.g., Marklund et al. 1994; Carlson et al. 1998; McFadden et al. 1999). Although such upwards-directed electrons are accelerated at high altitudes, they are believed associated with current closure in the lower ionosphere, where the current is carried by lower energy electrons. The manner in which the lower ionosphere controls this current closure remains one of the fundamental outstanding problems in auroral physics today.

Oscillating Auroral Electrons Associated with Alfvén Waves. Another means by which electrons are accelerated by electric fields and subsequently precipitate into the upper atmosphere to create aurora is their interactions with Alfvén waves. In this case, the oscillating electric fields in the wave accelerate field-aligned electrons in both directions along the magnetic field which are often referred to as "counter-streaming" accelerated electrons. An

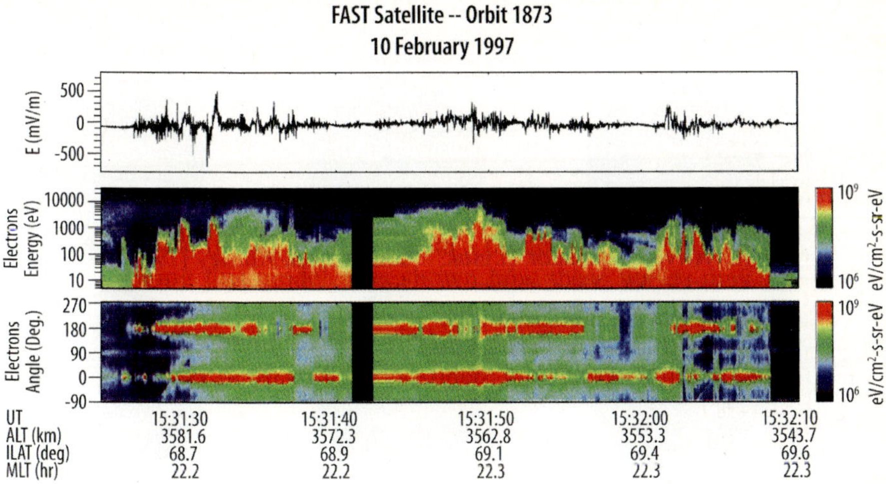

Fig. 38 Observations from the FAST satellite on the polar cap boundary. *Top*: the electric field along the spacecraft trajectory, or roughly north-south. *Middle*: the energetic electron spectra. *Bottom*: the pitch angle spectra of the energetic electrons with 0° pointing downwards (after Chaston et al. 2004)

example gathered by the FAST satellite is shown in Fig. 38 (Chaston et al. 2004). Notice that the pitch angle information shows the electrons as highly aligned along, and anti-parallel to, the magnetic field direction, although the energies themselves are spread-out over a large range of energy space, typically between 10 eV and 1 keV. Other detailed observations of Alfvén-wave driven field-aligned electrons, often within an inverted-V event characterized by higher energy electrons, have been reported in sounding rocket data sets (e.g., Chen et al. 2005).

Although not originally appreciated as an important source of aurora, recent observations by probes on the FAST satellite have shown that such Alfvén wave accelerated electrons can account for as much as 1/3 of the total energy deposited in the ionosphere during elevated auroral activity (Chaston et al. 2007). Wygant et al. (2000) showed how the Poynting flux associated with strong electric and magnetic signatures of Alfvén waves observed on the high-altitude Polar satellite corresponded to, and was sufficient to power, intense auroral optical emissions.

Diffuse and Pulsating Aurora. Diffuse aurorae are considerably fainter than the visual aurora associated with discrete auroral arcs and are best discerned with the aid of photometers. Although such aurorae are generally without distinct, sharp features, there is some structure that can be discerned from the measurements (Lui et al. 1973; Johnstone et al. 1993). In general, the diffuse aurora is most prevalent in the post-midnight sector and has wider latitudinal extent than the discrete aurora.

The diffuse aurora may include large scale pulsations in which large patches of aurora faintly turn on and off with periods typically lasting from 1–30 seconds. Correlated with the recovery phase of magnetic sub-storms, the pulsating aurora is also prevalent in the post-midnight hours and, as with the diffuse aurora, is believed caused by wave-particle interactions with whistler mode waves (e.g., Johnstone 1983; Davidson 1990; Nishimura et al. 2010). Recent observations suggest that ionosphere-magnetosphere coupling may play an important role in the generation mechanism (Sato et al. 2004).

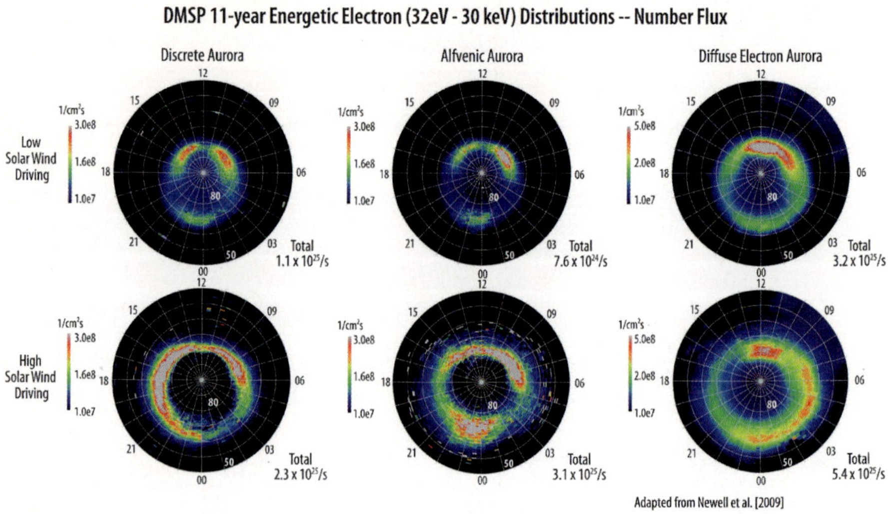

Fig. 39 Comparisons of average number fluxes of the discrete aurora (*left*), Alfvénic aurora (*middle*), and diffuse electron aurora (*right*), for low solar wind driving conditions (*top*) and high solar wind driving conditions (*bottom*). The data were recorded by the energetic particle instrument on the DMSP satellites over the interval of 1988 to 1998. The total number flux for each group is shown (after Newell et al. 2009). The *outermost circle* in each case corresponds to 50° and corrects an error in the original manuscript (Newell, personal communication, 2012)

Unlike the discrete and Alfvénic aurorae which are produced by energetic electrons accelerated by components of electric fields along the magnetic field direction, diffuse aurora are caused by pitch angle scattering with whistler mode chorus waves in the magnetosphere (Thorne et al. 2010), and then precipitate when their loss cones intersect the upper atmosphere. Despite the fact that the diffuse aurora does not create as strong an optical signature, it deposits considerably more energy than the other forms of aurora, as we discuss next. Diffuse ion aurora also exists, and precipitates via the same mechanism, as discussed, for example, by Newell et al. (2009).

Relative Energy Flux of Different Auroral Processes. Newell et al. (2009) have averaged energetic electron data from the polar-orbiting DMSP satellites over an 11 year time from 1988 to 1998. The data were binned according to the type of auroral acceleration (discrete, Alfvénic, or diffuse) and the results are shown in Figs. 39, 40 corresponding to averages of number flux (Fig. 39) and energy flux (Fig. 40). The data in each case were further binned by low and high solar wind driving conditions. As explained by those authors, the local time coverage of the satellites was fairly uniform, although the post-midnight region was covered only from 1992–1998.

The distributions shown in Figs. 39, 40 compiled by Newell et al. (2009) immediately show that discrete aurora are most prevalent on the dayside and pre-midnight portions of the auroral oval, depositing most of their energy in the post-noon, pre-midnight sector. The Alfvénic aurora is a dayside and midnight phenomenon, yet deposits most of its energy on the nightside, just prior to midnight. The diffuse aurora is primarily a post-midnight/morning phenomenon, although it is also the most prevalent form of aurora in the cusp centered at noon on the dayside. Of paramount importance is that the diffuse electron aurora is clearly the dominant source of particle energy flux into the high latitude ionosphere, constituting 63 % of the energy in the low solar wind driving conditions and 57 % during the

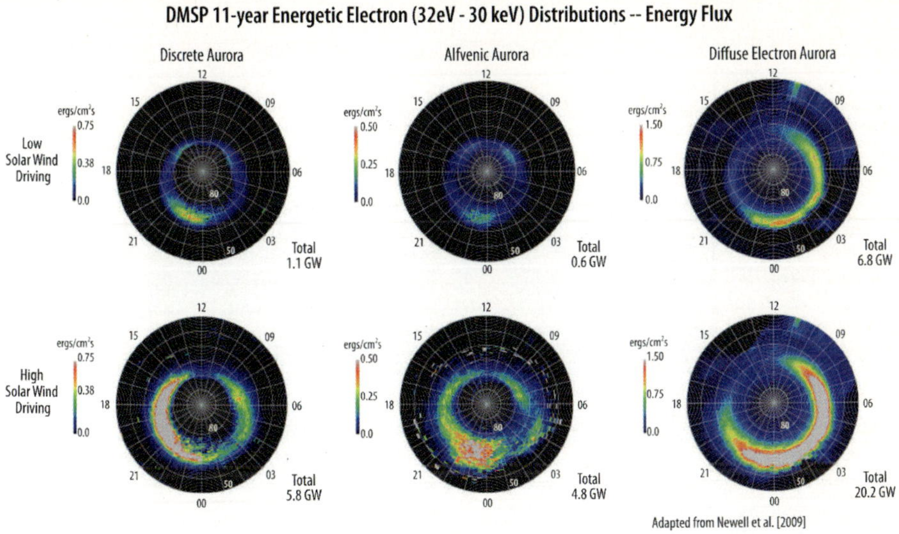

Fig. 40 Comparisons of average energy fluxes of the discrete aurora (*left*), Alfvénic aurora (*middle*), and diffuse electron aurora (*right*) for low solar wind driving conditions (*top*) and high solar wind driving conditions (*bottom*). The data were recorded by the energetic particle instrument on the DMSP satellites over the interval of 1988 to 1998. The total energy flux for each group is shown (after Newell et al. 2009). The outermost circle in each case corresponds to 50° and corrects an error in the original manuscript (Newell, personal communication, 2012)

high solar wind driving conditions, according to Newell et al. (2009). Note that the ion diffuse aurora were also included in the statistics presented by these authors but are not shown here. That the diffuse electron aurora is so prevalent has important consequences for the conductivity of the high latitude, post midnight ionosphere, as we discuss further on below.

Cusp Aurora. The cusps are funnel-shaped magnetic field regions in the northern and southern hemispheres on the dayside of the earth directly connected with the solar wind. As such, they permit direct entry of solar wind particles and energy into the high latitude ionosphere/upper atmosphere. The cusp aurora thus represents a distinct region of the near-earth plasma at high latitudes. Indeed, the ion diffuse aurora peaks in the cusp (Newell et al. 2009).

An example of data gathered by probes on the FAST satellite as it crossed the cusp region near its perigee at 348 km is shown in Fig. 41. The lower panel shows the downgoing, energetic ions of solar wind origin, exhibiting an energy dispersion indicative of a cusp crossing in the presence of the southward interplanetary magnetic field (e.g., Reiff et al. 1977). The precipitating electrons are organized in very narrow, inverted-V forms with modest peak energies of 100's of eV and thus deposit their energy and plasma at higher altitudes, in the 200–300 km region (as discussed in the next sub-section). The presence of such inverted-V signatures suggest that local electron acceleration processes in the cusp are similar to those in the auroral acceleration region on the nightside (e.g., Yamauchi et al. 1998; Pfaff et al. 1998). The upper panel shows the highly correlated electric and magnetic field signatures, indicative of field-aligned currents and downward directed Poynting Flux present in the cusp regions (e.g., Strangeway et al. 2000; Le et al. 2002), as we will discuss later.

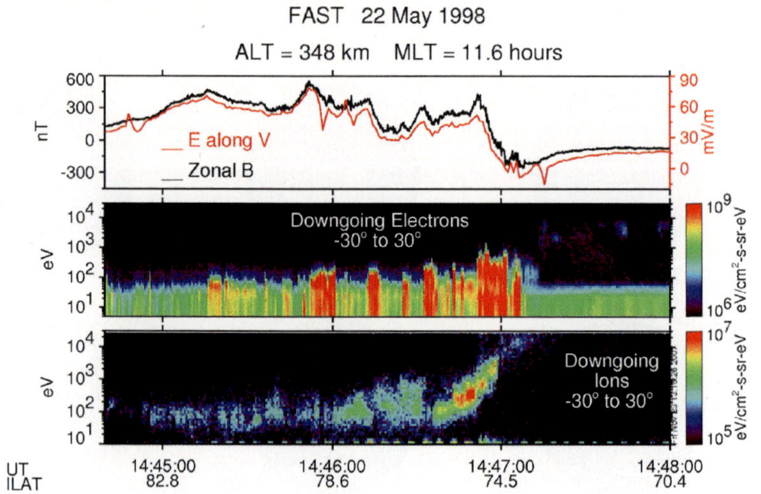

Fig. 41 Observations by instruments on the FAST satellite recorded in the cusp region near 348 km altitude and local magnetic noon. The bottom and middle panels show energy spectra of highly structured downgoing ions and electrons, respectively. The top panel shows the electric field component along the spacecraft velocity (*red*) and the zonal component of the magnetic field (*black*), which are highly correlated

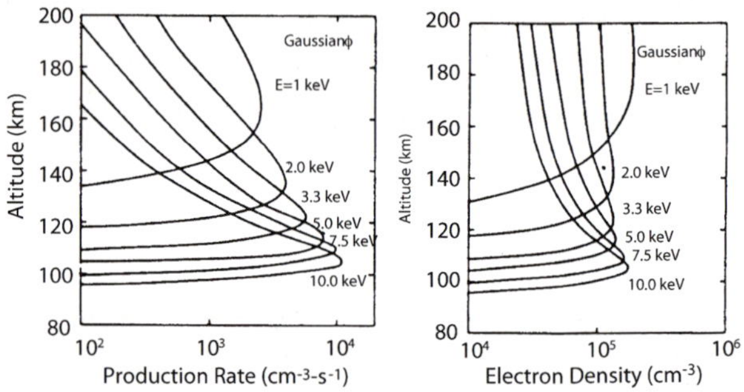

Fig. 42 Thermal plasma production rates (*left*) and resulting density profiles (*right*) as a function of incoming electron energy (Strickland et al. 1983)

5.2 Thermal Plasma Created by Precipitating Energetic Electrons

The interaction altitude of the energetic electrons with the atmosphere depends on both the energy of the precipitating electrons and on the neutral density and composition of the atmosphere. The effective penetration altitude of precipitating electrons as a function of energy is shown in Fig. 42 for a model atmosphere (Strickland et al. 1983). Alongside is a panel that shows the ion production rates associated with these interactions. As with the mid and low-latitude plasma discussed previously, the molecular ions (NO^+ and O_2^+) produced at the lower altitudes (100–130 km) recombine quickly. At higher altitudes (i.e., 200–300 km), the plasma created by the softer electrons (i.e., a few hundred eV) that ionize the O^+, have much longer lifetimes and hence are important in subsequent transport and dynamic coupling to the neutral gas, as we discuss later on below.

Another example of precipitating, energetic electrons measured by the FAST satellite at 437 km altitude and organized in inverted-V shaped energy functions with peak energies ranging from 100's of eV to several keV is shown in the upper panel of Fig. 43. As discussed above, the inverted-V shapes represent spatial variations of the electron energy that occur

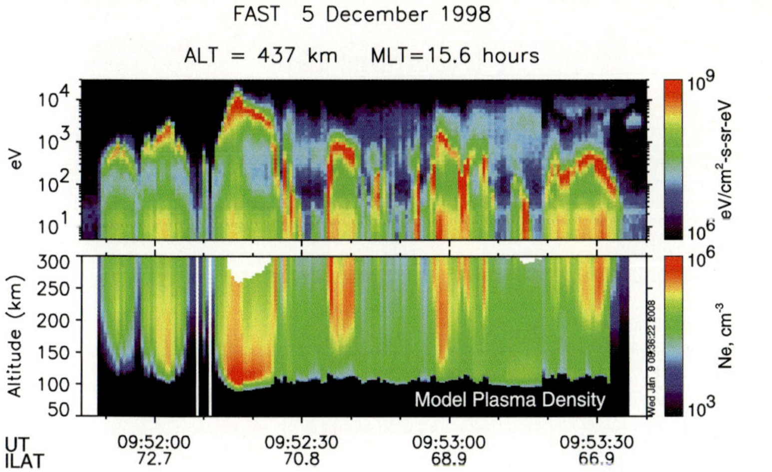

Fig. 43 Observations recorded on the FAST satellite near 437 km are shown in the top panel and reveal narrow beams of downgoing, energetic electron spectra. The bottom panel shows the corresponding thermal plasma created by these electrons, according to the electron transport model described in Strickland et al. (1993)

across auroral arcs. The lower panel shows the expected thermal plasma associated with these measured electron energies, according to the electron transport model described in Strickland et al. (1993). Note how the altitude of the model thermal plasma in the lower panel varies in accordance with the measured energy of the precipitating electrons.

Incoherent scatter radar observations reveal how thermal plasma created by keV electrons impinging on the atmosphere is indeed extended in altitude within the lower ionosphere. Figure 44 (figure courtesy I. Häggström) shows a time sequence of continuous data gathered with a fixed radar beam directed up the magnetic field line, and hence the same space-time cautions that were discussed with respect to the radar data presented in Fig. 24 must again be invoked here. These radar observations were gathered simultaneously with all-sky camera observations showing a series of "intense, high, rayed arcs", as reported by Kirkwood et al. (1988).

Diffuse aurora, which is characterized by precipitation of electrons with a broad energy spectrum (from 100's of eV to tens of keV) over a wide latitudinal extent, can last for hours, creating substantial ionization over a wide range of altitudes (e.g., 100–300 km). The ionization at the higher altitudes persists much longer due the much longer recombination times of the O^+ ions, as noted above and discussed in Sect. 2.

5.3 Characterizing the High Latitude Ionosphere Plasma

The high latitude plasma density is directly formed via: (1) precipitating energetic electrons that create thermal plasma as they are braked by the atmosphere (as discussed above) and (2) photo-ionization. We next discuss the contribution from photo-ionization at high latitudes. In subsequent sections, we discuss how other processes must be taken into account, particularly transport via $\mathbf{E} \times \mathbf{B}$ drifts and winds, in order to specify the state of the high latitude ionosphere at any given location and time.

Photo-Ionization at High Latitudes. In the preceding sub-sections we showed how the aurora is statistically distributed within the high latitude region and discussed how the resulting thermal plasma varies as a function of altitude, depending primarily on the energy of

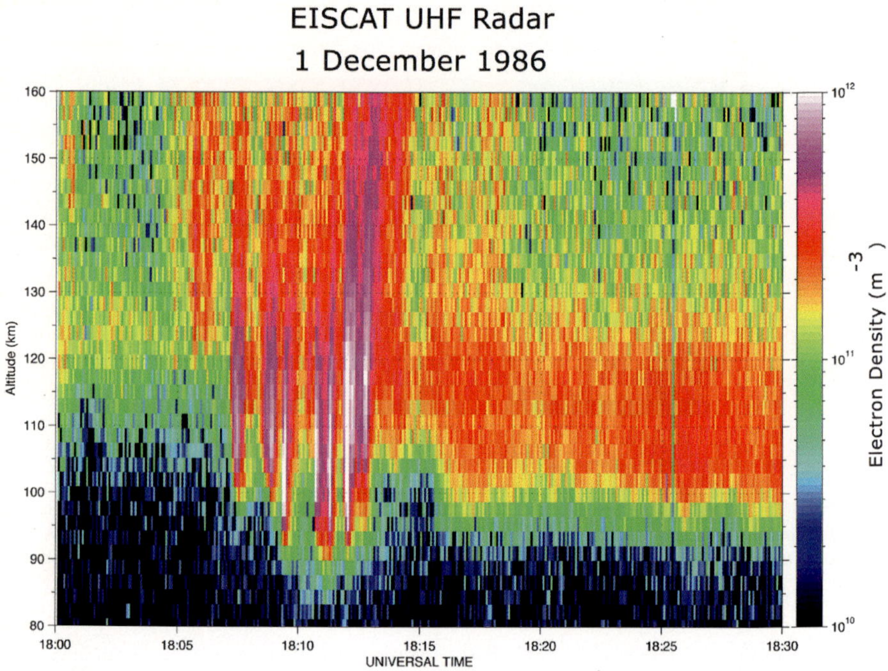

Fig. 44 EISCAT incoherent scatter radar observations of plasma density recorded with a beam oriented in a fixed position up, along the magnetic field direction (figure courtesy I. Häggström)

the precipitating electrons and the background neutral density and composition. Although ionization produced by such energetic auroral particles often dominates the local high latitude ionosphere, photo-ionization at high latitudes is nevertheless an extremely important, regular source of plasma in these regions as well.

Contributions to the plasma density from photo-ionization in the high latitude ionosphere calculated using the IRI model for summer and winter hemispheres at the December solstice are shown in Fig. 45. Although the solar elevation angle is low at high latitudes, the continuous sunlight during the summer creates a conductive E-region over the polar cap to 70 degrees on the midnight side (lower panel). During the winter, on the other hand, the opposite conditions occur, and the E-region is devoid of plasma from the midnight sector over the poles to 70 degrees on the noon side (upper panel). As with the mid- and low-latitude ionosphere, recombination efficiently diminishes the plasma population below about 200 km in the absence of sunlight, yet the F-region ionosphere is maintained, albeit at lower densities, even during the winter months at the pole.

Plasma Temperature and the Polar Wind. The plasma temperature in the high latitude ionosphere is highly variable due to the many different and highly dynamic processes at work in this region. Precipitating energetic electrons heat the ambient plasma to higher temperatures, although the temperature of the secondary electrons as well as heating contributions from associated local plasma waves remain controversial, as discussed in Wahlund et al. (1993). Those authors used EISCAT radar data to show that electron temperature enhancements associated with 100–500 eV electron precipitation correlated well with plasma density enhancements in the altitude region of 170–230 km. Heating via other mechanisms

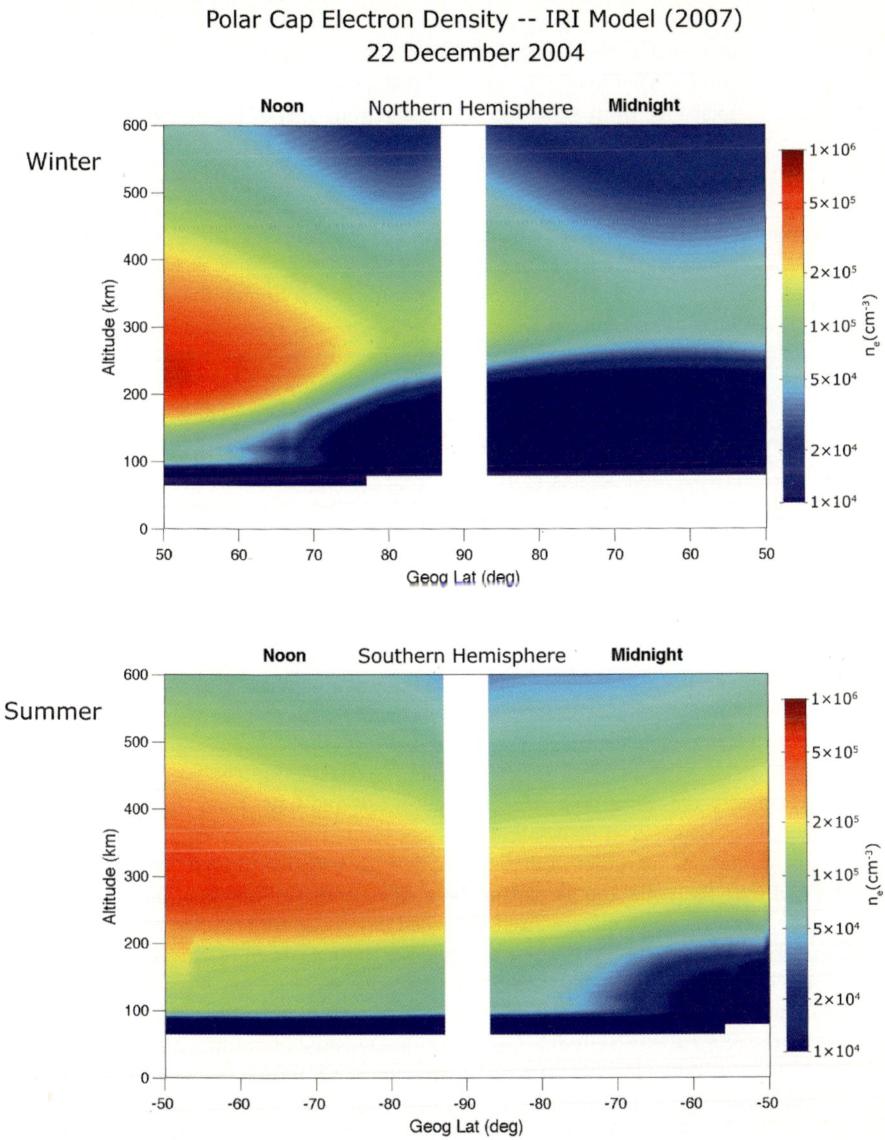

Fig. 45 IRI model calculations (for December 22, 2004) of electron density in the ionosphere starting at 50 degrees latitude, crossing the pole, and extending to 50 degrees latitude on the other side. *Top*: northern hemisphere. *Bottom*: southern. Evidence for numerical deviations in the model at the geographic pole precludes accurate estimates in this region

such as frictional Joule heating driven by perpendicular electric fields (discussed below), and plasma instabilities also create regions of increased plasma temperatures (e.g., Jones et al. 1992; Williams et al. 1992). Although this heating is initially localized, the increased pressure gradients subsequently affect neighboring regions. Despite the strong variability of the auroral precipitation within the high latitude ionosphere, the electron temperature has been modeled for different altitudes of this region by Schunk et al. (1986).

A consequence of the larger temperatures and the nearly vertical magnetic field direction in the polar region is that plasma flows outward along the magnetic field lines, often with significant velocities. This "polar wind" is a feature of the polar ionosphere that results from an ambipolar electric field set up as the electrons rise to higher altitudes along the magnetic field lines due to their higher temperature and lighter mass, as discussed in reviews by Ganguli (1996) and Schunk and Sojka (1997). The polar wind includes both O^+ and H^+ ions that eventually escape the ionosphere into the magnetosphere and distant tail regions. Observations of the polar wind have been measured by instruments on numerous spacecraft, including Dynamics Explorer-1 (Chappell et al. 1982, 1987; Nagai et al. 1984), Akebono (Abe et al. 1993), and Polar (Su et al. 1998). It is important to note that other forms of ion outflow, in which ions are *accelerated* to higher energies via wave-particle interactions, also help populate the magnetosphere with ions of ionospheric origin, as reviewed by André and Yau (1997) and others.

Structure in the High Latitude Ionosphere. The high latitude ionosphere is highly structured, which results from a variety of mechanisms too numerous to review here. (See, Kelley 2009, for a wide-ranging discussion.) The most obvious source of plasma structure is precipitating auroral particles, as illustrated in the auroral images in Fig. 33, 34, 35 and 36. These structures can involve scales sizes from 1000's of km to less than a km. Such density forms have well-defined plasma gradients and associated electric field and current structures that drive local instabilities and produce irregularities with associated scintillations.

An important feature of the structured high latitude ionosphere is the appearance of patches of enhanced plasma density in the F-region that convect over the polar cap in the anti-sunward direction. These patches have horizontal scales that range from 100–1000 km and have associated smaller scale structure and irregularities that also produce radio scintillation effects (e.g., Tsunoda 1988; Coley and Heelis 1995). The patches appear when the interplanetary magnetic field (IMF) is southward and have been shown by numerous researchers (e.g., Buchau et al. 1985; Valladares et al. 1996) to result not from energetic particle precipitation, but rather from detached dayside F-region plasma that is subsequently transported over the polar cap as part of the general anti-sunward flow pattern (discussed below). This narrow, sunward stream of enhanced dayside aurora is referred to as a "tongue of ionization" and is strongest when magnetic activity is high (e.g., Sojka et al. 1981; Middleton et al. 2005; Foster et al. 2005).

The ionospheric plasma is particularly structured within the cusp region. Using a survey of 18 months of ELF electric field data from the Dynamics Explorer-2 satellite, Heppner et al. (1993) showed that the most intense low frequency (4–8 Hz) structures at high latitudes clearly are those in the dayside cusp region. These authors interpreted their electric field variations as either static structures or Alfvén waves. Measurements gathered along vertical profiles provided by a sounding rocket, as shown in Fig. 46, show that the plasma density is highly structured along the portion of the trajectory within the cusp proper on open magnetic field lines, yet is smooth on that portion of the trajectory that is outside of the cusp measured on closed magnetic field lines. These data were gathered in conjunction with low frequency wave electric field irregularities and are interpreted primarily as static structures that created strong scintillations that were observed on the ground during the flight (Pfaff et al. 2003).

5.4 Large-Scale Electrodynamics of the High Latitude Ionosphere: Magnetospheric Electric Fields and Field-Aligned Currents

We now discuss the electrodynamics of the high latitude ionosphere, for which we focus on the coupling of this plasma to the magnetosphere via DC electric fields and field-aligned

Fig. 46 Plasma density along vertical profiles, measured by a sounding rocket in the cusp (*lower panels*). The *upper panel* shows how the rocket payload was launched in the cusp region from Ny-Alesund and remained there during the upleg portion of the trajectory. The rocket then crossed from open to closed magnetic fields lines during the downleg portion of the trajectory, where the plasma irregularities and structure are no longer observed

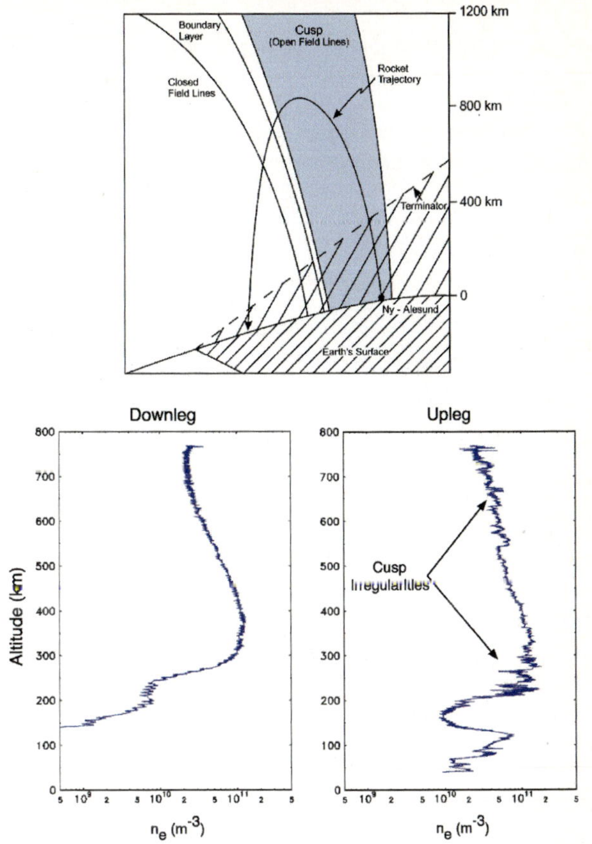

currents. Later, we introduce effects of the neutral atmosphere. Space limitations restrict our discussion here to descriptions of general features relevant to the dynamical and electrical coupling of the ionosphere and magnetosphere. For more detailed discussions, readers are referred to numerous publications on this subject, including Lyons (1992), Richmond and Thayer (2000), Cowley (2000), and Paschmann et al. (2003).

Overview of the Electrical Connection Between the Solar Wind, Magnetosphere, and Ionosphere. The solar wind is a magnetized plasma that flows around the earth's magnetic envelope, or magnetosphere. Figure 47 depicts the magnetosphere in the noon-midnight/north-south plane in which the earth is shown to scale in the center. For reference, the radius of the earth is ~6400 km. Although the near-earth plasma region that is the subject of this article would only extend a short distance on this diagram, the earth's magnetic field enables this region to communicate and exchange energy and momentum with the farthest reaches of the magnetosphere.

Since the solar wind is supersonic, a bow shock and underlying magnetosheath region is formed upstream of the compressed magnetosphere which then deflects impinging solar wind particles. The outer boundary of the magnetosphere is a current layer called the magnetopause and exists where the solar wind pressure and the earth's magnetic field pressure are equal. On the sunward side, the magnetopause is typically located at 8–12 earth radii (R_E), depending on the solar wind dynamic pressure.

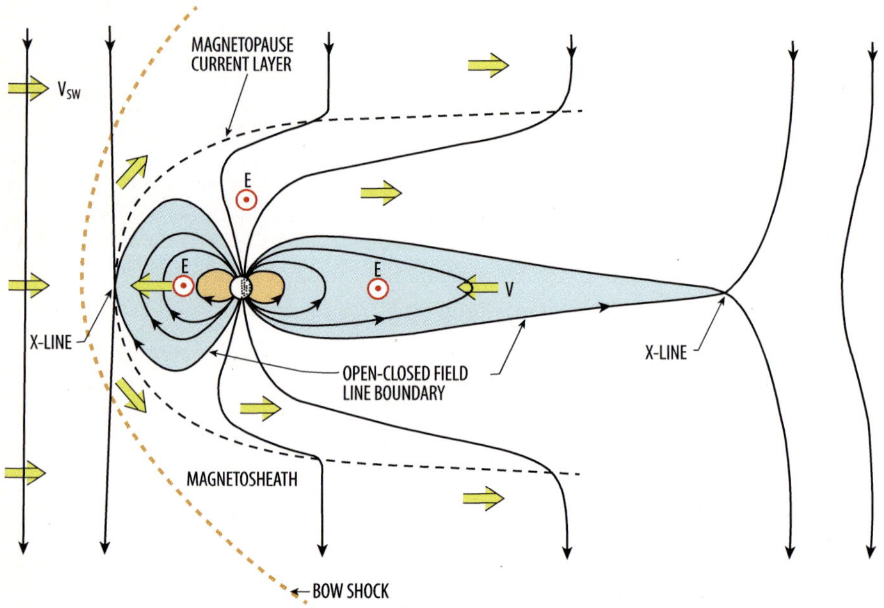

Fig. 47 The main characteristics of the magnetosphere and its interaction with a $B_z < 0$ interplanetary magnetic field. The dawn-dusk electric field is shown, along with the principal direction of the plasma flow. The toroidal volume near the earth (*shaded orange*) is the co-rotating plasmasphere (adapted from Lyons 1992)

Solar wind energy connects through the magnetopause to the magnetosphere via its magnetic field. The presentation shown in Fig. 47 depicts the standard Dungey-type "X-line" configuration of interaction of the earth's magnetosphere with the solar wind with the IMF $B_z < 0$ (Dungey 1961). (The standard coordinates used for describing the sun-magnetosphere system are: x, directed from the center of the earth towards the sun, y, directed from dawn to dusk, and z, directed upwards, perpendicular to the ecliptic.) The X-line represents a merging of the magnetic fields of the earth and of the solar wind and is referred to as magnetic "reconnection" (e.g., Priest and Forbes 2000). In cases where the IMF $B_z < 0$, significant energy is transferred from the solar wind to the magnetosphere, whereas, when IMF $B_z > 0$, this transfer is reduced.

An essential feature of solar wind/magnetosphere interaction described above is that a dawn-dusk electric field is set up as a result of the solar wind flow across the earth's magnetic field dipole, given by $\mathbf{E} = -\mathbf{V}sw \times \mathbf{B}$. This electric field creates anti-sunward $\mathbf{E} \times \mathbf{B}$ plasma flow in the polar cap regions. The polar cap regions are essentially characterized by "open" magnetic field lines that do not close in the conjugate hemisphere of the earth, but are swept back to form the earth's magnetotail. Equatorward of these open field lines is the plasma sheet, which is a region of trapped plasma on closed magnetic field lines that are stretched down the tail and yet whose "feet" map to the auroral oval regions of the high latitude ionosphere in each hemisphere. This region corresponds to the annular regions where auroral optical emissions are most common, as shown earlier in the images in Fig. 36. The direction of the electric field within this ring in the dawn-dusk direction is opposite that in the polar cap, which is key to understanding plasma flows in the high latitude ionosphere.

To illustrate this important feature, consider the magnetosphere depicted from the perspective of looking from the magnetotail towards the sun in Fig. 48 in which only the upper

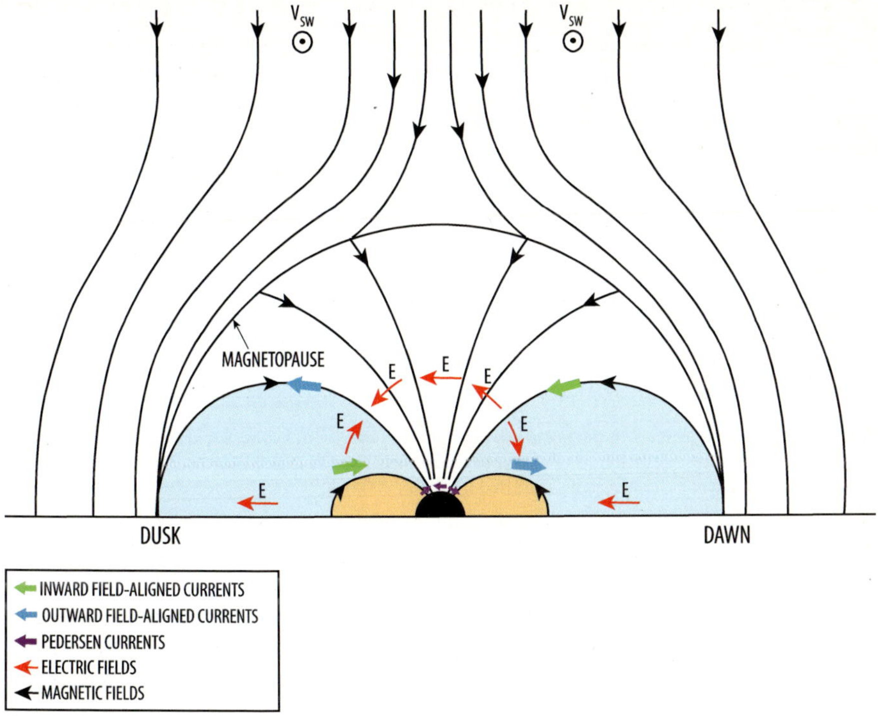

Fig. 48 The magnetosphere from the perspective of the tail, looking toward the sun. The principal current systems and electric fields are shown. The Pedersen currents with which the magnetospheric currents close in the ionosphere are generally much closer to the earth than shown (adapted from Lyons 1992)

half of the magnetosphere is shown. In the polar cap region above the earth at the center of the illustration where the magnetic field lines are essentially vertical, the dawn-dusk electric field extends directly to the earth's lower ionosphere. On the flanks of the polar cap adjacent to the magnetopause, the return plasma flow is sunward at the equator and the associated duskward-directed electric field on each side maps back (earthward) along the magnetic field lines to create an electric field oriented in the opposite direction to that in the polar cap. In this manner, we see that the electric field associated with the return, sunward flow, when mapped to regions closer to the earth, is oriented towards the pole on the dusk side and towards the equator on the dawn side. In each case, these electric fields drive sunward plasma flow in the auroral oval.

The electric field orientation and resulting plasma flow in the high latitude ionosphere is now shown from a polar perspective, in the left hand side of Fig. 49(a). The polar cap and auroral oval electric fields combine to drive a pair of vortices, or convection cells, that are responsible for the plasma circulation within the entire high latitude ionosphere. The shapes and orientations of the dual cell patterns are highly dependent on forcing by the interplanetary magnetic field (IMF), as discussed by many authors (e.g., Heelis 1988; Cowley 2000).

In addition to the patterns of high latitude electric fields, field-aligned currents of magnetospheric origin are also organized in a highly systematic manner. The principal field-aligned currents that connect the magnetosphere and ionosphere are those that flow at the boundaries of the auroral oval in each hemisphere. These currents are depicted in Fig. 49(b) where their

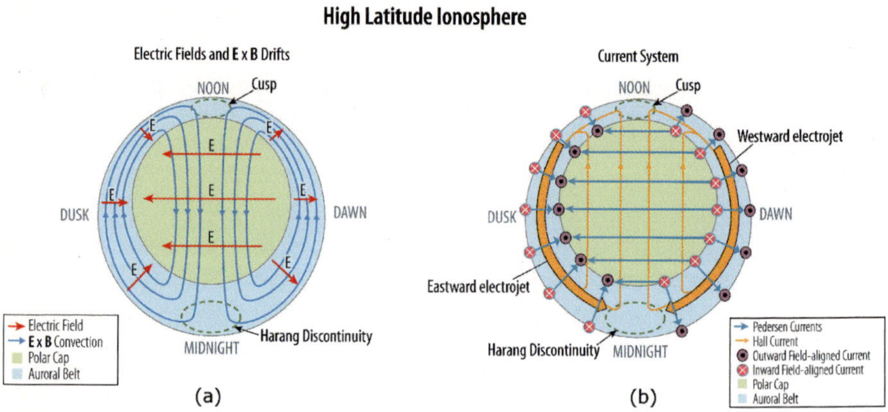

Fig. 49 The electrodynamics of the polar cap region for an idealized, symmetric coupling to the magnetosphere. The illustration on the left (**a**) shows the electric fields and $\mathbf{E} \times \mathbf{B}$ drifts; the illustration on the right (**b**) shows the principal current systems (inspired by Prölss 2004)

relation to the two-celled convection pattern and high latitude electric fields is apparent. The field-aligned currents on the poleward side (the so-called "region 1" currents) connect to the magnetopause, as shown in Fig. 48. Driven by stresses on the dayside magnetopause (e.g., Siscoe and Siebert 2006), the region 1 currents flow out of the ionosphere on the duskside and into the ionosphere on the dawn side. These currents then flow as Pedersen currents in the ionosphere and return to the magnetosphere on the equatorward side of the auroral zone, where they are referred to as "region 2" currents, flowing into the ionosphere on the dusk side and out of the ionosphere on the dawn side.

A second set of currents is driven within the high latitude ionosphere by the applied electric fields discussed above. These fields drive not only Hall and Pedersen currents, given by (6) in Sect. 3 and also shown in Fig. 49(b), but may also drive field aligned currents in the following way. If the conductivity of the ionosphere were uniform, the $\mathbf{E} \times \mathbf{B}$ plasma flows might exist without the need for a field aligned current system to maintain equilibrium. This is not the case in nature. Gradients in the conductivities force ionospheric currents to flow along field lines. Furthermore, as pointed out by Cowley (2000), the Pedersen currents flowing along the electric field direction in the ionosphere cannot close in the ionosphere, but rather a system of field-aligned currents is needed to accommodate their divergence. Note that these field-aligned currents are in addition to the principal, region-1 field-aligned currents driven by the stresses at the dayside magnetopause.

We now present experimental evidence that supports the main features of the electrical coupling between the high latitude ionosphere and magnetosphere described above. All of the data shown were gathered by probes flown on polar-orbiting satellites in the high latitude ionosphere.

High Latitude Convection. Measurements of DC electric fields and plasma drifts have been fundamental in establishing the plasma circulation in the high-latitude ionosphere. Among the earliest satellite measurements of the high latitude DC electric field are those from the OGO-6 satellite (Heppner 1972), for which an example is shown in Fig. 50. These data were gathered on a dusk-to-dawn pass of the satellite across the polar ionosphere. The large amplitude electric fields in the auroral oval which drive sunward flow are clearly visible. The polar cap electric fields are considerably weaker, but clearly correspond to anti-sunward $\mathbf{E} \times \mathbf{B}$ flow.

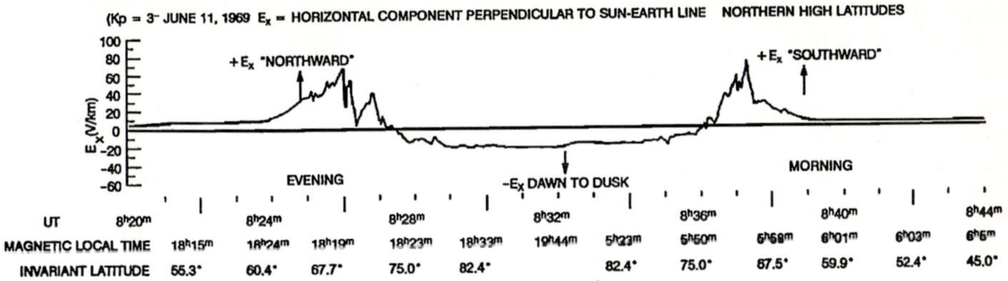

Fig. 50 An example of the horizontal DC electric field component perpendicular to the sun-earth line, observed along the dusk-dawn meridian from a low altitude pass of the OGO-6 satellite over the northern polar cap (Heppner 1972)

Direct measurements of the DC electric fields and plasma drifts in the high latitude ionosphere by numerous authors including Heelis (1984), Heppner and Maynard (1987), and others have shown how high latitude convection changes in strength and pattern as a function of the IMF. In the case where the IMF $B_z < 0$, the plasma drifts are not only stronger, but they also show a much more organized structure with two well-defined cells that may be skewed dawnward or duskward according to the direction of the IMF By component. In cases where IMF B_z is >0, the convection is not well organized, and the observed fields (drifts) are weaker. In fact, during such weakly driven conditions, the potential patterns are such that additional, isolated convection cells may be present (Heelis et al. 1986).

Models of the high latitude convection patterns, such as those based on ionospheric electric fields measured by the Dynamics Explorer-2 satellite (Weimer 1995) show how the two-celled patterns respond to the interplanetary magnetic field, as shown in Fig. 51. This particular presentation shows the potential patterns, which correspond to plasma convection flow lines. These patterns underscore how the general flow of the high latitude ionosphere is governed by the interplanetary magnetic field.

Field-Aligned Currents. In concert with large-scale patterns of magnetospheric electric fields, measurements of field-aligned currents have helped to confirm the basic magnetospheric energy flow processes discussed above. The charges carried by the precipitating energetic particles that create the aurora discussed in previous sub-sections have an associated current that flows along the direction of the magnetic field. Although such field-aligned currents were first proposed by Birkeland (1908) with respect to the aurora, their existence was not fully accepted until space-based magnetometer measurements showed their presence definitively (Zmuda 1966; Cummings and Dessler 1967).

The large-scale field-aligned currents that flow between the ionosphere and magnetosphere are not necessarily associated with optical aurora displays. In fact, these currents, which are almost always present, may be carried by low energy electrons that would not necessarily have sufficient energy to create the impact ionization or optical emissions discussed above. By convention, the field-aligned current is designated upward (downward) for downgoing (upgoing) electrons.

In situ magnetometer data from polar-orbiting satellites revealed that such field-aligned currents are typically observed as a pair of infinite sheets, providing important clues regarding how the majority of the currents might close. The upper portion of Fig. 52 depicts field-aligned currents as flowing vertically along an infinite current sheet that is traversed in the normal direction by a spacecraft carrying a vector magnetometer. The left hand illustration of a single, upward current would produce a step function in the east-west (or

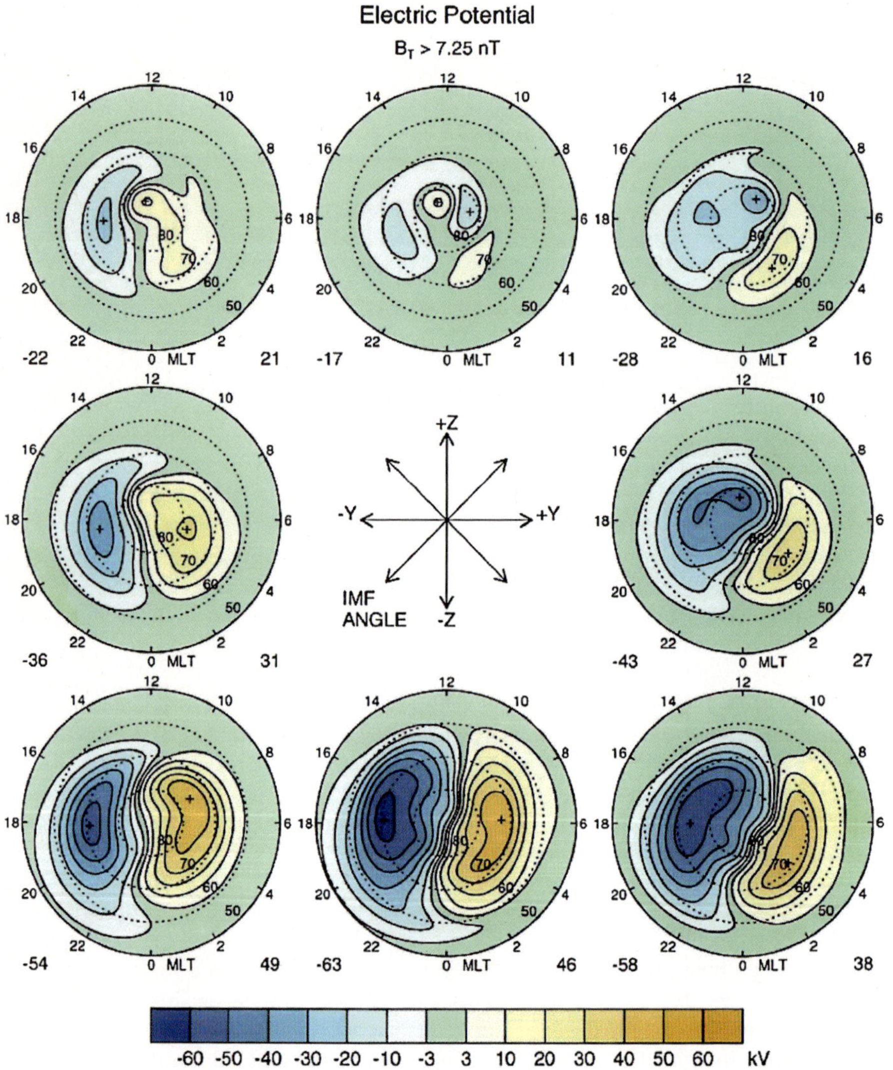

Fig. 51 Models of the high-latitude electric potential as a function of the interplanetary magnetic field (IMF), using spherical harmonic coefficients derived from measurements of the electric field by the Dynamics Explorer-2 satellite (Weimer 1995). The models shown here are for cases in which the IMF magnitude, B_T, was greater than 7.25 nT. See Weimer (1995) for an explanation of details on the figure panels

zonal) component of the satellite magnetometer data. The right hand side shows the magnetic perturbation from two anti-parallel sheets. Magnetometer data from the MAGSAT satellite gathered along a dusk-to-dawn polar orbit are shown in the lower portion of the figure (Zanetti et al. 1984). The observed zonal magnetometer component is consistent with two parallel current sheets of oppositely directed vertical currents encountered on each side of the polar cap. Note that the observed magnetometer signature appears of opposite sign when converted to east-west coordinates on the equatorward-bound satellite trajectory.

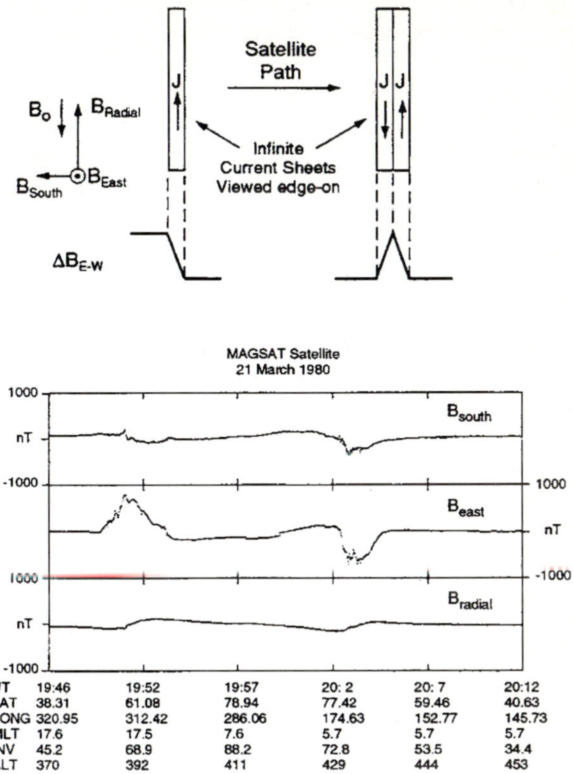

Fig. 52 (*Top*) How data from the east-west component of a vector magnetometer along a polar orbit can be used to deduce field-aligned currents in single and dual parallel current sheets (Pfaff 1996). (*Bottom*) Fluxgate magnetometer data recorded on a polar crossing by the MAGSAT satellite. The measurements show two regions of parallel current sheets, with field-aligned currents into and away from the ionosphere, encountered on each side of the polar cap (Zanetti et al. 1984)

Statistical studies of these events showed that the field-aligned currents form a characteristic pattern, arranged in a belt around the polar cap corresponding to where auroral optical emissions are customarily observed, as shown in Fig. 53 (Iijima and Potemra 1976). Furthermore, the patterns show that the currents on the poleward edge of the polar cap flow into the ionosphere on the dawn side and out on the dusk side, with opposite flows on the equatorward edge of the oval. These measurements were fundamental to demonstrating the patterns of Region 1/Region 2 currents about the auroral oval that connect the ionosphere and magnetosphere, as shown in Fig. 48. Similar patterns have been recorded in other satellite data, for example in the FAST data set by Peria et al. (2000).

In cases where the field-aligned currents are carried primarily by accelerated energetic particles, detailed comparisons of data such as those gathered on the FAST satellite show that zonal magnetometer perturbations match extremely well the field-aligned currents computed from the measured fluxes of field-aligned, energetic electrons (Elphic et al. 1998). For example, returning to Fig. 41 that shows FAST satellite observations within the cusp, notice that in each instance where downgoing packets of electrons were observed, the sign of the zonal magnetic field component is negative, indicative of upgoing currents. Such measurable variations, even for extremely small electron fluxes as shown in this figure, illustrate how commonplace field-aligned currents are in the earth's high latitude plasma.

The primary manner in which field-aligned currents are believed to "close" in the ionosphere is by horizontal Pedersen currents, flowing in the direction of the ambient DC electric field, as sketched in Fig. 48 and Fig. 49(b). Although it is expected that the majority of such currents would close via a neighboring current sheet as discussed above, these illustrations

Fig. 53 A summary of the distribution and flow directions of large-scale, field-aligned currents determined from data obtained on 493 passes of the Triad satellite during weakly disturbed conditions (|AL| < 100γ). The *hatched area* shown between 1130 and 1230 MLT in the polar cusp region indicates the current flow directions are not distinct (Iijima and Potemra 1976)

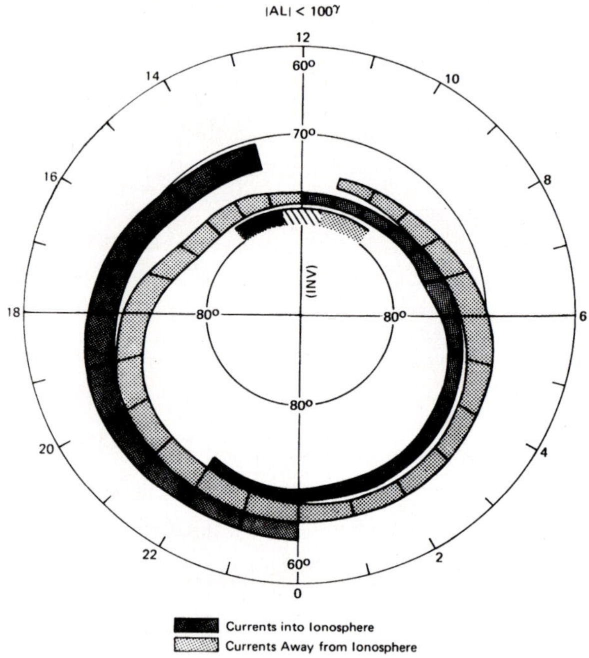

show that some of the field-aligned currents might close via Pedersen currents over the polar cap. In fact, Le et al. (2010), using ST-5 satellite data, demonstrated that such polar cap Pedersen currents were balanced by an observed asymmetry in the measured Region 1/Region 2 currents.

Closure of the field-aligned currents occurs in the lower ionosphere, where the Pedersen conductivity is largest. It is controlled both by the variations with altitude in the local thermal plasma (i.e., local conductivity), determined primarily by the spectra of precipitating energetic electrons, and by the Pedersen mobilities that depend on the charged-neutral collision frequencies and thus the properties of the atmospheric gas (see Sect. 3.2). The sunlit ionosphere on the dayside provides increased conductivity via photo-ionization to facilitate this closure.

Correlation of Electric and Magnetic Fields and Poynting Flux. Observations of DC electric fields in the high latitude ionosphere show a very strong correlation with the magnetic field variations, as shown in Fig. 54 (Sugiura et al. 1982). The strong correlation is generally present between the poleward component of the electric field, which we designate as ΔE_x, and the eastward component of the perturbation magnetic field (i.e., in which the earth's ambient magnetic field has been subtracted), which we designate as ΔB_y. These observations support the infinite current sheet approximation for the field-aligned currents, and is consistent with meridional closure, i.e., via Pedersen currents. Using this assumption and assuming that the Pedersen conductivity is uniform along the measurement track, the two quantities may be related by $\Delta B_y/\Delta E_x = \mu_o \Sigma_p$, where Σ_p is the height integrated Pedersen conductivity and μ_o is the permeability of free space (e.g., Smiddy et al. 1980).

Such combined electric field and magnetic field measurements are also important for calculating the Poynting flux into the ionosphere (e.g., Kelley et al. 1991). The Poynting flux is given by $\Delta \mathbf{E} \times \Delta \mathbf{B}$ and its divergence can be related directly to the energy input to the iono-

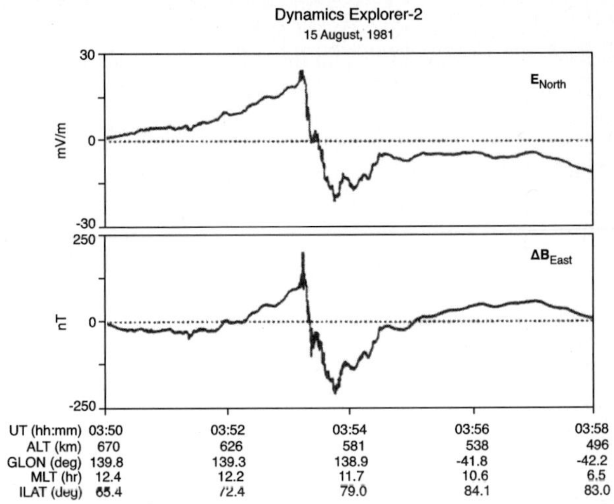

Fig. 54 Data from the east-west magnetic field component (after subtraction of the earth's ambient magnetic field) and the north-south electric field component measured on the polar-orbiting Dynamics Explorer-2 satellite (Sugiura et al. 1982)

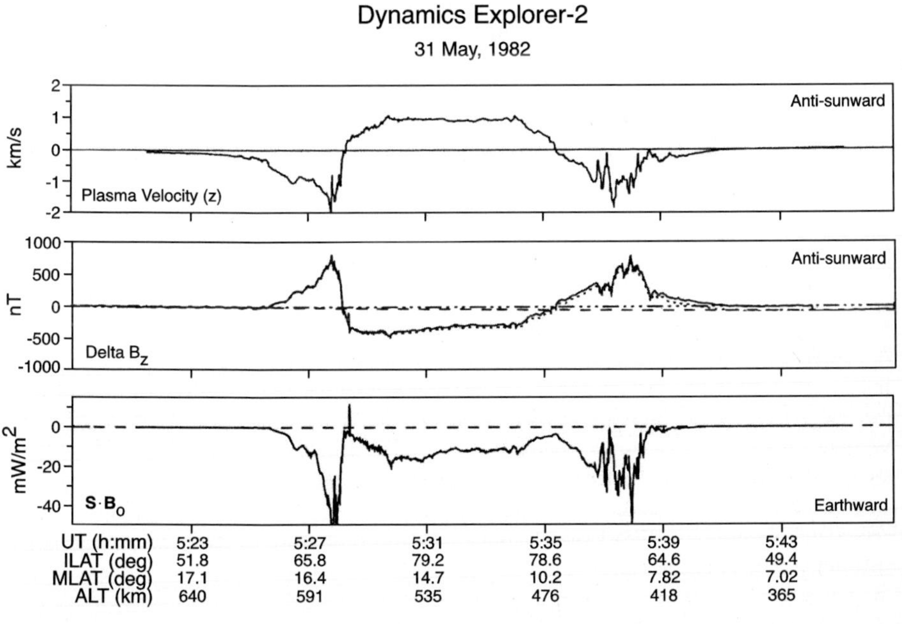

Fig. 55 Ion drift velocity (*upper panel*) and magnetic field (*middle panel*) components in the anti-sunward direction recorded by the Dynamics Explorer-2 satellite for a pass over the northern polar region. The *bottom panel* shows the field-aligned Poynting flux $S_\| = \mathbf{S} \cdot \mathbf{B}_o$, which is primarily from these two components. The data are from Gary et al. (1994), which shows all measured components

sphere that is dissipated as Joule heat, as discussed further in the next section. In Fig. 55, we present an example of field-aligned Poynting flux, $\mathbf{S}_\|$, measured by instruments on the Dynamics Explorer-2 satellite (Gary et al. 1994). In this presentation, the ion drift component is shown, which can be equated to the $\mathbf{E} \times \mathbf{B}$ component at these altitudes (>350 km). The data shown here were gathered on a dusk-to-dawn traversal of the polar cap in which the components of the ion drift and magnetic field were measured in the anti-sunward direction. (See Gary et al. 1994 for a detailed discussion as well as a presentation of all 6 of the measured components.) Not only does this figure again show the correlation of the crossed electric and magnetic field components, but it emphasizes how the electromagnetic energy is directly delivered to the high latitude ionosphere and upper atmosphere—i.e., downward, along the magnetic field direction.

Alfvén Waves. Alfvén waves are essentially ubiquitous in the high latitude ionosphere, as the medium is not in a steady state and motions of "slabs" of plasma (e.g., enhanced conductivity regions corresponding to auroral arcs), or shears in the electric field, will launch Alfvén waves which will carry field-aligned currents (e.g., Kelley 2009). Furthermore, Alfvén "resonators" have been proposed as regions between large gradients in the plasma density, such as between the F-region and the lower portion of the magnetosphere near (say) 3000 km (e.g., Lysak 1993), where Alfvén waves may be trapped. Note that such resonators need not be confined to the high latitude region.

From an experimental standpoint, following the discussion above concerning the high degree of correlation between electric and magnetic fields, we note that the electric and magnetic field signatures within Alfvén waves are also highly correlated and can mimic "static structures", particularly since the static structures can be easily Doppler-shifted by the satellite motion into the same frequency regime as the Alfvén waves. An examination of the ratios of the measured ΔE and ΔB components can be compared to the expected Alfvén velocity, V_A, which is given by $V_A = |B_G|/(4\pi\rho)^{1/2}$, where, ρ is the particle density which depends on the plasma number density and the ambient ion composition. A comparison of estimates of the Alfvén velocity and the integrated Pedersen conductivity (discussed above) can then be used to distinguish the origin of the correlated electric and magnetic fields (e.g., Knudsen et al. 1990). Such comparisons using *in situ* measurements have been discussed in detail by Gurnett et al. (1984), Berthelier et al. (1989), Knudsen et al. (1990), Heppner et al. (1993), and others.

5.5 Atmospheric Interactions with the Magnetospheric Energy and Momentum

In this sub-section, we introduce the role of the neutral atmosphere within the framework of the interactions of the high latitude ionosphere/upper atmosphere with the magnetosphere. We have already shown how the atmosphere brakes incoming charged particles, creating thermal plasma as well as auroral optical emissions, and have noted the atmosphere's role with respect to current closure of field-aligned currents. We now address additional processes that involve the high latitude upper atmosphere and its interactions with magnetospheric electric fields and particles:

1. $\mathbf{E} \times \mathbf{B}$ *Drift/Atmosphere Interactions*
   - Ion drift deceleration below 250 km
   - Winds driven by magnetospheric electric fields
   - Joule heating
   - Electric fields driven by winds and upwards Poynting flux
   - Two-stream waves driven by electrons streaming through collision-dominated ions
2. *Energetic Particle/Atmosphere Interactions*
   - Creation of thermal plasma by energetic particle precipitation
   - Heating due to particle precipitation
   - Formation of Hall and Pedersen currents (in conjunction with the electric field) and field-aligned current closure

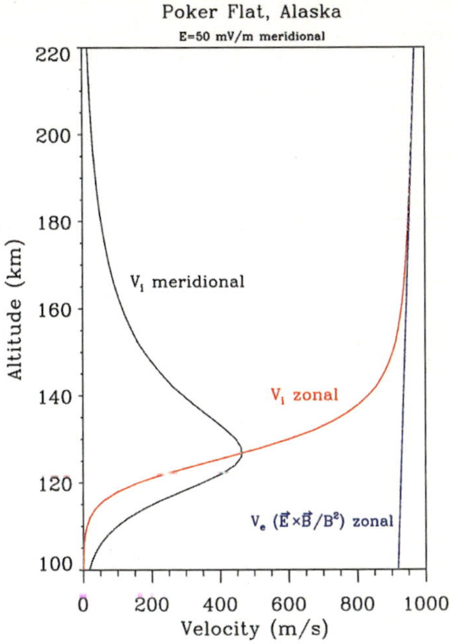

Fig. 56 Ion drift velocity perpendicular to the magnetic field in the zonal (*red*) and meridional (*black*) directions, and the $\mathbf{E} \times \mathbf{B}$ drift in the zonal direction (*blue*), as a function of altitude for a DC electric field of 50 mV/m applied in the meridional directional and assumed to be constant with altitude. The variations of the ion drift component with altitude are largely due to ion-neutral collisions, as explained in the text. The slight variation of the $\mathbf{E} \times \mathbf{B}$ magnitude with altitude is due to small changes of the magnetic field strength with altitude

Ion Drift Deceleration Below 250 km. As discussed above, the DC electric fields that map down from the magnetosphere set the high latitude plasma in motion via $\mathbf{E} \times \mathbf{B}$ drifts. At lower altitudes (below 250 km), however, the ion motion is appreciably affected by ion-neutral collisions such that the ion plasma drift is noticeably slowed and undergoes a change in direction. The electrons, on the other hand, remain fully "magnetized" and continue to execute $\mathbf{E} \times \mathbf{B}$ drifts at altitudes as low as 80–90 km. Below these altitudes, the electron-neutral collision frequency becomes significant when compared to the electron gyro frequency, as shown earlier in Sect. 3.

The ion and electron plasma drifts, calculated versus altitude for a 50 mV/m poleward (meridional) DC electric field at Poker Flat, Alaska ($|\mathbf{B}| = 0.54$ G), are plotted in Fig. 56. Ions and electrons both undergo an $\mathbf{E} \times \mathbf{B}$ drift of approximately 950 m/s at 250 km. However, at lower altitudes, where the ion-neutral collisions become increasingly substantial, the ion drift begins to both slow and change its orientation towards the direction of the DC electric field. At about 125 km altitude, where the Pedersen mobility peaks, the ion drift is one half the value given by $\mathbf{E} \times \mathbf{B}$ and is equally divided between flow along and perpendicular to the electric field direction. As the ion-neutral collisions slow the ion drifts, momentum is transferred to the neutral gas, as discussed next.

Winds Driven by the Magnetospheric Convection. The strong plasma motions in the high latitude ionosphere that are driven by the magnetospheric DC electric fields set the neutral upper atmosphere in motion via ion-neutral collisions. Thus, the electromagnetic energy of the magnetosphere is converted to mechanical energy. To illustrate this coupling, Fig. 57 shows simultaneous measurements of the ion drifts (associated with the $\mathbf{E} \times \mathbf{B}$ drift) and the neutral wind measured by the Dynamics Explorer-2 satellite as it transited the polar region near 350 km altitude (Killeen et al. 1984). The neutral winds essentially track the ion drifts in the evening sector and within the polar cap, yet diverge considerably in the morning sector. Where the two vectors are not aligned is where Joule heating and compositional changes occur, as discussed below.

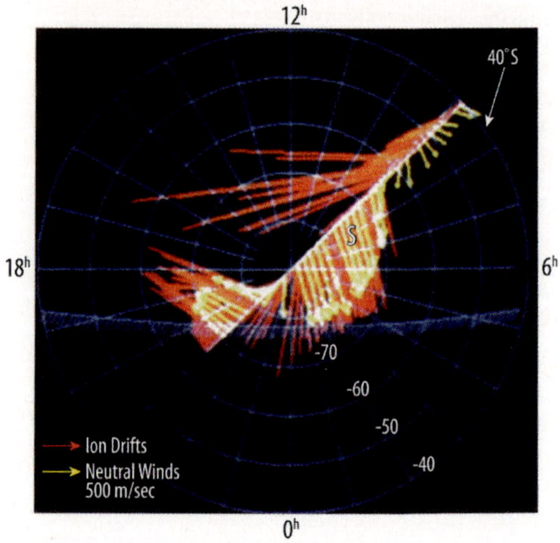

Fig. 57 The neutral wind and ion drift vectors measured by Dynamics Explorer-2 on a perigee pass over the southern polar region, plotted in geographic polar coordinates. The neutral winds are coded by the *yellow arrows*, and the ion drifts are coded by the *red* bars. The *curved line* represents the location of the solar terminator (90° solar zenith angle). The *symbol S* refers to the location of the invariant pole at the universal time of the given pass (Killeen et al. 1984)

As the high latitude neutral motions are influenced by both the amplitude and direction of the $\mathbf{E} \times \mathbf{B}$ drifts associated with the magnetospheric electric fields, the resulting neutral winds show a strong and remarkable IMF dependence in accordance with the $\mathbf{E} \times \mathbf{B}$ convection, as discussed by Killeen and Roble (1988) (see, in particular, their Fig. 18).

Although the neutral wind circulation is expected to follow that of the plasma convection at altitudes of 300 km and above, since the ion velocities decrease and rotate with decreasing altitude below 250 km (see Fig. 56), the neutral wind circulation at lower altitudes is not well established. The situation is further complicated by the fact that the neutral winds are also driven by tidal forcing from below. Furthermore, during auroral events, the plasma conductance is often increased within certain altitude intervals corresponding to thermal plasma created by energetic particles, altering the coupling efficiency of the ions and neutrals. For cases where steady plasma channels might be created via diffuse auroral precipitation, atmospheric models predict that neutral "jets" may form (Brinkman et al. 1995). The existence of such jets and the neutral circulation at low altitudes associated with magnetospheric electric fields are largely unexplored in any systematic way.

Joule Heating. The relative motions of the ion and neutral drifts have a profound effect on the temperature of the neutral gas via Joule heating. Joule heating may be considered the result of friction between the ion and neutral gases that occurs when the gases are not flowing at the same speed and direction. The Joule heating rate per unit volume, Q_j, may be represented as:

$$Q_j = \sigma_p \mathbf{E}'^2 \tag{7}$$

where $\mathbf{E}' = (\mathbf{E} + \mathbf{U} \times \mathbf{B})$ and σ_p is the Pedersen conductivity, discussed in Sect. 3.2. Clearly, the Joule heating depends linearly on the plasma density and on the square of the relative motions of the plasma drifts and neutral winds. Joule heating has a strong altitude dependence due to the variation of the Pedersen conductivity and of the winds with altitude. The neutral gas heating is also affected by the fact that cooling rates increase with decreasing altitude as the neutral density increases.

To illustrate local Joule heating, we return to the example in Fig. 57, but now present in Fig. 58 other quantitative measurements gathered *in situ* by instruments on the Dynamics Explorer-2 satellite during this same event. The neutral winds track the ion drifts throughout

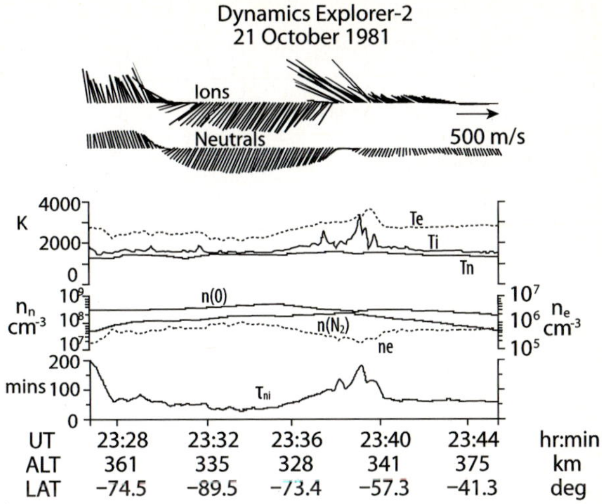

Fig. 58 Geophysical quantities measured along the track of Dynamics Explorer-2 during the orbit shown in Fig. 20(a). The ion drifts and the neutral winds are shown in the top two traces. plotted against time, altitude, and latitude of the spacecraft. The *second panel* shows the electron, ion, and neutral temperatures measured along the track, and the *third panel* shows the atomic oxygen and molecular nitrogen number densities (*left-hand scale*) and the electron density (*right-hand scale*). The bottom trace shows the ion-neutral coupling time constant measured along the track, as discussed in Killeen et al. (1984)

most of the satellite pass over the polar cap prior to 23:36 UT and at these times, there is little local Joule heating expected and hence there is little change in observed ion temperature. Notice, however, how the ion temperatures increase at later times when the drifts diverge (Killeen et al. 1984). This is also a region where compositional changes are observed caused by thermal expansion.

It is important to note that structured electric fields will produce significantly more Joule heating than smoothed fields that typically are used in model calculations (Codrescu and Fuller-Rowell 1995). Indeed, various researchers (e.g., Thayer 1998; Emery et al. 1999) have invoked short scale electric fields to account for a "missing energy" in circulation models and other related calculations. Joule heating represents a fundamental process whereby magnetospheric electromagnetic energy is deposited into the upper atmosphere in the form of heat, and, assuming a steady state and motionless neutral gas, can be shown to be equivalent to the Poynting flux (e.g., Thayer and Semeter 2004). Joule dissipation, frictional heating, and the related energy exchange in the ionosphere/thermosphere system have been discussed by numerous authors (e.g., Thayer et al. 1995; Thayer 1998; 2000; Strangeway 2012) and represent an important means by which the atmosphere acts as a reactive load on the magnetosphere.

Wind Driven Electric Fields and Upwards Poynting Flux at High Latitudes. The winds set up by the plasma drifts driven by magnetospheric electric fields have other profound consequences in the high latitude regions. The winds, unbound by the magnetic field, transport energy and momentum outside the regions of the plasma drifts. Furthermore, due to their inertia, they do not readily change direction as the applied $\mathbf{E} \times \mathbf{B}$ direction changes nor do they cease to flow when the driving DC electric fields are turned off. In other words, the high latitude neutral winds have persistent scales and time constants that are longer than those corresponding to the applied electric fields. Consider the case where the neutral gases are set in motion and then the driving magnetospheric electric fields are shut off abruptly. In this case, the neutral gases maintain their proper motion and generate an associated electric field that then acts back on the system. This "flywheel" effect then generates Poynting flux directed *upwards*, into the magnetosphere, opposite to the customary Poynting flux associated with the impinging magnetospheric energy (Kelley et al. 1991). Such "upwards" Poynting

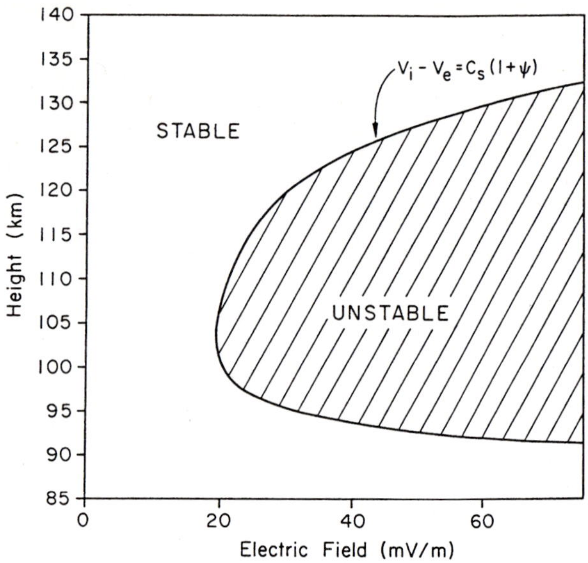

Fig. 59 Altitudes in the high latitude, lower ionosphere (Kiruna, Sweden) where the two-stream instability is unstable according to the linear dispersion relation, as a function of the driving DC electric field (Pfaff 1986)

flux has been observed on the Dynamics Explorer-2 spacecraft by Gary et al. (1994) during IMF $B_z > 0$ conditions.

Two-Stream Waves in the High Latitude, Lower Ionosphere Regions. The collisional two-stream instability represents another common mechanism in the high-latitude lower ionosphere in which energy is extracted from the moving plasma driven by magnetospheric electric fields. This process occurs in the lower ionosphere, where $\mathbf{E} \times \mathbf{B}$ driven electrons "stream" through the ions that are collisionally bound to the neutral gas. A two-stream instability occurs when the relative ion-electron velocity exceeds a threshold given approximately by the ion acoustic velocity, as discussed earlier with respect to the equatorial electrojet. In the auroral electrojet, such relative ion-electron drifts occur at altitudes between 95 and 125 km where the Hall current is dominant. For typical magnetic field strengths at these altitudes within the high latitude regions, the instability threshold is met for DC electric fields of approximately 20–25 mV/m, since the ion acoustic velocity here is about 350–400 m/sec. Although the two-stream instability is typically associated with the auroral electrojets, this condition may be met throughout the high latitude ionosphere, including the polar cap, provided the DC electric field is of sufficient magnitude.

The altitudes where the two-stream instability is generally met are shown in Fig. 59 as calculated for Kiruna, Sweden, by Pfaff (1986). The unstable region is given by the threshold condition of fluid theory (e.g., Sudan et al. 1973) given by $V_i - V_e = C_s(1 + \psi)$. Here, C_s is the acoustic velocity and ψ is the ratio of the product of the ion-neutral and electron-neutral collision frequencies to the product of the ion and electron gyro frequencies. Although ψ is about 0.1 at 105 km at high latitudes and decreases for higher altitudes, this variable grows significantly larger at lower altitudes and accounts for the sharp cut off of the instability at altitudes of about 95 km and below, as shown in the figure. At the higher altitudes, the instability shuts off as the ions become magnetized and the relative ion-electron velocity is reduced. Electric field observations of two stream waves measured on a sounding rocket launched from Esrange, Sweden are shown in Fig. 60 and are described in Pfaff et al. (1992). In general, the waves correspond to meter scales and typically display large

Fig. 60 Observations of DC electric field magnitude (*left*) and wave electric fields driven by the two-stream instability (*right*) measured on the downleg of a sounding rocket flight from Kiruna, Sweden. These data are described in Pfaff et al. (1992)

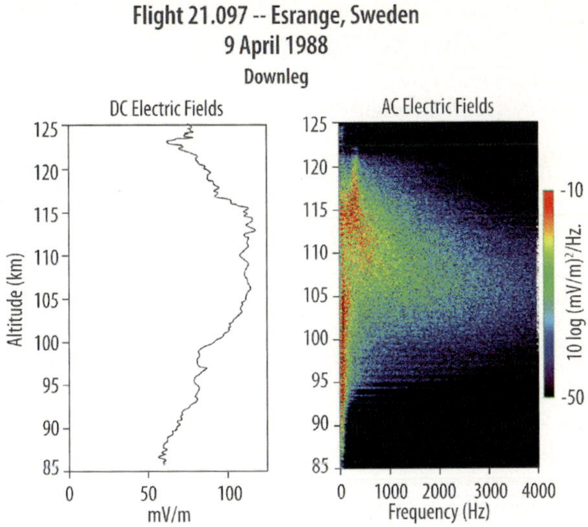

amplitudes in which the wave electric fields are a significant fraction of the driving DC electric field. Furthermore, the waves are known to heat the local electrons, raising the electron temperature by factors of 2–4 (e.g., Schlegel and St.-Maurice 1981; Jones et al. 1991; Williams et al. 1992). Although driven by DC electric fields set up by an entirely different process, the same instability is also common in the equatorial electrojet, as discussed in Sect. 4.

Creation of Thermal Plasma and Heating by Energetic Particle Precipitation. In Sect. 5.2, we discussed impact ionization and how the interactions of energetic particles of sufficient energy create thermal plasma as they are braked by the upper atmosphere. This mechanism represents a fundamental means by which the atmospheric gas and the magnetospheric plasma interact.

Absorbing the kinetic energy of incoming particles heats the neutral gas. On average, as reported by Lu et al. (1995), this heat source is about $\frac{1}{4}$ that of typical Joule heating, although it can be significant during large auroral events (e.g., Thayer and Semeter 2004). The heating caused by the mechanical energy of the precipitating particles is important for the high-latitude energy budget as well for the creation of localized effects. For example, Clemmons et al. (2008) have invoked heating by soft particle precipitation in the cusp to suggest that upwellings might explain the depletions in the neutral density measured by the Streak satellite at low altitudes.

Formation of Hall and Pedersen Currents and Field-Aligned Current Closure. In the lower ionosphere, where the ion drifts are slowed by collisions, differential drift motion of the ions and electrons drives a Hall current. As discussed with respect to the equatorial electrojet, this Hall current maximizes at the lower altitudes where the ions become essentially collision dominated. As shown in Fig. 49(b), on the nightside, the pre-midnight poleward DC electric field drives the electrons westward in the auroral oval, which results in an eastward electrojet, and the opposite is true for the equatorward DC electric field at post midnight, where a westward electrojet is driven. The electrojet currents close in the polar cap via the weaker Hall currents, which depend on both the available plasma density as well as the driving electric fields, and thus the current path is not straightforward, particularly where both of these quantities are highly variable. Where the currents converge in the midnight sector, known as the Harang Discontinuity, as well as in the cusp region on the dayside, field-aligned currents may be produced to keep the overall currents divergence free.

Pedersen currents are also driven by the applied electric fields, as discussed above with respect to the closure of the main Region 1/Region 2 field-aligned currents. In discussing the Pedersen currents it is important to realize that whereas the Pedersen mobility peaks near 125 km, the altitude profile of the thermal plasma on the nightside auroral zone is highly dependent on the energy of the incoming energetic particles, as shown earlier in Fig. 43, and hence the Pedersen conductivity may peak at a different altitude than the Pedersen mobility. In other words, as the Pedersen conductivity is dependent on the combined mobility and plasma density effects, we see that the efficiency of the closure currents for a given auroral precipitation event is thus intricately tied to both the energy spectrum of its energetic particles and the characteristics of the ambient atmosphere.

5.6 Auroral Arc Electrodynamics and Feedback to the Magnetosphere

We conclude this section by briefly returning to the subject of discrete auroral arcs and their associated electrodynamics. Given the relevance of the aurora to the physics of the high latitude ionosphere, a few comments regarding their inherent physical processes as well as their role in coupling the ionosphere and magnetosphere are appropriate.

We showed previously how energetic precipitating auroral particles impact the neutral atmosphere to form narrow channels of enhanced thermal plasma within a limited altitude range, generally below 300 km. In many ways, the creation of thermal plasma is simply one step in a chain of processes that includes the acceleration of the energetic particles and the feedback of the enhanced conductivity on the generation mechanism itself. The many related electrodynamic processes associated with discrete auroral arcs include: (1) the closure of the current associated with the precipitating particles, (2) the effect of the newly formed auroral arc plasma on the ambient DC electric field, (3) the enhancement of electric fields within the arc and along its sides, and (4) the feedback of the enhanced conductivity on the acceleration mechanism responsible for accelerating energetic particles. The reader is referred to the various chapters in the ISSI monograph edited by Paschmann et al. (2003) for a detailed discussion of many of these predominantly plasma processes.

Among the earliest observational results from *in situ* measurements gathered in the auroral ionosphere by sounding rockets is that of the anti-correlation of DC electric fields and plasma density within auroral arcs (e.g., Aggson 1969; Potter and Cahill 1969; Evans et al. 1977; Maynard et al. 1973; Mallinckrodt and Carlson 1985). To first order, such an anti-correlation is explained via Ohm's law, $\mathbf{J} = \overline{\overline{\sigma}} \cdot \mathbf{E}$. In order to maintain current continuity, the electric field is reduced where the conductivity (i.e., inside the arc) is increased. How the modified electric field effects the source potential in the magnetosphere remains an open question, particularly in cases where the conductivity of the entire auroral oval is significantly enhanced (e.g., Fig. 36). The picture is further complicated by the motions of the enhanced conducting "slabs" which affect electric fields on neighboring magnetic field lines. Several researchers have shown that auroral arcs have a proper motion that is different from that of the ambient $\mathbf{E} \times \mathbf{B}$ drifts (e.g., Haerendel et al. 1993; Frey et al. 1996). Indeed, Mallinckrodt and Carlson (1985) suggested that a larger collisional coupling with the neutrals may be necessary to explain the different velocities, and obtained a relationship between the neutral wind and the cross-arc electric field potential.

With respect to the electrodynamics within the arc, the observed electric fields are not always decreased (e.g., Marklund 1984; Paschmann et al. 2003). The creation of polarization electric fields *within* auroral arcs has been explained by invoking enhanced Cowling conductivities within the arc channel (e.g., Boström 1973; Paschmann et al. 2003). Other researchers have reported enhanced electric fields and associated velocity shears at the edges of auroral arcs (e.g., Kelley and Carlson 1977). Measurements with the EISCAT radar have shown that the plasma drifts at the edges of the arcs may be in opposite directions (e.g., Haerendel et al. 1996), which may eventually help explain auroral motions such as curls and folds (e.g., Hallinan and Davis 1970; Hallinan 1976; Davis 1978).

As discussed earlier, it is now fairly well established that electric field structures (sometimes referred to as "shocks") at altitudes of several thousand km and higher accelerate charged particles along magnetic field lines, and that electrons accelerated downward by these structures impact the atmosphere and create the discrete aurora. These electric field structures are set up to enable the ionospheric plasma to maintain continuity of the field-aligned currents. If the available conductivity in the volume in which the currents must flow is too small, as explained by Knight (1973), electric fields oriented parallel to the magnetic field are generated to accelerate the plasma along the magnetic field direction and hence greatly increase the available current in a given volume. (See, for example, Paschmann et al. 2003; Ergun et al. 2004; and references therein.)

Of particular importance is the discovery of "inverse" electrostatic structures of opposite polarity that account for the acceleration of upgoing electrons (e.g., Marklund et al. 1994; Carlson et al. 1998; McFadden et al. 1999). Besides direct measurements of these accelerated particles, as shown by these authors, the satellite probes have measured the perpendicular, U-shaped potentials that have the necessary reversed polarity to produce the oppositely-directed accelerated particles. The currents of these paired shocks appear to be associated with the field-aligned currents discussed above which close in the lower ionosphere. Furthermore, electrodynamic models which include low altitude acceleration of upward-directed electrons have been advanced by Doe et al. (1995) and by Marklund et al. (1997) to explain what appear to be auroral density cavities at altitudes as low as 150 km in the high latitude ionosphere (Doe et al. 1993). These and related observations underscore our limited knowledge of how the ionosphere participates in the acceleration of energetic particles and the closure of field-aligned currents.

To illustrate the importance of the ionosphere in understanding this fundamental mechanism of nature, consider the role of the ionospheric conductivity in the acceleration of energetic particles. Newell et al. (1996) have carried out statistical analysis of aurorae using direct measurements of energetic particle data to demonstrate that discrete aurorae are more likely to be generated in darkness than in sunlight, as shown in Fig. 61. These data support the concept that the lower conductivity of the dark ionosphere requires an auroral acceleration mechanism to be set up in order to produce the keV electrons to carry the current as required by the Knight relation. In a similar fashion, the abundance of the diffuse aurora in the post-midnight auroral zone may provide sufficient conductivity in the ionosphere such that the field-aligned currents are maintained without the need for potential structures to be set up to accelerate the particles. The diffuse aurora thus supplies the conductivity in the same manner as that provided by photo-ionization during sunlit conditions, in accordance with the fact that discrete arcs are not commonly observed during these post-midnight periods.

To summarize, since the ionosphere must support the field-aligned current circuit between the ionosphere and magnetosphere, its properties are essential to maintaining these

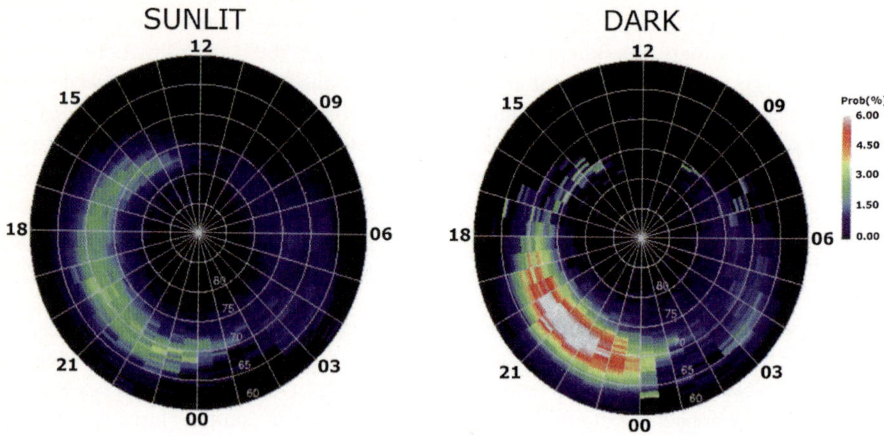

Fig. 61 Probability of observing discrete aurora when the polar ionosphere is sunlit (*left*) and dark (*right*) as a function of invariant latitude and magnetic local time, based on measurements of energetic particles recorded by the DMSP satellites (Newell et al. 1996)

currents and to the formation of the electric field structures that ultimately accelerate both the downgoing and upgoing electrons. In this fashion, the near-earth plasma plays a critical role in establishing the conditions to accelerate energetic particles which create the discrete aurora. Because the neutral atmosphere not only brakes the incoming energetic particles but also enables the high conductivity thermal plasma to form and the currents to close (via Pedersen mobilities), the critical role that the *neutral atmosphere* plays in the coupling of the high latitude ionosphere with the magnetosphere cannot be overstated.

6 Ionosphere Response to Magnetic Storms

We now provide a brief discussion of some of the ways in which the near-earth plasma responds to solar-induced magnetic activity and in particular to magnetic storms. As we shall show, the entire ionosphere including middle and low latitudes undergoes dramatic changes during the strongest magnetic storms. As we cannot describe here all of the ways in which the ionosphere responds to magnetic storms, we direct the reader to a number of reviews on this subject, including Prölss (1995), Forbes et al. (2000), and Mendillo (2006), as well as the articles in the AGU monograph on mid-latitude ionospheric disturbances edited by Kintner et al. (2008).

The near-earth plasma environment is continuously responding to two types of variations of the sun: (1) variations of the flux of solar EUV radiation, as discussed in Sect. 2, and (2) variations of the solar wind at the earth's location which subsequently influence the magnetosphere/ionosphere/thermosphere system. In this section, we focus on the second type, and in particular, the response of the near-earth plasma environment to magnetic storms. Magnetic storms are initiated by prolonged periods of the solar wind's southward ($B_z < 0$) magnetic field direction and are most severe when associated with a "cloud" of solar wind plasma (e.g., Gonzalez et al. 1994). Major storms may last 2–3 days and are typically monitored by the storm-time disturbance index, or Dst. Dst is calculated from magnetometer data gathered at low latitude stations that provide a measure of the magnetospheric ring current energy. Energy transfer from the solar wind to the magnetosphere results in an increase in the intensity of the ring current that corresponds to a negative Dst perturbation (e.g, Daglis et al. 1999). The main phase of a magnetic storm has an associated negative Dst excursion which lasts for several hours, with the largest storms displaying Dst values of <-100 nT.

The magnetic substorm is another important category of magnetic activity to which the near-earth plasma responds (Sergeev et al. 2012). Magnetic substorms are different than magnetic storms in that they are generally much briefer, and result from energy released from the tail of the magnetosphere which then propagates into the high latitude auroral oval. Substorms are characterized by brightenings and movements of auroral displays and have strong associated high latitude magnetic disturbances that result from increased currents. The appearance of substorms as well as magnetic storms are evident in an index of magnetic activity is known as K_p that monitors worldwide magnetic disturbances every 3-hours with an integer index from 0–9 corresponding to a logarithmic scale. A general discussion of geomagnetic indices is provided by McPherron (1995).

The influence of magnetic storms on the near-earth plasma may be considered from two general perspectives: (1) magnetic storm effects that originate at high latitudes, for example increased magnetospheric energy and momentum deposition in the high latitude ionosphere/upper atmosphere, which then have subsequent consequences at lower latitudes, and (2) global magnetic storm effects which are applied to the ionosphere/thermosphere system at all latitudes within a brief time interval after storm onset, for example those effects driven via prompt penetration electric fields. Below, we discuss examples of magnetic storm effects on the ionosphere/thermosphere system, corresponding to each category.

6.1 Magnetic Storm Effects that Originate at High Latitudes

Magnetospheric Energy and Particle Increases at High Latitudes. During a magnetic storm, the polar cap and auroral oval widen and the electromagnetic input and auroral emissions increase considerably. Although these changes are driven by the larger-scale magnetospheric activity, the increased energy and particle input to the ionosphere/thermosphere system is initially concentrated within the high latitude regions. For example, an estimate of increased heating at high latitudes during the magnetic storm of January 10–11, 1997, compared to quiet conditions, is shown in Fig. 62. This heating has been calculated by Richmond and Lu (2000) using a model that assimilates a variety of ground-based and space-based measurements. The upper panels show the energy input due to particle heating, whereas the lower panels show the energy input due to Joule heating. It is clear that the energy input to the high latitude region is significantly increased in both cases compared to quiet conditions and that this energy is greater for the Joule heating. As noted by these authors, the Joule heating has a greater effect on the thermospheric circulation not only because of its larger energy but also because the particle energy is deposited within an altitude region that extends as low as 90 km where the neutral gas, with its significantly larger density, is more difficult to perturb. The storm-driven energetic particles nevertheless provide extremely important conductivity changes via impact ionization, as discussed in the last section, as well as other effects.

Magnetospheric energy input at high latitudes during storms, such as shown in this example, can lead to significant effects at lower latitudes, as reviewed by Fuller-Rowell et al. (2008). For example, strong pressure gradients caused by the enhanced Joule heating in the high latitude regions during active periods drive strong equatorward neutral winds that are not constrained by the magnetic field. Neutral wind changes driven by a sudden increase in Joule heating at high latitudes have been computed via the CTIPe numerical simulation and are shown in Fig. 63 (Fuller-Rowell et al. 2008). The winds are shown at 15 UT for the time corresponding to three hours after the onset of the high latitude forcing. Note the wind

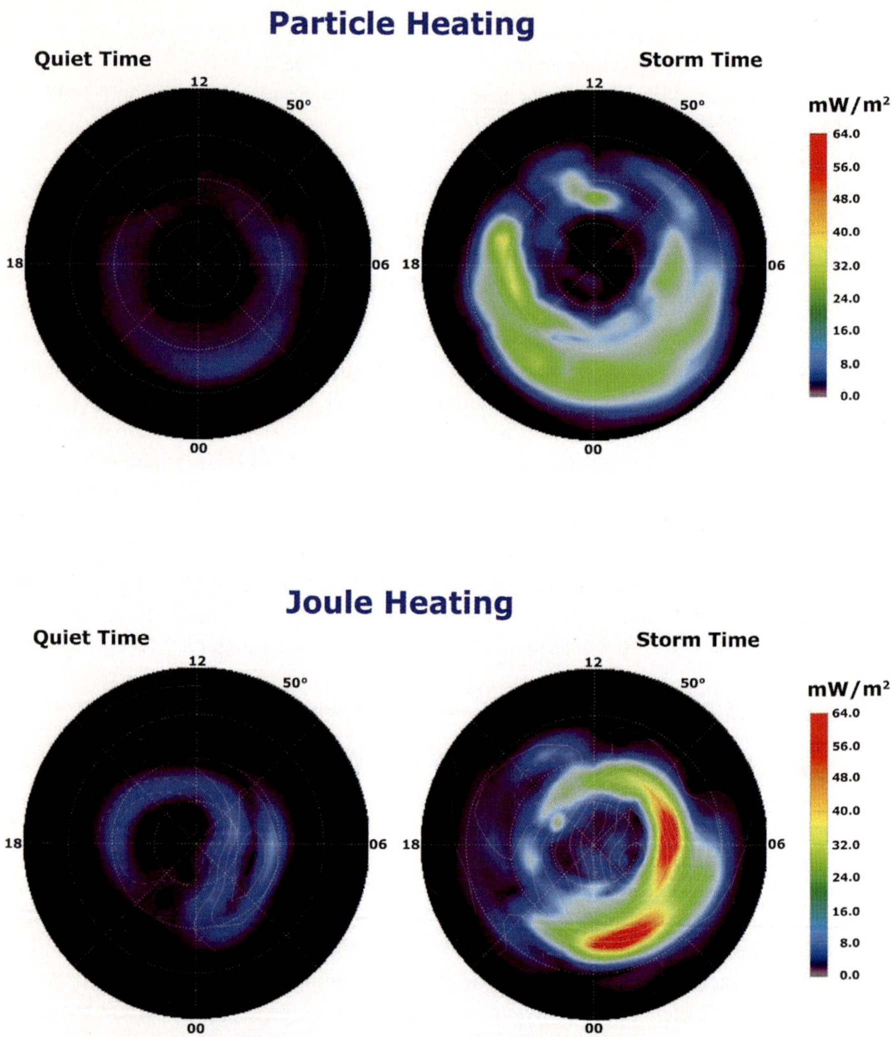

Fig. 62 Distributions in magnetic latitude and local time of the estimated auroral electron energy flux (*top*), the height-integrated Joule heating rate (*bottom*), and the electric potential (*bottom*, contours) above 50° magnetic latitude in the northern hemisphere for a quiet period on 1997 January 10: 0255 UT (*left*), and an active period at 1105 UT (*right*) on the same day. The electric-potential contour interval is 10 kV (Richmond and Lu 2000)

surges of 100–150 m/s which are predominantly on the nightside. Note further that by this time the winds have reached the equator where they will interact with the opposing winds from the other pole, producing an apparent "sloshing" of the winds between hemispheres, as discussed by these authors.

Large neutral winds driven by storms can subsequently lead to changes in the plasma density, electrodynamics, and composition. Indeed, Lu et al. (2009) showed that disturbance neutral winds driven by Joule heating associated with magnetic storms transfer significant energy into the mid and low latitudes in the form of gravity waves. These authors showed that such gravity waves travel with phase speeds close to the local acoustic speed (~500 m/s

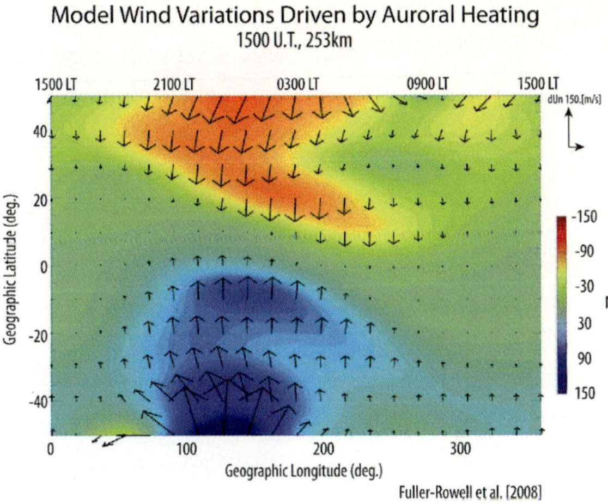

Fig. 63 Model neutral wind variations driven by auroral heating (Fuller-Rowell et al. 2008). The model calculations show the northern component of the wind variation in color, starting at 3 hours after the initiation of an impulse auroral heating event. The vectors (*arrows*) show the total horizontal component, including the zonal wind component, represented according to the scale at the *upper right*

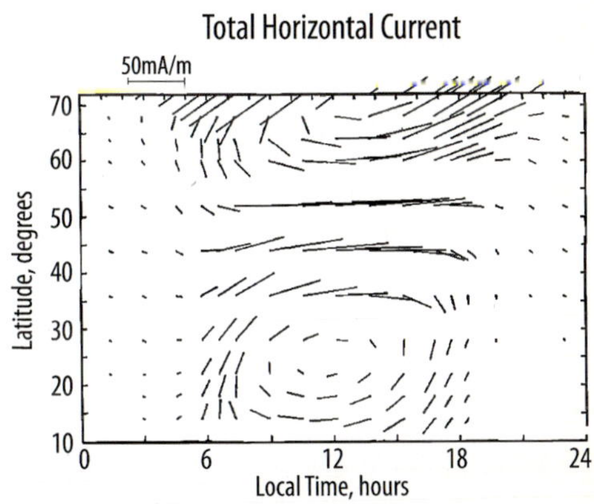

Fig. 64 Model current patterns in the northern hemisphere showing the disturbance dynamo 6 hours after the initiation of a magnetic storm (after Blanc and Richmond 1980)

or higher) and subsequently push plasma up the magnetic field lines, greatly altering the F-region plasma density profile along their track, as observed by these authors in data gathered at the Millstone Hill and Arecibo incoherent scatter radar observatories.

Disturbance Dynamo. Another consequence of the auroral heating and associated winds directed equatorward from high latitude regions is the creation of new global circulation patterns at lower latitudes, as the equatorward winds develop zonal components at mid-latitudes due to Coriolis forces. The resulting circulation creates a system of electric fields and currents known as the disturbance dynamo (Blanc and Richmond 1980). During active conditions, the Sq current patterns discussed earlier (see Fig. 16) actually split into two cells in the upper and lower latitudinal portions of the hemisphere as shown in Fig. 64. Driven by auroral heating at high latitudes, a second cell is introduced with an "anti-Sq" rotation in the upper middle latitudes. Here, strong poleward electric fields driven by strong poleward electric fields that produce a westward $\mathbf{E} \times \mathbf{B}$ drift or eastward currents centered near 45 degree latitude. At the equator, the onset of perturbed east-west electric fields has been shown by Fejer et al. (1983) and Scherliess and Fejer (1997) to be delayed by appropriate time scales, including short (1–12 hrs) and long (22–28 hrs) delays, implying a thermospheric circulation source and supporting the disturbance dynamo theory. Other electrodynamic contributions

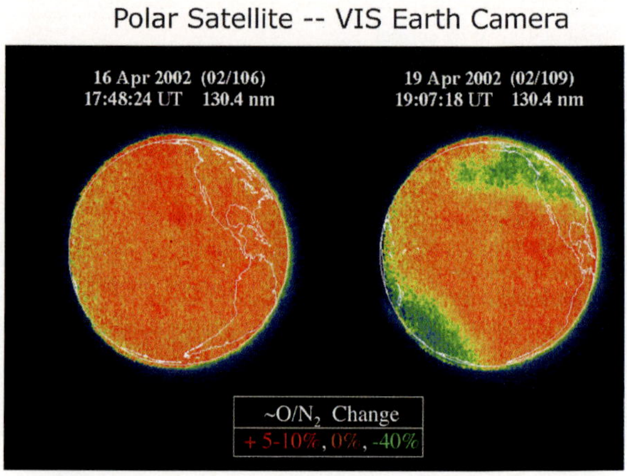

Fig. 65 Images of the earth at 130.4 nm during the daytime as observed by the VIS camera on the Polar satellite, showing the O/N_2 ratios during a quiet period before a storm on April 16, 2002 (*left*), and during a magnetic storm on April 19, 2002 (*right*). The data are provided by J. Sigwarth and J. Kozyra

resulting from magnetic storms include those from penetration electric fields and are discussed further on below.

Composition Changes Related to Magnetic Storms. Other major effects of magnetic storms that are quite spectacular are those associated with neutral composition changes that, in turn, affect the plasma density. Variations in the column-integrated O/N_2 ratio observed by the VIS camera on the Polar satellite are presented in Fig. 65 (J. Sigwarth and J. Kozyra, personal communication). The figures show global views of the 130.4 nm airglow on April 16, 2002 prior to the magnetic storm (left) and on April 19, 2002, during the storm interval (right). Such large-scale composition variations are a well-known consequence of magnetic storms (e.g., Rishbeth et al. 1987; Prölss 1997; Strickland et al. 2001) and are associated with high latitude heating of neutral gases that are then transferred to lower latitudes. The heating causes upwelling of molecules in the air (O_2, N_2) and drives changes in the diffusive equilibrium of the neutral species, enriching heavy species and depleting lighter ones. The molecular species become ionized via charge exchange and then undergo rapid loss as recombination is much faster for the molecular ions. Ultimately, the lower conductivity affects the electrical structure of the ionosphere, as well as its coupling with the magnetosphere. Other examples of changes in the column-integrated O/N_2 ratio have been observed in the GUVI instrument on the TIMED satellite, as shown in Fig. 66, and compare well with simulations using the TIMEGCM model (e.g., Meier et al. 2005). A review of current understanding of the variations of O/N_2 ratios during magnetic storms is provided by Crowley and Meier (2008).

Sub-auroral Polarization Streams. Another consequence of magnetic storms is that they create increased pressure gradients at the inner edge of the plasma sheet, as well as strong conductivity gradients in the ionosphere. This combination results in enhanced, poleward-directed electric fields in the pre-midnight sector at the equatorward edge of the auroral oval (Wygant et al. 1998). These electric fields map into the ionosphere and can become quite large, driving strong westward plasma drifts, particularly in the low conductivity region equatorward of the auroral oval in the pre-midnight sector. This phenomenon is known as sub-auroral polarization streams or SAPS, a term proposed by Foster and Burke (2002), that includes similar phenomena known as polarization jets (Galperin et al. 1973) and sub-auroral ion drifts or SAIDS (Spiro et al. 1979) which have been discussed by numerous

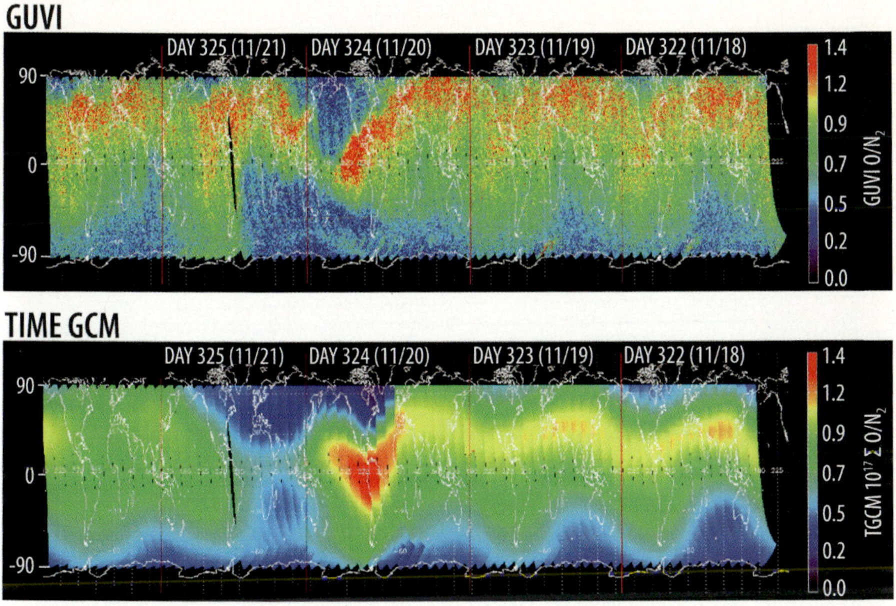

Fig. 66 Observations of O/N_2 ratios as observed by GUVI on the TIMED satellite for five days from November 18–22, 2003 (*upper panel*), and TIEGCM model calculations of this ratio (Meier et al. 2005). The data are essentially recorded at the same local time, in this case near noon. Note that the observing time (i.e., UT) runs from right to left within each day panel and the panel boundaries are near 0000 UT

authors (e.g., Smiddy et al. 1977; Anderson et al. 1991, 1993; Karlsson et al. 1998). Situated equatorward of the auroral precipitation (hence the term, "sub-auroral"), these streams are closely related to where region 2 current are located. In fact, the electric fields are believed to increase within the low conductivity regions to maintain Pedersen current closure of the field-aligned currents. The appearance of enhanced, latitudinally narrow SAIDS has been shown to be related to substorms (e.g., Anderson et al. 1991, 1993; Karlsson et al. 1998). The SAPS phenomenon is an important element in the electrodynamics of magnetic storms (e.g., Burke et al. 2000).

Examples of the enhanced electric fields associated with the sub-auroral polarization streams measured by the Dynamics Explorer-2 satellite are shown in Fig. 67 together with simultaneous measurements of the plasma density (Pfaff et al. 1988). The upper three rows show examples of large amplitude (some greater than 300 mV/m) enhancements with narrow latitudinal extent, whereas the bottom row shows examples of broader regions of enhanced electric fields with somewhat lower amplitudes. Examples of polarization streams that are wider and not peaked have also been reported by Foster and Vo (2002). Notice that the enhanced electric fields are often coincident with sharp decreases in the plasma density, though this is not always the case. The enhanced electric fields have been observed in conjugate hemispheres, as indicated by the pairs of data gathered on the same orbit. These observations are not magnetically conjugate as the satellite orbit did not intersect the same magnetic field line as it encountered the SAPS event in the northern and southern hemisphere. However, the data clearly show that the electric fields are enhanced over a wide range of magnetic local time. Simultaneous measurements of this phenomenon at high (9000 km) and low alti-

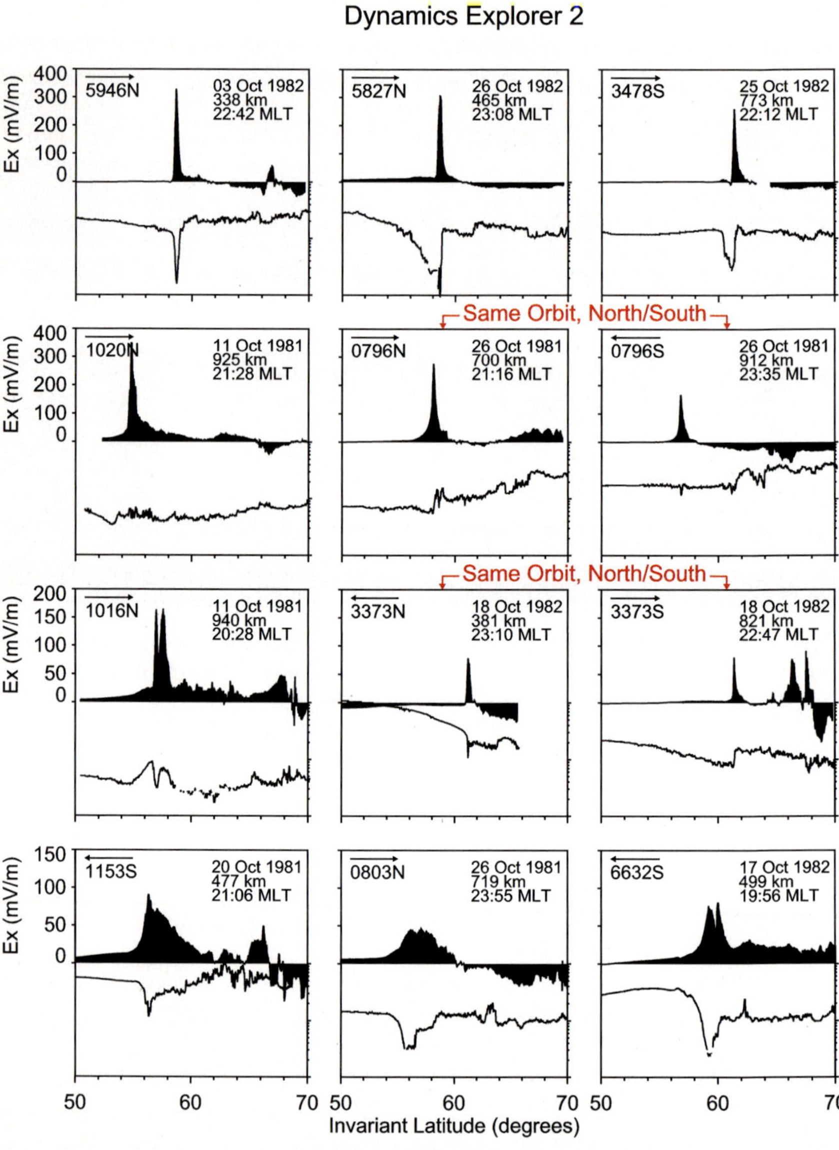

Fig. 67 Examples of large "sub-auroral" electric fields and concurrent plasma density observed by the Dynamics-Explorer-2 satellite. These electric fields are oriented toward the pole producing westward plasma drifts often referred to as SAPS (sub-auroral plasma streams). Two pairs of measurements shown here were recorded along the same orbit, and reveal similar electric fields in conjugate hemispheres

tudes (840 km) by probes on the Akebono and DMSP satellites, respectively, were reported by Anderson et al. (2001), and support the fact that these fields map between hemispheres along the magnetic field.

SAPS have profound effects on the local ionospheric conductivity and composition, as demonstrated in a series of papers by Anderson et al. (1991, 1993). Pointing out that the rate of the charge exchange reaction (discussed in Sect. 2):

$$O^+ + N_2 \rightarrow NO^+ + N$$

depends on the 4th power of the velocity differential between the ions and the neutrals (Schunk and Banks 1976), these authors showed that the NO^+ concentration will increase substantially in regions where strong SAPS are present. Since the recombination rate of NO^+ is much faster than O^+, the plasma number density will decrease further in regions of large electric fields. Anderson et al. (1991) also showed that large vertical ion flows that result from heating via ion-neutral collisions within SAPS will also contribute to the density depletions in the sub-auroral ionosphere.

6.2 Prompt "Global" Magnetic Storm Effects

Penetration Electric Fields and Their Effect on Mid- and Low Latitude Plasma Density. The broadening of the auroral oval and the transfer of the Joule heating energy to mid-latitudes is only part of the story of the upper atmosphere's response to magnetic storms. For the largest storms, the entire ionosphere may be subjected to global scale variations such as those driven by prompt penetration electric fields. These have profound effects on the entire ionosphere/upper atmosphere at all latitudes. Such penetration fields are associated with southward turnings of the IMF and are believed to represent interplanetary electric fields that have penetrated the inner magnetosphere to the mid- and low-latitude ionosphere. Examples of such fields are shown in Figs. 68a, 68b in which the zonal electric field measured at the Jicamarca Radio Observatory in Peru (at the magnetic equator) increases in concert with the southward turning of the IMF B_z component (Huang et al. 2005). (Note that these authors have included a time shift to allow for the solar wind to travel from the satellite measurement location in the solar wind upstream of the magnetosphere to the ionosphere, which obscures a detailed comparison of the timing of the solar wind events and the exact onset of the ionospheric effects.) Although the details of such penetration fields as well as the shielding efficiency and time constants of the inner magnetosphere are still being debated, these fields appear to have dramatic effects on the distribution of plasma in the low and mid-latitudes.

One of the most important responses of the global ionosphere to large magnetic storms is that the low latitude ionospheric plasma can become depleted within a significant latitudinal extent centered on the magnetic equator, at least for localized longitude regions. Plasma density measurements from consecutive passes of the DMSP satellite in a circular orbit at 840 km clearly show how the ionospheric plasma "disappeared" from $-20°$ to $+20°$ (magnetic latitude) during a large magnetic storm (at least for the longitudes sampled), as shown in Fig. 69 (Greenspan et al. 1991). The simultaneously measured upwards plasma drifts were in excess of 200 m/s and are shown by these authors to be consistent with plasma being elevated above the satellite altitude. In this manner the phenomenon is similar to the fountain effect (discussed in Sect. 4), driven by sustained, large amplitude east-west zonal electric fields. Discussed in terms of a so-called "super fountain" effect, Heelis (2008) argues that, taking into account neutral winds and other effects, most of the plasma transport under such conditions is poleward (i.e., horizontal). When such density "holes" are created during the daytime due to the large plasma uplift, note that photo-ionization continues at the lower altitudes and the creation of a new F-region proceeds.

An example of large increases in the plasma density in the upper F-region during the daytime occurring within a significant magnetic storm is provided by measurements of total electron content (TEC) above 400 km, as measured by the CHAMP satellite and shown in the lower panel of Fig. 70 (Mannucci et al. 2005). Here, data from three consecutive satellite orbits (spaced ~90 minutes apart) show how the TEC changes during the storm main

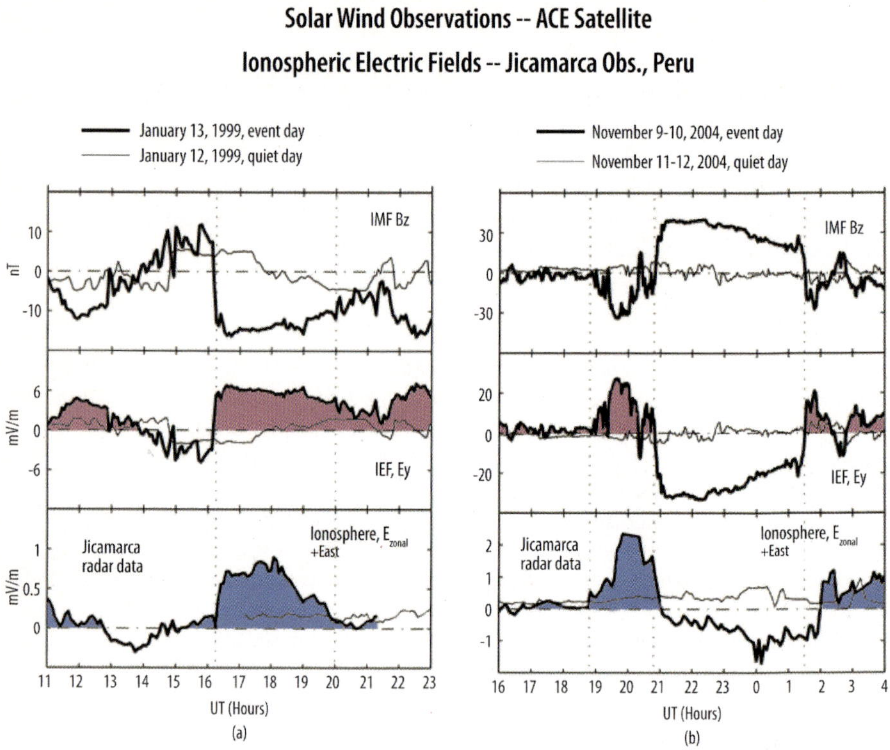

Fig. 68 Solar wind IMF B_z (*top*) measured by the ACE satellite with associated interplanetary electric field, E_y (*middle*) and the equatorial zonal ionospheric electric field (IEF) E_{zonal} measured by the Jicamarca radar (*bottom*). Data from two storm periods are shown, for January 13, 1999 (*left*) and November 9–10, 2004 (*right*). Also shown are data recorded during quiet conditions on neighboring days for comparison. The periods shaded in *red* and *blue* highlight the interval of positive IEF, corresponding to southward IMF, and enhanced eastward ionospheric electric fields, respectively. Notice the increased scales in the panels on the right, both for the solar wind measurements as well as for the electric fields measured at Jicamarca (adapted from Huang et al. 2005)

phase, indicated by the decrease in Dst, which is plotted in the upper panel. Notice how the maxima of the equatorial anomaly that are normally observed nearer to $\pm 15°$ (see blue trace corresponding to 19:00 UT) were "pushed" to values of $\pm 30°$ (see red and black traces) yet still retained their approximately symmetric, winged shape. These data were gathered near 1300 hrs SLT at low and middle latitude (A. Mannucci, personal communication, 2012) and show prompt dayside total electron content increases of $\sim$900 % above 400 km, suggesting that the normal, daytime ionosphere at lower altitudes had risen above the CHAMP satellite altitude. These authors also invoke prompt penetrating electric fields which create plasma uplift with associated diffusion along the magnetic field lines to higher latitudes to explain both the large TEC values and the observed, symmetric latitudinal spreading. A different interpretation for the storm-time density enhancements on the dayside is that due to high latitude convection electric fields expanded equatorward, as discussed by Heelis et al. (2009a, 2009b) and Heelis and Mohapatra (2009).

Examples of Low and Mid-Latitude Irregularities Associated with Magnetic Storms. To demonstrate further how the ionospheric storm response appears to be initiated at the equator

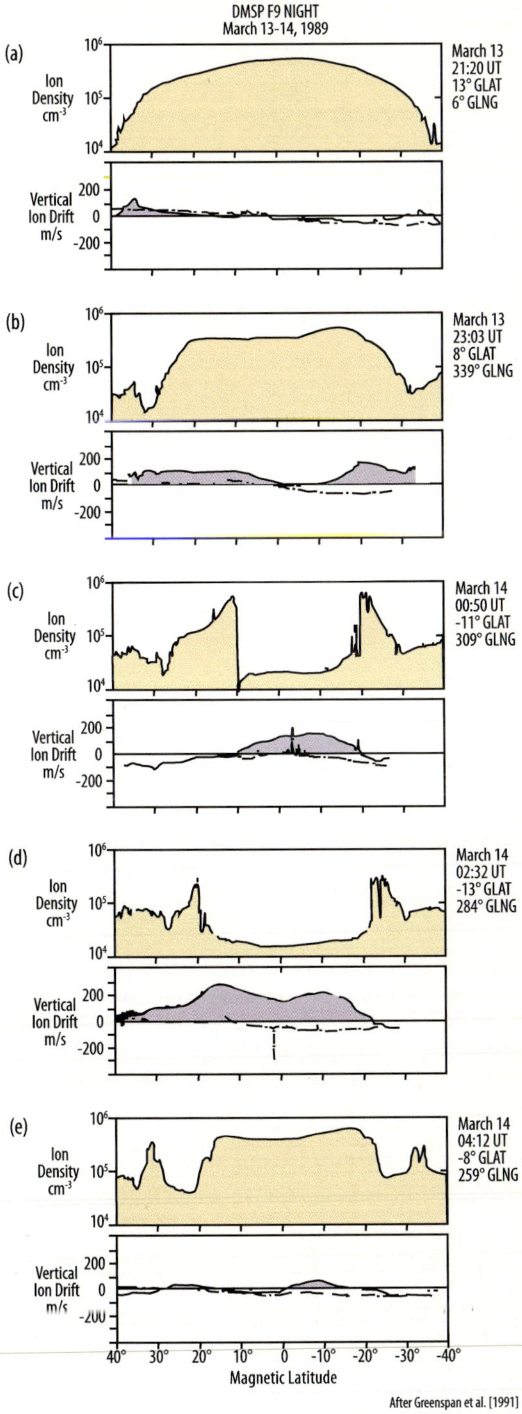

Fig. 69 Log of total ion density in cm^{-3} (*upper panels, light orange*) measured by the DMSP spacecraft plotted as a function of magnetic latitude for five successive passes, each near 2130 LT, during the magnetic storm of March 13–14, 1989. The *lower panels* show measurements of the vertical ion drifts in which the positive values are shaded *blue*. The labels to the right of each plot give the universal time and geographic latitude and longitude at which the spacecraft crossed the magnetic equator (adapted from Greenspan et al. 1991)

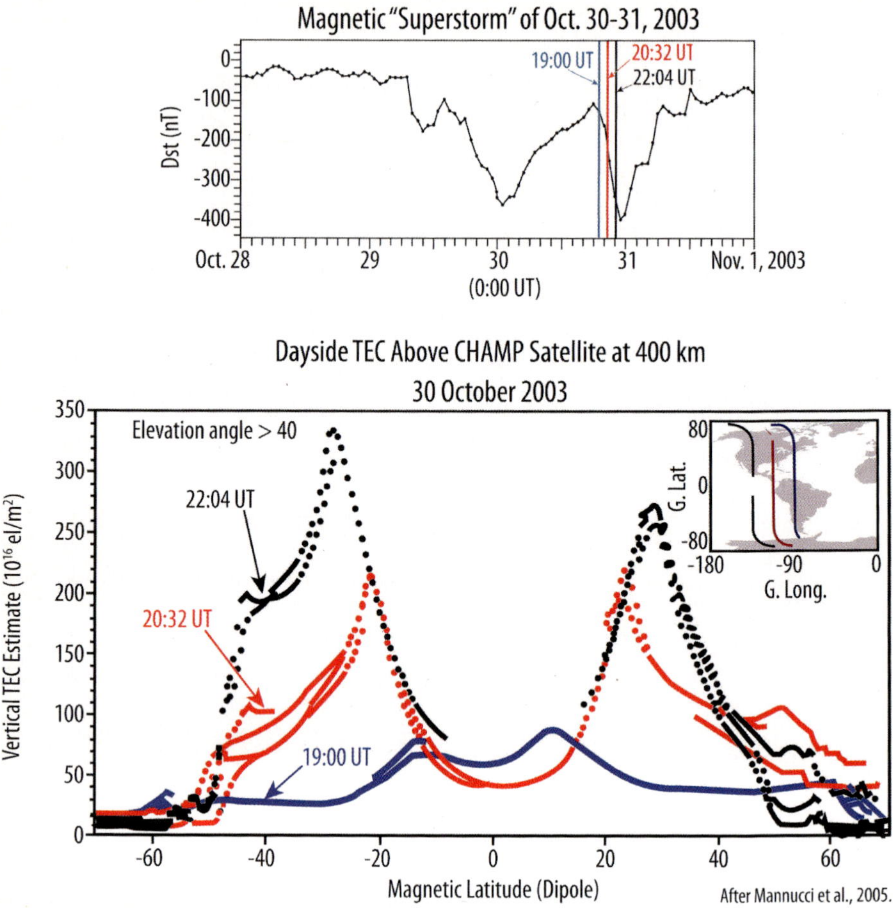

Fig. 70 The integrated electron content (corresponding to total electron content or TEC) as measured by the GPS instrument on the CHAMP satellite for altitudes above the CHAMP altitude of 400 km (*lower panel*). Data are shown for three orbits plotted as a function of magnetic latitude which correspond to times just prior to (*blue*) and after (*red* and *black*) the onset of a magnetic storm on October 30, 2003, as shown in the plot of Dst in the *upper panel*. The locations of the CHAMP orbits are shown in the upper right corner of the lower panel. The local times of these orbits range from 1230–1330 LT for latitudes within ±60 degrees. Points missing near the anomaly trough are due to the elevation angle cut-off (adapted from Mannucci et al. 2005)

and then move polewards simultaneously in both hemispheres, at least during the evening, we show DEMETER satellite data in Fig. 71. Consecutive orbits of plasma density and electric field irregularities are presented for two different magnetic storms, as shown in the Dst data in the upper panels (Pfaff et al. 2008). These data were gathered along circular orbits at 710 km in the evening (~22.5 LT) sector. At the storm commencement the first indication of both enhanced plasma density and associated irregularities at 710 km occurred at the equator, and then proceeded towards the poles in both hemispheres as the storm progressed. The irregularities associated with the inter-hemispheric, unstable flux tubes continue to rise in altitude and latitude, symmetrically in both hemispheres and are encountered by the consecutive DEMETER orbits at successively higher latitudes. These data reinforce the hypothesis that the storm response at low and mid-latitudes is initiated at the equator. At the same time,

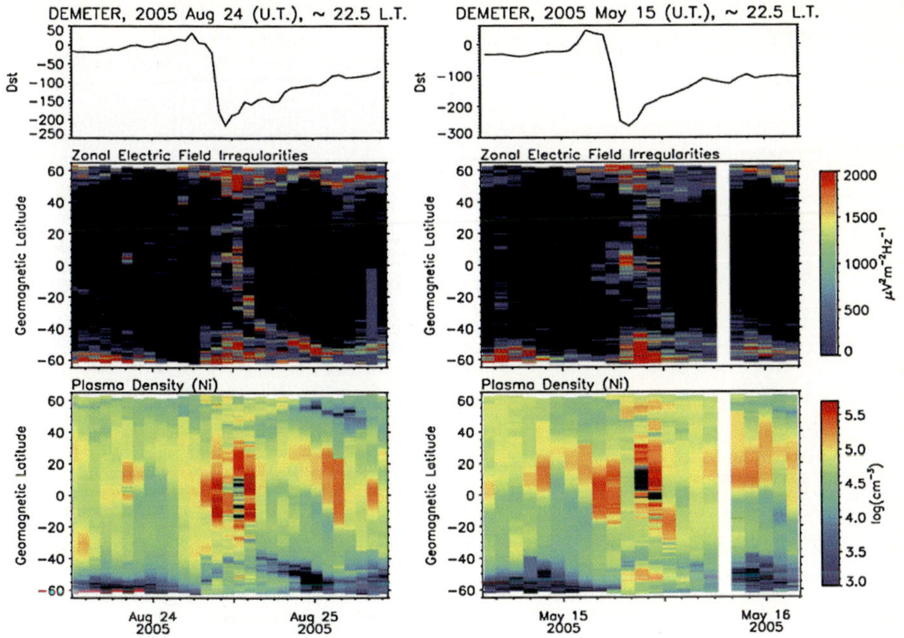

Fig. 71 DEMETER satellite data (near 22:30 LT) for the magnetic storm periods of August 24, 2005 (**a**) and May 15, 2005 (**b**). The *top panels* show the Dst estimates based on ground-based magnetic field measurements. The *middle* and *lower panels* show the DEMETER integrated electric field irregularity power from 10–100 Hz and the plasma number density along each orbit. The satellite data have been converted to color scales and mapped to bins of magnetic latitude between −65 and +65 degrees (Pfaff et al. 2008)

the polar cap high latitude ionosphere widens and expands equatorward, as evidenced in the lower panels in Fig. 71 where the plasma density decrease or trough near 60 degrees magnetic latitude prior to the storms expands equatorward towards ±40 degrees magnetic latitude in both the northern and southern hemispheres.

Ground-based observatories at mid latitudes have also observed ionospheric irregularities associated with magnetic storms that provide important clues concerning whether unstable flux tubes might be set up between hemispheres, as suggested by the DEMETER data above. Two examples of radar measurements gathered in Puerto Rico during magnetic storms are shown in Fig. 72. The upper right-hand panel shows data gathered with a VHF coherent scatter radar looking northward from Puerto Rico (figure courtesy E. Kudeki; see Swartz et al. 2000) and shows irregularities emerging from what is believed to be descending F-region ledge, similar to those associated with low latitude spread-F. The lower panel shows Arecibo incoherent scatter measurements during a similar magnetic storm (Kelley et al. 2002). Although the fixed radar beam cannot distinguish temporal and spatial variations that may be at work during this event, the data suggest the presence of depleted flux tubes that emerge from the F-region ledge and which may be expected to give rise to electrostatic irregularities. Other measurements (e.g., TEC and airglow depletions) during these and similar magnetically-disturbed events at mid latitudes have been reported by other researchers (e.g., Makela et al. 2000; Kelley et al. 2000; and Bust et al. 2000) and demonstrate that the mid-latitude ionosphere becomes highly structured at night during magnetically disturbed conditions.

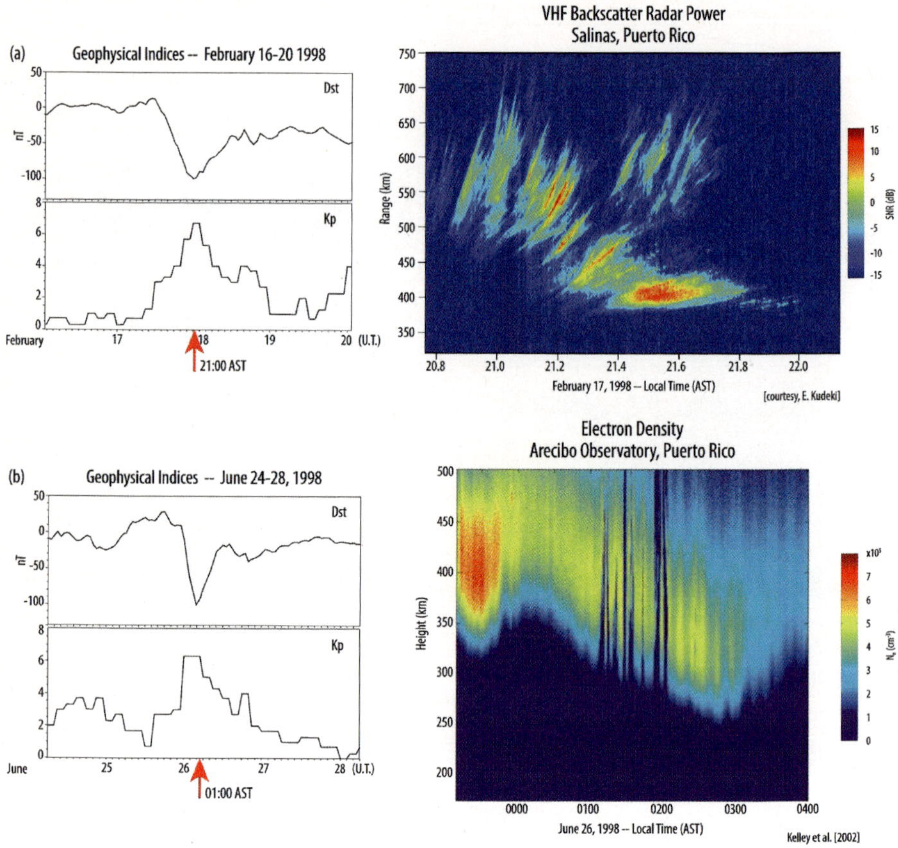

Fig. 72 Measurements by ground-based instruments in Puerto Rico of plasma density structures associated with two modest magnetic storms, whose Dst and Kp signatures are shown on the left hand side. The *panel on the upper right* shows 50 MHz backscatter radar measurements that resemble spread-F scatter that were coincident with the period of largest negative Dst (figure courtesy E. Kudeki; See also Swartz et al. 2000). The *panel on the lower right* shows plasma density measurements from the incoherent scatter radar at Arecibo that were also coincident with the period of the largest negative Dst (Kelley et al. 2002). The large plasma depletion "chimneys" that emerge from the lower edge of the F-region between 0100 and 0200 LT resemble spread-F depletions more commonly observed at lower latitudes

The panels showing the Dst data on the left hand side of Fig. 72 indicate that in each case, the radar irregularities and depletions were observed at the peak times of their respective storms. The fact that the radars in each case were situated on the nightside during each of these two storm events, in which the main phase lasted less than 12 hours, undoubtedly helped enable the observations of these particular storm signatures. These mid-latitude observations of F-region irregularities and depletions resemble spread-F processes known to occur at low latitudes at night.

Storm-Enhanced Density Structures. A final example of significant effects on the near-earth plasma from magnetic storms is planetary-scale plasma density structures that appear to break off from the enhanced mid-latitude plasma that accumulates during a large amplitude magnetic storm and then is transported poleward, as shown in Fig. 73 (figure courtesy of A. Coster). These plumes are referred to as storm enhanced density (SED) erosion plumes (see Foster et al. 2005, and references therein) and are evi-

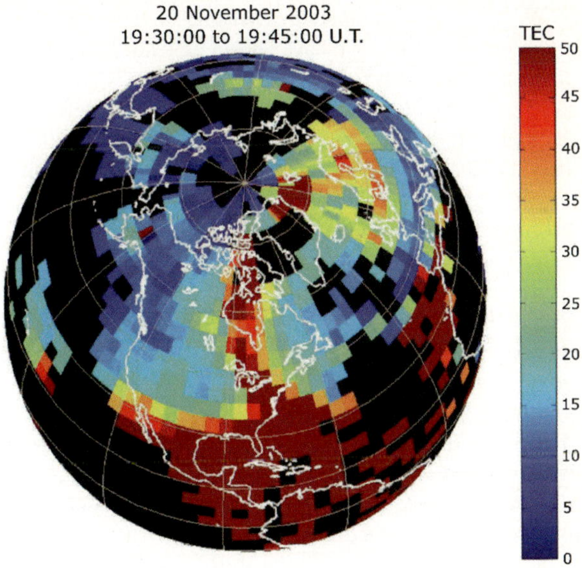

Fig. 73 Storm-enhanced plasma density (SED) signatures in total electron content (TEC) observed on November 20, 2003. These are believed to be connected to plasmasphere erosion and driven by sub-auroral electric fields from the inner magnetosphere. Strong plasma density gradients are observed by a network of ground-based GPS observatories. Figure courtesy of A. Coster

dent in widely spaced GPS measurements of total electron content. Foster et al. (2005) reported that the plume shown in the figure is carried through the cusp during the main phase of the magnetic storm of November 20, 2003. In such a scenario, the enhanced plasma density is driven upwards and poleward by the prompt penetration electric fields, then towards the dayside by storm-related SAPS, and finally by dayside, anti-sunward convection to form the plume across the polar cap at higher latitudes. Such plumes which arrive at the high latitudes may enhance ion outflow into the magnetosphere.

7 Summary

In this paper, we have reviewed the origins of the near-earth plasma created by photoionization due to solar EUV radiation, primarily at mid and low latitudes, and by impact ionization from precipitating energetic particles at higher latitudes. Not only is the near-earth plasma created directly from the earth's extended atmosphere, but the ionization that is formed remains coupled to the neutral gases and their dynamics. Clearly, to fully understand the near-earth space environment, we must consider both the neutral and ionized gases as an ensemble, forming one envelope that surrounds our planet.

We have examined many of the numerous processes and dynamics inherent in the coupled ion-neutral gas at low and mid latitudes. Throughout this region, the ionized gas component is largely controlled by the neutral dynamics. However, as we have shown, small changes between the ion and electron distributions, primarily set up as a result of collisions with the neutral particles, are sufficient to generate global systems of currents and electric fields. These currents map along the magnetic field lines to other regions of Geospace where their effects may be profound, for example near the magnetic equator just after sunset. Furthermore, the fact that the ionosphere/upper atmosphere is strongly influenced from forcing from below—i.e., from tides, gravity waves, and thunderstorm activity originating in the troposphere—underscores the controlling role of the atmosphere on the behavior of the ionized gases, particularly at mid and low latitudes.

We have also examined the dynamics and processes inherent in the coupled ion-neutral gas at high latitudes. We have shown how strong electric fields and currents originating in the magnetosphere often dominate the dynamics, "stirring the pot" of the polar regions with complex convection and field-aligned current patterns that together yield a self-consistent electrodynamic picture. As the magnetospheric electric fields drive neutral winds via ion-neutral collisions, the neutral atmosphere is shown to play a key role in the conversion of magnetospheric electromagnetic energy to mechanical energy in the high latitude upper atmosphere. Furthermore, the neutral atmosphere also plays a critical role in determining how the magnetospheric energy is dissipated, primarily through Joule heating which essentially results from friction between the differential motion of the neutral and ionized gases. Indeed, the response of the neutral gases includes large horizontal and vertical flows that are not confined by the magnetic field and which can subsequently convey the magnetospheric energy to locations far beyond where it was initially deposited. The neutral atmosphere also affects the closure, in the lower ionosphere, of magnetospheric field-aligned currents via Pedersen currents. Finally, the local conductivity in the ionosphere, which is also affected by the neutral atmosphere, has been shown to have a direct effect on the creation of electrostatic structures at higher altitudes that accelerate the energetic particles that ultimately precipitate to form the discrete aurora.

We have also examined many of the fundamental physical processes inherent in the near-earth plasma that result from magnetic storms, such as the effects of increased energy deposition at high latitudes, and have shown how such energy is then distributed away from the polar regions. We also showed evidence that the global ionosphere is subjected to large scale variations when the earth is impacted by magnetic storms of appreciable magnitude. Such effects include substantial changes in the ionospheric density, which may be attributable to prompt penetration electric fields which have been observed at the magnetic equator near the onset of magnetic storms. These observations demonstrate the global and rapid response of the nearearth plasma to solar wind forcing associated with significant magnetic storms.

Progress in our understanding of the ionosphere/thermosphere system requires adopting a systems approach with respect to interpreting the large scale dynamics associated with the energy and momentum exchanges within this gaseous envelope as it adjusts internally and responds to input from both the troposphere below and the magnetosphere above. Understanding such a self-consistent electrodynamic and hydrodynamic response to these drivers is a critical, long-standing objective which is at the forefront of future research in this field.

One area that is highly relevant to future progress in our understanding is that of the tremendous promise within the modeling community of incorporating both physics and observations in computational models of the upper atmosphere. Examples of such models that include first principles of ionosphere/thermosphere coupling are the Thermosphere-Ionosphere-Mesosphere-Electrodynamics General Circulation Model (TIMEGCM) (Roble et al. 1988) and the Coupled Thermosphere-Ionosphere-Plasmasphere-Electrodynamics (CTIPe) model (Fuller-Rowell et al. 1996). Furthermore, the Whole Atmosphere Model (WAM) (Fuller-Rowell et al. 2010) is an extension of the US National Weather Service Global Forecast System and includes the impact of tides, planetary waves, and gravity waves. Finally, the Assimilative Mapping of Ionospheric Electrodynamics (AMIE) model (Richmond et al. 1992) includes the most accurate high latitude specification, which is important for understanding the high latitude forcing.

While the models mentioned above rely on both physical understanding and observations, in some areas the models are considerably "data starved". For example, very few simultaneous measurements of the dynamics of neutral and ionized gases at the lower altitudes exist. Although targeted research rockets, satellites, and ground-based measurements are essential for advancing our understanding, the systems approach requires simultaneous measurements using a number of observational techniques, including space-based *in situ* and imaging experiments as well as arrays of ground-based radars and other instruments. The space-based direct measurements show particular promise when organized as "constellation" missions, particularly in configurations where data may be gathered simultaneously at multiple local times and longitudes. Such platforms promise to reveal how the neutral and ionized dynamics are organized, as well as show how energy and momentum is distributed, and re-distributed, globally. For example, such measurements would reveal the two-dimensional coupling of neutral and plasma gases and their dynamics across the entire high latitude ionosphere, show the global consequences of propagating tides, and demonstrate how magnetic storms promptly effect the entire ionosphere/thermosphere system as a function of latitude, local time, and longitude.

Finally, we emphasize that to be most effective, such measurement platforms need to journey where the most important physical processes are actually taking place. In other words, to examine how the ionized and neutral gases exchange energy and momentum, it is important that the instruments sample altitudes where these interactions are most profound. For example, simultaneous measurements of ions and neutrals below 400 km are essential to understand the coupled behavior of these gases. Focused missions that furthermore explore well below 300 km are expected to reveal important fundamental physical processes involving the ion-neutral gases where the coupling is maximized and where our knowledge of the resulting gas dynamics is rudimentary, at best.

Ultimately, the behavior of the Earth's ionized and neutral gases depends on non-linear coupling between complex, highly varying external sources. These external sources include the variable EUV radiation from the sun as well as intricate energy and momentum exchanges with the troposphere/stratosphere/mesosphere below and the magnetosphere and solar wind above. It is precisely because these coupled regions are exceedingly rich in physical processes that the near-earth plasma is one of the most exciting, and important, natural environments in which to focus scientific research and resources. The new knowledge promised from such future research is essential, not only for understanding the Earth, but also for understanding the upper atmospheres of other planets, both in our solar system and beyond.

Acknowledgements I am indebted to Dr. J. Klenzing for calculations of the model ionosphere parameters provided in this article. I thank Drs. L. Kepko, J. Grebowsky, J. Clemmons, R. Walterscheid, D. Rowland, and H. Freudenreich for very useful comments on the entire manuscript. I thank Drs. G. Le, P. Anderson, T. Fuller-Rowell, and J. Thayer, for valuable comments on individual sections or sub-sections. I thank Ms. H. Joliffe and Dr. S. Martin for providing many of the figures shown here, as well as other aspects of the manuscript preparation.

References

T. Abe, B.A. Whalen, A.W. Yau, R.E. Horita, S. Watanabe, E. Sagawa, EXOS-D (Akebono) suprathermal mass spectrometer observations of the polar wind. J. Geophys. Res. **98**, 11191–11203 (1993)

T.L. Aggson, Probe measurements of electric fields in space, in *Proc. NATO Adv. Study Inst* (1969), p. 305

T.L. Aggson, N.C. Maynard, W.B. Hanson, J.L. Saba, Electric-field observations of equatorial bubbles. J. Geophys. Res. **97**, 2997–3009 (1992)

P.C. Anderson, R.A. Heelis, W.B. Hanson, The ionospheric signatures of rapid subauroral ion drifts. J. Geophys. Res. **96**, 5785–5792 (1991)

P.C. Anderson, W.B. Hanson, R.A. Heelis, J.D. Craven, D.N. Baker, L.A. Frank, A proposed production model of rapid subauroral ion drifts and their relationship to substorm evolution. J. Geophys. Res. **98**, 6069–6078 (1993). doi:10.1029/92JA01975

P.C. Anderson, D.L. Carpenter, K. Tsuruda, T. Mukai, F.J. Rich, Multi-satellite observations of rapid subauroral ion drifts (SAID). J. Geophys. Res. **106**, 29585–29600 (2001)

M. André, A. Yau, Theories and observations of ion energization and outflow in the high latitude magnetosphere. Space Sci. Rev. **80**, 27–48 (1997)

E.V. Appleton, Two anomalies in the ionosphere. Nature **157**, 691 (1946)

P.M. Banks, G. Kockarts, *Aeronomy, Part A* (Academic Press, New York, 1973)

A. Berthelier, J.-C. Cerisier, J.-J. Berthelier, J.-M. Bosqued, R.A. Kovrazkhin, The electrodynamic signature of short scale field aligned currents, and associated turbulence in the cusp and dayside auroral zone, in *Electromagnetic Coupling in the Polar Clefts and Caps*, ed. by P.E. Sandholt, A. Egeland (Kluwer Academic, Dordrecht, 1989)

D. Bilitza, B.W. Reinisch, International reference ionosphere 2007: improvements and new parameters. Adv. Space Res. **43**, 599–609 (2008). doi:10.1016/j.asr.2007.07.048

D. Bilitza, L.-A. McKinnel, B. Reinisch, T. Fuller-Rowell, International reference, ionosphere today and in the future. J. Geod. **85**, 909–920 (2011). doi:10.1007/s00190-010-0427-x

K. Birkeland, On the cause of magnetic storms. C. R. Hebd. Séances l'Acad. Sci. **147**, 539–543 (1908)

M. Blanc, A.D. Richmond, The ionospheric disturbance dynamo. J. Geophys. Res. **85**, 1669–1686 (1980)

R. Boström, Electrodynamics of the ionosphere, in *Cosmical Geophysics*, ed. by A. Egeland, O. Holter, A. Omholt (Universitetsforlaget, Oslo, 1973)

D.G. Brinkman, R.L. Walterscheid, L.R. Lyons, D.C. Kayser, A.B. Christensen, *E* region neutral winds in the postmidnight diffuse aurora during the Atmospheric Response in Aurora 1 rocket campaign. J. Geophys. Res. **100**, 17,309–17,320 (1995)

J.E. Buchau, J. Weber, D.N. Anderson, H.C. Carlson Jr., J.G. Moore, B.W. Reinisch, R.C. Livingston, Ionospheric structures in the polar cap: their origin and relation to 250 MHz scintillation. Radio Sci. **20**, 325 (1985)

O. Buneman, Excitation of field aligned sound waves by electron streams. Phys. Rev. Lett. **10**, 285–287 (1963)

W.J. Burke, T.L. Aggson, N.C. Maynard, W.R. Hoegy, R.A. Hoffman, R.M. Candy, C. Liebrecht, E. Rodgers, Effects of a lightning discharge detected by the DE 2 satellite over Hurricane Debbie. J. Geophys. Res. **97**, 6359–6367 (1992). doi:10.1029/92JA00305

W.J. Burke, A.G. Rubin, N.C. Maynard, L.C. Gentile, P.J. Sultan, F.J. Rich, O. de La Beaujardière, C.Y. Huang, G.R. Wilson, Ionospheric disturbances observed by DMSP at middle to low latitudes during the magnetic storm of June 4–6, 1991. J. Geophys. Res. **105**, 18391–18405 (2000)

G.S. Bust, D. Coco, J.J. Makela, Combined ionospheric campaign. 1. Ionospheric tomography and GPS total electron count (TEC) depletions. Geophys. Res. Lett. **27**, 2849–2852 (2000)

C.W. Carlson, J.P. McFadden, R.E. Ergun, M. Temerin, W. Peria, F.S. Mozer, D.M. Klumpar, E.G. Shelley, W.K. Peterson, E. Moebius, R. Elphic, R. Strangeway, C. Cattell, R. Pfaff, FAST observations in the downward auroral current region: energetic upgoing electron beams, parallel potential drops, and ion heating. Geophys. Res. Lett. **25**, 2017–2020 (1998)

Z. Ceplecha, Influx of interplanetary bodies onto earth. Astron. Astrophys. **263**, 361–366 (1992)

E. Chapin, E. Kudeki, Plasma-wave excitation on meteor trails in the equatorial electrojet. Geophys. Res. Lett. **21**, 2433–2436 (1994)

S. Chapman, The absorption and dissociative or ionizing effect of monochromatic radiation in an atmosphere on a rotating earth. Proc. Phys. Soc. London **43**, 26–45 (1931)

C.R. Chappell, J.L. Green, J.F.E. Johnson, J.H. Waite Jr., Pitch angle variations in magnetospheric thermal plasma—initial observations from Dynamics Explorer-1. Geophys. Res. Lett. **9**, 933–936 (1982)

C.R. Chappell, T.E. Moore, J.H. Waite Jr., The Ionosphere as a fully adequate source of plasma for the earth's magnetosphere. J. Geophys. Res. **92**, 5896–5910 (1987)

C.C. Chaston, J.W. Bonnell, C.W. Carlson, J.P. McFadden, R.E. Ergun, R.J. Strangeway, E.J. Lund, Auroral ion acceleration in dispersive Alfvén waves. J. Geophys. Res. **109**, A04205 (2004). doi:10.1029/2003JA010053

C.C. Chaston, C.W. Carlson, J.P. McFadden, R.E. Ergun, R.J. Strangeway, How important are dispersive Alfvén waves for auroral particle acceleration? Geophys. Res. Lett. **34**, L07101 (2007). doi:10.1029/2006GL029144

L.-J. Chen, C.A. Kletzing, S. Hu, S.R. Bounds, Auroral electron dispersion below inverted-V energies: resonant deceleration and acceleration by Alfvén waves. J. Geophys. Res. **110**, A10S13 (2005). doi:10.1029/2005JA011168

J.H. Clemmons, J.H. Hecht, D.R. Salem, D.J. Strickland, Thermospheric density in the Earth's magnetic cusp as observed by the Streak mission. Geophys. Res. Lett. **35**, L24103 (2008). doi:10.1029/2008GL035972

S. Close, M. Oppenheim, S. Hunt, L. Dyrud, Scattering characteristics of high-resolution meteor head echoes detected at multiple frequencies. J. Geophys. Res. **107**(A10), 1295 (2002). doi:10.1029/2002JA009253

M.V. Codrescu, T.J. Fuller-Rowell, On the importance of E-field variability for Joule heating in the high-latitude thermosphere. Geophys. Res. Lett. **22**, 2393–2396 (1995)
W.R. Coley, R.A. Heelis, Adaptive identification and characterization of polar ionization patches. J. Geophys. Res. **100**(A12), 23819–23827 (1995)
S.W.H. Cowley, Magnetosphere-Ionosphere Interactions: a tutorial review, in *Magnetosphere Current Systems*, ed. by S. Ohtani et al. AGU Geophys. Monograph, vol. 118 (2000)
P.D. Craven, D.L. Gallagher, R.H. Comfort, Relative concentration of He^+ in the inner magnetosphere as observed by the DE 1 retarding ion mass spectrometer. J. Geophys. Res. **102**, 2279–2289 (1997)
G. Crowley, Dynamics of the Earth's thermosphere: a review. Rev. Geophys. **29**, 1143–1165 (1991)
G. Crowley, R.R. Meier, Disturbed O/N_2 ratios and their transport to middle and low latitudes, in *Midlatitude Ionospheric Dynamics and Disturbances*, ed. by P.M. Kintner et al. AGU Geophys. Monograph, vol. 181 (2008). doi:10.1029/181GM06
W.D. Cumming, A.J. Dessler, Field-aligned currents in the magnetosphere. J. Geophys. Res. **72**, 1007 (1967)
I. Daglis, R. Thorne, W. Baumjohann, S. Orsini, The terrestrial ring current: origin, formation, and decay. Rev. Geophys. **37**, 407 (1999)
G.T. Davidson, Pitch angle diffusion and the origin of temporal and spatial structures in pulsating aurorae. Space Sci. Rev. **53**, 45–82 (1990)
T.N. Davis, Observed characteristics of auroral forms. Space Sci. Rev. **22**, 77 (1978)
O. de La Beaujardière, C/NOFS: a mission to forecast scintillations. J. Atmos. Sol.-Terr. Phys. **66**, 1573–1591 (2004). doi:10.1016/j.jastp.2004.07.030
R. Doe, M. Mendillo, J. Vickrey, L. Zanetti, R. Eastes, Observations of nightside auroral cavities. J. Geophys. Res. **98**, 293–310 (1993). doi:10.1029/92JA02004
R. Doe, J.F. Vickrey, M. Mendillo, Electrodynamic model for the formation of auroral ionospheric cavities. J. Geophys. Res. **100**, 9683–9696 (1995)
J.W. Dungey, Interplanetary field and the auroral zones. Phys. Rev. Lett. **6**, 47 (1961)
L.P. Dyrud, M.M. Oppenheim, A.F. vom Endt, The anomalous diffusion of meteor trails. Geophys. Res. Lett. **28**(14), 2775–2778 (2001)
G.D. Earle, R.L. Bishop, Q. Zhou, S.P. Wallace, A comparative study of in-situ and remote intermediate layer measurements against wind model predictions of vertical ion drift. J. Atmos. Sol.-Terr. Phys. **60**, 1313 (1998)
G.D. Earle, T.J. Kane, R.F. Pfaff, S.R. Bounds, Ion layer separation and equilibrium zonal winds in midlatitude sporadic E. Geophys. Res. Lett. **27**, 461–464 (2000)
G.D. Earle, A.M. Musumba, S.L. Vadas, Satellite-based measurements of gravity wave-induced midlatitude plasma density perturbations. J. Geophys. Res. **113**, A03303 (2008). doi:10.10299/5117JA012766
J.V. Eccles, A simple model of low-latitude electric fields. J. Geophys. Res. **103**, 26699 (1998)
R.C. Elphic, J.W. Bonnell, R.J. Strangeway, L. Kepko, R.E. Ergun, J.P. McFadden, C.W. Carlson, W. Peria, C.A. Cattell, D. Klumpar, E. Shelley, W. Peterson, E. Moebius, L. Kistler, R. Pfaff, The auroral current circuit and field-aligned currents observed by FAST. Geophys. Res. Lett. **25**, 2033–2036 (1998)
B.A. Emery, C. Lathuillere, P.G. Richards, R.G. Roble, M.J. Buonsanto, D.J. Knipp, P. Wilkinson, D.P. Siplerd, R. Niciejewski, Time dependent thermospheric neutral response to the 2–11 November 1993 storm period. J. Atmos. Sol.-Terr. Phys. **61**, 329–350 (1999)
S.L. England, A review of the effects of non-migrating atmospheric tides on the earth's low-latitude ionosphere. Space Sci. Rev. (2011). doi:10.1007/s11214-011-9842-4
S.L. England, S. Maus, T.J. Immel, S.B. Mende, Longitudinal variation of the E-region electric fields caused by atmospheric tides. Geophys. Res. Lett. 33 (2006). doi:10.1029/2006GL027465
R.E. Ergun, L. Andersson, D. Main, Y.-J. Su, D.L. Newman, M.V. Goldman, C.W. Carlson, A.J. Hull, J.P. McFadden, F.S. Moser, Auroral particle acceleration by strong double layers: the upward current region. J. Geophys. Res. **109**, A12220 (2004). doi:10.1029/2004JA010545
D.S. Evans, The observations of a near monoenergetic flux of auroral electrons. J. Geophys. Res. **73**, 2315 (1968)
D.S. Evans, N.C. Maynard, J. Trøim, T. Jacobsen, A. Egeland, Auroral vector electric field and particle comparisons. 2. Electrodynamics of an arc. J. Geophys. Res. **82**, 2235–2259 (1977)
D.T. Farley Jr., Two-stream plasma instability as a source of irregularities in the ionosphere. Phys. Rev. Lett. **10**, 279–282 (1963)
D.T. Farley, E. Bonelli, B.G. Fejer, M.F. Larsen, The prereversal enhancement of the zonal electric field in the equatorial ionosphere. J. Geophys. Res. **91**, 13723 (1986)
W.M. Farrell, T.L. Aggson, E.B. Rodgers, W.B. Hanson, Observations of ionospheric electric fields above atmospheric weather systems. J. Geophys. Res. **99**, 19475–19483 (1994)
B.G. Fejer, D.T. Farley, B.B. Balsley, R.F. Woodman, Vertical structure of the VHF backscattering region in the equatorial electrojet and the gradient drift instability. J. Geophys. Res. **80**(10), 1313–1324 (1975)

B.G. Fejer, M.F. Larsen, D.T. Farley, Equatorial disturbance dynamo electric fields. Geophys. Res. Lett. **10**, 537–540 (1983)
B.G. Fejer, J.W. Jensen, S.-Y. Su, Quiet time equatorial *F* region vertical plasma drift model derived from ROCSAT-1 observations. J. Geophys. Res. **113**, A05304 (2008). doi:10.1029/2007JA012801
J.M. Forbes, S.E. Palo, X. Zhang, Variability of the ionosphere. J. Atmos. Sol.-Terr. Phys. **62**, 685–693 (2000)
J.C. Foster, W.J. Burke, SAPS: a new categorization for sub-auroral electric fields. Eos **83**(3), 393–394 (2002). doi:10.1029/2002EO000289
J.C. Foster, H.B. Vo, Average characteristics and activity dependence of the subauroral polarization stream. J. Geophys. Res. **107**(A12), 1475 (2002). doi:10.1029/2002JA009409
J.C. Foster, A.J. Coster, P.J. Erickson, J.M. Holt, F.D. Lind, W. Rideout, M. McCready, A. Van Eyken, R.J. Barnes, R.A. Greenwald, F.J. Rich, Multiradar observations of the polar tongue of ionization. J. Geophys. Res. **110**, A09S31 (2005). doi:10.1029/2004JA010928
L.A. Frank, K.L. Ackerson, Observations of charged particle precipitation into the auroral zone. J. Geophys. Res. **76**, 3612 (1971)
L.A. Frank, J.D. Craven, Imaging results from Dynamics Explorer 1. Rev. Geophys. **26**(2), 249–283 (1988)
H.U. Frey, G. Haerendel, D. Knudsen, S. Buchert, O.H. Bauer, Optical and radar observations of the motion of auroral arcs. J. Atmos. Sol.-Terr. Phys. **58**, 57–69 (1996)
T.J. Fuller-Rowell, D. Rees, S. Quegan, R.J. Moffett, M.V. Codrescu, G.H. Millward, A coupled thermosphere ionosphere model (CTIM), in *Handbook of Ionospheric Models*, STEP report, ed. by R.W. Schunk (1996), pp. 217–238
T.J. Fuller-Rowell, A.D. Richmond, N. Maruyama, Global modeling of storm-time thermospheric dynamics and electrodynamics, in *Midlatitude Ionospheric Dynamics and Disturbances*, ed. by P.M. Kintner et al. AGU Geophys. Monograph, vol. 181 (2008). doi:10.1029/181GM06
T.J. Fuller-Rowell, F. Wu, R. Akmaev, T.W. Fang, E. Araujo-Pradere, A whole atmosphere model simulation of the impact of a sudden stratospheric warming on thermosphere dynamics and electrodynamics. J. Geophys. Res. **115**, A00G08 (2010). doi:10.1029/2010JA015524
Y.L. Galperin, Y.N. Ponomarov, A.G. Zosinova, Direct measurement of ion drift velocity in the upper atmosphere during a magnetic storm. Kosm. Issled. **11**, 273 (1973)
S. Ganguli, The polar wind. Rev. Geophys. **34**(3), 311–348 (1996). doi:10.1029/96RG00497
J.B. Gary, R.A. Heelis, W.B. Hanson, J.A. Slavin, Field-aligned Poynting flux observations in the high-latitude ionosphere. J. Geophys. Res. **99**, 11417–11427 (1994)
W.D. Gonzalez, J.A. Joselyn, Y. Kamide, H.W. Kroehl, G. Rostoker, B.T. Tsurutani, V.M. Vasyliunas, What is a geomagnetic storm? J. Geophys. Res. **99**, 5771–5794 (1994)
J.M. Grebowsky, A.C. Aiken, In situ measurements of meteoric ions, in *Meteors in the Earth's Atmosphere*, ed. by E. Murad, I.P. Williams (Cambridge University Press, Cambridge, 2002)
M.E. Greenspan, C.E. Rasmussen, W.J. Burke, M.A. Abdu, Equatorial density depletions observed at 840 km during the great magnetic storm of March 1989. J. Geophys. Res. **96**, 13931–13942 (1991)
D.A. Gurnett, R.L. Huff, J.D. Menietti, J.L. Burch, J.D. Winningham, Correlated low-frequency electric and magnetic noise along the auroral field lines. J. Geophys. Res. **89**, 8971–8985 (1984)
G. Haerendel, J.V. Eccles, The role of the equatorial electrojet in the evening ionosphere. J. Geophys. Res. **97**, 1181 (1992)
G. Haerendel, S. Buchert, C. La Hoz, B. Raff, E. Rieger, On the proper motion of auroral arcs. J. Geophys. Res. **98**, 6087–6099 (1993)
G. Haerendel, B.U. Olipitz, S. Buchert, O.H. Bauer, E. Rieger, C. La Hoz, Optical and radar observations of auroral arcs with emphasis on small scale structures. J. Atmos. Sol.-Terr. Phys. **58**, 71–83 (1996)
C. Haldoupis, Midlatitude sporadic *E*: A typical paradigm of atmosphere-ionosphere coupling. Space Sci. Rev. (2011). doi:10.1007/11214-011-9786-8
T. Hallinan, Spiral-like auroral forms: observations and a proposed theory, Ph.D. thesis presented to the University of Alaska (1976)
T. Hallinan, T. Davis, Small-scale auroral arc distortions. Planet. Space Sci. **18**, 1735 (1970)
J.K. Hargreaves, *The Solar-Terrestrial Environment* (Cambridge University Press, Cambridge, 1992)
R.A. Heelis, The effects of interplanetary magnetic field orientation on dayside high-latitude ionospheric convection. J. Geophys. Res. **89**, 2873–2880 (1984)
R.A. Heelis, Studies of ionospheric plasma and electrodynamics and their application to ionosphere-magnetosphere coupling. Rev. Geophys. **26**, 317–328 (1988)
R.A. Heelis, Electrodynamics in the low and middle latitude ionosphere: a tutorial. J. Atmos. Sol.-Terr. Phys. **66**, 825–838 (2004)
R.A. Heelis, S. Mohapatra, Storm time signatures of the ionospheric zonal ion drift at middle latitudes. J. Geophys. Res. **114**, A02305 (2009). doi:10.1029/2008JA013620
R.A. Heelis, P.C. Kendall, R.J. Moffett, D.W. Windle, H. Rishbeth, Electrical coupling of the *E*-and *F*-regions and its effect on *F*-region drifts and winds. Planet. Space Sci. **22**, 743 (1974)

R.A. Heelis, P.H. Reiff, J.D. Winningham, W.B. Hanson, Ionospheric convection signatures observed by DE 2 during north-ward interplanetary magnetic field. J. Geophys. Res. **91**, 5817–5830 (1986)
R.A. Heelis, Low- and middle-latitude ionospheric dynamics associated with magnetic storms, in *Midlatitude Ionospheric Dynamics and Disturbances*, ed. by P.M. Kintner et al. AGU Geophys. Monograph, vol. 181 (2008). doi:10.1029/181GM06
R.A. Heelis, W.R. Coley, A.G. Burrell, M.R. Hairston, G.D. Earle, M.D. Perdue, R.A. Power, L.L. Harmon, B.J. Holt, C.R. Loppincott, Behaviour of the O^+/H^+ transition height during the extreme solar minimum of 2008. Geophys. Res. Lett. **36**, L00C03 (2009a). doi:10.1029/2009GL038652
R.A. Heelis, J.J. Sojka, M. David, R.W. Schunk, Storm time density enhancements in the middle-latitude dayside ionosphere. J. Geophys. Res. **114**, A03315 (2009b). doi:10.1029/2008JA013690
J.P. Heppner, Electric field variations durings aubstorms: OGO-6 measurements. Planet. Space Sci. **20**, 1475–1498 (1972)
J.P. Heppner, N.C. Maynard, Empirical high-latitude electric field models. J. Geophys. Res. **92**, 4467–4489 (1987)
J.P. Heppner, M.C. Liebrecht, N.C. Maynard, R.F. Pfaff, High-latitude distributions of plasma waves and spatial irregularities from DE 2 alternating current electric field observations. J. Geophys. Res. **98**, 1629–1652 (1993)
C.O. Hines, The formation of midlatitude sporadic *E* layers. J. Geophys. Res. **69**, 1018–1019 (1964)
C.-S. Huang, J.C. Foster, M.C. Kelley, Long duration penetration of the interplanetary electric field to the low-latitude ionosphere during the main phase of magnetic storms. J. Geophys. Res. **110**, A11309 (2005). doi:10.1029/2005JA011202
J.D. Huba, G. Joyce, J.A. Fedder, Sami2 is another model of the ionosphere (SAMI2): a new low-latitude ionosphere model. J. Geophys. Res. **105**, 23035–23053 (2000)
D.L. Hysell, E. Kudeki, Collisional shear instability in the equatorial F region ionosphere. J. Geophys. Res. **109**, A11301 (2004)
D. Hysell, M. Larsen, Q. Zhou, Common volume coherent and incoherent scatter radar observations of mid-latitude sporadic E-layer and QP echoes. Ann. Geophys. **22**, 3277–3290 (2004)
D.L. Hysell, M.F. Larsen, C.M. Swenson, A. Barjatya, T.F. Wheeler, T.W. Bullett, M.F. Sarango, R.F. Woodman, J.L. Chau, D. Sponseller, Rocket and radar investigation of background electrodynamics and bottom-type scattering layers at the onset of equatorial spread *F*. Ann. Geophys. **24**, 1387–1400 (2006)
T. Iijima, T.A. Potemra, Field-aligned currents in the dayside cusp observed by Triad. J. Geophys. Res. **34**, 5971 (1976)
T.J. Immel, E. Sagawa, S.L. England, S.B. Henderson, M.E. Hagan, S.B. Mende, H.U. Frey, C.M. Swenson, L.J. Paxton, Control of equatorial morphology by atmospheric tides. Geophys. Res. Lett. **33**, L15108 (2006). doi:10.1029/2006GL026161
D. Janches, L.P. Dyrud, S.L. Broadley, J.M.C. Plane, First observation of micrometeoroid differential ablation in the atmosphere. Geophys. Res. Lett. **36**, L06101 (2009). doi:10.1029/2009GL037389
A.D. Johnstone, The mechanism of pulsating aurora. Ann. Geophys. **1**, 397–410 (1983)
A.D. Johnstone, D.M. Walton, R. Liu, D.A. Hardy, Pitch angle diffusion of low-energy electrons by whistler mode waves. J. Geophys. Res. **98**, 5959–5967 (1993)
B. Jones, P.J.S. Williams, K. Schlegel, T. Robinson, I. Häggström, Interpretation of enhanced electron temperatures measured in the auroral *E*-region during the ERRRIS campaign. Ann. Geophys. **9**, 55–59 (1991)
G.O.L. Jones, P.J.S. Williams, K.J. Winser, Large ionospheric plasma flows along magnetic field lines: EISCAT observations and their interpretation. Adv. Space Res. **12**, 157–160 (1992)
T. Karlsson, G. Marklund, L. Blomberg, A. Malkki, Subauroral electric fields observed by the Freja satellite: a statistical study. J. Geophys. Res. **103**(A), 4327–4341 (1998). doi:10.1029/97JA00333
M.C. Kelley, *The Earth's Ionosphere: Plasma Physics and Electrodynamics*, 2nd edn. Int. Geophys. Ser, vol. 96 (Academic Press, Boston, 2009)
M.C. Kelley, C.W. Carlson, Observations of intense velocity shear and associated electrostatic waves near an auroral arc. J. Geophys. Res. **82**, 2343–2348 (1977)
M.C. Kelley, C.L. Siefring, R.F. Pfaff, P.M. Kintner, M. Larsen, R. Green, R.H. Holzworth, L.C. Hale, J.D. Mitchell, D. LeVine, Electrical measurements in the atmosphere and the ionosphere over an active thunderstorm. 1. Campaign overview and initial ionospheric results. J. Geophys. Res. **90**, 9815–9823 (1985). doi:10.1029/JA090iA10p09815
M.C. Kelley, J. LaBelle, E. Kudeki, B.G. Fejer, Sa. Basu, Su. Basu, K.D. Baker, C. Hanuise, P. Argo, R.F. Woodman, W.E. Swartz, D.T. Farley, J.W. Meriwether Jr., The Condor equatorial spread *F* campaign: overview and results of the large-scale measurements. J. Geophys. Res. **91**, 5487–5503 (1986). doi:10.1029/JA091iA05p05487
M.C. Kelley, D.J. Knudsen, J.F. Vickrey, Poynting flux measurements on a satellite: a diagnostic tool for space research. J. Geophys. Res. **96**(A1), 201–207 (1991)

M.C. Kelley, F. Garcia, J. Makela, T. Fan, E. Mak, C. Sia, D. Alcocer, Highly structured tropical airglow and TEC signatures during strong geomagnetic activity. Geophys. Res. Lett. **27**, 465–468 (2000)

M.C. Kelley, J.J. Makela, A. Saito, The mid-latitude *F* region at the mesoscale: some progress at last. J. Atmos. Sol.-Terr. Phys. **64**, 1525–1529 (2002)

M.C. Kelley, J.J. Makela, O. de La Beaujardière, J. Retterer, Convective ionospheric storms: a review. Rev. Geophys. **49**, RG2011 (2011). doi:10.1029/2010RG000340

T.L. Killeen, R.G. Roble, Thermospheric dynamics: contributions from the first 5 years of the Dynamics Explorer Program. Rev. Geophys. **26**, 329–367 (1988)

T.L. Killeen, P.B. Hays, G.R. Carignan, R.A. Heelis, W.B. Hanson, N.W. Spencer, L.H. Brace, Ion-neutral coupling in the high-latitude *F* region: evaluation of ion heating terms from Dynamics Explorer 2. J. Geophys. Res. **89**, 7495–7508 (1984). doi:10.1029/JA089iA09p07495

P.M. Kintner Jr., A.J. Coster, T. Fuller-Rowell, A.J. Mannucci, M. Mendillo, R. Heelis (eds), *Midlatitude Ionospheric Dynamics and Disturbances*, AGU Geophys. Monograph, vol. 181 (2008). doi:10.1029/181GM06

S. Kirkwood, H. Opgenoorth, J.S. Murphree, Ionospheric conductivities, electric fields and currents associated with auroral substorms measured by the EISCAT radar. Planet. Space Sci. **36**(12), 1359–1380 (1988)

J.H. Klenzing, D.E. Rowland, R.F. Pfaff, G. Le, H. Freudenreich, R.A. Haaser, A.G. Burrell, R.A. Stoneback, W.R. Coley, R.A. Heelis, Observations of low-latitude plasma density enhancements and their associated plasma drifts. J. Geophys. Res. **116**, A09324 (2011). doi:10.1029/2011JA016711

S. Knight, Parallel electric fields. Planet. Space Sci. **21**, 741 (1973)

D.J. Knudsen, M.C. Kelley, G.D. Earle, J.F. Vickrey, M. Boehm, Distinguishing Alfvén waves from quasi-static field structures associated with the discrete aurora: sounding rocket and HILAT satellite measurements. Geophys. Res. Lett. **17**(7), 921–924 (1990)

E. Kopp, On the abundance of metal ions in the lower ionosphere. J. Geophys. Res. **102**(A5), 9667–9674 (1997)

E. Kudeki, S. Bhattacharyya, Postsunset vortex in equatorial *F*-region plasma drifts and implications for bottomside spread-*F*. J. Geophys. Res. **104**(A), 28163–28170 (1999). doi:10.1029/1998JA900111

E. Kudeki, D.T. Farley, B.G. Fejer, Long wavelength irregularities in the equatorial electrojet. Geophys. Res. Lett. **9**, 684 (1982)

E. Kudeki, B.G. Fejer, D.T. Farley, C. Hanuise, The Condor equatorial electrojet campaign: radar results. J. Geophys. Res. **92**, 13,561–13,577 (1987)

E. Kudeki, A. Akgiray, M. Milla, J.L. Chau, D.L. Hysell, Equatorial spread-F initiation: post-sunset vortex, thermospheric winds, gravity waves. J. Atmos. Sol.-Terr. Phys. **69**, 2416–2427 (2007)

J.W. LaBelle III, *Ionospheric Turbulence: Case Studies in Equatorial Spread F and Development of a Rocket-Borne Interferometer*, Ph.D. thesis presented to Cornell University (1985)

M.F. Larsen, A shear instability seeding mechanism for quasi-periodic radar echoes. J. Geophys. Res. **105**, 24931–24940 (2000)

M.F. Larsen, S. Fukao, M. Yamamoto, R. Tsunoda, K. Igarashi, T. Ono, The SEEK chemical release experiment: observed neutral wind profile in a region of sporadic-E. Geophys. Res. Lett. **25**, 1789–1792 (1998)

M.F. Larsen, M. Yamamoto, S. Fukao, R. Tsunoda, A. Saito, Observations of neutral winds, winds shears, and wave structure during a sporadic-E/QP event. Ann. Geophys. **23**, 2369–2375 (2005)

G. Le, G. Lu, R.J. Strangeway, R.F. Pfaff Jr., Strong interplanetary magnetic field B_Y-related plasma convection in the ionosphere and cusp field-aligned currents under northward interplanetary magnetic field conditions. J. Geophys. Res. **107**, 1477 (2002). doi:10.1029/2001JA007546

G. Le, C.-S. Huang, R.F. Pfaff, S.-Y. Su, H.-C. Yeh, R.A. Heelis, F.J. Rich, M. Hairston, Plasma density enhancements associated with equatorial spread F: ROCSAT-1 and DMSP observations. J. Geophys. Res. **108**, 1318 (2003). doi:10.1029/2002JA009592

G. Le, J.A. Slavin, R.J. Strangeway, Space Technology 5 observations of the imbalance of regions 1 and 2 field-aligned currents and its implication to the cross-polar cap Pedersen currents. J. Geophys. Res. **115**, A07202 (2010). doi:10.1029/2009JA014979

C.H. Lin, C.C. Hsiao, J.Y. Liu, C.H. Liu, Longitudinal structure of the equatorial ionosphere: time evolution of the four-peaked EIA structure. J. Geophys. Res. **112**, A12305 (2007). doi:10.1029/2007JA012455

G. Lu, A. Richmond, B. Emery, R. Roble, Magnetosphere-ionosphere-thermosphere coupling: effect of neutral winds on energy transfer and field-aligned current. J. Geophys. Res. **100**, 19643–19659 (1995). doi:10.1029/95JA00766

G. Lu, L.P. Goncharenko, A.D. Richmond, R.G. Roble, N. Aponte, A dayside ionospheric positive storm phase driven by neutral winds. J. Geophys. Res. **113**, A08304 (2009). doi:10.1029/2007JA012895

A.T. Lui, Y.P. Perrault, S.-I. Akasofu, C.D. Anger, The diffuse aurora. Planet. Space Sci. **21**, 857 (1973)

D. Lummerzheim, Modeling and forecasting aurora. Computing in Sci. Eng. **9**(5), 53–61 (2007)

L.R. Lyons, Formation of auroral arcs via magnetosphere-ionosphere coupling. Rev. Geophys. **30**, 93–112 (1992)
R. Lysak, Generalized model of the ionospheric Alfvén resonator, in *Auroral Plasma Dynamics*, ed. by R. Lysak. AGU Geophys. Monograph, vol. 80 (1993)
J.J. Makela, Y. Otsuka, Overview of nighttime ionospheric instabilities at low- and mid-latitudes: coupling aspects resulting in structuring at the mesoscale. Space Sci. Rev. (2011). doi:10.1007/s11214-011-9816-6
J.J. Makela, S.A. Gonzalez, B. MacPherson, X. Pi, M.C. Kelley, P.J. Sultan, Intercomparisons of total electron content measurements using the Arecibo incoherent scatter radar and GPS. Geophys. Res. Lett. **27**(18), 2841–2844 (2000)
A.J. Mallinckrodt, C.W. Carlson, On the anticorrelation of the electric field and peak electron energy within an auroral arc. J. Geophys. Res. **90**, 399–408 (1985)
A.J. Mannucci, B.T. Tsurutani, B.A. Iijima, A. Komjathy, A. Saito, W.D. Gonzalez, F.L. Guarnieri, J.U. Kozyra, R. Skoug, Dayside global ionospheric response to the major interplanetary events of October 29–30, 2003 "Halloween Storms". Geophys. Res. Lett. **32**, L12S02 (2005). doi:10.1029/2004GL021467
G.T. Marklund, Auroral arc classification scheme based on the observed arc-associated electric field pattern. Planet. Space Sci. **32**, 193–211 (1984)
G.T. Marklund, Electric fields and plasma processes in the auroral downward current region, below, within, and above the acceleration region. Space Sci. Rev. **142**, 1–21 (2009). doi:10.1007/s11214-008-9373-9
G.T. Marklund, L. Blomberg, C.-G. Falthämmar, P.-A. Lindqvist, On intense diverging electric fields associated with black aurora. Geophys. Res. Lett. **21**, 1859–1862 (1994)
G. Marklund, T. Karlsson, J. Clemmons, On low-altitude particle acceleration and intense electric fields and their relationship to black aurora. J. Geophys. Res. **102**, 17509–17522 (1997). doi:10.1029/97JA00334
D.F. Martyn, Atmospheric tides in the ionosphere. I. Solar tides in the $F2$ region. Proc. R. Soc. Lond. Ser. A, Math. Phys. Sci. **189**, 241–260 (1947)
J.D. Mathews, Sporadic E: current views and recent progress. J. Atmos. Sol.-Terr. Phys. **60**, 413–435 (1998)
J.D. Matthews, S. Briczinski, D. Meisel, C. Heinselman, Radio and meteor science outcomes from comparisons of meteor radar observations at AMISR Poker Flat, Sondrestrom, and Arecibo. Earth Moon Planets **102**, 365–372 (2008). doi:10.1007/s11038-007-9168-0
S. Matsushita, Global presentation of the external S_q and L current systems. J. Geophys. Res. **70**, 4395 (1965)
N.C. Maynard, A. Bahnsen, P. Christophersen, A. Egeland, R. Lundin, An example of anticorrelation of auroral particles and electric fields. J. Geophys. Res. **78**, 3976–3980 (1973)
J.P. McFadden, C.W. Carlson, R.E. Ergun, Microstructure of the auroral acceleration region as observed by FAST. J. Geophys. Res. **104**, 14453–14480 (1999)
C. McIlwain, Direct measurements of particles producing visible aurora. J. Geophys. Res. **65**, 2727 (1960)
R.L. McPherron, Magnetosphere dynamics, in *Introduction to Space Physics*, ed. by M.G. Kivelson, C.T. Russell (Cambridge University Press, Cambridge, 1995)
R.R. Meier, G. Crowley, D.J. Strickland, A.B. Christensen, L.J. Paxton, D. Morrison, C.L. Hackert, First look at the 20 November 2003 superstorm with TIMED/GUVI: comparisons with a thermospheric global circulation model. J. Geophys. Res. 110 (2005). doi:10.1029/2004JA010990
M. Mendillo, Storms in the ionosphere: patterns and processes for total electron content. Rev. Geophys. **44**, RG4001 (2006). 1–47
H.R. Middleton, S.E. Pryse, L. Kersley, G.S. Bust, E.J. Fremouw, J.A. Secan, Evidence for the tongue of ionization under northward interplanetary magnetic field conditions. J. Geophys. Res. **110**, A07301 (2005). doi:10.1029/2004JA010800
T. Nagai, J.H. Waite Jr., J.L. Green, C.R. Chappell, First measurements of supersonic polar wind in the polar magnetosphere. Geophys. Res. Lett. **11**, 669–672 (1984)
P.T. Newell, C.-I. Meng, K.M. Lyons, Suppression of discrete aurorae by sunlight. Lett. Nat. **381**, 766–767 (1996)
P.T. Newell, T. Sotirelis, S. Wing, Diffuse, monoenergetic, and broadband aurora: the global precipitation budget. J. Geophys. Res. **114**, A09207 (2009). doi:10.1029/2009JA014326
Y. Nishimura, J. Bortnik, W. Li, R.M. Thorne, L.R. Lyons, V. Angelopoulos, S.B. Mende, J.W. Bonnell, O.L. Contel, C. Cully, R. Ergun, U. Auster, Identifying the driver of pulsating aurora. Science **330**, 81–84 (2010). doi:10.1126/science.1193186
Y. Otsuka, K. Shiokawa, T. Ogawa, P. Wilkinson, Geomagnetic conjugate observations of medium-scale traveling ionospheric disturbances at mid-latitude using all-sky airglow imagers. Geophys. Res. Lett. **31**, L15803 (2004). doi:10.1029/2004GL020262
G. Paschmann, S. Haaland, R. Treumann, *Auroral Plasma Physics* (Kluwer Academic, Dordrecht, 2003)
W.J. Peria, C.W. Carlson, R.E. Ergun, J.P. McFadden, J. Bonnell, R.C. Elphic, R.J. Strangeway, Characteristics of field-aligned currents near the auroral acceleration region: FAST observations, in *Magnetosphere Current Systems*, ed. by S. Ohtani et al. AGU Geophys. Monograph, vol. 118 (2000)

R.F. Pfaff, *Rocket Studies of Plasma Turbulence in the Equatorial and Auroral Electrojets*. Ph.D. thesis presented to Cornell University (1986)

R.F. Pfaff Jr., In-situ measurement techniques for ionospheric research, in *Modern Ionospheric Science*, ed. by H. Kohl et al. (Max-Planck-Institut für Aeronomie, Katlenburg-Lindau, 1996)

R.F. Pfaff, M.C. Kelley, E. Kudeki, B.G. Fejer, K.D. Baker, Electric field and plasma density measurements in the strongly driven daytime equatorial electrojet. 1. The unstable layer and gradient drift waves. J. Geophys. Res. **92**, 13,578–13,596 (1987)

R.F. Pfaff, J. Heppner, W. Hoegy, J. Slavin, N. Maynard, M.C. Liebrecht, A survey of enhanced sub-auroral electric fields observed by Dynamics Explorer-2, In: EOS Trans. Am. Geophys. Union, Fall Meeting (1988)

R.F. Pfaff, J. Sahr, J.F. Providakes, W.E. Swartz, D.T. Farley, P.M. Kintner, I. Haggstrom, A. Hedberg, H. Opgenoorth, G. Holmgren, The E-region Rocket/Radar Instability Study (ERRRIS)—scientific objectives and campaign overview. J. Atmos. Sol.-Terr. Phys. **54**, 779–808 (1992)

R.F. Pfaff, M.H. Acuna, P.A. Marionni, N.B. Trivedi, DC polarization electric field, current density, and plasma density measurements in the daytime equatorial electrojet. Geophys. Res. Lett. **24**, 1667–1670 (1997a)

R.F. Pfaff, P.A. Marionni, W.E. Swartz, Wavevector observations of the two-stream instability in the daytime equatorial electrojet. Geophys. Res. Lett. **24**, 1671–1674 (1997b)

R.F. Pfaff, J. Clemmons, C. Carlson, R. Ergun, J. McFadden, F. Mozer, M. Temerin, D. Klumpar, W. Peterson, E. Shelley, E. Moebius, L. Kistler, R. Strangeway, R. Elphic, C. Cattell, Initial FAST observations of acceleration processes in the cusp. Geophys. Res. Lett. **25**, 2037–2040 (1998)

R.F. Pfaff et al., Combined sounding rocket and SuperDarn/EISCAT radar observations of plasma convection, shear, irregularities, and other phenomena in the cusp and boundary layer during IMF B_z negative and By negative conditions. In: EOS Trans. Am. Geophys. Union, Fall Meeting (2003)

R.F. Pfaff, C. Liebrecht, J.-J. Berthelier, M. Malingre, M. Parrot, J.-P. Lebreton, DEMETER satellite observations of plasma irregularities in the topside ionosphere at low, middle, and sub-auroral latitudes and their dependence on magnetic storms, in *Midlatitude Ionospheric Dynamics and Disturbances*, ed. by P.M. Kitner et al. AGU Geophys. Monograph, vol. 181 (2008). doi:10.1029/181GM06

J.M. Picone, A.E. Hedin, D.P. Drob, A.C. Aikin, NRLMSISE-00 empirical model of the atmosphere: statistical comparisons and scientific issues. J. Geophys. Res. **107**(A12), 1468 (2002). doi:10.1029/2002JA009430

W.E. Potter, L.J. Cahill Jr., Electric and magnetic field measurements near an auroral electrojet. J. Geophys. Res. **74**, 5159–5160 (1969)

E. Priest, T. Forbes, *Magnetic Reconnection, MHD Theory and Applications* (Cambridge University Press, Cambridge, 2000)

G.W. Prölss, Ionospheric F-region storms, in *Handbook of Atmosp. Electrodynamics, 2*, ed. by H. Volland (CRC Press, Boca Raton, 1995), Chap. 8, pp. 195–248

G.W. Prölss, Magnetic storm associated perturbations of the upper atmosphere, in *Magnetic Storms*, ed. by B. Tsurutani et al. AGU Geophys. Monograph Series, vol. 98 (1997)

G. Prölss, *Physics of the Earth's Space Environment: an Introduction* (Springer, Heidelberg, 2004)

M.H. Rees, *Physics and Chemistry of the Upper Atmosphere*. Cambridge Atmos. and Space Sci. Series (Univ. Press, Cambridge, 1989)

P. Reiff, T. Hill, J. Burch, Solar wind plasma injection at the dayside magnetospheric cusp. J. Geophys. Res. **82**, 479 (1977)

A.D. Richmond, G. Lu, Upper-atmospheric effects of magnetic storms: a brief tutorial. J. Atmos. Sol.-Terr. Phys. **62**, 1115–1127 (2000)

A.D. Richmond, J.P. Thayer, Ionospheric electrodynamics: a tutorial, in *Magnetosphere Current Systems*, ed. by S. Ohtani et al. AGU Geophys. Monograph, vol. 118 (2000)

A.D. Richmond, E. Ridley, R. Roble, A thermosphere/ionosphere general-circulation model with coupled electrodynamics. Geophys. Res. Lett. **19**, 601–604 (1992). doi:10.1029/92GL00401

H. Rishbeth, The effect of winds on the ionospheric F2-peak. J. Atmos. Sol.-Terr. Phys. **29**, 225–288 (1967)

H. Rishbeth, The F-layer dynamo. Planet. Space Sci. **19**, 263–267 (1971)

H. Rishbeth, The ionospheric E-layer and F-layer dynamos-a tutorial review. J. Atmos. Sol.-Terr. Phys. **59**, 1873–1880 (1997)

H. Rishbeth, O.K. Garriot, *Introduction to Ionospheric Physics* (Academic Press, New York, 1969)

H. Rishbeth, T.J. Fuller-Rowell, A.S. Rodger, F-layer storms and thermospheric composition. Phys. Scr. **36**, 327–336 (1987)

R.G. Roble, E. Ridley, A. Richmond, R. Dickinson, A coupled thermosphere ionosphere general-circulation model. Geophys. Res. Lett. **15**, 1325–1328 (1988). doi:10.1029/GL015I012P01325

A. Rodger, The mid-latitude trough–revisited, in *Midlatitude Ionospheric Dynamics and Disturbances*, ed. by P.M. Kintner et al. AGU Geophys. Monograph, vol. 181 (2008). doi:10.1029/181GM06

A. Saito, M. Nishimura, M. Yamamoto, S. Fukao, M. Kubota, K. Shiokawa, Y. Otsuka, T. Tsugawa, T. Ogawa, M. Ishii, T. Sakanoi, S. Miyazaki, Traveling ionospheric disturbances detected in the FRONT campaign. Geophys. Res. Lett. **28**, 689–692 (2001)

N. Sato, D.M. Wright, C.W. Carlson, Y. Ebihara, M. Sato, T. Saemundsson, S. Milan, M. Lester, Generation region of pulsating aurora obtained simultaneously by the FAST satellite and a Syowa-Iceland conjugate pair of observations. J. Geophys. Res. **109**, A10201 (2004). doi:10.1029/2004JA010419

L. Scherliess, B.G. Fejer, Storm time dependence of equatorial disturbance dynamo zonal electric fields. J. Geophys. Res. **102**(A11), 24,037–24,046 (1997)

L. Scherliess, B.G. Fejer, Radar and satellite global equatorial *F* region vertical drift model. J. Geophys. Res. **104**, 6829–6842 (1999). doi:10.1029/1999JA900025

K. Schlegel, J.P. St.-Maurice, Anomalous heating of the polar *E* region by unstable plasma waves. 1. Observations. J. Geophys. Res. **86**(A3), 1447–1452 (1981)

R. Schunk, P.M. Banks, Effects of electric fields and other processes upon the nighttime high-latitude *F* layer. J. Geophys. Res. **81**, 3271–3282 (1976)

R. Schunk, A.F. Nagy, *Ionospheres: Physics, Plasma Physics, and Chemistry* (Cambridge Univ. Press, Cambridge, 2009)

R.W. Schunk, J.J. Sojka, Global ionosphere-polar wind system during changing magnetic activity. J. Geophys. Res. **102**, 11,625–11,651 (1997)

R.W. Schunk, J.J. Sojka, M.D. Bowline, Theoretical study of the electron temperature in the high-latitude ionosphere for solar maximum and winter conditions. J. Geophys. Res. **91**, 12,041–12,054 (1986)

V.A. Sergeev, V. Angelopoulos, R. Nakamura, Recent advances in understanding substorm dynamics. Geophys. Res. Lett. **39**(5), L05101 (2012) (ISSN 0094-8276). doi:10.1029/2012GL050859

E.G. Shelley, R.D. Sharp, R.G. Johnson, Satellite observations of an ionospheric acceleration mechanism. Geophys. Res. Lett. **3**, 654 (1976)

K. Shiokawa, Y. Otsuka, T. Ogawa, P. Wilkinson, Time evolution of high-altitude plasma bubbles imaged at geomagnetic conjugate points. Ann. Geophys. **22**, 3137–3143 (2004)

F. Simoes, R. Pfaff, J.-J. Berthelier, J. Klenzing, A review of low frequency electromagnetic wave phenomena related to tropospheric–ionospheric coupling mechanisms. Space Sci. Rev. (2011). doi:10.1007/s112140-011-9854-0

G.L. Siscoe, K.D. Siebert, Bimodal nature of solar wind–magnetosphere–ionosphere–thermosphere coupling. J. Atmos. Sol.-Terr. Phys. **68**, 911–920 (2006)

M. Smiddy, M.C. Kelley, W. Burke, F. Rich, R. Sagalyn, B. Shuman, R. Hays, S. Lai, Intense poleward directed electric fields near the ionospheric projection of the plasmapause. Geophys. Res. Lett. **4**, 543 (1977)

M. Smiddy, W.J. Burke, M.C. Kelley, N.A. Saflekos, M.S. Gussenhoven, D.A. Hardy, F.J. Rich, Effects of high-latitude conductivity on observed convection electric fields and Birkeland currents. J. Geophys. Res. **85**, 6811–6818 (1980)

J.J. Sojka, W.J. Raitt, R.W. Schunk, Plasma density features associated with strong convection in the winter high-latitude *F* region. J. Geophys. Res. **86**, 6916–6968 (1981)

R.W. Spiro, R.A. Heelis, W.B. Hanson, Rapid subauroral ion drifts observed by Atmosphere Explorer C. Geophys. Res. Lett. **6**, 657–660 (1979)

H. Stenbaek-Nielsen, T. Hallinan, D. Osborne, J. Kimball, C. Chaston, J. McFadden, G. Delory, M. Temerin, C. Carlson, Aircraft observations conjugate to FAST. Geophys. Res. Lett. **25**, 2073 (1998)

R.J. Strangeway, The equivalence of Joule dissipation and frictional heating in the collisional ionosphere. J. Geophys. Res. **117**, A02310 (2012). doi:10.1029/2011JA017302

R. Strangeway, C.T. Russell, C.W. Carlson, J.P. McFadden, R.E. Ergun, M. Temerin, D.M. Klumpar, W.K. Peterson, T.E. Moore, Cusp field-aligned currents and ion outflows. J. Geophys. Res. **105**(A9), 21129–21142 (2000)

D.J. Strickland, J.R. Jasperse, J.A. Whalen, Dependence of auroral FUV emissions on the incident electron spectrum and neutral atmosphere. J. Geophys. Res. **88**, 8051 (1983)

D.J. Strickland, R.E. Daniell Jr., J.R. Jasperse, B. Basu, Transport-theoretic model for the electron-proton-hydrogen atom aurora. 2. Model results. J. Geophys. Res. **98**, 21533–21548 (1993)

D.J. Strickland, R.E. Daniell, J.D. Craven, Negative ionospheric storm coincident with DE 1-observed thermospheric disturbance on October 14, 1981. J. Geophys. Res. **106**(A10), 21049–21062 (2001)

Y.-J. Su, J.L. Horwitz, T.E. Moore, B.L. Giles, M.O. Chandler, P.D. Craven, M. Hirahara, C.J. Pollock, Polar wind survey with the Thermal Ion Dynamics Experiment/Plasma Source Instrument suite aboard POLAR. J. Geophys. Res. **103**(A12), 29,305–29,337 (1998)

R.N. Sudan, J. Akinrimisi, D.T. Farley, Generation of small-scale irregularities in the equatorial electrojet. J. Geophys. Res. **78**(1), 240 (1973)

M. Sugiura, N.C. Maynard, W.H. Farthing, J.P. Heppner, B.G. Ledley, L.J. Cahill Jr., Initial results on the correlation between the magnetic and electric fields observed from the DE-2 satellite in the field-aligned current regions. Geophys. Res. Lett. **9**, 985–988 (1982)

W.E. Swartz, M.C. Kelley, J.J. Makela, S.C. Collins, E. Kudeki, S. Franke, J. Urbina, N. Aponte, M.P. Sulzer, S.A. González, Coherent and incoherent scatter radar observations during intense mid-latitude spread *F*. Geophys. Res. Lett. **27**(18), 2829–2832 (2000)

J.P. Thayer, Height-resolved Joule heating rates in the high-latitude *E* region and the influence of neutral winds. J. Geophys. Res. **103**, 471–487 (1998)

J.P. Thayer, High-latitude currents and their energy exchange with the ionosphere-thermosphere system. J. Geophys. Res. **105**, 23,015–23,024 (2000)

J.P. Thayer, J. Semeter, The convergence of magnetospheric energy flux in the polar atmosphere. J. Atmos. Sol.-Terr. Phys. **66**, 807–824 (2004)

J.P. Thayer, J.F. Vickrey, R.A. Heelis, J.B. Gary, Interpretation and modeling of the high-latitude electromagnetic energy flux. J. Geophys. Res. **100**, 19,715–19,728 (1995)

R.M. Thorne, B. Ni, X. Tao, R.B. Horne, N.P. Meredith, Scattering by chorus waves as the dominant cause of diffuse auroral precipitation. Nature **467**, 943–946 (2010)

R.T. Tsunoda, High-latitude *F* region irregularities: a review and synthesis. Rev. Geophys. **26**, 719–760 (1988). doi:10.1029/RG026i004p00719

R.T. Tsunoda, R.C. Livingston, J.P. McClure, W.B. Hanson, Equatorial plasma bubbles: vertically elongated wedges from the bottomside *F* layer. J. Geophys. Res. **87**, 9171–9180 (1982)

S.L. Vadas, D.C. Fritts, Thermospheric responses to gravity waves: Influences of increasing velocity and thermal diffusivity. J. Geophys. Res. **110**, D15103 (2005). doi:10.1029/2004JD005574

S.L. Vadas, Horizontal and vertical propagation and dissipation of gravity waves in the thermosphere from lower atmosphere and thermospheric sources. J. Geophys. Res. **112**, A06305 (2007). doi:10.1029/2006JA011845

C.E. Valladares, D.T. Decker, R. Sheehan, Modeling the formation of polar cap patches using large plasma flows. Radio Sci. **31**, 573–593 (1996)

T. Vondrak, J.M.C. Plane, S. Broadley, D. Janches, A chemical model of meteoric ablation. Atmos. Chem. Phys. **8**, 7015–7031 (2008)

J.-E. Wahlund, H.J. Opgenoorth, F.R.E. Forme, M.A.L. Persson, I. Häggström, J. Lilensten, Electron energization in the topside auroral ionosphere: on the importance of ion-acoustic turbulence. J. Atmos. Sol.-Terr. Phys. **55**, 623–645 (1993)

R.L. Walterscheid, Gravity wave transports and their effects on the large-scale circulation of the upper mesosphere and lower thermosphere. Adv. Space Res. **27**, 1713–1721 (2001)

D.R. Weimer, Models of high-latitude electric potentials derived with a least error fit of spherical harmonic coefficients. J. Geophys. Res. **100**, 19,595–19,607 (1995). doi:10.1029/95JA01755

J.D. Whitehead, Production and prediction of sporadic *E*. Rev. Geophys. **8**, 65–144 (1970)

P.J.S. Williams, B. Jones, G.O.L. Jones, The measured relationship between electric field strength and electron temperature in the auroral *E*-region. J. Atmos. Sol.-Terr. Phys. **54**, 741–748 (1992)

C.T.R. Wilson, Investigation on lightning discharges and on the electric field of thunderstorms. Philos. Trans. R. Soc. Lond. **A221**, 73–115 (1920)

R.F. Woodman, Spread *F*—an old equatorial aeronomy problem finally resolved? Ann. Geophys. **27**, 1915–1934 (2009)

R.F. Woodman, C. La Hoz, Radar observations of *F* region equatorial irregularities. J. Geophys. Res. **81**, 5447–5466 (1976)

J. Wygant, D. Rowland, H. Singer, M. Temerin, F. Mozer, M. Hudson, Experimental evidence on the role of the large spatial-scale electric field in creating the ring current. J. Geophys. Res. **103**, 29527 (1998)

J.R. Wygant, A. Keiling, C.A. Cattell, M. Johnson, R.L. Lysak, M. Temerin, F.S. Mozer, C.A. Kletzing, J.D. Scudder, W. Peterson, C.T. Russell, G. Parks, M. Brittnacher, G. Germany, J. Spann, Polar spacecraft based comparisons of intense electric fields and Poynting flux near and within the plasma sheet-tail lobe boundary to UVI images: an energy source for the aurora. J. Geophys. Res. **105**, 18,675–18,692 (2000)

M. Yamamoto, S. Fukao, R.F. Woodman, T. Ogawa, T. Tsuda, S. Kato, Mid-latitude *E* region field-aligned irregularities observed with the MU radar. J. Geophys. Res. **96**, 15,943–15,949 (1991). doi:10.1029/91JA01321

M. Yamauchi, R. Lundin, L. Eliasson, S. Ohtani, J. Clemmons, Relationship between large-, meso-, and small-scale field aligned currents and their current carriers, in *Polar Cap Boundary Phenomena*, ed. by J. Moen et al. (1998), pp. 173–188

T. Yokoyama, D.L. Hysell, Y. Otsuka, M. Yamamoto, Three-dimensional simulation of the coupled Perkins and E_s-layer instabilities in the nighttime midlatitude ionosphere. J. Geophys. Res. **114**, A03308 (2009). doi:10.1029/2008JA013789

L.J. Zanetti, W. Baumjohann, T.A. Potemra, P.F. Bythrow, Three-dimensional Birkeland-ionospheric current system, determined from MAGSAT, in *Magnetospheric Currents*, ed. by T.A. Potemra. AGU Monograph, vol. 28 (1984)

A.J. Zmuda, Ionization enhancement from Van Allen electrons in South Atlantic magnetic anomaly. J. Geophys. Res. **71**, 1911 (1966)

Space Sci Rev (2012) 168:113–145
DOI 10.1007/s11214-011-9799-3

Trends in the Neutral and Ionized Upper Atmosphere

Jan Laštovička · Stanley C. Solomon · Liying Qian

Received: 24 February 2011 / Accepted: 7 June 2011 / Published online: 30 June 2011

Abstract This article reviews our knowledge of long-term changes and trends in the upper atmosphere and ionosphere. These changes are part of complex and comprehensive pattern of long-term trends in the Earth's atmosphere. They also have practical impact. For example, decreasing thermospheric density causes the lifetime of orbiting space debris to increase, which is becoming a significant threat to important satellite technologies. Since the first paper on upper atmosphere trends was published in 1989, our knowledge has progressed considerably. Anthropogenic emissions of greenhouse gases affect the whole atmosphere, not only the troposphere. They cause warming in the troposphere but cooling in the upper atmosphere. Greenhouse gases such as carbon dioxide are not the only driver of long-term changes and trends in the upper atmosphere and ionosphere. Anthropogenic changes of stratospheric ozone, long-term changes of geomagnetic and solar activity, and other drivers play a role as well, although greenhouse gases appear to be the main driver of long-term trends. This makes the pattern of trends more complex and variable. A consistent, although incomplete, scenario of trends in the upper atmosphere and ionosphere is presented. Trends in F2-region ionosphere parameters, in mesosphere-lower thermosphere dynamics, and in noctilucent or polar mesospheric clouds, are discussed in more detail. Advances in observational and theoretical analysis have explained some previous discrepancies in this global trend scenario. An important role in trend investigations is played by model simulations, which facilitate understanding of the mechanisms behind the observed trends.

Keywords Global change · Long-term trends · Ionosphere · Upper atmosphere

J. Laštovička (✉)
Institute of Atmospheric Physics ASCR, Prague, Czech Republic
e-mail: jla@ufa.cas.cz

S.C. Solomon · L. Qian
High Altitude Observatory, National Center for Atmospheric Research, Boulder, CO, USA

1 Introduction

The anthropogenic emissions of greenhouse gases influence the troposphere, weather, and climate. They appear to be the primary driver of a significant increase of surface temperature in recent decades, thus affecting directly our life. Another anthropogenic pollution of the atmosphere, ozone depleting substances, is responsible for depletion of the Earth's stratospheric ozone layer, particularly in Antarctica, in the form of the spring ozone hole. Both greenhouse gases and ozone depleting substances thus affect the environment, and can adversely influence human health. However, these polluting substances affect not only the troposphere and stratosphere, their effect is observed to penetrate throughout the upper atmosphere, including the mesosphere (~50–90 km), the thermosphere (~90–1000 km), and the ionosphere, which is embedded in the upper atmosphere (e.g., Rishbeth and Roble 1992; Laštovička et al. 2006a). The thermosphere is the operating environment of many satellites, including the International Space Station, and thousands of pieces of space debris, the orbital lifetime of which depends on long-term changes of thermospheric density. Propagation of Global Positioning System (GPS) signals and radio communications are affected by the ionosphere, so anthropogenic changes of these high-altitude regions can affect also satellite-based technologies which are increasingly important to modern life. As with global change near the Earth's surface, the challenge facing upper atmosphere climate scientists is to detect long-term trends and understand their primary causes, so that society can mitigate against potential harmful changes.

The 0.6°C increase in global surface air temperature during the twentieth century (e.g., IPCC 2007) has been attributed predominantly to the increasing atmospheric concentration of greenhouse gases. However, the greenhouse gas increase has an opposite, cooling effect in the upper atmosphere. Greenhouse gases in the troposphere are optically thick to outgoing longwave (infrared) radiation, which they both absorb and reemit back to the surface to produce the heating effect. In contrast, greenhouse gases in the much lower density upper atmosphere are optically thin to outgoing infrared radiation. In-situ collisional excitation results in atmospheric thermal energy readily lost to space via outgoing infrared radiation, while the absorption of radiation emanating from the lower atmosphere plays only a secondary role in the energy balance. The net result is that the radiatively active greenhouse gases act as cooling agents, and their increasing concentrations enhance the cooling effect in the upper atmosphere. This effect of greenhouse gases may be called "greenhouse cooling" (Cicerone 1990).

The global atmosphere is close to a hydrostatic equilibrium, which means that the height of a given pressure surface is determined by the average atmospheric temperature below. The cooling is expected to result in thermal contraction of the upper atmosphere and we may expect a significant decline in thermospheric density at fixed heights, which was observed in long-term satellite drag data (e.g., Emmert et al. 2008). Downward displacement of ionospheric layers should accompany this contraction. The cooling will also affect chemical reaction rates and, thus, the chemistry of minor constituents, resulting in further changes to the ionosphere.

The combination of the lower atmosphere heating and the upper atmosphere cooling is supported by a much stronger greenhouse effect that is observed on Venus where the 96% concentration of carbon dioxide in the atmosphere and a more dense atmosphere result in a troposphere that is more than twice as warm as the Earth's troposphere and a thermosphere that is 4–5 times colder than the Earth's thermosphere (e.g., Bougher and Roble 1991). While the Earth's troposphere is much colder than the thermosphere, the Venusian troposphere is substantially warmer than the thermosphere

Long-term changes in the upper atmosphere and ionosphere have been of interest since pioneering study of Roble and Dickinson (1989). They suggested that global cooling will occur in the upper atmosphere in conjunction with global warming in the troposphere due to long-term increase of greenhouse gas concentrations, particularly carbon dioxide (CO_2). Modeling studies by Rishbeth (1990) and Rishbeth and Roble (1992) broadened these results to the thermosphere-ionosphere system. Investigation of long-term changes in the upper atmosphere has become a significant topic in global change investigation, and many results have been published in the last twenty years (e.g., Aikin et al. 1991; Akmaev 2003; Akmaev and Fomichev 2000; Akmaev et al. 2006; Alfonsi et al. 2002, 2008; Beig 2000, 2006; Beig et al. 2003; Bencze 2009; Bremer 1998, 2001, 2005, 2008; Bremer et al. 2004, 2006, 2009a; Bremer and Berger 2002; Bremer and Peters 2008; Cannon et al. 2004; Chandra et al. 1997; Clemesha et al. 1992; Clilverd et al. 2003; Cnossen and Richmond 2008; Danilov 1997, 2002, 2005, 2008a, 2008b, 2009; DeLand et al. 2006, 2007; Elias 2009; Emmert et al. 2004, 2008; Fomichev et al. 2007; Foppiano et al. 1999; Garcia et al. 2007; Gavrilov et al. 2002; Golitsyn et al. 2000; Gruzdev and Brasseur 2005; Hall et al. 2007a, 2007b; Holt and Zhang 2008; Keating et al. 2000; Jacobi et al. 2005, 2006, 2008, 2009; Jarvis 2006; Jarvis et al. 1998, 2002; Keuer et al. 2007; Kirkwood et al. 2008; Kubicky et al. 2008; Laštovička 2002, 2005, 2009; Laštovička et al. 1994, 2006a, 2006b, 2008a, 2008b; Laštovička and Bremer 2004; Laštovička and Pancheva 1991; Lübken 2000, 2001; Marcos et al. 2005; Marsh et al. 2003; Merzlyakov et al. 2009; Mikhailov 2002, 2006a, 2006b, 2008; Mikhailov and de la Morena 2003; Mikhailov and Marin 2000, 2001; Nedoluha et al. 2003; Offermann et al. 2004; Ortiz de Adler et al. 2002; Portnyagin et al. 2006; Qian et al. 2006, 2008, 2009; Remsberg 2007; Rishbeth 1997; Schmidt et al. 2006; Semenov et al. 2002; Sharma et al. 1999; Shettle et al. 2009; Smirnova et al. 2010; Sridharan et al. 2010; Ulich et al. 2003, 2007; Ulich and Turunen 1997; Xu et al. 2004; Yue et al. 2006, 2008).

In the early period of long-term trend studies in the upper atmosphere, results of individual studies were often inconsistent and sometimes controversial. However, with the increasing amount of observational and model results and findings, a global pattern of trend behavior began to emerge, and, in 2006, the first global scenario of trends in the upper atmosphere and ionosphere was constructed (Laštovička et al. 2006a, 2008a). However, at that time results concerning some variables were not consistent with the global scenario. Moreover, in recent years it became increasingly clear that other drivers play an important role in long-term trends in the upper atmosphere and ionosphere, in addition to the increasing atmospheric concentration of greenhouse gases, which is the dominant driver of upper atmosphere trends.

This paper provides contemporary information, reviews recent developments and suggests some directions of future research in the field of the long-term trends in the upper atmosphere and ionosphere. We focus primarily on the ionosphere and thermosphere, but long-term trends in the mesosphere are also treated. Observational and modeling results are mixed together; the structure of the paper is topical rather than methodological. Section 2 discusses the term "long-term trend" in order to clarify the topic of investigation. Section 3 describes difficulties and problems in trend investigations. Section 4 deals with other drivers of trends. Section 5 presents the global scenario of trends in the upper atmosphere and ionosphere and some results supporting this scenario. Sections 6–8 describe recent developments in three areas which were initially inconsistent with the global scenario of trends, namely the F2-region ionosphere, middle atmospheric dynamics, and noctilucent or polar mesospheric clouds. Section 9 treats aspects of the recent development of modeling of trends in the upper atmosphere and ionosphere, including current problems to be overcome by future development. Section 10 contains a brief discussion of possible or desirable future development of trend investigations. Section 11 closes the paper with a summary and conclusions.

2 What Does the Term "Long-Term Trend" Mean?

Most authors understand the term "trend" as a "long-term trend", i.e. as a long-term tendency of essentially monotonic or quasi-linear change, either an increase or decrease of the values of a particular variable with time. If the long-term behavior is substantially unstable on medium time-scales, or oscillatory, the term "long-term change" is usually used. Strictly speaking, trends are often not linear. However, in most cases the linear trend approximation is sufficient, and in other cases we use the linear approximation to facilitate comparison with trends in other parameters. Ideally, a trend continues to infinity. In reality, a period of quasi-stable trends begins and ends in concrete years/periods. An example would be the Antarctic ozone hole, which appeared in the late 1970s, has recently stabilized, and it is expected to disappear in the second half of this century. Since the mutual role of some trend drivers in the upper atmosphere and ionosphere, and in some cases even the sign of their effect, change with time (e.g. ozone), an overall linear trend cannot be stable in time, and in reality a quasi-stable trend should begin and end within a certain period. Such behavior may be characterized by a sophisticated mathematical curve, but may often be approximated by the so-called piecewise contiguous linear trend model (Weatherhead et al. 2000; Reinsel et al. 2005), wherein substantially different linear trends are fit to the data within different time intervals, the latter also being model parameters. Upper atmosphere and Ionosphere linear trends should be understood in such a way, even though for some parameters the period of quasi-stable trend may be identical with the period of available data for trend studies, i.e. such trends might appear to be monotonic.

It is also possible to use a slightly different and more mathematical approach. A truly linear trend in a time series, by definition, corresponds to the lowest frequencies that are not taken into account in a spectral decomposition. More precisely, quoting Kendall (1973) "The essential idea of the trend is that it shall be smooth", which is equivalent to a continuous slowly varying change in a time series over long time scales (Graigmile et al. 2004). Assuming the trend is not a small temporal part of a major very long period cycle or due to some temporally-coincident stochastic change, then it is possible to temporally and spectrally decompose the time series to separate the oscillatory and true trend elements using, for instance, the wavelet transform technique. However, practical application of such an approach has to be made with care, as the trend has a non-infinite period limited by the length of the dataset or by the limited length of the stable trend interval. This is probably the reason why the attempt of a Chilean team to use such an approach in foF2 trend analysis (Laštovička et al. 2006b) was unsuccessful (Laštovička et al. 2008b).

3 Difficulties and Problems—Methodology, Data Uncertainties, Natural Variability

Important points to be discussed in relation to long-term trend studies include various uncertainties in data, the long-term consistency of measurements, and effects of applications of different methods of trend determination. These issues are particularly important for detecting weak trends in noisy time series, for instance trends in foF2. Different methods have been developed for extracting long-term trends from particular data sets. These methods often include either averaging or interpolation procedures, and thus it is very difficult to compare trend estimates obtained by different authors. Ulich et al. (2003) discussed some practical problems of determining long-term change.

Long time series of measurements need to be consistent. Instrumental changes and malfunctions need to be well documented and corrected for, and subsequent instruments must be well inter-calibrated with previous ones, which is of particular importance for satellite-based data. Before attempting to compute any trend, one has to verify whether or not the time series in question includes instrumental effects. Inhomogeneous data series are one of the reasons of differences of results of different authors. Furthermore, historic time series usually contain data gaps, which have to be treated with care. Some methods of analysis require continuous data sets with constant sampling frequency, but interpolation can introduce incorrect data points, which might skew the results of the analysis.

In the F2-region of the ionosphere, studies of the F2-layer peak height, hmF2, suffer from an additional complication. The height cannot be directly scaled from ionograms; for historical data it was computed from the critical frequencies of the E and F2 layers as well as the propagation parameter M(3000)F2 using empirical formulae. These formulae were derived in the 1970s based on various regional data sets and to be simple and straightforward enough to be used by the computers of those times, but these formulae are by no means globally valid, and their use results in different trends. It is necessary to carefully verify which one is applicable to the area in question.

Atmospheric time series usually vary on diurnal and seasonal time scales. In order to see beyond this variability, cyclic influences on the measurements need to be properly removed and/or considered in interpretation of results. This is usually done by fitting multi-parameter models to the data. Care must be taken that these cyclic parameters are adequately removed and/or considered, especially if the diurnal and seasonal sampling is not uniform.

Solar activity can have a crucial impact on trend determination when data series are relatively short, or when trends in the ionosphere or thermosphere are estimated. However, the effect of the 11-year solar cycle can often be suppressed or minimized with proper selection of the analyzed period, with application of corrections for solar activity, or by studying the residuals from empirical models based upon solar (and geomagnetic) inputs, season, latitude and time. However, for variables with very strong solar cycle effects and weak trends, such as foF2, even relatively small "remnants" of solar cycle effects may influence determination of long-term trends and, thus, different methods of removal/suppression of the solar cycle effect may contribute to differences between various reported trends. Solar activity changes are usually represented by the F10.7 and E10.7 indices, or by sunspot numbers. Estimated trends can be sensitive to the choice of solar activity proxy; for example, the use of sunspot numbers instead of F10.7 or E10.7 produces less reliable trends for F2-layer parameters (Laštovička et al. 2006b).

The temporal extent necessary to obtain reliable trend estimates from a data set is not easy to determine (Weatherhead et al. 2002). For a time series beginning and ending at different phases of the solar cycle, even if it covers several solar cycles, residual solar cycle effects affect the estimated trend. If the trend estimate is plotted as a function of the length of the time series it is derived from, a damped oscillation results, the decay rate of which can be used to estimate the number of years of data required to produce a reliable trend (Jarvis et al. 2002; Clilverd et al. 2003).

One relatively little known factor, which affects the magnitude of determined trends, is data smoothing, e.g., in the form of running averages, which may artificially improve the statistical significance of results if the reduction in the number of degrees of freedom is not accounted for properly. This also tends to reduce the magnitude of trends, as illustrated by the behavior of trends in foF2 (Laštovička et al. 2006b).

Comparison of the results of different methods of trend determination applied to the same data set of foF2 values showed good agreement among some methods, but substantially different results were obtained by other methods (Laštovička et al. 2006b). This is the case for trends that are much weaker than overlapping natural variations. On the other hand, strong trends, as expected, appear to be essentially method-independent, as documented by Križan and Laštovička (2005) for laminae in ozone profiles.

Long-term changes of solar activity and/or an increase of geomagnetic activity throughout the twentieth century might play an important role in the observed long-term trends. This is treated in the next section about other drivers of trends.

4 Drivers of Long-Term Trends

Which agents are responsible for the observed long-term trends? The main reason for the observed long-term trends is considered to be the increasing concentration of greenhouse gases (predominantly CO_2) in the atmosphere, which affects the whole atmosphere from the surface up to heights of several hundred kilometers, both in the neutral and ionized component. However, there are also other drivers of long-term changes and trends in the upper atmosphere, which cannot be neglected (e.g., Laštovička 2009).

In the thermosphere and ionosphere, several drivers can play a role in long-term trends. Solar activity appears to have decreased during in the second half of the 20th century, and particularly in the beginning of the 21st century, which is a tendency opposite to what is required to explain the observed ionospheric trends in the E and F1 regions. Moreover, the effect of solar activity, on solar cycle time scales, can be reduced when long-term trends are computed both in the ionosphere and thermosphere. Different corrections to solar activity is one of the sources of differences between different trend results in F2-region parameters, foF2 and hmF2. Thus, solar activity itself has little or no direct effect on observed ionospheric trends.

However, it is necessary to mention that trends may be quantitatively different under solar activity maximum and minimum conditions, which is the case for thermospheric density (e.g., Emmert et al. 2008). For thermospheric density the reason is much larger relative role of the CO_2 radiative cooling compared to the NO radiative cooling under solar minimum conditions (Qian et al. 2006). The NO radiative power decreased by almost an order of magnitude from 2002 to 2009, whereas the CO_2 radiative power decreased only by $\sim$35% according to SABER/TIMED measurements (Mlynczak et al. 2010). This effect may be considered as an indirect effect of solar activity on trends in the thermosphere and ionosphere.

Another potentially important driver of trends is geomagnetic activity and its long-term changes, an increase of geomagnetic activity throughout the twentieth century (Clilverd et al. 1998; Stamper et al. 1999; Mursula and Martini 2006), even though due to some calibration/instrumental problems in the 1950s the increase is probably smaller than thought earlier (e.g., Martini and Mursula 2008). Geomagnetic activity might be even dominant in trends in foF2. Laštovička (2005) found that in the 20th century the role of geomagnetic activity in the observed long-term trends/changes in the atmosphere-ionosphere system was decreasing from its beginning towards its end. Mikhailov and de la Morena (2003) found that trends in foE were controlled by geomagnetic activity before about 1970, but not in more recent years. Bremer et al. (2009a, 2009b) found that the change of dependency of trends on long-term change of geomagnetic activity occurred around 1970 in the E region, in mid 1990s in the F1 region, and around 2000 in the F2 region. Quite recently, the geomagnetic

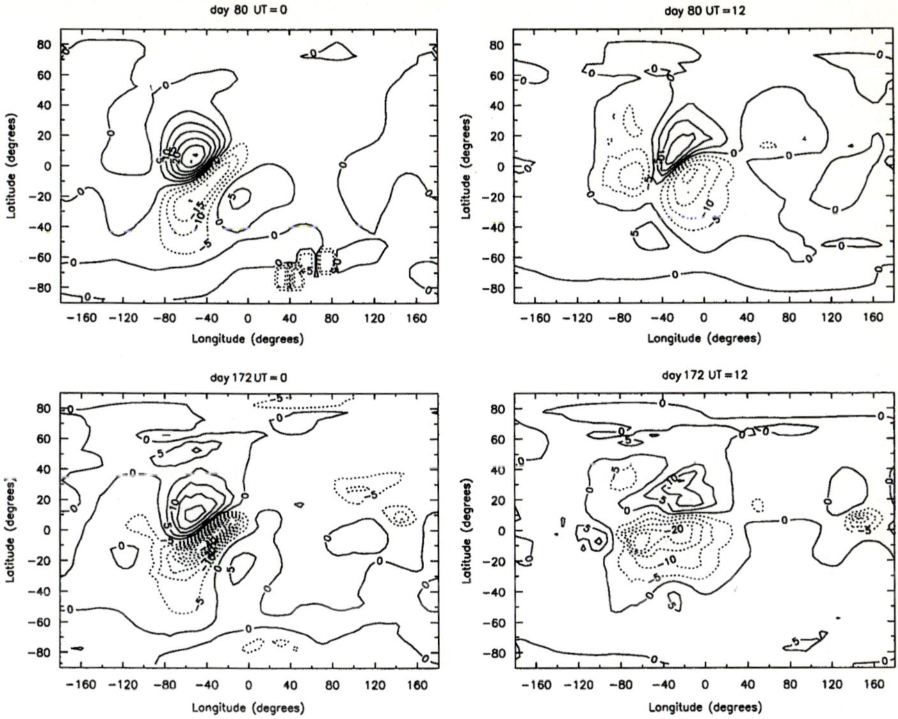

Fig. 1 TIE-GCM model simulation of effect of changes of the Earth's magnetic field in hmF2 (difference between 1997 and 1957 in km) at day 80 (*top*) and day 172 (*bottom*), 00 UT (*left*) and 12 UT (*right*). After Cnossen and Richmond (2008)

activity control of trends was lost even in the F2 region, as it will be discussed in more detail in Sect. 7. Thus it seems that currently the long-term changes in geomagnetic activity are not the main driver of trends in any variable, contrary to past years, when it probably controlled ionospheric trends.

A regionally important driver is the secular change of the main magnetic field of the Earth, which forces the International Association of Geomagnetism and Aeronomy (IAGA) to issue every five years a new reference map of the surface magnetic field (IGRF—International Geomagnetic Reference Field). Figure 1 shows the results of model computations using the NCAR Thermosphere-Ionosphere-Electrodynamics General Circulation Model (TIE-GCM) of the impact of long-term (secular) changes of the Earth's magnetic field on hmF2, the height of maximum of the ionosphere (fof2 and/or maximum electron density provide a very similar pattern). This effect is insignificant in most regions, but in the equatorial to mid-latitude Atlantic Ocean and in the South America it seems to play a substantial role in trends in hmF2 and foF2 (from Cnossen and Richmond 2008; c.f., Yue et al. 2008). Physical analysis by Elias and Ortiz de Adler (2006) demonstrates the important role of secular changes of the Earth's magnetic field in long-term trends in the F2 region parameters in the South America. This effect is dominated by changes of inclination and to some extent also declination of the Earth's magnetic field, but changes of field strength cannot be neglected, as well. Cnossen and Richmond (2008) found that most of the modeled changes in hmF2 and foF2 were due to plasma transport driven by neutral winds, which are mainly caused by changes in the inclination of the magnetic field.

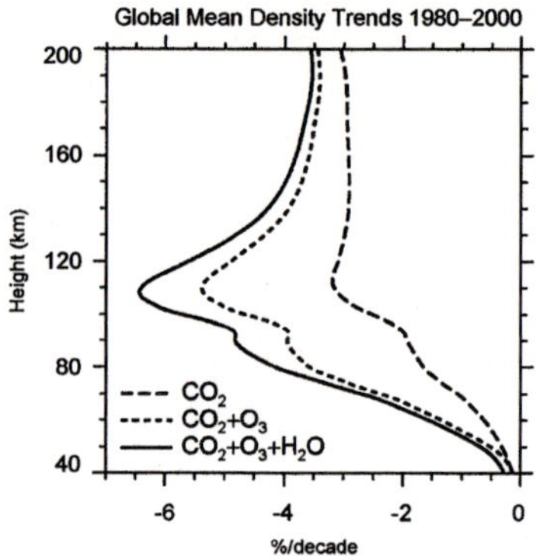

Fig. 2 Trends in global-mean atmospheric density at 40–200 km caused by changes of CO_2, ozone and water vapor concentrations over 1980–2000 according to model calculations by Akmaev et al. (2006)

In the mesosphere and lower thermosphere the most important additional driver seems to be the stratospheric ozone depletion, with an important but unclear role played by changes of the MLT region dynamics, including atmospheric wave activity, and possibly also by mesospheric water vapor concentration. Figure 2 shows the results of model calculations (Akmaev et al. 2006) of the impact of ozone and water vapor trends on trends in atmospheric density. The maximum temperature effect of stratospheric ozone depletion, which amplifies greenhouse thermal contraction of the upper atmosphere, occurs 110 km, i.e. in the lower thermosphere, and it seems to be detectable well into the thermosphere. Water vapor also seems to enhance the contraction of the atmosphere, but its role is less important. This model probably somewhat overestimates trends in atmospheric density (see Fig. 8), but the existence of some influence of these two drivers is indisputable. Unfortunately, there is limited observational information about trends in ozone concentration in the mesosphere and mesopause region, which could affect other trends in the MLT region. Analysis of ozone data from UARS/HALOE satellite observations, finds no detectable trend for sunrise ozone but a strong negative trend for sunset ozone in a narrow region near 80 km, in anti-correlation with water vapor trends (Marsh et al. 2003). However, due to the length of the analyzed interval and uncertainty in the present-day water vapor trend, the mesospheric ozone trend remains uncertain.

Additional support of the role of ozone in long-term trends in the MLT region is provided by ionospheric data analyses. Analysis of long-term variations of the low frequency reflection height near 81 km and the total ozone content from Arosa revealed similarity in trends in the sense that in the period of the midlatitude ozone decrease of 1980–1995 the trend in reflection heights was steeper (Bremer and Peters 2008). Figure 3 shows longitudinal variation of deviations of foE trends and total ozone trends from zonal mean trends (ionosondes and ERA-40 in 30–75°N). A clear anti-correlation between foE trend deviations and total ozone is evident both for winter and summer with slight phase shift, with more pronounced effect in winter. On the other hand, at F1 region heights in foF1 no such relation between trends in foF1 and total ozone has been observed (Bremer 2008).

The role of stratospheric ozone in long-term trends in the mesosphere, lower thermosphere and E-region ionosphere is crucial. Stratospheric O_3 concentration has been decreasing since about 1979, with a turnaround of the trend near 1995 at northern middle latitudes (e.g., Harris et al. 2008) at northern middle latitudes and later also in other regions. Thus now ozone changes affect trends in a way opposite to that in the 1980s. This should result in some modification of trends e.g. in mesospheric temperatures and as observations document, this is the case (see Sect. 5).

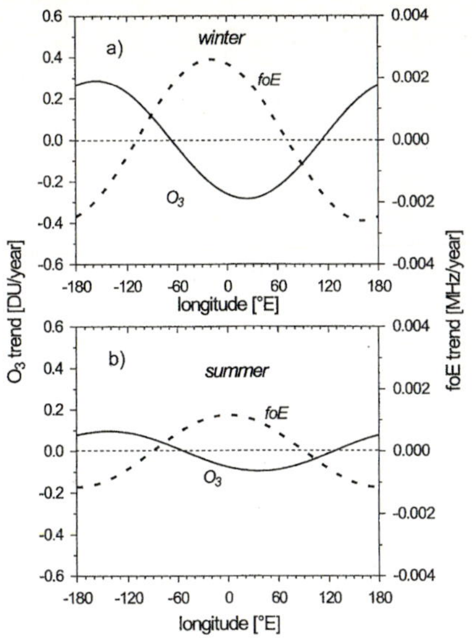

Fig. 3 Longitudinal variation of deviations of foE trends from the average trend in foE (ionosondes between 30–75°N) and of deviations of total ozone trends from the average trend (ERA40, 30–75°N); *top panel*—winter, *bottom panel*—summer. After Bremer and Peters (2008)

Another important factor is long-term trends in atmospheric dynamics and atmospheric wave activity. They are treated in more detail in Sect. 7. Here we only mention that these trends also seem to be non-stable and changing (Laštovička 2009). Trends in mesospheric water vapor are little known (see Sect. 8) but with respect to recent drops in the stratospheric water vapor concentration in 2001 and 2008 (e.g., Solomon et al. 2010a) we expect also rather non-stable trends in the mesospheric water vapor concentration.

Greenhouse gas concentration is steadily and more or less homogeneously increasing. However, other drivers change their behavior both in time and space. Stratospheric ozone concentration after a period of depletion reversed trend and began to recover; its changes are large at high latitudes but almost negligible at low latitudes. Geomagnetic activity is no more increasing and its effect is more important at high than at low latitudes. Secular changes of the main magnetic field of the Earth are very regional phenomenon introducing in various regions even opposite effects on trends. The mesospheric water vapor is uncertain but probably non-stable. Possible trends in atmospheric wave activity are very probably also non-stable. *Therefore we can hardly expect stable, non-changing trends and stable role of various factors responsible for trends—they will be changing both with time and space* (*location*). This phenomenon complicates trend investigations in the upper atmosphere and ionosphere, but on the other hand it contributes to or even explains substantial part of differences between various results of different authors, mainly observational results.

Recent model calculations illustrate the role of other trend drivers (see also Figs. 1 and 2). Figure 4 shows TIME-GCM simulated changes of temperature and neutral density due to estimated changes of trace gases other than CO_2 (Qian et al. 2011). Forcing from these trace gases caused additional cooling in the mesosphere, slight cooling in the upper thermosphere, but some warming in the altitude region around 100–120 km (in fixed height coordinates); it caused neutral density decrease in the mesosphere and thermosphere with maximum density reduction around 110 km. The results on atmospheric density changes are qualitatively consistent with the results by Akmaev et al. (2006) (Fig. 2).

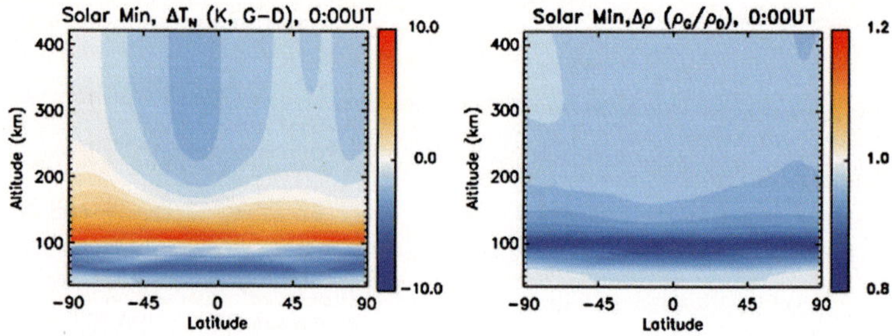

Fig. 4 Changes of neutral temperature and neutral density due to trace gases other than CO_2, such as O_3 depletion (*halved*), H_2O increase (*doubled*), and CH_4 increase (*doubled*), under solar minimum and geomagnetic quiet conditions, at 0:00UT, simulated by the NCAR TIME-GCM model. The difference between the simulation case with doubling of CO_2 only and the simulation case with doubling of CO_2 plus changes of concentrations of other gases is shown. After Qian et al. (2011)

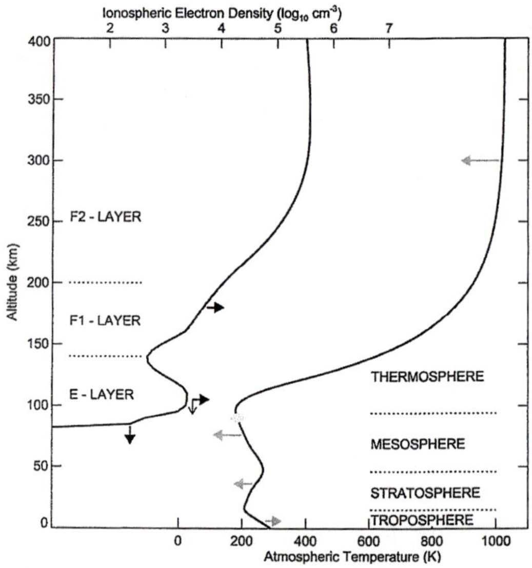

Fig. 5 The first scenario of global change in the atmosphere and ionosphere. Adopted from Laštovička et al. (2008a). Arrows indicate the direction of change. Temperature profile—*left* cooling, *right* heating, no change of temperature in the mesopause; electron density profile—changes in electron density (*horizontal*) and heights of ionospheric layers (*vertical*)

5 Global Scenario of Trends in the Upper Atmosphere and Ionosphere

Early observational results on long-term trends in the upper atmosphere and ionosphere were rather chaotic, partly contradictory and they did not provide a scenario of trends. However, with increasing number of observational and model results a systematic pattern of trends began to emerge. Then in 2006 the first global scenario of long-term trends in the upper atmosphere and ionosphere has been constructed (Laštovička et al. 2006a, 2008a) as it is shown in Fig. 5. The scenario is created by trends in mesospheric temperature (mesospheric cooling and no trend in mesopause region), electron concentration in the lower ionosphere below 100 km (increase at fixed heights), maximum electron concentration (slight increase) and height of this maximum (statistically rather insignificant decrease due to coarse resolution of height measurements by ionosondes) in the E region, electron concentration in the F1-region maximum (slight increase), thermospheric neutral density (moderate decrease) and F2-region ion temperature (some decrease—preliminary result). All these trends are qualitatively mutually consistent and qualitatively agree with model simulations of consequences of the enhanced greenhouse effect, and with the hypothesis of global cooling and contracting of the upper atmosphere. Changes in temperature influenced chemistry of minor constituents, which is probably the main reason of observed changes in the E and F1 region ionosphere.

As regards trends in mesospheric temperatures, the 2006 scenario was based on an extensive analysis by Beig et al. (2003) updated by Beig (2006). New results since 2006 essentially confirm previous results, i.e. a cooling by 2–3 K/decade for middle latitudes and maybe a little bit less at low latitudes for heights of about 50–80 km and essentially no trend in the mesopause region, and partly remove some discrepancies and/or "outlier" results. Remsberg (2007) re-analyzed temperature trends from HALOE measurements at latitudes 40°N–40°S and found a more pronounced cooling at midlatitudes compared to low latitudes. Kubicky et al. (2008) re-analyzed long-term trends in middle atmospheric temperature at northern polar latitudes from historical rocket measurements at Heiss Island. Cooling peaks with about −6 K/decade at 65 km both in winter and summer; however, in summer at 75 km a statistically insignificant heating is observed. Previous high estimates of trends are considerably reduced. Perminov and Semenov (2007) analyzed results of Russian rocket soundings over 1964–1994. Above 55 km (top height of measurements was 75 km) they found large seasonal variation of trends and their remarkable latitudinal dependence; trends were substantially (2–3 times) stronger in winter than in summer. Inclusion of recent data on low frequency phase reflection heights near 81 km confirmed previous results on mesospheric cooling based on older data (Bremer and Peters 2008).

The absence of temperature trends in the mesopause region appears surprising at first glance. However, Fig. 6 shows that all observational data provide for the Northern Hemisphere no statistically significant trend except for the results of Semenov et al. (2002). This data series is the longest, but with several two-year data gaps, which, makes reliability of the trend result quite questionable. Re-analyzed OH temperatures (~87 km, mesopause region) above Wuppertal (Germany, 51°N) give no detectable trend in annual average values over 1994–2004, but the annual amplitude shows a moderate increase before 2001 and substantial decline after 2001 (Offermann et al. 2006), which reflects seasonality in trends. In a longer data series of 1988–2008, a slight but statistically significant overall cooling trend was found, with substantially different trends in individual months, and systematic changes in the shape of seasonal variation of temperature (Offermann et al. 2010). Also, recent model simulations reveal a lack of any significant temperature trend near the mesopause (Schmidt et al. 2006; Garcia et al. 2007; Fomichev et al. 2007). Similarly predominantly no statistically significant trends were reported for the Southern Hemisphere (Beig et al. 2003; Beig 2006).

Changes in the stratospheric ozone concentration have some influence on mesospheric temperatures (e.g., Bremer and Peters 2008; see also Sect. 4) and as the ozone trend reversed in the mid-1990 or in the second half of the 1990s, an impact of this ozone trend reversal should be expected. Observational data confirm this. In the mesosphere, HALOE trends are weaker than older rocket and lidar-based trends, because the ozone effect over the period of HALOE observations was close to none (Remsberg 2009). In the mesopause region the turnaround of ozone trends resulted in a change from no trend to a mild negative trend, at least at low latitudes (e.g., Venkat Ratnam et al. 2010). Temperature trends in the mesopause region in general changed from no trend to a slight negative trend (Beig 2010). This all means that the global scenario of trends remains qualitatively the same but it is somewhat changed quantitatively. Further monitoring of trends in mesosphere and mesopause temperatures is needed.

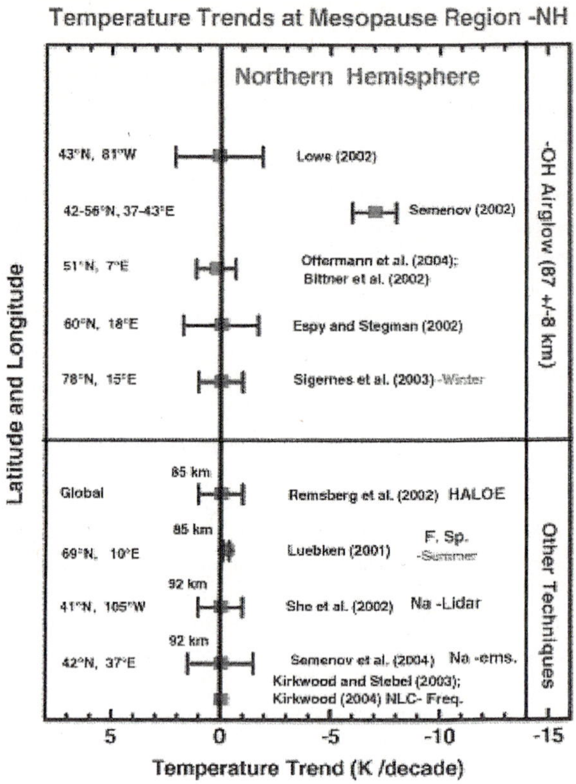

Fig. 6 Summary of trends in the northern hemisphere mesopause region temperatures. After Beig (2006)

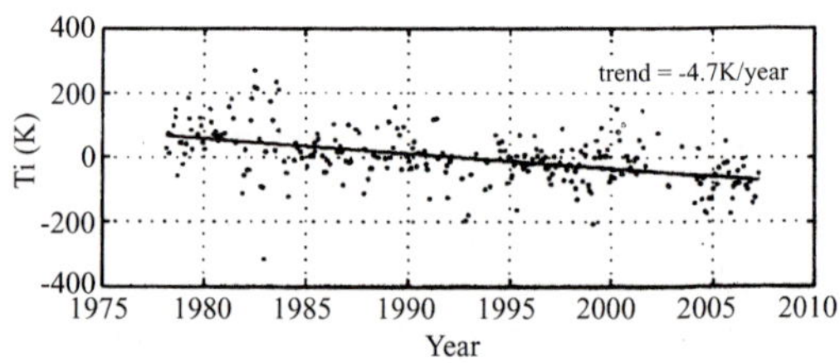

Fig. 7 Long-term trend in ion temperature, Ti, 1978–2007, noon, 350–450 km, Millstone Hill (Holt and Zhang 2008)

Above the mesopause region there are no observational data which would make it possible to calculate temperature trends, only model simulations, which indicate slight heating at fixed altitude just above 100 km, which is replaced by cooling at higher altitudes (e.g., Qian et al. 2011). However, there is some observational information about trends in ion temperature in the upper thermosphere, which is based on long-term measurements of the incoherent scatter radar at Millstone Hill (46.2°N, 288.5°E). The first paper on long-term trends in ion temperatures has been published by Holt and Zhang (2008). Figure 7 displays long-term trends in ion temperatures at 350–450 km. The trend is quite large, a cooling by 47 K/decade, even though it is substantially less than the solar cycle effect. Scatter of data is quite large but the cooling trend is strong enough to make cooling itself reliable. Quite recent results (Zhang et al. 2011) reveal that this trend depends strongly on height, as shown in Table 1. Trends are negative but they are strong and peak near 400 km, are still relatively strong and statistically significant in the exosphere, but they are weak and statistically insignificant below 250 km. The negative trend in the thermospheric ion temperatures together

Table 1 Dependence of the Millstone Hill ion temperature trends on height. After Zhang et al. (2011). Bold—statistically significant trends

Heights (km)	Trend (K/year)
200–250	−0.4
250–300	**−2.6**
300–350	**−4.1**
350–400	**−4.9**
Exosphere	**−2–3**

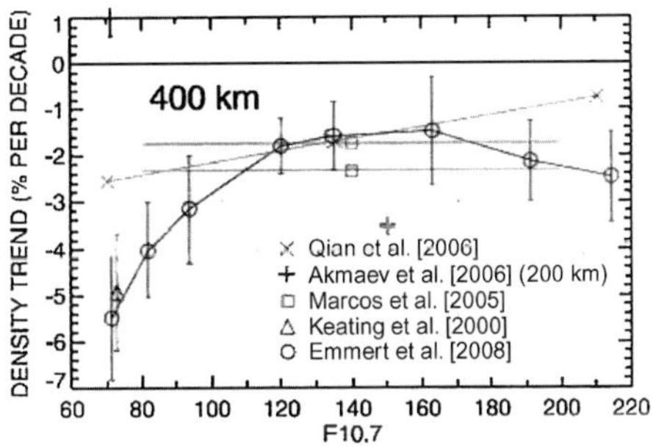

Fig. 8 Summary of various observational (Keating et al. 2000); (Marcos et al. 2005; Emmert et al. 2008) and model (Akmaev et al. 2006; Qian et al. 2006) results on long-term trends in the thermospheric neutral density (adapted from Emmert et al. 2008)

with a negative trend in the thermospheric neutral density indicates a negative trend in the thermospheric neutral temperature, as well. However, it is necessary to mention that model trends in ion temperatures display quite different height dependence and much weaker; the origin of this discrepancy is not known (Qian et al. 2011).

Trends in thermospheric neutral density are very important. They affect the ionosphere but moreover they influence orbital lifetime of low orbiting satellites and space debris. Figure 8 summarizes various modeling and observational thermospheric density trend results. The curve by Emmert et al. (2008) is based on thermospheric densities derived from an analysis of drag of about 5,000 space objects. Observational and model results display a general agreement. These results essentially confirm previous results with only slight quantitative corrections. Figure 8 reveals a clear negative trend for 400 km, typically a decrease by about 2%/decade for high and medium solar activity, which becomes much stronger for low solar activity, more than 5%/decade at solar minimum. Thermospheric density trends at 200–600 km are similar; they become somewhat stronger with increasing height. These trends not only support the idea of thermospheric cooling, they have also important practical impact. A decrease of neutral density results in longer orbital lifetime of space objects including space debris, which means larger than expected future concentration of dangerous space debris at satellite orbits.

Trends in the ionized component are substantial part of the global scenario of trends. Figure 9 shows a global trend in foE and a histogram of trends at 71 individual ionosonde stations (Bremer 2008). Positive trends dominate, but number of stations providing negative trends is larger than negligible. The average trend is $+0.013 \pm 0.005$ MHz/decade. Trends slightly weaken with increasing latitude; their longitudinal dependence seems to be affected, if not determined, by longitudinal dependence of ozone (Fig. 3). Model simulated trends of greenhouse origin are weaker than the observed trends; however, first, part of the difference is very probably attributable to ozone contribution to trends in foE (Sect. 4) and, second, both observational trends and particularly model trends of Rishbeth and Roble (1992) have some degree of uncertainty. Positive trends in foE are supported also by model calculations

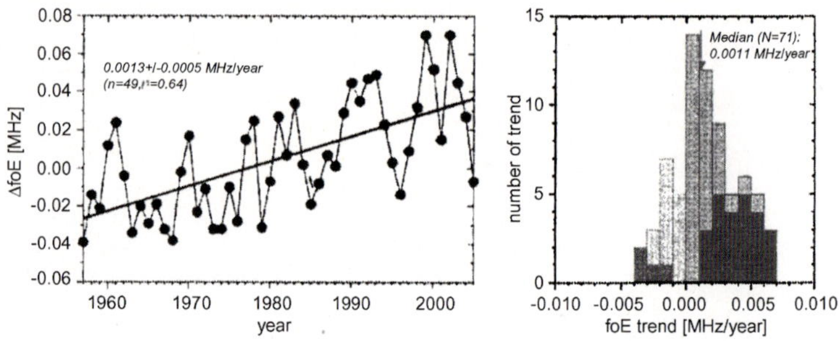

Fig. 9 Global mean foE trend (*left*) and histogram of individual foE trends at 71 ionosonde stations (*right—black* is statistically significant, *grey* insignificant) after Bremer (2008)

of Qian et al. (2008). Differences in trends obtained from data of individual stations displayed by histogram, may be partly of regional origin but mostly they do not seem to be a systematic function of any parameter, therefore they are of unclear origin. However, it is probable that data quality and/or long-term data homogeneity issues play a role in such differences, particularly for relatively close stations. Bremer (1998, 2008) made a similar trend study for hmE. The global trend is negative, -0.29 ± 0.20 km/decade, as expected consequence of mesospheric cooling and, thus, of thermal shrinking. Also histogram shows rather large scatter of trends obtained from data of individual stations. This all makes this trend rather insignificant and/or questionable, mainly due to data problems. First, historical hmE data have coarse height resolution of 5 km. To reach 5 km decrease of hmE, the trend should continue for 170 years. Second, evidently there are some data problems. The longest data series of h'E from Tromso (northern Norway, 1948–2006) reveals an evident decrease after 1975–1980 and no detectable trend before (Hall et al. 2007a).

Bremer (1998, 2008) analyzed foF1 in the same way as foE and obtained from global ionosonde network a weak positive trend, which is only slightly stronger than that for foE and reaches a value of 0.019 ± 0.011 MHz/year. However, there is one significant difference between trends in foE and foF1—no influence of ozone depletion on trends in foF1 was found contrary to that in foE trends.

Cooling and contraction of the upper atmosphere is changing the neutral and ion composition. The ionosphere in the E and F1 regions is approximately under photochemical equilibrium. Therefore, electron density in the E and F1 regions are determined by neutral and ion composition, and solar irradiance reaching to these regions. Cooling and contraction of the upper atmosphere has the effect of increasing ionization rate due to reduced absorption above these regions, enhancing O/N2, and reducing the ratio of NO^+ and O_2^+, the major ions in the E region (Qian et al. 2008). The net effect is an increase of electron density in the E and F1 regions.

Three main areas of discrepancies or unclear trends of the 2006 scenario have been established by Laštovička et al. (2008a). Their key words are: (1) F2-region ionosphere, (2) MLT (mesosphere/lower thermosphere) dynamics, (3) water vapor. These three topics are treated in Sects. 6–8. Item (1) includes contradiction among various results on trends in foF2 and hmF2 and their drivers. Item (2) consists of unstable and/or non-linear trends in MLT winds and largely unknown and/or uncertain trends in atmospheric wave activity in MLT. Item (3) means changes in behavior of water vapor concentration in the last ten years and discrepancies in ground-based observations of noctilucent clouds and satellite-based observations of polar mesospheric clouds. In Sect. 8 we shall deal only with noctilucent cloud trends, as the situation with mesospheric water vapor is not clear due to shortage of observational data. Moreover, we are missing observational information about trends in two important minor constituents of the MLT region, nitric oxide and ozone, except for analysis of insufficiently long ozone data series from limb measurements of HALOE (Marsh et al. 2003); more observations are needed.

6 F2 Region Ionosphere

The F2 region ionosphere main variables, critical frequency foF2 (equivalent to maximum electron density) and height of the F2 region peak, hmF2, did not fit the 2006 scenario of trends due to two problems: (1) Contradictions in magnitude and even sign/sense of trends. (2) Unclear origin of trends—dominantly anthropogenic or dominantly geomagnetic. Here we shall deal with progress since 2006.

In literature there are large discrepancies and controversies between the results of different authors and different methods as for trends in the F2 region parameters foF2 and hmF2. A trial to remove and/or explain some discrepancies between the results of six different groups on trends in foF2 was performed by Laštovička et al. (2006b) via comparison of the results of their different methods applied to a high-quality test data set (1976–1996, two comparable solar cycles) from Juliusruh (northern Germany). The necessity of a very careful correction for the solar cycle effect was pointed out. The correction for the solar cycle effect with F10.7 or E10.7 was evidently better than that with R or R12. Various regression-based methods used by five teams provided comparable weak negative trends in foF2. A Chilean team analyzed the data with regression-based and wavelet-based methods and got quite different results of opposite sign. As shown by Laštovička et al. (2008b), there was however an error in their approach. Also, a neural network-based method (Yue et al. 2006) gave results comparable with regression-based results (Laštovička et al. 2008b). Only the foF2 trends derived by Mikhailov's method, which uses 131-month smoothed values and a special way of eliminating geomagnetic activity effects, are substantially different from other results. However, the explanation has been made that without a special elimination of geomagnetic activity effects, these results are consistent with the others (A.V. Mikhailov, private communication, 2008). Thus, application of different methods can explain only a relatively small part of differences between the results of various authors

The main sources of majority of discrepancies in trend results among various authors seem to be different data sets, quality of data, different analyzed periods, and different way of eliminating the effect of solar and geomagnetic activity. These influences can explain a substantial part of discrepancies among trends obtained by various authors. Since long-term trends in foF2 are by almost two orders of magnitude weaker than the solar cycle change over a decade, the resulting trends are sensitive to selection of analyzed period with respect to solar cycle and to the method of solar cycle effect removal and/or suppression. Ulich et al. (2007) analyzed a 48-years long series of foF2 measurements from Sodankylä (northern Finland). They found that trends depended on the selection of data smoothing. Models based on monthly averages and 13-month running means yielded identical trend of -14 ± 4 kHz/year, while that based on 131-month running means provides weaker trend of -10 ± 4 kHz/year. It is consequence of known effects in statistics; too much smoothing of data to some extent reduces trends (E.C. Weatherhead, private communication, 2005). With trends in hmF2 there is one more problem with calculating hmF2 from M(3000)F2 (direct measurements by incoherent scatters or deduced from digisonde electron density profiles do not provide sufficiently long or continuous data sets), because various formulas used for re-calculating are applicable only under some conditions, and under specific conditions different formulas are the best (McNamara 2008).

Another controversy concerns origin of trends in foF2 and hmF2. The two main candidates are increasing concentration of greenhouse gases and long-term changes of geomagnetic activity (the role of secular changes of the Earth's magnetic field has been discussed in Sect. 4). Based on analyses of foF2 and hmF2 long-term variations of Eurasian ionosonde stations, the so-called geomagnetic control concept has been developed by Mikhailov (2002). Initially formulated for ionospheric F2-layer parameters, later it was extended for the foE (Mikhailov and de la Morena 2003; Mikhailov 2006b) and foF1 (Mikhailov 2008) long-term variations. According to this concept the observed long-term variations of electron concentration in the ionosphere are dominantly controlled by the geomagnetic activity long-term variations (11-year running means Ap indices Ap_{132} are used).

However, this geomagnetic activity control was fully valid only in the past; it is no more the case at present. In the E region the geomagnetic control broke down around 1970 (Mikhailov and de la Morena 2003). Figure 10 reveals years 1972 for Slough and 1967 for Rome. Figure 10 also shows continuous loss of geomagnetic control of foF1 in the 1990s, where the dominant geomagnetic control appears to be lost in the first half of the 1990s. Figure 10 also indicates for both stations possible beginning of the loss of the predominant geomagnetic activity control in foF2 around 2000. Two more data points available now confirm this tendency, so we may say that even in foF2 the predominant geomagnetic activity control of trends was lost at about 2000 and that at present long-term changes in the ionosphere are no more predominantly controlled by long-term changes of geomagnetic activity; the increasing concentration of greenhouse gases seem to be the main controlling factor. This all coincides with Laštovička's (2005) finding that the greenhouse gas control of long-term changes in the atmosphere-ionosphere system was increasing throughout the 20th century, while the solar and geomagnetic control was decreasing.

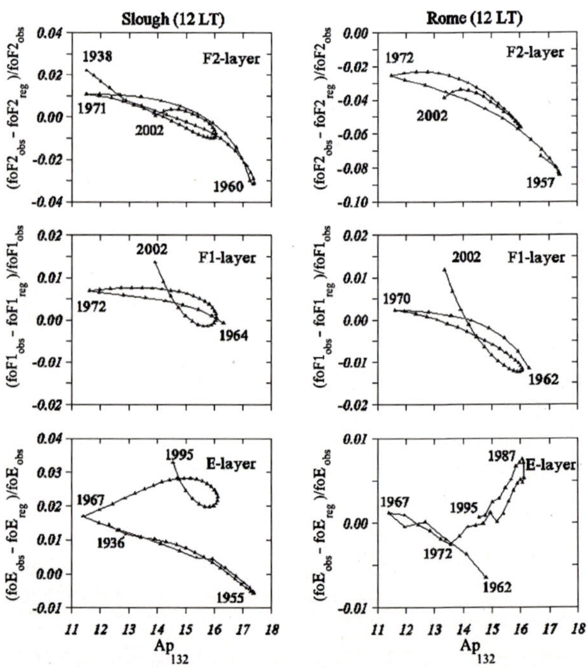

Fig. 10 Relationships between δfoF2 (*top panels*), δfoF1 (*middle panels*) and δfoE (*bottom panels*), and Ap_{132} variations for Slough (*left panels*) and Rome (*right panels*). The change in the type of the dependences appeared earlier in the E region, later in the F1 region, and eventually it began to appear in F2 region. After Bremer et al. (2009a)

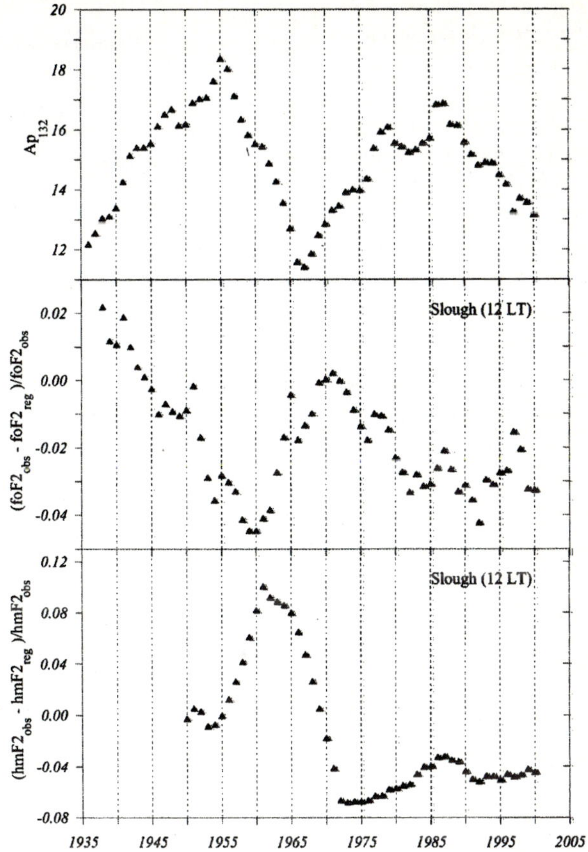

Fig. 11 11-year running means (132 months): Ap132 (*top panel*); ΔfoF2132 (*middle panel*) and ΔhmF2132 (*bottom panel*) long-term variations at noon for Slough (U.K.). After Mikhailov (2006a)

How is it with hmF2? Figure 11 illustrates development of foF2 and hmF2 in dependence on smoothed geomagnetic activity for Slough. Until 1972, foF2 is anti-correlated and hmF2 correlated (with time delay of about 4–5 years) with geomagnetic activity. However, since about 1973 the hmF2 response to geomagnetic activity becomes much smaller (almost negligible) and in the early 1990s hmF2 looses any relation to geomagnetic activity, i.e. we can assume the dominant greenhouse gas control to govern long-term changes of hmF2 at present.

Model calculations of the effect of CO_2 doubling on the ionosphere by Qian et al. (2008) qualitatively explain how such a different behavior of foF2 and hmF2 is possible. Figure 12 show that these calculations confirm observational findings of increasing electron densities at fixed heights in the lower ionosphere, increasing foE and foF1. In the F2 region, plasma transport becomes increasingly important. The combined effect of changes in photochemical and plasma transport processes together with thermal shrinking of the upper atmosphere result in an evident decrease of hmF2 (by more than 20 km), while decrease of foF2 is small, because the cross point between increasing and decreasing electron densities is only a little below the height of maximum electron density in the F2 region; hmF2 for doubled CO_2 concentration is located almost at line of no change in electron density. The electron density trend above the peak of the F2 region is negative. In addition, the interplay of photochemical and plasma transport processes causes complex regional, diurnal, seasonal, and solar cycle

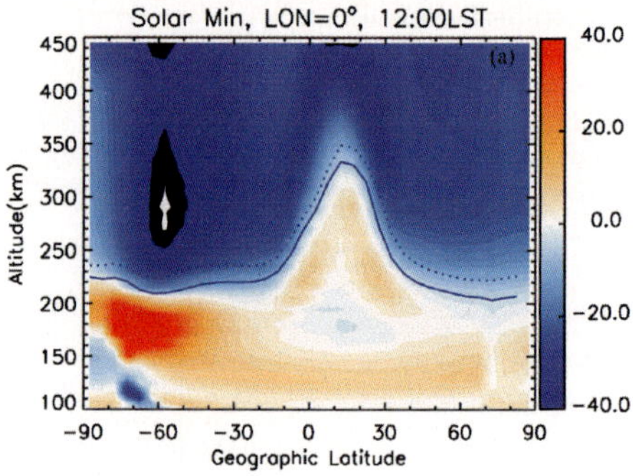

Fig. 12 Model simulation of trends in foF2 and hmF2 at noon, longitude 0°, as a difference between the basic state and the state with doubled CO_2 concentration. *Dashed curve*—hmF2 for basic state; *solid curve*—hmF2 for doubled CO_2. After Qian et al. (2008)

variations of trends in the upper ionosphere (Qian et al. 2009); e.g. the nighttime pattern is significantly different from the daytime pattern.

Thus *trends in the F2-region are at present already predominantly controlled by greenhouse gas increases and, thus, consistent with the presented scenario of trends in the ionosphere and upper atmosphere.* Now we also essentially understand the large scatter and even contradictions of the past reported observational trends in foF2. Factors responsible for this are as follows: small greenhouse gas effect (see Fig. 12), changing role of trend drivers (particularly geomagnetic activity) with time, regional influences of secular change of the geomagnetic field, difficulty in removing of solar cycle influence, local-time dependence of trends, and possibly other factors.

Bencze (2009) claims that the geographical distribution of trends in hmF2 can be explained in terms of the inclination and declination of geomagnetic field and nonmigrating tides. However, observational results supporting this statement are not very conclusive, and this hypothesis is preliminary, pending further research and clarification.

7 Atmospheric Dynamics and Trends

Changes in the temperature field should affect winds. Routine wind measurements in the MLT region at heights 80–100 km have been carried out for several decades, particularly at several stations at northern higher middle latitudes. The most prominent components of the mid-latitude MLT winds are the prevailing (mean) and tidal winds. Long-term trends of prevailing wind and semidiurnal tides have been examined and detected using long-term wind measurements at mid-high latitude stations in the northern hemisphere (e.g., Portnyagin et al. 2006).

Figure 13 shows time series of annual mean zonal and meridional prevailing winds over Obninsk (55°N, 37°E) and Collm (52°N, 15°E). In spite of differences in the measurement techniques and strong year-to-year wind variations, the general tendencies in the climatic MLT wind variations at both stations are similar. Before about 1990, both the annual prevailing zonal and meridional wind weakened, while after about 1990, zonal wind strengthened and the negative trend in meridional wind leveled off. It should be mentioned that lower stratospheric winds at 52.5°N change their trend also around 1990 (Laštovička et al. 2010). However, if look on Fig. 13 in more detail, we see some differences between Collm and Obninsk winds even for annual means. Other stations in the same latitudinal belt, especially Saskatoon in Canada, provide partly different results. Figure 14 gives an explanation why the observational trend results at different longitudes partly differ, even though these model calculations are made for somewhat lower latitude (41.25°N). This simulation is calculated

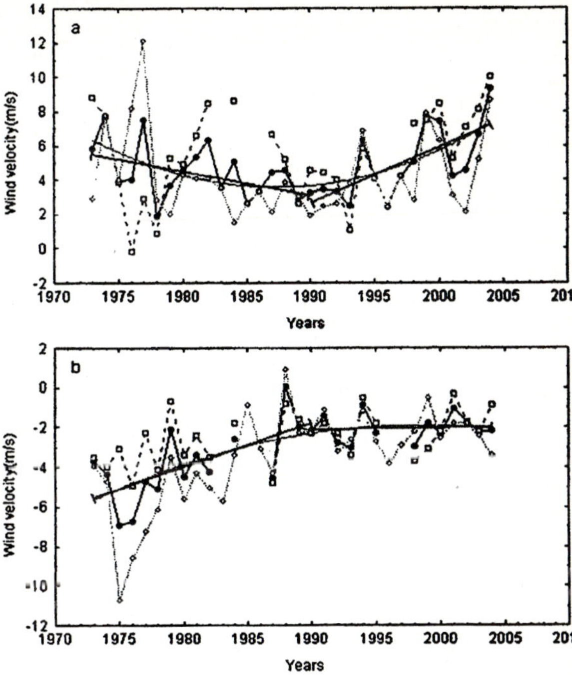

Fig. 13 MLT annual mean prevailing winds over Obninsk (55° N, 37° E) and Collm (52° N, 15° E), *dotted line*—Obninsk, *dashed line*—Collm, *full line*—mean values; *top panel*—zonal component, *bottom panel*—meridional component. From Laštovička et al. (2008a)

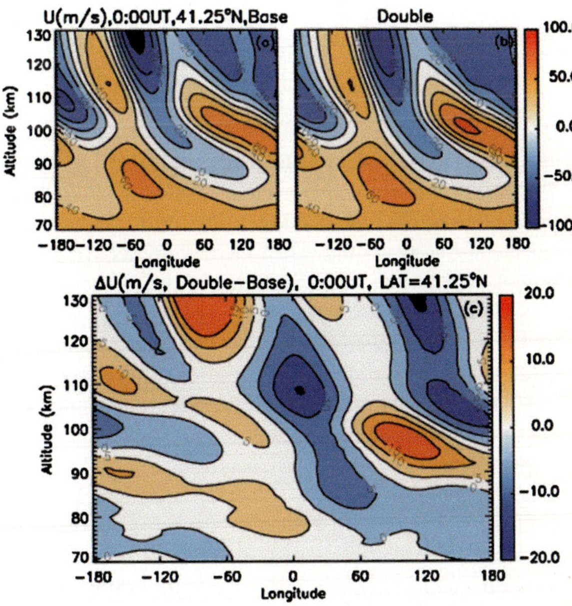

Fig. 14 Zonal wind and change (double CO_2–base CO_2) of the zonal wind at 41.25°N, under solar minimum and geomagnetic quiet conditions, at 0:00UT, simulated by the NCAR TIME-GCM. (**a**) Zonal wind for the base case; (**b**) zonal wind for the double CO_2 case; (**c**) change of the zonal wind between the two cases

for 00 UT, so a part of apparent longitudinal difference of trends is an effect of different day- and night-time trends but even for well daytime or well nighttime intervals the trends are somewhat different, i.e. they reveal some real longitudinal difference which is reflected in the observation-based trends. Another problem is that trends in prevailing wind are height dependent even as to their sign/sense; this is shown by Fig. 14 for simulations and confirmed also by observations (e.g., Keuer et al. 2007). However, simulations shown in Fig. 14 include only changes in the greenhouse gas concentration while ozone and water vapor effects are omitted therefore they illustrate only the greenhouse gas effect, not complete change. More global data coverage is also required in order to get more reliable information about trends in prevailing wind in the MLT region.

Atmospheric waves, namely planetary, tidal and gravity waves coming from below appear to be the main agent responsible for coupling of the lower atmosphere to the upper atmosphere. Very little is known about their long-term changes and trends. Their trends may partly be caused by changes in the source region, i.e. in the troposphere and, therefore, they may differ from the trends in other upper atmosphere variables. There are some indications of possible trends in gravity wave activity (e.g., Jacobi et al. 2006), tides (e.g., Portnyagin et al. 2006), and planetary waves (e.g., Jacobi et al. 2008). However, information on trends in atmospheric waves in the mesosphere and also in the thermosphere is very limited and not quite consistent. Infrasonic waves with periods of seconds to a few minutes (e.g., Blanc 1983; Krasnov et al. 2006) are very probably less important and no information is available about possible trends due to lack of measurements.

Information on trends in planetary wave activity as summarized by Laštovička et al. (2008a) does not provide clear pattern. Planetary wave activity in the MLT region appears to have increased, even though this increase seems to be intermittent. As for recent results, analysis of NCEP/NCAR reanalyses over the period 1960–2000 reveals some increase of the stationary planetary wave of the zonal wave numbers 1 and 2 activity in the stratosphere but a reduction of traveling stratospheric wave activity at typical periods ~5, ~10 and ~16 days (Pogorelcev et al. 2009). Jacobi et al. (2008) analyzed MLT wind measurements at Collm. They report trends in planetary wave activity, which are highly variable with period range and season, being generally positive in zonal wind and weaker and negative in meridional wind, and in total wind being in average slightly positive. The trends are more pronounced for short periods ($T < 7$ days) and for summer, whereas at longer periods and in other seasons they are very predominantly relatively small and statistically insignificant. Overall we can say that there is probably a tendency to strengthening planetary wave activity in the MLT region but more data are needed. Long-term changes in tropospheric/stratospheric stationary planetary waves could affect the occurrence frequency of (major) sudden stratospheric warming. Indeed, the occurrence rate of the major stratospheric warming in the 1990s was low, while in the 2000s it was high. The strong major warming affect the ionosphere at all latitudes down to equator in such a way that morning values of NmF2, hmF2 and ITEC considerably increase whereas their afternoon values considerably decrease (e.g., Yue et al. 2010). Thus long-term changes in the occurrence rate of major stratospheric warming may affect long-term trends of ionospheric parameters at fixed times, which is another potential way of influence of atmospheric wave activity on long-term trends in the ionosphere-upper atmosphere system.

Diurnal and semidiurnal tides are the most important tidal modes. In the mid- and high-latitude MLT region, the semidiurnal tide is dominant. The negative trend in semidiurnal tidal winds seems to cease after the mid 1980s or 1990 (e.g., Portnyagin et al. 2006), or may even reverse. Observations at Scott Base, Antarctica reveal a positive trend since the late 1980s (Baumgaertner et al. 2005). Portnyagin et al. (2006) and Merzlyakov et al. (2009) showed that the prevailing wind and semidiurnal tide in the mesopause region on the Northern Hemisphere as well as in Antarctica changed the trend (mostly to opposite one) near 1990. The most recent results from Collm, however, indicate an overall negative trend (weakening) of the semidiurnal tide (with large seasonal variation of trend) over 1979–2007.

Thus observational data do not provide a consistent pattern of trends in tidal activity in the MLT region. More observations are required.

There are very few studies of long-term trends in the MLT region gravity wave activity. Results of gravity wave analyses from ground-based systems can be strongly dependent on system characteristics. The wave periods seen in wind data measured by MF radars (>10 min), meteor radars (>1 h), and LF drifts (0.7–3 h) are very different and results are not easily comparable. Indirect evidence was rather in favor of some trends in gravity wave activity, but wind measurements themselves did not reveal a trend, as summarized by Laštovička et al. (2008a). There are some indirect indications of possible trends in gravity wave activity (e.g., Jacobi et al. 2006); direct investigation of gravity wave activity in MLT winds at Shigaraki (southern Japan) did not reveal any significant trend (Gavrilov et al. 2002); Hoffmann et al. (2010) found for Juliusruh (northernmost Germany) a weak positive trend at 80–90 km and they mention that some stations display trends, some not. Some mechanisms of trends in GW activity, like changes of storm tracks at midlatitudes, may result in regionally/locally different trends in GW activity even as to sign/sense, thus caution is necessary. Again, more information is very required.

A phenomenon closely related to dissipation of gravity waves is turbulence. Hall et al. (2007b) reported MLT turbulence measurement analysis for Tromso (70°N, 19°E), 1999–2006. Even though this data series is too short to provide reliable trend information, it gives a hint. Steadily positive trend was observed in the upper mesosphere and lower thermosphere and negative trend in the middle mesosphere with average transition height 80–81 km (varying with season). Solar and geomagnetic activities are unable to explain the observed changes therefore strong candidate is anthropogenic forcing. Kalgin (1998) also reported a positive trend in turbulence in the lower thermosphere.

There is no direct information about trends in atmospheric winds and waves in the middle and upper thermosphere. However, Danilov (2008b, 2009) and Danilov and Vanina-Dart (2010) proposed a new method how to estimate trend in horizontal winds based on analysis of foF2 and hmF2, namely foF2(night)/foF2(day) ratio, correlation between foF2(night) and foF2 (day) and scatter of hmF2 values. All these parameters indicate a change in trends in thermospheric winds in about 1980 from a no trend regime to a trend regime. The trend was different for different stations in dependence on station magnetic inclination and declination (Danilov and Vanina-Dart 2010). The trend was better pronounced in spring than in other seasons.

Summarizing, it may be said that there was some but limited progress in the area of long-term trends in the upper atmosphere dynamics and atmospheric wave activity in recent period and main open questions remain unanswered. The trends in atmospheric wave activity represent the key problem of trend investigations as they are capable to affect trends in majority of upper atmospheric and ionospheric variables.

8 Noctilucent Clouds Versus Polar Mesospheric Clouds

The highest atmospheric clouds, noctilucent clouds (NLC) observed by ground-based instruments, appear in extremely cold summer polar mesopause region at heights of about 82–85 km. The same phenomenon, observed from above by satellites, is called polar mesospheric clouds (PMC). Their appearance is controlled by temperature and water vapor content. Long-term trends in the occurrence frequency and brightness of NLCs observed from the ground, at latitudes below about 65°N in Europe do not display a detectable trend in the NLCs according to Kirkwood et al. (2008). Satellite observations cover a larger geographical

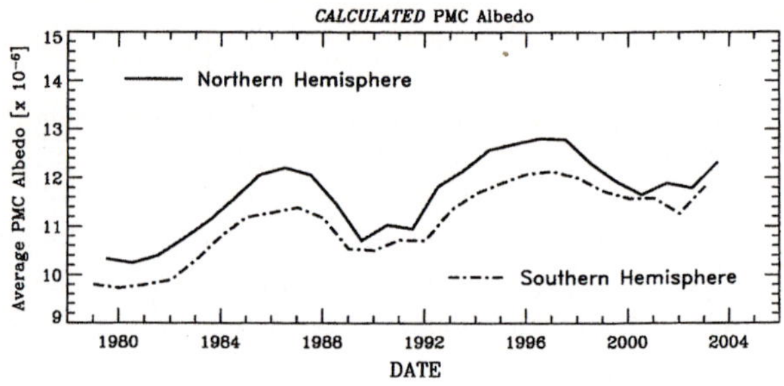

Fig. 15 Average PMC albedo calculated from a fit to the combined SBUV and SBUV/2 data: *Solid line*—Northern Hemisphere, *dot-dash line*—Southern Hemisphere. After DeLand et al. (2006)

Table 2 Latitudinal dependence of trends in the occurrence frequency of polar mesospheric clouds separately for individual latitudinal bands. The relative occurrence frequency in latitudinal bands is as follows: 54–64 : 64–74 : 74–82 = 1 : 2 : 4. Adopted from Shettle et al. (2009)

Latitudes (°N)	Trend coefficient t	Trend %/decade	95% confidence per decade	Statistical significance
54–64	0.017 ± 0.014	9.9	12.4	Insignificant
64–74	0.022 ± 0.018	6.7	9.2	Insignificant
74–82	$\mathbf{0.134 \pm 0.031}$	**19.7**	**9.1**	**Significant**
54–82	$\mathbf{0.069 \pm 0.021}$	**15.6**	**9.3**	**Significant**

area and the highest occurrence frequency of PMCs is around 80°N. The satellite observations indicate an increase in the PMC occurrence frequency and brightness, as illustrated by Fig. 15 for the PMC albedo. Thus NLC-based and PMC-based trends are quite different, which is evident contradiction as NLC and PMC is the same physical phenomenon, only seen from below and from above.

In order to solve this discrepancy, Shettle et al. (2009) analyzed 28 years of satellite PMC observations for latitudinal bands 50°N–64°N (NLC observation latitudes), 64°N–74°N, and 74°N–82°N (Table 2). They observed statistically significant increase only in 74°N–82°N, whereas in 50°N–64°N, it was half in magnitude and statistically insignificant. At NLC observational latitudes, the PMC trend of Shettle et al. (2009) is +9.9%/decade, the NLC trend of Kirkwood et al. (2008) is +4.4%/decade for moderate and bright NLCs, and ~14%/decade for all NLCs; all trends being statistically insignificant. This means that the PMC and NLC trends are comparable and do not differ within accuracy of their determination in spite of partly different analyzed periods. Relatively large differences between older results were strengthened by the fact that the occurrence frequency of PMC in the latitudinal band with the largest trend is four times as high as in the NLC latitudinal band (Table 2).

Simulations with the LIMA/ice model (Lübken et al. 2009) show that all three factors, temperature, water vapor and Lyman-α solar radiation are important and none of them is dominant in PMC/NLC behavior. Thus, the origin of trends in PMC/NLC remains to be discovered. However, Lübken et al. (2010) found that the occurrence frequency of NLC is determined mainly by water vapor concentration, whereas the NLC layer height is affected dominantly by temperature. The observed trends in PMC (no change of height but increasing occurrence frequency) would then indicate little change in temperature, in line with other mesopause region temperature trend estimates, (Sect. 5, Fig. 6) but increasing water vapor concentration.

Polar mesospheric summer echoes (PMSE), the anomalous radar echoes found between 80–90 km from May through early August in the Arctic, and from November to February in the Antarctic, is a phenomenon closely related to PMC. Bremer et al. (2009b) found a positive but insignificant trend in PMSE occurrence frequency and length of season for Andenes (69°N) over 1994–2008, whereas Smirnova et al. (2010) found insignificant negative and positive trends, respectively, for Kiruna (68°N) over 1997–2008. Longer data series are needed to get more reliable trend results.

9 Modeling of Trends—Recent Developments, Current Problems and Challenges

Since the original finding of Roble and Dickinson (1989) using a global mean model, advances in model development have gained considerable insight into the causes and expected nature of upper atmospheric trends. Ionospheric effects (Rishbeth and Roble 1992; Rishbeth 1997; Qian et al. 2008, 2009) have been modeled, and mesosphere and lower thermosphere responses to multiple trace-gas changes have been studied (Akmaev and Fomichev 2000; Akmaev et al. 2006; Gruzdev and Brasseur 2005). Recent advances in three-dimensional whole-atmosphere models that include both the tropospheric heating and upper-atmosphere cooling response to anthropogenic forcing, such as the Hamburg Model of the Neutral and Ionized Atmosphere (HAMMONIA) (Schmidt et al. 2006), the NCAR Whole Atmosphere Community Climate Model (WACCM) (Garcia et al. 2007) and the Canadian Middle Atmosphere Model (CMAM) (Fomichev et al. 2007) have enabled detailed examination of the stratosphere/mesosphere response to solar and anthropogenic forcing. Development of the NCAR Thermosphere-Ionosphere-Mesosphere-Electrodynamics General Circulation Model (TIME-GCM) has also continued, including its three-dimensional and global mean versions, and has been employed to understand the interaction of the solar cycle with thermosphere and ionosphere trends (Roble and Ridley 1994; Qian et al. 2006, 2011). Recent studies employing the TIME-GCM quantify the role CO_2 compared to other anthropogenic gases, as shown above in Fig. 4 (Qian et al. 2011).

Model simulations have enabled a better understanding of some of the specific aspects of the trend scenario described in Sect. 5 and illustrated by Fig. 5. The whole atmosphere models agree that transition from heating to cooling occurs above the tropopause, near 15 to 20 km, and that ozone changes dominate stratospheric cooling; Garcia et al. (2007) obtained reasonable agreement with observed stratospheric trends. In the mesosphere, the dominant cooling mechanism becomes CO_2 radiation, with ozone and water vapor playing additional roles (Akmaev et al. 2006) as shown in Fig. 2. The effects of stratospheric ozone depletion, causing cooling and contraction of the stratosphere, propagate even into the thermosphere.

All the large-scale models also agree that there should be little or no temperature trend near the mesopause. This appears to be in agreement with most of the observations. However, there is not yet any comprehensive explanation of what mechanism is responsible for this apparent paradox, since CO_2 cooling is large near the mesopause. Schmidt et al. (2006) speculate that since the mesopause region is dominated by dynamics, the models are insensitive to radiative changes. McLandress and Fomichev (2006) found amplification in tidal forcing using the CMAM, and proposed a secular increase in the amplitude of the migrating diurnal tide in the upper mesosphere. Garcia et al. (2007) and Fomichev et al. (2007) also found no significant temperature change near the mesopause. TIME-GCM simulations (Qian et al. 2011) show a small negative temperature trend at 90 km, but zero trend at 100 km, when analyzed in altitude coordinates, primarily due to the effect of the positive temperature gradient above the mesopause. Thus, there is a need for further analysis of model processes in order to understand the operative mechanisms in this region, since the results of modern fully-coupled three-dimensional models can be as difficult to understand as the atmosphere itself. Progress in modeling another important mesopause-region phenomenon, PMC/NLC, is described in Sect. 8 above. Lübken et al. (2010) found that the occurrence frequency of NLC is determined mainly by water vapor concentration, while temperature determines the layer height. Thus, anthropogenic changes to the mesospheric hydrogen budget, caused largely by methane increases, may be implicated in possible trends.

Returning to the thermosphere, where the clearest detection of upper atmosphere trends has been established through long-term satellite drag measurements (e.g., Keating et al. 2000; Emmert et al. 2004, 2008; Marcos et al. 2005), significant progress has been made in understanding the solar cycle dependence of temperature and density trends (see Fig. 8). Qian et al. (2006, 2011) showed that cooling trends should be larger at solar minimum, due to the larger relative role of CO_2 radiative cooling compared to the nitric oxide (NO) radiative cooling. This is confirmed by SABER measurements showing that global integrated NO radiative power decreased by almost an order of magnitude from 2002 to 2009, while CO_2 radiative power decreased only by $\sim$35% (Mlynczak et al. 2010). This may explain why early results by Keating et al. (2000) were higher than model estimates at that time, and subsequent analyses by Marcos et al. (2005). Keating et al. compared successive solar minima in order to eliminate the large confounding effect of the solar cycle, obtaining an estimate of 5%/decade decrease in density at 400 km. Emmert et al. (2004, 2008) also found solar minimum variation near 5%/decade, decreasing to about 2%/decade at higher solar activity. This is in reasonable agreement with models for moderate solar activity, but still larger than model simulations at solar minimum.

Modeling of ionospheric effects, starting with Rishbeth and Roble 1992, have been extended by Qian et al. (2008, 2009). The basic predictions, as described in Sections 5 and 6, are that the E and F1 regions should experience a small increase in density, due to ionization rate and composition changes, as the thermosphere cools and contracts, while the F2 region above the peak should decrease in density (see Fig. 12). The height of all layers is expected to decrease, due to thermospheric contraction. The net result is a very small decrease in foF2. Simulations of the effect of geomagnetic field evolution (Cnossen and Richmond 2008) show that the complicating effect of trends caused by these different effects, and their complex diurnal and spatial structure, may make F-region density trends very difficult to observe. F-region heights should be less ambiguous, but these are more difficult to measure.

Challenges for future model studies include resolving the remaining discrepancies with solar minimum trends in the thermosphere, understanding the reasons for the apparent lack of trends near the mesopause, and explicating the role of changes in atmospheric dynamics, especially tidal and wave forcing, expected from the combined effects of tropospheric heating and stratospheric cooling. For the thermosphere, there is still considerable uncertainty in a fundamental chemical kinetic parameter, the rate of CO_2 collisional activation-deactivation by atomic oxygen, as discussed by Akmaev (2003). Employing a higher rate in model calculations would increase the efficiency of CO_2 radiative cooling above the mesopause, which could increase model trend estimates, particularly at solar minimum. However, Akmaev found that the strong temperature dependence of the radiative forcing and molecular conduction offset changes to the collision rate, resulting in surprisingly low model sensitivity to this parameter. Additionally, atomic oxygen densities themselves are highly variable and not well known. Improved model quantification of lower thermosphere processes will be needed to resolve these issues.

An additional complexity has also been introduced by measurements during the recent minimum between solar cycles 23 and 24. In addition to the inter-cycle period being unusually long, the Sun was unusually quiet, certainly with respect to magnetic activity, and possibly also photon emissions (e.g., Russell et al., 2010). Emmert et al. (2010) found that the thermosphere reached record-low density levels during 2007–2009, 30% lower than the previous solar minimum, which has been explained by a combination of lower solar extreme-ultraviolet radiation, lower geomagnetic forcing, and increased CO_2 cooling (Solomon et al. 2010b, 2011). However, the solar extreme-ultraviolet decrease was the dominant effect. Thus, the simple method employed by Keating et al. (2000), comparing successive solar minima to make the first clear detection of upper atmosphere cooling, can not be considered reliable in the future, as we confront a new dimension in solar variability and upper atmosphere response.

10 Future Trend Investigations

Future investigations of trends in the upper atmosphere and ionosphere should proceed in four general directions, each of which includes both observational and modeling activities:

1. Investigations of trends in atmospheric wave activity (planetary, tidal and gravity waves). This is currently the key open problem of trend investigations in general. The origin of long-term changes is probably of dual nature, variability in the predominant tropospheric source of waves and variability in their filtering by middle (and upper) atmosphere. Both observation-based and model simulation results indicate that trends may vary regionally. Historical data mining could broaden the limited observational basis for trend investigations.
2. Investigations toward making the current global scenario of trends more complete and more quantified, as it is now essentially qualitative. Information about trends in variables not yet included into the scenario is desirable. Longer data series and more sophisticated and accurate models could help to quantify more accurately trends in various atmospheric and ionospheric variables.
3. Investigations of the impact of the changing roles of various trend drivers such as ozone and geomagnetic activity on trends in the upper atmosphere and ionosphere. This is a relatively new but important area in trend research particularly for predicting future trends in the upper atmosphere and ionosphere.
4. Investigations liking the trend scenario for the upper atmosphere and ionosphere with trends in the stratosphere. The impact of stratospheric trends on trends in the upper atmosphere and ionosphere, through changes in stratospheric ozone concentration, temperature, and possibly atmospheric waves and water vapor, is an important subject linking different aspects of global change.

11 Conclusions

The steady increase of greenhouse gas concentrations, primarily CO_2, has caused global warming in the troposphere, but global cooling in the upper atmosphere and ionosphere. In the mesosphere, the cooling trend is in the order of -2–3 K/decade from 50 km to 80 km, whereas no significant cooling in the mesopause region from 80 km to 100 km. In the upper thermosphere, the decreasing trend of neutral density at 400 km is in the order of ~ -1.7–-3.0%/decade for the past 3–4 decades. The cooling of the upper atmosphere causes contraction of the upper atmosphere, which lowers the altitude of the ionosphere, with a global average trend in the order of -0.29 ± 0.20 km/decade at E region peak.

In addition, a long-term global increase of electron density has been detected in these regions, in the order of $+0.013 \pm 0.005$ MHz/decade for foE and 0.019 ± 0.011 MHz/year for foF1. The dominant driver of these trends has been attributed to the increasing greenhouse gas forcing. Various model simulations on long-term trends of the upper atmosphere and ionosphere under greenhouse gas forcing at least qualitatively, sometimes quantitatively, agrees with these detected results. Other forcing mechanisms such as changes in other radiatively active trace gases (e.g., O_3 and H_2O), long-term change in geomagnetic activity, and long-term change in the Earth's magnetic field, also contributed to the observed trends. The combined effect of these drivers, and their changing roles, can cause complicated patterns of trends; this is particularly true for trends in the F2 region.

The internally consistent global pattern of trends in the upper atmosphere and ionosphere, formed by the above trends and trends in several other variables, emerged in 2006 from a chaos of individual trend results for individual variables. This pattern/scenario is formed by trends in mesospheric temperature (mesospheric cooling and no trend in mesopause region), electron concentration in the lower ionosphere below 100 km (increase at fixed heights), maximum electron concentration (slight increase) and height of this maximum (statistically rather insignificant decrease due to coarse resolution of height measurements) in the E region, electron concentration in the F1-region maximum (slight increase), thermospheric neutral density (moderate but statistically significant decrease) and F2-region ion temperature (cooling). All these trends are qualitatively mutually consistent and qualitatively agree with model simulations of consequences of the enhanced greenhouse effect. These trends are caused primarily by greenhouse cooling and thermal contraction of the atmosphere with related changes in minor constituents responsible for ionospheric changes.

Trends for the F2 region ionosphere, MLT dynamics and waves, and water vapor and water vapor related phenomena, are the three main areas that have had controversies or limited knowledge, consequently they were not included in the 2006 trend scenario. Progress has been made in the past four years in these areas. For the F2 region ionosphere, model simulations demonstrated that under greenhouse gas forcing, trend of electron density changes from being positive in the E and F1 regions to being negative in the F2 region; the F2 peak is slightly above this transitional altitude, consequently, the trend of NmF2 is small and negative on a global average basis; the trend of hmF2 is negative on global average basis, due to cooling and contraction; trends of both hmF2 and NmF2 show strong latitudinal, longitudinal, diurnal, and solar cycle variability; cooling and contraction of the upper atmosphere causes changes in neutral dynamics and electrodynamics, which in turn can cause regional positive trends of hmF2 and NmF2 at night time. Model simulations also showed that secular change of the Earth's magnetic field can cause significant regional trends of hmF2 and NmF2 in the equatorial to mid-latitude Atlantic Ocean and in the South American; the sign of the trends can be positive and negative depending on region; and these trends show diurnal and seasonal variations. Furthermore, the role of long-term changes of geomagnetic activity in the observed long-term trends was decreasing. While it was probably dominant in ionospheric trends in the past, since about 2000 trends in foF2 appear to be controlled predominantly by increasing concentration of greenhouse gases. The combined effects of these three drivers and the changing role of geomagnetic activity, as well as difficulty in removing solar cycle influence in trend analysis, have likely contributed to controversies and discrepancies on earlier reported the trends of hmF2 and foF2. In addition, progress has been made on trend of ion temperature in the F2 region. A negative trend of ion temperature was found in the incoherent scatter radar measurements. However, model simulations show a negative trend at lower altitude but positive trend at higher altitude ($>$350 km), and the simulated negative trend is much smaller than the detected negative trend. More investigations are needed.

There have been discrepancies between trends of NLC and PMC, even though the two are same phenomenon, excepting that NLC are observed from the ground and PMC are observed from space. It has been reported that there were no detectable trend in NLC, but satellite observations indicated an increase in PMC occurrence frequency and brightness over the past several decades. More recent studies found that PMC trends are only significant in the latitude band from 74°N to 82°N, but at NLC observational latitudes ($\sim$ 54°N–64°N), both PMC and NLC trends are statistically insignificant. These findings may explain and reconcile the discrepancies between PMC and NLC trends.

Waves dominate MLT dynamics and also affect the thermosphere. They perform vertical coupling between the lower and upper atmosphere, and are able to influence trends in the upper atmosphere and ionosphere. Limited observational studies indicate possible trends in wave activity, but understanding of this phenomenon is still very limited. Causes and effects of trends in wave activity, are key open questions regarding trends in the upper atmosphere.

An important finding of trend investigations in the last four years is realization of the importance of other trend drivers, and partial specification of their relative role. These drivers include anthropogenic changes of ozone, secular changes in geomagnetic activity and Earth's magnetic field, changes in atmospheric wave activity, and possibly changes of mesospheric water vapor and thermospheric atomic oxygen concentration. Changes of ozone concentration play a role in the mesosphere and lower thermosphere, while long-term changes of geomagnetic activity play a role in the ionosphere, mainly in its F region. Trends in both of these parameters are variable, which results in modification of overall long-term trends in the upper atmosphere and ionosphere, and even more in the relative role of various trend drivers.

The path to developing a comprehensive scenario of trends in the atmosphere-ionosphere system appears to be arduous, but part of it is accomplished. With continued advances in observational analysis and model simulation, a full understanding of the causes and consequences of global change throughout the Earth's atmosphere will be obtained.

Acknowledgements This work was supported by GACR grant P209/10/1792 to the Institute of Atmospheric Physics and by NASA grants NNX08AQ31G and NNX10AF21G to the National Center for Atmospheric Research. NCAR is sponsored by the National Science Foundation.

References

A.C. Aikin, M.L. Chanin, J. Nash, D.J. Kendig, Temperature trends in the lower mesosphere. Geophys. Res. Lett. **18**(3), 416–419 (1991)

R.A. Akmaev, Thermospheric resistance to "greenhouse cooling": effect of the collisional excitation rate by atomic oxygen on the thermal response to CO_2 forcing. J. Geophys. Res. **108**(A7), 1292 (2003). doi:10.1029/2003JA009896

R.A. Akmaev, V.I. Fomichev, A model estimate of cooling in the mesosphere and lower thermosphere due to the CO_2 increase over the last 3–4 decades. Geophys. Res. Lett. **27**, 2113–2116 (2000)

R.A. Akmaev, V.I. Fomichev, X. Zhu, Impact of middle-atmospheric composition changes on greenhouse cooling in the upper atmosphere. J. Atmos. Sol.-Terr. Phys. **68**, 1879–1889 (2006). doi:10.1016/j.jastp.2006.03.008

L. Alfonsi, G. DeFranceschi, L. Perrone, Long term trends of the critical frequency of the F2 layer at northern and southern high latitude regions. Phys. Chem. Earth **27**(6–8), 595–600 (2002)

L. Alfonsi, G. DeFranceschi, G. De Santi, Geomagnetic and ionospheric data analysis over Antarctica: a contribution to the long term trends investigation. Ann. Geophys. **26**, 1173–1179 (2008)

A.J.G. Baumgaertner, A.J. McDonald, G.J. Fraser, G.E. Plank, Long-term observations of mean winds and tides in the upper mesosphere and lower thermosphere above Scott Base, Antarctica. J. Atmos. Sol.-Terr. Phys. **67**, 1480–1496 (2005)

G. Beig, The relative importance of solar activity and anthropogenic influences on the ion composition, temperature, and associated neutrals of the middle atmosphere. J. Geophys. Res. **105**, 19841–19856 (2000)

G. Beig, Trends in the mesopause region temperature and our present understanding—an update. Phys. Chem. Earth **31**(1), 3–9 (2006). doi:10.1016/j.pce.2005.03.007

G. Beig, Recent advances in long-term trends of MLT-region, in *6th IAGA/ICMA/ CAWSES workshop "Long-Term Changes and Trends in the Atmosphere". HAO-NCAR*, Boulder (Book of abstracts, p. 12) (2010)

G. Beig, P. Keckhut, R.P. Lowe, R.G. Roble, M.G. Mlynczak, J. Scheer, V.I. Fomichev, D. Offermann, W.J.R. French, M.G. Shepherd, A.I. Semenov, E.E. Remsberg, C.-Y. She, F.-J. Lübken, J. Bremer, B.R. Clemesha, J. Stegman, F. Sigernes, S. Fadnavis, Review of mesospheric temperature trends. Rev. Geophys. **41**(4), 1015 (2003). doi:10.1029/2002RG000121

P. Bencze, Geographical distribution of long-term changes in the height of the maximum electron density of the F region: a nonmigrating tidal effect? J. Geophys. Res. **111**, A06304 (2009). doi:10.1029/2008JA013492

E. Blanc, Observations in the upper atmosphere of infrasonic waves from natural or artificial sources: a summary. Ann. Geophys. **3**, 673–688 (1983)

S.W. Bougher, R.G. Roble, Comparative terrestrial planet thermospheres. 1. Solar cycle variations of global mean temperatures. J. Geophys. Res. **96**, 11045–11055 (1991)

J. Bremer, Trends in the ionospheric E- and F-regions over Europe. Ann. Geophys. **16**, 986–996 (1998)

J. Bremer, Trends in the thermosphere derived from global ionosonde observations. Adv. Space Res. **28**(7), 997–1006 (2001)

J. Bremer, Detection of long-term trends in the mesosphere-lower thermosphere from ground-based radio propagation measurements. Adv. Space Res. **35**(8), 1398–1404 (2005)

J. Bremer, Long-term trends in the ionospheric E and F1 regions. Ann. Geophys. **26**, 1189–1197 (2008)

J. Bremer, U. Berger, Mesospheric temperature trends derived from ground-based LF phase-height observations at middle latitudes: Comparison with model simulations. J. Atmos. Sol.-Terr. Phys. **64**(7), 805–816 (2002)

J. Bremer, L. Alfonsi, P. Bencze, J. Laštovička, A.V. Mikhailov, N. Rogers, Long-term trends in the ionosphere and upper atmosphere parameters. Ann. Geophys. **47** (Supplement to Nos. 2–3), 1009–1029 (2004)

J. Bremer, P. Hoffmann, J. Höffner, R. Latteck, W. Singer, M. Zecha, O. Zeller, Long-term changes of mesospheric summer echoes at polar and middle latitudes. J. Atmos. Sol.-Terr. Phys. **68**(17), 1940–1951 (2006)

J. Bremer, P. Hoffmann, R. Latteck, W. Singer, M. Zecha, Long-term changes of (polar) mesosphere summer echoes. J. Atmos. Sol.-Terr. Phys. **71**, 1571–1576 (2009b). doi:10.1016/j.jastp.2009.03.010

J. Bremer, J. Laštovička, A.V. Mikhailov, D. Altadill, P. Bencze, D. Burešová, G. De Franceschi, C. Jacobi, S. Kouris, L. Perrone, E. Turunen, Climate of the upper atmosphere. Ann. Geophys. **52**(3/4), 273–299 (2009a)

J. Bremer, D. Peters, Influence of stratospheric ozone changes on long-term trends in the meso- and lower thermosphere. J. Atmos. Sol.-Terr. Phys. **70**(11-12), 1473–1481 (2008). doi:10.1016/j.jastp.2008.03.024

P.S. Cannon, N.C. Rogers, C.M. Hall, Trends in critical frequencies and layer heights over Tromso and their consequential impact for radio system modeling. Adv. Space Res. **34**(9), 2085–2091 (2004)

H. Chandra, G.D. Vyas, S. Sharma, Long-term changes in ionospheric parameters over Ahmedabad. Adv. Space Res. **20**, 2161–2164 (1997)

R.J. Cicerone, Greenhouse cooling up high. Nature **344**, 104–105 (1990)

B.R. Clemesha, D.M. Simonich, P.P. Batista, A long term trend in the height of the atmospheric sodium layer: possible evidence for global change. Geophys. Res. Lett. **19**, 457–460 (1992)

M.A. Clilverd, T.G.C. Clark, E. Clarke, H. Rishbeth, Increased magnetic storm activity from 1868 to 1995. J. Atmos. Sol.-Terr. Phys. **60**, 1047–1056 (1998)

M.A. Clilverd, T. Ulich, J.M. Jarvis, Residual solar cycle influence on trends in ionospheric F2-layer peak height. J. Geophys. Res., **108**(A12), 1450 (2003)

I. Cnossen, A.D. Richmond, Modeling the effect of changes in the Earth's magnetic field from 1957 to 1997 on the ionospheric hmF2 and foF2 parameters. J. Atmos. Sol.-Terr. Phys. **70**(11–12), 1512–1524 (2008)

A.D. Danilov, Long-term changes of the mesosphere and lower thermosphere temperature and composition. Adv. Space Res. **20**(11), 2137–2147 (1997)

A.D. Danilov, The method of determination of long-term trends in F2-region independent of geomagnetic activity. Ann. Geophys. **20**, 1–11 (2002)

A.D. Danilov, Long-term trends in F2-layer parameters and their relation to other trends. Adv. Space Res. **35**(8), 1405–1410 (2005)

A.D. Danilov, Time and spatial variations in the ratio of nighttime and daytime critical frequencies of the F2 layer. J. Atmos. Sol.-Terr. Phys. **70**(8–9), 1201–1212 (2008a)
A.D. Danilov, Long-term trends in the relation between daytime and nighttime values of foF2. Ann. Geophys. **26**(5), 1199–1206 (2008b)
A.D. Danilov, Scatter of hmF2 values as an indicator of trends in thermospheric dynamics. J. Atmos. Sol.-Terr. Phys. **71**, 1586–1591 (2009). doi:10.1016/j.jastp.2009.03.002
A.D. Danilov, L.B. Vanina-Dart, Parameters of the ionospheric F2 layer as a source of information on trends in thermospheric dynamics. Geomagn. Aeron. **50**(2), 187–200 (2010)
M.T. DeLand, E.P. Shettle, G.E. Thomas, J.J. Olivero, A quarter-century of satellite polar mesospheric cloud observations. J. Atmos. Sol.-Terr. Phys. **68**, 9–29 (2006)
M.T. DeLand, E.P. Shettle, G.E. Thomas, J.J. Olivero, Latitude-dependent long-term variations in polar mesospheric clouds from SBUV version 3 PMC data. J. Geophys. Res. **112**, D10315 (2007). doi:10.1029/2006JD007857
A.G. Elias, Trends in the F2 ionospheric layer due to long-term variations in the Earth's magnetic field. J. Atmos. Sol.-Terr. Phys. **71**, 1602–1609 (2009). doi:10.1016/j.jastp.2009.05.014
A.G. Elias, N. Ortiz de Adler, Earth magnetic field and geomagnetic activity effects on long-term trends in the F2 layer at mid-high latitudes. J. Atmos. Sol.-Terr. Phys. **68**(17), 1871–1878 (2006)
J.T. Emmert, J.M. Picone, J.L. Lean, S.H. Knowles, Global change in the thermosphere: compelling evidence of a secular decrease in density. J. Geophys. Res. **109**, A02301 (2004). doi:10.1029/2003JA010176
J.T. Emmert, J.M. Picone, R.R. Meier, Thermospheric global average density trends 1967-2007, derived from orbits of 5000 near-Earth objects. Geophys. Res. Lett. **35**, L05101 (2008). doi:10.1029/2007GL032809
J.T. Emmert, J.L. Lean, J.M. Picone, Record low thermospheric density during the 2008 solar minimum. Geophys. Res. Lett. **37**, L12102 (2010). doi:10.1029/2010GL043671
V.I. Fomichev, A.I. Jonsson, J. de Grandpré, S.R. Beagley, C. McLandress, K. Semeniuk, T.G. Shepherd, Response of the middle atmosphere to CO_2 doubling: results from the Canadian Middle Atmosphere Model. J. Climate **20**(7), 1255–1264 (2007)
A.J. Foppiano, L. Cid, V. Jara, Ionospheric long-term trends for South American mid-latitudes. J. Atmos. Sol.-Terr. Phys. **61**, 717–723 (1999)
R.R. Garcia, D.R. Marsh, D.E. Kinnison, B.A. Boville, F. Sassi, Simulation of secular trends in the middle atmosphere, 1950–2003. J. Geophys. Res. **112**, D09301 (2007). doi:10.1029/2006JD007485
N.M. Gavrilov, S. Fukao, T. Nakamura, Ch. Jacobi, D. Kürschner, A.H. Manson, C.E. Meek, Comparative study of interannual changes of the mean winds and gravity wave activity in the middle atmosphere over Japan, Central Europe and Canada. J. Atmos. Sol.-Terr. Phys. **64**, 1003–1010 (2002)
G.S. Golitsyn, A.I. Semenov, N.N. Shefov, Seasonal variations of the long-term temperature trend in the mesopause region. Geomagn. Aeron. **40**(2), 198–200 (2000)
P.F. Graigmile, P. Guttorp, D.B. Percival, Trend assessment in a long memory dependence model using the discrete wavelet transform. Environmetrics **15**(4), 313–315 (2004)
A.N. Gruzdev, G.P. Brasseur, Long-term changes in the mesosphere calculated by a two-dimensional model. J. Geophys. Res. **110**, D03304 (2005). doi:10.1029/2003JD004410
C.M. Hall, A. Brekke, P.S. Cannon, Climatic trends in E-region critical frequency and virtual height above Tromso (70°N, 19°E). Ann. Geophys. **25**, 2351–2357 (2007a)
C.M. Hall, A. Brekke, A.H. Manson, C.E. Meek, S. Nozawa, Trends in mesospheric turbulence at 70°N. Atmos. Sci. Lett. **8**, 80–84 (2007b)
N.R.P. Harris, E. Kyrö, J. Staehelin et al., Ozone trends at northern mid and high latitudes—a European perspective. Ann. Geophys. **26**(5), 1207–1220 (2008)
P. Hoffmann, E. Becker, M. Rapp, J. Bremer, W. Singer, M. Placke, Seasonal and interannual variation of mesospheric waves at middle and high latitudes, in *6th IAGA/ICMA/ CAWSES Workshop "Long-Term Changes and Trends in the Atmosphere", HAO-NCAR*, Boulder (Book of abstracts, p. 16) (2010)
J.M. Holt, S.-R. Zhang, Long-term temperature trends in the ionosphere above Millstone Hill. Geophys. Res. Lett. **35**, L05813 (2008). doi:10.1029/2007GL031148
IPCC (Intergovernmental Panel on Climate Change), in *Climate Change 2007: The Physical Science Basis*, ed. by S. Solomon (Cambridge Univ. Press, Cambridge, 2007)
C. Jacobi, Y.I. Portnyagin, E.G. Merzlyakov, T.V. Solovjova, N.A. Makarov, D. Kürschner, A long-term comparison of mesopause region wind measurements over Eastern and Central Europe. J. Atmos. Sol.-Terr. Phys. **67**, 227–240 (2005)
C. Jacobi, N.M. Gavrilov, D. Kürschner, K. Fröhlich, Gravity wave climatology and trends in the mesosphere/lower thermosphere region deduced from low-frequency drift measurements 1984–2003 (52.1°N, 13.2°E). J. Atmos. Sol.-Terr. Phys. **68**, 1913–1923 (2006)
C. Jacobi, P. Hoffmann, D. Kürschner, Trends in MLT region winds and planetary waves, Collm (52°N, 15°E). Ann. Geophys. **26**, 1221–1232 (2008)

C. Jacobi, P. Hoffmann, R.Q. Liu, P. Križan, J. Laštovička, E.G. Merzlyakov, T.V. Solovjova, Yu.I. Portnyagin, Midlatitude mesopause region winds and waves and comparison with stratospheric variability. J. Atmos. Sol.-Terr. Phys. **71**, 1540–1546 (2009)

M.J. Jarvis, Planetary wave trends in the lower thermosphere—evidence for 22-year solar modulation of the quasi 5-day wave. J. Atmos. Sol.-Terr. Phys. **68**(17), 1902–1912 (2006)

M.J. Jarvis, B. Jenkins, G.A. Rodgers, Southern hemisphere observations of a long-term decrease in F-region altitude and thermospheric wind providing possible evidence for global thermospheric cooling. J. Geophys. Res. **103**(A9), 20,774–20,787 (1998)

M.J. Jarvis, M.A. Clilverd, Th. Ulich, Methodological influences on F-region peak height trend analysis. Phys. Chem. Earth **27**, 589–594 (2002)

G.M. Keating, R.H. Tolson, M.S. Bradford, Evidence of long term global decline in the Earth's thermospheric densities apparently related to anthropogenic effects. Geophys. Res. Lett. **27**, 1523–1526 (2000)

Y.A. Kalgin, Dynamical aspect of long-term trend of the neutral atmosphere composition at turbopause region, in *Proc. Int. Workshop "Cooling and Sinking of the Middle and Upper Atmosphere"*, Moscow (1998), pp. 26–27

M. Kendall, *Time Series* (Charles Griffin, London, 1973)

D. Keuer, P. Hoffmann, W. Singer, J. Bremer, Long-term variations of the mesospheric wind field at midlatitudes. Ann. Geophys. **25**, 1779–1790 (2007)

S. Kirkwood, P. Dalin, A. Réchou, Noctilucent clouds observed from the UK and Denmark—trends and variations over 43 years. Ann. Geophys. **26**, 1243–1254 (2008)

V.M. Krasnov, Y.V. Drobzheva, J. Laštovička, Recent advances and difficulties of infrasonic wave investigation in the ionosphere. Surv. Geophys. **27**(2), 169–209 (2006)

P. Križan, J. Laštovička, Trends in positive and negative ozone laminae in the Northern Hemisphere. J. Geophys. Res. **110**, D10107 (2005). doi:10.1029/2004JD005477

A. Kubicky, P. Keckhut, M.-L. Chanin, G.S. Golitsyn, E. Lysenko, Temperature trends in the middle atmosphere as seen by historical Russian rocket launches. Part II. Heiss Island (80.6°N, 58°E). J. Atmos. Sol.-Terr. Phys. **70**(1), 145–155 (2008)

J. Laštovička, Long-term changes and trends in the lower ionosphere. Phys. Chem. Earth **27**, 497–507 (2002)

J. Laštovička, On the role of solar and geomagnetic activity in long-term trends in the atmosphere-ionosphere system. J. Atmos. Sol.-Terr. Phys. **67**(1–2), 83–92 (2005)

J. Laštovička, Global pattern of trends in the upper atmosphere and ionosphere: recent progress. J. Atmos. Sol.-Terr. Phys. **71**, 1514–1528 (2009). doi:10.1016/j.jastp.2009.01.010

J. Laštovička, J. Bremer, An overview of long-term trends in the lower ionosphere below 120 km. Surv. Geophys. **25**, 69–99 (2004)

J. Laštovička, D. Pancheva, Changes in characteristics of planetary waves at 80–100 km over Central and Southern Europe since 1980. Adv. Space Res. **11**(3), 31–34 (1991)

J. Laštovička, V. Fišer, D. Pancheva, Long-term trends in planetary wave activity (2–15 days) at 80–100 km inferred from radio wave absorption. J. Atmos. Sol.-Terr. Phys. **56**, 893–899 (1994)

J. Laštovička, R.A. Akmaev, G. Beig, J. Bremer, J.T. Emmert, Global change in the upper atmosphere. Science **314**(5803), 1253–1254 (2006a)

J. Laštovička, A.V. Mikhailov, Th. Ulich, J. Bremer, A.G. Elias, N. Ortiz de Adler, V. Jara, R. Abarca del Rio, A.J. Foppiano, E. Ovalle, A.D. Danilov, Long-term trends in foF2: a comparison of various methods. J. Atmos. Sol.-Terr. Phys. **68**(17), 1854–1870 (2006b)

J. Laštovička, R.A. Akmaev, G. Beig, J. Bremer, J.T. Emmert, C. Jacobi, M.J. Jarvis, G. Nedoluha, Y.I. Portnyagin, T. Ulich, Emerging pattern of global change in the upper atmosphere and ionosphere. Ann. Geophys. **26**(5), 1255–1268 (2008a)

J. Laštovička, X. Yue, W. Wan, Long-term trends in foF2: their estimating and origin. Ann. Geophys. **26**, 593–598 (2008b)

J. Laštovička, P. Križan, M. Kozubek, Long-term trends in the middle atmosphere dynamics at northern middle latitudes—one regime or two different regimes? Atmos. Chem. Phys. Discuss. **10**, 2631–2668 (2010)

F.-J. Lübken, Nearly zero temperature trend in the polar summer mesosphere. Geophys. Res. Lett. **27**(21), 3603–3606 (2000)

F.-J. Lübken, No long term change of the thermal structure in the mesosphere at high latitudes during summer. Adv. Space Res. **28**(7), 947–953 (2001)

F.-J. Lübken, U. Berger, G. Baumgartner, Stratospheric and solar cycle effects on long-term variability of mesospheric ice clouds. J. Geophys. Res. **114**(21), D00106 (2009). doi:10.1029/2009JD012377

F.-J. Lübken, U. Berger, J. Fiedler, G. Baumgartner, M. Gerding, Trends and solar cycle effects in mesospheric ice clouds, in *38th Sci. Ass. COSPAR, Symp. C2.1*, Bremen (2010)

F.A. Marcos, J.O. Wise, M.J. Kendra, N.J. Grossbard, B.R. Bowman, Detection of long-term decrease in thermospheric neutral density. Geophys. Res. Lett. **32**, L04103 (2005). doi:10.1029/2004GL021269

D. Marsh, A. Smith, E. Woble, Mesospheric ozone response to changes in water vapor. J. Geophys. Res. **108**(D3), 4109 (2003). doi:10.1029/2002JD002705

D. Martini, K. Mursula, Centennial geomagnetic activity studied by a new, reliable long-term index. J. Atmos. Sol.-Terr. Phys. **70**(7), 1074–1087 (2008)

C. McLandress, V.I. Fomichev, Amplification of the mesospheric diurnal tide in a doubled CO_2 atmosphere. Geophys. Res. Lett. **33**, L06808 (2006). doi:10.1029/2005GL025345

L.F. McNamara, Accuracy of models of hmF2 used for long-term trend analyses. Radio Sci. **43**, RS2002 (2008). doi:10.1029/2007RS003740

E. Merzlyakov, C. Jacobi, Y.I. Portnyagin, T.V. Solovjova, Structural changes in trend parameters of the MLT winds based on wind measurements at Obninsk (55°N, 37°E) and Collm (52°N, 15°E). J. Atmos. Solar-Terr. Phys. 71 (2009). doi:10.1016/j.jastp.2009.05.013

A.V. Mikhailov, The geomagnetic control concept of the F2-layer parameter long-term trends. Phys. Chem. Earth **27**, 595–606 (2002)

A.V. Mikhailov, Ionospheric long-term trends: can the geomagnetic control and the greenhouse hypothesis be reconciled? Ann. Geophys. **24**(10), 2533–2541 (2006a)

A.V. Mikhailov, Trends in the ionospheric E-region. Phys. Chem. Earth **31**, 22–23 (2006b)

A.V. Mikhailov, Ionospheric F1 layer long-term trends and the geomagnetic control concept. Ann. Geophys. **26**, 3793–3803 (2008)

A.V. Mikhailov, B.A. de la Morena, Long-term trends of foE and geomagnetic activity variations. Ann. Geophys. **21**, 751–760 (2003)

A.V. Mikhailov, D. Marin, Geomagnetic control of the foF2 long-term trends. Ann. Geophys. **18**, 653–665 (2000)

A.V. Mikhailov, D. Marin, An interpretation of the foF2 and hmF2 long-term trends in the framework of the geomagnetic control concept. Ann. Geophys. **19**, 733–748 (2001)

M. Mlynczak, L. Hunt, B.T. Marshall, F.J. Martin-Torres, C.J. Mertens, J.M. Russell III, E. Remsberg, M. Lopez-Puertas, R. Picard, J. Winick, P. Wintersteiner, R.E. Thompson, L.L. Gordley, Observations of infrared radiative cooling in the thermosphere on daily to multiyear timescales from the TIMED/SABER instrument. J. Geophys. Res. **115**, A03309 (2010). doi:10.1029/2009JA014713

K. Mursula, D. Martini, Centennial increase in geomagnetic activity: latitudinal difference and global estimates. J. Geophys. Res. **111**, A08209 (2006). doi:10.1029/2005JA011549

G.E. Nedoluha, R.M. Bevilacqua, R.M. Gomez, B.C. Hicks, J.M. Russell III, B.J. Connor, An evaluation of trends in middle atmospheric water vapor as measured by HALOE, WVMS, and POAM. J. Geophys. Res. **108**(D13), 4391 (2003). doi:10.1029/2002JD003332

D. Offermann, M. Donner, P. Knieling, B. Naujokat, Middle atmosphere temperature changes and the duration of summer. J. Atmos. Sol.-Terr. Phys. **66**, 437–450 (2004)

D. Offermann, M. Jarisch, M. Donner, W. Steinbrecht, A.I. Semenov, OH temperature re-analysis forced by recent variance increases. J. Atmos. Sol.-Terr. Phys. **68**(17), 1924–1933 (2006)

D. Offermann, P. Hoffmann, P. Knieling, R. Koppmann, J. Oberheide, W. Steinbrecht, Long-term-trends and solar cycle variations of mesospheric temperature and dynamics. J. Geophys. Res. **115**, D18127 (2010). doi:10.1029/2009JD013363

N. Ortiz de Adler, A.G. Elias, T. Heredia, Long-term trend of the ionospheric F2-layer peak height at a southern low latitude station. Phys. Chem. Earth **27**(6–8), 613–615 (2002)

V.I. Perminov, A.I. Semenov, Model of latitudinal, seasonal and altitudinal changes of long-term temperature trends in the middle atmosphere. Geomagn. Aeron. **47**(5), 685–691 (2007) (in Russian, abstract in English)

A.I. Pogorelcev, A.Yu. Kanukhina, E.V. Suvorova, E.N. Savenkova, Variability of planetary waves as a signature of possible climatic change. J. Atmos. Sol.-Terr. Phys. **71**, 1529–1539 (2009)

Y.I. Portnyagin, E.G. Merzlyakov, T.V. Solovjova, Ch. Jacobi, D. Kürschner, A. Manson, C. Meek, Long-term trends and year-to-year variability of mid-latitude mesosphere/lower thermosphere winds. J. Atmos. Sol.-Terr. Phys. **68**, 1890–1901 (2006). doi:10.1016/j.jastp.2006.04.004

L. Qian, R.G. Roble, S.C. Solomon, T.J. Kane, Calculated and observed climate change in the thermosphere, and a prediction for solar cycle 24. Geophys. Res. Lett. **33**, L23705 (2006). doi:10.1029/2006GL027185

L. Qian, S.C. Solomon, R.G. Roble, T.J. Kane, Model simulations of global change in the ionosphere. Geophys. Res. Lett. **35**, L07811 (2008). doi:10.1029/2007GL033156

L. Qian, A.G. Burns, S.C. Solomon, R.G. Roble, The effect of carbon dioxide cooling on trends in the F2 layer ionosphere. J. Atmos. Sol.-Terr. Phys. **71**, 1592–1601 (2009). doi:10.1016/j.jastp.2009.03.006

L. Qian, J. Laštovička, S.C. Solomon, R.G. Roble, Progress in observations and simulations of global change in the upper atmosphere. J. Geophys. Res. **111**(A4), A00H03, (2011) doi:10.1029/2010JA016317

G.C. Reinsel, A.J. Miller, E.C. Weatherhead, L.E. Flynn, R.M. Nagatani, G.C. Tiao, D.J. Wuebbles, Trend analysis of total ozone data for turnaround and dynamical contributions. J. Geophys. Res. **110**, D16306 (2005). doi:10.1029/2004JD004662

E.E. Remsberg, A reanalysis for the seasonal and longer-period cycles and the trends in middle atmosphere temperature from the Halogen Occultation Experiment. J. Geophys. Res. **112**, D09118 (2007). doi:10.1029/2006JD007489

E.E. Remsberg, Trends and solar cycle effects in temperatures versus altitude from the Halogen Occultation Experiment for the mesosphere and upper stratosphere. J. Geophys. Res. **114**, D12303 (2009). doi:10.1029/2009JD011897

H. Rishbeth, A greenhouse effect in the ionosphere? Planet. Space Sci. **38**, 945–948 (1990)

H. Rishbeth, Long-term changes in the ionosphere. Adv. Space Res. **20**(11), 2149–2155 (1997)

H. Rishbeth, R.G. Roble, Cooling of the upper atmosphere by enhanced greenhouse gases—modeling of thermospheric and ionospheric effects. Planet. Space Sci. **40**, 1011–1026 (1992)

R.G. Roble, R.E. Dickinson, How will changes in carbon dioxide and methane modify the mean structure of the mesosphere and lower thermosphere? Geophys. Res. Lett. **16**, 1441–1444 (1989)

R.G. Roble, E.C. Ridley, Thermosphere-ionospheremesosphere electrodynamics general circulation model (TIME-GCM): equinox solar min simulations, 30–500 km. Geophys. Res. Lett. **21**, 417–420 (1994)

H. Schmidt, G.P. Brasseur, M. Charron, E. Manzini, M.A. Giorgetta, T. Diehl, V.I. Fomichev, D. Kinnison, D. Marsh, S. Walters, The HAMMONIA chemistry climate model: sensitivity of the mesopause region to the 11-year solar cycle and CO_2 doubling. J. Climate **19**, 3903–3931 (2006)

A.I. Semenov, N.N. Shefov, E.V. Lysenko, G.V. Givishvili, A.V. Tikhonov, The seasonal peculiarities of behavior of the long-term temperature trends in the middle atmosphere at the mid-latitudes. Phys. Chem. Earth **27**, 529–534 (2002)

S.S. Sharma, H. Chandra, G.D. Vyas, Long-term ionospheric trends over Ahmedabad. Geophys. Res. Lett. **26**, 433–436 (1999)

P.E. Shettle, M.T. DeLand, G.E. Thomas, J.J. Olivero, Long term variations in the frequency of polar mesospheric clouds in the Northern Hemisphere from SBUV. Geophys. Res. Lett. **36**, L02803 (2009). doi:10.1029/2008GL036048

M. Smirnova, E. Belova, S. Kirkwood, N. Mitchell, Polar mesosphere summer echoes with ESRAD, Kiruna, Sweden: variations and trends over 1997–2008. J. Atmos. Solar-Terr. Phys. **72** (2010). doi:10.1016/j.jastp.2009.12.014

S. Solomon, K.H. Rosenlof, R.W. Portmann, J.S. Daniel, S.M. Davis, T.J. Sanford, G.-K. Plattner, Contributions of stratospheric water vapor to decadal changes in the rate of global warming. Science, **327**(5970), 1219–1223 (2010a)

S.C. Solomon, T.N. Woods, L.V. Didkovsky, J.T. Emmert, L. Qian, Anomalously low solar extreme-ultraviolet irradiance and thermospheric density during solar minimum. Geophys. Res. Lett. **37**, L16103 (2010b). doi:10.1029/2010GL044468

S.C. Solomon, L. Qian, L.V. Didkovsky, R.A. Viereck, T.N. Woods, Causes of low thermospheric density during the 2007–2009 solar minimum. J. Geophys. Res., submitted (2011), doi:10.1029/2011JA016508

S. Sridharan, T. Tsuda, S. Gurubaran, Long-term tendencies in the mesosphere/lower thermosphere mean winds and tides observed by medium frequency radar at Tirunaveli (8.7°N, 77.8°E). J. Geophys. Res. **115**, D08109 (2010). doi:10.1029/2008JD011609

R. Stamper, M. Lockwood, M.N. Wild, T.D.G. Clark, Solar causes of the long-term increase in geomagnetic activity. J. Geophys. Res. **104**, 28325–28342 (1999)

T. Ulich, E. Turunen, Evidence for long-term cooling of the upper atmosphere in ionosonde data. Geophys. Res. Lett. **24**, 1103–1106 (1997)

T. Ulich, M.A. Clilverd, H. Rishbeth, On determining long-term change in the ionosphere. EOS Trans. **84**(52), 581–585 (2003)

T. Ulich, M.A. Clilverd, M.J. Jarvis, H. Rishbeth, Unravelling signs of global change in the ionosphere, in *Space Weather*, ed. by J. Lilensten (Springer, Dordrecht, 2007), pp. 95–105

H.O. Upadhyay, E.E. Mahajan, Atmospheric greenhouse effect and ionospheric trends. Geophys. Res. Lett. **25**, 3375–3378 (1998)

M. Venkat Ratnam, A.K. Patra, B.V. Krishna Murthy, Tropical mesopause: is it always close to 100 km? J. Geophys. Res. **115**, D06106 (2010). doi:10.1029/2009JD012531

E.C. Weatherhead, G.C. Reinsel, G.C. Tiao et al., Detecting the recovery of total column ozone. J. Geophys. Res. **105**, 22201–22210 (2000)

E.C. Weatherhead, A.J. Stevermer, B.E. Schwartz, Detecting environmental changes and trends. Phys. Chem. Earth **27**, 399–403 (2002)

Z.-W. Xu, J. Wu, K. Igarashi, H. Kato, Z.-S. Wu, Long-term ionospheric trends based on ground-based ionosonde observations at Kokubunji, Japan. J. Geophys. Res. **109**, A09307 (2004)

X. Yue, W. Wan, L. Liu, B. Ning, B. Zhao, Applying artificial neural network to derive long-term foF2 trends in Asia/Pacific sector from ionosonde observations. J. Geophys. Res. **111**, D22307 (2006). doi:10.1029/2005JA011577

X. Yue, L. Liu, W. Wan, Y. Wei, Z. Ren, Modeling the effects of secular variation of geomagnetic field orientation on the ionospheric long term trend over the past century. J. Geophys. Res. **113**, A10301 (2008). doi:10.1029/2007JA012995

X. Yue, W.S. Schreiner, J. Lei, C. Rocken, D.G. Hunt, Y.-H. Kuo, W. Wan, Global ionospheric response observed by COSMIC satellites during the January 2009 stratospheric sudden warming event. J. Geophys. Res. **115**, A00G09 (2010). doi:10.1029/2010JA015466

S.-R. Zhang, J. Holt, J. Kurdzo, Millstone Hill ISR observations of upper atmospheric long-term changes: Height dependency. J. Geophys. Res. **116**, A00H05 (2011). doi:10.1029/2010JA016414

Space Sci Rev (2012) 168:147–173
DOI 10.1007/s11214-011-9810-z

Thermospheric Density: An Overview of Temporal and Spatial Variations

Liying Qian · Stanley C. Solomon

Received: 19 April 2011 / Accepted: 8 July 2011 / Published online: 5 August 2011

Abstract Neutral density shows complicated temporal and spatial variations driven by external forcing of the thermosphere/ionosphere system, internal dynamics, and thermosphere and ionosphere coupling. Temporal variations include abrupt changes with a time scale of minutes to hours, diurnal variation, multi-day variation, solar-rotational variation, annual/semiannual variation, solar-cycle variation, and long-term trends with a time scale of decades. Spatial variations include latitudinal and longitudinal variations, as well as variation with altitude. Atmospheric drag on satellites varies strongly as a function of thermospheric mass density. Errors in estimating density cause orbit prediction error, and impact satellite operations including accurate catalog maintenance, collision avoidance for manned and unmanned space flight, and re-entry prediction. In this paper, we summarize and discuss these density variations, their magnitudes, and their forcing mechanisms, using neutral density data sets and modeling results. The neutral density data sets include neutral density observed by the accelerometers onboard the Challenging Mini-satellite Payload (CHAMP), neutral density at satellite perigees, and global-mean neutral density derived from thousands of orbiting objects. Modeling results are from the National Center for Atmospheric Research (NCAR) thermosphere-ionosphere-electrodynamics general circulation model (TIE-GCM), and from the NRLMSISE-00 empirical model.

Keywords Thermosphere neutral density · Density variation · Satellite drag · Density data · Model simulation

1 Introduction

Although the atmosphere is nearly a vacuum in the thermosphere, perturbations on satellite orbits do occur, due to the cumulative effects of variable mass density. Errors in estimating thermospheric density cause differences between predicted and actual satellite orbits.

L. Qian (✉) · S.C. Solomon
High Altitude Observatory, National Center for Atmospheric Research, Boulder, CO, USA
e-mail: lqian@ucar.edu

On short time scales, these errors reduce the fidelity of catalogs of space object orbital elements, which degrades operational tracking, collision avoidance warnings, and re-entry predictions. On longer time scales, uncertainties in long-term density variation impacts satellite design in terms of lifetime, on-board fuel, and satellite attitude control. To address these problems, observational, empirical, and numerical modeling has been developed to analyze neutral density and its variations. The region of the Earth's atmosphere where satellite drag estimations are particularly important is between 150 km and 600 km. Since neutral density decreases exponentially with height, below 150 km, satellite drag is too large for satellites to maintain orbit without propulsion, whereas above 600 km, satellite drag is usually small and has less impact on satellite operations. This 150-km to 600-km region of the atmosphere, the thermosphere/ionosphere, is highly dynamic and variable, driven by energy and momentum forcing from above and from below, and responding to its complex internal dynamics.

The primary energy input to the thermosphere/ionosphere system is solar irradiance (Knipp et al. 2004). The thermosphere absorbs solar irradiance in XUV (the soft X-ray ultra-violet, 1–30 nm), EUV (extreme ultra-violet, 30–120 nm), and FUV (far ultra-violet, 120–200 nm). Although some XUV and FUV can penetrate to the mesosphere; solar EUV is completely absorbed in the thermosphere. The solar EUV ionizes, dissociates, and excites the thermospheric constituents, creates the ionosphere, and heats the thermosphere. The second most important energy input to the thermosphere/ionosphere is geomagnetic activity resulting from the interaction between the solar wind and the Earth's magnetic field, which amplifies the magnetospheric-ionospheric current system and causes energetic particle precipitation into the atmosphere. This current system drives plasma convection in the high-latitude ionosphere, transferring energy to the thermosphere through Joule heating. Energetic particles precipitate from the magnetosphere into the underlying thermosphere/ionosphere along the Earth's magnetic field lines into the auroral region, heating the thermosphere/ionosphere mainly through ionization of thermospheric constituents. Joule heating and energetic particle precipitation can become the primary energy source for the thermosphere and ionosphere system during major geomagnetic storms (Knipp et al. 2004), and energy input from geomagnetic activity is generally more dynamic and impulsive than solar irradiance. Thermospheric neutral density perturbation due to this second major energy input was recently reviewed by Prölss (2011). In addition to these two main external forcings, the thermosphere/ionosphere is also coupled to the mesosphere energetically, dynamically, and chemically, particularly with momentum and energy forcing from various waves in the mesopause region. Furthermore, infrared cooling by radiatively active trace gases, such as CO_2, NO, CH_4, H_2O, and O_3, are also important for the energy balance of the thermosphere and mesosphere, and concentrations of these trace gases have shown long-term changes due to anthropogenic effects.

Thermospheric dynamics, mainly driven by temporal variation and spatial distribution of solar irradiance and geomagnetic forcing, redistribute composition and energy within the thermosphere. In addition, the thermosphere and ionosphere are closely coupled energetically, dynamically, and chemically. For example, the ionospheric plasma convection at high latitudes transfers energy and momentum through collisions between the plasma and the neutral species, heats the neutral atmosphere (Joule heating), and accelerates the neutral atmosphere (ion drag); the thermospheric neutral wind moves the electrically conducting plasma through the Earth's geomagnetic field and generates electric fields (dynamo). The dynamo electric field and current then affect both the neutral and plasma dynamics; and the daytime ionosphere E-layer and F1 layer are in photochemical equilibrium between production of ions through ionization and loss of ions through ion-neutral and ion-electron chemical reactions.

Variations in the external forcing, internal dynamics of the system, and coupling between the thermosphere and ionosphere, can drive complicated neutral temperature and composition variations, which changes neutral density scale height and causes complicated density variations. Neutral density varies on time scales from minutes to decades, with variations on hourly, diurnal, multi-day, 27-day solar-rotational, semiannual/annual, and solar-cycle time scales in between, and significant variation in latitude, longitude, and altitude. The purpose of this paper is to summarize and discuss these variations and their driving mechanisms, using various data sets and modeling results. In the following sections, brief descriptions of data and models will be given, followed by discussions for the temporal variations of neutral density from abrupt changes to long-term trends, and spatial variations.

2 Data and Models

Neutral density data, measured by accelerometers onboard the Challenging Minisatellite Payload (CHAMP) satellite and derived from satellite drag data, will be used throughout the paper for discussions of density variations. The National Center for Atmospheric Research (NCAR) thermosphere-ionosphere-electrodynamics general circulation model (TIE-GCM) (Roble et al. 1988; Richmond et al. 1992) was used to simulate each type of density variation for model-data comparisons. In addition, neutral density calculated by the NRLMSISE-00 empirical model (Picone et al. 2002) was also used for discussions.

2.1 CHAMP Neutral Density Data

The CHAMP satellite was launched into a near-circular orbit with an inclination of 87.3° on 15 July 2000 (Reigber et al. 2002; Lühr et al. 2004). The high inclination ensures almost complete latitudinal coverage from pole to pole. Neutral density is obtained from accelerometer measurements of non-gravitational accelerations on the CHAMP satellite (e.g., Sutton et al. 2005). The measured densities are normalized to a constant altitude of 400 km using NRLMSISE-00 (Picone et al. 2002). The temporal resolution of CHAMP neutral density is 10 seconds and the spatial resolution is 75 km, so CHAMP density data can be used to investigate spatial gradients of neutral density with good temporal resolution.

2.2 Neutral Density Data at Satellite Perigees

These neutral density data are daily averaged density obtained at satellite perigees of several low-Earth orbiting satellites from 2002 to 2006 by Bowman et al. (2004). These satellites are defunct radar calibration spheres launched by the former Soviet Union during the 1990s: Cosmos 660 (07337), Cosmos 807 (08744), Cosmos 1236 (12388), and Cosmos 1238 (12138). They are all in moderately eccentric orbits with average perigee altitudes between 380 and 430 km, and the average apogee altitude from 1300 to 1650 km. The satellite perigees scan approximately three latitude cycles and five local time cycles in a year. Errors of these density data are ~2%–4% (Bowman et al. 2004).

2.3 Global-Mean Neutral Density from Satellite Drag Data

Emmert (2009) created a long-term data set of globally averaged thermosphere mass density, derived from drag data of ~5000 orbiting objects, for the period 1967–2009, in the altitude range from 200 to 600 km. The data have a temporal resolution of 3–6 days, a typical short-term precision of 2%, and a long-term accuracy of 5–10%.

2.4 NCAR TIE-GCM

The National Center for Atmospheric Research (NCAR) thermosphere-ionosphere-electrodynamics general circulation model (TIE-GCM) is a first principles upper atmospheric general circulation model that solves the Eulerian continuity, momentum, and energy equations for the coupled thermosphere/ionosphere system. It uses pressure surfaces as the vertical coordinate and extends in altitude from approximately 97 km to 600 km (Roble et al. 1988; Richmond et al. 1992). The default solar input is specified by the EUVAC solar proxy model (Richards et al. 1994) and solar energy deposition is calculated using the method described in Solomon and Qian (2005). In addition, solar spectral irradiance measured by the Thermosphere Ionosphere Mesosphere Energetics and Dynamics (TIMED) Solar EUV (Extreme Ultra-Violet) Experiment (SEE) (Woods et al. 2005) can be used as solar input (Qian et al. 2008, 2009), and flare spectra calculated by the Flare Irradiance Spectral Model (FISM) (Chamberlin et al. 2008) were used for flare simulations (Qian et al. 2010a). Magnetospheric energy inputs are parameterized using the 3-hour planetary Kp index. Ionospheric convection is specified by the empirical model of Heelis et al. (1982). Auroral particle precipitation and its ionization and dissociation are calculated by an analytical auroral model described by Roble and Ridley (1987). Migrating semidiurnal and diurnal tides are specified at the lower boundary using the global scale wave model (Hagan and Forbes 2002, 2003). The effect of gravity wave breaking in the mesosphere/lower thermosphere (MLT) region is included by specifying a seasonal variation of eddy diffusivity at the lower boundary that declines with altitude (Qian et al. 2009). Effects of planetary waves and nonmigrating tides are not considered in the simulations shown here.

3 Density Variations

The data sets and models described in the previous section are used in the following sections to discuss density variations. These density variations are: abrupt changes with time scales from minutes to hours, diurnal variation, multi-day variation, solar-rotational variation, annual/semiannual variation, solar-cycle variation, long-term trends with a time scale of decades, and geographic variations with latitude, longitude, and altitude.

3.1 Abrupt Changes

Abrupt changes of neutral density refer to density changes on the order of minutes to hours. Solar flares and geomagnetic storms associated with coronal mass ejections (CME) cause rapidly changing energy inputs, and thus can cause rapid changes in neutral density. Semi-empirical density models often do not adequately capture these abrupt changes in neutral density. Sudden changes seen in time series of derived ballistic coefficients for orbiting objects are often assumed to result from attitude changes, but could sometimes result from unexpected density variation. Therefore, abrupt density changes can be a major source of orbit prediction error and significantly impact satellite tracking and collision avoidance.

3.1.1 Response to Solar Flares

Solar flares can produce a large and rapid increase of solar irradiance in the X-ray and EUV ranges, causing immediate enhancement of ionization in the upper atmosphere. Most studies of terrestrial effects of solar flares have focused on the ionospheric aspects (e.g., Mitra

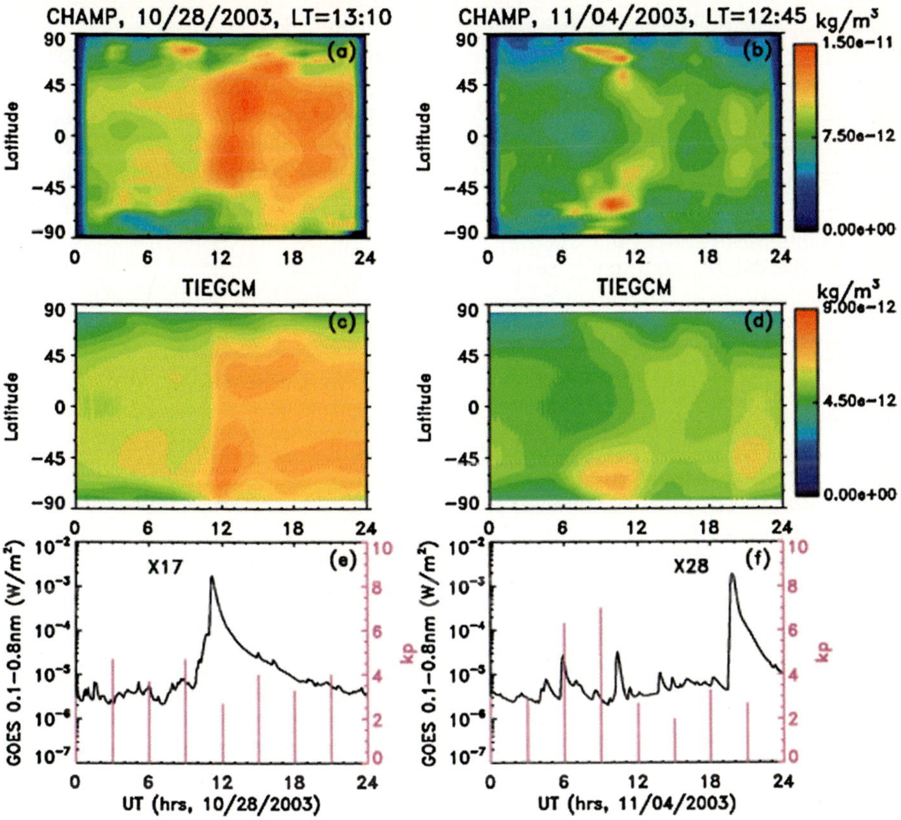

Fig. 1 Neutral density responses to an X17 flare occurred on October 28, 2003 and an X28 flare occurred on November 4, 2003. (**a**) Neutral density observed by CHAMP on October 28, 2003. The solar local time of CHAMP orbit was ~13:10. (**b**) Neutral density observed by CHAMP on November, 2003. The solar local time of CHAMP orbit was ~12:45. (**c**) Neutral density simulated by TIE-GCM for October 28, 2003, sampled along the CHAMP orbit. FISM flare spectra were used as solar input for the TIE-GCM. (**d**) Neutral density simulated by TIE-GCM for November, 2003, sampled along the CHAMP orbit. FISM flare spectra were used as solar input for the TIE-GCM; (**c**) GOES 0.1–0.8 nm solar irradiance and geomagnetic Kp index for October 28, 2003. (**d**) GOES 0.1–0.8 nm solar irradiance and geomagnetic Kp index for November 4, 2003

1974; Davies 1990; Mendillo et al. 1974; Zhang et al. 2005; Tsurutani et al. 2006), with relatively sparse research concerning the thermosphere (Sutton et al. 2005; Liu et al. 2007a; Pawlowski and Ridley 2008). The thermosphere, with its large mass and high heat capacity, is expected to be slower in responding to transient events such as solar flares. However, model simulations and observations show significant rapid density response to solar flares.

Figure 1a and 1b show neutral density at 400 km observed by the CHAMP satellite along its day-time orbit, for October 28, 2003, and November 4, 2003. Figure 1c and 1d are neutral density simulated by the TIE-GCM. Input flare spectra for the TIE-GCM were provided by the FISM (Chamberlin et al. 2008), which is based largely on data from TIMED/SEE and the X-ray monitors on the GOES satellites. TIE-GCM density was sampled along CHAMP orbits for comparison to the measured density. Figure 1e and 1f show solar flux in the wavelength range 0.1–0.8 nm measured by GOES 10, and geomagnetic Kp index. An X17 flare occurred at 11:10 UT on October 28, 2003, and an X28 flare occurred at 19:45 UT on November 4, 2003. For the X17 flare, the sudden increase of solar XUV and EUV drove a rapid increase in density. The magnitude of density enhancement and peak response time vary with latitude. The model simulated density enhancement is somewhat weaker than the observations. On average, peak density response is on the order of 30 to 40%. The peak response time varies from ~1 h to ~3 h, depending on latitude. These results are consistent with results by Sutton et al. (2005) and Liu et al. (2007a) for the same flare. By comparison, density enhancement during the X28 flare on November 4, 2003 is much weaker, as shown in Figs. 1b and 1d. Both model simulation and observations indicate a density enhancement on the order of ~10%. There is density enhancement earlier during the day between ~6:00 UT and ~15:00 UT, which is due to a geomagnetic disturbance, as indicated by the Ap index in Fig. 1f.

This seemingly contradictory relationship between the density response to the X28 flare and the X17 flare has to do with the spectral characteristics of the emission enhancement, which was due to their different locations on the solar disk. The X28 flare was a limb flare whereas the X17 flare occurred near the solar disk center. Qian et al. (2010a) investigated how the location of a flare on the solar disk affects thermosphere and ionosphere responses, finding that flare enhancements of XUV and EUV spectral irradiance are affected by disk location. Most emission lines in the XUV region (~0.1 to ~25 nm) are optically thin, and are weakly dependent on the location of a flare, but in the EUV region (~25 to ~120 nm), many important lines and continua are optically thick, so enhancements are weaker for flares located near the solar limb, due to absorption by the solar atmosphere. Consequently, in the E region ionosphere, where solar XUV dominates ionization, flare location does not affect thermosphere and ionosphere responses. However, flare response of neutral density in the upper thermosphere is 2 to 3 times stronger for a disk-center flare than for a limb flare, due to the fact that neutral gas heating through ionization by solar EUV is dominant in the 150–300 km region. Zhang et al. (2011) conducted statistical analysis and found that at the same X-ray class, flares near the solar disc center have much larger effects on the ionospheric TEC than those near the solar limb region.

The initial flare response occurs on the sunlit side of the Earth, so flare response is best observed on the dayside at low- to mid-latitudes. However, intense dayside heating can launch nightward propagating gravity waves that transport energy efficiently to the nightside at velocities near the local sound speed plus the bulk wind velocity in a matter of hours, with flare response occurring on the night side about 4 to 5 hours later. This nightside density enhancement was observed by CHAMP and simulated by upper atmospheric models (Pawlowski and Ridley 2008; Qian et al. 2010a).

Many questions concerning flare effects remain, for example, how other flare characteristics, such as flare rise time and decay time, affect thermosphere responses, whether latitudinal dependence of amplitude and density response time are determined by solar zenith angle, and how the thermosphere and ionosphere are coupled or decoupled in the phases of response and recovery (Qian et al. 2011). Combined modeling and data studies are needed to answer these questions and understand the underlying physical mechanisms.

3.1.2 Response to Geomagnetic Storms

Geomagnetic storms addressed in this section are those caused by CMEs, which occur as a result of a sudden increase of dynamic pressure in the solar wind, and changes in strength and orientation of the interplanetary magnetic field (IMF). These geomagnetic storms generate rapidly changing energy inputs, and occur predominantly at higher levels of solar activity. Solar wind energy is transferred from the magnetosphere to the ionosphere and the thermosphere through plasma convection and energetic particle precipitation; both are high latitude phenomena. The increased plasma convection causes a substantial increase in Joule heating, and energetic particle precipitation heats the thermosphere and the ionosphere in the auroral region through ionization, dissociation, and excitation of neutral constituents, mainly at E-layer altitudes. Joule heating usually dominates particle heating during large geomagnetic storms (Wilson et al. 2006). The initial storm response occurs at high latitudes, and neutral density responses to storms are best observed at high latitudes (e.g., Liu and Lühr 2005). The thermosphere responds to the heating with increased temperature, intensified large-scale circulation, redistribution of composition, and increased neutral mass density. The localized heating can then excite gravity waves at high latitude (e.g., Bruinsma et al. 2006). Disturbances at high latitude are then propagated to lower latitudes by large-scale circulation and waves, causing perturbation at lower latitudes with several hours delay (Bruinsma et al. 2006). As a result, geomagnetic storm-driven increases of density become a global phenomenon.

Neutral density response to a geomagnetic disturbance that occurred on November 4, 2003, is evident in Fig. 1b, 1d, and 1f. The 3-hour Kp index was 6–7 from 6:00–9:00 UT, and the neutral density was enhanced between ~6:00 and ~15:00 UT. Neutral density increased rapidly at high latitude, responding to the disturbance, and this density enhancement moved toward lower latitudes during the next several hours. The density enhancement was on the order of ~50%. Geomagnetic storms can last for several days. During the 2003 "Halloween storm" of October 28–November 4, (day 301–308) of 2003, the Kp index remained high for almost three days. Figure 2a and 2b show neutral density at perigees of satellites #12388 and #12138 from 2003270 to 2003350 (September 27, 2003–December 16, 2003), the period surrounding this storm, derived from satellite drag and simulated by the TIE-GCM. Figure 2c shows the daily Ap index for this period. These neutral density data are daily averaged density at the satellite perigees. As mentioned earlier, these satellites have moderately eccentric orbits with perigees between 380 km and 430 km and apogees near 1500 km. The TIE-GCM simulated neutral density was sampled at perigee local time, altitude, and latitude for comparison to the data. We would point out that the altitude, latitude, and local time of the perigees changed with time, which would change neutral density. However, this change of density is gradual compared to the impulsive change due to geomagnetic activity. Figure 2 shows the direct correspondences of each impulsive change in the line plot of the neutral density to the line plot of Ap index. This clearly demonstrates how geomagnetic forcing drives impulsive variation in neutral density. The "Halloween storm" caused neutral density enhancement on the order of 100 to 200%. The different morphology of neutral density for the two satellites following the "Halloween storm" is due to the changes of altitude, local time, and latitude at the perigees. Specifically, the perigee of satellite #12388 increased from ~390 km to ~415 km and the local time at perigee progressed from near local noon to night time, consequently, neutral density was low after the storm. By contrast, the perigee of satellite #12138 decreased from ~425 km to ~395 km and the local time at perigee progressed from night time to around local noon, therefore, overall neutral density remained high after the storm.

As with the response of neutral density to flares, the characteristics of an individual geomagnetic storm, such as the orientation, strength, and variability of the interplanetary magnetic field, and the solar wind speed and density, affect the magnitude, temporal, and spatial features of the resulting density changes (e.g., Liu et al. 2010, 2011). The level of solar activity, season, and UT at storm onset also has a significant effect (Burns et al. 2004), due largely to variations in ionospheric conductivity. Response timing and duration also vary strongly with altitude (Wang et al. 2008). The altitude distribution of Joule heating is very important

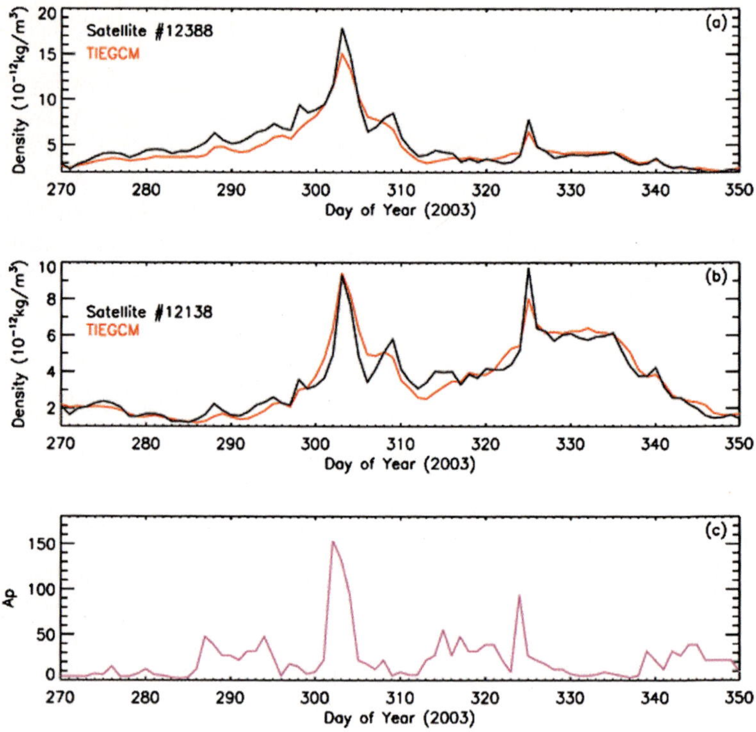

Fig. 2 The impulsive changes of neutral density responding to the geomagnetic storms in November, 2003. (**a**) Neutral density at the perigees of satellite #12388 from 2003270 to 2003350 (September 27, 2003–December 16, 2003). *Black*: derived from satellite drag; *red*: simulated by TIE-GCM. (**b**) Neutral density at the perigees of satellite #12138 from 2003270 to 2003350 (September 27, 2003–December 16, 2003). *Black*: derived from satellite drag; *red*: simulated by TIE-GCM. (**c**) Corresponding geomagnetic Ap index during the period. *Note*: the altitude, latitude, and local time of the satellite perigees changed with time through the period, which caused density changes. However, these density changes are gradual compared to the impulsive changes of neutral density directly corresponding to the impulsive changes in Ap index shown in this figure

to thermosphere and ionosphere responses. Compared to lower-altitude Joule heating, high-altitude heating produces a strong, immediate density response that rapidly decays when the heating ends (Deng et al. 2010). More investigation is needed to quantify the distribution of magnetospheric energy inputs to the thermosphere/ionosphere system during a storm, and how the energy partitioning and ionosphere conditions determine neutral density response.

3.2 Diurnal Variation

Since solar irradiance is the primary energy source for the thermosphere, thermospheric density exhibits large diurnal variation. Müller et al. (2009) conducted statistical analysis to CHAMP neutral density at low latitude (±30 geographical latitude) and found that the ratio between the density on the dayside (10:30–16:30 LT) and night side (22:30–04:30 LT) is 2 ± 0.1 during geomagnetic quite time, nearly independent of solar activity level and season. Figure 3a shows neutral density from day 270 to day 274 of 2007 observed by CHAMP along CHAMP's ascending orbit (~13:00LT) and descending orbit (~01:00LT). Figure 3c and 3d shows TIE-GCM simulated neutral density sampled along CHAMP's as-

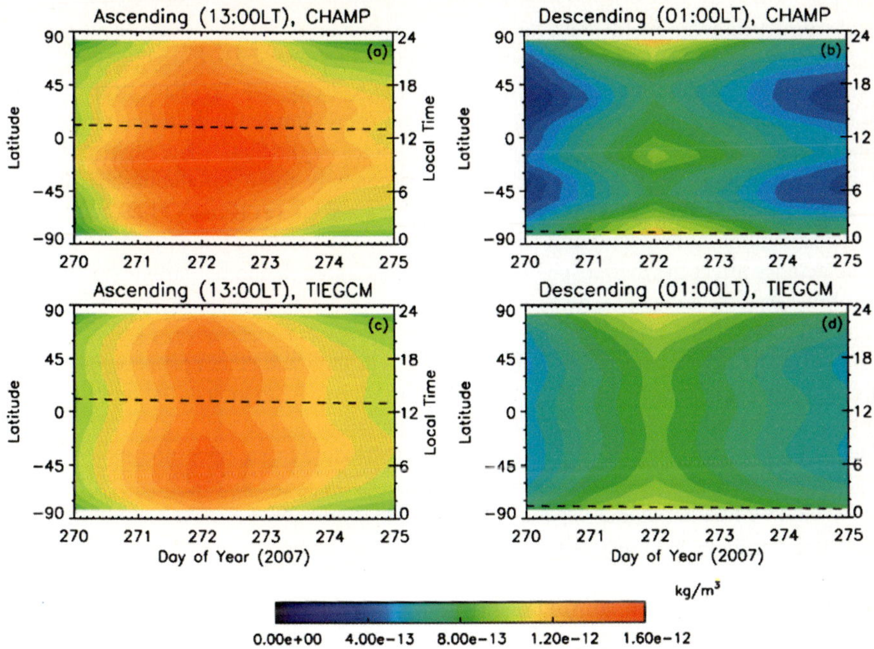

Fig. 3 Diurnal variation of neutral density. (**a**) Day time neutral density along CHAMP ascending phase of the orbit from day 270 to day 274, at ~13:00LT, observed by CHAMP. The *black dashed lines* shows local time of the ascending phase for each day. (**b**) Night time neutral density along CHAMP descending phase of the orbit from day 270 to day 274, at ~01:00LT, observed by CHAMP. The *black dashed lines* shows local time of the descending phase for each day. (**c**) Day time neutral density along CHAMP ascending phase of the orbit from day 270 to day 274, at ~13:00LT, simulated by TIE-GCM. The *black dashed lines* shows local time of the ascending phase for each day. (**d**) Night time neutral density along CHAMP descending phase of the orbit from day 270 to day 274, at ~01:00LT, simulated by TIE-GCM. The *black dashed lines* shows local time of the descending phase for each day

cending and descending orbits, respectively. The observed and simulated density difference between these two local times is ~80–100%.

These diurnal variations of neutral density in the upper thermosphere (around 400 km) are mainly caused by in-situ diurnal variation of solar irradiance. In the lower thermosphere, diurnal variation can be modulated by tides (e.g., Hagan et al. 1997; Forbes et al. 2003). These tides are originated in the troposphere and propagate upward into the lower thermosphere. Recent analysis of CHAMP and GRACE neutral density suggests that effects of these tides can reach the upper thermosphere (Forbes et al. 2009). Therefore, diurnal variation of neutral density, even in the upper thermosphere, may include contribution from the lower atmosphere tides, however, this contribution should be small compared to the contribution by the in-situ solar irradiance.

3.3 Multi-day Variation

Multi-day variations refer to density variation with a periodicity at sub-harmonics of the solar rotation period. The multi-day density variation is primarily caused by recurrent geomagnetic forcing. Compared to geomagnetic storms driven by CMEs discussed in Sect. 3.1, this recurrent geomagnetic forcing is at low to moderate geomagnetic activity levels associated with co-rotating interaction regions (CIR) (Tsurutani et al. 1995), that occur predominantly in the declining and minimum phases of a solar cycle. Multi-day periodicities at sub-harmonics of the solar rotation period (~9 day, ~7 day, ~5 day) have been observed in the interplanetary magnetic field and solar wind speed during the declining phase of solar cycle 23, related to the distribution of coronal holes on the Sun, where high-speed solar wind streams originate (Temmer et al. 2007). Consequently, these periodicities are also seen in auroral precipitation and various geomagnetic activity indices (Emery et al. 2009) as the solar wind interacts with the Earth's magnetosphere. The resulting periodicities in geomagnetic activity affect the thermosphere and ionosphere through auroral particle precipitation and Joule heating. Lei et al. (2008) found a strong (~30–50%) 9-day periodicity in neutral density observed by CHAMP during 2005.

Qian et al. (2010b) assessed modeling capability in simulating thermospheric response to this recurrent geomagnetic forcing, and investigated vertical variation of the multi-day variation. They found that the simulated multi-day variation agrees well with measurements. The multi-day oscillation of density is globally distributed in the upper thermosphere, but restricted to high latitudes in the lower thermosphere. The upper thermosphere density variation exhibits less latitudinal dependence than the temperature variation, because the latitudinal dependence of composition compensates for the temperature effect by reducing the density scale height. The differences between latitudinal variations of neutral density, neutral temperature, and composition will be further discussed in Sect. 3.8.

Figure 4a shows neutral density at 400 km from day 240 to day 300 of 2007, observed by CHAMP along its ascending orbit, and Fig. 4b shows TIE-GCM simulated density sampled along the CHAMP's ascending orbit. The local time changes from ~16:00 to ~10:45 during the 60 days. Figure 4c shows the corresponding F10.7 and Ap indices. Figure 4 clearly shows the observed and simulated multi-day variations of neutral density that are globally distributed at this altitude, and demonstrates the correlation between multi-day density variation and multi-day variation in Ap index.

Forcing from the lower atmosphere can also cause multi-day density variation. Although most planetary waves dissipate in the stratosphere and lower mesosphere, planetary waves with periods of 2, 5, 10, and 16 days have been observed in the mesosphere and lower thermosphere region (e.g., Jacobi et al. 1998; Clark et al. 2002). These waves, when present, may also contribute to multi-day periodicities observed in the thermosphere and ionosphere.

3.4 Solar-Rotational Variation

Solar-rotational variation of thermospheric density is caused by the appearance and disappearance of the Sun's active regions as the Sun rotates in an average 27-day period. The active regions can be seen through the Sun's hydrogen image that features bright regions identified as solar plages. Plages are localized regions with intense magnetic activities that are usually associated with sunspots. Plages are distributed non-uniformly and appear primarily within 30° latitude on each side of the equator of the Sun. These features usually appear and disappear over a period of several months. As the Sun rotates with an approximate 27-day period, the non-uniform active regions emerge and vanish to the Earth's atmosphere. This produces a rotational variability in solar irradiance with a magnitude proportional to the contrast and fractional areas between active regions and quiet regions, as well as a strong 27-day periodicity in geomagnetic activity. Since the Sun is very active under solar maximum conditions, solar-rotational variation is larger under solar maximum conditions compared to solar minimum conditions. Figure 5a and 5b shows global-mean neutral density at 400 km, derived from satellite drag (Emmert 2009) and simulated by the

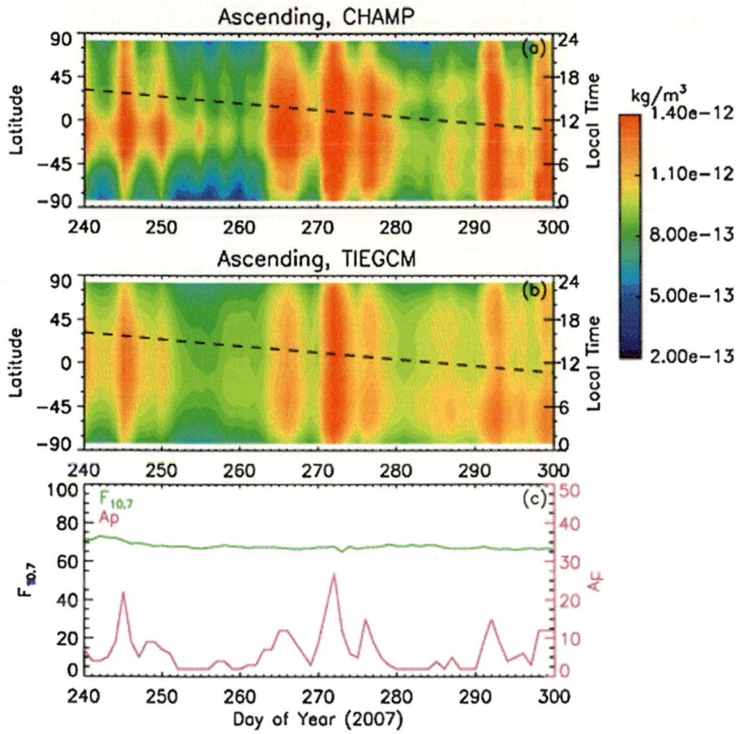

Fig. 4 Multi-day variation of neutral density. (**a**) Neutral density along CHAMP ascending phase of the orbit from day 240 to day 299 of 2007, observed by CHAMP. The *black dashed lines* shows local time of the ascending phase for each day. (**b**) Neutral density along CHAMP ascending phase of the orbit from day 240 to day 299 of 2007, simulated by TIE-GCM. (**c**) Corresponding F10.7 index Ap indices

TIE-GCM for 2003 (high solar activity) and 2008 (solar minimum), respectively. Figure 5c and 5d show the corresponding F10.7 and Ap indices for these two years. Solar-rotational variation of solar irradiance is strong in 2003 but very weak for 2008, as seen in the F10.7 index (Fig. 5c, 5d). Solar-rotational variations of neutral density are evident in the data and modeling results, and often exceed 100% for 2003. By comparison, solar-rotational variation is very minor during 2008; density during this year, a solar minimum year, exhibits a predominantly multi-day variation, primarily caused by the multi-day variation in geomagnetic activity, as we discussed in Sect. 3.3.

3.5 Annual/Semiannual Variation

Thermospheric mass density and composition exhibit a strong seasonal variation, with maxima near the equinoxes, a primary minimum during northern hemisphere summer, and a secondary minimum during southern hemisphere summer. The annual thermospheric density variation was first observed by Paetzold and Zschörner (1961) through analysis of satellite drag data. They found that neutral density has a minimum from May to August and a broad maximum during the rest of the year with a secondary minimum near January. The magnitude of the annual minimum to maximum variation is more than 100%. Using harmonic analysis, they further found that the annual variation is a semiannual variation superimposed

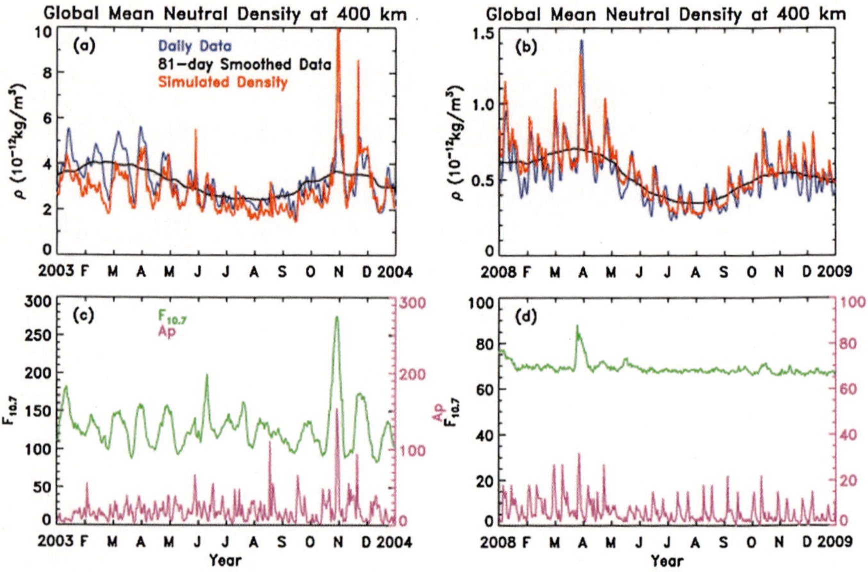

Fig. 5 Solar-rotational and annual/semiannual variations of neutral density. (**a**) Global-mean neutral density at 400 km for 2003. *Blue*: daily density data derived from satellite drag; *black*: 81-day running mean of the daily density data; *red*: daily density simulated by TIE-GCM. (**b**) Global mean neutral density at 400 km for 2008. *Blue*: daily density data derived from satellite drag; *black*: 81-day running mean of the daily density data; *red*: daily density simulated by TIE-GCM. (**c**) F10.7 and Ap indices for 2003. (**d**) F10.7 and Ap indices for 2008

on an annual variation, and the ratio of the amplitude of the annual to semiannual variation is 3:2.

Figure 6a shows global-mean neutral density at 400 km for 1979 derived from satellite drag data (Emmert 2009). Figure 6b shows Fourier decomposition of the density from 1 to 4 cycles per year. It is clear that on seasonal scale, main components of neutral density variation are annual and semiannual variation; the amplitude from annual minimum to annual maximum is ~140%, and the annual amplitude and semi-annual amplitude are comparable, on the order of ~100%.

Jacchia (1965) represented annual/semiannual density variation with temperature functions in his 1965 thermospheric empirical density model (J65). Jacchia (1971) then revised the J65 model approach, and represented annual/semiannual variation in the density formula with amplitude as a function of height. The MSIS series of models (Hedin 1991; Picone et al. 2002) represent annual/semiannual density variation with a combined contribution from temperature and composition variation. The composition function is imposed at 120 km and propagates to the upper thermosphere through molecular diffusion.

Bowman (2004) investigated the annual/semiannual density variation by analyzing drag data from 1970–2002 for 13 satellites with perigee heights between 200 km and 1100 km. He found that both the phase and amplitude of the annual/semiannual density variation change from year to year, with more complicated phase variation under high solar activity conditions. The amplitude increases with altitude from 200 km to 800 km and then declines at higher altitudes. The amplitude (yearly minimum to yearly maximum) can change from year to year from 30% to 250%, depending on solar activity and altitude. Bowman (2004) developed new empirical equations of annual/semiannual density variation based on the equation

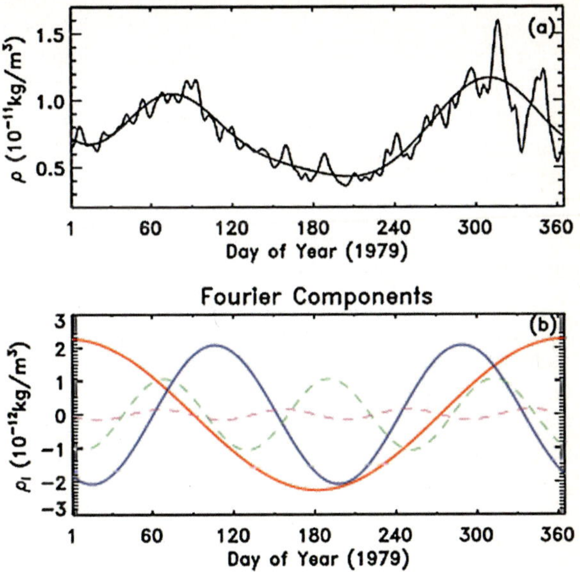

Fig. 6 Annual/semiannual variation of neutral density at 400 km. (**a**) Global-mean neutral density at 400 km for 1979, derived from satellite drag. The *smooth line* is Fourier fit of the data. (**b**) Components of Fourier decomposition of the neutral density data. *Red*: the annual component; *blue*: the semi-annual component; *green*: the three-cycle per year component; *magenta*: the four-cycle per year component

used in Jacchia71 model (Jacchia 1971); the amplitude and phase functions of the equation were based on data fitting and used F10.7 index. Bowman et al. (2007) updated the J71 empirical density model to Jacchia-Bowman 2006 model (JB2006), including the new annual/semiannual formula. Bowman et al. (2008) further investigated annual/semiannual density variation and found that the phase and amplitude of annual/semiannual density variation can be better parameterized using combined solar indices S_{10} and M_{10} as opposed to using F10.7 index alone. Bowman et al. (2008)) used this new parameterization of annual/semiannual variation in the JB2008 model, the revised version of JB2006.

Emmert and Picone (2010) developed an empirical Global Average Mass Density Model (GAMDM) using the global neutral density data set of Emmert (2009). The model represents density as a function of the F10.7 index, the day of year, and the Kp geomagnetic activity index. The model has two terms for annual/semiannual variation that include a core seasonal term and an F10.7 modulated seasonal term. They found that the amplitudes of the annual and semiannual harmonics increase with height and F10.7; the inter-annual variability of the amplitudes is comparable to their increase from solar minimum to solar maximum; and at 400 km, the average amplitude of the annual variation is $9.8 \pm 1.6\%$ at solar minimum and $15.1 \pm 1.8\%$ at solar maximum. The corresponding semiannual amplitudes are $16.4 \pm 1.1\%$ and $21.0 \pm 1.2\%$. We would note that these amplitudes are the values between the extremes and the mean values, therefore, we should double these amplitudes in comparison with the peak-to-peak amplitudes of Bowman (2004). The peak-to-peak amplitudes of Emmert and Picone (2010) are consistent with the amplitude of Bowman (2004) for solar minimum conditions, but are considerably smaller for solar maximum conditions.

Müller et al. (2009) examined the annual/semiannul variation of neutral density using CHAMP data. They used data at the low latitude, at fixed local time intervals (10:30–16:30 LT and 22:30–04.30 LT), and under geomagnetic quiet conditions ($Ap < 15$). In addition, they normalized the data to 400 km and a fixed solar activity level ($P_{10.7} = 130$, where $P_{10.7}$ is the average of $F_{10.7}$ and 81-day average $F_{10.7}$). The purpose of this data selection was to remove the effects of various forcing on the annual/semiannual variation, such as solar and geomagnetic activity, solar zenith angle, local time, and altitude. They found that the peak to peak amplitude of the annual/semiannual variation is 60% at $P_{10.7} = 130$, corresponding to the amplitude of Bowman (2004) for solar minimum conditions.

Empirical models can parameterize and reproduce the annual/semiannual density variation, but mechanisms for the variation are not addressed. The annual 7% variation of insolation due to variation of the Sun–Earth distance can result in an annual asymmetry, with terrestrial perihelion and hence maximum irradiance during January. However, Paetzold and Zschörner (1961) found that the resulting amplitude is too small compared to what was observed by satellite drag data. Sensitivity tests using the TIE-GCM showed that on global-average and under solar minimum conditions, the 7% variation in solar irradiance can cause a ~2%–5% annual asymmetry in neutral temperature and ~25% annual asymmetry in neutral density at 400 km. The observed annual asymmetry shown in Fig. 5b for 2008, a solar minimum year, is ~50–60%, therefore, the Sun–Earth distance alone cannot account for the observed annual asymmetry. Walterscheid (1982) suggested that there is a globally averaged temperature difference between solstice and equinox because of stronger geomagnetic activity at equinoxes than solstices due to the "Russell-McPherron effect" (Russell and McPherron 1973). This semiannual variation in geomagnetic activity can cause a small semiannual variation in neutral density (Detman 1996). Fuller-Rowell (1998) proposed the large-scale inter-hemispheric circulation as a mechanism for global semiannual density variation; the large-scale inter-hemispheric circulation acts as a "thermospheric spoon" to mix the atmosphere; since the circulation is stronger during solstice seasons, it causes a global semiannual variation in neutral density with maxima during equinox seasons and minima during solstices.

Recent model simulations (Qian et al. 2009) using the TIE-GCM indicate that these mechanisms do not fully account for the observed annual/semiannual amplitude, primarily because of the lack of a minimum during northern hemisphere summer. Qian et al. (2009) proposed forcing by the lower atmosphere as a mechanism for the annual/semiannual variation in the thermosphere: gravity wave breaking can cause a annual/semiannual variation in eddy mixing; increased eddy diffusion during northern hemisphere summer accelerates the downward transport of O to the mesopause where it recombines; composition change in the lower thermosphere affects the entire thermosphere through molecular diffusion. They derived an empirical variation of the eddy diffusion coefficient applied on a global basis to the lower boundary of the TIE-GCM (~97 km). The simulated global mean neutral density shown in Fig. 5 employed this annual/semiannual variation of eddy diffusion. The simulated neutral density agrees well with the data. Annual/semiannual density variation is evident in the observed and simulated neutral density for both the high solar activity year 2003 and the solar minimum year 2008. Amplitudes (between yearly minimum and yearly maximum) are ~60% and ~110% for 2003 and 2008, respectively. Combined data analysis using a thermosphere density data set, middle atmosphere temperature and wind data, and whole atmosphere model simulations, are needed to investigate this possible connection between the lower and upper atmosphere.

Annual/semiannual variation of neutral density is discussed again in Sect. 3.8 as it pertains to the latitudinal variation of neutral density for the four seasons. We will also discuss the annual asymmetry and possible equinoctial asymmetry components shown in the annual/semiannual variation of neutral density and composition.

3.6 Solar-Cycle Variation

Solar-cycle variation is driven by the intrinsic ~11-year variability of magnetic activity in the Sun. During a solar cycle, the plage fractional area over the solar disk varies from

0 to around 0.2. The variability of solar irradiance, mainly the EUV, drives a solar-cycle variability of neutral density in the upper thermosphere of more than an order of magnitude. Increased geomagnetic activity at solar maximum also accounts for part of this variation.

Figure 7 compares neutral density at perigee heights for satellite #12388, #07337, and #08744, derived from satellite drag data for these 3 satellites, to neutral density simulated

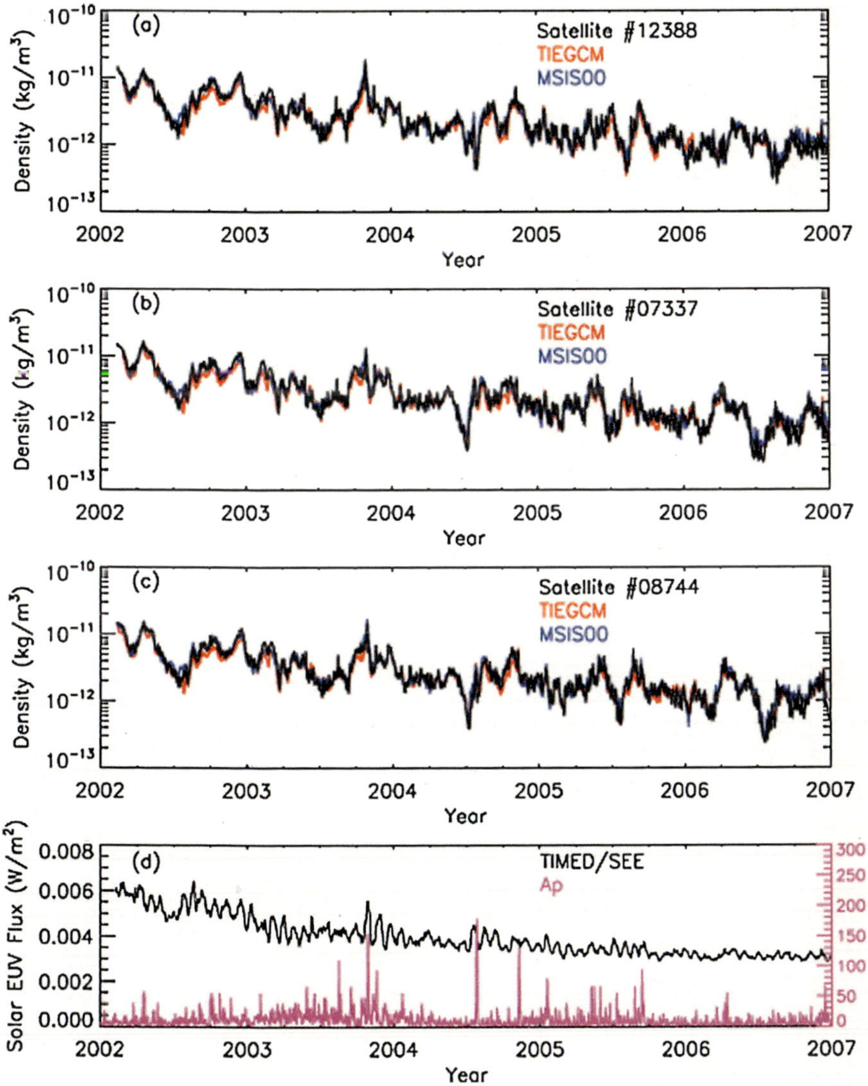

Fig. 7 Solar cycle variation of neutral density. For the simulated neutral density, TIMED/SEE measured spectral irradiance in the wavelength range 0–175 nm were used as solar input for TIE-GCM. (**a**) Neutral density at the perigees of satellite #12388 from 2002 to 2006. *Black*: derived from satellite drag; *red*: simulated by TIE-GCM; *blue*: calculated by MSIS00. (**b**) Neutral density at the perigees of satellite #07337 from 2002 to 2006. *Black*: derived from satellite drag; *red*: simulated by TIE-GCM; *blue*: calculated by MSIS00. (**c**) Neutral density at the perigees of satellite #08744 from 2002 to 2006. *Black*: derived from satellite drag; *red*: simulated by TIE-GCM; *blue*: calculated by MSIS00. (**d**) TIMED/SEE integrated solar EUV (5–105 nm) and geomagnetic Ap index from 2002 to 2006

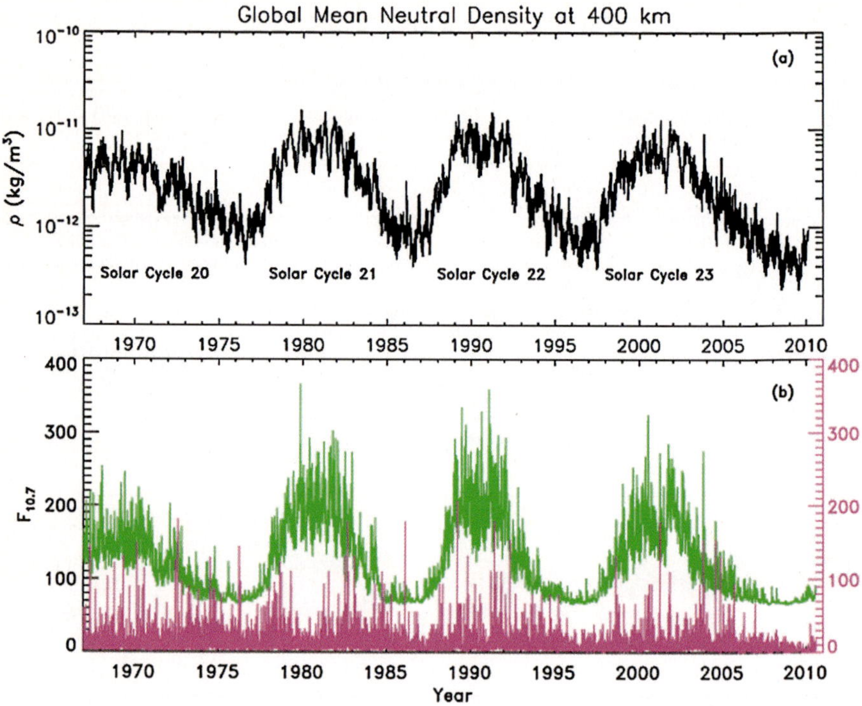

Fig. 8 Solar cycle variation and long-term trend of neutral density. (**a**) Global mean neutral density at 400 km derived from satellite drag data (Emmert 2009) from 1967 to 2010. (**b**) Corresponding F10.7 and Ap indices

by the TIE-GCM and calculated using MSIS00, from 2002 to the beginning of 2007. Neutral density from the TIE-GCM and MSIS00 was sampled at the perigees of each satellite. In these simulations, solar spectral irradiance in the range of 0–175 nm, observed by TIMED/SEE, was used as solar input for TIE-GCM. Figure 7d shows the corresponding integrated solar EUV irradiance (5–105 nm) measured by TIMED/SEE, and the Ap index. From 2002 to the beginning of 2007, the integrated solar EUV changed from $\sim$0.006 W/m^2 to $\sim$0.003 W/m^2; and neutral density changed about one order of magnitude.

Figure 8 shows global-mean neutral density at 400 km for nearly 4 solar cycles from the density data set of Emmert (2009), as well as the corresponding F10.7 and Ap indices. The amplitude of solar-cycle density variation for solar cycle 20 is about a factor of five; the amplitudes for solar cycles 21–23 are more than an order of magnitude. In addition, Fig. 8 demonstrates that solar irradiance is the primary forcing for the thermosphere that determines the baseline average neutral density, whereas geomagnetic forcing drives short-term variability.

3.7 Long-Term Trends

In Fig. 8, it is noticeable that neutral density at each solar minimum is lower than that at the previous solar minimum. This indicates a long-term trend of neutral density that we will discuss in this section. This long-term trend has been attributed primarily to the long-term changes of trace gases, mainly CO_2 (Laštovička et al. 2006).

Anthropogenic emissions of greenhouse gases are considered to be the primary cause of global warming in the troposphere during the past 50 years, due to the fact that these gases

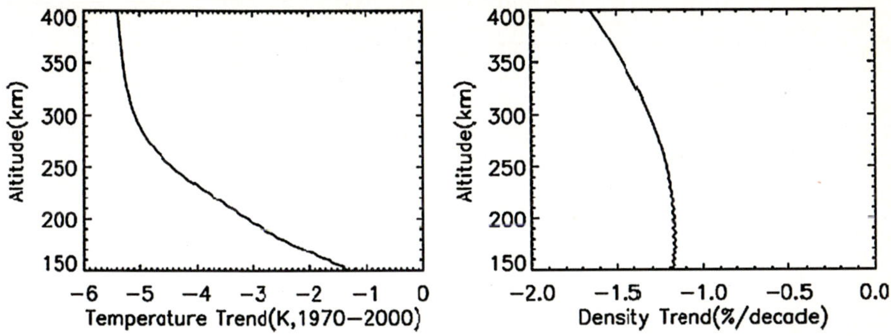

Fig. 9 Altitude profiles of long-term trends of neutral temperature and neutral density, simulated by the GLBMEAN, for the period from 1970–2000. CO_2 concentration from 1970 to 2000 observed at Mona Loa solar observatory was used as green house forcing for the model

are optically thick to outgoing infrared radiation and thus produce a heating effect. In the middle and upper atmosphere, however, greenhouse gases have an opposite, cooling effect. Greenhouse gases are optically thin to outgoing infrared radiation in the middle and upper atmosphere due to low density. Greenhouse gases absorb thermal energy through in-situ collisional excitation and then radiate to space in infrared, causing a cooling effect. Roble and Dickinson (1989) first suggested that global cooling would occur in the upper atmosphere in conjunction with global warming in the troposphere due to the long-term increase of greenhouse gas concentrations, particularly CO_2. Modeling studies by Rishbeth (1990) and Rishbeth and Roble (1992) broadened these results to the thermosphere-ionosphere system. Since the thermosphere is approximately under hydrostatic equilibrium, the height of a given pressure surface is determined by the average atmospheric temperature below. The cooling therefore results in thermal contraction of the upper atmosphere, and we may expect a significant decline in thermospheric density at fixed heights.

This long-term decreasing trend of neutral density has been observed using long-term satellite drag data (e.g., Keating et al. 2000; Emmert et al. 2004, 2008; Marcos et al. 2005). The estimated density trend at 400 km ranges from −1.7%/decade to −3.0%/decade, for the past 3 to 4 decades. The density trend anti-correlates with solar activity, with about −1%/decade to −2%/decade under solar maximum conditions and about −3%/decade to −5%/decade for solar minimum conditions. In addition, the density trend increases with altitude. The solar cycle dependence is due to the relative importance of CO_2 infrared cooling and NO infrared cooling under different solar activity conditions (Qian et al. 2006). These observational findings are qualitatively consistent with modeling results by Roble and Dickinson (1989). Using CO_2 concentrations measured at Mauna Loa, Qian et al. (2006) calculated neutral density trend for the period from 1970 to 2000, and found an average density trend of −1.7%/decade at 400 km. Model simulations also showed that this density trend corresponds to a neutral temperature trend of −5.4 K for this three decades at 400 km. Figure 9 shows altitude profiles of the temperature trend and neutral density trend from 1970 to 2000, from the model simulations of Qian et al. (2006).

As mentioned earlier, this long-term decreasing trend of neutral density is discernable in the long-term density data shown in Fig. 8, as indicated by the gradual decrease of density at each solar minimum. However, as can be seen in Fig. 8, density at the minimum between solar cycles 23 and 24 is ∼30% lower than the density at previous minimum (Emmert et al. 2010; Solomon et al. 2010), greatly exceeding the observed and modeled long-term trend. Model simulations using a global-mean version of TIE-GCM demonstrates that long-term

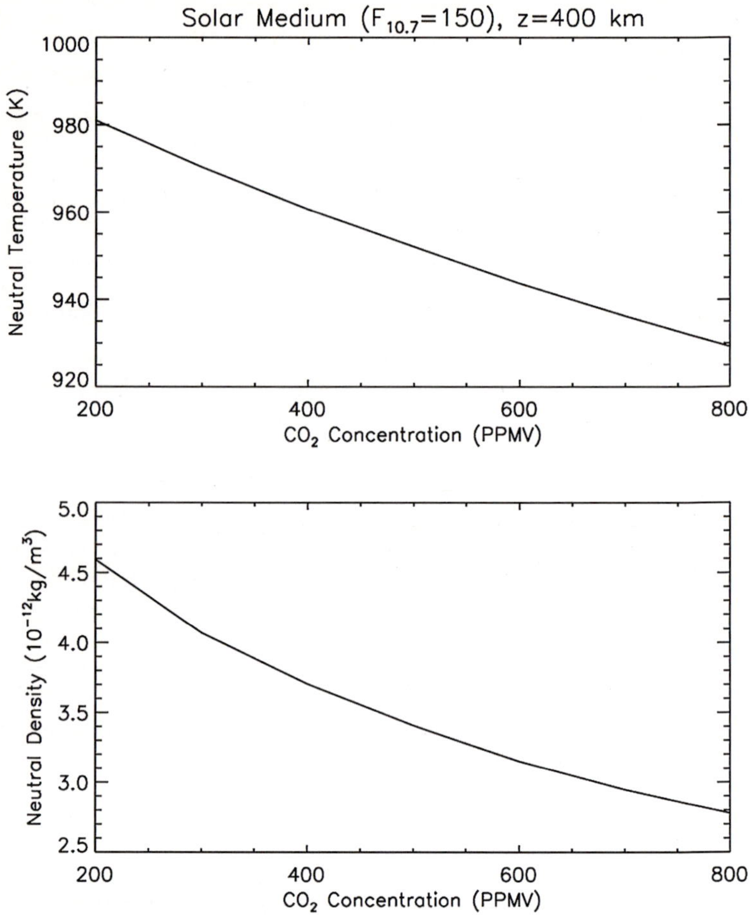

Fig. 10 Sensitivity of long-term trends of neutral temperature and neutral density to CO_2 concentration, simulated by the GLBMEAN, under solar medium condition

cooling due to CO_2 does not accelerate as CO_2 concentration increases. Figure 10 shows the simulated temperature and neutral density under solar medium ($F107 = 150$) as CO_2 concentration changes from 200 ppmv to 800 ppmv. Temperature variation with CO_2 concentration is quite linear in this large range of CO_2 concentration. Density shows only a slight non-linear relation with CO_2 concentration due to the exponential decrease of density with scale height. Therefore, the $\sim$30% decrease of neutral density at this solar minimum is not due to accelerated global cooling. Model simulations using SOHO SEM measurements (Solomon et al. 2010) and MgII core-to-wing ratio (Solomon et al. 2011) indicated that the primary cause of the anomalous low density at solar cycle 23 minimum is anomalous low solar irradiance at this solar minimum, with geomagnetic forcing and anthropogenic long-term trend playing small but significant roles. On the other hand, statistical analysis using global total electron content derived from multiple GPS stations suggested that the anomalously low solar irradiance for solar cycle 23 minimum measured by SOHO SEM is incompatible with the total electron content observations (Lean et al. 2011).

3.8 Spatial Variation

The density variations addressed so far are variations with time. Another aspect of density variations are their spatial distribution. The total mass density is determined by the densities of major species. The spatial distribution of external forcing, together with internal dynamics and thermosphere/ionosphere coupling, determines the spatial variation of neutral density. By comparison with horizontal variation, the larger spatial variation of neutral density is usually the variation with altitude. On a global mean basis, the thermosphere is in diffusive equilibrium state; therefore, thermospheric neutral density decreases exponentially with increasing altitude, according to the diffusive equilibrium density scale height, which is determined by temperature and composition. Since the middle and upper thermosphere (above ~200 km) is nearly isothermal, the exponential decrease of neutral density is mainly due to the exponential decrease of the neutral species, which are in diffusive separation according to the individual molecular mass of each species. However, thermospheric large-scale circulation, as well as other processes such as nonmigrating tides from the lower atmosphere, can significantly change the spatial distributions of temperature and composition, and thus the spatial distribution of neutral density. Vertically, molecular diffusion constantly works to restore the atmosphere to a new diffusive equilibrium, and neutral density by large decreases exponentially with increasing altitude. In this section, we will discuss spatial variation at a fixed altitude, including latitudinal and longitudinal variations. The modification of diffusive equilibrium distribution of neutral density will also be discussed below.

3.8.1 Latitudinal Variation

At a fixed altitude, the number densities of atmospheric constituents are thermally and dynamically governed. The solar irradiance that reaches each hemisphere changes with season, causing different latitudinal variations of temperature in each season, as well as driving changes in inter-hemisphere circulation. This large-scale inter-hemispheric circulation, together with the high latitude circulation driven by geomagnetic activity, transports lighter species to different latitudes and thus causes latitudinal variation of composition. Latitudinal variations of neutral temperature and composition then determine the latitudinal variation of neutral density. In addition, the latitudinal variation of neutral density may also be affected by thermosphere and ionosphere coupling through mechanisms such as ion drag and chemical heating (Liu et al. 2007b). Figure 11 shows latitudinal distributions of zonal-average neutral density, temperature, major species number densities at 350 km, under solar minimum conditions, for the two solstices and two equinoxes, simulated by TIE-GCM. At the solstices, neutral density (Fig. 11a) increases from the winter pole to the summer pole with amplitude ~40%; at equinoxes, the density exhibits little latitudinal variations. The large latitudinal gradient of neutral density at solstices and small latitudinal gradient of neutral density at equinoxes are consistent with the overall latitudinal distribution of neutral density observed by CHAMP (Liu et al. 2007b). However, Liu et al. (2007b) found the anomalous latitudinal variation of neutral density, the "equatorial mass anomaly" (EMA) in CHAMP density, with neutral mass density showing a dip at the magnetic equator and two crests on each side. This EMA feature is prominent during equinoxes, under high solar activity, and during 11:00LT–16:00LT. They suggested that the EMA was linked to the equatorial ionization anomaly (EIA) through neutral-ion coupling. The TIE-GCM simulated neutral density in Fig. 11a does not show the EMA feature. This is probably due to the fact that the model simulation was conducted for solar minimum conditions. In addition, Fig. 11a shows zonal-mean neutral density, which could also contribute to the lack of the EMA feature. Figure 11b

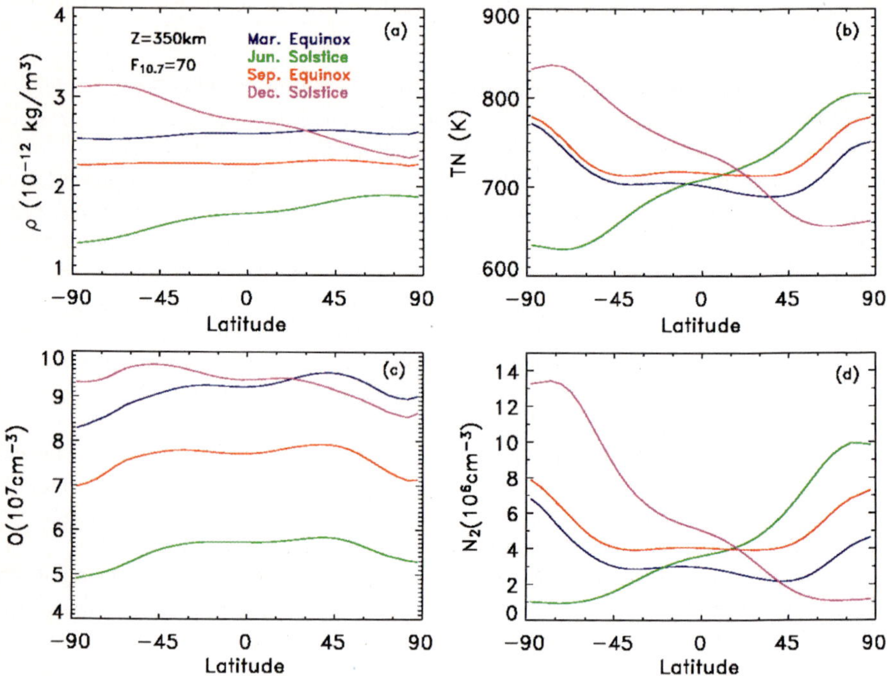

Fig. 11 Latitudinal variations of zonal-mean neutral density, neutral temperature, O number density, and N_2 number density at 350 km at equinoxes and solstices, under solar minimum and geomagnetic quiet conditions, simulated by TIE-GCM

shows latitudinal variation of zonal-mean temperature for each season. It is straightforward that temperature increases from the winter hemisphere toward the summer hemisphere. The temperature near summer pole is approximately 27% higher than near the winter pole. During equinoxes, temperature shows small latitudinal variation except at high latitudes where heating by geomagnetic activities contributes to higher temperatures in the regions.

There is a distinct difference in density annual variation (variation between the two solstices) between the Arctic and the Antarctic; the density annual variation in the southern hemisphere is much larger than that in the north (Fig. 11a). The explanation can be found by investigating the change of the Sun–Earth distance in a year. At June solstice, when the Arctic circle north of 66.5°N has 24 hours of daylight, the Earth is near the aphelion point where it is farthest from the Sun. At the December solstice, when the Arctic circle is having 24 hours nighttime, the Earth is near the perihelion point where it is closest to the Sun. This means that for the Arctic circle, the variation in insolation caused by the Sun–Earth distance is opposite to the annual variation, and thus partially cancels the annual temperature difference. Consequently, the annual temperature variation in the Arctic circle is smaller (~150 K) than that in the Antarctic circle (~200 K) (Fig. 11b). The smaller annual variation in density in the Arctic circle reflects the reduction in temperature annual variation in the region.

Figure 11c and 11d show the latitudinal distribution of the number densities of the two major species, O and N_2. There are two distinct features in O and N_2 latitudinal variations: (1) High latitude heating due to geomagnetic activity enhances N_2 number density whereas it decrease O number density at this fixed altitude (350 km); (2) O number density is nearly same in the winter and summer hemispheres whereas N_2 number density shows an even larger latitudinal gradient from the winter hemisphere to the summer hemisphere, compared to the temperature. This is because internal dynamics transport lighter species O (lighter than mean molecular weight) differently compared to the heavier species N_2 (heavier than mean molecular weight). There are two types of vertical motions; one is the thermal expansion and contraction and the other is due to horizontal divergence and convergence. Thermal expansion and contraction result in the rise and fall of a fixed pressure surface. In the process of expansion and contraction, different species move at the same speed, and therefore, thermal expansion and contraction does not change composition at a fixed pressure surface; Vertical motion associated with horizontal divergence and convergence, however, accumulates heavier species in the heated area but transports lighter species to the colder area. This is due to the fact that this type of closed circulation moves air parcels relative to a fixed pressure surface; energy is transported from the heated area to the colder area through adiabatic cooling in the upwelling and adiabatic heating in the downwelling. There is a return flow at lower altitude from the colder area to the heated area. Since density decreases with increasing altitude, by continuity of mass, the return flow is much slower than the flow at the greater altitude. Horizontally, due to the large difference in the speed of the two flows, lighter species that have larger scale height than the mean scale height will be effectively transported from the heated area to the colder area. The opposite is true for heavier species; the heavier species will accumulate in the heated area. Vertically, a heavier species, having smaller scale height than the mean scale height, is more abundant in the lower altitudes. The upward motion will increase its mixing ratio at higher altitude but decrease the mixing ratio of lighter species. The combined effect of thermal expansion/contraction and circulations determines the different latitudinal gradients of O and N_2 number densities shown in Fig. 11c and 11d. It is important to note that the transportation of lighter species from the summer hemisphere to the winter hemisphere by the inter-hemispheric large circulation is the mechanism of the winter anomaly (or winter bulge). Detection of winter Helium bulge by satellite drag observations (e.g., Keating and Prior 1968) and mass spectrometer data (e.g., Reber et al. 1968) revealed the winter anomaly phenomenon. Helium is considered as a minor species and is not included in the TIE-GCM. However, Helium can become a major species at ~500 km, especially in the Helium bulge in the mid- and high latitudes in the winter hemisphere, during solar minimum years.

It is important to point out that Fig. 11a again demonstrates the annual/semiannual variation of neutral density that we discussed in Sect. 3.5. Neutral density near June solstice is lower than the December solstice, and this annual asymmetry of neutral density is ~70% in this simulation which is under solar minimum condition, consistent with various observations (Bowman 2004; Müller et al. 2009; Emmert and Picone 2010). Neutral density also shows asymmetry between the two equinoxes (Fig. 11a). In this simulation with constant solar minimum forcing, neutral density near March equinox is higher than the September equinox by ~14%. This equinoctial asymmetry, with neutral density near March equinox being higher than that near the September equinox, is also observed by the CHAMP (Liu et al. 2007b; Müller et al. 2009). Furthermore, O number density shows similar annual/semiannual variations compared to the neutral density (Figs. 11a, 11c).

3.8.2 Longitudinal Variation

In order to show longitudinal variation of neutral density, we simulated neutral density under constant solar and geomagnetic forcing. We examined longitudinal variation by plotting variables at a fixed local time over a period of several days. Figure 12 shows TIE-GCM

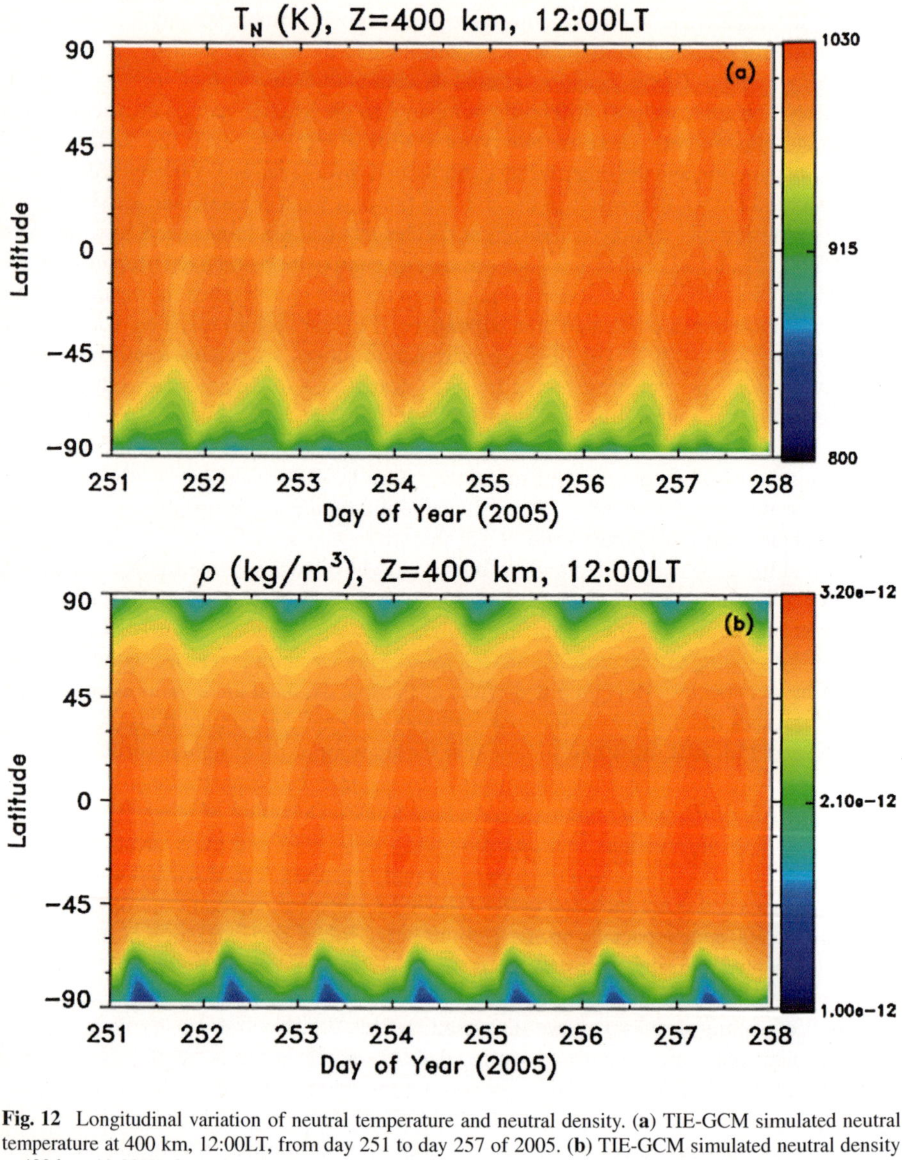

Fig. 12 Longitudinal variation of neutral temperature and neutral density. (**a**) TIE-GCM simulated neutral temperature at 400 km, 12:00LT, from day 251 to day 257 of 2005. (**b**) TIE-GCM simulated neutral density at 400 km, 12:00LT, from day 251 to day 257 of 2005

simulated neutral temperature and neutral density at 400 km, 12:00LT, from 2005251 to the beginning of 2005258 (September 8, 2005–September 16, 2005), assuming a constant solar and geomagnetic forcing. The real solar and geomagnetic forcing changed quite significantly in this period. F10.7 index was 94 on 2005251 (September 8, 2005), but was more than 110 through most of the period. In addition, there were several X-class flares on September 8 and September 9. On the other hand, Ap index was 8 on September 8, 2005, but increased to 105 on September 11, 2005. If we ran the model using the real forcing, then the longitudinal variation would be undistinguishable due to the larger density changes driven by the solar and geomagnetic forcing. We looked at CHAMP density in this period, and longitudinal variation is hard to discern due to the overwhelmingly larger density changes driven by the solar and geomagnetic forcing. In order to be able to show the longitudinal variation, we kept the solar and geomagnetic forcing to be the same as those on September 8, 2005 for the entire period. Both neutral temperature and neutral density show an apparent periodical variation, with amplitude $\sim$15%. Although the periodicity is one day, this is not the diurnal variation that we discussed in Sect. 3.3. The periodicity shown in Fig. 12 is purely a longitudinal variation. Diurnal variation of neutral density is primarily caused by diurnal variation of solar irradiance, whereas longitudinal variation of neutral density is likely due to displacement of the geomagnetic pole from the geophysical pole. It is well understood that this displacement of the two poles causes longitudinal variation in the ionosphere. Effect of this displacement on the thermosphere is not clear. Further investigations are needed to understand the possible connection between the displacement of the two poles and the longitudinal variations shown in Fig. 12. Since the ionosphere and thermosphere are closely coupled energetically and dynamically, it is likely that the longitudinal variation of energetics and dynamics, associated with longitudinal variations in electron density (conductance), Joule heating, and auroral heating, is the primary cause of the longitudinal variation in the thermosphere shown in Fig. 12.

In Sect. 3.2, we discussed the possible effect of nonmigrating tides on the diurnal variation of neutral density. The effect of nonmigrating tides can also cause longitudinal variation of neutral density. Forbes et al. (2009) analyzed CHAMP and GRACE neutral density data and found that in the low latitude band ($\pm 30°$), significant longitude variability (e.g., ± 25 K maximum to minimum over the equator) in geomagnetically-quiet exosphere temperatures exists, due to a spectrum of diurnal and semidiurnal thermal tides that are excited in the troposphere and strongly influenced by the global land-sea distribution. However, as we mentioned in Sect. 2, the effect of nonmigrating tides is not considered in the TIE-GCM, therefore, the longitudinal variation shown in Fig. 12 does not include contribution of nonmigrating tides.

There is a significant difference between latitudinal dependence of the longitudinal variation of temperature (Fig. 12a) and latitudinal dependence of longitudinal variation of neutral density (Fig. 12b). We have discussed differences in latitudinal variations of neutral temperature, neutral density, and neutral composition in previous subsection. Figure 12 again demonstrates that latitudinal gradient of composition, driven by different effect of circulations on lighter species O and heavier species N_2, compensates the latitudinal gradient of neutral temperature, and thus causes small latitudinal gradient in neutral density and its variability.

4 Summary

Determination and prediction of neutral density is critical for satellite operations including catalog maintenance, collision avoidance, and reentry prediction. Neutral density is governed by thermospheric external forcing, internal neutral dynamics, thermosphere and ionosphere coupling, and effects from the lower and middle atmosphere. The primary forcing for the thermosphere is solar irradiance ($<$175 nm). The secondary energy input for the thermosphere is geomagnetic activity through Joule heating and energetic particle precipitation. In addition to these two main external forcing, tides and waves propagating through the mesopause region provide significant forcing for the thermosphere. Variability in the external forcing, internal dynamics, and thermosphere and ionosphere coupling, drives complex density variation. This paper discussed temporal variations from abrupt changes to long-term trends, as well as spatial variations with latitude, longitude, and altitude. A summary of these variations, their amplitude (at ~400 km), and their driving forcing mechanisms are as follows:

- Abrupt changes (minutes to hours): driven by solar flares and geomagnetic storms (associated with CME and/or flares). A large (X17) centrally located solar flare can cause ~30–40% density enhancement on the day side. The flare response in neutral density can be best observed on the dayside at low to mid-latitudes. A large geomagnetic storm ($Kp = 9$) can cause density enhancement on the order of ~100–200%. Storm response is largest at high latitudes.
- Diurnal variation: driven by presence and absence of solar irradiance during day and night, and the effects of middle-atmosphere tides. The amplitude of the diurnal variation is on the order of ~100%.
- Multi-day variation: primarily driven by multi-day recurrent geomagnetic forcing associated with CIRs. The amplitude of the multi-day variation of density is on the order of ~30–50%.
- Solar-rotational variation: driven by the appearance and disappearance of the Sun's active regions as the Sun rotates in an average 27-day period. Solar-rotational variation is on the order of ~100% for high solar activity and decreases for lower solar activity.
- Annual/semiannual variation: driven by the annual variation of the Sun–Earth distance, semi-annual variation of inter-hemispheric large-scale circulation, semi-annual variation of geomagnetic activity, and likely lower atmospheric forcing through eddy diffusivity. The amplitude from annual minimum (~July) to annual maximum (equinoxes) is on the order of ~100%, but with strong inter-annual variability from ~30% to ~250%.
- Solar-cycle variation: driven by the intrinsic ~11-year variability of magnetic activity in the Sun. The amplitude of the solar-cycle variation is about one order of magnitude.
- Long-term trends: primarily driven by global cooling in the upper atmosphere due to increasing concentration of CO_2. The long-term trend of neutral density is approximately −1.7%/decade to −3.0%/decade for the past several decades.
- Latitudinal variation: driven by changes of solar irradiance in the two hemispheres associated with seasonal solar zenith angle changes, geomagnetic heating at high latitude, and internal dynamics. Neutral density shows little latitudinal variation during equinox seasons and ~40% increase from the winter pole to the summer pole.
- Longitudinal variation: likely driven by longitudinal variation of energetics and dynamics through thermosphere and ionosphere coupling, associated with longitudinal variations in electron density (conductance), Joule heating, and auroral heating. Amplitude of the longitudinal variation is on the order of ~15%. This result is from the model simulations that did not include nonmigrating tides effect. Nonmigrating tides may introduce additional longitudinal variation.
- Variation with altitude: on a global mean basis, the thermosphere is in diffusive equilibrium state; therefore, thermospheric neutral density decreases exponentially with increasing altitude, according to the diffusive equilibrium density scale height, which is determined by temperature and composition.

Understanding neutral density, its variability, and driving mechanisms provide the means to continually improve operational neutral density models for satellite operations. We have gained significant knowledge through observational and modeling studies as summarized and discussed in this paper. Examples of remaining challenges are to quantify the temporal and spatial characteristics of rapid density changes in relation to the characteristics of the solar flares and geomagnetic disturbances that drive them, to investigate how lower/middle atmosphere forcing contributes to the annual/semiannual variation of neutral density, and to determine whether the longitudinal variation of neutral density is due to magnetic field asymmetries affecting the ionosphere through thermosphere and ionosphere coupling.

Acknowledgements This research was supported by AFOSR contract FA9550-08-C0046 and NASA grants NNX08AQ31G, NNX10AF21G, and NNX07AC61G to the National Center for Atmospheric Research. NCAR is supported by the National Science Foundation. University of Colorado in Boulder supplied the CHAMP neutral density data that we used in this paper (http://sisko.colorado.edu/sutton/data.html). Bruce Bowman from the Air Force Space Command and John Emmert from the Naval Research Laboratory supplied the neutral density data at individual satellite perigees and global-mean neutral density data, respectively.

References

B.R. Bowman, The semiannual thermosphere density variation from 1970 to 2002 between 200–1100 km, AAS 2004-174, in *AAS/AIAA Spaceflight Mechanics Meeting*, Maui, Hawaii, 8–12 February (2004)

B.R. Bowman, F.A. Marcos, M. Kendra, A method for computing accurate daily atmospheric density values from satellite drag data, AAS 2004-173, in *AAS/AIAA Spaceflight Mechanics Meeting*, Maui, Hawaii, 8–12 February (2004)

B.R. Bowman, W.K. Tobiska, F.A. Marcos, C. Valladares, The JB2006 empirical thermospheric density model. J. Atmos. Sol.-Terr. Phys. **70**, 774–793 (2007). doi:10.1016/j.jastp.2007.10.002

B.R. Bowman et al., The thermospheric semiannual density response to solar EUV heating. J. Atmos. Sol.-Terr. Phys. (2008). doi:10.1016/j.jastp.2008.04.020

S. Bruinsma, J.M. Forbes, R.S. Nerem, X. Zhang, Thermosphere densit response to the 20–21 November 2003 solar and geomagnetic storm from CHAM and GRACE accelerometer data. J. Geophys. Res. **111**, A06303 (2006). doi:10.1029/2005JA011284

A.G. Burns, T.L. Killeen, W. Wang, R.G. Roble, The solar-cycle-dependent response of the thermosphere to geomagnetic storms. J. Atmos. Sol.-Terr. Phys. **66**, 14 (2004)

P.C. Chamberlin, T.N. Woods, F.G. Eparvier, Flare Irradiance Spectral Model (FISM): Flare component algorithms and results. Space Weather **6**, S05001 (2008). doi:10.1029/2007SW000372

R.R. Clark, M.D. Burrageb, S.J. Frankec, A.H. Mansond, C.E. Meekd, N.J. Mitchelle, H.G. Mullerf, Observations of 7-d planetary waves with MLT radars and the UARS-HRDI instrument. J. Atmos. Sol.-Terr. Phys. **64**, 1217–1228 (2002)

K. Davies, *Ionospheric Radio* (Peter Peregrinus, London, 1990)

Y. Deng, T.J. Fuller-Rowell, R.A. Akmaev, A.J. Ridley, Impact of the altitudinal Joule heating distribution on the thermosphere. J. Geophys. Res. (2010, submitted)

T.R. Detman, Cross validation comparisons of autonomous Ap predictions, in *Proceedings of Workshop on the Evaluation of Space Weather Forecasts*, ed. by K. Doggett (1996), p. 149, NOAA, ERL, Boulder, Colo

B.A. Emery, I.G. Richardson, D.S. Evans, F.J. Rich, Solar wind structure sources and periodicities of auroral electron power over three solar cycles, J. Atmos. Terr. Phys. **71** (2009). doi:10.1016/j.jastp.2008.08.005

J.T. Emmert, J.M. Picone, J.L. Lean, S.H. Knowles, Global change in the thermosphere: compelling evidence of a secular decrease in density. J. Geophys. Res. **109**, A02301 (2004)

J.T. Emmert, J.M. Picone, R.R. Meier, Thermospheric global average density trends 1967–2007, derived from orbits of 5000 near-Earth objects. Geophys. Res. Lett. **35**, L05101 (2008). doi:10.1029/2007GL032809

J.T. Emmert, A long-term data set of globally averaged thermospheric total mass density. J. Geophys. Res. **114**, A06315 (2009). doi:10.1029/2009JA014102

J.T. Emmert, J.L. Lean, J.M. Picone, Record-low thermospheric density during the 2008 solar minimum. Geophys. Res. Lett. **37**, L12102 (2010). doi:10.1029/2010GL043671

J.T. Emmert, J.M. Picone, Climatology of globally averaged thermospheric mass density. J. Geophys. Res. **115**, A09326 (2010). doi:10.1029/2010JA015298

J.M. Forbes, S.L. Bruinsma, X. Zhang, J. Oberheide, Surface-exosphere coupling due to thermal tides. Geophys. Res. Lett. **36**, L15812 (2009). doi:10.1029/2009GL038748

J.M. Forbes, X. Zhang, E.R. Talaat, W. Ward, Nonmigrating diurnal tides in the thermosphere. J. Geophys. Res. A **108**(1), 1033 (2003). doi:10.1029/2002JA009262

T.J. Fuller-Rowell, The "thermospheric spoon": A mechanism for the semiannual density variation. J. Geophys. Res. **103**, 3951–3956 (1998)

M.E. Hagan, C. McLandress, J.M. Forbes, Diurnal tidal variability in the upper mesosphere and lower thermosphere. Ann. Geophys. **15**, 1176–1186 (1997). doi:10.1007/s00585-997-1176-x

M.E. Hagan, J.M. Forbes, Migrating and nonmigrating diurnal tides in the middle and upper atmosphere excited by tropospheric latent heat release. J. Geophys. Res. D **107**(24), 4754 (2002). doi:10.1029/2001JD001236

M.E. Hagan, J.M. Forbes, Migrating and nonmigrating semidiurnal tides in the upper atmosphere excited by tropospheric latent heat release. J. Geophys. Res. A **108**(2), 1062 (2003). doi:10.1029/2002JA009466

A.E. Hedin, Extension of the MSIS thermosphere model into the middle and lower atmosphere. J. Geophys. Res. **96**, 1159–1172 (1991)

R.A. Heelis, J.K. Lowell, R.W. Spiro, A model of the highlatitude ionospheric convection pattern. J. Geophys. Res. **87**, 6339–6345 (1982)

L.G. Jacchia, Static diffusion models of the upper atmosphere with empirical temperature profiles. Smithson. Contrib. Astrophys. **8**, 215–257 (1965)

L.G. Jacchia, Semiannual variation in the heteorosphere: A reappraisal. J. Geophys. Res. **76**, 4602–4607 (1971)

C. Jacobi, R. Schminder, D. Kürschner, Planetary-wave activity obtained from long-period (2–18 days) variations of mesopause region winds over Central Europe (52°N, 15°E), J. Atmos. Sol.-Terr. Phys. **60**(1), 81–93 (1998)

G.M. Keating, E.J. Prior, The winter He bulge. Space Res. **8**, 982 (1968)

G.M. Keating, R.H. Tolson, M.S. Bradford, Evidence of long-term global decline in the Earth's thermospheric densities apparently related to anthropogenic effects. Geophys. Res. Lett. **27**, 1523–1526 (2000)

D.J. Knipp, W.K. Tobiska, B.A. Emery, Direct and indirect thermosphere heating sources for solar cycle 21–23. Sol. Phys. **224**, 495–505 (2004)

J. Laštovička, R.A. Akmaev, G. Beig, J. Bremer, J.T. Emmert, Global change in the upper atmosphere. Science **314**, 1253–1254 (2006)

J.L. Lean, J.T. Emmert, J.M. Picone, R.R. Meier, Global and regional trends in ionospheric total electron content. J. Geophys. Res. **116**, A00H04 (2011). doi:10.1029/2010JA016378

J. Lei, J.P. Thayer, J.M. Forbes, E.K. Sutton, R.S. Nerem, Rotating solar coronal holes and periodic modulation of the upper atmosphere. Geophys. Res. Lett. **35**, L10109 (2008). doi:10.1029/2008GL033875

H. Liu, H. Lühr, Strong disturbance of the upper thermospheric density due to magnetic storms: CHAMP observations. J. Geophys. Res. **110**, A09S29 (2005). doi:10.1029/2004JA010908

H. Liu, H. Lühr, S. Watanabe, W. Köhler, C. Manoj, Contrasting behavior of the thermosphere and ionosphere in response to the 28 October 2003 solar flare. J. Geophys. Res. **112**, A07305 (2007a). doi:10.1029/2007JA012313

H. Liu, H. Lühr, S. Watanabe, Climatology of the equatorial thermospheric mass density anomaly. J. Geophys. Res. **112**, A05305 (2007b). doi:10.1029/2006JA012199

R. Liu, H. Lühr, E. Doornbos, S.-Y. Ma, Thermospheric mass density variations during geomagnetic storms and a prediction model based on the merging electric field. Ann. Geophys. **28**, 1633–1645 (2010). doi:10.5194/angeo-28-1633-2010

R. Liu, S.-Y. Ma, H. Lühr, Predicting storm-time thermospheric mass density variations at CHAMP and GRACE altitudes. Ann. Geophys. **29**, 443–453 (2011). doi:10.5194/angeo-29-443-2011

H. Lühr, M. Rother, W. Köhler, P. Ritter, L. Grunwaldt, Thermospheric up-welling in the cusp region: Evidence from CHAMP observations. Geophys. Res. Lett. **31**, L06805 (2004). doi:10.1029/2003GL019314

F.A. Marcos, J.O. Wise, M.J. Kendra, N.J. Grossbard, B.R. Bowman, Detection of a long-term decrease in thermospheric neutral density. Geophys. Res. Lett. **32**, L04103 (2005). doi:10.1029/2004GL021269

M. Mendillo et al., Behavior of the ionospheric F region during the great solar flare of August 7, 1972. J. Geophys. Res. **79**(4), 665–672 (1974)

A.P. Mitra, *Ionospheric Effects of Solar Flares* (Reidel, Boston, 1974)

S. Müller, H. Lühr, S. Rentz, Solar and magnetospheric forcing of the low latitude thermospheric mass density as observed by CHAMP. Ann. Geophys. **27**, 2087–2099 (2009)

H.K. Paetzold, H. Zschörner, An annual and a semiannual variation of the upper air density. Pure Appl. Geophys. **48**, 85–92 (1961)

D.J. Pawlowski, A.J. Ridley, Modeling the thermospheric response to solar flares. J. Geophys. Res. **113**, A10309 (2008). doi:10.1029/2008JA013182

J.M. Picone, A.E. Hedin, D.P. Drob, A.C. Aikin, NRLMSISE-00 empirical model of the atmosphere: Statistical comparisons and scientific issues. J. Geophys. Res. A **107**(12), 1468 (2002). doi:10.1029/2002JA009430

G.W. Prölss, Density perturbations in the upper atmosphere caused by the dissipation of solar wind energy. Surv. Geophys. **32**, 101–195 (2011). doi:10.1007/s10712-010-9104-0

L. Qian, R.G. Roble, S.C. Solomon, T.J. Kane, Calculated and observed climate change in the thermosphere, and a prediction for solar cycle 24. Geophys. Res. Lett. **33**, L23705 (2006). doi:10.1029/2006GL027185

L. Qian, S.C. Solomon, R.G. Roble, B.R. Bowman, F.A. Marcos, Thermospheric neutral density response to solar forcing. Adv. Space Res. **42**(5), 926–932 (2008). doi:10.1016/j.asr.2007.10.019

L. Qian, S.C. Solomon, T.J. Kane, Seasonal variation of thermospheric density and composition. J. Geophys. Res. **114**, A01312 (2009). doi:10.1029/2008JA013643

L. Qian, A.G. Burns, P.C. Chamberlin, S.C. Solomon, Flare location on the solar disk: Modeling the thermosphere and ionosphere response. J. Geophys. Res. **115**, A09311 (2010a). doi:10.1029/2009JA015225

L. Qian, S.C. Solomon, M.G. Mlynczak, Model simulation of thermospheric response to recurrent geomagnetic forcing. J. Geophys. Res. **115**, A10301 (2010b). doi:10.1029/2010JA015309

L. Qian, A.G. Burns, P.C. Chamberlin, S.C. Solomon, Variability of thermosphere and ionosphere responses to solar flares. J. Grophys. Res. (2011, accepted)

C.A. Reber, J.E. Cooley, D.N. Harpold, Upper atmosphere hydrogen and helium measurements from the Explorer 32 satellite. Space Res. **8**, 993 (1968)

C. Reigber, H. Lühr, P. Schwintzer, CHAMP mission status. Adv. Space Res. **30**(2), 129–134 (2002)

P.G. Richards, J.A. Fennelly, D.G. Torr, EUVAC: A solar EUV flux model for aeronomic calculations. J. Geophys. Res. **99**, 8981–8992 (1994)

A.D. Richmond, E.C. Ridley, R.G. Roble, A thermosphere/ionosphere general circulation model with coupled electrodynamics. Geophys. Res. Lett. **19**, 601–604 (1992)

H. Rishbeth, A greenhouse effect in the ionosphere? Planet. Space Sci. **38**, 945–948 (1990)

H. Rishbeth, R.G. Roble, Cooling of the upper atmosphere by enhanced greenhouse gases—modelling of thermospheric and ionospheric effects. Planet. Space Sci. **40**, 1011–1026 (1992)

R.G. Roble, E.C. Ridley, An auroral model for the NCAR thermosphere general circulation model (TGCM). Ann. Geophys. A **5**(6), 369–382 (1987)

R.G. Roble, E.C. Ridley, A.D. Richmond, R.E. Dickinson, A coupled thermosphere/ionosphere general circulation model. Geophys. Res. Lett. **15**, 1325 (1988). doi:10.1029/GL015i012p01325

R.G. Roble, R.E. Dickinson, How will changes in carbon dioxide and methane modify the mean structure of the mesosphere and lower thermosphere? Geophys. Res. Lett. **16**, 1441–1444 (1989)

C.T. Russell, R.L. McPherron, Semiannual variation of geomagnetic activity. J. Geophys. Res. **78**, (1), 92–108 (1973)

S.C. Solomon, L. Qian, Solar extreme-ultraviolet irradiance for general circulation models. J. Geophys. Res. **110**, A10306 (2005). doi:10.1029/2005JA011160

S.C. Solomon, T.N. Woods, L.V. Didkovsky, J.T. Emmert, L. Qian, Anomalously low solar extreme-ultraviolet irradiance and thermospheric density during solar minimum. Geophys. Res. Lett. **37**, L16103 (2010). doi:10.1029/2010GL044468

S.C. Solomon, L. Qian, L.V. Didkovsky, R.A. Viereck, T.N. Woods, Causes of low thermospheric density during the 2007–2009 solar minimum. J. Geophys. Res. **116** (2011). doi:10.1029/2011JA016508

E.K. Sutton, J.M. Forbes, R.S. Nerem, Global thermospheric neutral density and wind response to the severe 2003 geomagnetic storms from CHAMP accelerometer data. J. Geophys. Res. **110**, A09S40 (2005). doi:10.1029/2004JA010985

M. Temmer, B. Vrsnak, A.M. Veronig, Periodic appearance of coronal holes and the related variation of solar wind parameters. Sol. Phys. **241**, 371–383 (2007)

B.T. Tsurutani, W.D. Gonzalez, A.L.C. Gonzalez, F. Tang, J.K. Arballo, M. Okada, Interplanetary origin of geomagnetic activity in the declining phase of the solar cycle. J. Geophys. Res. A **100**(11), 21717–21733 (1995)

B.T. Tsurutani et al., Extreme solar EUV flares and ICMEs and resultant extreme ionospheric effects: comparison of the Halloween 2003 and the Bastille Day events. Radio Sci. **41**, RS5S07 (2006). doi:10.1029/2005RS003331

R.L. Walterscheid, The semiannual oscillation in the thermosphere as a conduction mode. J. Geophys. Res. **87**, 10527–10535 (1982)

W. Wang, A.G. Burns, M. Wiltberger, S.C. Solomon, T.L. Killeen, Altitude variations of the horizontal thermospheric winds during geomagnetic storms. J. Geophys. Res. **113**, A02301 (2008). doi:10.1029/2007JA012374

G.R. Wilson, D.R. Weimer, J.O. Wise, F.A. Marcos, Response of the thermosphere to Joule heating and particle precipitation. J. Geophys. Res. **111**, A10314 (2006). doi:10.1029/2005JA011274

T.N. Woods, F.G. Eparvier, S.M. Bailey, P.C. Chamberlin, J. Lean, G.J. Rottman, S.C. Solomon, W.K. Tobiska, D.L. Woodraska, Solar EUV Experiment (SEE): Mission overview and first results. J. Geophys. Res. **110**, A01312 (2005). doi:10.1029/2004JA010765

D.H. Zhang, Z. Xiao, Study of ionospheric response to the 4B flare on 28 October 2003 using Internation GPS Service network data. J. Geophys. Res. **110**, A03307 (2005). doi:10.1029/2004JA010738

D.H. Zhang, X.H. Mo, L. Cai, W. Zhang, M. Feng, Y.Q. Hao, Z. Xiao, Impact factor for the ionospheric total electron content response to solar flare irradiation. J. Geophys. Res. **116**, A04311 (2011). doi:10.1029/2010JA016089

Space Sci Rev (2012) 168:175–209
DOI 10.1007/s11214-011-9837-1

Global Response of the Ionosphere to Atmospheric Tides Forced from Below: Recent Progress Based on Satellite Measurements

Global Tidal Response of the Ionosphere

Dora Pancheva · Plamen Mukhtarov

Received: 1 May 2011 / Accepted: 12 September 2011 / Published online: 11 October 2011

Abstract This paper provides an overview on the recent progress in studying the ionospheric response to atmospheric tides forced from below. The global spatial structure and temporal variability of the atmospheric temperature tides and their ionospheric responses are considered on the basis of modern satellite-board data (COSMIC and TIMED). The tidal waves from the two data sets have been extracted by one and the same data analysis method. The similarity between the lower thermospheric temperature tides and their ionospheric responses provides evidence for confirming the new paradigm of atmosphere-ionosphere coupling. This paper provides also new experimental results which give an explanation why the WN4 and partly WN3 longitude structures are so prominent pattern in the ionosphere. These results present evidence indicating that the WN4 (WN3) structure is not generated only by the DE3 (DE2) tide as it has been often assumed. The DE3 (DE2) tide remains the leading contributor, but the SPW4 and SE2 (SPW3, DW4 and SE1) waves have their effects as well in a way that the ionospheric response becomes almost double (1.5 time stronger). The paper presents also the global distribution and temporal variability of the sun-synchronous 24-h (DW1), 12-h (SW2) and 8-h (TW3) electron density oscillations. It has been shown that while the latitude and altitude structure of the ionospheric SW2 response is predominantly shaped by the migrating SW2 tide forced from below the DW1 response is mainly due to daily variability of the photo-ionization. The peculiar vertical structure of the ionospheric TW3 response, that shows downward/upward phase progression, calls for further study of the physical processes shaping this ionospheric response.

Keywords Nonmigrating tides · Ionospheric response · Modulated vertical plasma drift · "Fountain effect"

1 Introduction

The conventional sources of ionospheric structure and variability are related to changes in solar radiative output and geomagnetic activity, together with the subsequent response of

D. Pancheva (✉) · P. Mukhtarov
Geophysical Institute, Bulg. Acad. Sciences, Sofia, Bulgaria
e-mail: dpancheva@geophys.bas.bg

the thermosphere and ionosphere system and interaction between the components. The high sensitivity of the ionosphere to the external forcing causes its significant variability on time scales ranging from minutes to a solar cycle. To understand and forecast such variability is one of the main tasks of space weather research. Often, and especially recently when the level of solar activity is very low, quite large day-to-day changes of the ionosphere have been observed. With the recent accumulation of satellite measurements, attention is now being directed towards investigating the impact of the processes from below and particularly the wave forcing from the lower atmosphere.

The low-latitude ionospheric morphology is greatly affected by the electric fields and neutral winds which define its dependence on latitude, longitude and local time. The neutral wind and electric field effects on the ionosphere are variable with geomagnetic field configuration (magnetic field strength, magnetic declination, and displacement of the geomagnetic equator from the geographic equator) and this is the reason for many years the ionospheric longitudinal variations to be understood mainly in the context of geomagnetic field configuration (West and Heelis 1996; Thuillier et al. 2002; Kil et al. 2006). Recently, however, various ionospheric observations have shown the development of longitudinal wave-like patterns (as wave number three (WN3), or wave number four (WN4) structures) that cannot be explained by the geomagnetic field configuration. The wave-like longitudinal plasma density pattern was first reported by topside soundings from the Russian satellite Intercosmos-19 (Kochenova 1987, 1988; Karpachev 1988). Intercosmos-19 observed the formation of four peaks in plasma density (WN4 structure at a fixed local time frame) over the magnetic equator (Benkova et al. 1990), but the focus of this study was on the displacement of the geomagnetic equator from the geographic equator and the hemispheric asymmetry induced by neutral winds.

It has been mentioned above that the Earth's ionosphere is closely connected to atmospheric waves propagating from below. Among them are solar thermal tides which are global-scale atmospheric oscillations in all atmospheric fields, including wind, temperature, density and pressure, with periods that are subharmonics of a solar day. These waves profoundly affect the large-scale dynamics of the mesosphere and lower thermosphere (MLT) where they attain large amplitudes and dominate the large-scale atmospheric fields. Like other waves the tidal components grow in amplitude with increasing altitude since atmospheric density decreases and energy must be conserved. Generally, the tidal amplitudes are the largest of all the atmospheric oscillations in the MLT region, including also gravity and planetary waves. The major source of forcing at tidal periods in the lower and middle atmosphere is atmospheric heating due to absorption of solar radiation by zonally symmetric atmospheric components as stratospheric ozone and tropospheric water vapor. The resulting atmospheric variations are reflected in migrating tides which propagate with the apparent motion of the Sun to a ground-based observer, i.e. they are a function of local time alone (Chapman and Lindzen 1970). Other sources may contribute to tidal variations that depend explicitly on longitude and in this way resulting in nonmigrating tides (all tides propagating eastward, zonally symmetric tides and those propagating westward but with a different phase speed from that of the apparent motion of the Sun). It is now generally accepted that nonmigrating tides arise from at least two mechanisms: zonally asymmetric thermal forcing (surface topography, geographically varying heat sources, variation of solar heating with longitude) and nonlinear interactions between migrating tides and stationary planetary waves (SPWs) (Angelats i Coll and Forbes 2002; Pancheva et al. 2009a). For instance, the release of latent heat by deep convective clouds is dependent on universal time (UT), longitude, latitude and season hence it is a possible source that could generate a nonnegligible tidal migrating and nonmigrating response in the MLT (Hamilton 1981; Williams and Avery 1996;

Hagan 1996; Hagan and Forbes 2002, 2003; Oberheide et al. 2002). McLandress and Ward (1994) suggested that the interaction of a zonally asymmetric distribution of the gravity waves interacting with migrating tides could also generate nonmigrating tides.

Due to increasing eddy and molecular dissipation the lower thermosphere is a region where the atmospheric tides propagating from below usually damp. Dissipating tides deposit net momentum and energy into the mean flow and thus modify the mean circulation at these altitudes. They can also mix the atmospheric species there or nonlinearly interact with other waves generating additional wave activity (Forbes et al. 1993). Through impacting the wind fields and mixing, the tides can affect the density of different neutral species like NO, O, CO_2 and O_3 and in this way affect the energy budget either by radiating, or by absorbing solar radiation. The tides penetrating in the ionospheric E-region (∼100–170 km) can also affect the overlying ionosphere F-region by the so called "E-region wind dynamo". In this way the neutral dynamics renders the lower thermosphere the strongest driver from below of the ionosphere F-region. The "E-region wind dynamo" is composed by the following sequence of processes (Richmond 1995; Heelis 2004; Forbes et al. 2008): (i) the tidal winds move positive ions through collisions while electrons remain fixed to magnetic field lines due to their high gyrofrequency/collision frequency ratio; (ii) an electric current is thus induced; (iii) in order to maintain nondivergent flow of total electric current in accord with Maxwell's steady-state equations, polarization electric fields are set up almost instantaneously; (iv) these polarization electric fields are transmitted from the low- and middle-latitude E-region to the equatorial F-region along equipotential magnetic field lines and (v) there, they modulate the background zonal electric field and the accompanying $E \times B$ vertical drift that redistributes plasma upward and poleward to produce the equatorial ionization anomaly (EIA) near ±15–20° dip latitude. Additionally, recent investigations showed that the atmospheric tides can propagate also directly up to thermospheric heights (Oberheide et al. 2009). Therefore, either through complex coupling mechanisms or through direct propagation, the MLT tides can affect the whole ionosphere-thermosphere system.

Recent studies based on the observations made by the Sounding of the Atmosphere using Broadband Emission Radiometry (SABER) and the TIMED Doppler Interferometer (TIDI) instruments on the Thermosphere-Ionosphere-Mesosphere-Energetics and Dynamics (TIMED) satellite have provided new insight into tidal fields and revealed the global distribution and climatology of the most important tidal components in temperature and neutral winds respectively (e.g., Zhang et al. 2006; Huang et al. 2006a, 2006b; Forbes et al. 2006, 2008; Oberheide et al. 2007; Oberheide and Forbes 2008; Mukhtarov et al. 2009; Xu et al. 2009; Pancheva et al. 2009b, 2010a; Pancheva and Mukhtarov 2011). Some of the above mentioned studies clearly suggested that the nonmigrating tides are much larger than previous anticipated and often exceed the migrating tide counterparts in the lower thermosphere (e.g., Zhang et al. 2006; Forbes et al. 2006, 2008; Oberheide et al. 2006, 2007; Pancheva et al. 2010b). The nonmigrating tides give rise to longitudinal variations in local time structures of neutral winds and temperatures in the MLT region. Through coupling mechanisms (such as the wind dynamo), the nonmigrating tides can play an important role in generating the longitudinal variability in the ionosphere. In this way, a special attention has been paid recently on the nonmigrating tides as a possible link between the troposphere and ionosphere.

The last several years have seen an abundance of new observational evidence illustrating terrestrial weather impacts on the ionosphere. The evidence has emerged from different measurements and all of them unambiguously display manifestations of lower atmospheric dynamics on the upper atmosphere and ionosphere. Such illustrations are various ionospheric observations which show the development of longitudinal WN3 or WN4 structures which are suggested to be caused by different atmospheric tides. Sagawa et al. (2005) observed four enhanced longitude regions of the nighttime airglow intensity in the EIA and pointed out that the WN4 longitudinal structure of the plasma density is most probably forced by the atmospheric tides. The authors suggested that non-migrating tides that propagate from lower altitudes to the E region can modulate the E-region dynamo electric field and produce the observed WN4 density pattern in the F region. Later Immel et al. (2006) and England et al. (2006a) specified that the diurnal eastward propagating zonal wave number 3 (DE3, D corresponds to diurnal tide and E to eastward propagation) tide is responsible for the generation of the ionospheric WN4 density structure. We note that when viewed at a fixed local time the DE3 wave is seen as four maxima on longitude in all tidal fields. As evidence of modulation in the dynamo, a WN4 structure has been observed in $E \times B$ plasma drift (Hartman and Heelis 2007; Kil et al. 2007, 2008; Ren et al. 2008, 2009) and in the equatorial electrojet (EEJ) (England et al. 2006b; Lühr et al. 2008). The satellite observations of different ionospheric and thermospheric parameters, such as: total electron content (TEC), the electron temperature, F-region zonal neutral wind and neutral mass density, the ionospheric F2 layer height (h_mF2), also have shown WN4 longitude structures (e.g., Lin et al. 2007a, 2007b; Lühr et al. 2007; Häusler and Lühr 2009; Wan et al. 2008; Fejer et al. 2008; Bankov et al. 2009; Liu et al. 2009; Pancheva and Mukhtarov 2010).

Model simulations support the causal link between DE3 tide and WN4 ionospheric longitude structure. Hagan et al. (2007) used the GSWM tidal winds as input to the Thermosphere-Ionosphere-Mesosphere-Electrodynamics General Circulation Model (TIME-GCM) and identified the DE3 zonal wind as the cause of the formation of WN4 structure. Jin et al. (2008) and Ren et al. (2010) on the basis of electrodynamic simulations pointed out also that the charge separation induced by the Hall current driven by DE3 tide winds causes the WN4 structure in the E-region zonal electric field. With the influence of the coupling of ionospheric electric fields along the highly conducting magnetic field lines Ren et al. (2010) found that the symmetric wind component of DE3 tide plays a more important role in the WN4 structure in the vertical $E \times B$ plasma drift than does the antisymmetric component. For understanding the atmospheric vertical coupling, global first-principal models treating both atmospheric and ionospheric regions, like the above mentioned TIME-GCM (Roble 1996) are quite useful. Such models are: the TIME-GCM/CCM3, which couples TIME-GCM and a lower atmospheric model of the Climate Community Model version 3 (CCM3) (Roble 2000), the Whole Atmosphere Model and a Global Ionosphere Plasmosphere Model (Fuller-Rowell et al. 2008), and the model called GAIA which combines three models, that is, a whole atmospheric GCM, an ionospheric model and an electrodynamics model (Jin et al. 2011).

All the above mentioned studies clearly indicated that the WN4 longitude structure is a very prominent feature of the low-latitude ionosphere-thermosphere system that in general can be reproduced by the models. So far however, the ionospheric wave-like longitude structures have been investigated only at fixed local times (LT) most probably because they were observed by quasi-sun-synchronous satellites. In this case however the observed structure cannot be directly related to a certain tidal component. For example, if we consider only diurnal and semidiurnal tides, the WN4 structure can be forced by: DE3, DW5, SW6, SE2 and SPW4 (according to the standard nomenclature S corresponds to semidiurnal tide, while W to westward propagation; SPW means the stationary planetary wave) and the WN3 structure by: DW4, DE2, SW5, SE1 and SPW3 (Häusler and Lühr 2009) and any of these waves could in principal contribute to the ionosphere WN4/WN3 variability. Its relative importance, of course, depends on its magnitude in the E-region and characteristics such as latitudinal shape and vertical wavelength (Oberheide et al. 2011). Usually the annual and LT variations of the WN4 structure indicate the prevailing effect of the DE3 tide, but the contribution of the other aforementioned tides and SPW4 cannot be excluded. The formation mechanism of the WN3 structure, that is observed mainly during the Northern hemisphere (NH) winter (England et al. 2009; Pancheva and Mukhtarov 2010), has not yet been clarified. However Pedatella et al. (2008) have already made the point that the WN3 is mainly due to the DE2 by comparing the CHAMP CTR data to the TIDI neutral wind tides in the E-region. Similar to the WN4 structure, in this case also we cannot exclude the contribution of the other above mentioned tides and SPW3.

The basic aim of the present paper is to study the global distribution and temporal variability of the ionospheric response to atmospheric tides forced from below. Only when the ionospheric response to each wave forced from below is known then the relative contribution and importance of the above mentioned waves in generating the ionospheric longitudinal WN4 or WN3 structures can be studied. For this purpose the Constellation Observing System for Meteorology, Ionosphere, and Climate (COSMIC) electron densities at given heights are utilized in order to define the ionospheric wave response. The wave forcing from below, e.g. the lower thermospheric waves are defined by using the SABER/TIMED temperatures. The period of time 1 October 2007–31 March 2009 is considered in the present study; it covers solar minimum conditions when the tidal damping in the lower thermosphere should be weaker because of the reduced tidal dissipation.

2 Observations and Method for Data Analysis

2.1 COSMIC Electron Density Profiles Data

The electron density profiles employed in this study were retrieved by the advanced GPS receivers on board of COSMIC satellites using a radio occultation (RO) inversion technique (Cheng et al. 2006). Six identical microsatellites constitute the COSMIC constellation system. They were initially launched (on April 2006) into the same orbit, and then during the following 13 months, they were gradually spread to individual orbits at a higher altitude of about 800 km. Since late 2007, the satellites have operated on their final orbits with the inclination of 72° and 30° separation in longitude. Each COSMIC satellite has a GPS occultation experiment payload, performing the radio occultation observations in the ionosphere. The details of the inversion technique applied to invert the COSMIC occultation soundings to ionospheric electron density profiles are given by Schreiner et al. (1999) and Lei et al. (2007). The COSMIC satellites now provide approximately 24 h of LT coverage globally and about 2200 vertical electron density profiles per day. The ionospheric measurements of COSMIC have been assessed by comparing with ground-based measurements and model prediction (Lei et al. 2007). The advantages of making use of the COSMIC-measured data to study the F-region electron density variability are nearly uniformly global distribution of the electron density profiles with sufficiently fine height and horizontal resolutions, which cannot be achieved for the data taken from the conventional ground-based ionosonde networks. In this way the COSMIC GPS occultation experiment has proven to be a powerful tool in probing vertical profiles of the ionospheric electron density by its global-coverage observations.

The electron density data downloaded from the web site http://cosmic-io.cosmic.ucar.edu/cdaac/ had to be preprocessed before using them for the wave analysis. The preprocessing steps included removing: (i) negative values of the electron densities; (ii) electron densities corresponding to plasma frequencies larger than 20 MHz (in this study the electron density is measured by its plasma frequency in MHz); and (iii) spurious spikes. The electron density data from September 2007 to April 2009 are used in this study. The data were sorted out according to: (i) latitude, at each 10° between 80°S and 80°N; (ii) longitude, at each 15°; (iii) altitude, at each 25 km between 100 and 800 km; and (iv) time, for each UT hour. Finally, for each latitude and height the electron density data were arranged into a matrix with 24 columns (this is the longitude with a step of 15°) and a number of rows equal to the length of the considered period of time in hours. It has been mentioned that the ionospheric tidal response will be studied by considering the electron densities at fixed altitudes between 100 and 800 km. Besides them the F-region is characterized also by its maximum electron density (f_oF2), the h_mF2 and the bottom and top coefficients describing the respective vertical gradients of the electron density, called slope coefficients. The parameters f_oF2 and h_mF2 are defined by their real values from the profile, however to avoid the influence of the sporadic E_s layers they are found for altitudes higher than 175 km. The bottom and top slope coefficients are defined as scale heights B calculated for 50 km below and 100 km above the h_mF2 as the electron density profile is approximated by:

$$N(h) = N_m \sec h^2\left(\frac{h-h_m}{2B}\right) = \frac{4N_m}{(1+\exp(\frac{h-h_m}{B}))^2}\exp\left(\frac{h-h_m}{B}\right) \tag{1}$$

where B is a scale height and N and N_m are electron density at a given height h and at electron density peak height h_m respectively. The above approximation is used in modeling electron density profiles (Radicella and Leitinger 2001) or in the reconstruction of the electron density profiles in the topside ionosphere and plasmosphere (Stankov et al. 2003).

2.2 SABER Temperature Data

The TIMED satellite was launched on 7 December 2001 and the SABER instrument began making observations in January 2002. It measures CO_2 infrared limb radiance from approximately 20 km to 120 km altitude and kinetic temperature profiles are retrieved over these heights using LTE radiative transfer in the stratosphere and lowest part of the mesosphere (up to ~60 km), and a full non-LTE inversion in the MLT (Mertens et al. 2001, 2004). SABER views the atmosphere at an angle of 90° with respect to the satellite velocity vector in a 625 km height and 74° inclination orbit so that the latitude coverage on a given day extends from about 53° in one hemisphere to 83° in the other. About every 60 days, the latitude ranges flip as the spacecraft yaws to keep the instrument on the anti-sunward side of the spacecraft. In this way high latitude data are available only in 60-day segments, with no information for the 60 days preceding or following. This is the reason not to generate results poleward of 50°. The TIMED orbit precesses and covers 12 hours of LT in each 60-day yaw period, so that ascending and descending data together give almost 24 hours of LT sampling.

Our results are derived from the latest Version 1.07 of the SABER data, which were downloaded from the web site: http://saber.gats-inc.com. The details about the assessment of the SABER temperatures can be found in Remsberg et al. (2008). We use temperature data from September 2007 to April 2009 in order to study the temporal variability and global spatial structure of the atmospheric tides forced from below. The data were averaged into 5 km altitude and 10° latitude bins and each bin was independently fit. We work in UT and for each altitude (from 20 to 120 km in 5 km steps) and latitude (from 50°S to 50°N in 10° steps) the data were arranged into a matrix with 24 columns (this is the longitude with a step of 15°) and the number of rows is equal to the length of the considered period of time in hours.

2.3 Method for Extracting Waves from COSMIC Electron Density

The basic aim of this study is to obtain the global spatial structure and temporal variability of the ionospheric response to the known waves (defined by SABER temperature measurements) which propagate from the troposphere and reach the lower thermosphere. This means that we have to separate from the COSMIC electron density data the ionospheric oscillations with known periods and zonal structures. The known waves are solar tides with periods of 24, 12 and 8 hours and with zonal wavenumbers up to 4, as well as planetary waves (stationary and propagating) with zonal wavenumbers up to 4. The preliminary 2D spectral analysis of the COSMIC and SABER data revealed relatively weak zonally travelling planetary waves during the considered period of time. Only the zonally symmetric waves (zonal wavenumber zero) possessed large amplitudes, but it was found that they are due to external forcing (Lei et al. 2008). Therefore, the wanted tidal and planetary wave ionospheric responses are obtained by using the same method as that used for deriving the atmospheric waves from the SABER temperatures. This data analysis method is described in detail by Pancheva et al. (2009c). The advantages of this method are as follows: (i) to avoid the constraint of sampling uniformity, instead of a Fourier transform a linear two-dimensional (time-longitude) least squares fitting technique is used to extract the waves, (ii) to avoid a possible distortion of the weaker planetary waves by the stronger ones, all significant planetary waves are extracted simultaneously from the data, and (iii) to avoid the aliasing between the nonmigrating tides and the SPWs or between the migrating tides and the zonal mean, the tides (migrating and nonmigrating) are extracted simultaneously with the planetary waves.

Because the same method is used for the COSMIC and SABER data, and because it takes SABER 60 days to sample 24 hours in LT by combining ascending and descending data together, the length of the sliding window used for performing the 2D least-squares fitting procedure is 60 days. The 60-day window is then moved through the time series with steps of 24 hours in order to obtain the daily values of the wave characteristics for period of time October 2007–March 2009. Only continuous latitudinal results are considered in this study; for COSMIC electron density data, they are between 70°S and 70°N while for SABER temperature data, they are between 50°S and 50°N. It is worth noting that the use of a 60-day segment gives too smoothed ionospheric tidal response. However, because the emphasis is on the ionospheric response at low and middle latitudes, this helps to obtain reliable results as the number of the COSMIC measurements at low latitudes is smaller than that at high latitudes. The waves at given latitude and height are extracted by using the following expression:

$$\Theta(t,l) = \Theta_0 + \Theta_r t + \sum_{s=1}^{4} B_s \cos\left[\frac{2\pi}{360} sl - \frac{2\pi}{360}\psi_s\right] + \sum_{k=1}^{3}\sum_{s=-4}^{4} C_{ks} \cos\left[k\frac{2\pi}{24}t - \frac{2\pi}{360}sl - \frac{2\pi}{360}\gamma_{ks}\right] \tag{2}$$

where t is time in hours counted from 0 UT on 1 October, l—longitude in degrees, s—wavenumber, k—number of the diurnal harmonic. The eastward (westward) propagation corresponds to $s > 0$ ($s < 0$). The four terms on the right hand side of the above expression are as follows: (i) the mean ionospheric parameter Θ_0 (zonal and time mean), (ii) a linear term with coefficient Θ_r responsible for zonally symmetric variations with period longer than 60 days; this term includes a possible effect of the ionospheric semiannual variability,

(iii) stationary waves with amplitudes B_s and phases ψ_s, and (iv) 24-h, 12-h and 8-h tides with amplitudes C_{ks} and phases γ_{ks}. In the above expression the phases of the tides are defined as the time of the wave maxima at zero longitude and are presented in degrees, while the phase of the stationary wave—as the longitude of the wave maximum.

It is important to note that the results for the COSMIC electron density and SABER temperature sun-synchronous 24-h (DW1), 12-h (SW2) and 8-h (TW3, T corresponds to terdiurnal tide) tides have different nature. While in the SABER temperatures they are the migrating tides, those in the COSMIC electron density are produced by a combination of the diurnal variability of the photo ionization, which depends on the solar zenith angle, and the effect of the migrating tides forced from below. Therefore, by using the above mentioned data analysis method, the effect of the daily variability of the ionization cannot be separated from the effect of the migrating tides. In this study we will make an attempt to estimate which of the two effects is prevailing in the observed ionospheric DW1, SW2 and TW3 responses. The nonmigrating tidal effects in the COSMIC electron densities however, can be fully separated.

Because of the satellite data sampling pattern and the variability of the ionosphere particularly near the F-region maximum at low latitudes a very important problem in the ionospheric tidal response assessment is how to determine the reference noise level. For this purpose we use the same approach as that applied for the SABER satellite data and described in detail by Pancheva et al. (2009a). It is based on a method for generating of time series with a normal distribution by using of random number generator with an uniform probability distribution (Jeruchim et al. 2000). The generated in this way time series of random numbers will have a mean approaching zero and a standard deviation close to 1, i.e. white noise. This unit normal random time series can be further modified to have a particular mean and variance by multiplying with the given standard deviation and adding the mean.

As the WN4 ionospheric longitude structure related mainly to DE3 tide is particularly strong in August-September the results of the reference noise level analysis will be shown for this period of time. For each altitude and modified dip latitude (the clarification of the modified dip, or modip, latitude is given in the next section) 200 Gaussian hourly noise datasets with length of 61 days and with data sampling pattern, variance and mean as those of the true COSMIC electron density measurements were generated (Kong et al. 1998). Then the $2D$ spectral analysis in the periodic range of 3–30 hours was performed on the resulting noise datasets. The reference noise spectrum for each zonal wavenumber was calculated as an average spectrum. As an example for the reference noise spectra the result for the ionospheric parameter f_oF2 is shown in Fig. 1a, where the latitude structure of the noise spectrum for westward propagating waves with zonal wavenumber 2 (W2) is presented in the left plot while for eastward propagating waves with zonal wavenumber 2 (E2) is seen in the right plot. Both noise spectra do not reveal some particular peaks because of data sampling pattern and the ionospheric variability. An enhancement of the spectral power at modip latitude range of $\pm30°$ is seen but the amplitudes of the randomly distributed peaks do not exceed 0.016 MHz for W2 and 0.018 MHz for E2. Similar results are obtained for the other zonal wavenumber f_oF2 noise spectra. Figure 1b shows the altitude structure of the electron density noise spectra over the equator; the left plot presents the noise spectrum for eastward propagating waves with zonal wavenumber 3 (E3) while the right plot presents that for westward propagating waves with zonal wavenumber 1 (W1). Again both noise spectra do not reveal some particular peaks; only an enhancement of the spectral power is seen between altitudes of ∼250–350 km with peaks which do not exceed 0.017 MHz for E3 and 0.018 MHz for W1. The results of this reference noise spectral analysis clearly indicated that because of the COSMIC data sampling pattern and the ionospheric variability there aren't

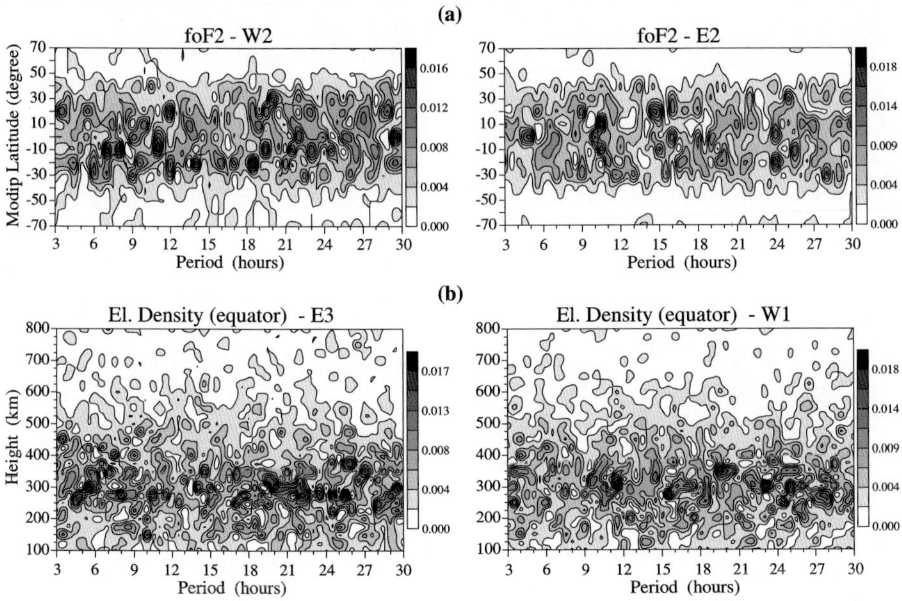

Fig. 1 (**a**) Latitude structure of the f_oF2 noise spectra in the periodic range of 3–30 hours for westward propagating waves with zonal wavenumber 2 (W2) (*left plot*) and for eastward propagating waves with zonal wavenumber 2 (E2) (*right plot*); (**b**) Altitude structure of the electron density noise spectra in the periodic range of 3–30 hours over the equator for eastward propagating waves with zonal wavenumber 3 (E3) (*left plot*) and for westward propagating waves with zonal wavenumber 1 (W1) (*right plot*)

revealed particular spectral peaks. The found low latitude amplification of the spectral f_oF2 noise power is related to the variability of the F-region maximum in the densest part of the ionosphere while the enhancement of the electron density noise spectral power between altitudes of ~250–350 km reflects the diurnal variability of the F-region over the equator.

Similar analysis of that shown in Fig. 1 was done on each 60-day window (because a 60-day window is used for extracting the waves) for the considered altitudes and latitudes. On the average the obtained noise amplitudes did not exceed 0.018 MHz. Therefore, all COSMIC tidal waves extracted from a given 60-day window with amplitudes larger than 0.02 MHz are accepted as statistically significant.

3 COSMIC Response to Nonmigrating Tides

As it was mentioned, the ionospheric WN4 longitudinal structures are thought to reflect the effect of the tropospheric weather on the ionosphere because of the nonmigrating DE3 tide which is mainly forced by deep tropical convection (Hagan and Forbes 2002). Similarly, the ionospheric WN3 structure, which is pronounced usually in winter, has been associated with DE2 tide. In view of this our analysis will start with the ionospheric response to the above mentioned nonmigrating tides seen in the ionospheric parameter h_mF2, which is closely related to the vertical plasma drift and in this way is more sensitive to dynamics.

If the ionospheric WN4 and WN3 structures are driven by tides, coming from below, then similar pictures have to be observed between the SABER lower thermospheric DE3 and DE2 tides and their ionospheric counterparts. The upper row of plots in Fig. 2a shows

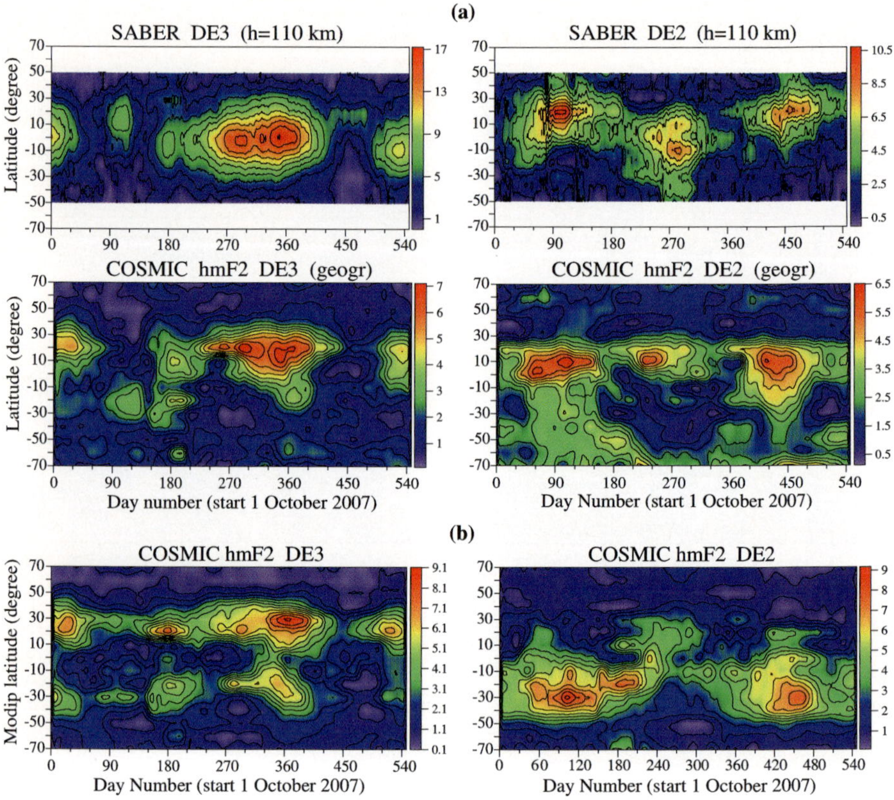

Fig. 2 (**a**) Latitude-time cross sections of the DE3 (*left column* of plots) and DE2 (*right column* of plots) tidal amplitudes seen in the SABER temperatures (*upper row* of plots, in Kelvin) and in the ionospheric h_mF2 (*bottom row* of plots, in km) for the period of time October 2007–March 2009; (**b**) The same as (**a**) but instead of geographic a modip latitude is used

the latitude (±50°) structures of the SABER temperature DE3 (left plot) and DE2 (right plot) tides at 110 km altitude, while the bottom row of plots shows the latitude (±70°) structures of their ionospheric responses seen in h_mF2. We note that both SABER and COSMIC results are in geographic latitudes. The similarity between the forcing from below (SABER tides, upper row of plots) and its ionospheric response (bottom row of plots) is remarkable. This strongly supports the causal link between the two phenomena. While the h_mF2 DE3 response prevails between June and October with maximum amplitude of ∼7 km that of DE2 response is pronounced mainly in winter (a secondary summer amplification as well) with amplitude of ∼6.5 km. The climatology of the SABER DE3 and DE2 tides (Pancheva et al. 2010b) indicated that while the DE3 is the predominant diurnal tide in the summer E-region (∼115 km height), the DE2 tide prevails during the winter. This defines their significant modulating effect on the vertical plasma drift/h_mF2 variability. The above result, reported by Pancheva and Mukhtarov (2010), is the first experimental evidence of the tidal coupling between lower atmosphere and ionosphere. It was possible because of using modern satellite-born (SABER and COSMIC) data effectively analyzed by one and the same method.

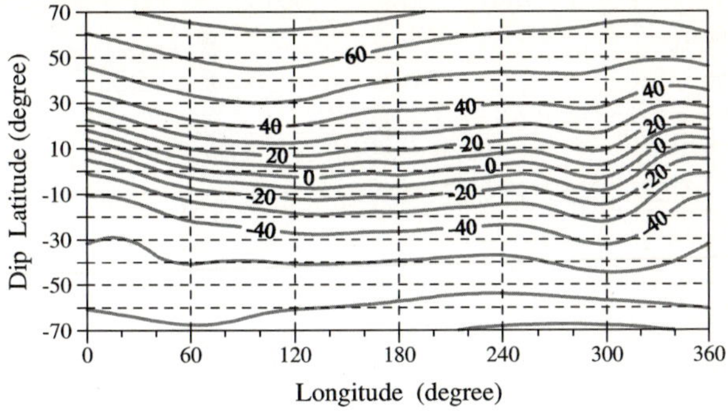

Fig. 3 (Color online) Relationship between dip (*black dash line*) and modip (*red solid line*) latitudes

It has already been mentioned that the neutral wind and electric field effects on the ionosphere are at variance with the geomagnetic field configuration. That is why the distribution of the ionospheric parameters is usually presented in geomagnetic latitude instead of geographic one. Early investigations (reported, e.g., in Rawer 1984) demonstrated the benefit of using the modified dip latitude (modip), introduced by Rawer (1963), to describe the variability of the densest part of the ionosphere, particularly at mid and low latitudes. The modified dip (modip) latitude which is adapted to the real magnetic field, e.g., to the magnetic inclination (dip), is defined by: $\tan\mu = I/\sqrt{\cos\Phi}$, where μ is the modip latitude, I is the true magnetic dip (usually at a height of 350 km), and Φ is the geographic latitude. Modip equator is the locus of points where the magnetic dip (or inclination) is 0. In the equatorial zone, the lines of constant modip are practically identical to those of the magnetic inclination but as latitude increases they deviate and come nearer to those of constant geographical latitude. The poles are identical to the geographic ones (Azpilicueta et al. 2006). Figure 3 shows the relationship between the frequently used dip (black dash lines) and modip (thick red lines) latitudes. The modip isolines are denser in the low latitudes; this allows a more detailed representation of the ionospheric tidal responses at low and middle latitudes.

Figure 2b displays the h_mF2 DE3 (left plot) and DE2 (right plot) responses in modip latitude frame. In general, the main features of the DE3 and DE2 responses from Fig. 1a remain, however the amplitudes are larger in modip frame; in this case they reach ~9 km. An important difference is seen however in modip frame; there is a splitting of the response at both sides of the dip equator particularly well evident for the DE3 response. The response is located near ±30° modip latitude, which is near ±10–20° dip latitude (see Fig. 3).

In order to study the spatial structure and particularly the vertical structure of the ionospheric DE3 and DE2 responses Mukhtarov and Pancheva (2011) utilized the COSMIC electron densities at given heights. Figure 4 shows the latitude (±70°) structure of the COSMIC DE3 (left column of plots) and DE2 (right column of plots) electron density responses at heights: 600 km (upper row of plots)), 400 km (middle row of plots) and 200 km (bottom row of plots). The latitude structure of the electron density responses show a splitting around a dip equator at 200 and 400 km heights, while at upper levels the splitting gradually disappears and the response moves toward the equator. This latitudinal structure indicates the role of the plasma drift in shaping the response: at 200 km and especially at 400 km the response is part of the "fountain effect" while at the upper levels, e.g. in the topside ionosphere, it is approximately confined over the dip equator. The latter feature is particularly well evident for the DE3 response. Figure 4 shows also that the strongest response for both tides produced by the "fountain effect" is observed near ±30° modip latitude. The response

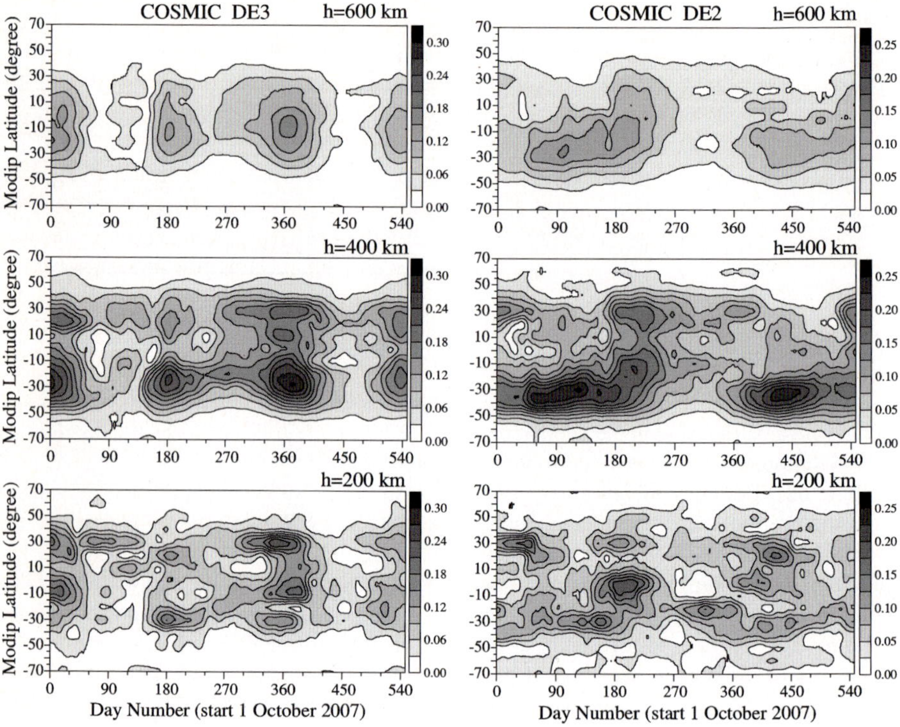

Fig. 4 Latitude structure (±70°) of the amplitudes in MHz of the electron density DE3 (*left column* of plots) and DE2 (*right column* of plots) responses at heights: 600 km (*upper row* of plots), 600 km (*middle row* of plots) and 200 km (*bottom row* of plots); all scales for DE3 and for DE2 are the same

at 200 km height is more complex; besides amplification near ±30° modip latitude it shows a response also close to the dip equator.

Figure 4 displays some difference between the COSMIC h_mF2 (Fig. 2a) and electron density DE3 responses: the electron density response in March, particularly up to 400 km height, is much stronger than that of the h_mF2 response. The reason is that while the h_mF2 response is mainly dependent on the dynamics of the thermosphere, the tidal response of the electron density depends on other factors as well. It is worth noting that, for example, the ratio of [O] and [N_2] that is controlled by the thermospheric circulation is the key parameter dominating the semiannual variation of the F-region anomaly at low latitudes with the greatest f_oF2 at equinox (Rishbeth et al. 2000). When the background electron density is higher then the DE3 electron density response also will be stronger.

Figure 5 shows the altitude structure of the amplitudes in MHz (upper row of plots) and phases in degrees (bottom row of plots) of the electron density DE3 response at 30°N (left column of plots) and DE2 response at 30°S (right column of plots), i.e. at these modip latitudes where the response is maximum. The main feature of the altitude structure at both responses is that there are upper level response (above ~300 km height) and bottom level response (below ~250 km height). The two regions of enhanced response are separated by a narrow altitude zone where the response is almost absent. The upper level response at ±30° modip latitude has maximum near 400–450 km height while the bottom level one near ~200–250 km. In the NH the upper and bottom level DE3 responses are similar while in the SH the upper level DE2 response is stronger than the bottom level one.

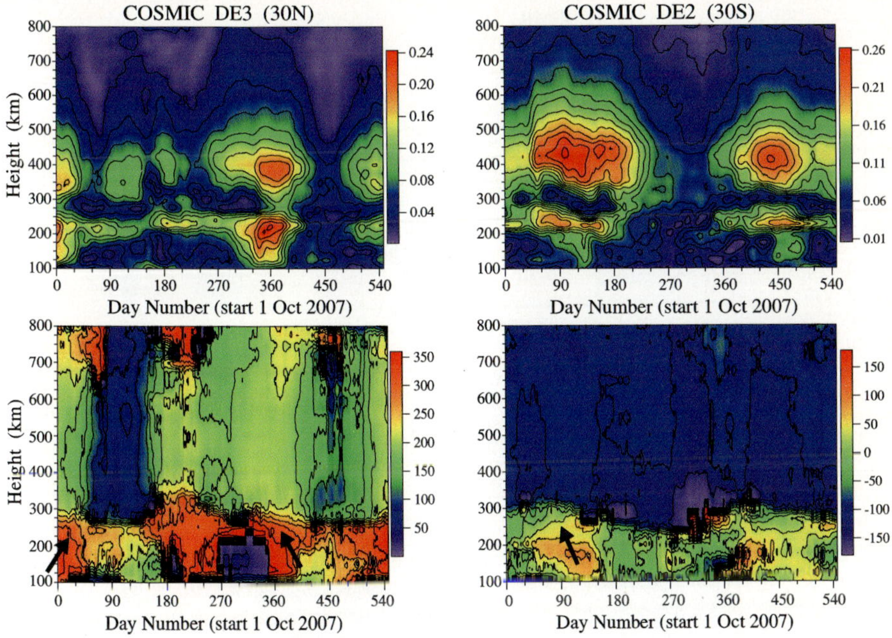

Fig. 5 Altitude structure of the amplitudes in MHz (*upper row* of plots) and phases in degrees (*bottom row* of plots) of the electron density DE3 response at 30°N (*left column* of plots) and DE2 response at 30°S (*right column* of plots); the *black arrows* at the bottom part of the phase plots display the vertical phase structure

The phase structures at both altitude regions are very different: while in the upper level response there is no any vertical phase structure in the bottom level there is such structure. The latter is well evident near the altitudes of enhanced response and is marked by black arrows in the bottom parts of the two phase plots. The absence of vertical phase structure in the upper level response indicates that the response is shaped mainly by the vertical plasma drift. The direct effect by the tides cannot be excluded also as the wind/temperature tides are almost trapped at these heights. The vertical phase structure in the bottom level response revealed upward propagation of the response on the average above 150–180 km altitude.

The above described important feature of the ionospheric response to nonmigrating tides forced from below, reported for the first time by Mukhtarov and Pancheva (2011), can be illustrated better in altitude-longitude frame. The upper plot of Fig. 6 shows the altitude-longitude structure of the monthly mean electron density DE3 variability in MHz at modip 30°N for September 2008 while the bottom plot shows the monthly mean DE2 variability at 30°S for January 2008; both electron density variabilities are at 00 UT. This type of presentation clearly distinguishes the two different types of response: the upper level response composed of trapped oscillations and bottom level response consisting of propagating oscillations. Generally, above ∼175 km altitude the slope of the bottom level response is eastward (as the DE3 and DE2 are zonally eastward propagating tides) which indicates an vertical upward propagation. The observed propagating tidal response at ±30° modip latitudes is a signature for a possible direct tidal effect in the considered altitude region. The tropical response has maximum near ∼200–230 km altitude. This could be an indication for a possible effect of the vertical wind DE3 tide (Oberheide et al. 2009), or a response shaped by a DE3/DE2 meridional wind modulated Pedersen currents. There is also

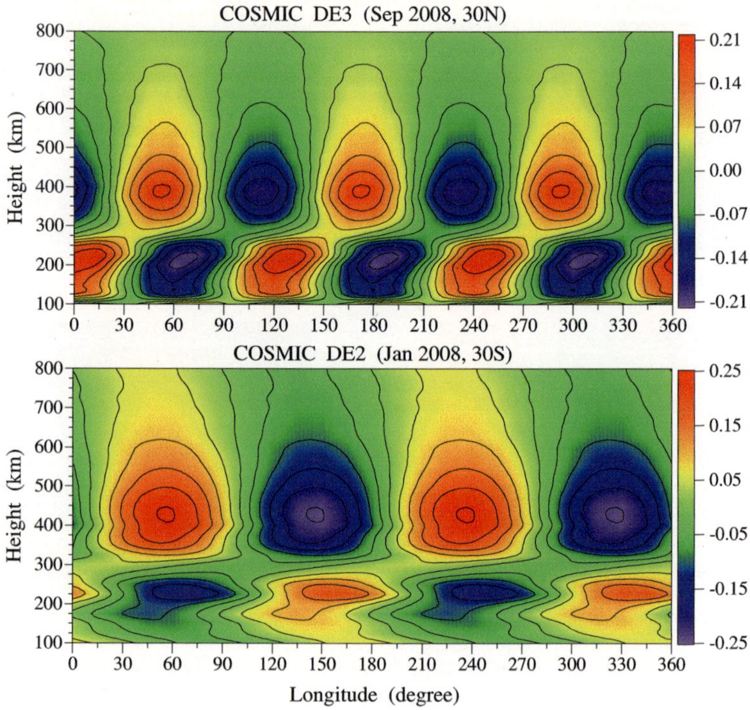

Fig. 6 Altitude-longitude structure of the electron density DE3 variability at 30°N in September 2008 (*upper plot*) and DE2 variability at 30°S in January 2008 (*bottom plot*); both plots present the tidal variability at 00 UT

a possibility of coupling between tides and the ionosphere via changes in neutral density and particularly a modification of [O], i.e. [O]/[N_2], which has effect on the recombination rate.

Figure 7a presents the electron density DE3 response over the equator; the amplitudes in MHz are shown in the upper plot while the phases in degrees can be seen in the bottom plot. There are also two altitude regions of enhanced response: the upper level response that is expanded from ∼400 km up to ∼700 km height (particularly that in September/October) and the bottom level one that maximizes near and below 150 km height. In this case however the phases at both regions are the same (particularly where the amplitudes are large and the phases can be defined with a high degree of confidence). Figure 7b displays the altitude-longitude structure of the monthly mean DE3 electron density variability over the equator for October 2007 at 00 UT. In this case the upper level and bottom level responses are in phase. Very recently Brahmanandam et al. (2011) have considered vertical and longitudinal electron density structures of the equatorial E- and F-regions in September–November 2007 and found very similar picture as that shown in Fig. 7b. The authors used also COSMIC electron density data but analyzed them in the traditional fixed LT frame (i.e. four structures are seen) by subtracting the background value from the original data. The obtaining of similar results by using two very different data analysis methods is a quite encouraging support of the result shown in Fig. 7b.

Figure 8 presents the latitude-altitude cross sections of the electron density DE3 response on September 26, 2008 (left plot) and DE2 response on March 25, 2008 (right plot), e.g.

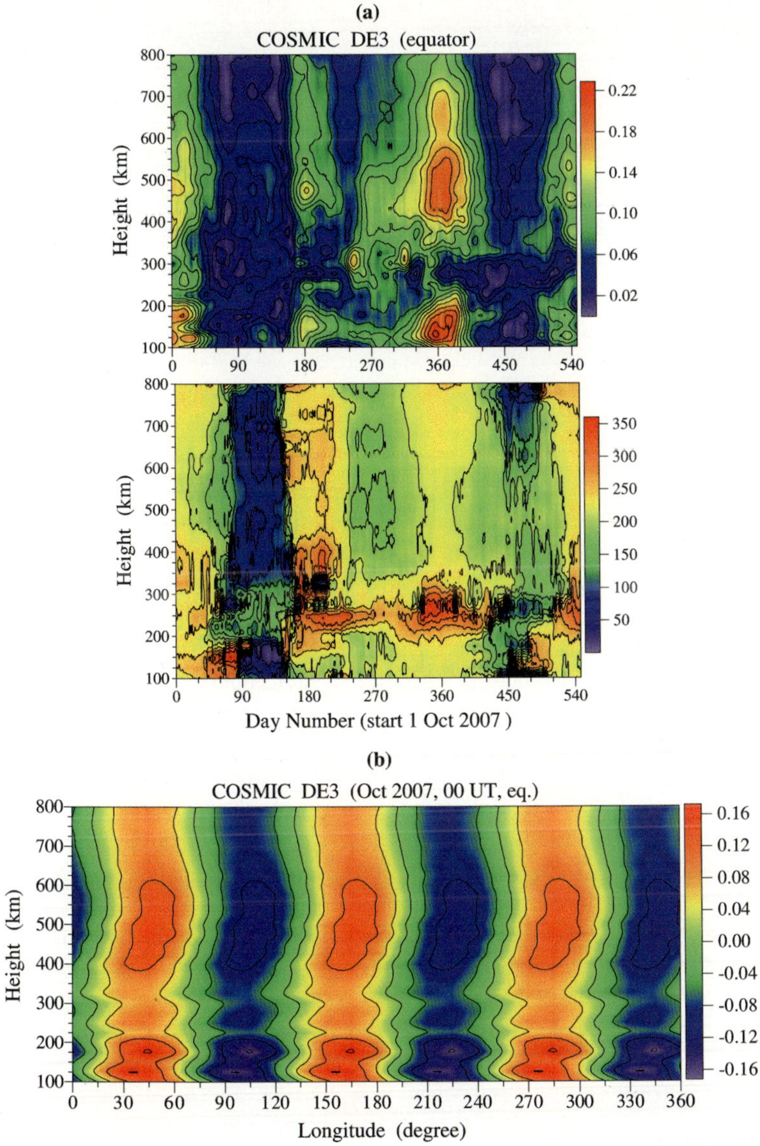

Fig. 7 (**a**) Altitude structure of the amplitude in MHz (*upper plot*) and phase in degrees (*bottom plot*) of the electron density DE3 response over the equator; (**b**) Altitude-longitude structure of the electron density DE3 variability over the equator in October 2007 at 00 UT

when the responses are strong (see Fig. 5). The both upper level responses clearly indicate the influence of the "fountain effect". Nevertheless that we consider equinoctial months the responses are stronger in the Southern Hemisphere (SH). It is worth noting that during the entire considered period of time (October 2007–March 2009) the ionospheric DE3 and DE2 responses at altitude of 400 km are stronger in the SH (see Fig. 4, 400 km height). This could be due to some equinoctial asymmetry of the vertical plasma drift found recently by

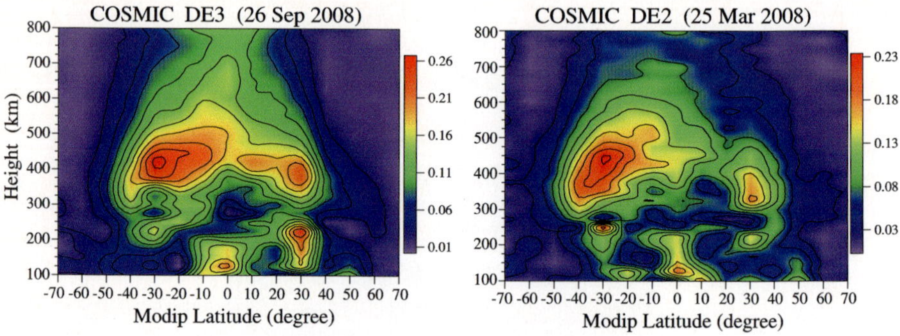

Fig. 8 Altitude-latitude structures of the electron density DE3 and DE2 responses observed on September 26, 2008 and March 25, 2008 respectively

Ren et al. (2011). The DE3 and DE2 bottom level responses are composed of two parts: a tropical response, near ±30° modip latitude, and over the dip equator. While the tropical response maximizes near 200–230 km height that over the equator amplifies at lower thermosphere, near 130–150 km. It should be noted that the bottom level tropical responses are just underneath the upper level ones.

It seems from Fig. 8 that the upper level DE3/DE2 response is shaped mainly by the DE3/DE2 modulated vertical plasma drifts. This is borne out by the strong DE3/DE2 h_mF2 response shown in Fig. 1 which reaches amplitudes of ~9 km.

England et al. (2010) found however that in addition to the E-region dynamo modulation other effects have the potential to contribute to the coupling between tides and the F-region ionosphere: neutral density variations, changes in thermospheric [O]/[N_2] ration, and meridional winds at F-region altitudes. In order to assess if these alternative effects contribute to the ionospheric DE3/DE2 response above ~300 km altitude we analyze both bottom and top slope coefficients described by (1), but our interest is directed mainly to the result for the top slope coefficient. We remind that these coefficients represent the bottom- and top-side vertical gradients of the electron density profile.

Figure 9 shows the DE3 (left column of plots) and DE2 (right column of plots) response in the top (upper row of plots) and bottom (bottom row of plots) slope coefficients. It is seen that both coefficients display clear response at the same time when the response is seen in the electron density (Figs. 4 and 5) and in h_mF2 (Fig. 2). The observed DE3 and DE2 variations have quite large amplitudes being near 10% of their mean values. This means that both (bottom- and top-side) slopes of the electron density profiles change during the coupling between tides and the ionosphere. This is an indication that the above mentioned additional affects contribute not only to shaping the bottom level response but to the upper level one as well.

The above obtained main features of the ionospheric DE3/DE2 response are seen in all nonmigrating responses considered in this study, i.e. nonmigrating diurnal, semidiurnal and terdiurnal tidal responses with zonal wavenumber up to 4. The SPWs with zonal wavenumber up to 4 are also considered in this study (see the third term on the right hand side of (2)). In general, the main features of the ionospheric SPW variability are similar to those of the tidal response, but the influence of the "fountain effect" in some cases is significantly weaker.

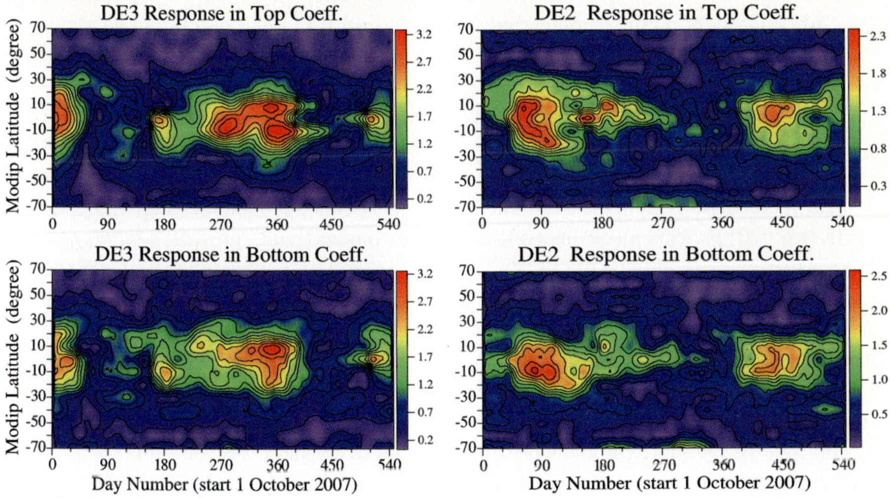

Fig. 9 Latitude-time cross sections of the DE3 (*left column* of plots) and DE2 (*right column* of plots) responses seen in the top (*upper row* of plots) and bottom (*bottom row* of plots) slope coefficients defining the electron density profiles

4 Contributors to the WN4 and WN3 Longitude Structures

It has been already mentioned that the WN4 longitude structure is the most prominent pattern observed by quasi-sun-synchronous satellites in a variety of ionospheric and thermospheric parameters. This structure has been often attributed to the DE3 tide based usually on the annual and LT variations of the WN4 structure. We noted also that all observational reports mentioned in the Introduction part were studied in a fixed LT. Hence, the observed structure cannot be directly related to a certain tidal component. Our approach however allows assessing the possible contributions of all waves that can force the WN4/WN3 structures and this is a significant advantage of the used in this study data analysis method.

As we noted before if only the diurnal and semidiurnal tides are considered, the WN4 structure can be forced by: DE3, DW5, SW6, SE2 and SPW4. The wave components with zonal wavenumber larger than 4 are not included in the decomposition procedure of both SABER and COSMIC data (see the formula (2)) hence the contribution of the DW5 and SW6 tides cannot be estimated. In general, the tidal components with high zonal wavenumbers have short vertical wavelengths hence they are more sensitive to the increasing dissipation in the lower thermosphere, i.e. they are susceptible to dissipation.

The climatology of the SABER eastward nonmigrating tides has been studied by Pancheva et al. (2010b) and found that the SE2 tide is one of the strongest semidiurnal nonmigrating tides with coherent latitude and altitude structures. It amplifies near tropics ($\pm 30°$) and reaches amplitudes of $\sim$7 K in the E-region of ionosphere (at $\sim$115 km altitude). The SE2 tide, particularly that in the NH, when propagating upward it propagates toward the equator as well. During the time period under consideration, the SABER SE2 tide shows some amplification in March and September–October 2008 with amplitudes of $\sim$5–5.5 K (not shown here). This tide is excited by latent heat release in the tropical troposphere, with some contributions from net radiative heating (Zhang et al. 2010). It can be produced also by a nonlinear interaction between the DE3 and migrating diurnal tide (DW1) tides (Hagan et al. 2009). The SE2 tide has relatively short vertical wavelength in winter,

~30 km, but longer one in summer and equinoxes, ~40–45 km, i.e. it might be efficient in the E-region dynamo modulation during the time when the ionospheric WN4 structure is observed.

The SABER SPW4 has relatively small amplitudes (~2.5–2.8 K) in the lower thermosphere over the equator in August–October 2008 (not shown here). However, very recently Oberheide et al. (2011) has reported quite large zonal wind SPW4 (~12 m/sec) with long vertical wavelength seen in the TIDI measurements over the equator (see Fig. 13 from Oberheide et al. 2011). This result fits well into recent model results presented by Hagan et al. (2009) where it is found that the SPW4 is excited by nonlinear interaction between the DE3 and the migrating diurnal tide (DW1). The observed strong zonal wind SPW4 with long vertical wavelength over the equatorial E-region indicates that this wave might efficiently modulate the E-region dynamo.

Figure 10 shows the latitude structure (±70°) of the amplitudes in MHz of the electron density DE3 (upper plot), SPW4 (middle plot) and SE2 (bottom plot) responses at altitude of 400 km. To facilitate the comparison the scales at all plots are the same. It is evident right away that the DE3 and SPW4 responses have very similar seasonal behaviour and similar amplitudes at 400 km height; while the largest amplitude of the DE3 response is ~0.30 MHz that of the SPW4 is 0.29 MHz. As it has been expected the SE2 response is much weaker, ~0.16 MHz. Similarly to the DE3 and SPW4 waves the SE2 response shows also amplification in July–October 2008.

If the wave responses seen in Fig. 10 contribute to the WN4 structure (seen in a LT fixed frame) then the ionospheric variability forced by the superposition of the DE3, SPW4 and SE2 waves should be similar to that of the DE3 tide alone, i.e. the superposition has to have similar seasonal behaviour and the same longitudinal structure as the DE3 tide. If each wave has a seasonal variability similar to that of the DE3, as it was shown in Fig. 10, then this will be valid for the superposition as well. Figure 11 presents the altitude-longitude structure of the electron density DE3 (upper plot) and the superposition of DE3, SE2 and SPW4 (bottom plot) variability at 30°S in October 2008 at 00 UT. The comparison between the two electron density variabilities reveals high degree of similarity: (i) the same longitude structure; (ii) two different types of response; while the upper level response is composed of trapped oscillations and bottom level one consists of propagating oscillations, and (iii) the longitude maxima of both upper and bottom level responses are situated at almost the same longitudes. There is however one important difference; while the maximum amplitude of the DE3 response in October 2008 is 0.27 MHz that of the superposition is ~0.5 MHz, i.e. almost two times stronger. This is very important fact that explains why the WN4 structure can be distinguished sometimes even in the raw data.

As we mentioned before the ionospheric WN3 structure can be forced by: DW4, DE2, SW5, SE1 and SPW3. All these waves, except the nonmigrating SW5 tide, are included in the decomposition scheme of SABER and COSMIC data and their contribution to the WN3 structure can be assessed. The analysis of the SABER temperatures revealed that the SW4, SE1 and SPW3 have amplitudes of ~5–6 K in the equatorial E-region of the ionosphere (~110–115 km altitude) and that they amplify usually in winter (not shown here). This result supports their contribution to the WN3 structure which is observed mainly in winter. The SPW3 penetrates to the lower thermosphere from below; it has stratospheric and mesospheric amplifications at altitudes of ~45 km and 85–90 km respectively. It might be the most efficient wave among the above mentioned ones in modulating the E-region dynamo because of having comparatively long vertical wavelength.

Figure 12 displays the latitude structure (±70°) of the amplitudes in MHz of the electron density DE2 (upper plot), SPW3 (second from above plot), DW4 (third from above plot)

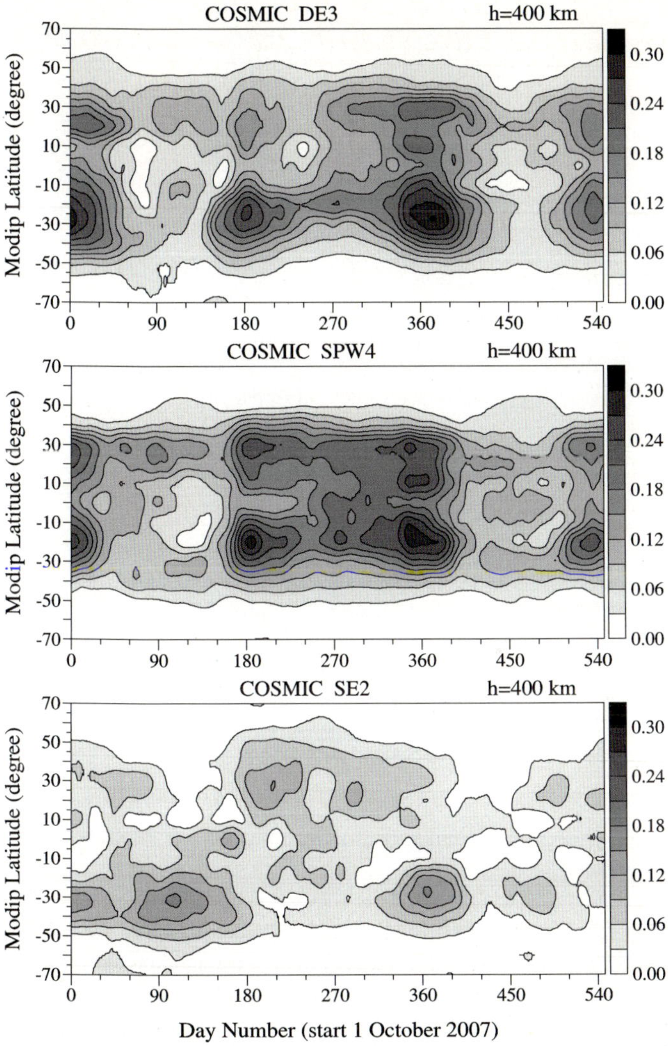

Fig. 10 Latitude structure (±70°) of the amplitudes in MHz of the electron density DE3 (*upper plot*), SPW4 (*middle plot*) and SE2 (*bottom plot*) responses at altitude of 400 km; the scales at all plots are the same

and SE1 (bottom plot) responses at altitude of 400 km. Again to facilitate the comparison the scales at all plots are the same. The vertical arrangement of the ionospheric responses is based also on their amplitudes as the strongest response is shown on the uppermost plot, i.e. DE2 response which is ∼0.25 MHz. Next to the DE2 are as follows: SPW3 (∼0.20 MHz), DW4 (∼0.15 MHz) and SE1 (∼0.13 MHz) responses. It is evident also that all these responses amplify during winter and spring, i.e. when the WN3 structure is often observed.

Figure 13 display the altitude-longitude structure of the electron density DE2 (upper plot) and the superposition of DE2, DW4, SE1 and SPW3 (bottom plot) variability at 30°S in January 2008 at 00 UT. Again the comparison between the both responses indicates high degree of similarity. All details valid for the comparison shown in Fig. 11 are applicable in

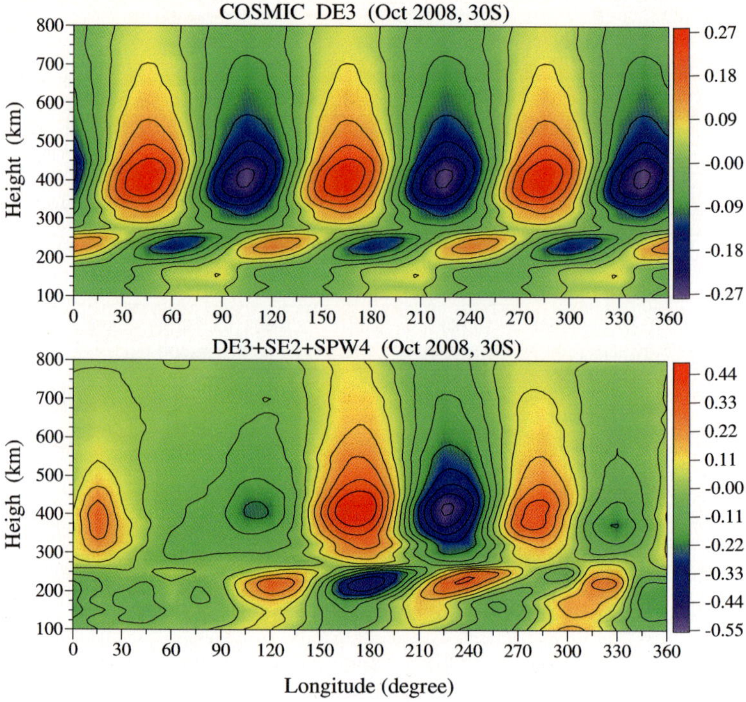

Fig. 11 Altitude-longitude structure of the electron density DE3 (*upper plot*) and the superposition of DE3, SE2 and SPW4 (*bottom plot*) variability at 30°S in October 2008 at 00 UT

this case as well. Here also the superposition response of ~0.36 MHz is 1.5 time stronger than the DE2 response (~0.25 MHz).

The results of this part are obtained for the first time from an analysis of the COSMIC electron density measurements where all forced from below waves are extracted simultaneously from the data. These new experimental results give an explanation why the WN4 and partly WN3 longitude structures are so prominent pattern in the ionosphere. We present experimental evidence which indicate that the WN4 (WN3) structure is not generated only by the DE3 (DE2) tide as it has been often assumed. The DE3 (DE2) tide remains the leading contributor, but the SPW4 and SE2 (SPW3, DW4 and SE1) waves have their effects as well in a way that the ionospheric response becomes almost double (1.5 time stronger).

5 Electron Density Sun-Synchronous 24-, 12- and 8-h Tidal Responses

We mentioned before that the sun-synchronous 24-h (DW1), 12-h (SW2) and 8-h (TW3) electron density oscillations are produced by a combination of the diurnal variability of the photo ionization which depends on the solar zenith angle and the effect of the migrating diurnal, semidiurnal and terdiurnal tides forced from below. It was mentioned also that the data analysis method used in this study cannot separate the two effects. Having in mind the main features of the latitude and altitude structures of the electron density tidal response described in Part 3, we will make an attempt to assess the effect of the migrating tides on the DW1, SW2 and TE3 electron density oscillations separated from the COSMIC data.

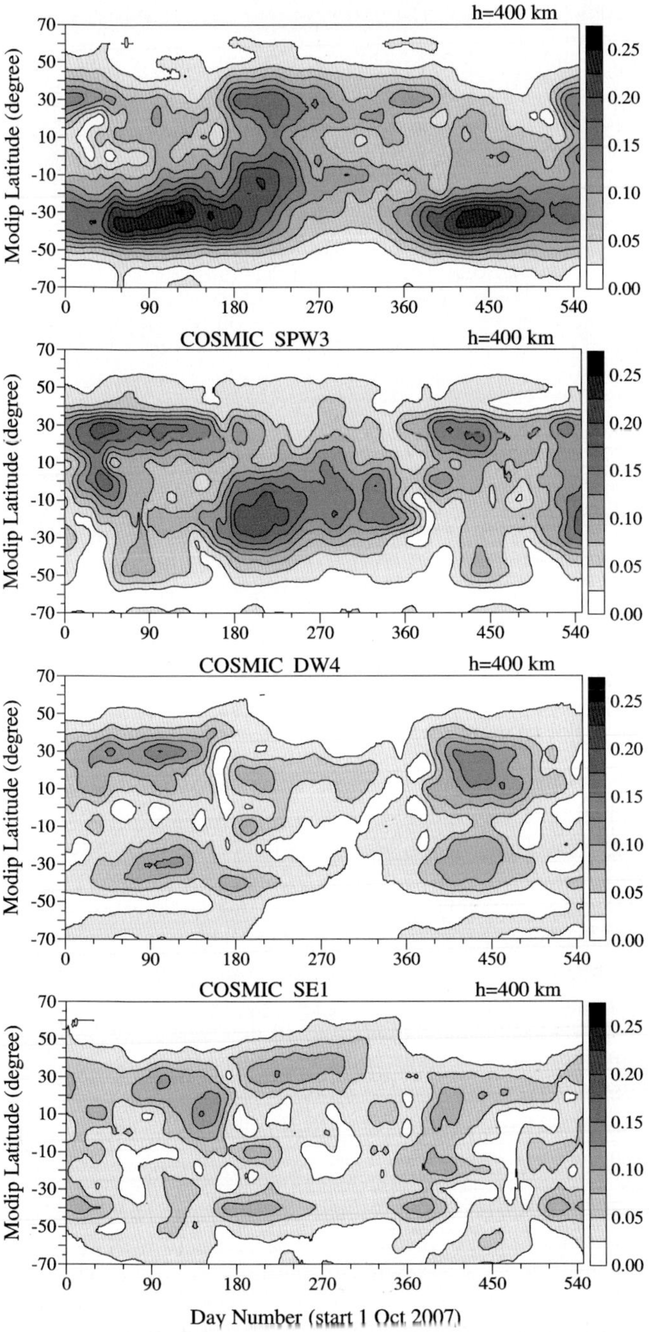

Fig. 12 Latitude structure (±70°) of the amplitudes in MHz of the electron density DE2 (*upper plot*), SPW3 (*second from above plot*), DW4 (*third from above plot*) and SE1 (*bottom plot*) responses at altitude of 400 km; the scales at all plots are the same

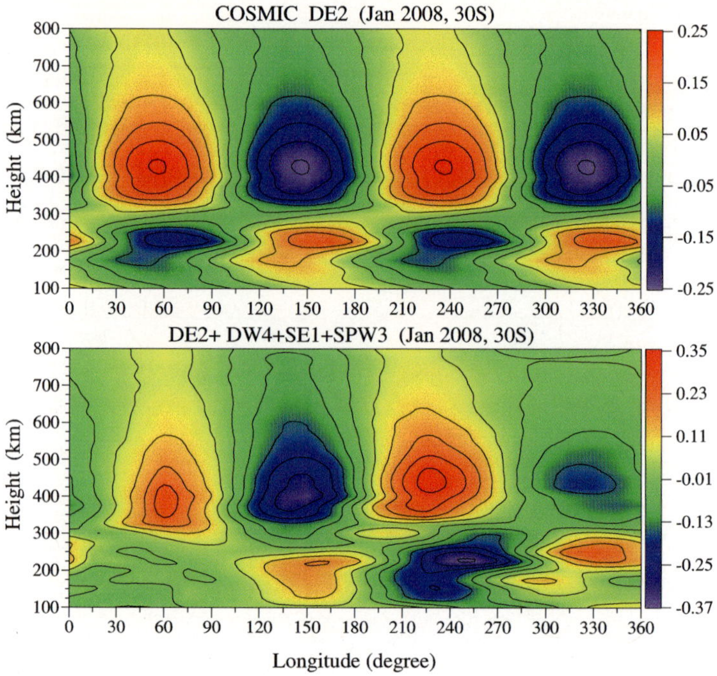

Fig. 13 Altitude-longitude structure of the electron density DE2 (*upper plot*) and the superposition of DE2, DW4, SE1 and SPW3 (*bottom plot*) variability at 30°S in January 2008 at 00 UT

Figure 14 shows the latitudinal structures of the sun-synchronous diurnal (upper plot), semidiurnal (middle plot) and terdiurnal (bottom plot) oscillations observed in the COSMIC electron density at 400 km height. We display the results at 400 km height because the splitting around the equator of the DE3 and DE2 electron density responses (Fig. 4) was most clearly evident at the height of 400 km. While the semidiurnal oscillation shows clear splitting around the equator with maxima near ±20–30° modip latitude the diurnal and terdiurnal ones does not show such splitting. It is worth noting that the used in this study a latitude resolution of 10° could screen some fine splitting located close to the equator (not far away from ±10°). As the latitude structure of the nonmigrating tidal responses clearly revealed maxima locations near ±30°, hence the latitude resolution of 10° is enough for the basic aims of this study.

Figure 15a presents the altitude structure of the sun-synchronous diurnal oscillation; the upper plot shows the amplitude in MHz, while the bottom plot shows the phase in degrees. There isn't any signature of the two altitude regions of ionospheric response which is typical for the nonmigrating tidal effect. The latitude (upper plot of Fig. 14) and altitude (Fig. 15a) structures of the COSMIC sun-synchronous 24-h electron density oscillation indicate that this response is shaped mainly by the photo ionization, i.e. it is due to a solar zenith angle effect. Figure 15b displays the altitude-longitude structure of the electron density DW1 variability over the equator in March 2008 at 00 UT. The maximum diurnal variability in March is seen at ~350 km height and the minimum electron density is seen between 02 and 03 am; the latter is supported by numerous ground-based ionosonde measurements.

Figure 16a presents the altitude structure of the sun-synchronous 12-h oscillation at latitude of 30°N (left column of plots) and 30°S (right column of plots); again the result for

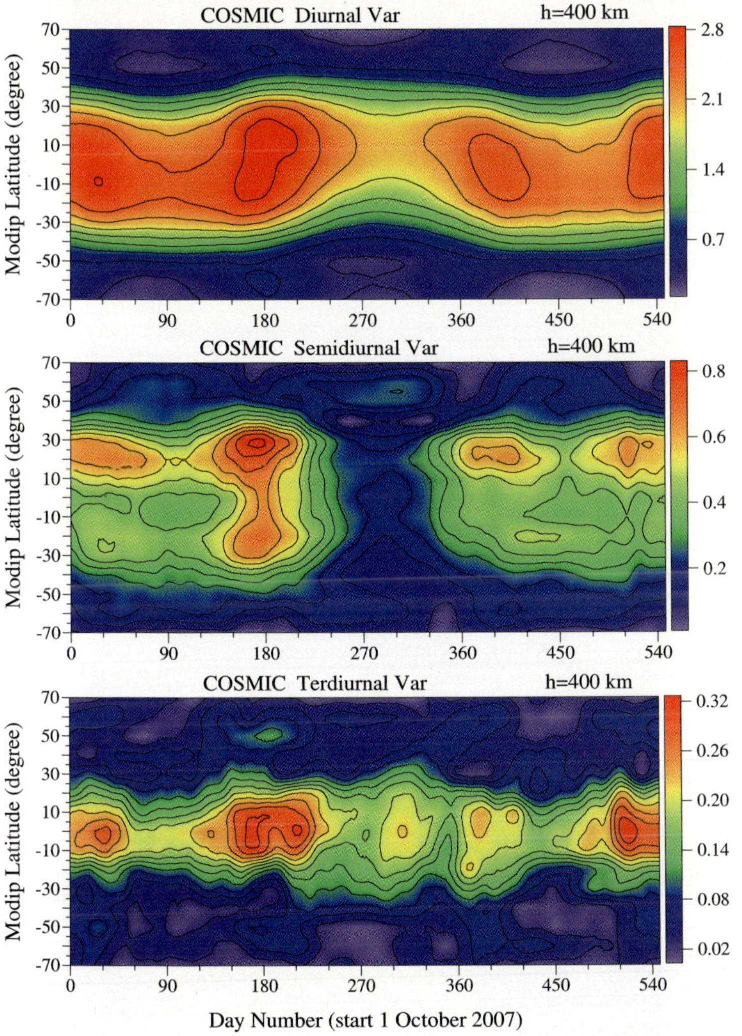

Fig. 14 Latitude structure (±70°) of the amplitudes in MHz of the electron density DW1 (*upper plot*), SW2 (*middle plot*) and TW3 (*bottom plot*) responses at altitude of 400 km

the amplitude is shown in the upper row of plots while for the phase is in the bottom row of plots. In this case the two altitude regions of response, separated by a narrow zone with no response, are clearly outlined. The SW2 amplitude particularly at the NH tropic reaches ∼0.9 MHz, as the SW2 maxima are situated near 350–400 km height. We remind that the SABER SW2 tide also maximizes at tropical lower thermosphere. The difference in the phase structure at the tropics is also well evident: no vertical phase structure in the upper level and the presence of phase structure at the bottom level; the latter is marked by the red arrows situated in the bottom side of the phase plots. The vertical phase structure indicates that the bottom level response at least above ∼150 km height is vertically upward propagating response.

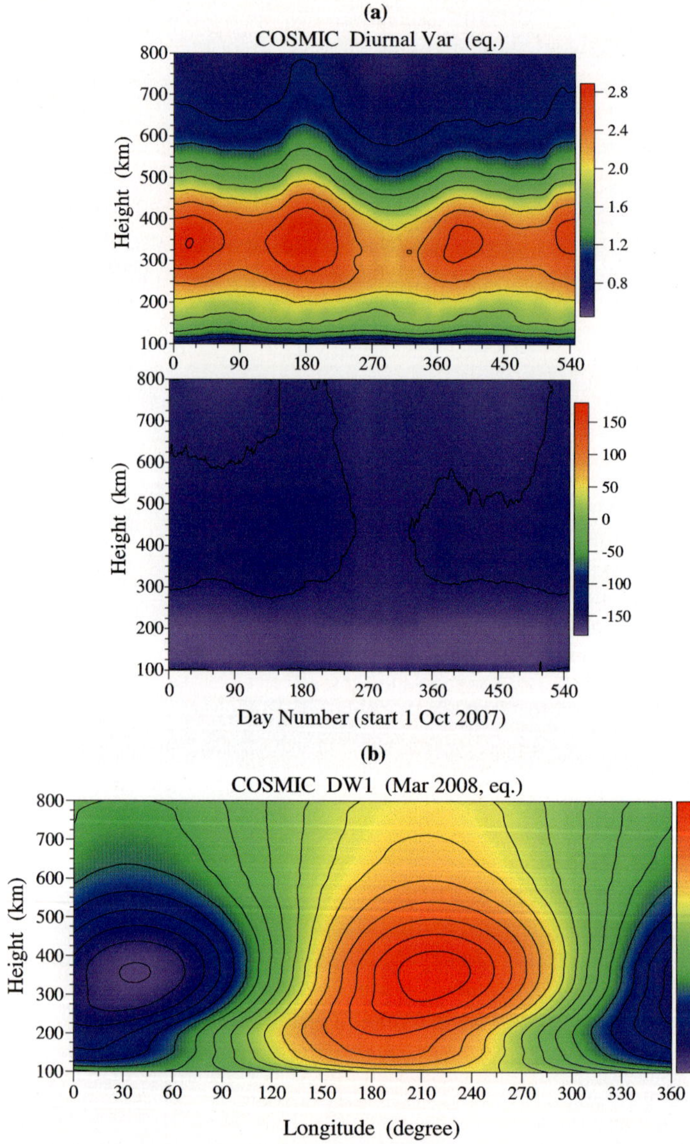

Fig. 15 (**a**) Altitude structure of the amplitude in MHz (*upper plot*) and phase in degrees (*bottom plot*) of the electron density DW1 response over the equator; (**b**) Altitude-longitude structure of the electron density DW1 variability over the equator in March 2008 at 00 UT

Figure 16b is the same as Fig. 16a, but for the SW2 response over the equator. Two levels of response are again clearly distinguished. Similarly to the DE3 response over the equator, shown in Fig. 7a, there is also no phase difference between the upper and bottom level phase structures. The SW2 response over the equator is weaker than those at the tropics and it reaches ~0.6 MHz.

Figure 17 presents the altitude-longitude structure of the electron density SW2 variability at 30°N (upper plot) and over the equator (bottom plot) in March 2008 at 00 UT. The SW2

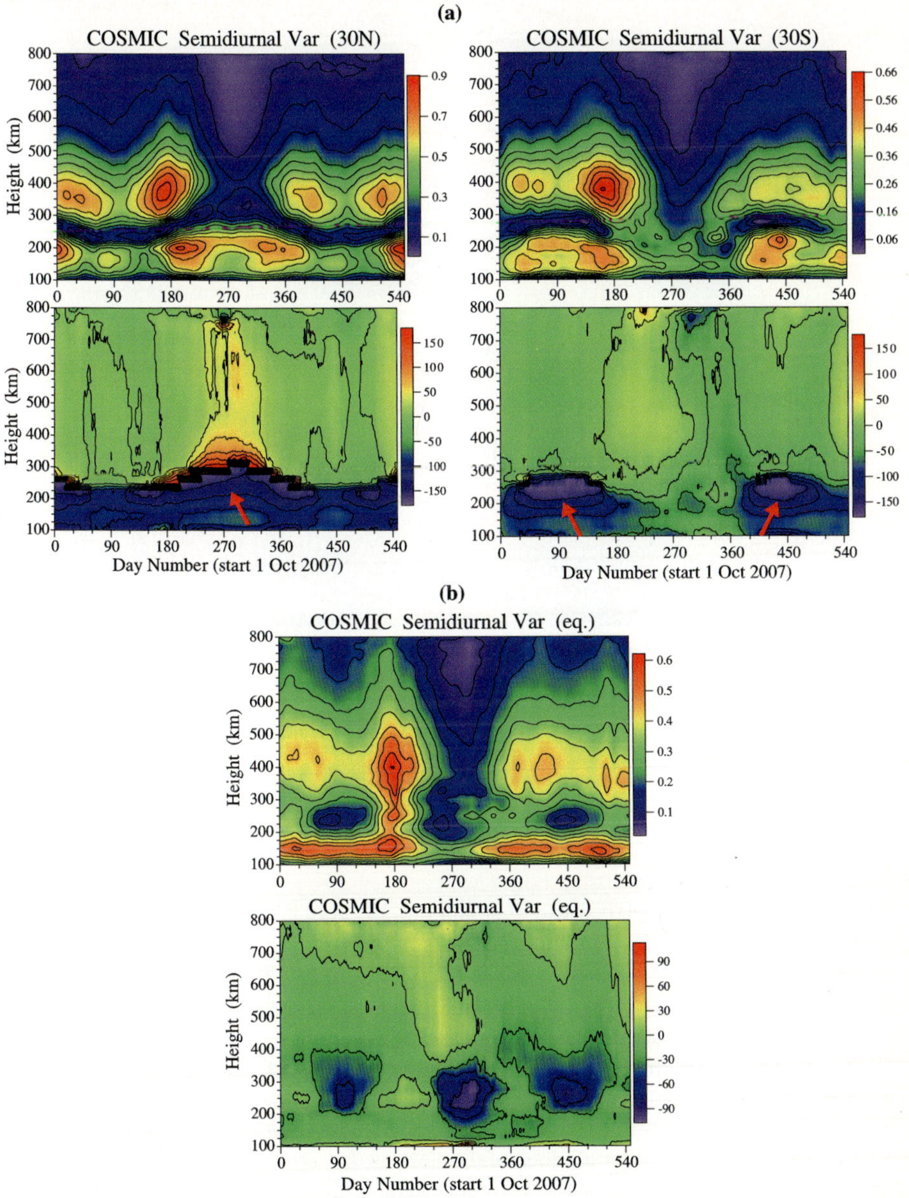

Fig. 16 (**a**) Altitude structure of the amplitudes in MHz (*upper row* of plots) and phases in degrees (*bottom row* of plots) of the electron density DW1 responses at 30°N (*left column* of plots) and at 30°S (*right column* of plots); the *red arrows* at the bottom part of the phase plots display the vertical phase structure; (**b**) The same as (**a**) but over the equator

response at tropics is similar to the tropical DE3/DE2 response shown in Fig. 6. The upper level response is composed of trapped oscillations and bottom level response consists of propagating oscillations. Generally, above ~150 km altitude the slope of the bottom level response is westward (as the SW2 is zonally westward propagating tide) which indicates a

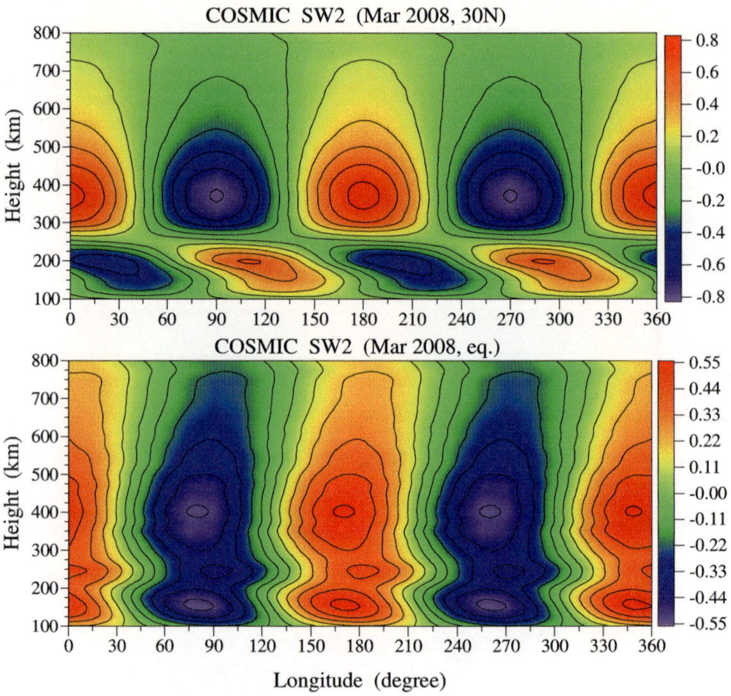

Fig. 17 Altitude-longitude structure of the electron density SW2 variability at 30°N (*upper plot*) and over the equator (*bottom plot*) in March 2008 at 00 UT

vertical upward propagation. The SW2 response over the equator (bottom plot) is similar to the equatorial DE3 response shown in Fig. 7b with no vertical phase structure in the entire altitude range.

The latitude (middle plot of Fig. 14) and altitude (Figs. 16 and 17) structures of the COSMIC sun-synchronous 12-h electron density oscillation indicate that this ionospheric oscillation is significantly affected by the migrating SW2 tide forced from below. We note that the climatology of the SABER SW2 tide (Pancheva et al. 2009b) revealed that this is the strongest semidiurnal tide observed in the low-latitude E-region of the ionosphere. The SW2 vertical wavelength increases during the equinoctial and summer months, hence its efficiency for electric field generation increases as well.

Figure 18 shows the altitude structure of the amplitudes in MHz (upper row of plots) and phases in degrees (bottom row of plots) of the electron density TW3 responses over the equator (left column of plots) and at 30°N (right column of plots). This ionospheric oscillation, similarly to the DW1 one, does not show two altitude levels of response. The largest TW3 response over the equator is near 250 km altitude, while at the tropics it is slightly lower, ~225 km. The response maximizes during the equinoxes, similarly to the SABER TW3 tide. The TW3 amplitude at the tropic is slightly larger than that over the equator, 0.73 MHz vs. 0.65 MHz. In general, the ionospheric TW3 response is very strong and it approaches the SW2 response. The TW3 phase structure is completely different from those of the DW1 and SW2 responses. At both latitudes there is an altitude region with a phase maximum; such region above the equator is located near 500–550 km while at 30°N it is near 400–450 km. The phase plots reveal also that below and above this region the phase decreases, i.e. there is upward/downward phase progression in electron density perturbation. This is an indication

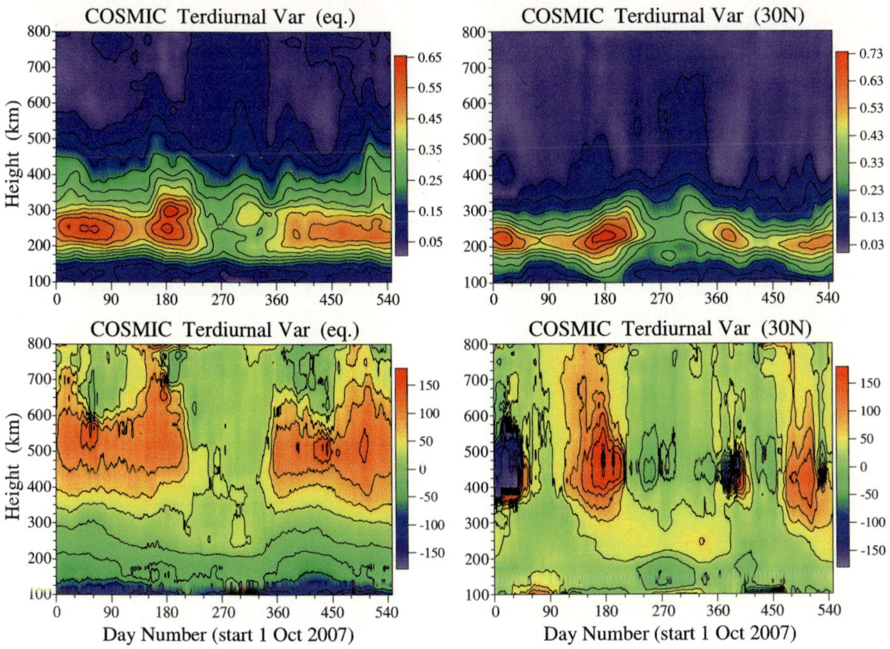

Fig. 18 Altitude structure of the amplitudes in MHz (*upper row* of plots) and phases in degrees (*bottom row* of plots) of the electron density TW3 responses over the equator (*left column* of plots) and at 30°N (*right column* of plots)

that the electron density TW3 response propagates downward/upward below and above the phase maximum region (i.e. there is downward/upward energy propagation).

The above described peculiar feature of the ionospheric TW3 response is illustrated better in the altitude-longitude frame. Figure 19 displays the altitude-longitude structure of the electron density TW3 variability over the equator (upper plot), at 30°N (middle plot) and at 50°N (bottom plot) in March 2008 at 00 UT. The main features of the TW3 response can be summarized as: (i) at all modip latitudes the slope of the largest oscillations is eastward, this indicates an downward propagation of the oscillations; (ii) there is lowering of the altitude with the largest response when the modip latitude increases; it is near 250 km over the equator while is near and even below 200 km at 50°N; (iii) the lowering of the altitude with the largest response is accompanied by a lowering of the altitude where the phase slope changes its direction. It is worth noting that the standard version of the TIME-GCM model (i.e. version where only the migrating tides from the GSWM are included) also produces downward/upward phase progression in the electron density TW3 perturbation at geographic latitude of 31°N (which is ~40–50°N modip latitude), i.e. the electron density TW3 response does not correspond to the neutral wind and temperature TW3 upward energy propagation in the lower-middle thermosphere (up to ~300 km altitude) (H.-L. Liu, personal communication, 2011).

The peculiar vertical structure of the ionospheric TW3 response hints at possible coupling between the SW2 tide forced from below and the diurnal variability of the electron density, i.e. at ion-neutral momentum coupling. The answer of this question cannot be found only from the analysis of COSMIC data; it needs further detailed numerical simulations as well.

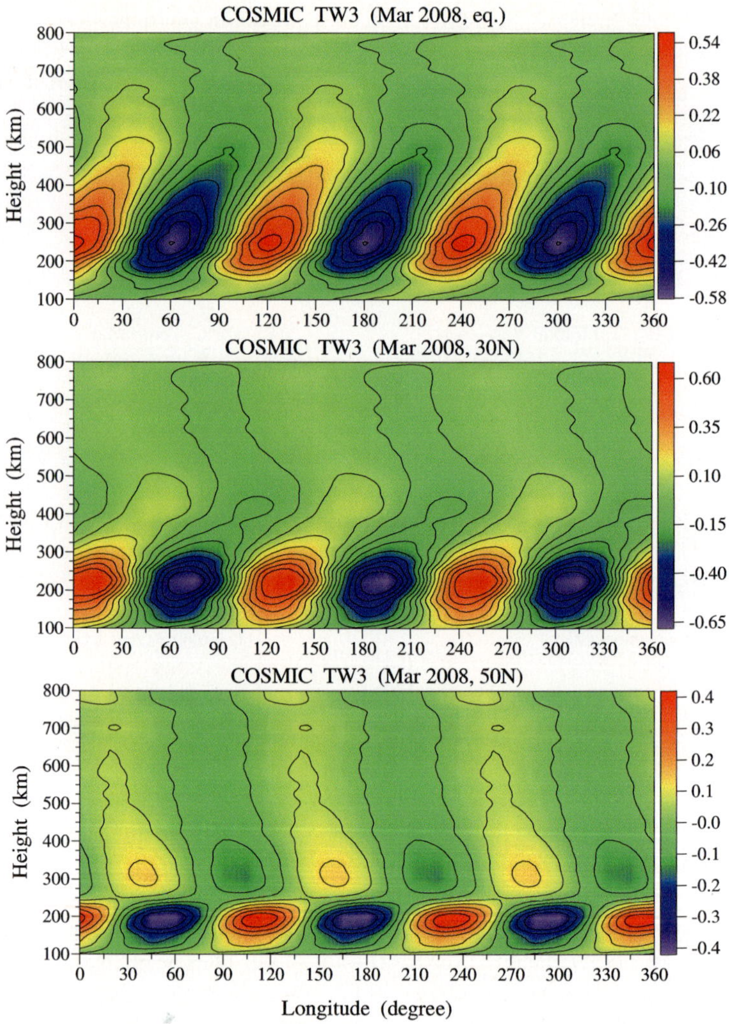

Fig. 19 Altitude-longitude structure of the electron density TW3 variability over the equator (*upper plot*), at 30°N (*middle plot*) and at 50°N (*bottom plot*) in March 2008 at 00 UT

6 Summary and Concluding Comments

The paper is an overview of the ionospheric response to atmospheric tides forced from below. Most of the results have been very recently produced and they provide strong observational support for tidal-ionosphere coupling processes beyond the lower thermosphere dynamo region. Moreover, not traditional data concerning of middle atmosphere and ionosphere but modern satellite-board data (COSMIC and SABER/TIMED) were effectively used for the analysis. The tidal waves from the two data sets have been extracted by the same data analysis method described in detail by Pancheva et al. (2009a). The similarity between the lower thermospheric temperature tides and their ionospheric electron density responses provides evidence for confirming the new paradigm of atmosphere-ionosphere coupling.

It was mentioned in the COSMIC data section that the electron density profiles employed in this study were retrieved by using a RO inversion technique. The electron density profiles obtained by this technique have been validated by comparing them with ground-based ionosonde and Incoherent Scatter Radar (ISR) observations (Lei et al. 2007; Kelley et al. 2009). The RO inversion approach relies on a few assumptions and approximations described in detail by Kuo et al. (2004) and Lei et al. (2007). All of them introduce some limitations and possible errors in determining the electron density profiles. Recently it has been found that particularly the spherical symmetry assumption used in Abel inversion is possibly the most significant error source (Schreiner et al. 1999; Lei et al. 2007; Kelley et al. 2009; Yue et al. 2010). In Yue et al. (2010) and Liu et al. (2010) an attempt has been made to simulate error distribution of RO electron density profiles from the Abel inversion by using some empirical ionospheric models. Their results show that the retrieved N_mF2 and h_mF2 are generally in good agreement with the true values, but the reliability of the retrieved electron density degrades at low latitudes and altitudes. Specifically the above mentioned authors found that the Abel retrieval method introduces artificial plasma caves underneath the EIA crests along with three density enhancements in adjacent latitudes. The artifact appears mainly below 250 km altitudes and becomes pronounced when the EIA is well developed. Liu et al. (2010) found also that the deviation of the retrieved electron density from the true one depends also on the local times.

The above mentioned simulations indicate that the error in defining the electron density below 250 km height depends on the latitudes where the EIA crests are located. In our case the EIA crests can be defined by the latitude structures of the electron density mean value (the first and second terms on the right hand side of (2) which is not used in this paper) and the sun-synchronized ionospheric tidal responses (DW1, SW2 and TW3). In general, the latitude maxima (i.e. EIA), obtained by the superposition of the above mentioned components, are situated near ±30° modip latitude (not shown here). We remind that the upper level COSMIC nonmigrating electron density responses were located also near ±30° modip latitude and that the bottom level responses are just underneath the upper level ones. This means that the bottom level responses are just underneath the EIA crests. We remind also that the bottom level response at tropics has vertically propagating structure. According to the above mentioned simulation results (Yue et al. 2010; Liu et al. 2010) the Abel retrieval method underestimates the electron density underneath the EIA crests; hence, it may degrade the electron density responses but cannot generate the vertically propagating response. Additionally, the LT dependence of the error in the retrieved electron density indicates that mainly the migrating ionospheric response could be affected. The nonmigrating electron density responses are however the main focus of this paper because they are responsible for the ionospheric WN4/WN3 longitude structures.

Recently Chu et al. (2010) have compared the N_mE electron density retrieved by COSMIC satellites with ground-based ionosonde peak values of E region in different latitudinal ranges and different seasons for 2007–2009. The authors found that no matter what season and latitudes are considered, the mean COSMIC N_mE tends to be constantly greater than the ionosonde measurements. This comparison indicates that there is a systematic error in retrieved E-region electron density. The data analysis method used in the paper however filters very efficiently each systematic error. The reason is that such type of errors will affect mainly the constant term (mean electron density) in the decomposition. This parameter is not considered in this study.

Very recently Perrone et al. (2011) have compared the COSMIC N_mE data with the ionosonde measurements at Rome and Gibilmanna (Sicily) for some months of 2006–2007. Italy is completely located at the artificially density enhancement area due to the use of the Abel retrieval method (Yue et al. 2010 and Liu et al. 2010). The authors found that particularly for the summer months the COSMIC N_mE data turn out to be close to the blanketing frequency f_bEs. Therefore, the sporadic E layers which permanently exist during the summer at middle latitudes strongly contribute to the COSMIC N_mE measurements. Having in mind the contribution of the tides for producing not only the tidal ion layers (Mathews and Sporadic 1998) but also the wind shear for the normal sporadic E layers (Pancheva et al. 2003; Haldoupis et al. 2004, 2006; Haldoupis and Pancheva 2006; Arras et al. 2009) then the ionospheric tidal response seen in the E-region heights could be related to the sporadic E-layer variability as well.

At this stage we may conclude with confidence that the bottom level ionospheric tidal response is real one but its quantitative description has to be accepted with some caution. It is worth noting that the retrieval error problem is too complex and needs to be seriously investigated in future.

This overview study has demonstrated that the application of one and the same data analysis method to the global atmospheric (SABER/TIMED temperatures) and ionospheric (COSMIC electron density) satellite measurements provides an opportunity for a detailed study of the atmosphere-ionosphere coupling by atmospheric waves. The main features of the latitude and altitude structures of the ionospheric nonmigrating tidal response can be summarized as follows:

- The latitude structure of the ionospheric response for heights up to 450–500 km is distributed at both sides of the dip equator at about ±30° modip latitude; above 500 km the response is approximately confined to the equator.
- Two altitude regions of enhanced electron density response are found: an upper level response (above 300 km), which usually maximizes near 400–450 km, and a bottom level one (at below 250 km); the two regions are separated by a narrow altitude zone with no tidal response.
- The bottom level response is composed of two parts: a tropical response, near ±30° modip latitude, e.g. just underneath the EIA, and a response over the dip equator (the second is not well evident in all nonmigrating tidal responses). While the tropical response maximizes near 200–230 km height that over the equator amplifies at lower levels, near 120–150 km.
- The phase structures at both altitude regions are very different in the tropics: while in the upper level response there is no vertical phase structure, at the bottom level one there is such a structure; over the dip equator however the phase structure at both altitude regions is the same.
- The different phase structures of the ionospheric response in both altitude regions suggests: (i) the upper level response is mainly due to the DE3/DE2 modulated vertical plasma drift, and (ii) the lower level response is complex and probably is caused by more than one mechanism.

The present paper provided also new experimental results which give an explanation why the WN4 and partly WN3 longitude structures are so prominent pattern in the ionosphere. We presented evidence indicating that:

- WN4 longitude structure is most probably caused at least by the combined action of DE3, SE2 and SPW4. The combined electron density variability is almost two times stronger than that generated only by DE3.
- WN3 longitude structure is caused at least by the combined action of DE2, DW4, SE1 and SPW3. The combined electron density variability is almost 1.5 times stronger than that generated only by DE2.

The paper presented also the global distribution and temporal variability of the sun-synchronous 24-h (DW1), 12- (SW2) and 8-h (TW3) electron density oscillations. They are produced by a combination of the diurnal variability (an its harmonics as well) of the photo ionization which depends on the solar zenith angle and the effect of the migrating diurnal, semidiurnal and terdiurnal tides forced from below. The main results can be summarized as follows:

- While the latitude and altitude structure of the ionospheric SW2 response is predominantly shaped by the migrating SW2 tide forced from below the DW1 response is mainly due to daily variability of the photo-ionization.
- The peculiar vertical structure of the ionospheric TW3 response, indicating downward/upward phase progression, hints at possible coupling between the SW2 tide forced from below and the diurnal variability of the electron density, i.e. at ion-neutral momentum coupling.

Despite some progress in investigating the ionospheric response to the atmospheric tides forced from below there are still observational results which need further explanations. For example, it is not clear why the bottom level response over the equator is not vertically propagating one, or why the electron density TW3 response does not correspond to the neutral wind and temperature TW3 upward energy propagation in the lower-middle thermosphere (up to ∼300 km altitude), etc. The answer of these and other questions cannot be found only from the analysis and combination of satellite and ground based measurements; it needs further detailed numerical simulations as well.

In conclusion, we note that by means of a detailed analysis of the SABER temperature tides and their global COSMIC electron density responses, we presented valuable and strong experimental evidence that confirms the new paradigm in upper atmospheric and ionospheric physics—that the entire ionosphere regularly responds to the troposphere and stratosphere. These results highlight the importance of understanding the variability of the lower atmospheric weather systems and the possible predictability of the ionospheric response to them. Additionally, the ionospheric tidal diagnostics presented in this overview paper provides the necessary observational constraints for the numerical simulations.

Acknowledgement We are grateful to the COSMIC and SABER teams for the access to the data respectively on http://cosmic-io.cosmic.ucar.edu/cdaac/ and http://saber.gats-inc.com.

References

M. Angelats i Coll, J.M. Forbes, Nonlinear interactions in the upper atmosphere: the $s = 1$ and $s = 3$ nonmigrating semidiurnal tides. J. Geophys. Res. (2002). doi:10.1029/2001JA900179

C. Arras, Ch. Jacobi, J. Wicker, Semidiurnal tidal signature in sporadic E occurrence rates derived from GPS radio occultation measurements at midlatitudes. Ann. Geophys. **27**, 2555–2563 (2009)

F. Azpilicueta, C. Brunini, S.M. Radicella, Global ionospheric maps from GPS observations using modip latitude. Adv. Space Res. **38**, 2324–2331 (2006)

L. Bankov, R. Heelis, M. Parrot, J.-J. Berthelier, P. Marinov, A. Vassileva, WN4 effect on longitudinal distribution of different ion species in the topside ionosphere at low latitudes by means of DEMETER, DMSP-F13 and DMSP-F15 data. Ann. Geophys. **27**, 1–10 (2009)

N.P. Benkova, M.G. Deminov, A.T. Karpachev, N.A. Kochenova, Yu.V. Kusnerevsky, V.V. Migulin, S.A. Pulinets, M.D. Fligel, Longitude features shown by topside sounder data and their importance in ionospheric mapping. Adv. Space Res. **10**, 857–866 (1990)

P.S. Brahmanandam, Y.-H. Chu, K.-H. Wu, H.-P. Hsia, C.-L. Su, G. Uma, Vertical and longitudinal electron density structures of equatorial E- and F-regions. Ann. Geophys. **29**, 81–89 (2011)

S. Chapman, R.S. Lindzen, *Atmospheric Tides: Thermal and Gravitational* (Gordon and Breach, New York, 1970), p. 200

C.-Z. Cheng, Y.-H. Kuo, R.A. Anthes, L. Wu, Satellite constellation monitors global and space weather. Eos Trans. AGU **87**, 166–167 (2006)

Y.-H. Chu, K.-H. Wu, C.-L. Su, Reply to comment by Lei et al. on "A new aspect of ionospheric E region electron density morphology". J. Geophys. Res. **115**, A07314 (2010). doi:10.1029/2010JA015334

S.L. England, T.J. Immel, E. Sagawa, S.B. Henderson, M.E. Hagan, S.B. Mende, H.U. Frey, C.M. Swenson, L.J. Paxton, The effect of atmospheric tides on the morphology of the quiet-time post-sunset equatorial ionospheric anomaly. J. Geophys. Res. **111**, A10S19 (2006a). doi:10.1029/2006JA011795

S.L. England, S. Maus, T.J. Immel, S.B. Mende, Longitudinal variation of the E-region electric fields caused by atmospheric tides. Geophys. Res. Lett. **33**, L21105 (2006b). doi:10.1029/2006GL027465

S.L. England, X. Zhang, T.J. Immel, J.M. Forbes, R. DeMajistre, The effect of non-migrating tides on the morphology of the equatorial ionospheric anomaly: seasonal variability. Earth Planets Space **61**, 493–503 (2009)

S.L. England, T.J. Immel, J.D. Huba, M.E. Hagan, A. Maute, R. DeMajistre, Modeling of multiple effects of atmospheric tides on the ionosphere: an examination of possible coupling mechanisms responsible for the longitudinal structure of the equatorial ionosphere. J. Geophys. Res. **115**, A05308 (2010). doi:10.1029/2009JA014894

B.G. Fejer, J.W. Jensen, S.-Y. Su, Quiet time equatorial F region vertical plasma drift model derived from ROCSAT-1 observations. J. Geophys. Res. **113**, A05304 (2008). doi:10.1029/2007JA012801

J.M. Forbes, R.G. Roble, C. Fesen, Accelerating, heating and compositional mixing of the thermosphere due to upward propagating tides. J. Geophys. Res. **98**(A1), 311–321 (1993). doi:10.1029/1992JA00442

J.M. Forbes, J. Russell, S. Miyahara, X. Zhang, S. Palo, M. Mlynczak, C.J. Mertens, M.E. Hagan, Troposphere-thermosphere tidal coupling as measure by the SABER instrument on TIMED during July-September 2002. J. Geophys. Res. **111**, A10S06 (2006). doi:10.1029/2005JA011492

J.M. Forbes, X. Zhang, S. Palo, J. Russell, C.J. Mertens, M. Mlynczak, Tidal variability in the ionospheric dynamo region. J. Geophys. Res. **113**, A02310 (2008). doi:10.1029/2007JA012737

T.J. Fuller-Rowell, R.A. Akmaev, F. Wu et al., Impact of terrestrial weather on the upper atmosphere. Geophys. Res. Lett. **35**, L09808 (2008). doi:10.1029/2007GL032911

M.E. Hagan, Comparative effects of migrating solar sources on tidal signatures in the middle and upper atmosphere. J. Geophys. Res. **101**(D16), 21213–21222 (1996)

M.E. Hagan, J.M. Forbes, Migrating and nonmigrating diurnal tides in the middle and upper atmosphere excited by tropospheric latent heat release. J. Geophys. Res. **107**, 4754 (2002). doi:10.1029/2001JD001236

M.E. Hagan, J.M. Forbes, Migrating and nonmigrating semidiurnal tides in the upper atmosphere excited by tropospheric latent heat release. J. Geophys. Res. **108**(A2), 1062 (2003). doi:10.1029/2002JA009466

M.E. Hagan, A. Maute, R.G. Roble, A.D. Richmond, T.J. Immel, S.L. England, Connections between deep tropical clouds and the Earth's ionosphere. Geophys. Res. Lett. **34**, L20109 (2007). doi:10.1029/2007GL030142

M.E. Hagan, A. Maute, R.G. Roble, Tropospheric tidal effects on the middle and upper atmosphere. J. Geophys. Res. **114**, A01302 (2009). doi:10.1029/2008JA013637

C. Haldoupis, D. Pancheva, N.J. Mitchell, A study of tidal and planetary wave periodicities present in midlatitude sporadic E layers. J. Geophys. Res. **109**, A02302 (2004). doi:10.1029/2003JA010253

C. Haldoupis, C. Meek, N. Christakis, D. Pancheva, A. Bourdillon, Ionogram height-time-intensity observations of descending sporadic E layers. J. Atmos. Sol.-Terr. Phys. **68**, 539–557 (2006)

C. Haldoupis, D. Pancheva, Terdiurnal tide-like variability in sporadic E layers. J. Geophys. Res. **111**, A07303 (2006). doi:10.1029/2005JA011522

K. Hamilton, Latent heat release as a possible forcing mechanism for atmospheric tides. Mon. Weather Rev. **109**, 3–17 (1981)

W.A. Hartman, R.A. Heelis, Longitudinal variations in the equatorial vertical drift in the topside ionosphere. J. Geophys. Res. **112**, A03305 (2007). doi:10.1029/2006JA011773

K. Häusler, H. Lühr, Nonmigrating tidal signals in the upper thermospheric zonal wind at equatorial latitudes as observed by CHAMP. Ann. Geophys. **27**, 2643–2652 (2009)

R.A. Heelis, Electrodynamics in the low and middle latitude ionosphere: a tutorial. J. Atmos. Sol.-Terr. Phys. **66**, 825–838 (2004)

F.T. Huang, H.G. Mayr, C.A. Reber, T. Killeen, J.M. Russell, M. Mlynczak, W. Skinner, J.G. Mengel, Diurnal variations of temperature and winds inferred from TIMED and UARS measurements. J. Geophys. Res. **111**, A10S04 (2006a). doi:10.1029/2005JA011426

F.T. Huang, H.G. Mayr, C.A. Reber, J.M. Russell, M. Mlynczak, J.G. Mengel, Stratospheric and mesospheric temperature variations for quasi-biennial and semiannual (QBO and SAO) oscillations based on measurements from SABER (TIMED) and MLS (UARS). Ann. Geophys. **24**, 2131–2149 (2006b)

T.J. Immel, E. Sagawa, S.L. England, S.B. Henderson, M.E. Hagan, S.B. Mende, H.U. Frey, C.M. Swenson, L.J. Paxton, Control of equatorial ionospheric morphology by atmospheric tides. Geophys. Res. Lett. **33**, L15108 (2006). doi:10.1029/2006GL026161

M.C. Jeruchim, P. Balaban, K.S. Shanmugan, *Simulation of Communication Systems: Modeling, Methodology and Techniques* (Kluwer Academic Plenum Publishers, Dordrecht, 2000), pp. 397–399
H. Jin, Y. Miyoshi, H. Fujiwara, H. Shinagawa, Electrodynamics of the formation of ionospheric wave number 4 longitudinal structure. J. Geophys. Res. **113**, A09307 (2008). doi:10.1029/2008JA013301
H. Jin, Y. Miyoshi, H. Fujiwara, H. Shinagawa, K. Terada, N. Terada, M. Ishii, Y. Otsuka, A. Saito, Vertical connection from the tropospheric activities to the ionospheric longitudinal structure simulated by a new Earth's whole atmosphere-ionosphere coupled model. J. Geophys. Res. **116**, A01316 (2011). doi:10.1029/2010JA015925
A.T. Karpachev, Characteristics of the global longitudinal effect in the night-time equatorial anomaly. Geomagn. Aeron. **28**(1), 46–49 (1988)
M.C. Kelley, V.K. Wong, N. Aponte, C. Coker, A.J. Mannucci, A. Komjathy, Comparison of COSMIC occultation-based electron density profiles and TIP observations with Arecibo incoherent scatter radar data. Radio Sci. **44**, RS4011 (2009). doi:10.1029/2008RS004087
H. Kil, R. DeMajistre, L.J. Paxton, Y. Zhang, Nighttime F-region morphology in the low and middle latitudes seen from DMSP F15 and TIMED/GUVI. J. Atmos. Sol.-Terr. Phys. **68**, 1672–1681 (2006)
H. Kil, S.-J. Oh, M. Kelley, L. Paxton, S. England, E. Talaat, K.-W. Min, S.-Y. Su, Longitudinal structure of the vertical $E \times B$ drift and ion density seen from ROCSAT-1. Geophys. Res. Lett. **34**, L14110 (2007). doi:10.1029/2007GL030018
H. Kil, E.R. Talaat, S.-J. Oh, L.J. Paxton, S.L. England, S.-Y. Su, The wave structures of the plasma density and vertical E × B drift in low-latitude F region. J. Geophys. Res. **113**, A09312 (2008). doi:10.1029/2008JA013106
N.A. Kochenova, Longitudinal variations of the equatorial ionosphere according to Intercosmos-19 data. Geomagn. Aeron. **21**(1), 142–144 (1987)
N.A. Kochenova, Longitudinal variations of $N(h)$ profiles at the magnetic equator. Geomagn. Aeron. **28**(1), 144–146 (1988)
A.K.H. Kong, P.A. Charles, E. Kuulkers, Long-term X-ray variability in GX 354-0. New Astron. **3**(5), 301–307 (1998)
Y.-H. Kuo, T.-K. Wee, S. Sokolovskiy, C. Rocken, W. Schreiner, D. Hunt, R.A. Anthes, Inversion and error estimation of GPS radio occultation data. J. Meteorol. Soc. Jpn. **82**(1B), 507–531 (2004)
J. Lei, S. Syndergaard, A.G. Burns et al., Comparison of COSMIC ionospheric measurements with ground-based observations and model predictions: Preliminary results. J. Geophys. Res. **112**, A07308 (2007). doi:10.1029/2006JA012240
J. Lei, J.P. Thayer, J.M. Forbes, Q. Wu, C. She, W. Wan, W. Wang, Ionosphere response to solar wind high-speed streams. Geophys. Res. Lett. **35**, L19105 (2008). doi:10.1029/2008GL035208
C.H. Lin, W. Wang, M.E. Hagan, C.C. Hsiao, T.J. Immel, M.L. Hsu, J.Y. Liu, L.J. Paxton, T.W. Fang, C.H. Liu, Plausible effect of atmospheric tides on the equatorial ionosphere observed by the FORMOSAT-3/COSMIC: three-dimensional electron density structures. Geophys. Res. Lett. **34**, L11112 (2007a). doi:10.1029/2007GL029265
C.H. Lin, C.C. Hsiao, J.Y. Liu, C.H. Liu, Longitudinal structure of the equatorial ionosphere: time evolution of the four-peaked EIA structure. J. Geophys. Res. **112**, A12305 (2007b). doi:10.1029/2007JA012455
H. Liu, M. Yamamoto, H. Lühr, Wave-4 pattern of the equatorial mass density anomaly: a thermospheric signature of tropical deep convection. Geophys. Res. Lett. **36**, L18104 (2009). doi:10.1029/2009GL039865
J.Y. Liu, C.Y. Lin, C.H. Lin, H.F. Tsai, S.C. Solomon, Y.Y. Sun, T. Lee, W.S. Schreiner, Y.H. Kuo, Artificial plasma caves in the low-latitude ionosphere results from the radio occultation inversion of the FORMOSAT-3/COSMIC. J. Geophys. Res. (2010). doi:10.1029/2009JA015079
H. Lühr, K. Häusler, C. Stolle, Longitudinal variation of F region electron density and thermospheric zonal wind caused by atmospheric tides. Geophys. Res. Lett. **34**, L16102 (2007). doi:10.1029/2007GL030639
H. Lühr, M. Rother, K. Häusler, P. Alken, S. Maus, Influence of nonmigrating tides on the longitudinal variations of the equatorial electrojet. J. Geophys. Res. **113**, A08313 (2008). doi:10.1029/2008JA013064
J.D. Mathews, E. Sporadic, Current views and recent progress. J. Atmos. Sol.-Terr. Phys. **60**, 413–435 (1998)
C. McLandress, W.E. Ward, Tidal/gravity wave interactions and their influence on the large scale dynamics of the middle atmosphere: model results. J. Geophys. Res. **99**, 8139–8156 (1994)
C.J. Mertens et al., Retrieval of mesospheric and lower thermospheric kinetic temperature from measurements of CO_2 15 μm earth limb emission under non-LTE conditions. Geophys. Res. Lett. **28**, 1391–1394 (2001)
C.J. Mertens et al., SABER observations of mesospheric temperature and comparisons with falling sphere measurements taken during the 2002 summer MaCWINE campaign. Geophys. Res. Lett. **31**, J03105 (2004). doi:10.1029/2003GL018605
P. Mukhtarov, D. Pancheva, B. Andonov, Global structure, seasonal and interannual variability of the migrating diurnal tide seen in the SABER/TIMED temperatures between 20 and 120 km. J. Geophys. Res. **114**, A02309 (2009). doi:10.1029/2008JA013759

P. Mukhtarov, D. Pancheva, Global ionospheric response to nonmigrating DE3 and DE2 tides forced from below. J. Geophys. Res. (2011). doi:10.1029/2010JA016099

J. Oberheide, M.E. Hagan, R.G. Roble, D. Offermann, Sources of nonmigrating tides in the tropical middle atmosphere. J. Geophys. Res. **107**, 4567 (2002). doi:10.1029/2002JD002220

J. Oberheide, Q. Wu, T.L. Killeen, M.E. Hagan, R.G. Roble, Diurnal nonmigrating tides from TIMED Doppler Interferometer wind data: monthly climatologies and seasonal variations. J. Geophys. Res. **111**, A10S03 (2006). doi:10.1029/2005JA011491

J. Oberheide, Q. Wu, T.L. Killeen, M.E. Hagan, R.G. Roble, A climatology of nonmigrating semidiurnal tides from TIMED Doppler Interferometer (TIDI) wind data. J. Atmos. Sol.-Terr. Phys. **69**, 2203–2218 (2007)

J. Oberheide, J.M. Forbes, Tidal propagation of deep tropical cloud signatures into the thermosphere from TIMED observations. Geophys. Res. Lett. **35**, L04816 (2008). doi:10.1029/2007GL032397

J. Oberheide, J.M. Forbes, K. Haüsler, Q. Wu, S.L. Bruinsma, Tropospheric tides from 80–400 km: propagation, inter-annual variability and solar cycle effects. J. Geophys. Res. **114**, D00I05 (2009). doi:10.1029/2009JD012388

J. Oberheide, J.M. Forbes, X. Zhang, S.L. Bruinsma, Wave-driven variability in the ionosphere-thermosphere-mesosphere system from TIMED observations: what contributes to the "wave4"? J. Geophys. Res. **116**, A01306 (2011). doi:10.1029/2009JA015911

D. Pancheva, C. Haldoupis, C. Meek, A. Manson, N. Mitchell, Evidence for a role of modulated atmospheric tides in the dependence of sporadic E layers on planetary waves. J. Geophys. Res. **108**(A5), 1176 (2003). doi:10.1029/2002JA009788

D. Pancheva, P. Mukhtarov, B. Andonov, Nonmigrating tidal activity related to the sudden stratospheric warming in the Arctic winter of 2003/2004. Ann. Geophys. **27**, 975–987 (2009a)

D. Pancheva, P. Mukhtarov, B. Andonov, Global structure, seasonal and interannual variability of the migrating semidiurnal tide seen in the SABER/TIMED temperatures (2002–2007). Ann. Geophys. **27**, 687–703 (2009b)

D. Pancheva, P. Mukhtarov, B. Andonov, N.J. Mitchell, J.M. Forbes, Planetary waves observed by TIMED/SABER in coupling the stratosphere-mesosphere-lower thermosphere during the winter of 2003/2004: Part 1, Comparison with the UKMO temperature results. J. Atmos. Sol.-Terr. Phys. **71**, 61–74 (2009c)

D. Pancheva, P. Mukhtarov, B. Andonov, Reply to Manson et al.'s comment on "Global structure, seasonal and interannual variability of the migrating semidiurnal tide seen in the SABER/TIMED temperatures (2002–2007)". Ann. Geophys. **28**, 677–685 (2010a)

D. Pancheva, P. Mukhtarov, B. Andonov, Global distribution, seasonal and interannual variability of the eastward propagating tides seen in the SABER/TIMED temperatures (2002–2007). Adv. Space Res. **46**, 257–274 (2010b). doi:10.1016/j.asr.2010.03.026

D. Pancheva, P. Mukhtarov, Strong evidence for the tidal control on the longitudinal structure of the ionospheric F-region. Geophys. Res. Lett. **37**, L14105 (2010). doi:10.1029/2010GL044039

D. Pancheva, P. Mukhtarov, Atmospheric tides and planetary waves: recent progress based on SABER/TIMED, in *IAGA Special Sopron Book Series 2, Aeronomy of the Earth's Atmosphere and Ionosphere*, ed. by M. Abdu, D. Pancheva (Springer, Berlin, 2011), pp. 19–56, doi:10.1007/978-94-007-0326-1

N.M. Pedatella, J.M. Forbes, J. Oberheide, Intra-annual variability of the low-latitude ionosphere due to nonmigrating tides. Geophys. Res. Lett. **35**, L18104 (2008). doi:10.1029/2008GL035332

L. Perrone, A.V. Mikhailov, L.P. Korsunova, FORMOSAT-3/COSMIC E region observations and daytime f_oE at middle latitudes. J. Geophys. Res. **116**, A06307 (2011). doi:10.1029/2010JA016411

S.M. Radicella, R. Leitinger, The evolution of the DGR approach to model electron density profiles. Adv. Space Res. **27**(1), 35–40 (2001)

K. Rawer, in *Meteorological and Astronomical Influences on Radio Wave Propagation*, ed. by B. Landmark (Oxford, Pergamon Press, 1963), pp. 221–250

K. Rawer (ed.), *Encyclopedia of Physics, Geophysics III, Part VII* (Springer, Berlin, 1984), pp. 389–391

E.E. Remsberg, B.T. Marshall, M. García-Comas, D. Krueger, G.S. Lingenfelser, J. Martin-Torres, M.G. Mlynczak, J.M. Russell, A.K. Smith, Y. Zhao, C. Brown, L.L. Gordley, M.J. Lopez-Gonzales, M. Lopez-Puertas, C.-Y. She, M.J. Taylor, R.E. Thompson, Assessment of the quality of the Version 1.07 temperature-versus-pressure profiles of the middle atmosphere from TIMED/SABER. J. Geophys. Res. **113**, D17101 (2008). doi:10.1029/2008JD0100113

Z. Ren, W. Wan, L. Liu, B. Zhao, Y. Wei, X. Yue, R.A. Heelis, Longitudinal variations of electron temperature and total ion density in the sunset equatorial topside ionosphere. Geophys. Res. Lett. **35**, L05108 (2008). doi:10.1029/2007GL032998

Z. Ren, W. Wan, L. Liu, J. Xiong, Intra-annual variation of wavenumber-4 structure of vertical $E \times B$ drifts in the equatorial ionosphere seen from ROCSAT-1. J. Geophys. Res. (2009). doi:10.1029/2009JA014060

Z. Ren, W. Wan, J. Xiong, L. Liu, Simulated wavenumber 4 structure in the equatorial F region vertical plasma drifts. J. Geophys. Res. **115**, A05301 (2010). doi:10.1029/2009JA014746

Z. Ren, W. Wan, L. Liu, Y. Cheng, H. Le, Equinoctial asymmetry of ionospheric vertical plasma drifts and its effect on F-region plasma density. J. Geophys. Res. **116**, A02308 (2011). doi:10.1029/2010JA016081

A.D. Richmond, The ionospheric wind dynamo: effects of its coupling with different atmospheric regions, in *The Upper Mesosphere and Lower Thermosphere: A Review of Experiments and Theory*, ed. by R.M. Johnson, T.L. Killeen. Geophys. Monogr. Ser., vol. 87 (AGU, Washington, 1995), pp. 49–65

H. Rishbeth, I.C.F. Müller-Wodarg, L. Zou, T.J. Fuller-Rowell, G.H. Millward, R.J. Moffett, D.W. Idenden, A.D. Aylward, Annual and semiannual variations in the ionospheric F2-layer: II. Physical discussion. Ann. Geophys. **18**, 945–956 (2000)

R.G. Roble, The NCAR thermosphere-ionosphere-mesosphere-electrodynamics general circulation model, in *STEP Handbook on Ionospheric Models*, ed. by R.W. Schunk (Utah State Univ., Logan, 1996), pp. 207–216

R.G. Roble, On the feasibility of developing a global atmospheric model extending from the ground to the exosphere, in *Atmospheric Science Across the Stratopause*, ed. by D.E. Siskind, S.D. Eckermann, M.E. Summers. Geophys. Monogr. Ser., vol. 123 (AGU, Washington, 2000), pp. 53–67

E. Sagawa, T.J. Immel, H.U. Frey, S.B. Mende, Longitudinal structure of the equatorial anomaly in the nighttime ionosphere observed by IMAGE/FUV. J. Geophys. Res. **110**, A11302 (2005). doi:10.1029/2004JA010848

W.S. Schreiner, S.V. Sokolovskiy, C. Rocken, D.C. Hunt, Analysis and validation of GPS/MET radio occultation data in the ionosphere. Radio Sci. **34**(4), 949–966 (1999)

S.M. Stankov, N. Jakowski, S. Heise, P. Muhtarov, I. Kutiev, R. Warnant, A new method for reconstruction of the vertical electron density distribution in the upper ionosphere and plasmosphere. J. Geophys. Res. **108**(A5), 1164 (2003). doi:10.1029/2002JA009570

G. Thuillier, J.R.H. Wiens, G.G. Shepherd, R.G. Roble, Photochemistry and dynamics in 566 thermospheric intertropical arcs measure by the WIND imaging interferometer on board UARS: 567 A comparison with TIE-GCM simulations. J. Atmos. Sol.-Terr. Phys. **64**, 405–415 (2002)

W. Wan, L. Liu, X. Pi, M.-L. Zhang, B. Ning, J. Xiong, F. Ding, Wavenumber-4 patterns of the total electron content over the low latitude ionosphere. Geophys. Res. Lett. **35**, L12104 (2008). doi:10.1029/2008GL033755

K.H. West, R.A. Heelis, Longitude variations in ion composition in the morning and evening 574 topside equatorial ionosphere near solar minimum. J. Geophys. Res. **101**, 7951–7960 (1996)

C.R. Williams, S.K. Avery, Diurnal nonmigrating tidal oscillations forced by deep convective clouds. J. Geophys. Res. **101**, 4079–4091 (1996)

J. Xu, A.K. Smith, H.-L. Liu, W. Yuan, Q. Wu, G. Jiang, M.G. Mlynczak, J.M. Russell III, S.J. Franke, Seasonal and quasi-biennial variations in the migrating diurnal tide observed by Thermosphere, Ionosphere, Mesosphere, Energetics and Dynamics (TIMED). J. Geophys. Res. **114**, D13107 (2009). doi:10.1029/2007JD011298

X. Zhang, J.M. Forbes, M.E. Hagan, J.M. Russell III, S.E. Palo, C.J. Mertens, M.G. Mlynczak, Monthly tidal temperatures 20–120 km from TIMED/SABER. J. Geophys. Res. **111**, A10S08 (2006). doi:10.1029/2005JA011504

X. Zhang, J.M. Forbes, M.E. Hagan, Longitudinal variation of tides in the MLT region: 2. Relative effects of solar radiative and latent heating. J. Geophys. Res. **115**, A06317 (2010). doi:10.1029/2009JA014898

X. Yue, W.S. Schreiner, J. Lei, S.V. Sokolovsky, C. Rocken, D.C. Hunt, Y.-H. Kuo, Error analysis of Abel retrieved electron density profiles from radio occultation measurements. Ann. Geophys. **28**, 217–222 (2010)

Space Sci Rev (2012) 168:211–236
DOI 10.1007/s11214-011-9842-4

A Review of the Effects of Non-migrating Atmospheric Tides on the Earth's Low-Latitude Ionosphere

S.L. England

Received: 1 April 2011 / Accepted: 20 September 2011 / Published online: 15 October 2011

Abstract Solar thermal tides are planetary-scale waves in the neutral atmosphere with periods that are harmonics of 24 hours. In the thermosphere, they can achieve significant amplitude and can be the dominant source of variation in the atmosphere. Through their modification of the neutral atmosphere, they can also significantly modify the ionosphere, especially at low-latitudes where the dynamics of the Earth's ionosphere is determined to a large extent by the neutral atmosphere. Much recent work has focused on characterizing and understanding the impact of one sub-group of tides, known as non-migrating tides, on the ionosphere. Whereas migrating tides are responsible for creating strong day-night variations in the ionosphere, non-migrating tides create longitudinal variations in the ionosphere, the signature of which can only be detected with distributed networks of ground-based observations or spacecraft. The present work reviews the recent observations and modeling efforts that have helped to characterize and explain this longitudinal variability. Emphasis is placed on the characteristics of tides throughout the thermosphere, their impacts on the chemical composition of the thermosphere, and impacts on the ionosphere.

Keywords Atmospheric tides · Earth's ionosphere

1 Introduction

Over half of all of the plasma in the Earth's ionosphere is found in the O^+ F-region at low latitudes. This plasma is produced from photoionization of O atoms in the thermosphere. This production is strongest close to the sub-solar point at a height that is close to the bottom of the F-region (~200 km). The daily pattern of neutral winds in the thermosphere, a combination of the background diurnal pressure gradient, the migrating diurnal tide and other tidal components (Tarpley 1970) drives a Pedersen current in the E-region ionosphere that produces a diurnal eastward electric field at low latitudes (Kato 1956, 1957). This dynamo electric field leads to an $\mathbf{E} \times \mathbf{B}$ drift of the plasma at low-latitudes. Close

S.L. England (✉)
Space Sciences Laboratory, University of California Berkeley, Berkeley, CA, USA
e-mail: england@ssl.berkeley.edu

to the magnetic equator at local noon, the direction of this drift is upwards. This drift removes O^+ ions from the production region, which allows for further O^+ production as the region returns to photochemical equilibrium. Thus, newly created O^+ ions are continuously lifted to higher altitudes, where recombination via collisions with neutral molecular species is greatly reduced and the lifetime of the O^+ increases. This leads to the accumulation of dense plasma at the F-region peak (~300–400 km close to the geomagnetic equator. This dense plasma moves along the magnetic field lines to higher latitudes and lower altitudes due to gravity and pressure gradients (Hanson and Moffett 1966). This leads to the formation of two bands of dense plasma, located ~10–20° either side of the geomagnetic equator and a local minimum at the magnetic equator (Namba and Maeda 1939; Appleton 1946). This is known as the equatorial ionospheric anomaly (EIA hereafter).

Early observations of variations in the EIA with longitude have been reviewed by Walker (1981). Variations with longitude are expected to arise solely from geometry of the migrating tide, geomagnetic field and ion production for four different reasons. (1) As the geomagnetic equator does not follow the geographic equator (see Fig. 1), the distance between the sub-solar point and the geomagnetic equator vary as a function of longitude. This can be expected to lead to longitudinal variations in the EIA as the distance between the O^+ production region and region of vertical plasma uplift varies. This effect should vary with season, with the largest effects expected at solstice. (2) The variation in the magnetic declination as a function of longitude changes the angle between the neutral winds and geomagnetic field. Thus, even if the tidal winds do not vary with longitude, the component of the wind parallel and perpendicular to the geomagnetic field will vary, thus affecting the way in which these winds move the ions (Challinor and Eccles 1971; Eccles et al. 1971; Thuilleir and Blamont 1973). (3) The strength of the horizontal component of the geomagnetic field varies by a factor of 1.5 from around South America to around India. While this may be expected to modify the $\mathbf{E} \times \mathbf{B}$ drift velocities and hence the EIA, early work by Walton and Bowhill (1979) appeared to show that this was not the case. (4) Winds associated with the migrating tide at F-region altitudes blow poleward from the sub-solar point. At longitudes where the geomagnetic equator is offset from this location, an asymmetric meridional wind pattern exists in geomagnetic coordinates. This can lead to asymmetric advection of plasma along the magnetic-field lines, producing asymmetries in the EIA that vary with longitude (Bramley and Young 1968; Lyon and Thomas 1963; Thuillier et al. 2002).

The present work is focused on reviewing the effects of non-migrating tides and specifically the longitudinal variations these wave create in the EIA, which are distinct and (sometimes) easy to separate from those induced by the geometric effects described above.

Solar thermal tides can be generated by the diabatic heating of the atmosphere, either directly by absorption of solar radiation, absorption of re-radiated light (especially infrared from the Earth's surface) or from latent heat release when water vapor condenses into water droplets during cloud formation. Alternatively, they can be generated by wave-wave interactions. Zonally symmetric forcing of the atmosphere produces the migrating tides with periods of 24, 12 and 8 hours and phase velocities that match the apparent motion of the sub-solar point. Zonally asymmetric forcing is able to produce non-migrating tides, which have the same periods, but different phase velocities to the migrating tides. These waves can propagate eastwards, westwards or have stationary phase relative to the ground. By convention, tides are named according to their period, propagation direction and zonal wavenumber; for example the DE2 tide is the diurnal (24 hour), eastward propagating tide with zonal wavenumber-2. An illustration of this tide in different reference frames is given in Fig. 1. This shows the wave when viewed as a snapshot in time (a), when its propagation in universal time (UT hereafter) is plotted in geographic coordinates (b) and when its propagation

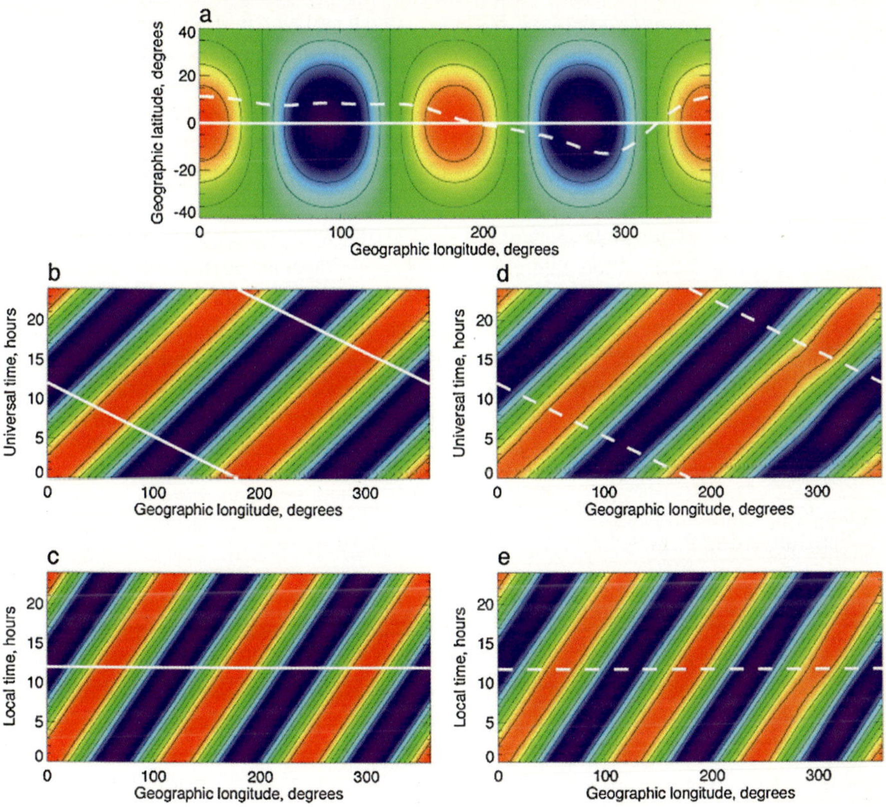

Fig. 1 Amplitude of an idealized diurnal eastward wavenumber 2 tide (DE2). (**a**) The amplitude as a function of geographic latitude and longitude at a snapshot in time. The geographic and geomagnetic equators are marked by the *thick solid* and *dashed lines* respectively. (**b**) The amplitude at the geographic equator as a function of geographic longitude and universal time. Local noon is marked by the *white line*. (**c**) The amplitude at the geographic equator as a function of local time and geographic longitude. Local noon is marked by the *white line*. (**d**) The amplitude at the geomagnetic equator as a function of universal time and geographical longitude. Local noon is marked by the *dotted line*. (**e**) The amplitude at the geomagnetic equator as a function of local time and geographic longitude. Local noon is marked by the *dotted line*. The units of amplitude are arbitrary

in local time (LT hereafter) is plotted in geographic coordinates (c). The shift in reference frame from UT to LT has the result of Doppler-shifting the wavenumber of the wave from 2 to 3 (the result of viewing an eastward propagating wave in a westward propagating reference frame). This can also be thought of as a rotation of the axes of panel b such that the dashed line becomes parallel with the bottom axis (panel c). As the ionosphere is so variable from day to night, with different processes dominating at different local times, many of the observations of the ionosphere that will be presented in the following sections are shown in the LT frame. Further, many are plotted in geomagnetic coordinates, rather than geographic. The effect of this shift in reference frame on the idealized wave considered here is shown in Fig. 1 panels (d) and (e).

Tides are forced at all altitudes within the atmosphere, but at low-latitudes perhaps the strongest source of zonal asymmetry is the land-sea distribution. The contrast between atmospheric conditions over land and sea strongly affects tropospheric dynamics and alters both radiative and latent heat patterns. Variations in the radiative heating were shown to give rise to non-migrating tides by Kato et al. (1982). Deep convection cells that form over tropical land masses are associated with intense thunderstorm activity in these regions (e.g. Kurz and Grewe 2002). The cloud formation associated with these deep convection cells deposits latent heat into the atmosphere over a ~15 km altitude range (Hong and Wang 1980). This, combined with longitudinal distribution of land-masses at low-latitudes makes this latent heat deposition another efficient source of the non-migrating atmospheric tides (Hagan and Forbes 2002, 2003). The effect of these tides on the EIA will be considered in this work. The present work will focus on the nature of non-migrating tides in the thermosphere and their effects on the EIA. Section 2 will outline observations and models of non-migrating tides at thermospheric altitudes. Section 3 will describe the subsequent impacts of these waves on thermospheric composition. The signature of these waves in the ionosphere will be reviewed in Sect. 4. Section 5 will describe the mechanisms by which non-migrating tides in the thermosphere may impose a variation on the ionosphere. Outstanding problems and topics for future research will be described in Sect. 6.

2 Non-migrating Atmospheric Tides in the Thermosphere

Tides can be forced at all altitudes throughout the atmosphere. Provided the wave energy is conserved, those that propagate vertically will increase (decrease) in amplitude as the atmospheric density decreases (increases). This means that those which propagate upwards tend to grow rapidly while those that propagate downwards diminish. Therefore, when considering the thermosphere, it is important to account for tides that originate at all altitudes throughout the atmosphere. Information about these waves in the thermosphere can be obtained from direct observations of the wave parameters (temperature, density, wind) or from observations of the subsequent effect these waves have on the atmosphere (changes in composition, airglow etc.). This section will describe the former—the wave properties that are directly linked to the propagation of the waves. Section 3 will describe the effects these waves have on thermospheric properties that are unrelated to the restoring forces that maintain them.

2.1 Tidal Winds

Hagan and Forbes (2002, 2003) used the Global Scale Wave Model (GSWM) to calculate which non-migrating tides were excited with the largest amplitudes by the latent-heat release in the tropics and radiative forcing throughout the atmosphere. They predicted that the largest amplitude non-migrating tides that should reach the upper mesosphere/lower thermosphere should be the DE3 and SW6. The authors also noted that the dominant tide would be a function of season as both the source in the troposphere and the propagation conditions throughout the middle atmosphere vary with season. They showed that the zonal winds and temperatures associated with the DE3 in the low-latitude lower thermosphere (below 30° latitude, 115 km altitude) could be expected to reach 45 m/s and 30 K amplitudes, exceeding the amplitude of the migrating diurnal tide in this location. With a vertical wavelength of around 30–40 km (large compared to the Hall conductivity layer) and symmetric latitudinal structure, this result suggested that the DE3 tide would be very effective at modifying the E-region dynamo. However, the model used by these authors did not account for dissipation effects that become important in the thermosphere, so could not provide accurate estimates of these tides throughout the thermosphere. This approach has been built

Fig. 2 (*Top*) Amplitude and (*bottom*) phase of the wavenumber-4 component in a LT frame of the (*left*) zonal and (*right*) meridional winds observed by WINDII. After Talaat and Lieberman (2010). Reproduced by permission of American Geophysical Union

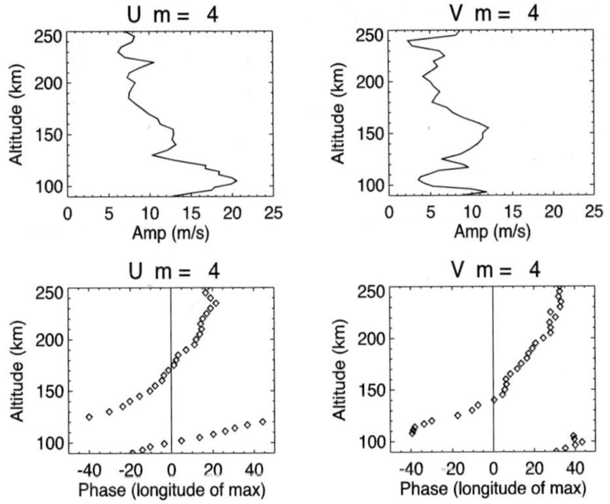

upon by Zhang et al. (2010), who have updated the GSWM model to include latent heat profiles from the Tropical Rainfall Measuring Mission (TRMM) and longitudinal asymmetries in the radiative forcing from the International Satellite Cloud Climatology Project (ISCCP). The updated GSWM results presented by these authors have shown that the radiate heating variations are important during all seasons and for both diurnal and semi-diurnal tides.

Continuous propagation of non-migrating tides throughout the thermosphere was identified by Talaat and Lieberman (2010), who used the wind observations from UARS-WINDII of the zonal and meridional wind components from 90–250 km altitude. In the lower thermosphere, where WINDII offers complete local time coverage of the winds, the authors were able to confirm that the zonal component of the non-migrating tidal winds are dominated by the DE3 tide (see their Fig. 3). They were then able to study the amplitude and phase of the wavenumber-4 component in a fixed LT frame (that would correspond to the DE3 tide) throughout the thermosphere (see Fig. 2). The continuous amplitude and phase structure appears to show that the DE3 tide can propagate at significant amplitude (above 5 m/s zonal winds) to heights above 250 km, although it should be noted that this analysis, which is based on daytime only data, cannot distinguish the DE3 tide from other components that have a wavenumber-4 structure in the LT frame (e.g. DW5, SE2 etc.).

It has been suggested by other studies that while the DE3 dominates the wavenumber-4 LT wind structure in the lower thermosphere, at higher altitudes a number of other waves may make a significant contribution. Hagan et al. (2009) used the TIME-GCM model to simulate the propagation of waves throughout the thermosphere. They found that a number of waves could contribute significantly to the amplitude of the wavenumber-4 structure in the zonal winds in a LT frame (see Fig. 3). Principle among these are the SE2 non-migrating tide and the stationary planetary wave 4. The stationary planetary wave is believed to be generated through a non-linear interaction of the DE3 and DW1 tides, and thus is expected to vary with season in the same manner as the DE3 tide, which could make it difficult to separate from the DE3 tide in observations that are only able to sample a limited range of LT (such as that shown in Fig. 2) and therefore deserves careful attention.

One important source of global-scale thermospheric wind data that covers all LT are the observations from the accelerometer onboard the CHAMP spacecraft. CHAMP was

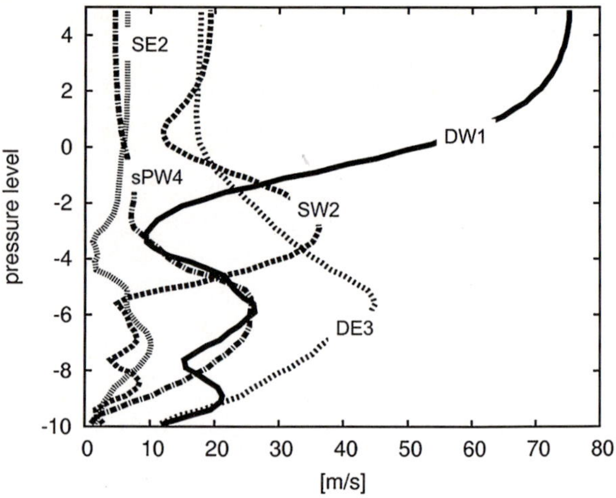

Fig. 3 Zonal wind amplitudes of the migrating DW1 tide, non-migrating DE3, SW2 and SE2 tides and the stationary planetary wave 4 (sPW4) simulated by the TIME-GCM model for September conditions. The *vertical scale* shows the TIME-GCM model pressure surfaces, where the range shown is equivalent to ~80–360 km altitude. After Hagan et al. (2009). Reproduced by permission of American Geophysical Union

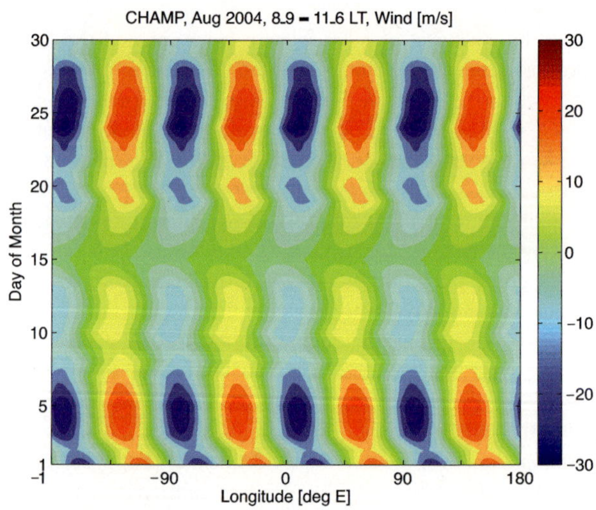

Fig. 4 Wavenumber-4 component of the zonal winds at ~400 km altitude for 8.9–11.6 hours LT inferred from the CHAMP accelerometer data during August of 2004. After Lühr et al. (2007). Reproduced by permission of American Geophysical Union

launched into a nearly polar orbit at ~400 km altitude, which allows the accelerometer data to be used to find the in situ zonal wind close to the equator. Lühr et al. (2007) showed that at this altitude, the zonal winds are dominated by a wavenumber-4 structure in LT during August–September (see Fig. 4), with wind amplitudes of order 30 m/s seen on some days. Häusler et al. (2010) performed a decomposition of the CHAMP data to determine which waves were contributing to these wavenumber-4 pattern throughout the year. The authors compared these data with the amplitudes predicted from the TIME-GCM model for March, June, September and December. The comparison for March is shown in Fig. 5. While the observations and model do not agree in detail, both show that several waves contribute to the various wavenumber components seen in a LT frame throughout the year. For the case of the wavenumber-4 component in March, the TIME-GCM predicted that this was dominated by the DE3 and to a lesser extent the SE2. The CHAMP observations appear to show that this is made up of a combination of the DE3 (with an amplitude of ~4 m/s) and the SE2, SW6 and sPW4 waves (each of which has an amplitude of ~2 m/s).

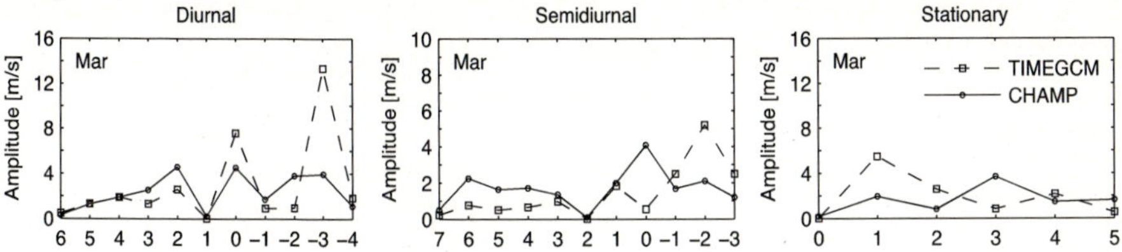

Fig. 5 Decomposition of diurnal, semi-diurnal and stationary waves that contribute to each wavenumber seen in the CHAMP and TIME-GCM zonal winds for March. After Häusler et al. (2010). Reproduced by permission of American Geophysical Union

2.2 Tidal Temperatures

Forbes et al. (2008) argued that while, for example the DE3 tide has the largest amplitude of those waves that produce a wavenumber-4 structure, it is the aggregate effect of the constructive/destructive interference of multiple waves that will be important for considering the E-region ionosphere. Therefore, in the case of the wavenumber-4 structure, we must also consider the impacts of the SW6, SE2, TE1 and TW7 tides as these all have moderate to significant amplitudes in the lower thermosphere. This combination of tides makes the seasonal and inter-annual variation of the wavenumber-4 structure of the atmosphere, and that can be expected to exist in the ionosphere, non-trivial. While the basic seasonal pattern repeats each year, it also responds to effects in the lower levels of the atmosphere such as the quasi-biennial and El-Niño oscillations (e.g. Oberheide et al. 2009). Their analysis was based on the amplitudes of the temperature perturbations created by these waves, as found from TIMED-SABER observations. These observations have been used to study the seasonal variations in non-migrating tides by a number of authors (Zhang et al. 2006; Kil et al. 2008; England et al. 2009) and for comparisons between these tides and the structure of the low-latitude ionosphere (England et al. 2009; Pancheva and Mukhtarov 2010).

More recently, Akmaev et al. (2008) have performed a detailed comparison of the non-migrating tidal temperature amplitudes observed by SABER in the lower thermosphere with those simulated using the Whole Atmosphere Model (WAM; see Fig. 6). While the TIME-GCM simulations cited above have non-migrating tides imposed at their lower boundary, the WAM model generates these waves internally. Thus, the relatively good comparison between the seasonal evolution of the temperature amplitudes simulated by WAM and those observed by SABER, as reported by the authors, demonstrates the ability of such models to simulate the generation and propagation of the key non-migrating tides from the lower atmosphere to the lower thermosphere in a self-consistent manner. Such results have also been demonstrated in other models such as the Whole Atmosphere Community Climate Model (WACCM) (Liu et al. 2010c). The temperature perturbations associated with non-migrating tides calculated from the GSWM were also used as the benchmark against which to compare the observed ionospheric signatures of these waves by Immel et al. (2006), England et al. (2006b) and Kil et al. (2007), demonstrating the important role that tidal temperature simulations have played in advancing understanding of the effects of non-migrating tides on the ionosphere.

2.3 Combined Wave Parameters

Using an analysis method known as Hough Mode Extension (HME hereafter; Oberheide and Forbes 2008b), it is possible to combine simultaneous observations of multiple tidal

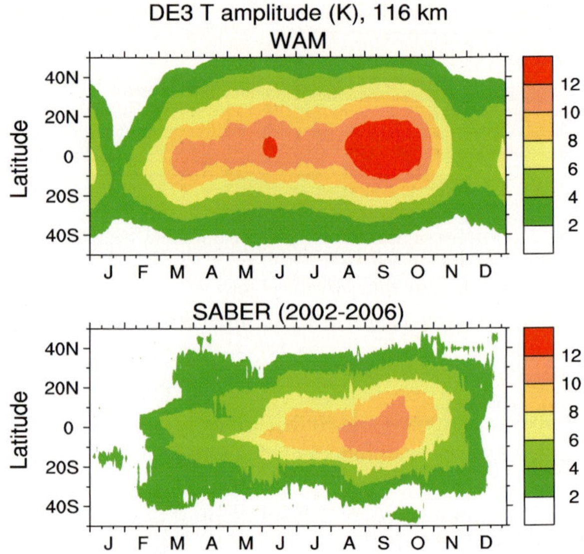

Fig. 6 Temperature amplitude associated with the DE3 tide as a function of latitude and month of the year at 116 km altitude, (*top*) as simulated by the WAM model and (*bottom*) from TIMED-SABER observations over 2002–2006. After Akmaev et al. (2008). Reproduced by permission of American Geophysical Union

parameters (temperatures and winds) and fit idealized tidal models (known as Hough Modes) to these parameters. Such fits can then be extended to higher altitudes, where fewer or no observations exist, given no further wave sources exist in that region. As observations of all atmospheric parameters do not exist at all altitudes, the HME method relies on some assumptions about the background atmosphere. In the case of Oberheide et al. (2009), the authors solved for the tidal propagation using the GSWM model with zero background zonal wind and which does not include a complete solution for the dissipation of these waves due to ion drag. Such simplifications do place limitations on the results of such an analysis. Nevertheless, these analyses have proved useful in climatological studies.

Oberheide et al. (2009) used a HME analysis constrained by TIMED-TIDI and SABER observations in the lower thermosphere in order to find the tidal amplitudes predicted throughout the thermosphere (see Fig. 7). For the case of the strong DE3 tide, they then compared the HME amplitudes to those observed in situ by the CHAMP spacecraft at ∼400 km altitude and found a relative good overall agreement on a season-to-season basis. Thus, while observations of all of the tidal parameters are not available at all altitudes throughout the thermosphere, the HME provides a useful tool for examining their structure throughout the entire thermosphere, at least for climatological studies.

3 The Thermospheric Response to Non-migrating Tides

3.1 Atomic Oxygen and Molecular Nitrogen

Panels (c–e) of Fig. 7 show that the typical vertical winds are around 2 orders of magnitude lower than the horizontal winds associated with atmospheric tides. However, given the relative scales of variations in the thermosphere (∼1000s km horizontally and ∼10s km vertically), all components of the wind can play an important role in advecting air parcels. In the lower thermosphere, strong vertical gradients in constituent mixing ratios exist. These result from altitude variations in production rates (from photodissociation etc.), recombination rates and scale heights that depend on molecular weight. These gradients are therefore

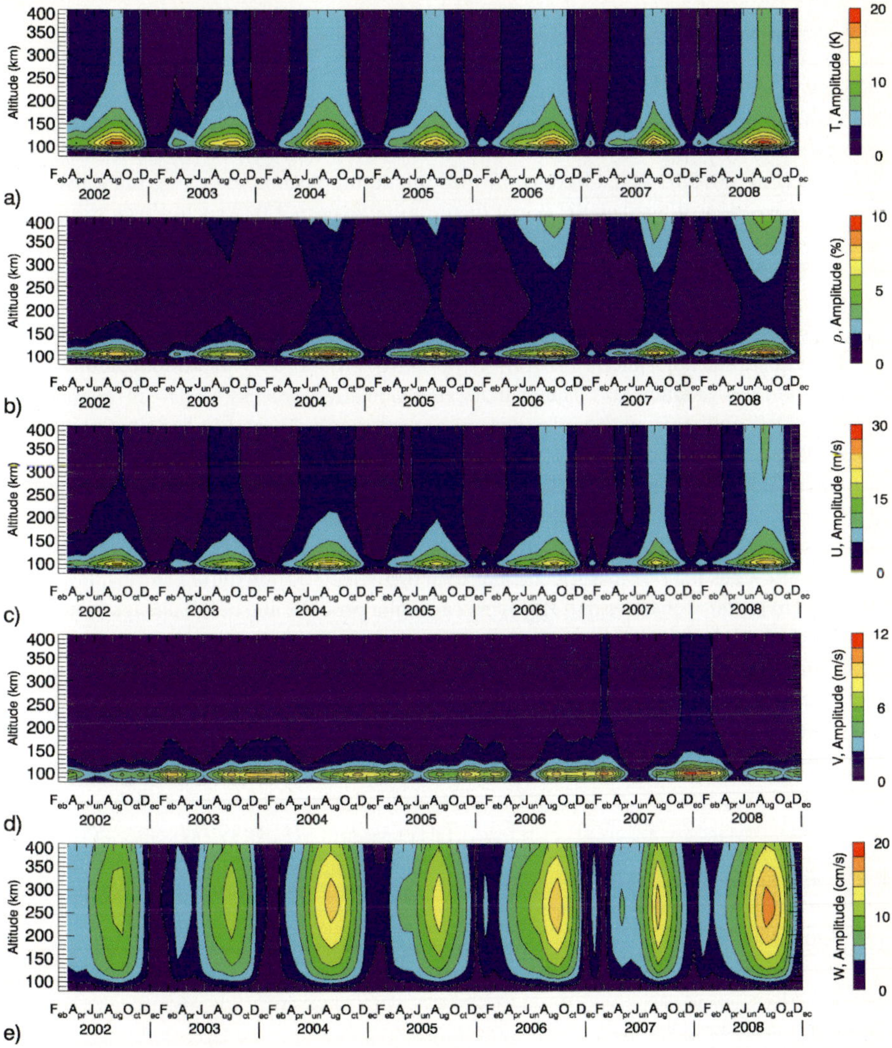

Fig. 7 DE3 tidal amplitudes averaged between 5° north and south of the geographic equator for (**a**) temperatures, (**b**) density and (**c**–**e**) neutral wind components (where U is the zonal, V is the meridional and W is the vertical component) given by the HME analysis constrained by TIMED-TIDI and TIME-SABER observations. After Oberheide et al. (2009). Reproduced by permission of American Geophysical Union

different for each species. The vertical displacement of an air parcel into a region where the background density is different modifies the collision rate for species within that parcel. For species that are in an excited state or are reactive on short time scales, such a vertical displacement can therefore alter the lifetime of species within that parcel. The combination of strong vertical gradients in composition, background density and species lifetimes means that as atmospheric tides cause vertical advection of air parcels, they can modify the composition of the atmosphere through which they propagate (Shepherd et al. 1997). This has been studied in detail for the case of the migrating diurnal tide at low-latitudes. Roble and Shepherd (1997) used a combination of both the TIME-GCM model and UARS-WINDII data to demonstrate that the vertical advection caused by this wave could create both a diurnal modulation of the density of O^1S and its associated 557.7 nm airglow in the lower thermosphere, and a net decrease in the total concentration of O^1S through the decreased species lifetime at lower altitudes. The impact of this wave on this excited state of atomic oxygen has been studied using other models (Harris et al. 2002; England et al. 2006a), which have confirmed this result. As non-migrating tides also cause vertical advection of air parcels, it can therefore be expected that such waves would have a similar impact on the composition of the thermosphere as the migrating tide described above. The effects of these waves should be separable from the migrating tide as the non-migrating tides would produce longitudinal variations in the composition in a LT frame, whereas the effects of the migrating tide ought to be approximately zonally symmetric.

The equatorial ionospheric F-region is primarily produced by photo-ionization of thermospheric O to produce O^+. The primary recombination process for O^+ involves charge exchange via collisions with molecular species (primarily N_2 and O_2) (Schunk and Nagy 2004). Thus, the ratio of $[O]/[N_2]$ is widely used as an indicator of when and where thermospheric composition may produce changes in the F-region ion density. During geomagnetic storms, strong Joule heating causes vertical uplift of air parcels which in turn alters thermospheric $[O]/[N_2]$ (Burns et al. 1995). While the vertical advection associated with atmospheric tides may be less spectacular than that which occurs during storms, it could be expected to produce changes in $[O]/[N_2]$ via a similar process, and hence change ionospheric density during quiet times.

Zhang et al. (2010) were the first to demonstrate that the quiet-time $[O]/[N_2]$ is dominated by wave-like structures in a LT frame that are indicative of non-migrating tides (see Fig. 8). The authors used observations from TIMED-GUVI to infer the vertical-column-integrated $[O]/[N_2]$. The authors found that throughout most of the year, the longitudinal pattern in

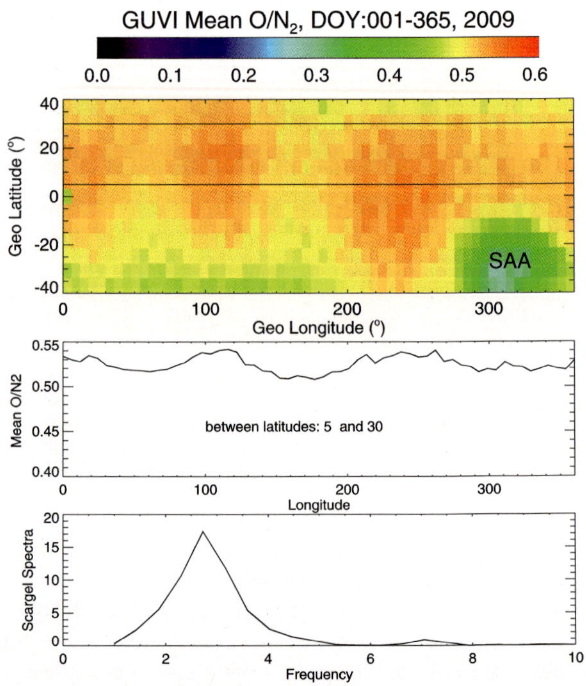

Fig. 8 (*Top*) Daytime average of the column $[O]/[N_2]$ during 2009, (*middle*) longitudinal variation in $[O]/[N_2]$ averaged over 5–30° north and (*bottom*) Scargel power spectrum of the longitudinal variation in $[O]/[N_2]$. The South Atlantic Anomaly is marked as SAA. After Zhang et al. (2010). Reproduced by permission of American Geophysical Union

daytime $[O]/[N_2]$ was dominated by a wavenumber-3 structure and only during summer is a wavenumber-4 structure strong. The authors attributed this pattern to the DE2 and DE3 tides and their associated wavenumber-three and -four structures. The DE2 tide has a longer vertical wavelength than the DE3 and so the authors argued that this might be more effective at creating longitudinal variations seen in this kind of vertical-column-integrated measure of $[O]/[N_2]$, but that during summer when the DE3 tide is especially strong, the wavenumber-four structure associated with this tide can also be seen. However, Kil et al. (2010) have examined the LT evolution of the wavenumber-3 structure and found it to be inconsistent with that expected if the DE2 tide was creating this pattern, creating some doubt as to its origin. Using the same type of GUVI data, He et al. (2010) performed a more detailed analysis of the wavenumber structure in thermospheric $[O]/[N_2]$ and its LT evolution. Using data from 2002–2007, they showed that the wavenumber-4 structure in $[O]/[N_2]$ follows the same seasonal pattern as the strength of the DE3 tide, but does not vary in LT as expected for the DE3 tide (does not move eastwards as in Fig. 1, panel c). The authors were not able to account for this from their observations alone.

England et al. (2010) used the TIME-GCM model to examine the wavenumber-4 structure in $[O]/[N_2]$ during equinox conditions. They found that the wavenumber-4 structure in $[O]/[N_2]$ in the model was the result of a combination of diurnal, semi-diurnal and stationary waves. This, combined with the large altitude integration of the GUVI observations described above, may account for the unexpected LT evolution of the wavenumber-4 $[O]/[N_2]$ structure that has been reported.

3.2 Nitric Oxide

The density distribution of nitric oxide (NO) in the lower thermosphere has been observed using the SNOE spacecraft. Oberheide and Forbes (2008a) used these observations from 2002–2006 in order to determine the impact of non-migrating tides on the distribution of NO. In the lower thermosphere, they found strong wavenumber-3 and -4 structures with up to 40% variations in NO seen from one longitude region to another (see Fig. 9). The wavenumber-4 signature in NO was shown to be dominant during September, while a strong wavenumber-3 signature was identified during December. By comparison to TIMED-SABER and TIDI observations, the authors determined that wavenumber-4 signature was likely the result of the DE3 tide, whereas both the DE2 and SE1 tides contributed strongly to the wavenumber-3 signature.

4 The Ionospheric Response to Non-migrating Tides

As atmospheric tides propagate through the thermosphere, they may modify the ionosphere through a variety of mechanisms that will be discussed in Sect. 5. Ultimately, through modification of the thermosphere, these waves are able to leave a wave-like signature in the ionosphere. While this wave-like signature itself is normally not thought of as a wave, it often has the same zonal wavenumber, LT evolution etc. as the atmospheric tide that is believed to have created it. The section will review the wide-range of observations and models of the ionosphere that have been used to define the low-latitude ionospheric response to non-migrating tides.

4.1 The Equatorial Electrojet

The equatorial electrojet (EEJ hereafter) is a thin band of intense current that flows close to the magnetic equator at a height of ~105 km and is strongest within the hours around local

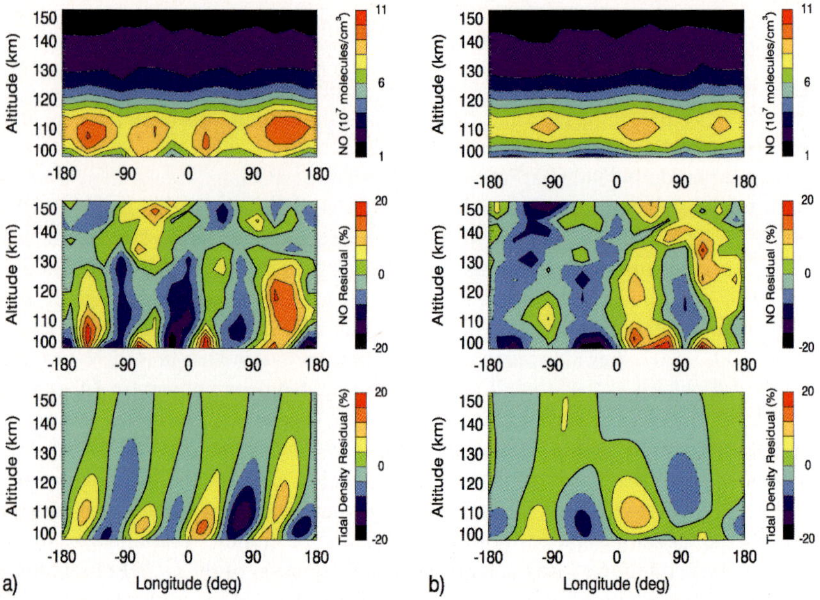

Fig. 9 Monthly-averaged (*top*) NO observations from SNOE, (*middle*) NO residuals from SNOE, and (*bottom*) tidal density residuals from HME based on TIMED-TIDI and SABER at 0° latitude for (**a**) September and (**b**) December for LT = 10.5 hours. After Oberheide and Forbes (2008b). Reproduced by permission of American Geophysical Union

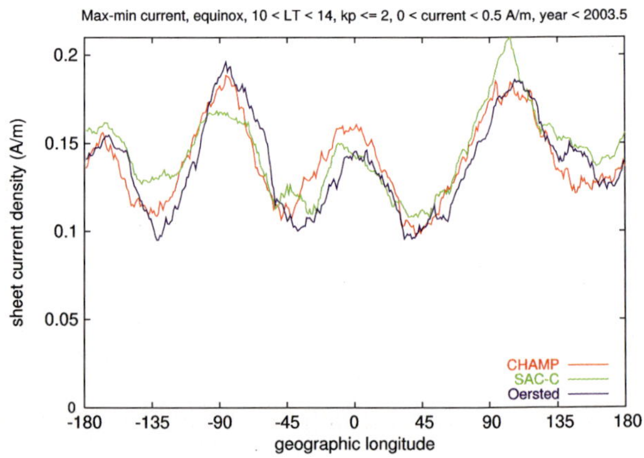

Fig. 10 Peak of the noontime electrojet sheet current density as a function of longitude for geomagnetically quiet periods ($Kp \leqslant 2$) around the March equinox derived from CHAMP, SAC-C and Ørsted magnetic field measurements. After England et al. (2006c). Reproduced by permission of American Geophysical Union

noon (Forbes 1981). The current in this region is particularly intense because the effective (Cowling) conductivity is very high in this limited region. The current density that flows in the EEJ responds to the strength of the electric field in the dayside ionosphere at these altitudes. England et al. (2006c) used observations of the current density in the EEJ derived from the magnetometer measurements from the CHAMP, SAC-C and Ørsted spacecraft to study the impact of non-migrating tides on this current system. They demonstrated that around March equinox, the EEJ displayed a strong wavenumber-4 signature in a fixed LT frame (see Fig. 10). This signature corresponded to, and was in-phase with the diurnal non-migrating tidal pattern calculated by the GSWM for the same season and thus the authors argued that this was evidence that the DE3 tide was modulating the E-region dynamo and hence the electric fields associated with the EEJ.

Alken et al. (2009) used the CHAMP, SAC-C and Ørsted spacecraft data to develop an empirical model of the EEJ. This model shows that the wavenumber-4 pattern is dominant during the northern spring, summer and autumn seasons, but the longitudinal structure of the EEJ is dominated by a wavenumber-3 pattern around the northern winter solstice. Lühr et al. (2008) examined the LT evolution of the wavenumber-4 strucutre in September and showed that this moved to the east with increasing LT at the same rate as the DE3 tide. This LT behavior and the correspondence of the seasonal behavior to the relative amplitude of the DE2 and DE3 tides, provides additional evidence that these waves are largely responsible for producing the longitudinal patterns seen in the EEJ.

4.2 Plasma Drifts and Currents

The observations of the EEJ appear to indicate that the noontime zonal electric field in the E-region ionosphere is varying spatially in response to the DE2 and DE3 tides. If this is true, a similar response should also be seen in the $\mathbf{E} \times \mathbf{B}$ drift of the ions in the equatorial ionosphere. As a result of the geomagnetic field lines being quasi-equipotentials, the $\mathbf{E} \times \mathbf{B}$ drift at F-region altitudes should also display the same longitudinal structure as the electric field in the E-region. It is therefore reasonable to expect this variation in the drift to be measured in situ by spacecraft at F-region altitudes. The ROCSAT-1 spacecraft measurements of these drifts at 600 km altitude have been used by a number of authors to study this. Kil et al. (2007) showed that the $\mathbf{E} \times \mathbf{B}$ drift around local noon (10–11 LT) averaged over March 1999–June 2004 have the same type of wavenumber-4 pattern as reported for the EEJ above. As both the EEJ and these $\mathbf{E} \times \mathbf{B}$ drift observations show the same feature at the same LT, that provides strong evidence to support the theory that the electric fields are varying in response to non-migrating tides. Kil et al. (2007) also examined the $\mathbf{E} \times \mathbf{B}$ drifts around 18–19 LT, when the F-region dynamo dominates these drifts and a strong uplift associated with the pre-reversal enhancement is observed. At this LT, the authors did not find any wavenumber-3 or wavenumber-4 pattern, but simply a variation in the drifts from a maximum around 0° longitude to a minimum around 60°–180° longitude. This suggests that either there is no variation in the F-region dynamo drifts, or that this variation is small compared with the other factors that affect the strength of the drifts at this LT.

Ren et al. (2009) studied the LT variations of the $\mathbf{E} \times \mathbf{B}$ drifts observed by ROCSAT-1 (see Fig. 11). They showed that the apparent shift in longitude with LT of the wavenumber-4 pattern matched what is expected for the DE3 tide. Further, they showed that the variation in the amplitude of the wavenumber-4 pattern with season matched that observed for the DE3 in the lower thermosphere by TIMED-SABER. From the modulation of the E-region dynamo by the DE3, it is expected that the $\mathbf{E} \times \mathbf{B}$ drift pattern should change sign at sunrise and sunset. However, as shown in Fig. 11, a more complex behavior was noted by Ren et al. (2009) at these LT. The authors argued that this was the result of the switch from the E- to F-region dynamos around these LT, each of which is driven by different wind patterns.

Hartman and Heelis (2007) used $\mathbf{E} \times \mathbf{B}$ drift measurements from the DMSP-15 spacecraft from 2001–2002 to study the longitudinal structure of these drifts at $\sim$830 km altitude. The DMSP 15 spacecraft samples these drifts at 9:30 LT, when the longitudinal variation in these drifts is not particularly strong (see Fig. 11). Nonetheless, the authors were able to show that the wavenumber-4 signature is seen even at these altitudes. Their observations also revealed a strong dependence on the declination of the geomagnetic field, which is less obvious in the observations from ROCSAT-1. This may be the result of the weaker

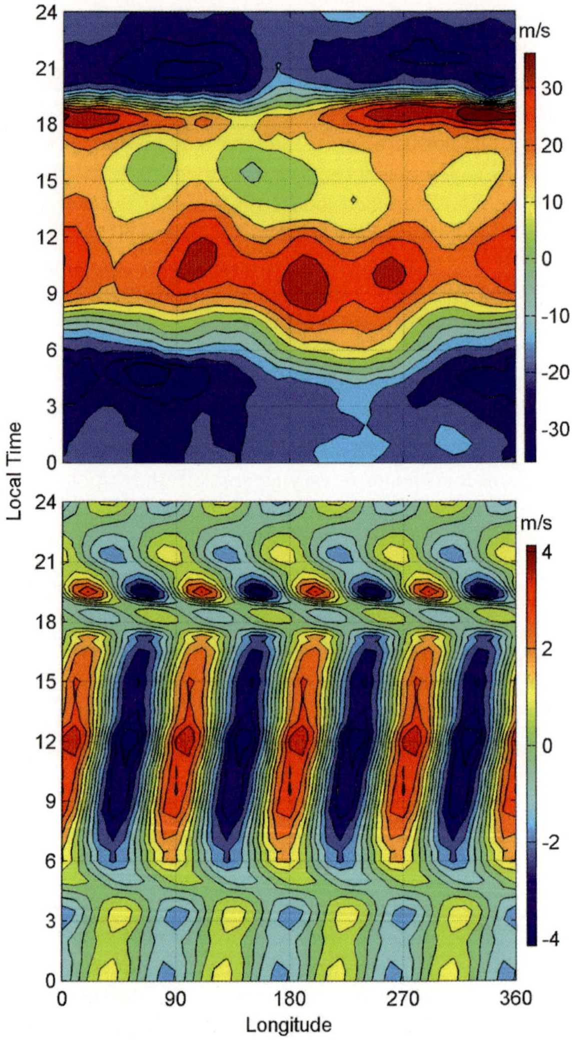

Fig. 11 (*Top*) Variation of the $\mathbf{E} \times \mathbf{B}$ drift observed by ROCSAT-1 as a function of LT and longitude, averaged over each October from 1999–2003, and (*bottom*) the wavenumber-4 component of the $\mathbf{E} \times \mathbf{B}$ drifts shown in the *top panel*. After Ren et al. (2009). Reproduced by permission of American Geophysical Union

wavenumber-4 at this LT, or that the wavenumber-4 is weaker at this altitude/on these higher L-value magnetic field lines.

Park et al. (2010) used the fluxgate magnetometer onboard the CHAMP spacecraft to derive the longitudinal structure of the currents that flow in the F-region at an altitude of ~400 km. The dynamo currents they found are primarily Pedersen currents that are generated by winds at F-region altitudes. The authors showed that around local noon, these currents display a wavenumber-4 signature, with a seasonal variation that matches that seen for the DE3 and the E-region electric fields described above. At dusk they found that this wavenumber-4 structure was significantly reduced, which is in agreement with the $\mathbf{E} \times \mathbf{B}$ drifts reported by Kil et al. (2007).

Several authors have calculated the $\mathbf{E} \times \mathbf{B}$ drifts induced by the DE3 tide in the dynamo region using different types of electrodynamic models. Ren et al. (2010) use an electrodynamic model (TIDM-IGGCAS II), driven by wind observations from TIMED-TIDI. These

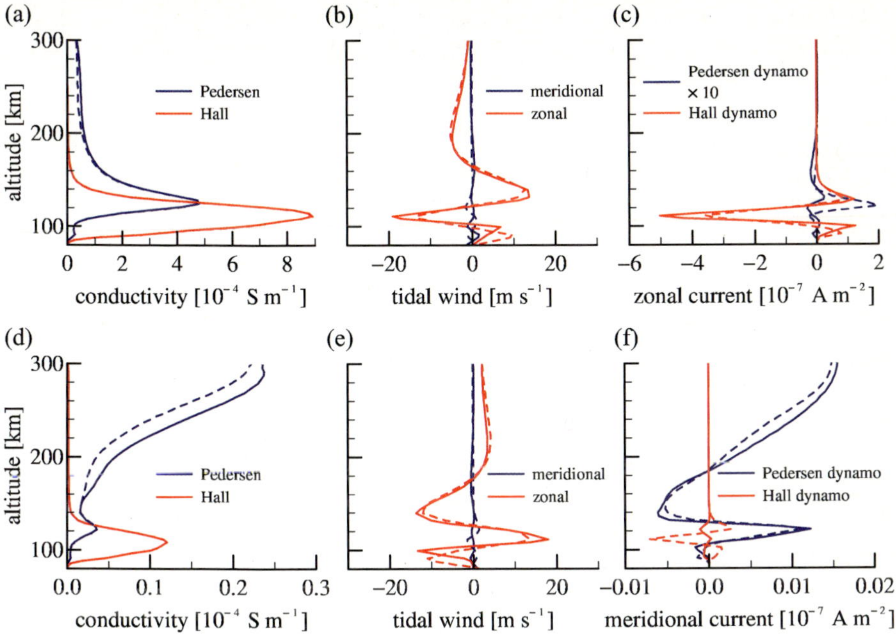

Fig. 12 Altitude profiles of the conductivities, DE3 tidal winds and dynamo currents along the field line with apex height of 300 km, calculated with the whole atmosphere model and electrodynamic model of Jin et al. (2008). Panels (**a**–**c**) are for 12 LT and (**d**–**f**) are for 20 LT. *Solid* and *dashed lines* represent the southern and northern hemispheres, respectively. In panels (**b**), (**c**), (**e**) and (**f**), positive values indicate eastward and northward. After Jin et al. (2008). Reproduced by permission of American Geophysical Union

observations covered the altitude range from 85–105 km. Above 105 km, the authors extrapolated these wind velocities with a linear decrease such that they reached 0 m/s at an altitude of 150 km. Their calculations showed that it was the symmetric, zonal wind component of the DE3 tide, driving a Hall current in the E-region ionosphere that was the most effective portion of the dynamo current system at creating the wavenumber-4 pattern in the $\mathbf{E} \times \mathbf{B}$ drifts in their model. Jin et al. (2008) performed a similar set of experiments with an electrodynamic model, but used a dynamical model of the atmosphere, rather than observed winds (see Fig. 12). In this model, the DE3 tide was seen to propagate throughout the thermosphere (panel b of Fig. 12) and created a strong zonal Hall current, similar to that seen in the Ren et al. (2010) model. England et al. (2010) used the TIME-GCM to examine the wavenumber-4 component of the equatorial $\mathbf{E} \times \mathbf{B}$ drift. Unlike both of the models described above, the TIME-GCM model has a fully coupled ionosphere and thermosphere. The authors found a strong wavenumber-4 signature in the daytime $\mathbf{E} \times \mathbf{B}$ drifts and a secondary peak around 18–19 LT, which is in agreement with the model results of Jin et al. (2008) (see their Fig. 3), but not in agreement with the observations of Kil et al. (2007). This suggests that some important component of either the propagation of non-migrating tides through the thermosphere, or their electrodynamic effects, is not well represented in these models. Millward et al. (2001) examined the impact of atmospheric tides on the vertical drifts in the equatorial F-region using the coupled thermosphere-ionosphere-plasmasphere model (CTIP). By changing the amplitude and phase of the thermospheric tidal forcing in the model, the authors showed that the vertical drifts around sunset where sensitive to changes in the semi-diurnal tide at low levels of solar activity, but not at high levels of solar activity.

While this alone may not explain the apparent disagreement between the various models and observations cited above, it is possible that through similar tests as those of Millward et al., the source of these discrepancies may be found.

4.3 Altitude Variations

As the equatorial **E** × **B** drifts show a wavenumber-4 pattern, and these drifts are responsible for the uplift of the O^+ in the F-region during the formation of the EIA, it can be expected that this will result in an in-phase variation of the altitude of the F-region at low-latitudes. This was shown in the observations reported by Thuillier and Blamont (1973), who used the scanning photometer onboard OGO-4 to study the spatial variation in the altitude of the 630.0 nm emission at ∼23 LT. The 630.0 nm emission involves both O^+ and O_2, where the density of O_2 is the limiting factor. The result of this is that the brightness decreases as the bottom side of the O^+ layer increases in altitude. The OGO-4 observations showed that during October 1967, the 630.0 nm brightness and emission height both varied in a wavenumber-4 pattern, but these were out of phase, which is consistent with a wavenumber-4 variation in the height of the F-region. England et al. (2006b) compared the 630.0 nm brightness observations of Thuillier and Blamont (1973) to the diurnal non-migrating tidal pattern from the GSWM and showed that these were out-of-phase, indicating that the bottom-side of the O^+ layer was being lifted higher in regions where the these tides were stronger at local noon.

Liu et al. (2010b, 2010a) have examined the variation in the peak of the F-region ($h_m F_2$) in a fixed LT frame using observations of the electron density from COSMIC-GPS. They showed that the height of the F-region peak also varies with a wavenumber-4 pattern, confirming that the variation in **E** × **B** drift described above causes a variation in the altitude of the F-region.

The 630.0 nm observations described above revealed a variation in the ion-density at the bottom of the F-region (∼200–250 km altitude). In situ measurements of the ion density can also be made on the topside of the F-region and these have been used to study the impacts of non-migrating tides on the density of the ionosphere at these altitudes. Lühr et al. (2007) and Kil et al. (2008) used in situ measurements of the ion density from the CHAMP and ROCSAT-1 spacecraft, respectively, to deduce the longitudinal variation in the ion density at ∼400 km and ∼600 km altitude, respectively. Lühr et al. (2007) reported a wavenumber-4 in the ion density, both pre-noon and post-sunset. The authors showed that the pattern in the ion-density was in-phase with the zonal-wind pattern in situ during the night, but out-of-phase during the day. Kil et al. (2008) compared the in situ ion densities observations to the in situ **E** × **B** drift observations during different seasons and found that these were in good agreement, suggesting that the variations in both **E** × **B** and ion density have the same source.

4.4 Total Electron Content

Given that both the height of the F-region and density in both the bottom- and top-sides display a response to non-migrating tides, a number of authors have looked at the total column electron density in the ionosphere (Scherliess et al. 2008), or the column density integrated over a range of altitudes (Lin et al. 2007). These data have been analyzed over each season, LT and altitude range, in order to reveal the dominant longitudinal structures. One example is shown in Fig. 13, adapted from Lin et al. (2007). This shows the strong diurnal variation in the electron density in the F-region that is driven by the diurnal cycles in

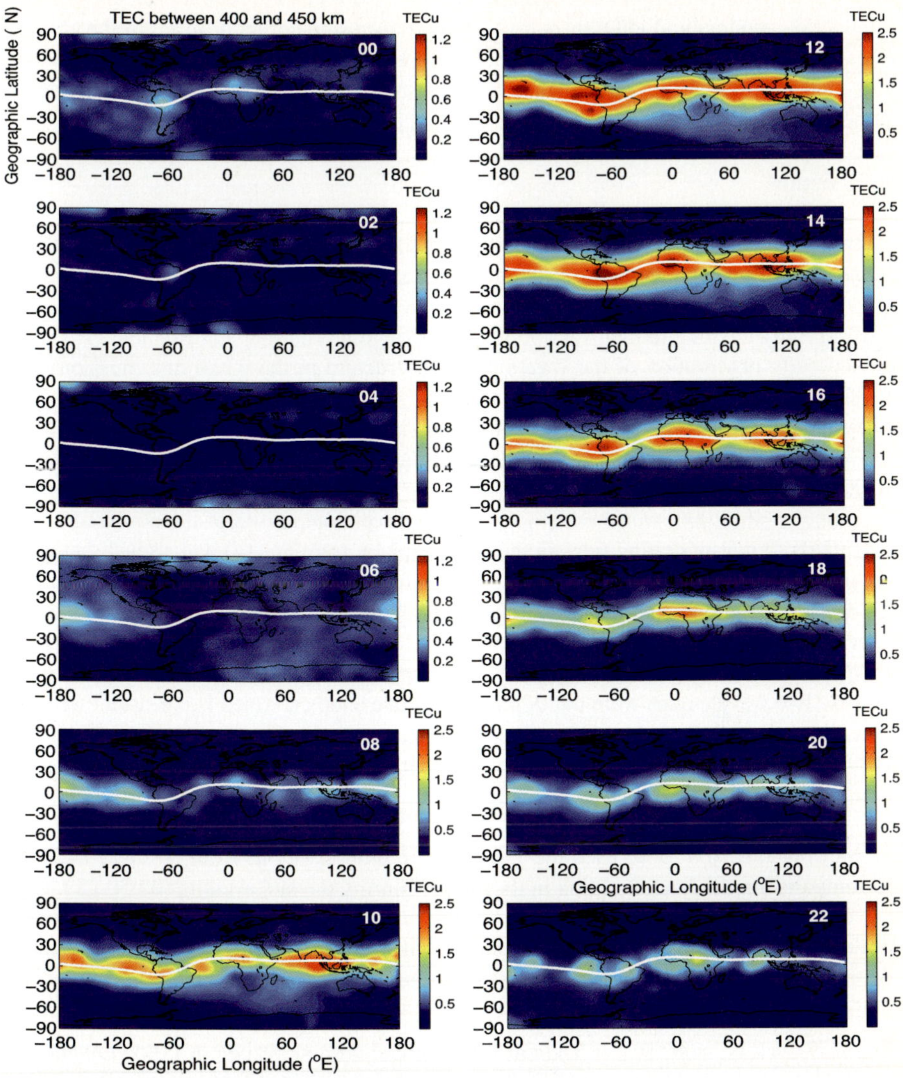

Fig. 13 Longitudinal structure of the TEC between 400–450 km altitude derived from COSMIC-GPS measurements averaged over September–October 2006 for different LT. LT is shown in the *upper right corner of each panel*. Note the *different colour scales* used at different LT. Adapted from Lin et al. (2007). Modified by permission of American Geophysical Union

plasma motion, production and recombination. On top of this diurnal cycle, for the example of September–October shown here, Fig. 13 shows a clear wavenumber-4 variation in these densities. This structure is seen to develop by ∼10 LT and last until at least ∼22 LT.

Hagan et al. (2007) used the TIME-GCM model to simulate the total column electron density for the March equinox. They examined the impact of both the non-migrating tides and the geomagnetic field (compared to a spin-axis aligned dipole field) on the longitudinal distribution of the plasma at low-mid latitudes. The authors showed that a self-consistent model of the atmosphere and ionosphere, including the photochemistry related to ion-production and loss and ion transport could reproduce some of the features of the observed wavenumber-pattern observed in the ionosphere when non-migrating tides were included in this model, providing strong evidence for a causal link between these waves in the neutral atmosphere and the wave-like response seen in the low-latitude ionosphere.

4.5 Airglow

Along with the changes in the ion density of the F-region are changes in the airglow emissions from this region. Both the 630.0 nm and 135.6 nm nighttime emissions are related to the recombination of O^+, the main species in the F-region ionosphere and both have been used to study the longitudinal variations of the F-region. Thuillier and Blamont (1973) presented early observation of the wavenumber-4 structure in the 630.0 nm emissions from OGO-4 during equinox conditions. Sagawa et al. (2005) and Henderson et al. (2005) found wavenumber-4 variations in the 135.6 nm emissions observed by IMAGE-FUV and TIMED-GUVI respectively. Sagawa et al. (2005) postulated that this structure in the airglow was related to non-migrating semi-diurnal tides.

Immel et al. (2006) wer the first to present a direct comparison of the observed airglow brightness to non-migrating tides in the thermosphere (see Fig. 14) and argued that in-fact the diurnal tide, DE3, was primarily responsible for creating the wavenumber-4 variation in the airglow brightness and location. England et al. (2006b) compared the 135.6 nm airglow observations from IMAGE-FUV and TIMED-GUVI with the much earlier OGO-4 630.0 nm observations, showing that for equinox conditions the longitudinal patterns in the airglow at these two wavelengths were out of phase with one another, which is consistent with both emissions responding to the same variation in the ionospheric dynamo (see Sect. 5). Kil et al. (2008) and England et al. (2009) both showed that the seasonal variations in the airglow patterns corresponded to the seasonal changes in atmospheric tides in the lower thermosphere observed by TIMED-SABER.

McDonald et al. (2008) used the 135.6 nm limb observations from ARGOS-LORAAS to study the longitudinal variation in the airglow during the late evening (02:30 LT). Using these limb data, the authors were able to determine the variation in brightness, F-region peak height and density. The authors reported that at this LT, the airglow pattern displayed a strong inter-hemispheric asymmetry on top of the wavenumber structure reported by other

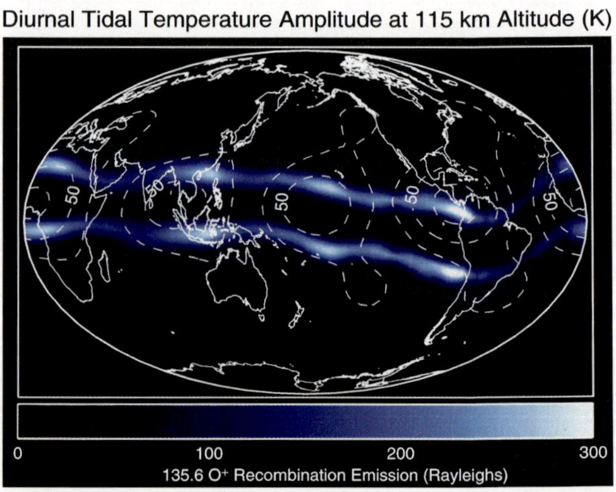

Fig. 14 Reconstruction of the 135.6 nm airglow brightness at 20 LT based on the IMAGE-FUV observations from March–April 2002. Overlaid are the temperature amplitudes of the diurnal migrating tide at 115 km from the GSWM (*white dashed line*). Due to poor sampling of the southern airglow band, the airglow band shown south of the geomagnetic equator are based on a mirror image of the northern band. After Immel et al. (2006). Reproduced by permission of American Geophysical Union

authors cited above. The authors went on to performed simulations using the SAMI2 model using different neutral wind patterns as the inputs to the model. These neutral wind inputs affect the advection of plasma within the SAMI2 model. By comparing their observations to these model outputs, the authors demonstrated that a combination of the F-region wind, geometry of the geomagnetic equator and forcing from non-migrating tides appeared to be necessary in order to interpret the longitudinal variation in the airglow observed.

5 The Coupling Mechanisms Between Non-migrating Tides in the Thermosphere and the Signature of These Waves in the Ionosphere

The mechanisms by which the thermosphere can impact the ionosphere have been studied extensively through both observation and simulation. Of the these, a sub-set appear to be significant in the coupling of non-migrating tides to the ionosphere. This section will describe these, and how they may be responsible for producing each of the wave-like structures seen in the ionosphere (described in Sect. 4).

5.1 Modification of the E-Region Dynamo

Perhaps the most direct coupling between atmospheric tides and the ionosphere occurs in the E-region, where frequent collisions make both the Hall and Pedersen conductivities significant. Strong horizontal winds associated with non-migrating tides shown in Sect. 2 can drive currents that modify the zonal electric field associated with the migrating diurnal tide, described in Sect. 1. Changes in this electric field may be seen in the E-region as changes in the strength of the EEJ. This mechanism would explain the longitudinal variations in the EEJ described in the Sect. 4.

Modifying the electric field in the E-region will in-turn modify the $\mathbf{E} \times \mathbf{B}$ drift in the F-region during both the day and night when the E-region dynamo dominates (Heelis 2004), but not (or to a far less extent) at sunrise and sunset when the F-region dynamo dominates. This is consistent with the observations described in Sect. 4. Several of the model studies described previously (Ren et al. 2009; Jin et al. 2008) have calculated the changes in the $\mathbf{E} \times \mathbf{B}$ drift, based only on changes in these E-region winds. Jin et al. (2008) showed that, for the case of the DE3 tide driving a wavenumber-4 variation in the $\mathbf{E} \times \mathbf{B}$ drift, the zonal winds in the Hall layer were of primary import. Ren et al. (2009) showed that the magnitude of the observed DE3 winds from TIMED-TIDI were sufficient to account for the variations in $\mathbf{E} \times \mathbf{B}$ drift observed.

A change in the vertical uplift of plasma during the day associated with changes in the $\mathbf{E} \times \mathbf{B}$ drifts can account for the variation in the height of the F-region reported by observations of changes in HmF2 and the brightness of the 630.0 nm airglow described in Sect. 4.

Figure 15 shows a depiction of the sequence of events proposed by Immel et al. (2006) to explain the impact of non-migrating tides on the F-region ionosphere, and specifically in the formation of the wavenumber-four signature observed in the 135.6 nm airglow. Step 1 shown is the modification of the E-region electric fields that have been inferred via their impact on the EEJ described above. Steps 2 and 3 are the modification of the F-region $\mathbf{E} \times \mathbf{B}$ drifts as described above. Step 4 is the uplift of plasma that may explain the HmF2 and 630.0 nm airglow variations. As the production of $[O^+]$ is determined by photochemical equilibrium, removal of $[O^+]$ from the peak $[O^+]$ production region via ion transport leads to enhanced $[O^+]$ production in the lower F-region. Over several hours, this leads to an increase in both the amount of plasma that builds up at higher altitudes (where the 135.6 nm emissions peak)

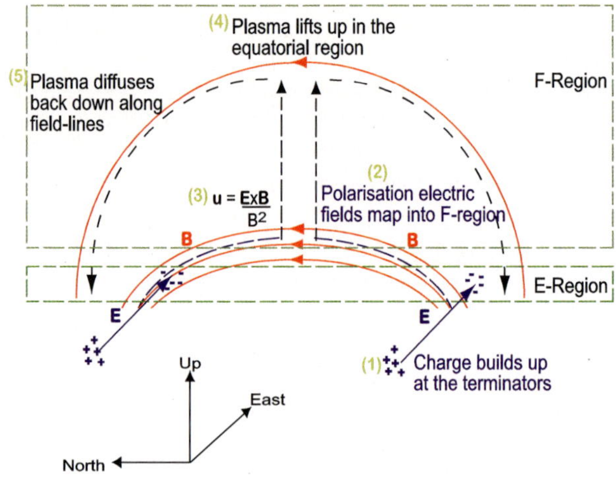

Fig. 15 Diagram describing the development of the EIA and the longitudinal wave-like structure therein. (**1**) The E-region dynamo, driven by neutral wind, produces an eastward electric field across the dayside. (**2**) These fields map upward along magnetic field lines into the F-region, causing the plasma to $\mathbf{E} \times \mathbf{B}$ drift upward (**3**, **4**) at the magnetic equator. (**5**) Through diffusion and gravitational sedimentation, the upward lifted plasma settles along the magnetic field to locations north and south of the equator. After Immel et al. (2006). Reproduced by permission of American Geophysical Union

and the total column ion density (TEC). Immel et al. (2006) proposed that this sequence of events, originating simply with a change in the wind at E-region altitude, could describe the observed variation in ionospheric density and 135.6 nm airglow.

England et al. (2006c) used the observed variation in E-region electric fields derived from observations of the EEJ to impose a modulation in the strength of the $\mathbf{E} \times \mathbf{B}$ drift in the SAMI2 model. They demonstrated that the magnitude of the variation in the E-region electric fields was sufficient to account for the observed variations in 135.6 nm airglow, when ion production, recombination and transport were all taken into account. Fang et al. (2009) performed a similar study with the GIP model, in which they used the observed variations in the $\mathbf{E} \times \mathbf{B}$ drift from ROCSAT-1 and demonstrated that these were sufficient to account for the observed wavenumber-four variation in the plasma density observed in situ by the same spacecraft. These studies provide a wealth of evidence that show that, at least in the case of the DE3 and wavenumber-4, the horizontal winds associated with non-migrating tides at E-region altitudes could be responsible for the wave-like signatures seen in the ionosphere.

5.2 Modification of Plasma Advection at F-Region Altitudes

Other authors have examined alternative mechanisms by which non-migrating tides may influence the ionosphere. Section 2 described the observational evidence for and modeling studies of the propagation of atmospheric tides throughout the entire altitude range of the thermosphere. These waves can still have significant horizontal winds at F-region altitudes. Meridional winds at F-region altitudes are efficient at transporting plasma along geomagnetic field lines, thus raising or lowering its altitude and altering its lifetime (Thuillier et al. 1976). At the altitudes close to the F-region peak, the meridional wind component associated with non-migrating tides has not been observed. However, observations of the companion zonal component of these winds (e.g. from the CHAMP observations discussed earlier) suggests that such meridional tidal winds should exist at these altitudes. This would provide

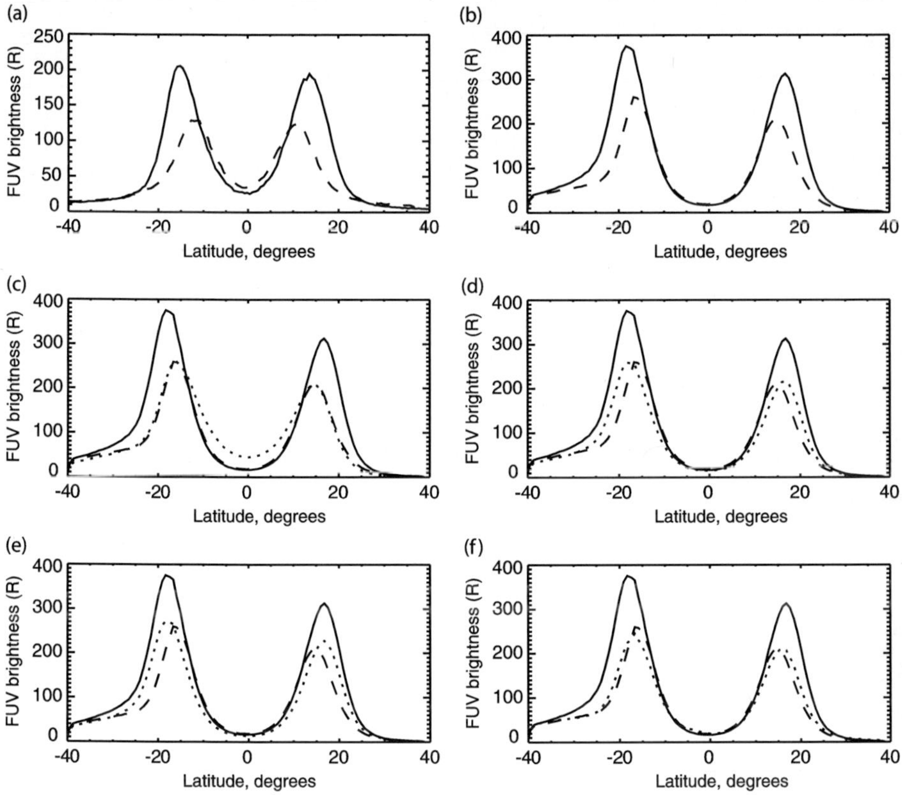

Fig. 16 Plots of the 135.6 nm emission brightness as a function of latitude relative to the equatorial trough at 21 LT for the vernal equinox from (**a**) observations by TIMED-GUVI from days 86–95 of 2002 at (*solid*) one maximum in the wavenumber-4 pattern (averaged over 180–210° longitude) and (*dashed*) the adjacent minimum (averaged over 135–165° longitude), (**b**) simulated with SAMI2 at 190° longitude for (*solid*) the maximum and (*dashed*) minimum E × B drifts associated with the E-region dynamo based on EEJ observations, (**c**) as (**b**) but including (*dotted*) the SAMI2 simulation of the change in E × B drifts around sunset, (**d**) as (**b**) but including (*dotted*) the SAMI2 simulation of the change in thermospheric neutral density, (**e**) as (**b**) but including (*dotted*) the SAMI2 simulation of the change in thermospheric O/N_2 ratio, (**f**) as (**b**) but including (*dotted*) the SAMI2 simulation of the change in thermospheric meridional winds. After England et al. (2010). Reproduced by permission of American Geophysical Union

them with a mechanism for affecting the F-region ion density that could act completely independently of the $\mathbf{E} \times \mathbf{B}$ drift modulation described above. England et al. (2010) used the SAMI2 model to perform a series of sensitivity tests and found that a diurnal meridional wind of 20 m/s could be as effective at modifying the 135.6 nm airglow as the E-region electric field modulations that have been observed (see Fig. 16, panel f).

In addition to their ability to move the plasma along the field lines, winds at F-region altitudes also drive the F-region dynamo. The winds associated with non-migrating tides can modify this current system, as reported by Park et al. (2010). As the F-region dynamo currents can be separated from the E-region dynamo, the observations reported by Park et al. (2010) provide direct evidence that the winds associated with non-migrating tides have a direct impact on the F-region ionosphere, in addition to any effect they may have on the plasma density.

5.3 Modification of Photochemical Equilibrium

There is at least one more mechanism by which non-migrating tides can modify the ionospheric density for which a significant amount of observational evidence exists. Changes in thermospheric [O]/[N_2] can strongly influence the density of O^+ in the ionosphere as increases in [O] allows for more O^+ production at any given altitude and decreases in [N_2] increase the lifetime of the O^+ produced. Non-migrating tides have been shown to modify thermospheric [O]/[N_2] as detailed in Sect. 3 and thus they may be able to modify the F-region [O^+] through their influence on the photochemical equilibrium of this region. England et al. (2010) showed that a 20% variation in [O]/[N_2] could create a change in the 135.6 nm airglow equivalent to that produced from the changes in the E-region electric fields described above (see Fig. 16, panel e). Based on the magnitude of the variations in [O]/[N_2] reported by Zhang et al. (2010), the authors showed that the observed variations in the composition of the thermosphere could account for at least ~30% of all of the variation in ionospheric [O^+] that is observed. Given that their observations were for a vertical column of [O]/[N_2], the variations at any one given altitude are likely to be even greater.

In summary, there are at least three mechanisms by which non-migrating tides can affect the ionospheric F-region density: via electrodynamic coupling to the E-region dynamo; via advection of the plasma along the geomagnetic field lines at F-region altitudes; and via modification of the photochemical equilibrium. There is strong observational evidence to support each of these three processes existing simultaneously. What is not known is how effective each of these are, or how each process might vary depending on which non-migrating tide is dominant.

6 Future Directions

Since the work by Immel et al. (2006) linked the global-scale observations of the structure of the low-latitude ionosphere to the GSWM model of non-migrating tides of tropospheric origin, there has been an extraordinary effort by a large number of researchers to characterize both of these. This review has summarized some of the key observations of these waves, along with the wealth of observations that describe how the ionosphere reacts to them. While this apparent cause and effect have been observed in detail, there is perhaps still some significant uncertainty about how these two are connected, as described in Sect. 5. In addition to this, below is a brief list of some of the other outstanding questions that, in this author's opinion, are particularly interesting avenues for further research in this area.

1. Upward propagating gravity waves carry energy and momentum from the lower and middle atmosphere into the thermosphere. The propagation of these waves is strongly influenced by the conditions of the atmosphere, which can affect their growth rate, affect their dissipation and filter out certain waves as these reach critical levels where their phase speed matches the background flow. As non-migrating tides can reach significant amplitudes in the mesosphere and lower thermosphere, even equalling the amplitude of the migrating tide, these should strongly influence the propagation of gravity waves into the thermosphere. Does this further impact the ionosphere as these gravity waves propagate to higher altitudes? If such an impact exists, is it sporadic and local, or does a coherent global-scale response observed? Recent work by Yigit et al. (2009) has demonstrated that, by incorporating the necessary physical processes relevant to gravity wave propagation and dissipation in the thermosphere, global-scale numerical models can include the effects of these waves throughout the atmosphere. This may provide the kind of tool necessary to address these questions.

2. The ionosphere has long been known to respond to the periodicities in the atmosphere associated with planetary waves. It has also recently been shown by Immel et al. (2009) and Liu et al. (2010b, 2010a), that the wavenumber-4 structure in the ionosphere also varies in strength in response to planetary waves observed in the mesosphere. How is the signature of planetary waves communicated to the ionosphere? How often is such a periodic variation in both the spatial and temporal structure of the ionosphere seen?
3. Pedatella et al. (2008) showed the response of the wavenumber structure in the EIA to a sequence of coronal mass ejections (CME) in July 2004. The authors showed that as the geomagnetic storm produced from these CMEs progressed, the wavenumber structure of the ionosphere was overwhelmed by perturbations associated with the storm. As geomagnetic storms vary greatly in their magnitude, duration and effects on the thermosphere and ionosphere, there are doubtless a wide range of interactions that can exist and the combination of the influence of the lower atmosphere (via atmospheric tides) and disturbances in the solar-wind on the longitudinal variations in the ionosphere has not been explored in detail.
4. This work has focused on the impact of atmospheric tides on the ionosphere, but at F-region altitudes, the longitudinal variations of the ionosphere have been seen to have a significant impact on the atmosphere. This feedback has been seen in both observations of the density of the neutral atmosphere (Liu et al. 2009) and zonal-winds (Lühr et al. 2007). How this feedback occurs is not well understood, especially considering that the variations in the neutral atmospheric density is orders of magnitude larger than the absolute variation in density of the ionosphere which is believed to cause it.

The significant progress that has been made in recent years has shown that studying how atmospheric tides impact the ionosphere can be a useful tool for understanding how the Earth's atmosphere and ionosphere interact as a whole, which remains an exciting frontier of research.

Acknowledgements S.L. England was supported by the National Science Foundation's CEDAR program through Award number AGS-1042261.

References

R.A. Akmaev, T.J. Fuller-Rowell, F. Wu, J.M. Forbes, X. Zhang, A.F. Anghel, M.D. Iredell, S. Moorthi, H. Juang, Tidal variability in the lower thermosphere: comparison of Whole Atmosphere Model (WAM) simulations with observations from TIMED. Geophys. Res. Lett. **35**, L03810 (2008). doi:10.1029/2007GL032584

P. Alken, Modeling equatorial ionospheric currents and electric fields from satellite magnetic field measurements. Ph.D. thesis, University of Colorado at Boulder (2009)

E.V. Appleton, Two anomalies in the ionosphere. Nature **157**, 691 (1946). doi:10.1038/157691a0

E.N. Bramley, M. Young, Winds and electromagnetic drifts in the equatorial F2-region. J. Atmos. Terr. Phys. **30**, 99–111 (1968). doi:10.1016/0021-9169(68)90044-5

A.G. Burns, T.L. Killeen, G.R. Carignan, R.G. Roble, Large enhancements in the O/N_2 ratio in the evening sector of the winter hemisphere during geomagnetic storms. J. Geophys. Res. **100**, 14,661–14,672 (1995). doi:10.1029/94JA03235

R.A. Challinor, D. Eccles, Longitudinal variations of the mid-latitude ionosphere produced by neutral-air winds-I. Neutral-air winds and ionospheric drifts in the northern and southern hemispheres. J. Atmos. Terr. Phys. **33**, 363 (1971)

D. Eccles, J.W. King, P. Rothwell, Longitudinal variations of the mid-latitude ionosphere produced by neutral-air winds-II. Comparisons of the calculated variations of electron concentration with data obtained from the Ariel I and Ariel III satellites. J. Atmos. Terr. Phys. **33**, 371 (1971)

S.L. England, A. Dobbin, M.J. Harris, N.F. Arnold, A.D. Aylward, A study into the effects of gravity wave activity on the diurnal tide and airglow emissions in the equatorial mesosphere and lower thermosphere using the Coupled Middle Atmosphere and Thermosphere (CMAT) general circulation model. J. Atmos. Sol.-Terr. Phys. **68**, 293–308 (2006a). doi:10.1016/j.jastp.2005.05.006

S.L. England, T.J. Immel, E. Sagawa, S.B. Henderson, M.E. Hagan, S.B. Mende, H.U. Frey, C.M. Swenson, L.J. Paxton, Effect of atmospheric tides on the morphology of the quiet time, postsunset equatorial ionospheric anomaly. J. Geophys. Res. **111**(A10) (2006b). doi:10.1029/2006JA011795

S.L. England, S. Maus, T.J. Immel, S.B. Mende, Longitudinal variation of the E-region electric fields caused by atmospheric tides. Geophys. Res. Lett. **33** (2006c). doi:10.1029/2006GL027465

S.L. England, X. Zhang, T.J. Immel, J. Forbes, R. DeMajistre, The effect of non-migrating tides on the morphology of the equatorial ionospheric anomaly: seasonal variability. Earth Planets Space **61**, 493–503 (2009)

S.L. England, T.J. Immel, J.D. Huba, M.E. Hagan, A. Maute, R. DeMajistre, Modeling of multiple effects of atmospheric tides on the ionosphere: an examination of possible coupling mechanisms responsible for the longitudinal structure of the equatorial ionosphere. J. Geophys. Res., Space Phys. **115**, A05308 (2010). doi:10.1029/2009JA014894

T. Fang, H. Kil, G. Millward, A.D. Richmond, J. Liu, S. Oh, Causal link of the wave-4 structures in plasma density and vertical plasma drift in the low-latitude ionosphere. J. Geophys. Res., Space Phys. **114**, A10315 (2009). doi:10.1029/2009JA014460

J.M. Forbes, The equatorial electrojet. Rev. Geophys. Space Phys. **19**, 469–504 (1981)

J.M. Forbes, X. Zhang, S. Palo, J. Russell, C.J. Mertens, M. Mlynczak, Tidal variability in the ionospheric dynamo region. J. Geophys. Res., Space Phys. **113**, 2310 (2008). doi:10.1029/2007JA012737

M.E. Hagan, J.M. Forbes, Migrating and nonmigrating diurnal tides in the middle and upper atmosphere excited by tropospheric latent heat release. J. Geophys. Res.-Atmos. **107**, 4754 (2002). doi:10.1029/2001JD001236

M. Hagan, J. Forbes, Migrating and nonmigrating semidiurnal tides in the upper atmosphere excited by tropospheric latent heat release. J. Geophys. Res. **108**(A2), 1062 (2003). doi:10.1029/2002JA009, 466

M.E. Hagan, A. Maute, R.G. Roble, A.D. Richmond, T.J. Immel, S.L. England, Connections between deep tropical clouds and the Earth's ionosphere. Geophys. Res. Lett. **34**, 20,109 (2007). doi: 10.1029/2007GL030142

M.E. Hagan, A. Maute, R.G. Roble, Tropospheric tidal effects on the middle and upper atmosphere. J. Geophys. Res., Space Phys. **114**, 1302 (2009). doi:10.1029/2008JA013637

W.B. Hanson, R.J. Moffett, Ionization transport effects in the equatorial F region. J. Geophys. Res. **71**, 5559 (1966)

M.J. Harris, N.F. Arnold, A.D. Aylward, A study into the effect of the diurnal tide on the structure of the background mesosphere and thermosphere using the new coupled middle atmosphere and thermosphere (CMAT) general circulation model. Ann. Geophys. **20**, 225–235 (2002). doi:10.5194/angeo-20-225-2002

W.A. Hartman, R.A. Heelis, Longitudinal variations in the equatorial vertical drift in the topside ionosphere. J. Geophys. Res., Space Phys. **112**, 3305 (2007). doi:10.1029/2006JA011773

K. Häusler, H. Lühr, M. Hagan, A. Maute, R. Roble, Comparison of CHAMP and TIME-GCM nonmigrating tidal signals in the thermospheric zonal wind. J. Geophys. Res., Space Phys. **115**, (2010). doi:10.1029/2009JD012394

M. He, L. Liu, W. Wan, J. Lei, B. Zhao, Longitudinal modulation of the O/N_2 column density retrieved from TIMED/GUVI measurement. Geophys. Res. Lett. **37**, L20108 (2010). doi:10.1029/2010GL045105

R.A. Heelis, Electrodynamics in the low and middle latitude ionosphere: a tutorial. J. Atmos. Sol.-Terr. Phys. **66**, 825–838 (2004). doi:10.1016/j.jastp.2004.01.034

S.B. Henderson, C.M. Swenson, J.H. Gunther, A.B. Christensen, L.J. Paxton, Method for characterization of the equatorial anomaly using image subspace analysis of Global Ultraviolet Imager data. J. Geophys. Res. **110**(A9) (2005). doi:10.1029/2004JA010830

S.S. Hong, P.H. Wang, On the thermal excitation of atmospheric tides. Bull. Geophys. Natl. Cent. Univ. Taiwan **19**, 56–83 (1980)

T.J. Immel, E. Sagawa, S.L. England, S.B. Henderson, M.E. Hagan, S.B. Mende, H.U. Frey, C.M. Swenson, L.J. Paxton, Control of equatorial ionospheric morphology by atmospheric tides. Geophys. Res. Lett. **33** (2006). doi:10.1029/2006GL026161

T.J. Immel, S.L. England, X. Zhang, J. Forbes, R. DeMajistre, Upward propagating tidal effects across the E- and F-regions of the ionosphere. Earth Planets Space **61**, 505–512 (2009)

H. Jin, Y. Miyoshi, H. Fujiwara, H. Shinagawa, Electrodynamics of the formation of ionospheric wave number 4 longitudinal structure. J. Geophys. Res., Space Phys. **113**, 9307 (2008). doi:10.1029/2008JA013301

S. Kato, Horizontal wind-systems in the ionospheric E region deduced from the dynamo theory of the geomagnetic S_q variation. Part II. Rotating Earth. J. Geomagn. Geoelectr. **8**, 24–37 (1956)

S. Kato, Horizontal wind-systems in the ionospheric E region deduced from the dynamo theory of the geomagnetic S_q variation. Part IV. J. Geomagn. Geolectr. **9**, 107–115 (1957)
S. Kato, T. Tsuda, F. Watanabe, Thermal excitation of non-migrating tides. J. Atmos. Terr. Phys. **44**(2), 131–146 (1982). doi:10.1016/0021-9169(82)90116-7
H. Kil, S.J. Oh, M.C. Kelley, L.J. Paxton, S.L. England, E. Talaat, K.W. Min, S.Y. Su, Longitudinal structure of the vertical E × B drift and ion density seen from ROCSAT-1. Geophys. Res. Lett. **34**, 14,110 (2007). doi:10.1029/2007GL030018
H. Kil, E.R. Talaat, S.J. Oh, L.J. Paxton, S.L. England, S.Y. Su, Wave structures of the plasma density and vertical E × B drift in low-latitude F region. J. Geophys. Res., Space Phys. **113**, 9312 (2008). doi:10.1029/2008JA013106
H. Kil, L.J. Paxton, W.K. Lee, Z. Ren, S. Oh, Y. Kwak, Is DE2 the source of the ionospheric wave number 3 longitudinal structure? J. Geophys. Res., Space Phys. **115**, A11319 (2010). doi:10.1029/2010JA015979
C. Kurz, V. Grewe, Lightning and thunderstorms, Part I: Observational data and model results. Meteorol. Z. **11**, 379–393 (2002). doi:10.1127/0941-2948/2002/0011-0379
C.H. Lin, C.C. Hsiao, J.Y. Liu, C.H. Liu, Longitudinal structure of the equatorial ionosphere: time evolution of the four-peaked EIA structure. J. Geophys. Res., Space Phys. **112**, 12,305 (2007). doi:10.1029/2007JA012455
H. Liu, M. Yamamoto, H. Lühr, Wave-4 pattern of the equatorial mass density anomaly: a thermospheric signature of tropical deep convection. Geophys. Res. Lett. **36**, L18104 (2009) doi:10.1029/2009GL039865
G. Liu, T.J. Immel, S.L. England, K.K. Kumar, G. Ramkumar, Temporal modulation of the four-peaked longitudinal structure of the equatorial ionosphere by the 2 day planetary wave. J. Geophys. Res., Space Phys. **115**, A12338 (2010a). doi:10.1029/2010JA016071
G. Liu, T.J. Immel, S.L. England, K.K. Kumar, G. Ramkumar, Temporal modulations of the longitudinal structure in F_2 peak height in the equatorial ionosphere as observed by COSMIC. J. Geophys. Res., Space Phys. **115**, A04303 (2010b). doi:10.1029/2009JA014829
H. Liu, B.T. Foster, M.E. Hagan, J.M. McInerney, A. Maute, L. Qian, A.D. Richmond, R.G. Roble, S.C. Solomon, R.R. Garcia, D. Kinnison, D.R. Marsh, A.K. Smith, J. Richter, F. Sassi, J. Oberheide, Thermosphere extension of the Whole Atmosphere Community Climate Model. J. Geophys. Res., Space Phys. **115**, A12302 (2010c). doi:10.1029/2010JA015586
H. Lühr, K. Häusler, C. Stolle, Longitudinal variation of F region electron density and thermospheric zonal wind caused by atmospheric tides. Geophys. Res. Lett. **34**, 16,102 (2007). doi:10.1029/2007GL030639
H. Lühr, M. Rother, K. Häusler, P. Alken, S. Maus, The influence of nonmigrating tides on the longitudinal variation of the equatorial electrojet. J. Geophys. Res., Space Phys. **113**(A12), A08313 (2008). doi:10.1029/2008JA013064
A.J. Lyon, L. Thomas, The F2-region equatorial anomaly in the African, American and East Asian sectors during sunspot maximum. J. Atmos. Terr. Phys. **25**, 373–386 (1963). doi:10.1016/0021-9169(63)90170-3
S.E. McDonald, K.F. Dymond, M.E. Summers, Hemispheric asymmetries in the longitudinal structure of the low-latitude nighttime ionosphere. J. Geophys. Res., Space Phys. **113**, A08308 (2008) doi:10.1029/2007JA012876
G.H. Millward, I.C.F. Müller-Wodarg, A.D. Aylward, T.J. Fuller-Rowell, A.D. Richmond, R.J. Moffett, An investigation into the influence of tidal forcing on F region equatorial vertical ion drift using a global ionosphere-thermosphere model with coupled electrodynamics. J. Geophys. Res. **106**, 24,733–24,744 (2001). doi:10.1029/2000JA000342
S. Namba, K.I. Maeda, in *Radio Wave Propagation* (Corona, Toyko, 1939)
J. Oberheide, J.M. Forbes, Thermospheric nitric oxide variability induced by nonmigrating tides. Geophys. Res. Lett. **35**, L16814 (2008a). doi:10.1029/2008GL034825
J. Oberheide, J.M. Forbes, Tidal propagation of deep tropical cloud signatures into the thermosphere from TIMED observations. Geophys. Res. Lett. **35**, 4816 (2008b). doi:10.1029/2007GL032397
J. Oberheide, J.M. Forbes, K. Häusler, Q. Wu, S.L. Bruinsma, Tropospheric tides from 80 to 400 km: propagation, interannual variability, and solar cycle effects. J. Geophys. Res., Space Phys. **114**, D00I05 (2009). doi:10.1029/2009JD012388
D. Pancheva, P. Mukhtarov, Strong evidence for the tidal control on the longitudinal structure of the ionospheric F-region. Geophys. Res. Lett. **37**, L14105 (2010). doi:10.1029/2010GL044039
J. Park, H. Lühr, K.W. Min, Characteristics of F-region dynamo currents deduced from CHAMP magnetic field measurements. J. Geophys. Res., Space Phys. **115**, A10302 (2010). doi:10.1029/2010JA015604
N.M. Pedatella, J.M. Forbes, J. Oberheide, Intra-annual variability of the low-latitude ionosphere due to nonmigrating tides. Geophys. Res. Lett. **35**, 18104 (2008). doi:10.1029/2008GL035332
Z. Ren, W. Wan, L. Liu, J. Xiong, Intra-annual variation of wave number 4 structure of vertical E × B drifts in the equatorial ionosphere seen from ROCSAT-1. J. Geophys. Res., Space Phys. **114**, A05308 (2009). doi:10.1029/2009JA014060

Z. Ren, W. Wan, J. Xiong, L. Liu, Simulated wave number 4 structure in equatorial F-region vertical plasma drifts. J. Geophys. Res., Space Phys. **115**, A05301 (2010). doi:10.1029/2009JA014746

R.G. Roble, G.G. Shepherd, An analysis of wind imaging interferometer observations of $O(^1S)$ equatorial emission rates using the thermosphere-ionosphere-mesosphere-electrodynamics general circulation model. J. Geophys. Res. **102**, 2467–2474 (1997). doi:10.1029/96JA02930

E. Sagawa, T.J. Immel, H.U. Frey, S.B. Mende, Longitudinal structure of the equatorial anomaly in the nighttime ionosphere observed by IMAGE/FUV. J. Geophys. Res. **110**(A9) (2005). doi:10.1029/2004JA010848

L. Scherliess, D.C. Thompson, R.W. Schunk, Longitudinal variability of low-latitude total electron content: tidal influences. J. Geophys. Res., Space Phys. **113**, 1311 (2008). doi:10.1029/2007JA012480

R.W. Schunk, A.F. Nagy, *Ionospheres Cambridge Atmospheric and Space Science Series* (Cambridge University Press, Cambridge, 2004)

G.G. Shepherd, R.G. Roble, C. McLandress, W.E. Ward, WINDII observations of the 558 nm emission in the lower thermosphere: the influence of dynamics on composition. J. Atmos. Sol.-Terr. Phys. **59**(6), 655–667 (1997). doi:10.1016/S1364-6826(96)00142-3

E.R. Talaat, R.S. Lieberman, Direct observations of nonmigrating diurnal tides in the equatorial thermosphere. Geophys. Res. Lett. **37**, L04803 (2010). doi:10.1029/2009GL041845

J.D. Tarpley, The ionospheric wind dynamo-II. Solar tides. Planet. Space Sci. **18**, 1091 (1970). doi:10.1016/0032-0633(70)90110-8

G. Thuilleir, J.E. Blamont, Vertical red line 6300 Å distribution and tropical nightglow morphology in quiet magnetic conditions, in *Physics and Chemistry of Upper Atmospheres*, ed. by B.M. McCormac. Astrophysics and Space Science Library, vol. 35 (1973), p. 219

G. Thuillier, J.W. King, A.J. Slater, An explanation of the longitudinal variation of the O1D /630 nm/ tropical nightglow intensity. J. Atmos. Terr. Phys. **38**, 155–158 (1976)

G. Thuillier, R.H. Wiens, G.G. Shepherd, R.G. Roble, Photochemistry and dynamics in thermospheric intertropical arcs measured by the WIND Imaging Interferometer on board UARS: a comparison with TIE-GCM simulations. J. Atmos. Sol.-Terr. Phys. **64**, 405–415 (2002). doi:10.1016/S1364-6826(01)00109-2

G.O. Walker, Longitudinal structure of the F-region equatorial anomaly—a review. J. Atmos. Terr. Phys. **43**, 763–774 (1981)

E.K. Walton, S.A. Bowhill, Seasonal variations in the low latitude dynamo current system near sunspot maximum. J. Atmos. Terr. Phys. **41**, 937–949 (1979)

E. Yiğit, A.S. Medvedev, A.D. Aylward, P. Hartogh, M.J. Harris, Modeling the effects of gravity wave momentum deposition on the general circulation above the turbopause. J. Geophys. Res.-Atmos. **114**, D07101 (2009). doi:10.1029/2008JD011132

X. Zhang, J.M. Forbes, M.E. Hagan, J.M. Russell, S.E. Palo, C.J. Mertens, M.G. Mlynczak, Monthly tidal temperatures 20–120 km from TIMED/SABER. J. Geophys. Res., Space Phys. **111**, A10S08 (2006). doi:10.1029/2005JA011504

Y. Zhang, S. England, L.J. Paxton, Thermospheric composition variations due to nonmigrating tides and their effect on ionosphere. Geophys. Res. Lett. **37**, L17103 (2010). doi:10.1029/2010GL044313

Space Sci Rev (2012) 168:237–260
DOI 10.1007/s11214-011-9798-4

In-Situ CHAMP Observation of Ionosphere-Thermosphere Coupling

Hermann Lühr · Jaeheung Park · Patricia Ritter · Huixin Liu

Received: 3 February 2011 / Accepted: 6 June 2011 / Published online: 5 July 2011

Abstract The coupling between the ionised plasma and the neutral thermospheric particles plays an important role for the dynamics of the upper atmosphere. Significant progress in understanding the related processes has been achieved thanks to the availability of continuous accurate measurements of thermospheric parameters like mass density and wind by high resolution accelerometers on board the satellites CHAMP and GRACE. Here we present some examples of ionosphere-thermosphere coupling where CHAMP observations contributed considerably to their interpretation. We start with the derived properties of the thermosphere at altitudes around 400 km. A new aspect is the significant control of the geomagnetic field geometry on thermospheric features. Phenomena discussed in some depths are the equatorial mass density anomaly, the cusp-related mass density enhancement and the thermospheric response to magnetospheric substorms. Here we consider both the effect on the density and on the wind. A long predicted process is the wind-driven ionospheric *F* region dynamo. The high-resolution magnetic field measurements of CHAMP enabled for the first time a systematic study of that phenomenon considering longitudinal, local time, seasonal and solar flux dependences. Some open issues that require further investigations are mentioned at the end.

Keywords Ionosphere-thermosphere coupling · Low-latitude thermosphere · High-latitude thermosphere · Substorm effects · Ionospheric *F* region dynamo

H. Lühr (✉) · J. Park · P. Ritter
Deutsches GeoForschungsZentrum, GFZ, Telegrafenberg, 14473 Potsdam, Germany
e-mail: hluehr@gfz-potsdam.de

J. Park
Department of Physics, Korea Advanced Institute of Science and Technology (KAIST), Daejeon, Republic of Korea

H. Liu
Research Institute for Sustainable Humanosphere, Kyoto University, Uji, Japan

1 Introduction

The Earth's atmosphere up to about 100 km altitude can be considered as well mixed. The relative abundance of the main constituents stays about the same as near the surface. Above 100 km the mean free path between collisions becomes increasingly larger. Therefore the constituents start to separate and decay with increasing height following their individual barometric law. The height range 100–1000 km is commonly termed thermosphere. Here the extreme ultraviolet (EUV) radiation of the sun (5∼100 nm wavelength) is almost completely absorbed. This causes the temperature to rise with increasing altitude from its minimum of about 180 K to above 1000 K. The solar EUV radiation causes also a dissociation of the gas molecules. In particular, atomic oxygen is created from O_2. Because of its lighter weight and fairly large number density, atomic oxygen is the most abundant species in the height range 200∼1000 km. A general introduction into the properties of the thermosphere is given, for example, by Prölss (2004).

Another effect of the high energetic solar radiation is the partial ionisation of atoms and molecules. The ionised component of the upper atmosphere is called ionosphere, and it is immerged in the thermosphere. The vertical structure of the electron density is determined by the balance between ionisation rate and recombination rate. Largest electron densities are typically found in the F region at altitudes between 250–300 km. Another layer forms during daytime around 115 km altitude. This so-called E region disappears during night due to the high recombination rate and the lack of solar insulation. In the F region the recombination rate is much lower. Therefore it persists throughout the night.

The thermosphere and the ionosphere particles show different dynamic behaviour. The neutrals are driven by hydrodynamic forces, but the ions are controlled in addition by electrodynamic forces. Due to frequent collisions between ions and neutrals the frictional force enters the momentum equations of both species as a further term. The coupling between the ionospheric plasma and the neutral thermosphere is particularly strong in the E-layer between 100 and 120 km altitude during day time. Here the horizontal conductivity peaks, and any electric field mapped into this layer will drive currents. Since the electrical conductivity along magnetic field lines is very high, the ionospheric E and F regions are electromagnetically coupled. During day time the highly conductive E region shorts out most of the F region voltages while at night the E region behaves as an open circuit. More details about the ionospheric electrodynamics can be found, for example, in Kelley (2009).

The mass density of the thermosphere is several orders of magnitude larger than that of the ionospheric plasma. Therefore, any heat generated by the dissipation of ionospheric currents is dumped into the thermosphere, and the thermospheric temperature determines largely the ion temperature, at least up to ∼400 km altitude.

In the subsequent sections we will present primarily CHAMP observations of coupling processes between ions and neutrals taking place at altitudes around 400 km. At this height range the thermosphere is largely dominated by atomic oxygen, and, in terms of the ionosphere, being generally above the F region density peak, the most abundant ion is single charged oxygen (O^+). The coupling between ions and neutrals can be considered weak at that altitude. Nevertheless, several important coupling processes are taking place there. In this article we will report about the most prominent ones.

2 CHAMP Satellite Measurements

The data presented here are primarily measured by the German satellite CHAMP (CHAllenging Minisatellite Payload). CHAMP was launched into a circular, near-polar (inclination: 87.2°) orbit at 456 km altitude (Reigber et al. 2002). Over its 10 years life time the

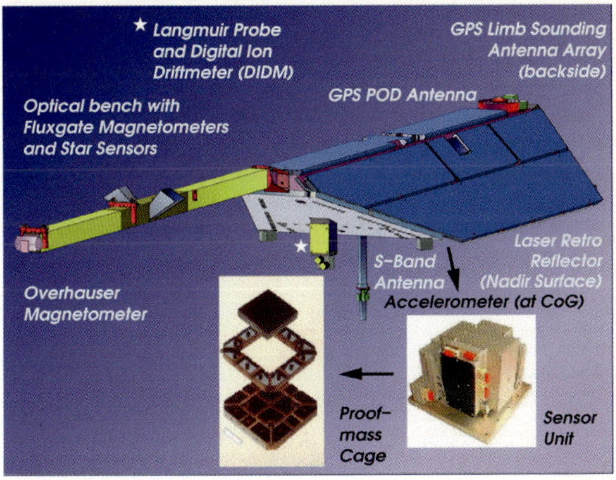

Fig. 1 Schematic illustration of the CHAMP satellite and the allocation of the scientific instruments

orbit slowly decayed, and reentry into the atmosphere occurred on 19 September 2010. Due to the chosen inclination, the orbital plane precessed through local time at a rate of 1 hour per 11 day. It thus took CHAMP 131 days to sample all local times on ascending and descending orbital arcs. When considering a time period of 5 years the local time distribution of CHAMP readings is evenly distributed over all seasons. For that reason a 5-year interval is most suitable for a statistical study.

A schematic picture of the CHAMP satellite is shown in Fig. 1. The spacecraft is 3-axes stabilised, pointing with the boom into flight direction. The scientific payload consists among others of a scalar magnetometer at the boom tip. This absolute instrument is used for magnetic calibration purposes. A fluxgate magnetometer is mounted at the centre of the boom, together with a pair of star cameras for precise attitude determination. Readings of the vector magnetometer are employed for the determination of ionospheric currents. A planar Langmuir probe (PLP) faces into the ram direction. From its readings the electron density and the electron temperature can be deduced. Of particular interest for the results presented here is the accelerometer accommodated at the satellite's center of gravity. At this position it senses only the non-gravitational forces acting on the spacecraft. The measurement principle is based on a proof-mass of about 100 g that is kept floating in the center of a vacuum cage by electrostatic forces. The acceleration acting on the spacecraft body is deduced from the restoring forces in the three spatial directions that are required to keep the proof-mass in the center.

From the air drag experienced by the satellite the thermospheric mass density and wind can be deduced. The basis equation is

$$\boldsymbol{a} = -\frac{1}{2}\rho\frac{\boldsymbol{C_d}}{m}A_{eff}v^2 \tag{1}$$

where $\boldsymbol{a}$ is the measured acceleration, ρ is the mass density, m is the satellite mass, A_{eff} denotes the effective cross-section area in ram direction and v the velocity of the satellite. $\boldsymbol{C_d}$ is the drag coefficient vector with different values for along-track and cross-track directions.

Equation (1) can be solved for the mass density, ρ, since the other quantities are known or measured. Further details about the interpretation of accelerometer data are given by Doornbos et al. (2010).

For the estimation of the thermospheric winds we make use of the measured acceleration components. Under the assumption that the experienced acceleration, $\boldsymbol{a}$, is aligned with the velocity, $\boldsymbol{v}$, with respect to the air rest frame we may write for the vector components:

$$\frac{a_y}{a_x} = \frac{v_x}{v_y} \tag{2}$$

where the x component is aligned with the spacecraft along-track axis and y with the cross-track axis. Vertical wind contributions are ignored because they are much weaker than the horizontal ones. Since CHAMP had a polar orbit, the zonal wind velocity, u, can be derived by

$$u = v_x \frac{a_y}{a_x} - v_{cor} \tag{3}$$

where v_x is the orbital velocity (7.6 km/s) and v_{cor} is the corotation velocity of the atmosphere (∼490 km/s at the equator). The approach actually used for deriving the zonal wind is more sophisticated and is described by Doornbos et al. (2010).

3 Gross Features of Thermospheric Mass Density

Before looking into the coupling between ionosphere and thermosphere we present gross features of the quiet time thermosphere. CHAMP crosses all latitudes on its orbit and covers all local times within 4 months. Thanks to the long and homogeneous set of accelerometer measurements the climatology of the mass density at an altitude of 400 km was compiled. Since the CHAMP orbit decays slowly, the actual density readings, $\rho(h)$, have been normalised to a common altitude of 400 km (ρ (400 km)). For this purpose we made use of collocated mass density predictions of the MSIS model.

$$\rho(400\text{ km}) = \rho(h)\frac{MSIS(400\text{ km})}{MSIS(h)} \tag{4}$$

This approach is regarded justified since the height change of CHAMP during the studied periods was less than one scale height (∼60 km).

Based on these density data important features were identified by Liu et al. (2005). A quantitative description of the low-latitude thermospheric density and its variation with local time and season was given later by Müller et al. (2009). They also looked into the response to changes in solar EUV flux and magnetic activity. One of their conclusions is that the effects of the four controlling parameters considered in the study, local time, season, solar flux, and magnetic activity, on the thermospheric density can be treated as a linear combination of the different influences. Non-linear coupling was not identified as an important process.

In a statistical survey of the thermospheric mass density, Liu et al. (2005) determined among others the mean diurnal variation. Opposed to previous studies, they binned the density readings by magnetic latitude rather than geographic latitude. As can be seen in Fig. 2, largest densities are observed about 1.5 hours after noon. Peak values, however, do not appear at the subsolar point, but at latitudes some 25° north and south of the magnetic equator. This is an unexpected phenomenon. For comparison, the thermospheric density distribution as predicted by the MSIS model (Hedin 1991) is shown below. Although gross features are well reflected by the model, there is no sign of a bifurcation at day time hours. Liu et al. (2007a) had a closer look at the so-called equatorial mass density anomaly (EMA). They find that, especially during equinox seasons, the two bands of enhanced density follow reasonably well magnetic latitudes, as can be seen in Fig. 3. In addition there is some

Fig. 2 Diurnal variation of the thermospheric mass density (in 10^{-12} kg/m^3). (*Top*) Latitudinal distribution as observed by CHAMP; (*bottom*) as predicted by the MSIS model. (After Fig. 2 of Liu et al. 2005)

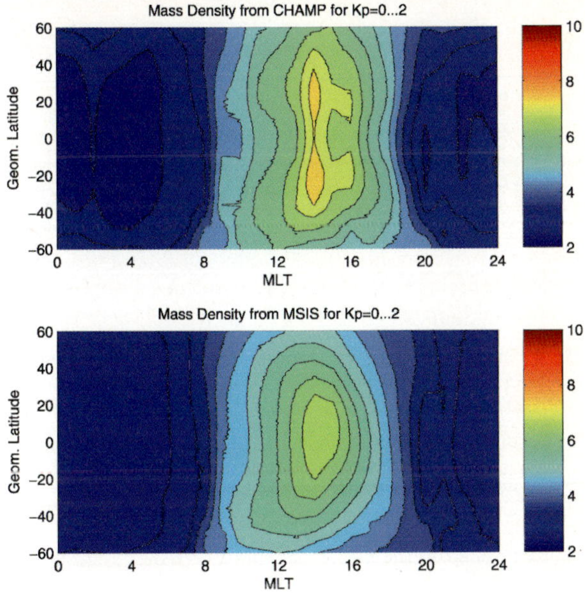

Fig. 3 Longitude/latitude distribution of the mass density (in 10^{-12} kg/m^3) on the dayside. The two bands of enhanced density are closely aligned to the magnetic dip-equator. (After Fig. 2 in Liu et al. 2007a)

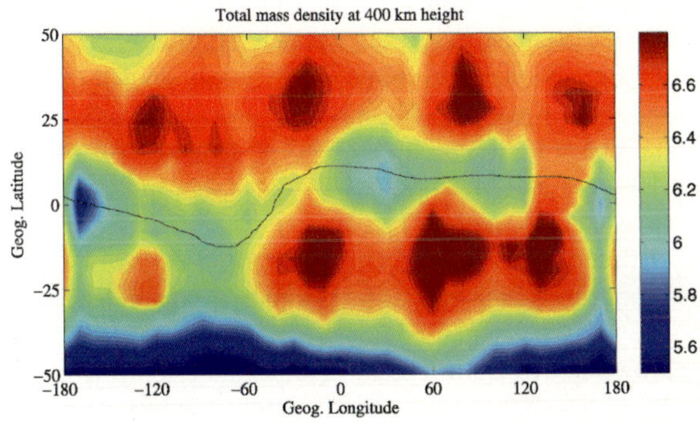

longitudinal structure visible. This can be related to tidal signatures, as was shown later by Liu et al. (2009b), but this topic will not be addressed here. During solstice seasons the EMA becomes asymmetric, favouring the summer hemisphere (Liu et al. 2007a). There have been attempts to link the EMA with the well-known equatorial ionisation anomaly (EIA). A possible mechanism for causing the density bulges at low latitudes is chemical heating. Following the suggestions of Fuller-Rowell et al. (1997), oxygen ions, O^+, are lifted up over the magnetic equator by the ion fountain effect, then diffuse down along field lines into the northern and southern hemisphere. At E region altitude, charge exchange, in particular between O^+ and O, takes place. The difference in ionisation energy (1.6 eV per nucleus) is emitted, thus heats and may lift the atmosphere at footprint latitudes of the EIA flux-tubes. However, there are also other processes discussed. A final verification of the EMA mechanism is still pending. In any case, the equatorial mass density anomaly is a product of ionosphere-thermosphere coupling. Plasma is the medium to communicate the magnetic geometry to the neutral gas.

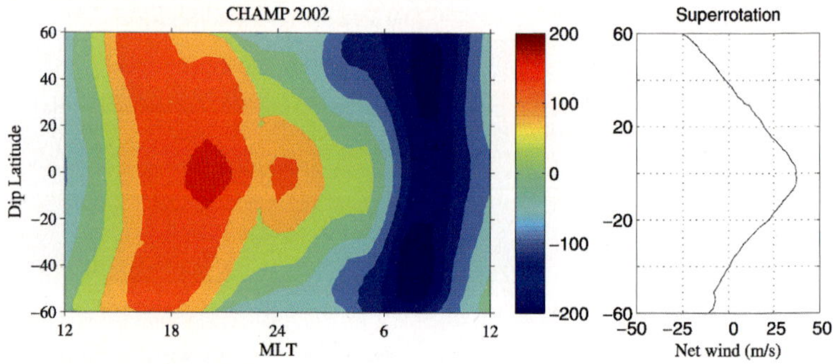

Fig. 4 Diurnal variation of the thermospheric zonal wind (in m/s). (*Left*) Latitudinal distribution as observed by CHAMP; (*right*) latitude variation of the longitudinally integrated zonal wind speed. (After Fig. 2 in Liu et al. 2009a)

4 Gross Features of Thermospheric Winds

As mentioned in the earlier sections, CHAMP measurements provide also information on thermospheric winds. For these studies altitude changes of the orbit have not been taken into account since it is known that the wind above 300 km depends little on altitude (Prölss 2004). A first global survey of CHAMP low-latitude zonal wind was presented by Liu et al. (2006). They determined the diurnal variation of wind speed and its dependence on solar flux level and magnetic activity. Previously known features could be confirmed: westward wind at the equator at day time and eastward winds during evening and night hours. In a follow-on study Liu et al. (2009a) investigated the zonal wind distribution within the latitude range $\pm 60^\circ$ of magnetic latitude. As can be seen in Fig. 4, highest wind speeds are observed at the equator. The reversals of wind direction do not take place simultaneously at all latitudes. The switch from westward to eastward in the afternoon occurs about an hour later at the equator than at 60° latitude. An even larger time difference of up to 7 hours is found in the early morning between equator and 60° latitude for the direction change from eastward to westward. The observed latitude dependence of the zonal wind direction reversal on local time can be explained in part by the influence of the Coriolis force. If an air parcel is moved from the equator to the pole, it is deflected eastward. This situation, valid on the day side, causes the earlier eastward turning at high latitudes around noon. Conversely, air moving from the poles to the equator, as happens on the night side, experiences a westward deflection. Since wind speeds are larger on the night side than on the day side, the Coriolis effect here causes a larger delay between high and low latitude reversals.

From Fig. 4 it is obvious that highest eastward wind speeds of about 150 m/s are observed at the night time magnetic equator. In order to investigate this situation further Liu et al. (2009a) have looked at the longitude distribution of the zonal wind velocity. As can be seen in Fig. 5, there is a jet stream of fast zonal wind along the magnetic equator at all longitudes. Liu et al. (2009a) have explained this high-speed channel by the reduced ion air drag within the equatorial ionisation trough. In that sense the wind distribution controlled by the geomagnetic field geometry is another form of ionosphere-thermosphere coupling.

When integrating the zonal wind velocity over all longitudes, as shown in Fig. 4, we obtain the net zonal motion of the air at given latitudes. From the right frame of Fig. 4 we see that the air rotates faster than the planet at low latitude and slower at high latitude. For an outside observer the Earth's upper atmosphere exhibits a differential rotation velocity,

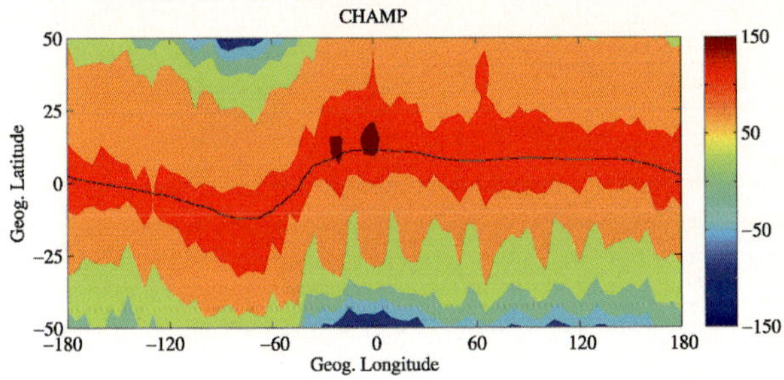

Fig. 5 Longitude/latitude distribution of the zonal wind (in m/s) on the night side. The channel of enhanced velocity follows the latitude variation of the magnetic dip-equator. (After Fig. 1 in Liu et al. 2009a)

fastest at the equator and progressively slower towards the poles. The excess rotation rate at the equator is 8% according to Fig. 4.

An even stronger interaction between ions and neutrals is observed at high magnetic latitudes. The wind distribution in the polar region, as observed by CHAMP, was first studied by Lühr et al. (2007). A similar study based on DE-2 satellite data was presented by Thayer et al. (1987). Due to the short DE-2 mission duration and sparse sampling, detailed features could not be resolved. A common result of both studies is the strong day-to-night wind over the polar cap with speeds in excess of 600 m/s. In this region hydrodynamic forces and plasma drifts are well aligned. From comparison with the low-latitude evening terminator region we can expect velocities up to 150 m/s due to neutral forcing (see Fig. 4), such as the pressure gradient. A much larger effect has to come from the cross-polar cap potential difference which drives the plasma into the anti-sunward direction at speeds of more than 1 km/s. The plasma drift velocity, $\boldsymbol{v}$, at CHAMP altitude is described by $\boldsymbol{v} = \boldsymbol{E} \times \boldsymbol{B}/B^2$. For a typical polar cap electric field strength of $E = 50$ mV/m and an ambient magnetic field, $B = 50000$ nT, we obtain a plasma velocity, $v = 1$ km/s. Plasma transfers momentum to the neutral particles by collisions. The relevant momentum equation can be written as

$$\rho \frac{d\boldsymbol{u}}{dt} = -\nabla P + \rho \nu_{i,n}(\boldsymbol{v} - \boldsymbol{u}) \tag{5}$$

where ρ is the mass density, u is the wind velocity, P is the thermal pressure and $\nu_{i,n}$ is the ion/neutral collision frequency. This equation tells us that the acceleration of wind speed by the plasma is proportional to the collision frequency and to the velocity difference between ions and neutrals. Due to the large inertia of the thermosphere the wind speed will respond to changes in plasma drift with a time delay of tens of minutes.

The average polar region wind distribution for the four months around June solstice 2003 is shown in Fig. 6. At both hemispheres we observe similar distributions. Signatures are clearer and winds are stronger in the northern (summer) hemisphere. This can be attributed to the higher plasma and neutral densities during that season, giving rise to a higher collision frequency and thus causing a closer coupling between ions and neutrals. There is a clear asymmetry between the dawn and dusk sides. Lühr et al. (2007) report fast winds towards the night time sector on the dawn side, but stagnant flow on the dusk side. There are two effects contributing to this difference. One is the plasma flow along the auroral oval streaming generally sunward. This moves in opposite direction to the prevailing thermospheric wind. The other effect is caused by a combined action of Coriolis and centrifugal forces. For an explanation of the fast day-to-night winds on the dawn side, Lühr et al. (2007) refer to the arguments of Fuller-Rowell and Rees (1984) who state that particles starting at noon and moving westward experience an equatorward centrifugal force, but at the same time a poleward Coriolis force. The velocity at which the two forces cancel each other depends on

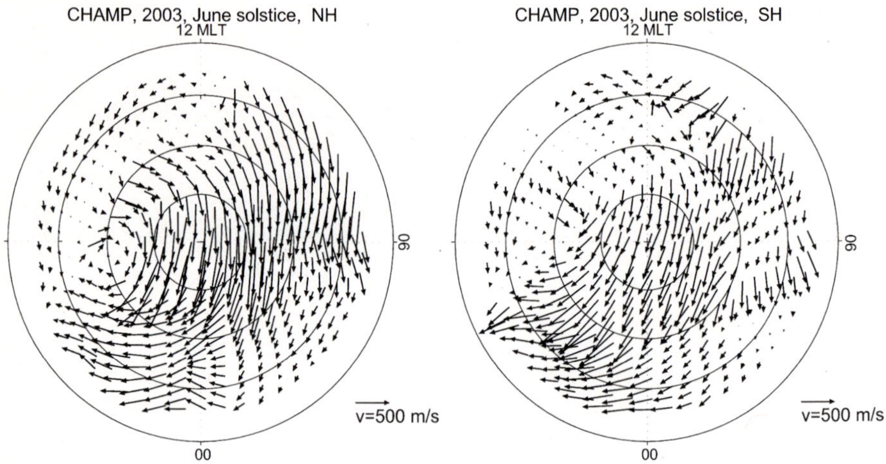

Fig. 6 Distribution of mean thermospheric wind vectors in the Northern (summer) (*left*) and Southern (winter) (*right*) hemispheres for June solstice 2003. Concentric circles mark magnetic latitude at 10° spacing (80°, 70°, 60°, and 50°). (After Fig. 4 in Lühr et al. 2007)

the polar distance, θ

$$\sin\theta = v/985\ [\mathrm{m/s}] \tag{6}$$

At a latitude of 60° ($\theta = 30°$) we obtain $v = 498$ m/s for the critical velocity. This value is close to the wind speed observed on the dawn side (cf. Fig. 6). Obviously, many air particles move along the stable paths and the counter-streaming plasma has little effect. On the dusk side the Coriolis and centrifugal forces act into the same direction deflecting the particles into an anti-cyclonic spiral motion. Such a signature can be seen in Fig. 6 near 70° of magnetic latitude at 18:00 local time. The additional effect of a sunward plasma flow along the auroral oval causes wind stagnation on the afternoon side.

Within the night sector the equatorward wind is deflected westward, as expected from the action of the Coriolis force. This observation is consistent with our interpretation of the latitude dependent shift in local time, where the zonal wind switches direction, as evident from Fig. 4.

The effect of the high-latitude plasma drift pattern on thermospheric winds was investigated in more details by Förster et al. (2008). The authors clearly show that the wind direction over the polar regions depends on the interplanetary magnetic field (IMF) components By and Bz. Their results provide further evidence for the close control of the plasma dynamics on the thermospheric wind. Over the northern hemisphere polar cap highest wind speeds are observed for a combination of negative IMF By and Bz components. In the southern hemisphere wind speeds maximise for positive IMF By and negative Bz. The clockwise wind vortex on the dusk side increases in the northern hemisphere when IMF By is positive. The opposite By polarity causes the vortex to grow in the south. This behaviour is consistent with the response of the plasma convection cells to IMF conditions. Interestingly, on the dawn side the wind velocity shows little dependence on IMF orientation. Obviously, the coupling between ions and neutrals is less efficient here. For a dedicated investigation of the interaction between neutral air and plasma in the *F* region, it would be desirable to measure wind and ion drift by the same spacecraft. ESA's upcoming constellation mission "Swarm" will provide this capability; it carries both an accelerometer and an ion driftmeter on each of the three spacecraft.

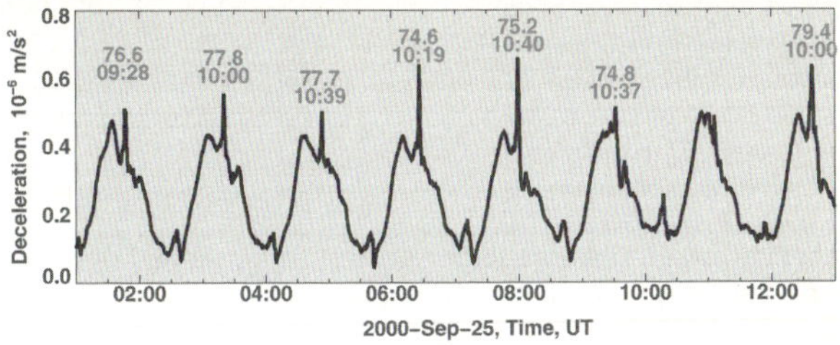

Fig. 7 Deceleration of CHAMP by air drag. Local peaks in air drag are experienced in the cusp region. (After Fig. 1 in Lühr et al. 2004)

5 The Cusp-Related Mass Density Anomaly

In Sect. 3 we had presented the average thermospheric mass density at low and mid latitudes for altitudes around 400 km. Here we focus on local phenomena at high magnetic latitudes. A surprising observation, made already early in the CHAMP mission, was the appearance of rather localized large peaks in air drag. As an example, acceleration measurements over the first 8 orbits on 25 September 2000 are shown in Fig. 7. Clearly visible are the harmonic variations of air drag over an orbit. Near the orbital maxima there are large small-scale spikes in deceleration. All these peaks tend to appear at magnetic latitudes near 75° and within the time sector around 10:30 of magnetic local time (MLT). This confinement in local occurrence suggests a relation to the ionospheric cusp. Lühr et al. (2004) were the first to look into this phenomenon in some detail. They related the deduced local mass density enhancement to concurrent peaks in ionospheric current observations. During the prominent event shown in Fig. 8 the air density almost doubles in the cusp region with respect to the background. Collocated with the density peak intense field-aligned currents (FACs) are observed. A particularly outstanding feature is the strong burst of kilometre-scale FACs. Also the Hall current in the electrojet region is enhanced. Both FAC and electrojet intensities are deduced from the CHAMP high-resolution magnetic field measurements. The approach applied for deriving these currents is described by Wang et al. (2005) and Ritter et al. (2004) for the FAC and electrojet, respectively. After CHAMP has crossed the polar cap we find on the same pass comparably strong Hall currents in the evening auroral region, but they are not accompanied by a density anomaly. Lühr et al. (2004) provided additional evidence that intense small-scale FACs play an important role for the formation of cusp-related density anomalies.

A more extended study of this density phenomenon was performed by Schlegel et al. (2005), but no conclusive explanations for a generation mechanism could be offered. Demars and Schunk (2007) made an attempt to reproduce the density anomaly by means of their high-resolution thermosphere model. For achieving a local doubling of the mass density near the cusp, as reported by Lühr et al. (2004), they had to increase the heating in the *E* region by an unrealistic factor of greater than 100.

Motivated by these open issues Rentz and Lühr (2008) performed a systematic study of the cusp-related density anomaly, based on CHAMP data from the years 2002–2005. An anomaly is defined as an enhancement above the large-scale background density. It is quantified either as an absolute increase above background or as a relative enhancement. The region scanned comprises the magnetic local time sector 08 to 16 MLT and the latitude range of 60° to 80° of magnetic latitude in both hemispheres. The peak value of density enhancement from each overflight was sorted into the appropriate equal-area bin of a latitude-by-local-time grid. Figure 9 shows the resulting average amplitude distribution on a dial plot separately for every year. Density anomalies cluster around the nominal cusp location in all cases. The amplitudes get smaller each year and decrease by a factor of 4 from

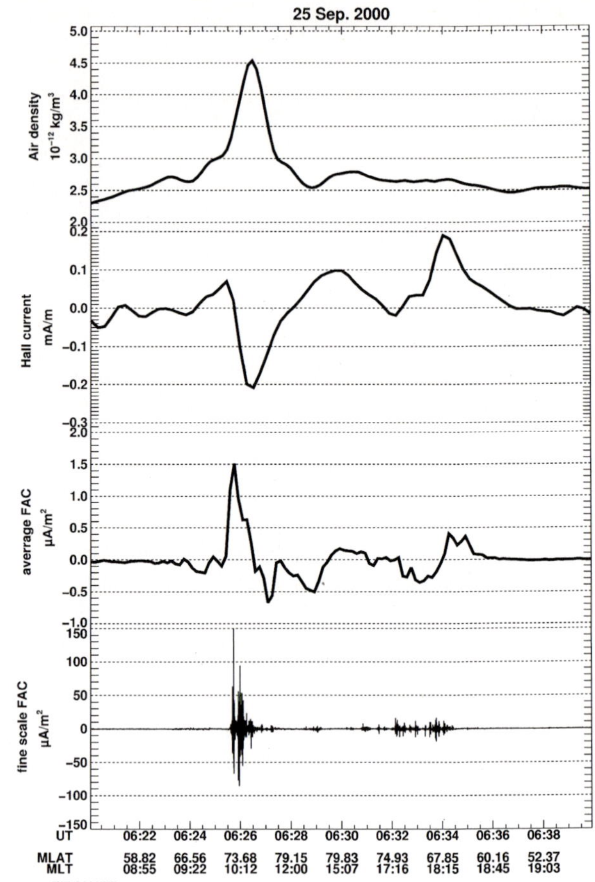

Fig. 8 Synoptic view of the cusp-related density anomaly together with collocated ionospheric currents. (*From top to bottom*) (1) mass density, (2) electrojet current density, (3) large-scale field-aligned currents, and (4) kilometre-scale field-aligned currents. (After Fig. 2 in Lühr et al. 2004)

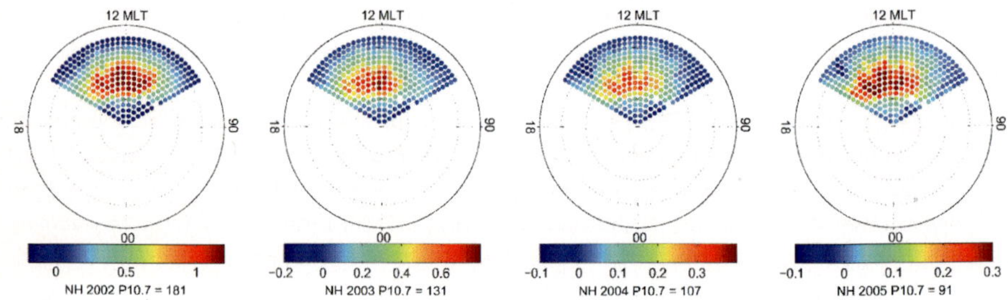

Fig. 9 Average distribution of the cusp-related density anomaly amplitude (in 10^{-12} kg/m^3) in the northern hemisphere for four successive years. Note the different density scales. Yearly averaged solar flux values, P10.7, are listed below each plot. (After Fig. 4 in Rentz and Lühr 2008)

2002 to 2005. This is accompanied by a reduction of solar EUV flux. The mean F10.7 index declines from 181 sfu in 2002 to 91 sfu in 2005. This suggests a significant dependence of the density anomaly's magnitude on solar activity. Similar results, as shown for the northern hemisphere, are observed in the south, but there the amplitudes are smaller by a factor of 1.7 on average.

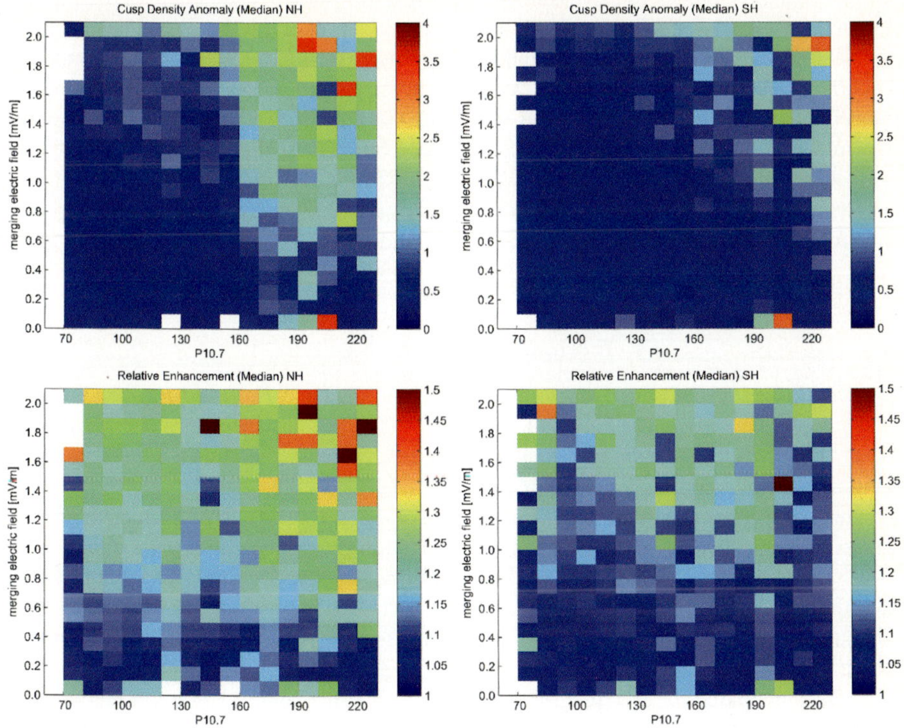

Fig. 10 Dependence of the cusp-related density anomaly on solar flux level, P10.7 and on merging electric field, separately for both hemispheres. (*Top*) Amplitude of anomaly (in 10^{-12} kg/m^3), (*bottom*) relative amplitude of anomaly with respect to background density

In the quest for the mechanism that causes the mass density enhancement near the cusp Rentz and Lühr (2008) correlated the anomalies with solar flux and solar wind parameters. The solar wind input is characterised by the merging electric field, E_m, as defined by Kan and Lee (1979)

$$E_m = v_{sw}\sqrt{B_y^2 + B_z^2}\,\sin^2\left(\frac{\theta}{2}\right) \tag{7}$$

where B_y and B_z are the IMF components, v_{sw} is the solar wind speed, and θ is the IMF clock angle ($\tan\theta = B_y/B_z$). The solar EUV input is quantified by P10.7 = 0.5 (F10.7 + aF10.7) where aF10.7 is the average of F10.7 over 81 days. Figure 10 shows the dependence of the anomaly amplitude on P10.7 and E_m. From the top two frames we may conclude that a combination of high solar flux with strong solar wind input favours the formation of large density peaks. For this reason we find anomalies with appreciable amplitude only in the upper right corner. The reduced amplitude of the anomalies in the southern hemisphere, as quantified above, is clearly visible in this figure.

Beside the absolute amplitude, also the relative amplitude of the anomaly with respect to the background density is shown in Fig. 10 (lower frames). In that case the dependence on solar EUV is largely reduced but not fully removed. In any case, it demonstrates the dominant role of the solar wind input for the formation of the anomalies.

A prime question that remains after all is, what causes the confinement of the anomalies to the ionospheric cusp region? For part of the answer we can refer to the association with the intense bursts of small-scale FACs as shown by Lühr et al. (2004). It had been reported by Neubert and Christensen (2003) and later by Rother et al. (2007) that there exists a clear occurrence maximum of kilometre-scale FACs in the polar cusp region. With respect to the model results obtained by Demars and Schunk (2007), however, it was questionable whether the amount of Joule heating in the E region caused by small-scale FACs would be sufficient to fuel the observed amount of air up-welling.

Rentz and Lühr (2008) suggested that particle precipitation may also play an important role for the formation of the density anomaly. In order to test the hypothesis we may have a look at the important quantities. The effective amount of Joule heating can be expressed as function of altitude

$$\boldsymbol{j} \cdot \boldsymbol{E} = \sigma_P(h) E^2 = \sigma_P(h)(E_{DC} + E_{AC})^2 \tag{8}$$

where $\boldsymbol{j}$ is the local current density, $\boldsymbol{E}$ is the electric field and E_{DC}, and E_{AC} are the large- and small-scale parts of the electric field, $\sigma_P(h)$ is the height-dependent Pedersen conductivity. The E-field depends only weakly on altitude because the almost vertical magnetic field lines can be considered as equipotential lines. For a given electric field the Pedersen conductivity determines the amount of Joule heating and the affected altitude range. For assessing the effect of precipitation on the conductivity we make use of the CTIP model (Millward et al. 1999), since there are no particle measurements from CHAMP. It is known that the cusp is characterised by high fluxes of soft particle precipitation. Typical energies for particles have been used, Maxwellian distributions peaking at 50 eV for electrons and 500 eV for ions. Geophysical conditions were chosen as those on 14 December 2002. CTIP runs were compared for precipitation switched on and off. Figure 11 presents results of the two runs. Height profiles of the Pedersen conductivity show a clear response to the precipitating particles. For altitudes between 130 and 400 km the conductivity is significantly enhanced by precipitation. When looking at the relative increase of conductivity (plotted in the right frame), which is proportional to the ratio of the Joule heating rate with and without precipitation, we find the largest effects up to a factor of 9 above 150 km altitude, i.e. in the $F1$ region. This uplift of the conductive and current-carrying layer is caused by high fluxes of low energetic particles which is a typical feature of the cusp. Particle precipitation in the cusp is primarily caused by reconnection at the dayside magnetopause. The strength of the merging electric field, E_m, is commonly seen as an indicator of an enhanced reconnection rate. In this respect E_m seems to indicate a two-fold effect on the strength of the cusp-related density anomalies: (1) reconnection on the dayside controls the flux of precipitating soft particles, (2) merging enhances the convection electric field which is partly mapped into the cusp region. It has been noticed earlier from Freja and FAST satellite data that high fluxes of cusp-type soft precipitation are commonly accompanied by intense small-scale FACs. Watermann et al. (2009) provided observational evidence for this by comparing DMSP particle precipitation measurements with FAC estimates from Ørsted and CHAMP. Another particular feature is the strong conductivity enhancement above 150 km of altitude. This causes a large fraction of currents to flow in the $F1$ region. In this altitude range the thermospheric density is about two orders of magnitude lower than in the E region at 110 km. This means, a given amount of Joule heating will cause a much larger temperature enhancement in the $F1$-layer than in the E-layer.

By summarizing the observations made in the context of mass density anomalies we may offer a suggestion why prominent density peaks occur preferably in association with the ionospheric cusp. The intense precipitation of low energy particles is regarded as the key process for determining the location of the anomaly. Some researchers even claim that the energy deposited by the incident particles may already account for the major part of heating

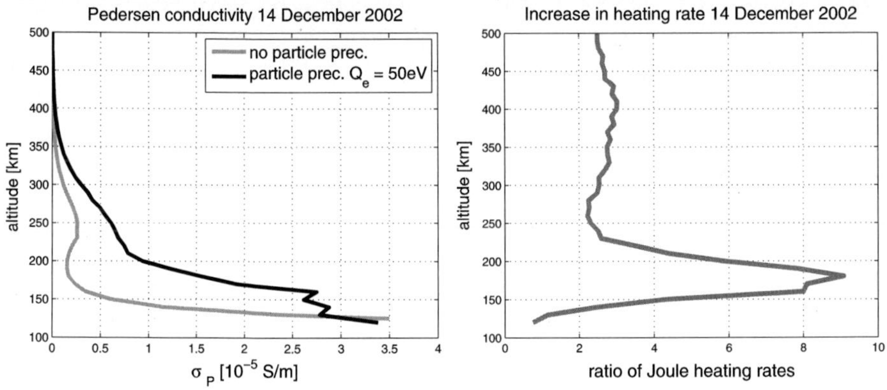

Fig. 11 Effect of soft cusp-like particle precipitation on the conductivity and the Joule heating in the ionosphere, as derived from the Sheffield High Latitude model. (*Left*) Vertical profiles of the Pedersen conductivity, (*right*) height profile of the relative Joule heating increase

(Clemmons et al. 2008). Soft precipitation occurs frequently in the cusp region. The largely enhanced conductivity in the $F1$ region due to precipitation places the heating region to 150 km and higher. At this altitude predominantly molecules are heated and subsequently uplifted. Crowley et al. (2008) presented observational evidence for a reduced O/N_2 ratio in the cusp/cleft region during enhanced magnetic activity. This increased abundance of heavier particles in the upper thermosphere adds to the density anomaly detected by CHAMP. Our suggested mechanism is also consistent with the strong dependence of the anomaly amplitude on the merging electric field. The merging rate controls both the precipitating particle flux and the cross-polar cap potential. A verification of this chain of arguments by physics-based numerical modelling would be highly desirable.

6 Substorm Effects on the Thermosphere

In their early paper Liu et al. (2005) reported about a mass density anomaly observed by CHAMP at auroral latitudes in the time sector before midnight. The authors found that the anomaly intensifies during magnetically active periods. From that dependence they suggested a relation between the enhanced air density and substorm processes. In a dedicated statistical study Ritter et al. (2010) investigated the thermospheric response to the substorm onset. CHAMP mass density and wind measurements were investigated for a large number of substorms. The survey was performed using onsets from the catalogue published by Frey and Mende (2006), containing more than 4000 entries. Events were selected based on the criteria that CHAMP passed the onset location sufficiently close in space and time. By stacking the results of many substorm events significant structures in mass density and wind could be deduced.

In order to determine the spatial and temporal evolution of substorm-related thermospheric effects CHAMP data were binned according to certain criteria. Figure 12 shows a local time dial plot with a schematic illustration of the auroral oval. The substorm onset location is marked by a yellow/orange star. There are three local time bins defined, each one is 30° in longitude wide (2 h in LT). When CHAMP passed the actual onset location to the west (0°–30° in longitude) data are sorted into bin 1, when it passed to the east (0°–30° in

Fig. 12 Scheme of binning CHAMP data by local time. Observations from passes to the west (0°–30°) of the substorm onset are sorted into bin 1, passes to the east (0°–30° in long.) into bin 2, and (30°–60° in long.) into bin 3

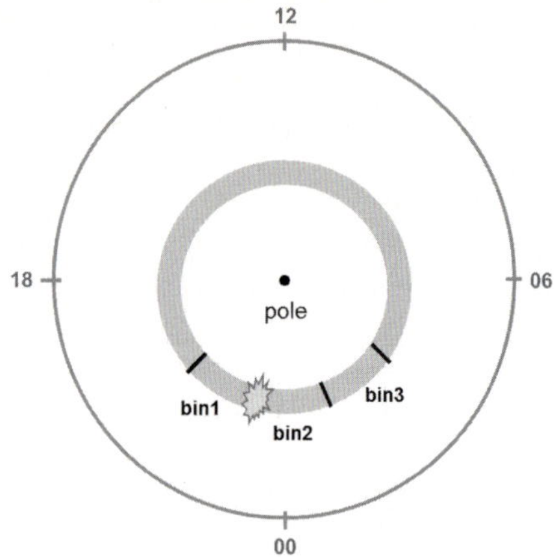

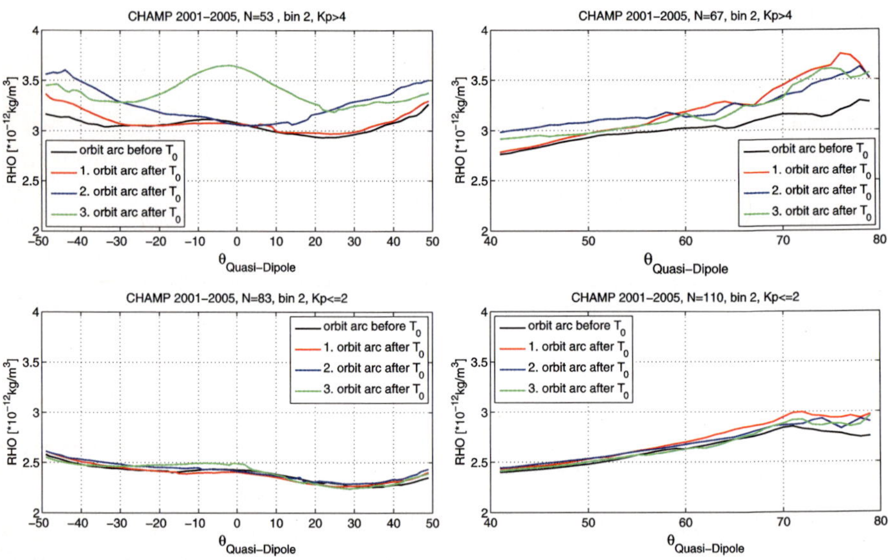

Fig. 13 Latitude profiles of mean thermospheric density distribution before and after a substorm; (*top row*) events from active period, (*bottom row*) events from quiet period, (*left frames*) low and mid latitudes, (*right frame*) high latitudes. The colours of the curves mark different temporal offsets to the substorm onset, T0. *Black line*: pass before T0; *red line*: 0.25–1.25 h after T0; *blue line*: 1.75–2.75 h after T0; *green line*: 3.25–4.25 h after T0. (After Figs. 3 and 4 in Ritter et al. 2010)

longitude) data went into bin 2 and further east (30°–60° in longitude) into bin 3. For tracking the temporal evolution of the disturbance the satellite pass before the substorm is taken as reference and is used for comparison with the passes subsequent to the onset. The orbital period of CHAMP is about 93 minutes.

Figure 13 shows the mean thermospheric density response to a substorm. Presented are observations of bin 2, around midnight. The left frames contain variations at middle latitudes, and on the right side the high-latitude response is shown. Substorms have a significant effect on the thermosphere during magnetically active periods, as can be seen in the top panels ($Kp \geq 4$). Conversely, for substorms during quiet times ($Kp \leq 2$), we hardly see

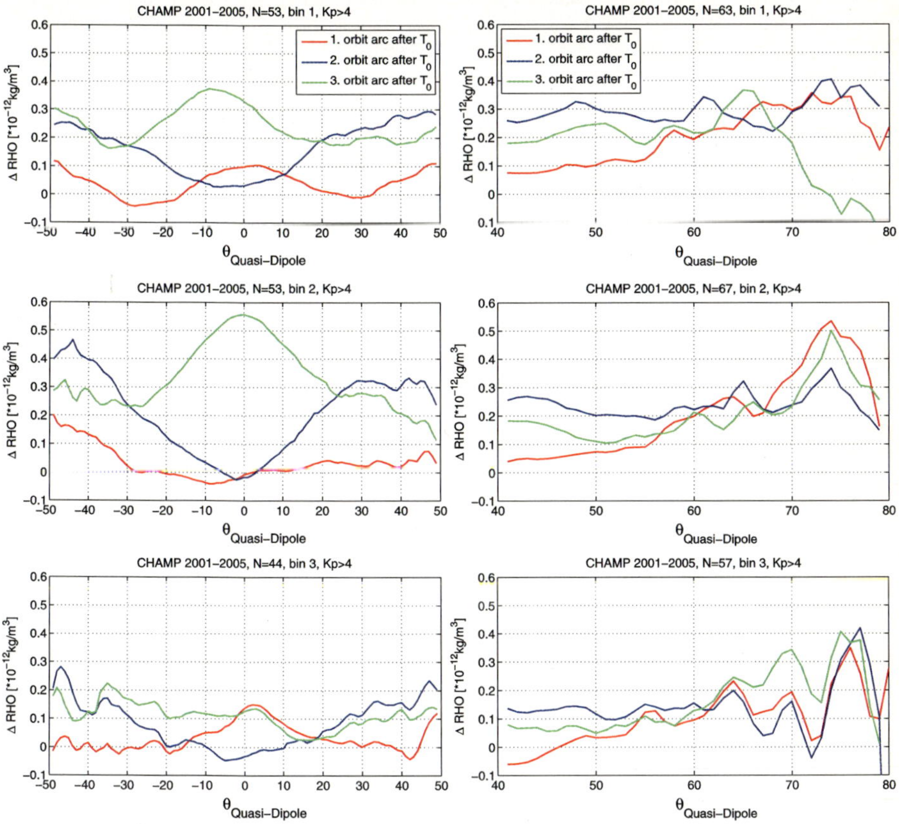

Fig. 14 Temporal evolution of thermospheric density disturbance caused by a substorm during magnetically active periods. Separate results for the three local time bins are shown. The sequences of orbits and the colour-coded time offsets are the same as in Fig. 13. (After Fig. 5 in Ritter et al. 2010)

any difference between the before-substorm curve (black) and the later passes. The amount of energy accumulated in the magnetospheric tail and released into the thermosphere during quiet-time isolated events is obviously too small. For that reason we focus in the subsequent paragraphs on active periods. Figure 14 shows mass density changes with respect to the before-substorm measurements. Plotted are mean latitude profiles derived from subsequent orbits. Measurements of the three consecutive passes are taken on average 45 min, 2.4 h and 4 h after the onset. Results of all three local time bins are presented. An outstanding feature is the mass density bulge appearing at the magnetic equator 4 h after the onset. Density changes are largest in bin 2, around midnight. First effects are observed at high latitudes between 70° and 80° of magnetic latitude (red curves). During the second pass (blue curve) the density enhancement has already expanded to mid latitudes, both from the northern and southern auroral regions. By the time of the third pass (green curves), about 4 hours after onset, the substorm-related density enhancements from both hemispheres reach the equator and pile up there. All these effects are largest in the time sector around midnight (bin 2). Here the density bulge at the equator surmounts the background density by 20%. At locations to the west of the onset (bin 1) the same density features are observed, but with smaller amplitude. The apparent minor density bulge of the red curve at the equator in bin 1 is a remnant

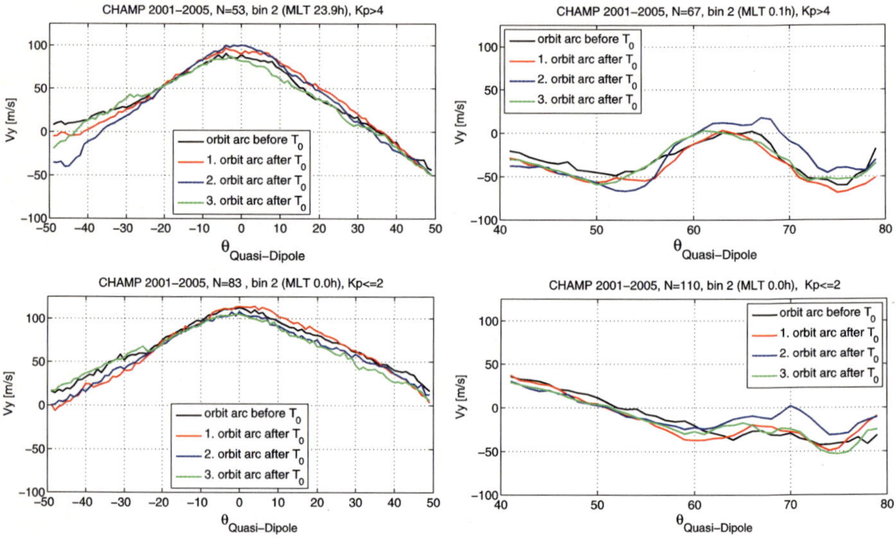

Fig. 15 Latitude profiles of the mean thermospheric zonal wind velocity around midnight; (*top row*) for high magnetic activity, (*bottom row*) for low activity. The colour-coding of the curves is the same as described in Fig. 13. (After Figs. 7 and 8 in Ritter et al. 2010)

of a substorm taking place about 3 hours earlier. More details of this study can be found in Ritter et al. (2010). Interestingly, in the post-midnight sector (bin 3) only marginal substorm-related thermospheric variations are recorded. The amplitudes are in the range of uncertainties ($\pm 0.1 \times 10^{-12}$ kg/m^3). This asymmetry between pre- and post-midnight effects supports the notion of the widely accepted current wedge model (e.g. Clauer and McPherron 1974; Ritter and Lühr 2008) according to which most of the energy injected from the tail is deposited in the vicinity of the upward directed field-aligned current region in the pre-midnight sector.

The density enhancement caused by the substorm in the auroral region propagates as travelling atmospheric disturbance (TAD) equatorward. Ritter et al. (2010) have deduced a mean propagation speed of 650 m/s from their observations. Furthermore they report a westward deflection of the travel path. This can largely be explained by the Coriolis force acting on the air parcel which moves equatorward, but an additional westward velocity component is needed for the full explanation of the observations. From these results we see that auroral phenomena like substorms can affect the thermosphere at low and mid latitude significantly.

Substorms are expected to have also an influence on thermospheric winds. In order to check that, Ritter et al. (2010) had a look at zonal winds which are detectable by CHAMP. The same procedure as used for the density study was applied to the cross-track wind component. Figure 15 shows the obtained average zonal wind distribution observed in association with substorms. In general, there is little difference between the various passes before and after the onset both during magnetically active periods (top frames) and quiet periods (bottom frames). This suggests that substorms do not influence the zonal wind on the night side significantly. However, we have no information about effects on the meridional wind. Based on simulation results Fujiwara and Miyoshi (2006) predict a strong equatorward meridional wind surge around midnight which is related to the equatorward TAD initiated by the substorm. For zonal winds their model also predicts only small effects.

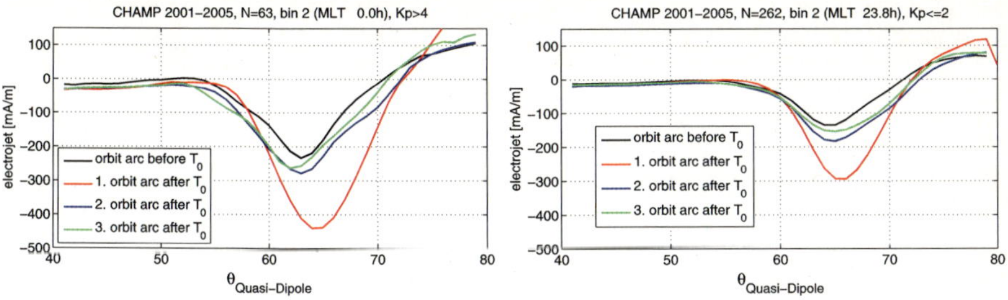

Fig. 16 Current density latitude profiles of the auroral electrojet in the midnight sector before and after a substorm; (*left frame*) high magnetic activity, (*right frame*) low activity. Negative values signify westward currents. The colour-coding of the curves is the same as described in Fig. 13

For both cases of high and low magnetic activity we find peak eastward zonal wind speeds at the magnetic equator and lower velocities at higher latitudes. This signature is consistent with the global zonal wind distribution shown in Fig. 4. Interestingly, peak wind velocities are higher during quiet times. This observation confirms the prediction of the DWM07 model that there is a westward directed disturbance wind during active periods at low and mid latitude on the night side (Emmert et al. 2008). The disturbance wind amounts to 20 m/s at the equator and increases monotonically to 50 m/s at 50° of magnetic latitude. At high latitudes, Ritter et al. (2010) report a reduction of the westward wind to almost stagnation around 65° of magnetic latitude during high magnetic activity. In order to find an explanation for the wind speed reduction we estimated the electrojet strength in this region from CHAMP magnetic field data. Ritter et al. (2004) described and validated in a systematic study their approach for determining source-free currents (largely identical with the auroral electrojet) from CHAMP total magnetic field recordings. We applied this technique to magnetic field data from the same set of CHAMP passes as used in connection with our substorms study. Figure 16 shows the latitudinal distribution of the electrojet current density as derived from a superposed epoch analysis applied to many substorms. Again magnetically active and quiet periods were considered separately. Also the colours of the curves mark the same succession of orbits after onset. In both frames we find negative (westward) electrojet currents at latitudes where the westward wind is diminished. Since the plasma drifts in the opposite direction as the electrojet current, we can imply eastward flowing plasma in the latitude range of electrojet. This can explain neatly the retarding effect on the westward wind by ion friction through eastward drifting plasma.

When looking at the evolution of the electrojet strength over the course of a substorm we find a persistent westward electrojet before and well after the substorm. This seems to be driven by the prevailing convection electric field. An outstanding feature can be observed at the first orbit (~45 min) after the onset (red curve). There we find a current strength increase on average by a factor of 1.5, which is attributed to the substorm-related current wedge (see Ritter and Lühr 2008). Interestingly, this additional current does not seem to have an influence on the zonal wind. This observation implies that there is no enhancement of the electric field (plasma drift) by the substorm, just an increase in conductivity. The highly conductive channel appears a few degrees further poleward of the prevailing electrojet. This interesting behaviour of the substorm electrojet justifies further investigations including concurrent plasma drift measurements.

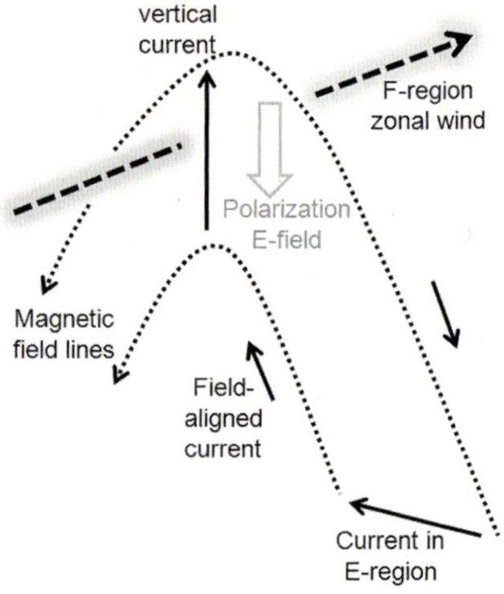

Fig. 17 Schematic drawing of the F region dynamo concept. Shown is the configuration for eastward zonal wind (evening time sector)

7 The Equatorial F Region Dynamo

In the previous sections we presented examples for thermospheric signatures caused by plasma processes. Here we show a case of an ionospheric process driven by the thermosphere. The concept of an F region dynamo was presented already several decades ago by Rishbeth (1971). He suggested a wind-driven dynamo in the low-latitude ionospheric F region. Zonal wind blowing across the geomagnetic field will cause a charge separation by interaction with the plasma. This sets up a polarisation electric field in the meridional plane. The strength of the E-field, $|\boldsymbol{E}|$, driven by the F region dynamo depends on the ratio between the conductance in the F and E regions. According to Kelley (2009) it can be expressed as

$$\boldsymbol{E} = \frac{\Sigma_P^F}{\Sigma_P^{EN} + \Sigma_P^{ES} + \Sigma_P^F} \boldsymbol{u} \times \boldsymbol{B} \tag{9}$$

where Σ_P^{EN} and Σ_P^{ES} are E region height-integrated Pedersen conductivities at the northern and southern ionospheric footprints, Σ_P^F is the fluxtube-integrated conductivity in the F region, $\boldsymbol{u}$ is the wind velocity, and $\boldsymbol{B}$ is the ambient magnetic field. The F region electric field drives toroidal current systems in both hemispheres. Figure 17 shows schematically the important elements of the F region dynamo in the evening sector.

First observational evidence of the F region dynamo was presented by Maeda et al. (1982) based on Magsat data. Since that satellite was in orbit only for a few months and sampled only the dawn/dusk sector, many issues were left unresolved. New opportunities for studying this phenomenon arose with the CHAMP mission lasting for almost one solar cycle. The diurnal variation of the F region dynamo current strength was first presented by Lühr and Maus (2006). They found downward currents in the F region which peak around noon and upward currents in the evening that are largest between 18 and 19 local time. This notion is consistent with prevailing westward zonal winds during daytime and eastward winds in the evening and night time hours (see Sect. 4).

A detailed statistical study based on CHAMP data for determining the climatological characteristics of the F region dynamo was conducted by Park et al. (2010a). The authors sorted the data by various parameters and thus could determine the dependence of the current strength on local time, season, longitude and solar flux. Figure 18 shows the local time

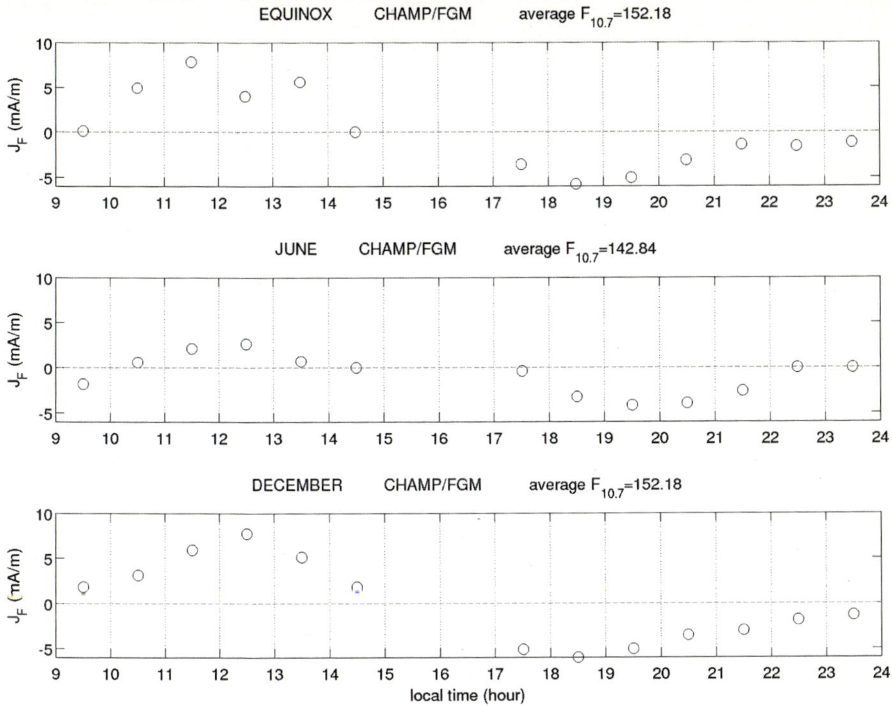

Fig. 18 Diurnal variation of the F region dynamo current zonal mean separately for the seasons. Downward currents at the equator are denoted positive. Average solar flux levels (F10.7 index) for the seasons are given in the heading of each panel. (After Fig. 3 in Park et al. 2010a)

variation of the sheet current density generated by the F region dynamo. Downward vertical currents are denoted positive. The diurnal variations are determined separately for each season. During post-midnight and morning hours no clear current signal could be deduced from the magnetic field measurements. Consistent with the results of Lühr and Maus (2006), Park et al. (2010a) find downward (positive) currents in all seasons around noon and upward (negative) current densities around evening hours. In the vicinity of the current reversal the automatic processing does not work properly. Therefore, the unreliable results in this area have been omitted in Fig. 18. Nevertheless, a zero crossing of the current between 15 and 16 LT can be found for all seasons. This local time is consistent with the reversal in zonal wind direction from westward to eastward (see Fig. 4). Smallest current densities are attainted around June solstice, while during equinox and December solstice months the diurnal variations are comparable and amplitudes are about twice as large as in June.

The vertical current in the F region is diverted and flows along field lines into the E regions of both hemispheres. For the closure currents, J_P, in the E-layer we obtain

$$J_P = \Sigma_P^E E \tag{10}$$

where E in this case is the sum of E and F region electric fields. When inserting the electric field from (9) into (10) and assuming that the conductance, Σ_P^E, is the same in the two E regions and that the conductance in the E region is much larger than that in the F region, which is reasonable during day time, we obtain

$$J_P = \frac{1}{2}\Sigma_P^F [uB]^F - \Sigma_P^E [uB_z]^E \tag{11}$$

Here the terms in brackets with superscript F denote the zonal wind velocity and the field strength in the equatorial F region, whereas in the brackets with superscript E we consider the zonal wind and vertical magnetic field at the E region footprint of the fluxtube. Equation (11) reveals that the sheet current density in the E region depends on the zonal winds in the E and F regions. Around noon the zonal winds at the two layers have, according to the wind model HWM07, opposite directions, and the product $(\boldsymbol{u x B})$ is about 5 times larger in the F region than in the E-layer. Thus, J_P is the sum of two almost equally strong contributions from both layers. However, the vertical F region dynamo current, J_z, is less influenced due to the small E region electric field contribution.

$$J_z \approx \Sigma_P^F ([uB]^F - [uB_z]^E) \tag{12}$$

From (12) we can see that the current of the F region dynamo is strongly controlled by the fluxtube-integrated conductivity in the F region. This quantity is highly dependent on the electron and neutral particle density. For the interpretation of the current strength minimum around June solstice one has to look into the electron density (Liu et al. 2007b) and neutral density (Müller et al. 2009). Both exhibit their annual minimum during that season. The reported density depletions are sufficient to explain the June depression of the F region dynamo currents.

A detailed investigation of the longitudinal variation of the F region dynamo current was performed by Park et al. (2010a). Extracted from that study we show the longitude dependence of the noon-time F region dynamo currents in Fig. 19, separately for the seasons. Quite obvious are again the small amplitudes at times around June solstice. During equinoxes there is a remarkable longitudinal variation of the current strength. Values vary between 2 and 15 mA/m. Rather obvious is a four-peaked longitudinal pattern. This is indicative of forcing by non-migrating tides. We will not go into the details of that topic, but refer to Park et al. (2010a) who discussed the tidal aspect in some depth. During the months around December solstice the current intensity varies not much with longitude. A wavenumber-2 structure seems to dominate. During all seasons an outstanding current feature is observed around 300°CE longitude. We relate this prominent peak in F region dynamo current to the South Atlantic Anomaly (SAA). Takeda (1996) predicted that in a region of low ambient magnetic field the F region is lifted higher up. As a consequence, it populates a longer fluxtubes and the field-aligned integrated Pedersen conductivity become larger. The resulting positive effect on the vertical current density (cf. (12)) is expected to surmount that of the reduced polarisation field, $u \times B$. Our observations are in support of Takeda's (1996) results.

Quite different longitudinal variations were observed during evening hours. In this local time sector several features of the noon-time recordings are missing. From Fig. 20 it can be deduced that the current distribution is less structured. There is no outstanding current density peak near the SAA (around 300°E longitude). Also the wavenumber-4 longitudinal pattern has disappeared. In particular, during equinoxes there is only little variation with longitude. For June solstice months we observe very week F region dynamo currents over the longitude sector 300–108°E. In other regions they attain appreciable values, peaking around 230°E of longitude. Similarly, during December solstice months we find a large-scale variation with minimum around 150°E and maximum at 320°E of longitude. The observed

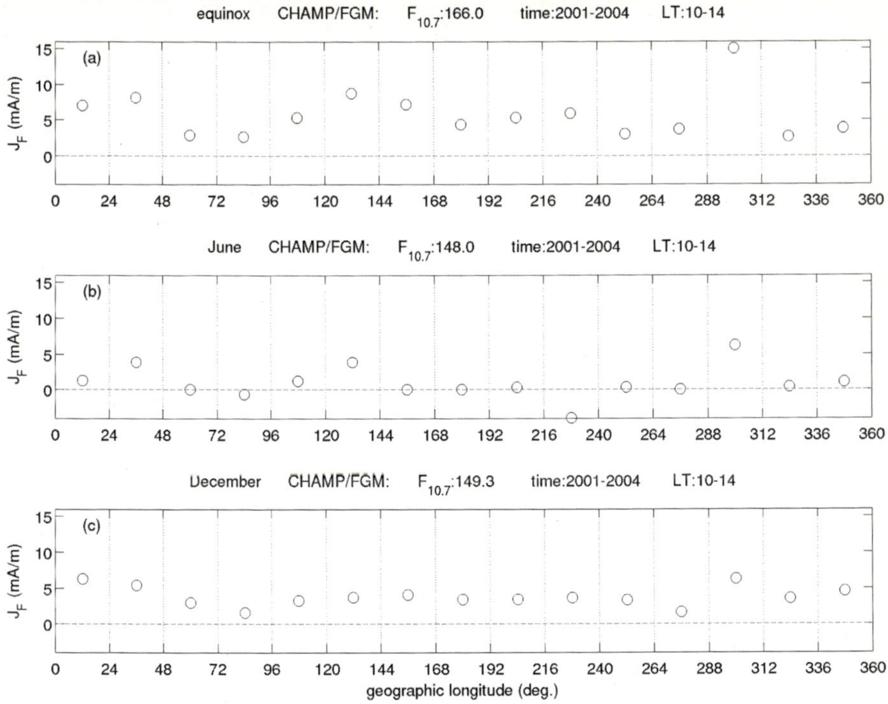

Fig. 19 Longitude distribution of *F* region dynamo currents around the noon-time sector separatly for each season, (**a**) equinoxes, (**b**) June solstice, and (**c**) December solstice. (After Fig. 4 in Park et al. 2010a)

phase difference of the dominating wavenumber-1 longitudinal variation between the two solstices is, according to Park et al. (2010a), 115°E in longitude. This large-scale variation of the vertical *F* region current density can be attributed to the combined effects of the longitude-dependent displacement of the dip-equator from the geographical equator and the seasonal variation of upper atmospheric conditions.

In general, we can state that the *F* region dynamo currents around noon reflect many features of the *E* region dynamics, e.g. wavenumber-4 longitudinal pattern and SAA enhancement. All these do not appear in the recordings from the evening hours' *F* region dynamo. This confirms the strong coupling between *E* and *F* region during day time and the disconnection of the regions at sunset. Further studies are required to resolve all the different longitudinal variations of the *F* region dynamo.

Another influence that has been investigated by Park et al. (2010a) is the dependence on the solar flux index, F10.7. According to (12) the *F* region dynamo current is directly proportional to the conductance in the *F* region. The conductance depends both on electron density and neutral particle density. These two quantities increase with the solar flux at *F* region heights, as has been shown for the neutral density, e.g. by Müller et al. (2009) and for the electron density in the equatorial ionisation anomaly, e.g. by Stolle et al. (2008). Figure 21 presents the dependence of the *F* region dynamo current density on the solar flux level, as observed by CHAMP, separately for data taken around noon and during evening hours. For both time sectors we find a clear linear relation. Interestingly, the fitted regression lines have practically the same slopes. The noon values, however, are consistently higher by about 1 mA/m. Such an amplitude difference is also evident in Fig. 18. The similarity in

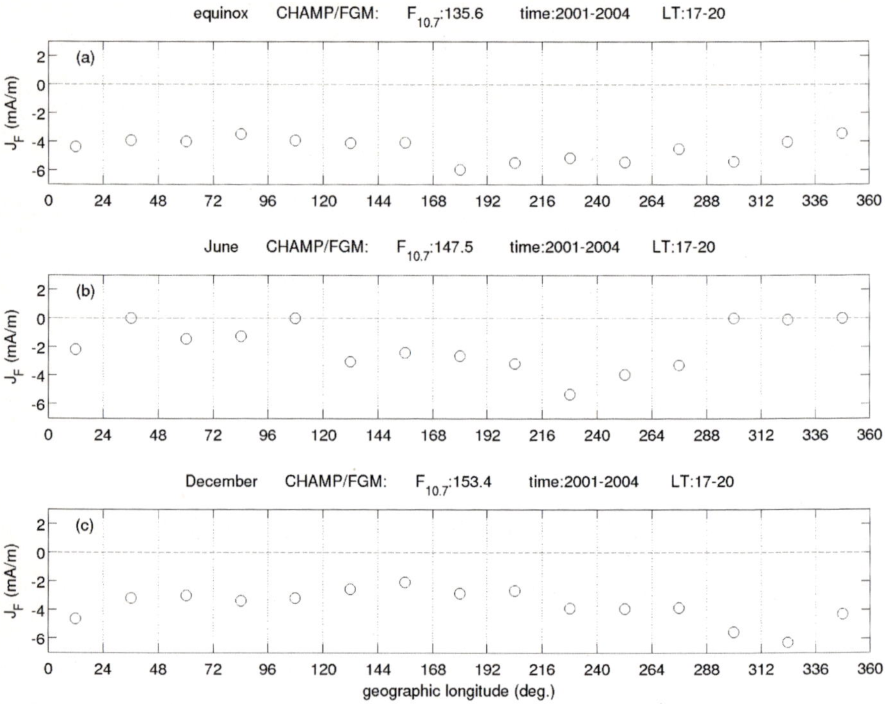

Fig. 20 The same as Fig. 19, but for the evening sector. (After Fig. 5 in Park et al. 2010a)

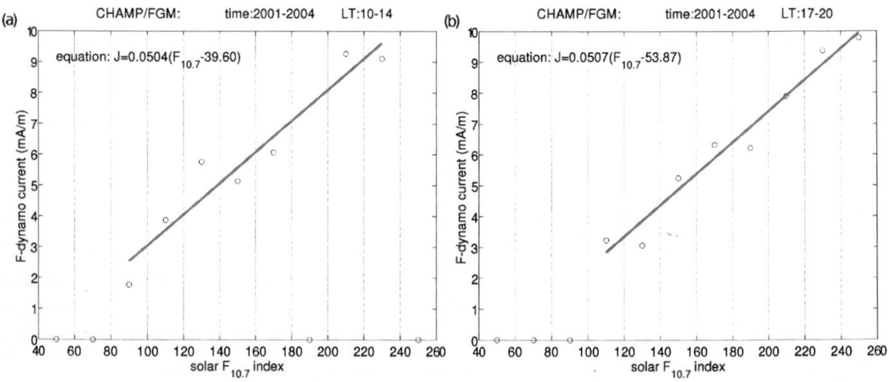

Fig. 21 Dependence of *F* region dynamo current density on solar flux level (F10.7), separately for (**a**) noon sector and (**b**) dusk sector. (After Fig. 6 in Park et al. 2010a)

slope around noon and dusk suggest that the same mechanisms cause the *F* region dynamo currents to increase in response to higher solar flux. As mentioned before, a more conductive *F* region is our preferred candidate.

It has been suggested for quite some time that the *F* region dynamo in the dusk sector might be the main driver for the peak in vertical plasma drift around sunset. Park et al. (2010b) were the first to provide observational evidence, based in CHAMP data, for the close relationship between the *F* region dynamo current strength and the peak value of the vertical plasma drift during the pre-reversal enhancement (PRE). They also show that the azimuthal conductivity gradient of the E-layer along the terminator is an important factor in this relationship. Since the peak plasma drift during PRE plays a crucial role for initiating ionospheric instabilities during pre-midnight hours, a better understanding of the dusk-sector current generator is of great interest. Further studies on the relation between F region dynamo and PRE are thus highly recommended.

8 Conclusion and Outlook

The previous sections gave an idea of the large variety of processes coupling the ionosphere with the thermosphere. The German satellite, CHAMP, with its continuous measurements over almost one solar cycle helped to advance our understanding of many phenomena. In the past, progress in research was often hampered by the lack of sufficient observations of thermospheric quantities. This has changed drastically since satellites like CHAMP and GRACE carry sensitive accelerometers for recording the air drag. Only now it has become clear that thermospheric density distribution is significantly controlled by the shape of the geomagnetic field. Examples for that are found at low latitudes, e.g. the equatorial mass density anomaly, and at high latitudes, e.g. the cusp-related density enhancement. Both these features are not reproduced by any of the thermospheric models. Since these are systematic features, their consideration in models is highly recommended.

Ionospheric processes at auroral latitudes have been shown to have strong effects on the thermospheric dynamics. This connection is known qualitatively since the early days of satellite missions. Only now, statistically significant quantitative relations between plasma dynamics and thermospheric response have been deduced. An example presented here is the thermospheric wind at polar regions. It is driven strongly by plasma convection, but shows also a clear response to Coriolis force and centrifugal force. Therefore, plasma convection and polar region wind show rather different patterns. Also in this case, empirical models provide only a rather poor representation of wind patterns at high latitudes. Another example is the thermospheric disturbance caused by substorms. CHAMP measurements revealed that air parcels lifted up at auroral latitudes propagate as density bulges equatorward and cause large transient density enhancements at low latitude. Zonal winds are only marginally influenced by substorms.

The equatorial F region dynamo is presented, as an example of a ionospheric process driven by the thermosphere. Here again, CHAMP measurements provide the first comprehensive overview of its properties. Features of the F region dynamo in the noon sector reflect many properties of the E region dynamics. This is indicative of a close connection between E and F regions during that time of the day. Conversely, a decoupling of the two regions is deduced from the F region dynamo current features in the dusk sector. On the other hand, we found a close relation between the F region current strength in that sector with the vertical plasma drift velocity during the pre-reversal enhancement (PRE). This result suggests that the size of the PRE, which plays an important role for post-sunset ionospheric instabilities, is controlled by a wind-driven dynamo. In particular, this coupling process between neutral particles and plasma calls for further detailed investigations.

ESA's upcoming constellation mission *Swarm*, consisting of 3 CHAMP-like spacecraft, is well-equipped to advance the studies of ionosphere-thermosphere coupling. With its instrumentation package involving accelerometer, ion driftmeter and magnetometers it will record the ion and neutral dynamics simultaneously. The foreseen orbit configuration of the constellation enables the employment of spatial gradients in the data interpretation. The launch of *Swarm* is scheduled for July 2012.

Acknowledgements The CHAMP mission was sponsored by the Space Agency of the German Aerospace Center (DLR) through funds of the Federal Ministry of Economics and Technology, following a decision of the German Federal Parliament (grant code 50EE0944). The data retrieval and operation of the CHAMP satellite was performed by the German Space Operations Center (GSOC).

References

C.R. Clauer, R.L. McPherron, J. Geophys. Res. **79**, 2812–2820 (1974)
J.H. Clemmons, J.H. Hecht, D.R. Salem, D.J. Strickland, Geophys. Res. Lett. **35**, L24103 (2008). doi:10.1029/2008GL035972
G. Crowley, A. Reynolds, J.P. Thayer, J. Lei, L.J. Paxton, A.B. Christensen, Y. Zhang, R.R. Meier, D.J. Strickland, Geophys. Res. Lett. **35**, L21106 (2008). doi:10.1029/2008GL035745
H.G. Demars, R.W. Schunk, J. Atmos. Sol.-Terr. Phys. **69**, 649–660 (2007)
E. Doornbos, J. van den Ijssel, H. Lühr, M. Förster, G. Koppenwallner, J. Spacecr. Rockets **47**, 580–589 (2010)
J.T. Emmert, D.P. Drob, G.G. Shepherd, G. Hernandez, M.J. Jarvis, J.W. Meriwether, R.J. Niciejewski, D.P. Sipler, C.A. Tepley, J. Geophys. Res. **113**, A11319 (2008). doi:10.1029/2008JA013541
M. Förster, S. Rentz, W. Köhler, H. Liu, S.E. Haaland, Ann. Geophys. **26**, 1581–1595 (2008)
H.U. Frey, S.B. Mende, in *Proceedings of the 8th International Conference on Substorms*, ed. by M. Syrjasuo, E. Donovan (University of Calgary, Calgary, 2006), pp. 71–76
H. Fujiwara, Y. Miyoshi, Geophys. Res. Lett. **33**, L20108 (2006). doi:10.1029/2006GL027103
T.J. Fuller-Rowell, D. Rees, Planet. Space Sci. **32**, 69–86 (1984)
T.J. Fuller-Rowell, M.V. Codrescu, B.G. Fejer, W. Borer, F. Marcos, D.N. Anderson, J. Atmos. Sol.-Terr. Phys. **61**, 1533–1540 (1997)
A.E. Hedin, J. Geophys. Res. **96**, 1159–1172 (1991)
J.R. Kan, L.C. Lee, Geophys. Res. Lett. **6**, 577–580 (1979)
M.C. Kelley, *The Earth's Ionosphere*, 2nd edn. (Academic Press, San Diego, 2009)
H. Liu, H. Lühr, V. Henize, W. Köhler, J. Geophys. Res. **110**, A04301 (2005). doi:10.1029/2004JA010741
H. Liu, H. Lühr, S. Watanabe, W. Köhler, V. Henize, P. Visser, J. Geophys. Res. **111**, A07307 (2006). doi:10.1029/2005JA011415
H. Liu, H. Lühr, S. Watanabe, J. Geophys. Res. **112**, A05305 (2007a). doi:10.1029/2006JA012199
L. Liu, B. Zhao, W. Wan, S. Venkartraman, M.-L. Zhang, X. Yue, J. Geophys. Res. **112**, A07303 (2007b). doi:10.1029/2007JA012283
H. Liu, S. Watanabe, T. Kondo, Geophys. Res. Lett. **36**, L08103 (2009a). doi:10.1029/2009037377
H. Liu, M. Yamamoto, H. Lühr, Geophys. Res. Lett. **36**, L18104 (2009b). doi:10.1029/2009GL039865
H. Lühr, M. Rother, W. Köhler, P. Ritter, L. Grunwaldt, Geophys. Res. Lett. **31**, L06805 (2004). doi:10.1029/2003GL019314
H. Lühr, S. Maus, Geophys. Res. Lett. **33**, L24102 (2006). doi:10.1029/2006GL028374
H. Lühr, S. Rentz, P. Ritter, H. Liu, K. Häusler, Ann. Geophys. **25**, 1093–1101 (2007)
H. Maeda, T. Iyemori, T. Araki, T. Kamei, Geophys. Res. Lett. **9**, 337–340 (1982)
G. Millward, R. Moffett, H. Balmforth, A. Rodger, J. Geophys. Res. **104**(A11), 24603–24612 (1999)
S. Müller, H. Lühr, S. Rentz, Ann. Geophys. **27**, 2087–2099 (2009)
T. Neubert, F. Christensen, Geophys. Res. Lett. **30**(19) (2003). doi:10.1029/2003GL017808
J. Park, H. Lühr, K.W. Min, J. Geophys. Res. **115**, A10302 (2010a). doi:10.1029/2010JA015604
J. Park, H. Lühr, B.G. Fejer, K.W. Min, Ann. Geophys. **28**, 2097–2101 (2010b)
G.W. Prölss, *Physics of the Earth's Space Environment* (Springer, Berlin, 2004)
C. Reigber, H. Lühr, P. Schwintzer, Adv. Space Res. **30**(2), 129–134 (2002)
S. Rentz, H. Lühr, Ann. Geophys. **26**, 2807–2823 (2008)
H. Rishbeth, Planet. Space Sci. **19**, 263–267 (1971)
P. Ritter, H. Lühr, A. Viljanen, O. Amm, A. Pulkkinen, Sillanpää, Ann. Geophys. **22**, 417–430 (2004)
P. Ritter, H. Lühr, Ann. Geophys. **26**, 2781–2793 (2008)
P. Ritter, H. Lühr, E. Doornbos, Ann. Geophys. **28**, 1207–1220 (2010)
M. Rother, K. Schlegel, H. Lühr, Ann. Geophys. **25**, 1603–1615 (2007)
K. Schlegel, H. Lühr, J.-P. St.-Maurice, G. Crowley, C. Hackert, Ann. Geophys. **23**, 1659–1672 (2005)
C. Stolle, C. Manoj, H. Lühr, S. Maus, P. Alken, J. Geophys. Res. **113**, A09310 (2008). doi:10.1029/2007JA012781
M. Takeda, J. Geophys. Res. **101**(A4), 7875–7880 (1996)
J.P. Thayer, T.L. Killeen, F.G. McCormac, C.R. Tschan, J.-J. Ponthieu, N.W. Spencer, Ann. Geophys. **5**, 363–368 (1987)
H. Wang, H. Lühr, S.-Y. Ma, J. Geophys. Res. **110**, A03306 (2005). doi:10.1029/2004JA010530
J. Watermann, P. Stauning, H. Lühr, P.T. Newell, F. Christiansen, K. Schlegel, Adv. Space Res. **43**, 41–46 (2009)

Space Sci Rev (2012) 168:261–282
DOI 10.1007/s11214-011-9829-1

Observations of Stratosphere-Troposphere Coupling During Major Solar Eclipses from FORMOSAT-3/COSMIC Constellation

Kuo-Ying Wang · Chao-Han Liu · Lou-Chuang Lee · Peter Braesicke

Received: 30 May 2011 / Accepted: 27 August 2011 / Published online: 28 October 2011

Abstract Sudden tropospheric cooling and induced stratospheric warming were found during the 22 July 2009 total solar eclipse. Can the 22 July 2009 hallmark also be seen in other major solar eclipses? Here we hypothesize that the tropospheric cooling and the stratospheric warming can be predicted to occur during a major solar eclipse event. In this work we use the FORMOSAT-3/COSMIC (F3C) Global Positioning System (GPS) radio occultation (RO) data to construct eclipse-time temperature profiles before, during, and after the passages of major solar eclipses for the years 2006–2010. We use four times a day of meteorological analysis from the European Centre for Medium Range Weather Forecast (ECMWF) global meteorological analysis to construct non-eclipse effect temperature profiles for the same eclipse passages. The eclipse effects were calculated based on the difference between F3C and ECMWF profiles. A total of five eclipse cases and thirteen non-eclipse cases were analyzed and compared. We found that eclipses cause direct thermal cooling in the troposphere and indirect dynamic warming in the stratosphere. These results are statistically significant. Our results show -0.6 to -1.2°C cooling in the troposphere and 0.4 to 1.3°C warming in the middle to lower stratosphere during the eclipses. This characteristic stratosphere-troposphere coupling in temperature profiles represent a distinctive atmospheric responses to the solar eclipses.

Keywords Solar eclipse · Stratosphere-troposphere coupling · FORMOSAT-3/COSMIC · ECMWF

K.-Y. Wang (✉)
Department of Atmospheric Sciences, National Central University, Chung-Li, Taiwan
e-mail: kuoying@mail.atm.ncu.edu.tw

C. H. Liu · L. C. Lee
Institute of Space Science, National Central University, Chung-Li, Taiwan

P. Braesicke
Centre for Atmospheric Science, Department of Chemistry, Cambridge University, Lensfield Road, Cambridge, CB2 1EW, UK

1 Introduction

Major solar eclipses such as total solar eclipse and annular solar eclipse arise from direct block out of incoming solar radiation by the moon to the atmosphere (Zirker 1980; Lindsey et al. 1992; Pasachoff 2009). This sudden drop in the incoming solar radiation causes the cooling of the surface and higher up, resulting in the changes of atmospheric temperatures. These temperature variations drive changes in pressures and winds (Ballard et al. 1969; Anderson et al. 1972; Founda et al. 2007; Gerasopoulos et al. 2007; Kameda et al. 2009; Wang and Liu 2010), and induces vertically propagating gravity waves (e.g., Chimonas 1970; Chimonas and Hines 1971; Seykora et al. 1985; Zerefos et al. 2007). Solar eclipse also changes the ionosphere total electron density (Le et al. 2009). Treumann et al. (2008) described the physics of electric discharges in the atmospheric gases, and average altitude profiles of temperature and mass density from the surface to 150-km altitude.

Though the troposphere and the stratosphere are regarded as a coupled system (Holton et al. 1995), the effects of the eclipse on this coupled system is not well known, due to the lack of observational data from the troposphere to the stratosphere when the lunar shadows move through the atmosphere.

Wang and Liu (2010) used F3C data to study the atmospheric effect of the 22 July 2010 total solar eclipse. They found a significant cooling through the troposphere and a distinctive warming in the lower stratosphere. This sudden tropospheric cooling and stratospheric warming appeared as a hallmark in the coupled stratosphere-troposphere system during the 22 July 2009 total solar eclipse. Can the 22 July 2009 hallmark also be seen in other major solar eclipses? Here we hypothesize that the tropospheric cooling and the stratospheric warming can be predicted to occur during a major solar eclipse event. The motivation for this study is to use F3C GPS RO data, together with the daily ECMWF global meteorological analysis, to demonstrate that the tropospheric cooling and stratospheric warming feature is indeed a hallmark during a major solar eclipse.

2 Data and Methods

2.1 The 2006–2010 Solar Eclipses

Table 1 shows a list of five major solar eclipse events that had occurred between September 2006 and January 2010. Figure 1 shows spatial distribution of the 15 January 2010, 26 January 2009, 1 August 2008, and 22 September 2006 solar eclipse paths, and distribution of the GPS RO profiles. Here the green dots indicate all the GPS RO profiles that are located within the 3-hour time window of the 1-s eclipse centers, the red dots indicate all the GPS RO profiles that are within 2250-km radius of the eclipse path, and the blue dots indicate

Table 1 Major solar eclipses analyzed in this work (Espenak and Meeus 2006; Espenak and Anderson 2008)

Time	Event	Region
22 September 2006	Annular solar eclipse	South Atlantic Ocean
1 August 2008	Total solar eclipse	Northern Polar regions-Central Asia
26 January 2009	Annular solar eclipse	South Indian Ocean
22 July 2009	Total solar eclipse	South Asia-North Pacific Ocean
15 January 2010	Annular solar eclipse	East Africa-North Indian Ocean

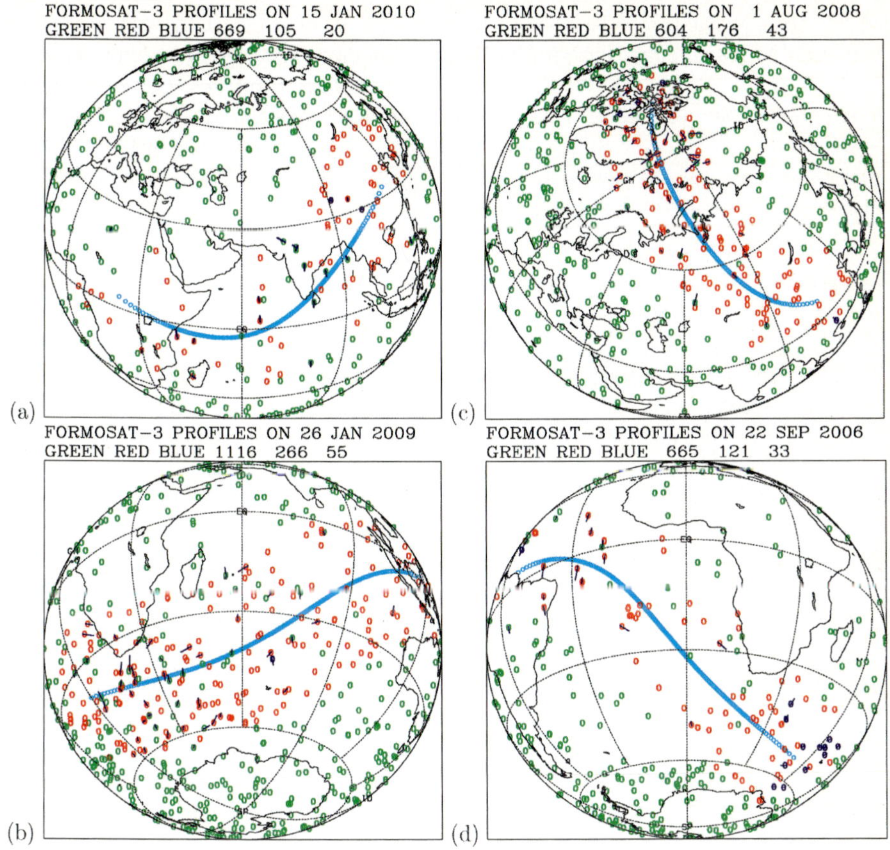

Fig. 1 Spatial distribution of F3C GPS RO profiles and path of the total eclipse (*sky-blue circles*) on (**a**) 15 January 2010, (**b**) 26 January 2009, (**c**) 1 August 2008, and (**d**) 22 September 2006. RO profiles located within 3-hour time window of the 1-s eclipse centers are shown in *green dots*, within a distance of 2250-km radius are shown in *red circles*, within both 3-hour time window and 2250-km radius are shown in *blue circles* and its ray path in *blue lines*

all the GPS RO profiles that are within the 3-hour time window and 2250-km radius of the eclipse centers calculated by Espenak and Meeus (2006). Only the blue dots were selected for analysis. The locations of these blue dots were also used to select ECMWF grids that coincide with the GPS RO locations. The path of the 22 July 2009 eclipse and the distribution of the GPS RO profiles were shown in Wang and Liu (2010).

In this work we select only major eclipse events and the periods of the events that were covered by the availability of the F3C data. Two total solar eclipses occurred in this period. One occurred on 1 August 2008, during the Northern Hemisphere (NH) summer, covering polar regions of Greenland, North Atlantic, and mid-latitude of Central Asia (Fig. 1(c)). The second one occurred on 22 July 2009 also during the NH summer, covering latitudinal bands between equator and 30° N of the Asian and Pacific regions (see Fig. 1 of Wang and Liu 2010). The rest three are annular solar eclipses. The 22 September 2006 annular solar eclipse occurred over the Southern Hemisphere (SH), covering latitudes from equator to 60°S, and regions of South Atlantic Ocean (Fig. 1(d)). The 26 January 2009 annular solar eclipse occurred over the SH as well, covering latitudes between 45°S to equator and regions of

South Indian Ocean (Fig. 1(b)). The 15 January 2010 annular solar eclipse covered regions from East Africa, through North Indian Ocean to Southeast Asia and China (Fig. 1(a)).

The eclipse paths shown in Fig. 1 represent centers of lunar shadows as the lunar shadow moves through the atmosphere. These are the locations where the maximum effect of moon's block out of solar radiation occurred. For example, during a total solar eclipse, the center of the eclipse represents 100% block out of the sun. The percentage of block out decreases as distance to the center of the eclipse increases.

2.2 Eclipse-Effect Recording GPS RO Profiles: The Eclipse-Time Mean Temperature Differences dT(F3C)

The successful launch of the F3C satellite on 14 April 2006 provides a new way to measure temperature profiles in the troposphere and the stratosphere on a global domain (Anthes et al. 2008). The GPS RO method provides highly accurate temperature profiles from the middle troposphere to the stratosphere (Kursinski et al. 1997; Wickert et al. 2001; Hajj et al. 2004; Kuo et al. 2005). The high spatial and temporal coverage of the GPS RO profiles enable us to study the atmospheric effects of prominent events that were previously difficult to investigate. For example, effects of the volcanic plumes following volcano eruption (Wang et al. 2009), thermal effects of the Saharan dust over the tropical North Atlantic (Wang 2010), high resolution vertical temperature structure inside the Antarctic polar vortex (Wang and Lin 2007), large-scale planetary waves (Shepherd and Tsuda 2008), and variability of the upper troposphere and lower stratosphere (e.g., Schmidt et al. 2010).

The F3C data set used in this work are the same as those described in Wang and Liu (2010). This work used Level-2 post-processed atmospheric temperature profiles obtained from F3C data center at Taiwan Analysis Center for COSMIC (TACC; http://tacc.cwb.gov.tw). Each F3C RO profile contains a vertical 1-km resolution, extending from the surface to 40-km altitude.

In order to estimate the effect of the major eclipses on the stratosphere-troposphere coupled system, we calculate and compare temperature profiles before, during, and after the eclipse. We note that the selections of the FORMOSAT-3/COSMIC data along the vicinity of the eclipse paths were followed the method described in Wang and Liu (2010). Basically, the F3C GPS RO profiles were selected when if it occurred within a 3-hour time window, and 2250-km distance in space between the RO points and the centers of the eclipse. For simplicity of following discussion, we call these profiles as the eclipse-effect recording RO profiles.

Firstly, we take profiles of eclipse-time GPS RO temperature measurements inside the lunar shadow as it moves through the atmosphere. These represent temperature profiles inside the real lunar shadows. Secondly, we take temperature profiles following the same eclipse path but on non-eclipse days. These temperatures are calculated as follows. Let $T_{i,k,l}^{\text{F3C}}$ denotes temperature profiles from F3C. Here i indicates GPS RO profile index, and k indicates a vertical level index for each GPS RO profile. Here l indicates day index for each eclipse event, including 10 days before a major eclipse, during the eclipse day, and 10 days after the eclipse day. Details of these days for each major eclipse event are shown in Table 2. Hence, the mean temperature profiles for each day in the investigated period (10 days before and 10 days after an eclipse) is given by

$$\overline{T_{k,l}^{\text{F3C}}} = \frac{\sum_{i=1}^{M} T_{i,k,l}^{\text{F3C}}}{M} \tag{1}$$

Table 2 Analysis periods

Pre-eclipse days	Eclipse day	Post-eclipse days	Eclipse year	Comparison years
12 Sep–21 Sep	09:52–13:27 UT, 22 Sep	23 Sep–2 Oct	2006	2007, 2008, 2009
22 Jul–31 Jul	09:23–11:18 UT 1 Aug	2 Aug–11 Aug	2008	2007, 2009
16 Jan–25 Jan	06:07–09:50 UT, 26 Jan	27 Jan–5 Feb	2009	2007, 2008, 2010
12 Jul–21 Jul	00:53–04:17 UT, 22 Jul	23 Jul–1 Aug	2009	2007, 2008
5 Jan–14 Jan	05:18–08:54 UT, 15 Jan	16 Jan–25 Jan	2010	2009, 2008, 2007

Here M is the total number of GPS RO profiles that had occurred in the 3-h time-window and the 2250-km vicinity of the eclipse centers. These are the eclipse-effect recording RO profiles. In the comparing non-eclipse years, no eclipse had occurred but we still take temperature profiles inside a virtual lunar shadow as it moves through the atmosphere. Note that the virtual lunar shadows and the real lunar shadows are exactly the same, except that the block out of solar radiation occurred only inside the real lunar shadows.

Equation (1) was applied to both the eclipse year and the comparing non-eclipse years for each of the five major solar eclipses shown in Table 2. Based on (1), we can calculate the mean eclipse-time temperature difference profiles between the eclipse day and the non-eclipse days as:

$$\overline{dT_k^{\text{F3C}}} = \frac{\sum_{l=1}^{L} \overline{T_{k,\text{eclipse day}}^{\text{F3C}}} - \overline{T_{k,l}^{\text{F3C}}}}{L} \tag{2}$$

Here L is equal to 20 (10 days before and after the eclipse day). These profiles of temperature differences are also denoted as dT(F3C). Hence,

$$dT(\text{F3C}) = \overline{dT_k^{\text{F3C}}} \tag{3}$$

The GPS RO data from F3C provides temperature profiles inside the lunar shadows at eclipse days ($\overline{T_{k,\text{eclipse day}}^{\text{F3C}}}$), and inside the virtual lunar shadows at the non-eclipse days ($\overline{T_{k,l}^{\text{F3C}}}$). Since all the GPS RO data were taken during the day time when eclipse occurred, these profiles represent eclipse-time mean temperatures (dT(F3C)).

Table 2 shows a list of days for calculating the F3C-based eclipse-time mean temperatures inside the lunar shadow (eclipse day), and inside the virtual lunar shadows at the non-eclipse days (pre-eclipse days and post-eclipse days) for the eclipse year. For the comparison purpose, we also calculate the temperature difference along the same eclipse path for the same period but in non-eclipse years (comparison years). Note that an eclipse occurred only in the eclipse year and not in the non-eclipse years. The eclipse-time mean temperatures were calculated following the eclipse path and vicinity (Fig. 1) for each day of both the eclipse year and the comparison years shown in Table 2.

2.3 Non-eclipse Effect Analysis Profiles: The Daily Mean Temperature Differences dT(ECMWF)

We note that temperature profiles are also influenced by the atmospheric dynamic processes associated with the synoptic scale weather activity (e.g., advection of warm and cold airs), and the atmospheric physical processes such as the diabatic cooling/warming associated with the presence of the aerosols (e.g., Wang et al. 2009).

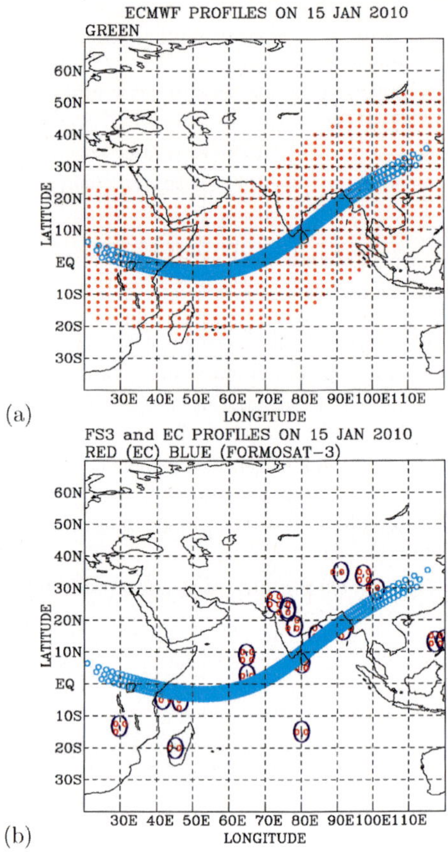

Fig. 2 (**a**) Spatial distribution of ECMWF grid profiles (*red dots*) and path of the total eclipse (*sky-blue circles*) on 15 January 2010. (**b**) The same as in (**a**) but also for F3C GPS RO profiles (*big blue circles*)

In order to quantify these non-eclipse effects, we use four times (at 00, 06, 12, and 18 UT) a day of meteorological analysis taken from the ECMWF (Buizza et al. 2005) to help define the variations of the daily average (basic state) of the atmosphere. The current operational ECMWF forecast model contains a horizontal resolution of 0.125° × 0.125° longitude/latitude, and a vertical resolution of 91 layers, extending from the surface up to 0.01 hPa (about 80 km). In this work we use four times daily global gridded data product, which contains horizontally a 2.5° × 2.5° longitude-latitude grid resolution, and at levels 1000, 950, 925, 900, 850, 800, 700, 600, 500, 400, 300, 250, 200, 150, 100, 70, 50, 30, 20, 10, 7, 5, 3, 2, and 1 hPa. More detailed information on the evolution of the ECMWF forecast system can be found at ECMWF (see http://www.ecmwf.int). These 6-hourly analysis data are vertically interpolated to the 1-km vertical resolution used in the F3C data for intercomparison.

The daily average temperature profiles from ECMWF were taken at eclipse day (from inside the lunar shadows) and also within the 250-km radius of the F3C GPS RO profiles. For example, Fig. 2(a) shows spatial distribution of the ECMWF grids that are close to within 2250 km search radius of the 15 Jan 2010 solar eclipse path. However, only those grids in Fig. 2(b) that are within 250-km radius of the eclipse-effect recording RO profiles were selected for analysis.

Similarly, the daily average temperature profiles were also calculated for the non-eclipse days (from inside the virtual lunar shadows). The daily mean temperature profiles ($\overline{T_{k,l}^{\text{ECMWF}}}$) were calculated for each day of the eclipse year and the non-eclipse years as shown in Table 2:

$$\overline{T_{k,l}^{\text{ECMWF}}} = \frac{\sum_{i=1}^{N} T_{i,k,l}^{\text{ECMWF}}}{N} \tag{4}$$

Here N represents the total number of the ECMWF grid profiles (shown from a horizontal 2.5° longitude-latitude grid resolution) that are located within the 250-km radius of each eclipse-recording GPS RO profiles.

Based on these daily average temperature profiles, we can calculate the profiles of difference in daily average temperatures between the eclipse day and each day of the non-eclipse days in the eclipse year and the non-eclipse years, respectively:

$$\overline{dT_k^{\text{ECMWF}}} = \frac{\sum_{l=1}^{L} \overline{T_{k,\text{eclipse day}}^{\text{ECMWF}}} - \overline{T_{k,l}^{\text{ECMWF}}}}{L} \tag{5}$$

These profiles of daily mean temperature difference are also denoted as dT(ECMWF). Hence,

$$dT(\text{ECMWF}) = \overline{dT_k^{\text{ECMWF}}} \tag{6}$$

2.4 The Eclipse Tropospheric Cooling and Stratospheric Warming Hypothesis: dT(F3C)–dT(ECMWF)

Due to the block out of direct solar radiation compared with no block out of solar radiation on the non-eclipse days, $\overline{T_{k,\text{eclipse day}}^{\text{F3C}}}$ reduced, and $\overline{T_{k,\text{eclipse day}}^{\text{F3C}}} - \overline{T_{k,l}^{\text{F3C}}}$ becomes more negative. This leads $dT(\text{F3C}) = \overline{dT_k^{\text{F3C}}}$ to become more negative. Hence, profiles of temperature differences between the eclipse day and the non-eclipse days in the eclipse years should become more negative than the differences from the non-eclipse years.

The block out effect appears on the F3C data but not fully represented on the operational meteorological analysis data such as ECMWF used here. Hence, the tropospheric cooling hypothesis predicts that in the troposphere of the eclipse years, dT(ECMWF) changes little, while dT(F3C) becomes more negative. This leads $dT_{\text{eclipse effect}}^{\text{TROPO}} = dT(\text{F3C}) - dT(\text{ECMWF})$ to become more negative. In the stratosphere of the eclipse years, dT(ECMWF) changes little, and dT(F3C) becomes more positive (due to the eclipse-induced dynamic warming effect). This leads $dT_{\text{eclipse effect}}^{\text{STRAT}} = dT(\text{F3C}) - dT(\text{ECMWF})$ to become more positive.

We expect that (1) profile of temperature differences between the eclipse day and the non-eclipse days in the non-eclipse years should be small for both the eclipse-time average data (dT(F3C)) and the daily average data (dT(ECMWF)); (2) profiles of temperature differences between the eclipse day and the non-eclipse days in the eclipse years should be large from the eclipse-time average data, due to the eclipse block out effect observed by the F3C; (3) profiles of temperature differences in the eclipse years from F3C should be larger than those from the ECMWF, due to the eclipse effect observed by F3C but not fully expressed in the ECMWF data.

3 Results

3.1 Spatial Distribution of Eclipse Paths and GPS RO profiles

Figure 3 shows latitudinal distribution of the GPS RO profiles for the eclipse years and the comparison years (Table 2). These eclipses occurred in latitudes ranging from the NH polar regions, the NH mid-latitudes, tropical regions of both hemispheres, the SH mid-latitudes, to the SH high latitudes. Two total eclipses occurred in the NH summer seasons: the 22 July 2009 eclipse over the NH tropics to mid-latitudes (Fig. 3(b)), and the 1 August 2008 eclipse over the NH high to polar latitudes (Fig. 3(a)). The 15 Jan 2010 annular solar eclipse occurred during the NH winter season, with the GPS RO profiles distributed from the SH tropical region to close to the NH mid-latitudes (Fig. 3(c)). The 22 September 2009 annular solar eclipse occurred during the SH spring season, and in the SH mid-latitudes (Fig. 3(d)). The 22 September 2006 annular solar eclipse occurred in the SH spring season, and in latitudes extend from the NH tropical region to the SH high latitudes (Fig. 3(e)).

Figure 4 shows mean temperature profiles for the eclipse day and the non-eclipse days of each eclipse event. These profiles are shown following the occurrence of the eclipse latitudes from the NH polar to high latitudes (Fig. 4(a)), the NH mid-latitudes to the SH tropical latitudes (Fig. 4(b)), the SH mid to high latitudes (Fig. 4(c)), to the NH tropical latitudes to the SH high latitudes (Fig. 4(d)). We can clearly see the different tropopause height at high and low latitudes, varying between 11-km altitudes over the high latitudes and 17–19 km over the low latitudes.

3.2 The NH Polar to High Latitudes Eclipse: 1 August 2008

Figure 5(a) compares mean temperature differences during the eclipse occurred on 1 August 2008, and those mean temperature differences from the non-eclipse years. The eclipse-time mean temperature differences (dT(F3C)) show pronounced negative values in the troposphere (4–10 km altitudes), and positive values in the stratosphere (10–16 km altitudes). This is a clear indication of eclipse-induced cooling in the troposphere and warming in the lower stratosphere. These results are consistent with what we observed in Wang and Liu (2010).

Figure 5(a) shows that daily mean temperature differences (dT(ECMWF)) in the troposphere (below 8-km altitude) were positive, indicating that temperatures in the troposphere during the post-eclipse period (2–11 August) are generally warmer than temperature on the eclipse day (1 August) and in pre-eclipse days (22–31). This feature also appear in the non-eclipse years of 2007 and 2009 (Fig. 5(b)). Hence, the daily mean temperature differences are predominantly positive for these 3 years based on the ECMWF analysis.

For the non-eclipse years, Fig. 5(b) shows strong positive temperature differences from the eclipse-time means, and these values are larger than those from the daily means. Positive temperature differences appear in the troposphere, and these values are larger than the daily means. Hence, the non-eclipse years show that eclipse-time mean temperature differences are much warmer than the daily mean temperature differences. This indicates more warming of the troposphere in post-eclipse periods than the pre-eclipse periods. This feature also reveals that the cooling in the troposphere during the eclipse year of 2008 is indeed caused by the eclipse effect, as mean temperature differences (dT(F3C)–dT(ECMWF)) in the troposphere are distinctively negative compared with the predominantly positive values from the other two non-eclipse years.

Fig. 3 Latitudinal distribution of F3C GPS RO profiles for (**a**) 1 August 2008, (**b**) 22 July 2009, (**c**) 15 January 2010, (**d**) 26 January 2009, and (**e**) 22 September 2006 eclipse cases. *Solid lines* indicate GPS RO profile counts in the eclipse years, while the *dashed lined* indicate GPS RO counts in the comparison years

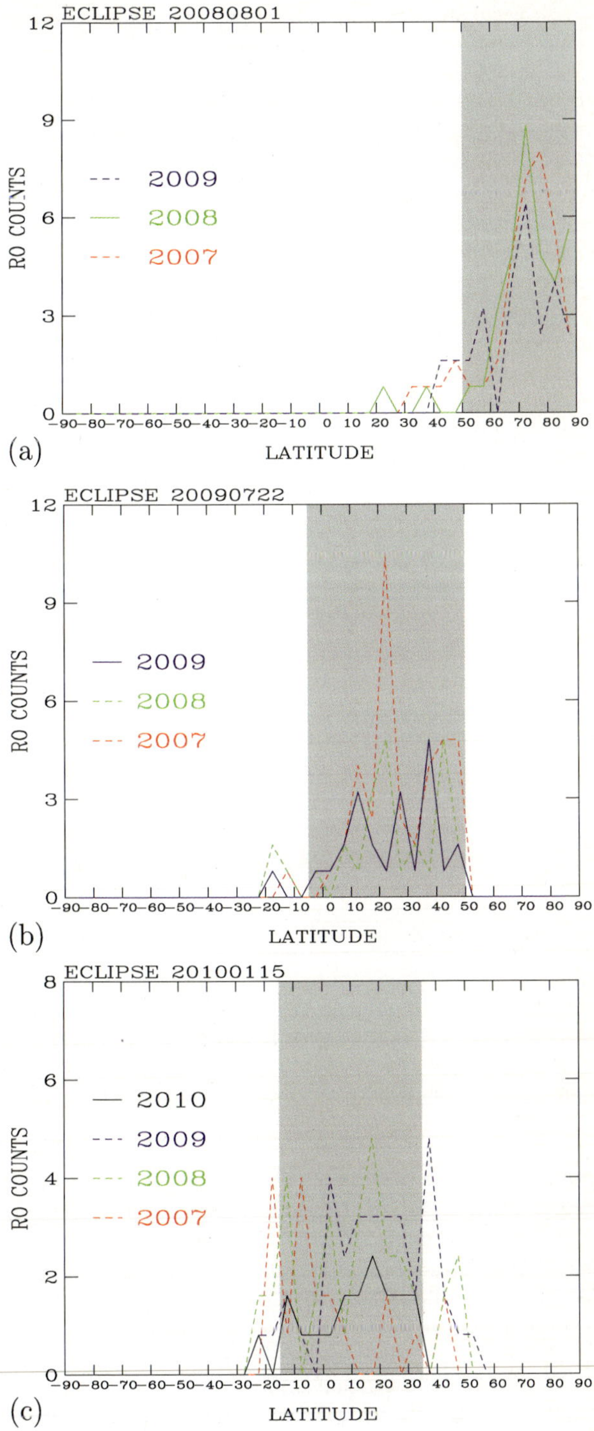

Fig. 3 (*Continued*)

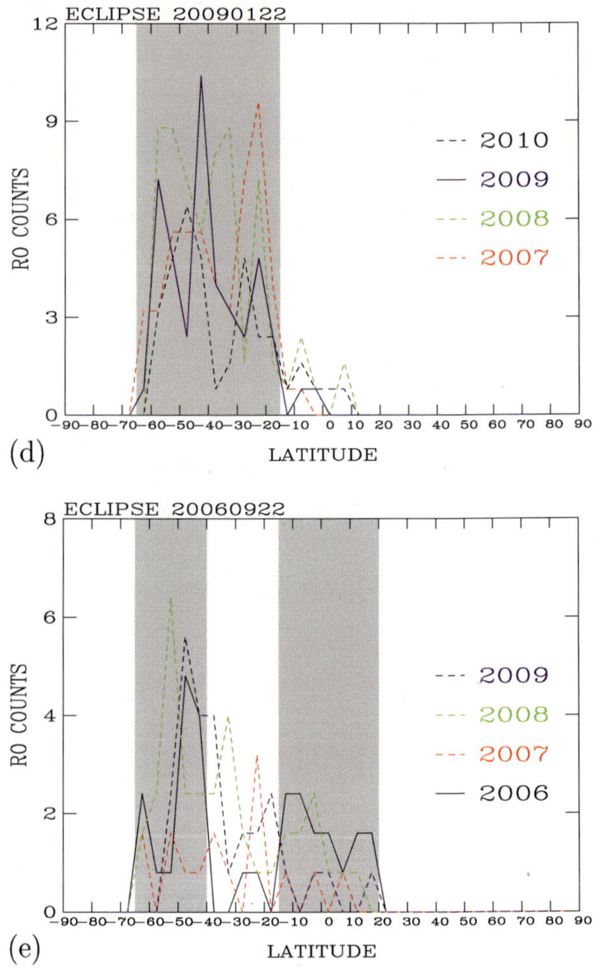

3.3 The NH Mid to the SH Tropical Latitudes Eclipse: 22 July 2009

Figure 6(a) compares average of profiles of temperature differences between the eclipse day (22 July) and the non-eclipse days (12–21 July, and 23 July–1 August) in 2009. The daily average profiles of temperature differences (dT(ECMWF)) show negative daily mean temperature differences of about -1°C exist throughout the troposphere (below 10 km altitude), while positive daily mean temperature differences of up to 3°C appear in the stratospheric region of 12 to 30 km. The eclipse-time mean temperature differences (dT(F3C)) shows similar pattern compared with the daily mean temperature differences, but with a large negative mean eclipse-time temperature differences of -2°C at altitudes below 10 km altitudes and positive eclipse-time mean temperature differences of up to 6°C in the stratospheric region of 12 to 30 km.

Hence, the mean temperature differences (dT(F3C)–dT(ECMWF)) on the eclipse day of 22 Jul 2009 are about 1°C cooler in the troposphere, and about 0–3°C warmer in the lower stratosphere than those daily mean temperature differences of the non-eclipse days of 12–21 July 2009 and 23 July–1 August 2009. The warming in the stratosphere occurred in

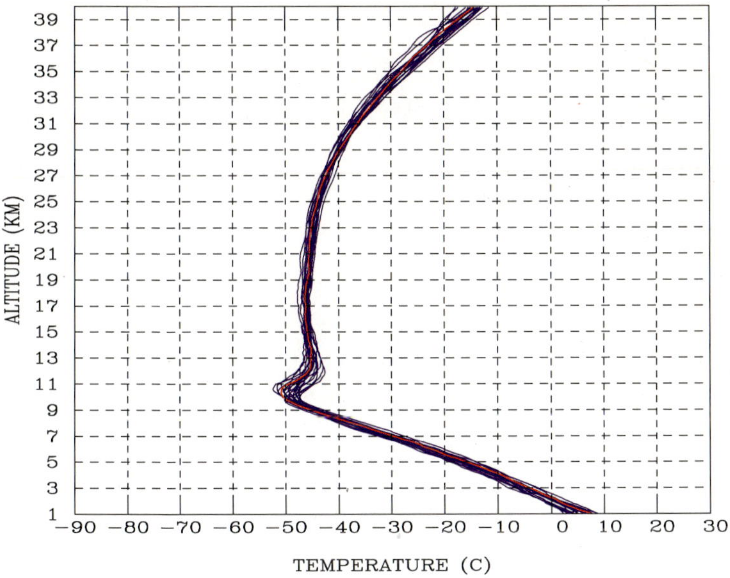

(a) The 200808 T. Eclipse

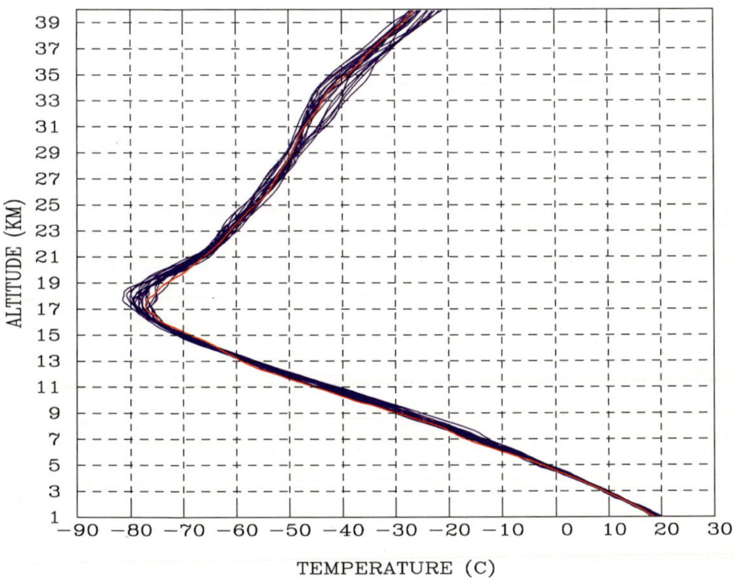

(b) The 201001 A. Eclipse

Fig. 4 Profiles of temperatures from F3C for the (**a**) 1 August 2008, (**b**) 15 January 2010, (**c**) 26 January 2009, and (**d**) 22 September 2006 (occurred between 20°N–15°S) eclipses. *Red curves* indicate temperature profile occurred in the eclipse days, while *blue curves* indicates temperature profiles occurred 10-day before and after the eclipse day of each event

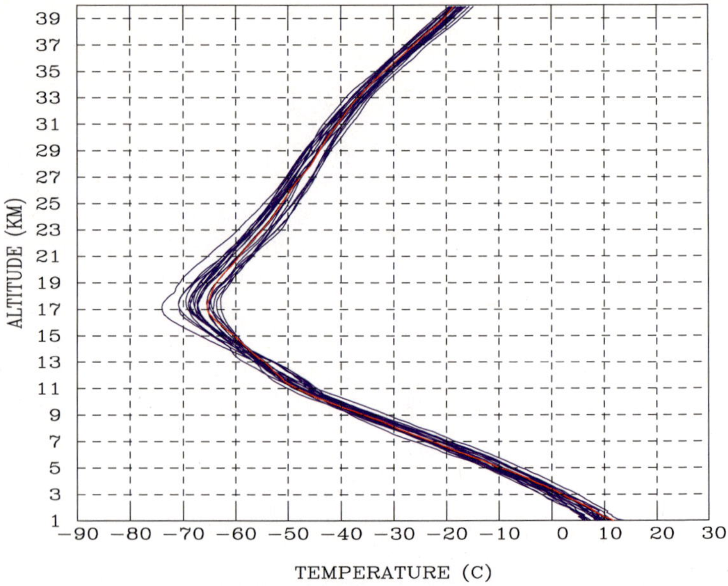

(c) The 200901 A. Eclipse

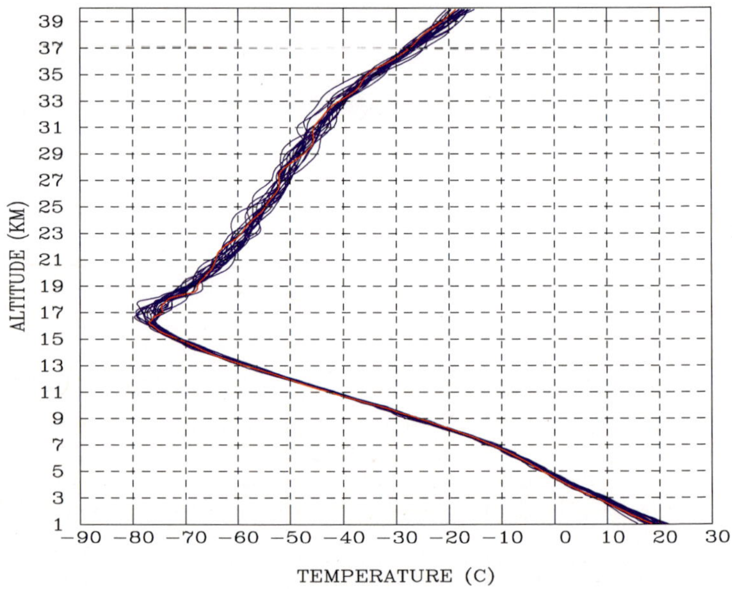

(d) The 200609 A. Eclipse

Fig. 4 (*Continued*)

the regions between 13 and 22 km altitudes, with a peak of 3°C warming centered around 17 km altitude.

Figure 6(b) compares mean profiles of temperature differences occurred in the same period but for non-eclipse years of 2007 and 2008. The calculations show that the mean

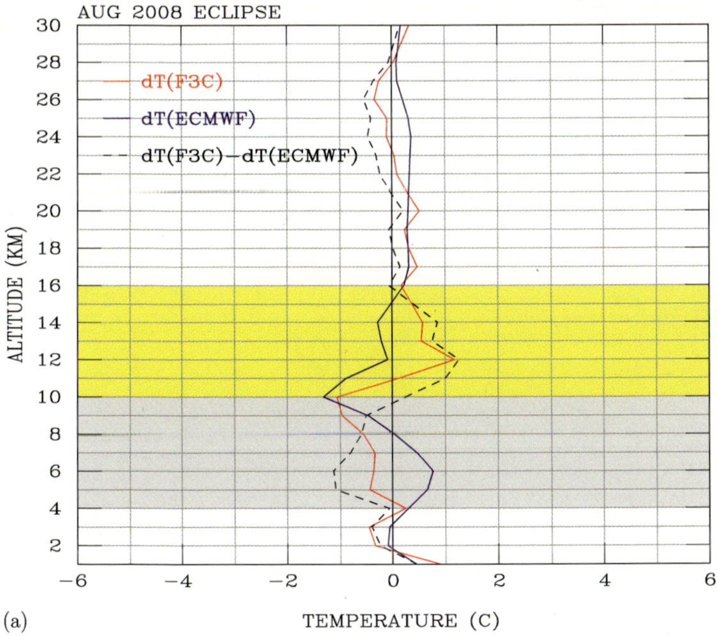

(a)

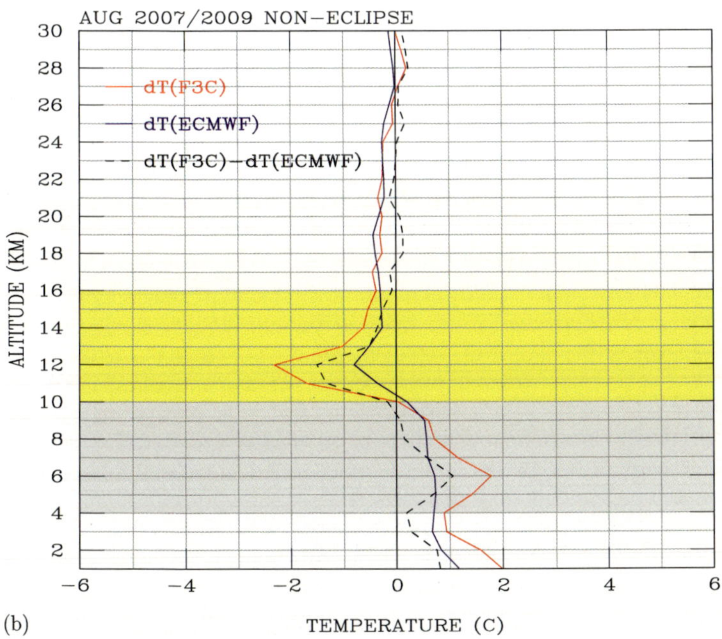

(b)

Fig. 5 (**a**) The 1 August 2008 total solar eclipse. Mean profiles of temperature differences between eclipse and non-eclipse days from F3C (*red curves*) and ECMWF (*blue curves*), respectively; and the difference of mean profiles between F3C and ECMWF (*black dashed curves*). (**b**) Average results from the non-eclipse years of 2007 and 2009

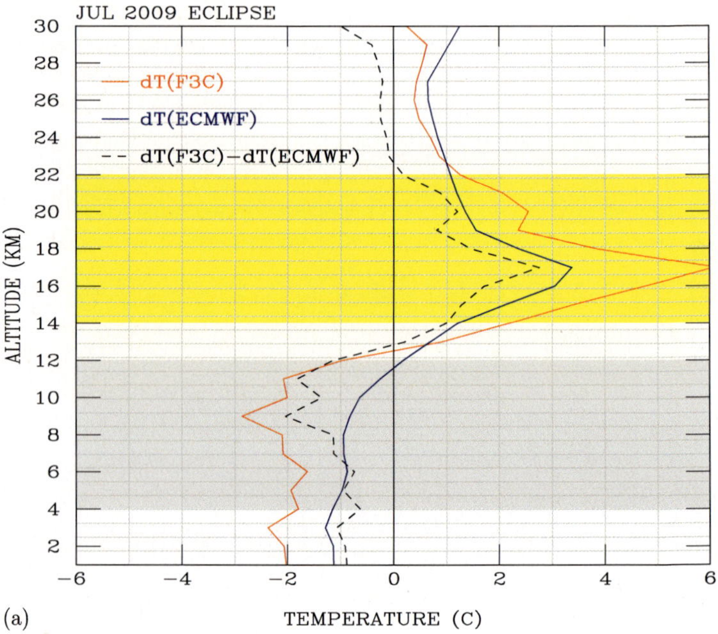

(a)

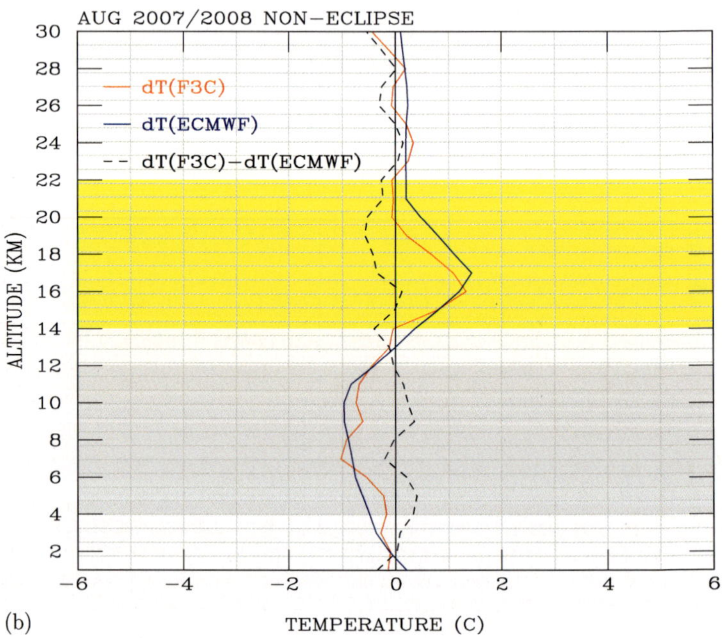

(b)

Fig. 6 (**a**) The 22 July 2009 total solar eclipse. Mean profiles of temperature differences between eclipse and non-eclipse days from F3C (*red curves*) and ECMWF (*blue curves*), respectively; and the difference of mean profiles between F3C and ECMWF (*black dashed curves*). (**b**) Average results from the non-eclipse years of 2007 and 2008 eclipse-time temperature differences and the mean daily temperature differences are very close to each other. The difference between the mean eclipse-time profiles and the mean daily profiles (dT(F3C)–dT(ECMWF)) are small, within 0.5 and −0.5°C through most of the troposphere and the stratosphere.

Figures 6(b) demonstrates that the temperature differences from the eclipse-time means are close to the daily means in the troposphere. These results validate our hypothesis shown in Sect. 2.4, that the eclipse-time mean temperatures are generally similar to the daily mean temperatures in the non-eclipse years of 2007 and 2008. When compared with these two non-eclipse years, the eclipse year of 2009 show very distinctive cooling in the troposphere and warming in the middle to lower stratosphere.

Hence, we further prove that the 22 July 2009 eclipse do indeed produced significant cooling in the troposphere and warming in the lower stratosphere when compared with the non-eclipse years. The results shown here are consistent with those reported in Wang and Liu (2010).

3.4 The NH Sub-tropical to the SH Tropical Latitudes Eclipse: 15 January 2010

Figure 7(a) shows comparison of mean profiles of temperature differences during the 15 January 2010 annular solar eclipse. The mean daily temperature differences (dT(ECMWF)) are negative in the troposphere and positive in the lower stratosphere. The mean eclipse-time temperature differences (dT(F3C)) show similar pattern to those of mean daily temperature differences. However, the mean temperature differences (dT(F3C)–dT(ECMWF)) are more negative in the troposphere and more positive in the stratosphere than the mean daily temperature differences. This elevated cooling in the troposphere and warming in the stratosphere (as shown in dT(F3C)–dT(ECMWF)) indicates the effect of the eclipse. These results are consistent with those shown during the 22 July 2009 total solar eclipse (Fig. 6(a)), and the 1 August 2008 total solar eclipse (Fig. 5(a)).

Further comparisons were made to compare mean temperature differences between the daily and the eclipse-time values following the 15 January 2010 eclipse path but in the non-eclipse year of 2007, 2008, and 2009. For comparisons in these non-eclipse years (Fig. 7(b)), we find that mean temperature differences (dT(F3C)–dT(ECMWF)) in the troposphere and the lower stratosphere are small compared with those shown in the eclipse year. These comparisons again demonstrate the eclipse effect. We note that the wavy structure of temperature differences occurs in the stratosphere looks like gravity waves. This will require further study to demonstrate the vertical propagating gravity waves associated with this case.

3.5 The SH Mid to High Latitudes Eclipse: 26 January 2009

Figure 8(a) compares mean temperature differences between the eclipse-time and the daily mean values during the annular solar eclipse that had occurred on 26 January 2009, and those occurred in the non-eclipse years of 2007, 2008, and 2010 (Fig. 8(db)). For the mean temperature differences in the eclipse year of 2009, eclipse-time mean temperature differences show clear cooling in the middle to upper troposphere and warming in the lower stratosphere (14–22 km altitudes, Fig. 8(a)). The daily average mean temperature differences show only clear cooling in the upper troposphere (10–12 km altitudes) and moderate warming in the stratosphere. As such, the mean temperature differences (dT(F3C)–dT(ECMWF)) are negative in the troposphere and positive in the lower stratosphere. These results are consistent with the pattern of the eclipse-induced tropospheric cooling and stratospheric warming shown during the 22 July 2009 total solar eclipse.

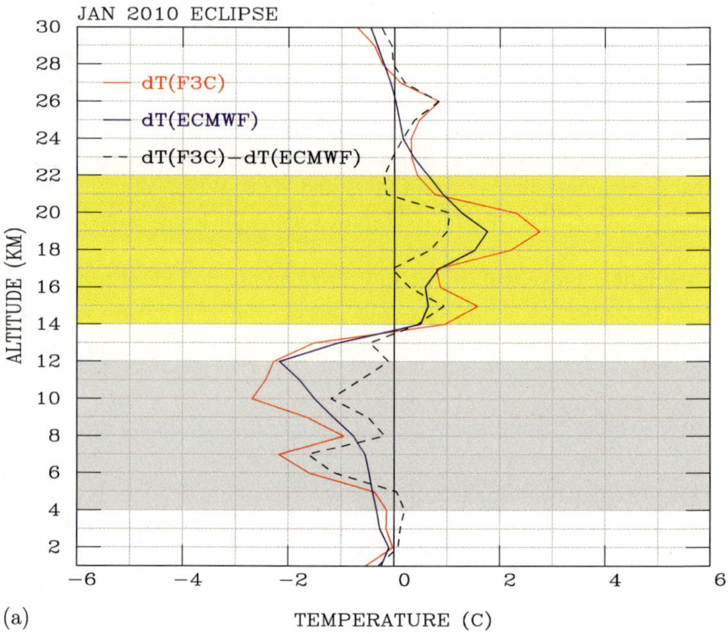

(a)

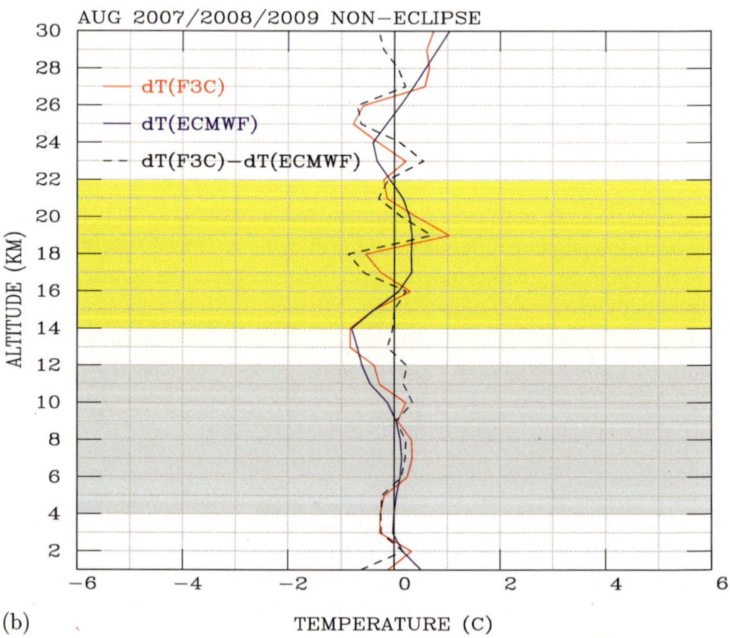

(b)

Fig. 7 (**a**) The 15 January 2010 annular solar eclipse. *Red curves* show mean profiles of temperature differences between eclipse and non-eclipse days from F3C. Mean temperature differences from ECMWF are shown in *blue curves*. The difference of mean profiles between F3C and ECMWF are shown in *black dashed curves*. (**b**) Average results from the non-eclipse years of 2007, 2008, and 2009

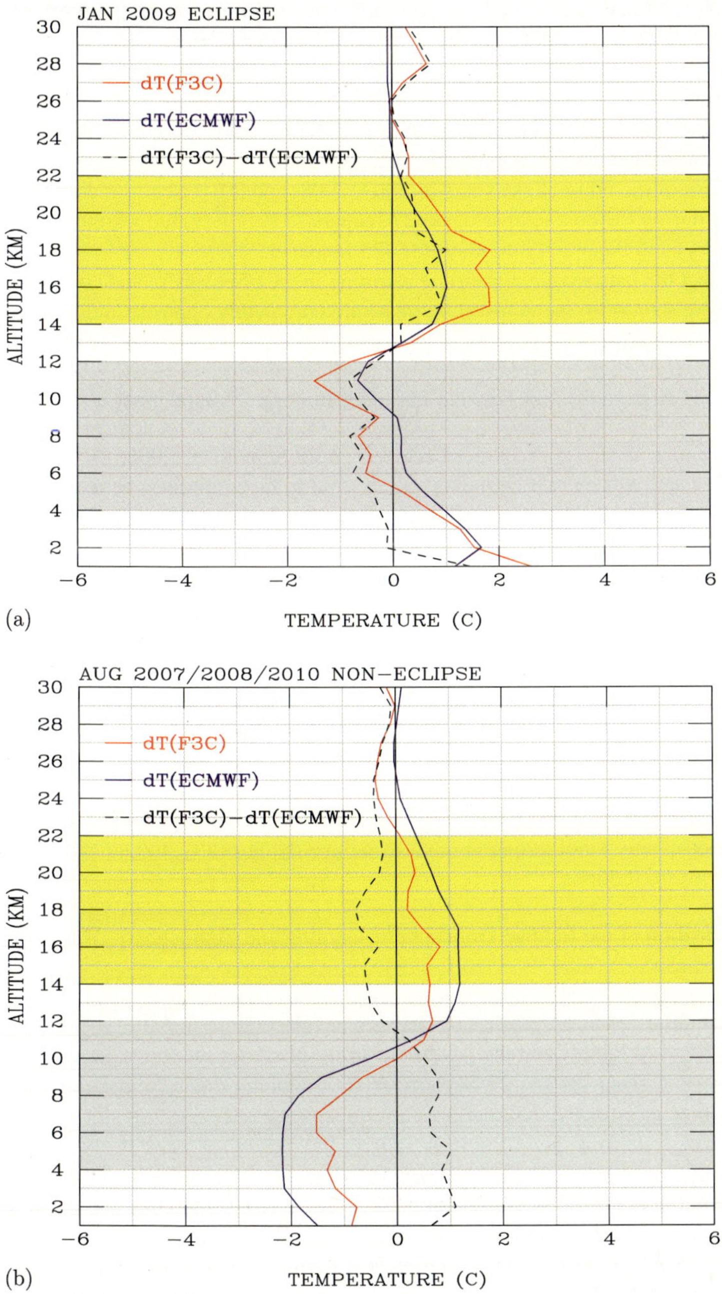

Fig. 8 (**a**) The 26 January 2009 annular solar eclipse. Mean profiles of temperature differences between eclipse and non-eclipse days from F3C are shown in *red curves* and from ECMWF are shown in *blue curves*. The difference of mean profiles between F3C and ECMWF are shown in *black dashed curves*. (**b**) Average results from the non-eclipse years of 2007, 2008, and 2010

The non-eclipse years (Fig. 8(b)) show more negative mean temperature differences in the troposphere and more positive mean temperature differences in the lower stratosphere from the daily average values than from the eclipse-time average values. This results in the warming effect in the eclipse-time troposphere and cooling effect in the eclipse-time lower stratosphere.

The 26 January 2009 eclipse tracked a path that mostly remains in the summer of the South Hemisphere. The predominantly negative daily mean temperature differences in the troposphere of the non-eclipse years of 2007, 2008, and 2010 clearly indicate that the temperatures during the post-eclipse days (27 January–5 February) are warmer than the eclipse (26 January) and the pre-eclipse days (16–25 January).

Note that the pattern of eclipse-time warming in the troposphere and cooling in the lower stratosphere seen from the non-eclipse years are completely opposite to the eclipse-induced pattern of the eclipse-time cooling in the troposphere and warming in the stratosphere.

3.6 The SH High to the NH Tropical Latitudes Eclipse: 22 September 2006

For the annular solar eclipse event occurred on 22 September 2006, the F3C RO profiles mainly concentrates on two latitudinal bands: mid to high latitudes, between 40° and 65°S; and tropical latitudes, between 20°N and 15°S. For the RO profiles in the mid to high latitudes (40° to 65°S), we find they exhibit strong year to year variability (not shown here). This may be due to the presence of the Antarctic polar vortex during this time of the year. Here we analyze profiles of temperature differences in the lower latitudes.

Figure 9 compares profiles of mean temperature differences (dT(F3C)–dT(ECMWF)) between the eclipse and the non-eclipse days on the latitudes in the tropical latitudes. In the lower troposphere (below 6-km altitude), the general cooling from the eclipse year in 2006 (Fig. 9(a)) is more pronounced than the minor warming from the non-eclipse years of 2007, 2008, and 2009 (Fig. 9(b)). This characteristic cooling in the lower troposphere is consistent with those shown previously. However, the warming in the lower stratosphere in the 2006 eclipse year is less clear when compared with other non-eclipse years. It may be the cooling in the troposphere is not strong enough to induce dynamically and adiabatically warming from downward motions.

3.7 Verification of the Eclipse Tropospheric Cooling and Stratospheric Warming Hypothesis

Table 3 summarizes mean profiles of changes of temperature (dT(F3C)–dT(ECMWF)) in the 4–10 km troposphere and the 14–20 km stratosphere for the five solar eclipses analyzed in this work. Except for the 22 September 2006 annular solar eclipse, all the other four eclipses show a distinctive hallmark of the atmospheric response to major solar eclipses.

For these four eclipses in the troposphere, the values of dT(F3C)–dT(ECMWF) are consistently negative, with values varying between −0.6 and −1.2°C for the eclipse years. On the contrary, the non-eclipse comparing years show that the values of dT(F3C)–dT(ECMWF) are either greater or equal to zero. The value of dT(F3C)–dT(ECMWF) in the troposphere for 2006 is zero, which is still lower than those of the non-eclipse years. This is consistent with the eclipse cooling in the troposphere. In the stratosphere of these four eclipses, the values of dT(F3C)–dT(ECMWF) are consistently positive, with values varying between 0.4 and 1.3°C. The values of dT(F3C)–dT(ECMWF) are mostly negative, three positive values, and one zero for other non-eclipse comparing years.

For a total of five eclipse cases shown in Table 3, four cases show tropospheric cooling and stratospheric warming results while one case (22 September 2006) does not show this

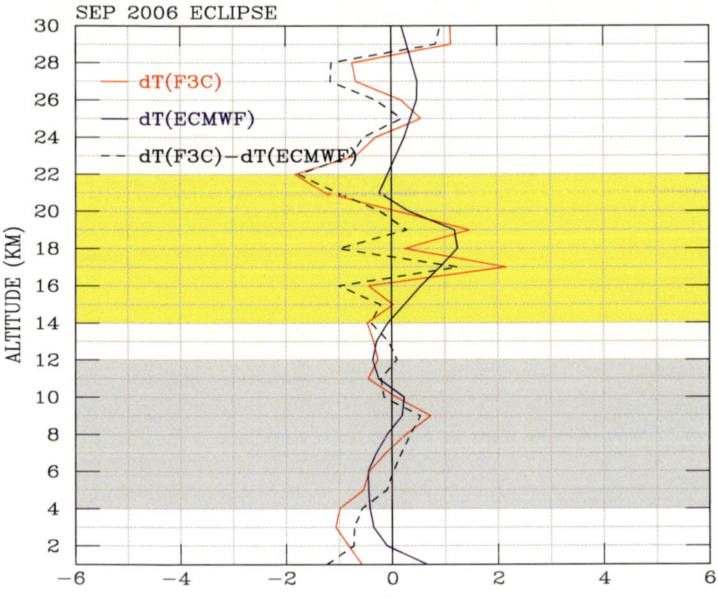

(a)

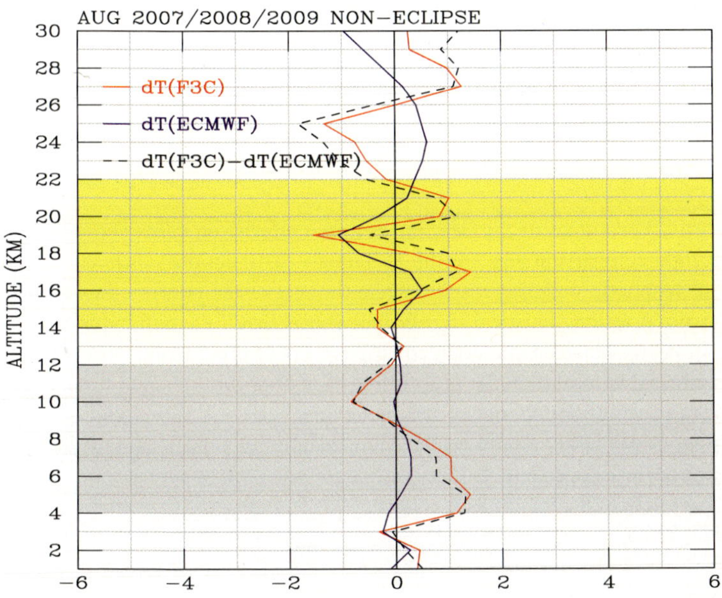

(b)

Fig. 9 (**a**) The 22 September 2006 annular solar eclipse. Mean profiles of temperature differences between eclipse and non-eclipse days from F3C (*red curves*) and ECMWF (*blue curves*), respectively; and the difference of mean profiles between F3C and ECMWF (*black dashed curves*). (**b**) Average results from the non-eclipse years of 2007, 2008, and 2009. These analysis are based on GPS and ECMWF profiles occurred between 20°N and 15°S

Table 3 Mean temperature differences (dT(F3C)–dT(ECMWF)) for the eclipse and non-eclipse cases

Latitudinal Extent	Eclipse events	Comparison Years	Strato-sphere	Tropo-sphere
NH high latitudes	1 Aug 2008[a]		0.6	−0.6
		2009	−0.7	0.7
$\delta T = 1:55$		2007	−0.5	0.0
NH mid latitudes	22 Jul 2009[b]		1.3	−1.2
		2008	0.1	0.1
$\delta T = 3:24$		2007	−0.7	0.2
NH low lat.–SH tropics	15 Jan 2010[b]		0.4	−0.6
		2009	−0.4	0.3
$\delta T = 3:36$		2008	−0.2	0.0
		2007	0.3	−0.1
SH mid latitudes	26 Jan 2009[b]		0.5	−0.5
		2010	−1.0	1.1
$\delta T = 3:43$		2008	−0.5	0.5
		2007	0.0	0.0
NH tropics.–SH mid-high lat.	22 Sep 2006[b]		−0.4	0.0
		2009	0.6	0.7
$\delta T = 3:35$		2008	0.2	0.5
		2007	0.1	0.0

[a]The tropospheric mean is calculated for the 4–10 km altitudes, while the stratospheric mean is calculated for the 10–16 km altitudes

[b]The tropospheric mean is calculated for the 4–12 km altitudes, while the stratospheric mean is calculated for the 14–22 km altitudes

δT: Eclipse duration

characteristic. For a total of thirteen non-eclipse comparison cases shown in Table 3, one case (the 2009 comparison year for the 15 January 2010 eclipse) shows tropospheric cooling and stratospheric warming results, while the rest of the twelve cases do not show this characteristic. A test of statistical significance shows that if the null hypothesis ($H_0 =$ there is no relationship between the coupled troposphere-cooling-stratosphere-warming events and the eclipses) is true, then the probability of getting four coupled troposphere-cooling-stratosphere-warming events out of the five eclipse cases is very small ($P = 0.002$, Chi-Square Test) (More 2001). These results show that our hypothesis is statistically significant. The cooling of the troposphere and the warming of the stratosphere is a predictable response of the atmosphere to the major solar eclipse.

4 Summary

The block out of short wave solar radiation by major eclipse events provides great opportunities to study the coupling between the troposphere and the stratosphere. In this work we used the F3C RO profiles to obtain temperature profiles inside the lunar shadow, and the ECMWF analysis to obtain background non-eclipse temperature profiles. The eclipse effects were calculated based on the difference of F3C and ECMWF profiles. We have analyzed five eclipse cases, and thirteen non-eclipse comparison cases. Our results show −0.6 to −1.2°C cooling in the troposphere and 0.4 to 1.3°C warming in the middle to lower stratosphere during the eclipses. These results are statistically significant ($P = 0.002$ with a Chi-Square Test).

In this work we showed that the direct cooling in troposphere and induced warming in the middle to lower stratosphere occurs during the major solar eclipses. The mechanism for the cooling in the troposphere is thermally driven while the warming in the stratosphere is dynamically controlled. As discussed in Wang and Liu (2010), the cooling produces contraction of air in the troposphere, which induces downward movement in the middle to lower stratosphere. The downward motion adiabatically heats the middle to lower stratosphere, resulting in the warming in that region.

Our analyses of the stratosphere-troposphere coupling during the 2006–2010 major solar eclipses shows that the troposphere is the region in the atmosphere that responses quickly to the shield of solar radiation. This direct thermal effect in the troposphere then propagates upward through the stratosphere-troposphere coupling into the stratosphere, which not only responses to the shield of solar radiation but also to the dynamic forcing from below. The resulting warming indicates that dynamically driven warming dominates over the thermally cooling in the middle to lower stratosphere.

Acknowledgements We thank NSPO for providing F3C data, NASA for the eclipse path coordinates (eclipse predictions by Fred Espenak at NASA's GSFC), and discussions with the GPSARC team in Taiwan. We are very grateful to the anonymous reviewer for insightful comments that greatly improved the clarity of this paper. We are also very grateful to Rudolf Treumann for inviting us to present this work in the ISSI-Coupling 2010 workshop in Bern, Switzerland. This work is supported by NSPO-S-099010(E), NSPO-S-100011, and NSC-100-2119-M-008-020.

References

R.C. Anderson, D.R. Keefer, O.E. Myers, Atmospheric pressure and temperature changes during the 7 March 1970 solar eclipse. J. Atmos. Sci. **29**, 583–587 (1972)

R. Anthes et al., The COSMIC/FORMOSAT-3 mission: early results. Bull. Am. Meteorol. Soc. **89**, 313–333 (2008)

H.N. Ballard, R. Valenzuela, M. Izquierdo, J.S. Randhawa, R. Morla, J.F. Bettle, Solar eclipse: temperature, wind, and ozone in the stratosphere. J. Geophys. Res. **74**, 711–712 (1969)

R. Buizza, P.L. Houtekamer, G. Pellerin, Z. Toth, Y. Zhu, M. Wei, A comparison of the ECMWF, MSC, and NCEP global ensemble prediction systems. Mon. Weather Rev. **133**, 1076–1097 (2005)

G. Chimonas, Internal gravity-wave motions induced in the Earth's atmosphere by a solar eclipse. J. Geophys. Res. **75**, 5545–5551 (1970)

G. Chimonas, C.O. Hines, Atmospheric gravity waves induced by a solar eclipse 2. J. Geophys. Res. **76**, 7003–7005 (1971)

F. Espenak, J. Anderson, Total solar eclipse of 2009 July 22. NASA/TP-2008-214169 (2008)

F. Espenak, J. Meeus, Five Milliennium canon of solar eclipses: −1999 to +3000 (2000 BCE to 3000 CE). NASA/TP-2006-214141 (2006)

D. Founda et al., The effect of the total solar eclipse of 29 March 2006 on meteorological variables in Greece. Atmos. Chem. Phys. **7**, 5543–5553 (2007)

E. Gerasopoulos et al., The total solar eclipse of March 2006: overview. Atmos. Chem. Phys. **7**, 5205–5220 (2007)

G.A. Hajj et al., CHAMP and SAC-C atmospheric occultation results and intercomparisons. J. Geophys. Res. **109**, D06109 (2004). doi:10.1029/2003JD003909

J.R. Holton, P.H. Haynes, M.E. McIntyre, A.R. Douglass, R.B. Rood, L. Pfister, Stratosphere–troposphere exchange. Rev. Geophys. Space Phys. **33**, 403–439 (1995)

T. Kameda, K. Fujita, O. Sugita, N. Hirasawa, S. Takahashi, Total solar eclipse over Antarctica on 23 November 2003 and its effects on the atmosphere and snow near the ice sheet surface at Dome Fuji. J. Geophys. Res. **114**, D18115 (2009). doi:10.1029/2009JD011886

Y.-H. Kuo, W.S. Schreiner, J. Wang, D.L. Rossiter, Y. Zhang, Comparison of GPS radio occultation soundings with radiosondes. Geophys. Res. Lett. **32**, L05817 (2005). doi:10.1029/2004GL021443

E.R. Kursinski, G.A. Hajj, J.T. Schofield, R.P. Linfield, K.R. Hardy, Observing Earth's atmosphere with radio occultation measurements using the Global Positioning System. J. Geophys. Res. **102**(D19), 23429–23465 (1997)

H. Le, L. Liu, X. Yue, W. Wan, B. Ning, Latitudinal dependence of the ionospheric response to solar eclipses. J. Geophys. Res. **114**, A07308 (2009). doi:10.1029/2009JA014072

C. Lindsey et al., Extreme-infrared brightness profile of the solar chromospheres obtained during the total eclipse of 1991. Nature **358**, 308–310 (1992)

D.S. More, *Statistics: Concepts and Controversies*, 5th edn. (Freeman, New York, 2001), p. 557

J.M. Pasachoff, Solar eclipses as an astrophysical laboratory. Nature **459**, 789–795 (2009)

T. Schmidt, J. Wickert, A. Haser, Variability of the upper troposphere and lower stratosphere observed with GPS radio occultation bending angles and temperatures. Adv. Space Res. **46**, 150–161 (2010)

E.J. Seykora, A. Bhatnagar, R.M. Jain, J.L. Streete, Evidence of atmospheric gravity waves produced during the 11 June 1983 total solar eclipse. Nature **313**, 124–125 (1985)

M.G. Shepherd, T. Tsuda, Large-scale planetary disturbances in stratospheric temperature at high-latitudes in the southern summer hemisphere. Atmos. Chem. Phys. **8**, 7557–7570 (2008)

R.A. Treumann, Z. Klos, M. Parrot, Physics of electric discharges in atmospheric gases: An informal introduction. Space Sci. Rev. **137**, 133–148 (2008)

K.-Y. Wang, Profiles of the atmospheric temperature response to the Saharan dust outbreaks derived from FORMOSAT-3/COSMIC and OMI AI. Atmos. Res. **96**, 110–121 (2010)

K.-Y. Wang, S.-C. Lin, First continuous GPS soundings of temperature structure over Antarctic winter from FORMOSAT-3/COSMIC constellation. Geophys. Res. Lett. **34**, L12805 (2007). doi:10.1029/2007GL030159

K.-Y. Wang, C.-H. Liu, Profiles of temperature responses to the 22 July 2009 total solar eclipse from FORMOSAT-3/COSMIC constellation. Geophys. Res. Lett. **37**, L01804 (2010). doi:10.1029/2009GL040968

K.-Y. Wang, S.-C. Lin, L.-C. Lee, Immediate impact of the Mt. Chaiten eruption on atmosphere from FORMOSAT-3/COSMIC constellation. Geophys. Res. Lett. **36**, L03808 (2009). doi:10.1029/2008GL036802

J. Wickert et al., Atmospheric sounding by GPS radio occultation: First results from CHAMP. Geophys. Res. Lett. **28**(17), 3263–3266 (2001)

C.S. Zerefos et al., Evidence of gravity waves into the atmosphere during the March 2006 total solar eclipse. Atmos. Chem. Phys. **7**, 4943–4951 (2007)

J.B. Zirker, Total eclipse of the sun. Science **210**, 1313–1319 (1980)

Space Sci Rev (2012) 168:283–314
DOI 10.1007/s11214-011-9841-5

Dynamical Control of the Middle Atmosphere

Erich Becker

Received: 2 April 2011 / Accepted: 20 September 2011 / Published online: 29 October 2011

Abstract This review recapitulates the concept of the wave-driven residual circulation in the stratosphere and mesosphere. The residual circulation is defined as the conventional mean meridional circulation corrected by the quasi-linear Stokes drift due to atmospheric waves. Only when the zonal-mean primitive equations are transformed using the residual circulation, they reflect the causality arising from the Eliassen-Palm (EP) theorem. The EP theorem states that the proper wave-mean flow interaction, defined as the EP flux divergence, vanishes for waves that are linear, conservative, and steady. In the real atmosphere, this theorem is violated mainly due to wave breaking and turbulence. The resulting EP flux divergence then drives a residual circulation which causes the observed substantial deviations from some hypothetical radiatively determined state. With regard to this dynamical control we discuss the different contributions of Rossby waves and gravity waves. Recapitulation of Lindzen's theory of gravity-wave saturation allows us to interpret various phenomena in the upper mesosphere such as interhemispheric coupling or modulations of the gravity-wave driven branch of the residual circulation by solar proton effects and thermal tides. In addition we discuss the relative importance of changes in radiative transfer and tropospheric gravity-wave sources on the long-term temperature trends in the summer mesosphere.

Keywords Residual circulation · Eliassen-Palm theorem · Rossby waves · Gravity waves · Variability of the summer mesosphere

1 Introduction

The general circulation of the atmosphere is defined as the zonally and temporally averaged wind and temperature fields. Scientific knowledge of the general circulation has mainly evolved over the last six decades. It is appropriate to distinguish between height regions and their particular physical and dynamical properties as discussed in the review by Smith (this

E. Becker (✉)
Leibniz Institute of Atmospheric Physics at the University of Rostock, Schlossstr. 6,
18225 Kühlungsborn, Germany
e-mail: becker@iap-kborn.de

issue). In the present review we focus on the dynamical concepts that allow us to understand the general circulation in the middle atmosphere.

A comprehensive historical review of the explanations for the prevailing zonal-mean wind patterns in the troposphere can be found in Lorenz (1967). The main notion is that the equatorial zonal-mean circulation in the tropics can at least qualitatively be understood in terms of a thermally driven Hadley circulation (e.g., Held and Hou 1980), while in middle and high latitudes the strong influences of Rossby waves must be taken into account. The proper physical picture to explain the zonal-mean circulation in the extratropics was developed in the 1960s and 1970s (e.g., Charney and Drazin 1961; Eliassen and Palm 1960; Dickinson 1969; Matsuno 1971; Boyd 1976; Andrews and McIntyre 1976; Dunkerton 1978) and is nowadays expressed in terms of the transformed Eulerian mean equations (hereafter: TEM equations). The fundamental advantage of the TEM equations over the conventional zonal-mean picture is that it provides a clear causality of the wave-mean flow interaction and the so-induced deviations from some radiatively (and convectively) determined state via the wave-driven residual circulation. In Sect. 2.2 we recapitulate the TEM equations in quasi-geostrophic scaling and their application to the troposphere and stratosphere.

Lorenz (1967) also gave a complete picture about the energy conversions in the troposphere, known as the Lorenz energy cycle. This cycle describes the conversion of large-scale available potential energy into kinetic energy by baroclinic Rossby waves (Lorenz 1955; Orrt 1964) and its subsequent dissipation, which in the real atmosphere is due to a forward energy cascade that is ultimately balanced by molecular friction. When viewed in a broader context, the Lorenz energy cycle also includes the generation of kinetic energy by mesoscale weather systems and convection (Haimberger and Hantel 2000). Though the energy cycle is basically restricted to the troposphere,[1] it is important for the middle atmosphere through the vertical coupling by internal gravity waves (hereafter: GWs). GWs are mainly generated in the troposphere by various processes such as convection, shear instability, geostrophic adjustment, and flow over topography. While the latter mechanism gives rise to so-called mountain waves, which are relevant for the upper troposphere and the stratosphere (Alexander et al. 2010), convection and shear instability typically generate GWs that propagate up to the mesosphere and lower thermosphere (height region from about 50 to 110 km, hereafter: MLT), or even higher. The averaged intensity by which these waves are generated is presumably linked to the overall strength of the tropospheric energy cycle (Becker 2009). A comprehensive review about our present knowledge of atmospheric GWs, including in particular the various mechanisms for wave generation, wave instability, and transition to turbulence can be found in Fritts and Alexander (2003).

Hines (1960) first argued that GWs play a fundamental role in determining the large-scale flow in the MLT. It took another twenty years until a quantitative physical picture of the dynamical control of the MLT by GWs was provided with the classical paper of Lindzen (1981). This picture is based on the linear saturation theory for GWs (e.g., Fritts 1984) and has since then frequently been used as a basis for parameterizing the effects of GWs in comprehensive general circulation models (e.g., Garcia et al. 2007). The fact that GWs are required for a realistic description of the general circulation in climate models that include the MLT has stimulated various alternative parameterization schemes. Among those the methods of Hines (Hines 1997; Fomichev et al. 2002) and of Medvedev and Klaassen (Medvedev and Klaassen 2000; Yiğit et al. 2008) should be mentioned in first place. An

[1] The reason is that baroclinic and convective instabilities are absent in the stratosphere and that conversion rates higher up are negligible when the whole atmosphere is considered.

instructive comparison of the performance of different schemes can be found in McLandress and Scinocca (2005).

The assumptions underlying any conventional GW parameterization must be questioned in many respects. This holds in particular with regard to the breaking mechanism (see Sect. 6 in Fritts and Alexander 2003). Nevertheless, the arguments underlying Lindzen's theory provide a helpful physical way of thinking in order to interpret the dynamical control of the MLT by GWs, as well as vertical coupling mechanisms and variability phenomena. We recapitulate the main arguments underlying Lindzen's saturation theory in Sect. 2.3. This allows us to extend the quasi-geostrophic TEM equations into the MLT.

The extended TEM picture of the dynamical control by Rossby waves and GWs is limited in several respects. In Sect. 2.5 we shall list these limitations and mention the generalized TEM equations according to the text of Andrews et al. (1987). Unlike the TEM equations in quasi-geostrophic scaling, their generalized form describes also the thermally driven branches of the residual circulation in the equatorial regions. Moreover, the large-scale sensible heat budget is fundamentally incomplete in quasi-geostrophic scaling. This is mainly because the direct heating rates due to GW breaking, like the turbulent dissipation (Lübken 1997), are substantial in the MLT. The situation is further complicated by the interaction of GWs and Rossby waves (Sect. 2.6).

The wave driving in the MLT is subject to changes in the underlying stratosphere and troposphere. Even if we assume fixed tropospheric GW sources, changes in the propagation conditions can efficiently modulate the wave driving in the MLT. In Sect. 3 we consider several examples, such as the interhemispheric coupling (Karlsson et al. 2008) and the possible modulations of GW breaking by solar proton events and thermal tides. In addition, we briefly speculate about the effects of climate change in the summer MLT (Sect. 3.4). Again, the arguments from Lindzen (1981) are advantageous for these considerations. We close this review by a summary of the concept of dynamical control and the aforementioned examples, and by mentioning some issues in current research (Sect. 4).

Throughout the paper we take advantage of simulations based on the Kühlungsborn Mechanistic general Circulation Model (herafter: KMCM). A detailed model description is given in Körnich et al. (2006) and Becker (2009). A particular feature of the KMCM is that it can be run at very high resolutions such as to simulate the dynamical control of the MLT by GWs explicitly in a general circulation model (hereafter: GCM). A prerequisite for this feature is an advanced parameterization of turbulent diffusion. Even though such an approach is limited by the restriction of GW scales to the available resolution, some of the strong assumptions made in GW parameterizations are relaxed.

2 Wave Driving and TEM Equations

2.1 Main Problem

The basic question we have to answer is comprised in Fig. 1: How do we understand the differences between the upper and lower panels. Figure 1a represents some radiatively determined state for January (Shine 1987). One may obtain such a temperature distribution by running a radiation model in an annual cycle mode and then extract the averaged temperature for January. The present T_E was constructed empirically. The corresponding zonal wind structure (Fig. 1b) may be inferred from gradient-wind balance, i.e., by the meridional momentum equation for a stationary zonal flow. It is clear that these fields bear at most a very poor similarity with the observed climatology (see for instance Figs. 1 and 2 in Richter et al. 2008). Moreover, gradient-wind balance even leads

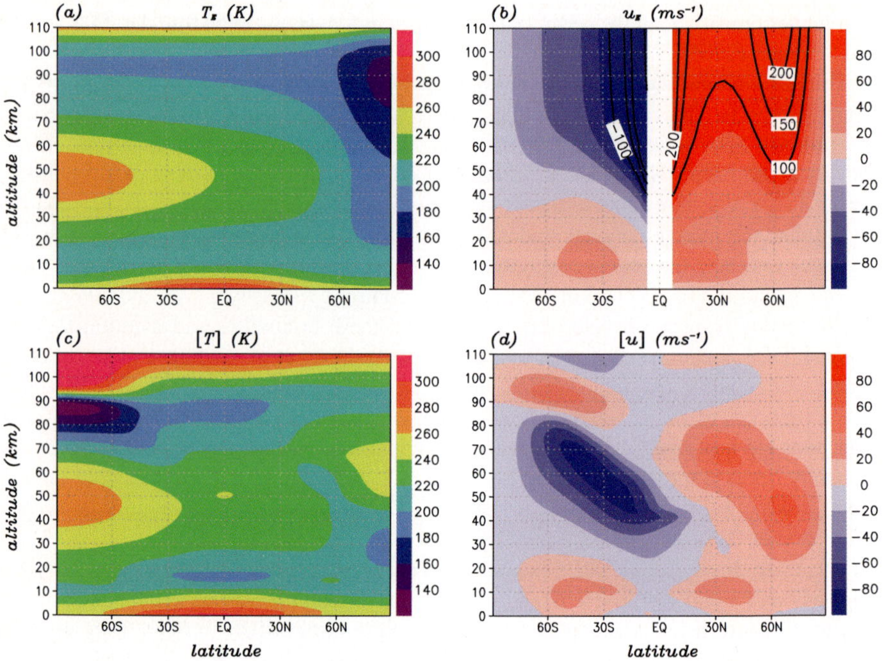

Fig. 1 Comparison of a synthetic radiatively determined state (*upper panels*) and the simulated zonal-mean climatology (*lower panels*) during January for the temperature (*left*) and zonal wind (*right*). See text for further details

to singularities in the vicinity of the equator for a cross-equatorial temperature gradient. Hence, the observed climatological zonal-mean state must be strongly controlled by dynamical processes at all latitudes. The main features not captured by the radiatively determined state are the cold summer mesopause, the warm winter stratopause, and reversals of the zonal wind in the upper mesosphere. These features are reproduced in any middle atmosphere GCM that properly describes the dynamical control (e.g., Fomichev et al. 2002).

Figures 1c and d show the corresponding results for the January climatology simulated with the KMCM. These temperature and zonal wind structures reflect the dynamical control when compared to the radiatively determined state. This holds in particular because the radiative heating in the KMCM is parameterized in a simplistic way by temperature relaxation toward the T_E shown in Fig. 1a. Hence, the substantial deviations of the temperature and zonal wind fields in the lower panels from those in the upper panels of Fig. 1 are entirely dynamically determined. As we shall see farther below, these deviations are closely related to a wave-driven residual circulation.

In order to recapitulate the wave driving concept we start with the primitive equations (hereafter: PE). These equations form the basis of dynamic meteorology, as well as of numerical models used for weather forecasting and climate studies. For the pedagogical purpose of the present review it is convenient to use the PE in the anelastic approximation:

$$\partial_t \mathbf{v} = \mathbf{v} \times (f + \xi)\mathbf{e}_z - w\partial_z \mathbf{v} - \nabla\big(p/\rho_r + \mathbf{v}^2/2\big) + \mathbf{R}, \tag{1}$$

$$\partial_z p = -g\rho, \tag{2}$$

$$0 = \rho_r \nabla \cdot \mathbf{v} + \partial_z(\rho_r w), \tag{3}$$

$$c_p d_t T = c_p(\partial_t + \mathbf{v} \cdot \nabla + w\partial_z)T = \rho_r^{-1}(d_t p + g(\rho - \rho_r)w) + Q. \tag{4}$$

Here, $\mathbf{v}$ and w denote the horizontal and vertical wind and f is the Coriolis parameter. The horizontal vorticity is defined as $\xi = \mathbf{e}_z \cdot (\nabla \times \mathbf{v})$ with ∇ being the horizontal gradient operator and $\mathbf{e}_z$ the unit vector in the vertical direction on the sphere. The thermodynamic variables T, p, and ρ have their usual meanings with c_p being the heat capacity per unit mass at constant pressure. $\rho_r(z)$ is a reference density profile according to the anelastic approximation.[2] The symbols $\mathbf{R}$ and Q represent turbulent momentum diffusion and the overall diabatic heating, respectively.

The PE describe the large-scale dynamics in the tropics, as well as the quasi-geostrophic dynamics in the extratropics. These motions are resolved in state-of-the-art GCMs. The PE describe also low- and mid-frequency GWs. Mid-frequency GWs are not affected by the Coriolis force, have typical horizontal scales of about 100 to 500 km, and are typically parameterized in GCMs.

2.2 Dynamical Control by Quasi-Geostrophic Waves

Let us follow the nomenclature of Lorenz (1967) and denote the zonal or temporal mean of some quantity X by $[X]$ or $\overline{X}$. Deviations from the zonal mean are written as X^* such that $[X^*] = 0$. We now ask for the zonal and temporal-mean mass circulation in a latitude-height cross-section. As a first attempt we solve the averaged continuity equation (3),

$$(a \cos\phi)^{-1}\partial_\phi(\cos\phi \rho_r[\overline{v}]) + \partial_z(\rho_r[\overline{w}]) = 0, \tag{5}$$

by the Eulerian mass streamfunction $\Psi(\phi, z)$ according to

$$\rho_r[\overline{v}] = \frac{-\partial_z \Psi}{2\pi a \cos\phi} \quad \text{and} \quad \rho_r[\overline{w}] = \frac{\partial_\phi \Psi / a}{2\pi a \cos\phi}. \tag{6}$$

Here, v is the meridional wind component, ϕ is latitude and a the Earth's radius. In practise, Ψ can be computed by numerically evaluating the integral

$$\Psi = 2\pi a \cos\phi \int_z^\infty \rho_r[\overline{v}]dz' = 2\pi a \cos\phi \int_0^p [\overline{v}]\frac{dp'}{g}. \tag{7}$$

Figure 2a shows Ψ and $[\overline{v}]$ from the January simulation with the KMCM. There is the well-known 3-cell structure in each hemisphere of the troposphere: A tropical Hadley cell that is strongest in winter, a thermally indirect Ferrel cell at middle latitudes, and a weak thermally direct polar cell. We recognize a 2-cell structure in the winter stratosphere and

---

[2]The main purpose of the anelastic approximation is to filter out sound waves. This is done by introducing a (hydrostatic) reference state $(\rho_r(z), p_r(z))$ and neglecting the deviations from ρ_r in the continuity equation. The momentum and thermodynamic equations are linear with regard to the deviations from ρ_r, p_r, or T_r while the hydrostatic approximation and the equation of state $p = \rho RT$ are as usual. The anelastic approximation is useful in our context since it allows for a straight-forward formulation of the generalized TEM equations (Sect. 2.5) without using pressure (or log-pressure) as vertical coordinate. Note also that for vanishing $\mathbf{R}$ and Q, (1)–(4) fulfill the energy conservation law

$$\partial_t(\rho_r\epsilon) + \nabla \cdot (\rho_r \mathbf{v}\epsilon) + \partial_z(\rho_r w\epsilon) = \nabla \cdot (p\mathbf{v}) - \partial_z(pw),$$

where $\epsilon = c_v T + \mathbf{v}^2/2 + gz$ is sum of internal, kinetic, and geopotential energy per unit mass.

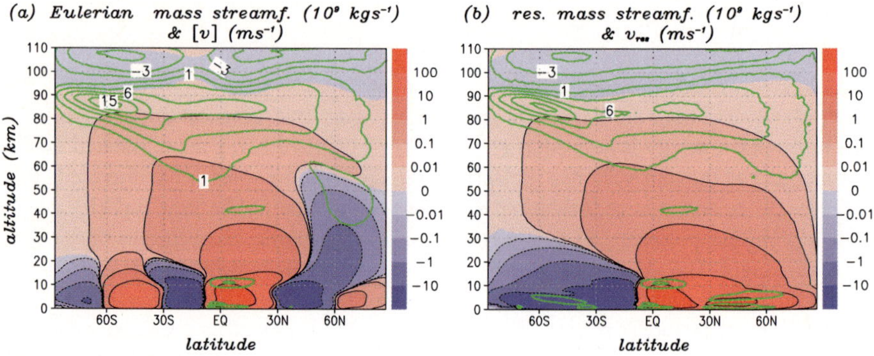

Fig. 2 (**a**) Eulerian mass streamfunction (*colours and thin black contours*) and zonal-mean meridional wind (*green contours drawn* for $\pm 1, 3, 6, 9, 12, 15\ \mathrm{m\,s^{-1}}$). (**b**) Same as (**a**) but for the residual mass streamfunction and the residual meridional wind

lower mesosphere with the stratospheric branch of the Ferrel extending to the pole. Such a mean meridional circulation for the middle atmosphere was anticipated by Dunkerton (1978). He also pointed out the paradoxical behavior of Ψ when compared to the Brewer-Dobson circulation. The latter implies that stratospheric mass is transported from low to high latitudes with downwelling over the polar cap (see review of Smith, this issue). It is obvious that Ψ does not describe this Lagrangian mean circulation. A formal reason is that the streamlines of an averaged flow do usually not describe averaged parcel trajectories. On fluid mechanical grounds, Ψ would describe the Lagrangian mean mass transport only if the atmospheric circulation was stationary and zonally symmetric, which is not the case at all.

The main physical reason why Ψ does not describe averaged parcel trajectories is that atmospheric waves possess meridional and vertical Stokes drifts, even in the limit of linear, frictionless, isentropic, and amplitude conserving waves (hereafter: ideal waves). For this reason, ideal Rossby waves are always accompanied by some mean meridional circulation that balances the Stokes drifts. To illustrate this point we consider the zonally averaged zonal momentum and thermodynamic equations in quasi-geostrophic (herafter: qg) scaling,[3]

$$\partial_t[u] = f[v] + [\xi^* v^*] + [R_x] = f[v] - (a\cos^2\phi)^{-1}\partial_\phi(\cos^2\phi[u^*v^*]) + [R_x], \qquad (8)$$

$$\partial_t T = -(a\cos\phi)^{-1}\partial_\phi(\cos\phi[T^*v^*]) - \frac{g}{c_p}[w] + [Q]. \qquad (9)$$

In the temporal mean without friction and diabatic heating, the momentum and heat flux terms on the right-hand side of (8) and (9) must be balanced by the Coriolis force and the adiabatic heating term, respectively. On the other hand, when considering the qg potential vorticity equation, ideal qg waves do not account for any wave-mean flow interaction (see Sect. 3.6 in the text of Andrews et al. 1987). The seemingly paradoxical behavior of (8) and (9) is resolved by the Eliassen-Palm (hereafter: EP) theorem for Rossby waves. Here we shall motivate the EP theorem with the help of the residual circulation.

Any linear atmospheric wave can be shown to possess meridional and vertical Stokes drifts that scale with the meridional wave heat flux. As pointed out by Dunkerton (1978),

[3] In qg scaling the wave velocities u^* and v^* are due to the geostrophic flow and can be represented by a streamfunction. Applying the horizontal Laplacian to that streamfunction yields the wave horizontal vorticity ξ^*.

the residual velocities are given by the conventional (Eulerian) mean velocities corrected by the Stokes drifts:

$$\overline{v}_{res} = [\overline{v}] - \frac{c_p}{g\rho_r}\partial_z\left(\rho_r[\overline{T^*v^*}]\right) = [\overline{v}] + (\text{merid. Stokes drift})/\rho_r, \tag{10}$$

$$\overline{w}_{res} = [\overline{w}] + \frac{c_p}{g}\frac{\partial_\phi}{a\cos\phi}\left(\cos\phi[\overline{T^*v^*}]\right) = [\overline{w}] + (\text{vert. Stokes drift})/\rho_r. \tag{11}$$

Eliminating the mean meridional and vertical velocities in (8) and (9) with the help of (10) and (11), the corresponding TEM equations are obtained:

$$\partial_t[u] = f v_{res} - \frac{\partial_\phi}{a\cos^2\phi}(\cos^2\phi[u^*v^*]) + \frac{c_p f}{g\rho_r}\partial_z(\rho_r[T^*v^*]) + [R_x], \tag{12}$$

$$0 = (a\cos\phi)^{-1}\partial_\phi(\cos\phi\rho_r v_{res}) + \partial_z(\rho_r w_{res}), \tag{13}$$

$$\partial_t T = -\frac{g}{c_p}w_{res} + [Q], \tag{14}$$

$$f\partial_z[u] = -\frac{g}{T_r}\frac{\partial_\phi}{a\cos\phi}[T]. \tag{15}$$

The zonal momentum and thermodynamic equations (12) and (14) are completed by the continuity equation for the residual winds (13) and by the thermal wind relation (15).[4] The EP theorem states that

$$-\frac{\partial_\phi}{a\cos^2\phi}(\cos^2\phi[u^*v^*]) + \frac{c_p f}{g\rho_r}\partial_z(\rho_r[T^*v^*]) = 0 \quad \text{for ideal qg waves.} \tag{16}$$

Moreover, if in addition $[Q]=0$ and $[R_x]=0$, also the residual circulation vanishes. Other frequently used terms for the EP theorem are the *Charney-Drazin* or *non-acceleration* theorem. The EP theorem is often also written as

$$(\rho_r a\cos\phi)^{-1}\text{div}(\mathbf{EPF}_{qg}) = 0 \quad \text{for ideal qg waves,} \tag{17}$$

where

$$\mathbf{EPF}_{qg} = a\cos\phi\rho_r\left(-[u^*v^*]\mathbf{e}_y + \frac{c_p f}{g}[T^*v^*]\mathbf{e}_z\right) \tag{18}$$

is the qg EP flux and the divergence operator in (17) has to be taken with regard to the meridional and vertical directions in spherical geometry. The EP flux corresponds to a negative angular momentum flux density and describes how the wave action is transported in a latitude-height cross-section (Edmon Jr. et al. 1980). The main implications from the EP theorem are as follows:

(1) An ideal qg wave is always accompanied by a mean meridional circulation ($[\overline{v}], [\overline{w}]$). This circulation balances the Stokes drifts, as well as the meridional momentum and heat fluxes associated with the wave. The Ferrel cells in Fig. 2a reflect this balance, indicating that qg waves in the real atmosphere are on average close to being ideal.

[4]Note that only the temporal mean residual circulation fully includes the Stokes drifts. Our assumption here is that this holds also for time scales associated with slow variations of the zonal-mean flow. For this reason we retain the time dependence in (12)–(15).

(2) The left-hand side of (16) (or (17)) describes the proper wave-mean flow interaction when the waves are no longer ideal. The resulting residual circulation $(\overline{v}_{res}, \overline{w}_{res})$ describes the averaged parcel trajectories of air parcels (Dunkerton 1978) and therefore corresponds to the Brewer-Dobson circulation.

(3) The TEM equations provide a clear causality for the residual circulation in the extratropics. This is most evident when we specify Q as temperature relaxation, take the temporal average of (12)–(15), and neglect $[\overline{R_x}]$. The TEM equations can then be written as:

$$0 = f\overline{v}_{res} + (\rho_r a \cos\phi)^{-1}\mathrm{div}(\overline{\mathbf{EPF}}_{qg}), \tag{19}$$

$$0 = (a\cos\phi)^{-1}\partial_\phi(\cos\phi\rho_r\overline{v}_{res}) + \partial_z(\rho_r\overline{w}_{res}), \tag{20}$$

$$0 = -\frac{g}{c_p}\overline{w}_{res} - \frac{[\overline{T}] - T_E}{\tau}, \tag{21}$$

$$f\partial_z[\overline{u}] = -\frac{g}{T_r}\frac{\partial_\phi}{a\cos\phi}[\overline{T}]. \tag{22}$$

From (19) we see that in the climatological mean, any residual meridional flow is *driven* by the EP flux divergence. Due to continuity (20) this flow is balanced by some residual upwelling or downwelling, which in turn yields adiabatic cooling or heating rates. These induce deviations from the radiatively determined state such that the resulting diabatic heating balances the residual vertical motion according to (21). Hence, within the range of the idealizations made, the residual circulation corresponds to the diabatic circulation. Finally, the mean zonal wind follows via the thermal wind relation (22) from the temperature distribution.

The residual mass streamfunction is defined analogously to (6) and can be computed from

$$\Psi_{res} = 2\pi a\cos\phi\int_z^\infty \rho_r\overline{v}_{res}dz' = 2\pi a\cos\phi\int_0^p \overline{v}_{res}\frac{dp'}{g}. \tag{23}$$

As a typical example of its structure we show in Fig. 2b the result from the KMCM using the same dynamical data as in the previous figures. In the troposphere and lowermost stratosphere, the three-cell structure is replaced by a single, thermally direct cell in each hemisphere. This is consistent with an exchange of heat and constituents between low and high latitudes due to the residual circulation. There are only slight differences between the two circulation systems in the tropical troposphere, where the Hadley circulation prevails. This is because the Hadley circulation is, to a first approximation, determined by differential heating and the nonlinear advection terms not retained in the qg TEM equations (Held and Hou 1980). Nevertheless, extratropical Rossby waves may extend into low latitudes and modulate the overall shape of the Hadley circulation (Becker and Schmitz 2001).

In the mid stratosphere and lower mesosphere, the pronounced winter-hemispheric two-cell structure (Fig. 2a) is merged into one single residual circulation cell with upwelling in the tropics and downwelling in the polar-night region (Fig. 2b). This is consistent with the Brewer-Dobson circulation.

As first noted by Dunkerton (1989) and further analyzed by Seminiuk and Shepherd (2001), there is an additional Hadley circulation branch around the equatorial stratopause. This is indicated in Figs. 2 by a corresponding local maximum of $[\overline{v}]$ or $\overline{v}_{res}$. The associated cross-equatorial flow toward the winter pole accounts for the strong westward zonal flow in this region due to nonlinear advection of angular momentum. It is not clear to date to what

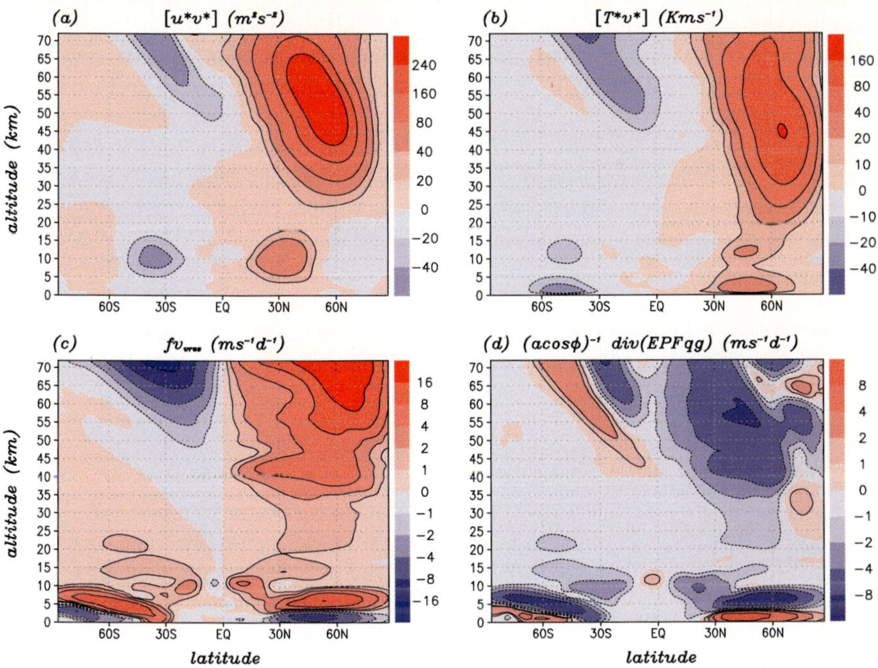

Fig. 3 Simulated January climatologies of (**a**), (**b**) the meridional momentum and temperature fluxes due to synoptic-scale and planetary waves, (**c**) the Coriolis force due to the residual circulation, and (**d**) the quasi-geostrophic Eliassen-Palm flux divergence

extent this Hadley circulation in the equatorial stratopause is affected by waves, in particular by GWs, or how it interacts with the quasi-biennial and the semi-annual oscillations.

In the upper mesosphere the flow is from the summer pole to high winter latitudes, either for the Eulerian and the residual circulation. Nevertheless, for the residual circulation the equatorward branch in summer is weaker and the poleward branch in winter is stronger than for the Eulerian circulation. We discuss this behavior farther below.

Figure 3 shows the meridional momentum and temperature fluxes, as well as the Coriolis force associated with the residual meridional wind and the qg EP flux divergence. The extratropical momentum flux is mainly poleward in the troposphere and, hence, for ideal qg waves it must be balanced by some equatorward flow between the subtropics and high latitudes. The meridional heat flux is generally poleward and concentrated at middle latitudes in the troposphere. For ideal qg waves this must be compensated by downwelling at subtropical latitudes and upwelling at high latitudes. These balances give rise to the tropospheric Ferrel cells visible in Fig. 2a. In the winter stratosphere and lower mesosphere, the meridional momentum and heat flux maxima shift to higher latitudes such that the winter Ferrel cell extends to the pole. In summer, on the other hand, quasi-geostrophic waves cannot propagate from the troposphere into the stratosphere and lower mesosphere by virtue of the Charney Drazin criterion (see Sect. 4.5 in the text of Andrews et al. 1987). As we shall discuss in Sect. 2.6, qg waves do nevertheless exist in the summer mesopause region.

The qg EP flux divergence (Fig. 3d) is basically negative above the boundary layer in the summer troposphere and in winter up to the mesosphere. Hence it drives a poleward flow in these regions, which is consistent with the residual circulation shown in Fig. 2b. Note also that the extratropical balance between the EP flux divergence and the Coriolis force associated with the residual circulation is not perfect. The reason is that also ageostrophic terms play a significant role. In particular, GWs become dominant in the upper mesosphere in both hemispheres.

It is remarkable that Dunkerton (1978) inferred the structures of the Eulerian and residual circulation in the winter stratosphere (Fig. 2) solely by reasoning about the Stokes drifts and EP flux divergence due to planetary Rossby waves, without having the corresponding data analysis from observations or from some middle atmospheric GCM available. Moreover, based on calculations of the net radiative heating rates constrained by available temperature measurements, he proposed that there should be a pole-to-pole circulation in the upper mesosphere. As is known nowadays, the summer-to-winter-pole branch of the circulation in the upper mesosphere is dynamically driven by GWs. We now extend the TEM equations by the corresponding GW drag.

2.3 Dynamical Control by Internal Gravity Waves (GWs)

The first quantitative explanation of the wave driving for the pole-to-pole circulation in the upper mesosphere was provided by Lindzen (1981). He applied one-dimensional WKB theory for GWs in mid-frequency approximation and included a weak damping by means of vertical diffusion. According to this analysis, which is summarized in Sect. 3 of Becker (2004), the vertical flux of zonal momentum for a monochromatic GW with zonal phase speed c and wavenumber k can be written as

$$F(z) = F(z_0)\exp\left(-\int_{z_0}^{z} \frac{2DN^3}{k(u-c)^4}dz'\right). \tag{24}$$

Here, the zonal background flow, u, the background buoyancy frequency, N, and the diffusion coefficient, D, can all be slowly varying functions of height. z_0 denotes the launch level where the momentum flux $F(z_0)$ must be prescribed. With regard to the planetary scale, all quantities may also depend on longitude, latitude, and time.

The integrant in (24) is proportional to the diffusion coefficient. This reflects the non-acceleration theorem for monochromatic mid-frequency GWs (also known as the second theorem of Eliassen and Palm): In the conservative case ($D = 0$) the momentum flux is independent of height. The integrant in (24) is furthermore inversely proportional to the difference between the background wind and the phase speed to the power of 4. Hence, the GW will be strongly damped for any finite D when the background wind approaches the phase speed, i.e., close to a critical level. In this case the mid-frequency assumption breaks down since the intrinsic wave frequency

$$\omega_I = \omega - ku = k(c-u) \tag{25}$$

approaches zero. The physical behavior of (24) is nevertheless quite reasonable: Due to the compression of the vertical wavelength beneath a critical level by virtue of the dispersion relation for the real part of the vertical wavenumber,

$$m_r = N/(u-c), \tag{26}$$

the wave becomes dynamically unstable, which is expected to trigger the generation of turbulence and the subsequent damping of the wave. This reasoning forms the basis of most GW parameterizations currently used in middle atmosphere climate models.

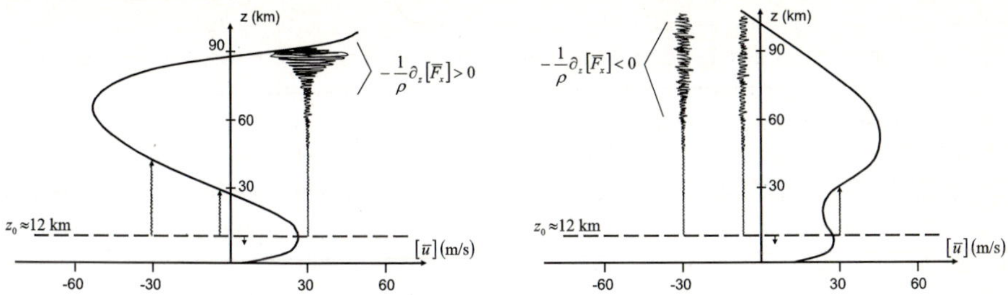

Fig. 4 Schematics to illustrate the GW filtering and breaking for typical extratropical zonal wind profiles during summer (*left*) and winter (*right*). See text for further details

It is, however, important to note that wave instability can occur well below a critical level. This is included in GW schemes by different means. For instance, the method developed by Medvedev and Klaassen (1995, 2000) explicitly considers nonlinear wave instability in order to derive a continuous WKB expression analogous to (24). Other spectral methods do not resort to WKB solutions and specify some criterion for the truncation of a broad GW spectrum with increasing height (e.g., Hines 1997).

Lindzen's saturation theory assumes that a monochromatic GW propagates conservatively as long as the mean flow plus the wave is statically stable. When the background wind gets close to the phase speed and/or the wave amplitude is large enough such that the stratification of the mean flow plus the wave becomes statically unstable, D becomes positive such that neutral static instability is maintained with increasing height. The level above which $D > 0$ is called the saturation or breaking level. Lindzen (1981) motivated his assumption with rocket-borne measurements of GW-induced temperature fluctuations which indicated that the wave amplitude did no longer increase with height above a certain level. Within his theoretical approach, the corresponding D and the resulting GW drag can be expressed analytically.

In general, we consider the GW drag,

$$\text{GW drag} = -\rho_r^{-1}\partial_z F, \tag{27}$$

as an additional wave driving, where F now includes the zonal projections from several monochromatic GWs having different horizontal phase speeds, propagation directions, and launch level fluxes (e.g., Becker 2004; Garcia et al. 2007). The extension of the qg TEM equations into the upper mesosphere then simply consists of completing (12) with the zonal-mean GW drag:

$$\partial_t[u] = f v_{res} + (\rho_r a\cos\phi)^{-1}\text{div}(\mathbf{EPF}_{qg} - a\cos\phi[F]\mathbf{e}_z). \tag{28}$$

Lindzen's theory provides an intimate relation between GW breakdown and turbulence and it reflects the generic mechanisms for GW instability, namely strong vertical shear due to $[u] \to c$ and/or growth of wave amplitude with increasing height leading to turbulent damping of the wave. The usefulness of these arguments can be illustrated with the schematics in Fig. 4. Let us consider mean zonal wind profiles typical for the summer or winter hemisphere. We assume that GWs with small and large zonal phase speeds of ± 5 and ± 30 m s^{-1} are launched in the upper troposphere, and that the complete GW drag exerted on the mean flow consists of a linear superposition of the contributions from these waves. In summer, all GWs except for the fast eastward propagating one encounter critical levels below the mesosphere and can thus not contribute to the GW drag in the upper mesosphere. The fast eastward propagating wave, on the other hand, propagates nearly conservatively to the upper mesosphere since the background wind and the phase speed differ strongly, i.e., since the wave has a long vertical wavelength and weak vertical shear according to (26). However, in the mesopause region the zonal wind reverses from westward to eastward. This leads to a sudden onset of saturation and wave damping, resulting in a concentrated and strongly positive GW drag. In winter, on the other hand, the fast and the slow westward waves can propagate into the middle atmosphere. Therefore, the GW activity in the stratosphere and lower mesosphere is generally expected to be stronger than in summer. On the other hand, while the winter mean zonal wind reverses smoothly from eastward to westward flow in the upper mesosphere and is, at least during northern winter, much more variable than during summer, the breaking levels of the westward GWs are on average lower in altitude and distributed over a deeper altitude range compared to the situation in summer. As a result, the maximum negative GW drag per unit mass is weaker, less concentrated, and on average located at lower altitudes compared to the eastward GW drag in summer.

Lindzen (1981) used observational indications of the saturation levels occurring around 70 and 50 km in summer and winter, respectively, to compute the GW drag just from the saturation assumption. Assuming typical values for the horizontal wavenumber he inferred values of about $+130\ \mathrm{m\,s^{-1}\,d^{-1}}$ in summer versus $-100\ \mathrm{m\,s^{-1}\,d^{-1}}$ in winter. These values provided an approximate balance between the GW drag and the Coriolis force associated with observational estimates of the mean meridional flow. Moreover, the theory predicts the main charateristics of the summer-winter differences of GW activity and GW-induced turbulence, with a sharp onset of strong turbulent dissipation in the summer upper mesosphere and weaker values in winter (Lübken 1997; Becker 2004). Corresponding distributions of GW-induced wind and temperature variations as a function of season have recently also been inferred on the basis of satellite or radar measurements and model simulations (e.g., Preusse et al. 2006; Hoffmann et al. 2010). In addition, the study of Hoffmann et al. (2010) has revealed a semi-annual component in the seasonal cycle of the upper mesospheric GW activity with absolute minima during the equinoxes. This feature can be attributed to the very weak mean zonal wind during the transition from summer to winter conditions and vice versa, giving rise to breaking levels of most GWs below the upper mesosphere.

It is worth noting that the values predicted by Lindzen (1981) for the summer and winter mesospheric GW drag have later on been confirmed in all GCMs that resolve the mesopause region. Early applications of linear saturation theory for parameterizing the effects of GWs in GCMs were presented by Holton (1982) or Garcia and Solomon (1985). In particular, the three currently existing comprehensive GCMs that extend from the surface up to at least the mesopause region, i.e., the Whole Atmosphere Community Climate Model (WACCM, Richter et al. 2008), the Canadian Middle Atmosphere Model (CMAM, Fomichev et al. 2002); and the HAMburg MOdel of the Ionized and neutral Atmosphere (HAMMONIA, Schmidt et al. 2006), agree on these values for the GW drag (e.g., see Fig. 10c in Fomichev et al. 2002 or Figs. 7g and 7h in Richter et al. 2008). Also GCMs that focus on the upper atmosphere and have lower boundary conditions in the lower stratosphere, like the University College London CMAT2 (Middle Atmosphere Thermosphere-2) model for instance, yield a mean solstice GW drag in the MLT that compares well to other model results (see Figs. 3 and 4 in the paper of Yiğit et al. 2009). While the WACCM includes a Lindzen-type GW parameterization, the CMAM and HAMMONIA use the Doppler-spread parameterization (DSP) proposed by Hines (1997) and the CMAT2 model employs the method of Medvedev and Klaassen (2000). Also our present January simulation using the KMCM with resolved GWs yields a GW drag that is quantitatively consistent with the other model estimates (see Sect. 2.5 and Fig. 5c below).

2.4 Steady Downward Control

The TEM equations as derived in the last subsections give rise to a theorem known as the *downward control principle* (Haynes et al. 1991). The downward control principle (hereafter: DCP) basically reflects angular momentum conservation of any wave-mean flow interaction. For example, a conventional Rayleigh friction as a surrogate for the GW drag violates the DCP and generates spurious feedbacks in climate sensitivity experiments (Shepherd et al. 1996; Shaw et al. 2009). In order to derive the DCP we must combine the temporal mean of (28) with the continuity equation (20) and apply several algebraic manipulations. As a result, we can express the residual vertical mass flux poleward of some latitude ϕ as

$$a\int_{\phi}^{\text{pole}} \rho_r \overline{w}_{res} \cos\phi' d\phi' = a\cos\phi \left(\frac{1}{f}[\overline{F}] - \frac{c_p \rho_r}{g}[\overline{T^* v^*}] \right) - \int_z^{\infty} \frac{\rho_r}{f} \frac{\partial_\phi}{\cos\phi} \left(\cos^2\phi [\overline{u^* v^*}] \right) dz'. \tag{29}$$

The momentum flux $[\overline{u^* v^*}]$ is negligible at polar latitudes. Therefore, the residual vertical mass flux at height z poleward of some middle or high latitude ϕ is mainly determined by the qg vertical EP flux completed by the GW momentum flux at (ϕ, z). In other words, the deposition of momentum at ϕ above z controls the residual vertical mass flux (and thus also $[\overline{T}] - T_E$) poleward of ϕ at z. This relation gave rise to the term *downward control*. For instance, the warm polar winter stratopause can be understood as a result of the downward control of westward propagating GWs breaking in the mesosphere. In the same fashion, lower stratospheric polar temperatures are closely related to the strengths of planetary Rossby waves as measured by their (pseudo) vertical momentum flux proportional to $-[\overline{T^* v^*}]$ (e.g., Eyring et al. 2006). Note that the wave-mean flow interaction above z is determined by the EP flux generated below. In this respect, the DCP does not represent a *causal* relationship by which upper levels would dynamically control the flow below.

2.5 The Generalized TEM Equations and the Complete Sensible Heat Budget

Quasi-geostrophic (qg) scaling does not hold in the tropics, and it does in particular not apply to GWs. On the other hand, the Stokes-drift corrections to the zonal-mean meridional and vertical winds hold for any linear waves. We may therefore use the residual velocities (10) and (11) to transform the zonal-mean zonal momentum equation without an additional scaling assumption. A corresponding transformation is given in the text of Andrews et al. (1987) for the log-p vertical coordinate system and using potential temperature as prognostic thermodynamic variable. An equivalent transformation with pressure as vertical coordinate and enthalpy as thermodynamic variable is given in Becker (2009).[5] Here, we stay with the PE in the anelastic approximation (1)–(4). Inserting the residual velocities (10) and (11) into the zonal-mean zonal component of (1) and using the continuity equations (3) and (14), the transformed zonal momentum equation can be written as

$$\partial_t [u] = v_{res}(f + [\xi]) - w_{res}\partial_z [u] + (\rho_r a \cos\phi)^{-1} \mathrm{div}(\mathbf{EPF}) + [R_\lambda], \tag{30}$$

[5]There is a typo in the first line of (A4) of Becker (2009): A factor $\cos^2\phi$ is missing in front of the second term proportional to $[T^* v^*]$ under the y-derivative. The definition of the EP flux in Becker (2009) also differs by the factor of $\rho_r a \cos\phi$ from the present review or the text of Andrews et al. (1987).

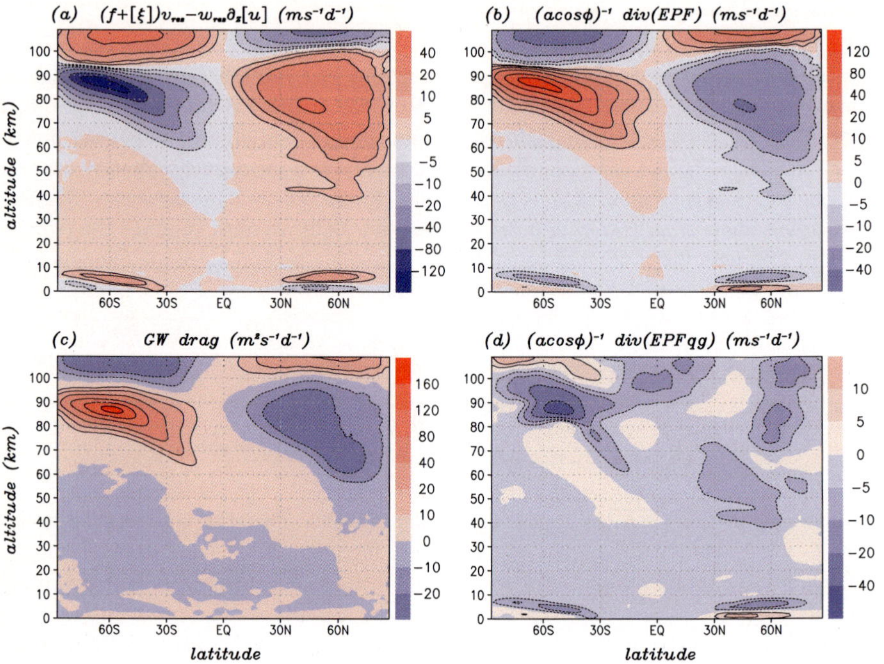

Fig. 5 Complete climatological zonal-mean momentum budget from the January simulation with the KMCM: (**a**) Coriolis force plus advection by the residual circulation, (**b**) generalized EP flux divergence, (**c**) GW drag, and (**d**) quasi-geostrophic EP flux divergence

where the generalized EP flux is defined as

$$\mathbf{EPF} = \rho_r a \cos\phi \left(\mathbf{e}_y \left(-[u^* v^*] + \partial_z [u] \frac{c_p}{g} [T^* v^*] \right) + \mathbf{e}_z \left((f + [\xi]) \frac{c_p}{g} [T^* v^*] - [u^* w^*] \right) \right). \tag{31}$$

Concerning the differences to the qg system extended by the GW drag we must make the following remarks:

(1) The GW drag is included in (30) completed by (31) as

$$\text{GW drag} = -\rho_r^{-1} \partial_z (\rho_r [u^* w^*]). \tag{32}$$

When computing this term from high-resolution data, it may be interpreted as GW drag only if it is dominated by the mesoscale wind perturbations in u^* and w^*. This is the case in our diagnostics of the KMCM simulation for the mesosphere (Fig. 5c).

(2) A generalized non-acceleration theorem exists for (30) and (31) according to the works mainly of Eliassen and Palm (1960), Boyd (1976), and Andrews and McIntyre (1976), as described in Sect. 3.6 of the text of Andrews et al. (1987). Hence, we have $(\rho_r a \cos\phi)^{-1} \text{div}(\mathbf{EPF}) = 0$ in general for ideal waves.

(3) In the temporal mean, the generalized EP flux divergence drives the residual circulation not just by a balance with the Coriolis force. Rather, the material rate of change of zonal

momentum of an air parcel transported by the residual circulation, $-\overline{v}_{res}[\xi]+\overline{w}_{res}\partial_z[\overline{u}]$, is important as well. This is essential in the tropics. For instance, in the poleward branch of an ideal Hadley circulation we have the balance $\overline{v}_{res}[\overline{\xi}]+\overline{v}_{res}f=0$ (Held and Hou 1980). Furthermore, as the diagnostic relation between the zonal wind and temperature, thermal wind balance must be substituted by gradient wind balance,

$$\big(f+2\text{tg}\phi[u]/a\big)\partial_z[u] = -\frac{g}{T_r}\frac{\partial_\phi}{a\cos\phi}[T], \tag{33}$$

which is a good approximation in the troposphere and stratosphere but less accurate in the mesosphere (McLandress 2006).

(4) Combining the temporal means of (30) and (31) with the continuity equation (14) and the definition of the residual mass streamfunction (23), one can derive also a generalized version or the downward-control principle. The resulting expression can be used to attribute residual circulation changes in the stratosphere to specific changes in the wave driving (e.g., McLandress and Shepherd 2009).

(5) The upper panels in Fig. 5 show the complete zonal momentum budget from the January simulation with the KMCM. We infer a precise balance between the EP flux divergence and the acceleration due to the residual circulation (Coriolis force plus advection). That is, even though the subgrid-scale turbulent diffusion scheme applied in the KMCM gives rise to violation of the EP theorem and thereby induces the wave-mean flow interaction, its contribution to the momentum equation averages out in the climatological zonal mean, as it should be according to the wave-driving concept.[6]

(6) The zonal-mean thermodynamic equation is much more complicated when qg scaling is relaxed, particularly because irreversible internal heating rates that are due to mesoscale motions and small-scale turbulence become substantial in the MLT. Nevertheless, the large-scale dynamic heating, defined as the adiabatic heating plus advection due to the large scales (mainly the zonal-mean flow and Rossby waves), still represents the dynamic heating by the residual circulation. This large-scale dynamic heating is then balanced by the external diabatic heating (radiative and chemical heating in comprehensive models), as well as by the internal irreversible heating rates. Figure 6 shows a corresponding estimate of the so-defined zonal-mean sensible heat budget in the MLT from the KMCM (for details see Becker and von Savigny 2010). First of all we note the qualitative balance between the temperature relaxation (Q_{rad}, panel a) and the large-scale dynamic heating (panel b). The latter shows dynamic cooling below the summer mesospause and heating below the winter mesopause as expected. The reversed signs above the mesopause are caused by a reversed residual circulation in the lower thermosphere. In fact, Fig. 5 indicates a change in sign of the EP flux divergence across the mesopause—an effect that is also seen in comprehensive models with GW parameterizations (Richter et al. 2008) and presumably caused by very fast GWs not filtered by the stratospheric mean winds. When focusing on the polar summer mesopause region and comparing the absolute numbers in Figs. 6a and b, we see that the maximum large-scale dynamic cooling/heating rates are about twice as strong as the temperature relaxation. The reason is that the dynamic heating rates due to the mesoscale flow, as well due to turbulent diffusion and dissipation (frictional heating) are substantial. The sum of

[6]In comprehensive climate models with parameterized GWs, strong vertical diffusion coefficients are usually predicted by the GW scheme and applied to the resolved large-scale flow. This can give notable contributions to the mean momentum budget (see Fig. 10 in Fomichev et al. 2002).

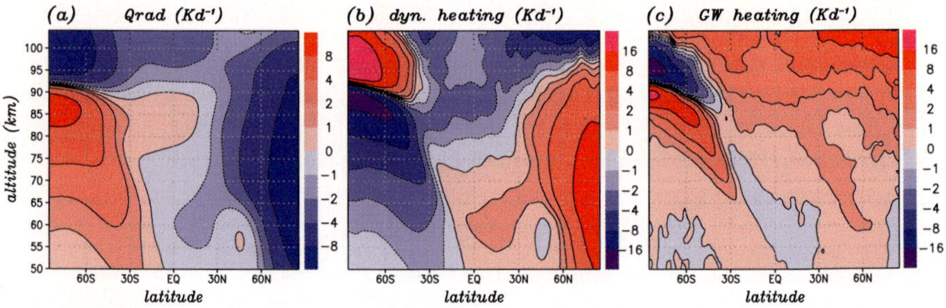

Fig. 6 Complete climatological zonal-mean sensible heat budget in units of Kelvin per day from the January simulation with the KMCM: (**a**) Temperature relaxation (Q_{rad}), (**b**) large-scale dynamic heating, and (**c**) direct GW heating (or irreversible internal heating) defined as the mesoscale dynamic heating plus diffusion and frictional heating

these terms (Fig. 6c) may be interpreted as the direct GW heating and exhibits a heating/cooling dipole pattern similar to the temperature relaxation signal. Such a structure was first estimated by Liu (2000) and later found in circulation models with parameterized GWs (e.g., Fig. 9 in Fomichev et al. 2002 or Fig. 2d in Becker 2004). In particular, the lower thermospheric cooling in polar summer results from a downward potential temperature flux due to GWs and GW-induced turbulent vertical diffusion acting on the large-scale stratification. The heating farther below represents partly the convergence of the downward heat flux and, in the main, the dissipation of GW kinetic energy. Note that the present model estimate of the dissipation (see Fig. 5b in Becker 2009) is based on resolved GWs combined with an advanced parameterization of turbulence. Like former estimates from GW parameterizations it reflects the summer-winter-asymmetry that is expected from the simple arguments underlying the saturation theory. The maximum dissipation rate of up to $\sim$10 K d^{-1} in the summer MLT between 80 and 90 km (not shown) corresponds quantitatively well to the results of Lübken (1997) based on rocket-borne in-situ measurements. General theoretical accounts on how the sensible heating rates due to GWs and turbulence must be formulated in a model with parameterized GWs can be found for instance in Becker (2004), Akmaev (2007), Yiğit et al. (2008), Becker and McLandress (2009), and Shaw and Shepherd (2009).

2.6 Wave Driving by Gravity Waves and Rossby Waves in the MLT

The wave driving in the upper mesosphere by GWs is complicated due to thermal tides and Rossby waves. While tides give notable contributions to the EP flux divergence at most at low latitudes, Rossby waves are quite important in the extratropics of either hemisphere. A thorough analysis of this subject based on the CMAM can be found in McLandress (2006).

In Northern Hemisphere winter, quasi-stationary planetary Rossby waves forced in the troposphere can propagate up to the mesosphere where they contribute to the negative wave drag. The question how the nonlinear interaction between GWs and quasi-stationary Rossby waves may affect the occurrence of the latter in the winter MLT was investigated by Smith (2003) (see also references therein). We furthermore note that the polar winter stratopause does not exist in the radiatively determined state and is generated by the dynamic heating associated with the GW-driven winter mesospheric branch of the residual circulation. The resulting temperature gradient toward the pole in the winter lower mesosphere can even

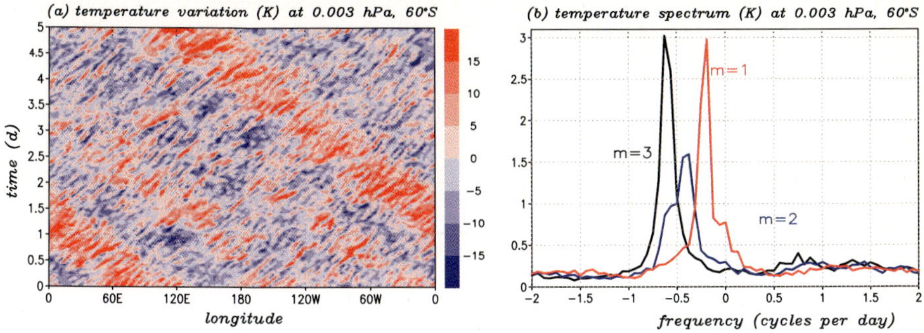

Fig. 7 (**a**) Typical longitude-time cross-section of the temperature variation in the summer upper mesosphere for an arbitrary 5-day time interval of the January simulation with the KMCM. (**b**) Corresponding frequency spectra for planetary waves numbers $m = 1, 2, 3$ computed from the whole time series (240 days, snapshot interval 45 minutes)

allow the in-situ generation of an eastward propagating 4-day wave, which gives rise to an equatorward heat flux that offsets the effect of the westward GW drag to some extent (McLandress 2006).

The reversed mechanism is well known to apply in the polar summer mesopause region (Norton and Thuburn 1996; McLandress 2006; Lieberman 1999, 2002; Lieberman et al. 2003). Let us reconsider the individual contributions to the momentum budget in Fig. 5. We see that the GW drag in the southern summer mesopause region is significantly larger than the complete EP flux divergence. The reason is a strong negative qg EP divergence in the summer mesopause region (Fig. 5d). This deceleration results from westward propagating planetary Rossby waves, such as the 2-day and 5-day waves. These waves add a poleward component to the equatorward residual circulation and thereby reduce the cold temperatures and strong mean wind shear that would develop due to GWs alone. The qg EP flux divergence also exhibits a dipole structure with deceleration located upon a region of acceleration, which is analogous to the pattern seen in the troposphere. In both height regions baroclinic Rossby waves develop because of a strong decrease of the temperature toward the pole and a corresponding vertical wind shear. The baroclinicity of the troposphere is maintained by the greenhouse effect together with the finite surface heat capacity and the annual-mean solar insolation. The resulting baroclinic Rossby waves generate GWs in the storm tracks (O'Sullivan and Dunkerton 1995). In turn, these GWs drive a strong equatorward circulation in the summer mesopause region and thereby set the stage for baroclinic Rossby waves (Plumb 1983). In the KMCM, the overall reduction of the EP flux divergence in the summer MLT by Rossby waves is about 25%.

Figure 7a shows a longitude-time cross-section of the temperature variation for an arbitrary 5-day time interval from the January simulation. This Hovmöller diagram illustrates how the eastward propagating mesoscale GWs and the westward propagating planetary Rossby waves in the summer MLT (around 60°S) are simultaneously resolved in the KMCM. From the eastward phase propagation of the GWs we can estimate zonal phase speeds and wavelengths of about 40 $\mathrm{m\,s^{-1}}$ and 500 km. These values depend on the model resolution, which appears to be a general property of GW-resolving large-scale circulation models (Becker 2009). The westward propagating planetary waves have zonal wave numbers of 1–3 (corresponding to wavelengths of ~7000 to 20000 km) and show periods of a few days. Though both wave types are clearly visible in this representation, the temperature variations are dominated by the planetary waves. This is obvious from the frequency power spectra for individual zonal wavenumbers (Fig. 7b). The main contributions are those of the quasi 5-day and 2-day waves with zonal wavenumbers 1 and 3, respectively. Note that if the GW drag was not be present, also the westward propagating baroclinic waves would be absent in the summer MLT as was shown by a sensitivity experiment with the CMAM by McLandress (2006).

3 Mechanism for Modulations of the Wave Driving in the MLT

The GW drag in the winter upper mesosphere strongly depends on variations of the polar night jet below. The most prominent example in this context are sudden stratospheric warmings, during which the wind reversal in the stratosphere leads to a temporary cessation of the westward GW drag in the mesosphere. This in turn is accompanied by mesospheric cooling due to the reduction of the GW-driven branch of the residual circulation (Labitzke 1972; Holton 1983; Liu and Roble 2002; Siskind et al. 2005). Such an intra-hemispheric vertical coupling induced by internal variability of planetary Rossby waves forced in the troposphere is not possible in summer by virtue of the Charney-Drazin criterion. Nonetheless, the summer mesospheric dynamical control by GWs can be affected by various mechanisms. In the following we consider some prominent examples.

3.1 Interhemispheric Coupling

The dominant zonal-mean internal variability patterns in the extratropics are the northern and southern annular modes (hereafter: NAM and SAM), with the lower tropospheric branch of the NAM being closely related to the Arctic oscillation (Wallace, 2000). The wintertime NAM and SAM are usually considered for the troposphere and stratosphere (Baldwin and Dunkerton 2001) while their role in the mesosphere has long been ignored. Sudden stratospheric warmings can be interpreted as extreme NAM events. Since they have strong signals also in the mesosphere and the thermosphere, it is tempting to argue that the MLT responds to the NAM and SAM in general. Moreover, given the summer-to-winter pole extent of the upper mesospheric branch of the residual circulation one may even speculate about some influence of the winter annular mode on the summer mesopause region. Such a link has recently been found in various modeling studies and observational data and is known as interhemispheric coupling (hereafter: IHC).

First indications of IHC can be found in studies based on global circulation models. Volodin and Schmitz (2001) computed the first empirical orthogonal function of the zonal-mean temperature during northern hemispheric winter and detected IHC as a mode of internal variability. Becker and Schmitz (2003) studied the model sensitivity to the tropospheric forcing of quasi-stationary Rossby waves during January and found the IHC signature in the temperature, zonal wind, and residual circulation response. The main connection that can be drawn from these results is that a warmer and weaker polar night jet in the stratosphere leads to a colder winter mesosphere, as well as a warmer summer upper mesosphere. While the first effect is similar to that during a sudden stratospheric warming, the link to the summer mesopause region was unclear and has later been interpreted by Becker and Fritts (2006) and Karlsson et al. (2008). Before mentioning the observational evidence for IHC we shall recapitulate the mechanism as proposed by Karlsson et al. (2008) and recently confirmed within a highly idealized model by Körnich and Becker (2010).

Karlsson et al. (2008) analyzed the internal variability in long-term simulations with the CMAM and showed that the IHC is a generic mode of internal variability in a complex

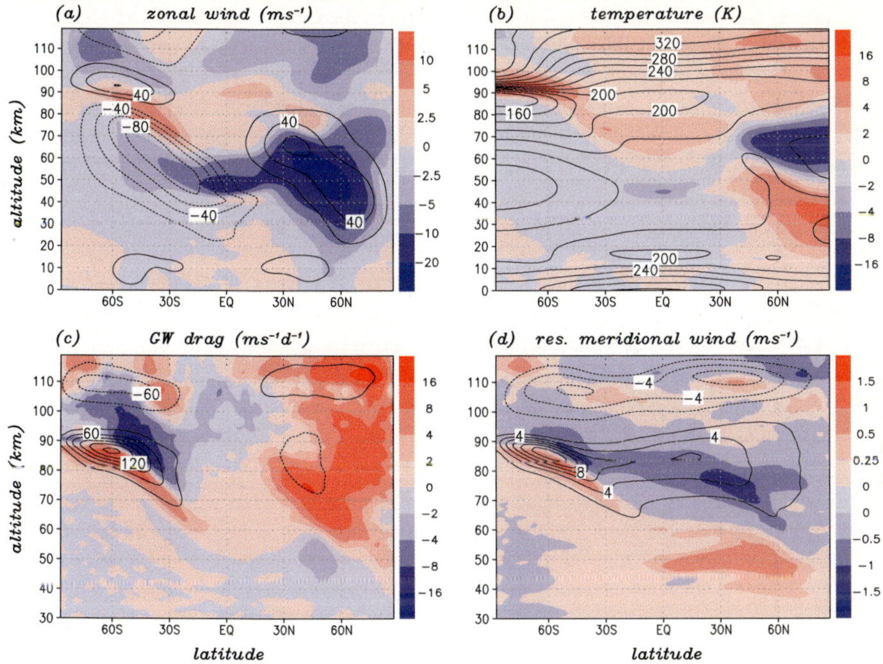

Fig. 8 Variability pattern of interhemispheric coupling (IHC) illustrated by the differences (*colors*) computed from a composite of sudden stratospheric warming events simulated with the KMCM minus the corresponding January climatology. (**a**), (**b**) Zonal-mean temperature and zonal wind from 0 to ~120 km, and (**c**), (**d**) GW drag and $\overline{v}_{res}$ from 30 to ~120 km. The control climatology is indicated by *black contours* for reference (intervals 20 m s^{-1}, 20 K, 30 m s^{-1} d^{-1}, and 2 m s^{-1} with the zero contours omitted)

climate model. Alternatively, we can illustrate the IHC variability pattern by the difference between a composite of sudden stratospheric warming events that occur during the present January simulation with the KMCM and the January climatology. Figure 8 shows this difference for the temperature, zonal wind, GW drag, and residual meridional wind. Inspecting the northern winter stratosphere we see that the poleward (residual) flow is strongly enhanced for the sudden warming composite, as expected. As a result of the so-induced warmer temperature and weaker zonal wind, the westward GWs break at lower altitudes or encounter critical levels such the westward (negative) GW drag in the winter mesosphere is strongly reduced. This causes a weakening of the winter mesospheric branch of the residual circulation, which is accompanied by mesospheric cooling. Karlsson et al. (2008) pointed out that the polar winter temperature changes are balanced by opposite temperature signals in the tropics. In particular, the warming in the polar stratosphere is balanced by enhanced upwelling and cooling in the tropical stratosphere. Likewise, polar mesospheric cooling is balanced by some downwelling superposed on the cross-equatorial flow and hence warming of the tropical mesosphere. Such a seesaw was first mentioned by Yulaeva et al. (1994) for the winter lower stratosphere. The fact that it holds also higher up is crucial for IHC. Mainly because the warmer equatorial mesosphere corresponds to a stronger temperature decrease toward the summer pole and hence will induce reduced westward flow in the summer upper mesosphere according to gradient wind balance. Referring to Sect. 2.3 we conclude that the altered wind will shift the GW breaking levels and hence the equatorward residual circulation to lower altitudes. This results in higher temperatures around and above the summer mesopause. Of course, the chain of events also works the other way round. The temperature signal requires at most a few days to propagate from the winter lower stratosphere all the way to the summer mesopause.

One important implication from the IHC is that the breaking levels of GWs in the summer MLT are not defined by just the conditions in the summer hemisphere itself. Rather, the breaking levels are subject to modulation by variability in the winter middle atmosphere, which ultimately is controlled by planetary Rossby waves forced in the winter troposphere. In this respect the IHC can be understood as an extension of the annular mode into the mesosphere on a global scale.

First observational evidence for IHC emerged from an international measurement campaign that took place at the site of Andøya in Norway (69°N) in June and July 2002 (Goldberg et al. 2004). When comparing the summer-MLT measurements of Rapp et al. (2004) and Singer et al. (2005) during June and July in 2002 to corresponding data from previous years, a systematic pattern of anomalies in winds, temperature, and turbulence could be inferred which fitted qualitatively well to IHC as described above (with reversed latitude). That is, the summer mesosphere in June and July 2002 was presumably subject to IHC because of the strong planetary Rossby-wave activity in Austral winter at the same time (Becker et al. 2004). In fact, the year 2002 was exceptional because of high amplitudes of planetary Rossby waves in the Austral winter stratosphere which culminated in the first observed southern major sudden stratospheric warming in September 2002 (e.g., Harnik et al. 2005). Becker and Fritts (2006) noted that also stronger tropospheric GW sources are required for a more quantitative interpretation of the unusual state of the northern polar summer MLT in 2002.

Meanwhile, IHC has been detected in the interannual and intraseasonal variations of noctilucent clouds (Karlsson et al. 2007, 2009), which can be used as a proxy of upper mesospheric temperatures. A recent observational analysis by Espy et al. (2011) reveals a correlation between the summer MLT temperatures from airglow measurements in the northern hemisphere to the phase of the quasi-biennial oscillation (QBO). Here, IHC comes into play since the QBO modulates the vertical propagation of synoptic-scale Rossby waves at middle latitudes in Austral winter via the Holton-Tan mechanism (Baldwin and Dunkerton 1998). This leads to mean wind variations in the Austral winter stratosphere at middle latitudes due to the QBO. Hence, the mesospheric residual circulation is presumably modulated by the QBO via a combination of the Holton-Tan and IHC mechanisms.

3.2 Solar Proton Events

Protons emitted by the sun can temporally reach very high energies during solar flares or when they are accelerated by an interstellar shock front induced by a coronal mass ejection. Such a *solar proton event* (hereafter: SPE) typically lasts for several days. The main effect of the energetic solar protons precipitating into the polar middle atmosphere is a temporary ozone depletion in the lower mesosphere due to odd hydrogen and odd nitrogen production as a result of the ionization (Rohen et al. 2006). We therefore expect a temporary anomalous diabatic cooling in the lower polar mesosphere in summer because of reduced short-wave radiative heating. The question then is how this primary thermal effect of an SPE can be accompanied by a warming of more than 10 K around the mesopause and a depletion of noctilucent clouds, as was observed by von Savigny et al. (2007). A warming due to Joule heating in the mesopause region, which also occurs during an SPE (Roble et al. 1987), cannot explain the observations. Jackman et al. (2007) argued that rather changes in the wave driving might be important.

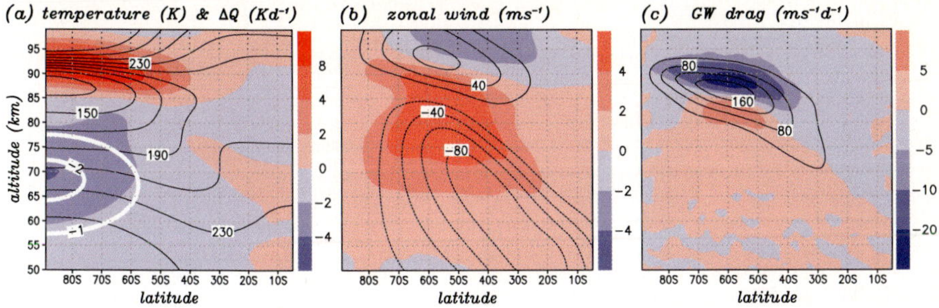

Fig. 9 Model response in the summer MLT to an SPE perturbation. Differences (*colors*) in the SPE composite from the unperturbed January composite for (**a**) temperature, (**b**) zonal wind, and (**c**) GW drag. The control composite is indicated by *black contours* (intervals 30 K, 20 m s^{-1}, and 40 m s^{-1} d^{-1} with the zero contours omitted). The *white contours in panel* a display the diabatic cooling perturbation in the SPE case (contours −2 and −1 K d^{-1})

In a time-slice sensitivity experiment with the present version of the KMCM, Becker and von Savigny (2010) have identified a straight forward dynamical mechanism by which the diabatic cooling in the lower summer polar mesopause can lead to a strong dynamical warming of the mesospause region above. This sensitivity experiment was designed as follows. An arbitrary snapshot from the January simulation was extracted as initial condition and the model was integrated forward in time with an additional diabatic cooling concentrated in the lower mesosphere above the summer pole. After five model days, this diabatic cooling perturbation was switched off and the integration extended for another 10 days. The resulting 15 day time series represented an SPE-related perturbation simulation when compared to the corresponding 15 day time series of the unperturbed January simulation. This procedure was repeated 5 times for other arbitrarily picked initial conditions. Averaging all 15 day series for the unperturbed and the SPE cases then yielded statistically averaged composite time series.

Figure 9 shows the zonal and temporal mean differences of the SPE case from the unperturbed case by colors. For reference, the black contours show the unperturbed composite. In addition, the white contours in panel a show the diabatic cooling perturbation applied in the SPE case. A maximum value of about −2.5 K d^{-1} above the summer pole is consistent with the estimate for equinox given by Jackman et al. (2007). From Fig. 9a it is clear that a heating around the mesopause induced by the SPE is well simulated. All three panels together reveal the following mechanism: As a result of the temporary diabatic cooling in the lower polar mesosphere in the SPE composite, the averaged temperature response amounts to about −3 K in this region. The zonal wind response (Fig. 9b) reflects this cooling by increased vertical shear below 80 km and a resulting positive anomaly in the summer MLT up to 90 km. This gives rise to shorter vertical wavelengths of the eastward propagating GWs and hence induces dynamical instability at lower altitudes, as is reflected by a downward shift of the GW drag (Fig. 9c). Becker and von Savigny (2010) proved that this downward shift is indeed mainly balanced by a downward shift of the residual circulation and that it therefore induces a large-scale relative dynamic warming (less large-scale dynamic cooling) around and below the mesopause which causes the higher temperatures. They furthermore estimated how these arguments must be quantitatively modified by a downward shift of the 2-day and 5-day waves, as well as by changes in the direct GW heating.

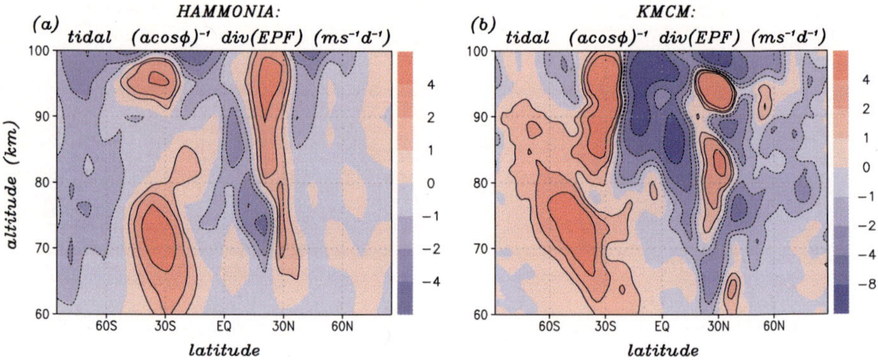

Fig. 10 Generalized EP flux divergence of thermal tides during January (**a**) in the HAMMONIA and (**b**) in the KMCM when completed by the thermal forcing of tides extracted from the HAMMONIA. See text for further details

3.3 Thermal Tides

Thermal tides manifest themselves as ubiquitous, regular, and strong variations of winds and temperature in the upper mesosphere and higher up in the thermosphere. They are mainly forced by the diurnal cycle of solar insolation via absorption of visible light by tropospheric water vapor and of UV radiation by stratospheric ozone. In addition, thermal tides are forced by the daily cycle of cumulus convection in the tropics. Though thermal tides have their strongest amplitudes in the tropics and subtropics (e.g., McLandress 2002; Achatz et al. 2008), where the daily cycle is strongest, they may nevertheless attain notable amplitudes even at high and polar latitudes. Here we ask for the role of tides in contributing to or modulating the EP flux divergence.

Figure 10a shows an evaluation of the generalized EP flux divergence (31) for the thermal tides during January from the HAMMONIA (Schmidt et al. 2006). Note the different color shading when comparing this representation to the momentum budget in Fig. 5. It follows that the contribution of thermal tides to the generalized EP flux divergence is negligible, except for their role in modulating the cross-equatorial flow.

The thermal forcing of tides was extracted from the HAMMONIA from a control climatology in terms of the averaged daily cycles of the diabatic heating for each month. This tidal diabatic heating is defined such that it vanishes when averaged over one day. It has already been used in a linear tidal model (Achatz et al. 2008) and can be included in the present version of the KMCM as well. The simulated thermal tides are then similar to those in the HAMMONIA (not shown). Also the tidal EP flux divergence (Fig. 10b) is roughly similar to that from the HAMMONIA. The advantage of having a mechanistic GCM with tidal forcing is that we can consistently compare the model climatologies with and without tides.

Figure 11 shows the climatological zonal-mean changes of the temperature, zonal wind, and GW drag in the simulation with tides relative to our control January simulation. We infer a significant warming around the polar summer mesopause that is obviously caused by a downward shift of the GW drag. The change of the GW drag can be understood in terms of a modulation of the GW propagation in the summer mesosphere in the following way: During the positive tidal phase of the zonal wind, the eastward propagating GWs are Doppler-shifted to smaller intrinsic frequencies (shorter vertical wavelengths) and hence are subject to enhanced damping by turbulent diffusion, while the opposite effect during the

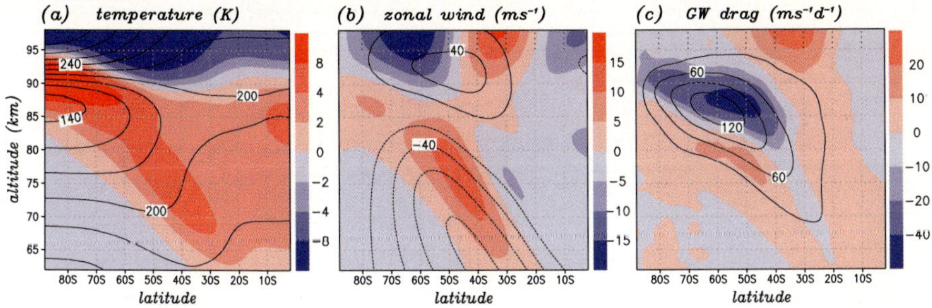

Fig. 11 Model response to the inclusion of thermal tides. Differences (*colours*) in the January simulation with tides from the control simulation for (**a**) temperature, (**b**) zonal wind, and (**c**) GW drag. The *black contours* show the climatology of the simulation with tides for reference (intervals 20 K, 20 $m s^{-1}$, and 30 $m s^{-1} d^{-1}$ with the zero contours omitted)

westward tidal phase of the zonal wind is less efficient. Therefore, the zonal wind variations due to thermal tides in the summer mesosphere give rise to a net reduction of the GW drag in the mesopause region. Such a modulation is likely also effective in comprehensive models with parameterized GWs. Further details of this mechanism will be presented elsewhere.

3.4 Climate Change: Radiatively Induced Changes Versus Altering Tropospheric GW Sources

The observed negative temperature trends in the middle atmosphere are much stronger than the global warming trends in the troposphere (Keckhut et al. 1995; Ramaswamy et al. 2001; Bremer 2005). Comprehensive climate models generally fail to quantitatively describe these long-term cooling trends. The situation is particularly complicated in northern winter since also the possible changes in the stationary Rossby waves must be taken into account (Gabriel et al. 2007). In the following we consider the summer middle atmosphere, where we can assume that temporal averages are presentative of the zonal mean.

The temperature trend during summer from 1950 to 2003 as simulated with the WACCM by Garcia et al. (2007) shows maximum cooling around the summer stratopause and a slight warming around the summer mesopause.[7] However, the magnitude of the simulated temperature change does not explain the trend in phase-height measurements over Europe (Bremer 2005). This is illustrated in Fig. 12. The red curves correspond to the temperature and pressure-height changes simulated with the WACCM, which are mimicked here using a simple radiative-convective equilibrium model that has the same vertical discretization as the KMCM. The resulting change in pressure height yields a value of about −0.6 km at 0.005 hPa. The corresponding result of the phase-height measurements is indicated by the green dot and amounts to about −1.3 km, which is about twice as strong as the model result.

In order to seek for some explanation we note that GWs are parameterized in a comprehensive chemistry-climate model (hereafter: CCM) like the WACCM. This widely prevents

[7]The mesopause warming likely results from a stronger long-wave heating in the CO_2 15 μm band in the vicinity of the cold summer mesopause, where thermal emission does not exceed the absorption of upward radiation. Everywhere else in the middle atmosphere, there is generally radiative cooling for this band (e.g., Fomichev et al. 2002).

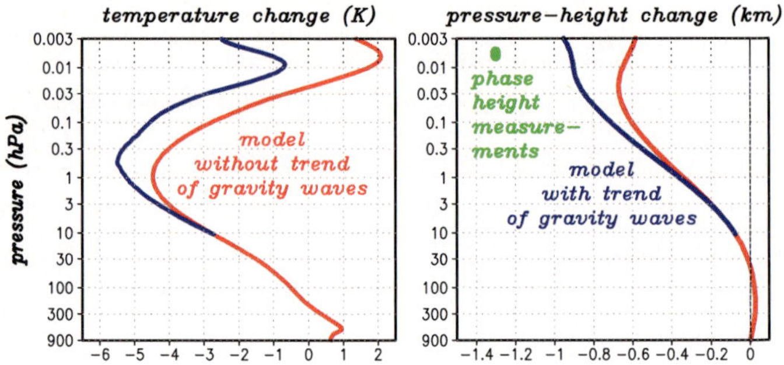

Fig. 12 Long-term changes of temperature and pressure heights in northern summer at 55°N. The simulation without trend of GWs (*red curves*) corresponds to results obtained with the WACCM (Garcia et al. 2007, data for 55°N provided by Rolando Garcia) for the period from 1950 to 2003 and has been mimicked here using a one-dimensional radiative-convective-equilibrium model. For the *blue curves*, the temperature change above 10 hPa (30 km) is completed by the effect of stronger GW activity according to Becker (2009). The *green dot* indicates phase-height measurements at the IAP for the period from 1959 to 2003 according to Bremer (2005)

that the possible changes in the extratropical tropospheric GW sources due to global warming are fully taken into account. Recently Becker (2009) performed a sensitivity experiment with the KMCM in order to estimate such an effect. For this purpose, the zonal-mean warming pattern in the troposphere was prescribed by changes in the equilibrium temperature T_E, the surface temperature, and the tropical latent heating rates. All other model parameters were unchanged; in particular, the T_E above 200 hPa ($\sim$10 km) was the same in both the control and the simplistic global warming simulation. The model showed that, despite a reduced equator-to-pole temperature difference in the lower troposphere, the tropospheric Lorenz energy cycle becomes more intense by about 10–20%. The main reason is that the increased surface temperatures result in a decreased tropospheric static stability and thereby enhanced available potential energy. A stronger energy cycle reflects stronger baroclinic Rossby wave activity which, as a side effect in a high-resolution GCM, yields stronger GW sources. This change in tropospheric dynamics has a pronounced effect in the polar summer mesosphere of the KMCM: Due to the stronger GW momentum flux, the residual circulation intensifies such that the temperature decreases by a few degrees up to the mesopause region in accordance with the downward control principle (Sect. 2.4). If we add this purely dynamical effect to the red curve in Fig. 10a, we obtain a synthetic temperature signal that includes the changes due to radiation and chemistry, and in addition the likely changes due to stronger tropospheric GW sources. The result is indicated by the blue curves and leads to a stronger mesospheric cooling such that the pressure-height change is in better agreement with the measurements. Moreover, the blue curve shows a zero temperature trend around the mesopause which is consistent with long-term ground-based temperature measurements (Lübken 2000).

4 Summary and Outlook

The first part of this review (Sect. 2) has recapitulated the causality and consequences of the wave-driven residual circulation. The main notion is that, except for the Hadley circulations in the tropical troposphere and tropical stratopause region, any mean-meridional circulation is caused by waves. However, we must distinguish between the residual (Lagrangian mean) circulation and any secondary meridional circulation. A secondary circulation already occurs for ideal (linear, steady, and conservative) waves and neutralizes their meridional and vertical Stokes drifts, as well as their momentum and heat fluxes. The Ferrel cell in middle latitudes of the troposphere and in high latitudes of the winter stratosphere is such a secondary circulation that accompanies Rossby waves. Secondary circulations occur as soon as the waves are affected by the Coriolis force, which is the case for Rossby waves, inertia gravity waves, thermal tides, and tropical Rossby-gravity waves.

The residual circulation is obtained by adding the quasi-linear Stokes drifts to the mean meridional and vertical velocities (Dunkerton 1978). This gives rise to the so-called transformed Eulerian mean (TEM) equations (see Sect. 3.5 in the text of Andrews et al. 1987 and references therein). The fundamental importance of the TEM equations derives from the Eliassen-Palm (EP) theorem: The wave-mean flow interaction in the transformed zonal-mean zonal momentum equation, also known as the Eliassen-Palm (EP) flux divergence or wave drag, vanishes for ideal waves. Therefore, any EP flux divergence results from a violation of the EP theorem.

In the temporal mean (at least several days or longer), the EP theorem is mainly violated due to irreversible processes such as wave breaking and the so-induced turbulent diffusion. The resulting wave drag must be balanced by the Coriolis force plus advection of zonal momentum associated with the residual circulation. In other words, the extratropical residual circulation along with the associated deviations from some hypothetical radiative-convective equilibrium are caused by the EP flux divergence. The residual circulation is representative of, and under the assumption of quasi-geostrophic scaling even identical with, the diabatic circulation, with the diabatic heating being the result, not the cause, of the residual circulation.

While Rossby waves drive the extratropical residual circulation in the troposphere and (winter) stratosphere, mesoscale gravity waves (GWs) are responsible for the summer-to-winter-pole residual circulation in the upper mesosphere. In addition, orographic GWs contribute to the deceleration in the upper troposphere and lower stratosphere (Alexander et al. 2010). Furthermore, the positive GW drag in the summer mesopause region is opposed by the EP flux divergence of in-situ generated baroclinic planetary waves by about one quarter (Norton and Thuburn 1996). Rossby waves in the winter mesosphere are either forced in the troposphere or generated by the zonally varying GW drag (Smith 2003). In addition, also eastward propagating planetary waves can develop in the polar winter mesosphere (McLandress 2006). The contributions of the different types of waves to the overall wave driving is less certain in winter than in summer.

The sensible heat budget in the mesosphere is strongly affected by GW breakdown and turbulence. For instance, though the turbulent dissipation (frictional heating) is usually very small in the troposphere and stratosphere, it becomes substantial in the mesosphere and lower thermosphere (Lübken 1997), where it represents the dissipation of GW kinetic energy. Particularly in the summer mesopause region, the turbulent dissipation is on average as strong as the net radiative plus chemical heating. This poses strong constraints on the representation of GWs in general circulation models that resolve the mesopause region (Fomichev et al. 2002; Becker 2004, 2009; Akmaev 2007; Yiğit et al. 2009).

In the second part of this review (Sect. 3) we have discussed several mechanisms by which the wave driving in the mesopause region can be modulated. Such modulations are well known for northern winter and may be understood as a vertical extent of the annular mode. In particular, using arguments from Lindzen (1981) we can anticipate that a weaker polar night jet will cause a reduction and downward shift of the westward GW drag per unit mass and vice versa. The most obvious example for this vertical coupling is found during sudden stratospheric warmings (Holton 1983).

Modulations of the GW drag in summer are less obvious. They typically result from some changes of the background zonal wind, giving rise to net vertical shifts and changes of the drag generated by eastward propagating GWs. In this context we have discussed the examples of interhemispheric coupling, solar proton events, and thermal tides. Interhemispheric coupling can be understood as an extension of the winter hemispheric annular mode all the way to the polar summer mesopause region through changes in the tropical mesospheric temperature and hence of the zonal wind in the summer mesosphere. Through this mechanism (Karlsson et al. 2008), a weaker polar night jet will result in a warmer summer mesopause since the eastward GW drag is shifted to lower altitudes. A similar pattern is found when we consider the simulated response to solar proton events (Becker and von Savigny 2010), where the primary perturbation is a diabatic cooling in the lower polar summer mesosphere and hence a reduction of the mesospheric westward zonal wind. Also thermal tides may modulate the GW drag in summer: During the eastward tidal phase of the zonal wind, the eastward propagating GWs already break well below the mesopause region, while the westward tidal phase is less effective in shifting GW breakdown to higher altitudes. The net effect of thermal tides in the summer MLT is thus a significant weakening and downward shift of the GW drag. The wave driving in the summer mesopause region may also be affected by other vertical coupling mechanisms within the summer hemisphere itself. Smith et al. (2010) have recently shown that the ozone hole in early Austral summer causes an anomalous diabatic cooling in the lower polar stratosphere and thus induces a significant eastward zonal wind even in early December. This filters some of the eastward propagating gravity waves and thereby reduces the GW drag beneath the polar summer mesopause, leading to a significant positive temperature anomaly around 80 km.

A significant response of the GW drag in the summer MLT may furthermore be caused by tropospheric global warming, which according to the mechanistic high-resolution sensitivity experiment of Becker (2009) is likely associated with intensifications of the Lorenz energy cycle and the GW sources in the tropospheric storm tracks. In line with the downward control principle (Haynes et al. 1991, see also Sect. 2.4), the stronger GW momentum flux entering the summer mesosphere induces a stronger upwelling and thereby a cooling of the entire polar summer mesosphere by a few degrees. Such a mechanism may explain part of the differences between the observed trends in the northern summer meosphere (Bremer 2005; Keckhut et al. 1995) and estimates from comprehensive climate models with parameterized GWs (Garcia et al. 2007).

This review has concentrated on the general circulation of the middle atmosphere and the causality that allows to interpret dynamically equilibrated states. It is not meant to be comprehensive, but rather as an introduction to the dynamical control of the middle atmosphere during the solstices. In closing we list some issues concerning the subject that may attract particular scientific interest in the future:

(1) In order to better distinguish between natural long-term variability and anthropogenic climate change it is necessary to improve our understanding of the various variability patterns, like the annular modes (Baldwin and Dunkerton 2001), the QBO (Baldwin et al. 2001), or the semi-annual oscillation (Richter and Garcia 2006), as well as their interactions and dependencies on solar activity (Schmidt et al. 2010). For this purpose it is important to know how the individual contributions from different types of waves contribute to the wave driving and how these contributions change over a longer period. Our knowledge in this respect is still limited.

(2) Model estimates of long-term variability should preferably include interactive chemistry for the lower and middle atmosphere. Mainly because dynamically induced changes of the transport of minor constituents may easily feed back on the radiative heating and thus on the propagation conditions for waves (Shepherd 2008). While various middle atmosphere GCMs were developed in the past, only a few with interactive chemistry (so-called chemistry climate models, hereafter: CCMs) that extend from the surface to the lower thermosphere currently exist (Fomichev et al. 2007; Garcia et al. 2007; Schmidt et al. 2006). Even these models are still limited with regard to several dynamical aspects, some of which are mentioned below.
(3) Dunkerton's insight that the residual circulation represents an approximate description of the mean Lagrangian motion in the zonal-mean frame must be modified by an effect known as wave mixing. That is, since the waves are neither entirely linear nor conservative, the corresponding wave motions of a passive tracer will not be ideal too. The interaction of the wave perturbation in a tracer field with the small-scale turbulence will then lead some mixing of the tracer in addition to the transport by the residual circulation and the turbulent diffusion (Shepherd 2007). Whereas meridional mixing by Rossby waves is important in the lower stratosphere (e.g., Shepherd and Koshyk 2000), vertical mixing by GWs must be considered in the MLT (Grygalashvyly et al. 2011). In current CCMs the representation of vertical mixing by GWs is represented in terms of the GW-induced turbulent diffusivity, which is likely incomplete.
(4) The fact that the transport in a CCM depends not only on the resolved flow, but is sensitive to the parameterized scales as well, couples the various uncertainties in subgrid-scale parameterizations to the transport problem. Moreover, even the consistent numerical representation of the resolved transport remains an issue (Shepherd 2007). In addition to these complications in formulating models, a detailed understanding of the different transport ways in the middle atmosphere is required in order to properly validate simulations against observations (Birner 2010; Birner and Bönisch 2011).
(5) With regard to vertical coupling of the lower and middle atmosphere it would be desirable to represent all important waves explicitly. This will not be technically feasible in the foreseeable future in CCMs with regard to GWs. Nevertheless, high-resolution estimates of GW dynamics from more or less idealized middle atmosphere GCMs already exist (e.g., Watanabe et al. 2008; Becker 2009). As more and more computer resources become available in the future, the challenge will be to run also CCMs with very high resolutions, including the proper modification of all kind of parameterizations such as to be physically consistent with the scales resolved. Only such high-resolution comprehensive CCMs will allow to study the climate sensitivity of the MLT to the combined changes in radiation and chemistry, in the tropospheric forcing of planetary Rossby waves, and in the extratropical dynamical GW sources that ultimately give rise to the pole-to-pole residual circulation in the mesosphere. Even in such a model, the proper representation of orographic GWs and tropical GWs generated by convection will most likely remain an issue.
(6) The GW problem has long been neglected for the thermosphere. Estimates of GW effects in the thermosphere have been provided for instance by Vadas and Fritts (2005), Fritts and Vadas (2008), and Yiğit et al. (2008). The first parameterization of GWs in a thermospheric GCM has been achieved by Yiğit et al. (2009) and Yiğit and Medvedev (2009), who found strong influences of GWs on the large-scale momentum and heat budgets. The general notion is that GW drag, GW heating, and presumably also GW mixing are even more important above the mesopause than below (see Fritts and Lund 2011). Moreover, thermosphere-mesosphere coupling is important for minor

constituents in the middle atmosphere (Smith et al. 2011) and this will require particular attention in future studies focussing on the MLT.

Acknowledgements I am indebted to Erdal Yiğit for his thorough review and many helpful comments on the manuscript, as well as to Norbert Grieger and Christoph Zülicke for preparing the data analyses on tides and sudden stratospheric warmings used in this paper. Other valuable comments by Fabian Senf, Thomas Birner, and Corinna Schütt are gratefully acknowledged. Last not least I thank Rudolf Treumann for initiating this review.

References

U. Achatz, N. Grieger, H. Schmidt, Mechanism controlling the diurnal solar tide: analysis using a GCM and a linear model. J. Geophys. Res. **113** (2008). doi:10.1029/2007JA012967

R.A. Akmaev, On the energetics of mean-flow interactions with thermally dissipating gravity waves. J. Geophys. Res. **112** (2007). doi:10.1029/2006JD007908

M.J. Alexander, M. Geller, C. McLandress, S. Polavarapu, P. Preusse, F. Sassi, K. Sato, S. Eckermann, M. Ern, A. Hertzog, Y. Kawatani, M. Pulido, T. Shaw, M. Sigmond, R. Vincent, S. Watanabe, Recent developments in gravity-wave effects in climate models and the global distribution of gravity-wave momentum flux from observations and models. Q. J. R. Meteorol. Soc. **136**, 1103–1124 (2010)

D.G. Andrews, M.E. McIntyre, Planetary waves in horizontal and vertical shear: the generalized Eliassen-Palm relation and the mean zonal acceleration. J. Atmos. Sci. **33**, 2031–2048 (1976)

D.G. Andrews, J.R. Holton, C.B. Leovy, *Middle Atmosphere Dynamics* (Academic Press, San Diego, 1987), p. 489

M.P. Baldwin, T.J. Dunkerton, Quasi-biennial modulation of the southern hemisphere stratospheric polar vortex. Geophys. Res. Lett. **25**, 3343–3346 (1998)

M.P. Baldwin, T.J. Dunkerton, Stratospheric Harbingers of anomalous weather regimes. Science 581–584 (2001)

M.P. Baldwin, L.J. Gray, T.J. Dunkerton, K. Hamilton, P.H. Haynes, W.J. Randel, J.R. Holton, M.J. Alexander, I. Hirota, T. Horinouchi, D.B.A. Jones, J.S. Kinnersley, C. Marquardt, K. Sato, M. Takahashi, The quasi-biennial oscillation. Rev. Geophys. **39**, 179–229 (2001)

E. Becker, Direct heating rates associated with gravity wave saturation. J. Atmos. Sol.-Terr. Phys. **66**, 683–696 (2004)

E. Becker, Sensitivity of the upper mesosphere to the Lorenz energy cycle of the troposphere. J. Atmos. Sci. **66**, 647–666 (2009)

E. Becker, D.C. Fritts, Enhanced gravity-wave activity and interhemispheric coupling during the MaCWAVE/MIDAS northern summer program 2002. Ann. Geophys. **24**, 1175–1188 (2006)

E. Becker, C. McLandress, Consistent scale interaction of gravity waves in the Doppler spread parameterization. J. Atmos. Sci. **66**, 1434–1449 (2009)

E. Becker, G. Schmitz, Interaction between extratropical stationary waves and the zonal mean circulation. J. Atmos. Sci. **58**, 462–480 (2001)

E. Becker, G. Schmitz, Climatological effects of orography and land-sea heating contrasts on the gravity-wave driven circulation of the mesosphere. J. Atmos. Sci. **60**, 103–118 (2003)

E. Becker, C. von Savigny, Dynamical heating of the polar summer mesopause induced by solar proton events. J. Geophys. Res. **115** (2010). doi:10.1029/2009JD012561

E. Becker, E.A. Müllemann, F.-J. Lübken, H. Körnich, P. Hoffmann, M. Rapp, High Rossby-wave activity in austral winter 2002: modulation of the general circulation of the MLT during the MaCWAVE/MIDAS northern summer program. Geophys. Res. Lett. **31** (2004). doi:10.1029/2004GL019615

T. Birner, Recent widening of the tropical belt from global tropopause statistics: sensitivities. J. Geophys. Res. **115** (2010). doi:10.1029/2010JD014664

T. Birner, H. Bönisch, Residual circulation trajectories and transit times into the extratropical lowermost stratosphere. Atmos. Chem. Phys. **11**, 817–827 (2011). doi:10.5194/acp-11-817-2011

J. Boyd, The noninteraction of waves with the zonally-averaged flow on a spherical earth and the interrelationships of eddy fluxes of energy, heat and momentum. J. Atmos. Sci. **33**, 2285–2291 (1976)

J. Bremer, Detection of long-term trends in the mesosphere/lower thermosphere from ground-based radio propagation measurements. Adv. Space Res. **35**, 1398–1404 (2005)

J.G. Charney, P.G. Drazin, Propagation of planetary-scale disturbances from the lower into the upper atmosphere. J. Geophys. Res. **66**, 83–109 (1961)

R.E. Dickinson, Theory of planetary wave-zonal flow interaction. J. Atmos. Sci. **26**, 73–81 (1969)

T.J. Dunkerton, On the mean meridional mass motions of the stratosphere and mesosphere. J. Atmos. Sci. **35**, 2325–2333 (1978)

T.J. Dunkerton, Nonlinear Hadley circulation driven by asymmetric differential heating. J. Atmos. Sci. **46**, 956–974 (1989)

H.J. Edmon Jr., B.J. Hoskins, M.E. McIntyre, Eliassen-Palm cross sections for the troposphere. J. Atmos. Sci. **37**, 2600–2616 (1980)

A. Eliassen, E. Palm, On the transfer of energy in stationary mountain waves. Geophys. Norv. **22**(3), 1–23 (1960)

P.J. Espy, S. Ochoa Fernández, P. Forkman, D. Murtagh, J. Stegman, The role of the QBO in the inter-hemispheric coupling of summer mesospheric temperatures. Atmos. Chem. Phys. **11**, 495–502 (2011). doi:10.5194/acp-11-495-2011

V. Eyring, N. Butchart, D.W. Waugh, et al., Assessment of temperature, trace species, and ozone in chemistry-climate model simulations of the recent past. J. Geophys. Res. **11** (2006). doi:10.1029/2006JD007327

V.I. Fomichev, W.E. Ward, S.R. Beagley, C. McLandress, J.C. McConnell, N.A. McFarlane, T.G. Shepherd, Extended Canadian Middle Atmosphere Model: zonal-mean climatology and physical parameterizations. J. Geophys. Res. **107** (2002). doi:10.1029/2001JD000479

V.I. Fomichev, W.E. Ward, S.R. Beagley, C. McLandress, J.C. McConnell, N.A. McFarlane, T.G. Shepherd, Extended Canadian Middle Atmosphere Model: zonal-mean climatology and physical parameterizations. J. Climate **107** (2007). doi:10.1029/2001JD000479

D.C. Fritts, Gravity saturation in the middle atmosphere: a review of theory and observations. Rev. Geophys. **2**, 275–308 (1984)

D.C. Fritts, M.J. Alexander, Gravity wave dynamics and effects in the middle atmosphere. Rev. Geophys. **41** (2003). doi:10.1029/2001/RG000106

D.C. Fritts, T. Lund, Gravity wave influences in the thermosphere and ionosphere: observations and recent modeling, in *Aeronomy of the Earth's Atmosphere and Ionosphere*, ed. by M. Abdu, D. Pancheva (2011), pp. 141–162

D.C. Fritts, S.L. Vadas, Gravity wave penetration into the thermosphere: sensitivity to solar cycle variations and mean winds. Ann. Geophys. **26**, 3841–3861 (2008)

A. Gabriel, D. Peters, I. Kirchner, H.-F. Graf, Effect of zonally asymmetric ozone on stratospheric temperature and planetary wave propagation. Geophys. Res. Lett. **34** (2007). doi:10.1029/2006GL028998

R.R. Garcia, S. Solomon, The effect of breaking gravity waves on the dynamics and chemical composition of the mesosphere and lower thermosphere. J. Geophys. Res. **90**, 3850–3868 (1985)

R.R. Garcia, D.R. Marsh, D.E. Kinnison, B.A. Boville, F. Sassi, Simulation of secular trends in the middle atmosphere, 1950–2003. J. Geophys. Res. **112** (2007). doi:10.1029/2006JD007485

R.A. Goldberg, D.C. Fritts, B.P. Williams, F.-J. Lübken, M. Rapp, W. Singer, R. Latteck, P. Hoffmann, A. Müllemann, G. Baumgarten, F.J. Schmidlin, C.-Y. She, D.A. Krueger, The MaCWAVE/MIDAS rocket and groundbased measurements of polar summer dynamics: overview and mean state structure. Geophys. Res. Lett. **31** (2004). doi:10.1029/2004GL019411

M. Grygalashvyly, E. Becker, G.R. Sonnemann, Wave mixing effects on minor chemical constituents in the MLT region: results from a global CTM driven by high-resolution dynamics. J. Geophys. Res. **116** (2011). doi:10.1029/2010JD015518

L. Haimberger, M. Hantel, Implementing convection into Lorenz's global cycle. Part II. A new estimate of the conversion rate into kinetic energy. Tellus **52** (2000)

N. Harnik, R.K. Scott, J. Perlwitz, Wave reflection and focusing prior to the major stratospheric warming of September 2002. J. Atmos. Sci. **62**, 640–650 (2005)

P.H. Haynes, H.J. Marks, M.E. McIntyre, T.G. Shepherd, K.P. Shine, On the "downward control" of extratropical diabatic circulations by eddy-induced mean zonal forces. J. Atmos. Sci. **48**, 651–678 (1991)

I.M. Held, A.Y. Hou, Nonlinear axially symmetric circulations in a nearly inviscid atmosphere. J. Atmos. Sci. **37**, 515–533 (1980)

C.O. Hines, Internal atmospheric gravity waves at ionospheric heights. Can. J. Phys. **38**, 1441–1481 (1960)

C.O. Hines, Doppler-spread parameterization of gravity-wave momentum deposition in the middle atmosphere. Part 1: Basic formulation. J. Atmos. Sol.-Terr. Phys. **59**, 371–386 (1997)

P. Hoffmann, E. Becker, W. Singer, M. Placke, Seasonal variation of mesospheric waves at northern middle and high latitudes. J. Atmos. Sol.-Terr. Phys. **72**, 1068–1079 (2010)

J.R. Holton, The role of gravity wave induced drag and diffusion in the momentum budget of the mesosphere. J. Atmos. Sci. **39**, 791–799 (1982)

J.R. Holton, The influence of gravity wave breaking on the general circulation of the middle atmosphere. J. Atmos. Sci. **40**, 2497–2507 (1983)

C.H. Jackman, R.G. Roble, E.L. Fleming, Mesospheric dynamical changes induced by the solar proton events in October-November 2003. Geophys. Res. Lett. **34** (2007). doi:10.1029/2006GL028328

B. Karlsson, H. Körnich, J. Gumbel, Evidence for interhemispheric stratosphere-mesosphere coupling derived from noctilucent cloud properties. Geophys. Res. Lett. **34** (2007). doi:10.1029/2007GL030282

B. Karlsson, C. McLandress, T.G. Shepherd, Inter-hemispheric mesospheric coupling in a comprehensive middle atmosphere model. J. Atmos. Sol.-Terr. Phys. **71**, 518–530 (2008). doi:10.1016/j.jastp.2008.08.006

B. Karlsson, C.E. Randall, S. Benze, M. Mills, V.L. Harvey, S.M. Bailey, J.M. Russel, Intra-seasonal variability of polar mesospheric clouds due to inter-hemispheric coupling. Geophys. Res. Lett. **36** (2009). doi:10.1029/2009GL040348

P. Keckhut, A. Hauchcorne, M.-L. Chanin, Midlatitude long-term variability of the middle atmosphere: trends and cyclic and episodic changes. J. Geophys. Res. **100**, 18887–18897 (1995)

K. Körnich, E. Becker, A simple model for the interhemispheric coupling of the middle atmosphere circulation. Adv. Space Res. **45**, 661–668 (2010). doi:10.1016/j.asr.2009.11.001

K. Körnich, G. Schmitz, E. Becker, The role of stationary waves in the maintenance of the northern annular mode as deduced from model experiments. J. Atmos. Sci. **63**, 2931–2947 (2006)

K. Labitzke, The interaction between stratosphere and mesosphere in winter. J. Atmos. Sci. **29**, 1395–1399 (1972)

R.S. Lieberman, Eliassen-Palm fluxes of the 2-day wave. J. Atmos. Sci. **56**, 2846–2861 (1999)

R.S. Lieberman, Corrigendum. J. Atmos. Sci. **59**, 2625–2627 (2002)

R.S. Lieberman, D.M. Riggin, S.J. Franke, A.H. Hanson, C. Meek, T. Nakamura, T. Tsuda, R.A. Vincent, I. Reid, The 6.5-day wave in the mesosphere and lower thermosphere: evidence for baroclinic/barotropic instability. J. Geophys. Res. **108** (2003). doi:10.1029/2002JD003349

R.S. Lindzen, Turbulence and stress owing to gravity wave and tidal breakdown. J. Geophys. Res. **86**, 9707–9714 (1981)

H.L. Liu, Temperature changes due to gravity wave saturation. J. Geophys. Res. **105**, 12329–12336 (2000)

H.L. Liu, R.G. Roble, A study of a self-generated stratospheric sudden warming and its mesospheric-lower thermospheric impacts using the coupled TIME-GCM/CCM3. J. Geophys. Res. **107** (2002). doi:10.1029/2001JD001533

E.N. Lorenz, Available potential energy and the maintenance of the general circulation. Tellus **7**, 157–167 (1955)

E.N. Lorenz, The nature and theory of the general circulation of the atmosphere. World Meteorol. Organ. Monogr. **218**, 1–161 (1967)

F.-J. Lübken, Seasonal variation of turbulent energy dissipation rates at high latitudes as determined by in situ measurements of neutral density fluctuations. J. Geophys. Res. **102**, 13441–13456 (1997)

F.-J. Lübken, Nearly zero temperature trend in the polar summer mesosphere. Geophys. Res. Lett. **21**, 3603–3606 (2000)

T. Matsuno, A dynamical model of the stratospheric sudden warming. J. Atmos. Sci. **28**, 1479–1494 (1971)

C. McLandress, The seasonal variation of the propagating diurnal tide in the mesosphere and lower thermosphere. Part II: The role of tidal heating and zonal mean winds. J. Atmos. Sci. **59**, 907–922 (2002)

C. McLandress, Large-scale dynamics of the mesosphere and lower thermosphere: an analysis using the extended Canadian Middle Atmosphere Model. J. Geophys. Res. **111** (2006). doi:10.1029/2005JD006776

C. McLandress, J.F. Scinocca, The GCM response to current parameterizations of non-orographic gravity wave drag. J. Atmos. Sci. **62**, 2394–2413 (2005)

C. McLandress, T.G. Shepherd, Simulated anthropogenic changes in the Brewer-Dobson circulation, including its extension to high latitudes. J. Climate **22**, 1516–1540 (2009)

A.S. Medvedev, G.P. Klaassen, J. Geophys. Res. **100**, 25841–25853 (1995)

A.S. Medvedev, G.P. Klaassen, J. Atmos. Sol.-Terr. Phys. **62**, 1015–1033 (2000)

W.A. Norton, J. Thuburn, The two-day wave in a middle atmosphere GCM. Geophys. Res. Lett. **23**(16), 2113–2116 (1996)

A.H. Orrt, On estimates of the atmospheric energy cycle. Mon. Weather Rev. **92**, 483–493 (1964)

D. O'Sullivan, T.J. Dunkerton, Generation of inertia-gravity waves in a simulated life-cycle of baroclinic instability. J. Atmos. Sci. **52**, 3695–3716 (1995)

R.A. Plumb, Baroclinic instability of the summer mesosphere: a mechanism for the quasi-two-day wave? J. Atmos. Sci. **40**, 262–270 (1983)

P. Preusse, M. Ern, S.D. Eckermann, C.D. Warner, R.H. Picard, P. Knieling, M. Krebsbach, J.M.I. Russel, M.G. Mlynczak, C.J. Mertens, M. Riese, Tropopause to mesopause gravity waves in August: measurement and modeling. J. Atmos. Sol.-Terr. Phys. **68**, 1730–1751 (2006)

V. Ramaswamy, M.-L. Chanin, J. Angell, J. Barnett, D. Gaffen, M. Gelman, P. Keckhut, Y. Koshelkov, K. Labitzke, J.-J.R. Lin, A.O. O'Neill, J. Nash, W. Randel, R. Rood, K. Shine, M. Shiotani, R. Swinbank, Stratospheric temperature trends: observations and model simulations. Rev. Geophys. **39**, 71–122 (2001)

M. Rapp, B. Strelnikov, A. Müllemann, F.-J. Lübken, D.C. Fritts, Turbulence measurements and implications for gravity wave dissipation during the MaCWAVE/MIDAS rocket program. Geophys. Res. Lett. **31** (2004). doi:10.1029/2003GL019325

J.H. Richter, R.R. Garcia, On the forcing of the mesospheric semi-annual oscillation in the Whole Atmosphere Community Climate Model. Geophys. Res. Lett. **33** (2006). doi:10.1029/2005GL024378

J.H. Richter, F. Sassi, R.R. Garcia, K. Matthes, C.A. Fischer, Dynamics of the middle atmosphere as simulated by the Whole Atmosphere Community Climate Model, version 3 (WACCM3). J. Geophys. Res. **113** (2008). doi:10.1029/2007JD009269

R.G. Roble, B.A. Emery, R.R. Garcia, T.L. Killeen, P.B. Hays, G.C. Reid, S. Solomon, D.S. Evan, N.W. Spencer, L.H. Brace, Joule heating in the mesosphere and thermosphere during the 13 July 1982, solar proton event. J. Geophys. Res. **92**, 6083–6090 (1987)

G.J. Rohen, C. von Savigny, M. Sinnhuber, E.J. Llewellyn, J.W. Kaiser, C.H. Jackman, M.B. Kallenrode, J. Schröter, K.-U. Eichmann, H. Bovensmann, J.P. Burrows, Solar Proton Events on Mesospheric Ozone: SCIAMACHY Measurements and Model Results. J. Geophys. Res. **110** (2006). doi:10.1029/2004JA010984

H. Schmidt, G.P. Brasseur, M. Charron, E. Manzini, M.A. Giorgetta, T. Diehl, V.I. Fomichev, D. Kinnison, D. Marsh, S. Walters, The HAMMONIA chemistry climate model: sensitivity of the mesopause region to the 11 year solar cycle and CO_2 doubling. J. Climate **19**, 3903–3931 (2006)

H. Schmidt, G.P. Brasseur, M.A. Giorgetta, The solar cycle signal in a general circulation and chemistry model with internally generated QBO. J. Geophys. Res. **115** (2010). doi:10.1029/2009JD012542

K. Seminiuk, T.G. Shepherd, The middle-atmosphere Hadley circulation and equatorial inertial adjustment. J. Atmos. Sci. **58**, 3077–3096 (2001)

T.A. Shaw, T.G. Shepherd, Theoretical framework for energy and momentum consistency in subgrid-scale parameterization for climate models. J. Atmos. Sci. **66**, 3095–3114 (2009)

T.A. Shaw, M. Sigmond, T.G. Shepherd, J.F. Scinocca, Sensitivity of simulated climate to conservation of momentum in gravity wave drag parameterization. J. Climate **22**, 2726–2742 (2009)

T.G. Shepherd, Transport in the middle atmosphere. J. Meteorol. Soc. Jpn. **85** (2007)

T.G. Shepherd, Dynamics, stratospheric ozone, and climate change. Atmos.-Ocean **46**, 117–138 (2008)

T.G. Shepherd, J.N. Koshyk, On the nature of large-scale mixing in the stratosphere and mesosphere. J. Geophys. Res. **105**, 12433–12446 (2000)

T.G. Shepherd, K. Seminiuk, J.N. Koshyk, Sponge layer feedbacks in middle-atmosphere models. J. Geophys. Res. **101**, 23447–23464 (1996)

K.P. Shine, The middle atmosphere in the absence of dynamical heat fluxes. Q. J. R. Meteorol. Soc. **113**, 603–633 (1987)

W. Singer, R. Latteck, P. Hoffmann, B.P. Williams, D.C. Fritts, Y. Murayama, K. Sakanoi, Tides near the Arctic summer mesopause during the MaCWAVE/MIDAS summer program. Geophys. Res. Lett. **32** (2005). doi:10.1029/2004GL021607

D.E. Siskind, L. Coy, P. Espy, Observations of stratospheric warmings and mesospheric coolings by the TIMED SABER instrument. Geophys. Res. Lett. **32** (2005). doi:10.1029/2005GL022399

A.K. Smith, The origin of stationary planetary waves in the upper mesosphere. J. Atmos. Sci. **60**, 3033–3041 (2003)

A.K. Smith, Interactions between the lower, middle and upper atmosphere. Space Sci. Rev. (2011). doi: 10.1007/s11214-011-9791-y

A.K. Smith, R.R. Garcia, D.R. Marsh, D.E. Kinnison, J.H. Richter, Simulations of the response of mesospheric circulation and temperature to the Antarctic ozone hole. Geophys. Res. Lett. **37** (2010)

A.K. Smith, R.R. Garcia, D.R. Marsh, J.H. Richter, WACCM simulations of the mean circulation and trace species transport in the winter mesosphere. J. Geophys. Res. (2011)

S.L. Vadas, D.C. Fritts, Thermospheric responses to gravity waves: influences of increasing viscosity and thermal diffusivity. J. Geophys. Res. **110** (2005). doi:10.1029/2004JD005574

E.M. Volodin, G. Schmitz, A troposphere-stratosphere-mesosphere general circulation model with parameterization of gravity waves: climatology and sensitivity studies. Tellus A **53**, 300–316 (2001)

C. von Savigny, M. Sinnhuber, H. Bovensmann, J.P. Burrows, M.B. Kallenrode, M. Schwartz, On the disappearance of noctilucent clouds during the January 2005 solar proton events. Geophys. Res. Lett. **134** (2007). doi:10.1029/2006GL028106

S. Watanabe, Y. Kawatani, Y. Tomikawa, K. Miyazaki, M. Takahashi, K. Sato, General aspects of a T213L256 middle atmosphere general circulation model. J. Geophys. Res. **113** (2008). doi:10.1029/2008JD010026

E. Yiğit, A.S. Medvedev, Heating and cooling of the thermosphere by internal gravity waves. Geophys. Res. Lett. **36** (2009). doi:10.1029/2009GL038507

E. Yiğit, A.D. Aylward, A.S. Medvedev, Parameterization of the effects of vertically propagating gravity waves for thermosphere general circulation models: sensitivity study. J. Geophys. Res. **113** (2008). doi:10.1029/2008JD010135

E. Yiğit, A.S. Medvedev, A.D. Aylward, P. Hartogh, M.J. Harris, Modeling the effects of gravity wave momentum deposition on the general circulation above the turbopause. J. Geophys. Res. **114** (2009). doi:10.1029/2008JD011132

E. Yulaeva, J.R. Holton, J.M. Wallace, On the cause of the annual cycle in tropical lower-stratospheric temperatures. J. Atmos. Sci. **51**, 169–174 (1994)

Space Sci Rev (2012) 168:315–332
DOI 10.1007/s11214-011-9861-1

Stratosphere NO_y Species Measured by MIPAS and GOMOS Onboard ENVISAT During 2002–2010: Influence of Plasma Processes onto the Observed Distribution and Variability

Enrico Arnone · Alain Hauchecorne

Received: 31 July 2011 / Accepted: 21 November 2011 / Published online: 9 December 2011

Abstract We present observations of stratosphere NO_y species from 2002 to 2010 taken by the Michelson Interferometer for Passive Atmospheric Sounding (MIPAS) and by the Global Ozone Monitoring by Occultation of Stars (GOMOS) instruments on board ENVISAT. We jointly used observations of MIPAS NO_2, HNO_3, N_2O_5, $ClONO_2$ and N_2O, and GOMOS NO_2 and NO_3. MIPAS results are part of the MIPAS2D database retrieved adopting a full 2D tomographic approach. We describe the mean distribution and variability of NO_y species in the stratosphere, identifying changes induced by plasma processes. Beside enhancements due to sporadic solar proton events, we show that winter polar NO_2 has an almost linear relationship with the geomagnetic activity index Ap down to about 10 hPa. This indicates a dominant role of energetic precipitating particles in the production of upper atmosphere NO_y. The correlation has clear signatures extending to mid latitudes. Partitioning of the NO_y reservoir species are also traced, with HNO_3 and N_2O_5 showing a correlation with Ap extending to lower altitude within the polar regions. We found no large signatures of an impact of thunderstorm-induced plasma processes onto monthly means of NO_y species in the stratosphere.

Keywords Stratosphere · NO_y · NO_x · Energetic particle precipitation · Plasma processes

1 Introduction

The largest fraction of the Earth's atmosphere is composed by inactive molecular nitrogen. Active nitrogen is present in the form of odd nitrogen NO_x ($N + NO + NO_2$), which is a key family involved in the regulation of stratospheric ozone through catalytic cycles (Crutzen 1971). NO_x in the stratosphere is due primarily to oxidation of N_2O which is injected from

E. Arnone (✉)
Istituto di Scienze dell'Atmosfera e del Clima, CNR, Bologna, Italy
e-mail: e.arnone@isac.cnr.it

A. Hauchecorne
LATMOS/IPSL, UVSQ, CNRS-INSU, Guyancourt, France

the tropical troposphere and transported at higher altitude and latitude by the stratospheric meridional circulation. An overview of the atmosphere and the interaction of its different layers was given by Smith (2011). Stratospheric nitrogen can also be under the form of less reactive nitrogen compounds and reservoirs, such as NO_3, N_2O_5, HNO_3, HO_2NO_2, or as halogenated nitrates $ClONO_2$ and $BrONO_2$. Together with NO_x, these nitrogen containing species form the NO_y family (see e.g. Brasseur and Solomon 2005). Throughout the stratosphere, the lifetime of NO_x is of the order of 10^6 s while that of NO_y is orders of magnitude larger. Within the NO_x family, a marked diurnal cycle exists with NO being converted into NO_2 at nighttime on timescales of minutes, which is then largely converted back to NO during daytime (Crutzen 1971). Together with N_2O_5 which is produced from NO_2 during nighttime, NO_x should be therefore considered to achieve a cycling equilibrium over a 24 hours period (e.g. Solomon and Garcia 1983). In the lower stratosphere, NO_x and HNO_3 lifetimes become comparable to transport timescales so that their distribution is due to the competing action of transport and photochemistry. This is also the case of the winter polar regions. NO_x is generally the main NO_y component above about 30 km altitude, HNO_3 below this altitude. In the upper stratosphere and above, NO_y is destroyed by the $N + NO$ reaction, leading to re-formation of molecular nitrogen and to the mesospheric minimum of NO_y in its vertical profile. When transported to the upper troposphere, HNO_3 is readily incorporated in rain droplets leading to scavenging and a net loss of nitrogen oxides. A second pathway for loss is through sedimentation of HNO_3 through polar stratospheric clouds in the polar vortex, a process referred to as denitrification and fundamental in the occurrence of major ozone depletion in winter (Fahey et al. 1990). For further details see e.g. Brasseur and Solomon (2005).

Nitrogen compounds are also produced by solar shortwave radiation and energetic particle precipitation (EPP) in the form of auroral zonal electrons, solar proton events (SPEs) and galactic cosmic rays (GCRs) (see e.g. Semeniuk et al. 2011). These processes lead to ionization, dissociation and excitation of atmospheric constituents in the stratosphere and above. In particular, charged particles can enter the atmosphere at high geomagnetic latitude, where the Earth's geomagnetic field lines are open. The altitude of penetration and cut-off latitude depend on the energy of the precipitating particle: 10 keV–1 MeV electrons penetrate to 110–60 km altitude and down to 86–84 degrees latitude, whereas 1–100 MeV protons reach 90–35 km altitude and 75–65 degree latitude (Hargreaves 1979). Energetic protons (up to tens of MeV) precipitating in the polar cap during SPEs cause dissociation of molecular nitrogen via charged particle impact (Crutzen et al. 1975; Rusch et al. 1981). Large SPEs lead to sporadic hundreds of percent increases of odd nitrogen and tens of percent decreases in ozone, as modeled by several authors (Rusch et al. 1981; Reid et al. 1991; Jackman et al. 2000; Verronen et al. 2002) and observed in great detail from space. This was the case of the July 2000 SPE showing a simultaneous NO_x increase and O_3 depletion in the upper stratosphere and mesosphere in the polar regions (Jackman et al. 2001) and more recently confirmed by observations of the Halloween October 2003 SPE (e.g. Seppälä et al. 2004; López-Puertas et al. 2005; Orsolini et al. 2005). Precipitating protons produce also large amounts of odd hydrogen HO_x ($H + OH + OH_2$), although its short lifetime generally leads only to short term perturbations. See Jackman et al. (2008) for an overview of the atmospheric effects of large SPEs. NO is produced by auroral activity generally above the stratopause, due to direct particle impact and heating due to collisions of ions and neutral species (Barth et al. 1999; Seppälä et al. 2007; Sinnhuber et al. 2011). Electrons precipitating from the Earth's radiation belt are also observed to cause ionization at high latitude, causing sporadic changes in NO_x typically at the stratopause and above (Callis et al. 1991). High energy particles produce also X-ray emissions through bremsstrahlung and large amounts of secondary electrons, both able to penetrate deeper into the stratosphere. These effects become important to the stratosphere when enhanced NO_x is transported downward through diabatic descent in the polar night where its lifetime is long (Siskind et al. 1997; Callis and Lambeth 1998). The downward transported NO_x can reach up to 10% of the yearly N_2O source (Funke et al. 2005). Galactic cosmic rays penetrate as low as to the troposphere, with maximum ionization rates at about 15 km altitude (Usoskin and Kovaltsov 2006). NO_x production by cosmic rays may reach as much as 20% at the polar tropopause regions (Calisto et al. 2011). These processes are greatly modulated by solar variability, primarily on the 11 year cycle timescale. As geomagnetic activity and particle precipitation peak on the declining phase of the 11 year solar cycle, there is typically a time lag of 2 years between the maximum of respective indices such as Ap (geomagnetic activity) and F10.7 (solar radio flux) (Verbanac et al. 2011). GCRs, on the contrary, are shielded by the solar wind and anticorrelate with the 11 year solar cycle. Peak ionization rates for these processes range around 10^1–10^3 ions $cm^{-3}\,s^{-1}$ at 77 km altitude for auroral zone electrons, 10^{-1}–10^4 ions $cm^{-3}\,s^{-1}$ at 50 km altitude for sporadic SPEs, 25–45 ions $cm^{-3}\,s^{-1}$ at 12 km altitude for GCRs (Semeniuk et al. 2011, and references therein). As the significant impact of these processes on the atmospheric content of nitrogen oxides at high latitude is becoming evident, recent studies have included these sources in global circulation models (e.g. Marsh et al. 2007, Semeniuk et al. 2011; Funke et al. 2011). Solar variability indeed acts also through a series of coupling mechanisms which indirectly impact NO_x, e.g. through changes in temperature, ozone, atmospheric circulation etc. An extensive literature exists on the response of the atmosphere to solar variability, as recently reviewed by Gray et al. (2010).

Among plasma processes, thunderstorms also have an influence on NO_x through lightning. In the upper troposphere, lightning is a dominant source of nitrogen oxides (see the review by Schumann and Huntrieser 2007), which may consequently be transported into the lower stratosphere by the slow meridional circulation or directly injected by thunderstorm intrusion (see e.g. Ko et al. 1986; Kotamarthi et al. 1994; Baehr et al. 2003). Despite the large number of lightning-NO_x estimates (Schumann and Huntrieser 2007), observation of lightning NO_x, and in particular of its transport into the stratosphere, has proven to be a challenge. Further limitations arise from the reconversion of lightning NO_x into HNO_3 with consequent loss through scavenging, from the high dynamical variability of the upper troposphere-lower stratosphere region, and from the large variability among thunderstorm events (see e.g. Beirle et al. 2010). Thunderstorms also affect directly the stratosphere through upper atmospheric lightning processes collectively called transient luminous events (TLEs, see e.g. Neubert et al. 2008; Pasko et al. 2011). Recent ion-neutral chemistry model studies focused on sprites, huge discharges occurring in the upper stratosphere and mesosphere. Modeling of the chemistry occurring within sprite streamers showed local enhancements of NO_x up to orders of magnitude induced by electron impact on molecular nitrogen (Enell et al. 2008; Sentman et al. 2008; Gordillo-Vázquez 2008; Hiraki et al. 2008). Enell et al. (2008) estimated the sprite-NO_x source could be irrelevant at global scale but significant at regional scale. Observations seem to confirm a lack of signature at global scales and possible impact above individual thunderstorms (Arnone et al. 2008, 2009; Rodger et al. 2008). The effects of transport that were not considered by these studies in detail will need to be addressed using global circulation models to correctly estimate their source. Detecting and quantifying thunderstorm driven perturbations onto stratospheric NO_x and ozone remain an open scientific challenge.

In this paper, we present observations of stratospheric NO_y species taken in 2002–2010 by the MIPAS and GOMOS instruments flying onboard ESA ENVISAT satellite. Section 2 recalls the main characteristics of two instruments and adopted data. We used these observations to describe the mean distribution and variability of these NO_y species throughout the stratosphere, and identify large changes that can be related to plasma processes. The influence of plasma processes at high latitude is investigated in Sect. 4, and its effects at mid to low latitude in Sect. 5. Conclusions are given in Sect. 6.

2 Observations

2.1 MIPAS

The Michelson Interferometer for Passive Atmospheric Sounding (MIPAS) is part of Europe's environmental monitoring satellite, ENVISAT, launched on 1 March 2002 (Fischer et al. 2008). ENVISAT has a polar orbit with altitude of 800 km and orbital period of about 101 minutes. The equator is crossed at about 10:00 local time (LT) during the descending node and at 22:00 LT during the ascending node. We refer to these measurements respectively as am and pm. Global coverage is assured by 14.3 orbits per day. MIPAS is an actively cooled atmospheric thermal emission limb sounder, working in the mid-infrared recording high resolution spectra (0.035 cm^{-1} full width half maximum, unapodized) in the spectral range 685–2410 cm^{-1} (14.6–4.15 μm). The across track field-of-view at tangent point is approximately 3 km (vertical) × 30 km (horizontal). The complete limb scan sequence of the original Full Resolution (FR) mission consists of 17 spectra with nominal tangent points at 68, 60, 52, 47 and 42 km, and then down to 6 km at 3 km steps. The measurement time for a complete elevation scan is 70–80 seconds which corresponds to a horizontal separation of 500–600 km between adjacent scans. The FR observation modes cover the time period from July 2002 to March 2004 when MIPAS operations in the original configuration were suspended due to the deterioration of the interferometric slides. Starting from January 2005 all MIPAS observation modes have been redefined for a new configuration in which the instrument is operated at 41% of its maximum spectral resolution, with consequent increase in the spatial resolution of the measurements. The new configuration was called optimized resolution (OR). See further details in Dinelli et al. (2010). We retrieved MIPAS observations with the Geo-fit Multi-Target Retrieval (GMTR) 2D tomographic analysis tool (Carlotti et al. 2006). GMTR is based on the Geo-fit approach (Carlotti et al. 2001) upgraded with the Multi-Target Retrieval (MTR) functionality (Dinelli et al. 2004). The Geo-fit approach (2D retrieval of the whole orbit) enables to take into account the horizontal atmospheric inhomogeneities that are modeled using a 2D discretization of the atmosphere. The MTR approach, by simultaneously fitting pressure (p), temperature (T) and the interfering species, eliminates the systematic error components due to the propagation of the uncertainties on p, T and on the amount of molecules that generate interfering spectral features. A horizontal retrieval grid fixed at steps of five degrees of latitude was adopted. The database of results obtained with the GMTR 2D approach was called MIPAS2D and is thoroughly described in Dinelli et al. (2010). Results presented here were obtained with the recent reprocessing IPF5 of ESA Level 1b data. An upgraded version of the GMTR tool has been used to determine atmospheric fields of p, T, and volume mixing ratio (vmr) of target species relevant for a description of NO_y. Pressure, temperature, H_2O and O_3 profiles have been retrieved simultaneously performing a MTR analysis; the other targets were then retrieved in cascade, exploiting the atmospheric fields previously determined. The GMTR Level-2 data results have been computed in the altitude range from 6 to 68 km over a fixed vertical grid that coincides with the MIPAS nominal mode altitudes with no regularization and using optimal estimation (in order to handle observations affected by clouds). The analyzed dataset and further details can be found at http://www.mbf.fci.unibo.it/mipas2d.html, http://www.isac.cnr.it/~rss/mipas2d.htm. The webpage includes also details of the retrieval of the additional N_2O_5 and $ClONO_2$ molecules that were not previously published. ESA's MIPAS operational NO_2 data was validated by Wetzel et al. (2007) finding a general good agreement within 10–20% with NO_2 from other instruments. A noticeable discrepancy arose in MIPAS NO_2 retrieved in the stratosphere when large amounts of mesospheric NO_2 are encountered by MIPAS lines of sight.

2.2 GOMOS

The Global Ozone Monitoring by Occultation of Stars (GOMOS), also on board ESA's ENVISAT, is a stellar occultation spectrometer aimed at building a global climatology of ozone and related species in the middle atmosphere (15 to 100 km) with a very high accuracy using the technique of stellar occultation (Bertaux et al. 2004). Bertaux et al. (2010) give an overview of the scientific results. Each constituent can be identified by its absorption spectrum. The atmospheric transmission spectrum is equal to the ratio between the stellar spectrum absorbed and scattered by the atmosphere and the reference stellar spectrum measured outside the atmosphere. As the reference spectrum is measured at the beginning of each occultation, we can consider that GOMOS is a self-calibrated instrument, independent of any radiometric calibration. Furthermore the stellar occultation technique gives an excellent knowledge of the tangent altitude, depending only on the geometry of the light path between the star and the satellite. The 250–680 nm spectral domain is used for the determination of O_3, NO_2, NO_3 and aerosols from the upper troposphere to the mesosphere. In addition, two high spectral resolution channels centered at 760 and 940 nm allow the measurement of O_2 and H_2O and two fast photometers (1 kHz sampling rate) are used to correct stellar scintillation perturbations and to determine high vertical resolution temperature profiles. Global latitude coverage is obtained with up to 40 stellar occultations per orbit from South Pole to North Pole. Results presented here are obtained using version IPF 5.0 of the algorithms in the operational processor. This version has been implemented in April 2004 for the full reprocessing of 2003 data. The main characteristics of the data inversion are summarized below. More details are given in Kyrölä et al. (2010). The inversion of spectrometer data from Level 1b (calibrated transmission spectra) to Level 2 (vertical profiles of constituents) is made in two steps, the spectral inversion and the vertical inversion. The spectral inversion allows determining the number of molecules along the horizontally integrated line of sight densities (slant densities) of each absorber from the transmission spectrum. It is made globally on all retrieved species using a Levenberg-Marquardt method except for NO_2 and NO_3 for which a DOAS (differential optical absorption spectroscopy) method is applied. The temperature dependence of O_3, NO_2 and NO_3 cross-sections is taken into account using ECMWF temperature along the line of sight. The vertical inversion is made using a Tikhonov regularization with an imposed vertical resolution depending on the species. It is fixed to a constant value equal to 4 km for NO_2 and NO_3 and it increases from 2 km below 20 km to 4 km above 30 km for O_3. The data quality depends on illumination conditions and on stellar brightness and temperature. Data acquired on dark limb (nighttime) are of better quality than for the bright limb (daytime) because of a smaller perturbation by background light. Hot stars have a strong UV emission and allow ozone profiles to be obtained up to the lower thermosphere using UV absorption bands. Cool-star ozone profiles are limited to the stratosphere using the visible Chappuis band. NO_2 profiles extend from 20 to 50 km in typical non-polar conditions and up to 70 km during polar winter when a strong NO_2 enhancement occurs. The typical accuracy is 5% for O_3 (Kyrölä et al. 2006) and 15–20% for

NO_2 (Hauchecorne et al. 2005). A comparison of NO_2 from MIPAS and GOMOS was performed by Verronen et al. (2009) who found a good agreement of the measurements from the two instruments in the stratosphere. The difference in NO_2 from the two instruments deteriorated above 0.5 hPa at low-to-mid latitude, but was smaller than 35% in the polar mesosphere. In the polar winter stratosphere the difference exceeded the systematic error due to MIPAS reduced sensitivity under the presence of a large mesospheric NO_2 load. For this study, MIPAS monthly mean pressure measurements were adopted to convert GOMOS data from their altitude grid.

3 NO_y Species in the Stratosphere

We present MIPAS and GOMOS observations of NO_y species in terms of monthly averages over a common pressure vertical grid and within 15 degree latitude bands. Because of the diurnal cycle of NO_y species, we averaged separately MIPAS am and pm observations, respectively from the descending and ascending parts of ENVISAT orbits. The data are here considered in terms of MIPAS2D orbital coordinate (OC, see Dinelli et al. 2010) which spans from 0 at the North Pole (NP) to 180 at the South Pole (SP) along the descending satellite track, and up to 360 back to the NP along the ascending satellite track. The Equator is crossed respectively at OC = 90 at about 10:00 LT and at OC = 270 at about 22:00 LT. For GOMOS, measurements are taken only during nighttime so that observations are largely from the ascending part of the orbit. Gaps in the monthly mean data are mostly due to sunlit regions. To ease comparison between the two dataset also GOMOS data are reported in terms of OC.

Figure 1 reports maps of MIPAS monthly mean vertical profiles of NO_y species (NO_2, HNO_3, N_2O_5 and $ClONO_2$) in January 2008 along the satellite orbit. Also shown are N_2O and ozone as a reference of the overall behavior of the stratosphere. Figure 2 reports GOMOS NO_2 and NO_3 in January 2008 in the night side only. The figures show that NO_y in the stratosphere is mostly in the form of reactive NO_2 (above about 10 hPa) and HNO_3 (below 10 hPa). The two species have peak values respectively above 15 and 10 ppbv, whereas N_2O_5, $ClONO_2$ and NO_3 reach respectively at most 2, 1.5 and 0.3 ppbv at 10, 10 and 3 hPa. In the night side (OC from 180 to 360), NO_2 is a good proxy for NO_x so that the reported species correctly describe the overall NO_y distribution. In the day side, NO_x is mostly in the form of NO and therefore the main NO_y component is missing. The day side should therefore be considered to understand the magnitude of the diurnal cycle of the NO_x family and not as overall NO_y.

As discussed in Sect. 1, the main source of NO_y in the stratosphere is oxidation of N_2O. Its maximum production rate at 10 hPa (30 km altitude) explains the peak values of NO_2 in the middle stratosphere between about 10 and 1 hPa in the vertical direction, and within middle to low latitudes, horizontally matching the distribution of N_2O. Below 10 hPa, conversion of NO_x into HNO_3 occurs which then shows relevant values down to about 50 hPa (20 km altitude) where significant N_2O oxidation halts because of low amounts of $O(^1D)$ (Crutzen and Schmailzl 1983). The diurnal NO_x cycle reduces NO_2 from its peak values exceeding 15 ppbv at nighttime to less than 5 ppbv during daytime. On the contrary, during nighttime NO density drops by several orders of magnitude (Brasseur and Solomon 2005). High latitude regions with typically low NO_2 vmr values switch from daytime to nighttime conditions according to the season, and the extension of the two regions is tracked by the strong horizontal gradients occurring at the day-night terminator in the two hemispheres. The strong horizontal gradients are evident in the summer hemisphere (at about OC = 200

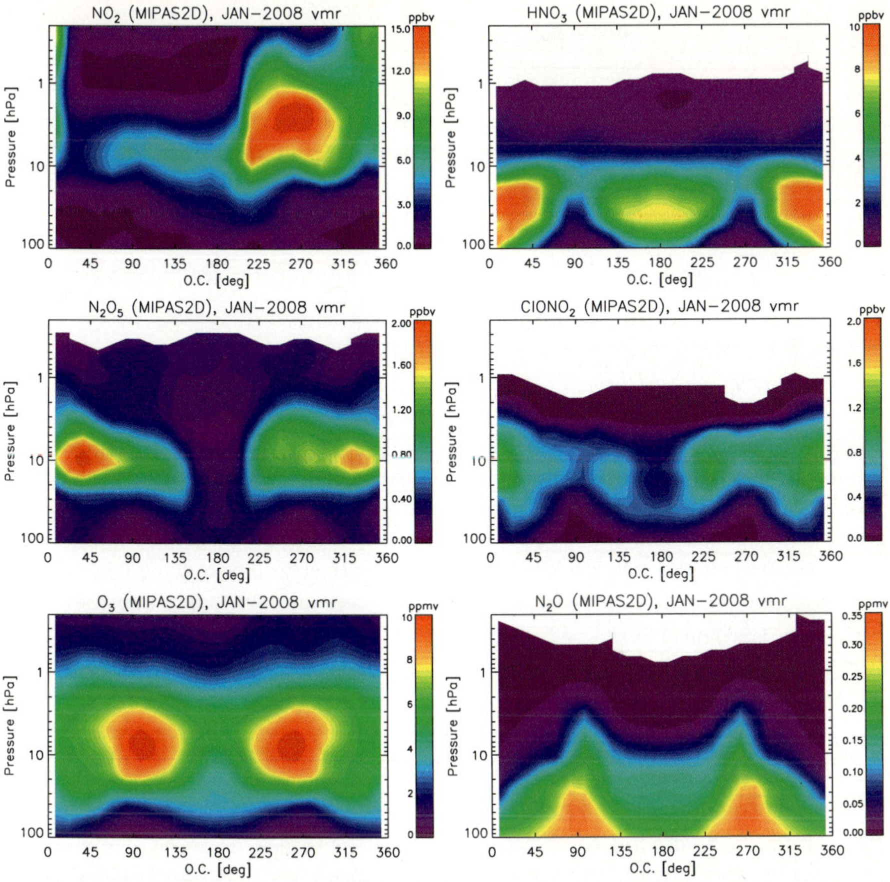

Fig. 1 MIPAS2D monthly mean NO_2 and HNO_3 (*top panels*), N_2O_5 and $ClONO_2$ (*middle panels*), O_3 and N_2O (*bottom panels*) for January 2008 in terms of pressure levels vs. orbital coordinate (OC). The OC spans from 0 at the NP to 180 at the SP along the day side descending part of the orbit, and back to 360 at the NP along the night side ascending part of the orbit

in the figure) where prompt conversion of stratospheric NO_2 into NO occurs. Towards high latitudes, NO_y tends to be more in the form of HNO_3, N_2O_5 and $ClONO_2$ as these less reactive species are favored on the long timescales needed by meridional transport. The conversion of NO_x into less reactive species is particularly enhanced in the darker winter pole (NP at OC = 0 and 360 in January) and much less pronounced in the summer pole (SP at OC = 180 in January). Below 50 hPa at low-to-mid latitudes, NO_y vmr are very low, pointing to a secondary order contribution from lightning to stratospheric NO_x in the above picture.

Beside the core abundances in the tropical stratosphere, NO_2 is strongly enhanced by descending mesospheric NO_x into the winter polar region together with sporadic polar night build up of NO_2 due to insitu production by precipitating particles. Such an effect in January 2008 is visible towards OC = 360 above 1 hPa. The evolution in time of these effects is discussed in the following section.

The distribution of NO_2 in January 2008 seen by GOMOS (Fig. 2) shows a consistent maximum at low latitudes (OC between 225 and 315) around 2–3 hPa with values up to

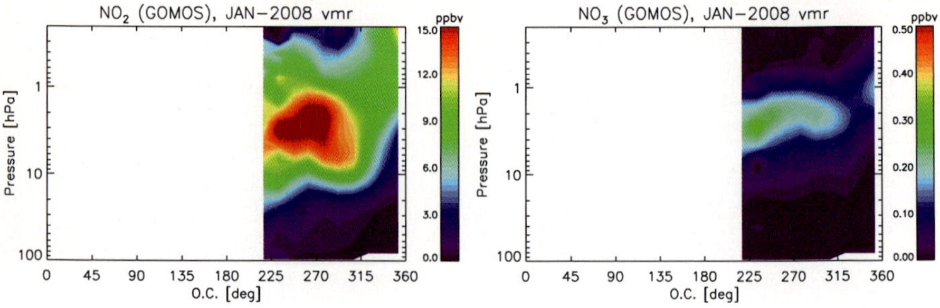

Fig. 2 As in Fig. 1 but for GOMOS monthly mean NO_2 (*left*) and NO_3 (*right*). Note that GOMOS data were averaged over nighttime observations only and are reported in terms of OC to ease comparison with MIPAS data

15 ppbv. At high Northern latitudes, the maximum in the lower mesosphere due to in situ production and descent of NO_x in the winter polar vortex is also consistent with MIPAS results. NO_3 exhibits a maximum around 2–3 hPa at low and mid-latitudes with mixing ratio 0.3 ppbv, which makes it a rather low abundant NO_y species. A maximum is observed at high Northern latitudes around 1 hPa. As explained by Hauchecorne et al. (2005), nighttime NO_3 is in rapid equilibrium with ozone and is highly correlated with temperature. The polar maximum at 1 hPa is probably due to the increase of temperature in this region in relation with planetary wave activity.

4 High Latitude Plasma Influence

Figure 3 reports timeseries of monthly mean vertical profiles at the North Pole (NP, averaged over 75 to 90 N) of NO_y species and N_2O. Top panels report NO_2 from MIPAS (left) and GOMOS (right). The figure also reports timeseries of HNO_3, N_2O_5, $ClONO_2$, and N_2O. NO_3 is not reported because of the extremely low values in the polar region (see Fig. 2). The limited GOMOS NO_2 data highlight the short period of polar night available at GOMOS measurement time. Besides some differences, MIPAS and GOMOS data are consistently showing recurrent downward propagation of NO_2 in the winter mesosphere with a strong interannual variability in the amplitude of the NO_2 peaks at the stratopause. Large enhancements of these downward propagating features are seen in coincidence with known SPEs such as in October 2003 and January 2005 (see Sect. 1). A trend is found in the maximum NO_2 values reached in the lower mesosphere and upper stratosphere, which tend to weaken up to 2009 and then increase again. The corresponding timeseries at the stratopause are reported in Fig. 4 for both NP and South Pole (SP, averaged over 75 to 90 S). The figure shows monthly means of NO_2 at 1 hPa for MIPAS2D (red) and GOMOS (blue). MIPAS data are reported with their standard deviation about the average (red error bars). Both MIPAS2D and GOMOS confirm the trend is not instrumental dependent, and is consistent with the downward phase of the 11-years solar cycle from 2004 to 2009 and increasing phase from 2010. Since the enhanced NO_2 displaying the trend occurs consistently during both polar winters, we build a new timeseries extracting only nighttime winter polar NO_2 from both hemispheres: December, January and February (DJF) for NP, and June, July and August (JJA) for SP. Due to consistency of the two datasets we adopt MIPAS2D NO_2. The winter polar NO_2 timeseries at various pressure levels is shown in Fig. 5 together with the Ap and F10.7 indices (respectively black, red and blue lines). All

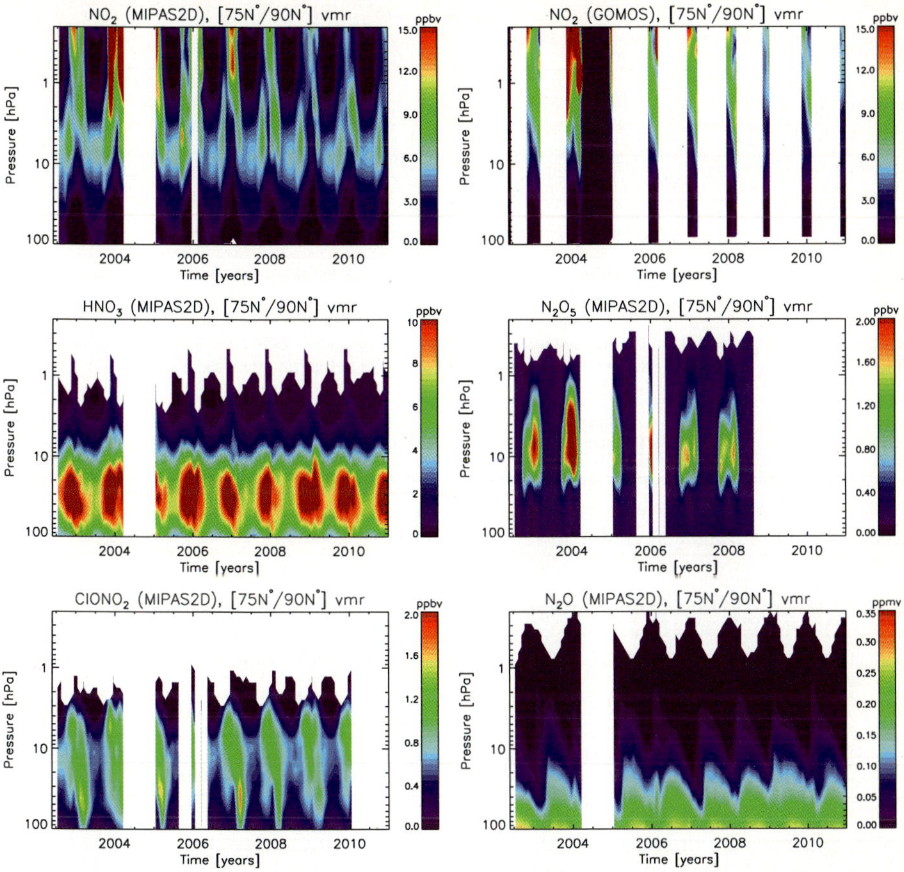

Fig. 3 Timeseries of monthly mean vertical profiles of NO_y species at the North Pole (75N–90N) during 2002–2010. *Top panels*: NO_2 from MIPAS2D (*left*) and GOMOS (*right*); *Middle panels*: MIPAS2D HNO_3 (*left*) and N_2O_5 (*right*); *Bottom panels*: MIPAS2D $ClONO_2$ (*left*) and N_2O (*right*). White *vertical lines* indicate missing data

indices were normalized and smoothed with a running average over a 6 month time window to remove high frequency variability. Since we extract only DJF and JJA, the running average corresponds to a solar year. The low NO_2 values observed in 2002–2003 clearly point to a correlation of NO_2 with the geomagnetic Ap index rather than with the solar irradiance F10.7 index. The NO_2–Ap correlation is shown by the scatter plots in the right column of Fig. 5. Pearson linear correlation coefficients for the 6 pressure levels 0.24, 1, 5, 9, 26 and 50 hPa (about 55 to 19 km altitude) are 0.93, 0.94, 0.75, 0.33, 0.28 and 0.92. The corresponding correlation coefficients with the F10.7 index (or sunspot number, not shown) range between 0.15 and 0.62. Despite the simple correlation analysis we applied to the data, the correlation in the upper stratosphere is remarkable and consistent with what reported for the previous solar cycles for the Southern Hemisphere (SH) (Siskind 2000; Randall et al. 2007). Although sporadic large SPEs have greatly contributed to the averages in 2003–2005, the average NO_2 in the upper stratosphere has a linear relationship with geomagnetic activity, pointing to a dominant role of the modulation of energetic electrons in the upper layers. In the middle stratosphere, between roughly 40 and 5 hPa (about 22 to 33 km

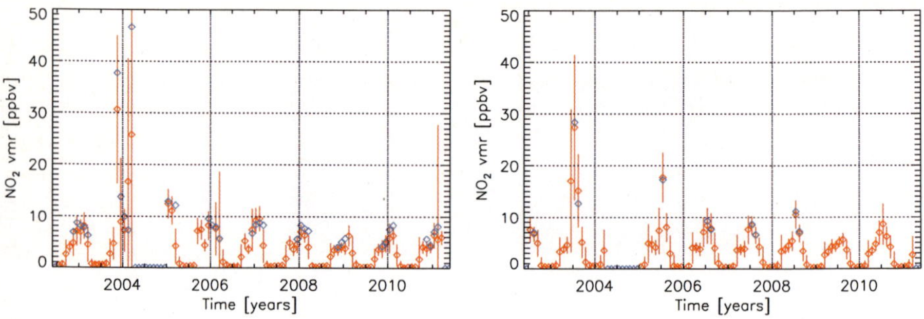

Fig. 4 Timeseries of NO_2 at the stratopause North Pole (*left*) and South Pole (*right*) for MIPAS2D (*red*) and GOMOS (*blue*). MIPAS2D data are reported with their standard deviation about the average

altitude), the correlation is lost. In situ production by SPEs cannot reach this altitude, and downward transport of upper atmospheric NO_x is too slow to reach this altitude range during winter (see e.g. GOMOS data in Fig. 3 and see e.g. Funke et al. 2005). Surprisingly, the correlation is strong again in the lower stratosphere at 50–80 hPa (19–16 km), accompanied by a negative correlation of N_2O with the Ap index (with a correlation coefficient of -0.80 at 50 hPa). The latter is due to changes in dynamics as confirmed by analogous behavior of the CH_4 passive tracer (not shown), likely induced by indirect effects of the solar cycle onto the troposphere (Gray et al. 2010). The correlation with Ap of the winter polar NO_2 at 60 to 75 N and S latitude bands is very consistent with the 75 to 90 latitude band, a part from the 50 hPa pressure level where the correlation is lost. N_2O is shown to trace downward transport at the pole (see bottom right panel in Fig. 3) and the influence of dynamics on the observed timeseries. In particular, the data show a quasi-biennial oscillation (QBO) modulation in the middle and lower stratosphere, especially during low solar and geomagnetic activity. Above the stratopause (at 0.24 hPa), an N_2O positive correlation with the Ap index is not reproduced by CH_4, pointing to additional production of N_2O as discussed by Funke et al. (2008). Despite changes occurring in all NO_y species, other tracers do not show clear trends. The most evident changes are related to the NO_2 sudden increase with the late 2003 SPEs, accompanied by clear N_2O_5 enhancement and $ClONO_2$ decrease (see Fig. 3).

5 Effects at Low and Mid Latitude

Figure 6 shows timeseries of monthly mean profiles within the 45N–60N latitude band. The figure is consistent with Fig. 3, apart for NO_3 which has higher values at mid latitude and is now shown instead of N_2O (see bottom right panel). Some downward propagating features in the timeseries of Fig. 6 resemble those seen in Fig. 3, especially if comparing the more disturbed years around 2004–2006 with 2009. In particular, the October 2003 Halloween SPE perturbation is evident in NO_2 from both instruments, and in N_2O_5 at 10 hPa. The influence at mid-latitude is more easily traced by timeseries at constant pressure level reported in Fig. 7. The figure is now correlating the temporal evolution of NO_2 at 1 and 4 hPa (i.e., close to its maximum vmr) to that of N_2O, in terms of orbital coordinate (as discussed in Fig. 1) versus time. To ease the comparison between the two species, N_2O contour lines (white) were plotted over the NO_2 timeseries. Most of the NO_2 variability at mid and low latitude is superimposed to the natural N_2O source, which gets seasonally modulated by the meridional circulation. High latitude plasma effects on NO_2 are clearly visible at the South

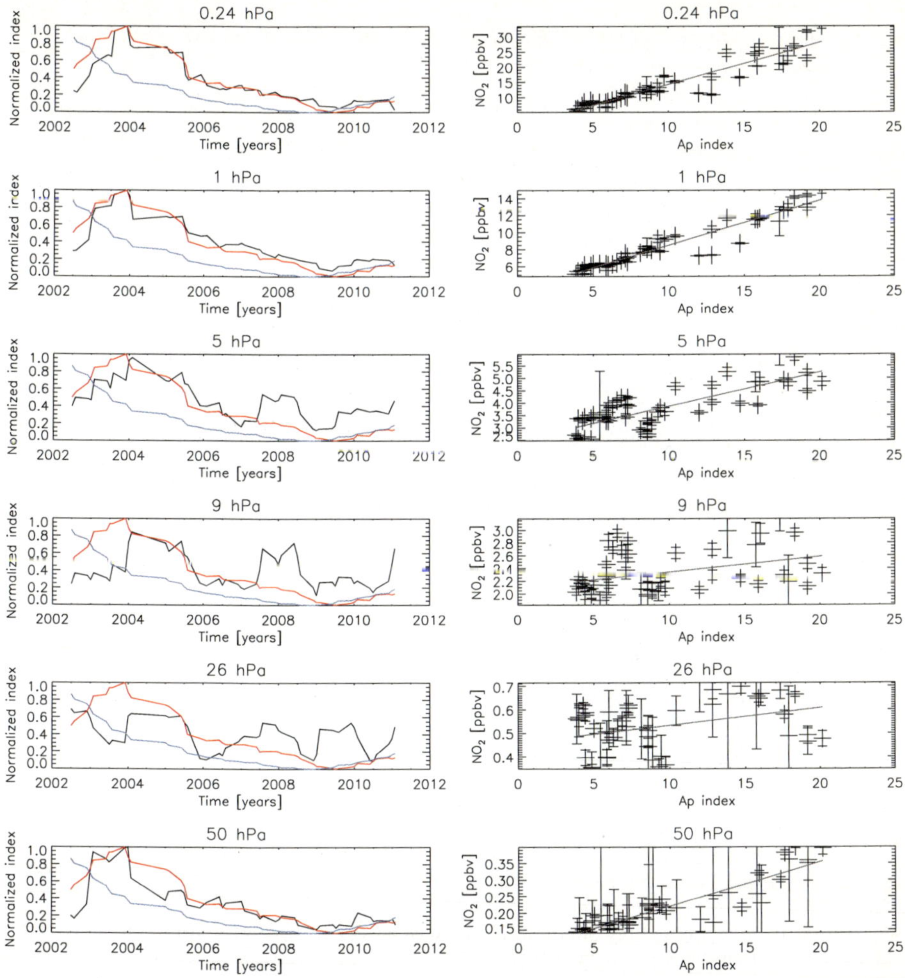

Fig. 5 *Left*: Timeseries of polar winter NO_2 (*black*), Ap (*red*) and F10.7 (*blue*) indices at various pressure levels (0.24 to 50 hPa, top to bottom). All indices were normalized and smoothed over a 6 month time window. *Left*: scatter plots of polar winter NO_2 and the Ap index. NO_2 data are reported with their standard error. See text for further details

Pole as high values around OC = 180. Less evidently, high NO_2 values in the winter polar night are visible also for the North Pole at OC = 360 and OC = 0. The previously noticed downward trend in NO_2 at 1 hPa (Fig. 4) is remarkably visible. Its maximum at the South Pole during Southern Hemisphere winter clearly extended to latitudes below 45S (OC = 225) in 2003. In the following years, the extension of these NO_2 perturbed regions became weaker down to a minimum in 2009. A sign of increase is then visible again in 2010 consistently with the upward trend in the Ap index. A clear influence of the QBO on transport seen at high latitude makes the attribution of this trend more difficult. The QBO influence is visible for example in the high N_2O values observed at low-to-mid latitudes reaching higher peak values approximately every two years (see the evolution of peak values in red around OC = 270 in the N_2O timeseries at 1 hPa). The N_2O modulation by the QBO is then reproduced in the observed NO_2 (Hauchecorne et al. 2010).

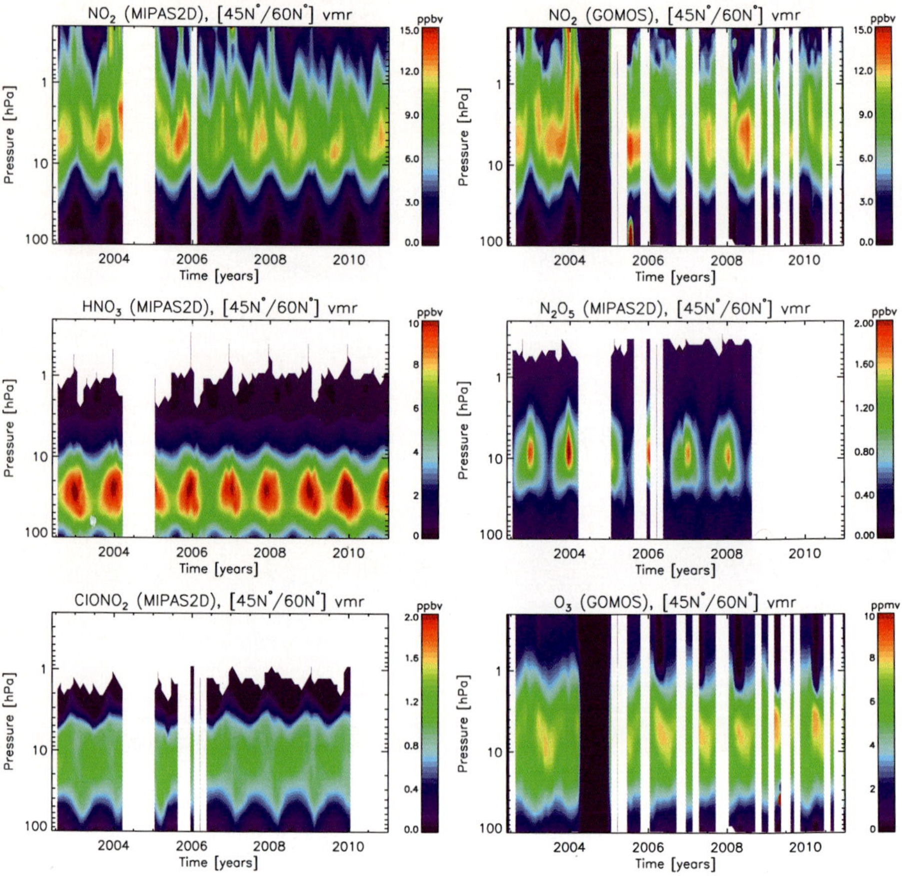

Fig. 6 Time series of monthly mean vertical profiles of NO_y species at Northern mid-latitude during 2002–2010. *Top panels*: NO_2 from MIPAS2D (*left*) and GOMOS (*right*); *Middle panels*: MIPAS2D HNO_3 (*left*) and N_2O_5 (*right*); *Bottom panels*: MIPAS2D $ClONO_2$ (*left*) and GOMOS NO_3 (*right*)

A correlation analysis confirms the impact of geomagnetic activity also at mid latitude. Also in this case we applied a simple linear correlation. Maps of the linear Pearson coefficient of the correlation of NO_y species with the Ap index are reported in Fig. 8, for Northern hemisphere summer (JJA, left) and winter (DJF, right). The highest correlation of NO_2 and Ap occurs in the high latitude regions during the respective winters, but extends to mid latitude in the upper stratosphere and above. Above 20 hPa, both HNO_3 and N_2O_5 have a consistent positive correlation in the winter polar region. This is consistent with enhanced HNO_3 observed at high latitude following descend of upper atmospheric NO_x (Orsolini et al. 2005). $ClONO_2$ show a tendency to increase with the Ap index below 10 hPa, and decrease above that level. This tendency is partially replicated by HNO_3 and N_2O_5. The different patterns at low and mid latitude, and in the summer polar region, suggest an impact on the partitioning of NO_y. Below 20 hPa, similar patterns were found correlating NO_y species with the solar F10.7 index and with a simple linear downward trend. Therefore the correlation with solar plasma processes in the lower stratosphere, either directly or indirectly through changes in the troposphere dynamics (Gray et al. 2010) will need to be further investigated with more advanced statistics.

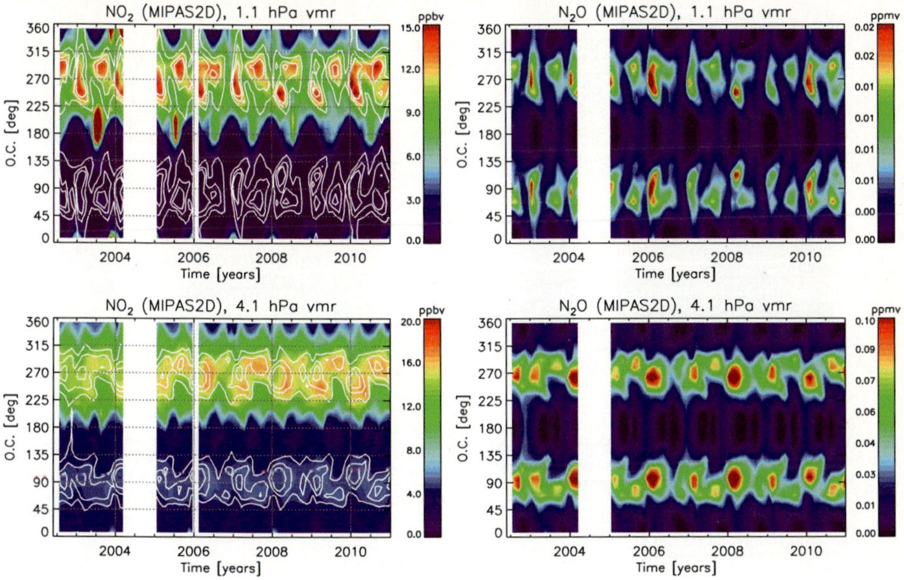

Fig. 7 Time series of MIPAS2D monthly mean NO_2 and N_2O at 1 hPa (*top panels*) and at 4 hPa (*bottom panels*) during 2002–2010. N_2O vmr main contour levels (*white*) are replicated over the NO_2 timeseries to ease the comparison

The overall behavior is largely consistent with the results by Semeniuk et al. (2011) who modeled the impact of EPP from auroral zonal electrons, SPEs and GCRs onto the middle atmosphere with a varying solar irradiance. Their results show that winter NO_y is directly controlled by auroral zonal electrons down to about 40 km altitude at mid to high latitude (see their Fig. 6, top panels). At high latitude, auroral EEP induces changes by tens of percent down to about 30 km. They found few to a few percent enhancement in NO_y also in the summer middle and lower stratosphere, possibly consistently with the positive correlation we find in HNO_3 in the Southern high latitude (Fig. 8 left, second panel from the top). The $ClONO_2$ positive correlation we found resembles the response of their model to GCRs, due either to direct changes in NO_y or through changes in HO_x. The more robust consistency of our finding with their results is given by the different behavior of the two hemispheres, with steeper gradient of the high correlation region in the winter Southern high latitude as compared to the Northern high latitude (see their Fig. 11).

We eventually investigated possible evidence of enhancements in the monthly means of NO_y species due to thunderstorm-induced plasma processes. Much of the variability in the tropical stratosphere is dominated by the QBO signal (Hauchecorne et al. 2010). In the stratosphere below 50 hPa, at low to mid latitudes, vmr of NO_y species are very low (see Fig. 1), typically below 1 ppbv. Considering a build up of lightning NO_x in the upper troposphere which gets either transported or directly injected into the lower stratosphere (Schumann and Huntrieser 2007), the observed vmr points to a secondary order contribution from lightning to monthly mean stratospheric NO_y. In MIPAS2D data, we found recurrent signs of regional enhancements of NO_2 at 100 to 60 hPa over the tropics, with monthly mean values around 1.6 ppbv in the nightside and 0.6 in the dayside (such as those seen at the bottom levels of the NO_2 panel in Fig. 1). Some of these enhancements are found also in GOMOS data (Fig. 2), although often inconsistent in latitude with those found in

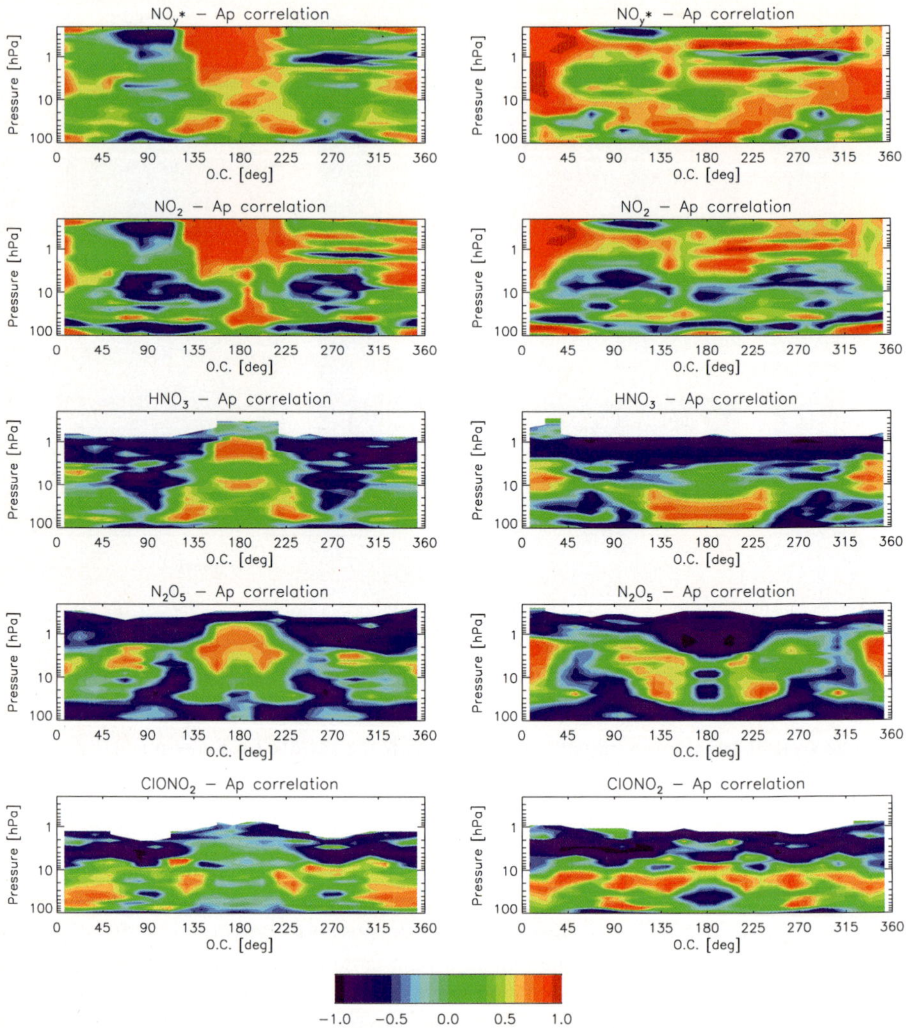

Fig. 8 Correlation coefficient of MIPAS2D NO_y species (see labels) with the Ap index during Northern hemisphere summer (JJA, *left panels*) and winter (DJF, *right panels*) over 2002–2010. NO_y^* (*top panel*) is the sum $NO_2 + HNO_3 + N_2O_5 + ClONO_2$ reported in the panels below

MIPAS2D data. HNO_3 which should be enhanced from lightning NO_x also does not show any consistent localized change, with average vmr below 0.2 ppbv at the tropics at and below 100 hPa. The sensitivity of both MIPAS and GOMOS to NO_y species over the tropics around 100 hPa and below is greatly challenged, so that spurious oscillations in the profile may be causing the observed enhancements. A possible detection of lightning induced changes will therefore need a further dedicated investigation, although the average contribution can be considered to have an upper limit around 1 ppbv. Similar considerations can be made for thunderstorm induced transient luminous events, such as sprites occurring at 40–90 km altitude mostly over the tropics (see e.g. Pasko et al. 2011). The expected maximum sprite-NO_x production is over the summer tropics around 60–80 km altitude (e.g. Enell et al. 2008), corresponding to about 0.2 hPa and above. Peak values in NO_2 at 0.2 hPa are around 2 ppbv (see Figs. 1 and 2), although these peak values are achieved in coincidence with the maxima in N_2O in a very similar fashion as occurring at 1 hPa (see Fig. 7). Peak values in NO_2 at 1 hPa and above show maxima in the autumn tropics following the N_2O source, rather than in the summer tropics as it would be expected for sprite perturbations. Also in this case, above 0.2 hPa over the tropics the quality of MIPAS and GOMOS measurements deteriorate quickly so that signs of enhanced NO_2 found in the data are largely due to retrieval artifacts. The sprite-NO_x source is therefore below the N_2O source which leads to peak values of 2 ppbv. Its magnitude will need to be investigated with dedicated studies involving the effects of mixing and transport in the atmosphere.

6 Conclusions

The time evolution of NO_y species measured by MIPAS and GOMOS between 2002 and 2010 have been presented. A general agreement is found in the main characteristics of the NO_2 evolution during the periods. At high latitudes, both instruments observe recurrent downward propagation of NO_2 in the winter mesosphere with a strong interannual variability in the amplitude of the NO_2 peaks at the stratopause. Beside sporadic very large enhancements of upper stratosphere polar NO_2 due to sporadic SPEs, an almost linear relationship of NO_2 with the Ap index is found in the winter polar upper stratosphere for both poles. These enhancements represent the most evident feature of plasma impact onto the stratosphere. This confirms previously reported nearly linear relationships over 2002–2006 (Seppälä et al. 2007), extending it to the complete downward phase of the Ap cycle. Indeed, the further 2007–2010 years added to the sample improved the correlation. The linear relationship was shown to affect also HNO_3 and N_2O_5 in the middle stratosphere. Our study indicates a clear propagation toward mid-latitudes of these NO_y enhancements in the middle and upper stratosphere. Below about 20 hPa, vast regions of positive correlation of NO_y species with the Ap index were found, although a similar and stronger correlation could be replicated with a simple linear trend. Eventually, we point to a secondary role of thunderstorm-induced plasma processes to the distribution of NO_y species in the stratosphere and above, with a first order upper limit of 1 ppbv in the lower stratosphere for the lightning source, and below 2 ppbv above 1 hPa for sprites. A more detailed investigation will be conducted to identify solar-modulated plasma effects in the lower stratosphere, and contributions from thunderstorm-driven processes.

Acknowledgements E.A. acknowledges the support by ESA CHIMTEA project within the framework of the Changing Earth Science Network Initiative.

References

E. Arnone, A. Kero, B.M. Dinelli, C.-F. Enell, N.F. Arnold, E. Papandrea, C.J. Rodger, M. Carlotti, M. Ridolfi, E. Turunen, Seeking sprite-induced signatures in remotely sensed middle atmosphere NO_2. Geophys. Res. Lett. **35**, 5807 (2008). doi:10.1029/2007GL031791

E. Arnone, A. Kero, C.-F. Enell, M. Carlotti, C.J. Rodger, E. Papandrea, N.F. Arnold, B.M. Dinelli, M. Ridolfi, E. Turunen, Seeking sprite-induced signatures in remotely sensed middle atmosphere NO_2: latitude and time variations. Plasma Sources Sci. Technol. **18**(3), 034014 (2009). doi:10.1088/0963-0252/18/3/034014

J. Baehr, H. Schlager, H. Ziereis, P. Stock, P. van Velthoven, R. Busen, J. Ström, U. Schumann, Aircraft observations of NO, NO_y, CO, and O_3 in the upper troposphere from 60N to 60S—interhemispheric differences at mitlatitudes. Geophys. Res. Lett. **30**(11), 1598 (2003). doi:10.1029/2003GL016935

C.A. Barth, S.M. Bailey, S.C. Solomon, Solar-terrestrial coupling: solar soft X-rays and thermospheric nitric oxide. Geophys. Res. Lett. **26**, 1251–1254 (1999). doi:10.1029/1999GL900237

S. Beirle, H. Huntrieser, T. Wagner, Direct satellite observation of lightning-produced NO_x. Atmos. Chem. Phys. **10**, 10965–10986 (2010). doi:10.5194/acp-10-10965-2010

J.L. Bertaux, A. Hauchecorne, F. Dalaudier, C. Cot, E. Kyrölä, D. Fussen, J. Tamminen, G.W. Leppelmeier, V. Sofieva, S. Hassinen, O. Fanton D'Andon, G. Barrot, A. Mangin, B. Théodore, M. Guirlet, O. Korablev, P. Snoeij, R. Koopman, R. Fraisse, First results on GOMOS/ENVISAT. Adv. Space Res. **33**, 1029–1035 (2004). doi:10.1016/j.asr.2003.09.037

J.L. Bertaux, E. Kyrölä, D. Fussen, A. Hauchecorne, F. Dalaudier, V. Sofieva, J. Tamminen, F. Vanhellemont, O. Fanton D'Andon, G. Barrot, A. Mangin, L. Blanot, J.C. Lebrun, K. Pérot, T. Fehr, L. Saavedra, G.W. Leppelmeier, R. Fraisse, Global ozone monitoring by occultation of stars: an overview of GOMOS measurements on ENVISAT. Atmos. Chem. Phys. **10**, 12091–12148 (2010). doi:10.5194/acp-10-12091-2010

G.P. Brasseur, S. Solomon, *Aeronomy of the Middle Atmosphere: Chemistry and Physics of the Stratosphere and Mesosphere* (Springer, Berlin, 2005)

M. Calisto, I. Usoskin, E. Rozanov, T. Peter, Influence of galactic cosmic rays on atmospheric composition and dynamics. Atmos. Chem. Phys. **11**, 4547–4556 (2011). doi:10.5194/acp-11-4547-2011

L.B. Callis, J.D. Lambeth, NO_y formed by precipitating electron events in 1991 and 1992: Descent into the stratosphere as observed by ISAMS. Geophys. Res. Lett. **25**, 1875–1878 (1998). doi:10.1029/98GL01219

L.B. Callis, R.E. Boughner, D.N. Baker, J.B. Blake, J.D. Lambeth, Precipitating relativistic electrons—their long-term effect on stratospheric odd nitrogen levels. J. Geophys. Res. **96**, 2939–2976 (1991). doi:10.1029/90JD02184

M. Carlotti, B.M. Dinelli, P. Raspollini, M. Ridolfi, Geo-fit approach to the analysis of limb-scanning satellite measurements. Appl. Opt. **40**, 1872–1885 (2001). doi:10.1364/AO.40.001872

M. Carlotti, G. Brizzi, E. Papandrea, M. Prevedelli, M. Ridolfi, B.M. Dinelli, L. Magnani, GMTR: Two-dimensional geo-fit multitarget retrieval model for Michelson Interferometer for Passive Atmospheric Sounding/Environmental Satellite observations. Appl. Opt. **45**, 716–727 (2006). doi:10.1364/AO.45.000716

P.J. Crutzen, Ozone production rates in an oxygen-hydrogen-nitrogen oxide atmosphere. J. Geophys. Res. **76**, 7311–7327 (1971). doi:10.1029/JC076i030p07311

P.J. Crutzen, U. Schmailzl, Chemical budgets of the stratosphere. Planet. Space Sci. **31**, 1009–1032 (1983). doi:10.1016/0032-0633(83)90092-2

P.J. Crutzen, I.S.A. Isaksen, G.C. Reid, Solar proton events—stratospheric sources of nitric oxide. Science **189**, 457–459 (1975). doi:10.1126/science.189.4201.457

B.M. Dinelli, D. Alpaslan, M. Carlotti, L. Magnani, M. Ridolfi, Multi-target retrieval (MTR): the simultaneous retrieval of pressure, temperature and volume mixing ratio profiles from limb-scanning atmospheric measurements. J. Quant. Spectrosc. Radiat. Transf. **84**, 141–157 (2004). doi:10.1016/S0022-4073(03)00137-7

B.M. Dinelli, E. Arnone, G. Brizzi, M. Carlotti, E. Castelli, L. Magnani, E. Papandrea, M. Prevedelli, M. Ridolfi, The MIPAS2D database of MIPAS/ENVISAT measurements retrieved with a multi-target 2-dimensional tomographic approach. Atmosp. Meas. Techn. **3**, 355–374 (2010)

C.-F. Enell, E. Arnone, O. Chanrion, T. Adachi, P.T. Verronen, A. Seppälä, T. Neubert, T. Ulich, E. Turunen, Y. Takahashi, R.-R. Hsu, Parameterisation of the chemical effect of sprites in the middle atmosphere. Ann. Geophys. **26**, 13–27 (2008)

D.W. Fahey, K.K. Kelly, S.R. Kawa, A.F. Tuck, M. Loewenstein, Observations of denitrification and dehydration in the winter polar stratospheres. Nature **344**, 321–324 (1990). doi:10.1038/344321a0

H. Fischer, M. Birk, C. Blom, B. Carli, M. Carlotti, T. von Clarmann, L. Delbouille, A. Dudhia, D. Ehhalt, M. Endemann, J.M. Flaud, R. Gessner, A. Kleinert, R. Koopman, J. Langen, M. López-Puertas, P. Mosner, H. Nett, H. Oelhaf, G. Perron, J. Remedios, M. Ridolfi, G. Stiller, R. Zander, MIPAS: an instrument for atmospheric and climate research. Atmos. Chem. Phys. **8**, 2151–2188 (2008)

B. Funke, M. López-Puertas, S. Gil-López, T. von Clarmann, G.P. Stiller, H. Fischer, S. Kellmann, Downward transport of upper atmospheric NO_x into the polar stratosphere and lower mesosphere during the Antarctic 2003 and Arctic 2002/2003 winters. J. Geophys. Res. **110**, 24308 (2005). doi:10.1029/2005JD006463

B. Funke, M. López-Puertas, M. Garcia-Comas, G.P. Stiller, T. von Clarmann, N. Glatthor, Mesospheric N_2O enhancements as observed by MIPAS on Envisat during the polar winters in 2002–2004. Atmos. Chem. Phys. **8**, 5787–5800 (2008)

B. Funke, A. Baumgaertner, M. Calisto, T. Egorova, C.H. Jackman, J. Kieser, A. Krivolutsky, M. López-Puertas, D.R. Marsh, T. Reddmann, E. Rozanov, S.-M. Salmi, M. Sinnhuber, G.P. Stiller, P.T. Verronen, S. Versick, T. von Clarmann, T.Y. Vyushkova, N. Wieters, J.M. Wissing, Composition changes after the "Halloween" solar proton event: the High Energy Particle Precipitation in the Atmosphere (HEPPA) model versus MIPAS data intercomparison study. Atmos. Chem. Phys. **11**, 9089–9139 (2011). doi:10.5194/acp-11-9089-2011

F.J. Gordillo-Vázquez, Air plasma kinetics under the influence of sprites. J. Phys. D, Appl. Phys. **41**(23), 234016 (2008). doi:10.1088/0022-3727/41/23/234016

L.J. Gray, J. Beer, M. Geller, J.D. Haigh, M. Lockwood, K. Matthes, U. Cubasch, D. Fleitmann, G. Harrison, L. Hood, J. Luterbacher, G.A. Meehl, D. Shindell, B. van Geel, W. White, Solar influences on climate. Rev. Geophys. **48**, 4001 (2010). doi:10.1029/2009RG000282

J.K. Hargreaves, *The Upper Atmosphere and Solar-Terrestrial Relations—An Introduction to the Aerospace Environment* (Van Nostrand-Reinhold, New York, 1979)

A. Hauchecorne, J.-L. Bertaux, F. Dalaudier, C. Cot, J.-C. Lebrun, S. Bekki, M. Marchand, E. Kyrölä, J. Tamminen, V. Sofieva, D. Fussen, F. Vanhellemont, O. Fanton D'Andon, G. Barrot, A. Mangin, B. Théodore, M. Guirlet, P. Snoeij, R. Koopman, L. Saavedra de Miguel, R. Fraisse, J.-B. Renard, First simultaneous global measurements of nighttime stratospheric NO_2 and NO_3 observed by Global Ozone Monitoring by Occultation of Stars (GOMOS)/Envisat in 2003. J. Geophys. Res. **110**, 18301 (2005). doi:10.1029/2004JD005711

A. Hauchecorne, J.L. Bertaux, F. Dalaudier, P. Keckhut, P. Lemennais, S. Bekki, M. Marchand, J.C. Lebrun, E. Kyrölä, J. Tamminen, V. Sofieva, D. Fussen, F. Vanhellemont, O. Fanton D'Andon, G. Barrot, L. Blanot, T. Fehr, L. Saavedra de Miguel, Response of tropical stratospheric O_3, NO_2 and NO_3 to the equatorial quasi-biennial oscillation and to temperature as seen from GOMOS/ENVISAT. Atmos. Chem. Phys. **10**, 8873–8879 (2010). doi:10.5194/acp-10-8873-2010

Y. Hiraki, Y. Kasai, H. Fukunishi, Chemistry of sprite discharges through ion-neutral reactions. Atmos. Chem. Phys. **8**, 3919–3928 (2008)

C.H. Jackman, E.L. Fleming, F.M. Vitt, Influence of extremely large solar proton events in a changing stratosphere. J. Geophys. Res. **105**, 11659–11670 (2000). doi:10.1029/2000JD900010

C.H. Jackman, R.D. McPeters, G.J. Labow, E.L. Fleming, C.J. Praderas, J.M. Russell, Northern hemisphere atmospheric effects due to the July 2000 solar proton event. Geophys. Res. Lett. **28**, 2883–2886 (2001). doi:10.1029/2001GL013221

C.H. Jackman, D.R. Marsh, F.M. Vitt, R.R. Garcia, E.L. Fleming, G.J. Labow, C.E. Randall, M. López-Puertas, B. Funke, T. von Clarmann, G.P. Stiller, Short- and medium-term atmospheric constituent effects of very large solar proton events. Atmos. Chem. Phys. **8**, 765–785 (2008)

M.K.W. Ko, D.K. Weisenstein, N.D. Sze, M.B. McElroy, Lightning—a possible source of stratospheric odd nitrogen. J. Geophys. Res. **91**, 5395–5404 (1986). doi:10.1029/JD091iD05p05395

V.R. Kotamarthi, M.K.W. Ko, D.K. Weisenstein, J.M. Rodriguez, N.D. Sze, Effect of lightning on the concentration of odd nitrogen species in the lower stratosphere: an update. J. Geophys. Res. **99**, 8167–8174 (1994). doi:10.1029/93JD03477

E. Kyrölä, J. Tamminen, G.W. Leppelmeier, V. Sofieva, S. Hassinen, A. Seppälä, P.T. Verronen, J.L. Bertaux, A. Hauchecorne, F. Dalaudier, D. Fussen, F. Vanhellemont, O.F. d'Andon, G. Barrot, A. Mangin, B. Theodore, M. Guirlet, R. Koopman, L.S. de Miguel, P. Snoeij, T. Fehr, Y. Meijer, R. Fraisse, Nighttime ozone profiles in the stratosphere and mesosphere by the global ozone monitoring by occultation of stars on Envisat. J. Geophys. Res. **111**, 24306 (2006). doi:10.1029/2006JD007193

E. Kyrölä, J. Tamminen, V. Sofieva, J.L. Bertaux, A. Hauchecorne, F. Dalaudier, D. Fussen, F. Vanhellemont, O. Fanton D'Andon, G. Barrot, M. Guirlet, A. Mangin, L. Blanot, T. Fehr, L. Saavedra de Miguel, R. Fraisse, Retrieval of atmospheric parameters from GOMOS data. Atmos. Chem. Phys. **10**, 11881–11903 (2010). doi:10.5194/acp-10-11881-2010

M. López-Puertas, B. Funke, S. Gil-López, T. von Clarmann, G.P. Stiller, M. Höpfner, S. Kellmann, H. Fischer, C.H. Jackman, Observation of NO_x enhancement and ozone depletion in the Northern and Southern hemispheres after the October–November 2003 solar proton events. J. Geophys. Res. **110**, 09-43 (2005). doi:10.1029/2005JA011050

D.R. Marsh, R.R. Garcia, D.E. Kinnison, B.A. Boville, F. Sassi, S.C. Solomon, K. Matthes, Modeling the whole atmosphere response to solar cycle changes in radiative and geomagnetic forcing. J. Geophys. Res. **112**, 23306 (2007). doi:10.1029/2006JD008306

T. Neubert, M. Rycroft, T. Farges, E. Blanc, O. Chanrion, E. Arnone, A. Odzimek, N. Arnold, C.-F. Enell, E. Turunen, T. Bösinger, Á. Mika, C. Haldoupis, R.J. Steiner, O. van der Velde, S. Soula, P. Berg, F. Boberg, P. Thejll, B. Christiansen, M. Ignaccolo, M. Füllekrug, P.T. Verronen, J. Montanya, N. Crosby, Recent results from studies of electric discharges in the mesosphere. Surveys in Geophysics **14** (2008). doi:10.1007/s10712-008-9043-1

Y.J. Orsolini, G.L. Manney, M.L. Santee, C.E. Randall, An upper stratospheric layer of enhanced HNO_3 following exceptional solar storms. Geophys. Res. Lett. **32**, 12-01 (2005). doi:10.1029/2004GL021588

V.P. Pasko, Y. Yair, C.-L. Kuo, Lightning related transient luminous events at high altitude in the Earth's atmosphere: phenomenology, mechanisms and effects. Space Science Rev., 287 (2011). doi:10.1007/s11214-011-9813-9

C.E. Randall, V.L. Harvey, C.S. Singleton, S.M. Bailey, P.F. Bernath, M. Codrescu, H. Nakajima, J.M. Russell, Energetic particle precipitation effects on the Southern Hemisphere stratosphere in 1992–2005. J. Geophys. Res. **112**, 8308 (2007). doi:10.1029/2006JD007696

G.C. Reid, S. Solomon, R.R. Garcia, Response of the middle atmosphere to the solar proton events of August–December, 1989. Geophys. Res. Lett. **18**, 1019–1022 (1991). doi:10.1029/91GL01049

C.J. Rodger, A. Seppälä, M.A. Clilverd, Significance of transient luminous events to neutral chemistry: experimental measurements. Geophys. Res. Lett. **35**, 7803 (2008). doi:10.1029/2008GL033221

D.W. Rusch, J.-C. Gerard, S. Solomon, P.J. Crutzen, G.C. Reid, The effect of particle precipitation events on the neutral and ion chemistry of the middle atmosphere. I. Odd nitrogen. Planet. Space Sci. **29**, 767–774 (1981). doi:10.1016/0032-0633(81)90048-9

U. Schumann, H. Huntrieser, The global lightning-induced nitrogen oxides source. Atmos. Chem. Phys. **7**, 3823–3907 (2007)

K. Semeniuk, V.I. Fomichev, J.C. McConnell, C. Fu, S.M.L. Melo, I.G. Usoskin, Middle atmosphere response to the solar cycle in irradiance and ionizing particle precipitation. Atmos. Chem. Phys. **11**, 5045–5077 (2011). doi:10.5194/acp-11-5045-2011

D.D. Sentman, H.C. Stenbaek-Nielsen, M.G. McHarg, J.S. Morrill, Plasma chemistry of sprite streamers. J. Geophys. Res. **113**, 11112 (2008). doi:10.1029/2007JD008941

A. Seppälä, P.T. Verronen, E. Kyrölä, S. Hassinen, L. Backman, A. Hauchecorne, J.L. Bertaux, D. Fussen, Solar proton events of October-November 2003: ozone depletion in the Northern hemisphere polar winter as seen by GOMOS/Envisat. Geophys. Res. Lett. **31**, 19107 (2004). doi:10.1029/2004GL021042

A. Seppälä, P.T. Verronen, M.A. Clilverd, C.E. Randall, J. Tamminen, V. Sofieva, L. Backman, E. Kyrölä, Arctic and Antarctic polar winter NO_x and energetic particle precipitation in 2002–2006. Geophys. Res. Lett. **341**, 12810 (2007). doi:10.1029/2007GL029733

M. Sinnhuber, S. Kazeminejad, J.M. Wissing, Interannual variation of NO_x from the lower thermosphere to the upper stratosphere in the years 1991–2005. J. Geophys. Res. **116**, 2312 (2011). doi:10.1029/2010JA015825

D.E. Siskind, An assessment of Southern Hemisphere stratospheric NO_x enhancements due to transport from the upper atmosphere. Geophys. Res. Lett. **27**, 329–332 (2000). doi:10.1029/1999GL010940

D.E. Siskind, J.T. Bacmeister, M.E. Summers, J.M. Russell III, Two-dimensional model calculations of nitric oxide transport in the middle atmosphere and comparison with Halogen Occultation Experiment data. J. Geophys. Res. **102**, 3527–3546 (1997). doi:10.1029/96JD02970

A.K. Smith, Interactions between the lower, middle and upper atmosphere. Space Science Reviews, 76 (2011). doi:10.1007/s11214-011-9791-y

S. Solomon, R.R. Garcia, Simulation of NO_x partitioning along isobaric parcel trajectories. Geophys. Res. Lett. **88**, 5497–5501 (1983). doi:10.1029/JC088iC09p05497

I.G. Usoskin, G.A. Kovaltsov, Cosmic ray induced ionization in the atmosphere: full modeling and practical applications. J. Geophys. Res. **111**, 21206 (2006). doi:10.1029/2006JD007150

G. Verbanac, M. Mandea, B. Vršnak, S. Sentic, Evolution of solar and geomagnetic activity indices, and their relationship: 1960–2001. Sol. Phys. **271**, 183–195 (2011). doi:10.1007/s11207-011-9801-y

P.T. Verronen, E. Turunen, T. Ulich, E. Kyrölä, Modelling the effects of the October 1989 solar proton event on mesospheric odd nitrogen using a detailed ion and neutral chemistry model. Ann. Geophys. **20**, 1967–1976 (2002). doi:10.5194/angeo-20-1967-2002

P.T. Verronen, S. Ceccherini, U. Cortesi, E. Kyrölä, J. Tamminen, Statistical comparison of night-time NO_2 observations in 2003–2006 from GOMOS and MIPAS instruments. Adv. Space Res. **43**, 1918–1925 (2009). doi:10.1016/j.asr.2009.01.027

G. Wetzel, A. Bracher, B. Funke, F. Goutail, F. Hendrick, J.-C. Lambert, S. Mikuteit, C. Piccolo, M. Pirre, A. Bazureau, C. Belotti, T. Blumenstock, M. de Mazière, H. Fischer, N. Huret, D. Ionov, M. López-Puertas, G. Maucher, H. Oelhaf, J.-P. Pommereau, R. Ruhnke, M. Sinnhuber, G. Stiller, M. van Roozendael, G. Zhang, Validation of MIPAS-ENVISAT NO_2 operational data. Atmos. Chem. Phys. **7**, 3261–3284 (2007)

Space Sci Rev (2012) 168:333–362
DOI 10.1007/s11214-011-9857-x

Gravity Wave Mixing and Effective Diffusivity for Minor Chemical Constituents in the Mesosphere/Lower Thermosphere

M. Grygalashvyly · E. Becker · G.R. Sonnemann

Received: 16 July 2011 / Accepted: 5 November 2011 / Published online: 29 November 2011

Abstract The influence of gravity waves (GWs) on the distributions of minor chemical constituents in the mesosphere-lower thermosphere (MLT) is studied on the basis of the effective diffusivity concept. The mixing ratios of chemical species used for calculations of the effective diffusivity are obtained from numerical experiments with an off-line coupled model of the dynamics and chemistry abbreviated as KMCM-MECTM (Kuehlungsborn Mechanistic general Circulation Model—MEsospheric Chemistry-Transport Model). In our control simulation the MECTM is driven with the full dynamical fields from an annual cycle simulation with the KMCM, where mid-frequency GWs down to horizontal wavelengths of 350 km are resolved and their wave-mean flow interaction is self-consistently induced by an advanced turbulence model. A perturbation simulation with the MECTM is defined by eliminating all meso-scale variations with horizontal wavelengths shorter than 1000 km from the dynamical fields by means of spectral filtering before running the MECTM.

The response of the MECTM to GWs perturbations reveals strong effects on the minor chemical constituents. We show by theoretical arguments and numerical diagnostics that GWs have direct, down-gradient mixing effects on all long-lived minor chemical species that possess a mean vertical gradient in the MLT. Introducing the term wave diffusion (WD) and showing that wave mixing yields approximately the same WD coefficient for different chemical constituents, we argue that it is a useful tool for diagnostic irreversible transport processes. We also present a detailed discussion of the gravity-wave mixing effects on the photochemistry and highlight the consequences for the general circulation of the MLT.

Keywords Wave diffusion · Gravity waves · Wave mixing · Effective diffusivity · Atmospheric chemistry · MLT

M. Grygalashvyly (✉) · E. Becker · G.R. Sonnemann
Leibniz-Institute of Atmospheric Physics at the University Rostock in Kühlungsborn, Schloss-Str. 6, 18225 Ostseebad Kühlungsborn, Germany
e-mail: gryga@iap-kborn.de

1 Introduction

The concept of effective diffusivity was independently introduced by Winters and D'Asaro (1996) and Nakamura (1996) for the Lagrangian formalism and, later on, it was developed by Nakamura (2001) for the Eulerian-mean formalism. This concept has widely been applied for diagnostics of transport in the middle atmosphere (Haynes and Shuckburgh 2000a, 2000b; Allen and Nakamura 2001; Shuckburgh et al. 2001; Kostrykin and Schmitz 2006), and in oceanic flows as well (Eden et al. 2007, 2009; Shuckburgh et al. 2009a, 2009b). In atmospheric dynamics the concept has mainly been utilized for analyses of meridional mixing by Rossby-waves and planetary-scale disturbances, but not for vertical mixing. Understanding the large-scale mixing processes in the stratosphere and mesosphere was advanced in a paper by Shepherd et al. (2000). The authors noted that the transport in the mesosphere is strongly controlled by gravity waves (hereafter: GWs). A comprehensive overview of the transport and mixing in the middle atmosphere is given in Shepherd (2007).

The fact that GWs play an important role in determining the distributions of minor chemical constituents in the mesosphere has been emphasized in various studies (e.g. Frederick 1979; Hickey et al. 2000, 2003; Xu et al. 2000a, 2000b, 2003; Huang and Hickey 2007). Thus it would be natural to apply the concept of effective diffusivity to GWs (Medvedev and Greatbatch 2004). The primary importance of GWs in the middle atmosphere is due to their contributions to the energy and momentum budgets and the driving of the residual circulation (Lindzen 1981; Garcia and Solomon 1985). A comprehensive review about the current knowledge of the generation and propagation of GWs, as well as dynamic instability and transition to turbulence has been provided by Fritts and Alexander (2003).

Previous researches with respect to the GWs and chemistry in the middle atmosphere have mainly focused on variations of airglow due to relaxation of excited hydroxyl OH* (Walterscheid et al. 1987; Taylor et al. 1991, 1995a, 1995b, 1997, 1998; Makhlouf et al. 1995, 1998; Wang et al. 2001a, 2001b; Huang et al. 2003; Huang 2006; Nielsen et al. 2006) and nightglow from O_2 (Weinstock 1978; Takahashi et al. 1992; Zhang et al. 1993a, 1993b; Leko et al. 2002; Belyaev 2009), and on constituents involved in airglow processes. There are only a few studies of the influence of GWs on the other chemical constituents in the mesosphere-lower thermosphere (MLT) (Frederick 1979; Hickey et al. 2000, 2003; Xu et al. 2000a, 2000b, 2003; Huang and Hickey 2007). All of them employ 1D or 2D models. Some are furthermore constrained by linear steady-state chemistry (Walterscheid et al. 1987; Walterscheid and Schubert 1989; Hickey et al. 1992; Hickey and Walterscheid 1994) and others by linear parameterization of GWs (e.g., Hickey et al. 1993, 1997, 2003; Xu et al. 2000a, 2000b, 2003; Huang et al. 2003; Huang and Hickey 2007). In these studies, the strongest effects due to GWs (in terms of redistributions of chemical species and a resultant modulation of exothermic chemical heating) showed up around the mesopause. Nevertheless, it has been noted that linear steady-state models do not take into account wave-wave interactions, non-linear effects, and wave transience, and that a time-dependent coupled non-linear model of dynamics and chemistry is required in order to completely describe the influence of GWs on minor chemical constituents in the MLT (Hickey and Walterscheid 1994; Xu et al. 2000b; Huang and Hickey 2007).

The purpose of the present study is to employ such a model, albeit with off-line photochemistry, in order to estimate the effects of mesoscale GWs on minor constituents of the MLT on a global scale. This is done in terms a sensitivity experiment where a chemistry-transport model (CTM) is driven with the same large-scale dynamical fields but two different filtering of GW motions. Comparing the two simulations allows (1) to infer the changes in the photochemical system that arise directly from the resolved GW motions and (2) to apply the concept of effective diffusivity to the vertical mixing by GWs.

This paper is structured as follows. The model and numerical experiments are described in Sect. 2. The concept of wave mixing and effective diffusivity along with the definition of the wave-driven residual circulation is outlined in Sect. 3. In Sect. 4 we show that GW mixing plays an important role for minor chemical constituent distributions and estimate the consequences of such an influence. The impact on the thermal regime is discussed in Sect. 5. In Sect. 6 we discuss the influences of GWs on minor constituents using the concept of effective diffusivity. In the last section we summarize our results and present some further conclusions.

2 Model and Numerical Experiment

2.1 Dynamical Part of the Model

The dynamical part is the Kühlungsborn Mechanistic general Circulation Model (KMCM) and has widely been applied in previous studies (Körnich et al. 2006; Becker 2009; Becker and von Savigny 2010). It solves the primitive equations like other so-called spectral models according to the method of Simmons and Burridge (1981). The KMCM is distinguished from conventional general circulation models including the MLT because GWs are simulated explicitly. The GWs are generated dynamically within the extratropical storm tracks mainly by fronts and Rossby-wave breaking. The vertical propagation and subsequent turbulent damping of the resolved GWs is simulated by combining a high spatial resolution with an advanced parameterization of turbulent diffusion (Becker 2009). On the other hand, the radiative heating in the middle atmosphere must be represented by some mechanistic means for which we use temperature relaxation. The overall thermal forcing of the model was previously applied only for permanent January or July conditions. Recently, also the annual cycle has been simulated (Hoffmann et al. 2010), yielding a quite reasonable representation of the observed mean flow, planetary waves, and GW activity. The current resolution of the KMCM is spectral truncation at total wavenumber 120 and 190 hybrid levels up to ~125 km. In the middle atmosphere these levels corresponds to pressure levels. Model results are saved with a snapshot interval of 45 minutes. Here we utilize the dynamical fields of the annual cycle simulated with the KMCM (Hoffmann et al. 2010) to the drive the MEsospheric Chemistry Transport Model (MECTM), which has also been used previously in various studies (e.g., Grygalashvyly et al. 2009). Since the KMCM is a spectral model, the simulated dynamical fields can easily be filtered with regard to the horizontal scales. This allows for a straight-forward sensitivity experiment with the MECTM in order isolate the effects of GWs.

2.2 Chemical Part of the Model

The dynamical fields calculated with the KMCM are used in the MECTM. The MECTM includes a chemical, radiation, and transport code. A high horizontal resolution is essential to reproduce the GWs; thus, simulations presented here were performed on a grid with 288 longitudinal and 144 latitudinal grid points, as well as 118 vertical levels from the ground to approximately 120 km. The time step of integration is 20 s and the dynamical fields are updated every 45 minutes. Note that the dynamics drives the chemical part but not vice versa. Thus, in the present study, the model is coupled only in the direction from dynamics to chemistry.

The chemistry module consists of 19 constituents. The reaction scheme includes 49 chemical reactions and 14 photodissociation processes. The reaction rates used in the model are taken from Sander et al. (2006). For the first time in 3D modeling, the temperature-dependent reaction rates are calculated on-line and are thus sensitive to small fluctuations of the temperature. The chemistry is based on the family concept (Shimazaki 1985), including in particular the odd hydrogen (H, OH, HO_2, H_2O_2), the odd oxygen (O, $O(^1D)$, O_3), and the odd nitrogen (NO, NO_2, $N(^4S)$, $N(^2D)$) families. Time integration is performed using the implicit Euler method with a quadratic loss term. The integration of long-lived constituents is done separately while the short-lived constituents are included in chemical families, which are solved as fully implicit subsystems.

The diurnal variation in chemistry results from the diurnal variation of the solar insolation. The corresponding radiation code used in the MECTM is described in Sonnemann et al. (1998a). This code was primarily developed by Fichtelmann and Sonnemann (1989) for mesospheric and thermospheric modeling and considers the EUV and X-ray spectrum of the solar radiation. The lower mesospheric photodissociation rates were revised by Kremp et al. (1999) by implementing a code developed by Röth (1992) for stratospheric modeling. The annual variation due to the eccentricity of the Earth's orbit is approximated by a sine function with an amplitude of 3.5% of the calculated photodissociation rates. The radiation code runs off-line, meaning that it uses fixed absorber concentrations for the calculation of the photodissociation rates.

The transport code considers both the advective and diffusive transport. A newly developed advective transport scheme (Walcek and Aleksic 1998; Walcek 2000) that is almost free of numerical diffusion is applied in the model. The scheme is mass conserving, monotonic, and positive-definite. It utilizes a second-order upstream approximation with monotone limiters and adjusts the fluxes at cell edges around local extremes.

The diffusion code includes turbulent and molecular diffusion and calculates vertical fluxes of the individual gas species according to Colegrove et al. (1965) with the implicit Thomas algorithm as described in Morton and Mayers (1994). The species-dependent molecular diffusion coefficients depend on temperature and number density of the air, and they are calculated at each time step. The turbulent diffusion coefficient is based on the results of Lübken (1997), using a fixed and horizontally uniform vertical profile that corresponds to the annual mean at high latitudes.

Calculations of exothermic chemical heating are based on the model described in the paper of Sonnemann et al. (1997), which is extended here to include the 7 most significant chemical reactions with exothermicity according to the work of Mlynczak and Solomon (1993). As the CTM runs in off-line mode, the chemical heating rates should be considered as diagnostic estimates only. Further information about the CTM can be found in the papers referred to above and in other more recent works (Grygalashvyly et al. 2009, 2011; Hartogh et al. 2010).

2.3 Description of the Numerical Experiment

We performed two numerical experiments with the MECTM driven with the dynamical fields from the KMCM. In our control simulation (case 1) the full dynamical fields are applied. For this purpose, the grid of the MECTM is adjusted to resemble the spectral resolution of the circulation model (see previous subsections). A conventional simulation (case 2) is defined in such a way that GW motions are filtered out when driving the MECTM. This is achieved by truncating the spectral expansion coefficients computed with the KMCM at a total wavenumber of 42. This corresponds to a minimum horizontal wavelength of

1000 km in the conventional simulation compared to 350 km in the control case with the full spectral resolution up to wavenumber 120. It was shown previously that the resolved GW variations and GW effects on the large-scale flow lie between these horizontal wavelengths for the present setup of the KMCM (Becker 2009; Becker and von Savigny 2010; Hoffmann et al. 2010). Note that the large-scale flow (horizontal wavelengths longer than 1000 km) is identical in both simulations with the MECTM. We also employ the same spatial resolution and time step in each case in order to ensure a numerically consistent sensitivity experiment with the MECTM. The deviations between the control and conventional simulation will then mirror the influence of the GWs resolved in the KMCM on the photochemistry in the MECTM.

Even though the limited spectral resolution of the GCM excludes GWs with horizontal wavelength shorter than 350 km, the simulated GWs effects on the large-scale circulation (GW drag and dissipative heating) have realistic magnitudes (Becker 2009). The reason is that the resolved GW activity is biased toward the resolvable scales. We therefore expect that the simulated wave mixing will be comparable for higher resolution, since the solved GWs will have higher frequencies but similar vertical winds. A detailed discussion of the resolved GW scales and the parameterized turbulence is given in Becker (2009). Additional information is given in Becker and von Savigny (2010) and in Hoffmann et al. (2010). To get an impression of the simulated GW dynamics, Fig. 1 shows instantaneous longitude-height cross section at 64°S during January 30 for the zonal and vertical wind. The GWs have typical horizontal wavelengths of about 500 km and periods of a few hours. Their zonal phase speed is about 40 m/s. In the region of the zonal wind reversal (80–90 km) the vertical wavelength is compressed, as expected. This induces strong vertical shear and turbulent diffusion. The latter is evident from a minimum in the vertical wind variations around 85–90 km. In this region we expect that the vertical mixing due to GWs will have a minimum too (see Sect. 6).

3 Concept of Effective Diffusivity

If chemical processes, advective transport, and turbulent and molecular vertical diffusion are taken into account, the continuity equation for a single species with mixing ratio c is given by

$$\frac{\partial c}{\partial t} = \frac{1}{\rho_0}(P(T) - L(T)) - \vec{v} \cdot \nabla c + \frac{1}{\rho_0}\frac{\partial}{\partial z}\left(\rho_0 K_{turb}\frac{\partial}{\partial z}c\right) + \frac{1}{\rho_0}\frac{\partial}{\partial z}\left(\rho_0 K_{mol}(T)\frac{\partial}{\partial z}c\right). \quad (1)$$

Here, P and L are chemical production and loss terms, respectively, which depend on temperature due to their dependencies on the chemical reaction rates. $\vec{v}$ is the 3D wind vector and ρ_0 the background density. K_{turb} is the turbulent diffusion coefficient, and is a function only of height z in the MECTM (see Fig. 1 of Sonnemann and Körner 2003 for the vertical profile). K_{mol} is the molecular diffusion coefficient and depends on temperature and the sort of molecule. Thus, in our model, GWs can influence the distributions of minor constituents by chemical reactions, as represented by the 1st term on the right hand side (rhs) of (1), by transport processes (including resolved wave mixing, 2nd term on the rhs), and via variations in the molecular diffusion (last term on the rhs).

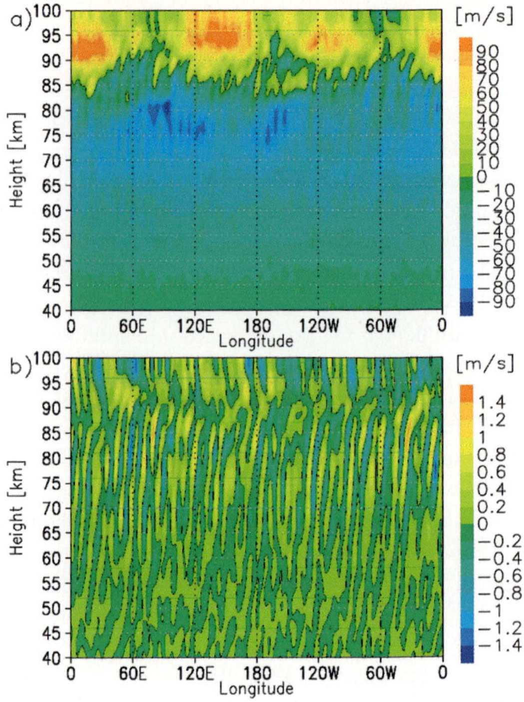

Fig. 1 Longitude-height sections of (**a**) the zonal wind and (**b**) the vertical wind at 64°S for southern summer (January 30) from the KMCM when all scales are retained (case 1)

In the absence of sources and sinks the continuity equation for a tracer with vertical diffusion is

$$\frac{dc}{dt} = \frac{\partial c}{\partial t} + \vec{v} \cdot \nabla c = \rho_0^{-1} \partial_z (\rho_0 K_z \partial_z c), \tag{2}$$

where K_z denotes the sum of the turbulent and molecular diffusion coefficients. Any flow variable can be decomposed into a zonal mean and a zonally varying component as $X = [X] + X^*$ with $[X^*] = 0$. Temporal averaging of the flow variable will be indicated as $\bar{X}$. Applying the zonal and temporal average to (2) yields

$$\left([\overline{v}]\frac{\partial_\phi}{a} + [\overline{w}]\partial_z\right)[\overline{c}] = -\frac{\partial_\phi}{a\cos\phi}(\cos\phi[\overline{c^* v^*}]) - \frac{\partial_z(\rho_0[\overline{c^* w^*}])}{\rho_0} + \frac{\partial_z(\rho_0 K_z \partial_z [\overline{c}])}{\rho_0}, \tag{3}$$

where a denotes the radius of Earth, ϕ is latitude, v and w are the meridional and vertical wind components. Using the primitive equations in the anelastic approximation and sensible heat as thermodynamic variable, the residual velocities are defined as (see Becker 2011, this issue)

$$\begin{aligned} v_{res} &= \rho_0[\overline{v}] - \frac{c_p}{g}\partial_z(\rho_0[\overline{T^* v^*}]), \\ w_{res} &= \rho_0[\overline{w}] + \frac{c_p \rho_0}{g}\frac{\partial_\phi}{a\cos\phi}(\cos\phi[\overline{T^* v^*}]). \end{aligned} \tag{4}$$

Here, g is the gravity acceleration and c_p the specific heat capacity at constant pressure.

Transforming (3) into the residual frame with the help of (4) we get

$$\left(v_{res}\frac{\partial_\phi}{a}+w_{res}\partial_z\right)[\overline{c}]$$
$$=-\frac{\partial_\phi}{\rho_0 a\cos\phi}\left(\cos\phi\left(\rho_0[\overline{c^*v^*}]-\frac{c_p\rho_0}{g}[\overline{T^*v^*}]\partial_z[\overline{c}]\right)\right)$$
$$-\frac{\partial_z}{\rho_0}\left(\rho_0[\overline{c^*w^*}]+\frac{c_p\rho_0}{g}[\overline{T^*v^*}]\frac{\partial_\phi[\overline{c}]}{a}\right)+\rho_0^{-1}\partial_z(\rho_0 K_z\partial_z[\overline{c}]). \tag{5}$$

Note that the left-hand side of (5) represents the material rate of change in mixing ratio for an air parcel that is transported by the residual circulation. For a uniform mean tracer distribution this rate of change vanishes. However, when the mean tracer distribution has gradients, the mixing ratio within the air parcel will change according to (1) wave mixing associated with the wave tracer flux convergence, (2) Stokes corrections associated with the meridional heat flux and the mean tracer gradient, and (3) vertical diffusion.

Assuming that only vertical variations of the mean flow enter the wave equation of the tracer (single-column approximation, e.g., Lindzen 1981), that the vertical wavelength is smaller than the scale height, and that the wave propagates in zonal direction ($v^*=0$), linear wave theory can be used to eliminate the vertical wave mixing as follows. The linear wave equation for the tracer yields:

$$\frac{\partial c^*}{\partial t}+U\frac{\partial_\lambda}{a\cos\phi}c^*+w^*\partial_z[\overline{c}]=K_z\partial_z^2c^*, \tag{6}$$

where U is the mean zonal flow and λ is longitude. Multiplying (6) by c^* and averaging over time and longitude leads to

$$[\overline{c^*w^*}]\partial_z[\overline{c}]=K_z[\overline{c^*\partial_z^2c^*}]=K_z\partial_z[\overline{c^*\partial_zc^*}]-K_z[\overline{(\partial_zc^*)^2}]. \tag{7}$$

For conservative waves we have $[\overline{c^*\partial_zc^*}]=0$. This is still a good approximation in the case of weak diffusion. Hence, (7) can be written as

$$[\overline{c^*w^*}]=-K_{waves}\partial_z[\overline{c}], \tag{8}$$

where

$$K_{waves}=K_z[\overline{(\partial_zc^*)^2}]/(\partial_z[\overline{c}])^2. \tag{9}$$

Equation (8) shows that mean gradients in the tracer distribution lead to a local down-gradient wave mixing of the tracer if some finite small-scale diffusivity is present. This wave mixing is mainly opposite to the large-scale vertical gradient of the mean mixing ratio and, in particular, independent of the direction of the residual circulation.

Neglecting the meridional wave mixing and using (8), (5) can be rewritten as

$$\left(v_{res}\frac{\partial_\phi}{a}+w_{res}\partial_z\right)[\overline{c}]=-\frac{\partial_\phi}{\rho_0 a\cos\phi}\left(\cos\phi\left(-\frac{c_p\rho_0}{g}[\overline{T^*v^*}]\partial_z[\overline{c}]\right)\right)$$
$$-\frac{\partial_z}{\rho_0}\left(\frac{c_p\rho_0}{g}[\overline{T^*v^*}]\frac{\partial_\phi[\overline{c}]}{a}\right)+\rho_0^{-1}\partial_z(\rho_0K_{eff}\partial_z[\overline{c}]), \tag{10}$$

where K_{eff} is the effective diffusivity

$$K_{eff}=K_z+K_{waves}=K_z+K_z[\overline{(\partial_zc^*)^2}]/(\partial_z[\overline{c}])^2=K_z(1+[\overline{(\partial_zc^*)^2}]/(\partial_z[\overline{c}])^2)$$
$$=K_z\cdot[\overline{(\partial_zc)^2}]/(\partial_z[\overline{c}])^2. \tag{11}$$

Thus, K_{eff} describes a local mixing analogously to the K_m in work of Nakamura (2001). It is the result of diffusivity at unresolved scales K_z (in our case K_z is the sum of turbulent and molecular diffusion), multiplied by the factor which is known as the mixing efficiency or as the Osborne-Cox relation (Osborn and Cox 1972). We interpret the effective wave diffusion, or simply wave diffusion (WD) with coefficient K_{waves} as an analog of the Winters-D'Asaro-Nakamura effective diffusion, which in our case is due to resolved GWs. Note, that the WD coefficient maintains all the properties of K_m as defined by Nakamura (2001).

Hereafter we will denote the net advection due to GWs as wave mixing. To the extent that the quasi-linear theory described above applies, the wave mixing can be described as wave diffusion (WD) in terms of K_{waves}. Note, that in our nomenclature eddy diffusion (or effective diffusion according to the nomenclature of Nakamura 2001, Haynes and Shuckburgh 2000a, 2000b) means the combined action of turbulent diffusion and mixing by GWs.

From our sensitivity experiment we can infer a vertical flux due to all resolved waves $\rho_0 \cdot [\overline{w^* c^*}]$. The difference of $\rho_0 \cdot [\overline{w^* c^*}]$ between case 1 (including all wave lengths >350 km) and case 2 (including only wave lengths > 1000 km) gives us the flux due to the resolved GWs. In complex chemistry-climate models with coarse resolution and parameterized GWs (e.g. WACCM), the eddy diffusion is usually just the turbulent diffusion provided by the GW scheme. We shall show in the present study that the GW mixing can be approximately parameterized by an additional vertical diffusion that is similar for different constituents, though being quite different from the turbulent diffusivity profile assumed in the CTM.

4 Influence of GW Mixing on Minor Chemical Constituents

We distinguish three cases to characterize the influence of dynamics and chemistry on the distribution of a tracer by considering the photochemical lifetime (Brasseur and Solomon 1986). If the photochemical lifetime is smaller than the characteristic transport time (advective or diffusive), the species will be in photochemical equilibrium and transport has no direct influence on the distribution. Nevertheless, an indirect influence by the interaction with long-lived constituents is possible. If the photochemical lifetime is longer than the characteristic transport time, transport processes strongly determine the distribution of the constituent. In the last case, if the characteristic photochemical and transport times are comparable, the distribution depends on both dynamics and chemistry. Below we analyze the modulation of minor chemical constituents by GWs by using the lifetime concept and considering the different processes according to (1).

From our sensitivity experiment with the MECTM we cannot separate direct effects of GW mixing from the effects due to the temperature dependencies of the molecular diffusion coefficients. However, the molecular diffusion becomes larger than the assumed K_{turb} only well above ~100 km (see Fig. 1 and discussion of Sonnemann and Körner 2003). Thus, the influence of GWs that comes into play through the last term on the rhs of (1) is not expected to be significant for the mesosphere and mesopause region.

We define absolute deviations as differences between case 1 and 2. When zonally and diurnally averaged, the difference of a quantity X is denoted as $[\overline{X}]_{L>350\ \mathrm{km}} - [\overline{X}]_{L>1000\ \mathrm{km}}$ indicating that the mixing due to waves with horizontal wavelengths between 350 and 1000 km determines the absolute deviation. Note that both simulations with the MECTM (case 1 and 2) are subject to the same large-scale ($L > 1000$ km) fields. This holds in particular for the residual (Lagrangian mean) circulation, which implicitly includes the wave driving of all waves resolved in the KMCM.

4.1 Water Vapor

Correct modeling of the water vapor distribution is the one of key points of photochemical models of the MLT. Mainly because water vapor is the source gas for the hydrogen radicals (H, OH, HO_2), which determine the composition of the chemical active odd oxygen constituents (O, O_3, $O(^1D)$). The water vapor mixing ratio (hereafter mixing ratio means volume mixing ratio) influences such phenomena as noctilucent clouds (NLC) and polar mesospheric summer echoes (PMSE) at high latitudes in summer. The maximum water vapor mixing ratio in the middle atmosphere is found at high summer latitudes between 40 and 50 km and it is formed due to oxidation of methane.

The residual circulation is characterized by upward transport in the summer mesosphere and downward transport in the winter middle atmosphere. Thus, the water vapor mixing ratio at middle and high latitudes of the MLT is typically higher during summer than during winter. The formation of a strong meridional gradient of water vapor at the edge of the winter polar jet has recently been observed by the satellite and rockct measurements (Nassar et al. 2005; Lossow et al. 2009a, 2009b). Our simulation (Fig. 2a) reproduces the observed water vapor distribution in the middle atmosphere quite well.

Figure 2b depicts a latitude-height section of the absolute deviation of the zonally and diurnally averaged water vapor mixing ratio, i.e., the impact of GW mixing for the 30th of January. The maximum positive deviation occurs above ~80 km at high summer latitudes and amounts to 1.2 ppmv. At mesopause altitudes, water vapor still has a photochemical lifetime of ~10 days and is thus still subject to transport processes. Therefore, the rise of water vapor in the mesopause region is mainly determined by wave mixing. A similar argument can be applied to the negative deviation which occurs at low and equatorial latitudes from 55 to 70 km height, reaching ~−1.2 ppmv. Here, the photochemical lifetime of water

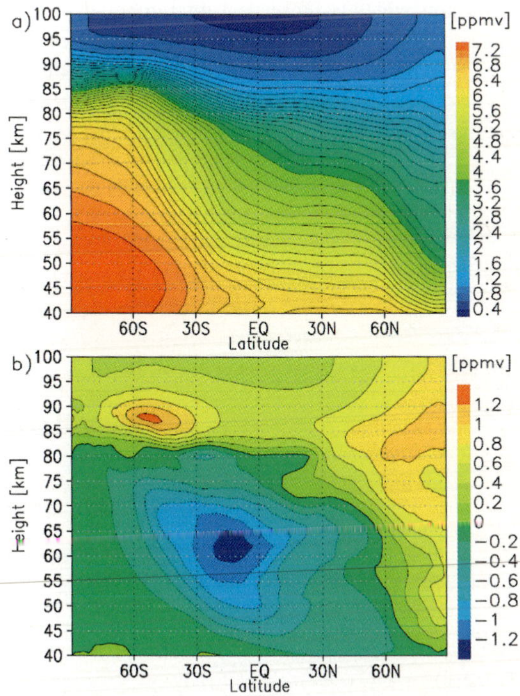

Fig. 2 Latitude-height section of the zonally and diurnally averaged water vapor mixing ratio (**a**) and its absolute deviation (**b**) for January 30

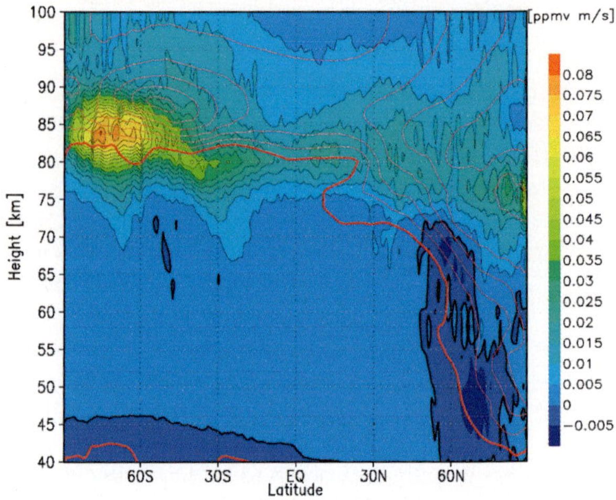

Fig. 3 GW flux of the water vapor mixing ratio for January 30. The *thin red lines* correspond to Fig. 2b and indicate the region of positive absolute deviation with the *bold red line* being the zero contour

vapor is longer than 1 month and molecular diffusion plays no role. Hence, the reduction of water vapor below the mesopause results from wave mixing too.

In order to prove that wave mixing directly determines the redistribution of water vapor, we calculate the flux due to GWs (Fig. 3), i.e. $[\overline{w^*c^*}]_{L>350\text{ km}} - [\overline{w^*c^*}]_{L>1000\text{ km}}$. Since the water vapor mixing ratio has a negative vertical gradient (Fig. 2a), its GW flux is positive (upward). This flux has a maximum in the summer mesopause region below the altitude of the maximum deviation (indicated by thin red lines in Fig. 3).

With increasing altitude the dissociation rates of water vapor grow. Water vapor dissociates into atomic hydrogen and hydroxyl. These are converted into the more stable molecular hydrogen, which is a reservoir of hydrogen in the upper MLT and becomes more abound, being comparable with water vapor, above ~70 km. The partitioning of the total H_2-equivalent between these two reservoirs depends on chemical processes, and particularly on autocatalytic cycles. Two autocatalytic cycles were identified in which hydroxyl plays the role of a catalyst (Sonnemann et al. 2005). In the first cycle,

$$\begin{aligned}
&H_2 + OH \rightarrow H_2O + H,\\
&H + O_2 + M \rightarrow HO_2 + M,\\
&HO_2 + O \rightarrow OH + O_2,\\
&\text{net: } H_2 + O \rightarrow H_2O,
\end{aligned}$$

atomic oxygen is the oxidizing agent. In the second cycle,

$$\begin{aligned}
&H_2 + OH \rightarrow H_2O + H,\\
&H + O_3 \rightarrow OH + O_2,\\
&\text{net: } H_2 + O_3 \rightarrow H_2O + O_2,
\end{aligned}$$

ozone plays the same role. In both cycles molecular hydrogen is oxidized into water vapor. Thus, GWs may modulate the catalytic cycles of water vapor formation in the mesopause region by altering the mixing ratios of the oxidants. Consequently, in addition to the direct effect of wave mixing, the redistribution of water vapor could be amplified by photochemical processes.

Fig. 4 Latitude-height section of the zonally and diurnally averaged molecular hydrogen mixing ratio (**a**) and its absolute deviation (**b**) for January 30

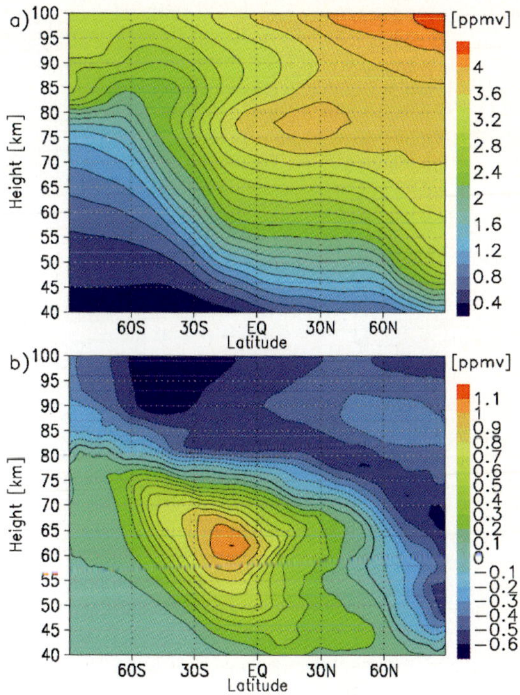

Furthermore, due to redistribution of water vapor we can expect an influence on short-lived constituents such as O_3, H, OH, and indirectly on chemical heating, radiative cooling, and radiative heating by the ozone. A direct impact on the cooling rate through infrared emission at 80 μm is negligible in the mesosphere since the contribution of water vapor is relatively weak and reaches only ~0.5 K/day at the water vapor maximum near 50 km (London 1980).

4.2 Molecular Hydrogen

Analogously to water vapor, one can expect strong influences of GW mixing on other long-lived constituents. Molecular hydrogen, for example, has a photochemical lifetime of more than 1 month throughout the entire MLT. Figure 4 displays latitude-height cross-sections of the absolute distribution and GW-induced deviation of the diurnally and zonally averaged molecular hydrogen mixing ratio at January 30. Since molecular hydrogen is a product of water vapor dissociation, it possesses a positive vertical gradient (Fig. 4a). Thus we can expect a downward transport by wave mixing. The maximum of the molecular hydrogen deviation (Fig. 4b) occurs at low latitudes between 55 and 65 km with values up to ~1 ppmv. On the other hand, GWs reduce molecular hydrogen above ~75–80 km, except for the polar night area where the reduction extends into the stratosphere. The largest reduction occurs in the summer hemisphere with a maximum decrease of ~0.6 ppmv, placed around 15°–60°S, 89–100 km height. The GW flux for molecular hydrogen (Fig. 5) is downward, except for the positive area around ~80–90 km at equatorial and middle latitudes where the vertical gradient of the mean mixing ratio has an opposite sign. The strongest downward flux occurs in the summer mesopause region at high latitudes and in the winter hemisphere around ~100 km.

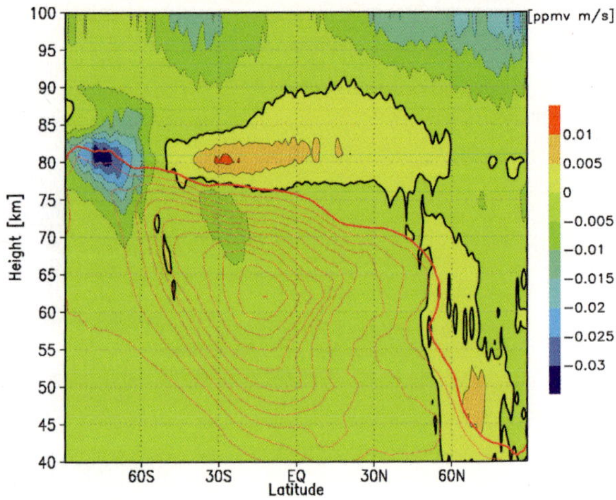

Fig. 5 GW flux of the molecular hydrogen mixing ratio. The *thin red lines* correspond to Fig. 4b and indicate the region of positive absolute deviation with the *bold red line* being the zero contour

4.3 Odd Hydrogen Family

Odd hydrogen constituents like hydroxyl (OH), hydroperoxy radicals (HO_2), and atomic hydrogen (H) mainly play a role in catalytic processes of ozone destruction (discussed below) and autocatalytic processes of water vapor formation (discussed above). Also, they are involved in exothermic chemical reactions. Moreover, monitoring the OH layer is widely used to measure physical quantities around the mesopause. For example, the emission from OH^* allows to infer temperatures (von Zahn et al. 1987; Offermann and Gerndt 1990; She and Lowe 1998; Bittner et al. 2002). Furthermore, important information about GWs (Tarasick and Shepherd 1992; Makhlouf et al. 1995; Taylor et al. 1995a, 1995b, 1997, 1998; Nielsen et al. 2006) and traveling planetary waves (Offermann et al. 2011) is obtained. Hydroxyl observations can be used to deduce atomic oxygen and atomic hydrogen mixing ratios (Thomas 1990; McDade and Llewellyn 1987; Takahashi et al. 1996). OH may also be used as a tracer for chemical processes that form or destroy this constituent. The hydroxyl radical is involved in the oxidation of methane and molecular hydrogen and it controls the odd oxygen chemistry of ozone.

The odd-hydrogens have short photochemical lifetimes, thus they can be directly influenced by GW mixing only in the frame of the odd-hydrogen family, or, indirectly, by redistribution of long-lived species such as water vapor and molecular hydrogen. Figures 6a and 6b illustrate the absolute distribution and deviation of odd hydrogen, respectively. The maximum odd-hydrogen mixing ratio is placed at high summer latitudes above ~90 km. As will be illustrated below, its distribution reflects the distribution of atomic hydrogen, which is the family forming constituent. The absolute deviation of odd-hydrogen (Fig. 6b) maximizes in the equatorial mesopause region, reaching ~1.2 ppmv.

The entire odd-hydrogen family has a photochemical lifetime of several hours in the mid mesosphere, but longer than 1 day above ~75 km (Brasseur and Solomon 1986). Hence, we expect a strong direct impact of GW mixing in this region, which moreover exhibits the largest vertical gradient of the absolute distribution and where the photochemical lifetime is longer than GW periods. Figure 7 shows the vertical flux of odd-hydrogen due to GW mixing. Since the vertical gradient of the absolute distribution is positive (Fig. 6a), the GW flux is negative (downward). It maximizes around 75–90 km at high summer latitudes, as is expected from Fig. 6.

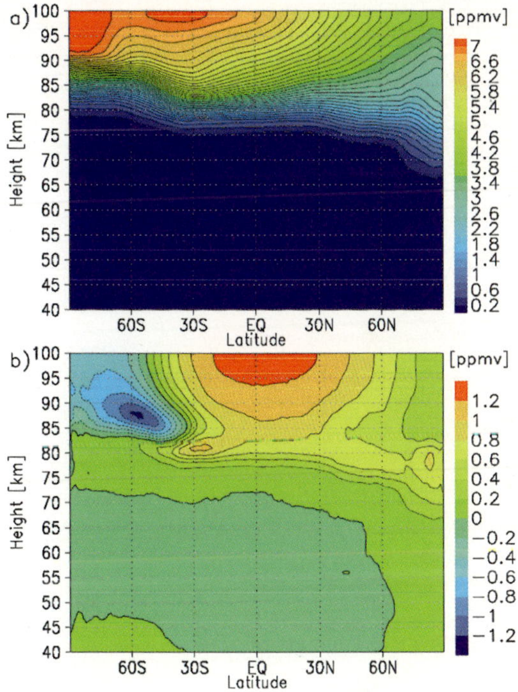

Fig. 6 Latitude-height section of the zonally and diurnally averaged odd-hydrogen mixing ratio (**a**) and its absolute deviation (**b**) for January 30

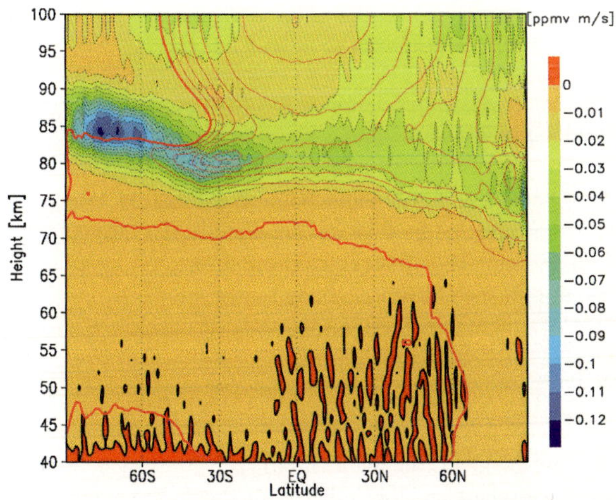

Fig. 7 GW flux of the odd-hydrogen mixing ratio. The *thin red lines* correspond to Fig. 6b and indicate the region of positive absolute deviation with the *bold red line* being the zero contour

The most important species among the odd-hydrogens are hydroxyl (for the reasons mentioned above) and atomic hydrogen. These constituents are most abundant above ∼70 km. Figure 8a shows the zonally and diurnally averaged hydroxyl mixing ratio. We infer the hydroxyl layer at ∼75–85 km with its summer-winter gradient and typical values of ∼8–12 ppbv. Note that the layer of vibrationally excited hydroxyl is placed ∼5 km above the layer of maximum hydroxyl concentration (Adler-Golden 1997). Hence, the signal from airglow measurements would, in fact, be expected to come from ∼80–90 km according to our model result. The strongest reduction of hydroxyl due to GWs (Fig. 8b) occurs precisely in

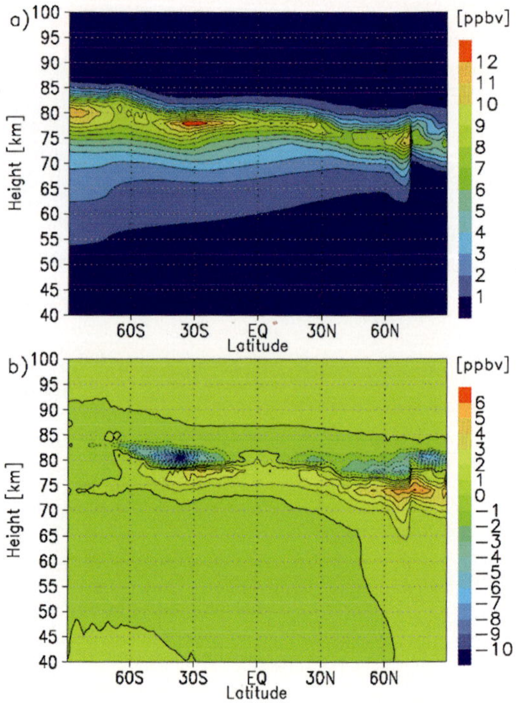

Fig. 8 Latitude-height section of the zonally and diurnally averaged hydroxyl mixing ratio (**a**) and its absolute deviation (**b**) for January 30

the region of the mesospheric hydroxyl layer. The decrease in the mixing ratio amounts to ~5–10 ppbv. This reduces the emission in the Meinel bands. The hydroxyl is only indirectly affected by GWs through the redistribution of atomic hydrogen and ozone, and through the modulation of reaction rates by GW-induced temperature variations.

The distribution of atomic hydrogen has a positive vertical gradient (Fig. 9a). In the mesopause region its photochemical lifetime is comparable to the periods of GWs. Thus, it can be affected by GW mixing directly. The structure and values of the atomic hydrogen redistribution (Fig. 9b) reflect those of the odd hydrogen redistribution (Fig. 6b). Thus, the direct influence of wave mixing on the atomic hydrogen is responsible for most of the odd hydrogen family redistribution.

4.4 Odd Oxygen Family

Atomic oxygen is the main odd oxygen constituent above ~70 km. It is produced in the upper mesosphere and lower thermosphere during daytime due to the dissociation of molecular oxygen by UV radiation. During night, atomic oxygen is converted into ozone by the 3-body reaction $O + O_2 + M \rightarrow O_3 + M$, forming a distinct nighttime secondary ozone maximum. Being a main source gas for exothermic chemical heating due to the three-body reaction $O + O + M \rightarrow O_2 + M$, atomic oxygen may modify the thermal regime in the mesopause region and is likely involved in feedback processes between dynamics and chemistry (Xu et al. 2001).

Figure 10a shows the zonally and diurnally averaged latitude-height cross-section of the atomic oxygen concentration at January 30. The absolute maximum of atomic oxygen is located around 80–100 km. The maximum in summer is stronger and located at somewhat

Fig. 9 Latitude-height section of the zonally and diurnally averaged atomic hydrogen mixing ratio (**a**) and its absolute deviation (**b**) for January 30

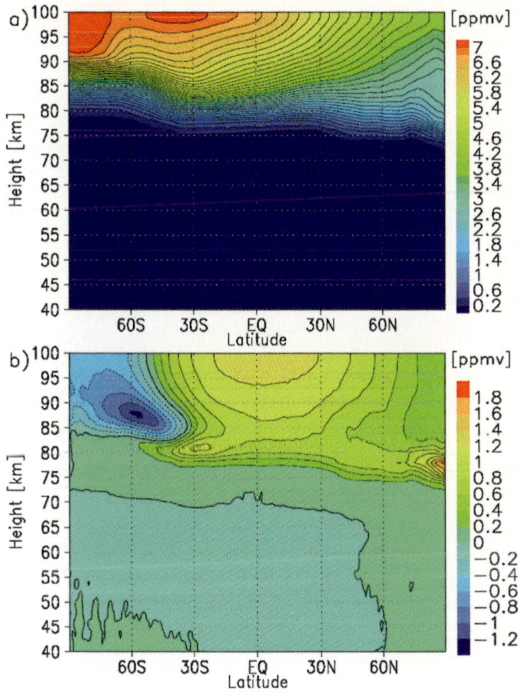

Fig. 10 Latitude-height section of the zonally and diurnally averaged atomic oxygen number density (**a**) and its absolute deviation (**b**) for January 30

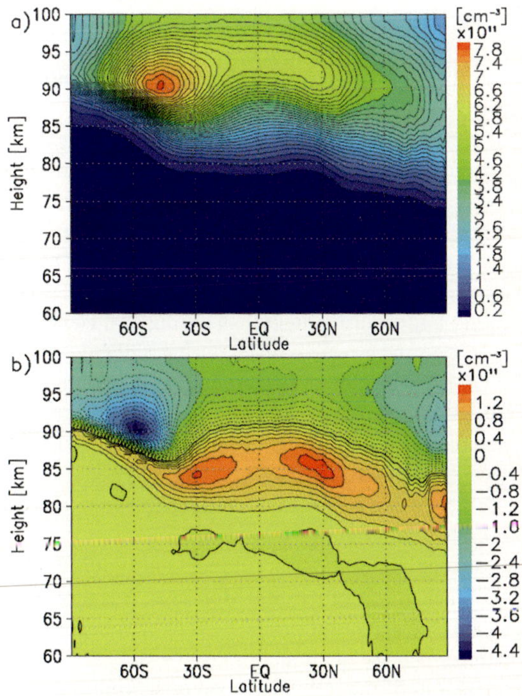

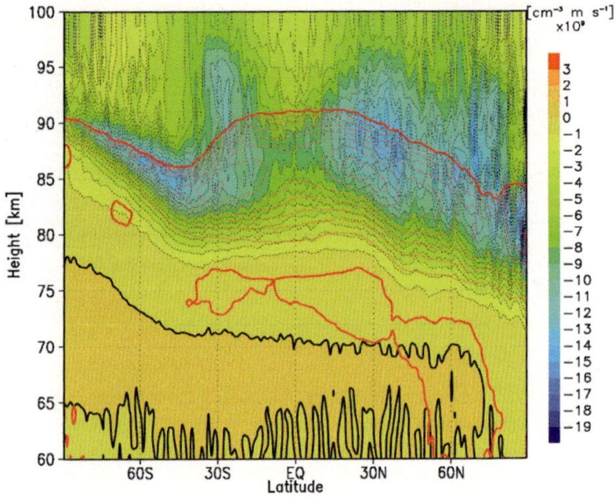

Fig. 11 GW flux of the atomic oxygen concentration. The *thin red lines* correspond to Fig. 10b and indicate the region of positive absolute deviation with the *bold red line* being the zero contour

higher altitudes than in winter. The maximum values vary in the range of 10^{11}–10^{12} cm^{-3} which is in good agreement with measurements (Russell et al. 2005; Smith et al. 2010).

The photochemical lifetime of atomic oxygen increases with altitude from several seconds in the lower mesosphere to about one hour beneath the mesopause, which is still lower than the characteristic transport time. Thus atomic oxygen is mainly in photochemical equilibrium throughout the mesosphere. However, around the mesopause and in the lower thermosphere its photochemical lifetime increases and becomes comparable to the characteristic large-scale transport time or even larger. Consequently, the strongest impact of GWs on the atomic oxygen distribution is expected at 80–100 km. The simulated atomic oxygen absolute deviations (Fig. 10b) illustrates that this is indeed the case. Owing to the influence of GWs, the concentration of atomic oxygen rises by $5 \cdot 10^{10}$–$1.5 \cdot 10^{11}$ cm^{-3} beneath the mesopause. The corresponding reduction of atomic oxygen by $2.5 \cdot 10^{11}$–$5 \cdot 10^{11}$ cm^{-3} occurs at 87–100 km.

Between 70 and 95 km the vertical gradient of atomic oxygen is large, which implies a large GW flux. Figure 11 depicts the atomic oxygen flux due to GWs together with the region of the positive absolute deviation (red lines). The maximum downward flux occurs at the location of the maximum absolute deviation and somewhat farther above. This finding provides strong evidence that GW mixing plays an important role for the distribution of atomic oxygen. Note that the vertical gradient of atomic oxygen around 70–85 km is significantly larger during night than during day. Hence, we expect the strongest influence of GWs on ozone and atomic oxygen during night.

It is well known that atomic oxygen induces nightglow of exited molecular oxygen ($O_2(b^1\Sigma_g^+)$ and $O_2(c^1\Sigma_u^-)$) through the Barth (two-stage population) or Chapman (direct population) mechanisms (Solheim and Llewellyn 1979; Murtagh et al. 1990; Hickey et al. 1993; Leko et al. 2002). Therefore, any reduction of atomic oxygen in the nightglow layer at ∼85–95 km will result in reduced volume emission. The effects on exothermic chemical heating will be addressed in Sect. 5.2.

Ozone is an important minor constituent in the MLT since it is involved in radiation and in chemical processes. Moreover, it is a reservoir of odd oxygen during nighttime. The balance between the atomic oxygen number density, which increases with altitude, and the number density of air and molecular oxygen, which both decrease with altitude, determines the height of the secondary ozone maximum which is typically placed around 85–95 km.

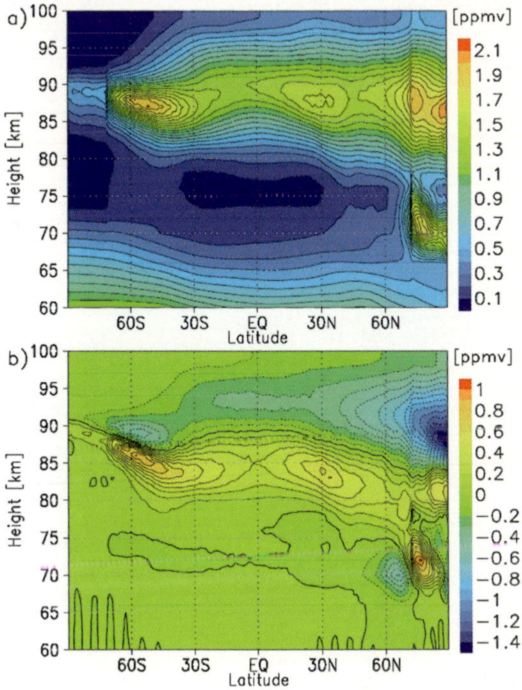

Fig. 12 Latitude-height section of the zonally and diurnally averaged ozone mixing ratio (**a**) and its absolute deviation (**b**) for January 30

Figure 12a exhibits the secondary ozone maximum at 85–95 km with zonally and diurnally averaged values up to ~2 ppmv which is consistent with recent satellite observations (Smith et al. 2009). The model also simulates the tertiary ozone maximum around ~70–75 km. This feature was discovered at high winter latitudes by observations (Marsh et al. 2001) and thoroughly investigated by modeling (Hartogh et al. 2004; Sonnemann et al. 2007).

The only reaction for ozone formation in the MLT is the three-body reaction $O + O_2 + M \rightarrow O_3 + M$. Therefore, it is not surprising that the absolute deviation of ozone reflects the corresponding distributions of atomic oxygen. Figure 12b represents the absolute deviation of the ozone mixing ratio due to the influence of GWs for January 30. The reduction of ozone occurs in the region of the secondary ozone layer by ~0.5 ppmv and maximizes near the winter pole up to ~1 ppmv. At 75–85 km, one can see a growth of the ozone mixing ratio by ~0.5–0.7 ppmv. Due to the very fast photolysis of ozone in the MLT its photochemical lifetime is only 1–2 minutes during daytime; hence, it is determined by photochemistry. During nighttime, however, the photochemical lifetime increases to several years below ~70 km, it decreases drastically to several minutes at ~70–80 km, and approaches 1 hour above ~80 km. Hence, ozone is directly affected by GWs during night below ~70 km.

5 Influence on the Thermal Regime

5.1 Microwave Cooling (CO_2)

Carbon dioxide is the main radiative cooling agent in the middle atmosphere. Figures 13a and 13b show height-latitude cross-sections of the distribution of carbon dioxide and its absolute deviation during January 30. A positive deviation due to GWs is found above ~85–90 km and amounts to ~120 ppmv. The corresponding reduction of carbon dioxide in the

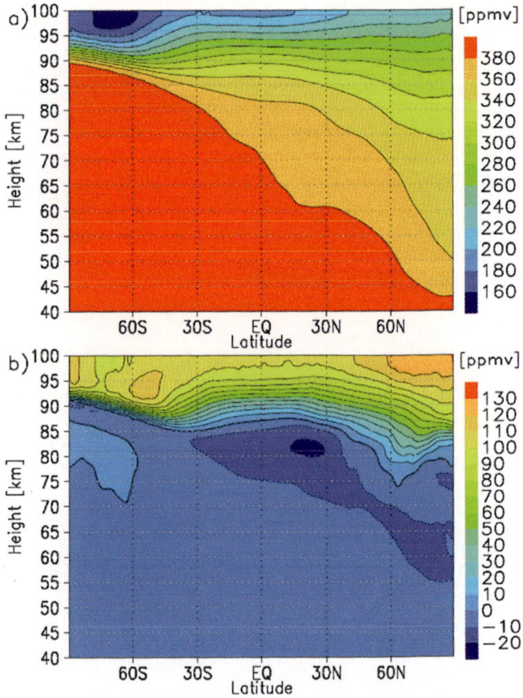

Fig. 13 Latitude-height section of the zonally and diurnally averaged carbon dioxide mixing ratio (**a**) and its absolute deviation (**b**) for January 30

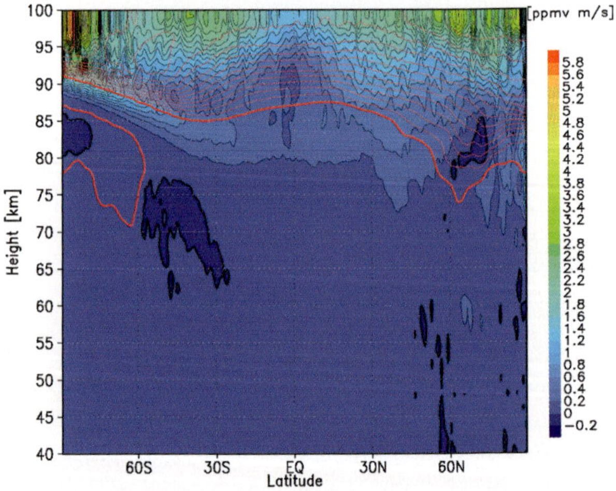

Fig. 14 GW flux of the carbon dioxide mixing ratio. The *thin red lines* correspond to Fig. 13b and indicate the region of positive absolute deviation with the *bold red line* being the zero contour

mesosphere does not exceed 20 ppmv. Since the photochemical life-time of carbon dioxide amounts to several years throughout the middle atmosphere, its redistribution can only be explained by GW mixing. Figure 14 shows the GW flux of carbon dioxide. Since the vertical gradient of carbon dioxide is negative, this flux is upward. As expected, it maximizes beneath the strongest absolute deviation, which is indicated in Fig. 14 by red contours.

Radiative transfer in the 15 μm band of carbon dioxide under non-LTE conditions is the main diabatic cooling mechanism in the MLT. Large cooling rates on average of about 13 K/d are found between 80 and 100 km K/day (London 1980). The strong dissociation of

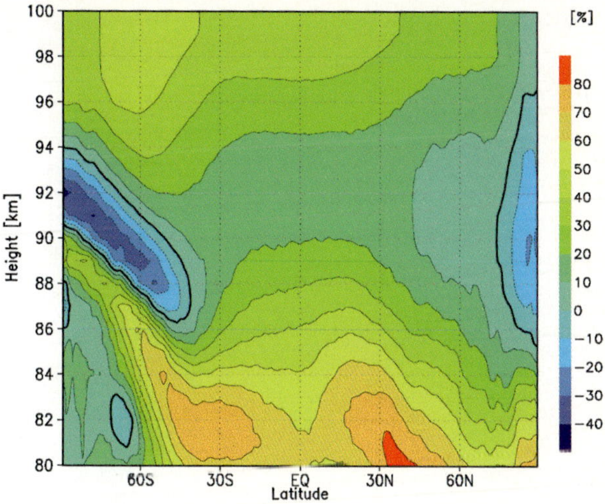

Fig. 15 Relative deviation of cooling rate [%] due to redistribution of carbon dioxide and atomic oxygen by GWs

molecular oxygen above the mesopause generates atomic oxygen. As a result, the anelastic collisions between atomic oxygen and carbon dioxide reduce non-LTE, and this becomes the dominant pathway for radiative cooling in this region. Assuming that the cooling rate (CR) due to infrared emission by carbon dioxide is proportional to the number density of carbon dioxide, $[CO_2]$, and the number density of atomic oxygen, [O], we can estimate the relative deviation of the cooling rate at about 80–100 km that is due to the redistribution of carbon dioxide by GWs as

$$RD_{CR} = \frac{([CO_2][O])|_{L>350\ \mathrm{km}} - ([CO_2][O])|_{L>1000\ \mathrm{km}}}{([CO_2][O])|_{L>350\ \mathrm{km}}}.$$

Figure 15 presents the corresponding estimate. The cooling rate increases at high and middle latitudes around the summer mesopause by up to 80%. There is a region of decrease by 20–40% around 86–92 km at high latitudes during the summer. During winter a reduction by 20% occurs at high latitudes centered round 90 km. Below 70 km we can apply the "cooling to space" approximation (Brasseur and Solomon 1986), yielding that the relative cooling rate deviation does not exceed 5.5% (not shown).

5.2 Chemical Heating Rates

Figure 16a displays a latitude-height cross-section of the zonally and diurnally averaged total chemical heating rate. The maximum is placed at 80–95 km with values up to ~27 K/day around ~88 km at middle latitudes in summer which is in agreement with results of Xu et al. (2000a). It is general wisdom that the strongest chemical heating results from the recombination of atomic oxygen (Mlynczak and Solomon 1993; Sonnemann et al. 1997; Xu et al. 2000a).

As indicated in Fig. 16b, the maximum chemical heating at summer middle latitudes around the mesopause is strongly reduced due to GWs. Nevertheless, it is still very strong in the control simulation (Fig. 16a). However, the K_{turb} in the MECTM is likely too weak around the summer mesopause since it does not reflect the pronounced summer-winter asymmetry known from measurements and models (Lübken 1997; Becker 2009). With regard to (11) we therefore expect that a more consistent K_{turb} would give rise to a much weaker chemical heating maximum in summer.

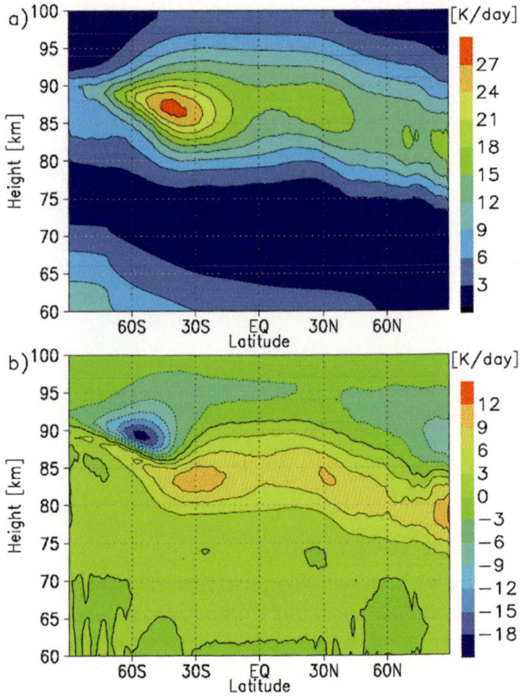

Fig. 16 Latitude-height section of the zonally and diurnally averaged exothermic chemical heating (**a**) and its absolute deviation (**b**) for January 30

The reduction of chemical heating above ~85 km (Fig. 16b) induced by GWs mainly results from the decrease of the reaction yield of $O + O + M \rightarrow O_2 + M$, and is determined by the decrease of atomic oxygen in this region (Fig. 10b). While this explanation for GW-induced changes of chemical heating holds at high and middle summer latitudes around 85–90 km, the situation at high winter latitudes around 90–95 km is different. Here, the change in atomic oxygen recombination contributes only by ~50%, while the rest of the decrease in chemical heating results from reduced rates of the following reactions (in the order of significance): $H + O_3 \rightarrow OH + O_2$, $O + O_2 + M \rightarrow O_3 + M$, $O + OH \rightarrow H + O_2$, and $O + O_3 \rightarrow 2O_2$. GW mixing brings additional atomic oxygen into the height region beneath 85 km, where it contributes to the exothermic formation of ozone due to the reaction $O + O_2 + M \rightarrow O_3 + M$. A more specific analysis of our model data showed that the increased yield of this reaction contributes by about ~40% to the chemical heating increase beneath 85 km. In turn, the increased production of ozone enhances the production from its exothermic reaction with atomic hydrogen, $H + O_3 \rightarrow OH + O_2$, which yields another ~40% to the chemical heating increase beneath 85 km. The so-produced hydroxyl reacts with atomic oxygen according to $O + OH \rightarrow H + O_2$, which accounts for ~20% of the relative increase in chemical heating. The combination of these three reactions modifies the first catalytic cycle of odd oxygen destruction:

$$O + O_2 + M \rightarrow O_3 + M$$
$$H + O_3 \rightarrow OH + O_2$$
$$O + OH \rightarrow H + O_2$$
$$\text{Net: } O + O + M \rightarrow O_2 + M$$

Here, hydroxyl and atomic hydrogen play the role of catalysts.

The change of the thermal regime by the exothermic chemical heating due to redistribution of minor chemical constituents induced by GWs is expected to feed back on the dynamical system; for example, it may provide destabilization of GWs (Leovy 1966; Xu et al. 2001). A thorough modeling of such effects would require a GW resolving global chemistry climate model for the MLT, which has not been developed yet.

6 Effective Diffusivity

Here we utilize the concept of effective diffusivity (see Sect. 3) for an analysis of GW mixing in the MLT. We can characterize the mixing processes by estimating the vertical wave diffusion (WD) coefficient K_{waves} due to the resolved GWs from the simulated distributions of long-lived chemical constituents. Figure 17 shows this coefficient, i.e., $K_{waves}|_{L>350\text{ km}} - K_{waves}|_{L>1000\text{ km}}$ calculated using (8) for six long-lived chemical constituents that can be considered as passive tracers in the MLT: Water vapor, methane, molecular hydrogen, atomic oxygen, nitric oxide, and nitrous oxide. Similar distributions occur for K_{waves} when calculated for different chemical constituents in regions where the given species are passive tracers. In particular, all calculated fields display a local maximum at high summer latitudes between ~75–90 km, which is the region of maximum vertical wind variations due to GWs (Fig. 1). The height of the maximum shifts downward toward the middle latitudes consistent with the distribution of the GW kinetic energy (see Fig. 8b in Becker 2009). A local minimum occurs at ~90–95 km in summer, representing some kind of a vertical mixing barrier. Secondary maxima are found in the lower thermosphere at middle and high latitudes in each hemisphere. These maxima are possibly induced by those GWs not being subject to turbulent damping around and below the mesopause. The WD coefficients minimize in the equatorial MLT which is at least partly caused by the lack tropical GW sources in the present mechanistic GCM.

The differences in K_{waves} for the different constituents, including the occurrence of singularities, have several reasons. The weak negative region below ~65 km around 60°N in K_{waves} as calculated for water vapor, methane, molecular hydrogen, and nitrous oxide (Figs. 17a, b, c, and f, respectively) occurs due to the polar vortex and the consequent formation of strong meridional gradients. In this case, the assumption of a dominant vertical mixing as applied in Sect. 3 does not hold, i.e., the GW tracer is dominated by meridional mixing due to low-frequency GWs which have smaller vertical and larger horizontal winds compared to the mid-frequency GWs that dominate K_{waves} in the summer mesopause region. Indeed, water vapor and molecular hydrogen both exhibit strong meridional gradients near the polar night jet (Figs. 2a and 4a, respectively). In this case a more complicated, 3-dimensional wave mixing concept as outlined by Medvedev and Greatbatch (2004) should be employed. This is, however, beyond the scope of the present study.

If, on the other hand, $\partial_z[\overline{c}]$ tends to zero for a particular constituent, (8) yields a singularity in K_{waves}. This is the case for the two layers of a diverging K_{waves} around 80 and 90 km at low latitudes for molecular hydrogen (Fig. 17c). In fact, Fig. 4a indicates changes in sign for the vertical derivative of the mean molecular hydrogen mixing ratio in these regions.

If in some region chemical reactions actively influence the distribution of species, we cannot consider the given constituent as a passive tracer. This is the case for atomic oxygen below ~70 km (Fig. 17d), where its photochemical lifetime is even shorter than a typical GW period (a few hours). The transition region toward long photochemical lifetimes is characterized in Fig. 17d by a line of singularities around 70 km which probably arises because of a resonance-like effect when the photochemical lifetime is equal to the GW period such as to generate $\partial_z[\overline{c}] \sim 0$ for atomic oxygen.

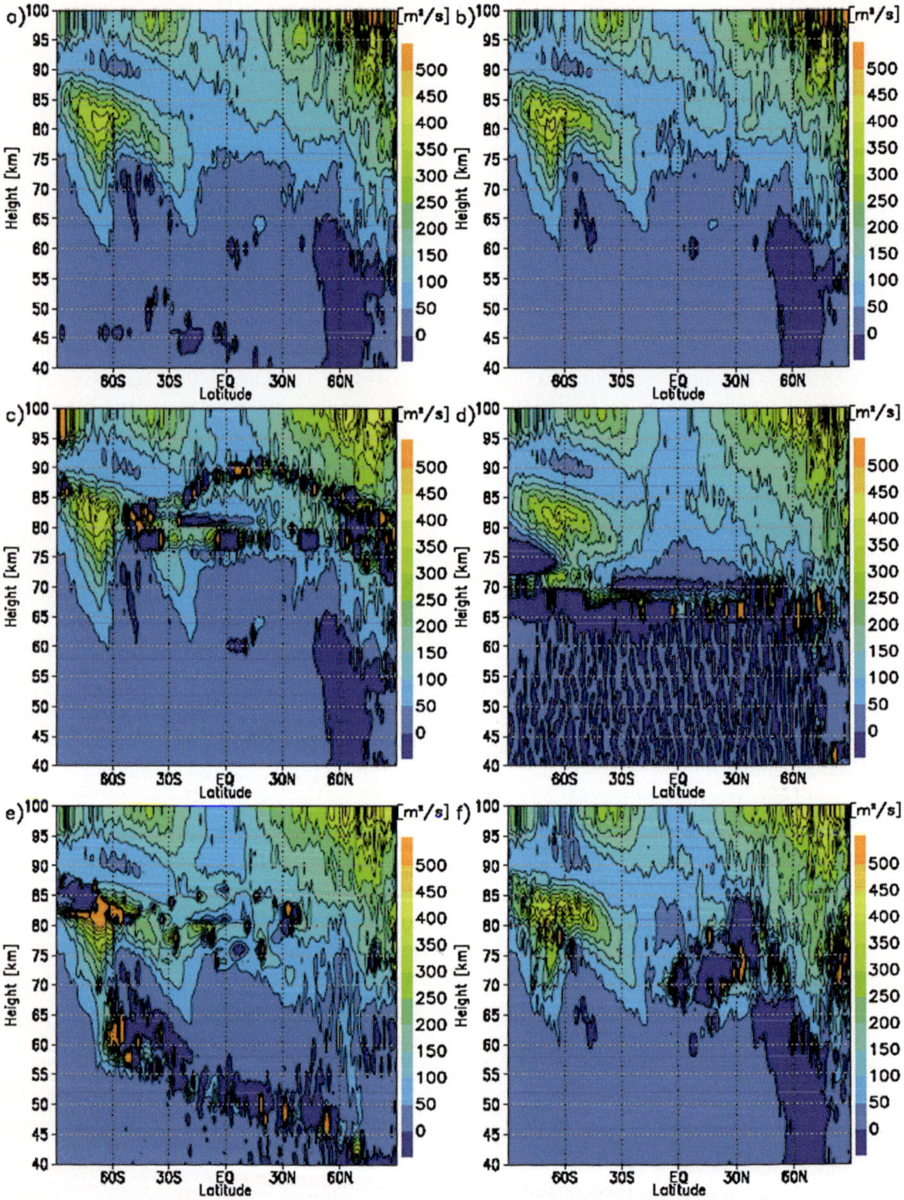

Fig. 17 Latitude-height section of the wave diffusion coefficient ($K_{waves|L>350\ \mathrm{km}} - K_{waves|L>1000\ \mathrm{km}}$) for (**a**) water vapor, (**b**) methane, (**c**) molecular hydrogen, (**d**) atomic oxygen, (**e**) nitric oxide, and (**f**) nitrous oxide for January 30

Figure 18 presents the WD coefficients as computed from case 1 ($K_{waves|L>350\ \mathrm{km}}$) and case 2 ($K_{waves|L>1000\ \mathrm{km}}$) separately for water vapor (Fig. 18a, b) and methane (Fig. 18c, d). These results reveal that the WD coefficients are clearly dominated by the mesoscales, i.e., the resolved GWs. Nonetheless, larger-scale GWs motions are significant in the polar night stratopause region and in the lower thermosphere. Like in Fig. 17 the WD coefficients are

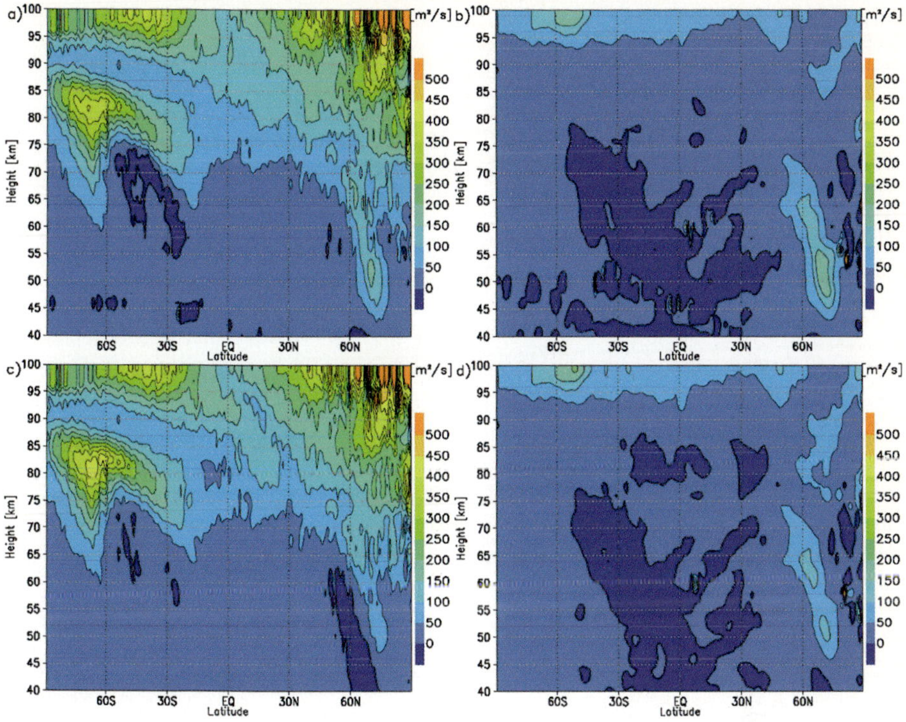

Fig. 18 Latitude-height section of the wave diffusion coefficient $K_{waves|L>350\text{ km}}$ for the control run (case 1, *left*) and $K_{waves|L>1000\text{ km}}$ for the conventional run (case 2, *right*) calculated for water vapor (**a**, **b**) and methane (**c**, **d**) for January 30

quite similar for water vapor and methane. This holds also for the other constituents (not shown).

Figure 19 illustrates that K_{waves} is almost identical for several long-lived minor chemical constituents above about 75 km, except for the regions where a given species can not be considered as a passive tracer or where $\partial_z[\overline{c}] \sim 0$. In general, such a result is not surprising, as was noted by Nakamura (2001) for the upper troposphere-lower stratosphere (UTLS). Haynes and Shuckburgh (2000a, 2000b) found that the meridional effective diffusivity for planetary-scale disturbances is independent of the tracer initial distribution in the stratosphere. They argue that passive tracers form linear relationships according to $c^i = a_{ij} + b_{ij}c^j$, where i and j enumerate the constituents and the coefficients a_{ij} and b_{ij} are constants; then one trivially obtains $K^i_{eff} = K^j_{eff}$. The tendency of long-lived tracers to develop linear relationships was shown for the stratosphere based on measurements (Plumb and Ko 1992; Plumb 2007) and on simulations (Ma 1999). To the best of our knowledge, there is no confirmation that this tendency toward compact linear relationships extends into the MLT. For instance it is quite obvious that for the height region from 70 to 100 km, the mean water vapor distribution (Fig. 2a) can hardly be converted into the mean atomic oxygen distribution (Fig. 10a) by a linear transformation than is independent of latitude and height. Nevertheless the corresponding wave diffusion coefficients are more or less the same (Figs. 17a and d).

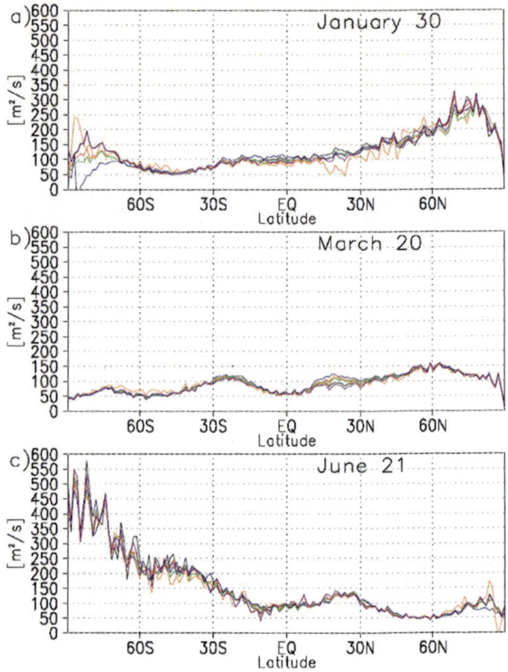

Fig. 19 Wave diffusion coefficient ($K_{waves|L>350\ \mathrm{km}} - K_{waves|L>1000\ \mathrm{km}}$) calculated for distributions of water vapor (*black*), methane (*violet*), carbon dioxide (*green*), atomic oxygen (*red*), nitric oxide (*blue*), and nitrous oxide (*orange*), at 88 km for January 30 (**a**), March 20 (**b**) and June 21 (**c**)

Alternatively to the assumption of compact linear relationships, we argue that for GW mixing in the MLT we can roughly assume that

$$-i\omega_I c^* \sim w^* \partial_z[\overline{c}], \tag{12}$$

where ω_I is the intrinsic frequency of a monochromatic mid-frequency GW. Denoting its vertical and horizontal wave numbers by m and k, respectively, and the Brunt-Väisällä frequency as N, we use the GW dispersion relation, $m = kN/\omega_I$, as well as $\partial_z c^* \sim imc^*$ in order to transform (12) into

$$[\overline{(\partial_z c^*)^2}] \sim \frac{m^4}{N^2k^2}[\overline{w^{*2}}](\partial_z[\overline{c}])^2. \tag{13}$$

Consequently, from (8) and the GW continuity equation $ku^* + mw^* = 0$ we get

$$K_{waves} \sim K_z \frac{m^4}{N^2k^2}[\overline{w^{*2}}] \sim K_z \frac{m^2}{N^2}[\overline{u^{*2}}], \tag{14}$$

where u^* is the GW horizontal wind perturbation. This expression does not depend on a particular tracer distribution. It may furthermore be written as a sum over a discrete GW spectrum. Hence, K_{waves} can approximately be considered as an intrinsic property of the GW activity in the MLT. In fact, the spatial distribution of K_{waves} is strongly reminiscent of the simulated GW kinetic energy during January in the KMCM (see Fig. 20).

7 Summary and Conclusions

From a sensitivity experiment with an advanced global chemistry-transport model MECTM for the MLT we have inferred the effects of resolved GWs on minor chemical constituents.

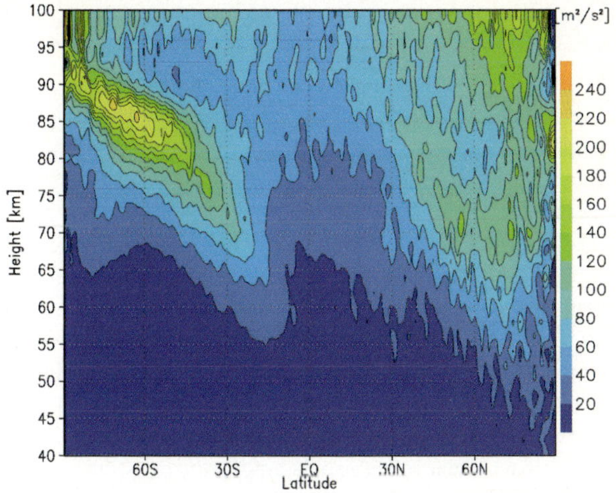

Fig. 20 Kinetic energy contained in horizontal wavelengths between 350 and 1000 km, zonally and diurnally averaged for January 30

For this purpose we have used the dynamical fields from a high-resolution mechanistic GCM (KMCM). This model includes an advanced parameterization of turbulent diffusion, giving rise to a self-consistent and explicit simulation of GWs and their effects on the large-scale flow. An annual cycle from the KMCM has been utilized with different spectral filtering in order to drive the MECTM: A control simulation with the full dynamical fields including meso-scale GW motions (case 1) and a conventional simulation where the MECTM is driven just by the large-scale flow from the KMCM (case 2). The differences between the 2 simulations with the MECTM reveal substantial effects of GWs on the distributions of chemical species, as well as on chemical and radiative heating rates. The general implication of these results is that, in the temporal and zonal average, minor constituents in the MLT are not simply transported by the wave-driven residual circulation and mixed by the turbulent diffusion induced by GW breakdown and by molecular diffusion, as is assumed in comprehensive modeling of the photochemistry of the MLT (e.g., Schmidt et al. 2006; Garcia et al. 2007; Megner et al. 2008). Rather, both the turbulent diffusion implied by GW breakdown and the molecular diffusion can be substantially amplified due to an effect that is known as wave mixing (e.g., Medvedev and Greatbatch 2004; Osborn and Cox 1972). According to quasi linear theory, the additional diffusivity is proportional to the squared GW variations of the tracer field divided by the squared large-scale tracer gradient. A rigorous description of GW mixing is only achieved when the GWs are part of the resolved flow. The present sensitivity experiment allows estimate the effects of GW mixing in relation to the conventional modeling approach.

Our results show that GWs have direct, down-gradient mixing effects on minor chemical constituents such as water vapor, molecular hydrogen, carbon dioxide, methane, atomic oxygen, and nitrous oxide, all of which possess a mean vertical gradient in the MLT and have photochemical lifetime larger than a typical GW period. An additional influence arises from the GW-induced temperature variations that modulate the molecular diffusion coefficient. Strong GW influences on carbon dioxide, which is mixed upwards across the mesopause, and on atomic oxygen, which is mixed downward, lead to modulations of radiative cooling in the mesopause region.

The redistribution of long-lived species leads in addition to redistribution of short-lived chemical constituents like odd-hydrogen radicals and ozone. Hence, short-lived constituents are indirectly affected by GWs. The temperature variations also affect species with short photochemical lifetime via the temperature-dependence of reaction rates. However, the strongest effects on the photo-chemistry of short-lived constituents are induced by the changes in the long-lived constituents. The so-induced redistributions of short-lived species modulate hydroxyl emission, catalytic processes of odd oxygen destruction, and autocatalytic cycles of water vapor production. Furthermore, chemical heating rates increase at ~75–85 km and decrease above. Thus GW mixing modulates the thermal regime of mesopause region and the lower thermosphere quite effectively, and this may induce further feedback mechanisms.

Utilizing the effective diffusivity concept, we have calculated the wave diffusion coefficient, which represents the contribution to the effective diffusion coefficient due to mixing by resolved GWs for a particular constituent. We have found that the wave diffusion coefficient is approximately the same for all passive tracers in the MLT. We have then argued that, unlike to the meridional mixing by planetary waves in the stratosphere, the similarity of the vertical GW mixing for different constituents in the MLT is not necessarily a consequence of compact linear relationships among the constituents. Rather, the effective diffusivity due to GWs may represent intrinsic properties of the GW field. It is not necessarily proportional to the turbulent plus molecular diffusion assumed in the CTM.

A corresponding estimate of the wave diffusion (WD) coefficient for passive tracers has been given (see (14)). Turning this argument the other way round, estimates of the GW mixing from mean and variable tracer fields allow to estimate the GW activity in terms of the GW kinetic energy times the corresponding squared vertical wave number. Indeed, our numerical results for the WD coefficient resemble the distribution of the simulated GW kinetic energy. It should thus be possible to use estimates of the WD coefficients to study long-term variability of the GW activity in the MLT, particularly in the summer mesopause region.

References

S. Adler-Golden, Kinetic parameters for OH nightglow modeling consistent with recent laboratory measurements. J. Geophys. Res. **102**(A9), 19,969–19,976 (1997). doi:10.1029/97JA01622

D.R. Allen, N. Nakamura, A seasonal climatology of effective diffusivity in the stratosphere. J. Geophys. Res. **106**(D8), 7917–7935 (2001)

E. Becker, Sensitivity of the upper mesosphere to the Lorenz Energy Cycle of the Troposphere. J. Atmos. Sci. **66**, 647–666 (2009). doi:10.1175/2008JAS2735.1

E. Becker, Dynamical control of the middle atmosphere. Space Sci. Rev. (2011). doi:10.1007/s11214-011-9841-5

E. Becker, C. von Savigny, Dynamical heating of the polar summer mesopause induced by solar proton events. J. Geophys. Res. **115**, D00I18 (2010). doi:10.1029/2009JD012561

A.N. Belyaev, An emission layer as a gravity wave detector. J. Atmos. Sol.-Terr. Phys. **71**, 1974–1981 (2009). doi:10.1016/j.jastp.2009.09.001

M. Bittner, D. Offermann, H.-H. Graef, M. Donner, K. Hamilton, An 18 year time series of OH rotational temperatures and middle atmosphere decadal variations. J. Atmos. Sol.-Terr. Phys. **64**, 1147–1166 (2002)

G. Brasseur, S. Solomon, *Aeronomy of the Middle Atmosphere: Chemistry and Physics of the Stratosphere and Mesosphere*, 2nd edn. (Reidel, Dordrecht, 1986)

F.D. Colegrove, W.B. Hanson, F.S. Johnson, Eddy diffusion and oxygen transport in the lower thermosphere. J. Geophys. Res. **70**, 4931 (1965)

C. Eden, R. Greatbatch, D. Olbers, Interpreting eddy fluxes. J. Phys. Oceanogr. **37**, 1282–1296 (2007). doi:10.1175/JPO3050.1

C. Eden, D. Olbers, R. Greatbatch, A generalised Osborn-Cox relation. J. Fluid Mech. **632**, 457–474 (2009). doi:10.1017/S0022112009007484

B. Fichtelmann, G. Sonnemann, On the variation of ozone in the upper mesosphere and lower thermosphere: a comparison between theory and observation. Z. Meteorol. **9**(6), 297–308 (1989)

J.E. Frederick, Influence of gravity wave activity on lower thermospheric photochemistry and composition. Planet. Space Sci. **27**, 1469–1477 (1979)
D.C. Fritts, M.J. Alexander, Gravity wave dynamics and effects in the middle atmosphere. Rev. Geophys. **41**(1), 1003 (2003). doi:10.1029/2001RG000106
R.R. Garcia, S. Solomon, The effect of breaking gravity waves on the dynamics and chemical composition of the mesosphere and lower thermosphere. J. Geophys. Res. **90**, 3850–3868 (1985)
R.R. Garcia, D.R. Marsh, D.E. Kinnison, B.A. Boville, F. Sassi, Simulation of secular trends in the middle atmosphere, 1950–2003. J. Geophys. Res. **112**, D09301 (2007). doi:10.1029/2006JD007485
M. Grygalashvyly, G.R. Sonnemann, P. Hartogh, Long-term behavior of the concentration of the minor constituents in the mesosphere—a model study. Atmos. Chem. Phys. **9**, 2779–2792 (2009)
M. Grygalashvyly, E. Becker, G.R. Sonnemann, Wave mixing effects on minor chemical constituents in the MLT-region: results from a global CTM driven by high-resolution dynamics. J. Geophys. Res. **116**, D18302 (2011). doi:10.1029/2010JD015518
P. Haynes, E. Shuckburgh, Effective diffusivity as a diagnostic of atmospheric transport, 1. Stratosphere. J. Geophys. Res. **105**(D18), 22,777–22,794 (2000a)
P. Haynes, E. Shuckburgh, Effective diffusivity as a diagnostic of atmospheric transport, 2. Troposphere and lower stratosphere. J. Geophys. Res. **105**(D18), 22,795–22,810 (2000b)
P. Hartogh, C. Jarchow, G.R. Sonnemann, M. Grygalashvyly, On the spatiotemporal behavior of ozone within the mesosphere/mesopause region under nearly polar night conditions. J. Geophys. Res. **109**, D18303 (2004). doi:10.1029/2004JD004576
P. Hartogh, G.R. Sonnemann, M. Grygalashvyly, S. Li, U. Berger, F.-J. Luebken, Water vapor measurements at ALOMAR over a solar cycle compared with model calculations by LIMA. J. Geophys. Res. **115**, D00I17 (2010). doi:10.1029/2009JD012364
M.P. Hickey, R.L. Walterscheid, Wave-modified mean exothermic heating in the mesopause region. Geophys. Res. Lett. **21**, 2413–2416 (1994)
M.P. Hickey, G. Schubert, R.L. Walterscheid, Seasonal and latitudinal variations of gravity wave-driven fluctuations in OH nightglow. J. Geophys. Res. **97**, 14,911–14,922 (1992)
M.P. Hickey, G. Schubert, R.L. Walterscheid, Gravity wave-driven fluctuations in the O_2 atmospheric (0–1) nightglow from an extended, dissipative emission region. J. Geophys. Res. **98**, 13717–13730 (1993)
M.P. Hickey, R.L. Walterscheid, M.J. Taylor, W. Ward, G. Schubert, Q. Zhou, F. Garcia, M.C. Kelly, G.G. Shepherd, A numerical calculations of gravity waves imaged over Arecibo during the 10-day January 1993 campaign. J. Geophys. Res. **102**, 11,475 (1997)
M.P. Hickey, R.L. Walterscheid, P.G. Richards, Secular variations of atomic oxygen in the mesopause region induced by transient gravity wave packets. Geophys. Res. Lett. **27**, 3599–3602 (2000)
M.P. Hickey, T.-Y. Huang, R.L. Walterscheid, Gravity wave packet effects on chemical exothermic heating in the mesopause region. J. Geophys. Res. **108**, 1448–1454 (2003). doi:10.1029/2002JA009363
P. Hoffmann, E. Becker, W. Singer, M. Placke, Seasonal variation of mesospheric waves at northern middle and high latitudes. J. Atmos. Sol.-Terr. Phys. **72**, 1068–1079 (2010). doi:10.1016/j.jastp.2010.07.002
T.-Y. Huang, Simulations of OH nightglow emission in the occurrence of sprites. J. Geophys. Res. **111**, A11311 (2006)
T.-Y. Huang, M.P. Hickey, On the latitudinal variations of the non-periodic response of minor species induced by a dissipative gravity-wave packet in the MLT region. J. Atmos. Sol.-Terr. Phys. **69**, 741–757 (2007). doi:10.1016/j.jastp.2007.01.011
T.-Y. Huang, M.P. Hickey, T.-F. Tuan, On nonlinear response of minor species with a layered structure to gravity waves. J. Geophys. Res. **108**, 1173–1184 (2003). doi:10.1029/2002JA009497
H. Körnich, G. Schmitz, E. Becker, The role of stationary waves in the maintenance of the Northern Annular Mode as deduced from model experiments. J. Atmos. Sci. **63**, 2931–2947 (2006)
S.V. Kostrykin, G. Schmitz, Effective diffusivity in the middle atmosphere based on general circulation model winds. J. Geophys. Res. **111**, D02304 (2006). doi:10.1029/2004JD005472
Ch. Kremp, U. Berger, P. Hoffmann, D. Keuer, G.R. Sonnemann, Seasonal variation of middle latitude wind fields of the mesopause region—a comparison between observation and model calculation. Geophys. Res. Lett. **26**, 1279–1282 (1999)
J.J. Leko, M.P. Hickey, P.G. Richards, Comparison of simulated gravity wave-driven mesospheric airglow fluctuations observed from the ground and space. J. Atmos. Sol.-Terr. Phys. **64**, 397–403 (2002)
C.B. Leovy, Photochemical destabilization of gravity wave near the mesopause. J. Atmos. Sci. **23**, 223–232 (1966)
R.S. Lindzen, Turbulence and stress owing to gravity wave and tidal breakdown. J. Geophys. Res. **86**, 9707–9714 (1981)
J. London, Radiative energy sources and sinks in the stratosphere and mesosphere, in *Proc. NATO Advanced Study Institute on Atmospheric Ozone: Its Variation and Human Influences, Algarve, Portugal*, Report FAAEE-80-20, 703 (1980)

S. Lossow, J. Urban, H. Schmidt, D.R. Marsh, J. Gumbel, P. Eriksson, D. Murtagh, Wintertime water vapor in the polar upper mesosphere and lower thermosphere: first satellite observations by Odin submillimeter radiometer. J. Geophys. Res. **114**, D10304 (2009a). doi:10.1029/2008JD011462
S. Lossow, M. Khaplanov, J. Gumbel, J. Stegman, G. Witt, P. Dalin, S. Kirkwood, F.J. Schmidlin, K.H. Fricke, U. Blum, Middle atmospheric water vapour and dynamics in the vicinity of the polar vortex during the Hygrosonde-2 campaign. Atmos. Chem. Phys. **9**, 4407–4417 (2009b)
F.-J. Lübken, Seasonal variation of turbulent energy dissipation rates at high latitudes as determined by in situ measurements of neutral density fluctuations. J. Geophys. Res. **102**, 13,441–13,456 (1997)
J. Ma, The modified Lagrangian-mean diagnostics of the stratospheric transport and chemistry, Ph.D. dissertation, University of Chicago, 130 pp., 1999
U.B. Makhlouf, R.H. Picard, J.R. Winick, Photochemical-dynamical modeling of the measured response of airglow to gravity waves. 1. Basic model for OH airglow. J. Geophys. Res. **100**, 11289–11311 (1995)
U.B. Makhlouf, R.H. Picard, J.R. Winick, T.F. Tuan, A model for the response of the atomic oxygen 557.7 nm and the OH Meinel airglow to atmospheric gravity waves in a realistic atmosphere. J. Geophys. Res. **103**, 6261–6269 (1998)
D. Marsh, A. Smith, G. Brasseur, M. Kaufmann, K. Grossmann, The existence of a tertiary ozone maximum in the high latitude middle mesosphere. Geophys. Res. Lett. **28**, 4531–4534 (2001)
I.C. McDade, E.J. Llewellyn, Kinetic parameters related to sources and sinks of vibrationally excited OH in the nightglow. J. Geophys. Res. **92**, 7643–7650 (1987)
A.S. Medvedev, R.J. Greatbatch, On advection and diffusion in the mesosphere and lower thermosphere: the role of rotational fluxes. J. Geophys. Res. **109**, D07104 (2004). doi:10.1029/2003JD003931
L. Megner, D.E. Siskind, M. Rapp, J. Gumbel, Global and temporal distribution of meteoric smoke: a two dimensional simulation study. J. Geophys. Res. **113**, D03202 (2008). doi:10.1029/2007JD009054
M.G. Mlynczak, S. Solomon, A detailed evaluation of the heating efficiency in the middle Atmosphere. J. Geophys. Res. **98**, 10,517–10,541 (1993)
K.W. Morton, D.F. Mayers, *Numerical Solution of Partial Differential Equations* (Cambridge University Press, Cambridge, 1994)
D.P. Murtagh, G. Witt, J. Stegman, I.C. McDade, E.J. Llewellyn, F. Harris, R.G.H. Greer, An assessment of proposed $O(^1S)$ and $O_2(b^1\Sigma_g{}^1)$ nightglow excitation parameters. Planet. Space Sci. **38**, 43–53 (1990)
N. Nakamura, Two-dimensional mixing, edge formation, and permeability diagnosed in an area coordinate. J. Atmos. Sci. **53**, 1524–1537 (1996)
N. Nakamura, A new look at eddy diffusivity as a mixing diagnostic. J. Atmos. Sci. **58**, 3685–3701 (2001)
R. Nassar, P.F. Bernath, C.D. Boone, G.L. Manney, S.D. McLeod, C.P. Rinsland, R. Skelton, K.A. Walker, ACE-FTS measurements across the edge of the winter 2004 Arctic vortex. Geophys. Res. Lett. **32**, L15S05 (2005). doi:10.1029/2005GL022671
K. Nielsen, M.J. Taylor, P.-D. Pautet, D.C. Fritts, N. Mitchell, C. Beldon, B.P. Williams, W. Singer, F.J. Schmidlin, R.A. Goldberg, Propagation of short-period gravity waves at high-latitudes during the MaCWAVE winter campaign. Ann. Geophys. **24**, 1227–1243 (2006)
D. Offermann, R. Gerndt, Upper mesospheric temperature from OH* emissions, in CIRA 1986. Adv. Space Res. **10**(12), 217–221 (1990)
D. Offermann, P. Hoffmann, P. Knieling, R. Koppmann, J. Oberheide, D.M. Riggin, V.M. Tunbridge, W. Steinbrecht, Quasi-two day waves in the summer mesosphere: triple structure of amplitudes and long-term development. J. Geophys. Res. **116**, D00P02 (2011). doi:10.1029/2010JD015051
T.R. Osborn, C.S. Cox, Oceanic fine structure. Geophys. Astrophys. Fluid Dyn. **3**(1), 321–345 (1972)
R.A. Plumb, Tracer interrelationships in the stratosphere. Rev. Geophys. **45**, RG4005 (2007). doi:10.1029/2005RG000179
R.A. Plumb, M.K.W. Ko, Interrelationships between mixing ratios of long-lived stratospheric constituents. J. Geophys. Res. **97**, 10,145–10,156 (1992)
E.-P. Röth, Fast algorithm to calculate the photon flux in optically dense media for use in photochemical models. Ber. Bunsenges. Phys. Chem. **96**, 417–420 (1992)
J.P. Russell, W.E. Ward, R.P. Lowe, R.G. Roble, G.G. Shepherd, B. Solheim, Atomic oxygen profiles (80 to 115 km) derived from Wind Imaging Interferometer/Upper Atmospheric Research Satellite measurements of the hydroxyl and greenline airglow: local time–latitude dependence. J. Geophys. Res. **110**, D15305 (2005). doi:10.1029/2004JD005570
S.P. Sander, R.R. Friedl, D.M. Golden, M.J. Kurylo, R.E. Huie, V.L. Orkin, G.K. Moortgat, P.H. Wine, A.R. Ravishankara, C.E. Kolb, M.J. Molina, B.J. Finlayson-Pitts, Chemical kinetics and photochemical data for use in stratospheric modeling, JPL Publication 06-2, Evaluation Number 15, California Institute of Technology, Pasadena, CA, USA (2006)
G.A. Schmidt, R. Ruedy, J.E. Hansen, I. Aleinov, N. Bell, M. Bauer, S. Bauer, B. Cairns, V. Canuto, Y. Cheng, A. Del Genio, G. Faluvegi, A.D. Friend, T.M. Hall, Y. Hu, M. Kelley, N.Y. Kiang, D. Koch, A.A. Lacis, J. Lerner, K.K. Lo, R.L. Miller, L. Nazarenko, V. Oinas, J. Perlwitz, J. Perlwitz, D. Rind, A. Romanou,

G.L. Russell, M. Sato, D.T. Shindell, P.H. Stone, S. Sun, N. Tausnev, D. Thresher, M.-S. Yao, Present day atmospheric simulations using GISS ModelE: comparison to in-situ, satellite and reanalysis data. J. Climate **19**, 153–192 (2006). doi:10.1175/JCLI3612.1

C.-Y. She, R.P. Lowe, Seasonal temperature variations in the mesopause region at mid-latitude: comparison of lidar and hydroxyl rotational temperatures using WINDII/UARS height profiles. J. Atmos. Sol.-Terr. Phys. **60**, 1573–1583 (1998)

T.G. Shepherd, Transport in the middle atmosphere. J. Meteorol. Soc. Jpn. **85B**, 165–191 (2007)

T.G. Shepherd, J.N. Koshyk, K. Ngan, On the nature of large-scale mixing in the stratosphere and mesosphere. J. Geophys. Res. **105**(D10), 12,433–12,446 (2000)

T. Shimazaki, *Minor Constituents in the Middle Atmosphere* (Reidel, Dordrecht, 1985)

E. Shuckburgh, W. Norton, A. Iwi, P. Haynes, Influence of the quasi-biennial oscillation on isentropic transport and mixing in the tropics and subtropics. J. Geophys. Res. **106**(D13), 14,327–14,337 (2001)

E. Shuckburgh, H. Jones, J. Marshall, C. Hill, Robustness of an effective diffusivity diagnostic in oceanic flows. J. Phys. Oceanogr. **39**, 1993–2009 (2009a). doi:10.1175/2009JPO4122.1

E. Shuckburgh, H. Jones, J. Marshall, C. Hill, Understanding the regional variability of eddy diffusivity in the Pacific sector of the southern ocean. J. Phys. Oceanogr. **39**, 2011–2023 (2009b). doi:10.1175/2009JPO4115.1

A.J. Simmons, D.M. Burridge, An energy and angular momentum conserving vertical finite-difference scheme and hybrid vertical coordinates. Mon. Weather Rev. **109**, 758–766 (1981)

A.K. Smith, M. Lopez-Puertas, M. Garcia-Comas, S. Tukiainen, SABER observations of mesospheric ozone during NH late winter 2002–2009. Geophys. Res. Lett. **36**, L23804 (2009). doi:10.1029/2009GL040942

A.K. Smith, D.R. Marsh, M.G. Mlynczak, J.C. Mast, Temporal variation of atomic oxygen in the upper mesosphere from SABER. J. Geophys. Res. **115**, D18309 (2010). doi:10.1029/2009JD013434

B.H. Solheim, E.J. Llewellyn, An indirect mechanism for the production of $O(^1S)$ in the aurora. Planet. Space Sci. **27**, 473–479 (1979)

G.R. Sonnemann, U. Körner, Total hydrogen mixing ratio anomaly around the mesopause region. J. Geophys. Res. **108**, 4692–4702 (2003). doi:10.1029/2002JD003015

G. Sonnemann, Ch. Kremp, A. Ebel, U. Berger, Calculation of the global chemical heating rates by means of a 3d-model of dynamics and chemistry. Adv. Space Res. **20**(6), 1153–1156 (1997)

G. Sonnemann, Ch. Kremp, A. Ebel, U. Berger, A three-dimensional dynamic model of minor constituents of the mesosphere. Atmos. Environ. **32**, 3157–3172 (1998a)

G.R. Sonnemann, M. Grygalashvyly, U. Berger, Autocatalytic water vapor production as a source of large mixing ratios within the middle to upper mesosphere. J. Geophys. Res. **110**, D15303 (2005). doi:10.1029/2004JD005593

G.R. Sonnemann, M. Grygalashvyly, P. Hartogh, C. Jarchow, Behavior of mesospheric ozone under nearly polar night conditions. Adv. Space Res. **40**, 846–854 (2007)

H. Takahashi, Y. Sahai, P.P. Batista, B.R. Clemesha, Atmospheric gravity wave effect on the airglow O (0, 1) and OH (9, 4) band intensity and temperature variations observed from a low latitude station. Adv. Space Res. **12**, 131–134 (1992)

H. Takahashi, M.L. Stella, B.R. Clemesha, D.M. Simonich, Atomic hydrogen and ozone concentration derived from simultaneous lidar and rocket airglow measurements in the equatorial region. J. Geophys. Res. **101**, 4033 (1996)

D.W. Tarasick, G.G. Shepherd, Effects of gravity waves on complex airglow chemistries. 2. OH emission. J. Geophys. Res. **97**, 3195–3208 (1992)

M.J. Taylor, P.J. Espy, D.J. Baker, R.J. Sica, P.C. Neal, W.R. Pendleton Jr., Simultaneous intensity, temperature and imaging measurements of short period wave structure in the OH nightglow emission. Planet. Space Sci. **39**, 1171–1188 (1991)

M.J. Taylor, D.C. Fritts, J.R. Isler, Determination of horizontal and vertical structure of an unusual pattern of short period gravity waves imaged during ALOHA-93. Geophys. Res. Lett. **22**, 2837–2840 (1995a)

M.J. Taylor, Y.Y. Gu, X. Tao, C.S. Gardner, M.B. Bishop, An investigation of intrinsic gravity wave signatures using coordinated lidar and nightglow image measurements. Geophys. Res. Lett. **22**, 2853–2856 (1995b)

M.J. Taylor, W.R. Pendleton Jr., S. Clark, H. Takahashi, D. Gobbi, R.A. Goldberg, Image measurements of short-period gravity waves at equatorial latitudes. J. Geophys. Res. **102**, 26,283–26,299 (1997)

M.J. Taylor, S.H. Seo, T. Nakamura, T. Tsuda, H. Fukunishi, Y. Takahashi, Long base-line measurements of short-period mesospheric gravity waves during the SEEK campaign. Geophys. Res. Lett. **25**, 1797–1800 (1998)

R.J. Thomas, Atomic hydrogen and atomic oxygen density in the mesopause region: global and seasonal variations deduced from Solar Mesosphere Explorer near-infrared emissions. J. Geophys. Res. **95**, 16,457–16,476 (1990)

U. von Zahn, K.H. Fricke, R. Gerndt, T. Blix, Mesospheric temperatures and the OH layer height as derived from ground-based lidar and OH* spectrometry. J. Atmos. Sol.-Terr. Phys. **49**, 863–869 (1987)
C.J. Walcek, Minor flux adjustment near mixing ratio extremes for simplified yet highly accurate monotonic calculation of tracer advection. J. Geophys. Res. **105**, 9335–9348 (2000)
C.J. Walcek, N.M. Aleksic, A simple but accurate mass conservative, peak preserving, mixing ratio bounded advection algorithm with Fortran code. Atmos. Environ. **32**, 3863–3880 (1998)
R.L. Walterscheid, G. Schubert, Gravity wave fluxes of O_3 and OH at the nightside mesopause. Geophys. Res. Lett. **16**, 719–722 (1989)
R.L. Walterscheid, G. Schubert, J.M. Straus, A dynamicalchemical model of wave-driven fluctuations in OH nightglow. J. Geophys. Res. **92**, 1241–1254 (1987)
D.Y. Wang, W.E. Ward, Y.J. Rochon, G.G. Shepherd, Airglow intensity variations induced by gravity waves. Part 1: Generalization of the Hines-Narasick theory. J. Atmos. Sol.-Terr. Phys. **63**, 35–46 (2001a)
D.Y. Wang, Y.J. Rochon, S.P. Zhang, W.E. Ward, R.H. Wiens, D.Y. Liang, W.A. Gault, B.H. Solheim, G.G. Shepherd, Airglow intensity variations induced by gravity waves. Part 2: Comparisons with observations. J. Atmos. Sol.-Terr. Phys. **63**, 47–60 (2001b)
J. Weinstock, Theory of the interaction of gravity waves with $O_2(^1\Sigma)$ airglow. J. Geophys. Res. **83**, 5175 (1978)
K.B. Winters, E.A. D'Asaro, Diascalar flux and the rate of fluid mixing. J. Fluid Mech. **317**, 179–193 (1996)
J. Xu, Y. Wang, Y. Wang, The loss of photochemical heating caused by gravity waves in the mesopause region. J. Atmos. Sol.-Terr. Phys. **62**, 37–45 (2000a)
J. Xu, A.K. Smith, G.P. Brasseur, The effects of gravity waves on distributions of chemically active constituents in the mesopause region. J. Geophys. Res. **105**, 26,593–26,602 (2000b)
J. Xu, A.K. Smith, G.P. Brasseur, Conditions for the photochemical destabilization of gravity waves in the mesopause region. J. Atmos. Sol.-Terr. Phys. **63**, 1821–1829 (2001)
J. Xu, A.K. Smith, R. Ma, A numerical study of the effect of gravity-wave propagation on minor species distributions in the mesopause region. J. Geophys. Res. **108**, 4119–4130 (2003). doi:10.1029/2001JD001570
S.P. Zhang, R.H. Wiens, G.G. Shepherd, Gravity waves from O_2 nightglow during the AIDA '89 campaign I: emission rate/temperature observations. J. Atmos. Terr. Phys. **55**, 355 (1993a)
S.P. Zhang, R.H. Wiens, G.G. Shepherd, Gravity waves from O_2 nightglow during the AIDA '89 campaign II: a theory of the emission rate/temperature ratio, η. J. Atmos. Terr. Phys. **55**, 377 (1993b)

Space Sci Rev (2012) 168:363–384
DOI 10.1007/s11214-011-9830-8

Electromagnetic Atmosphere-Plasma Coupling: The Global Atmospheric Electric Circuit

Michael J. Rycroft · R. Giles Harrison

Received: 17 June 2011 / Accepted: 31 August 2011 / Published online: 28 October 2011

Abstract A description is given of the global atmospheric electric circuit operating between the Earth's surface and the ionosphere. Attention is drawn to the huge range of horizontal and vertical spatial scales, ranging from 10^{-9} m to 10^{12} m, concerned with the many important processes at work. A similarly enormous range of time scales is involved from 10^{-6} s to 10^{9} s, in the physical effects and different phenomena that need to be considered. The current flowing in the global circuit is generated by disturbed weather such as thunderstorms and electrified rain/shower clouds, mostly occurring over the Earth's land surface. The profile of electrical conductivity up through the atmosphere, determined mainly by galactic cosmic ray ionization, is a crucial parameter of the circuit. Model simulation results on the variation of the ionospheric potential, ~250 kV positive with respect to the Earth's potential, following lightning discharges and sprites are summarized. Experimental results comparing global circuit variations with the neutron rate recorded at Climax, Colorado, are then discussed. Within the return (load) part of the circuit in the fair weather regions remote from the generators, charge layers exist on the upper and lower edges of extensive layer clouds; new experimental evidence for these charge layers is also reviewed. Finally, some directions for future research in the subject are suggested.

Keywords Global circuit · Thunderstorms · Electrified cloud generators · Cosmic rays · Atmospheric electrical conductivity profile · Fair weather regions · Ionospheric potential · Lightning · Sprites · Layer cloud electrification

M.J. Rycroft (✉)
CAESAR Consultancy, Cambridge CB3 9HW, UK
e-mail: michaelrycroft@btinternet.com

R.G. Harrison
University of Reading, Reading RG6 6BB, UK
e-mail: r.g.harrison@reading.ac.uk

1 Introduction

This paper is concerned with atmospheric electrical coupling from near the Earth's surface up into—and down from—the ionosphere at ~80 km altitude and higher. This coupling takes place rapidly, at, or close to, the speed of light c (Rycroft 2006), as opposed to coupling mechanisms involving mechanical waves of one type or the other which propagate at speeds much slower than c, and which are discussed in other papers in this volume. As outlined by Aplin et al. (2008), the subject of atmospheric electricity had its origins in the eighteenth century, grew into the concept of the global atmospheric electric circuit in the early twentieth century with the seminal papers of Wilson (1921, 1929, 1956), and matured considerably in the first decade of the twenty first century.

The global circuit is formed between the surface of the Earth and the ionosphere, both of which are good electrical conductors in comparison with the insulating atmosphere between them. D.C. and A.C. electric generators exist in the atmosphere, for example in thunderstorms, creating currents flowing up to the ionosphere. The current circuit closes through fair weather parts of the atmosphere that are remote from the generators. Useful background papers on different aspects of the global atmospheric electric circuit have been written by Vonnegut (1973), Hays and Roble (1979), Roble and Hays (1979), Makino and Ogawa (1984, 1985), Roble and Tzur (1986), Volland (1987), Roble (1991), Bering et al. (1998), Rycroft et al. (2000, 2007, 2008), Williams (2002, 2009), Harrison (2004), Siingh et al. (2005, 2007, 2011), Aplin (2006), Markson (2007), Tinsley et al. (2007), Harrison et al. (2008) and Tinsley (2008). The present paper complements these papers and those which they cite. The D.C. global atmospheric electric circuit has been considered in the context of the changing climate of planet Earth by Williams (1992), Price (1993), Tinsley et al. (1994), Gray et al. (2010) and Siingh et al. (2011).

When considering a subject in physics, and its mathematical representation, the treatment conventionally begins simply, using linear theory. However, as the subject develops, it is usually appreciated that a more complex representation is appropriate. At larger amplitudes nonlinearities can arise and the system may even become chaotic on the small scale, or on the local, moderate (regional) or largest (global) scales. An important characteristic of such complex systems is that they simultaneously need to consider a wide range of spatial scales and temporal scales. Up to the present, the treatment of atmospheric electricity has remained linear, whereas nowadays climate studies often involve multi-scale and nonlinear behaviour (Donner et al. 2009; Slingo et al. 2009; Palmer and Williams 2010). The global electric circuit may be involved in climate change via nonlinear electrical effects on cloud microphysical processes (Aplin et al. 2008; Carslaw 2009; Harrison and Ambaum 2008, 2010; Nicoll and Harrison 2010), as discussed later in this paper.

Figure 1 shows the huge range of horizontal and vertical scales involved in the diverse phenomena and processes of interest which occur in the atmosphere and in the near-Earth space environment (Rycroft 2010). The horizontal scale extends over 18 orders of magnitude, and the vertical scale over 12 orders. At the Earth's surface, point discharge currents (Chalmers 1962; Ette and Utah 1973; Märcz and Bencze 1998) emanate from sharp-ended vertical conductors, such as grass and the spiky needles of coniferous trees, which have scales of millimetre size. Pointed hills and ridges have scales of kilometres to hundreds of kilometres, and the oceans longer scales, although sea spray has dimensions of millimetres.

In the lowest part of the atmosphere over continental surfaces, ionisation is generated from the escape of radon isotopes (Harrison et al. 2010) and by galactic cosmic rays arriving from beyond the solar system (Bazilevskaya et al. 2000, 2008; Velinov et al. 2009). In clean

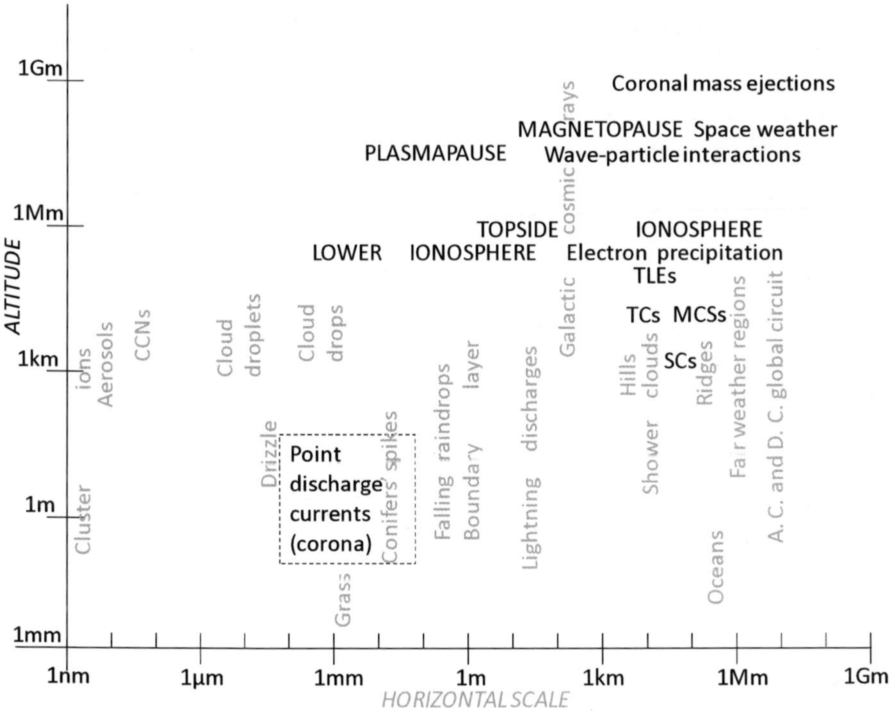

Fig. 1 Diagram indicating the very large range of horizontal and vertical scales involved in various electrical phenomena of interest in the Earth's atmosphere and near-Earth space environment. *Words appearing horizontally* show different regions at different altitudes where different physical processes make their presence felt and *Words appearing vertically* show features with a certain horizontal scale which are important over the altitude range indicated. (Here, TCs is an abbreviation for thunderclouds—other such terms are explained in the text)

air, water vapour condenses onto these ions to form cluster ions ~1 nm in size (Aplin et al. 2008; Rycroft et al. 2008). Some cluster ions are removed by ion-ion recombination and others are lost to aerosol particles (Harrison and Tammet 2008; Hirsikko et al. 2011). In some circumstances in which condensible vapours such as sulphuric acid are abundant, the cluster ions may eventually grow into ~100 nm sized cloud condensation nuclei (CCN) onto which cloud droplets can form; raindrops, which are up to three orders of magnitude larger, result from coalescence of the cloud droplets. It is worth mentioning here that Enghoff et al. (2011) have recently studied sulphuric acid aerosol nucleation in an atmospheric pressure reaction vessel where a 580 MeV electron beam has ionised the gas. They found evidence for an ion-induced effect on aerosol nucleation under conditions which resemble those of the Earth's atmosphere. Kirkby et al. (2011) have shown that "ground-level galactic cosmic ray ionisation substantially increases the nucleation rate of sulphuric acid and sulphuric acid-ammonia particles, by between twofold and tenfold or more, provided that the nucleation rate lies below the limiting ion-pair production rate".

Also shown in Fig. 1 are low level stratiform clouds (SCs) such as stratocumulus, where the electric charges at the cloud edges are important (Nicoll and Harrison 2009, 2010), and thunderstorm cells (TCs) which can grow into large thunderstorms (Williams and Yair 2006), termed mesoscale convective systems (MCSs). Thunderstorms produce lightning dis-

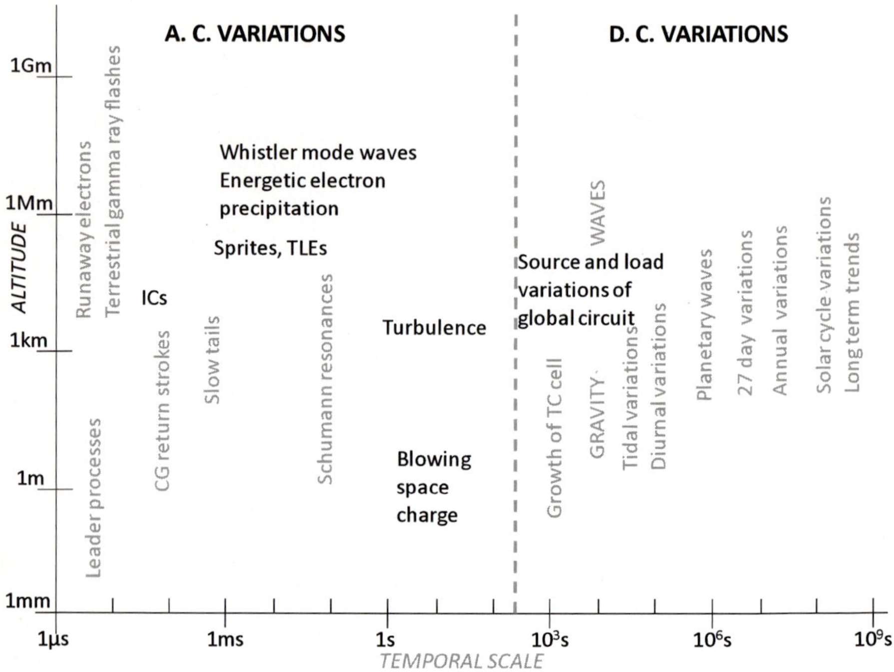

Fig. 2 Diagram showing the enormous range of time scales involved in important processes (*horizontal words*) occurring at different altitudes. *Vertical words* represent different phenomena which occur on different temporal scales over different altitude ranges—see the text for a fuller discussion. The boundary between A.C. and D.C. variations is placed at ~200 s, the *RC* time constant of the poorly conducting atmospheric region between the good conducting Earth and the ionosphere

charges which radiate electromagnetic waves across a broad range of frequencies; these constitute the A.C. part of the global electric circuit, discussed later.

Above large thunderstorms transient luminous events (TLEs), such as sprites, elves and blue jets (Füllekrug et al. 2006), may occur below and in the ionosphere. The lower ionosphere responds to activity from above, in the form of wave-particle interactions between whistler-mode waves from lightning and energetic electrons trapped in the magnetosphere (Rycroft 2010); extra ionisation is then produced in the lowest ionosphere (Rodger et al. 2001). The magnetosphere is stimulated by activity on the Sun, that information travelling through interplanetary space in the form of coronal mass ejections (CMEs); these phenomena are generically termed space weather (Bothmer and Daglis 2007). Such phenomena (Rycroft 2010) are important in terms of possible damage to satellites and other assets in space and to humans aboard spacecraft.

Figure 2 shows the broad range of temporal scales that are involved in the many phenomena of importance. On the shortest time scales of microseconds are electrical discharge phenomena. These are leader processes which occur as a lightning discharge progresses in steps from a thundercloud towards the ground, the cloud-to-ground (CG) return stroke which is a large (~30 kA) current to the cloud, and intra-cloud (IC) discharges (Rakov and Uman 2003). Lightning radiates all radio frequencies from MHz (associated with leader processes) to ~10 kHz (where the spectrum peaks, Smith et al. 2010a) to "slow tails" (~100 Hz, Mullayarov et al. 2010), and to the longest wavelength electromagnetic waves occurring in the Earth's environment (∼10 Hz). These latter waves excite Schumann resonances of the spherical shell cavity between the good conducting Earth and ionosphere, the fundamental of which is at 8 Hz (Williams 1992; Price et al. 2007; Simoes et al. 2008; Yeng et al. 2009; Shvets et al. 2010; Nickolaenko et al. 2010; Golkowski et al. 2011).

A few stations around the world can record the radiation of various frequencies generated by lightning and by sprites in order to find their location. Williams et al. (2010) did this for radio signals produced over Africa, investigating their dependence on the charge moment changes of the parent lightning. Whitley et al. (2011) have recently shown that with four stations around the world sources can be geolocated to an accuracy of ∼100 km, which is a remarkable achievement. Shvets et al. (2009, 2010), Nakamura et al. (2010b), Shvets and Hayakawa (2011) and Yamashita et al. (2011) have also presented several interesting results on these topics.

Runaway breakdown (Roussel-Dupré et al. 2008; Fishman and Chilingarian 2010) is believed to occur for some lightning discharges, possibly intra-cloud discharges (Williams et al. 2006), or active leader channels (Carlson et al. 2010). This runaway breakdown produces upward-going beams of runaway relativistic electrons which may be accelerated to a high enough energy to create a terrestrial gamma ray flash (TGF—see Fig. 2; Miller 2011). This is believed to be created as upward-going Bremsstrahlung radiation when the electrons collide with the nuclei of atmospheric atoms.

Figure 2 also shows that, on the time scale from seconds to hundreds of seconds, blowing space charge associated with turbulence in the boundary layer from the Earth's surface up to ∼2 km altitude generates electrical fluctuations. The space charge can be primarily ionic, in clean air, or particulate, in polluted air. Particulate space charge is subject to turbulent motions, which result in electrical fluctuations. All the processes shown in Fig. 2 ultimately contribute to the generation of A.C. variations in the global circuit.

The electromagnetic pulse from a very large CG discharge, which transfers positive charge to ground (termed a +CG discharge), deposits energy in the mesosphere and creates a rapidly expanding ring of light at ∼90 km altitude, termed an elve. The consequent large electrostatic field above the thundercloud exceeds the conventional threshold field for electrical breakdown, creating a sprite from about 80 km at the base of the ionosphere down to about 55 km (Füllekrug et al. 2006; Rycroft and Odzimek 2009, 2010; Pasko 2010; Pasko et al. 2011). Rarely, blue jets appear from the top of the thundercloud. Extremely rarely, gigantic jets reach from the top of the thundercloud right up to the ionosphere—they will short circuit the gap between the thundercloud and the ionosphere over some area as yet to be determined.

Also shown in Fig. 2 are D.C. variations; on time scales of a fraction of an hour, these are linked to the development and evolution of thunderstorm cells. Mechanical waves occur in the form of upward propagating gravity waves, tidal variations (e.g., semi-diurnal, diurnal and two day waves), and longer period planetary waves. At even longer time scales are variations associated with the changing energy input to the stratosphere and thermosphere over the 27 day solar rotation period, annual variations, variations over the 11 year solar cycle and long term trends. The boundary between D.C. and A.C. phenomena is here shown at 200 s, as this is the electrical time constant of the global atmospheric electric circuit, namely its resistance R multiplied by its capacitance C. The capacitance is that of the spherical capacitor formed between the Earth and the ionosphere, with the dielectric atmosphere being mostly concentrated near the Earth's surface, due to gravity (Rycroft et al. 2000).

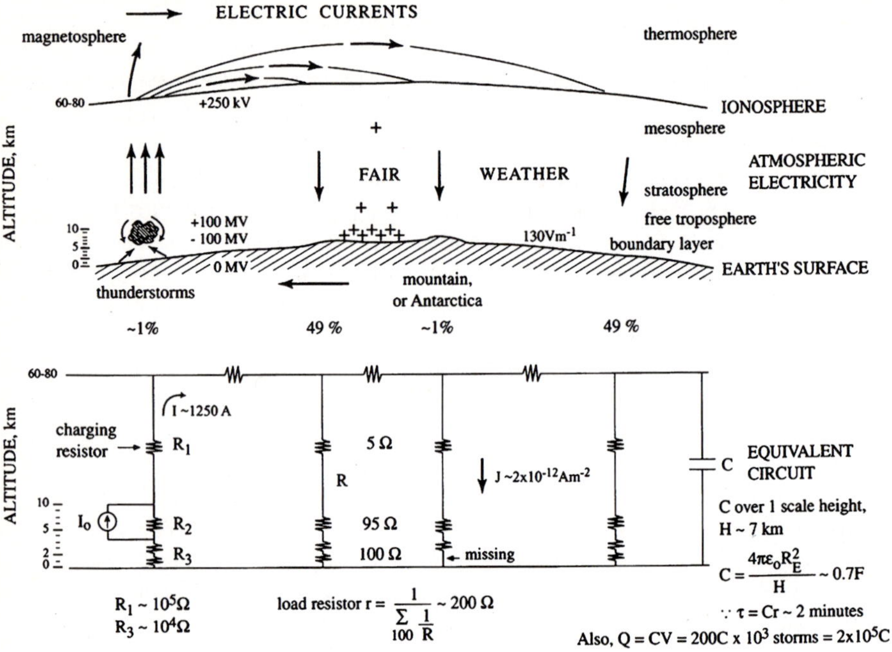

Fig. 3 Taken from Rycroft et al. (2000), the *upper panel* illustrates a part of the global atmospheric electric circuit. *On the left* is shown a representative thunderstorm, one of the ~ thousand which are active at all times and which occur over only a small part of the Earth's surface. Huge potential differences are generated inside thunderclouds; acting as giant batteries, they drive an upward (Wilson) current to the ionosphere at an altitude of ~80 km. The ionosphere is essentially an equipotential surface at ~+250 kV with respect to the Earth's surface, but small currents reach up into the magnetosphere. In fair weather regions remote from thunderstorms, over most of the Earth's surface, downward currents ~2 pA/m^2 flow vertically (radially) to the ground. Much more positive charge resides near the Earth's surface than in the stratosphere or mesosphere, as indicated by the density of the + symbols. The *lower panel* outlines a simple electrical engineering representation of the circuit. Over mountainous regions, the atmospheric columnar resistance is much less than it is elsewhere. Values for the *RC* (or *Cr*) time constant are shown at the *lower right* (Reprinted with permission from Elsevier)

2 Properties of the Global Electric Circuit

2.1 Generator and Load Regions

Figure 3 is a schematic representation of the circuit taken from Rycroft et al. (2000). In the upper part the bold arrows show the flow of electric current generated by thunderstorms (on the left), and occurring over $\lesssim 1\%$ of the Earth's surface, up to the ionosphere. A tiny part of the electric current flows up into the magnetosphere along geomagnetic field line, up to the apex of each particular field line. In the fair weather part of the circuit remote from thunderstorms, the current flows down to the Earth's surface (with negative ions moving upwards). The circuit is completed by the good conductivity of land and sea surfaces and up to the base of thunderstorms by point discharge (corona) currents from sharp-tipped objects in the boundary layer. It is evident that electric charge is distributed throughout the insulating atmosphere, but most of the electric charge resides near the surface where, as is apparent from the density profile, most of the atmosphere is.

The lower part of Fig. 3 shows the equivalent circuit. It is generally believed that about a thousand thunderstorms are active globally at any one time; these are represented by a current generator (of ~1 to 2 kA) which acts as a D.C. battery in the circuit. The electrical resistance of the atmosphere is appreciable, especially below the thunderstorms, whereas the ionospheric resistances shown are by comparison extremely small, effectively ensuring that the ionosphere is an equipotential; its potential, V_I, is ~250 kV positive with respect to the Earth's surface.

As well as thunderstorms, there is another significant generator in the global circuit. This is due to electrified rain/shower clouds which generally bring negative charge to the ground on raindrops (Liu et al. 2010), as was first discussed by Wilson (1921). Rycroft et al. (2007) estimated that thunderstorms contribute about 60% of the upward current to the ionosphere, with electrified rain/shower clouds being responsible for ~40%. Odzimek et al. (2010) deduced that these percentages are 80% and 20%, respectively. A very recent paper by Mach et al. (2011) has, for the first time, used experimental data from aircraft and satellites to deduce that thunderstorms over the land contribute 1.1 kA to the global circuit and, over the oceans, 0.7 kA. The contributions to the global circuit made by rain/shower clouds are 0.22 kA for ocean storms and 0.04 kA for storms over the land. Thus, Mach et al. (2011) consider that thunderstorms contribute 90% and rain/shower clouds 10% to the total ~2 kA flowing in the global circuit.

In the return (or load) part of the circuit in the fair weather region (over ~98% of the Earth's surface in area) the downward current density J_z flowing is ~2 pAm^{-2} (Wilson 1921). Most of the resistance is near the surface, due to the exponential distribution of the atmospheric density with a scale height H of ~7 km. The equivalent load resistor of the entire circuit, R, is ~200 Ω. The lower right corner shows that the capacitance C of the circuit is the value calculated by the standard electrostatics method. Here, R_E is the Earth's radius, and the separation between the two conductors is taken as the scale height H of the dielectric atmosphere rather than their physical separation, which is almost 12 times greater. The negative charge on the Earth's surface is -2×10^5 C, and the electric energy stored in the circuit is $\sim 2 \times 10^{10}$ J.

The conductivity and the electric current density J_z flowing in the fair weather regions determine the vertical electric field E. Near the Earth's surface away from aerosol pollution in fair weather $E = -130$ V/m; the minus sign indicates that the electric field is directed downwards. Meteorologists refer to this as the potential gradient (PG) rather the electric field; the PG and E have the same magnitude, but by convention the PG is positive in fair weather.

2.2 Vertical Variations

The cosmic ray fluxes at different altitudes and for different rigidities (i.e. for different momenta) have been reported by Bazilevskaya et al. (2000, 2008) and Stozhkov et al. (2001). In the presence of the geomagnetic field, protons with rigidities <0.6 GV can only gain access to the Earth's environment in the polar regions; those with rigidities between 2.5 and 6 GV are found at middle latitudes, and only those with rigidities of 13.5 GV or more can enter the magnetosphere at equatorial latitudes. Figure 4, from Bazilevskaya et al. (2000), shows, in panel (a), long term observations of the fluxes (J) of cosmic rays of different rigidities (R_c) at 33 degrees magnetic latitude (6.7 GV), 51 degrees magnetic latitude (2.4 GV) and 64 degrees (0.6 GV) at 25 to 30 km altitude in the middle atmosphere; panel (b) shows similar data in the upper troposphere between 6 and 12 km altitude.

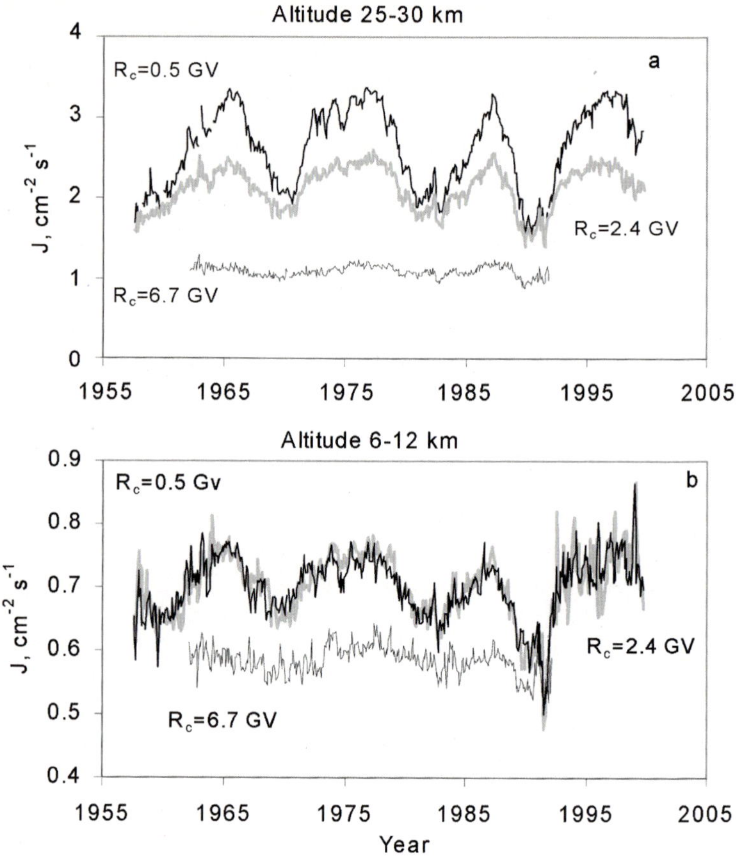

Fig. 4 Taken from Bazilevskaya et al. (2000), the *upper* (*lower*) *panel plots* the monthly average values of the omnidirectional flux of cosmic rays J at altitudes between 25 and 30 km (6 to 12 km) at geomagnetic latitudes of 33 degrees (cut-off rigidity $R_C = 6.7$ GV), 51 degrees (2.4 GV) and 64 degrees (0.6 GV), showing a clear 11 year solar cycle variation (Reprinted with permission from Elsevier)

There is a strong solar cycle variation, with large fluxes near the solar minima around 1965, 1976, 1987 and approaching 1998. In the troposphere the cosmic ray flux at high latitudes is typically ~20% larger in solar minimum conditions than near solar maximum; it is ~10% larger at 33 degrees magnetic latitude. The ion-pair production rate at different altitudes varies by ~2.5 as one moves from the geomagnetic equator to the magnetic poles. The principal nuclear species contributing to the cosmic ray spectrum have been presented recently by Nakamura et al. (2010a) and Schwarzchild (2011).

Because the conductivity of the atmosphere is least near the Earth's surface, most of the electrical resistance of the circuit lies there. Harrison and Bennett (2007) and Rycroft et al. (2008) have published diagrams showing that ~95% of the atmospheric columnar resistance R_c lies at altitudes below 10 km, with half of the columnar resistance being within the lowest 1.6 km. Figure 5 displays a graph of the percentage of the columnar resistance calculated at height increments of 1 km from the surface up to 10 km height; its value is 156 P Ω m^2. The value of R_c all the way up to the ionosphere is 167 P Ω m^2. In the fair weather part of the circuit, it is obvious from Ohm's Law that $J_z = V_I/R_c$. Inserting the model values which

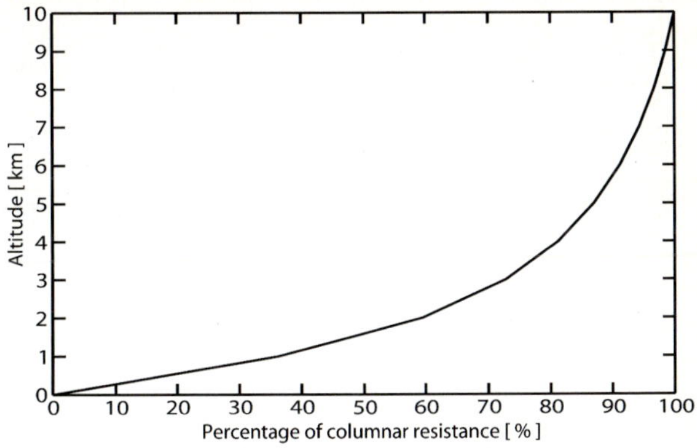

Fig. 5 The variation of the percentage of the columnar resistance up to 10 km altitude in fair weather regions, for the model atmospheric conductivity profile given by Rycroft et al. (2008). About 93% of the total resistance from the ground up to the ionosphere at 80 km altitude is at altitudes below 10 km (taken from Rycroft et al. 2008)

we have presented, J_z is found to be 250 kV/167 P $\Omega\,m^2 = 1.5$ pAm^{-2}, which is consistent with many experimental determinations, such as the original direct measurements of Wilson (1906).

Thus, it is evident that the atmospheric (or planetary) boundary layer is the most important load in the fair weather part of the global circuit. When considering how the global circuit operates, it is most important to realise this fact. Also, as mentioned earlier, most of the positive charge distributed through the atmosphere resides near the surface; it is not on the ionospheric upper conductor of the capacitor. Nogueira (2009) has constructed a finite element model of the fair weather electric circuit.

In order to perform model calculations, Rycroft et al. (2007) and Rycroft and Odzimek (2010) plotted a realistic average model of the electrical conductivity profile through the Earth's atmosphere. The conductivity increases by seven orders of magnitude from the surface, where the conductivity is $\sim 10^{-14}$ S/m, to $\sim 10^{-7}$ S/m at ~80 km altitude; Odzimek et al. (2010) have presented more complicated models displaying geographic variations. Using the simpler conductivity model, Rycroft et al. (2007) and Rycroft and Odzimek (2009, 2010) constructed an electrical engineering model of the global atmospheric electric circuit, which they used to evaluate the effects of lightning having different charge moment changes and of sprites on the potential of the ionosphere and also to show how sprites develop.

Rycroft et al. (2007, 2008) and Rycroft and Odzimek (2010) briefly discussed some relevant experimental observations of atmospheric electric fields, currents and air conductivities made from different balloons at altitudes up to 35 km. Thomas et al. (2009) reviewed the literature on these topics. Further, they presented the results of making such observations in the stratosphere (between 30 and 35 km) above an active thunderstorm in southeastern Brazil. They found that the conductivity was about 8.5 pS/m; for their model thunderstorm of radius ~60 km, they estimated that the upward current above this active thunderstorm was ~2.5 A. In relation to the results presented in Sect. 2.1, they reported that "the overall charge removal contribution of large +CG flashes to the global circuit was small". And, moreover, since the –CG flashes provide charging and +CG flashes provide discharging of nearly equal magnitudes, we conjecture that the combined contribution due to all lightning could be very small. These results are in good agreement with those presented by Rycroft et al. (2007) and Rycroft and Odzimek (2010).

Inside an active thundercloud, the electrical conductivity is not well-constrained, but observations discussed by Rycroft et al. (2007) show that it is at least a factor of six less than its value in the clear air surrounding the thundercloud; Rycroft et al. (2007, 2008) showed values for a model thundercloud. Figure 3 of the Rycroft and Odzimek (2010) model considered the effect on the global circuit of reducing the conductivity within the thundercloud by a factor ranging from 2 to 29. They reported that the ionospheric potential would then increase from ~150 kV to ~415 kV.

Similarly, but inside a stratiform cloud in the fair weather region of the global circuit, the conductivity is less that just outside the cloud. By Gauss' law of electrostatics, there has to be an electric charge at the top and bottom edges of a uniform stratiform cloud, and the current flows through the cloud with the same current density value as above and below it. This topic is the subject of recent experimental studies by Nicoll and Harrison (2010) which are discussed in Sect. 3.4.

With the increase in conductivity with height, the vertical electric field becomes so small with increasing height that the potential in the model atmosphere at 60 km is only 24 V less than the 250,000 V ionospheric potential (Rycroft et al. 2007). Since these potentials are the same to within one part in 10^4, we may say that the 60 km level is the height of the electrosphere. Chalmers (1967) defined the electrosphere as being at the level at which, horizontally, essentially the same electric potential exists.

However, this concept does not allow for the presence of electric fields arising from magnetospheric sources (Tinsley 2008), such as the flow of the solar wind past the magnetosphere (Rycroft 2010). The value of this additional electric field depends strongly on the value (~ some nT) of the southward component of the interplanetary magnetic field (often called the IMF). It also depends, to a lesser extent, on the sign and magnitude of the dawn-to-dusk component of the interplanetary magnetic field. Tinsley (2008) splits the atmospheric columnar resistance into two parts—a tropospheric part and a stratospheric part—the latter is much smaller than the former. The magnitude of the stratospheric part varies markedly with latitude, because of the latitudinal variation of the ionisation source. Tinsley et al. (2007) and Zhou and Tinsley (2010) emphasise the generator provided by coupling from the solar wind.

Electric currents flowing in the atmosphere find it easier to continue upwards into a region of ever-increasing conductivity than to flow horizontally. This statement is true, above the electrosphere, up to and through the lower ionosphere. It is even true through the ionospheric dynamo field region at ~100 to ~130 km altitude (Rishbeth and Garriott 1969; Kelley 2009). Thus, small currents flow up to the apex of the geomagnetic field line at that particular L-value (McIlwain 1961). (The McIlwain parameter L is, essentially, the distance measured (in Earth radii) from the centre of the Earth to where that magnetic field line crosses the equatorial plane. The geomagnetic latitude of the foot of the field line on the Earth's surface, λ, is related to L by the equation $L \cos^2 \lambda = 1$.) Stansbery et al. (1993) estimate that half of the current that reaches the ionosphere flows into the geomagnetically conjugate hemisphere.

Figure 6 plots some results of the Rycroft et al. (2007) model on properties of the fair weather field. Panel (a) on the left shows the height variation of the electric potential, from which the vertical electric field is calculated; that is shown in panel (b). The consequent electric charge density is found from Gauss' law, and plotted in panel (c) on the right hand side of the figure. In the free atmosphere, at a few km altitude, the charge density is only ~0.1 pC/m^3.

3 Recent Findings Concerning the Global Circuit

3.1 Source Term

Figure 7 shows the position of the RHESSI satellite when it detected a TGF (taken from Smith 2009; more results are given by Smith et al. 2010b). The TGFs occur over land

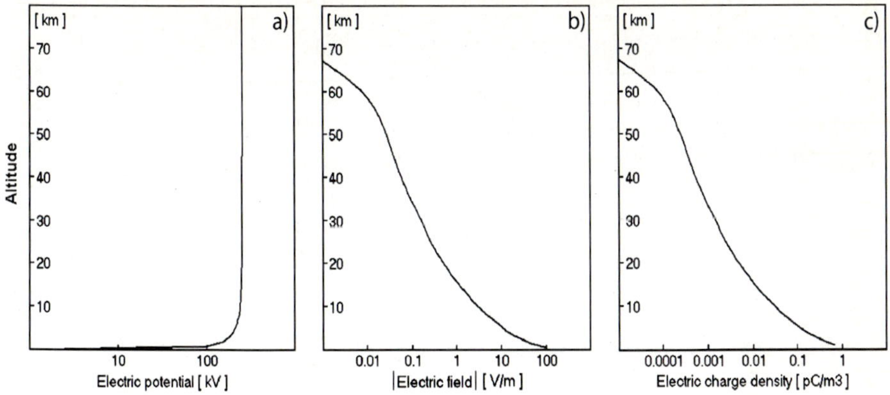

Fig. 6 In fair weather regions of the Earth's atmosphere are plotted the variation with altitude of (**a**) the electric potential, (**b**) the downward electric field, termed the potential gradient (*PG*), and (**c**) the electric charge density, taken from the model of Rycroft et al. (2007) (Reprinted with permission from Elsevier)

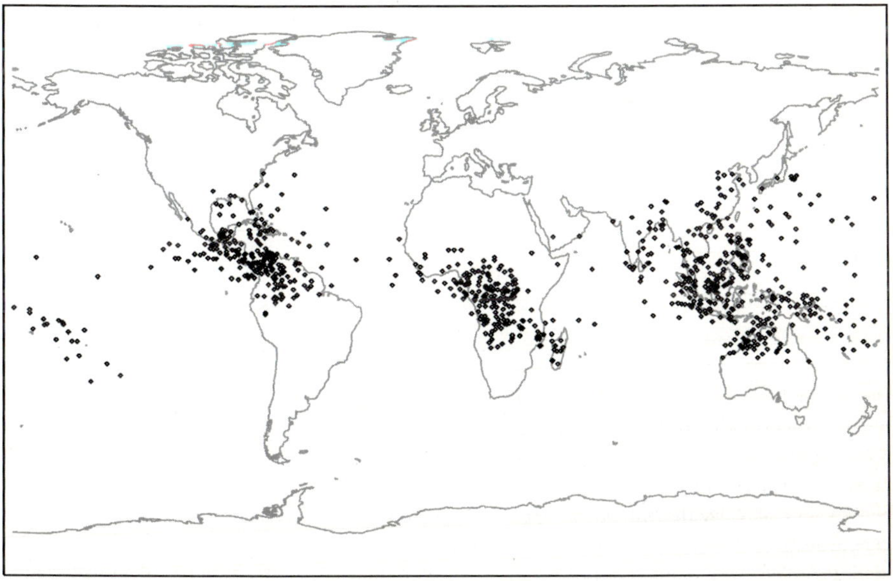

Fig. 7 Map showing the positions of the RHESSI satellite when its instrument observed terrestrial gamma-ray flashes (TGFs)—the TGF itself may occur a few hundred km from the *dot* shown. There are no *dots* at latitudes above 38 degrees, the orbital inclination of the satellite. The *dots* are concentrated in the tropics, but not over Brazil, the region of the South Atlantic Geomagnetic Anomaly (SAGA), where the instrument is disabled due to the presence of a high radiation dose from the inner Van Allen belt (taken from Smith 2009) (Reprinted with permission from David M. Smith, AIP Conference Proceedings, Vol. 1118, p. 40 (2008). Copyright 2008, American Institute of Physics)

and sea, preferentially over tropical thunderstorms where the troposphere can be as high as 18 km. This indicates that the TGFs may well originate in or above IC discharges up to the positive charge at the top of an energetic tropical thundercloud (Stanley et al. 2006; Williams et al. 2006; Lu et al. 2010). The lack of events observed over Brazil and to the

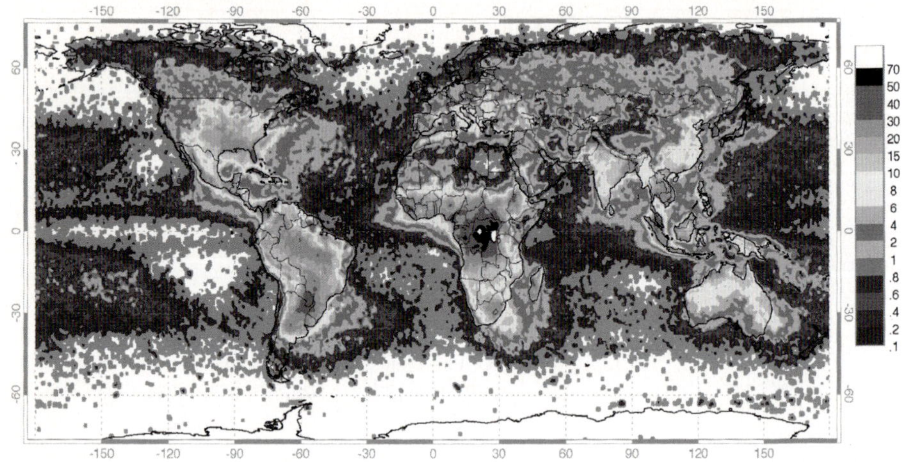

Fig. 8 Map showing the positions of lightning discharges, in number of flashes per km^2 per year, as observed in the infrared part of the spectrum by the Optical Transient Detector (OTD) and Lightning Imaging Sensor (LIS) instruments on two Earth-orbiting satellites. The lightning is concentrated over the three tropical land masses, with rather few occurring over the tropical oceans, taken from National Space Science and Technology Center (NSSTC), Huntsville, Alabama, USA, data (taken from http://www.science.nasa.gov)

East is due to the instrument being switched off as the satellite travels through the inner van Allen radiation belt (Rycroft 2010) in the vicinity of the South Atlantic Geomagnetic Anomaly (SAGA); here, the intense radiation may cause the satellite's electronic systems to fail.

Whilst it is not proven that TGFs play a role in maintaining the global circuit, it is important to consider the impact of relativistic processes taking place above thunderstorms on the global circuit. Carlson et al. (2009, 2010) suggest that TGF production is associated with current pulses ($\sim$1 ms) in lightning leader channels and runaway processes, and they have produced promising simulations. The three brightest TGFs observed aboard the Fermi satellite (Briggs et al. 2011) have relativistic electron-positron beams associated with them. Fishman (2011) has discussed the behaviour of these electrons and positrons moving along geomagnetic field lines at low L-shells. Füllekrug et al. (2011) have simulated the production of relativistic electrons above a thundercloud and below a sprite, which can radiate radio waves with frequencies up to 400 kHz. They note that such "relativistic electron beams are a new form of impulsive energy transfer between thunderclouds and the middle atmosphere which need to be considered as a novel element in the global atmospheric electric circuit".

The long term global distribution of lightning discharges, derived at latitudes up to 38 degrees North and South from data from the Lightning Imaging Sensor (LIS) on a satellite, is shown in Fig. 8. Three centres of activity lie over South East Asia (sometimes called the maritime continent), central Africa and South America. At these three different meridians, their activity peaks daily at around 1500 local time, as a result of daytime solar heating of the Earth's surface which stimulates enhanced atmospheric convection. It is clear that over the land the number of lightning discharges is typically a hundred times greater than over the tropical oceans. Earlier satellite data (Christian et al. 2003) gave the number of flashes per square km per year. Globally, there are about 44 flashes per second, of which less than one per second is over the oceans. For these discharges, in one second, 33 are likely to be intra-cloud (IC) discharges, 10 −CG discharges and 0.7 a +CG discharge (Christian et al. 2003).

Price (2006) plotted (his Fig. 8) the Universal Time variation of the thunderstorm area of the three tropical continental regions. The sum of these three UT curves gives the global thunderstorm activity as a function of UT. It has a minimum at 03 UT and a broad maximum from 14 to 19 UT. Measurements made during the 1920s aboard the research ship Carnegie showed that the mean PG under fair weather conditions of clear sky varied in a very similar way to the global thunderstorm activity, thereby demonstrating that thunderstorms are an important driver of the global atmospheric electric circuit. The UT variation of thunderstorm activity is called the Carnegie Curve, and shown as Fig. 8(b) of Price (2006).

3.2 Current Flow and Ionospheric Potential

Integrating the vertical electric field profile measured aboard aircraft or balloons up to the troposphere essentially determines the ionospheric potential. At 12 UT this has the value 250 kV, and it rises to 310 kV at 16 and 20 UT; it then falls to 200 kV at 24 UT. The curve closely resembles the Carnegie Curve (Markson 1986), which is one of the "confirming ideas" (Aplin et al. 2008) supporting the behaviour of the Earth-atmosphere global circuit system. A significant generator of the conduction currents up to the ionosphere (also known as Wilson currents) is due to the action of thunderstorms which act as batteries.

Harrison and Bennett (2007) considered the observations made on some days, in different years from 1966 to 1971, using electric field sensors carried on balloons launched from Weissenau, Germany, from which the ionospheric potential was derived by vertical integration. The results were compared with fair weather observations made on the same days at the Kew observatory, on the outskirts of London. A good fit to a linear model, plotting the ionospheric potential normalised by the PG at Kew against the derived surface air conductivity at Kew, was found. This indicates that global circuit concept holds over an area at least as great as the size of Europe.

3.3 Global Circuit Modulation

Next, we investigate studies of the response of the global atmospheric electric circuit to changes of the flux of cosmic rays in order to test our understanding of the global circuit. Measurements of the conduction current show a positive response to cosmic ray changes, driven by the solar cycle (Markson 1986; Harrison and Usoskin 2010). Panel (a), from Harrison and Usoskin (2010), demonstrates the response observed in V_I. It shows as individual symbols the ionospheric potential observed from several different investigators on specific days; this usually lies between 150 and 300 kV.

At the bottom of the panel is plotted the daily neutron count rate as observed at Climax, Colorado, USA. This indicates the flux of galactic cosmic rays with rigidities >3 GV. At solar maximum, around the year 1970, the cosmic ray flux is $\sim$12% less than nearer solar minimum (1966 and 1972–1973). This is because the scattering of cosmic rays by irregularities in the solar wind and the interplanetary magnetic field which it carries out into the heliosphere is enhanced at solar maximum. If the conductivity varies as the square root of the cosmic ray ion production rate (Rycroft et al. 2008), as expected in marine air where there is no radon contribution, nor appreciable ion removal by aerosol, it will be $\sim$6% less near solar maximum.

The straight line fit presented in panel (b) of Fig. 9 (Harrison and Usoskin 2010) shows that the ionospheric potential is $\sim$17% less at solar maximum (when the Climax neutron rate is $\sim$4100 ($\times 10^2$ s^{-1})) than at solar minimum (count rate $\sim$3600 ($\times 10^2$ s^{-1})). The atmospheric conductivity is less at solar maximum than at solar minimum and the ionospheric potential is accordingly less. Also the fair weather current density is $\sim$23% less at

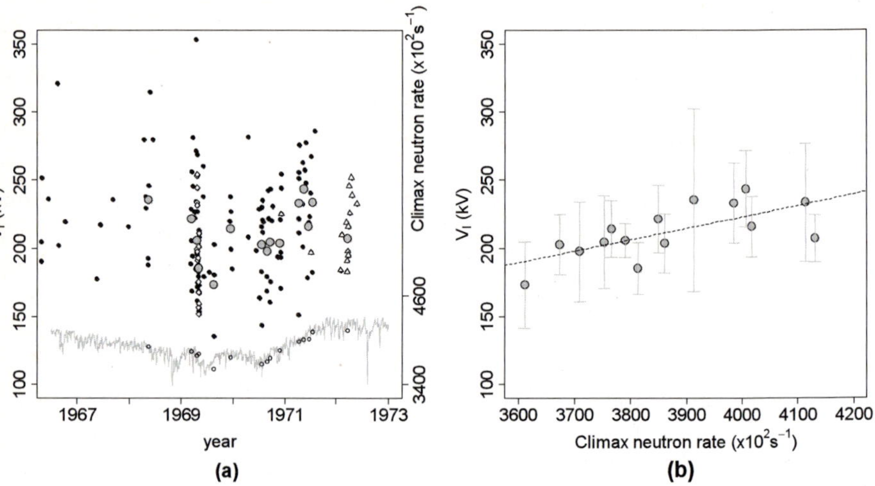

Fig. 9 Taken from Harrison and Usoskin (2010), the *left panel* (**a**) shows the ionospheric potential (V_I) from various data sets, with *larger circles* showing monthly averages for months having four or more V_I values, and the neutron count rate at Climax, Colorado, as the *grey line*, observed from 1966 to 1972. The *right panel* (**b**) plots the monthly-averaged V_I values against the monthly average neutron count rate; the *error bars* show $\pm$ two standard errors (Reprinted with permission from Elsevier)

solar minimum than at solar maximum (Harrison and Usoskin 2010); that result is in good agreement with the 6% conductivity change and the 17% potential change. The conclusion that can be drawn without doubt from these figures is that the flux of cosmic rays affects the global circuit.

3.4 Global Circuit Cloud Coupling

This modulation of the fair weather current density by solar activity and associated cosmic ray changes provides a potential mechanism whereby the properties of clouds at low heights in fair weather regions could be changed by the currents passing through them, with implications, currently not quantified, for changes in weather and climate as a result. Theory indicates that, at the edge of a horizontal layer cloud, the transition from low conductivity air within a cloud to air of greater conductivity outside the cloud will be accompanied by a region of enhanced space charge, when the current flows vertically through the cloud boundary. Nicoll and Harrison (2009) have confirmed empirically that current flow passes through layer clouds.

Nicoll (2010) and Nicoll and Harrison (2010) have described an ingenious cloud edge charge detector (CECD) carried aloft by a meteorological balloon to investigate layer cloud edges at high vertical resolution. At the bottom of a low level cloud, an instrumented aircraft detected cloud droplets, as indicated in panel (a) of Fig. 10. From the balloon data, in a region where J_z is $\sim$2 pAm^{-2}, the air conductivity was calculated using the observed droplet properties. This is shown in panel (c) of Fig. 10 as the dark curve, with the lower scale. In this stratiform cloud in the fair weather part of the circuit, the conductivity is found to be three times less than that outside it. At the cloud edge, the negative space charge as a result of Gauss' law is up to 35 pCm^{-3}. The equivalent mean charge per droplet would be that of a few elementary electronic charges, but this would vary with droplet size. It is believed that

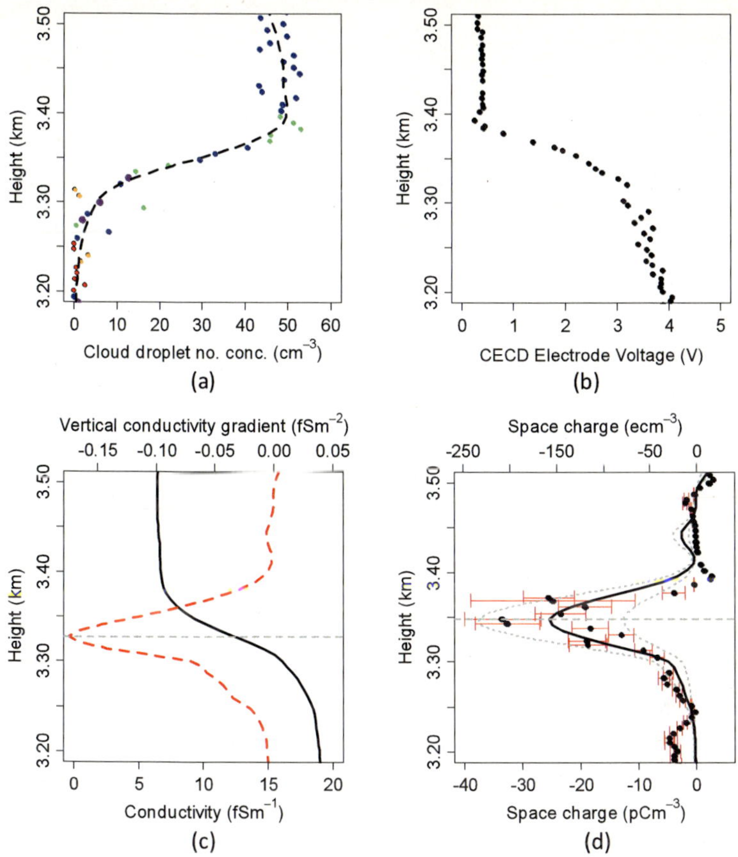

Fig. 10 Observations are shown of atmospheric electricity parameters and deduced electric charges at the lower edge of a low-level stratiform cloud at heights from 3.2 to 3.5 km, taken from Nicoll and Harrison (2010). Panel (**a**) shows the number concentration of cloud droplets, with an interpolating spline curve. Panel (**b**) plots the electrode voltage of a balloon-borne cloud edge charge detector (CECD); panel (**c**) shows the calculated conductivity (*solid curve*) in fSm^{-1} (*lower scale*) from the droplet concentrations, and the vertical conductivity gradient (*dashed line curve*, *upper scale*). Panel (**d**) shows the space charge density derived from the sensor voltages (*points*, with uncertainty), and the calculated space charge density, for a range of conduction current density assumptions (*dashed line*, *solid line* and *dotted lines*). The altitude of the space charge maximum, at 3.35 km, is shown by the *grey dashed horizontal line* (Copyright 2011 American Geophysical Union, reproduced with permission)

these charges are sufficient to influence collision processes between small cloud droplets in stratiform clouds.

The difficulty in determining the sensitivity of clouds to such changes, empirically at least, is the need to remove the substantial natural variability commonly present in cloud and clearly evident from satellite images of planet Earth. In one approach, Kniveton et al. (2008) investigated changes of the cloud cover at Vostok, Antarctica, and extreme increases of the vertical electric field there. Harrison and Ambaum (2010) reported a median 10% reduction in cloud amount at Lerwick Observatory, on the Shetland Islands to the north-east of Scotland, at $L \sim 4$, during Climax neutron rate reductions of at least 10% (Fig. 11). However, at a single site there is always substantial variability in the cloud data, which, even with averaging over multiple events, usually dominates. This result is not inconsistent with

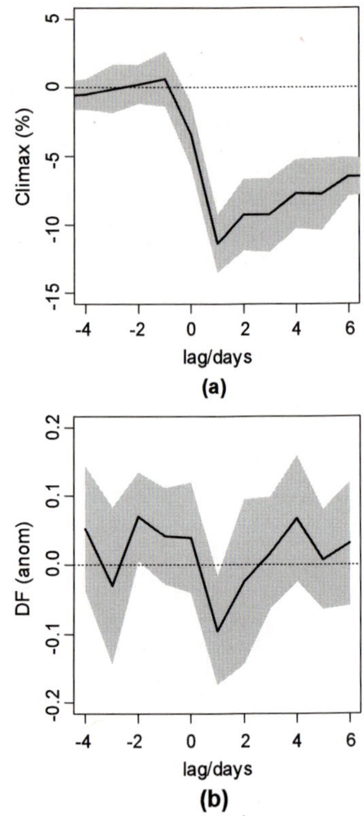

Fig. 11 Changes in (de-seasonalised) Lerwick diffuse fraction (DF, a measure proportional to cloud) around Climax neutron counter decreases of 10% or greater. (**a**) Median of neutron counter changes of 10% or greater (*black line*). (**b**) Median anomalies in diffuse fraction (*black line*) for the same times. For both (**a**) and (**b**), the 95% confidence range has been *shaded*

a cosmic ray effect on the global circuit which also influences clouds through a conduction current mechanism, and the analysis indicates rapid time scales of ~1 day or less.

Other than cosmic ray step changes, an alternative method of identifying cloud and global circuit responses to cosmic rays is to use spectral analysis methods to identify periodicities which are unique to cosmic rays. For example, a 1.68 year quasi-periodicity is known to occur from time to time in cosmic ray data (Valdés-Galicia et al. 1996), which is generated in the heliosphere rather than the photosphere (Rouillard and Lockwood 2004). A similar periodicity is apparent in surface PG data at Nagycenk Observatory, Hungary (Harrison and Märcz 2007), during fair weather conditions but absent during disturbed weather when global circuit influences would be masked.

Applying this spectral approach to a long series of cloud data from Lerwick, Harrison (2008) showed that a 1.68 year quasi-periodicity was also present during 1978–1990 when the periodicity was strong in neutron counter data. Specifically, this periodicity was found on days which were overcast. Hence, if the signal detected in the clouds originated in cosmic rays, this would not be inconsistent with a cloud edge response to modulation of the conduction current.

As common periodicities can arise by chance, a further requirement is for phase matching of the similar periodicities observed. Figure 12 shows a combination of neutrons (a) and Lerwick cloud data (c), filtered as in Harrison (2008), but with the short period of conduction current data available from Lerwick around the same time (Harrison and Nicoll 2008), panel (b). As must be expected with an intermittent and short data series, the periodicity after filtering the conduction current is well within the range expected to be produced by

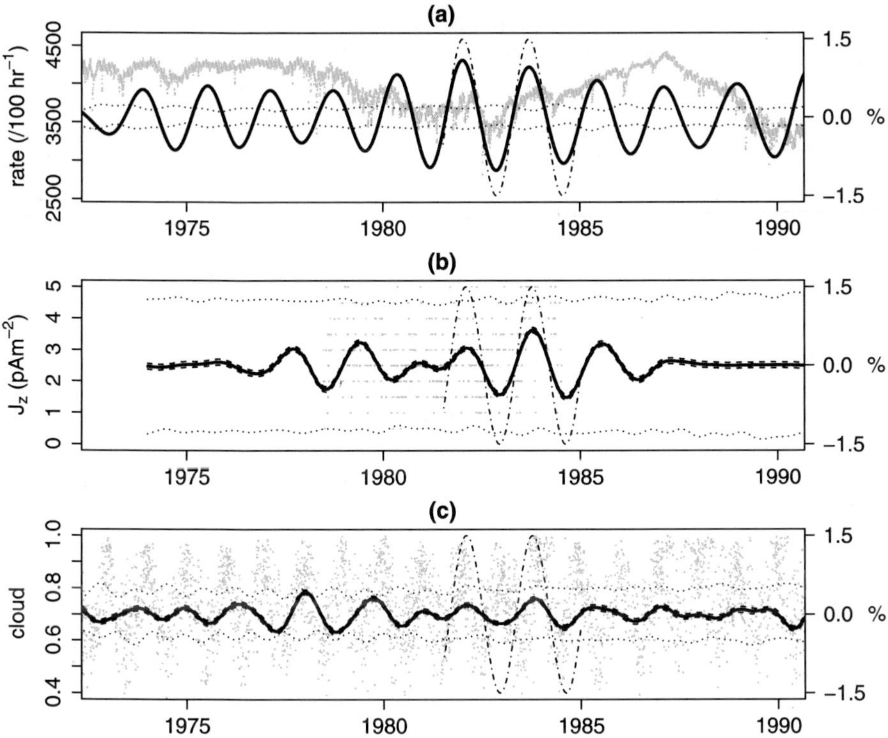

Fig. 12 Daily raw (*grey points*) and filtered (*black lines*) data for (**a**) Climax neutrons, (**b**) Lerwick conduction current density (**c**) Lerwick cloud "thickness" (opaqueness) on overcast days. The *black lines* represent the bandpass filtered versions of the raw data, passband 1.55 to 1.81 years with phase relationships preserved, as a percentage of the mean value of the raw data. Randomly selected points from the relevant series have been substituted for missing data, and the mean filtered version (*thick line*) is determined from multiple realisations of the random data replacements, with 95% confidence limits shown (*dashed lines*). *Dotted thin lines* show the 95% confidence limits on the variability in multiple realisations obtained by passing randomly chosen points through the bandpass filter. *Dashed-dotted lines* show a fitted sine wave with period 1.68 years, beginning at 1981.5. The median phase angles for the duration of the fitted sine wave are (**a**) 166 ± 0.2 degrees, (**b**) 153 ± 6 degrees and (**c**) 160 ± 6 degrees, with the 95% confidence range found from fitting to each of the multiple realisations

noise. However, a similar phase response is apparent in each of the neutron counter, conduction current and cloud data series, which, at the periodicity considered, is not inconsistent with a heliospheric signal propagating through the global circuit into cloud. Further work (Harrison et al. 2011) has shown that the cloud base at Lerwick also varies with cosmic rays, as might be expected from changes in the lower edge of clouds. This response was apparent in clouds identified by a trained meteorological observer as layer clouds.

3.5 Applications of Global Atmospheric Electricity

The potential gradient (PG) in surface air is sensitive to aerosol pollution, because of the removal of ions by aerosol particles. In severely polluted air, the PG can be substantially raised and, since historically many measurements were made in urban regions, this provides a method by which past urban pollution information can be reconstructed (Harrison and Aplin 2002, 2003; Harrison 2006, 2009).

At a regional scale Harrison et al. (2010) have discussed how radon emanating from the land (but not the oceans) in larger amounts than usual before a major earthquake increases the atmospheric conductivity in the surface layer, the lowest 250 m of the atmosphere. This reduces the columnar resistance to the ionosphere and, in a fair weather region, the current to the ionosphere increases. The PG before a major earthquake should therefore be reduced, as has previously been observed (Kondo 1968). The associated increased current, i.e. the increased upwards flow of negative ions, has the effect of lowering the ionosphere and so increasing the cut-off frequency ($\sim$1.7 kHz) for radio waves propagating in the Earth-ionosphere waveguide.

This mechanism accounts for the observations of the changed radio noise observed by the Demeter satellite over regions before major earthquakes occur (Nemec et al. 2009). It would be interesting to observe the cut-off frequency of "tweeks" (Reeve and Rycroft 1972), signals from distant lightning propagating at night over the earthquake-affected region in order to test the prediction of this mechanism where precursors to major earthquakes occurring over land affect the ionosphere. Pulinets and Ouzounov (2010) have presented a more complex mechanism relating these observable phenomena, and others. Mechanical (i.e. acoustic or gravity wave, or tidal, mechanisms to account for such effects have been considered by Hayakawa 2011). Hayakawa et al. (2005, 2008) have reported some anomalous third and fourth harmonic Schumann resonance effects associated with earthquakes.

4 Concluding Remarks

In this paper we have outlined the physical processes operating in the D.C. and A.C. global atmospheric electric circuit which rapidly couple phenomena occurring near the Earth's surface with the ionosphere and the near-Earth space environment.

Finally, we suggest some studies which could advance the subjects discussed further. It seems to be desirable to:

(i) investigate in more detail the relative contributions made by thunderstorm generators and by rain/shower cloud generators as drivers of the global electric circuit (Liu et al. 2010; Mach et al. 2011)
(ii) study the effects of cosmic rays (especially Forbush decreases) on lightning (Chronis 2009), low level clouds (Harrison et al. 2011), and their effects in the fair weather (load) part of the global circuit
(iii) investigate land/ocean differences in greater detail (Kulkarni 2009)
(iv) study the energy densities of the many different physical processes involved (see Feldstein et al. 2003)
(v) continue the search for signatures in the vertical electric field observed near the Earth's surface and throughout the atmosphere due to
  (a) solar flares (see Cobb 1967; Holzworth 1981; Kasatkina et al. 2010a, 2010b) or coronal mass ejections (see Kumar et al. 2008)
  (b) Forbush decreases (see Roble 1985; Sapkota and Varshneya 1990; Märcz 1997)
  (c) solar proton events (see Willett 1979; Farrell and Desch 2002)
  (d) auroral activity (see Hale and Croskey 1979; Belova et al. 2001; Kleimenova et al. 2010), and
  (e) gigantic jets (see Riousset et al. 2010)

(vi) conduct a variety of novel observations in space (ASIM, Chibis, Firefly, GLIMS, Sprite-sat, and TARANIS), and
(vii) test experimentally the hypothesis that radon emanating from the land before a major earthquake lowers the ionosphere.

Acknowledgements We thank many colleagues around the world for discussions over a good number of years on the topics presented here. This work partly results from an ISSI team project *Study of cosmic ray influence upon atmospheric processes*.

References

K.L. Aplin, Surv. Geophys. **27**, 63 (2006)
K.L. Aplin, R.G. Harrison, M.J. Rycroft, Space Sci. Rev. **137**, 107 (2008). doi:10.1007/978-0-387-87664-1_3
G.A. Bazilevskaya, M.B. Krainev, V.S. Makhmutov, J. Atmos. Sol.-Terr. Phys. **62**, 1577 (2000)
G.A. Bazilevskaya, I.G. Usoskin, E.O. Fluckiger, R.G. Harrison, L. Desorgher, R. Butikofer, M.B. Krainev, Y.I. Stozhlov, A.K. Svirzhevskaya, N.S. Svirzhevsky, G.A. Kovaltsov, Space Sci. Rev. **137**, 149 (2008)
E. Belova, S. Kirkwood, H. Tammet, Ann. Geophys. **18**, 1623 (2001)
E.A. Bering, A.A. Few, J.R. Benbrook, Phys. Today **51**, 24 (November 1998)
V. Bothmer, I.A. Daglis, *Space Weather—Physics and Effects* (Springer Praxis, Berlin, 2007)
M.S. Briggs, V. Connaughton, C. Wilson-Hodge, R.D. Preece, G.J. Fishman, R.M. Kippen, P.N. Bhat, W.S. Paciesas, V.L. Chaplin, C.A. Meegan, A. von Kienlin, J. Greiner, J.R. Dwyer, D.M. Smith, Geophys. Res. Lett. **38**, L02808 (2011). doi:10.1029/2010GL046259
B.E. Carlson, N.G. Lehtinen, U.S. Inan, J. Geophys. Res. **114**, 8 (2009). doi:10.1029/2009JA014531
B.E. Carlson, N.G. Lehtinen, U.S. Inan, J. Geophys. Res. **115**, A10324 (2010). doi:10.1029/2010JA015647
K. Carslaw, Nature **460**, 332 (2009)
J.A. Chalmers, J. Atmos. Terr. Phys. **24**, 1059 (1962)
J.A. Chalmers, *Atmospheric Electricity*, 2nd edn. (Pergamon, Oxford, 1967)
H.J. Christian, R.J. Blakeslee, D.J. Boccippio, W.L. Boeck, D.E. Buechler, K.T. Driscoll, S.J. Goodman, J.M. Hall, W.J. Koshak, D.M. Mach, M.F. Stewart, J. Geophys. Res. D **108**(1), 4005 (2003)
T.G. Chronis, J. Climate **22**, 5748 (2009)
W.E. Cobb, Mon. Weather Rev. **95**, 905 (1967)
R. Donner, S. Barbosa, J. Kurths, N. Marwan, Eur. Phys. J., Spec. Top. **174**, 1 (2009)
M.B. Enghoff, J.O.P. Pedersen, U.J. Uggerhoj, S.M. Paling, H. Svensmark, Geophys. Res. Lett. **38**, L09805 (2011). doi:10.1029/2011GL047036
A.I.I. Ette, U. Utah, J. Atmos. Terr. Phys. **35**, 785 (1973)
W.M. Farrell, M.D. Desch, Geophys. Res. Lett. **29**, 1323 (2002)
Y.I. Feldstein, I.A. Dremukhina, A.E. Levitin, U. Mall, I.I. Alexeev, V.V. Kalegaev, J. Atmos. Sol.-Terr. Phys. **65**, 429 (2003)
G. Fishman, A. Chilingarian, Eos **91**, 446 (2010)
G.J. Fishman, Eos **92**, 185 (2011). doi:10.1029/2011EO220001
M. Füllekrug, E.A. Mareev, M.J. Rycroft (eds.), *Sprites, Elves and Intense Lightning Discharges* (Springer, Dordrecht, 2006)
M. Füllekrug, R. Roussel-Dupré, E.M.D. Symbalisty, J.J. Colman, O. Chanrion, S. Soula, O. van der Velde, A. Odzimek, A.J. Bennett, V.P. Pasko, T. Neubert, Atmos. Chem. Phys. **11**, 7747 (2011)
M. Golkowski, M. Kubicki, M. Cohen, A. Kulak, U.S. Inan, Acta Geophys. **59**, 183 (2011)
L. Gray, J. Beer, M. Geller, J.D. Haigh, M. Lockwood, K. Matthes, U. Kubasch, D. Fleitmann, G. Harrison, L. Hood, J. Luterbacher, G.A. Meehl, D. Shindell, B. van Geel, W. White, Rev. Geophys. **48** (2010). doi:10.1029/2009RG000282
L.C. Hale, C.L. Croskey, Nature **278**, 239 (1979)
R.G. Harrison, Surv. Geophys. **25**, 441 (2004)
R.G. Harrison, Atmos. Environ. **40**(18), 3327 (2006). doi:10.1016/j.atmosenv.2006.01.042
R.G. Harrison, Proc. R. Soc. Lond. A **454**, 2575 (2008)
R.G. Harrison, Atmos. Environ. **43**, 1364 (2009). http://dx.doi.org/10.1016/j.atmosenv.2008.11.034
R.G. Harrison, K.L. Aplin, Atmos. Environ. **36**(25), 4037 (2002)
R.G. Harrison, K.L. Aplin, Atmos. Environ. **37**, 5319 (2003). doi:10.1016/j.atmosenv.2003.09.042
R.G. Harrison, A.J. Bennett, Adv. Geosci. **13**, 17 (2007)
R.G. Harrison, F. Märcz, Geophys. Res. Lett. **34**, L23816 (2007). doi:10.1029/2007GL031714
R.G. Harrison, K.L. Aplin, F. Leblanc, Y. Yair, Space Sci. Rev. **137**, 107 (2008). doi:10.1007/9787664-1_2

R.G. Harrison, M.H.P. Ambaum, Proc. R. Soc. Lond. Ser. A **464**, 2561 (2008)
R.G. Harrison, H. Tammet, Space Sci. Rev. **137**, 107 (2008). doi:10.1007/9787664-1_7
R.G. Harrison, K.A. Nicoll, Atmos. Res. **89**, 181 (2008). doi:10.1016/j.atmosres.2008.01.008
R.G. Harrison, M.H.P. Ambaum, J. Atmos. Sol.-Terr. Phys. **72**, 1408 (2010)
R.G. Harrison, I. Usoskin, J. Atmos. Sol.-Terr. Phys. **72**, 176 (2010)
R.G. Harrison, K. Aplin, M. Rycroft, J. Atmos. Sol.-Terr. Phys. **72**, 376 (2010)
R.G. Harrison, M.H.P. Ambaum, M. Lockwood, Proc. R. Soc. Lond. A **464**, 2561 (2011). doi:10.1098/rspa2011.0040
M. Hayakawa, K. Ohta, A.P. Nickolaenko, Y. Ando, Ann. Geophys. **23**, 1335 (2005)
M. Hayakawa, A.P. Nickolaenko, M. Sekiguchi, K. Yamashita, Y. Ida, M. Yano, Nat. Hazards Earth Syst. Sci. **8**, 1309 (2008)
M. Hayakawa, Nat. Hazards Earth Syst. Sci. **11**, 301 (2011)
P.B. Hays, R.G. Roble, J. Geophys. Res. **84A**, 3205 (1979)
A. Hirsikko, T. Niemenen, S. Gagne, K. Lehtipalo, H.E. Manninen, M. Ehn, U. Horrak, V.-M. Kerninen, L. Laakso, P.H. McMurry, A. Mirme, T. Petaja, H. Tammet, V. Vakkari, M. Vana, M. Kulmala, Atmos. Chem. Phys. **11**, 767 (2011)
R.H. Holzworth, J. Atmos. Sol.-Terr. Phys. **43**, 1115 (1981)
E.A. Kasatkina, O.I. Shumilov, M.J. Rycroft, F. Märcz, A.V. Frank-Kamanetsky, Atmos. Chem. Phys. Discuss. **9**, 21941 (2010a)
E.A. Kasatkina, O.I. Shumilov, M.J. Rycroft, F. Märcz, A.V. Frank-Kamanetsky, Atmos. Chem. Phys. Discuss. **10**, 24015 (2010b)
M.C. Kelley, *The Earth's Ionosphere: Plasma Physics and Electrodynamics*, 2nd edn. (Elsevier, Amsterdam, 2009)
J. Kirkby, J. Curtius, J. Almeida, et al., Nature **476**, 429 (2011). doi:10.1038/nature10343
N.G. Kleimenova, O.V. Kozyreva, M. Kubicki, S. Michnowski, Geomagn. Aeron. **50**, 48 (2010)
D.R. Kniveton, B.A. Tinsley, G.B. Burns, E.A. Bering, O.A. Troshichev, J. Atmos. Sol.-Terr. Phys. **70**, 1633 (2008)
G. Kondo, Mem. Kakioka Magn. Obs. **13**, 11 (1968)
M.N. Kulkarni, J. Geophys. Res. **114**, D06107 (2009). doi:10.1029/2008JD010009
C.P.A. Kumar, C. Panneerselvam, K.U. Nair, K. Jeeva, C. Selvaraj, S. Gurubaran, R. Rajaram, Indian J. Radio Space Res. **37**, 39 (2008)
C. Liu, E.R. Williams, E.J. Zipser, G. Burns, J. Atmos. Sci. **67**, 309 (2010). doi:10.1175/2009JAS3248.1
G. Lu, R.J. Blakeslee, J. Li, D.M. Smith, X.-M. Shao, E.W. McCaul, D.E. Buechler, H.J. Christian, J.M. Hall, S.A. Cummer, Geophys. Res. Lett. **37**, L11806 (2010). doi:10.1029/2010GL043494
D.M. Mach, R.J. Blakeslee, M.G. Bateman, J. Geophys. Res. **116**, D05201 (2011). doi:10.1029/2010JD014462
M. Makino, T. Ogawa, J. Atmos. Sol.-Terr. Phys. **46**, 431 (1984)
M. Makino, T. Ogawa, J. Geophys. Res. **90**, 5961 (1985)
F. Märcz, J. Atmos. Sol.-Terr. Phys. **59**, 975 (1997)
F. Märcz, P. Bencze, J. Atmos. Sol.-Terr. Phys. **60**, 1435 (1998)
R. Markson, Nature **320**, 588 (1986)
R. Markson, Bull. Am. Meteorol. Soc. **88**, 223 (2007)
C.E. McIlwain, J. Geophys. Res. **66**, 3681 (1961)
J. Miller, Phys. Today **64**, 16 (2011)
V.A. Mullayarov, V.I. Kozlov, A.A. Toropov, R.R. Karimov, J. Atmos. Sol.-Terr. Phys. **72**, 409 (2010)
K. Nakamura, et al. (Particle Data Group), J. Phys. G, Nucl. Part. Phys. **37**, 075021 (2010a). doi:10.1088/0954-3899/37/7A/075021
T. Nakamura, M. Sekiguchi, Y. Hobara, M. Hayakawa, J. Geophys. Res. **115**, A00E39 (2010b). doi:10.1029/2009JA014992
F. Nemec, O. Santolik, M. Parrot, J. Geophys. Res. **114**, A04303 (2009). doi:10.1029/2008JA013972
A.P. Nickolaenko, M. Hayakawa, Y. Hobara, Surv. Geophys. **31**, 409 (2010)
K.A. Nicoll, PhD thesis, University of Reading, 2010
K.A. Nicoll, R.G. Harrison, J. Atmos. Sol.-Terr. Phys. **71**, 2040 (2009)
K.A. Nicoll, R.G. Harrison, Geophys. Res. Lett. **37**, L13802 (2010). doi:10.1029/2010GL043605
A.F.L. Nogueira, IEEE Lat. Am. Trans. **7**, 223 (2009)
A. Odzimek, M. Lester, M. Kubici, J. Geophys. Res. **115**, D18207 (2010). doi:10.1029/2009JD013341
T. Palmer, P. Williams (eds.), *Stochastic Physics and Climate Modelling* (Cambridge, 2010)
V.P. Pasko, J. Geophys. Res. **115**, A00E35 (2010). doi:10.1029/2009JA014860
V.P. Pasko, Y. Yair, C.-L. Kuo, This issue (2011). doi:10.1007/s11214-011-9813-9
C. Price, Geophys. Res. Lett. **20**, 1363 (1993)
C. Price, in M. Füllekrug, E.A. Mareev, M.J. Rycroft (eds.), p. 85 (2006)

C. Price, O. Pechony, E. Greenberg, J. Light. Res. **1**, 1 (2007)
S. Pulinets, D. Ouzounov, J. Asian Earth Sci. **41**, 371 (2010)
V.A. Rakov, M.A. Uman, *Lightning: Physics and Effects* (Cambridge, 2003)
C.D. Reeve, M.J. Rycroft, J. Atmos. Terr. Phys. **34**, 667 (1972)
H. Rishbeth, O.K. Garriott, *Introduction to Ionospheric Physics* (Academic Press, New York, 1969)
J.A. Riousset, V.P. Pasko, P.R. Krehbiel, W. Rison, M.A. Stanley, J. Geophys. Res. **115**, A00E10 (2010). doi:10.1029/2009JA014286
R.G. Roble, J. Geophys. Res. **90**, 6000 (1985)
R.G. Roble, P.B. Hays, J. Geophys. Res. **84**, 7247 (1979)
R.G. Roble, I. Tzur, in *The Earth's Electrical Environment* (National Academy Press, Washington, 1986), p. 206
R.G. Roble, J. Atmos. Terr. Phys. **53**, 831 (1991)
C.J. Rodger, M. Cho, M.A. Clilverd, M.J. Rycroft, Geophys. Res. Lett. **28**, 199 (2001)
A. Rouillard, M. Lockwood, Ann. Geophys. **22**, 4381 (2004)
R. Roussel-Dupré, J.J. Colman, E. Symbalisty, D. Sentman, V.P. Pasko, Space Sci. Rev. **137**, 107 (2008). doi:10.1007/9787664-1_5
M.J. Rycroft, S. Israelsson, C. Price, J. Atmos. Sol.-Terr. Phys. **62**, 1563 (2000)
M.J. Rycroft, J. Atmos. Sol.-Terr. Phys. **68**, 445 (2006)
M.J. Rycroft, A. Odzimek, N.F. Arnold, M. Füllekrug, A. Kulak, T. Neubert, J. Atmos. Sol.-Terr. Phys. **69**, 2485 (2007)
M.J. Rycroft, R.G. Harrison, K.A. Nicoll, E.A. Mareev, Space Sci. Rev. **137**, 107 (2008). doi:10.1007/9787664-1_6
M.J. Rycroft, A. Odzimek, AIP Conf. Proc. **1118**, 124 (2009)
M.J. Rycroft, A. Odzimek, J. Geophys. Res. **115**, A00E37 (2010). doi:10.1029/2009JA014758
M.J. Rycroft, The plasma and radiation environment in Earth orbit, in *Encyclopedia of Aerospace Engineering*, vol. 6, eae323, ed. by R. Blockley, W. Shyy (Wiley, Chichester, 2010). doi:10.1002/978040686652
B.K. Sapkota, N.C. Varshneya, J. Atmos. Terr. Phys. **52**, 1 (1990)
B. Schwarzchild, Phys. Today **64**, 10 (May 2011)
A.V. Shvets, M. Hayakawa, Surv. Geophys. **32** (2011, in press). doi:10.1007/s10712-011-9135-1
A.V. Shvets, M. Hayakawa, M. Sekiguchi, Y. Ando, J. Atmos. Sol.-Terr. Phys. **71**, 1405 (2009)
A.V. Shvets, Y. Hobara, M. Hayakawa, J. Geophys. Res. **115**, A12316 (2010). doi:10.1029/2010JA015851
D. Siingh, R.P. Singh, A.K. Kamra, P.N. Gupta, R. Singh, V. Gopalakrishnan, A.K. Singh, J. Atmos. Sol.-Terr. Phys. **67**, 637 (2005)
D. Siingh, V. Gopalakrishnan, R.P. Singh, A.K. Kamra, S. Singh, V. Pant, R. Singh, A.K. Singh, Atmos. Res. **84**, 91 (2007)
D. Siingh, R.P. Singh, A.K. Singh, M.N. Kulkarni, A.S. Gautam, A.K. Singh, Surv. Geophys. **32** (2011, in press). doi:10.1007/s10712-011-9127-1
F. Simoes, M. Rycroft, N. Renno, Y. Yair, K.L. Aplin, Y. Takahashi, Space Sci. Rev. **137**, 107 (2008). doi:10.1007/9787664-1_28
J. Slingo, K. Bates, N. Nikiforakis, M. Piggott, M. Roberts, L. Shaffrey, I. Stevens, P.L. Vidale, H. Weller, Philos. Trans. R. Soc. Lond. A **367**, 815 (2009)
D.M. Smith, AIP Conf. Proc. **1118**, 34 (2009)
D.M. Smith, B.J. Hazelton, B.W. Grefenstette, J.R. Dwyer, R.H. Holzworth, E.H. Lay, J. Geophys. Res. **115**, A00E49 (2010a). doi:10.1029/2009JA014853
A.J. Smith, R.B. Horne, N.P. Meredith, J. Atmos. Sol.-Terr. Phys. **72**, 463 (2010b)
M.A. Stanley, X.-M. Shao, D.M. Smith, L.I. Lopez, M.B. Pongratz, J.D. Harlin, M. Stock, A. Regan, Geophys. Res. Lett. **33**, L06803 (2006). doi:10.1020/2005GL025537
E.K. Stansbery, A.A. Few, P.B. Geis, J. Geophys. Res. **98**, 16,591 (1993)
Y.I. Stozhkov, V.I. Ermakov, V.S. Makhmutov, in *Proc. ICRC* 2001 (2001), p. 4157
J.N. Thomas, R.H. Holzworth, M.P. McCarthy, Atmos. Res. **91**, 153 (2009). doi:10.1016/j_atmosres.2008.03.026
B.A. Tinsley, J.T. Hoeksema, D.N. Baker, J. Geophys. Res. **99**(D8), 16,805 (1994)
B.A. Tinsley, G.B. Burns, L. Zhou, Adv. Space Res. **40**, 1126 (2007)
B.A. Tinsley, Rep. Prog. Phys. **71**, 066801 (2008). doi:10.1088/0034-4885/71/6/066801
J.F. Valdés-Galicia, R. Pérez-Enríquez, J.A. Otaoloa, Sol. Phys. **167**, 409 (1996). doi:10.1007/BF00146349
P.I.Y. Velinov, A. Mishev, L. Mateev, Adv. Space Res. **44**, 1002 (2009)
H. Volland, Phys. Scr. **T18**, 289 (1987)
B. Vonnegut, Annu. Rev. Earth Planet. Sci. **1**, 297 (1973)
T. Whitely, M. Füllekrug, M. Rycroft, A. Bennett, F. Wyatt, D. Elliott, G. Heinson, A. Lewis, R. Sefako, P. Fourie, J. Dyers, A. Thomson, S. Flower, Radio Sci. **46**, RS4007 (2011). doi:10.1029/2010RS004523
J.C. Willett, J. Geophys. Res. **84**(C8), 4999 (1979)

E.R. Williams, Science **256**, 1184 (1992)
E.R. Williams, *Encyclopedia of Atmospheric Sciences*, ed. by J.R. Holton, J.A. Pyle, J.A. Curry, p. 724 (Academic Press, San Diego, 2002)
E.R. Williams, Y. Yair, ed. by M. Füllekrug, E.A. Mareev, M.J. Rycroft, p. 57 (2006)
E.R. Williams, Atmos. Res. **91**, 140 (2009)
E.R. Williams, R. Boldi, J. Bor, G. Satori, C. Price, E. Greenberg, Y. Takahashi, K. Yamamoto, Y. Matsudo, Y. Hobara, M. Hayakawa, T. Chronis, E. Anagnostou, D.M. Smith, L. Lopez, J. Geophys. Res. **111**, D16209 (2006). doi:10.1029/2005JD006447
E.R. Williams, W.A. Lyons, Y. Hobara, V.C. Mushtak, N. Ascencio, R. Boldi, J. Bor, S.A. Cummer, E. Greenberg, M. Hayakawa, R.H. Holzworth, V. Kotroni, J. Li, C. Morales, T.E. Nelson, C. Price, B. Russell, M. Sato, G. Satori, K. Shirahata, Y. Takahashi, K. Yamashita, Q. J. R. Meteorol. Soc. **136**(S1), 257 (2010). doi:10.1002/qj.489
C.T.R. Wilson, Proc. Camb. Philos. Soc. **13**, 363 (1906)
C.T.R. Wilson, Philos. Trans. R. Soc. Lond. A **221**, 73 (1921)
C.T.R. Wilson, J. Franklin Inst. **208**, 1 (1929)
C.T.R. Wilson, Proc. R. Soc. Lond. Ser. A **236**, 297 (1956)
K. Yamashita, Y. Takahashi, M. Sato, H. Kase, J. Geophys. Res. **116**, A00E61 (2011). doi:10.1029/2009JA014792
H. Yeng, V.P. Pasko, G. Satori, J. Geophys. Res. **114**, D01103 (2009). doi:10.1029/2008JD009961
L. Zhou, B.A. Tinsley, J. Atmos. Sci. **67**, 1143 (2010)

Space Sci Rev (2012) 168:385–417
DOI 10.1007/s11214-011-9797-5

Equatorial and Low Latitude Ionospheric Effects During Sudden Stratospheric Warming Events

Ionospheric Effects During SSW Events

Jorge L. Chau · Larisa P. Goncharenko · Bela G. Fejer · Han-Li Liu

Received: 21 March 2011 / Accepted: 6 June 2011 / Published online: 14 July 2011

Abstract There are several external sources of ionospheric forcing, including these are solar wind-magnetospheric processes and lower atmospheric winds and waves. In this work we review the observed ion-neutral coupling effects at equatorial and low latitudes during large meteorological events called sudden stratospheric warming (SSW). Research in this direction has been accelerated in recent years mainly due to: (1) extensive observing campaigns, and (2) solar minimum conditions. The former has been instrumental to capture the events before, during, and after the peak SSW temperatures and wind perturbations. The latter has permitted a reduced forcing contribution from solar wind-magnetospheric processes. The main ionospheric effects are clearly observed in the zonal electric fields (or vertical $\mathbf{E} \times \mathbf{B}$ drifts), total electron content, and electron and neutral densities. We include results from different ground- and satellite-based observations, covering different longitudes and years. We also present and discuss the modeling efforts that support most of the observations. Given that SSW can be forecasted with a few days in advance, there is potential for using the connection with the ionosphere for forecasting the occurrence and evolution of electrodynamic perturbations at low latitudes, and sometimes also mid latitudes, during arctic winter warmings.

Keywords Atmosphere-ionosphere coupling · Equatorial aeronomy · Low latitude electrodynamics · Stratospheric warming

J.L. Chau (✉)
Radio Observatorio de Jicamarca, Instituto Geofisico del Peru, Lima, Peru
e-mail: jorge.chau@jro.igp.gob.pe

L.P. Goncharenko
Massachusetts Institute of Technology, Haystack Observatory, Westford, MA, USA

B.G. Fejer
Center for Atmospheric and Space Sciences, Utah State University, Logan, UT, USA

H.-L. Liu
High Altitude Observatory, National Center for Atmospheric Research, Boulder, CO, USA

1 Introduction and Scope

Forecasting the ionospheric weather is very challenging. A significant portion of the day-to-day variability of the ionospheric parameters cannot be explained by the major drivers coming from above, i.e., solar ionizing flux and geomagnetic activity. The sources of such variability have been associated for many years with lower atmospheric processes (e.g., Rishbeth 2006). However the understanding of these links has been difficult due to the rapid response of the ionosphere to solar and magnetospheric drivers, and previously limited databases.

In this work we review the systematic studies of one of these links, i.e., ionospheric variability during sudden stratospheric warming (SSW) events and in particular at equatorial and low latitudes. A SSW is a large meteorological event where westerly winds in the Northern winter stratosphere abruptly (i.e., in a few days time) slow down or even reverses direction, accompanied by a break-down of the polar vortex and a rise of stratospheric temperature by several tens of degrees. As a large-scale meteorological process in the winter polar region, it strongly affects vertical coupling in a large range of altitudes, from the stratosphere to the thermosphere (Liu and Roble 2002). It has been under extensive observational and theoretical investigation for many years (e.g., Andrews et al. 1987). The key mechanism for SSW events is the growth of planetary waves propagating upward from the troposphere and their non-linear interaction with the zonal mean flow (Matsuno 1971). The beauty of this connection is that SSWs are predicted a few days in advance (e.g., Kim and Flatau 2010), and therefore understanding this connection could help in forecasting the ionospheric weather.

Research studying the lower atmosphere-ionosphere coupling during SSWs has been catalyzed after the observations reported by Goncharenko and Zhang (2008) and Chau et al. (2009). They clearly showed that the ion temperature at mid-latitudes and vertical $\mathbf{E} \times \mathbf{B}$ plasma drifts at equatorial latitudes, respectively, were correlated with the January 2008 SSW event. Those observations and results were a consequence of an unprecedented observing campaigns using all the incoherent scatter radars and other ground-based instruments. Since then, a significant number of upper atmospheric effects have been associated with this coupling, e.g.:

- Multiple studies have demonstrated SSW effects in various ionospheric parameters: total electron content (TEC) (e.g., Goncharenko et al. 2010a, 2010b; Pedatella and Forbes 2010a; Chau et al. 2010), peak electron density (f_oF_2) (e.g., Yue et al. 2010; Pancheva and Mukhtarov 2011), electric fields (e.g., Anderson and Araujo-Pradere 2010; Fejer et al. 2010).
- Differences in longitudinal response have been documented with observations over the Philippines (e.g., Anderson and Araujo-Pradere 2010; Fejer et al. 2010) and Indian (Vineeth et al. 2009; Sridharan et al. 2009) sectors.
- Global character of the observed phenomena have been studied with satellite observations using the Challenging Minisatellite Payload (CHAMP) (e.g., Fejer et al. 2010), the Constellation Observing System for Meteorology, Ionosphere, and Climate (COSMIC) (Yue et al. 2010; Pancheva and Mukhtarov 2011).
- The coupling mechanisms were investigated with modeling studies using the NCAR Thermosphere–Ionosphere–Mesosphere Electrodynamics General Circulation Model (TIMEGCM) (e.g., Liu et al. 2010) and the Whole Atmosphere Model (WAM) (e.g., Fuller-Rowell et al. 2010).

Our paper is organized as follows. We first start with background material related to: (a) the ionospheric variability (Sect. 2), (b) the quiet-time equatorial vertical $\mathbf{E} \times \mathbf{B}$ drifts including their statistics and measurement techniques (Sect. 3), and (c) sudden stratospheric warming events (Sect. 4). We put special emphasis on the vertical $\mathbf{E} \times \mathbf{B}$ drifts since, as we will explain later, it is one of the most important parameters that controls the electrodynamics at equatorial and low latitudes. Then, we review the effects on ionospheric parameters that have been associated with the SSWs, in particular, the vertical $\mathbf{E} \times \mathbf{B}$ drifts, TEC, and f_oF_2 (Sect. 5). Finally, we discuss the possible scenarios including those supported by current numerical modeling efforts.

2 Ionospheric Variability

There is a very good understanding of the F-region ionosphere climatology, i.e., mean values for different seasons, solar conditions, and degrees of magnetic activity, as should be expected after more than 70 years of ionospheric observations. However, the understanding of its weather and thus its predictive capability is very poor. This variability displays oscillations over a wide range of timescales (from minutes to severals days). During geomagnetic storms and substorms, there are significant ionospheric changes which can last for hours and it can take days for the ionosphere to recover. Even during nonstorm periods, geomagnetic perturbations associated with solar coronal holes and high-speed solar wind streams can cause periodic ionosphere and thermosphere oscillations with the solar rotation period and their subharmonics of this period (Lei et al. 2008). Ionospheric variability has been also correlated on a statistical basis with the K_p or A_p index (e.g., Rishbeth and Mendillo 2001; Pancheva et al. 2002; Altadill and Apostolov 2001; Xiong et al. 2006).

During quiet times, i.e., under geomagnetically undisturbed conditions, significant ionospheric day-to-day variability is also observed. For example, it has long been recognized that there is large day-to-day variability in the vertical plasma drift observed at the magnetic equator, with this variability being the largest during solar minimum conditions (e.g., Fejer et al. 1991; Fejer and Scherliess 2001). Similarly, changes of up to 20% in daytime F-region maximum electron density that cannot be associated with solar or magnetic drivers have been reported by Rishbeth and Mendillo (2001) and Mendillo et al. (2002). These variations are comparable to ionospheric changes related to geomagnetic activity (e.g., Fuller-Rowell et al. 2000; Rishbeth and Mendillo 2001; Forbes et al. 2000). In the equatorial ionosphere, the monthly ionospheric variability (i.e., monthly standard deviation over monthly mean) reaches 30–40% at sunrise and 15–30% at sunset (e.g., Bilitza et al. 2004), while quiet-time variations in daytime electron density can be as high as 200% (Zhao et al. 2008). Mikhailov et al. (2007) also showed that there exist synchronous quiet day electron density variations in both the E and F regions. Recent observations by C/NOFS have revealed very large equatorial ionospheric depletion near dawn under geomagnetically quiet conditions (De la Beaujardière et al. 2009).

The electrodynamic plasma ($\mathbf{E} \times \mathbf{B}$) drifts, and in particular the vertical component, play important roles on the plasma distribution and dynamics in the low latitude ionosphere. In the next section (Sect. 3), we summarize the main characteristics of these drifts under geomagnetically quiet-time conditions. These characteristics will be later put into context with the observations under similar quiet-time conditions but during SSW periods.

3 Quiet-Time Equatorial Vertical E x B Drifts

The quiet-time equatorial $\mathbf{E} \times \mathbf{B}$ drifts are driven by neutral wind generated E-region dynamo electric fields (e.g., Richmond et al. 1976; Stening 1981) and by F-region polarization electric fields (e.g., Rishbeth 1971; Heelis et al. 1974). The thermospheric winds are highly variable as a result of changes in the global tidal forcing, and effects of irregular winds, planetary, and gravity waves. Previous studies suggest that planetary waves and tides play important roles in the electrodynamics of the lower thermosphere (e.g., Chen 1992; Forbes and Leveroni 1992; Parish et al. 1994; Immel et al. 2009). In addition, the relative efficiencies of the E- and F-region dynamos change significantly throughout the day, and also with season and solar activity.

For a comprehensive review on the physical principles underlying the electrodynamic properties of the equatorial and low-latitude ionosphere, we recommend reading Heelis (2004). Moreover, Fejer (2010) has recently reviewed the ionospheric electrodynamics at low latitudes. In this section we focus on the statistics and measurement techniques for the equatorial vertical $\mathbf{E} \times \mathbf{B}$ drifts.

3.1 Equatorial Vertical $\mathbf{E} \times \mathbf{B}$ Drift Statistics

The equatorial vertical $\mathbf{E} \times \mathbf{B}$ plasma drifts have been extensively studied using ground-based instruments, i.e., coherent and incoherent scatter radars, ionosondes and magnetometers. These ground-based measurements started in 1968 using the Jicamarca Incoherent Scatter radar (Woodman and Hagfors 1969; Woodman 1970). The characteristics of the Jicamarca quiet-time vertical drifts have been described in several publications (e.g., Fejer et al. 1991; Scherliess and Fejer 1999; Fejer and Scherliess 2001). Figure 1 shows scatter plots of these drifts for periods around equinox and June solstice for two ranges of solar flux index. The December solstice results are not shown since there were not enough measurements for the period studied, i.e., between 1968 and 1992. These drifts show a large day-to-day variability at all local times. They are typically upward during the day ($\sim$20 m/s) and downward at night ($\sim$10–40 m/s) and do not change much with season and solar flux, except near sunset and sunrise. Amplitudes of the pre-reversal enhancement (around sunset) increase strongly with solar activity (e.g., Scherliess and Fejer 1999). The variability of these vertical drifts decreases from low to high solar flux periods for all seasons.

The quiet-time variability of the Jicamarca vertical drifts is local time, seasonal, and solar cycle dependent. Figure 2 shows the mean values and standard deviations of these drifts for morning and afternoon periods as function of day of the year and two solar flux indexes. This variability is largest in the dawn–noon sector and during March equinox solar minimum periods, when the midday average upward drift velocity from consecutive magnetically quiet days can often change by more than 10 m/s. The day-to-day variability of the vertical drifts decreases in the afternoon sector and with the increase of solar activity for all seasons. For the results we present in later sections, it is important to take note of the December solstice daytime values, in particular: (1) the standard deviation is larger at lower solar flux indexes with a value of $\sim$7 m/s, and (2) the afternoon mean value is usually downward, particularly at low solar flux values.

Combining Jicamarca vertical drifts and AE-E (Atmosphere-Explorer 2) satellite measurements, Scherliess and Fejer (1999) developed a quiet-time empirical F region vertical drift global model for different seasons and solar conditions. This model showed large longitudinal variations, particularly, for the evening upward drifts. Using drift measurements from the Republic of China Satellite (ROCSAT-1), Fejer et al. (2008) has developed a new empirical global model for moderate and high solar conditions for three seasons. They showed that the longitudinal dependence of the daytime and nighttime vertical drifts is much stronger than reported earlier, especially during December and June solstice. Among other important results, the morning and afternoon December solstice drifts have significantly different longitudinal dependence, and the daytime upward drifts have strong wave number-four signatures during equinox and June solstice. Using the same ROCSAT-1 data, Kil et al. (2009)

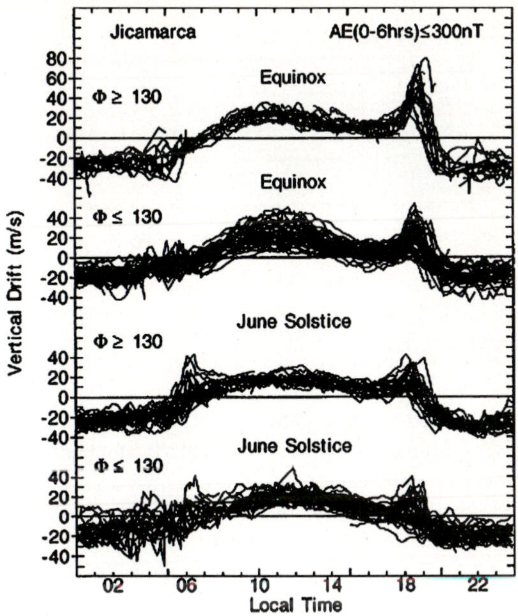

Fig. 1 Scatter plots of quiet-time Jicamarca F-region vertical plasma drifts (positive upward) during equinox (March-April, September-October) and June solstice (May-August) for two ranges of the solar flux index (Φ) (from Fejer and Scherliess 2001, Fig. 1)

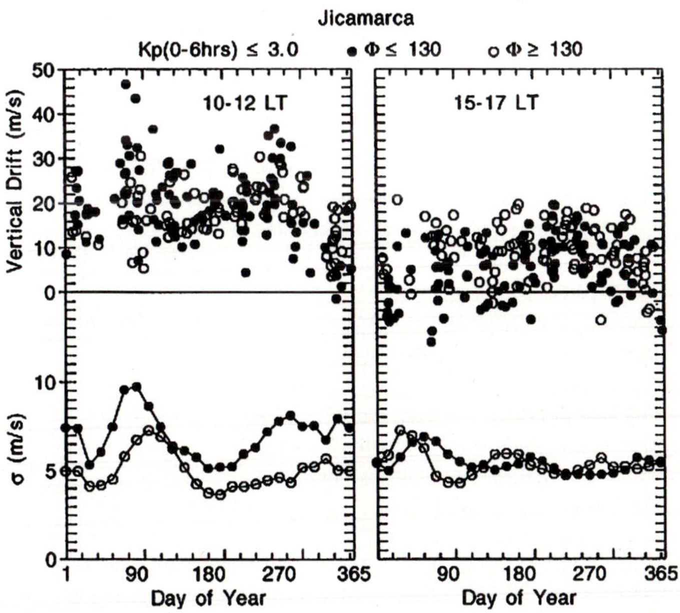

Fig. 2 Midday and afternoon quiet-time vertical plasma drift velocities for two ranges of the solar flux index (*top panels*). Standard deviations of the vertical drifts (*bottom panels*) (from Fejer and Scherliess 2001, Fig. 2)

introduced a similar empirical drift model but for every month instead of three seasons. Regional models at other longitude sectors have been also developed using ground-based ionosonde measurements in the Brazilian (Batista et al. 1996) and Indian (Sastri 1996) sectors.

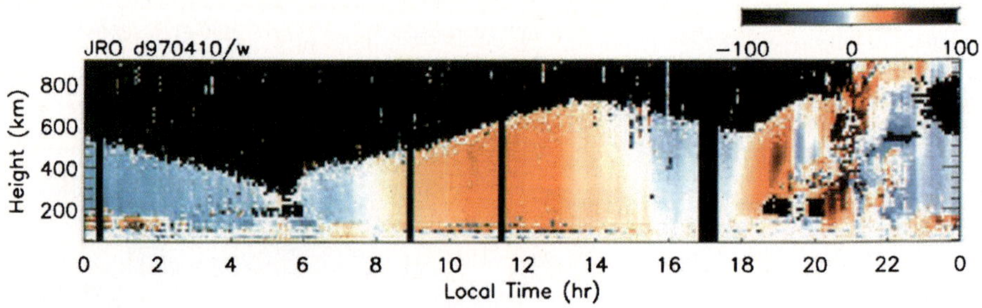

Fig. 3 Range-time plot of vertical drifts measured at Jicamarca on April 10, 1997. *Red* indicates upward drift while *blue* indicates downward drift (from Kudeki et al. 1999, Plate 1)

Other indirect measurements of daytime vertical F region drifts (or zonal electric fields) at different longitudes, particularly at equatorial latitudes, have been provided by the CHAMP satellite using magnetic field measurements of the equatorial electrojet (Lühr et al. 2008). These measurements show a strong longitudinal modulation of the equatorial daytime eastward electric fields associated with the wavenumber-four structure in agreement with the ROCSAT-1 measurements. These ionospheric signatures have been identified before in measurements of the nighttime equatorial ionization anomaly (e.g., Immel et al. 2006) and noontime equatorial electrojet (England et al. 2006) using far ultraviolet images. The reported wavenumber-four structure represents another evidence of atmosphere-ionosphere coupling, but this connections is not part of this review.

3.2 Equatorial Vertical $\mathbf{E} \times \mathbf{B}$ Measurements

Most of the equatorial drift measurements have been made with the Jicamarca Incoherent Scatter radar (ISR). Here we describe those measurements and also other ground-based measurements with a particular emphasis on the daytime capabilities. As we will see later, so far most of the ionospheric effects associated with the SSW events have been identified during the day.

Since 1968 (Woodman and Hagfors 1969), very precise measurements of vertical plasma drifts have been made at the Jicamarca Radio Observatory (JRO) from ISR echoes. By using antenna beams pointing perpendicular to the magnetic field, the correlation times of the received signals are very long, allowing the use of pulse-to-pulse techniques. From 1968 to 1994, daytime and nighttime measurements with 1–2 m s^{-1} precision have been obtained few days per year. Using a spectral analysis and inversion techniques, Kudeki et al. (1999) introduced a new mode that improved the precision further and allowed higher altitude resolution. Measurements using this new approach started in 1994. Figure 3 shows a typical example of the type of vertical drifts that are measured at Jicamarca. Those measurements are obtained with an altitude resolution of 15 km and time resolution that varies between 1 and 5 min. The altitudinal coverage varies depending on the electron density profile.

Since 2001 equatorial daytime vertical drifts have been routinely derived from the Doppler shift measurements of coherent echoes occurring in the equatorial F_1 region, the so-called 150-km echoes (Kudeki and Fawcett 1993). Although the physics behind these echoes is not known yet, their vertical velocities are in excellent agreement with the F region vertical drifts (Chau and Woodman 2004). Given that 150-km echoes are much stronger than incoherent scatter echoes, such measurements are done with the JULIA (Jicamarca Unattended Long-term Investigations of the Atmosphere) system that uses the large Jicamarca

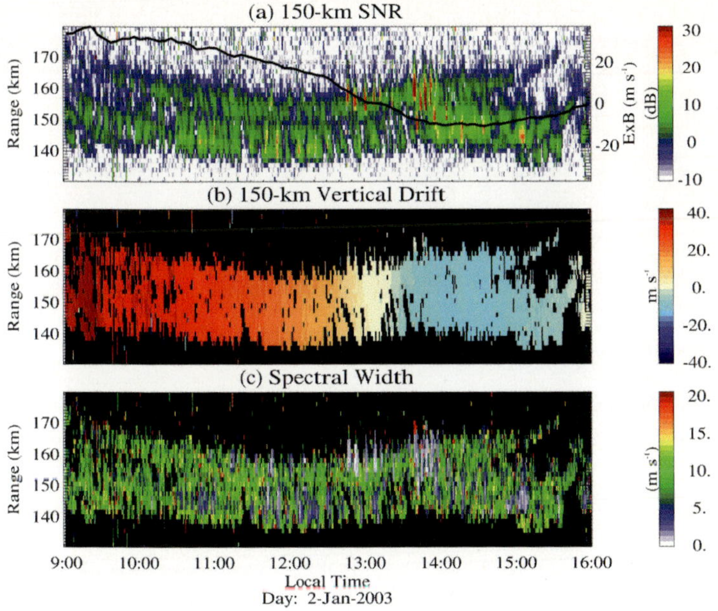

Fig. 4 Example of 150-km parameters obtained with the JULIA radar on January 3, 2003: (**a**) Signal-to-noise-ratio (dB), (**b**) vertical drift (m/s), and (**c**) spectral width (m/s). The *solid* in the *top panel* represents the average (in range) drifts (from Chau and Kudeki 2006, Fig. 1)

array but low power transmitters, allowing a better time coverage (between 125–180 days per year). The precision of these velocities is comparable to the ISR drifts, however they provide only mean values for the F region. Figure 4 shows an example of typical radar parameters from 150-km echoes. These measurements are only possible during the day and in the case of Jicamarca, the echoes do not show a strong seasonal or solar dependence (e.g., Chau and Kudeki 2006). Similar measurements are now routinely made at other longitudinal sectors: Brazil (de Paula and Hysell 2004), India (Patra and Rao 2007), and Pohnpei (Tsunoda and Ecklund 2004). It should be noted that the 150-km and F region vertical drifts are quite similar but not identical, otherwise there would be a violation of the curl-free conditions. The vertical drifts increase with height in the morning and decrease in the afternoon as shown by Pingree and Fejer (1987).

Recently, Alken (2009) presented a model of vertical plasma drifts from 150-km echoes observed by the JULIA system. The model includes a climatology of the 150 km equatorial vertical drifts as well as an estimate of the day-to-day variability, which can be significant, for different seasons, solar conditions and daytime (0800 to 1600 LT) hours in the Peruvian sector. Figure 5 shows the drift values as function of day of the year and daytime hour for low solar activity conditions: (a) the Scherliess/Fejer model (Scherliess and Fejer 1999), (b) the JULIA data, and (c) the JULIA Vertical Drift model (JVDM). There is very good agreement between the JVDM model and the data. All major features of the vertical plasma drifts have been reproduced. The drift maxima during March and September equinox and minima during June and December solstice agree well. An important improvement in the JVDM compared to the Scherliess and Fejer model is the allowance of differences between March and September equinox. Again, for the purpose of this review, the most important results are the December solstice values, i.e., positive values in the morning and negative values in the afternoon.

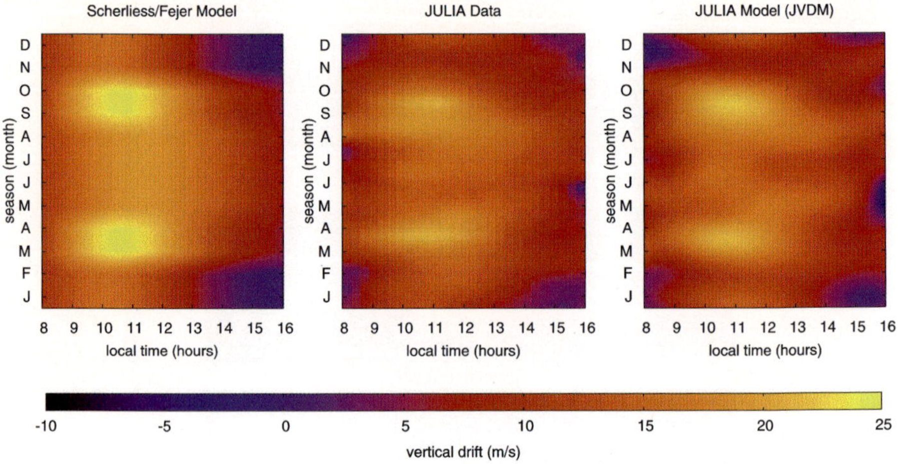

Fig. 5 Vertical drifts values for low solar conditions: (*left*) Scherliess and Fejer model output, (*middle*) raw JULIA vertical drift data, and (*right*) JULIA Vertical Drift Model (from Alken 2009, Fig. 4)

Equatorial daytime vertical drifts are also now routinely derived from the difference of horizontal geomagnetic fields measured at pairs of stations, one very near the magnetic equator and another a few degrees off at about the same longitude (e.g., Anderson et al. 2002). Vertical drifts derived using magnetic field data from Jicamarca and Piura ($\sim$6.8° magnetic) are generally in good agreement with radar measured drifts from both 150-km echoes and Incoherent scatter signals, but their accuracy can sometimes be affected by magnetic field changes produced by variable low latitude E region winds (e.g., Fang et al. 2008) or unexpected changes in the E region conductivity (e.g., during solar flares). Anderson et al. (2006) has used this technique at other longitudes.

In Fig. 6 we show examples of daytime vertical $\mathbf{E} \times \mathbf{B}$ drifts using (a) ISR, (b)150-km echoes, and (c) magnetometer ΔH, at Jicamarca for similar dates. The bottom panels shows the values during the January 2009 event, and the top panel shows the control days (similar season and solar conditions but without SSW) for all three techniques. In general, all three techniques are in very good agreement, particularly around noontime hours. In Table 1 we summarize the main parameters for all three techniques.

Recently, equatorial zonal electric fields have been derived from equatorial magnetic field measurements by the CHAMP satellite (Alken and Maus 2010). Although we are focusing on daytime vertical drift measurements, evening and early night equatorial F region vertical plasma drifts are routinely derived from ionosonde observations (e.g., Abdu et al. 2007). The accuracy of the ionosonde derived nighttime vertical drifts is improved when production and recombination effects are properly taken into account using theoretical models (e.g., Bertoni et al. 2006). On the other hand the ionosonde derived vertical velocity during the day does not represent the F region vertical $\mathbf{E} \times \mathbf{B}$ drift, particularly at the equator (Woodman et al. 2006).

4 Sudden Stratospheric Warming Events

In this section we describe the large SSW meteorological phenomena. Then, we summarize some of the effects that are not part of this review, i.e., they occur outside the equatorial/low-

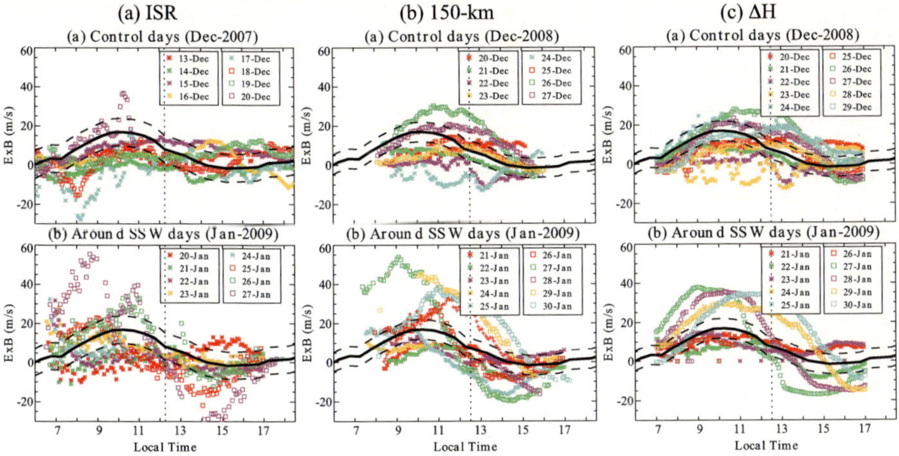

Fig. 6 Examples of daytime **E** × **B** vertical drifts using: (**a**) ISR, (**b**) 150-km echoes, (**c**) ΔH. The *bottom panels* show days (in different colors) around the Jan 2009 SSW event, and *top panels* show control days for a similar season and solar conditions. The *black solid line* and the *dashed lines* represent the expected average and standard deviations values, respectively. (Adapted from Chau et al. 2009, Fig. 2)

Table 1 **E** × **B** vertical drift measurements at Jicamarca using different techniques: (a) ISR, (b) 150-km echoes, (c) Magnetometer ΔH

Value	ISR	150-km echoes	ΔH
Instrument	Jicamarca ISR	JULIA	Mag. at JRO and Piura
Annual coverage	~20 days	>150 days	>300 days
Since	1968	2001	2000
Time coverage	All day*	07-18 LT	09-17 LT
Time res.	5 min	1–5 min	1 min
Accuracy	<1 m/s	<1 m/s	<3 m/s
Altitudes	150–1000 km	Mean *F* region	Mean *F* region
Reference	Kudeki et al. (1999)	Chau and Woodman (2004)	Anderson et al. (2002)

latitude ionospheric region, and present the main features of the January 2008 and January 2009 SSW campaigns.

4.1 SSW: A Large Meteorological Phenomena

The stratospheric sudden warming events represent the most spectacular meteorological fluctuations in the wintertime polar stratosphere and serve as the strongest manifestation of dynamical coupling in the troposphere-stratosphere-mesosphere system. During these events, stratospheric temperatures abruptly increase by tens of degrees, the normal winter polar vortex changes its position and shape (minor warming) or breaks up (major warming), and zonal mean zonal winds become weaker or even change direction (O'Neill 2003).

Following initial studies by Matsuno (1971), it is generally accepted that SSWs result from the interaction between the enhanced planetary waves and the zonal mean flow. These planetary waves are generated in the troposphere, propagate upward, and deposit the westward (easterly) momentum which slows down or even reverses the stratospheric winter jet.

The associated Coriolis force strengthens the poleward circulation (Brewer-Dobson circulation). As a result, the downwelling in the polar region is enhanced, which causes the enhanced adiabatic warming. However, the cause of the rapid amplification of planetary waves is not well understood. Suggested explanations include resonant growth of planetary waves (Tung and Lindzen 1979; Plumb 1981) and resonant enhancement of a barotropic mode of the polar vortex (Esler and Scott 2005; Matthewman et al. 2009).

On the basis of changes in the stratospheric circulation and timing, SSW events are classified as major and minor. Although there is an ambiguity in definitions used by different studies (Andrews et al. 1987; Labitzke and Naujokat 2000; Limpasuvan et al. 2004; Kruger et al. 2005; Charlton and Polvani 2007), the main criterion is related to the dynamics. The SSW is considered to be major when the mean zonal wind at 60°N and 10 hPa becomes easterly. In the minor warming, the mean zonal wind is decelerated, or even reversed, although the reversal level does not reach down to 10 hPa at 60°N. Following the World Meteorological Organization (WMO) definition, a "minor warming" is identified when polar temperatures increase by 25 K or more within a week at any stratospheric level.

Upward propagating planetary waves consist mostly of zonal wavenumbers 1 and 2, and traditionally SSWs were classified into wave 1 or wave 2 events depending on the amplitude of the dominant wave. Recently, Charlton and Polvani (2007) and Matthewman et al. (2009) suggested a vortex-oriented classification as a more appropriate approach and separated SSW events into those which split or do not split the stratospheric polar vortex. In the vortex split event, the polar vortex breaks up into two pieces of comparable size, generating multiple areas of colder and warmer temperatures. In the vortex displacement event, the polar vortex is shifted off the pole, generally leading to one area of colder temperature and one area of warmer temperature. The two types of events are dynamically different at tropospheric and stratospheric altitudes, and associated differences at the altitudes of mesosphere and thermosphere are unknown. Vortex splits occur mostly in January and February, while vortex displacements occur during all winter months (Charlton and Polvani 2007). The occurrence rate is almost the same, with 46% of major SSWs classified as vortex splits. Although wave 1 is often associated with vortex displacement and wave 2 is associated with vortex split, their identification is not straightforward due to the presence of both waves and nonlinearity of the flow.

Climatological studies of major SSW events reported on average 6.2 SSW events per decade (Charlton and Polvani 2007), and their distribution is affected by multiple external factors. The SSW events occur more frequently in the east phase of quasi-biannual oscillation (QBO) (Holton and Tan 1980; Labitzke 1982). Later studies (Labitzke and van Loon 1988; Gray et al. 2001) have shown that the relationship between SSW and QBO is affected by solar activity: SSW is most likely to occur at QBO-East at solar minimum years and at QBO-West at solar maximum years. On a longer time scales, discernible trends in the occurrence of major SSWs could not be identified reliably. As shown in Fig. 7, there were no major SSW events between February 1989 and December 1998 (Charlton and Polvani 2007; Taguchi 2008), and the reasons remain unknown. The early 1970s and the late 1990s were particularly active. The last several years were marked by unusually high SSW activity, when SSW events occurred more frequently (4 events in the winter of 2007–2008), reached high stratospheric temperatures (winters of 2007–2008, 2008–2009), or lasted unusually long time (2005–2006, 2008–2009) (Manney et al. 2009; Labitzke and Kunze 2009).

4.2 Global Effects Due to SSW Events

Although SSWS have been historically seen as a stratospheric phenomenon, the evidence collected within the last decade points to numerous associated phenomena at both lower and

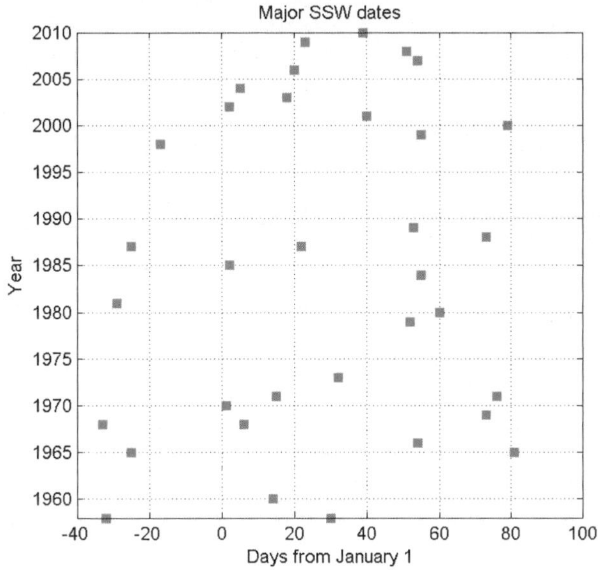

Fig. 7 Occurrence of SSW events as function of day of the year and year, since 1958 (after Taguchi 2008)

higher altitudes. A distinct tropospheric response to SSW was found in both experimental (e.g., Baldwin and Dunkerton 2001; Thompson et al. 2002; Thompson and Solomon 2002) and modeling (Sexton 2001; Polvani and Kushner 2002; Charlton et al. 2004) studies. Stratospheric anomalies propagate downward on timescales of several weeks, and the impact of SSW on the circulation in the troposphere can last for many weeks (Baldwin and Dunkerton 2001; Polvani and Waugh 2004). This influence has become accepted by the research community only recently.

In the mesosphere-lower thermosphere (MLT) region, large variability in association with SSW events is also evident in both observational studies and simulations. It is well known that stratospheric warmings lead to a cooling at mesospheric altitudes (Labitzke 1972, 1981). Recent progress in experimental techniques has resulted in a number of studies reporting effects of SSW at the altitudes of the upper mesosphere and lower thermosphere, with mesospheric cooling at 70–90 km altitude found for example in OH airglow temperature (Walterscheid et al. 2000; Azeem et al. 2005), lidar, meteor radar (Hoffmann et al. 2002, 2007) and SABER (Siskind et al. 2005) data. Cooling in the mesosphere reportedly precedes warming in the stratosphere. Cooling is also associated with large dynamical changes in the mesosphere. These include reversal of zonal and meridional prevailing winds, enhanced energy dissipation rates, increases in planetary wave activity and changes in the gravity wave activity (i.e., Hoffmann et al. 2007; Dowdy et al. 2007; Pancheva et al. 2009; Yamashita et al. 2010a). Other effects include the extension of the polar vortex into the mesosphere and the displacement of the temperature peak associated with the stratopause up to 80 km (Siskind et al. 2007; Manney et al. 2009). Unusually bright OH airglow was reported at altitudes 5–8 km lower than normal and interpreted as an evidence of enhanced downward transport (Winick et al. 2009). Of a particular interest for ionospheric research, the semidiurnal tidal amplitudes in the MLT region can increase or decrease, while the phases are shifted in comparison with undisturbed winter conditions (see review by Kazimirovsky and Vergasova 2009 and references therein). These mesospheric changes result from both planetary wave and gravity wave forcing (Siskind et al. 2010; Yamashita et al. 2010a, 2010b).

Recently Vineeth et al. (2010) used reanalysis of NCEP/NCAR stratospheric wind and temperature to suggest that the zero wind line (critical layer for the quasi-stationary planetary waves where the zonal wind amplitude is zero) appear over the tropics about 60 days prior to major SSWs and moves toward the pole. They also suggest that enhanced planetary wave activity with a quasi periodicity of 16-days, which occurs nearly simultaneously with the zero line, also propagates from the equator to the pole.

In general, the strength of anomalous variations associated with SSW decreases with latitude in the stratospheric and MLT regions. While the effects of stratospheric warming are better known at high latitudes, only few studies have been devoted to processes in the low-latitude and tropical regions (e.g., Pancheva et al. 2008, and references therein). Experimental data indicate that SSW at high latitudes are accompanied by cooling in the low-latitude upper troposphere and lower stratosphere (Fritz and Soules 1970; Kodera 2006; Mukhtarov et al. 2007) and cooling in the tropical mesosphere (Shepherd et al. 2007). An SSW-associated enhancement in the gravity wave perturbations at the MLT heights was reported in low-latitude data (Sridharan and Sathishkumar 2008; Sathishkumar and Sridharan 2009), similar to enhancements found at high latitudes (Hoffmann et al. 2007). SSW manifestations in the MLT region remain an active area of research, as they are not well characterized and understood. In contrast, prior to 2008 any effects at higher altitudes (above ~100 km) were practically unknown in observations.

Following the series of multi-instrument observational campaigns in 2008–2010, significant variations in the lower and upper thermosphere have been reported. In particular, alternating regions of ~50 K warming in the lower thermosphere (120–140 km) and ~70 K cooling in the upper thermosphere (above 150 km) were first reported at middle latitude location (Millstone Hill ISR, 42.6°N) (Goncharenko and Zhang 2008). SSW-associated temperature disturbances were recently identified at high latitudes in PFISR and Poker Flat FPI temperature (Conde and Nicolls 2010), EISCAT ion temperatures (Kurihara et al. 2010), MIPAS instrument on board ESA's Envisat satellite (Funke et al. 2010), and Arecibo ISR electron temperatures (Chau et al. 2010). The above mentioned observations point to a warming in the lower thermosphere above the region of mesospheric cooling, in consistency with model predictions (Liu and Roble 2002). In the upper thermosphere, recent works report evidence of cooling of the order of 50–100 (Goncharenko and Zhang 2008; Conde and Nicolls 2010), which was not expected from the simulations. The analysis of the zonal temperature structure during the SSW demonstrates a wave 1 pattern in the lower thermosphere (Funke et al. 2010), indicating that either warming or cooling can be expected for the observer at a single ground-based location.

4.3 SSW: The 2008 and 2009 Events

The evolution of the stratospheric (10 hPa) polar (60–90°N) parameters for both the January-February 2008 (red) and January-February 2009 (blue) periods are shown in Fig. 8: (a) Temperature (K), (b) Zonal wind (m/s), (c) Amplitude (m) of planetary wave 1 (PW1), and (d) Amplitude (m) of PW2. These data have been obtained from the National Center for Environmental Prediction (NCEP). The solid black line represents a 30-year median value.

The January-February 2008 SSW event is one of the strongest recorded events. It represents a minor warming because the zonal neutral wind reduced its magnitude but did not reverse sign at 10 hPa and 60°N. Note that after the initial warming, there were three more warming periods in the following weeks. The largest warmings were the first and fourth, respectively. On the other hand, the January-February 2009 event is the strongest and the longest lasting (more than three weeks) recorded event (e.g., Manney et al. 2009). This is

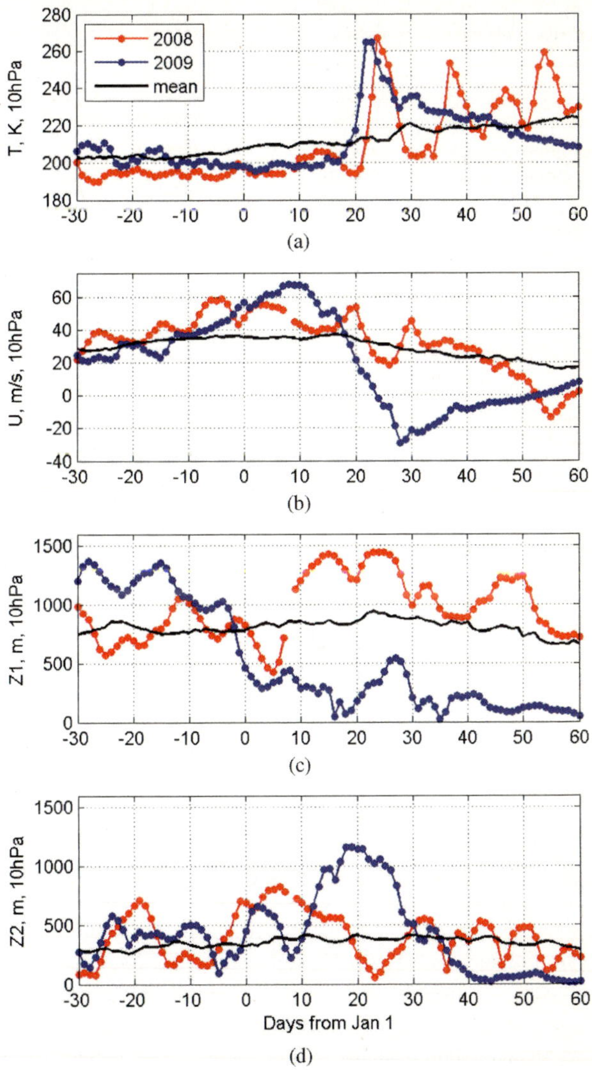

Fig. 8 Summary of stratospheric (10 hPa) polar (60–90° N) parameter for the January 2008 (*red*) and January 2009 (*blue*) SSW events (**a**) Temperature, (**b**) Zonal wind, (**c**) Amplitude PW1, and (**d**) Amplitude of PW2. The *black solid lines* indicate 30-year mean values (after Goncharenko et al. 2010a)

considered a major warming event since the zonal neutral wind reversed sign. More details on the January 2008 event, the solar and magnetic indexes, and the ionospheric signatures observed at Millstone Hill and Jicamarca ISRs are given in the following studies: Goncharenko and Zhang (2008) and Chau et al. (2009). Similarly, the details of the January-February 2009 SSW event, including other SSW parameters, are shown in Goncharenko et al. (2010a). In general, both periods were characterized by a very low solar flux (F10.7 < 80) and mostly quiet time conditions ($K_p < 3$).

5 Equatorial and Low-Latitude Ionospheric Effects During SSWs

Although lower atmospheric effects in the ionosphere have been suggested for many years, the identification of such connection has not been easy. In June 2007, one of us (LPG)

Fig. 9 Daytime vertical **E** × **B** drifts measured over Jicamarca on (**a**) December 13–20, 2007 (control days) and (**b**) January 17–26, 2008 (around SSW days), as function of local time. Data for different days are indicated with different symbols and colors. The expected average and standard deviation quiet-time values are indicated with *black solid* and *dashed curves*, respectively (from Chau et al. 2009, Fig. 2)

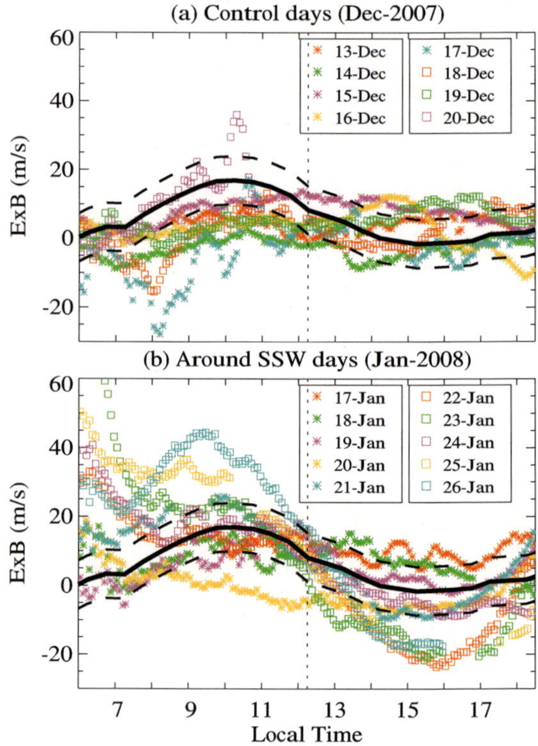

proposed an ambitious World-day campaign to run all the Incoherent Scatter Radars in the World for at least 10 days around a possible SSW in 2008. In a way the request was unusual (a) such long runs are very difficult to perform, (b) the exact dates were going to be determined one or two weeks before the SSW event (for some of the facilities this is not an easy task), and (c) most of the facility scientists and users were skeptical about finding any easy connection between the ionosphere and the polar stratosphere (including two of us). In the case of Jicamarca, we were even more skeptical to see a connection between an event occurring in the polar stratosphere and the equatorial ionosphere! As we will see later, we should have not been too skeptical since such connection was suggested few decades ago (e.g., Stening 1977).

5.1 Equatorial Vertical **E** × **B** Drifts

Figure 9 shows the daytime vertical drift measurements at Jicamarca during the January 2008 SSW event. The top panel shows the drifts for the control days (a month before) while the bottom panels shows the measurements around the SSW days. The solid black line represents the expected mean quiet-time value for the same season and similar solar conditions using the Scherliess-Fejer model, and the dashed line shows the expected standard deviation (see Fig. 2). Chau et al. (2009) found out that these measurements presented peculiar features that were not usually observed before, particularly in ISR measurements: (a) there was a clear signature of a semidiurnal wave with amplitudes much larger (more than 2 times) than the usual standard deviation (σ), (b) the pattern persisted for few days, and (c) the drift difference was the largest around sunrise (more than 6σ).

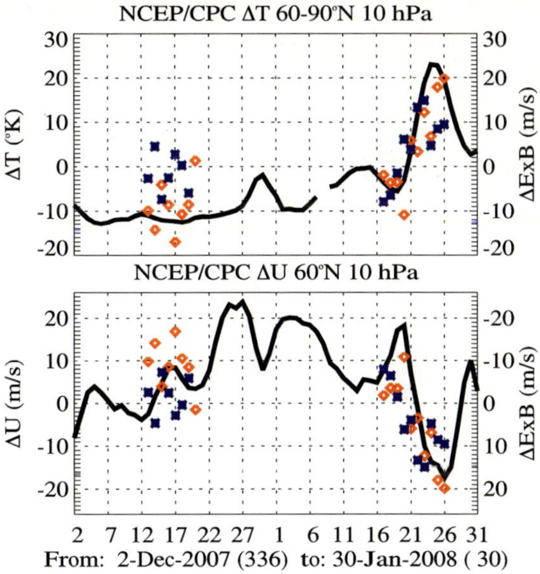

Fig. 10 Stratospheric polar differences with respect to the 30-year median values around the January 2008 SSW: (*top*) stratospheric temperature difference at 60–90°N, and (*bottom*) stratospheric zonal wind difference at 60–90°N. The morning/afternoon averaged vertical **E** × **B** differences are overplotted in *red/blue* (see scale in the right *y*-axis). Note that signs have been reversed for the zonal wind plot (from Chau et al. 2009, Fig. 4)

But, how were these results related to the SSW event? Once a semidiurnal pattern was identified, Chau et al. (2009) used average morning (and afternoon) drift differences (measurement—expected mean value) to characterize the drift patterns for different days. Figure 10 shows the difference of the SSW parameters (ΔT and ΔU) using the 30-year median values as a reference, and the morning/afternoon drift differences. This result clearly showed that the SSW ΔT/ΔU was correlated/anticorrelated with the drift-difference amplitude, i.e., the drift difference increases as the SSW temperature/zonal wind increases/decreases.

After the 2008 peculiar results, we decided to: (1) look into previous drift measurements, and (2) perform new measurements in conjunction with new SSW events. To our fortunate surprise, we found similar drift signatures in previous measurements (using all the three drift techniques mentioned before) that were associated with SSWs (e.g., March 1988, January 2003). Moreover, instead of looking into dates of previous SSW events, one of us (BGF) found drift patterns similar to the 2008 results, and to his big surprise there were SSW events around those periods. This finding gave us more confidence that somehow the two observations (SSW and equatorial vertical drifts) were correlated.

In January 2009 the largest SSW event occurred and the Jicamarca ISR operated around that period. This time the expected signature was not observed in conjunction with the peak SSW temperature, instead it occurred 4–5 days later. The drift results of the January 2009 are presented and discussed later along with the TEC results.

Using data from magnetometers and the CHAMP satellite, Fejer et al. (2010) studied the January 2003 SSW event. Figure 11 shows the magnetometer data for the Peruvian sector (Jicamarca–Piura). As in the case of January 2008, there was a clear signature of a semidiurnal pattern. Moreover such pattern was shifting with time. However, such pattern started few days (5–6) (see Fig. 18 below) after the SSW peak temperature. Using magnetometer data at other longitudes (India and Indonesia) and the CHAMP measurements, Fejer et al. (2010) found that such pattern repeats at other longitudes but occurs at different local times. Anderson and Araujo-Pradere (2010) using magnetometer drift data in both the Peruvian and the Philippines longitude sectors, studied the January 2003 and January 2004 SSWs

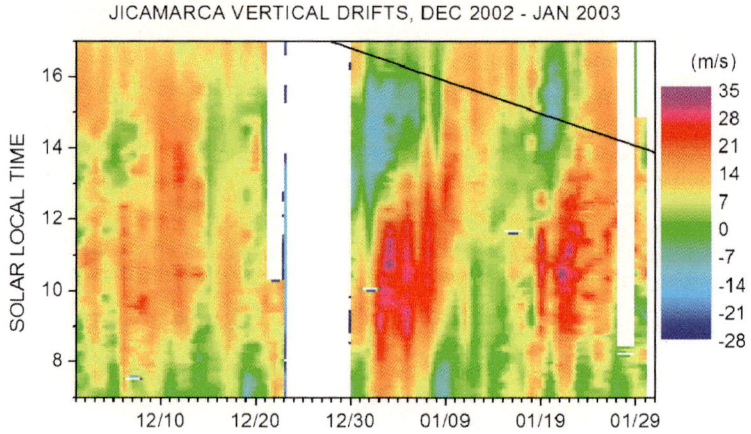

Fig. 11 Vertical plasma drift velocities over Jicamarca derived from magnetic field observations. The *slant line* indicates the local time coverage of the CHAMP satellite (from Fejer et al. 2010, Fig. 3)

events. They found that the semidiurnal shaped signature first appears in the Peruvian sector and 3 days later appears in the Philippines sector. In both sectors, the $\mathbf{E} \times \mathbf{B}$ drift signature lasted for approximately 5 days. Fejer et al. (2010) has associated the observed delays and the longitudinal differences to the phases of the Moon. We will discuss these results later.

Indirectly, similar features in the equatorial vertical drifts (or equatorial *E* region zonal electric fields) has been associated with SSW events. Stening (1977) suggested an association of counter equatorial electrojets (CEEJ) and stratospheric warmings. Similarly, Stening et al. (1996) pointed out that counterelectrojet events in the Indian sector during northern winter were accompanied by eastward-to-westward upper atmosphere wind reversals at a height of 99 km at Saskatoon, Canada (35°S, 107°W), and that they generally occurred during SSW periods. Stening (1989) pointed out that CEE events could result from the dynamo effects of global wind changes. Stening et al. (1997) used numerical simulations to show that lunar tidal amplitudes and phases undergo large changes during stratospheric warmings.

Rastogi (1999) suggested that quiet time CEEJ events during low solar activity northern winter months result from changes in the ionospheric wind system associated with SSWs. Sridharan et al. (2009) used magnetic field and mesospheric wind measurements in the Indian region to show that SSW events affect the day-to-day variability of equatorial CEEJ through the enhancement of semidiurnal tides. Vineeth et al. (2009) examined equatorial counter-electrojet occurrences over Trivandrum (8.5°N, 76.5°E, 0.5°N diplat) with quasi-periodicities of 16 days associated with SSW events. They report that prior to the SSWs there is a sudden cooling at about 30 km over Trivandrum and counterelectrojet maximum intensities. They also pointed out that during the SSWs the stratospheric zonal mean wind also show a significant pattern. They conclude that the lower atmospheric waves interact with tidal modes to give rise to counter electrojet occurrence with planetary wave periods.

5.2 Total Electron Content Results

At low latitudes, the ionospheric plasma is extremely sensitive to the strength of the zonal electric field. The upward vertical drift that accompanies eastward zonal electric field lifts ionospheric plasma to the altitudes of lower recombination rates, thus increasing plasma density. Due to pressure gradients and gravity forces, plasma diffuses along the geomagnetic field lines and downward from higher altitudes, forming peaks of equatorial ionization anomaly at $\sim 15°$ of magnetic latitude. Alternatively, westward (or less eastward) electric field is accompanied by a downward vertical drift (or decreased upward drift) which is expected to decrease ionospheric electron density via transport to lower altitudes. According to these general considerations, strong variations in the daytime low-latitude vertical drifts reported by Chau et al. (2009) and discussed in Sect. 5.1 are expected to result in significant variations in low-latitude electron density.

Such expectations were confirmed by analysis of GPS TEC data during several SSW events (Goncharenko et al. 2010a, 2010b). Goncharenko et al. (2010b) analyzed the January 2009 SSW event focusing on the Western hemisphere sector, where the good coverage of GPS TEC receivers and Jicamarca vertical drift data are available. Figures 12a and 12b show maps of 10-day mean TEC in the morning sector (15 UT) and afternoon sector (21 UT) prior to SSW to demonstrate average wintertime behavior in TEC. Figures 12c and 12d show TEC maps on January 27, 2009 (during SSW) for the morning and afternoon sectors, respectively; and Figure 12d presents vertical drifts over Jicamarca, in comparison with the average vertical drift for winter season and low solar activity from the Scherliess-Fejer model. The maps clearly demonstrate a strong increase in TEC during SSW (up to 150%) in the morning sector in response to the large upward drift, and a large decrease (50%) in TEC in the afternoon sector in response to the downward drift. This variation is observed in a large range of longitudes and extended up to middle latitudes. The magnitude of the variation in the low-latitude TEC is comparable to TEC variations observed during major geomagnetic storms (Mannucci et al. 2005). As the solar ionizing flux and geomagnetic activity were at extremely low and stable levels during the winter of 2008–2009, they could not explain the observed ionospheric features, in particularly the semidiurnal variation in both vertical drifts and TEC and persistence of the observed disturbances for several days after the peak of high-latitude stratospheric temperature.

The subsequent detailed study of this and other SSW events (Goncharenko et al. 2010a) confirmed that the observed phenomenon is repeatable for other SSW cases, and revealed several important features of low-latitude ionospheric response to SSW event. The TEC variation is most pronounced in daytime hours and has a clear semidiurnal character. The semidiurnal signature in TEC perturbation develops in a matter of several days, as illustrated in Fig. 13, and maximizes 3–6 days after the peak in high-latitude stratospheric temperature. This signature persists for several days. An important aspect of this TEC variation is a progressive shift to later local times, as seen in Fig. 13. Both the large magnitude of semidiurnal variation in TEC and a rapid phase change contribute to the extreme day-to-day variability in the equatorial ionization anomaly.

Chau et al. (2010) analyzed the GPS TEC observations from a single location at Arecibo for SSW events of 2008 and 2009 and compared them to the observations of electron density by ISR and vertical drifts from Jicamarca ISR, magnetometer, and 150-km echoes. They concluded that all parameters exhibit consistent variations during SSW which are similar to those reported with GPS TEC maps. Figure 14a presents variations in GPS TEC difference from a single receiver as a function of day and local time, i.e. ΔTEC is the difference between observed TEC and a median TEC. Figure 14b shows difference in vertical drifts at Jicamarca obtained from 150-km echoes. The black lines on both plots indicate the difference in stratospheric temperature at 10 hPa level to describe the development of stratospheric warming. The ΔTEC peaks 4–5 days after the peak in stratospheric temperature. The difference in equatorial vertical drift exhibits the already familiar semidiurnal pattern which begins few days before the peak in stratospheric temperatures, reaches the maximum 3–4 days after the peak, and shifts to later local times.

The observed TEC variations during SSW events are fully consistent with observations of variations in vertical drift described in Sect. 5.1. The semidiurnal character and the observed

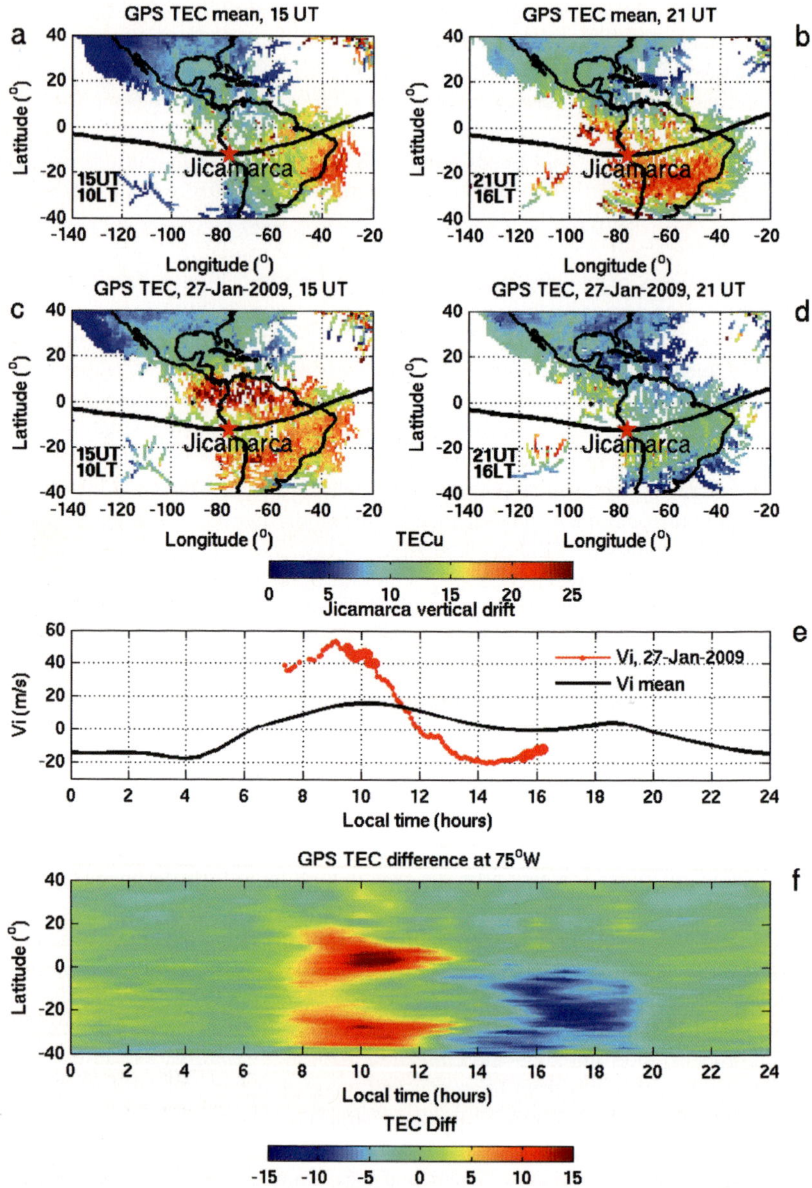

Fig. 12 Observations of ionospheric behavior during stratospheric warming: (**a**) Typical distribution of total electron content (TEC) in the western hemisphere at 15 UT (morning sector, 10 LT at 75-W), the *black line* shows magnetic equator and the star shows the location of the Jicamarca radar; (**b**) Same as (**a**), but for 21 UT (afternoon sector, 16 LT at 75-W); (**c**) TEC in the morning sector (15 UT) on January 27, 2009, during SSW; (**d**) TEC in the afternoon sector (21 UT) on January 27, 2009; (**e**) Vertical drift observations by the Jicamarca radar at 200–500 km above ground. *Red line* presents observations on Jan 27, 2009, during SSW. *Black line* presents average behavior for winter season and low solar activity. (**f**) Change in TEC at 75-W during SSW as function of local time and latitude (from Goncharenko et al. 2010b, Fig. 1)

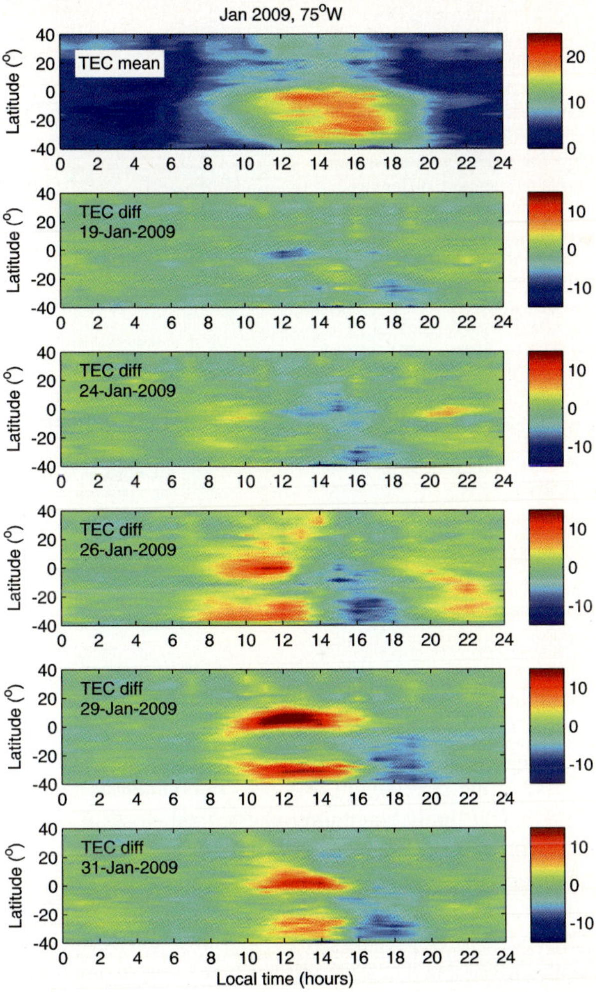

Fig. 13 TEC variation at 75° W in local time and latitude during the January 2009 SSW. (*Top*) The 10 day mean TEC prior to SSW. (*Lower*) Differences in TEC from the mean state during the SSW period (from Goncharenko et al. 2010a)

phase change with local time lend support to the idea of strong non-migrating semidiurnal tide resulting from non-linear interaction of planetary waves with migrating tides, which will be discussed in detail in Sect. 6. Pedatella and Forbes (2010a) analyzed GPS TEC observations from the International GNSS Service (IGS) Global Ionosphere Maps (GIMs) during the 2009 SSW and identified both migrating and nonmigrating perturbations to the semidiurnal tide in the equatorial ionization anomaly crest region. In particular, a significant enhancement was observed in the nonmigrating semidiurnal westward propagating tide with zonal wavenumber 1 (SW1).

5.3 Electron and Neutral Density

The global ionospheric response to SSW was recently investigated with FORMOSTAT3/COSMIC data by Yue et al. (2010) and Pancheva and Mukhtarov (2011). The study by Yue et al. (2010) focused on the January 2009 SSW event. Due to the limitations of data coverage, 3-day averages were used to delineate the SSW signature. Figure 15 shows examples

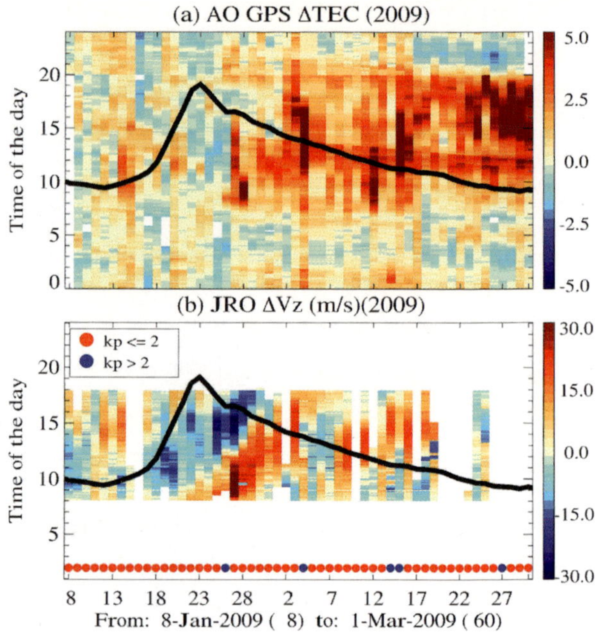

Fig. 14 (**a**) Arecibo Observatory (AO) TEC difference (ΔTEC) using the median values as a reference and (**b**) JRO vertical drift differences (ΔV_Z), both as function of time of the day and day for 2009. The stratospheric polar temperature difference (*black solid line*) is shown as a reference. The *red/blue circles* represent magnetically quiet/active days (from Chau et al. 2010)

of eight electron density profiles over the American (60°W) and Atlantic (0°) sectors. In the morning, electron density for all the cases increases significantly during the SSW days in comparison with that of non-SSW days except for one case. At the same time, the increase of the peak height of the F_2 region can be seen clearly. The peak density (NmF_2), peak height (hmF_2), and ionospheric total electron content (TEC) were found, in comparison with those during non-SSW days, to increase in the morning hours and decrease in the afternoon globally for 80% of the cases where SSW and non-SSW data were available for the same location. NmF_2, hmF_2, and TEC during SSW days, on average, increase 42%, 33 km, and 33%, respectively, in the morning; and decrease 29%, 24 km, and 28%, respectively, in the afternoon. The COSMIC observations also revealed global ionospheric response, including middle and high latitudes, though the character of this response is more complex than at low latitudes.

In addition to multiple evidence of enhanced semidiurnal behavior in ionospheric electron density or TEC in response to SSW, Pancheva and Mukhtarov (2011) presented evidence of changes in zonal mean electron density and in the diurnal component of electron density from the analysis of COSMIC data around the December solstices of 2007–2008 and 2008–2009. They found negative response to SSW in both zonal mean f_oF_2 and hmF_2. This response is strongest at low latitudes, but is extended up to middle latitudes. Maximum decrease in NmF_2 and hmF_2 is observed 2–4 days after the peak in stratospheric temperature. Although the temporal resolution was limited due to the 16-day window used to determine the full spectrum of the components (waves with periods of 24, 12 and 8 hours with zonal wavenumbers up to 4), the results show that there is no significant time delay in the appearance of disturbances at middle and low latitudes. Additional features revealed by Pancheva and Mukhtarov (2011) include ionospheric response to SSW primarily in the northern hemisphere, and most pronounced response above the NmF_2 peak, i.e., above 300–350 km.

Analysis of thermospheric density observed by the CHAMP and GRACE satellites (Liu et al. 2011) has revealed a density decrease by ~30% at 325 km (CHAMP data) and by

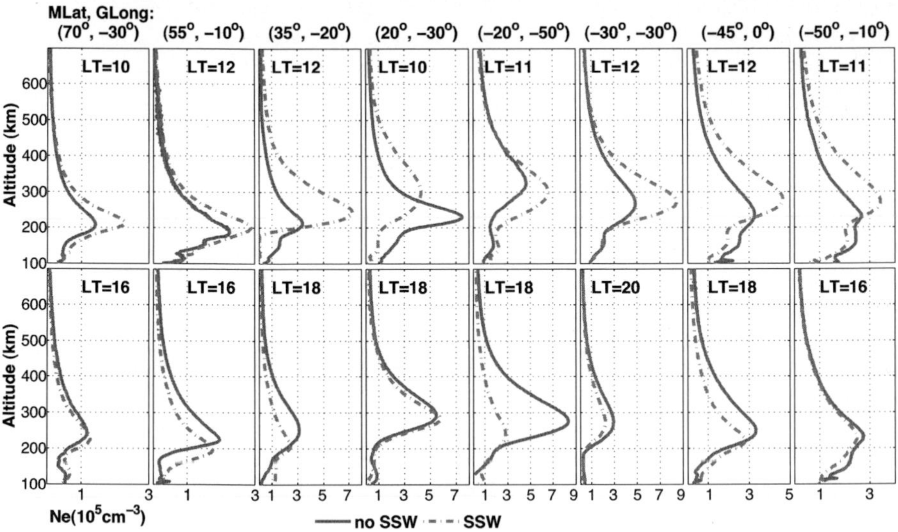

Fig. 15 Electron density profiles during SSW (*dashed lines*) and non-SSW days (*solid lines*) in both morning (*top*) and afternoon (*bottom*) sectors around the same location and local time bins over American and Atlantic sectors. The corresponding geomagnetic latitude and geographic longitude and the local time are indicated (from Yue et al. 2010, Fig. 2)

∼45% at 475 km (GRACE data) during the SSW of January 2009. This decrease is observed in both afternoon and pre-dawn sectors and maximizes on January 24, 2009, e.g., after the peak in high-latitude stratospheric temperature. The southern hemisphere experiences a larger relative density decrease with a peak at ∼20°S, while density decrease in the northern hemisphere peaks at ∼40°N. Since thermospheric density is closely controlled by thermospheric temperature, Liu et al. (2011) estimated that the observed reduction in thermospheric density corresponds to ∼50 K cooling in the upper thermosphere. Observations of electron density at 325 km altitude from CHAMP during the 2009 SSW also exhibits a significant depletion, which closely follows the thermospheric temperature with a delay of 1–2 days and is consistent with observations of depletion in Ne and TEC already described in this section.

6 Atmosphere-Ionosphere Coupling: Possible Scenarios

From the results presented in the previous sections, it is clear that there is a connection between the equatorial and low latitude ionosphere and the high latitude SSW. The equatorial and low latitude ionospheric results can be summarized as follows:

- There is a clear semidiurnal pattern in the daytime equatorial vertical drifts.
- The largest perturbation is observed around sunrise times.
- The semidiurnal pattern persists for several days.
- The daytime TEC observed at low latitudes is consistent with the equatorial vertical drifts and the electron density profiles.
- Similar effects are shown at different longitudes, although they occur at different local times.

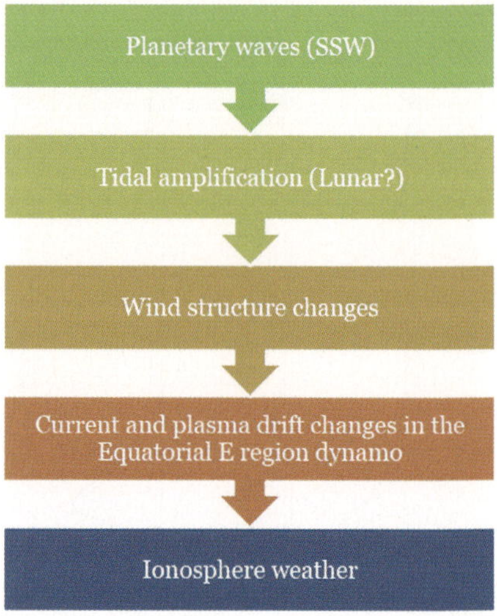

Fig. 16 A simple block diagram of a possible connection between SSW and the weather of the equatorial and low latitude ionosphere

– The equatorial ionospheric disturbances are observed few days (1 to 6) later than the peak stratospheric temperature.

The key problem is to explain how such geographically separated regions are connected with each other in altitude and latitude. Since the occurrence of SSW is related to the growth of quasi-stationary planetary waves (PWs), these waves might also played an important role in the equatorial ionospheric disturbances. Moreover, since the main ionospheric disturbance shows a clear semidiurnal pattern, lower atmospheric tides should also be included. Then the atmospheric-ionospheric coupling at equatorial and low latitudes during SSW events, might be either (1) direct (e.g., vertical propagation of PWs) or (2) indirect (e.g., through the *E* region dynamo). The former is difficult to conceive since quasi-stationary PWs are mostly limited to high and middle latitudes (e.g., Pogoreltsev et al. 2007) and their vertical propagation is confined below the critical layer near the mesopause where the zonal wind reverses (e.g., Sassi et al. 2002).

A number of authors have invoked the indirect coupling via the *E* region dynamo, particularly to explain the observed daytime effects. Figure 16 shows a block diagram with the main ideas of this indirect coupling. Although generally trapped below the mesopause, somehow quasi-stationary PWs (QSPWs) interact with the tides at lower altitudes causing large changes in the tides (solar and/or lunar). The tidal winds can modulate electric fields through the *E* region dynamo (at $\sim$115 km). At low latitudes, these modulated electric fields map along magnetic field lines to higher altitudes ($\sim$250–400 km) and produce tidal variations in vertical plasma drifts, therefore changing the ionosphere weather. However there are still many unanswered questions, for example, (1) how is the interaction of PWs and tides?, (2) which tides are the ones that amplified the most?, ... In the next paragraphs we review the recent modeling efforts that explain most of the observed ionospheric parameters by including planetary waves and solar tides. In addition, we discuss the possible role of lunar tides and summarize other mechanisms that have been proposed.

6.1 Planetary Waves and Solar Tides: Modeling Results

As mentioned earlier, numerical modeling studies have been conducted to understand the connection between the SSW and the ionospheric variability. Because QSPWs are the primary drivers of SSWs, an idealized wavenumber 1 QSPW was prescribed at the lower boundary of the NCAR TIME-GCM (at 10 hPa) for control simulations to understand its role in the SSW-ionosphere connection under solar minimum conditions (Liu et al. 2010). In particular, Liu et al. (2010) examined the spatial and temporal patterns of the ionospheric plasma drift changes, their relationship to the wind dynamo perturbations due to changes of migrating and non-migrating tides, and the dependence of these changes on the QSPW phase.

Compared with base case simulation without QSPW, the control TIME-GCM simulation does show that the introduction of a QSPW can cause large ionospheric changes, including changes in the ion drift, F_2 peak height and electron density. The QSPW, however, does not seem to directly affect the ionosphere at low latitudes because (1) it is mostly confined to the winter hemisphere at high latitudes below the mesopause, due to their interaction with critical layers and the increasing molecular dissipation, and (2) QSPW decays rapidly in the middle and upper atmosphere after the peak warming due to their interaction with the mean flow and the resultant jet reversal. On the other hand, both migrating and non-migrating tides show significant changes in ionospheric E and/or F regions at lower latitudes in the control simulation with QSPW. The changes of the migrating tides are likely caused by the nonlinear interaction with QSPW and the mean wind change during SSW. The non-migrating tides in the simulation are excited by nonlinear interaction between QSPW and migrating tides, since no non-migrating component is specified at the lower boundary.

In the model, these changes introduce perturbations to the neutral winds and their zonal gradients in the E-region. Along with large zonal gradient of electric conductivities at dawn, these perturbations can produce large convergence/divergence of Hall and Pedersen currents and thus change the polarization electric field and ion drift. The numerical results also show that the changes in E-region wind dynamo are longitude and local time dependent, and display complex spatio-temporal patterns even though the QSPW introduced is already idealized compared with realistic atmospheric conditions. This complexity is probably related to (1) the superposition of both migrating and non-migrating tides, (2) convolution with the variation of electric conductivity, and (3) the variation of the geomagnetic field with respective to the geographic locations. Ionospheric changes also depend on the phase of the QSPW, according to the control simulations. Therefore, quantitative knowledge of the tides and planetary waves is important for the correct interpretation of the observed ionospheric variability.

It is conceivable that changes of various wave components are more complex during an actual SSW event. Fuller-Rowell et al. (2010) demonstrate the changes of migrating and non-migrating waves during a minor warming event naturally generated in a NOAA Whole Atmosphere Model (WAM) simulation. These wave components show large natural variability in the model, and it is not evident if all are related to SSW. One component that does clearly correspond with the warming event is the migrating terdiurnal tide (TW3), whose amplitude increases by about 50% a few days after the peak warming. This delay may indicate that the TW3 change results from nonlinear interaction between the migrating diurnal and semi-diurnal component. The amplitude and phase changes of TW3 are expected to affect the wind dynamo.

The TIME-GCM simulations produce some features that are qualitatively similar to observations, such as the enhancement of upward/downward ion drift in the morning/afternoon, and corresponding TEC changes (e.g., Chau et al. 2009; Goncharenko et al. 2010a, 2010b), but the magnitude and timing of the changes differ. Figure 17 shows the vertical ion drifts near F_2 peak at 1000 UT on day 31 for the cases without and with QSPW. The upward enhancement of the vertical drifts is evident around 70°W around sunrise, in

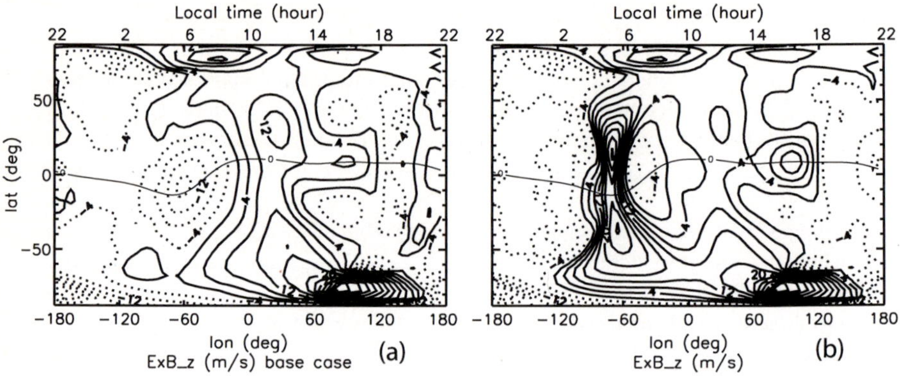

Fig. 17 *F* region vertical ion drift at 1000 UT, day 31 from (**a**) the case without QSPW and (**b**) the case with QSPW. Contour intervals are 4 m $s^{?1}$ (solid is for upward) (from Liu et al. 2010, Fig. 3)

very good qualitative agreement with Jicamarca vertical drift results (see Fig. 9). The resulting vertical drifts shows significant changes in latitude as well as longitudes and local times. More recently, WAM has been combined with the operational data assimilation scheme to produce realistic neutral dynamics (Wang et al. 2011), which has been used to drive the Coupled Thermosphere Ionosphere Plasmasphere electrodynamics (CTIPe) model and obtained ionospheric responses consistent with observations Fuller-Rowell et al. (2011). In particular, the WAM simulation predicted a substantial increase in the amplitude of the 8-hour ter-diurnal tide in the lower thermosphere dynamo in response to the warming, which occurred at the expense of the more typical semi-diurnal tides.

To test this hypothesis of PWs-tide interaction, Pedatella and Forbes (2010a) analyzed TEC data from Global Ionospheric Maps during the January 2009 SSW event. The study obtained a full spectrum of migrating and non-migrating tides and found increase in both migrating semidiurnal and non-migrating semidiurnal westward propagating (SW1) components during the SSW. In addition, the study also revealed a correlation between a PW1 activity in the high-latitude stratosphere and SW1 magnitude in the low-latitude ionosphere, despite very low levels of PW1 during this SSW event. The similarity in the oscillations in PW1 and SW1 provides strong observational evidence that the nonlinear interaction between tides and planetary waves occurs during SSWs, and furthermore, that global tidal modulation due to PWs is the driving mechanism for ionospheric perturbations during SSWs.

6.2 Lunar Effects

The modulation of the equatorial electrojet intensity by the lunar tide has been discussed by several authors following initial equatorial magnetic field observations (e.g., Bartels and Johnson 1940; Hutton and Oyinloye 1970; Mayaud 1977) and radar electron drift measurements (e.g., Tarpley and Balsley 1972). Lunar tidal effects have been clearly associated with the occurrence of daytime reversals of the equatorial electrojet current during geomagnetic quiet times (e.g., Rastogi 1974). The occurrence of these events with lunar age varies with longitude (Rastogi 1974). Over Peru, afternoon (morning) counter electrojets occur around lunar ages 1.4 and 13.4 (4.8). This modulation of the ionospheric current is due mostly to superposition in the ionospheric *E* region of the vertical propagation of the lunar semidiurnal tide on the generally stronger solar quiet current system (Vial and Forbes 1994).

Stening and Fejer (2001) studied the modulation of the lunar semi-diurnal tide on the equatorial vertical plasma drift measured by the Jicamarca radar. Fejer et al. (2010) showed

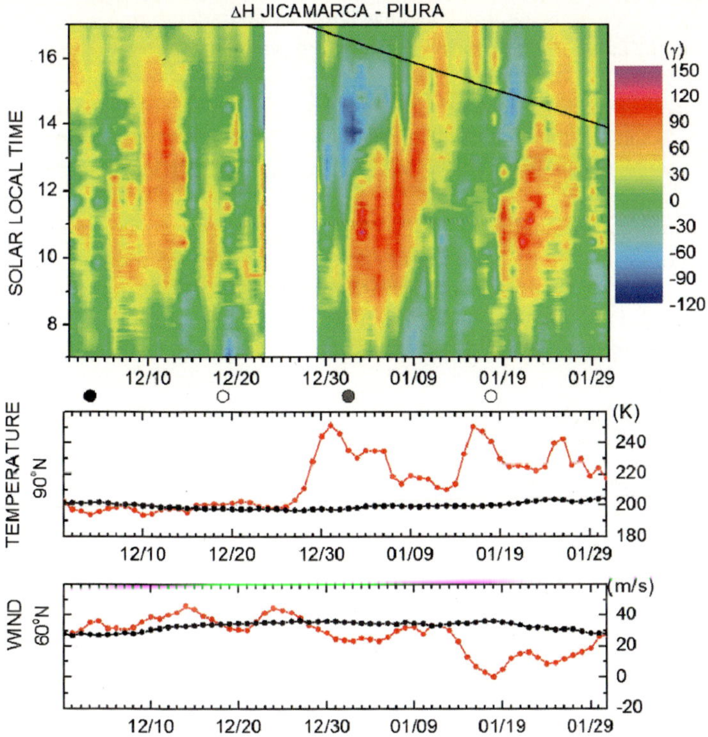

Fig. 18 Equatorial horizontal magnetic field components in the Peruvian sector and high latitude zonally averaged stratospheric temperatures and zonal winds during December 2002 and January 2003. The days of new and full moons are indicated by *open* and *solid circles*, respectively (after Fejer et al. 2010, Fig. 4)

that large equatorial electrodynamic perturbations during SSW periods were associated with strongly enhanced lunar semidiurnal wave effects, nonetheless, additional tidal modes could also play an important role during SSW events. They reported that the longitude dependent onsets of these events during northern hemisphere winter were observed near new and full Moons. Figure 18 shows the intensity of the equatorial electrojet in the Peruvian sector between December 1 2002 and January 29 2003. In this period, new Moons (solid circles) occurred on December 4th and January 2nd, and full Moons (open circles) occurred on December 19 and January 18. After the late December warming onset, the electrojet intensities first weakened for about 2 days and then, around new moon, suddenly developed very similar multiday perturbation patterns with largely enhanced morning eastward and afternoon westward currents that systematically shifted to later local times. Strong morning counter-electrojets were seen around the first and last quarters (January 10 and 25). The morning and afternoon current perturbations occurred first and were strongest in the American sector and developed last and were weakest (particularly in the afternoon) in the Pacific sector (see results of the Indian and Pacific sectors in Fejer et al. 2010, Fig. 4). Similar, but weaker, current perturbations developed near full moon after the second warming, when the high-latitude wind perturbations were even larger. These perturbation events are about 14 days apart, as expected for the lunar semimonthly tide (Stening 1989).

Pedatella and Forbes (2010b) used global GPS TEC data to study the solar local time, seasonal and longitudinal variability of the ionospheric lunar tide. They reported that the semidiurnal component is dominant and has largest amplitudes during northern hemisphere winter and in the equatorial ionization anomaly region. They pointed out that the significant longitudinal variability of the semidiurnal tide indicates the existence of non migrating lunar tidal components in addition to the stronger migrating tide. Using Huancayo ionosonde data, Bartels (1950) showed that amplitude of daytime foF_2 lunar tide in local summer (northern hemisphere winter) is about five times larger than that in local winter (northern hemisphere summer).

Some features associated to the lunar effects, e.g., gradual time shift to later times of the peak morning amplitude as time progresses and gradual decrease in magnitude, have been reproduced for the January 2009 using more realistic WAM simulations (e.g., Fuller-Rowell et al. 2011), but without including lunar effects. So far the phase of the lunar semi diurnal tide has not been included in the numerical simulations using WAM (Fuller-Rowell et al. 2010) or TIME-CGM (Liu et al. 2010). The inclusion of these tides as well as other waves (e.g., traveling planetary waves) are worth further investigation to better understand the ionospheric variability.

6.3 Other Possible Scenarios

Although the modeling work supports the hypothesis that PWs play a key role in SSW-Equatorial Ionosphere coupling, there is still significant work to do to understand how exactly the coupling is done. Several mechanisms involving planetary waves, tides, and gravity waves can lead to the enhancement of tidal amplitudes in the lower thermosphere. It is very likely that they complement each other, producing complex variations in interacting tidal modes and, through E-region dynamo, in low-latitude electron density. In all these mechanisms, the presence of planetary waves is a crucial element in the process.

One of such mechanisms was discussed by Goncharenko et al. (2010a) and includes changes in the background stratospheric and mesospheric parameters at low latitudes. Although most studies emphasize stratospheric and mesospheric anomalies occurring during SSW events at high latitudes, there is a sufficient experimental evidence of SSW effects in the tropical stratosphere and mesosphere, as summarized in the Sect. 4.2. Numerical simulations have demonstrated that changes in the background winds and temperatures can introduce enhancements of tidal amplitudes in the lower thermosphere (e.g., Hagan et al. 1992; Stening et al. 1997). Hagan et al. (1992) investigated the sensitivity of the upward propagating semidiurnal tide to the variability in the lower atmosphere. They modified the zonal mean temperature and wind profiles to include QBO variations at equatorial latitudes and demonstrated a significant (5–20 m s^{-1}) variability in the semidiurnal tide in the MLT region, although such signature was largely evident at middle and high latitudes. Stening et al. (1997) used a lunar tidal model to investigate the effects of changes in the background zonal wind on the structure of lunar tide and concluded that wind and temperature changes typical for SSW events can produce profound effects on lunar tide in the mesosphere. Although Stening et al. (1997) model considered only lunar tides, similar effects could be expected for solar tides. Thus, it is reasonable to expect changes in tidal amplitudes related to upward propagation of tidal modes through the modified atmosphere.

Recently Pancheva and Mukhtarov (2011) has suggested a different mechanism via the so called “disturbance dynamo” (Blanc and Richmond 1980). The increase of the zonal mean lower thermospheric temperature at high latitudes (for example those shown by Funke et al. 2010 at 70–90°N) could change the observed plasma drifts at low latitudes in a manner similar to the disturbed dynamo. In this case however, the disturbance dynamo is a current system driven by the equatorward lower thermospheric winds forced at high latitudes because of thermospheric heating. As the gas move toward the equator it must also move to the west (in NH) to preserve the original angular momentum. In the low and middle latitudes the wind surge has to have a significant zonal (westward) component that would affect the daily mean vertical plasma drift; it has to become predominantly negative (downward) or less positive. In this way the mean ionospheric plasma will be moved generally to the lower altitudes where the recombination is stronger, i.e. the mean ionospheric response to SSW events will be negative. This hypothesis needs to be tested with observations and numerical models.

7 Summary

In this paper we have reviewed the observations and current understanding related to the equatorial and low latitude ionospheric effects during SSWs. We have seen that by modifying the equatorial electric fields (in particular the zonal component) significant changes in TEC and electron density profiles are observed at low and sometimes midlatitudes. Moreover, the effects are observed at all longitudes. The main effect is a well-defined semidiurnal pattern with amplitudes above the expected variability that persists for few days. However, waves with other periods have also been identified in this connection. Many of the observed results can be partially reproduced by numerical models by including gravity waves. Moreover, it appears that the most efficient way to couple the lower atmosphere to the equatorial ionosphere is through the *E* region dynamo.

Despite this progress we are in the early stages of understanding the morphology, causes and consequences of this connection. For example, it is not clear if the observed semidiurnal variation is related to solar or lunar tidal modes. Though most studies assume solar tidal forcing, the analysis of Fejer et al. (2010) suggests enhancement in lunar tides. This suggestion is broadly consistent with a study of lunar tides in TEC by Pedatella and Forbes (2010b).

Temporal development of variations is another major unknown. While several studies reveal that mesospheric changes occur prior to SSW events (Azeem et al. 2005; Hoffmann et al. 2007), maximum ionospheric disturbances are reported within several days after the peak in SSW (e.g., Anderson and Araujo-Pradere 2010; Chau et al. 2009, 2010; Goncharenko et al. 2010a, 2010b; Fejer et al. 2010). Major discrepancies also exist between model simulations (e.g. Chang et al. 2009; Liu et al. 2010; Fuller-Rowell et al. 2010) and observations. These discrepancies include important differences in upper atmospheric response to SSW, including temporal development, different tidal features, and the magnitude of response. More work is needed for accurate specification of the tidal modes enhanced during SSW (e.g. solar or lunar, migrating or non-migrating, semidiurnal or other modes), together with the relative importance of different components. Questions arising from initial studies demonstrate a critical need for gathering accurate and detailed ionospheric observations prior, during, and after SSW events, and identification and specification of ionospheric disturbances related to SSW.

Acknowledgements The Jicamarca Radio Observatory is a facility of the Instituto Geofísico del Perú operated with support from the NSF AGS-0905448 through Cornell University. LPG's effort at the MIT Haystack Observatory has been supported by NSF Cooperative Agreement ATM-0733510 with the Massachusetts Institute of Technology. HLL's effort is partly supported by National Science Foundation grants ATM-0535466 and ATM-0836386, the NASA LWS NNX08AQ91G, and the Office of Naval Research (N00014-07-C-0209). The National Center for Atmospheric Research is sponsored by the National Science Foundation. The authors thank the International Space Science Institute (ISSI) for organizing the workshop on "Coupling between Earth's atmosphere and its plasma environment", and for sponsoring the team meetings on this research topic.

References

M.A. Abdu, T. Maruyama, I.S. Batista, S. Saito, M. Nakamura, Ionospheric responses to the October 2003 superstorm: Longitude/local time effects over equatorial low and middle latitudes. J. Geophys. Res. **112**(A10306) (2007). doi:10.1029/2006JA012228

P. Alken, A quiet time empirical model of equatorial vertical plasma drift in the Peruvian sector based on 150 km echoes. J. Geophys. Res. **114**(A2), 1–6 (2009). doi:10.1029/2008JA013751

D. Altadill, E.M. Apostolov, Vertical propagating signatures of wave-type oscillations (2 and 6.5 days) in the ionosphere obtained from electron-density profiles. J. Atmos. Sol.-Terr. Phys. **63**, 823–834 (2001)

D. Anderson, A. Anghel, K. Yumoto, M. Ishitsuka, E. Kudeki, Estimating daytime vertical $E \times B$ drift velocities in the equatorial F-region using ground-based magnetometers observations. Geophys. Res. Lett. **29** (2002). doi:10.1029/2001GL014562

D. Anderson, A. Anghel, J.L. Chau, K. Yumoto, Global, low-latitude, vertical $E \times B$ drift velocities inferred from daytime magnetometer observations. Space Weather **4**(S08003) (2006). doi:10.1029/2005SW000193

D. Anderson, E.A. Araujo-Pradere, Sudden stratospheric warming event signatures in daytime exb drift velocities in the Peruvian and Philippine longitude sectors for January 2003 and 2004. J. Geophys. Res. **115**, 1–7 (2010). doi:10.1029/2010JA015337

D.G. Andrews, J.R. Holton, C.B. Leovy, *Middle Atmosphere Dynamics* (Academic Press, London, San Diego, 1987)

S.M.I. Azeem, E.R. Talaat, G.G. Sivjee, H.-L. Liu, R.G. Roble, Observational study of the 4-day wave in the mesosphere preceding the sudden stratospheric warming events during 1995 and 2002. Geophys. Res. Lett. **32**(L15804) (2005). doi:10.1029/2005GL023393

M.P. Baldwin, T.J. Dunkerton, Stratospheric harbingers of anomalous weather regimes. Science **294**, 581–584 (2001)

J. Bartels, 27-day variations in f_2 layer critical frequencies at Huancayo. J. Atmos. Sol.-Terr. Phys. **1**, 2–12 (1950)

J. Bartels, H.F. Johnson, Geomagnetic tides in horizontal intensity at Huancayo. Terr. Magn. Atmos. Electr. **45**, 269–308 (1940)

I. Batista, R. de Medeiros, M. Abdu, J. de Souza, G. Bailey, E. de Paula, Equatorial ionospheric vertical plasma drift model over the Brazilian region. J. Geophys. Res. **101**(A5), 10887–10892 (1996)

F. Bertoni, I.S. Batista, M.A. Abdu, B.W. Reinisch, E.A. Kherani, A comparison of ionospheric vertical drift velocities measured by digisonde and incoherent scatter radar at the magnetic equator. J. Atmos. Sol.-Terr. Phys. **68**, 669–678 (2006)

D. Bilitza, O.K. Obrou, J.O. Adeniyi, O. Oladipo, Variability of fof2 in the equatorial ionosphere. Adv. Space Res. **34**(9), 1901–1906 (2004)

M. Blanc, A. Richmond, The ionospheric disturbance dynamo. J. Geophys. Res. **85**(A4), 1669–1686 (1980)

L.C. Chang, S.E. Palo, H.-L. Liu, Short-term variation of the $s = 1$ nonmigrating semidiurnal tide during the 2002 stratospheric sudden warming. J. Geophys. Res. **114**(D03109) (2009). doi:10.1029/2008JD010886

A.J. Charlton, L.M. Polvani, A new look at stratospheric sudden warmings. Part I: climatology and modeling benchmarks. J. Climate **20**, 449–469 (2007)

A.J. Charlton, A. O'Neill, A.C. Massacand, W.A. Lahoz, Sensitivity of tropospheric forecasts to stratospheric initial conditions. Q. J. R. Meteorol. Soc. **130**, 1771–1792 (2004)

J.L. Chau, E. Kudeki, Statistics of 150-km echoes over Jicamarca based on low-power VHF observations. Ann. Geophys. **24**, 1305–1310 (2006)

J.L. Chau, R.F. Woodman, Daytime vertical and zonal velocities from 150-km echoes: their relevance to F-region dynamics. Geophys. Res. Lett. **31**(L17801) (2004). doi:10.1029/2004GL020800

J.L. Chau, B.G. Fejer, L.P. Goncharenko, Quiet variability of equatorial $E \times B$ drifts during a sudden stratospheric warming event. Geophys. Res. Lett. **36**(5), 1–4 (2009). doi:10.1029/2008GL036785

J.L. Chau, N.A. Aponte, E. Cabassa, M.P. Sulzer, L.P. Goncharenko, S.A. González, Quiet time ionospheric variability over arecibo during sudden stratospheric warming events. J. Geophys. Res. **115**, 1–8 (2010). doi:10.1029/2010JA015378

P.-R. Chen, Two-day oscillation of the equatorial ionization anomaly. J. Geophys. Res. **97**(A5), 6343–6357 (1992)

M.G. Conde, M.J. Nicolls, Thermospheric temperatures above Poker Flat, Alaska, during the stratospheric warming event of January and February 2009. J. Geophys. Res. **115**(D00N05) (2010). 10–10292010014280

O. De la Beaujardière, J.M. Retterer, R.F. Pfaff, P.A. Roddy, C. Roth, W.J. Burke, Y.J. Su, M.C. Kelley, R.R. Ilma, G.R. Wilson, L.C. Gentile, D.E. Hunton, D.L. Cooke, C/nofs observations of deep plasma depletions at dawn. Geophys. Res. Lett. **36**, 1–4 (2009). doi:10.1029/2009GL038884

E.R. de Paula, D.L. Hysell, The Sao Luis 30 MHz coherent scatter ionospheric radar: system description and initial results. Radio Sci. **39**(RS1014) (2004). doi:10.1029/2003RS002914

A.J. Dowdy, R.A. Vincent, M. Tsutsumi, K. Igarashi, Y. Murayama, W. Singer, D.J. Murphy, D.M. Riggin, Polar mesosphere and lower thermosphere dynamics: 2. Response to sudden stratospheric warmings. J. Geophys. Res. **112**(D17105) (2007). doi:10.1029/2006JD008127

S.L. England, S. Maus, T.J. Immel, S.B. Mende, Longitudinal variation of the E-region electric fields caused by atmospheric tides. Geophys. Res. Lett. **33**(L21105) (2006). doi:10.1029/2006GL027465

J.G. Esler, R.K. Scott, Excitation of transient Rossby waves on the stratospheric polar vortex and the barotropic sudden warming. J. Atmos. Sci. **62**, 3661–3682 (2005)

T.W. Fang, A.D. Richmond, J.Y. Liu, A. Maute, Wind dynamo effects on ground magnetic perturbations and vertical drifts. J. Geophys. Res. **113**(A11313) (2008). doi:10.1029/2008JA013513

B.G. Fejer, L. Scherliess, On the variability of *F*-region vertical plasma drifts. J. Atmos. Sol.-Terr. Phys. **63**, 893–897 (2001)

B.G. Fejer, E.R. de Paula, S.A. Gonzalez, R.F. Woodman, Average vertical and zonal *F* region plasma drifts over Jicamarca. J. Geophys. Res. **96**, 13901–13906 (1991)

B.G. Fejer, M.E. Olson, J.L. Chau, C. Stolle, H. Lühr, L.P. Goncharenko, K. Yumoto, T. Nagatsuma, Lunar-dependent equatorial ionospheric electrodynamic effects during sudden stratospheric warmings. J. Geophys. Res. **115**, 1–9 (2010). doi:10.1029/2010JA015273

B.G. Fejer, Low latitude ionospheric electrodynamics. Space Sci Rev, 115 (2010). doi:10.1007/s11214-010-9690-7

B.G. Fejer, J.W. Jensen, S.-Y. Su, Quiet time equatorial *F* region vertical plasma drift model derived from rocsat-1 observations. J. Geophys. Res. **113**(A5), 10 (2008). doi:10.1029/2007JA012801

J.M. Forbes, S. Leveroni, Quasi 16-day oscillation in the ionosphere. Geophys. Res. Lett. **19**(10), 981–984 (1992)

J.M. Forbes, S.E. Palo, X. Zhang, Variability of the ionosphere. J. Atmos. Sol.-Terr. Phys. **62**, 685–693 (2000)

S. Fritz, S.D. Soules, Large-scale temperature changes in the stratosphere observed from Nimbus III. J. Atmos. Sci. **27**, 1091–1097 (1970)

T.J. Fuller-Rowell, M. Codrescu, P. Wilkinson, Quantitative modelling of the ionospheric response to geomagnetic activity. Ann. Geophys. **18**, 766–781 (2000)

T.J. Fuller-Rowell, H. Wang, R. Akmaev, F. Wu, T.W.F.M. Iredell, A.D. Richmond, Forecasting the dynamic and electrodynamic response to the January 2009 sudden stratospheric warming. Geophys. Res. Lett. (2011). doi:10.1029/2011GL047732

T. Fuller-Rowell, F. Wu, R. Akmaev, T.-W. Fang, E. Araujo-Pradere, A whole atmosphere model simulation of the impact of a sudden stratospheric warming on thermosphere dynamics and electrodynamics. J. Geophys. Res. **115**, 1–13 (2010). doi:10.1029/2010JA015524

B. Funke, M. López-Puertas, D. Bermejo-Pantaleón, M. García-Comas, G.P. Stiller, T.V. Clarmann, M. Kiefer, A. Linden, Evidence for dynamical coupling from the lower atmosphere to the thermosphere during a major stratospheric warming. Geophys. Res. Lett. **37**(13), 1–5 (2010). doi:10.1029/2010GL043619

L.P. Goncharenko, S.R. Zhang, Ionospheric signatures of sudden stratospheric warming: Ion temperature at middle latitude. Geophys. Res. Lett. **35**(L21103) (2008). doi:10.1029/2008GL035684

L.P. Goncharenko, A.J. Coster, J.L. Chau, C.E. Valladares, Impact of sudden stratospheric warmings on equatorial ionization anomaly. J. Geophys. Res. **115**, 1–11 (2010a). doi:10.1029/2010JA015400

L.P. Goncharenko, J.L. Chau, H.-L. Liu, A.J. Coster, Unexpected connections between the stratosphere and ionosphere. Geophys. Res. Lett. **37**(10), 1–6 (2010b). doi:10.1029/2010GL043125

L.J. Gray, E.F. Drysdale, T.J. Dunkerton, B.N. Lawrence, Model study of the inter annual variability of the northern hemisphere stratospheric winter circulation: the role of the quasi-biennial oscillation. Q. J. R. Meteorol. Soc. **127**, 1413–1432 (2001)

M.E. Hagan, F. Vial, J.M. Forbes, Variability in the upward propagating semidiurnal tide due to effects of QBO in the lower atmosphere. J. Atmos. Sol.-Terr. Phys. **54**, 1465–1474 (1992)

R.A. Heelis, Electrodynamics in the low and middle latitude ionosphere: a tutorial. J. Atmos. Sol.-Terr. Phys. **66**, 825–838 (2004)

R.A. Heelis, P.C. Kendall, R.J. Moffet, D.W. Windle, H. Rishbeth, Electrical coupling of the e- and f-region and its effects on F-region drifts and winds. Planet. Space Sci. **22**(5), 743–756 (1974)

P. Hoffmann, W. Singer, D. Keuer, Variability of the mesospheric wind field at middle and arctic latitudes in winter and its relation to stratospheric circulation disturbances. J. Atmos. Sol.-Terr. Phys. **64**, 1229–1240 (2002)

P. Hoffmann, W. Singer, D. Keuer, W.K. Hocking, M. Kunze, Y. Murayama, Latitudinal and longitudinal variability of mesospheric winds and temperatures during stratospheric warming events. J. Atmos. Sol.-Terr. Phys. **69**, 2355–2366 (2007)

J.R. Holton, H.-C. Tan, The influence of the equatorial quasi-biennial oscillation on the global circulation at 50 mb. J. Atmos. Sci. **37**, 2200–2208 (1980)

R. Hutton, J.O. Oyinloye, The counter-electrojet in Nigeria. Ann. Geophys. **26**, 921–926 (1970)

T.J. Immel, E. Sagawa, S.L. England, S.B. Henderson, M.E. Hagan, S.B. Mende, H.U. Frey, C.M. Swenson, L.J. Paxton, The control of equatorial ionospheric morphology by atmospheric tides. Geophys. Res. Lett. **33**(L15108) (2006). doi:10.1029/2006GL026161

T.J. Immel, S.B. Mende, M.E. Hagan, P.M. Kintner, S.L. England, Evidence of tropospheric effects on the ionosphere. Eos **90**, 69 (2009). doi:10.1029/2009EO090001

E.S. Kazimirovsky, G.V. Vergasova, Mesospheric, lower thermospheric dynamics and external forcing effects: a review. Indian J. Radio & Space Phys. **38**, 7–36 (2009)

H. Kil, R.A. Heelis, L.J. Paxton, S.-J. Oh, Formation of a plasma depletion shell in the equatorial ionosphere. J. Geophys. Res. **114**(A11), 1–7 (2009). doi:10.1029/2009JA014369

Y.-J. Kim, M. Flatau, Hindcasting the January 2009 arctic sudden stratospheric warming and its influence on the arctic oscillation with unified parameterization of orographic drag in NOGAPS. Part I: extended-range stand-alone forecast. Weather Forecast. **25**, 1628–1644 (2010)

K. Kodera, Influence of stratospheric sudden warming on the equatorial troposphere. Geophys. Res. Lett. **33**(L06804) (2006). doi:10.1029/2005GL024510

K. Kruger, B. Naujokat, K. Labitzke, The unusual midwinter warming in the southern hemisphere stratosphere 2002: a comparison to northern hemisphere phenomena. J. Atmos. Sci. **62**, 603–613 (2005)

E. Kudeki, C.D. Fawcett, High resolution observations of 150 km echoes at Jicamarca. Geophys. Res. Lett. **20**(18), 1987–1990 (1993)

E. Kudeki, S. Bhattacharyya, R.F. Woodman, A new approach in incoherent scatter F region $E \times B$ drift measurements at Jicamarca. J. Geophys. Res. **104**, 28145–28162 (1999)

J. Kurihara, Y. Ogawa, S. Oyama, S. Nozawa, M. Tsutsumi, C.M. Hall, Y. Tomikawa, R. Fujii, Links between a stratospheric sudden warming and thermal structures and dynamics in the high-latitude mesosphere, lower thermosphere, and ionosphere. Geophys. Res. Lett. **37**(L13806) (2010). doi:10.1029/2010GL043643

K. Labitzke, Temperature changes in the mesosphere and stratosphere connected with circulation changes in winter. J. Atmos. Sci. **29**, 756–766 (1972)

K. Labitzke, Stratospheric-mesospheric midwinter disturbances—a summary of observed characteristics. International Symposium on Middle Atmosphere Dynamics and Transport, vol. **86**, p. 9665 (1981). doi:10.1029/JC086iC10p09665

K. Labitzke, On the interannual variability of the middle stratosphere during the northern winters. J. Meteorol. Soc. Jpn. **60**, 124–139 (1982)

K. Labitzke, M. Kunze, On the remarkable Arctic winter in 2008/2009. J. Geophys. Res. **114**(D00I02) (2009). doi:10.1029/2009JD012273

K. Labitzke, B. Naujokat, The lower arctic stratosphere since 1952, in *SPARC Newsletter*, vol. 15 (SPARC Office, Toronto, 2000)

K. Labitzke, H. van Loon, Associations between the 11-year solar cycle, the QBO and the atmosphere. Part I: the troposphere and stratosphere in the northern hemisphere winter. J. Atmos. Sol.-Terr. Phys. **50**, 197–206 (1988)

J. Lei, J.P. Thayer, J.M. Forbes, E.K. Sutton, R.S. Nerem, M. Temmer, A.M. Veronig, Global thermospheric density variations caused by high-speed solar wind streams during the declining phase of solar cycle 23. J. Geophys. Res. **113**(A11303) (2008). doi:10.1029/2008JA013433

V. Limpasuvan, D.W.J. Thompson, D.L. Hartmann, The life cycle of the northern hemisphere sudden stratospheric warmings. J. Climate **17**, 2584–2596 (2004)

H.L. Liu, R.G. Roble, A study of a self-generated stratospheric sudden warming and its mesospheric/lower thermospheric impacts using coupled TIME-GCM/CCM3. J. Geophys. Res. **107**(4695) (2002). doi:10.1029/2001JD001533

H.-L. Liu, W. Wang, A.D. Richmond, R.G. Roble, Ionospheric variability due to planetary waves and tides for solar minimum conditions. J. Geophys. Res. **115**, 1–13 (2010). doi:10.1029/2009JA015188

H. Liu, E. Doornbos, M. Yamamoto, S.T. Ram, Strong thermospheric cooling during the 2009 major stratosphere warming. Geophys. Res. Lett. (2011). doi:10.1029/2011GL047898

H. Lühr, M. Rother, K. Häusler, P. Alken, S. Maus, The influence of nonmigrating tides on the longitudinal variation of the equatorial electrojet. J. Geophys. Res. **113**(A08313) (2008). doi:10.1029/2008JA013064

G.L. Manney, M.J. Schwartz, K. Kruger, M.L. Santee, S. Pawson, J.N. Lee, W.H. Daffer, R.A. Fuller, N.J. Livesey, Aura microwave limb sounder observations of dynamics and transport during the record-breaking 2009 arctic stratospheric major warming. Geophys. Res. Lett. **36**(L12815) (2009). doi:10.1029/2009GL038586

A.J. Mannucci, B.T. Tsurutani, B.A. Iijima, A. Komjathy, A. Saito, W.D. Gonzalez, F.L. Guarnieri, J.U. Kozyra, R. Skoug, Dayside global ionospheric response to the major interplanetary events of October 29-30, 2003, Halloween storms. Geophys. Res. Lett. **32**(L12S02) (2005)

T. Matsuno, A dynamical model of the stratospheric sudden warming. J. Atmos. Sci. **28**, 1479–1494 (1971)

N.J. Matthewman, J.G. Esler, A.J. Charlton-Perez, L.M. Polvani, A new look at stratospheric sudden warmings. Part III: Polar vortex evolution and vertical structure. J. Climate **22**, 1566–1585 (2009)

P.N. Mayaud, The equatorial counter-electrojet: a review of its geomagnetic aspects. J. Atmos. Sol.-Terr. Phys. **39**, 1055–1070 (1977)

M. Mendillo, H. Rishbeth, R.G. Roble, J. Wroten, Modelling F2-layer seasonal trends and day-to-day variability driven by coupling with the lower atmosphere. J. Atmos. Sol.-Terr. Phys. **64**, 1911–1931 (2002)

A.V. Mikhailov, V.H. Depuev, A.H. Depueva, Synchronous NmF2 and NmE daytime variations as a key to the mechanism of quiettime F2-layer disturbances. Ann. Geophys. **25**, 483–493 (2007)

P. Mukhtarov, D. Pancheva, B. Andonov, N.J. Mitchell, E. Merzlyakov, W. Singer, W. Hocking, C. Meek, A. Manson, Y. Murayama, Large-scale thermodynamics of the stratosphere and mesosphere during the major stratospheric warming in 2003/2004. J. Atmos. Sol. Terr. Phys. **69** (2007). doi:10.1016/j.jastp.2007.07.012

A. O'Neill, Stratospheric Sudden Warmings, in *Encyclopedia of Atmospheric Sciences* (Elsevier Science, San Diego, 2003), pp. 1342–1353

D. Pancheva, N. Mitchell, R.R. Clark, J. Drobjeva, J. Lastovicka, Variability in the maximum height of the ionospheric F2-layer over millstone hill (September 1998 to march 2000); influence from below and above. Ann. Geophys. **20**, 1807–1819 (2002)

D. Pancheva, P. Mukhtarov, N.J. Mitchell, B. Andonov, E. Merzlyakov, W. Singer, Y. Murayama, S. Kawamura, J. Xiong, W. Wan, W. Hocking, D. Fritts, D. Riggin, C. Meek, A. Manson, Latitudinal wave coupling of the stratosphere and mesosphere during the major stratospheric warming in 2003/2004. Ann. Geophys. **26**, 467–483 (2008)

D. Pancheva, P. Mukhtarov, B. Andonov, N.J. Mitchell, J.M. Forbes, Planetary waves observed by TIMED/SABER in coupling the stratosphere–mesosphere–lower thermosphere during the winter of 2003/2004: Part 2—altitude and latitude planetary wave structure. J. Atmos. Sol.-Terr. Phys. **71**(1), 75–87 (2009)

D. Pancheva, P. Mukhtarov, Stratospheric warmings: the atmosphere-ionosphere coupling paradigm. J. Atmos. Sol. Terr. Phys. **73** (2011). doi:10.1016/j.jastp.2011.03.066

H.F. Parish, J.M. Forbes, F. Kamalabadi, Planetary wave and solar emission signatures in the equatorial electrojet. J. Geophys. Res. **99**(A1), 355–368 (1994)

A.K. Patra, N.V. Rao, Further investigations of 150-km echoing riddle using simultaneous observations of 150-km and *E* region echoes from off-electrojet location Gadanki. J. Geophys. Res. **112**(A09301) (2007). doi:10.1029/2006JA012204

N.M. Pedatella, J.M. Forbes, Evidence for stratosphere sudden warming-ionosphere coupling due to vertically propagating tides. Geophys. Res. Lett. **37**(11), 1–5 (2010a). doi:10.1029/2010GL043560

N.M. Pedatella, J.M. Forbes, Global structure of the lunar tide in ionospheric total electron content. Geophys. Res. Lett. **37**, 06103 (2010b). doi:10.1029/2010GL042781

J.E. Pingree, B.G. Fejer, On the height variation of the equatorial *F* region vertical plasma drifts. J. Geophys. Res. **92**, 4763–4766 (1987)

R.A. Plumb, Instability of the distorted polar night vortex: a theory of stratospheric warmings. J. Atmos. Sci. **38**, 2514–2531 (1981)

A.I. Pogoreltsev, A.A. Vlasov, K. Fröhlich, C. Jacobi, Planetary waves in coupling the lower and upper atmosphere. J. Atmos. Sol.-Terr. Phys. **69**, 2083 (2007). doi:10.1016/j.jastp.2007.05.014

L.M. Polvani, P.J. Kushner, Tropospheric response to stratospheric perturbations in a relatively simple general circulation model. Geophys. Res. Lett. **29**(7) (2002). doi:10.1029/2001GL014284

L.M. Polvani, D.W. Waugh, Upward wave activity flux as precursor to extreme stratospheric events and subsequent weather regimes. J. Climate **17**, 3548–3554 (2004)

R.G. Rastogi, Lunar effects in the counter-electrojet near the magnetic equator. J. Atmos. Sol.-Terr. Phys. **36**, 167–170 (1974)

R.G. Rastogi, Morphological aspects of a new type of counter electrojet event. Ann. Geophys. **17**, 210 (1999). doi:10.1007/s00585-999-0210-6

A.D. Richmond, S. Matsushita, J.D. Tarpley, On the production mechanisms of electric currents and fields in the ionosphere. J. Geophys. Res. **81**(4), 547–555 (1976)

H. Rishbeth, The F-region dynamo. Planet. Space Sci. **19**(263) (1971)

H. Rishbeth, *F* region links with the lower atmosphere? J. Atmos. Sol.-Terr. Phys. **68**, 469–478 (2006)

H. Rishbeth, M. Mendillo, Patterns of F2-layer variability. J. Atmos. Sol.-Terr. Phys. **63**, 1661–1680 (2001)

F. Sassi, R.R. Garcia, B.A. Boville, H. Liu, On temperature inversions and the mesospheric surf zone. J. Geophys. Res. **107**(D19) (2002). doi:10.1029/2001JD001525

J.H. Sastri, Longitudinal dependence of equatorial *F* region vertical plasma drifts in the dusk sector. J. Geophys. Res. **101**(A2), 2445–2452 (1996)

S. Sathishkumar, S. Sridharan, Planetary and gravity waves in the mesosphere and lower thermosphere region over Tirunelveli (8.7°N, 77.8°E) during stratospheric warming events. Geophys. Res. Lett. **36**(7), 1–5 (2009). doi:10.1029/2008GL037081

L. Scherliess, B.G. Fejer, Radar and satellite global equatorial *F*-region vertical drift model. J. Geophys. Res. **104**, 6829–6842 (1999)

D.M.H. Sexton, The effect of stratospheric ozone depletion on the phase of the Antarctic oscillation. Geophys. Res. Lett. **28**, 3697–3700 (2001)

M.G. Shepherd, D.L. Wu, I.N. Fedulina, S. Gurubaran, J.M. Russell, M.G. Mlynczak, G.G. Shepherd, Stratospheric warming effects on the tropical mesospheric temperature field. J. Atmos. Sol.-Terr. Phys. **69**, 2309–2337 (2007)

D.E. Siskind, L. Coy, P. Espy, Observations of stratospheric warmings and mesospheric coolings by the TIMED SABER instrument. Geophys. Res. Lett. **32**(L09804) (2005). doi:10.1029/2005GL022399

D.E. Siskind, S.D. Eckermann, L. Coy, J.P. McCormack, C.E. Randall, On recent interannual variability of the arctic winter mesosphere: Implications for tracer descent. Geophys. Res. Lett. **34**(L09806) (2007). doi:10.1029/2007GL029293

D.E. Siskind, S.D. Eckermann, J.P. Mccormack, L. Coy, K.W. Hoppel, N.L. Baker, Case studies of the mesospheric response to recent minor, major, and extended stratospheric warmings. J. Geophys. Res. **115**, 1–16 (2010). doi:10.1029/2010JD014114

S. Sridharan, S. Sathishkumar, Seasonal and interannual variations of gravity wave activity in the low-latitude mesosphere and lower thermosphere over tirunelveli (8.7°N, 77.8°E). Ann. Geophys. **26**, 3215 (2008). doi:10.5194/angeo-26-3215-2008

S. Sridharan, S. Sathishkumar, S. Gurubaran, Variabilities of mesospheric tides and equatorial electrojet strength during major stratospheric warming events. Ann. Geophys. **27**, 4125 (2009)

R.J. Stening, Electron density profile changes associated with the equatorial electrojet. J. Atmos. Terr. Phys. **39**, 157 (1977)

R.J. Stening, A diurnal modulation of the lunar tide in the upper atmosphere. Geophys. Res. Lett. **16**, 307 (1989). doi:10.1029/GL016i004p00307 (ISSN 0094-8276)

R.J. Stening, B.G. Fejer, Lunar tide in the equatorial *F* region vertical ion drift velocity. J. Geophys. Res. **106**, 221–226 (2001)

R.J. Stening, C.E. Meek, A.H. Manson, Upper atmosphere wind systems during reverse equatorial electrojet events. Geophys. Res. Lett. **23**, 3243 (1996). doi:10.1029/96GL02611

R.J. Stening, J.M. Forbes, M.E. Hagan, A.D. Richmond, Experiments with a lunar atmospheric tidal model. J. Geophys. Res. **102**, 13465 (1997). doi:10.1029/97JD00778

R.J. Stening, A two-layer ionospheric dynamo calculation. J. Geophys. Res. **86**(A5), 3543–3550 (1981)

M. Taguchi, Is there a statistical connection between stratospheric sudden warming and tropospheric blocking events? J. Atmos. Sci. **65**, 1442 (2008). doi:10.1175/2007JAS2363.1

J.D. Tarpley, B.B. Balsley, Lunar variations in the Peruvian electrojet. J. Geophys. Res. **77**, 1951 (1972). doi:10.1029/JA077i010p01951

D.W.J. Thompson, S. Solomon, Interpretation of recent southern hemisphere climate change. Science **296**, 895–899 (2002)

D.W.J. Thompson, M.P. Baldwin, J.M. Wallace, Stratospheric connection to northern hemisphere wintertime weather: Implications for predictions. J. Climate **15**, 1421–1428 (2002)

R.T. Tsunoda, W.L. Ecklund, On a summer maximum in the occurrence frequency of 150 km (F_1) radar echoes over Pohnpei. Geophys. Res. Lett. **31**(L06810) (2004). doi:10.1029/2003GL018704

K.K. Tung, R.S. Lindzen, A theory of stationary long waves. Part I: a simple theory of blocking. Mon. Weather Rev. **107**, 714–734 (1979)

F. Vial, J.M. Forbes, Monthly simulations of the lunar semi-diurnal tide. J. Atmos. Terr. Phys. **56**, 1591–1607 (1994)

C. Vineeth, T.K. Pant, R. Sridharan, Equatorial counter electrojets and polar stratospheric sudden warmings—a classical example of high latitude-low latitude coupling? Ann. Geophys. **27**, 3147 (2009)

C. Vineeth, T.K. Pant, K.K. Kumar, S.G. Sumod, Tropical connection to the polar stratospheric sudden warming through quasi 16-day planetary wave. Ann. Geophys. **28**(11), 2007–2013 (2010). doi:10.5194/angeo-28-2007-2010

R.L. Walterscheid, G.G. Sivjee, R.G. Roble, Mesospheric and lower thermospheric manifestations of a stratospheric warming event over Eureka, Canada (80°N). Geophys. Res. Lett. **27**(18), 2897–2900 (2000)

H. Wang, T.J. Fuller-Rowell, R.A. Akmaev, M. Hu, D.T. Kleist, M.D. Iredell, Simulation of the January 2009 sudden stratospheric 1 warming with a whole atmosphere model. Geophys. Res. Lett. (2011)

J.R. Winick, P.P. Wintersteiner, R.H. Picard, D. Esplin, M.G. Mlynczak, J.M. Russell III, L.L.G. (2009), OH layer characteristics during unusual boreal winters of 2004 and 2006. J. Geophys. Res. **114**(A02303) (2009). doi:10.1029/2008JA013688

R.F. Woodman, Vertical drift velocities and east-west electric fields at the magnetic equator. J. Geophys. Res. **75**(31), 6249–6259 (1970)
R.F. Woodman, T. Hagfors, Methods for the measurement of vertical ionospheric motions near the magnetic equator by incoherent scattering. J. Geophys. Res. **74**, 1205–1212 (1969)
R.F. Woodman, J.L. Chau, R.R. Ilma, Comparison of ionosonde and incoherent scatter drift measurements at the magnetic equator. Geophys. Res. Lett. **33**(L01103) (2006). doi:10.1029/2005GL023692
J. Xiong, W. Wan, B. Ning, L. Liu, Y. Gao, Planetary wave-type oscillations in the ionosphere and their relationship to mesospheric/lower thermospheric and geomagnetic disturbances at Wuhan (30.6°N, 114.5°E). J. Atmos. Sol.-Terr. Phys. **68**, 498 (2006). doi:10.1016/j.jastp.2005.03.018
C. Yamashita, H.-L. Liu, X. Chu, Gravity wave variations during the 2009 stratospheric sudden warming as revealed by ECMWF-T799 and observations. Geophys. Res. Lett. **37**(L22806) (2010a). doi:10.1029/2010GL045437
C. Yamashita, H.-L. Liu, X. Chu, Responses of mesosphere and lower thermosphere temperatures to gravity wave forcing during stratospheric sudden warming. Geophys. Res. Lett. **37**(L09803) (2010b). doi:10.1029/2009GL042351
X. Yue, W.S. Schreiner, J. Lei, C. Rocken, D.C. Hunt, Y.-H. Kuo, W. Wan, Global ionospheric response observed by cosmic satellites during the January 2009 stratospheric sudden warming event. J. Geophys. Res. **115** (2010). doi:10.1029/2010JA015466
B. Zhao, W. Wan, L. Liu, K. Igarashi, M. Nakamura, L.J. Paxton, S.-Y. Su, G. Li, Z. Ren, Anomalous enhancement of ionospheric electron content in the Asian-Australian region during a geomagnetically quiet day. J. Geophys. Res. **113**(A11302) (2008). doi:10.1029/2007JA012987

Space Sci Rev (2012) 168:419–440
DOI 10.1007/s11214-011-9816-6

Overview of Nighttime Ionospheric Instabilities at Low- and Mid-Latitudes: Coupling Aspects Resulting in Structuring at the Mesoscale

Jonathan J. Makela · Yuichi Otsuka

Received: 23 March 2011 / Accepted: 26 July 2011 / Published online: 27 August 2011

Abstract We present a review of the current state of understanding regarding two classes of irregularities causing mesoscale structuring (hundreds of kilometers) in the nighttime ionosphere at low- and mid-latitudes. Additionally, current state of understanding of equatorial plasma bubbles at low latitudes, and medium-scale traveling ionospheric disturbances at mid latitudes and their relationship to possible seeding from lower altitudes are described. In each case, well-developed linear theories exist to explain the general properties of the irregularities. However, these linear theories have growth rates too low to explain the actual observations, giving rise to the need to invoke seeding mechanisms. We describe the observational databases that have been compiled over the decades and discuss possible coupling and seeding mechanisms that would overcome the low growth rate and explain the observed structuring at the mesoscale. Future research directions are also briefly discussed.

Keywords Ionospheric irregularities · Low-latitude ionosphere · Mid-latitude ionosphere · Rayleigh-Taylor instability · Perkins instability

1 Introduction

The Earth's ionosphere is primarily controlled by a complex interplay between the predominant forcing from above, controlled by solar influences, and what is increasingly being understood to be important influences from below. Our understanding of the basic daily, seasonal, and yearly trends has reached sufficient maturity that first-principle models do a reasonable job of predicting the state of the ionosphere for a wide variety of conditions. However, superposed on this well-understood and specified state, irregularities can develop

J.J. Makela (✉)
University of Illinois at Urbana-Champaign, Urbana, IL, USA
e-mail: jmakela@illinois.edu

Y. Otsuka
Solar-Terrestrial Environment Laboratory, Nagoya University, Nagoya, Japan
e-mail: otsuka@stelab.nagoya-u.ac.jp

internal to the system. Over the years, significant effort has been made to understand the physical mechanisms responsible for the growth of these irregularities. With our technological society's increasing reliance on trans-ionospheric radio wave propagation, understanding and (more importantly) predicting these irregularities has taken on renewed interest.

Although there are a wide variety of structures that can be considered irregularities, we concentrate here on two specific irregularity classes, both of which occur in the nighttime ionosphere, one at low-latitudes and the other at mid-latitudes that result in structuring at the mesoscale (hundreds of kilometers). These irregularity classes share a relatively long observational history using similar techniques, well-developed linear theories describing their development, and discrepancies between the linear theories and the observational databases used to study them. Although independent from one another, recent work has indicated that they may, at times, be coupled (e.g., Miller et al. 2009; Krall et al. 2011).

Observations of nighttime low-latitude irregularities date back to the 1930s (Booker and Wells 1938) and have a long history of being studied through a variety of techniques (see reviews by Hysell 2000; Makela 2006; Woodman 2009; Kelley et al. 2011). Their development is generally understood in terms of the gravity-driven Rayleigh-Taylor instability mechanism acting at the magnetic equator, where the magnetic field lines are nearly horizontal and can support the ionospheric plasma against gravity. In this instability mechanism, small perturbations in the bottomside of the ionosphere can evolve into vertically elongated regions of depleted electron density. These depletions also contain strong electric fields at a variety of scale sizes, the larger of which can map along the highly conducting magnetic field lines creating a "wedge" extending into the low- and tropical-latitude ionosphere. It is this mapping along the magnetic field lines that allows the effects of these irregularities to be seen over a large spatial region. It also allows a variety of techniques to be brought to bear on measuring the irregularity structures. Radar observations show large plumes of increased coherent backscatter, sometimes towering well into the topside ionosphere, indicating structures in electric field and electron density on the order of meters (e.g., Woodman and LaHoz 1976; Tsunoda et al. 1979; Patra et al. 1997; Hysell and Burcham 2002; de Paula and Hysell 2004; Yokoyama and Fukao 2006). Rocket (e.g., Kelley et al. 1976; Rino et al. 1981; Hysell et al. 1994) and satellite (e.g., McClure et al. 1977; Tsunoda et al. 1982; Huang et al. 2001; Su et al. 2001; Hysell et al. 2009) measurements have elucidated the relationship between structures over a range of scale sizes to those observed using ground-based instrumentation. Meanwhile, ground-based imaging techniques provide two-dimensional observations at the mesoscale showing that the structures are elongated along the magnetic field lines, sometimes impinging into the mid-latitudes (e.g., Kelley et al. 2002; Martinis and Mendillo 2007), have widths in the zonal direction significantly narrower than their lengths in the meridional direction (e.g., Mendillo and Tyler 1983; Kelley et al. 2002; Otsuka et al. 2002; Mukherjee 2003), and typically drift in the direction of the background ionospheric plasma (e.g., Mendillo and Baumgardner 1982; Taylor et al. 1997; Sinha and Raizada 2000; Pimenta et al. 2003; Yao and Makela 2007).

At mid-latitudes, traveling ionospheric disturbances (TIDs) are oftentimes observed as wavy structures in the plasma density at F-region heights having horizontal wavelengths of 100–1,000 km. Both nighttime and daytime TIDs have been observed by using various techniques, such as ionosondes, HF Doppler sounders, satellite beacons, and incoherent scatter radars (e.g., Hunsucker 1982; Hocke and Schlegel 1996). They are thought to be a manifestation of gravity waves that propagate upward from the lower atmosphere or auroral regions (Hines 1960; Hooke 1968). Here, we concentrate on a specific class of nighttime, electrified medium-scale traveling ionospheric disturbances (MSTIDs), described below.

The nighttime ionosphere at mid-latitudes is typically dynamically stable as the upward $\mathbf{E} \times \mathbf{B}$ drift due to the eastward electric field and/or equatorward neutral wind supports the F-region plasma against downward, gravity-driven diffusion (e.g., Perkins 1973; Kelley et al. 2003b). However, under certain conditions of the driving electric field and/or neutral winds, perturbations in the ionospheric conductivity can allow polarization electric fields to form, generating an instability leading to the development of MSTIDs. In contrast to the low-latitude structures described above, which tend to be aligned along the magnetic meridians, MSTIDs have been found to have a phase front aligned at a significant angle from the magnetic field lines (e.g., Garcia et al. 2000; Shiokawa et al. 2003a; Martinis et al. 2010). The Perkins (1973) instability is often cited as the cause of the development of MSTIDs, since this instability mechanism correctly predicts the anomalous alignment of the phase fronts in conductivity variations. Observations of MSTIDs using optical (e.g., Garcia et al. 2000; Shiokawa et al. 2003a) and GPS (e.g., Kotake et al. 2007) techniques have elucidated the structures' typical two-dimensional features and occurrence statistics. Observations made by incoherent scatter radar (e.g. Behnke 1979; Kelley et al. 2000; Swartz et al. 2002) and in-situ techniques confirm the electrified nature of MSTIDs (e.g., Saito et al. 1995; Shiokawa et al. 2003b; Park et al. 2009), which leads to the features being conjugate in nature (Otsuka et al. 2004). F-region Field-Aligned Irregularity (FAI) measurements with HF/VHF radars have been conducted simultaneously with optical observations of MSTIDs (Saito et al. 2008; Otsuka et al. 2009), showing that the FAIs are embedded in the MSTIDs and propagate with the same velocity with MSTIDs.

Despite the large observational database, general understanding of the occurrence statistics, and well-developed linear theories, there are still outstanding questions pertaining to each of these irregularity types. For example, in both cases, the growth rates of the associated instability processes are small, and significant research has been conducted to understand coupling and seeding processes that could increase the growth rate to appreciable levels. Here, we provide an overview of the nighttime instabilities at low- and mid-latitudes based on observations at the mesoscale. We begin by reviewing low-latitude instabilities, commonly referred to as equatorial spread-F or equatorial plasma bubbles. The Rayleigh-Taylor instability is presented as the likely generative mechanism. This is followed by a discussion of potential gravity wave seeding of the instability. We then turn to a review of mid-latitude instabilities, commonly referred to as medium scale traveling ionospheric disturbances. The Perkins instability is presented as the likely generative mechanism. A review of observations of the electric field perturbations associated with these structures as well as possible E- and F-region coupling is provided. We conclude with brief suggestions for future work to be performed to further our understanding of these structures.

2 Low-Latitude Instabilities

A fairly consistent observational database for structures associated with low-latitude instabilities, which we will call equatorial plasma bubbles (EPBs), has been obtained over the years using many different techniques from a wide variety of locations. From long-term observations of EPBs, the seasonal and solar-cycle dependence of these structures has been extensively studied (e.g., Aarons 1993; Abdu et al. 1998; Sahai et al. 2000; Hysell and Burcham 2002; Burke et al. 2004; Miller et al. 2010). In general, the occurrence rates of EPBs described in these studies follow the hypothesis of Tsunoda (1985), in which the seasonality of nighttime equatorial irregularities is controlled by the need to have near-simultaneous E-region sunset at the footprint of the equatorial F-region magnetic field lines. This requirement is fulfilled when the sunset terminator is aligned with the magnetic field. Maruyama and Matuura (1984) also suggested the contribution of the trans-equatorial neutral winds in defining the seasonal variation of EPBs.

Although the basic understanding of the phenomenology and occurrence of EPBs is well constrained by the observational database, there are still several outstanding questions which keep this field of scientific inquiry vibrant. First, the linear growth rate of the Rayleigh-Taylor instability is dreadfully low and cannot explain the growth times observed. Second, and perhaps more important, is the fact that there is significant day-to-day variability in the occurrence of EPBs at a given location. Although the sunset alignment requirement provides a general guideline for when and where EPBs are expected to occur, EPBs do not occur on every night during a location's "spread-F season", nor do they never occur outside of this season. These two facets of EPB development often lead to invoking the need for a seeding mechanism. The seed would increase the growth rate of the instability, addressing the first concern, and would have a variability which could account for the second. An oftentimes cited seeding mechanism is the atmospheric gravity wave. However, direct measurements of gravity-wave seeding of EPBs has been lacking, although there is tantalizing evidence from a variety of experiments that suggest that gravity waves may indeed play a role in EPB development.

2.1 Brief Overview of the Rayleigh-Taylor Instability

Dungey (1956) originally proposed the Rayleigh-Taylor instability (RTI) as the physical mechanism responsible for the growth of EPBs. Briefly, the conditions for the RTI are typically established in the post-sunset ionosphere at the geomagnetic equator. At this time, the bottomside of the F layer has recombined with the dense neutral atmosphere, while the F layer itself has been raised to high altitudes by the pre-reversal enhancement in the zonal electric field. This results in a sharp vertical gradient in the electron density. In an analogy to a fluid dynamic instability, the dense plasma in the F layer represents a heavy fluid, supported against gravity by the magnetic field. This situation is unstable to vertical perturbations in the ionospheric plasma, which can release the gravitational energy stored in the height of the ionosphere. The vertical perturbations are subject to polarization caused by gravity driven currents across the zonal variations in the plasma density, resulting in a zonally eastward electric field which further enhances the perturbation and the instability grows, as shown in Fig. 1. The Rayleigh-Taylor instability growth rate can be written as (e.g., Martinis et al. 2005):

$$\gamma_{RT} = \frac{\Sigma_P^F}{\Sigma_P^E + \Sigma_P^F} \frac{\nabla N}{N} \left[\frac{\mathbf{E} \times \mathbf{B}}{B^2} - \frac{\mathbf{g}}{\nu_{in}} - U_m \right], \qquad (1)$$

where Σ_P^E and Σ_P^F are the field-line integrated Pedersen conductivities in the E and F region, respectively, N is the field-line integrated electron density, $\mathbf{E}$ is the electric field, $\mathbf{B}$ is the magnetic field, $\mathbf{g}$ is gravity, ν_{in} is the ion-neutral collision frequency, and U_m is the meridional neutral wind.

The source of the perturbation that grows into the full EPB is still much debated. Two potential candidates include a collisional shear instability acting on the bottomside of the ionosphere (Hysell and Kudeki 2004) or gravity waves (e.g., Kelley et al. 1981; Hysell et al. 1990; Singh et al. 1997). Kudeki and Bhattacharyya (1999) reported on incoherent scatter radar measurements showing a post-sunset vortex below the F peak related to the development of bottomside spread-F layers. The enhanced growth rate of the wind-driven gradient drift instabilities driven by this vortex are sufficient to enhance the growth of

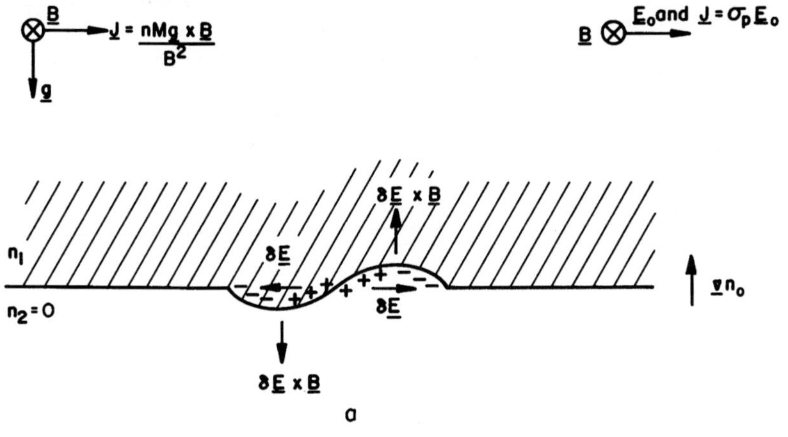

Fig. 1 Schematic showing an idealized case of the Rayleigh-Taylor instability at the magnetic equator (after Kelley 2009)

the larger-scale RTI. Recent combined rocket and radar measurements (Hysell et al. 2006) have studied these bottomside layers in detail. Using a radar imaging technique, Hysell et al. (2004) presented results indicating that the irregularities driven in this fashion could often times be found to occur in a periodic fashion along the zonal direction, with a separation of approximately 30 km, indicating a relationship between the shear flow represented by the vortex and a large-scale seed wave. Hysell and Kudeki (2004) presented a non-local instability analysis indicating that a collisional shear instability could explain these observations. More recently, Kudeki et al. (2007) presented modeling results indicating that a zonal neutral wind in the bottomside could further augment the collisional shear instability resulting in fast-growing modes with a horizontal scale on the order of tens of kilometers.

Another oft-cited potential mechanism is provided by gravity waves (e.g., Kelley et al. 1981; Hysell et al. 1990; Singh et al. 1997). Gravity waves could effect the Rayleigh-Taylor instability growth rate in several ways, thus affecting the day-to-day occurrence of EPBs. The first way a gravity wave could affect the growth of EPBs is through direct modulation of the neutral winds, which could increase the growth rate or increase the pre-reversal enhancement in the zonal electric field, which would push the F-layer to higher altitudes where it would be more conducive to instability growth. Second, gravity waves could perturb the background electron density, simply providing the initial perturbation in the layer's vertical density profile. Finally, the gravity waves could generate polarization electric fields in the E-region, which would map to the bottom side of the ionosphere, enhancing the growth rate through the $\mathbf{E} \times \mathbf{B}$ term.

The periodicity of gravity waves reaching the lower thermosphere have wavelengths larger than those indicated in the collisional shear studies mentioned above, on the order of hundreds of kilometers (e.g., Vadas 2007) and there is ample evidence for multiple scales existing within the unstable nighttime equatorial ionosphere. It is probable, then, that multiple mechanisms may be at play on a given night, making diagnostics of a specific seeding mechanism difficult. As Kudeki et al. (2007) point out, even though gravity waves are not required to enhance the RTI growth rate, once the wind-driven collisional shear instability theory is invoked, gravity waves "may still play a role in the variability of spread-F through their contributions to the thermospheric zonal winds at sunset." Thus, we should expect to see certain aspects of gravity wave phenomenology echoed in the larger-scale (hundreds of kilometer) EPB phenomenology. First, gravity waves are periodic structures. We should, therefore, expect that EPBs seeded by gravity waves would be periodic in nature. The spacing between isolated EPBs should echo the spacing of the underlying gravity wave. Second, there should be a relationship between gravity wave activity and the occurrence of EPBs. That is, there should be a measurable correlation between gravity wave and EPB occurrences. Finally, we might expect to see a relationship between variations in thermospheric winds, related to the passage of gravity waves, and the occurrence of EPBs. These variations, measurable using a Fabry-Perot interferometer, should be in line with gravity wave theory. Below, we review results that speak to each of these three points.

2.2 Periodic Spacing

Perhaps the relationship between gravity waves and EPBs that has received the most attention, historically, is related to the periodic spacing of irregularity structures. Observations have been made using a variety of modalities and, overall, a consistent picture supporting the plausibility of gravity-wave seeding of EPBs has emerged.

One of the earliest observations in which the spacing of EPBs was attributed to gravity waves was made by Röttger (1973) in an HF propagation experiment. Based on measurements of the non-great circle propagation paths of a trans-equatorial path (Lindau, West Germany to Tsumeb, South West Africa), irregularity patches associated with equatorial spread-F were observed with spacings between 210–640 km. The results were interpreted in terms of spatial resonance between gravity waves and the background $\mathbf{E} \times \mathbf{B}$ drift. Although later radar measurements using the Jicamarca radar (e.g., Kelley et al. 1981) relaxed the need for spatial resonance, these observations still represent some of the earliest linking gravity waves and the spacing of equatorial irregularities.

Radar measurements made at the Jicamarca Radio Observatory in the 1970s and 1980s showed large-scale undulations of the ionospheric layer, with large-scale backscatter plumes (which are known to be associated with EPBs) erupting from the westward edges of the undulations (e.g., Kelley et al. 1981). Although observations from the Jicamarca radar must be interpreted using a "slit camera" approximation, the scanning ALTAIR radar also showed similar undulations, termed "large scale wave structure" (LSWS) in the layer height. The undulation periodicity was similar to the periodicity of resultant backscatter plumes (e.g., Tsunoda and White 1981). The LSWS is plausibly linked to gravity waves under certain conditions (Tsunoda 2010). These early radar results of Kelley et al. (1981) suggested spacings ranging from 400–700 km. More recent results using a variety of techniques support the importance of the LSWS in the formation of EPBs (e.g., Fukao et al. 2006; Thampi et al. 2009; Tsunoda et al. 2010), and show scale sizes on the order of 200–1000 km.

Satellite measurements from the AE-E satellite show undulations in the in-situ electron density that, in the subsequent satellite pass, grow into large-scale density depletions spaced with the same periodicity as the undulations (Singh et al. 1997). More importantly, perhaps, Earle et al. (2008), using observations from the DE-2 satellite showed that the undulations in electron density are collocated with undulations in the neutral density, suggesting that they could be caused by a gravity wave.

From ground-based optics, studies have shown that the spacing of EPBs also falls into the same range as that observed by radio and satellite techniques. Takahashi et al. (2009) presented data taken using a wide-angle imaging system to study EPBs over Brazil. On 17 nights, multiple EPB structures were observed, with spacings ranging from approximately 100–300 km. In a similar study, Makela et al. (2010b) presented a study of EPB spacing using four years of data obtained from a field-aligned narrow-field imaging system located in

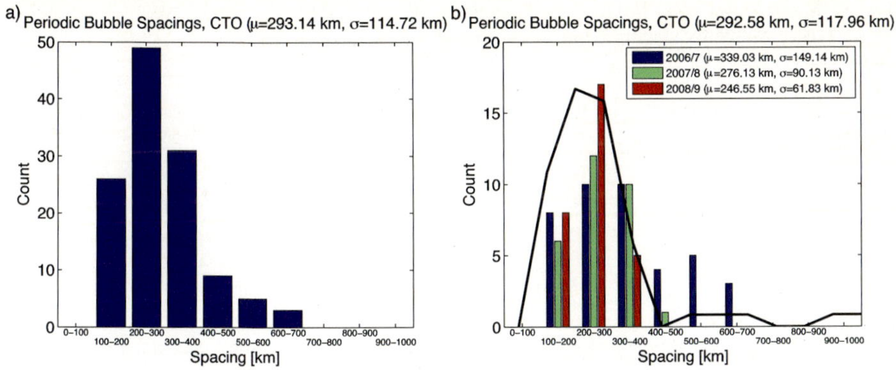

Fig. 2 Histogram of EPBs spacings measured using a narrow-field imaging system locate in Chile. (**a**) Spacings for all data collected between 22 Aug 2006 and 21 Aug 2009. (**b**) Spacings for data collected during the "spread-F season" (September through April), broken out by year. The *dark line* shows the distribution of horizontal wavelengths associated with secondary gravity waves measured by Vadas and Crowley (2010) (after Makela et al. 2010b)

Table 1 Observations of EPB spacing

Study	Technique	Spacing
Röttger (1973)	HF propagation	210–640 km
Kelley et al. (1981)	Radar backscatter	~600–700 km
Tsunoda and White (1981)	Radar backscatter	~400 km
Singh et al. (1997)	Satellite in-situ	150–800 km
Hysell et al. (2006)	Radar backscatter	~200 km
Fukao et al. (2006)	Radar backscatter	370–1000 km
Takahashi et al. (2009)	Imaging	100–300 km
Makela et al. (2010b)	Imaging	100–500 km

Chile. In this study, the spacing between EPBs ranged from 100–500 km, as shown in Fig. 2. The distribution of spacings agreed with the gravity wave spectrum measured by Vadas and Crowley (2010). The distribution also showed a solar-cycle dependence, which followed the expected dependence of the spectrum of gravity waves that could reach the lower thermosphere during different background conditions associated with the solar cycle. Importantly, this study showed that, even if gravity waves are a plausible seeding mechanism, they are likely not the only mechanism, as of the 317 nights with the presence of EPBs, only 123 showed periodic structure.

Interestingly, combined rocket and radar experiments have shown that in addition to the larger scale undulations seen in previous results, a smaller scale is also present, on the order of 30 km. These smaller scales, as argued by Kudeki et al. (2007) may be related to the evening vortex periods and have been shown in simulations to rapidly develop into larger-scale irregularity regions. However, the resultant spread-F structure is ordered at the larger scale size, rather than the 30-km size suggesting that a gravity wave, or some other process, must be involved.

Table 1 shows a summary of the studies mentioned above and demonstrates that a fairly consistent picture of periodic structuring emerges from a variety of instrumental techniques.

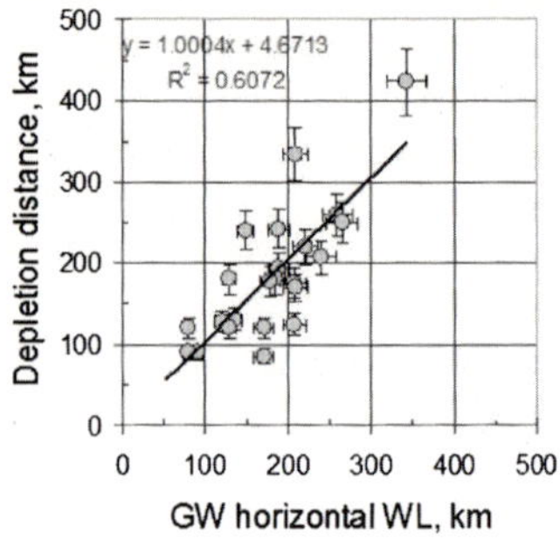

Fig. 3 Comparison of horizontal wavelength measured for gravity waves observed in the mesosphere and EPB spacings observed in the thermosphere. The observations were carried out in Brazil (after Takahashi et al. 2009)

2.3 Relationship Between Observed Gravity Wave Activity and EPB Occurrence

In contrast to the observations of EPB periodicity detailed above, observations of the relationship between gravity wave activity and EPB occurrence are not as prevalent. This is primarily due to the need for coordinated measurements over a large spatial region covering both the mesosphere (where GWs are significantly easier to observe) and the thermosphere. There was also a need for theory to be developed in order to link observed gravity wave properties in the mesosphere with the expected properties in the thermosphere, theory that was only recently developed (e.g., Vadas and Fritts 2004; Fritts and Vadas 2008). Thus, in comparison to the relative wealth of information on EPB spacing, coordinated measurements of gravity wave activity and EPB occurrence seem quite minimal.

Despite the difficulties, several important results warrant mentioning. Using data from a mesospheric temperature mapper and a narrow-field imaging system in the Pacific sector, Taori et al. (2010) presented two case studies using data from consecutive nights, one exhibiting the occurrence of EPBs and one lacking EPBs. They examined mesospheric variability to investigate the gravity wave properties, finding that growth for short-period waves was prominent between the OH (at ~86 km altitude) and O_2 (at ~94 km altitude) emission layers on the nights with EPB development. On the nights without EPBs, growth was not evident. This suggested that wave growth in the mesosphere could be an important indicator for the eventual development of EPBs at higher altitude.

Although the results above were obtained over a very limited time period, a more extensive campaign was carried out in Brazil in 2005 to investigate possible linkages between mesospheric gravity wave activity and the development of EPBs. The SpreadFEx campaign (Fritts et al. 2009) brought together a complement of instruments to make the required measurements from multiple field sites. Based on collocated OH and 630.0-nm imagery, Takahashi et al. (2009) presented a near one-to-one relationship between the horizontal wavelength of gravity waves observed in the mesosphere and the spacing of EPBs in the thermosphere, shown in Fig. 3. Vadas et al. (2009) used a ray tracing routine to track the waves seen in the mesosphere back to a convective source region, further suggesting that the waves seen in the mesosphere were gravity waves.

From these observations, it is evident that there is a correlative relationship between gravity wave activity and gravity wave parameters observed in the mesosphere and the occurrence and properties of EPBs in the thermosphere. However, a complete chain of direct measurements is still lacking, primarily due to the difficulty of measuring any useful parameters between the mesosphere/*E* region and the lower thermosphere.

2.4 Thermospheric Variations Due to Gravity Waves

Direct measurements of gravity wave activity in the thermosphere to address the possible relationship between gravity waves are quite rare to find in the literature. This is primarily a result of the relative lack of thermospheric neutral wind measurements with sufficient temporal resolution and accuracy to deduce potentially small gravity wave perturbations in the mean wind flow. It is also due to the difficulty of observing the signature of gravity waves in the 630.0-nm intensity. Compounding this is the fact that oftentimes the time period of most interest, prior to EPB onset in the sunset period, is a time when traditional nighttime Fabry-Perot interferometers cannot produce reliable results due to the bright continuum emission. Advanced modern dayglow instruments may provide important measurements of winds in the sunset period that may shed important light on the occurrence of gravity waves in the thermosphere and their possible relationship to EPB development, yet results are still forthcoming from such instruments

Fortunately, useful information may still be obtained from traditional nighttime FPIs, given sufficient accuracy, to deduce gravity wave parameters in the thermosphere, albeit only in the post-sunset time period. However, as EPBs do indeed develop at times later in the night, it may still be possible to observe a direct relationship at times. Several new FPI instruments have been installed in recent years in the S. American sector (Meriwether et al. 2011) and are operating in conjunction with a variety of additional instruments to hopefully provide direct measurements of gravity waves in the thermosphere and study their relationships to EPBs.

One such experiment is currently being conducted in northeastern Brazil as part of a follow-on experiment to the SpreadFEx campaign discussed above. The instrument complement is identical to that described in Makela et al. (2009). Two FPI instruments, capable of measuring the neutral winds with accuracies of ∼3–5 m/s have been deployed in a configuration that allows them to observe common volumes at the altitude of the 630.0-nm emission. This configuration provides two separate measurements of the total horizontal vector neutral wind, as well as three estimates of the vertical wind, through observations in the direction directly between the two observing sites as well as the zenith direction above each field site. Although the vertical wind at low-latitudes is expected to be minimal during nighttime conditions, a gravity wave propagating through the region would produce measurable vertical wind perturbations. Analysis of the data obtained from this experiment is just now getting underway. However, initial results have provided tantalizing measurements of wind perturbations that could be associated with the passage of gravity waves into the thermosphere. An example is shown in Fig. 4 in which periodic vertical wind fluctuations are observed in the three vertical wind measurements provided by the experiment. Collocated measurements from imaging and GPS systems would be able to investigate the relationship between these gravity waves and the occurrence of EPBs in a more complete fashion than was possible previously.

3 Mid-Latitude Instabilities

In contrast to the rich observational history for low-latitude irregularities, dating back to the 1930s, observations of irregularities at mid-latitudes are relatively recent. Despite this brief history, a fairly consistent climatological understanding of their seasonal and longitudinal occurrence as well as their general phenomenology has emerged from a variety of observational techniques.

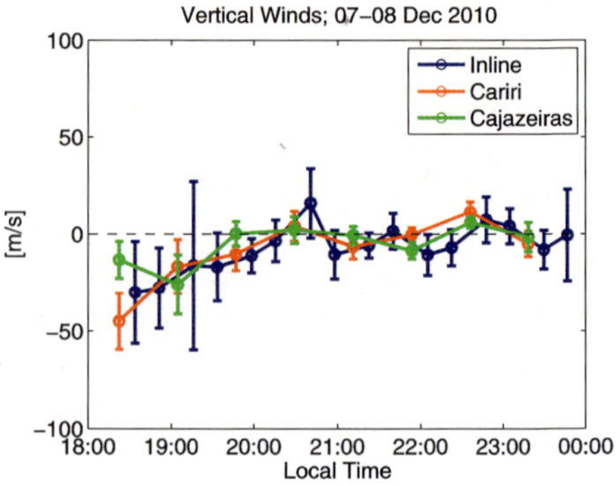

Fig. 4 Example of vertical wind measurements made by a dual-FPI experiment in northeastern Brazil. Wave structures of approximately the same frequency, but offset in phase, are evident in the three measurements suggesting the presence of a gravity wave

Mendillo et al. (1997) showed two-dimensional structures of the nighttime MSTIDs in nighttime airglow images obtained with highly sensitive cooled-CCD cameras. Garcia et al. (2000) and Shiokawa et al. (2003a) investigated statistical characteristics of the nighttime MSTIDs using 630.0-nm emission data taken from Arecibo (geographic: 18.3°N, 66.7°W) in Puerto Rico, and Rikubetsu (geographic: 43.5°N, 143.8°E) and Shigaraki (geographic: 34.8°N, 136.1°E) in Japan, respectively. They reported that the horizontal wavelength, period, and phase velocity of the nighttime MSTIDs are 100–300 km, 0.5–1.5 hour, and 50–120 m/s, respectively, and that most of the MSTIDs propagate southwestward. These results are consistent with those obtained from the GPS-TEC maps (e.g., Kotake et al. 2007; Otsuka et al. 2011). Ogawa et al. (2002) have shown fairly good correspondence of the MSTID structures and propagation direction in the TEC map and 630.0-nm airglow images.

In the Japanese and Australian longitudinal sector, the MSTID occurrence rate shows a semiannual variation with a major peak around the June solstice and a minor peak around the December solstice (Shiokawa et al. 2003a; Kotake et al. 2006). Unlike the Japanese and Australian longitudinal sector, the MSTID occurrence rate at Arecibo, Puerto Rico (18.3°N, 66.8°E) has a major peak around the December solstice (Garcia et al. 2000) as well as the June solstice (Martinis et al. 2010). Analyzing plasma density data obtained by satellites (CHAMP, KOMPSAT-1, and DMSP F15), Park et al. (2010) investigated the global distribution of plasma density variations caused by MSTIDs (Fig. 5). Their results show that the occurrence rate in the Asian/Oceanian (eastern Pacific/American) region is higher during June (December) solstice. Kotake et al. (2006) have pointed out that the major peak of the MSTID occurrence in summer or winter solstice could be explained by the linear growth rate of the Perkins instability, which is inversely proportional to the neutral density, as shown in Sect. 3.1. The neutral density shows semiannual variation with maxima at equinoxes and minima at solstices.

These seasonal and longitudinal variations of the MSTID occurrence rate are consistent with those of mid-latitude spread-F occurrence rates (Bowman 1992). Bowman (1992) pointed out that spread-F occurrence is attributed to a decrease in the neutral density in the thermosphere, although he also considered that MSTIDs were caused by atmospheric gravity waves and that the amplitude of gravity waves becomes larger for a smaller neutral density (e.g., Hines 1960).

Regarding the solar activity dependence of the MSTIDs, Kotake et al. (2006) have shown that MSTID activity, defined as $\delta I/\overline{I}$, where δI is the standard deviation of the TEC perturbations within one hour, and $\overline{I}$ is the background TEC, shows a clear anti-correlation with solar activity. Martinis et al. (2010) also report that MSTIDs observed in the airglow images show anti-correlation with the solar activity. As Kelley and Fukao (1991) have sug-

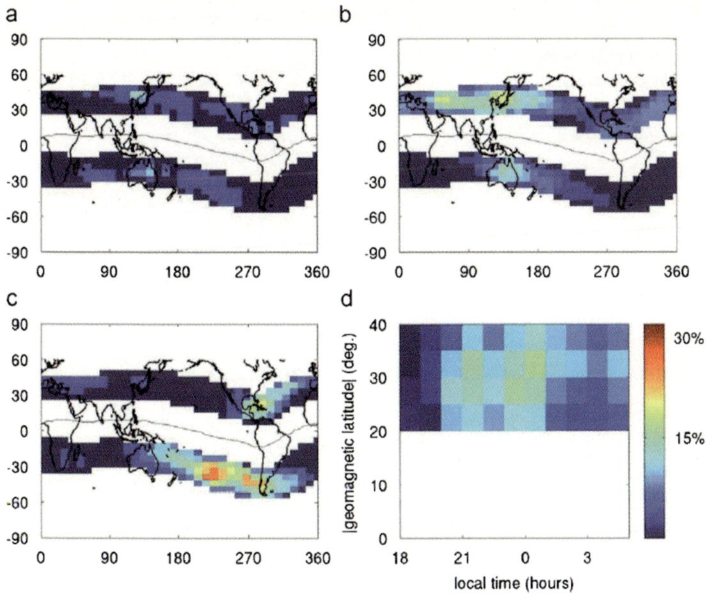

Fig. 5 Seasonal and global occurrence rate of mid-latitude density undulations as observed by CHAMP. The distribution is presented for each season centered on: (**a**) equinoxes, (**b**) June solstice, and (**c**) December solstice. Panel (**d**) shows the occurrence rate as a function of invariant latitude versus local time (after Park et al. 2010). Reprinted with permission from Elsevier

gested, the anti-correlation with the solar activity is predicted by the linear growth rate of the Perkins instability, which is discussed in detail below.

3.1 Brief Overview of the Perkins Instability

At mid-latitudes, the altitude of the nighttime F layer is in equilibrium, with a balance existing between the upward velocity due to the eastward electric fields and/or equatorward winds and the downward velocity due to gravitationally-induced ion diffusion. This equilibrium can be written by using an effective electric field $\mathbf{E}_0^* = \mathbf{E} + \mathbf{U} \times \mathbf{B}$, where $\mathbf{E}$ is electric field, $\mathbf{U}$ is neutral wind, and $\mathbf{B}$ is magnetic field (e.g., Kelley et al. 2003b):

$$\frac{|\mathbf{E}_0^*| \cos\theta}{B} \cos D = \frac{g}{\langle \nu_{in} \rangle} \sin^2 D \tag{2}$$

where θ is an angle between $\mathbf{E}_0^*$ and geomagnetic east, D is the dip angle of the magnetic field, g is the magnitude of the gravitational acceleration at the F layer, and $\langle \nu_{in} \rangle$ is the density-weighted collision frequency. The left and right sides of (2) express the upward drift caused by the electric field and/or neutral winds, and downward diffusion, respectively. They are balanced in the steady state.

Below, we assume a condition that $\mathbf{E}_0^*$ is northeastward and that field line integrated Pedersen conductivity (Σ_P) is reduced at a region with a phasefront elongated from NW to SE, as shown in Fig. 6. When the Pedersen current, $\Sigma_P \mathbf{E}_0^*$, traverses inhomogeneity of Σ_P, polarized electric fields (E_1) perpendicular to the phasefront of Σ_P could be generated at the region of low Σ_P to maintain the continuity of the current. Since Σ_P is proportional to $\langle \nu_{in} \rangle$,

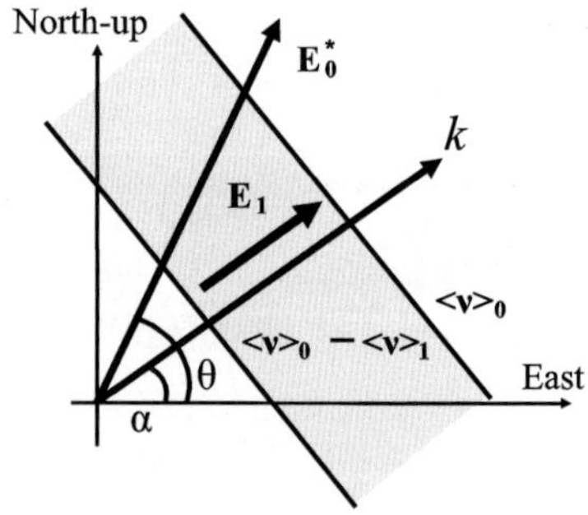

Fig. 6 Generation of the polarization electric fields (E_1) when the Pedersen conductivity perturbation have a band-like structure elongated from NW to SE in the northern hemisphere

current continuity of the Pedersen current in the direction perpendicular to the wavefront requires

$$\langle\nu\rangle_0 E_0^* \cos(\theta-\alpha) = (\langle\nu\rangle_0 - \langle\nu\rangle_1)\left\{E_0^*\cos(\theta-\alpha)+E_1\right\} \tag{3}$$

where α is an angle between geomagnetic east and the wave vector ($\mathbf{k}$), and $\langle\nu\rangle_0$ and $\langle\nu\rangle_1$ are background and perturbation of $\langle\nu_{in}\rangle$, respectively. Due to the eastward component of E_1, the F layer is lifted up to higher altitude by upward velocity of $\frac{E_1\cos\alpha}{B}\cos D$, so that downward velocity also increases to $\frac{g}{\langle\nu\rangle_0-\langle\nu\rangle_1}\sin^2 D$. The net upward plasma velocity (V_u) is

$$V_u = \frac{E_0^*\cos\theta + E_1\cos\alpha}{B}\cos D - \frac{g}{\langle\nu\rangle_0-\langle\nu\rangle_1}\sin^2 D \tag{4}$$

$$= \frac{\langle\nu\rangle_1}{\langle\nu\rangle_0-\langle\nu\rangle_1}\frac{E_0^*}{B}\cos D\sin(\theta-\alpha)\sin\alpha. \tag{5}$$

Equation (3) is used to obtain (5). When $\theta > \alpha$, V_u is positive. For this case, the F-region plasma moves to higher altitudes, causing further decrease in $\langle\nu_{in}\rangle$ because of the decrease in the collision frequency with altitude. The increase of the upward velocity is not compensated with the increase of the downward diffusion velocity, so that the F-region altitude is unstable. The unstable condition is satisfied when $\mathbf{k}$ lies between $\mathbf{E_0^*}$ and the eastward direction.

The most simple form of the linear growth rate (γ) for the Perkins instability (e.g., Perkins 1973; Hamza 1999) is given as:

$$\gamma = \frac{E_0^*\cos D}{BH}\sin(\theta-\alpha)\sin\alpha \tag{6}$$

$$= \frac{g\sin^2 D}{\langle\nu\rangle_0 H}\frac{\sin(\theta-\alpha)\sin\alpha}{\cos\theta}, \tag{7}$$

where H is the neutral scale height. γ is proportional to the effective electric fields and $\sin(\theta-\alpha)\sin\alpha$. Figure 7 shows unstable directions of $\mathbf{k}$. When $\mathbf{k}$ lies between $\mathbf{E_0^*}$ and zonal direction, the Perkins instability operates (e.g., Huang et al. 1994). $\mathbf{E}$ is considered to be smaller than and anti-parallel to $\mathbf{U}\times\mathbf{B}$ because $\mathbf{E}$ is generally generated by the F-region dynamo (Rishbeth 1981). Consequently, the direction of $\mathbf{E_0^*} = \mathbf{E}+\mathbf{U}\times\mathbf{B}$ is close to that of $\mathbf{U}\times\mathbf{B}$. Since $\mathbf{U}$ blows southeastward during the nighttime in the northern hemisphere due to a diurnal tide (HWM model; Hedin et al. 1996), $\mathbf{E_0^*}$ would typically be northeastward. For this condition, perturbations with phasefronts aligned from NW to SE are unstable for the Perkins instability. This feature is consistent with the phase alignment of the observed

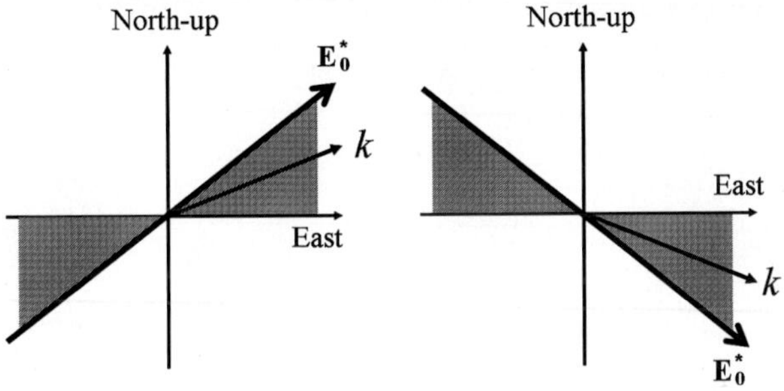

Fig. 7 Regions of unstable wave vectors k for the Perkins instability when effective electric field $\mathbf{E}_0^*$ lies (*left*) in the NE quadrant and (*right*) in the SE quadrant in the northern hemisphere. When k lies between **E** and zonal direction, the instability operates

nighttime MSTIDs. This is one of the major reasons why the Perkins instability is considered to be responsible for the nighttime MSTID generation.

3.2 Electric Field Perturbations Associated with MSTIDs

Using incoherent scatter (IS) measurements made at the Arecibo Observatory, Behnke (1979) reported remarkable height variations observed in the F layer, which had band-like structures elongated from the northwest to the southeast, traveling to the southwest. The observed height variations exceeded 50 km and were collocated with large electric fields (∼17 mV/m). He pointed out that these structures could be caused by the Perkins instability, an interpretation that remains commonplace to this day.

Fukao et al. (1991) observed intense backscatter echoes from field-aligned irregularities (FAIs) in the F region using the middle and upper atmosphere (MU) radar. The Doppler velocity of these echoes indicated that an intense polarization field is generated inside the FAIs. The Perkins instability is considered to play an important role in the evolution of these nighttime F-region phenomena (Kelley and Fukao 1991). Combining the incoherent scatter measurements by the Arecibo radar with airglow imaging observations, Kelley et al. (2000) have shown that polarization electric fields larger than the background field exist inside the depleted airglow regions.

The electric fields associated with nighttime MSTIDs have also been observed by in-situ measurements. Saito et al. (1995) presented results from the DE-2 satellite showing Mid-latitude Electric field Fluctuations (MEFs) having amplitudes comparable to that of the background electric field. Saito et al. (1998) showed similar MEFs detected by the Freja satellite (Lundin et al. 1994) that were associated with MSTIDs detected by the MU radar. Shiokawa et al. (2003b) compared the electric field variations observed with DMSP satellite and 630.0-nm airglow variations obtained with an all-sky imager at Shigaraki, Japan (Fig. 8), showing that the electric fields are southwestward (northeastward) in the region where the 630.0-nm airglow intensity increases (decreases), as shown in Fig. 8. As discussed in Makela and Kelley (2003), the field-line integrated Pedersen conductivity, Σ_P is related to the 630.0-nm intensity. Thus, the eastward component of the polarization electric field in the region of low Σ_P (low airglow intensity) pushes the F-region plasma to higher altitudes by $\mathbf{E} \times \mathbf{B}$ drift, causing further decrease in Σ_P (airglow intensity) because of the decrease in the ion-neutral collision frequency with altitude. The relationship between perturbations of Σ_P and the polarization electric fields is consistent with the theory for the Perkins instability.

By using the middle and upper atmosphere (MU) radar in Japan, Saito et al. (2008) have compared the F-region field-aligned irregularity (FAI) structures with MSTIDs observed with GPS-TEC measurements, showing that the 10-km scale FAIs move northwestward (southeastward) within the TEC depletion (enhancement). Furthermore, Otsuka et al. (2009)

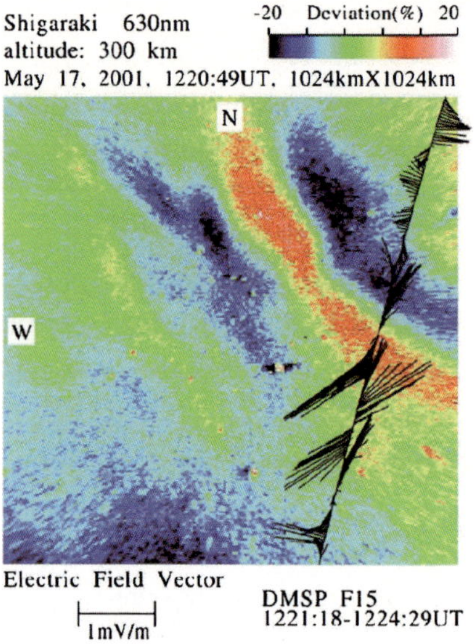

Fig. 8 Electric field vector measured by the DMSP F15 satellite along the footprint on the 630.0-nm image of the MSTID observed at Shigaraki at 1220:49 UT (2120:49 LT) on 17 May 2001. The images are deviations in percent from each 1-hour average image, in geographical coordinates of 1024 km × 1024 km at an altitude of 300 km. Only the variation component of the electric field are indicated by subtracting a running average of 1 min for the DMSP data. The vectors of electric field are indicated on the frame perpendicular to the local magnetic field. The satellite track was mapped from 846 km to 300 km along the local magnetic field line (after Shiokawa et al. 2003b)

have shown FAI velocity obtained from a combination of the Doppler velocities on the three radar beams oscillating in the northwest-southeast direction and that the FAI velocity was northwestward (southeastward) in the airglow depletion (enhancement). Using the Hokkaido Rikubetsu SuperDARN radar, Suzuki et al. (2009) and Koustov et al. (2009) have shown the same result. The directions of the FAI velocity were consistent with those of the $\mathbf{E} \times \mathbf{B}$ drifts caused by the polarized electric fields associated with the MSTIDs.

Saito et al. (1995) have shown that electric fields with spatial scale size exceeding about ten kilometers map almost without attenuation between both hemispheres. Otsuka et al. (2004) and Shiokawa et al. (2004) conducted geomagnetic conjugate observations of 630.0-nm airglow in Japan and Australia with two all-sky airglow imagers, showing that the observed MSTIDs have structures mirrored in the northern and southern hemispheres connected by geomagnetic field lines (Fig. 9). This result also indicates that polarization electric fields play an important role in generation of MSTIDs.

3.3 *E*- and *F*-Region Coupling

Kelley and Makela (2001) have pointed out two major discrepancies between the observations described above and theory for the Perkins instability. First, the growth rate predicted by the theory is too small (on the order of $10^{-4}\ \mathrm{s}^{-1}$) to explain the observations. Second, the instability theory cannot explain the equatorward and westward propagation of the MSTIDs. According to the Perkins instability theory, the structures generated by this instability should propagate at the same velocity as the background $\mathbf{E} \times \mathbf{B}$ drift, which is almost always to the east (Garcia et al. 2000). Consequently, the MSTIDs with phasefronts aligned from NW to SE should propagate northeastward. However, this propagation direction is opposite to that of the observed MSTIDs, which propagate southwestward in the northern hemisphere, as described above. These two discrepancies may be resolved by considering *E*- and *F*-region coupling processes.

Tsunoda and Cosgrove (2001) and Kelley et al. (2003a) have suggested that the electrodynamical coupling between the *E* and *F* regions could cause positive feedback, enhancing the polarization electric fields and plasma density perturbations that lead to the development

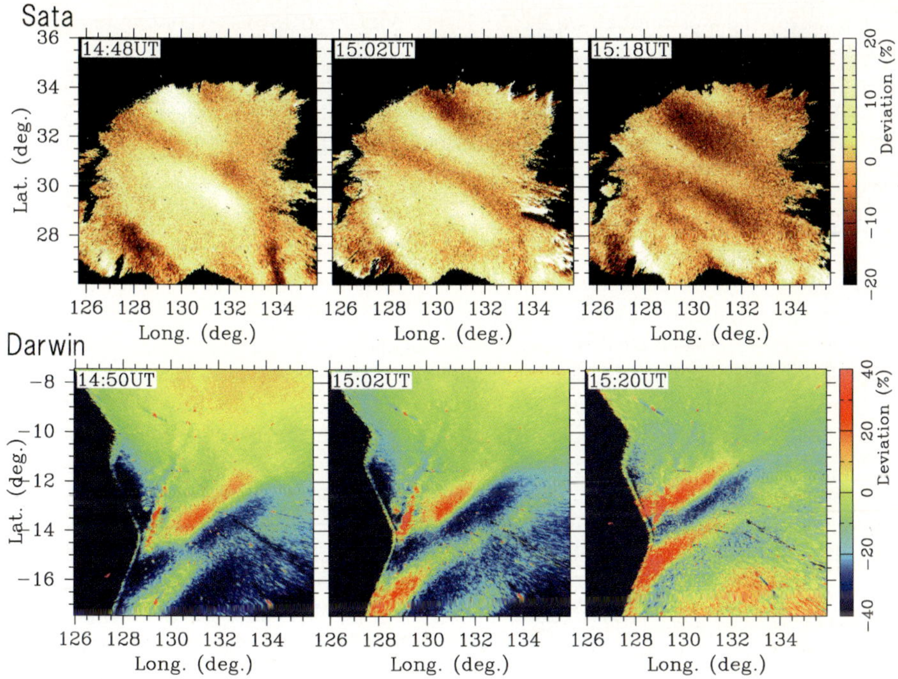

Fig. 9 Two-dimensional maps of 630-nm airglow intensity observed at (**a**) Sata and (**b**) Darwin for 1448–1520 UT (2348–2420 LT) on August 9, 2002 (LT = UT + 9 hours). The observed all-sky images are converted to geographical coordinates assuming that the emission layer exists at an altitude of 250 km. *Color levels* in each image show percentage of the airglow intensity deviations from 1-hour average to the background (after Otsuka et al. 2004)

of MSTIDs. Haldoupis et al. (1996) and Tsunoda (1998) have shown that Hall polarization processes, which drive the equatorial electrojet, can operate even in the mid-latitude *Es* layer, producing intense electric fields. Cosgrove and Tsunoda (2002, 2003) have proposed an *Es*-layer instability mechanism, in which the Hall polarization process is operated by a zonal wind-shear in the *Es* layer. The growth rate of the instability depends on the azimuthal alignment of the wavefront in the plasma density perturbations. The maximum growth rate occurs when a wavefront is aligned in the northwest to southeast direction in the northern hemisphere. This alignment is consistent with that of the *E*-region FAI structures observed at mid-latitudes (Yamamoto et al. 1994). Furthermore, the *Es*-layer instability mechanism predicts polarization electric field oscillation in the NE-SW direction ($\mathbf{E} \times \mathbf{B}$ drift oscillation in the NW-SE direction). Such $\mathbf{E} \times \mathbf{B}$ drift oscillations have been observed (Saito et al. 2007; Otsuka et al. 2007; Ogawa et al. 2009). Tsunoda and Cosgrove (2001) have suggested that electric fields generated by the Hall polarization process in an *Es* layer could map to the *F* region, causing a positive feedback process of the polarization electric fields. Cosgrove and Tsunoda (2004) and Cosgrove et al. (2004) have shown that the growth rate of the instability in the coupling system between the *Es* layer and *F* region exceeds that of the Perkins instability acting alone.

Performing three-dimensional simulation of the *E*- and *F*-region instabilities, Yokoyama et al. (2009) have explained the southwestward preference of the MSTID propagation in terms of southward neutral winds in the *Es* layer. The *Es* layer could be formed by the convergence of metallic ions due to the vertical shear in zonal wind which is eastward below the Es layer and westward above the Es layer (Whitehead 1961). According to tidal theory, the wind direction in the Es layer would be southward. Therefore, through ion-neutral collision, the Es layer would move southward. When the electric fields map between the E and F regions, the F-region plasma structures follow the Es layer movement so that the MSTIDs with NW-SE phasefront propagate southwestward.

4 Discussion

Considerable effort is currently being exerted to increase our understanding of the occurrence of ionospheric irregularities and structures, such as those described in this work. Although we have concentrated on observations of the mesoscale structure related to these irregularities, a comprehensive approach across decades of scale size is required to advance our understanding. A primary goal of ongoing work is understanding the generative mechanisms to a level of detail that would make prediction possible. In the cases of the instabilities described here, the low growth rate of the underlying irregularity process is often speculated to be overcome through a seeding mechanism. For equatorial plasma bubbles, as discussed above, there is tantalizing evidence that, at least in some cases, gravity waves can act as an effective seeding mechanism related to coupling from below. However, much of this evidence is indirect, with actual coincident measurements of gravity wave parameters and the resulting equatorial plasma bubbles still eluding observation. In any case, as described above and detailed in studies such as Kudeki et al. (2007), it is unlikely that gravity waves, alone, are the only seeding mechanism. Other mechanisms can likely trigger irregularity development. Coordinated measurements covering a large region using a multitude of observational modalities is likely the only way significant progress on this question will be made.

The Communication/Navigation Outage Forecasting System (C/NOFS; de La Beaujardière 2004) is one such program that has recently shown considerable progress in studying the equatorial irregularities described above. In addition to the suite of in-situ measurements made on-board the satellite, flying in a 13° inclination, elliptical orbit, the C/NOFS program involves advanced assimilative modeling techniques (e.g., Retterer 2005) with the aim of demonstrating the ability to predict the development of equatorial irregularities. Many interesting and informative results have come from the C/NOFS program (see, for example, the overview of initial results in Kelley et al. 2011). In addition to providing unprecedented measurements of the plasma, neutral, and electrodynamic conditions in the unstable equatorial ionosphere, the C/NOFS observations have identified a new class of large-scale structure immediately before sunrise (e.g., de La Beaujardière et al. 2009; Yizengaw et al. 2009). An unexpected result, these deep plasma depletions have yet to be seen in ground-based imaging observations, however, this may be due to the fact that most ground-based imaging systems turn off slightly before dawn, due to the increased sky intensity caused by the rising sun. This discovery, though, emphasizes the need for continued observational campaigns and the development of new experimental techniques.

Although the effects of gravity waves on the Rayleigh-Taylor instability and plasma bubble generation have received considerable attention, their effects on the development of nighttime MSTIDs have not been studied in much detail. Gravity waves, which perturb the plasma density, may seed the Perkins instability in the same way as is speculated for the Rayleigh-Taylor instability. Although the linear growth rate of the Perkins instability does not depend on spatial scale-size, most of the observed nighttime MSTIDs have horizontal wavelength of several hundred kilometers. The horizontal wavelength of the MSTIDs may correspond to that of gravity waves that seed the Perkins instability. However, direct measurements of the neutral wind oscillation caused by the gravity waves in the thermosphere are difficult and, consequently, comparisons between gravity waves and MSTIDs have not been done yet. On the other hand, the MSTID could be generated by the coupling processes between the Perkins and E_s layer instabilities, as described in Sect. 3.3, with the MSTIDs appearing in both the northern and southern hemispheres connected by the magnetic fields. Therefore, gravity waves in the E and F regions in both hemispheres need to be investigated in order to understand the effects of the gravity waves on MSTIDs.

Shiokawa et al. (2002) have suggested the low-latitude boundary of the MSTID occurrence exists around the crest region of the equatorial anomaly. However, Makela et al. (2010a) have reported that MSTIDs can occur closer to the magnetic equator during very low solar activity period, and pointed out that further theoretical and modeling work is needed to pursue the mechanism of MSTID occurrence close to the magnetic equator. Linear theory of the Perkins instability cannot account for the latitudinal dependence of the MSTIDs. The E- and F-region coupling processes and/or gravity wave seedings may be necessary to be considered in order to understand MSTID occurrence near the magnetic equator.

Yokoyama et al. (2009) and Yokoyama and Hysell (2010) have developed a new three-dimensional numerical model which simulates the coupling process between the E and F regions to produce MSITDs at mid-latitudes, and suggested that the polarization electric fields driven by the neutral winds in the E region are essentially important for the development of the Perkins instability in the F region. Furthermore, they have pointed out that gravity waves in the E region can seed the instability coupling between E and F regions through the modulation of the plasma density in the sporadic E layer. Yokoyama et al. (2004) show that equatorward-propagating gravity waves effectively generate polarization electric fields in the E_s layer. However, there are no observations showing that the equatorward-propagating gravity waves in the E region are associated with the MSTIDs because it is difficult to measure the gravity waves in the E region. For further studies, as Yokoyama and Hysell (2010) have suggested, gravity waves simulated in a global circulation model need to be used as an input for the numerical model that simulate the plasma instabilities with coupling processes between the E and F regions.

Acknowledgements JJM was supported by through the National Science Foundation grant ATM-09-40253. A part of the work done by YO was supported by Grant-in-Aid for Young Scientists A (20684021) of the Ministry of Education, Culture, Sports, Science and Technology of Japan. The authors thank the organizers of the International Space Science Institute workshop on "Coupling between Earth's atmosphere and its plasma environment."

References

J. Aarons, The longitudinal morphology of equatorial F-layer irregularities relevant to their occurrence. Space Sci. Rev. **63**, 209–243 (1993)

M.A. Abdu, J.H.A. Sobral, I.S. Batista, V.H. Rios, C. Medina, Equatorial spread-F occurrence statistics in the American longitudes: diurnal, seasonal and solar cycle variations. Adv. Space Res. **22**(6), 851–854 (1998). doi:10.1016/S0273-1177(98)00111-2

R. Behnke, F-layer height bands in the nocturnal ionosphere over Arecibo. J. Geophys. Res. **84**(A3), 974–978 (1979)

H.G. Booker, H.W. Wells, Scatter of radio waves by the F-region of the ionosphere. Terr. Magn. Atmos. Electr. **43**(3), 249–256 (1938)

G. Bowman, Upper atmosphere neutral-particle density variations compared with spread-f occurrence rates at locations around the world. Ann. Geophys. **10**(9), 676–682 (1992)

W. Burke, L. Gentile, C. Huang, C. Valladares, S. Su, Longitudinal variability of equatorial plasma bubbles observed by DMSP and ROCSAT-1. J. Geophys. Res. **109**(A12), A12301 (2004)

R. Cosgrove, R. Tsunoda, Simulation of the nonlinear evolution of the sporadic-e layer instability in the nighttime midlatitude ionosphere. J. Geophys. Res. **108**(A7), 1283 (2003)

R.B. Cosgrove, R.T. Tsunoda, A direction-dependent instability of sporadic-E layers in the nighttime midlatitude ionosphere. Geophys. Res. Lett. **29**(18), 1864 (2002)
R.B. Cosgrove, R.T. Tsunoda, Instability of the E-F coupled nighttime midlatitude ionosphere. J. Geophys. Res. **109**, A04305 (2004)
R.B. Cosgrove, R.T. Tsunoda, S. Fukao, M. Yamamoto, Coupling of the Perkins instability and the sporadic E layer instability derived from physical arguments. J. Geophys. Res. **109**, A06301 (2004)
J.W. Dungey, Convective diffusion in the equatorial F-region. J. Atmos. Terr. Phys. **9**(5–6), 304–310 (1956)
G.D. Earle, A.M. Masumba, S.L. Vadas, Satellite-based measurements of gravity wave-induced midlatitude plasma density perturbations. J. Geophys. Res. **113**, A03303 (2008)
D. Fritts, M. Abdu, B. Batista, I. Batista, P. Batista, R. Buriti, B. Clemesha, T. Dautermann, E. de Paula, B. Fechine et al., Overview and summary of the spread f experiment (spreadfex). Ann. Geophys. **27**, 2141–2155 (2009)
D.C. Fritts, S.L. Vadas, Gravity wave penetration into the thermosphere: sensitivity to solar cycle variations and mean winds. Ann. Geophys. **26**, 3841–3861 (2008)
S. Fukao, M.C. Kelley, T. Shirakawa, T. Takami, M. Yamamoto, T. Tsuda, S. Kato, Turbulent upwelling of the midlatitude ionosphere. 1. Observational results by the MU radar. J. Geophys. Res. **96**(A3), 3725–3746 (1991)
S. Fukao, T. Yokoyama, T. Tayama, M. Yamamoto, T. Maruyama, S. Saito, Eastward traverse of equatorial plasma plumes observed with the Equatorial Atmosphere Radar in Indonesia. Ann. Geophys. **24**, 1411–1418 (2006)
F.J. Garcia, M.C. Kelley, J.J. Makela, C.S. Huang, Airglow observations of mesoscale low-velocity traveling ionospheric disturbances at midlatitudes. J. Geophys. Res. **105**(A8), 18,407–18,415 (2000)
C. Haldoupis, K. Schlegel, D.T. Farley, An explanation for type 1 radar echoes from the midlatitude E-region ionosphere. Geophys. Res. Lett. **23**, 97–100 (1996)
A.M. Hamza, Perkins instability revisited. J. Geophys. Res. **104**(A10), 22,567–22,575 (1999)
A. Hedin, E. Fleming, A. Manson, F. Schmidlin, S. Avery, R. Clark, S. Franke, G. Fraser, T. Tsuda, F. Vial et al., Empirical wind model for the upper, middle and lower atmosphere. J. Atmos. Terr. Phys. **58**(13), 1421–1447 (1996)
C.O. Hines, Internal atmospheric gravity waves at ionospheric heights. Can. J. Phys. **38**(11), 1441–1481 (1960)
K. Hocke, K. Schlegel, A review of atmospheric gravity waves and traveling ionospheric disturbances: 1982–1995. Ann. Geophys. **14**, 14,917–14,940 (1996)
W.H. Hooke, Ionospheric irregularities produced by internal atmospheric gravity waves. JATP **38**, 1441–1481 (1968)
C.S. Huang, C. Miller, M. Kelley, Basic properties and gravity wave initiation of the midlatitude F region instability. Radio Sci. **29**(1), 395–405 (1994)
C.Y. Huang, W.J. Burke, J.S. Machuzak, L.C. Gentile, P.J. Sultan, DMSP observations of equatorial plasma bubbles in the topside ionosphere near solar maximum. J. Geophys. Res. **106**(A5), 8131–8142 (2001)
R. Hunsucker, Atmospheric gravity waves generated in the highlatitude ionosphere: a review. Rev. Geophys. Space Phys. **20**, 293–315 (1982)
D. Hysell, M. Larsen, C. Swenson, A. Barjatya, T. Wheeler, T. Bullett, M. Sarango, R. Woodman, Rocket and radar investigation of background electrodynamics and bottom-type scattering layers at the onset of equatorial spread F. Ann. Geophys. **24**(5), 1387–1400 (2006)
D.L. Hysell, An overview and synthesis of plasma irregularities in equatorial spread F. J. Atmos. Sol.-Terr. Phys. **62**(12), 1037–1056 (2000)
D.L. Hysell, J.D. Burcham, Long term studies of equatorial spread F using the JULIA radar at Jicamarca. J. Atmos. Sol.-Terr. Phys. **64**(12–14), 1531–1543 (2002)
D.L. Hysell, E. Kudeki, Collisional shear instability in the equatorial F region ionosphere. J. Geophys. Res. **109**(A11), A11301 (2004)
D.L. Hysell, M.C. Kelley, W.E. Swartz, R.F. Woodman, Seeding and layering of equatorial spread F by gravity waves. J. Geophys. Res. **95**(A10), 17,253–17,260 (1990)
D.L. Hysell, M.C. Kelley, W.E. Swartz, D.T. Farley, VHF radar and rocket observations of equatorial spread F on Kwajalein. J. Geophys. Res. **99**(A8), 15,065–15,085 (1994). doi:10.1029/94JA00476
D.L. Hysell, J. Chun, J.L. Chau, Bottom-type scattering layers and equatorial spread F. Ann. Geophys. **22**, 4061–4069 (2004)
D.L. Hysell, R.B. Hedden, J.L. Chau, F.R. Galindo, P.A. Roddy, R.F. Pfaff, Comparing F region ionospheric irregularity observations from C/NOFS and Jicamarca. Geophys. Res. Lett. **36**, L00C01 (2009). doi:10.1029/2009GL038983
M. Kelley, *The Earth's Ionosphere: Plasma Physics and Electrodynamics* (Academic Press, San Diego, 2009)
M. Kelley, C. Haldoupis, M.J. Nicolls, J.J. Makela, A. Belehaki, S. Shalimov, V.K. Wong, Case studies of coupling between the E and F regions during unstable sporadic-E conditions. J. Geophys. Res. **108**(A12) (2003a)

M. Kelley, J.J. Makela, M.N. Vlasov, A. Sur, Further studies of the Perkins stability during Space Weather Month. J. Atmos. Sol.-Terr. Phys. **65**(10), 1071–1075 (2003b)

M.C. Kelley, S. Fukao, Turbulent upwelling of the midlatitude ionosphere. 2. Theoretical framework. J. Geophys. Res. **96**(A3), 3747–3753 (1991)

M.C. Kelley, J.J. Makela, Resolution of the discrepancy between experiment and theory of midlatitude F-region structures. Geophys. Res. Lett. **28**(13), 2589–2592 (2001)

M.C. Kelley, G. Haerendel, H. Kappler, A. Valenzuela, B.B. Balsley, D.A. Carter, W.L. Ecklund, C.W. Carlson, B. Hausler, R. Torbert, Evidence for a Rayleigh-Taylor type instability and upwelling of depleted density regions during equatorial spread-F. Geophys. Res. Lett. **3**(8), 448–450 (1976)

M.C. Kelley, M.F. Larsen, C. Lahoz, J.P. McClure, Gravity-wave initiation of equatorial spread-F—a case-study. J. Geophys. Res. **86**(A11), 9087–9100 (1981)

M.C. Kelley, J.J. Makela, A. Saito, N. Aponte, M. Sulzer, S.A. González, On the electrical structure of airglow depletion/height layer bands over Arecibo. Geophys. Res. Lett. **27**(18), 2837–2840 (2000)

M.C. Kelley, J.J. Makela, B.M. Ledvina, P.M. Kintner, Observations of equatorial spread-F from Haleakala, Hawaii. Geophys. Res. Lett. **29**(20), 2003 (2002). doi:10.1029/2002GL015509

M.C. Kelley, J.J. Makela, O. de La Beaujardière, J. Retterer, Convective ionospheric storms: a review. Rev. Geophys. Space Phys. **49**, RG2003 (2011). doi:10.1029/2010RG000340

N. Kotake, Y. Otsuka, T. Tsugawa, T. Ogawa, A. Saito, Climatological study of GPS total electron content variations caused by medium-scale traveling ionospheric disturbances. J. Geophys. Res. **111**, A04306 (2006)

N. Kotake, Y. Otsuka, T. Tsugawa, T. Ogawa, A. Saito, Statistical study of medium-scale traveling ionospheric disturbances observed with the GPS networks in Southern California. Earth Planets Space **59**, 95–102 (2007)

A. Koustov, N. Nishitani, P.V. Panomarenko, K. Shiokawa, S. Suzuki, B.M. Shevtsov, J.W. MacDougall, Joint observations of a traveling ionospheric disturbance with the Paratunka OMTI camera and the Hokkaido HF radar. Ann. Geophys. **27**, 2399–2406 (2009)

J. Krall, J. Huba, S. Ossakow, G. Joyce, J. Makela, E. Miller, M. Kelley, Modeling of equatorial plasma bubbles triggered by non-equatorial traveling ionospheric disturbances. Geophys. Res. Lett. **38**, L08,103 (2011)

E. Kudeki, S. Bhattacharyya, Postsunset vortex in equatorial F-region plasma drifts and implications for bottomside spread-F. J. Geophys. Res. **104**(A12), 28,163–28,170 (1999)

E. Kudeki, A. Akgiray, M. Milla, J. Chau, D. Hysell, Equatorial spread-F initiation: post-sunsent vortex, thermospheric winds, gravity waves. J. Atmos. Sol.-Terr. Phys. **69**, 17–18 (2007)

O. de La Beaujardière, The C/NOFS Science Definition Team, C/NOFS: a mission to forecast scintillations. J. Atmos. Sol.-Terr. Phys. **66**(17), 1573–1591 (2004). doi:10.1016/j.jastp.2004.07.030

O. de La Beaujardière, J.M. Retterer, R.F. Pfaff, C. Roth, W.J. Burke, Y.J. Su, M.C. Kelley, R.R. Ilma, G.R. Wilson, L.C. Gentile, D.E. Hunton, D.L. Cooke, C/NOFS observations of deep plasma depletions at dawn. Geophys. Res. Lett. **36**, L00C06 (2009). doi:10.1029/2009GL038884

R. Lundin, G. Haerendel, S. Grahn, The Freja project. Geophys. Res. Lett. **21**(17), 1823–1826 (1994)

J.J. Makela, A review of imaging low-latitude ionospheric irregularity processes. J. Atmos. Sol.-Terr. Phys. **68**(13), 1441–1458 (2006)

J.J. Makela, M.C. Kelley, Using the 630.0-nm nightglow emission as a surrogate for the ionospheric Pedersen conductivity. J. Geophys. Res. **108**(A6), 1253 (2003)

J.J. Makela, J.W. Meriwether, J.P. Lima, E.S. Miller, S.J. Armstrong, The remote equatorial nighttime observatory of ionospheric regions project and the international heliospherical year. Earth Moon Planets **104**(1), 211–226 (2009)

J.J. Makela, E.S. Miller, E.R. Talaat, Nighttime medium-scale traveling ionospheric disturbances at low geomagnetic latitudes. Geophys. Res. Lett. **37**(24), L24,104 (2010a)

J.J. Makela, S.L. Vadas, R. Muryanto, T. Duly, G. Crowley, Periodic spacing between consecutive equatorial plasma bubbles. Geophys. Res. Lett. **37**, L14103 (2010b)

C. Martinis, M. Mendillo, Equatorial spread F-related airglow depletions at Arecibo and conjugate observations. J. Geophys. Res. **112**, A10310 (2007)

C. Martinis, J. Baumgardner, J. Wroten, M. Mendillo, Seasonal dependence of MSTIDs obtained from 630.0 nm airglow imaging at Arecibo. Geophys. Res. Lett. **37**(11), L11,103 (2010)

C.R. Martinis, M.J. Mendillo, J. Aarons, Toward a synthesis of equatorial spread F onset and suppression during geomagnetic storms. J. Geophys. Res. **110**, A07306 (2005)

T. Maruyama, N. Matuura, Longitudinal variability of annual changes in activity of equatorial spread F and plasma bubbles. J. Geophys. Res. **89**(A12), 10,903–10,912 (1984)

J.P. McClure, W.B. Hanson, J.H. Hoffman, Plasma bubbles and irregularities in the equatorial ionosphere. J. Geophys. Res. **82**(19), 2650–2656 (1977)

M. Mendillo, J. Baumgardner, Airglow characteristics of equatorial plasma depletions. J. Geophys. Res. **87**(A9), 7641–7652 (1982)

M. Mendillo, A. Tyler, Geometry of depleted plasma regions in the equatorial ionosphere. J. Geophys. Res. **88**(A7), 5778–5782 (1983)

M. Mendillo, J. Baumgardner, D. Nottingham, J. Aarons, B. Reinisch, J. Scali, M. Kelley, Investigations of thermospheric ionospheric dynamics with 6300-Å images from the Arecibo Observatory. J. Geophys. Res. **102**(A4), 7331–7343 (1997)

J. Meriwether, J.J. Makela, Y. Huang, D.J. Fisher, R.A. Buriti, A.F. Medeiros, H. Takahashi, Climatology of the nighttime equatorial thermospheric winds and temperatures over Brazil near solar minimum. J. Geophys. Res. **116**, A04322 (2011). doi:10.1029/2011JA016477

E.S. Miller, J.J. Makela, M.C. Kelley, Seeding of equatorial plasma depletions by polarization electric fields from middle latitudes: experimental evidence. Geophys. Res. Lett. **36**(18), L18105 (2009). doi:10.1029/2009GL039695

E.S. Miller, J.J. Makela, K.M. Groves, M.C. Kelley, R.T. Tsunoda, Coordinated study of coherent radar backscatter and optical airglow depletions in the central pacific. J. Geophys. Res. **115**, A06307 (2010)

G.K. Mukherjee, Studies of the equatorial F-region depletions and dynamics using multiple wavelength nightglow imaging. J. Atmos. Sol.-Terr. Phys. **65**(3), 379–390 (2003)

T. Ogawa, N. Balan, Y. Otsuka, K. Shiokawa, C. Ihara, T. Shimomai, A. Saito, Observations and modeling of 630 nm airglow and total electron content associated with traveling ionospheric disturbances over Shigaraki, Japan. Earth Planets Space **54**, 45–56 (2002)

T. Ogawa, N. Nishitani, Y. Otsuka, K. Shiokawa, T. Tsugawa, K. Hosokawa, Medium-scale traveling ionospheric disturbances observed with the SuperDARN Hokkaido radar, all-sky imager, and GPS network and their relation to concurrent sporadic E irregularities. J. Geophys. Res. **114**, A03316 (2009)

Y. Otsuka, K. Shiokawa, T. Ogawa, P. Wilkinson, Geomagnetic conjugate observations of equatorial airglow depletions. Geophys. Res. Lett. **29**(15), 1753 (2002)

Y. Otsuka, K. Shiokawa, T. Ogawa, P. Wilkinson, Geomagnetic conjugate observations of medium-scale traveling ionospheric disturbances at midlatitude using all-aky airglow imagers. Geophys. Res. Lett. **31**(15), L15803 (2004)

Y. Otsuka, F. Onoma, K. Shiokawa, T. Ogawa, M. Yamamoto, S. Fukao, Simultaneous observations of nighttime medium-scale traveling ionospheric disturbances and E-region field-aligned irregularities at midlatitude. J. Geophys. Res. **112**, A06317 (2007)

Y. Otsuka, K. Shiokawa, T. Ogawa, T. Yokoyama, M. Yamamoto, Spatial relationship of nighttime medium-scale traveling ionospheric disturbances and F region field-aligned irregularities observed with two spaced all-sky airglow imagers and the middle and upper atmosphere radar. J. Geophys. Res. **114**, A05302 (2009)

Y. Otsuka, N. Kotake, K. Shiokawa, T. Ogawa, T. Tsugawa, A. Saito, Statistical study of medium-scale traveling ionospheric disturbances observed with a GPS receiver network in Japan, in *Aeronomy of the Earth's Atmosphere and Ionosphere*, ed. by M.A. Abdu, D. Pancheva (Springer, Berlin, 2011)

J. Park, H. Lühr, C. Stolle, M. Rother, K. Min, J. Chung, Y. Kim, I. Michaelis, M. Noja, Magnetic signatures of medium-scale traveling ionospheric disturbances as observed by champ. J. Geophys. Res. **114**(A3), A03,307 (2009)

J. Park, H. Lühr, K.W. Min, J.J. Lee, Plasma density undulations in the nighttime mid-latitude F-region as observed by CHAMP, KOMPSAT-1, and DMSP F15. J. Atmos. Sol.-Terr. Phys. **72**(2–3), 183–192 (2010)

A.K. Patra, P.B. Rao, V.K. Anandan, A.R. Jain, Radar observations of 2.8 m equatorial spread-F irregularities. J. Atmos. Sol.-Terr. Phys. **59**(13), 1633–1641 (1997). doi:10.1016/S1364-6826(96)00162-9

E.R. de Paula, D.L. Hysell, The São Luís 30 MHz coherent scatter ionospheric radar: system description and initial results. Radio Sci. **39**, RS1014 (2004). doi:10.1029/2003RS002914

F. Perkins, Spread F and ionospheric currents. J. Geophys. Res. **78**(1), 218–226 (1973)

A.A. Pimenta, J.A. Bittencourt, P.R. Fagundes, Y. Sahai, R.A. Buriti, H. Takahashi, M.J. Taylor, Ionospheric plasma bubble zonal drifts over the tropical region: a study using OI 630 nm emission all-sky images. J. Atmos. Sol.-Terr. Phys. **65**(10), 1117–1126 (2003)

J.M. Retterer, Physics-based forecasts of equatorial radio scintillation for the Communication and Navigation Outage Forecasting System (C/NOFS). Space Weather **3**, S12C03 (2005)

C.L. Rino, R.T. Tsunoda, J. Petriceks, R.C. Livingston, M.C. Kelley, K.D. Baker, Simultaneous rocket-borne beacon and in situ measurements of equatorial spread F—intermediate wavelength results. J. Geophys. Res. **86**(A4), 2411–2420 (1981)

H. Rishbeth, The F-region dynamo. J. Atmos. Terr. Phys. **43**(5–6), 387–392 (1981)

J. Röttger, Wave-like structures of large-scale equatorial spread-F irregularities. J. Atmos. Terr. Phys. **35**(6), 1195–1206 (1973)

Y. Sahai, P.R. Fagundes, J.A. Bittencourt, Transequatorial F-region ionospheric plasma bubbles: solar cycle effects. J. Atmos. Sol.-Terr. Phys. **62**(15), 1377–1383 (2000)
A. Saito, T. Iyemori, M. Sugiura, N.C. Maynard, T.L. Aggson, L.H. Brace, M. Takeda, M. Yamamoto, Conjugate occurrence of the electric field fluctuations in the nighttime midlatitude ionosphere. J. Geophys. Res. **100**(A11), 21,439–21,451 (1995)
A. Saito, T. Iyemori, L.G. Blomberg, M. Yamamoto, M. Takeda, Conjugate observations of the mid-latitude electric field fluctuations with the MU radar and the Freja satellite. J. Atmos. Sol.-Terr. Phys. **60**, 129–140 (1998)
S. Saito, M. Yamamoto, H. Hashiguchi, A. Maegawa, A. Saito, Observational evidence of coupling between quasi-periodic echoes and medium scale traveling ionospheric disturbances. Ann. Geophys. **25**, 2185–2194 (2007)
S. Saito, M. Yamamoto, H. Hashiguchi, Imaging observations of nighttime mid-latitude f-region field-aligned irregularities by an mu radar ultra-multi-channel system. Ann. Geophys. **26**, 2345–2352 (2008)
K. Shiokawa, Y. Otsuka, M.K. Ejiri, Y. Sahai, T. Kadota, C. Ihara, T. Ogawa, K. Igarashi, S. Miyazaki, A. Saito, Imaging observations of the equatorward limit of midlatitude traveling ionospheric distrubances. Earth Planets Space **54**, 57–62 (2002)
K. Shiokawa, C. Ihara, Y. Otsuka, T. Ogawa, Statistical study of nighttime medium-scale traveling ionospheric disturbances using midlatitude airglow images. J. Geophys. Res. **108**(A1), 1052 (2003a)
K. Shiokawa, Y. Otsuka, C. Ihara, T. Ogawa, F.J. Rich, Ground and satellite observations of nighttime medium scale traveling ionospheric disturbance at midlatitude. J. Geophys. Res. **108**(A4), 1145 (2003b)
K. Shiokawa, Y. Otsuka, T. Ogawa, P. Wilkinson, Time evolution of high-altitude plasma bubbles imaged at geomagnetic conjugate points. Ann. Geophys. **22**, 3137–3143 (2004)
S. Singh, F.S. Johnson, R.A. Power, Gravity wave seeding of equatorial plasma bubbles. J. Geophys. Res. **102**(A4), 7399–7410 (1997)
H.S.S. Sinha, S. Raizada, Some new features of ionospheric plasma depletions over the Indian zone using all sky optical imaging. Earth Planets Space **52**(8), 549–559 (2000)
S.Y. Su, H.C. Yeh, R.A. Heelis, ROCSAT 1 ionospheric plasma and electrodynamics instrument observations of equatorial spread F: an early transitional scale result. J. Geophys. Res. **106**(A12), 29,153–29,159 (2001)
S. Suzuki, K. Hosokawa, Y. Otsuka, K. Shiokawa, T. Ogawa, N. Nishitani, T.F. Shibata, A.V. Koustov, B.M. Shevtsov, Coordinated observations of nighttime mediumscale traveling ionospheric disturbances in 630-nm airglow and HF radar echoes at midlatitudes. J. Geophys. Res. **114**, A07312 (2009)
W. Swartz, S. Collins, M. Kelley, J. Makela, E. Kudeki, S. Franke, J. Urbina, N. Aponte, S. González, M. Sulzer et al., First observations of an F-region turbulent upwelling coincident with severe E-region plasma and neutral atmosphere perturbations. J. Atmos. Sol.-Terr. Phys. **64**(12–14), 1545–1556 (2002)
H. Takahashi, M.J. Taylor, P.D. Pautet, A.F. Medeiros, D. Gobbi, C.M. Wrasse, J. Fechine, M.A. Abdu, I.S. Batista, E. Paula, J.H.A. Sobral, D. Arruda, S.L. Vadas, F.S. Sabbas, D.C. Fritts, Simultaneous observation of ionospheric plasma bubbles and mesospheric gravity waves during the SpreadFEx Campaign. Ann. Geophys. **27**, 1477–1487 (2009)
A. Taori, J.J. Makela, M. Taylor, Mesospheric wave signatures and equatorial plasma bubbles: a case study. J. Geophys. Res. **115**, A06302 (2010)
M.J. Taylor, J.V. Eccles, J. LaBelle, J.H.A. Sobral, High resolution OI (630 nm) image measurements of F-region depletion drifts during the Guará campaign. Geophys. Res. Lett. **24**(13), 1699–1702 (1997)
S. Thampi, M. Yamamoto, R. Tsunoda, Y. Otsuka, T. Tsugawa, J. Uemoto, M. Ishii, First observations of large-scale wave structure and equatorial spread F using CERTO radio beacon on the C/NOFS satellite. Geophys. Res. Lett. **36**(18), L18,111 (2009)
R. Tsunoda, R. Cosgrove, Coupled electrodynamics in the nighttime midlatitude ionosphere. Geophys. Res. Lett. **28**(22), 4171–4174 (2001)
R. Tsunoda, M. Baron, J. Owev, D. Towle, Altair: an incoherent scatter radar for equatorial spread F studies. Radio Sci. **14**(6), 1111–1119 (1979)
R.T. Tsunoda, Control of the seasonal and longitudinal occurrence of equatorial scintillations by the longitudinal gradient in integrated E region Pedersen conductivity. J. Geophys. Res. **90**(A1), 447–456 (1985)
R.T. Tsunoda, On polarized frontal structures, type-1 and quasi-periodic echoes in midlatitude sporadic E. Geophys. Res. Lett. **25**, 2614–2644 (1998)
R.T. Tsunoda, On seeding equatorial spread f during solstices. Geophys. Res. Lett. **37**(5), L05,102 (2010)
R.T. Tsunoda, B.R. White, On the generation and growth of equatorial backscatter plumes: 1. Wave structure in the bottomside F layer. J. Geophys. Res. **86**, 3610 (1981)
R.T. Tsunoda, R.C. Livingston, J.P. McClure, W.B. Hanson, Equatorial plasma bubbles: vertically elongated wedges from the bottomside F layer. J. Geophys. Res. **87**(A11), 9171–9180 (1982)
R.T. Tsunoda, D.M. Bubenik, S.V. Thampi, M. Yamamoto, On large-scale wave structure and equatorial spread F without a post-sunset rise of the F layer. Geophys. Res. Lett. **37**, L07105 (2010)

S. Vadas, D. Fritts, Thermospheric responses to gravity waves arising from mesoscale convective complexes. J. Atmos. Sol.-Terr. Phys. **66**(6–9), 781–804 (2004)

S.L. Vadas, Horizontal and vertical propagation and dissipation of gravity waves in the thermosphere from lower atmospheric and thermospheric sources. J. Geophys. Res. **112**, A06305 (2007)

S.L. Vadas, G. Crowley, Sources of the traveling ionospheric distrubances observed by the ionospheric TIDDBIT sounder near Wallops Island on October 30, 2007. J. Geophys. Res. **115**, A07324 (2010)

S.L. Vadas, M.J. Taylor, P.D. Pautet, P.A. Stamus, D.C. Fritts, H.L. Liu, F.T. Sao Sabbas, V.T. Rampinelli, P. Batista, H. Takahashi, Convection: the likely source of the medium-scale gravity waves observed in the OH airglow layer near Brasilia, Brazil, during the SpreadFEx campaign. Ann. Geophys. **27**, 231–259 (2009)

J.D. Whitehead, The formation of the sporadic-E layer in the temperate zones. J. Atmos. Terr. Phys. **20**, 49–58 (1961)

R.F. Woodman, Spread F—an old equatorial aeronomy problem finally resolved? Ann. Geophys. **27**, 1915–1934 (2009)

R.F. Woodman, C. LaHoz, Radar observations of F-region equatorial irregularities. J. Geophys. Res. **81**(31), 5447–5466 (1976)

M. Yamamoto, N. Komoda, S. Fukao, R.T. Tsunoda, T. Ogawa, T. Tsuda, Spatial structure of the E region field-aligned irregularities revealed by the MU radar. Radio Sci. **29**(1), 337–347 (1994)

D. Yao, J.J. Makela, Analysis of equatorial plasma bubble zonal drift velocities in the Pacific sector by imaging techniques. Ann. Geophys. **25**(3), 701–709 (2007)

E. Yizengaw, M.B. Moldwin, Y. Sahai, R. de Jesus, Strong postmidnight equatorial ionospheric anomaly observations during magnetically quiet periods. J. Geophys. Res. **114**, A12308 (2009). doi:10.1029/2009JA014603

T. Yokoyama, S. Fukao, Upwelling backscatter plumes in growth phase of equatorial spread F observed with the Equatorial Atmosphere Radar. Geophys. Res. Lett. **33**, L08104 (2006)

T. Yokoyama, D.L. Hysell, A new midlatitude ionosphere electrodynamics coupling model (MIECO): latitudinal dependence and propagation of mediumscale traveling ionospheric disturbances. Geophys. Res. Lett. **37**, L08105 (2010). doi:10.1029/2010GL042598

T. Yokoyama, T. Horinouchi, M. Yamamoto, S. Fukao, Modulation of the midlatitude ionospheric E region by atmospheric gravity waves through polarization electric field. J. Geophys. Res. **109**, A12307 (2004). doi:10.1029/2004JA010508

T. Yokoyama, D. Hysell, Y. Otsuka, M. Yamamoto, Three-dimensional simulation of the coupled Perkins and es-layer instabilities in the nighttime midlatitude ionosphere. J. Geophys. Res. **114**(A3), A03,308 (2009)

Space Sci Rev (2012) 168:441–461
DOI 10.1007/s11214-011-9786-8

Midlatitude Sporadic *E*. A Typical Paradigm of Atmosphere-Ionosphere Coupling

Christos Haldoupis

Received: 25 January 2011 / Accepted: 11 May 2011 / Published online: 10 June 2011

Abstract This paper provides a comprehensive update on sporadic *E* layers that is placed in the context of atmosphere-ionosphere coupling, exemplified here by the fundamental wind-shear theory processes that govern sporadic *E* layer formation and variability. Some basics of *windshear* theory are provided first, followed by a summary of key experimental results, their interpretation and physical understanding. The emphasis is placed on the wind shear control of the diurnal and sub-diurnal variability and altitude descent of sporadic *E* layers and the key role behind these properties of the diurnal and semidiurnal tides. Furthermore, the paper summarizes recent observations that establish a role also for the planetary waves in sporadic *E* layer occurrence and long-term variability. The possible mechanisms behind this interaction are examined and evidence is presented which shows that planetary waves affect sporadic *E* layers indirectly though the amplitude modulation of tides at lower altitudes in the MLT region. Only a brief mention is made about gravity wave effects on sporadic *E*, which apparently exist but cannot be as crucial in layer forming as thought in the past. There is now enough evidence to suggest that mid- and low-latitude sporadic *E* is not as "sporadic" as the name implies but a regularly occurring ionospheric phenomenon. This may suggest that the sporadic *E* layer physics can be incorporated in large-scale atmosphere-ionosphere coupling models.

Keywords Sporadic E layers · Windshear theory · Atmosphere-ionosphere coupling · Atmospheric tides · Planetary waves · Gravity waves

1 Introduction

The name *Sporadic E* and its abbreviation E_s refer to thin layers of metallic ion plasma which form in the dynamo region of the Earth's ionosphere, mostly between 100 and 125 km, where ion motion is controlled mainly by collisions with the neutrals, thus the

C. Haldoupis (✉)
Department of Physics, University of Crete, Heraklion, Greece
e-mail: chald@physics.uoc.gr

ions move with the local winds while electrons remain strongly magnetized. Since the invention of ionosonde in the nineteen thirties, sporadic E has been investigated extensively from ground, by means of analyzing ionosonde and incoherent scatter radar observations (e.g., see reviews by Whitehead 1989, and Mathews 1998 and textbook by Kelley 2009), and from in situ measurements made with rocket-borne sensors (e.g., see an early paper by Smith 1970, and more recent ones, for example by Fukao et al. 1998; Pfaff et al. 1998; Earle et al. 2000, and more references therein).

The physics of Sporadic E relies on the so called *windshear* theory, in which vertical shears in the horizontal wind play a key role in forming these layers from long-lived metallic ions through ion-neutral collisional coupling and geomagnetic Lorentz forcing. Once formed, the layers tend to descend with time downwards below 100 km where they eventually disappear because three-body collisional recombination becomes increasingly effective in depleting them of their metallic ion content. At present, the consensus is that the layer formation and altitude descent is driven mainly by the global system of tidal winds in the lower thermosphere. They provide the vertical windshear convergence nodes for the layers to form at and then to be dragged downwards through their tidal phase velocity propagation. In this sense, Sporadic E constitutes a prominent paradigm of energetic coupling between the neutral atmosphere and the ionospheric magnetoplasma. The purpose here is to substantiate this property by presenting a comprehensive summary of observations showing that sporadic E is a regularly occurring phenomenon controlled mainly by large scale neutral atmosphere waves.

The present review paper deals only with long-term properties of sporadic E without covering fully the great diversity of E_s characteristics and short-term variability, which are dealt with in more detail elsewhere, e.g., see review papers by Whitehead (1989) and Mathews (1998) and numerous references cited therein. Before going to the main part of the paper, which reviews the observational effects of tidal and planetary waves on sporadic E_s, a few theoretical elements are presented first that summarize the present knowledge and understanding of the E_s phenomenon.

2 Elements of Sporadic *E* Theory and Understanding

The physics of sporadic E layer formation is dealt with in the "windshear theory", conceived and formulated by Whitehead (1961), and Axford (1963). It relies on the concept of ionization layer-forming in the dynamo region of the ionosphere through altitudinal ion convergence forced by vertical shears in the horizontal wind acting in the presence of the Earth's magnetic field. The material in this section is based on a recent tutorial review paper by Haldoupis (2011) and is included here for the purpose of reference and completeness.

The layer formation is controlled by ion dynamics expressed through the ion momentum equation. Following Chimonas and Axford (1968) and neglecting pressure gradient forces at E region heights as well as electric field forces at middle and low latitudes, the ion momentum equation at steady state is simplified to include only the ion-neutral collisional and geomagnetic forces:

$$m_i \nu_i (\mathbf{v}_i - \mathbf{U}_n) - e\mathbf{v}_i \times \mathbf{B} = 0, \tag{1}$$

where m_i and ν_i are the ion mass and ion-neutral collision frequency, $\mathbf{v}_i$ and $\mathbf{U}_n$ are the ion drift and neutral wind velocity vectors respectively, e is the electronic charge, and $\mathbf{B}$ the geomagnetic field. By using a geomagnetic south, geomagnetic east and vertically up Cartesian (x, y, z) coordinate system for the northern hemisphere, and using the notations of Mathews (1998) for the vectors $\mathbf{v}_i(u, v, w)$, $\mathbf{U}_n(U, V, W)$ and $\mathbf{B}(-B\cos I, 0, -B\sin I)$ and taking the vertical wind component $W \simeq 0$, Equation (1) can be solved for the vertical ion drift:

$$w = f_z V + f_m U, \tag{2}$$

where the dimensionless parameters f_z and f_m are the zonal and merdional "ion drift factors":

$$f_z = \frac{(\nu_i/\omega_i)\cos I}{1 + (\nu_i/\omega_i)^2}, \qquad f_m = \frac{\cos I \sin I}{1 + (\nu_i/\omega_i)^2}, \tag{3}$$

with I denoting the magnetic dip angle and ν_i/ω_i the ratio of ion-neutral collision frequency to ion gyrofrequency.

2.1 Plasma Convergence Mechanisms

The equations (3) for f_z and f_m define two different processes of vertical ion convergence that associate with the vertical shears in the zonal and meridional winds, respectively. These two mechanisms are illustrated in Fig. 1 for the northern hemisphere magnetic field geometry. The zonal wind shear (dV/dz) mechanism is sketched in Fig. 1a. It involves the horizontal component of the magnetic field $B_H = B\cos I$ and a vertical wind shear characterized by a westward wind above and an eastward wind below. As the ions drift with the wind to the west above (east below) with V_{west} (V_{east}) they are Lorentz-forced $eV_{west}B_H$ ($eV_{east}B_H$) also to move downwards (upwards), gradually converging at an angle to the wind shear null, where $V = 0$, to form a layer. Figure 1b to the right sketches the meridional wind shear (dU/dz) mechanism which requires a northward wind above and a southward wind below. Here, the ions are wind-driven horizontally while at the same time are constrained by the Lorentz force to gyrate perpendicularly to the inclined magnetic field lines. As a result, the ions finally are forced to move along the magnetic field lines with the wind velocity

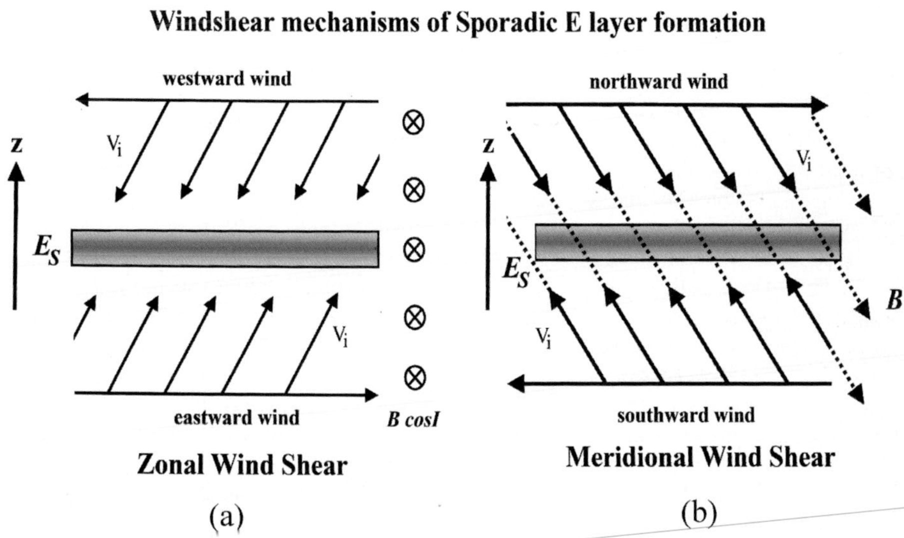

Fig. 1 Sketches illustrating the zonal (*left*) and meridional (*right*) windshear theory mechanisms of sporadic E layer formation. See text for more details

component $U_{north} \cos I$ ($U_{south} \cos I$), converging towards the wind shear null at $U = 0$ to form a layer. Note that the electrons are strongly magnetized ($\omega_e \gg \nu_e$), thus they are not affected directly by neutral winds. In the ion convergence process electrons simply follow the ions by moving along the magnetic field lines in order to maintain plasma neutrality.

Sporadic E layers live long times (from a few to several hours), implying that the windshear ion convergence process cannot rely on ambient molecular ion build up, because these ions have short lifetimes as they neutralize quickly at E region altitudes via dissociative recombination. Therefore, the E_s layers are attributed to metallic atomic ions of meteoric origin undergoing slow radiative recombination and thus having long lifetimes, ranging from a few days at ~120 km to a few hours at ~95 km (e.g., see MacDougall et al. 2000). For details on the relationship of sporadic E and metallic ions see Haldoupis et al. (2007).

It is important to stress that neither of the two ion-convergence mechanisms work efficiently either at the magnetic equator or the auroral zones. At the equator the magnetic field is horizontal, thus in the zonal wind shear process the ions are Lorentz-forced to move vertically but they cannot converge into a layer because are kept back by the magnetized electrons in order for the plasma to remain locally neutral. In the case of an equatorial meridional wind shear, the ions cannot converge vertically to a layer because no Lorentz force acts upon them since they move along the magnetic field lines. On the other hand, both windshear mechanisms are inefficient also at high magnetic latitudes (auroral zones). There the magnetic dip angles are large, thus the horizontal magnetic field component which is involved in the zonal windshear process is small, whereas in the meridional windshear process the driving wind component along the magnetic field is small. As a result of these geometrical constraints, the strongest and more frequent layers occur at midlatitudes. As for the sporadic E layers seen occasionally at high latitudes, these are attributed mostly to the relatively large auroral electric fields there, which in this case do contribute significantly in ion motion and thus cannot be excluded from (1).

2.2 Plasma Convergence Times

The layer-forming efficiency of the zonal and meridional windshear mechanisms differ at a given altitude because of different ion drift factor dependences on ν_i/ω_i. As it turns out (e.g., see Haldoupis 2011), both mechanisms are equally effective at around 125 km, where $\nu_i \sim \omega_i$, while at lower (upper) altitudes the zonal (meridional) windshear mechanism becomes dominant. At upper heights the ion-neutral collision frequency is reduced which makes the ions more magnetized, therefore the ability of a zonal wind to move them across the magnetic field is weakened. On the other hand, the meridional windshear process, which moves the ions along the inclined magnetic field direction, works better when the ions are strongly magnetized therefore it dominates at upper heights where $\nu_i \ll \omega_i$. At lower E region altitudes the ions become less magnetized, which reduces the action of the Lorentz force. This affects both windshear mechanisms but the effect is more severe in the meridional rather than the zonal windshear process. It turns out that the meridional windshear effects on vertical ion motion become minimal below 115 km, thus E_s layer-forming at altitudes below 115 km is governed largely by vertical shears in the zonal wind. Observations, which are to be shown later in the paper, show that the zonal wind shear process can still affect the layers down to altitudes as low as 90 km, and even lower.

Equations (2) and (3) can be used to estimate the convergence times t_z and t_m, needed for ions to move vertically a distance Δz, as a function of altitude and neutral wind velocity for the zonal and meridional wind shear mechanisms, respectively:

$$t_z = \frac{1 + (\nu_i/\omega_i)^2}{V(\nu_i/\omega_i)\cos I}\Delta z, \qquad t_m = \frac{1 + (\nu_i/\omega_i)^2}{U \cos I \sin I}\Delta z. \tag{4}$$

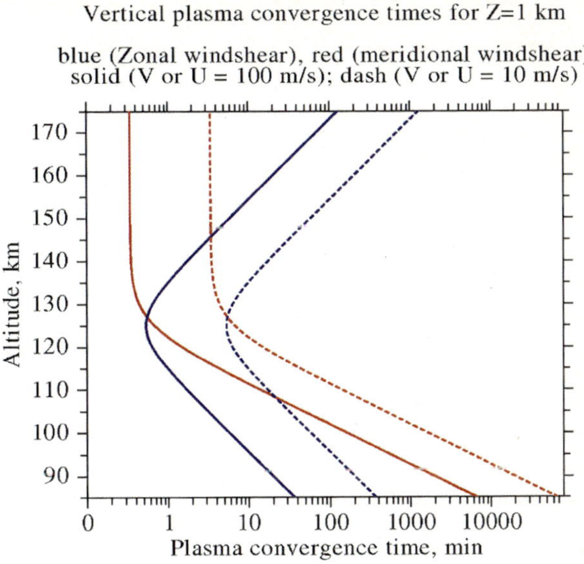

Fig. 2 Estimates of ion convergence times at different altitudes in the *E* region ionosphere for both the zonal and meridional wind shear mechanisms, for strong (*solid*) and weak (*dash*) zonal and meridional wind driving conditions

Such time estimates are useful not only in assessing the relative importance of the ion-convergence mechanisms with altitude but also in understanding the altitudinal layer descent, to be presented and discussed later.

Figure 2 shows estimates of vertical ion convergence times for a typical midlatitude location and an assumed ν_i/ω_i altitude profile. The latter was deduced from estimates given by Bishop and Earle (2003) which referred to metallic ion plasma with a mean ionic mass of 40 AMU (atomic mass units). The ion convergence times were computed for $\Delta z = 1$ km for two cases; a strong and a weak zonal and meridional wind components of 100 m/s and 10 m/s, respectively. As one may infer, a typical meridional wind of 50 m/s can force the ions to drift 1.0 km vertically in a few minutes at altitudes above 120 km, whereas below 100 km this requires times of a few hours. On the other hand, a zonal wind is much more efficient in forcing the ions to converge faster at lower altitudes compared to an equally strong meridional wind there. For example, at 110 km the ions need ~10 min to move one kilometer vertically under the forcing of a 50 m/s zonal wind, but ~100 min for a meridional wind of equal strength. Note that both converging processes act to transport vertically metallic ions by sweeping them in altitude and accumulating them into a layer that keeps re-forming and enhancing itself as it descents following the downward propagation of the wind shear convergence node.

It is important to realize that vertical ion-convergence becomes vertical ion-divergence for a windshear polarity opposite of that forming a layer, for instance in the case of a zonal (meridional) wind shear with an eastward (southward) wind above and a westward (northward) wind below. In these cases, an existing layer can be broadened and possibly be broken by a vertical wind shear forcing the ions vertically apart, therefore acting as a "de-layering" process. As expected, the times required for windshear de-layering are about the same as for layer forming, which means that the layers can disperse faster (slower) at altitudes where they form faster (slower). This is the reason why layers remain fairly stable for many hours at lower heights, especially below 100 km where ion-convergence, or divergence, times are long. On the other hand, at upper heights near 125 to 130 km, layers are more variable and shorter-lived since ion-convergence (divergence) is much faster there.

Finally a few words are provided on ambipolar plasma diffusion, a process that acts against plasma convergence in the ionosphere. Plasma diffusion is ignored in the sporadic E forming process because the ambipolar diffusion coefficient $D_a = k_B(T_i + T_e)/m_i \nu_i$ is rather small at lower E region altitudes. The role of plasma diffusion becomes increasingly important in the upper E region where D_a increases because ion-neutral collisions decrease (e.g., D_a is typically $\sim$1200 m^2/s at 150 km as compared to $\sim$50 m^2/s at 100 km).

Plasma diffusion prevents a layer from forming when the characteristic diffusion time is smaller than the zonal, or meridional, windshear plasma convergence time. As shown by Haldoupis (2011), who compared typical estimates of plasma diffusion versus plasma convergence times, plasma diffusion is effective in opposing E_s layer formation above $\sim$150 km and $\sim$180 km for the zonal and meridional windshear processes, respectively. Thus, plasma diffusion becomes unimportant in layer forming at lower heights, say below 130 km, where narrow sporadic E layers are located. On the other hand, ambipolar plasma diffusion is the physical reason why sporadic E layers are wider at higher E region altitudes (as seen for example in the Arecibo incoherent scatter radar observations), despite the fact that meridional wind shear convergence there is rather fast in forming a layer (e.g., see Fig. 2). This also explains why strong shears in the F region meridional wind cannot form there narrow layers.

3 Atmospheric Wave Effects in Sporadic E

The validity of the windshear theory is supported by various results obtained during the long period of E_s research (e.g., see reviews by Whitehead 1989, and Mathews 1998). This implies that neutral wind shears in the lower thermosphere are the key driving force behind the sporadic E phenomenon, which makes E_s a good paradigm to be evoked when it comes to emphasize the importance of atmosphere-ionosphere coupling. Furthermore, results obtained over the last several years made it clear that the wind shears required for E_s formation and variability are relating with atmospheric waves, and the purpose of the present paper is to review the role of these large scale waves and their importance.

The involvement of atmospheric waves and their dynamics in E_s formation and variability is implied by the periodicities imprinted in ionosonde E_s measurements. An example is illustrated in Fig. 3, which is based in Figs. 1 and 3 of Haldoupis et al. (2004). The upper panel displays time series of hourly means of sporadic E critical frequency $foEs$, obtained from a fast sampling (2 and 5 minute runs) ionosonde, operated in the Aegean island of Milos continuously from June 27 to September 4, 1996. The periodogram spectra of this ionosonde time series (shown in Fig. 1 of Haldoupis et al. 2004 but not here), disclosed that the amplitude spectrum in the 3 to 36 hours period was dominated by peaks at 24, 12, and 8 hours, which were identified, with additional analysis, to associate with the diurnal, semidiurnal and terdiurnal atmospheric tides. On the other hand the 1.5 to 20 days amplitude spectrum of the same $foEs$ time series (also shown in Fig. 1 of Haldoupis et al. but not here) was characterized by periodicities which had no relationship to geomagnetic activity but compared well with atmospheric normal Rossby wave modes predicted by theory, that is, planetary waves (PW) with periods of about 2, 5, 10 and 16 days (e.g., see Forbes 1994).

The tidal- and PW-like periodicities present in the $foEs$ times series in Fig. 3 are also identified in the wavelet period-time-spectrograms shown here in the middle and bottom panels, corresponding to the 3 to 36 hour (tidal) band and the 1.5 to 20 days (PW) band, respectively. The amplitudes of the tidal periodicities in the middle panels, mostly around

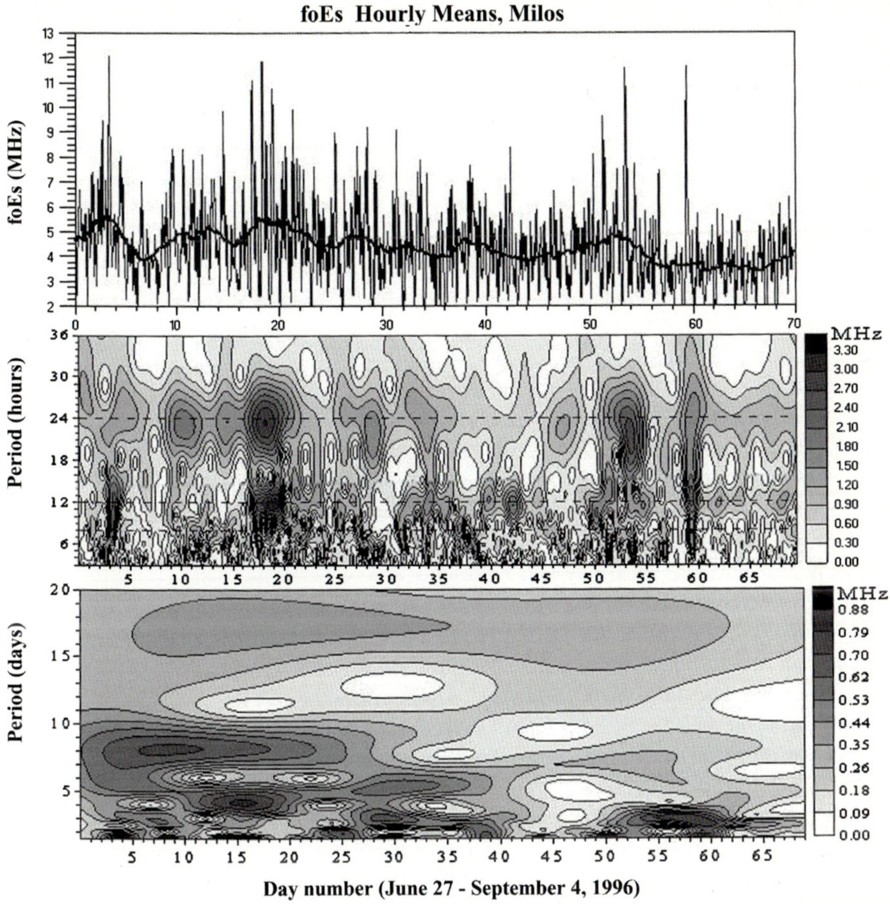

Fig. 3 The *top panel* displays time series of $foEs$ hourly means measured by an ionosonde in the Aegean island of Milos continuously for about 70 days. The *middle* and *bottom panels* show wavelet spectrograms corresponding to the $foEs$ time series, computed for the period intervals of 3 to 36 hours and 1.5 to 20 days, respectively. This figure illustrates that $foEs$ contains periodicities that compare well with atmospheric tidal and planetary wave periods, and possibly also with gravity waves

24- and 12-hour periods, were found to be modulated in time with periods of a few to several days which are approximately in accord with PW-like variations seen in the bottom panel (for details see Haldoupis et al. 2004). Shown also in the middle panel spectrogram are many short-lived spectral bursts corresponding to $foEs$ oscillations with periods shorter than 6 hours. Such shorter periods compare well with atmospheric gravity wave (GW) periods, although the likely association of these periodicities with GW cannot be certified from these ionosonde data alone.

In the following we review observations and coupling processes which are involved in the relationship between atmospheric waves and sporadic *E* layers. The emphasis is placed on the effects of tidal and planetary waves, for which there is strong supporting evidence and reasonable physical understanding.

4 Tidal Effects on Sporadic *E* Layers

Sporadic *E* observations, made with the powerful Arecibo incoherent scatter (ISR) radar and regular ionosonde measurements at midlatitudes, established that atmospheric tides are fundamental in the day to day formation and altitudinal motions of E_s (e.g., see review by Mathews 1998 and references therein). In this section, the basic role of tides and their effects on E_s will be summarized and discussed, with the emphasis placed on some key observations and our present physical understanding, examined within the framework of the windshear theory discussed earlier.

4.1 Arecibo ISR Observations

The Arecibo ISR studies, which refer at lower midlatitudes (Geog. Lat. ~18° N; Magnetic dip ~50°), revealed a key role for the diurnal and semidiurnal tides in the generation and altitude descent of E_s. This led Mathews to introduce the term "tidal ion layers" as more proper for the description of sporadic *E* layers. Figure 4 shows typical examples of diurnal Arecibo Sporadic *E* traces showing that E_s follows regular daily patterns governed by the dominating 24- and 12-hour tidal modes in the lower thermosphere. Each panel in Fig. 4 corresponds to a 24-h local day and represents a height-time-intensity (HTI) plot, where "intensity" here refers to the logarithm of the (positive) vertical electron density gradient, $\log(dNe/dz)$. This type of display is useful in detecting the altitude location of the layers quite accurately during the course of a 24-hour local day, and thus can help identify the tidal variability and descent of layers with time. The upper and lower parts in Fig. 4 correspond to two different periods of observations, each made of two consecutive days of continuous Arecibo ISR operation. As seen, there is a daily trace pattern in E_s that exhibits diurnal and semidiurnal variability in both time and altitude.

The tidal variability and descent of E_s layers at Arecibo were studied statistically by Christakis et al. (2009), using a large ISR data set of ~140 days of radar operation, distributed over several years and covering evenly all seasons. The findings show on the average a three layer daily pattern prevailing, with some differences, in all seasons, that is, for a given 24-hour local day there is usually a set of three separate E_s traces. These include a diurnal trace at lower altitudes below about 110 km and two much weaker semidiurnal-like upper altitude traces, a daytime trace and a less frequent and weaker nighttime trace, which appear at about 140 km prior to local noon and midnight, respectively. The lower altitude diurnal layer trace descends with a speed of ~1.0 km/h, which agrees reasonably well with the phase velocity of the theoretical diurnal tidal mode S(1, 1) of vertical wavelength $\lambda_z \simeq 28$ km, which was shown (e.g., see Harper 1977) to dominate the winds below 110 km at lower midlatitudes. The upper altitude semidiurnal-like layers descend faster, with speeds of ~2–4 km/h, towards lower heights where they often merge there with the slowly descending diurnal layer below. This repeating E_s trace pattern shows that sporadic *E* exhibits a regularity in both formation and descent which is in agreement with tidal windshear dynamics.

In line with the windshear theory outlined earlier, the Arecibo observations show that it is the vertical tidal wind profiles that supply the wind shears for the layers to form first at higher altitudes, where vertical ion convergence is much faster, and then to be dragged down by the wind shear that moves downwards with the phase velocity of the tidal waves. Christakis et al. (2009) performed numerical simulations by solving the windshear theory equation (2) for a simplified diurnal and semidiurnal zonal tidal wind of a given amplitude and vertical wavelength λ_z. The results suggested that the S(1, 1) 24-h tide with a vertical

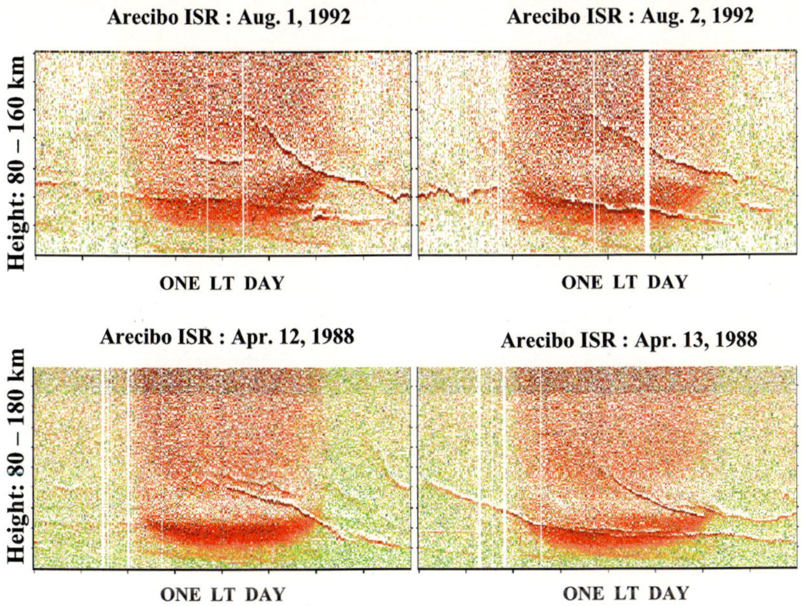

Fig. 4 Arecibo incoherent scatter radar HTI (height-local day time-log(dNe/dz) intensity) plots for 2 time intervals, each made of two consecutive day continuous radar runs. The local day E_s traces display a regular and repeating pattern in layer occurrence and altitude descent that follows diurnal and semidiurnal tidal-like periodicities

wavelength $\lambda_z \sim 25$ km controls fully the formation and descent of metallic E_s layers at low altitudes below $\sim$110 km. On the other hand, the higher altitude layers and their altitude descent was approximately accounted for by using the semidiurnal tidal mode S(2,6) with $\lambda_z \sim 35$ km.

4.2 Ionosonde Observations

The tidal effects on E_s were recognized also in ionosonde observations first by MacDougall (1974) and later by Wilkinson et al. (1992), and Szuszczewicz et al. (1995). More recently, Haldoupis et al. (2006) have studied these effects in detail by applying a new methodology of ionosonde height-time-intensity (HTI) analysis, which allowed the investigation of sporadic E layer tidal vertical motion and variability at different ionosonde frequency bands. The ionogram HTI display has the merits of the common RTI (range-time-intensity) radar plot which monitors signal dynamics over a given time interval. To study tidal variations and periodicites, the HTI plot was computed for a 24-hour local day for an ionosonde frequency band below the critical E_s frequency. Here, only the main findings of this work are summarized, whereas the interested reader can access the details in Haldoupis et al. (2006). The results of this study are exemplified in Fig. 5. The upper panels shows HTI plots based on ionograms measured in the island of Milos during the summer of 1996, while the lower panel shows E_s simulations obtained by applying basic windshear theory, using an over-simplified tidal wind system in the lower thermosphere.

The Milos ionogram HTI analysis revealed on the average a pronounced semidiurnal periodicity in layer descent and occurrence. This is characterized by a daytime layer starting at $\sim$115 km near 06 hours LT that moves downwards to altitudes below 100 km at $\sim$18

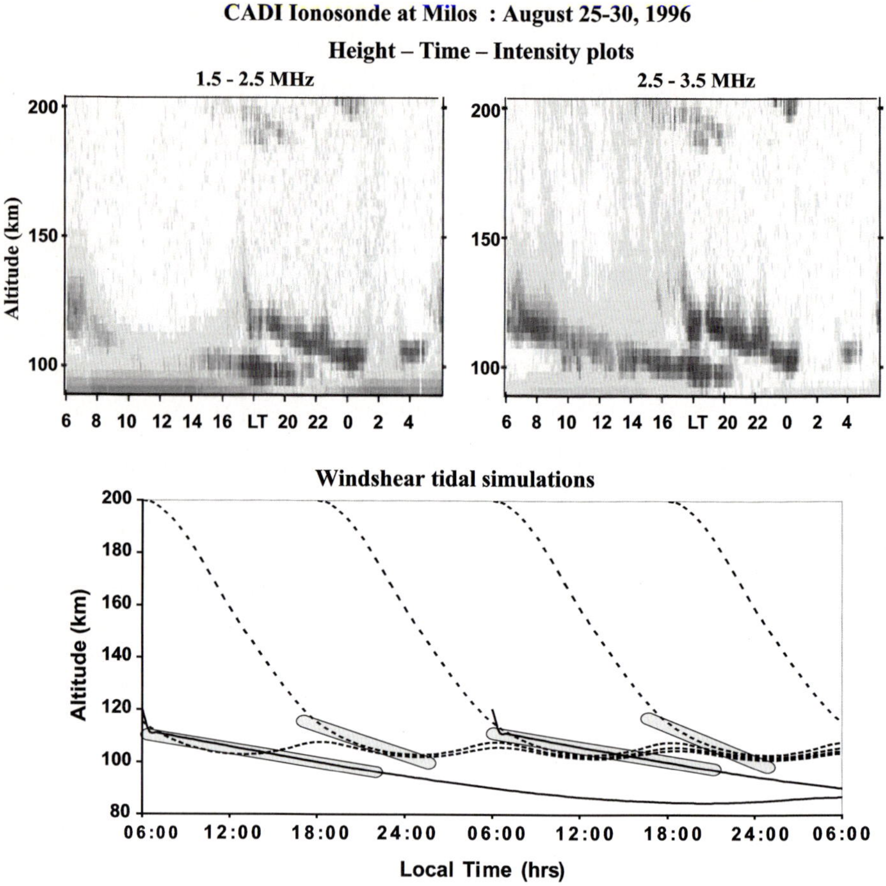

Fig. 5 The *top panels* show typical ionosonde HTI plots averaged here over the same 6 consecutive day interval from August 25 to 30, 1996. The descending E_s traces below ~125 km have a repeating daily pattern. There is a longer lasting daytime E_s trace starting near 120 km at about 06 h LT, followed by a nighttime one 12 h later. Both have a downward slope corresponding to E_s descent rates of about 1.0 and 2.0 km/h, respectively. The *bottom panel* shows simulation results based on windshear theory and a simple tidal wind system made of a S(2,3) semidiurnal tide in the upper E region and a S(1,1) diurnal tide in the lower E region, below 120 km. The *solid* and *dashed curves* correspond to E_s layer trajectories that form inside convergent nodes of vertical wind shears in the tidal winds that move downwards with the phase velocities of the tides. The *inclined shaded bars* represent the anticipated traces of forming E_s, which suggests a good agreement with the ionosonde observations in the upper HTI plots. For more details see text and paper by Haldoupis et al. (2006)

hours LT, at a time when a nighttime layer appears in the ionograms above at ~125 km. The nighttime layer moves also down but at a higher descent rate (1.8 to 2.2 km/h) than the daytime layer (0.8 to 1.2 km/h). In line also with the Arecibo observations, the nighttime E_s is weaker (in terms of critical frequencies $foEs$) and intermittent, while it has an overall shorter duration. The lower altitude layer is dominated by a diurnal periodicity. The HTI plots often show the daytime and nighttime E_s to connect in the upper E region with weak traces which occur with a semidiurnal, and at times a terdiurnal, periodicity. These upper E region layers, which identify with 12-h tides in the Arecibo ISR and classify as "descending intermediate layers" (DIL), e.g., see Mathews and Bekeny (1979), play a parenting role in reinforcing the narrow sporadic E layers at lower altitudes below about 115 km.

The key HTI ionosonde analysis findings were interpreted by applying a methodology, introduced first by Chimonas and Axford (1968), which relies on the windshear theory and a wind system that includes the sum of a semidiurnal and diurnal tide in the E region. To numerically simulate the intermediate and sporadic E layer vertical motion, (2), which is a first order differential equation, $dz/dt = w(z,t)$ is solved by using a harmonic tidal zonal wind profile $V(z,t)$ set equal to

$$V = V_0 \exp\frac{z - z_0}{2H} \cos\left[\frac{2\pi}{\lambda_z}(z - z_0) + \frac{2\pi}{T}(t - t_0)\right], \quad (5)$$

for both the 24-and 12-h tides. Here, T and λ_z are the period and vertical wavelength of the tidal wind, H is the scale height in the lower thermosphere, z_0 is a lower altitude boundary and t_0 is a free parameter representing the initial phase. The ion-neutral collision frequency profile used in this simulation is the same as the one used also to produce Fig. 2, that is, it is based on ν_i/ω_i altitude data given by Bishop and Earle (2003).

The results of the simulation, which are reproduced from Fig. 9 of Haldoupis et al. (2006), are shown in the lower panel of Fig. 5. The simulated E_s traces show that the observations, examples of which are shown in the upper panels, can be interpreted by considering the downward propagation of wind shear convergent nodes that associate with the theoretical S(1,1) diurnal tide in the lower E region and S(2,3) semidiurnal tide in the upper E region, which have a vertical wavelength λ_z of 25 km and 90 km, respectively. The daytime E_s is attributed to the confluence of semidiurnal and diurnal tidal convergence nodes, while the short duration nighttime layer is associated with the intrusion of the daytime DIL into lower E region heights trapped at a vertical wind shear convergence node as it descends with the phase velocity of the semidiurnal tide. Note that the shaded bars in the bottom panel of Fig. 5, which are sketched to represent the observed E_s traces with time, have slopes near 1.0 km/h and 1.8 km/h for the longer daytime and shorter nighttime sectors respectively, which are in fair agreement with the measured HTI trace descent rates.

Despite the severe limitations of the ionosonde technique, the findings outlined above are in reasonable agreement with several of the Arecibo ISR properties of layer descent and occurrence. Also, they compare well with recent findings on tidal effects on higher midlatitude E_s obtained by Arras et al. (2009) by means of using a new technique applied on low earth orbiting (LEO) satellite GPS (global positioning system) radio occultation data.

4.3 Physical Interpretation

Taking into consideration the observations and theoretical predictions discussed previously, a picture emerges which is summarized below.

First, the upper altitude daytime and nighttime weaker E_s layers are taken to be controlled by semidiurnal tides and to associate with the intermediate descending layers, which often are seen to detach from the F region bottomside (Mathews 1998). These layers form at shear convergence nulls at higher E region altitudes (say, 150 to 180 km) but become narrow only below about 135 km because plasma diffusion, which counteracts ion-convergence and plasma accumulation, becomes rather ineffective below 140 km. As a wind shear convergence node propagates downwards with the phase velocity of the tide, a layer "sticks" with a shear convergence null only if it forms fast enough compared to the time required for the null to phase-propagate downward a distance less or equal to the layer's width. As seen in Fig. 2, The layers from rapidly in the upper E region and therefore tend to constantly form at a wind shear convergence null as it moves downwards. This is manifested by the observed steady negative slopes seen in E_s traces above about 115 to 120 km (see examples in Fig. 4). Below these heights however, the situation changes gradually because ion-neutral collisions become increasingly effective in opposing/delaying ion convergence. As a result, layer descent slows down (trace slopes in Figs. 4 and 5 start curving) because ion convergence time becomes increasingly larger (Fig. 2), therefore a layer cannot form fast enough to remain inside a convergence null, as it does at upper E region heights. In fact, the layers lag steadily behind the downward propagating tidal convergence null in their downward motion, therefore they keep descending at rates progressively smaller than the vertical tidal phase velocity. This slow descent continues till a divergent tidal node that follows an ion convergence node catches up with E_s to impose a possible de-layering effect because the converged ions tend now to disperse vertically. Depending upon the amplitude and phase velocity of the semidiurnal tide, there is finally a lower altitude near 110 km at which the ion neutral collision frequency is high enough so there is not enough time for the 12-h tide to affect severely the layer. At these altitudes and below, the semidiurnal tides and their associated ion convergence (or divergence) wind shears "pass through" a stagnating layer without impacting much on it.

The upper altitude semidiurnal tidal layers may merge with lower altitude metallic ionization to contribute in the formation of the prominent 24-h sporadic E layer that is controlled by the diurnal S(1, 1) tide. Relative to the semidiurnal tides, the 24-h S(1, 1) tide has larger (wind speed) amplitudes (e.g., see Harper 1977) and a shorter vertical wavelength so that it phase-propagates downwards slower than the 12-h tides, therefore it has the time needed to fully control E_s layer formation and descent from about 110 km down to below 90 km. Although, the E_s picture can at a given day become more complicated relative to the one just described, because of the complexity of atmospheric dynamics and the confluence of coexisting tidal modes, as well as possible gravity wave and neutral wind instability effects, the overall pattern of sporadic E formation and descent remains well defined and predictable.

5 Planetary Wave Effects on Sporadic *E* Layers

Besides the undisputed role of atmospheric tides on sporadic E layer formation and variability, which is reviewed in the previous section, there is also now clear evidence in favor of a planetary wave role as well, a fact that escaped attention for many years in the long-going research of sporadic E. Note that planetary waves (PW) are very large horizontal scale oscillations in neutral density, pressure, and temperature, which propagate zonally and vertically in the mesosphere-lower thermosphere (MLT) region, having long quasi-periods near the so called normal Rossby wave modes with periods 2, 5, 10 and 16 days (e.g., see Forbes 1994).

5.1 Observational Evidence

The first indirect clue for a link between E_s and PW was provided by midlatitude field-aligned backscatter observations near 50 MHz, obtained with the MU (Mesosphere and Upper atmosphere) radar in Japan by Tsunoda et al. (1998), and the SESCAT (Sporadic E SCATter experiment) in Greece, as reported by Voiculescu et al. (1999, 2000). In these studies, the PW-like periodicities are observed in backscatter occurrence which are known to associate closely with strong sporadic E layers at midlatitude (e.g., see Hussey et al. 1998), thus they have been attributed to PW modulatory effects on sporadic E layer intensities.

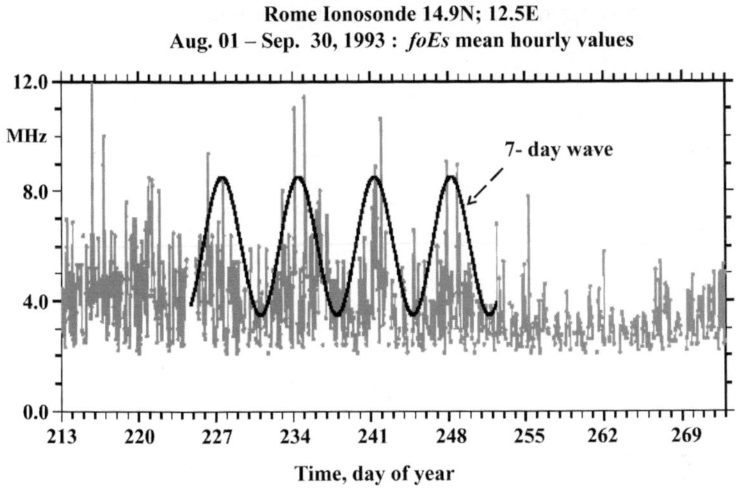

Fig. 6 Sporadic *E* critical frequency *foEs* time series measured by the Rome ionosonde during August–September 1993. A long period deep modulatory oscillation is evident in the E_s data from about August 15 to September 15, as emphasized by the superposition in the plot of a 7-day period pure sine wave. This periodicity occurred at the same time when a conspicuous 7-day period PW event was observed in MLT winds

The first strong evidence which established the existence of a PW role on E_s variability was obtained by Haldoupis and Pancheva (2002), who analyzed ionosonde data during the occurrence of a conspicuous, truly global, PW event in order to identify if concurrent PW-like periodicities are present in sporadic *E* as well. This event was a large amplitude 7-day period westward propagating PW that occurred during the second half of August and well into September 1993. It was detected by several ground radar and satellite wind observations in the MLT region and studied extensively (e.g., see Meyer and Forbes 1997 and Clark et al. 2002 and more references therein).

The first inspection of the ionosonde E_s data during this 7-day PW event period was revealing. A deep 7-day modulation in the magnitude of the E_s critical frequency *foEs* times series was present in the data recorded by several ionosonde stations in the northern hemisphere. This reassuring fact is illustrated here in Fig. 6 for the Rome ionosonde *foEs* times series from August 1 to September 30, 1993. As seen, a pronounced 7-day period modulation is prevailing in E_s intensity from middle of August to mid-September, 1993.

Haldoupis and Pancheva (2002) analyzed the *foEs* time series from eight midlatitude ionosonde stations, stretching over a large longitudinal band from ~50° E to ~160° W, to find a strong 7-day periodicity in the ionosonde data occurring simultaneously with the 7-day PW reported elsewhere. By using independent methods for times series signal processing, they computed identical estimates for the propagation direction, zonal wave number and the phase velocity of the 7-day wave that were all in agreement with those reported from radar and satellite neutral wind MLT observations. These findings provided the first definitive evidence proving that planetary waves play an important role in the physics of midlatitude sporadic *E* layers.

5.2 A Direct PW-E_s Mechanism: PW Horizontal Plasma Convergence

In an effort to provide a physical understanding for the association between PW and E_s, Shalimov et al. (1999) developed a new mechanism and a model based on the concept of

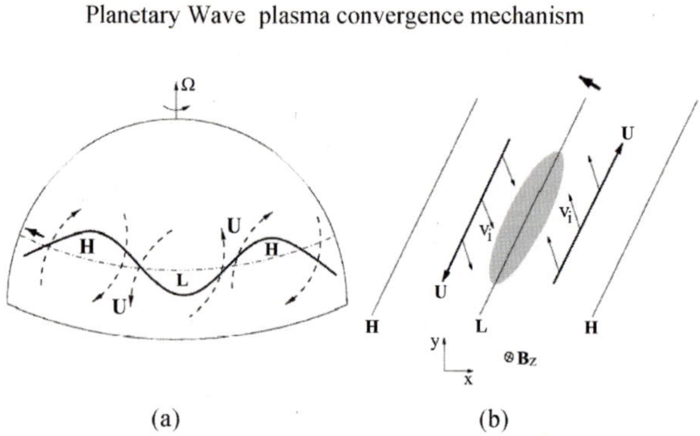

Fig. 7 (**a**): Visualization of a westward propagating planetary wave in the Northern hemisphere (adapted from Holton 1982). (**b**): Horizontal plasma convergence and accumulation in regions of wind shears produced by cyclonic winds in the propagating planetary wave to the left. This occurs in the E region where ion motion is controlled by the neutrals and the geomagnetic Lorentz force that deflects them inward

metallic E region plasma accumulation driven by planetary wave horizontal wind shears. In this process, the plasma is forced to converge horizontally and accumulate inside huge areas of positive vorticity set up by cyclonic neutral wind shears within a PW in the lower thermosphere. The mechanism and its effects are described here only in brief. One can find more details in Shalimov et al. (1999) and a rigorous theoretical treatment in Shalimov and Haldoupis (2002).

The mechanism is sketched in Fig. 7. Figure 7a to the left shows a traveling PW in the lower thermosphere, depicted from the PW visualization provided by Holton (1982). It is characterized by alternating areas of low and high pressure, which, under conditions of geostrophic balance, are accompanied by cyclonic and anticyclonic vortices in the wind field and thus large-scale wind shears in the horizontal plane. Figure 7b to the right shows a horizontal wind shear associated with a cyclonic wind vortex and its accumulating effect on the ambient plasma. Since the ion-neutral collision frequency, ν_{in} exceeds the ion gyrofrequency, ω_i at lower E region heights, the ions drift along with the neutral wind velocity $\mathbf{U}$ but in this way also experience a Lorentz force which deflects them at an angle to the wind velocity. In this way, the vertical magnetic field component $B_z = -B \sin I$ at a midlatitude location of magnetic Dip angle I in the northern hemisphere, will force the ions to move horizontally and converge inwards (outwards) in areas of cyclonic (anticyclonic) motion where vorticity is positive (negative). Under these conditions, the horizontal motion of the ions in the PW formation shown in Fig. 7a leads to plasma accumulation inside cyclonic vortices and plasma depletion inside anticyclonic wind motion.

The PW accumulation mechanism is conceptually similar as the vertical zonal windshear mechanism of E_s formation in Fig. 1a but with the geometry "turned on its side". As shown quantitatively by Shalimov et al. (1999), the large scales associated with planetary waves reduce plasma diffusion, which in turn reinforces the PW plasma accumulation process. This mechanism can lead to significant long lifetime (metallic) plasma accumulation inside large areas in the midlatitude E region, thus it complements the vertical wind shear process so that strong E_s layers can form more efficiently and frequently, alternating with weak (or no) layers inside areas of anticyclonic PW winds of plasma depletion, as the PW propagates westwards. This model provides a physical base for understanding the pronounced long-term variability (with periods of days) in occurrence and intensity of sporadic E layers. The validity of this mechanism however requires that the PW are present in higher than usual MLT altitudes, that is up to 120 km, and that the metallic ions there have lifetimes comparable with the periods of the driving PW.

5.3 An Indirect PW-E_s Mechanism: PW-modulation of Tides

Pancheva et al. (2003) continued the work started by Haldoupis and Pancheva (2002) in an effort to understand the physical reason behind the PW-E_s relationship. Towards this objective, these authors correlated two concurrent data sets which included radar-measured MLT neutral winds and ionosonde E_s recordings for the August–September 1993 strong 7-day PW event. The neutral wind data were hourly means of zonal and meridional winds measured by the medium frequency (MF) radar in Saskatoon, Canada (52°N, 107°W), and the meteor radar (MR) near Sheffield, UK (52°N, 2°W). The wind measurements were compared with $foEs$ observations made with ionosondes located as near to the radars as possible, that is, Boulder, Colorado (40°N, 105°W) $foEs$ data for the American sector (Saskatoon), and Lannion, France (49°N, 3°W) data for the West European sector (Sheffield).

Time and spectral analysis techniques verified that the 7-day wavelike variation was present in all data sets, that is, both the meridional neutral wind and the sporadic E critical frequencies $foEs$. Also, correlative analysis provided no convincing evidence in favor of a direct PW role in E_s generation. Next, and given that tides play a decisive role in forming E_s layers, the authors examined the possibility of an indirect PW mechanism acting upon sporadic E, probably through the modulation of atmospheric tides. This was reinforced as an option also by the fact that interactions between tides and PW are well known to take place in the middle atmosphere and that PW mediate their signatures in the ionosphere by modulating the tides which, contrary to planetary waves, penetrate deep into the upper ionosphere (e.g., Teitelbaum and Vial 1991; and Forbes et al. 1995; Pancheva 2000, and more references therein). These considerations led to the invention by Pancheva et al. (2003) of a mechanism that appears capable of explaining the observed relationship between planetary waves and sporadic E layers. Next, the evidence in favor of this mechanism is described in brief.

The interaction between PW and atmospheric tidal waves is a nonlinear process which resembles, for example, amplitude modulation (AM) in communication systems. If this is happening in the 1993 7-day PW event, then one would have expected to find two secondary waves having frequencies respectively equal to the sum and difference of the frequencies of the interacting primary (tidal and planetary) waves. In terms of periods, this means for the 7-day PW that the secondary waves must have periods: (a) for the 12-h tide near 11.1–11.2 hours and 12.9–13.0 hours, and (b) for the 24-h tide near 21.0 and 28.0 hours. This scenario was clearly verified by Pancheva et al. (2003), who used wave decomposition and correlation analyses to find that such secondary waves were indeed imbedded not only in the Saskatoon and Sheffield semidiurnal and diurnal tides but also in the 24- and 12-h tidal-like periodicities present also in the sporadic E critical frequencies $foEs$.

The key findings of Pancheva et al. (2003) are exemplified in Fig. 8. As shown in Fig. 8a, which includes the graphs in the left and middle panels, a 7-day tidal modulation was found to be present in both, the amplitude spectra of the zonal (left graphs) and meridional (middle graphs) winds measured in Saskatoon (upper panels) and UK (bottom panels) for the 30 day period, from day 235 to 265, 1993 when the 7-day PW was strongest. The two arrow pairs around the 12-h and 24-h tidal peaks indicate the position in the spectrum of the secondary peaks mentioned above. As evidenced in both the Saskatoon and UK longitudinal sectors, the measured tidal winds are 7-day amplitude modulated, with the secondary waves being stronger in the zonal component of the semidiurnal tide and the meridional component of

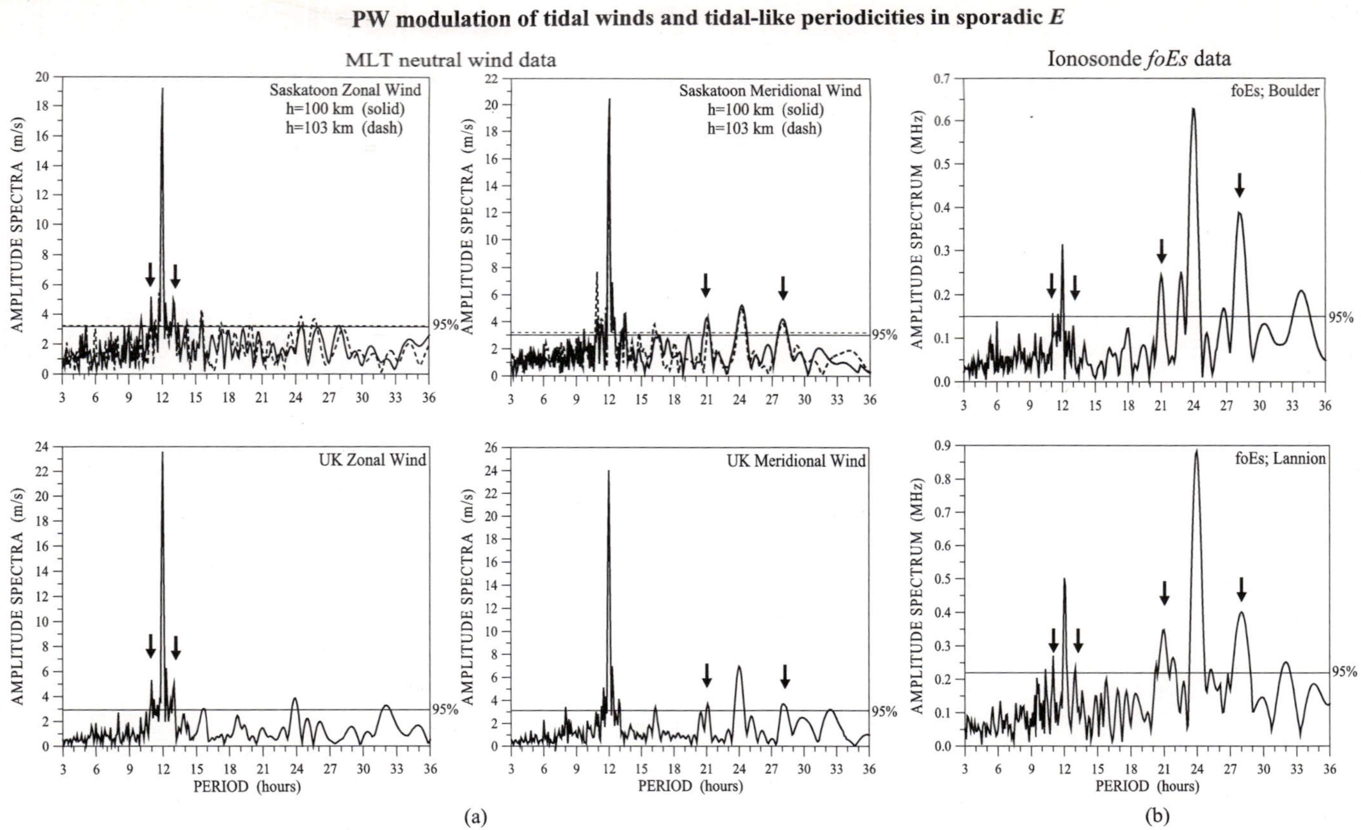

Fig. 8 The *upper* (*lower*) *panels* include tidal wind and sporadic E $foEs$ amplitude spectra obtained from time series data measured in the American (European) sector, respectively, during the August–September 1993 strong 7-day PW event. (**a**): The *arrows* indicate secondary wave peaks caused by a 7-day amplitude modulation of the diurnal and semidiurnal tides in the neutral wind. (**b**): The same 7-day modulation is also present in the 24-h and 12-h periodicities present in the $foEs$. See text for more details

the diurnal tide. The tidal modulations are found to be present also in time-period-amplitude spectrograms (not shown here) obtained by wavelet transform analysis of the tidal time series decomposed from the raw data after filtering the 7-day PW in order to avoid any artificial coupling between the tidal amplitudes and the mean wind (for more details and results see Pancheva et al. 2003).

Next, the same methodology was used in the concurrent $foEs$ time series to reveal, interestingly, that the same 7-day modulation found in the tidal winds was also clearly evident in the $foEs$ 12- and 24-hour tidal-like periodicities. This is illustrated in Fig. 8b which shows the $foEs$ amplitude spectra for Boulder (top panel) and Lannion (bottom panel) ionosondes. As seen, the secondary peaks were present about the 12-h and 24-h periodicities in $foEs$ at both stations, been much stronger in the diurnal periodicity. This suggested strongly that the 24- and 12-hour periodicities in $foEs$ were both 7-day amplitude modulated, in the same way as the diurnal and semidiurnal tidal winds, a fact that was confirmed by spectrogram wavelet transform analysis as well.

Although the study of Pancheva et al. (2003) cannot exclude a direct PW role on E_s, like the one discussed previously and depicted in Fig. 7, it shows clearly that E_s layers can be affected indirectly through the action of the tidal winds whose amplitudes are strongly modulated by PW, apparently through a nonlinear PW-tidal interaction process at the MLT region. The PW tidal modulation mechanism, as the base behind a physical understanding for the relation between planetary waves and sporadic E layers, was tested and confirmed in a follow up study by Haldoupis et al. (2004), by applying data analysis techniques used in neutral atmospheric dynamics on long times series of summertime ionosonde data. Also more evidence in favor of the results by Haldoupis and Pancheva (2002) and Pancheva et al. (2003) were published next by Haldoupis and Pancheva (2006) and other researchers (e.g., see Zuo and Wan 2008 and more relevant references therein).

6 Short Term Variability in Sporadic E Layers

The present paper is meant to be an overview of atmospheric wave effects on Sporadic E layers at large-scale, focusing on tidal and planetary wave influences. Variability of sporadic E at short scales with periods ranging from minutes to a few hours and spatial scales from meters to tens of kilometers, is a broad topic of active research that includes various phenomena and numerous observations (e.g., see Miller and Smith 1978 for a detailed Arecibo ISR study of short-scale irregular structure in sporadic E, the review of Mathews 1998 and more references therein). To review the subject of short term E_s variability is a major task that is well beyond the scope of, and the space available for, the present paper. Here, we include only a few general remarks on shorter scale E_s variations, mostly in relation with the role of gravity waves and instabilities.

There exist a wealth of gravity waves in the lower thermosphere, with periods from a few hours down to the Brunt-Vaissala period of 5 min, which may also act as a source of modulatory effects on E_s occurrence and intensity. If gravity waves have sufficiently large amplitudes and wind shears then they can affect the vertical motion of plasma locally and even increase/decrease horizontal transport within the layer itself, as argued for example theoretically by Chimonas (1971). In principle, gravity waves can intervene and alter the regular tidal forcing of E_s-forming process in reinforcing or disrupting its course and therefore could be responsible for the irregular variability in sporadic E, which is present at times in ionosonde (Spread E_s) and ISR observations. However, the unpredictable occurrence of gravity waves as well as their dynamic evolution and breaking at lower thermospheric heights makes it difficult for their effects on E_s to be either measured or modeled and thus be quantified.

On the other hand, filtering techniques applied on Arecibo ISR observations (e.g., see Djuth et al. 2004; Livneh et al. 2007 and more references therein) revealed the presence of continuous quasiperiodic perturbations of a few percent in thermospheric electron densities, having gravity wave-like periods in the 20 to 60 min range. HTI traces of these waves show up mostly in the upper E region and the F region having large negative dh/dt slopes due to fast downward phase propagation with high descent speeds, say higher than 100 km/h. Although the wind shears associated with such waves may affect at times upper altitude sporadic E layers (e.g., see Mathews et al. 1997), their effect is expected to be marginal on E_s below about 120 km, apparently because the waves propagate quickly through a layer, thus there is not enough time for their wind shears to impact much of vertical ion convergence or divergence effects on it.

Finally, there are also other forms of complex variability and wave-like structuring in sporadic E which are attributed to neutral density and/or plasma density instabilities, as well as localized electrodynamic processes. A neutral instability mechanism which can perturb severely E_s layers is the wind shear or Kelvin-Helmholtz instability which can produce short scale overturning structures, called Kelvin-Helmholtz billows (e.g., see Larsen 2000). Also short (meter to decameter) scale plasma instabilities, like the two-stream and gradient-drift instabilities (e.g., Haldoupis et al. 1997; Hysell et al. 2002) can lead to the generation to electrostatic plasma irregularities which are field aligned and can cause strong backscatter of radar signals incident perpendicular to the earth's magnetic field. The short scale neutral and/or plasma inhomogeneities in sporadic E is a complex topic of neutral and/or plasma turbulence, which is also beyond our scope here.

7 Summary and Concluding Comments

The present paper is an overview of sporadic E layer properties that originate within the general concept of neutral atmosphere-ionosphere coupling. First, a brief summary on E_s windshear theory basics is given. There, the emphasis is placed on the mechanisms of neutral wind shear vertical ion convergence that operate in the dynamo E region of the ionosphere. They may form narrow ionization layers under the action of a wind-driven geomagnetic Lorentz force in the presence of metallic ions provided by continuous meteoric bombardment and ablation.

Next, incoherent scatter radar and ionosonde observations are summarized, which establish that the vertical wind shears needed for E_s formation are mostly provided by the solar tidal winds, particularly the diurnal and semidiurnal tides. In general, a daily sporadic E layer pattern prevails in the form of: (1) a strong and slowly descending layer at lower E region altitudes, below 110 km, which is mastered by the global 24-hour period solar diurnal tide, and (2) two weaker layers at higher altitudes, a daytime and nighttime one, which descend faster with time and merge below with the stronger lower altitude layer. The upper altitude layers are controlled mainly by the semidiurnal tides which are known to dominate in the upper E and F region ionosphere. The characteristic E_s layer descent with time is driven mainly by the vertical motion of the windshear ion convergence nodes, where a layer forms, which move downwards with the phase velocity propagation of the tides.

Although there are more causes which can combine to affect the overall layer occurrence and intensity, the role of the diurnal and semidiurnal, and at times the terdiurnal, tides though their vertical wind shears remains dominant and is of fundamental importance in sporadic

E layer physics. The observations, especially those made by the powerful Arecibo ISR, support the notion that E_s is variable but "non-sporadic" phenomenon, a fact that has so far been overlooked. The regular or non-sporadic character of E_s implies that the physics of the layers should, in principle, be also included in atmosphere-ionosphere coupling models.

In addition, this paper reviews resent observations revealing that, besides tidal waves, planetary waves play also a role in E_s occurrence and long-term variability, a fact that went relatively un-noticed in the long-going research of sporadic *E*. Two possible mechanisms, which refer to direct and indirect PW forcing on sporadic *E* layer plasma are reviewed and discussed. The emphasis is placed on experimental evidence showing that E_s layers are likely affected indirectly by planetary waves. This is done through the usual action of tides on E_s which now are amplitude-modulated by planetary waves through a non-linear interaction process acting at altitudes below 100 km, in the mesosphere-lower-thermosphere (MLT) region.

Finally, the paper touches briefly and generally on short-term variability and dynamics of sporadic *E*, caused for example by shorter scale processes such as gravity waves and neutral and/or plasma instability processes. This is a major topic relating with various phenomena of E_s microphysics which was outside the scope of the present paper and needs to be dealt with in a separate review.

Acknowledgements I thank all my co-authors for their valuable contributions in our joint sporadic *E* research over the last several years. Especially, I wish to thank Dora Pancheva for a long-lasted fruitful collaboration and interaction. The completion of this paper was supported by ELKE, University of Crete, grant nr 3116.

References

W.I. Axford, The formation and vertical movement of dense ionized layers in the ionosphere. J. Geophys. Res. **68**, 769 (1963)

C. Arras, C. Jacobi, J. Wickert, Semidiurnal tidal signature in sporadic E occurrence rates derived from GPS radio occultation measurements at higher midlatitudes. Ann. Geophys. **27**, 2555–2563 (2009)

R.L. Bishop, G.D. Earle, Metallic ion transport associated with midlatitude intermediate layer development. J. Geophys. Res. **108**(A1), 1019 (2003)

G. Chimonas, W.I. Axford, Vertical movement of temperate zone sporadic E layer. J. Geophys. Res. **73**, 111 (1968)

G. Chimonas, Enhancement of sporadic E by horizontal transport within the layer. J. Geophys. Res. **76**, 4578 (1971)

N. Christakis, C. Haldoupis, Q. Zhou, C. Meek, Seasonal variability and descent of mid-latitude sporadic E layers at Arecibo. Ann. Geophys. **27**, 923–931 (2009)

R.R. Clark, M.D. Burrage, S.J. Franke, A.H. Manson, C.E. Meek, N.J. Mitchel, H.G. Muller, Observations of 7-day planetary waves with MLT radars and UARS/HRDI instrument. J. Atmos. Sol.-Terr. Phys. **64**, 1217 (2002)

F.T. Djuth, M.P. Sulzer, S.A. Gonzales, J.D. Mathews, J.H. Elder, A continuum of gravity waves in the Arecibo thermosphere. Geophys. Res. Lett. **31**, L16801 (2004). doi:10.1029/2003GL019376

G.D. Earle, T.J. Kane, R.F. Pfaff, S.R. Bounds, Ion layer separation and equilibrium zonal winds in midlatitude sporadic E. Geophys. Res. Lett. **27**(4), 461–464 (2000). doi:10.1029/1999GL900572

J.M. Forbes, Tidal and planetary waves, in *The Upper Mesosphere and Lower Thermosphere, a Review of Experiment and Theory*, ed. by R.M. Johnson, T.L. Killen (1994), p. 67

J.M. Forbes, M.E. Hagan, S. Miyahara, F. Vial, A.H. Manson, C.E. Meek, Yu. Portnyagin, Quasi-16-day oscillation in the mesosphere and lower thermosphere. J. Geophys. Res. **100**, 9149 (1995)

S. Fukao, M. Yamamoto, R.T. Tsunoda, H. Hayakawa, T. Mukai, The SEEK (Sporadic-E Experiment from Kyushu) campaign. Geophys. Res. Lett. **25**, 1761–1764 (1998)

C. Haldoupis, D.T. Farley, K. Schlegel, Type 1 echoes from the midlatitude E region ionosphere. Ann. Geophys. **15**, 908–917 (1997)

C. Haldoupis, D. Pancheva, Planetary waves and midlatitude sporadic E layers: Strong experimental evidence for a close relationship. J. Geophys. Res. **107** (2002). doi:10.1029/2001JA000212

C. Haldoupis, D. Pancheva, N.J. Mitchell, A study of tidal and planetary wave periodicities present in midlatitude sporadic E layers. J. Geophys. Res. **109**, A02302 (2004). doi:10.1029/2003JA010253
C. Haldoupis, C. Meek, N. Christakis, D. Pancheva, A. Bourdillon, Ionogram height-time intensity observations of descending sporadic E layers at mid-latitude. J. Atmos. Sol.-Terr. Phys. **68**, 539 (2006)
C. Haldoupis, D. Pancheva, Terdiurnal tidelike variability in sporadic E layers. J. Geophys. Res. **111**, A07303 (2006). doi:10.1029/2005JA011522
C. Haldoupis, D. Pancheva, W. Singer, C. Meek, J. MacDougall, An explanation for the seasonal dependence of midlatitude sporadic E layers. J. Geophys. Res. **112**, A06315 (2007). doi:10.1029/2007JA012322
C. Haldoupis, A tutorial review on sporadic E layers, in *Aeronomy of the Earth's Atmosphere and Ionosphere*. IAGA Special Sopron Book series, doi:10.1007/978-94-007-0326-1-29 (Springer, Berlin, 2011)
R.M. Harper, Tidal winds in the 100- to 200-km region at Arecibo. J. Geophys. Res. **82**, 3243 (1977)
J.R. Holton, *An Introduction to Dynamic Meteorology* (Academic Press, San Diego, 1982)
G.G. Hussey, K. Schlegel, C. Haldoupis, Simultaneous 50-MHz coherent backscatter and digital ionosonde observations in the midlatitude E region. J. Geophys. Res. **103**, 6991 (1998)
D.L. Hysell, M. Yamamoto, S. Fukao, Imaging radar observations and theory of type I and type II quasiperiodic echoes. J. Geophys. Res. **107**, 1360 (2002)
M.F. Larsen, A shear instability seeding mechanism for quasiperiodic echoes. J. Geophys. Res. **105**, 24931 (2000)
M.C. Kelley, *The Earth's Ionosphere: Plasma Physics and Electrodynamics*, 2nd edn. (Academic Press, San Diego, 2009)
D.J. Livneh, I. Seker, F.T. Djuth, J.D. Mathews, Continuous quasiperiodic thermospheric waves over Arecibo. J. Geophys. Res. **112**, A07313 (2007). doi:10.1029/2006JA012225
J.W. MacDougall, 110 km neutral zonal wind patterns. Planet. Space Sci. **22**, 545 (1974)
J.W. MacDougall, J.M. Plane, P.T. Jayachandran, Polar cap Sporadic E: part 2, modeling. J. Atmos. Sol.-Terr. Phys. **62**, 1169–1176 (2000)
J.D. Mathews, F.S. Bekeny, Upper atmosphere tides and the vertical motion of ionospheric sporadic layers at Arecibo. J. Geophys. Res. **84**, 2743–2750 (1979)
J.D. Mathews, M.P. Sulzer, P. Perillat, Aspects of layer electrodynamics from high-resolution ISR observations of the 80–270 km ionosphere. Geophys. Res. Lett. **24**(11), 1411–1414 (1997)
J.D. Mathews, Sporadic E: current views and recent progress. J. Atmos. Sol.-Terr. Phys. **60**, 413 (1998)
C.K. Meyer, J.M. Forbes, A 6.5-day westward propagating planetary wave: Origin and characteristics. J. Geophys. Res. **102**, 26173 (1997)
K.L. Miller, L.G. Smith, Incoherent scatter radar observations of irregular structure in mid-latitude sporadic E layers. J. Geophys. Res. **83**, 3761–3775 (1978)
D. Pancheva, Evidence for non-linear coupling of planetary waves and tides in the lower thermosphere over Bulgaria. J. Atmos. Sol.-Terr. Phys. **62**, 115 (2000)
D. Pancheva, C. Haldoupis, C.E. Meek, A.H. Manson, N.J. Mitchell, Evidence of a role for modulated atmospheric tides in the dependence of sporadic E on planetary waves, J. Geophys. Res. **108** (2003). doi:10.1029/2002JA009788
R.F. Pfaff, M. Yamamoto, P. Marioni, H. Mori, S. Fukao, Electric field measurements above and within a sporadic-E layer. Geophys. Res. Lett. **25**, 1769–1772 (1998)
S. Shalimov, C. Haldoupis, M. Voiculescu, K. Schlegel, Midlatitude E region plasma accumulation driven by planetary wave horizontal wind shears. J. Geophys. Res. **104**, 28207 (1999)
S. Shalimov, C. Haldoupis, A model of midlatitude E region plasma convergence inside a planetary wave cyclonic vortex. Ann. Geophys. **20**, 1193 (2002)
L.G. Smith, A sequence of rocket observations of nighttime sporadic E. J. Atmos. Sol.-Terr. Phys. **32**, 1427 (1970)
E.P. Szuszczewicz, R.G. Roble, P.J. Wilkinson, R. Hanbaba, Coupling mechanisms in the lower ionospheric-thermospheric system and manifestations in the formation and dynamics of intermediate descending layers. J. Atmos. Sol.-Terr. Phys. **57**, 1483 (1995)
H. Teitelbaum, F. Vial, On the tidal variability induced by non-linear interaction with planetary waves. J. Geophys. Res. **96**, 14169 (1991)
R. Tsunoda, M. Yamamoto, K. Igarashi, K. Hocke, S. Fukao, Quasiperiodic radar echoes from midlatitude sporadic E and role of the 5-day planetary wave. Geophys. Res. Lett. **25**, 951 (1998)
M. Voiculescu, C. Haldoupis, K. Schlegel, Evidence for planetary wave effects on midlatitude backscatter and sporadic *E* layer occurrence. Geophys. Res. Lett. **26**, 1105 (1999)
M. Voiculescu, C. Haldoupis, D. Pancheva, M. Ignat, K. Schlegel, S. Shalimov, More evidence for a planetary wave link with midlatitude E region coherent backscatter and sporadic E layers. Ann. Geophys. **18**, 1182 (2000)
J.D. Whitehead, Recent work on midlatitude and equatorial sporadic E. J. Atmos. Sol.-Terr. Phys. **51**, 401 (1989)

J.D. Whitehead, The formation of the sporadic E layer in the temperate zones. J. Atmos. Sol.-Terr. Phys. **20**, 49 (1961)
P.J. Wilkinson, E.P. Szuszczewicz, R.G. Roble, Measurements and modelling of intermediate, descending, and sporadic layers in the lower ionosphere: Results and implications for global-scale ionospheric-thermospheric studies. Geophys. Res. Lett. **19**, 95 (1992)
X.M. Zuo, W.X. Wan, Planetary wave oscillations in sporadic E layer occurrence at Wuhan. Earth Planets Space **60**, 647–652 (2008)

Space Sci Rev (2012) 168:463–473
DOI 10.1007/s11214-011-9847-z

Generation of Atmospheric Gravity Waves in the Polar Thermosphere in Response to Auroral Activity

S. Oyama · B.J. Watkins

Received: 29 June 2011 / Accepted: 21 October 2011 / Published online: 12 November 2011

Abstract Atmospheric gravity wave (AGW) is a typical phenomenon in the upper atmosphere. At mid/low latitudes, climatological sources such as unstable barometric activity in the troposphere play an important role to generate AGWs in the thermosphere. While these sources are also important at high latitudes, energy input from the magnetosphere has additional large contributions to AGW generation. This paper reviews previous studies of AGWs associated with auroral activity at high latitudes. Theoretical studies have indicated that Joule/particle heating and the Lorentz force are major processes for generating AGWs in the thermosphere. Many observations show that AGWs can propagate horizontally for thousands of km from the source region. The paper summarizes equations regarding AGW generation by Joule/particle heating and the Lorentz force, and discusses the relative importance of these two processes.

Keywords Atmospheric gravity wave · Aurora · High latitude · Thermosphere · Ionosphere

1 Introduction

Magnetic storm, ionospheric disturbance, and aurora are ionospheric signatures of disturbance in the magnetosphere. Electromagnetic and particle energy is associated with the solar wind that can directly cause such disturbances. However, some of the energy is transiently stored in the tail of magnetosphere and intermittently emitted, resulting in a magnetospheric substorm obtained mainly at high latitudes. The total energy causing individual substorms is around 10^{15}–10^{16} W s. Three important processes of the energy dissipation are: development of the ring current, auroral particle acceleration, and heating and plasma convection in the polar ionosphere. Approximately 10^{11}–10^{12} W of energy is directed to individual processes.

S. Oyama (✉)
Solar-Terrestrial Environment Laboratory, Nagoya University, F3-3(250), Furo Chikusa, Nagoya, Japan
e-mail: soyama@stelab.nagoya-u.ac.jp

B.J. Watkins
Geophysical Institute, University of Alaska Fairbanks, 903 Koyukuk Dr., Fairbanks, AK, 99775-7320, USA

If we assume that the heating/plasma convection energy is dissipated into the polar ionosphere at latitudes higher than 60°, the mean energy is of the order of 10^{-3} W/m^2, which is consistent with simulation results (Lu et al. 1995).

The polar ionosphere and thermosphere act as an energy and particle sink. The energy is used to heat and accelerate plasmas and neutral particles. These processes cause, for example, wind acceleration and generation of atmospheric gravity waves (AGWs) in the polar thermosphere with various spatiotemporal scales (Francis 1974; Kato et al. 1977; Hunsucker 1982; Hajkowicz 1991; Millward et al. 1993; Fujiwara et al. 1996; Fuller-Rowell et al. 1996; Hocke and Schlegel 1996; Richards and Wilkinson 1998; Smith 2000; Aruliah and Griffin 2001; Ishii et al. 2001, 2004; Oyama et al. 2001, 2008, 2010; Tsugawa et al. 2003, 2004; Aruliah et al. 2005; Nicolls and Heinselman 2007). AGWs generated in the auroral region can propagate horizontally for a long distance as detected with HF radars (Samson et al. 1990; Bristow et al. 1994) and GPS receiver networks (Ho et al. 1996, 1998; Tsugawa et al. 2003) as plasma manifestations of AGWs.

Theoretical calculations have indicated that there are at least three mechanisms in the polar thermosphere connected with the generation of AGWs: (1) perturbations in the auroral electrojet current, (2) thermospheric heating caused by highly energetic particles from the magnetosphere, and (3) rapid motions of the aurora (Chimonas and Hines 1970; Francis 1974; Brekke 1979; Williams et al. 1988; Walterscheid and Lyons 1992; Jing and Hunsucker 1993; Sun et al. 1995; Balthazor et al. 1997). The process (1) is in general more important than the process (2) (Hunsucker 1982; Hocke and Schlegel 1996). A theoretical investigation suggests that the Lorentz force tends to generate medium-scale AGWs and that Joule heating is more effective in generating large-scale AGWs (Jing and Hunsucker 1993). However, the relative contributions are not yet fully understood (Chimonas and Hines 1970; Williams et al. 1988). This paper reviews (i) effects of the Joule/particle heating and Lorentz force on AGWs' generation in the auroral thermosphere and (ii) horizontal propagation from the source.

2 AGW Generation

2.1 Particle Heating and Joule Heating

The ion production rate p_a due to auroral particle precipitation is related to the electron density by the equation (Banks 1977)

$$p_a = \alpha n_e^2 + \frac{dn_e}{dt} \tag{1}$$

where α is the dissociative recombination rate at E-region altitudes, and n_e is the electron density (m^{-3}). The time differential term can be neglect if the loss rate αn_e^2 is longer than the time scale of interest, which is for example a time resolution of the electron density variations. Since the profile of the rate of energy deposition from auroral electrons q_a is proportional to p_a, q_a, also known as the particle heating rate, this can be rewritten as (Wickwar et al. 1975; Vickrey et al. 1982)

$$q_a = 5.6 \times 10^{-6} \alpha_{eff} n_e^2 \tag{2}$$

where $\alpha_{eff} = 2.5 \times 10^{-6} \exp(-z/51.2)$ (z is height in km). The particle heating rate can temporarily exceed the Joule heating rate, particularly in an auroral arc; but on average the Joule heating rate is larger than the particle heating rate because of its extent over larger areas of the ionosphere and longer duration.

Energy transfer between the ionosphere and the thermosphere in the collisional process can be expressed by comparing the electromagnetic energy flux $\mathbf{j} \cdot \mathbf{E}$ and the Joule heating rate $\mathbf{j} \cdot \mathbf{E}'$ (Brekke and Rino 1978; Thayer et al. 1995; Fujii et al. 1998).

$$\mathbf{j} \cdot \mathbf{E} - \mathbf{j} \cdot \mathbf{E}' = \mathbf{u} \cdot (\mathbf{j} \times \mathbf{B}) \tag{3}$$

where **j**, **E**, **u**, and **B** are vectors of the current density, the electric field, the neutral wind velocity, and the magnetic field, respectively. $\mathbf{E}' = \mathbf{E} + \mathbf{u} \times \mathbf{B}$ is the total electric field as measured in a reference frame moving with the neutral wind. Since the current density perpendicular to **B** is given by

$$\mathbf{j}_{\perp} = \sigma_P \mathbf{E}' - \sigma_H \mathbf{E}' \times \mathbf{B}/B \tag{4}$$

where σ_P and σ_H are the Pedersen and Hall conductivities, respectively. The Joule heating rate $\mathbf{j} \cdot \mathbf{E}'$ can be written as

$$\mathbf{j} \cdot \mathbf{E}' = \sigma_P (\mathbf{E} + \mathbf{u} \times \mathbf{B})^2 = \sigma_P E^2 + \sigma_P |\mathbf{u} \times \mathbf{B}|^2 - 2\sigma_P \mathbf{u} \cdot (\mathbf{E} \times \mathbf{B}) \tag{5}$$

If the neutral wind blows in the same direction as the ion velocity but with a smaller speed, the sum of the second and the third terms in the right-hand-side of (5) can be negative. In this case $\sigma_P E^2$ is larger than the Joule heating rate (i.e. $\mathbf{j} \cdot \mathbf{E}' < \sigma_P E^2$). On the other hand, if the neutral-wind speed is considerably smaller than the ion speed, $\sigma_P E^2$ is almost equal to the Joule heating rate (i.e. $\mathbf{j} \cdot \mathbf{E}' \approx \sigma_P E^2$), which is likely in the polar ionosphere.

Joule heating of the ions, which is initiated by an electric field, may be computed by applying Ohm's law as indicated above. Subsequent increases of ion thermal energy are then transferred to the neutral gas. Since the ion density is much smaller than the neutral density, the ion temperature increases more rapidly than the neutral temperature at the onset of Joule heating. The neutral temperature increases relatively slowly compared to ion temperature increases. The neutral temperature increases then results in modulations of the pressure or the isobar. The first law of thermodynamics gives the rate of change of internal energy, η, as

$$\eta = \frac{dQ}{dt} - \frac{d\bar{W}}{dt} \tag{6}$$

Height-resolved data from the incoherent-scatter radar are available to calculate η from (2) and (5). dQ/dt, the rate of heat gained per unit volume, can be written as

$$\frac{dQ}{dt} = N_m C_p \frac{dT}{dt} = N_m C_p \frac{dT}{dl}\frac{dl}{dt} = N_m C_p \frac{dT}{dl} u_l \tag{7}$$

where N_m is the mass density of neutrals, C_p is the specific heat at constant pressure, T is the temperature of neutrals, and u_l is the wind flowing on the isobar in a direction l.

$$\frac{d\bar{W}}{dt} = \frac{dP}{dt} = \frac{dP}{dl}\frac{dl}{dt} = \frac{dP}{dl} u_l \tag{8}$$

is the rate at which work is done per unit volume, and P is the pressure. Substituting (7) and (8) into (6) gives u_l as

$$u_l = \frac{\eta}{N_m (C_p - k/(\langle m \rangle m_p)) dT/dl} \tag{9}$$

assuming $P = (N_m kT)/(\langle m \rangle m_p)$, where $\langle m \rangle$ and m_p are the mean neutral mass and the proton mass, respectively, and k is the Boltzmann constant. The gradient of N_m along l is assumed to be zero ($dN_m/dl = 0$). For example, horizontal winds of 100–300 m/s may be generated in the E region in association with the temperature gradient of $\sim 2 \times 10^{-4}$ K/m (Oyama et al. 2008).

Equation (9) represents wind along the isobar or l in association with natural ionospheric/thermospheric heating. Precisely speaking the pressure P also has some gradients in the direction perpendicular to l. However, the wind flux across the isobar is considered to be much smaller than the l component. Since l is almost parallel to the earth surface, u_l can be regarded as the geographically horizontal component of the wind. The horizontal wind acceleration contributes to equatorward propagation of AGW and TID (traveling ionospheric disturbance) from auroral latitudes (Fujiwara et al. 1996).

An assumption of no vertical wind is not always appropriate, for example in the vicinity of an auroral arc. The geographically vertical component of the wind flowing along the isobar may be comparable to the horizontal component (Oyama et al. 2008; Kurihara et al. 2009). To a reasonable approximation, the atmosphere is balanced in equilibrium between gravity and the vertical pressure gradient. Vertical displacement of an air parcel tends to be compensated by the pressure, temporarily destroying the balance with gravity; then, if no other force comes into play, the air parcel seeks to return to equilibrium by oscillating at a certain unique frequency, which is called the Brunt-Väisälä or buoyancy frequency. This frequency is a function of temperature, decreasing with height in the thermosphere: 3×10^{-2} and 1×10^{-2} Hz (corresponding to ~4 and ~10 minutes, respectively) at ~120 km and ~200 km height, respectively (Walterscheid et al. 2001). According to the dispersion relation proposed by Hines (1960), AGW frequency should be lower than the Brunt-Väisälä frequency. While the quantitative understanding of the generation mechanism is not well developed, there is experimental evidence to show thermospheric wind perturbations near the discrete auroral arc (Oyama et al. 2008, and references therein). If one calculates a spectrum from temporal and/or spatial data taken near the source region, the spectrum should not have specific peaks. However, some of the wave packets may grow to a monochromatic wave observed as an AGW propagating horizontally from the source region.

2.2 Ion Drag and Relative Importance to Joule Heating

The fundamental mechanism of the Joule heating process is collisions between ions and neutral particles initiated by an electric field. For ion drag, the collision process is also essential. Collisions result in increase of thermal and kinetic energy for Joule heating and ion drag, respectively. To be more specific regarding neutral particle motion, plasma energy is used to increase random velocity (as thermal motion) and flow velocity (as bulk motion with an average direction) for Joule heating and ion drag, respectively.

The Lorentz force can be written as

$$\mathbf{f}_L = \frac{\mathbf{j} \times \mathbf{B}}{\rho} \tag{10}$$

where ρ is the neutral mass density. While both the Lorentz force and Joule heating come into play to generate AGWs in the polar thermosphere, the relative importance (denoted as L/J in this paper) is not simply derived as the ratio of $\mathbf{f}_L$ to $\mathbf{j} \cdot \mathbf{E}'$ because we should take into account the ion-neutral collision frequency in a more precise manner. In the work by Chimonas and Hines (1970) L/J is expressed as

$$\frac{L}{J} = \frac{gHB}{C_L} \frac{\sigma_c}{j} \tag{11}$$

where σ_c and j are the Cowling conductivity and current density, respectively, and C_L is the phase speed of AGW. Brekke (1979) derives dependency of L/J on ion-neutral collision frequency ν_{in} and ion gyrofrequency Ω_i as shown in (12).

$$\frac{L}{J} \approx 15 \sqrt{\frac{\Omega_i^2 + \nu_{in}^2}{\Omega_i^2} \frac{1}{E}} \tag{12}$$

Since $\sqrt{(\Omega_i^2 + \nu_{in}^2)/\Omega_i^2}$ decreases with height, the relative importance L/J also decreases with height. This means that the contribution of Joule heating becomes larger with height. However, many observation and simulation studies suggest that the ion drag can modulate the horizontal wind pattern at polar F-region heights although the response is relatively sluggish at E-region heights (Conde and Smith 1998, and references therein). To clarify this issue, a consideration of the temperature dependence on L/J is helpful, although our understanding is not yet fully developed.

Richmond (1978) calculates individual contributions of the Lorentz force and Joule heating for generating AGWs:

$$L = \frac{\sigma_P B}{\rho} E_x u_y, \tag{13}$$

and

$$J = \frac{\sigma_P}{\rho} E_x^2 \frac{H'}{H\zeta}, \tag{14}$$

where E_x and u_y are one of the horizontal components of electric field and neutral wind velocity, respectively, H' is perturbation of the scale height H, and ζ is dimensionless parameter representing static stability. For simplicity one of the horizontal components is considered for the Lorentz force contribution. The relative importance L/J can be estimated in the first approximation form as

$$\frac{L}{J} = \frac{Bu_y}{E_x} \bigg/ \frac{H'}{H\zeta} \propto \frac{Bu_y}{E_x} \bigg/ \frac{T'}{T} \tag{15}$$

E_x is not equivalent to Bu_y in the lower thermosphere (below ~120 km) because of collisions between ions and neutral particles. There is a tendency for the wind velocity to be correlated with the Lorentz force and for the temperature perturbation to be correlated with Joule heating. Thus both terms come into play for generating AGWs in the lower thermosphere. However, in the upper thermosphere (corresponding to F-region heights), (15) may be more simplified as $L/J \propto T/T'$ because the horizontal neutral wind is more effectively accelerated along the ion velocity by the ion drag, that is, acceptable for assumption that Bu_y more quickly follows E_x variations. The neutral temperature can be perturbed by Joule heating but not by the Lorentz force. Therefore it is considered that Joule heating becomes more important for AGW generation in the upper thermosphere than the Lorentz force. The theoretical prediction of L/J made of data from the Chatanika incoherent-scatter radar was 0.1–10 for periods of moderately geomagnetic disturbance, and the average value of L/J was 2.37 for a mean electric-field magnitude of ~43 mV/m (Hunsucker 1982). The theoretical calculation by Brekke (1979) showed that the value of L/J at 140 km height was smaller than that at 100 km height by about two orders.

If we can estimate T'/T from measurements, one might think that T'/T increases with height because (12) suggests that L/J decreases with height. It is generally difficult to obtain the height-resolved neutral temperature from measurements; therefore ion temperature perturbations measured with incoherent-scatter radar have been studied as a tracer (Shibata and Schlegel 1993; Kirchengast et al. 1996). The results show that T_i'/T_i (T_i: ion temperature) decreases with height in the upper ionosphere. This trend is the opposite sense of the theoretical prediction by (12). However, T_i'/T_i increases with height for the case of lower-period band (14–60 minutes) in the lower thermosphere (Shibata and Schlegel 1993), which is consistent with (12).

3 AGW/TID Propagation

Energy input from the magnetosphere into the polar ionosphere/thermosphere has various temporal and spatial evolutions in the vicinity of the auroral arc much like complex arc structures. While the thermosphere results in temperature increases and wind acceleration from the energy input, the time constant of the thermospheric change depends on the energy transfer process. Furthermore, inertia, which plays an important role in the thermospheric wind dynamics below ~200 km (Lindzen 1967), delays the thermospheric response. It is

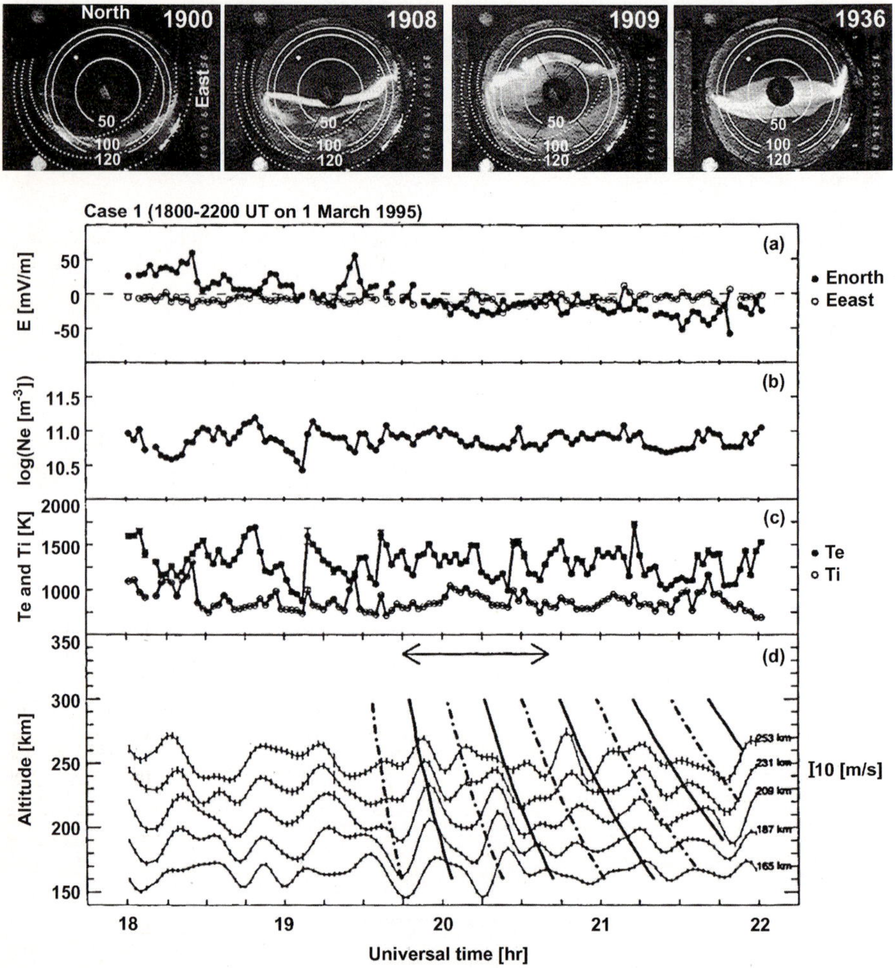

Fig. 1 *Top panel* shows all-sky camera images taken at Kilpisjärvi, Finland at 19:00, 19:08, 19:09, and 19:36 UT on 1 March 1995. The *three circles* shown with *solid lines* indicate distances of 50, 100, and 120 km from the zenith at Kilpisjärvi. The *dotted curves* indicate distances of 140, 200, and 290 km from the magnetic field line at Tromsø, Norway at an altitude of 110 km. *Four panels below* present temporal variations of the EISCAT Tromsø UHF radar data of (**a**) electric field, (**b**) electron density at 209 km, (**c**) electron and ion temperatures at 209 km, and (**d**) field-aligned component of the neutral wind velocity from 165 to 253 km heights. The modeled phase lines with minimum and maximum values are plotted with *thick dash-dotted* and *thick solid lines*, respectively. This study concluded that observed oscillations in the field-aligned neutral wind velocity was generated by poleward expansion of the auroral arc (after Oyama et al. 2001)

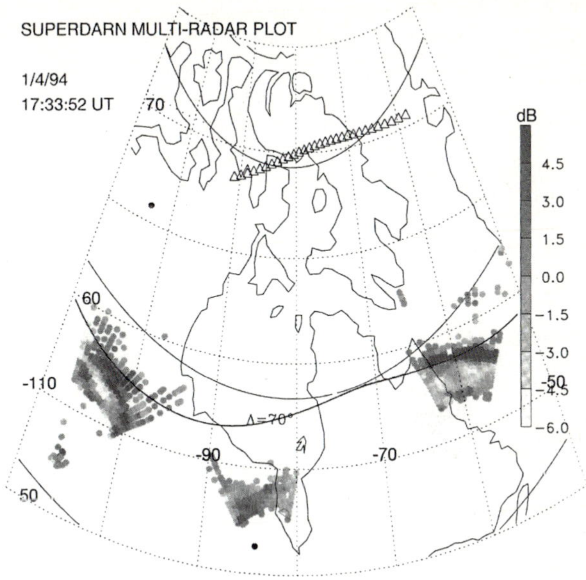

Fig. 2 A backscatter-echo map of SuperDARN radar data from 17:33 UT on January 4, 1994, showing the location of a gravity wave phase front and the estimated source region (marked by *triangles*) (after Bristow and Greenwald 1996)

thus generally difficult to find a source of the observed AGWs. However, some studies succeeded in identifying the source region of AGWs. Oyama et al. (2001) reported that AGWs at F-region heights were measured with the incoherent-scatter radar coincident with poleward expansion of the auroral arc (see Fig. 1). They estimated the horizontal phase speed, and compared with the distance to the arc measured with an all-sky camera in order to determine the source region. The horizontal wavelength of AGWs was comparable with the distance to the source (about 200 km) in their study. Thus in this case the waves propagated for only one wavelength until detected by the radar.

It is known that auroral AGWs can propagate horizontally for great distances. To detect them we need multiple instruments spaced horizontally or an instrument capable of observing over a wide area. The former examples are total electron content maps derived with GPS receiver networks (Tsugawa et al. 2003, 2004) and ionosonde networks (Maeda and Handa 1980; Hajkowicz and Hunsucker 1987; Hajkowicz 1990, 1991, 1999), and the latter example is an HF radar (Bristow et al. 1994, 1996; Bristow and Greenwald 1996). A case study of LSTIDs (large scale TIDs) measured with ionosondes in both hemispheres shows equatorward propagation of the wave, and its source region has been located along an L-shell with L-value of 4–5 (Hajkowicz and Hunsucker 1987). Statistics of TEC data in Japan for 45 months show that the occurrence rate of equatorward propagating LSTIDs increases with Kp index (Tsugawa et al. 2004). A horizontal map of the ground-backscatter power of the HF radar can be used to find the wave front (Bristow and Greenwald 1996) (see Fig. 2). Bristow et al. (1994) calculate propagation direction, speed, horizontal wavelength, and oscillation period from the HF radar data, and suggest that multiple waves from multiple directions are a common feature of the observations. While results of ionosonde and TEC present equatorward propagation of TIDs during periods of geomagnetic disturbance, statistics of the HF radar data suggest that TIDs can also propagate poleward from the auroral oval although there has been no observation of this yet.

Another important issue that requires more research is the damping rate, which is related to the energy dissipation rate. Tsugawa et al. (2003) calculated spatial variations of the LSTIDs' amplitude ratio along the propagation direction (see Fig 3). The e-folding distance

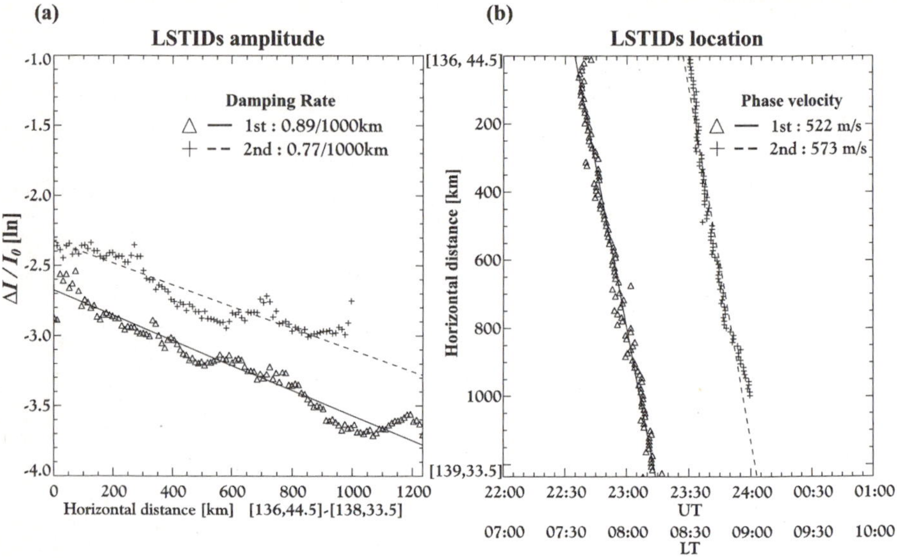

Fig. 3 Spatial variations of (**a**) the AGW-amplitude maxima and (**b**) temporal variations of their locations for the two LSTIDs. These maximum values and least-square fitted lines are represented by *open triangles* and a *solid line* for the first LSTID, and *crosses* and a *broken line* for the second LSTID, respectively (after Tsugawa et al. 2003)

was 961 km in the afternoon sector but smaller in the night sector. They concluded that the ion drag played an important role in damping LSTIDs.

4 Summary and Outstanding Questions

Atmospheric gravity wave at high latitudes is one of the thermospheric responses to the energy input from the magnetosphere. Joule heating, auroral particle heating, and ion drag play important roles in generating AGWs. However, debate remains about how much energy the thermosphere needs to generate AGWs, especially in the lower thermosphere (90–150 km heights). Previous results of numerical simulations provide strong evidence that it is difficult to reproduce wind and temperature fluctuations that simulate experimental values (observed with Fabry-Perot interferometer, incoherent-scatter radar, etc.) in the lower thermosphere unless the heating rate used in the model code is considerably larger and/or the heating duration is longer than a typical measured values (Sun et al. 1995; Shinagawa and Oyama 2006). Propagation distance of the AGW/TID from the source is also a question that remains to be answered. Since it may be difficult to identify the start/end point of a wave packet, the damping rate is more helpful for estimating the propagation distance. The damping rate is a key parameter for understanding the momentum dissipation process during wave propagation from high to mid/low latitudes.

Atmospheric gravity waves can be seen in the thermosphere even during geomagnetically quiet periods. Figure 4 shows a height-time profile of the field-aligned ion velocity measured with the European Incoherent Scatter (EISCAT) Svalbard radar. Periodic variations (at oscillation period of ~1 hour) are clearly seen from 00 to 04 UT on 14 August 2007 with changing sign. This is a very clear example of the AGW/TID in the thermosphere/ionosphere.

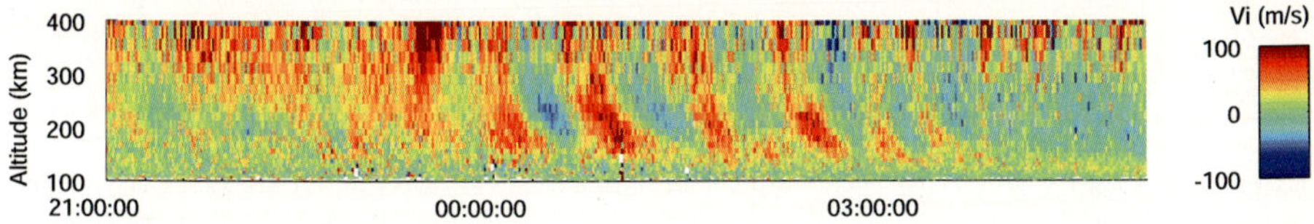

Fig. 4 Height-time profile of the field-aligned ion velocity measured with the EISCAT Svalbard radar from 21 UT on 13 August to 05 UT on 14 August 2007. Positive is upward along the magnetic field line. Wave patterns are clearly seen from 00 to 04 UT

Since magnetometer data suggest that the geomagnetic activity was quiet for this interval (not shown here), it is considered that the wave structure must have been generated independent of electromagnetic energy. The generation mechanism of this example is not clear but the effect of the lower atmosphere is an important issue to be understood separate from auroral effects.

Acknowledgements This research has been supported by a Grant-in-Aid for Scientific Research (2240-3010) from MEXT, Japan. We are indebted to the director and staff of EISCAT for operating the facility and supplying the data. EISCAT is an international association supported by research organizations in China (CRIPR), Finland (SA), Germany (DFG), Japan (STEL and NIPR), Norway (NFR), Sweden (VR), and the United Kingdom (STFC).

References

A.L. Aruliah, E.M. Griffin, Evidence of meso-scale structure in the high-latitude thermosphere. Ann. Geophys. **19**, 36–46 (2001)

A.L. Aruliah, E.M. Griffin, A.D. Aylward, E.A.K. Ford, M.J. Kosch, C.J. Davis, V.S.C. Howells, S.E. Pryse, H.R. Middleton, J. Jussila, First direct evidence of meso-scale variability on ion-neutral dynamics using co-located tristatic FPIs and EISCAT radar in Northern Scandinavia. Ann. Geophys. **23**, 147–162 (2005)

R.L. Balthazor, R.J. Moffett, G.H. Millward, A study of the Joule and Lorentz inputs in the production of atmospheric gravity waves in the upper thermosphere. Ann. Geophys. **15**, 779–785 (1997)

P.M. Banks, Observations of Joule and particle heating in the auroral zone. J. Atmos. Terr. Phys. **39**, 179–193 (1977)

A. Brekke, On the relative importance of Joule heating and the Lorentz force in generating atmospheric gravity waves and infrasound waves in the auroral electrojets. J. Atmos. Terr. Phys. **41**, 475–479 (1979)

A. Brekke, C.L. Rino, High-resolution altitude profiles of the auroral zone energy dissipation due to ionospheric currents. J. Geophys. Res. **83**(A6), 2517–2524 (1978)

W.A. Bristow, R.A. Greenwald, Multiradar observations of medium-scale acoustic gravity waves using the Super Dual Auroral Radar Network. J. Geophys. Res. **101**(A11), 24,499–24,511 (1996)

W.A. Bristow, R.A. Greenwald, J.C. Samson, Identification of high-latitude acoustic gravity wave sources using the Goose Bay HF radar. J. Geophys. Res. **99**(A1), 319–331 (1994)

W.A. Bristow, R.A. Greenwald, J.P. Villain, On the seasonal dependence of medium-scale atmospheric gravity waves in the upper atmosphere at high latitudes. J. Geophys. Res. **101**(A7), 15,685–15,699 (1996)

G. Chimonas, C.O. Hines, Atmospheric gravity waves launched by auroral currents. Planet. Space Sci. **18**, 565–582 (1970)

M. Conde, R.W. Smith, Spatial structure in the thermospheric horizontal wind above Poker Flat, Alaska, during solar minimum. J. Geophys. Res. **103**(A5), 9449–9471 (1998)

S.H. Francis, A theory of medium-scale travelling ionospheric disturbances. J. Geophys. Res. **79**, 5245–5260 (1974)

R. Fujii, S. Nozawa, N. Matuura, A. Brekke, Study on neutral wind contribution to the electrodynamics in the polar ionosphere using EISCAT CP-1 data. J. Geophys. Res. **103**(A7), 14,731–14,739 (1998)

H. Fujiwara, S. Maeda, H. Fukunishi, T.J. Fuller-Rowell, D.S. Evans, Global variations of thermospheric winds and temperatures caused by substorm energy injection. J. Geophys. Res. **101**, 225–239 (1996)

T.J. Fuller-Rowell, M.V. Codrescu, H. Rishbeth, R.J. Moffett, S. Quegan, On the seasonal response of the thermosphere and ionosphere to geomagnetic storms. J. Geophys. Res. **101**(A2), 2343–2353 (1996)

L.A. Hajkowicz, A global study of large scale travelling ionospheric disturbances (TIDs) following a step-like onset of auroral substorms in both hemispheres. Planet. Space Sci. **38**, 913–923 (1990)
L.A. Hajkowicz, Auroral electrojet effect on the global occurrence pattern of large scale travelling ionospheric disturbances. Planet. Space Sci. **39**, 1189–1196 (1991)
L.A. Hajkowicz, Monitoring ionospheric response to auroral electrojet activity from sub-auroral to equatorial latitudes in the East Asian-Australian longitudinal sector over a solar cycle (1978–1986). J. Atmos. Sol.-Terr. Phys. **61**, 857–866 (1999)
L.A. Hajkowicz, R.D. Hunsucker, A simultaneous observation of large-scale periodic TIDs in both hemispheres following an onset of auroral disturbances. Planet. Space Sci. **35**, 785–791 (1987)
C.O. Hines, Internal atmospheric gravity waves at ionospheric heights. Can. J. Phys. **38**, 1441–1481 (1960)
C.M. Ho, A.J. Mannucci, U.J. Lindqwister, X. Pi, B.T. Tsurutani, Global ionosphere perturbations monitored by the Worldwide GPS Network. Geophys. Res. Lett. **23**(22), 3219–3222 (1996)
C.M. Ho, A.J. Mannucci, L. Sparks, X. Pi, U.J. Lindqwister, B.D. Wilson, B.A. Iijima, M.J. Reyes, Ionospheric total electron content perturbations monitored by the GPS global network during two northern hemisphere winter storms. J. Geophys. Res. **103**(A11), 26,409–26,420 (1998)
K. Hocke, K. Schlegel, A review of atmospheric gravity waves and travelling ionospheric disturbances: 1982–1995. Ann. Geophys. **14**, 917–940 (1996)
R.D. Hunsucker, Atmospheric gravity waves generated in the high-latitude ionosphere: a review. Rev. Geophys. **20**, 293–315 (1982)
M. Ishii, M. Conde, R.W. Smith, M. Krynicki, E. Sagawa, S. Watari, Vertical wind observations with two Fabry-Perot interferometers at Poker Flat, Alaska. J. Geophys. Res. **106**, 10,537–10,551 (2001)
M. Ishii, M. Kubota, M. Conde, R.W. Smith, M. Krynicki, Vertical wind distribution in the polar thermosphere during Horizontal *E* Region Experiment (HEX) campaign. J. Geophys. Res. **109**, A12311 (2004). doi:10.1029/2004JA010657
J. Jing, R.D. Hunsucker, A theoretical investigation of sources of large and medium scale atmospheric gravity waves in the auroral oval. J. Atmos. Terr. Phys. **55**, 1667–1679 (1993)
S. Kato, T. Kawakami, D. St. John, Theory of gravity wave emission from moving sources in the upper atmosphere. J. Atmos. Terr. Phys. **39**, 581–588 (1977)
G. Kirchengast, K. Hocke, K. Schlegel, The gravity wave-TID relationship: insight via theoretical model-EISCAT data comparison. J. Atmos. Terr. Phys. **58**(1–4), 233–243 (1996)
J. Kurihara, S. Oyama, S. Nozawa, T.T. Tsuda, R. Fujii, Y. Ogawa, H. Miyaoka, N. Iwagami, T. Abe, K.-I. Oyama, M.J. Kosch, A. Aruliah, E. Griffin, K. Kauristie, Temperature enhancements and vertical winds in the lower thermosphere associated with auroral heating during the DELTA campaign. J. Geophys. Res. **114**, A12306 (2009). doi:10.1029/2009JA014392
R.S. Lindzen, Reconsideration of diurnal velocity oscillation in the thermosphere. J. Geophys. Res. **72**(5), 1591–1598 (1967)
G. Lu, A.D. Richmond, B.A. Emery, R.G. Roble, Magnetosphere-ionosphere-thermosphere coupling: effect of neutral winds on energy transfer and field-aligned current. J. Geophys. Res. **100**(A10), 19,643–19,659 (1995)
S. Maeda, S. Handa, Transmission of large-scale TIDs in the ionospheric F2-region. J. Atmos. Terr. Phys. **42**, 853–859 (1980)
G.H. Millward, S. Quegan, R.J. Moffett, T.J. Fuller-Rowell, D. Rees, A modeling study of the coupled ionospheric and thermospheric response to an enhanced high-latitude electric field event. Planet. Space Sci. **41**, 45–56 (1993)
M.J. Nicolls, C.J. Heinselman, Three-dimensional measurements of travelling ionospheric disturbances with the Poker Flat incoherent scatter radar. Geophys. Res. Lett. **34**, L21104 (2007). doi:10.1029/2007GL031506
S. Oyama, M. Ishii, Y. Murayama, H. Shinagawa, S.C. Buchert, R. Fujii, W. Kofman, Generation of atmospheric gravity waves associated with auroral activity in the polar *F* region. J. Geophys. Res. **106**(A9), 18,543–18,554 (2001)
S. Oyama, K. Shiokawa, J. Kurihara, T.T. Tsuda, S. Nozawa, Y. Ogawa, Y. Otsuka, B.J. Watkins, Lower-thermospheric wind fluctuations measured with an FPI during pulsating aurora at Tromsø, Norway. Ann. Geophys. **28**, 1847–1857 (2010)
S. Oyama, B.J. Watkins, S. Maeda, H. Shinagawa, S. Nozawa, Y. Ogawa, A. Brekke, C. Lathuillere, W. Kofman, Generation of the lower-thermospheric vertical wind estimated with the EISCAT KST radar at high latitudes during periods of moderate geomagnetic disturbance. Ann. Geophys. **26**, 1491–1505 (2008)
P.G. Richards, P.J. Wilkinson, The ionosphere and thermosphere at southern midlatitudes during the November 1993 ionospheric storm: a comparison of measurement and modeling. J. Geophys. Res. **103**(A5), 9373–9389 (1998)
A. Richmond, Gravity wave generation, propagation, and dissipation in the thermosphere. J. Geophys. Res. **83**(A9), 4131–4145 (1978)

J.C. Samson, R.A. Greenwald, J.M. Ruohoniemi, A. Frey, K.B. Baker, Goose Bay radar observations of earth-reflected, atmospheric gravity waves in the high-latitude ionosphere. J. Geophys. Res. **95**(A6), 7693–7709 (1990)

T. Shibata, K. Schlegel, Vertical structure of AGW associated ionospheric fluctuations in the E- and lower F-region observed with EISCAT—a case study. J. Atmos. Terr. Phys. **55**(4/5), 739–749 (1993)

H. Shinagawa, S. Oyama, A two-dimensional simulation of thermospheric vertical winds in the vicinity of an auroral arc. Earth Planets Space **58**, 1173–1181 (2006)

R.W. Smith, The global-scale effect of small-scale thermospheric disturbances. J. Atmos. Sol.-Terr. Phys. **62**, 1623–1628 (2000)

Z.P. Sun, R.P. Turco, R.L. Walterscheid, S.V. Venkateswaran, P.W. Jones, Thermospheric response to morningside diffuse aurora: high-resolution three-dimensional simulations. J. Geophys. Res. **100**, 23,779–23,793 (1995)

J.P. Thayer, J.F. Vickrey, R.A. Heelis, J.B. Gary, Interpretation and modeling of the high-latitude electromagnetic energy flux. J. Geophys. Res. **100**(A10), 19,715–19,728 (1995)

T. Tsugawa, A. Saito, Y. Otsuka, A statistical study of large-scale traveling ionospheric disturbances using the GPS network in Japan. J. Geophys. Res. **109**, A06302 (2004). doi:10.1029/2003JA010302

T. Tsugawa, A. Saito, Y. Otsuka, M. Yamamoto, Damping of large-scale traveling ionospheric disturbances detected with GPS networks during the geomagnetic storm. J. Geophys. Res. **108**(A3), 1127 (2003). doi:10.1029/2002JA009433

J. Vickrey, R. Vondrak, S. Matthews, Energy deposition by precipitating particles and Joule dissipation in the auroral ionosphere. J. Geophys. Res. **87**(A7), 5184–5196 (1982)

R.L. Walterscheid, L.R. Lyons, The neutral circulation in the vicinity of a stable auroral arc. J. Geophys. Res. **97**, A12 (1992). doi:10.1029/92JA01730

R. Walterscheid, G. Schubert, D. Brinkman, Small-scale gravity waves in the upper mesosphere and lower thermosphere generated by deep tropical convection. J. Geophys. Res. **106**(D23), 31825–31832 (2001)

V.B. Wickwar, M.J. Baron, R.D. Sears, Auroral energy input from energetic electrons and Joule heating at Chatanika. J. Geophys. Eng. **80**(31), 4364–4367 (1975)

P.J.S. Williams, G. Crowley, K. Schlegel, T.S. Virdi, I. McCrea, G. Watkins, N. Wade, J.K. Hargreaves, T. Lachlan-Cope, H. Muller, J.E. Baldwin, P. Warner, A.P. van Eyken, M.A. Hapgood, A.S. Rodger, The generation and propagation of atmospheric gravity waves observed during the Worldwide Atmospheric Gravity-wave Study (WAGS). J. Atmos. Terr. Phys. **50**(4/5), 323–338 (1988)

Space Sci Rev (2012) 168:475–516
DOI 10.1007/s11214-011-9813-9

Lightning Related Transient Luminous Events at High Altitude in the Earth's Atmosphere: Phenomenology, Mechanisms and Effects

Victor P. Pasko · Yoav Yair · Cheng-Ling Kuo

Received: 19 April 2011 / Accepted: 14 July 2011 / Published online: 13 September 2011

Abstract This paper presents a literature survey on the recent developments related to experimental and modeling studies of transient luminous events (TLEs) in the middle atmosphere termed elves, sprites and jets that are produced in association with thunderstorm activity at tropospheric altitudes. The primary emphasis is placed on publications that appeared in refereed literature starting from year 2008 and up to the present date. The survey covers general phenomenology of TLEs and their relationships to characteristics of individual thunderstorms and lightning, physical mechanisms and modeling of TLEs, past, present and future orbital observations of TLEs, and their chemical, energetic and electric effects on local and global scales.

Keywords Atmospheric electricity · Lightning · Sprites · Sprite halos · Elves · Blue jets · Gigantic jets

1 Introduction

Transient Luminous Events (TLEs) is the collective name given to a very wide variety of optical phenomena occurring over a wide range of altitudes in the stratosphere,

V.P. Pasko (✉)
Department of Electrical Engineering, Communications and Space Sciences Laboratory (CSSL),
The Pennsylvania State University, 211B Electrical Engineering East, University Park, PA 16802-2706, USA
e-mail: vpasko@psu.edu

Y. Yair
Department of Life and Natural Sciences, The Open University of Israel, 1 University Road, Ra'anana 43107, Israel
e-mail: yoavya@openu.ac.il

C.-L. Kuo
Department of Physics, National Cheng Kung University, No. 1, University Road, Tainan 70101, Taiwan
e-mail: johnny@phys.ncku.edu.tw

and the mesosphere and lower thermosphere, from about 15 to 110 km, in close conjunction with tropospheric electrical activity (e.g., Sentman et al. 1995; Neubert 2003; Pasko 2003, and references cited therein). Although eyewitness reports of TLEs above thunderstorms have been recorded for more than a century, the first image of one was captured in 1989, serendipitously during a test of a low-light television camera (Franz et al. 1990). Since then, several different types of TLEs above thunderstorms have been documented and classified. These include 'elves', which are lightning induced flashes that can spread over 300 km laterally (Boeck et al. 1992; Fukunishi et al. 1996; Inan et al. 1991, 1997; Mende et al. 2005; Cheng et al. 2007; Frey et al. 2005; Kuo et al. 2007b); 'sprites' that develop at the base of the ionosphere and move rapidly downwards at speeds up to 10,000 km/s (Sentman et al. 1995; Lyons 1996; Stanley et al. 1999; Gerken et al. 2000; Cummer et al. 2006b; McHarg et al. 2007; Stenbaek-Nielsen et al. 2007; Stenbaek-Nielsen and McHarg 2008); 'halos', which are brief descending glows with lateral extent 40-70 km, which sometimes (but not always) observed to accompany or precede more structured sprites (Barrington-Leigh et al. 2001; Frey et al. 2007); relatively slow-moving fountains of blue light, known as 'blue jets', that emanate from the top of thunderclouds up to an altitude of 40 km (Wescott et al. 1995; Boeck et al. 1995; Lyons et al. 2003); and upward moving 'gigantic jets', which establish a direct path of electrical contact between thundercloud tops and the lower ionosphere (Pasko et al. 2002; Su et al. 2003; Pasko 2003; Hsu et al. 2005; Fukunishi et al. 2005; van der Velde et al. 2007, 2010a; Kuo et al. 2007a, 2009; Cummer et al. 2009).

During the two decades elapsed since the original discovery in 1989 several hundred papers have been published in refereed scientific literature reflecting advances in this new and exciting research area. The research activities related to TLEs culminated in a first book, that have appeared in print in 2006 (Fullekrug et al. 2006). Planetary atmospheric electricity aspects relevant to TLEs have been recently covered in the book edited by Leblanc et al. (2008). Early theories of TLEs have been reviewed by Rowland (1998), Sukhorukov and Stubbe (1998) and Wescott et al. (1998). The experimental and theoretical findings related to TLEs have been summarized in several extensive review articles (Boeck et al. 1998; Rodger 1999; Inan 2002; Inan et al. 2010; Lyons et al. 2003; Pasko 2006, 2007, 2008, 2010; Raizer et al. 2010; Neubert et al. 2008; Roussel-Dupre et al. 2008; Mishin and Milikh 2008; Ebert and Sentman 2008; Ebert et al. 2010; Siingh et al. 2008). The goal of the present paper is to provide a limited overview of some of the most recent experimental and modeling developments in studies of TLEs putting emphasis on publications that appeared in refereed literature starting from year 2008 and up to the present date.

2 Phenomenology of TLEs and Their Relation to Properties of Lightning and Thunderstorms

The exact nature of the relationship between the properties of the parent thunderstorm that harbored the specific causative lightning flash of the TLE, and the properties of the observed optical emissions have been the basis for numerous works and several comprehensive reviews have already been published (Lyons 2006; Neubert et al. 2008; Pasko 2010). Sprites represent one of the most morphologically complex types of TLEs and in this section we mostly focus on phenomenology of sprites by describing the latest observational results, also pointing to open issues and outstanding questions. Additional information on phenomenology of other types of TLEs is included in Sects. 3 and 4 that follow.

Clearly, not only the source of the rapid change in the above-cloud electrical field plays a part in the formation of sprites: the "initial conditions" existing in the mesosphere and the ionosphere are undoubtedly important as well (density fluctuations, gravity waves, meteoritic dust), thus leading to an intricate and complicated interplay between the parent flash and the response of the atmospheric medium. Additionally, it is already recognized that sprites sometimes tend to re-appear in regions that had already been occupied by previous sprites, hinting at a local modification of the electrical and chemical properties of the mesosphere at sub-ionospheric altitudes (Haldoupis et al. 2010), thus pre-conditioning and facilitating the emergence of new sprites, pending the occurrence of lightning discharges with sufficient power. Significant progress had been made in quantifying and establishing concrete relations, though as can be expected, due the large variety of thunderstorms and of their lightning flashes, any generalization is bound to be limited and outstanding cases will exist.

2.1 Thunderstorms and Lightning Properties Leading to Sprite Formation

2.1.1 Meteorology of Sprite-Producing Thunderstorms

Not every thunderstorm produces sprites: rather, it is just a minority of storms that are accompanied by observable optical effects in the upper atmosphere. While the average global flash rate is approximately 44 flashes per second (Christian et al. 2003), the observed global sprite rate based on satellite observations is roughly 1 per minute (Kuo et al. 2008); and see Sect. 4 for a detailed overview). It has long been recognized that positive cloud-to-ground flashes (hereto after +CGs) are responsible for 99% of observed sprites, while negative cloud-to ground flashes (−CGs) only rarely produce sprites. Since −CGs are overwhelmingly more abundant than +CGs, it is no wonder that sprites are rather scarce, even though lightning is abundant. Another facet to explain this disparity is based on the nature of sprite-producing storms: these are most often the trailing stratiform regions (but not always) of large convective weather systems, which are typically continental in nature. Williams and Yair (2006) have mapped the types of sprite-producing weather systems in summer and winter storms, and noted the differences in size and in charge structure of the thunderstorms in various regions and conditions. Lyons et al. (2009) review the probability of different types of convective storms to generate sprite-halos, sprites and elves. Figure 17.1 in their book-chapter summarizes the different types of thunderstorms. Mesoscale Convective Complexes (MCC), with overall area of cold cloud tops exceeding 100,000 square kilometers and persisting for more than six hours, posses a "very high" likelihood to generate TLEs, while Mesoscale Convective Systems (MCS) rate second with a "high" likelihood. Winter convection, tropical storms and squall-lines have variable to low probability, while air-mass thunderstorms and super-cells produce sprites only rarely.

2.1.2 Parent Lightning Parameters

There are several different parameters which may be used to describe a lightning flash. The most well known are the polarity of the flash, its peak current and multiplicity, the existence of M-components, the magnitude and duration of the continuing current and the resultant charge moment change. It had been recognized that the decisive factor for the generation of sprites is the magnitude of the vertical charge-moment-change (CMC), defined as the product of the charge (Q) and the altitude from which it was lowered (h_Q) (Cummer et al. 2006b). A similar metric is the impulsive charge-moment-change (iCMC) defined as the total charge moment change within the first 2 ms after the lightning return stroke. Lyons et al. (2009) show that the magnitude of the iCMC for sprite-producing +CGs in the June 20

2007 MCC was always larger than the most commonly occurring +CG, which was typically smaller than 50 C km. Events which had an instantaneous charge moment change larger than 300 C km had a 75% likelihood to generate "an optically detected TLE" (Fig. 17.12, ibid). Hu et al. (2002) found that there is a 90% probability for sprite generation by a CMC of 1000 C km (or greater) caused within 6 ms of the parent +CG, and only 10% probability for a CMC of 600 C km or lower. Cummer and Lyons (2005) established a value of CMC of 500 C km as a limit for the generation of short-delayed sprites. Based on ELF data, Greenberg et al. (2007) found CMC values for sprite-producing +CG in eastern Mediterranean winter storms to range from 600 to 2800 C km, with ~1000 C km the frequent value. They noted that no significant difference in the ELF signatures and the computed CMCs was found for sprite- and non-sprite producing +CGs, suggesting that other factors play a role in sprite initiation. Lu et al. (2009) analyzed very strong +CG, and showed that even though they had large CMC (1500–3200 C km) they were not accompanied by any TLE. (Li et al. 2008) analyzed summer storms in central US and found that the total CMC ranged from 600 to 18,600 C km. While these values certainly match the concept of removal of charge from a wide area of the storm (Williams 2006), there are cases when even flashes with a CMC as low as 120 C km can trigger sprites (Hu et al. 2002). For winter thunderstorms, sprite producing +CG were found to have average CMC values of 1400 ± 600 C km, with some extreme events exceeding 3500 C km (Yair et al. 2009).

In addition to the CMC of the parent flash, it appears that in-cloud discharge processes preceding the actual attachment of the lightning channel to the ground, and some immediately following the return stroke, also play a role in the formation of sprites and in determining their properties. van der Velde et al. (2006) found distinct differences between in-cloud activity (as indicated by VHF bursts) prior to the generative +CG for column and carrot sprites. While column sprites appear with little or no VHF activity of the parent +CG, carrot sprites are usually associated with a burst of VHF activity, preceding the parent flash by 25–75 ms. Carrot sprites are also found to be associated with intense spherics caused by in-cloud lightning, lasting for 50 to 250 ms after the +CG, hinting at the existence of continuing current in the discharge channel, transferring the cloud charge to the ground. Yashunin et al. (2007) had suggested that the lightning M-components (these are transient surges or fluctuations in the continuing current stage after the first return stroke) may shift the electric field maximum from the location of the vertical channel, facilitating the appearance of delayed and offset sprites. The source of the lateral offset of sprite elements relative to the location of the +CG is still an open question for research. Recently, van der Velde et al. (2010b) analyzed the in-cloud component of the sprite-associated lightning flashes and those of other flashes. They showed that sprite development coincided with VHF bursts of the +CG, and is well correlated with the longest-lasting and largest lateral discharge channel dimensions.

2.2 Sprite Morphology

2.2.1 Shapes and Sizes of Sprites

There are many types and variations in the shapes and sizes of sprites. Indeed, due to the introduction of high-speed imagery, global satellite observations and advanced computer modeling and laboratory experiments, significant advances in understanding of sprite formation and propagation had been achieved. However, in any discussion of sprite properties, it should be born in mind that although in normal-rate video (which is most commonly used and the major source for the existing data) sprites appear like giant structures spanning large volumes and areas of the atmosphere, in reality such images are merely a "frozen

Fig. 1 A typical sprite event with multiple elements, with tendrils and bright cores. A large sprite-halo is clearly observed at the upper part. The image was taken by Michal Ganot with a Watec 902H camera at regular video speed, as part of the Eurosprite 2007 campaign in France. Courtesy of the ILAN team

history" of the fast movement of very small and bright volumes of intense ionization in streamer heads (Marshall and Inan 2006). One should also be aware that using 2D images (which in reality are projections of a 3D phenomenon) poses a severe limitation on our ability to fully describe sprites. That handicap can be partly overcome by multiple coordinated observations (e.g., Stenbaek-Nielsen et al. 2010) and using analogies and similarity rules between laboratory streamers and sprites, especially with the latest advances in stereographic imagery of streamers (Nijdam et al. 2008). Sprite shapes can be as simple as a single short vertical column, or can be composed of many distinct, spatially separated elements with a complex mixture of sizes, orientations and shapes. There exists a variety of sprite morphologies resembling straight columns, carrots, jelly-fish or fireworks. This variability necessitated inventing sub-classifications: (i) Columniform sprites (C-sprites) which appear as distinct, thin, vertical and elongated columns of light, either single or multiple (ii) Carrot sprites, which are composed of a heart-shaped central body with numerous protruding convulsed tendrils, resembling leaves or hairs (iii) Jellyfish sprites which are much larger in volume and area, with a big bright body and an overlaying canopy above numerous downward propagating tendrils and (iv) Angle sprites which are bifurcated columns with bright channels extending diagonally. A typical carrot sprite with a halo is given in Fig. 1. In general, one can distinguish between three distinct regions in sprites; the lower streamer region, the bright transition region, and an upper diffuse region (Pasko et al. 1998a; Pasko and Stenbaek-Nielsen 2002). The relative size, brightness and shape of these three regions are manifestations of the discharge process occurring at these altitudes, and can be attributed to the physical mechanisms of streamer generation and propagation that are a function of the reduced electric field, defined as a ratio of electric field magnitude E to the air density N, E/N (Ebert and Sentman 2008). Review of recent work on sprite streamers is presented in Sect. 3.2.

2.2.2 Time Delays of Sprite from the Parent +CG

Sprites appear in short conjunction after the causative +CG, typically of the order of several milliseconds. The difficulty in establishing the exact time delay is twofold: determining the exact sprite beginning time within the first video frame showing luminosity (typically inducing a 17 ms uncertainty in normal rate cameras), and the relative accuracy of the lightning detection network (usually GPS time-based). Observations in numerous campaigns in the US, Europe, Japan and Israel showed that a dividing time of approximately 10 ms can be defined to distinguish between short- (i.e., immediate, occurring within $\leq$10 ms) and long-delayed ($\geq$10 ms) sprites. Lyons et al. (2008) state that approximately half of observed sprites are delayed by more than 10 ms after their parent +CG, a fact that exemplifies the role of the continuing current for sprites initiation. Indeed, Li et al. (2008) found that out of 83 sprites they observed, 46% were long-delayed, and analyzed a typical long-delayed event: it was preceded by a +CG with an iCMC of 193 C km, which is below the usual threshold to ignite a sprite. The return stroke was followed by a 144 ms delay until the first sprite luminosity was detected, during which time a continuing current with an average 16 kA flowed, generating an integrated total charge moment change of 2650 C km. This event shows that it is not necessarily the iCMC which is important, but rather, the total CMC. In summer storms in Europe, van der Velde et al. (2006) reported that column sprites occur with a short time delay ($<$30 ms) from the causative +CG, while carrots are usually delayed up to $\sim$200 ms relative to their causative +CG stroke. Similar results were reported by Yair et al. (2009) for winter thunderstorms in the Eastern Mediterranean: the average time delay between the ELF transient of the parent +CG and the observed sprites was 55 ms, with shorter delays for column sprites (42 $\pm$ 17 ms) compared to carrot sprites (68 $\pm$ 17). Matsudo et al. (2009) compared time delays of sprites in winter storms in the Hokoriku region of western Japan. The most common sprite shape in this type of storms is simple vertical columns (68%) and the average delay time was found to be 90 ms. Similar winter storms over the Pacific Ocean east of Japan show delayed sprites by 43 ms on average.

2.2.3 Onset Altitudes

From the considerations of Wilson's theory (Wilson 1925), there exists an altitude in the atmosphere where the thunderstorm-induced electric field should surpass the conventional breakdown field (see Fig. 2 in Pasko 2010). Since the breakdown field is proportional to air density, it follows that sprites will initiate at altitudes where electric field can be sustained against fast relaxation in the highly conducting ionosphere and where the density of air is sufficiently low: between 75 and 80 km for typical conditions. Most of existing observations conducted with regular speed video have difficulty in determining the exact onset altitudes of sprites because of inaccuracies in range determination to the observed emissions, whose initial stages may be missed due to sensitivity issues. Range inaccuracy is inherent in single-station observations (even with star-field alignment) and in part stems also from the fact that sprites are often offset with respect to the ground location of the parent flash (which is precisely known based on lightning detection network data). In spite of these limitations, the determined altitudes range between 75 and 85 km agree with Wilson's predictions, where sprites seem to initiate and propagate. Figure 2 shows a sprite initiating from

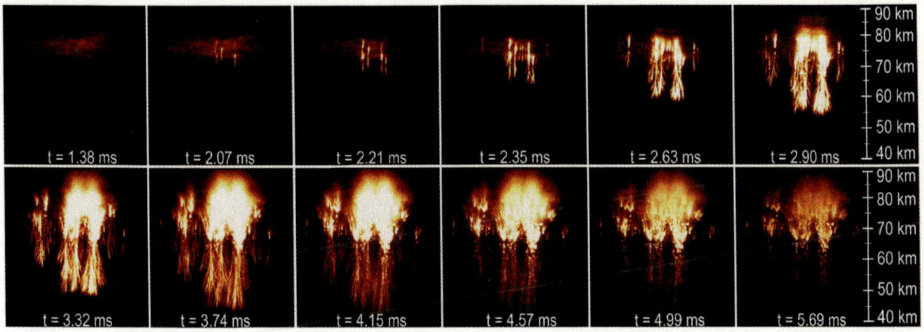

Fig. 2 The evolution of sprite streamers as depicted by high-speed imagery. The sprite emerges from the bottom of a diffuse glow (halo) and develops downward in visible complex tendrils. The uppermost main body of the sprite obtains a diffuse appearance. Courtesy of Steve Cummer, Duke University

a dim halo at 80 km, and propagates in several separate tendrils extending all the way down to 45 km. This observation of a dim halo at ~85 km preceding the appearance of streamers by Cummer et al. (2006b) was later modeled by Luque and Ebert (2009), as discussed in Sect. 3.2. High-speed (10,000 Hz) triangulated imagery by Stenbaek-Nielsen et al. (2010) showed that inception altitudes of the downward motion of sprite streamers is between 66 and 89 km, while in those cases where an upward streamer motion is found to occur (with a slight delay after the initial downward phase), it starts at lower altitudes: 64 to 78 km. This initiation may be related to negative charging of sprite elements (Luque and Ebert 2010; Li and Cummer 2011), as will be further discussed in Sect. 3.2. Gamerota et al. (2011) used a similar technique to compare observed and modeled sprite initiation altitudes. They analyzed 20 discrete sprite events and found that long-delayed sprites initiate at lower altitude. The average discrepancy between the measurements and their model simulation results was 0.35 km with a standard deviation of 3.6 km.

2.2.4 Sizes of Sprite Elements

The length of sprites is a measure of the overall duration of the movement of the streamer heads in the ambient electric field. High-speed imagery (Marshall and Inan 2006; McHarg et al. 2007; Stenbaek-Nielsen and McHarg 2008; Montanya et al. 2010) shows that sprites are composed of numerous compact, very fast ($\sim 10^7$ m/s) downward moving elements, that become brighter as they descent. The sizes of each individual streamers head are between tens to hundreds of meters and they leave a narrower trail in their wake. Telescopic observations by Gerken and Inan (2003) reported decametric structures in sprites. As mentioned above, the movement of streamer heads (sometimes referred to as "tendrils") as recorded by regular-speed video creates the illusion of elongated, vertical columns of light, and so any discussion of the vertical dimension of sprites actually reflects the time-integrated motion of these elements, clearly exemplified by Ebert et al. (2006) in the analogy to laboratory streamers at different time-exposures. Stenbaek-Nielsen and McHarg (2008) use the term "confetti falling from the sky" to describe this motion, as there are numerous such elements forming a single sprite. Relevant modeling results are described in Sect. 3.2. It should also be pointed out that in many cases, an upward movement of streamers is observed, 0.5–2 ms after the onset of the downward motion, with speeds slightly faster than the downward motion ($\sim 6 \times 10^7$ m/s). These upward streamers also possess a horizontal velocity component which leads to spreading out on a large scale, which creates the typical shape of carrot sprites. In column sprites, such upward streamers are not observed, and a simpler shape is maintained.

2.2.5 Spatial Organization of Sprite Elements

Sprites can span a large horizontal dimension, and impact a considerable volume of the mesosphere. Since almost all existing images offer only a 2-dimensional view of sprites, we lack a full mapping in 3 dimensions of their organization. In this regard we note that the first color imagery of sprites and their unambiguous physical dimensions and heights were obtained using triangulation from two jet aircraft (Sentman et al. 1995). Recent triangulation studies provided valuable information about initiation altitudes of downward (66 to 89 km) and upward (64 to 78 km) streamers, that associate with formation of columnar (C sprites) and carrot sprites, respectively (Stenbaek-Nielsen et al. 2010). Stanley et al. (2007) mapped sprite locations and compared them to LMA-derived flash dimensions and reported that sprites were arranged preferably on the periphery of the discharge. Vadislavsky et al. (2009) showed that in some cases, sprite elements appear to be organized in a circular shape above, or slightly offset, to the +CG location. Inan (personal communication, 2008) suggested that previous IC flashes lead to the creation of an attachment dominated "hole" of electron concentration at the region above the +CG, which prevents sprite formation directly above the +CG. Another possibility was suggested by Adachi et al. (2004), based on the circular shape of the lightning-induced EMP, that acts to pre-empt the mesosphere before the ensuing quasi-static electric (QE) field. van der Velde et al. (2010a) found that the propagation of the lightning channel during a VHF burst dictated the order of appearance of sprite elements, much like the early report by Stanley (2000). The cause of the arrangement of the different elements in sprites remains unclear.

3 Physical Mechanisms and Modeling TLEs

3.1 Interaction of Lightning Electromagnetic Pulses with Ionosphere and Elves

Inan et al. (1991) were first to predict that lightning might directly affect the lower ionosphere, producing both ionization and optical emissions which we know as elves phenomenon. After their initial discovery from the Space Shuttle (Boeck et al. 1992) the first ground observations of elves were made by Fukunishi et al. (1996) and Inan et al. (1997) using photometers. The summary of important historical facts and processes related to research on interaction of lightning induced electromagnetic pulses (EMPs) with the lower ionosphere have been given in recent review by Inan et al. (2010).

Inan et al. (1997) predicted that based on radiation pattern of a vertical cloud to ground lightning discharge, considered to be acting as a linear antenna element, the optical luminosity of elves should have a donut shape with a distinct minimum right above the causative discharge. Kuo et al. (2007b) recently performed detailed in-depth study of elves recorded by the ISUAL instrument on FORMOSAT-2 satellite. The modeling component of that study included validation of elves modeling indicating good agreement for: direct comparison of photon fluxes calculated by the model and observed by ISUAL spectrophotometers, direct comparison of modeled and observed morphologies of elves, and direct comparison of the calculated photon fluxes using peak currents for two elve-associated NLDN cloud to ground lightning discharges with those recorded by the ISUAL Imager (Kuo et al. 2007b). Figure 3 illustrates modeling results obtained in Kuo et al. (2007b) that are in excellent agreement

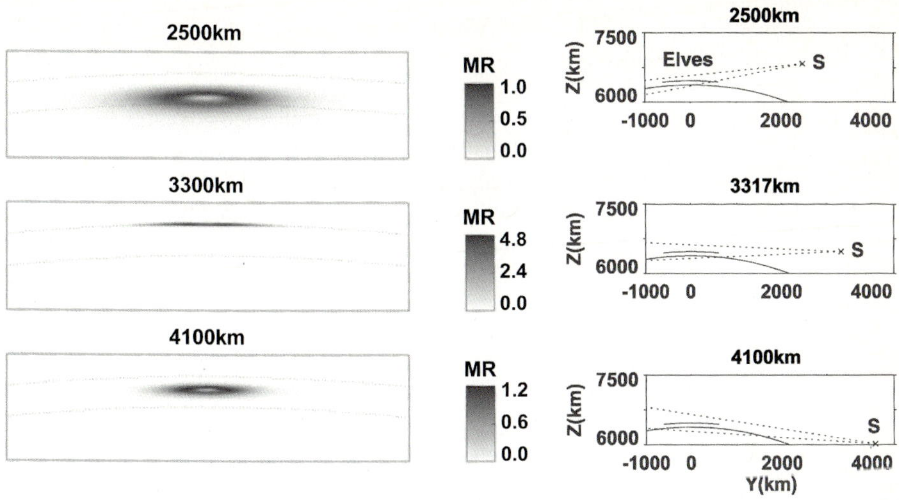

Fig. 3 Optical emissions of first positive band system of N_2 of a modeled elve (peak current 280 kA) in front (2500 km), at (3300 km) and behind (4100 km) the limb (Kuo et al. 2007b). Reprinted by permission from American Geophysical Union

with observed morphological features of elves recorded by the ISUAL under different viewing conditions, when elves appear in front, at, or behind the limb.

Marshall et al. (2010) have recently presented a three-dimensional finite difference time domain model for simulation of the lightning electromagnetic pulse and its interaction with the lower ionosphere. The results agree with the frequently observed, doughnut-shaped optical signature of elves, as illustrated by Fig. 3, but also show that the structure exhibits asymmetry due to the presence of Earth's ambient magnetic field. This prediction has not yet been confirmed by experiment. Additional quantitative results presented in Marshall et al. (2010) indicate that in-cloud (horizontal) lightning channels produce observable optical emissions without the doughnut shape and, in fact, produce a much stronger optical output for the same lightning current. Results of modeling of a realistic burst of in-cloud lightning activity, that involved series of pulses with realistic pulse amplitudes and random altitudes and orientations, produced an electron density increase of over 200% in the overlying ionosphere. Marshall and Inan (2010) further indicated that these factor of two increases in local density associated with elves yield small but detectable perturbations to the sub-ionospherically propagating very low frequency (VLF) transmitter signals and may explain at least some observed perturbations produced by direct interaction of lightning electromagnetic emissions with the lower ionosphere known as early/slow VLF events (see also recent work by Haldoupis et al. (2009) and Haldoupis et al. (2010) that include extensive list of references documenting the phenomenology of early VLF events in connection with sprites and elves, as reported by Mika et al. (2006)).

Lay et al. (2010) have recently performed a modeling study of cumulative effects of a sequence of strong lightning discharges on the lower ionosphere. The study involved a sequential application of ten lightning discharges with peak current 150 kA and rise time 20 μs. The stroke rates were based on World-Wide Lightning Location Network measurements. The study employed a two-dimensional axisymmetric finite-difference time-domain model that describes the effect of lightning electromagnetic pulses on the ionosphere. The authors noted an interesting effect of successive electromagnetic pulses in which each pulse interacts with a modified background ionosphere due to the previous pulses. In reported results a formation of nonlinear electron density perturbation over time, that eventually reached a limiting value, was observed. The limiting maximum electron density was reached earlier in time for higher altitudes, and the most significant effect occurred at 88 km. The limiting modeled electron density profile in the 83–91 km altitude range did not depend on the initial electron density (Lay et al. 2010). The effect of steepening of the ambient electron density profile reported by Lay et al. (2010) may have a direct facilitating effect on initiation of sprite streamers as discussed recently in Qin et al. (2011, see Fig. 9 and associated text).

3.2 Modeling of Sprites

3.2.1 Modeling of Large Scale Electrodynamics of Sprites

The charge moment change Qh_Q (i.e., charge removed by lightning Q times the altitude from which it was removed h_Q) represents the key parameter which is used in current sprite literature to measure the strength of lightning in terms of sprite production potential (Cummer et al. 1998; Hu et al. 2002, 2007; Li et al. 2008; Cummer 2003; Cummer and Lyons 2004, 2005; Cummer et al. 2006a). Computer simulations of Hiraki and Fukunishi (2006) and Asano et al. (2008) also indicate importance of the product of the current (Ih_Q) and charge (Qh_Q) moments for sprite initiation and suggest the threshold value: $(I\ h_Q) \times (Q\ h_Q) = 1.6 \times 10^6$ (A km) (C km). The results of Hiraki and Fukunishi (2006) and Asano et al. (2008) emphasize importance of the timescale of charge removal by lightning for initiation of sprites. These conclusions are generally in agreement with analysis reported in Barrington-Leigh et al. (2001) indicating that a lightning discharge with a fast (<1 ms) charge moment change may be sufficient to cause diffuse emissions at higher altitudes, where the threshold for ionization and optical excitation is lower. If lightning currents do not continue to flow, however, there may not be sufficient electric field to initiate streamers below ~75 km. Conversely, slow continuing currents may cause a (delayed) sprite without a significant initial flash in the diffuse (halo) region (Barrington-Leigh et al. 2001).

In recent years the ELF measurements of the time dynamics of the charge moment changes in sprite-producing lightning have progressed to the point allowing detailed event based testing of sprite theory (Hu et al. 2007; Li et al. 2008). The modeling analysis reported by Hu et al. (2007) provided quantitative information on time variation of lightning current moment, charge moment and electric field driving sprite phenomena and demonstrated that for bright, short-delayed sprites, the measurement-inferred mesospheric electric field agrees within 20% with the threshold electric field for conventional breakdown E_k. However, for long delayed sprite events and dimmer sprites, the measurement-inferred mesospheric electric field for sprite initiation is somehow below E_k values (Hu et al. 2007). Figure 4 shows measured current and charge moments and the simulated electric fields between 50 and 90 km altitude (normalized by E_k) for the typical long-delayed sprite (Li et al. 2008). The sprite is initiated at 144.4 ms after the lightning return stroke. The electric field increased before the sprite onset due to slow intensification of the continuing current. The peak normalized electric field in Fig. 4b producing the long-delayed sprite is $E/E_k = 0.45$ at 72 km altitude, which is similar to those for typical short-delayed sprites that are neither remarkably bright or dim (Hu et al. 2007). The modeled initiation altitude of 72 km falls in the 70–80 km range estimated from high-speed video images (Li et al. 2008). Comparison of modeling and observations indicate that the long-delayed sprites initiate 5 km lower than short-delayed sprites (Li et al. 2008). The results of Li et al. (2008) have recently been confirmed using analysis of 20 additional discrete sprite events (Gamerota et al. 2011). The

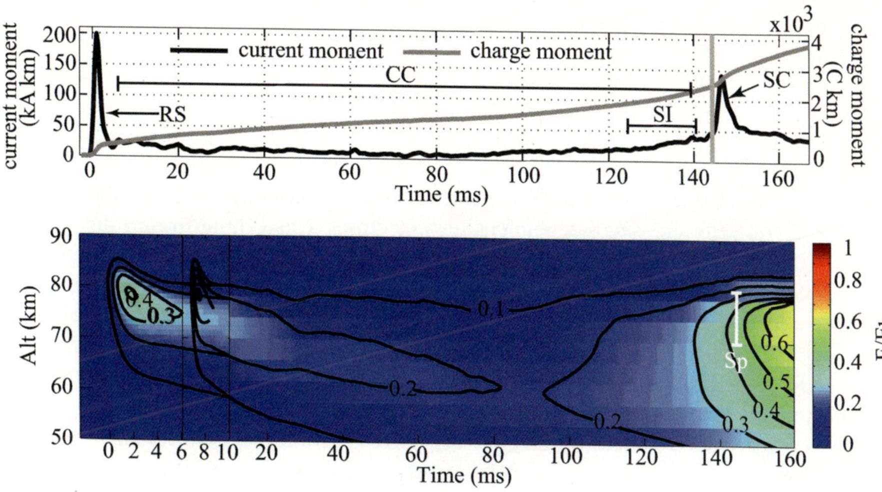

Fig. 4 Finite difference time domain (FDTD) simulation results for the typical long-delayed sprite event (Li et al. 2008). (**a**) Estimated current moment and total charge moment change. (**b**) Simulated electric fields above the lightning discharge. Reprinted from Li et al. (2008) by permission from American Geophysical Union

comparisons of 2D FDTD model results and observations indicate good agreement between measured and modeled initiation altitudes (defined on the basis of peak E/E_k values), with average discrepancy of 0.35 km with a standard deviation of 3.6 km.

Adachi et al. (2008) combined array photometer data obtained by ISUAL instrument on FORMOSAT-2 satellite with the ELF magnetic field data to deduce temporal evolutions of lightning charge moment changes driving sprites and halos. It was found that the lightning discharge producing a halo without streamers has a short time scale of $\sim$1 ms with a moderate charge moment of $Qh_Q \sim 400$ C km while that producing streamers as well as a halo have a similar time scale of $\sim$1 ms but a large charge moment of $Qh_Q \sim 1300$ C km. On the other hand, a lightning discharge producing sprite streamers without a discernible halo has a long time scale of $\sim$10 ms and a large charge moment of $\sim$1300 C km. These conclusions are in agreement with results reported in Barrington-Leigh et al. (2001), Hiraki and Fukunishi (2006), Asano et al. (2008).

Lightning M components are perturbations or transient enhancements in the continuing current (Rakov et al. 2001). Rakov et al. (2001) speculated that M-component type processes in positive lightning may play a role in the initiation of so-called delayed sprites that occur many tens of milliseconds after the return stroke. Modeling analysis of Yashunin et al. (2007) indicates that the occurrence of an M component shifts the electric field maximum from the vertical axis of the lightning channel and increases the likelihood of initiation of sprites spatially displaced with respect to the lightning channel axis. Asano et al. (2009b) established through computer simulations that M components in the continuing current with small amplitudes but fast time variations can initiate or enhance the occurrence of sprites. Asano et al. (2009a) utilized three-dimensional fully electromagnetic simulations to investigate effects of horizontal components of currents in lightning discharges in addition to the conventional vertical channel. It has been established that the position of sprite can be significantly shifted in response to the length of the horizontal lightning channel due to additional ionization effects created by electromagnetic pulse radiation from the horizontal channel (Asano et al. 2009a).

3.2.2 *Modeling of Sprite Streamers*

Significant efforts in recent years have been devoted to modeling of fine streamer structures in sprites and their interpretation. Streamers are needle-shaped filaments of ionization embedded in originally cold (near room temperature) air and driven by strong fields due to charge separation in their heads (Raizer 1991, p. 334). The streamer polarity is defined by the sign of the charge in its head. The positive streamer propagates against the direction of the electron drift and requires ambient seed electrons avalanching toward the streamer head for the spatial advancement (Dhali and Williams 1987). The negative streamer is generally able to propagate without the seed electrons since electron avalanches originating from the streamer head propagate in the same direction as the streamer (Vitello et al. 1994; Rocco et al. 2002). A detailed review of various properties of streamers, including fields required for their initiation and propagation, and similarity relationships allowing scaling of streamer parameters as a function of gas pressure for the purposes of interpretation of sprite discharges, is provided in Pasko et al. (1998a), Pasko (2006, 2007, 2008). Below we review related recent work directed to the modeling interpretation of sprite observations.

First particle simulations of sprite streamers have been reported by Chanrion and Neubert (2008). The code is in 2D axi-symmetric coordinates with charged particles followed in a Cartesian mesh and the electric field updated with Poisson's equation from the charged particle densities. Collisional processes between electrons and air molecules are simulated with a Monte Carlo technique, according to cross section probabilities. The code also includes photoionization processes of air molecules by photons emitted by excited constituents. Chanrion and Neubert (2008) presented detailed comparisons of results from the new particle model with previous fluid model of Liu and Pasko (2004) for the case of double headed streamers developing at 70 km altitude in air using identical initial plasma, electric field and neutral atmosphere parameters. The results demonstrated excellent agreement with the previous fluid modeling. It is shown that at 1 atm pressure the electric field must exceed similar to 7.5 times the breakdown field to observe runaway electrons in a constant electric field. It is also found that this value is reached in a negative streamer tip at 10 km altitude when the background electric field equals similar to 3 times the breakdown field. It is shown that the energetic runaway electrons produced by streamer tips create enhanced ionization in front of negative streamers. The simulations suggest that the thermal runaway mechanism may operate at lower altitudes and be associated with lightning and thundercloud electrification while the mechanism is unlikely to be important in sprite generation at higher altitudes in the mesosphere (Chanrion and Neubert 2008). The model of Chanrion and Neubert (2008) has recently been extended with an improved description of high-energy electrons in Chanrion and Neubert (2010).

The electric fields associated with streamer discharges in TLEs are now routinely measured considering the ratios of spatially integrated radiation intensities of band systems with different energy excitation thresholds (e.g., Kuo et al. 2005, 2009; Liu et al. 2006, 2009b; Adachi et al. 2006). In a streamer, the head is usually responsible for the most part of ionization and excitation of species, and therefore is responsible for the most part of emission. However, the spatial non-uniformity of streamer discharges is such that the maximum excitation rates are not exactly located at the spatial position of the maximum electric field (Naidis 2009). Following recent work by Naidis (2009), Celestin and Pasko (2010) demonstrated that due to strong spatial variations of the electron density and electric field in streamer heads previous measurements may lead to a significant underestimation of the peak values of electric fields in TLEs. The modeling analysis of streamers as a function of altitude, applied reduced electric field, and streamer polarity conducted by Celestin and Pasko (2010) indicated

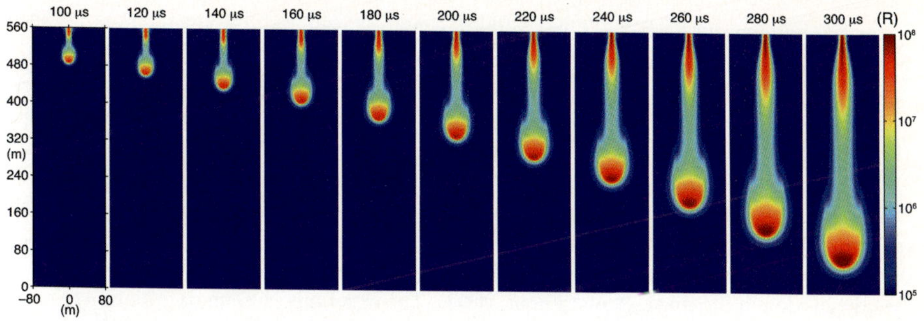

Fig. 5 A time sequence of intensity distributions of first positive band system of N_2 for a downward propagating model positive streamer at 75 km altitude (Liu et al. 2009a). The sequence of images is shown with 20-μs interval, starting at 100 μs and ending at 300 μs. The formatting is consistent with that of Stenbaek Nielsen et al. (2007, Fig. 2). Reprinted by permission from American Geophysical Union

that the ratios based electric fields derived from spectrophotometric data need to be multiplied by a corrective factor >1.4 for positive streamers and >1.5 for negative streamers, to obtain the true peak values of electric fields. The study of Celestin and Pasko (2010) brings the values of electric field previously deduced from observations (e.g., Kuo et al. 2005; Adachi et al. 2006) much closer to streamer simulations. Moreover, one can note that the value $5.5E_k$ documented in Kuo et al. (2009) for negative gigantic jet discharge would lead to the very high field $>1.5 \times 5.5E_k \simeq 8E_k$ approaching or exceeding magnitudes for which thermal electron runaway process becomes possible (e.g., Moss et al. 2006; Li et al. 2009; Chanrion and Neubert 2008, 2010; Celestin and Pasko 2011). These results, therefore, have direct implications for production of runaway electrons in jet discharges.

Liu et al. (2009a) compared sprite streamer modeling results with high-speed video recordings of sprites made with 50 μs temporal resolution (McHarg et al. 2007; Stenbaek-Nielsen et al. 2007). Both the modeling results and the sprite videos show that sprite streamers propagate with acceleration and expansion during the initial stage of sprite development. The acceleration computed from the modeling for the applied electric fields close to the conventional breakdown threshold field E_k is on the order of $(0.5\text{–}1) \times 10^{10}\ \text{m}\,\text{s}^{-2}$ and is in good agreement with the peak values observed experimentally. Mainly due to the increasing radius of the streamer head of an accelerating streamer, the brightness of the streamer head increases as well (see Fig. 5). The results reported in Liu et al. (2009a) demonstrate that the brightness of a sprite streamer head increases exponentially with time and can span more than 4 orders of magnitude in a very short period of about 1 ms. The rate of increase depends on the magnitude of the applied electric field. Liu et al. (2009a) proposed a method for remote sensing of the sprite-driving electric field in the mesospheric and lower ionospheric region by measuring the rate of the change of the brightness. In particular, Liu et al. (2009a) report that the sprite event presented in McHarg et al. (2007), Stenbaek-Nielsen et al. (2007) was initiated by fields close to the conventional breakdown threshold E_k.

We note that the exponential expansion and acceleration of streamers is an important effect that was recently identified to be also accompanied by exponential growth of electric potential differences in the streamer heads (Celestin and Pasko 2011). These electric potential differences have significant implications for the energy that thermal runaway electrons can gain once created and quantitative results of Celestin and Pasko (2011) indicate that under realistic conditions associated with lightning leaders streamers can create ~100 keV electrons capable of further acceleration to several MeVs energies needed for explanation of terrestrial gamma ray flashes (TGFs) in the Earth's atmosphere (e.g., Fishman et al. 1994; Smith et al. 2005). We note that the energy of photons in TGFs can reach up to 100 MeV (Tavani et al. 2011), and the most recent observations indicate that many TGFs are produced during upward negative leader progression prevalent in normal polarity intra-cloud flashes (Lu et al. 2010, 2011a).

Although it has been suggested that sprites may be initiated through simultaneous up and down propagating streamers (Liu and Pasko 2004), the recent observational evidence indicates that the preferential form of sprite initiation is through downward development of positive streamers launched in the region of the lower ledge of the Earth's ionosphere (see discussion in Stenbaek-Nielsen et al. 2007). The exact mechanism of initiation of sprite streamers is not well understood and related recent studies are reviewed below. The mechanism may be related to formation of upwardly concave ionization regions near the lower ionospheric boundary associated with sprite halos (Barrington-Leigh et al. 2001). As was discussed recently in Pasko (2010) the double headed streamers may be involved in initiation of spites, however, due to the fast exponential increate in streamer brightness in time (see review of Liu et al. 2009a above) the initial appearance of positive streamers is likely realized because of a relatively slow application of the electric field at sprite altitudes (1 ms) (Marshall and Inan 2006; Hu et al. 2007) coupled with lower propagation threshold for positive streamers in comparison with negative ones (Pasko et al. 2000), which creates asymmetric conditions with predominant initial propagation of positive streamers. Other factors have also been proposed and review of recent research work on this subject is provided below.

Luque and Ebert (2009) have recently reported a high resolution modeling of inception of sprite streamers as a result of sharpening and collapse of screening-ionization wave associated with sprite halo. The streamer emerged as a result of halo dynamics on the axis of symmetry in a two-dimensional axisymmetric plasma fluid model that employed a non-uniform, dynamically adapted computational grid (Luque and Ebert 2009). The transverse dimension of the modeled streamer was on the order of 1 km, and its time dynamics was resolved on a fine grid with 3 m spatial resolution (Luque and Ebert 2009). This result is qualitatively similar to the dynamics reported in early low resolution sprite modeling in which columns of ionization with transverse extent on the order of 10 km were observed to emerge from upwardly concave regions of sprites and to propagate down to altitudes ~45 km (Pasko et al. 1996). We note that although the upwardly concave regions of luminosity forming during sprite development were fully resolved and quantitatively described in early papers (Pasko 1995, 1996, 1997; Veronis et al. 2001) the name sprite halo, that is now used in association with these events, was introduced several years later when these events, driven primarily by quasi-static electric fields, were discovered, clearly identified and separated from elves phenomenon using comparisons of high speed video observations and fully electromagnetic modeling (see discussion in Barrington-Leigh et al. 2001, and references cited therein). A significant difficulty of both the early (Pasko et al. 1996) and more recent (Luque and Ebert 2009) modeling is that streamers appear in these models as a continuous process of halo development, while in many existing high resolution photometric and video records the sprite streamers exhibit significant spatial (both vertical and horizontal) as well as temporal separations with respect to position and timing of the sprite halo (Qin et al. 2011, and extensive list of references therein). The asymmetry in initiation of sprite streamers by cloud-to-ground lightning discharges with different polarities, and sprite streamers triggered by cloud-to-ground lightning with very low charge moment changes are also not reproduced by models reported in Pasko et al. (1996), Luque and Ebert (2009). Additionally, Qin et al. (2011) recently indicated that the collapse of sprite halo reported in Pasko et al. (1996), Luque and Ebert (2009) is due to numerical instability, that can be somewhat postponed but not eliminated by increasing resolution of numerical grid.

Liu (2010) has recently discussed mechanisms for the luminous trail or 'afterglow' region often observed at higher altitudes in downward propagating sprite streamers (McHarg et al. 2007; Stenbaek-Nielsen et al. 2007). The physical mechanisms of the observed phenomenon discussed in the existing literature include emissions produced due to energy transfer from N_2 and O_2 metastable electronic states, energy pooling between low energy metastable states, and energy pooling between the metastable states and the vibrationally excited ground electronic state of N_2 (e.g., Morrill et al. 1998; Bucsela et al. 2003; Kanmae et al. 2007; Pasko 2007; Sentman et al. 2008; Sentman and Stenbaek-Nielsen 2009). Liu (2010) demonstrates that the luminous trail also naturally appears in numerical modeling of sprite streamers in which the above-mentioned chemical processes are not taken into account. The effect is due to increase in current that flows in the sprite streamer body when streamer expands, accelerates and brightens. The increasing current self-consistently leads to the rise of the electric field in the streamer channel far behind the streamer head, which leads to effective production of N_2 excited states by electron impact excitation and then to the glowing trail (Liu 2010).

The results of Liu (2010) concerning the luminous trail are in agreement with more recent work (Luque and Ebert 2010) whose authors applied a model with adaptively refined grids to study propagation of long sprite streamers with transverse dimension on the order of 1 km through altitude range with significant variation of ambient air density. Luque and Ebert (2010) observed that in their model the optical emission intensity varied proportionally to air density and therefore exponentially in time for the streamer with velocity and radius that remained relatively stable. The authors also observed a negative charging of the upper section of the positive streamer that propagated downward, and suggested that this negative charging may facilitate emergence of upward negative streamers and attraction of positive streamer heads to previously formed channels as observed experimentally (Luque and Ebert 2010, and references therein).

Li and Cummer (2011) have recently used lightning-driven background electric fields at mesospheric altitudes deduced from ELF electromagnetic measurements in combination with high-speed images of sprite streamers to estimate charges present in bright columnar cores appearing at upper extremities of sprites. The analysis assumed that the observed negative streamers that emerge from the bright negatively charged cores propagate along the directions of the local electric field lines, and the amount and vertical distributions of charge were iterated until good agreement was achieved with high-speed video images in terms of angles of emergence of negative streamers with respect to the vertical cores. The authors were able to establish a lower bound on the electric charge in six observed sprites driven by positive lightning. They found, in particular, that individual bright sprite cores contain significant negative space charge between −0.01 C and −0.03 C. Given the significant negative charge, Li and Cummer (2011) interpreted the sprite core region as the partial and perhaps dominant sink of the negative charge created by the downward positive polarity streamers. This further suggests that when downward streamers supply more charge than can be absorbed by the sprite core, slightly delayed upward negative streamers initiate from the sprite core to disperse this charge (Li and Cummer 2011). This is consistent with observations that show that the subsequent upward streamers are not always present, especially in smaller sprites Li and Cummer (2011). The findings of Li and Cummer (2011) that the emergence of secondary negative streamers may be related to the negative charging of the upper section of the downward positive streamer are consistent with similar ideas advanced in Luque and Ebert (2010). An important original finding of the analysis of Li and Cummer (2011) is that it allowed for the first time to use sprite imagery to find if approximately 10 km long bright sprite core regions were electrically attached to the high conductivity regions of the lower ionosphere. The analysis indicated a weak or nonexistent electrical connection between the ionosphere and the downward streamer (Li and Cummer 2011). In addition to other important factors discussed above in the context of halo-streamer transition (Pasko et al. 1996; Luque and Ebert 2009), the new findings of Li and Cummer (2011) also support the idea that sprite streamers do not represent a continuous process of sprite halo development as they appear to be electrically disconnected from the upper high conductivity regions of the Earth ionosphere.

In order to monitor the inception of sprite streamers, that cannot be modeled with present computer resources in the framework of fluid models, Qin et al. (2011) have used an improved avalanche-to-streamer transition criterion, and have investigated the response of the lower ionosphere to the charge moment changes induced by lightning discharges as a system of avalanches. Qin et al. (2011) established a direct link between the charge moment change and the ambient electron density profile as required for triggering of sprite streamers. The related condition is defined as $h_{\text{tran}} \geq h_{\text{cr}}$, where h_{tran} represents the altitude at which the electron density is low enough so that the distance between electrons is equal to the radius of the electron avalanche at the moment of avalanche-to-streamer transition, and h_{cr} represents the altitude at which the ambient electric field E_{amb} created by the charge moment change in the thunderstorm is equal to the conventional breakdown field E_k defined by the equality of ionization and dissociative attachment frequencies in air. An analysis of the origin of polarity asymmetry between +CGs and −CGs in triggering of sprite streamers conducted by Qin et al. (2011) indicated that the vertical extent of streamer initiation region (SIR) created by a −CG was smaller than the SIR created by the opposite +CG that corresponded to the same charge moment change. This is due to the fact that, the positive SIR is basically equivalent to the high field $E > E_k$ region created by the halo because upward avalanches generated in a low electron density region can penetrate deep into the high electron density region of the halo. In contrast, the negative SIR is limited to the high field region where the electron density is low enough for avalanches to develop without overlapping. Qin et al. (2011) also demonstrated as part of the asymmetry study that the triggering of long-delayed sprites is a unique property of halos produced by positive cloud-to-ground lightning discharges due to the formation of a long-lasting high field region, that can be significantly enlarged by the lightning continuing current. Qin et al. (2011) demonstrate importance of initial inhomogeneities in the lower ionosphere for initiation of sprite streamers and indicate that a large number of initial seed electrons distributed in a compact region of space N_{0e} creates more favorable conditions for initiation of sprite streamers. It is important to emphasize that the initial inhomogeneities were discussed in many early papers on sprite modeling (Pasko et al. 1996, 1997; Raizer et al. 1998, 2010; Zabotin and Wright 2001). Raizer et al. (1998, 2010), for example, introduce a seed inhomogeneity with radius 60 m and electron density 150 cm^{-3} (corresponding to $N_{0e} \sim 10^{14}$) at 80 km altitude. The approach of Qin et al. (2011) is different as it considers the seed electrons in the framework of competition of many avalanches in which only a perturbation with a larger number of initial electrons reaches the avalanche-to-streamer transition, becomes dominant in defining the macroscopic electric field, and also eventually becomes observable. Qin et al. (2011) quantitatively demonstrate that in the framework of avalanche-to-streamer transition even values twelve orders of magnitude lower than those employed by Raizer et al. (1998, 2010) (i.e., $N_{0e} \sim 10^2$) play a definitive role in initiation of sprite streamers. The possible role of inhomogeneities in electron density in formation of bright beads and branching of sprite streamers has recently been discussed by Luque and Gordillo-Vazquez (2011).

3.3 Modeling of Jets

A summary of phenomenological features of blue jets and gigantic jets and their proposed relationship to different polarity leader processes has been presented recently in Pasko (2008). Assuming that both types of jets originate from streamer zones of conventional lightning leaders propagating upward from thundercloud tops, the continuous positive leader-like propagation of optically observed blue jets should be contrasted with the impulsive rebrightening of gigantic jets, resembling negative leader processes (Krehbiel et al. 2008). The polarity itself is not sufficient to explain all the morphological differences in jet events observed to date. In particular, Krehbiel et al. (2008) emphasize that the location of jet initiation and charge configuration in a thundercloud are also defining parameters for jet development. Although the discussion presented in recent reviews by Pasko (2008) and Pasko (2010) suggests association of blue jets with positive leaders and gigantic jets with negative leaders, the recent observations of a positive gigantic jet exhibiting similar morphological features to those in previously documented negative gigantic jets (van der Velde et al. 2010a) indicate that the situation is more complicated than originally thought. Extensive discussion on possible classification schemes of different jet events is given in Krehbiel et al. (2008) (see also Supplementary Information for Krehbiel et al. 2008 available on *Nature Geoscience* web site). We note that van der Velde et al. (2010a) indicated that there was an evidence that the positive gigantic jet was produced by an inverted polarity storm. This appears to be in excellent agreement with predictions of Krehbiel et al. (2008) of positive gigantic jets produced by inverted polarity storms (i.e., storms in which main negative charge is located above the main positive charge).

Raizer et al. (2006, 2007, 2010) associated both blue jets and gigantic jets with the streamer zone of positive leader, which is postulated to be initiated above the positive charge center positioned at altitude 12 km. Raizer et al. (2006, 2007, 2010) demonstrated that upward transfer of the high thundercloud potential by leader channel to lower air density regions with proportionally lower electric field threshold for propagation of streamers E_{cr}^{+}, allowed the sustainment of blue jet streamers by relatively moderate cloud charge of 50 C with a radius of 3 km.

Milikh and Shneider (2008) utilized the modeling approach proposed in Raizer et al. (2007) to study 300–400 nm emissions produced by a jet discharge in order to interpret flashes with 1–64 ms duration observed in equatorial regions by UV instrument on board of the microsatellite 'Tatiana'. The modeling considered the second positive band system of N_2 and the first negative band system of N_2^+ as primary emissions responsible for the observed flashes and accounted for excitation as well as for pressure dependent quenching effects of the responsible electronic states $C^3\Pi_u$ of N_2 and $B^2\Sigma_u^+$ of N_2^+, respectively. The model calculated UV fluxes are in agreement with those observed in terms of UV pulse duration and magnitude (Milikh and Shneider 2008).

In addition to an obvious disagreement with the inferred negative polarity of some of the gigantic jets observed to date, the limitation of models proposed in Pasko et al. (1999), Pasko and George (2002), Raizer et al. (2006, 2007, 2010) in support of the original idea expressed by Petrov and Petrova (1999) that jets correspond to the upward development of the pressure scaled streamer zone of a conventional leader is that they all postulate presence of a leader near the cloud top. These models do not provide a link to experimentally documented charge distributions and lightning phenomenology in thunderstorms leading to the initiation and upward escape of the leader process from the thundercloud top. There is experimental evidence, for example, that some of the gigantic jets are initiated as normal polarity intra-cloud lightning discharges between upper positive and lower negative charge centers (Mathews et al. 2002; Krehbiel et al. 2008) and the models advanced in Pasko et al. (1999), Pasko and George (2002), Raizer et al. (2006, 2007, 2010) do not reflect related scenarios.

Krehbiel et al. (2008) discussed the charge imbalances in thunderstorms as a fundamental condition allowing propagation of leaders downward as cloud-to-ground lightning or upward as jet discharges. This demonstrated that upward discharges are analogous to cloud-to-ground lightning and providing a unified view on how lightning escapes from a thundercloud. Cummer et al. (2009) recently reported measurement of total charge of −144 C transferred by a gigantic jet to the lower ionosphere, which is in the range of charge transfers to ground by strong negative cloud-to-ground discharges (−CGs), however, associated with current rise time in gigantic jet (∼30 ms), which is four orders of magnitude slower than in conventional −CGs (∼5 μs, Rakov and Uman 2003, pp. 7, 146, 154). Krehbiel et al. (2008) note that in accordance with existing experimental evidence the lightning initiation usually happens between adjacent charge regions of different polarity where the electric field is maximum. If the negative and positive charge centers are approximately equal in magnitude the bi-directional discharge propagates in the form of positive leaders inside of negative charge region and in the form of negative leaders inside of the positive charge region (e.g., Riousset et al. 2007). In this situation the leader system, which is assumed to be overall equipotential and neutral, remains at nearly zero potential (e.g., Riousset et al. 2007). Krehbiel et al. (2008) demonstrate that when the two charges are not balanced the leader potential can be significantly shifted in the direction defined by the charge with dominant magnitude and the propagation of the leader becomes essentially independent from the weaker charge center, allowing it to penetrate through the weaker charge center and to escape from the thundercloud. Krehbiel et al. (2008) presented a combination of observational and modeling results that indicate two principal ways in which upward discharges can be produced. The modeling presented in Krehbiel et al. (2008) indicates that blue jets occur as a result of electrical breakdown between the upper storm charge and screening charge attracted to the cloud top; they are predicted to occur 5–10 s or less after a cloud-to-ground or intracloud discharge produces a sudden charge imbalance in the storm. Gigantic jets are indicated to begin as a normal intracloud discharge between dominant mid-level charge and a screening-depleted upper level charge, that continues to propagate out the top of the storm (Krehbiel et al. 2008).

Figure 6 summarizes the results of simulating different types of discharges in normally electrified storms from Krehbiel et al. (2008). The lightning model employed to produce results shown in Fig. 6 uses a Lightning-Mapping-Array-Inferred multilayered charge structure positioned above a perfectly conducting flat ground plane (Riousset et al. 2007). The thundercloud and lightning discharge are modeled in a 3-D Cartesian domain using equidistant grids. The model uses a fractal approach to introduce stochasticity in a self-consistent model of the lightning channel, which fully satisfies Kasemir's hypothesis of equipotentiality and overall neutrality of the discharge (Kasemir 1960; Mazur and Ruhnke 1998; Riousset et al. 2007). In all cases shown in Fig. 6, the type of discharge results from a competition as to where breakdown is triggered first. Intracloud discharges usually win this competition because they occur between the two strongest charge regions during a storm's convective stages (Fig. 6a) (Krehbiel et al. 2008). The negative cloud-to-ground lightning discharges (−CGs, Figs. 6b, 6e) occur as descending precipitation generates lower positive charge (Williams 1989) or as the storm accumulates net negative charge, and can go either directly to ground or indirectly as a bolt-from-the-blue discharge (see discussion in Krehbiel et al. 2008). The negative gigantic jets (−GJs, Fig. 6f) provide an alternate way of relieving the mid-level negative charge, by discharging it to the upper atmosphere rather

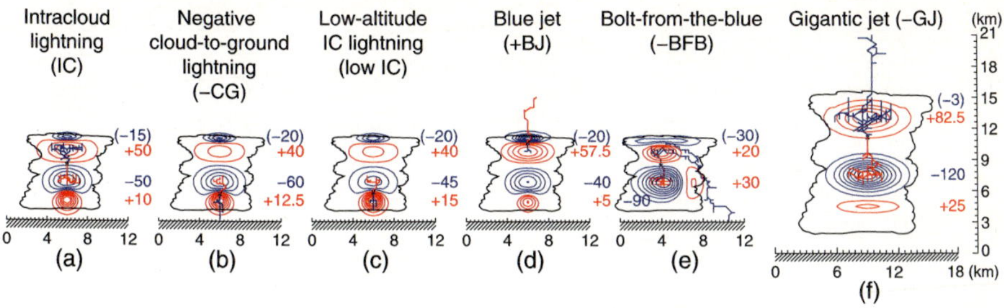

Fig. 6 Simulated discharges illustrating the different known and postulated lightning types in a normally electrified storm (Krehbiel et al. 2008). (**a–f**), *Blue and red contours* and numbers indicate negative and positive charge regions and charge amounts (**c**), respectively, each assumed to have a Gaussian spatial distribution. A partially analogous set of discharges occurs or would be predicted to occur in storms having inverted electrical structures (see Fig. S5 in Krehbiel et al. (2008, Supplementary Information)). Reprinted from Krehbiel et al. (2008) by permission from *Nature Geoscience*

than to ground (Krehbiel et al. 2008). The positive blue jets (+BJs) do the opposite, namely transport positive charge upward (Fig. 6d; Krehbiel et al. 2008).

Recent confirmation of the theory advanced by Krehbiel et al. (2008) was provided by Lu et al. (2011b). These authors reported observations of two negative polarity gigantic jets. In both cases the gigantic jet producing flash began as ordinary intra-cloud lightning with upper level channels attempting to exit the cloud, and then produced the upward gigantic jet. Neither flash had developed channels to ground, confirming that the major charge transfer during gigantic jets occurred between the cloud and ionosphere (Lu et al. 2011b).

Recently Riousset et al. (2010b) introduced a two-dimensional axisymmetric model of charge relaxation in the conducting atmosphere and applied this model in conjunction with three-dimensional lightning model proposed in Riousset et al. (2007) to illustrate how blue and gigantic jet discharges are produced above cloud tops. The model reported by Riousset et al. (2010b) accounts for the time dependent conduction currents and screening charges formed under the influence of the thundercloud charge sources and gives particular attention to realistic simulation of the dynamics of the screening charges near the cloud boundaries. The results demonstrate how the prior occurrence of intra-cloud discharges can prevent the development of a blue jet until a cloud-to-ground discharge enhances the excess of positive charge in the cloud by bringing negative charge to ground. The screening charge gradually developing at the cloud top leads to breakdown initiation near the cloud upper boundary, but is insufficient to contain the lightning leader channel within the cloud resulting in occurrence of upward propagating blue jet events. Furthermore, in thunderstorms where convective overturning near the cloud top is sufficiently strong, the screening layer that allowed for blue jet initiation, gets mixed with the storm's upper positive charge region, reducing the net positive charge in this region and causing a substantial charge imbalance between the two main layers of the thundercloud. Quantitative modeling of resulting discharge presented in Riousset et al. (2010b) reveals that the leader channels cannot be contained in the volume enclosed within the cloud boundary and eventually escape upward to form a gigantic jet, consistent with the ideas first expressed by Krehbiel et al. (2008).

Results presented in Krehbiel et al. (2008), Riousset et al. (2010b) provide experimentally substantiated mechanisms of escape of lightning leaders from cloud tops complementing the previous theoretical work (Petrov and Petrova 1999; Pasko et al. 1999; Pasko and George 2002; Tong et al. 2005; Raizer et al. 2006, 2007, 2010). The application of ideas advanced in Raizer et al. (2006, 2007, 2010) concerning possibility to map high potentials at the cloud top to higher altitudes using conducting leaders depends on the possibility to sustain the leader process at low air pressures at high altitude. The understanding of the streamer-to-leader transition and the development of accurate numerical parameterizations of streamer zones of lightning leaders of different polarities, especially under low air pressure conditions, represent an important problem in current research on transient luminous events (e.g., Pasko 2006; Pasko and Bourdon 2007) and relevant recent work on this subject is reviewed in Sect. 5.2 below.

4 Past, Present and Future Orbital Observation of TLEs

The first observation of a TLE from space was in a form of lightning associated emission layer at airglow altitude (~95 km) (Boeck et al. 1992), latter designated as elves (Fukunishi et al. 1996). The enhanced airglow events were recorded from space shuttle payload bay television cameras in a horizontal view. This first finding of lighting-induced airglow enhancement provided a new evidence of direct coupling between atmospheric lighting and enhanced airglow emission at the bottom of the ionosphere (Boeck et al. 1992). Elves, an acronym for Emission of Light and VLF perturbations due to EMP Source, were confirmed in ground observations using a multichannel high-speed photometer and image intensified CCD cameras at Yucca Ridge Field Station, Colorado as part of SPRITES'95 campaign in 1995 (Fukunishi et al. 1996).

During the STS-107 mission in January 2003, the Mediterranean Israeli Dust Experiment (MEIDEX) sprite campaign was conducted on board the space shuttle *Columbia* (Yair et al. 2003, 2004; Yair 2006). The total MEIDEX observation time was ~7 hours. In ~51 minutes of data collected over thunderstorms, 17 TLEs were identified including 7 sprites and 10 elves. The brightness of sprites was in the range of 0.3–1.7 MR in the 665 nm filter and 1.44–1.7 MR in the 860-nm filter. The estimated sprite detection rate was 0.13 sprites per minute (Yair et al. 2004). The observed elves were also analyzed, and the total emitted energy in the wavelength range 615–715 nm was estimated to be several to tens of kJ (Israelevich et al. 2004).

The LSO (Lightning and Sprite Observations) on board of ISS (International Space Station) is a first experiment dedicated to nadir observation of sprites in Earth orbit (Blanc et al. 2004, 2006, 2007). The first LSO measurements were conducted on the ISS in October of 2001. The LSO experiment used two digital space microcameras. One was equipped with a 762 nm narrowband filter targeting the specified $1PN_2$ (3–1) band of sprites, and the other camera covered a visible wavelength range (400–1000 nm, with maximum sensitivity at 690 nm) targeting lightning and sprite emissions. The major challenge to observe TLEs in a nadir view was to identify sprite emissions in the presence of overlapping lightning emissions. Blanc et al. (2004, 2007) designed a novel technique to separate the sprite events from lightning. Specifically, the LSO experiment used a method of spectral differentiation between sprites and lightning based on sprite $1PN_2$ (3–1) emission falling within the atmospheric absorption bands, $O_2(b^1\sum_g^+ - X^3\sum_g^+)$ (0–0) around 761.9 nm, in which lightning emissions are significantly attenuated. The LSO was a pioneer experiment to prepare the TARANIS satellite mission. The nadir observation is necessary to simultaneously study the optical emissions, relativistic runaway electrons and associated X-ray and γ-ray emissions, that is one of the primary goals of the TARANIS mission (see further discussion below).

During 19 hours of the LSO observations, 40 events were registered, total 17 sprites, 3 halos and 9 superbolts. The sprite observation frequency was about 1.7 per hour (Blanc et al. 2006). The detection rate of sprites was 0.033 sprites per minute. In the respective observation areas of LSO, the Lightning Imaging Sensor (LIS) estimated the number of lightning flashes to be about 1100 (Christian et al. 2003). Based on the fact that LSO observed only over the continents where 88% of the lighting occur, the occurrence of sprites number with respect to lightning was $17 \times 88\%/1100 = 1.4\%$. Having considered that the global number of lightning determined from LIS observation was 44 flashes per second (Christian et al. 2003) or 2640 flashes per minute, the global occurrence rate of sprites was estimated to be $2640 \times 1.4\% \sim 37$ sprites per minute (Blanc et al. 2006).

Present missions include ISUAL (2004) (Chern et al. 2003; Mende et al. 2006; Chen et al. 2008), Tatiana-1 (2005–2007) (Garipov et al. 2006), Tatiana-2 (2009) (Panasyuk et al. 2009; Garipov et al. 2010), and Sprite-Sat mission (2010) (Takahashi et al. 2006).

ISUAL (Imager of Sprites and Upper Atmospheric Lightnings) on board FORMOSAT-2 is the first satellite payload dedicated to the survey of the upper atmospheric discharge phenomena (Chern et al. 2003; Mende et al. 2006; Chen et al. 2008). The ISUAL project is an international collaboration between the National Cheng Kung University, Taiwan, Tohoku University, Japan and the instrument development team from the University of California, Berkeley. The ISUAL consists of three sensor packages including an intensified CCD imager, a six-channel spectrophotometer and a dual-band array photometer. The ISUAL observations have been successfully conducted during the first five years (2004–2009). Due to the significant scientific achievements during the first five years the ISUAL observations have been extended for three additional years (2010–2014) using funding from National Space Organization in Taiwan.

First sprite was recorded by ISUAL on July 4, 2004/21:31:15.451 (Mende et al. 2006). Furthermore, Mende et al. (2005) analyzed elves where the source lightning occurred beyond the solid Earth limb, and found that the elves contained significant 391.4 nm emission of the first negative N_2^+ band system ($1NN_2^+$). Mende et al. (2005) estimated that reduced electric field was > 200 Td by comparing the ratio of ISUAL spectrophotometer measurements of TLE produced emissions to theoretically derived emission intensities. Using estimated reduced electric field and the measured and total ionization rates derived from $1NN_2^+$ emission, an average electron density of 210 electrons cm^{-3} was estimated as being produced in the elves. Besides, the first FUV emissions (Lyman-Birge-Hopfield bands) were also detected in the ISUAL recorded elves (Mende et al. 2005). For sprites, Kuo et al. (2005) and Liu et al. (2006) estimated the intensities of electric field at the streamer tip to be 2–4E_k using the ISUAL spectrophotometric measurements, while Adachi et al. (2006) analyzed ISUAL array photometer data and obtained the electric field of 1–2E_k in the streamer region. As discussed in Sect. 3.2 the electric fields in sprite streamers are likely a factor of 1.5 higher (Celestin and Pasko 2010).

Chou et al. (2010) categorized ISUAL gigantic jets (GJs) into three types based on their generation dynamics and spectral properties. Type I GJs resemble those reported previously in Su et al. (2003), Kuo et al. (2009): after the fully developed jet (FDJ) established the discharge channel, the ISUAL photometers registered a peak that was from a return stroke-like-process. The associated ULF (ultra low frequency) spherics indicate that type I GJs are negative cloud-to-ionosphere discharges (−CIs). Type II GJs begin as blue jets and then develop into GJs in ∼100 ms. Blue jets also frequently occurred in the same region before and after the type II GJs. No identifiable ULF spherics for the type II GJs were found, though an extra event reported in Chou et al. (2010) with +CI ULF was likely a type II GJ. Thus the energy and the charge in these events may not accumulate to high enough levels to initiate a bright gigantic jet. Type III GJs were preceded by a lightning, and a GJ occurred near this preceding lightning. The spectral data of the type III GJs are dominated by lightning signals and the ULF data have a high background noise. The average brightness of the type III GJs falls between those of the other two types of GJs. It was proposed that the discharge polarity of the type III GJs could be either negative or positive, depending on the type of the charge imbalance left by the trigger lightning (Chou et al. 2010).

After analyzing the 1PN$_2$ brightness of the ISUAL elves and their FUV intensity and using modeling work for elves, Chang et al. (2010) demonstrated that ISUAL-FUV intensity of an elve could be used to estimate the peak current of the causative CG lightning. The ISUAL detection ratio of elves was improved since sensitivity of ISUAL FUV photometer is 16 times higher than that of ISUAL 1PN$_2$-filtered Imager. Hence, FUV photometer can be used to perform survey of elves to infer the peak current of the elve-producing lighting and other salient parameters. Besides, existences of multielves, which are FUV events from the M-components or the multiple strokes in lighting flashes were also found. Lee et al. (2010) analyzed the distribution of the TLEs registered by ISUAL, and deduced the synoptic-scale factors that control the occurrence of TLEs. Two different distribution patterns were analyzed. For the low latitude tropical regions (25°S $\sim$ 25°N), 84% of the TLEs were found to occur over the Intertropical Convergence Zone (ITCZ) and the South Pacific Convergence Zone. The distribution of TLEs exhibited a seasonal variation that migrates North and South with respect to the equator. For the mid-latitude regions (latitudes beyond $\pm$30°), 88% of the Northern winter TLEs and 72% of the Southern winter TLEs occurred near the mid-latitude cyclones. The winter TLE occurrence density and the storm track frequency share similar trends, with the distribution of the winter TLEs offset by 10°–15°.

Tatiana-1 is a Moscow State University's research educational microsatellite. Measurements with this satellite were conducted during the period from January 2005 to March 2007 (Garipov et al. 2005b, 2005a, 2006). The instruments on board Tatiana-1 consisted of photomultiplier tubes (PMTs) for UV detection with wavelength band 300–400 nm and the detectors of extremely high-energy cosmic rays (EHECRs). The EHECRs were dedicated to observe Cherenkov atmospheric radiation, the secondary radiation characteristic of the electron passage through the atmosphere. Two years of UV data recorded by Tatiana-1 were analyzed, and the brightness (energy) and occurrence rate of UV events were compared with the lunar phase. The correlation between UV events and lunar phase was found. Moreover, the rate of UV flashes with energy release in the atmosphere greater than 50 kJ was four times higher at full moon night than at other moon phase (Garipov et al. 2008).

Shneider and Milikh (2010) studied the atmospheric electricity phenomena that could serve as a source for short millisecond range UV flashes. It was shown that UV flashes in the millisecond range detected by Tatiana-1 could be originating from gigantic blue jets (GBJ). The influence of an assumed self-consistent governing electric field in the GBJ streamer zone on the UV pulse shape and duration was revealed. It was also shown that red sprites can be a source for UV flashes with similar temporal profiles but at much lower intensity (Shneider and Milikh 2010).

Tatiana-2 satellite was launched on 17 September 2009 with a new set of instruments that includes UV (300–400 nm)- and red (600–700 nm)-filtered photomultiplier tube (PMT) micro-electro-mechanical telescope for extreme lighting (MTEL), photo spectrometer and electron flux detector (Panasyuk et al. 2009; Garipov et al. 2010). The electron flux detector can detect the fluorescence light by penetrating electron with minimum energy greater than 1 MeV.

SPRITE-SAT is a micro satellite designed and developed by Tohoku University, Japan (Takahashi et al. 2006). The main scientific goal of SPRITE-SAT satellite is to simultaneously observe TLEs and terrestrial gamma-ray flashes (TGFs) in nadir direction and to study the relationship and generation mechanisms of TLEs and TGFs. The SPRITE-SAT has been successfully launched on January 23 2009 and operated by Tohoku University.

Future missions include Chibis-M (2010-) (Klimov et al. 2009), JEM-GLIMS (2011-) (Sato et al. 2009), ASIM (2014-) (Neubert et al. 2008), TARANIS (2016-) (Blanc et al. 2006; Lefeuvre et al. 2009).

Chibis-M mission (Klimov et al. 2009) (see http://chibis.cosmos.ru/) is a micro-satellite mission designed for studies of TLEs and TGFs. In this mission, scientific instruments include X-ray and γ-ray detectors with energy range 50–500 keV and time resolution 30 ns, UV detector sensitive in wavelength range of 180–800 nm, digital photo camera with exposure time 0.2 second, radio frequency sensor with frequency range 20–50 Hz, and ULF-VLF antennas. Micro-satellite is to be delivered to ISS by a transport vehicle.

Global Lightning and sprIte MeasurementS on JEM-EF (JEM-GLIMS) is a space mission to observe lightning and TLEs from the Exposure Facility (EF) of the Japanese Experiment Module (JEM) at International Space Station (ISS) (Sato et al. 2009). The scientific goal of the JEM-GLIMS mission is to study the generation mechanism of TLEs and to identify the relationship between lightning, TLEs and TGFs (TGFs). The scientific instruments of the JEM-GLIMS mission consist of two CMOS cameras, two photometers, one spectro-imager, and two VHF receivers. JEM-GLIMS is dedicated to collaborate with the ground-based observations and with other space missions, such as TARANIS and ASIM (Sato et al. 2009).

ASIM (Atmopshere-Space Interactions Monitor) (Neubert 2009) is an instrument suite, mounted on an external platform of the European Columbus module for the International Space Station (ISS). The ASIM will be coordinated with ground EuroSprite campaigns (Neubert et al. 2008). The ASIM payload is equipped with Modular Multi-spectral Imaging Array (MMIA) and Modular X- and Gamma-ray Sensor (MXGS). The MMIA consist of cameras and photometers where cameras provide spatial resolution and photometers temporal resolution. The scientific goals of ASIM are (1) to find the source region of TLEs and TGFs, (2) to study the physics of TLEs and TGFs, (3) to study effects of water vapor and aerosols on cloud formation and atmospheric electricity, and (4) to capture meteor events (Neubert 2009).

TARANIS (Tool for the Analysis of Radiations from lightNIngs and Sprites) (Blanc et al. 2006; Lefeuvre et al. 2009) is a CNES satellite project to study impulsive transfers of energy between the Earth atmosphere and the space environment. TARANIS will employ the micro-cameras, photometers, X-ray, γ-ray detectors, energetic electron detectors, and radio receiver and antenna. The goals of the mission are: (1) to advance the physical understanding of the links between TLEs and TGFs, (2) to clarify the potential signatures of impulsive transfers of energy, verified by physical mechanism, and (3) to characterize the physical parameters in TLEs and TGFs (Blanc et al. 2006; Lefeuvre et al. 2009).

5 Chemical, Energetic and Electric Effects of TLEs on Local and Global Scales

5.1 Chemical Effects of TLEs

Atmospheric discharge phenomena (corona discharge, lighting and TLEs) are important sources of NO_x (NO and NO_2, summarized as NO_x) in the troposphere and middle atmosphere. NO_x plays an important role in atmospheric chemistry since NO_x facilitates chemical reactions that influence the concentrations of ozone (O_3) and hydroxyl radical (OH) (Rakov and Uman 2003, p. 507). Appropriate quantification of lighting-produced NO_x and TLE-produced NO_x has become an important issue in studies of NO_x budget. The knowledge of occurrence rate and deposited energy, and understanding of triggered chemical reaction chain associated with lighting and TLEs are necessary to advance studies of NO_x budget.

However, the estimates of lightning induced NO_x (LNO_x) vary by two orders of magnitude and range from 1 to 200 Tg(N) yr^{-1}, where 1 Tg(N) = 10^{12} grams of nitrogen. The estimates of TLEs induced NO_x (TLE-NO_x) is limited because the types, occurrence rate and energy budget of TLEs are not well-studied (Rakov and Uman 2003, p. 508). The most recent accepted estimates of global LNO_x production range from 2 to 8 Tg(N) yr^{-1} (Schumann and Huntrieser 2007). In other publications Cooray et al. (2009) used the knowledge on the electrical discharges to extrapolate the measurements of NO_x yield in laboratory spark experiments to that in lighting, and derived the global annual production of NO_x of about 4 Tg(N). Beirle et al. (2004) used Global Ozone Monitoring Instrument (GOME) NO_2 column density and Lightning Imaging Sensor (LIS) to estimate 2.8 Tg(N) yr^{-1}. Beirle et al. (2010) noted that uncertainty of LNO_x estimates might be biased due to a high variability of NO_2 measurement for coincided lighting events and due to the fact that many LNO_x measurements are limited to the US region, not representative for global lighting.

Using in-situ NO_2 measurements from the DC-8 aircraft missions and lightning data from National Lightning Detection Network, Bucsela et al. (2010) estimated the LNO_x production to be in the range of 6–15 $\times$ 10^{25} molecules per flash. To date, there are no direct measurements of TLE induced NO_x production. Peterson et al. (2009) estimated TLE-NO_x production of blue jets and red sprites using the results of laboratory discharges. The blue jets were calculated to produce 1.7×10^{22}–7.4×10^{26} molecules of NO_x per event while corresponding numbers for sprites were estimated to be 6.8×10^{23}–6.3×10^{27} molecules of NO_x per event. On the basis of global TLEs occurrence rate (Ignaccolo et al. 2006), global annual NO_x production is estimated to be between 7×10^{23}–2×10^{28} molecules per second (Peterson et al. 2009) with five order of magnitude uncertainty, corresponding to global TLE-NO_x production of 5.2×10^{-4} to 14.7 Tg(N) yr^{-1}. The above mentioned TLE-NO_x estimates are still under debate (de Urquijo and Gordillo-Vazquez 2010; Nijdam et al. 2010; Peterson et al. 2010a, 2010b).

During the daytime, the NO_x gases are in photochemical balance in the stratosphere. After sunset NO is quickly converted to NO_2 in reaction with O_3, and the nighttime NO_2 measurements provide a good presentation of stratospheric NO_x (Arnone et al. 2008, 2009; Rodger et al. 2008). Arnone et al. (2008) investigated the NO_x production at sprite altitudes using the MIPAS (Michelson Interferometer for Passive Atmospheric Sounding) instrument on board European Space Agency's (ESA) ENVIromental SATellite (ENVISAT). The ENVISAT is an advanced polar-orbiting Earth observation satellite launched by ESA, which provides measurements of the atmosphere, ocean, land, and ice [see also http://envisat.esa.int]. MIPAS is a limb-scanning Fourier Transform spectrometer recording emission spectra in the mid-infrared 680–2410 cm^{-1} (4.15–14.6 μm) with spectral resolution of 0035 cm^{-1}. The MIPAS scan across the horizon detecting atmospheric spectral radiances, which can be inverted to vertical temperature and trace species (H_2O, O_3, CH_4, N_2O, HNO_3 and NO_2 profiles) (Fischer et al. 2008). Arnone et al. (2008) adopted the Geo-fit Multi-Target Retrieval (GMTR) algorithm (Carlotti et al. 2006) to retrieve the two dimensional emission profile of NO_2 from limb-scan data of MIPAS/GMTR on board of ENVISAT satellite. Arnone et al. (2008) used WWLLN (World Wide Lightning Location Network) lighting data as a proxy of sprite activity, and studied the possible sprite-induced NO_2 perturbation, derived from MIPAS/GMTR data, over the region with intense WWLLN lightning activity during the period of August to December 2003.

Arnone et al. (2008) examined the mean nighttime WWLLN-NO_2 relationship at an altitude of 52 km, where MIPAS NO_2 measurement coincided with high WWLLN lightning activity region, and suggested no significant impact on global scale of sprite perturbation for the time period August to December 2003. For a localized region measurements of

WWLLN-NO_2 in a 60-minute time window prior to MIPAS observations, the WWLLN-NO_2 data showed a positive displacement of about +1 ppbV (~10%) above the NO_2 background measurements (10 ± 1.5 ppbV). Arnone et al. (2008) indicated the possible observational evidence of ~10% NO_2 enhancement at sprite altitude of 52 km above the active thunderstorms, but provided no evidence of any significant increase on a global scale. Arnone et al. (2009) presented further analysis of an NO_2 enhancement as being dominated by the contribution from a region with low background winds north of the Equator (5°N to 20°N) during the first 30 to 40 days of the sample (i.e., the tail of Northern Hemisphere summer).

Rodger et al. (2008) inspected the possible sprite-induced nighttime NO_x using GOMOS NO_2 measurements at altitudes of 20–70 km. The GOMOS (Global Ozone Monitoring by Occultation of Stars) on board ENVISAT satellite measures vertical profiles of minor gases using >600 occultations per day with global coverage. Rodger et al. (2008) selected four land/ocean regions from LIS/OTD five year lighting distribution data with strong regional and seasonal variations in lighting activities. The two to three orders of magnitude peak differences between lightning activity in the selected land/ocean regions were used to evaluate the corresponding variation of GOMOS nighttime NO_2 measurements at blue jet altitudes of 30–50 km and at sprite altitudes of 50–70 km. No significant variations coincided with lightning rate changes over land/ocean contrast, or seasonal variations shown in LIS/OTD distribution data. Rodger et al. (2008) concluded that TLE may cause local variations in NO_x, but these do not appear to be significant on regional scales.

Huang et al. (2009) investigated the lightning-induced sudden brightening in the OH airglow layer using ISUAL 630-nm-filtered Imager. Huang et al. (2009) also proposed the conceptual mechanism of the induced OH nightglow enhancement. Huang et al. (2010) further analyzed 630-nm-filtered Imager recorded elves, and found that there is considerable intensity enhancement (~1.25 kR) unaccounted for after the $N_2$1P contribution has been removed. Huang et al. (2010) suggested that there might be OH emissions in elves and that OH species might also be involved in the lightning-induced processes and contribute to the intensity enhancements that were observed.

Enell et al. (2008) studied the chemical effects of the electrical discharge phenomena in the middle atmosphere using an ion-neutral chemical model, also including the ionization, excitation, dissociation and chemical reactions of active species. The model considered the chemical production and loss with the electric discharge associated reactions and important ion-neutral chemical reactions. The reaction rates of the model were derived from Bolsig solver (Pitchford et al. 1981) and Siglo database [CPAT and Kinema Software]. Enell et al. (2008) used the derived E-field ($\sim$ 0.5–2E_k) from ISUAL recorded TLE events at altitudes of 53, 63, 73, 83 km (Adachi et al. 2006) as input electric fields in the model. The modeling results show that the NO_x enhancements are at most one order of magnitude in the streamers, corresponding to at most 10 mol per event. Given the estimated global occurrence rate (3 per minute), the global production of NO_x by sprites is 150–1500 kg/day, i.e., 5.5×10^{-5}–5.5×10^{-4} Tg(N).

Sentman et al. (2008) utilized a simplified sprite streamer model coupled with the full kinetic modeling including 80+ species and 800+ chemical reactions. The species in their comprehensive kinetic model included background neutrals and atoms and their different electronic states, negative/positive ions and ion clusters. The reaction set considered the reactions accounting for cosmic ray background, electric field driven electron impact processes, electron-ion recombination, attachment-detachment processes, ground/active state chemistry, collisional and radiative deactivation, ion conversion and ion-ion recombination, odd hydrogen/nitrogen reactions, and positive/negative ion/chlorine chemistry. The sprite streamer model at an altitude of 70 km (Sentman et al. 2008) utilized an impulsive electric field at the streamer tip. The electric field was modeled as $E_h(t) = E_0 e^{-(t-t_0)^2/\Delta t^2}$ where $E_0 = 5E_k$, five times of the magnitude of breakdown electric field E_k, and Δt is the characteristic time duration (6 μs) of the impulse centered at time t_0. In the streamer head, the modeled impulse electric field triggered a series of chemical reaction chains. The principal chemical reactions were electron impact processes at streamer tip. The produced active chemical species remained in the trailing column of the streamer, and the chemiluminescent emissions persisted in that afterglow region for the time duration greater than tens of seconds (Sentman et al. 2008).

The simulation results of Sentman et al. (2008) predicted that the weak (~1 kR), but possibly detectable, OI 557.7 nm and O_2 atmospheric airglow emissions and OH Meinel emissions (~1–10 R). The derived sprite parameters (emission rate and brightness) in the plasma chemistry streamer model were consistent with ground observation of sprites (Stenbaek-Nielsen and McHarg 2008). Sentman and Stenbaek-Nielsen (2009) simulated the streamer plasma chemistry accounting for non-zero field ($<0.5E_k$) in the trailing region of the streamer channel. The results showed that the electron densities in the trailing region were slightly reduced due to enhanced dissociative attachment in the undervoltage environment. The densities of $O(^3P)$ and the metastable species $O(^1D)$ and $N_2(A^3\Sigma_u^+)$ were enhanced by a modest factor of 2 or less compared with a field-free tail. The $O_2(a^1\Delta_g)$ species showed slightly greater enhancements, by up to a factor of 3, and O showed enhancements by up to a factor of 5.

Gordillo-Vazquez (2008) developed a full time-dependent kinetic model to consider the microscopic collisional and radiative processes underlying the optical flashes associated with an impulsive discharge in the single sprite streamer passing through the air region in the mesosphere at altitudes, 63, 68 and 78 km. The kinetic model for air plasma included more than 75 species and almost 500 reactions. The reaction set in the kinetic model took into account the possible impact of H_2O (humid chemistry). The modeling results predicted the radiative emissions associated with vibrational kinetics of N_2 and CO_2, and local enhancement of NO_x at different altitudes. All calculations were conducted for midnight conditions in mid-latitude regions (+38N°) and 0° latitude. The initial values of the neutral species were taken from the Whole Atmosphere Community Climate Model (WACCM). The kinetic model of Gordillo-Vazquez (2008) assumed that a sprite streamer had a step-like impulse electric field with a peak value of $3.3E_k$ and time duration 5 μs in the streamer head, and weak electric field in the afterglow region in the streamer body. The characteristic value of electric field in the streamer body was 1 Td. The results demonstrated that the impact of 4 ppm of H_2O was only slightly visible in O_3^- at 68 and 78 km while it strongly affected the behavior of the anion CO_4^- at all the altitudes investigated. At 68 km, the concentrations of NO and NO_2 increased by about one order of magnitude while that of NO_3 exhibited a remarkable growth of up to almost three orders of magnitude. Gordillo-Vazquez (2008) investigated the brightness of the radiative transitions associated with sprite in the visible, IR and UV band emissions. The reported brightness of LBH band was always higher than the NO-γ band system. The vibrationally excited CO_2 emission bands (4.26, 9.4, 13.9 and 14.9 μm) exhibited noticeable enhancements of radiative emissions.

Liu and Pasko (2007) developed a model to study NO-γ emissions ($NO(A^2\Sigma^+) \rightarrow NO(X^2\Pi) + h\nu$) associated with streamer discharges in air at different pressures. The modeling results indicate that the $NO(A^2\Sigma^+)$ species in sprite streamers at 70 km altitude are mostly produced by interaction of $N_2(A^3\Sigma_u^+)$ metastable species with high-density ambient $NO(X^2\Pi_r)$, molecules via reaction $N_2(A^3\Sigma_u^+, w) + NO(X^2\Pi_r, v) \rightarrow NO(A^2\Sigma^+, v') + N_2(X^1\Sigma_g^+, w')$. Analysis of the production and loss mechanisms for the upper excited states leading to NO-γ and N_2 Lyman-Birge-Hopfield (LBH) emissions demonstrates that the total intensity of NO-γ emissions associated with sprites is substantially weaker than that of the N_2 LBH emissions. Campbell et al. (2007) recently emphasized importance of $N_2(A^3\Sigma_u^+) + O \rightarrow NO + N(^2D)$ reaction for production of NO in the upper atmosphere. Gordillo-Vazquez (2008) indicated possible importance of the direct electron impact excitation of $NO(A^2\Sigma^+)$ in sprite discharges as source of NO-γ emissions. The significance of the direct excitation is determined by the cross section of the related process, which varies by two orders of magnitude in the available literature (see Gordillo-Vazquez 2008, and references therein). Liu and Pasko (2010) have recently reviewed publications with relevant experimental data, and utilized streamer modeling to conclude that the direct electron impact excitation is not the dominant production mechanism for $NO(A^2\Sigma^+)$ species.

Analysis of ISUAL spectrophotometric data in the context of sprite streamer modeling has been reported in Liu et al. (2009b). These authors simulated two positive model streamers developing in strong and weak applied electric fields (with respect to E_k) at 70 km latitude. The intensity ratio of the second positive band system of N_2 to the first negative system of N_2^+ was obtained separately from the modeling and the ISUAL measurements. The comparison results indicate that the ratio obtained for the streamer developing in an electric field close to the conventional breakdown threshold field E_k agrees with the ISUAL measurements at the very early stage of the sprite development better than for the streamer developing in a field much lower than E_k. The authors utilized the strong field streamer case in conjunction with the ISUAL data to gain additional information on the poorly known quenching altitude of the $N_2(a^1\Pi_g)$ state, which is responsible for N_2 LBH band system. The results supported the conclusions of the previous study by Liu and Pasko (2005) that the 77 km is a good estimate for the quenching altitude of $N_2(a^1\Pi_g)$.

Gordillo-Vazquez and Donko (2009) considered the effect of relative humidity on plasma chemistry model. The relative humidity of the atmospheric air between 0 and 15 km can change between 15% and 100% depending on the altitude investigated and the ambient temperature. The reported calculations suggested that the relative humidity had a clear impact on the behavior of the electron energy distribution function (EEDF) and magnitude of the transport coefficients of air plasmas at ground (0 km) and room temperature conditions (293 K). At higher altitudes (11 and 15 km), the influence of the relative humidity was negligible when the values of the gas temperature were assumed to be the 'natural' ones corresponding to those altitudes, that is, ~215 K (at 11 km) and ~198 K (at 15 km). However, it was found that a small enhancement (of maximum 100 K) in the background gas temperature (that could be reasonably associated with the TLE activity) would lead to a remarkable impact of the relative humidity on the EEDF and transport coefficients of air plasmas under the conditions of blue jets, blue starters and gigantic jets at 11 and 15 km. The latter effects were visible for relatively low reduced electric fields ($E/N > 25$ Td) that could be controlling the afterglow kinetics of the air plasmas generated by TLEs.

Recently, Hiraki et al. (2008) and Hiraki (2009) used an ion-neutral chemical model to calculate the chemical effects under two situations. The first one corresponded to the impulsive electric field case. Hiraki et al. (2008) assumed a streamer tip with an impulsive electric field in the discharge region of TLEs. The tip region had a uniform magnitude of electric field (150 kV/cm) and moved at streamer speed $\sim 10^7$ cm/s. Nighttime condition results demonstrated that the densities of NO, O_3, H, and OH increased suddenly through reactions triggered by the initial atomic nitrogen and oxygen products and electrons just after streamer initiation. The NO and NO_2 species still remained for 1 hour, predominantly around an altitude of 60 km. Other species (O_3, OH, HO_2 and H_2O_2) remained in the altitude range between 40 and 70 km. The second considered situation corresponded to the quasi-electrostatic case. Hiraki (2009) calculated the chemical perturbations due to the lightning-induced electric field. Hiraki (2009) found that the recovery timescale of the electron density was much longer than the interval of the lightning stroke (~10 ms). The subsequent sprite events as well as the related field dynamics could be well affected by the residual effects of the primary sprite event (Hiraki 2009).

Haldoupis et al. (2010) reported narrowband VLF observations in which multiple transmitter-receiver VLF pairs with great circle paths passing near a sprite-producing thunderstorm were available, and presented an evidence that visible sprite occurrences are accompanied by early VLF perturbations in a one-to-one correspondence. The early VLF events are abrupt perturbations of sub-ionospherically propagating VLF waves (including both perturbations in amplitude and phase) that occur within 20 ms of a lightning event. These events have been discussed in Sect. 3.1 of this review with relationship to elves. The observations of Haldoupis et al. (2010) imply that the sprite generation mechanism may cause also sub-ionospheric conductivity disturbances that produce early VLF events. However, the one-to-one visible sprite to early VLF event correspondence, if viewed conversely, appears not to be always reciprocal (Haldoupis et al. 2010). This is because the number of early events detected in some case studies was considerably larger than the number of visible sprites. Since the great majority of the early events not accompanied by visible sprites appeared to be caused by positive cloud to ground (+CG) lightning discharges, it is possible that sprites or sprite halos were concurrently present in these events as well but were missed by the sprite-watch camera detection system (Haldoupis et al. 2010). In terms of general discussion of conductivity/chemical perturbations associated with sprites the average conductivity in sprite volume deduced from ELF radiation emitted by sprite current is $\sim 10^{-7}$ S/m (Pasko et al. 1998b). A recent study of Gordillo-Vazquez and Luque (2010) indicates that larger conductivity changes are possible due to associative detachment of electrons from O^- ions in the trailing regions of long sprite streamers when electric field in these regions approaches the conventional breakdown threshold field during the later stages of sprite evolution (Liu 2010; Luque and Ebert 2010).

5.2 Energetic Effects of TLEs

TLEs are discharge phenomenon, representing the mechanism for energy transfer from the tropospheric charge separation and lightning to the middle atmosphere and the ionosphere. The physical mechanism of TLEs is similar to charge/discharge or dielectric breakdown in a capacitor between two conducting planes. The redistribution of charge in the thunderstorm can generate strong electrostatic fields at high altitudes and the current in a lightning stroke can radiate the electromagnetic field. The leader process in lighting channels may produce runaway electrons leading to X-ray and gamma-ray emissions. The interactions between electric fields and particles in the middle atmosphere cause energy pooling or air breakdown in large volumes between the ground and the bottom of ionosphere. The transferred energy is released in the form of radiated photons (UV, visible, infrared emissions, X-rays and gamma-rays), stored in the medium in the form of vibrationally or electronically excited metastable species, or deposited as heat, leading to generation of infrasound waves. Review of recent literature relevant to infrasound emissions and heating effects of TLEs is included below in this subsection.

The global occurrence rates of TLEs are important to understand in order to estimate their effects on NO_x budget and energy budget in the mesosphere. First such estimations were conducted using ELF emissions from lightning discharges during a time period between June 19, 2001 and January 20, 2002. Based on the empirical sprite initiation probability with critical charge moment change, the global occurrence rate of sprites was estimated to be 0.5 sprites per minute (Sato and Fukunishi 2003). In MEIDEX mission, the average detection rate of TLEs during the MEIDEX was 0.13 sprites per minute (Yair et al. 2004). Multiplying the detection rate of sprites (0.13) by a conservative estimate of simultaneous tropical thunderstorms (100), we get a global occurrence rate of 13 sprites per minute in the tropics. In LSO mission, the detection rate of sprites was 0.3 sprites per minute and estimated global occurrence rate of sprites was ~37 sprites per minute (Blanc et al. 2004). Using ground observations, (Ignaccolo et al. 2006) estimated the occurrence rate of sprites to be ~2.8 events per minute. Chen et al. (2008) and Hsu et al. (2009) analyzed ISUAL data during an extended time period from 2004 to 2009 and reported the occurrence rate of TLEs of different types: 72, 3.7 and 1 per minute for elves, halos and sprites, respectively. The average observed $1PN_2$ energies in elves, halos and sprites are 1.9, 1.4 and 2.2 MJ, respectively (Kuo et al. 2008). Having considered the ratio of $1PN_2$ excitation rate to the total collision rates for the electron impact processes in the 78.1% N_2 and 20.9% O_2 air mixture, the ratio of total energy deposition to the $1PN_2$ energy is about 10. Therefore, the average energy deposition of elves, halos and sprites are 19, 14 and 22 MJ, respectively (Kuo et al. 2008). Therefore, elves, halos and sprites deposit 1370, 52 and 22 MJ of tropospheric energy in the middle atmosphere (Hsu et al. 2009).

In comparison to a typical total energy of 1–10 GJ in the cloud-to-ground (CG) lightning, the average total energy deposition (22 MJ) per sprite estimated by Kuo et al. (2008) is consistent with the theoretical calculation 1–10 MJ by Sentman et al. (2003). Other work based on scaling of observed optical emission line ratios had suggested that the total energy deposited in the mesosphere by sprites could be as large as 1 GJ (Heavner et al. 2000). Using infrasound detection with a method independent of an optical measurement, Farges et al. (2005) estimated a range of total energy between 0.4 and 40 GJ. Recently, Takahashi et al. (2010) estimated the optical energy of sprites using ISUAL array photometer. The averages of the time-integrated optical energies are 176 kJ and 119 kJ for the $1PN_2$ and second positive N_2 ($2PN_2$) band systems, respectively (Takahashi et al. 2010). Hence, the presently estimated total energy deposition by sprites varies by many orders of magnitude (1 MJ to 40 GJ) and depends on particular measurement technique employed. Both, the global occurrence rates of different types of TLEs and their energy budget are still poorly quantified and require further investigation.

Farges and Blanc (2010) reported inverted-chirp infrasonic signals with high frequencies arriving before low frequencies, possibly emitted by sprite discharges and observed on the ground at close range (<100 km) from the source. Following this report de Larquier and Pasko (2010) utilized a parallel version of a two-dimensional finite-difference time-domain model of infrasound propagation in a realistic atmosphere to demonstrate that the observed morphology of infrasound signals is consistent with general scaling of diameters of sprite streamers inversely proportionally to the air density. The smaller structures at lower altitudes radiate higher infrasonic frequencies that arrive first at the observational point on the ground, while the low frequency components are delayed because they originate at lower air densities at higher altitudes. Results of Farges and Blanc (2010) suggest that the observed infrasound signal is strongly related to the sprite vertical extent. The absorption of infrasonic waves increases with altitude and is generally proportional to the square of frequency (e.g., Sutherland and Bass 2004; de Groot-Hedlin 2008). Therefore the absorption scaling through the sprite vertical extent may be at least partly responsible for the observed infrasonic signal. Specifically, it was noted by de Larquier and Pasko (2010) that the signal arriving from higher altitudes would naturally have a depleted high frequency content. Also, geometrical considerations indicate that such a signal would arrive to an observer positioned several tens of kilometers horizontal distance from the sprite with some delay with respect to the signal emitted from the lower portions of the sprite. The quantitative results of de Larquier and

Pasko (2010) demonstrate that strong absorption of high frequency infrasonic components at high altitudes (i.e., similar to 0.2 dB/km for 8 Hz at 70 km) may also contribute to formation of inverted-chirp signals observed on the ground at close range.

The general scaling of the Joule heating time scale in the streamer channels as a function of air density can be quite easily deduced from basic similarity analysis of gas discharges to be inversely proportional to square of gas density (e.g., Achat et al. 1992; Tardiveau et al. 2001; Pasko 2006). Therefore it is generally expected that the heating processes and resulting streamer-to-leader transition should be delayed with reduction of air pressure (i.e., at higher altitudes in the Earth atmosphere) and it should be possible to define a set of specific conditions (i.e., altitude range, reduced electric field E/N, etc.) in the Earth atmosphere for which the transition becomes impossible. The visual appearance of some blue jets and gigantic jets [see for example Fig. 2 in Pasko et al. (2002), Fig. 1 in Wescott et al. (2001) and Fig. 10 in Pasko and George (2002)] is suggestive of a transition from hot lightning like channels to a cold streamer dominated region at higher altitudes. In terms of negative leader phenomenology, the moment of attachment of gigantic jet to the lower ionospheric boundary, can be interpreted as the "final jump stage", when the leader streamer zone makes contact with the opposite electrode (Bazelyan and Raizer 1998, p. 212). This stage may also have some resemblance to the negative corona flash stage of negative leader development (Bazelyan and Raizer 2000, p. 199). The high speeds during the final jump, 5×10^4 m/s to 10^6 m/s (Bazelyan and Raizer 1998, p. 212), are consistent with the range of speeds, from 5×10^4 m/s to more than 2×10^6 m/s, reported in Pasko et al. (2002). Kuo et al. (2009) recently reported first photometric measurements of the fast $\sim10^7$ m/s upward propagation during the initial stage (also referred to as fully developed stage in Su et al. 2003) of a gigantic jet probably indicating highly overvolted conditions, which streamers experienced as they moved to low air density (high reduced electric field E/N) regions at high altitude. The relatively bright persistent channel below ~40–50 km altitude observed in gigantic jets (Pasko et al. 2002; Su et al. 2003) and designated as trailing jet in Su et al. (2003), can be interpreted as an attempt of the negative leader to form a next step, which has not succeeded (possibly due to a fast dielectric relaxation response of a conducting atmosphere or a significant lengthening of the heating time scales at low air pressures).

As discussed recently in Riousset et al. (2010a, see also extensive list of references cited therein) there are two principal approaches that are used in the existing literature for modeling studies of air heating effects in streamer discharges. The first one usually postulates either a time-dependent or a stationary current and derives time dynamics of the electric field, conductivity and other parameters of the channel assuming continuity of the current along the channel. The second approach assumes a constant value of the electric field in the channel and then self-consistently evaluates dynamics of other discharge parameters (i.e., conductivity, current, gas temperature, etc). The second approach describes very well a situation after the streamer bridges a relatively short gap with a constant applied voltage, and there are several related experiments conducted under controlled laboratory conditions at ground and near ground air pressures in which streamer-to-spark transition times have been accurately measured (e.g., Černák et al. 1995; Larsson 1998). Riousset et al. (2010a) have recently reported modeling results relevant to the second approach. The related model set up represents a first step toward development of more general and scalable with air density leader models, and is motivated by the possibility to carefully test the model calculated streamer-to-spark transition times in comparison with experimental data. Riousset et al. (2010a), in particular, presented a formulation of streamer-to-spark transition model that allowed studies of gas dynamics and chemical kinetics involved in heating of air in streamer channels for a given air density N under assumption of constant applied electric field E. The model accounts for the dynamic expansion of the heated air in the streamer channel and resultant effects of E/N variations on plasma kinetics, the vibrational excitation of nitrogen molecules $N_2(v)$, effects of gains in electron energy in collisions with $N_2(v)$, and associative ionization processes involving $N_2(A^3\Sigma_u^+)$ and $N_2(a'^1\Sigma_u^-)$ species. The results are in excellent agreement with available experimental data at ground and near ground air pressures, and demonstrate that for the air densities corresponding to 0–70 km altitudes the kinetic effects lead to a significant acceleration of the heating, with effective heating times scaling closer to $1/N$ than to $1/N^2$ predicted on the basis of similarity laws for Joule heating (Riousset et al. 2010a). This acceleration is attributed to a strong reduction in electron losses due to three-body attachment and electron–ion recombination processes with reduction of air pressure (Riousset et al. 2010a).

5.3 Global Electric Circuit Effects of TLEs

As part of this review we also make some remarks as to recent research on global electric circuit processes as relevant to TLEs. It is well understood that electrified shower clouds, thunderstorms and lightning provide contributions to the total generator current in the global electric circuit with well defined, and also relatively well understood diurnal and annual variability (e.g., Mareev et al. 2008; Mareev 2010; Williams 2009; Liu et al. 2010, and references therein). Recent research indicates that in addition to above mentioned components there are several other types of electrical processes that may contribute to the global electric circuit (these include both charging and discharging effects). In particular, modeling studies reported in Krehbiel et al. (2008) emphasize that the recently discovered transient luminous events called gigantic jets (Pasko and Stenbaek-Nielsen 2002; Su et al. 2003) are highly energetic discharges forming a direct electrical path for transferring main negative charge of a thundercloud to the lower ionosphere (discharging the global electric circuit). These events are developing on time scales measured in hundreds of milliseconds. The charge transfers have only recently been measured and can exceed 100 C in one event (for example 144 C were reported in Cummer et al. 2009). Also, it is now known that sprite type of transient luminous events is driven by cloud-to-ground lightning discharges that possess significant continuing current. These discharges can move hundreds of Coulombs of positive charge to ground on a time scale of hundreds of ms. Cummer and Fullekrug (2001) reported an example of such discharge in which 850 C of charge were transferred in 150 ms. Given approximately 200,000 C residing in the global electric circuit, this indicates that on the order of 0.1% of that total can be moved in a single event. Although these events do not occur often (Williams 2009), their transient effects can be measurable (Fullekrug and Rycroft 2006; Rycroft and Odzimek 2010). Rycroft and Odzimek (2010) summarize recent parameterizations that are used to model thunderclouds and discuss an engineering model of global electric circuit including transient aspects related to lightning and sprites. This model is one of few current models that treats transient global electric circuit solutions. Rycroft and Odzimek (2010) point, in particular, that the role of high altitude intra-cloud lightning discharges is not fully understood and requires further investigation.

6 Summary and Concluding Remarks

In this review article we introduced transient luminous events of different types, including elves, sprite halos, sprites, blue jets and gigantic jets. Then, we described phenomenological features of sprites, including meteorological conditions leading to their production, parent lightning parameters, and their shapes, sizes, and dynamical characteristics of their formation. The most recent advances in modeling of elves have been presented, including event based finite-difference time-domain modeling allowing direct comparison with satellite observations. The review of theory on sprites and sprite halos focused on most recent literature devoted to initiation of sprite streamers, their observed exponential expansion, and their electrical connection to the lower ionosphere. A recently proposed view of mechanisms of blue and gigantic jets has been presented that emphasizes charge imbalances in thunderstorms and analogies with conventional cloud to ground discharges to articulate conditions leading to upward escape of lightning leaders and their propagation toward the ionosphere. We then reviewed various local and global effects of TLEs. In this context existing and planned orbital observations have been discussed in detail. That discussion followed by presentation of the summary of the most recent literature on chemical, energetic and global electric circuit effects of TLEs.

There are many experimental as well as theoretical problems that remain to be solved in TLE studies. Examples in theoretical field include exact mechanisms of initiation and propagation of sprite streamers in low applied electric fields, role of thermal electron runaway phenomena in lightning and different types of TLEs, and development of self-consistent quantitative theory of transition of electrical discharge from thermal (leader dominated) forms at low altitudes to non-thermal (streamer dominated) forms at lower air densities at high altitudes in the Earth's atmosphere (e.g. Pasko 2006, and references therein). Examples in experimental field include a definite need for higher resolution spectral measurements of rotational intensity distributions of N_2 molecular bands that would allow accurate determination of rotational temperatures in sprites (e.g. Pasko 2007, and references therein), and a definite need for at least 100 times better time resolution to resolve many missing details in blue and gigantic jets (e.g. Pasko 2010, and references therein).

Acknowledgement Participation of V.P. Pasko was supported by the United States National Science Foundation under AGS-0734083 and AGS-0836391 grants to Penn State University.

References

S. Achat, Y. Teisseyre, E. Marode, The scaling of the streamer-to-arc transition in a positive point-to-plane gap with pressure. J. Phys. D, Appl. Phys. **25**(4), 661–668 (1992)

T. Adachi, H. Fukunishi, Y. Takahashi, M. Sato, Roles of the EMP and QE field in the generation of columniform sprites. Geophys. Res. Lett. **31**(4), L04107 (2004). doi:10.1029/2003GL019081

T. Adachi, H. Fukunishi, Y. Takahashi, Y. Hiraki, R.R. Hsu, H.T. Su, A.B. Chen, S.B. Mende, H.U. Frey, L.C. Lee, Electric field transition between the diffuse and streamer regions of sprites estimated from ISUAL/array photometer measurements. Geophys. Res. Lett. **33**(17), L17803 (2006)

T. Adachi, Y. Hiraki, K. Yamamoto, Y. Takahashi, H. Fukunishi, R.R. Hsu, H.T. Su, A.B. Chen, S.B. Mende, H.U. Frey, L.C. Lee, Electric fields and electron energies in sprites and temporal evolutions of lightning charge momentmeasurements. J. Phys. D, Appl. Phys. **41**(23), 234010 (2008)

E. Arnone, A. Kero, B.M. Dinelli, C.F. Enell, N.F. Arnold, E. Papandera, C.J. Rodger, M. Carlotti, M. Ridolfi, E. Turunen, Seeking sprite-induced signatures in remotely sensed middle atmosphere NO_2. Geophys. Res. Lett. **35**, L05807 (2008)

E. Arnone, A. Kero, C.F. Enell, M. Carlotti, C.J. Rodger, E. Papandera, N.F. Arnold, B.M. Dinelli, M. Ridolfi, E. Turunen, Seeking sprite-induced signatures in remotely sensed middle atmosphere NO_2: latitude and time variations. Plasma Sources Sci. Technol. **18**, 034014 (2009)

T. Asano, M. Hayakawa, M. Cho, T. Suzuki, Computer simulations on the initiation and morphological difference of Japan winter and summer sprites. J. Geophys. Res. **113**(A2), A02308 (2008). doi:10.1029/2007JA012528

T. Asano, T. Suzuki, M. Hayakawa, M.G. Cho, Three-dimensional EM computer simulation on sprite initiation above a horizontal lightning discharge. J. Atmos. Sol.-Terr. Phys. **71**, 983–990 (2009a)

T. Asano, T. Suzuki, Y. Hiraki, E. Mareev, M.G. Cho, M. Hayakawa, Computer simulations on sprite initiation for realistic lightning models with higher-frequency surges. J. Geophys. Res. **114**, A02310 (2009b)

C.P. Barrington-Leigh, U.S. Inan, M. Stanley, Identification of sprites and elves with intensified video and broadband array photometry. J. Geophys. Res. **106**(A2), 1741–1750 (2001). doi:10.1029/2000JA000073

E.M. Bazelyan, Y.P. Raizer, *Spark Discharge* (Chemical Rubber Company Press, New York, 1998)

E.M. Bazelyan, Y.P. Raizer, *Lightning Physics and Lightning Protection* (IoP Publishing, Bristol, 2000),

S. Beirle, U. Platt, M. Wenig, T. Wagner, NO_x production by lightning estimated with GOME. Adv. Space Res. **34**(4) Sp. Iss., 793–797 (2004). doi:10.1016/j.asr.2003.07.069

S. Beirle, H. Huntrieser, T. Wagner, Direct satellite observation of lightning-produced NO_x. Atmos. Chem. Phys. **10**(22), 10965–10986 (2010). doi:10.5194/acp-10-10965-2010

E. Blanc, T. Farges, R. Roche, D. Brebion, T. Hua, A. Labarthe, V. Melnikov, Nadir observations of sprites from the international space station. J. Geophys. Res. **109**(A2), A02306 (2004). doi:10.1029/2003JA009972

E. Blanc, T. Farges, D. Brebion, A. Labarthe, V. Melnikov, Observations of sprites from space at the nadir: the LSO (lightning and sprite observations) experiment on board of the international space station, in *Sprites, Elves and Intense Lightning Discharges*, ed. by M. Füllekrug, E.A. Mareev, M.J. Rycroft. NATO Science Series II: Mathematics, Physics and Chemistry, vol. 225 (Springer, Heidelberg, 2006), pp. 151–166

E. Blanc, T. Farges, A.N. Belyaev, V.V. Alpatov, D. Brebion, A. Labarthe, V. Melnikov, Main results of LSO (Lightning and Sprite Observations) on board of the International Space Station. Microgravity Sci. Technol. **19**(5–6), 80–84 (2007)

W.L. Boeck, O.H. Vaughan, R.J. Blakeslee, B. Vonnegut, M. Brook, Lightning induced brightening in the airglow layer. Geophys. Res. Lett. **19**, 99–102 (1992)

W.L. Boeck, O.H. Vaughan, R.J. Blakeslee, B. Vonnegut, M. Brook, J. McKune, Observations of lightning in the stratosphere. J. Geophys. Res. **100**, 1465–1475 (1995)

W.L. Boeck, O.H. Vaughan, R.J. Blakeslee, B. Vonnegut, M. Brook, The role of the space shuttle videotapes in the discovery of sprites, jets and elves. J. Atmos. Sol.-Terr. Phys. **60**, 669–677 (1998)

E. Bucsela, J. Morrill, M. Heavner, C. Siefring, S. Berg, D. Hampton, D. Moudry, E. Wescott, D. Sentman, $N_2(B^3\Pi_g)$ and $N_2^+(A^2\Pi_u)$ vibrational distributions observed in sprites. J. Atmos. Sol.-Terr. Phys. **65**, 583–590 (2003)

E.J. Bucsela, K.E. Pickering, T.L. Huntemann, R.C. Cohen, A. Perring, J.F. Gleason, R.J. Blakeslee, R.I. Albrecht, R. Holzworth, J.P. Cipriani, D. Vargas-Navarro, I. Mora-Segura, A. Pacheco-Hernandez, S. Laporte-Molina, Lightning-generated NO_x seen by the Ozone Monitoring Instrument during NASA's Tropical Composition, Cloud and Climate Coupling Experiment (TC4). J. Geophys. Res. **115**, D00J10 (2010). doi:10.1029/2009JD013118

L. Campbell, D.C. Cartwright, M.J. Brunger, Role of excited N_2 in the production of nitric oxide. J. Geophys. Res. **112**, A08303 (2007). doi:10.1029/2007JA012337

M. Carlotti, G. Brizzi, E. Papandrea, M. Prevedelli, M. Ridolfi, B. Dinelli, L. Magnani, GMTR: Two-dimensional geo-fit multitarget retrieval model for Michelson interferometer for passive atmospheric sounding/environmental satellite observations. Appl. Opt. **45**(4), 716–727 (2006)

S. Celestin, V.P. Pasko, Effects of spatial non-uniformity of streamer discharges on spectroscopic diagnostics of peak electric fields in transient luminous events. Geophys. Res. Lett. **37**, L07804 (2010)

S. Celestin, V.P. Pasko, Energy and fluxes of thermal runaway electrons produced by exponential growth of streamers during the stepping of lightning leaders and in transient luminous events. J. Geophys. Res. **116**, A03315 (2011)

M. Černák, E.M. van Veldhuizen, I. Morva, W.R. Rutgers, Effect of cathode surface properties on glow-to-arc transition in a short positive corona gap in ambient air. J. Phys. D, Appl. Phys. **28**(6), 1126–1132 (1995)

S.C. Chang, C.L. Kuo, L.J. Lee, A.B. Chen, H.T. Su, R.R. Hsu, H.U. Frey, S.B. Mende, Y. Takahashi, L.C. Lee, ISUAL far-ultraviolet events, elves, and lightning current. J. Geophys. Res. **115**, A00E46 (2010). doi:10.1029/2009JA014861

O. Chanrion, T. Neubert, A PIC-MCC code for simulation of streamer propagation in air. J. Comput. Phys. **227**(15), 7222–7245 (2008)

O. Chanrion, T. Neubert, Production of runaway electrons by negative streamer discharges. J. Geophys. Res. **115**, A00E32 (2010). doi:10.1029/2009JA014774

A.B. Chen, C.L. Kuo, Y.J. Lee, H.T. Su, R.R. Hsu, J.L. Chern, H.U. Frey, S.B. Mende, Y. Takahashi, H. Fukunishi, Y.S. Chang, T.Y. Liu, L.C. Lee, Global distributions and occurrence rates of transient luminous events. J. Geophys. Res. **113**(A8), A08306 (2008)

Z. Cheng, S.A. Cummer, H.T. Su, R.R. Hsu, Broadband very low frequency measurement of D region ionospheric perturbations caused by lightning electromagnetic pulses. J. Geophys. Res. **112**(A6), A06318 (2007)

J.L. Chern, R.R. Hsu, H.T. Su, S.B. Mende, H. Fukunishi, Y. Takahashi, L.C. Lee, Global survey of upper atmospheric transient luminous events on the ROCSAT-2 satellite. J. Atmos. Sol.-Terr. Phys. **65**, 647–659 (2003). doi:10.1016/S1364-6826(02)00317-6
J.K. Chou, C.L. Kuo, L.Y. Tsai, A.B. Chen, H.T. Su, R.R. Hsu, S.A. Cummer, J. Li, H.U. Frey, S.B. Mende, Y. Takahashi, L.C. Lee, Gigantic jets with negative and positive polarity streamers. J. Geophys. Res. **115**, A00E45 (2010). doi:10.1029/2009JA014831
H. Christian, R. Blakeslee, D. Boccippio, W. Boeck, D. Buechler, K. Driscoll, S. Goodman, J. Hall, W. Koshak, D. Mach, M. Stewart, Global frequency and distribution of lightning as observed from space by the Optical Transient Detector. J. Geophys. Res. **108**(D1), 4005 (2003). doi:10.1029/2002JD002347
V. Cooray, M. Rahman, V. Rakov, On the NO_x production by laboratory electrical discharges and lightning. J. Atmos. Sol.-Terr. Phys. **71**(17–18), 1877–1889 (2009). doi:10.1016/j.jastp.2009.07.009
S.A. Cummer, Current moment in sprite-producing lightning. J. Atmos. Sol.-Terr. Phys. **65**, 499–508 (2003)
S.A. Cummer, M. Fullekrug, Unusually intense continuing current in lightning produces delayed mesospheric breakdown. Geophys. Res. Lett. **28**, 495–498 (2001)
S.A. Cummer, W.A. Lyons, Lightning charge moment changes in U.S. high plains thunderstorms. Geophys. Res. Lett. **31**(5), L05114 (2004). doi:10.1029/2003GL019043
S.A. Cummer, W.A. Lyons, Implication of lightning charge moment changes for sprite initiation. J. Geophys. Res. **110**, A04304 (2005). doi:10.1029/2004JA010812
S.A. Cummer, U.S. Inan, T.F. Bell, C.P. Barrington-Leigh, ELF radiation produced by electrical currents in sprites. Geophys. Res. Lett. **25**, 1281–1284 (1998)
S.A. Cummer, H.U. Frey, S.B. Mende, R.R. Hsu, H.T. Su, A.B. Chen, H. Fukunishi, Y. Takahashi, Simultaneous radio and satellite optical measurements of high-altitude sprite current and lightning continuing current. J. Geophys. Res. **111**(A10), A10315 (2006a)
S.A. Cummer, N.C. Jaugey, J.B. Li, W.A. Lyons, T.E. Nelson, E.A. Gerken, Submillisecond imaging of sprite development and structure. Geophys. Res. Lett. **33**, L04104 (2006b). doi:10.1029/2005GL024969
S.A. Cummer, J. Li, F. Han, G. Lu, N. Jaugey, W.A. Lyons, T.E. Nelson, Quantification of the troposphere-to-ionosphere charge transfer in a gigantic jet. Nat. Geosci. **2**, 1–4 (2009). doi:10.1038/NGEO607
C. de Groot-Hedlin, Finite-difference time-domain synthesis of infrasound propagation through an absorbing atmosphere. J. Acoust. Soc. Am. **124**(3, Part 1), 1430–1441 (2008). doi:10.1121/1.2959736
S. de Larquier, V.P. Pasko, Mechanism of inverted-chirp infrasonic radiation from sprites. Geophys. Res. Lett. **37**, L24803 (2010)
J. de Urquijo, F.J. Gordillo-Vazquez, Comment on "NO_x production in laboratory discharges simulating blue jets and red sprites" by Harold Peterson et al. J. Geophys. Res. **115**, A12319 (2010). doi:10.1029/2010JA015966
S.K. Dhali, P.F. Williams, Two-dimensional studies of streamers in gases. J. Appl. Phys. **62**, 4696–4707 (1987)
U. Ebert, D. Sentman, Editorial Review: Streamers, sprites, leaders, lightning: from micro- to macroscales. J. Phys. D, Appl. Phys. **41**, 230301 (2008)
U. Ebert, C. Montijn, T.M.P. Briels, W. Hundsdorfer, B. Meulenbroek, A. Rocco, E.M. van Veldhuizen, The multiscale nature of streamers. Plasma Sources Sci. Technol. **15**, S118–S129 (2006)
U. Ebert, S. Nijdam, C. Li, A. Luque, T. Briels, E. van Veldhuizen, Review of recent results on streamer discharges and discussion of their relevance for sprites and lightning. J. Geophys. Res. **115**, A00E43 (2010)
C.F. Enell, A. Arnone, T. Adachi, O. Chanrion, P.T. Verronen, A. Seppala, T. Neubert, T. Ulich, E. Turunen, Y. Takahashi, R.R. Hsu, Parameterisation of the chemical effect of sprites in the middle atmosphere. Ann. Geophys. **26**, 13–27 (2008)
T. Farges, E. Blanc, Characteristics of infrasound from lightning and sprites near thunderstorm areas. J. Geophys. Res. **115**, A00E31 (2010). doi:10.1029/2009JA014700
T. Farges, E. Blanc, A.L. Pichon, T. Neubert, T.H. Allin, Identification of infrasound produced by sprites during the Sprite2003 campaign. Geophys. Res. Lett. **32**(1), L01813 (2005). doi:10.1029/2004GL021212
H. Fischer, M. Birk, C. Blom, B. Carli, M. Carlotti, T. von Clarmann, L. Delbouille, A. Dudhia, D. Ehhalt, M. Endemann, J.M. Flaud, R. Gessner, A. Kleinert, R. Koopman, J. Langen, M. Lopez-Puertas, P. Mosner, H. Nett, H. Oelhaf, G. Perron, J. Remedios, M. Ridolfi, G. Stiller, R. Zander, MIPAS: an instrument for atmospheric and climate research. Atmos. Chem. Phys. **8**(8), 2151–2188 (2008)
G.J. Fishman, P.N. Bhat, R. Mallozzi, J.M. Horack, T. Koshut, C. Kouveliotou, G.N. Pendleton, C.A. Meegan, R.B. Wilson, W.S. Paciesas, S.J. Goodman, H.J. Christian, Discovery of intense gamma-ray flashes of atmospheric origin. Science **264**(5163), 1313–1316 (1994)
R.C. Franz, R.J. Nemzek, J.R. Winckler, Television image of a large upward electric discharge above a thunderstorm system. Science **249**, 48–51 (1990)
H.U. Frey, S.B. Mende, S.A. Cummer, A.B. Chen, R.R. Hsu, H.T. Su, Y.S. Chang, T. Adachi, H. Fukunishi, Y. Takahashi, Beta-type stepped leader of elve-producing lightning. Geophys. Res. Lett. **32**, L13824 (2005). doi:10.1029/2005GL023080

H.U. Frey, S.B. Mende, S.A. Cummer, J. Li, T. Adachi, H. Fukunishi, Y. Takahashi, A.B. Chen, R.R. Hsu, H.T. Su, Y.S. Chang, Halos generated by negative cloud-to-ground lightning. Geophys. Res. Lett. **34**, L18801 (2007)

H. Fukunishi, Y. Takahashi, M. Kubota, K. Sakanoi, U.S. Inan, W.A. Lyons, Elves: Lightning-induced transient luminous events in the lower ionosphere. Geophys. Res. Lett. **23**(16), 2157–2160 (1996)

H. Fukunishi, Y. Hiraki, T. Adachi, L. Tong, K. Nanbu, Occurrence conditions for gigantic jets connecting the thundercloud and the ionosphere. Eos Trans. AGU **86**(52), AE11A-02 (2005). Fall Meet. Suppl., Abstract AE11A-02

M. Fullekrug, M.J. Rycroft, The contribution of sprites to the global atmospheric electric circuit. Earth Planets Space **58**(9), 1193–1196 (2006)

M. Fullekrug, E.A. Mareev, M.J. Rycroft, *Sprites, Elves and Intense Lightning Discharges*. NATO Science Series II: Mathematics, Physics and Chemistry, vol. 225 (Springer, Heidelberg, 2006)

W.R. Gamerota, S.A. Cummer, J. Li, H.C. Stenbaek-Nielsen, R.K. Haaland, M.G. McHarg, Comparison of sprite initiation altitudes between observations and models. J. Geophys. Res. **116**, A02317 (2011). doi:10.1029/2010JA016095

G. Garipov, B. Khrenov, M. Panasyuk, V. Tulupov, A. Shirokov, I. Yashin, H. Salazar, UV radiation from the atmosphere: results of the MSU "Tatiana" satellite measurements. Astropart. Phys. **24**(4–5), 400–408 (2005a). doi:10.1016/j.astropartphys.2005.09.001

G. Garipov, M. Panasyuk, V. Tulupov, B. Khrenov, A. Shirokov, I. Yashin, H. Salazar, Ultraviolet flashes in the equatorial region of the earth. JETP Lett. **82**(4), 185–187 (2005b)

G. Garipov, M. Panasyuk, I. Rubinshtein, V. Tulupov, B. Khrenov, A. Shirokov, I. Yashin, H. Salazar, Ultraviolet radiation detector of the MSU research educational microsatellite Universitetskii-Tat'yana. Instrum. Exp. Tech. **49**(1), 126–131 (2006). doi:10.1134/S0020441206010180

G.K. Garipov, B.A. Khrenov, M.I. Panasyuk, Correlation of atmospheric UV transient events with lunar phase. Geophys. Res. Lett. **35**(10), L10807 (2008). doi:10.1029/2007GL032679

G.K. Garipov, B.A. Khrenov, P.A. Klimov, V.S. Morozenko, M.I. Panasyuk, S.N. Petrova, V.I. Tulupov, V.M. Shahparonov, S.I. Svertilov, N.N. Vedenkin, I.V. Yashin, J.A. Jeon, S.M. Jeong, A.R. Jung, J.E. Kim, J. Lee, H.Y. Lee, G.W. Na, J.W. Nam, S. Nam, I.H. Park, J.E. Suh, J.Y. Jin, M. Kim, Y.K. Kim, B.W. Yoo, Y.S. Park, H.J. Yu, C.H. Lee, J.H. Park, H.I. Salazar, O.B. Martinez, E.L. Ponce, J.P. Cotsomi, Program of transient UV event research at Tatiana-2 satellite. J. Geophys. Res. **115**, A00E24 (2010). doi:10.1029/2009JA014765

E.A. Gerken, U.S. Inan, Observations of decameter-scale morphologies in sprites. J. Atmos. Sol.-Terr. Phys. **65**, 567–572 (2003). doi:10.1016/S1364-6826(02)00333-4

E.A. Gerken, U.S. Inan, C.P. Barrington-Leigh, Telescopic imaging of sprites. Geophys. Res. Lett. **27**, 2637–2640 (2000)

F.J. Gordillo-Vazquez, Air plasma kinetics under the influence of sprites. J. Phys. D, Appl. Phys. **41**(23), 234016 (2008)

F.J. Gordillo-Vazquez, Z. Donko, Electron energy distribution functions and transport coefficients relevant for air plasmas in the troposphere: impact of humidity and gas temperature. Plasma Sources Sci. Technol. **18**, 034021 (2009)

F.J. Gordillo-Vazquez, A. Luque, Electrical conductivity in sprite streamer channels. Geophys. Res. Lett. **37**, L16809 (2010). doi:10.1029/2010GL044349

E. Greenberg, C. Price, Y. Yair, M. Ganot, J. Bor, G. Satori, ELF transients associated with sprites and elves in eastern Mediterranean winter thunderstorms. J. Atmos. Sol.-Terr. Phys. **69**(13), 1569–1586 (2007). doi:10.1016/j.jastp.2007.06.002

C. Haldoupis, A. Mika, S. Shalimov, Modeling the relaxation of early VLF perturbations associated with transient luminous events. J. Geophys. Res. **114**, A00E04 (2009)

C. Haldoupis, N. Amvrosialdi, B.R.T. Cotts, O.A. van der Velde, O. Chanrion, T. Neubert, More evidence for a one-to-one correlation between sprites and early VLF perturbations. J. Geophys. Res. **115**, A07304 (2010)

M.J. Heavner, D.D. Sentman, D.R. Moudry, E.M. Wescott, Sprites, blue jets, and elves: optical evidence of energy transport across the stratopause, in *Atmospheric Science Across the Stratopause*, ed. by D.E. Siskind, S.D. Eckermann, M.E. Summers. Geophysical Monograph Series, vol. 123 (American Geophysical Union, Washington, 2000), pp. 69–82

Y. Hiraki, Effects of ion-neutral chemical reactions on dynamics of lightning-induced electric field. Plasma Sources Sci. Technol. **18**(3), 034020 (2009)

Y. Hiraki, H. Fukunishi, Theoretical criterion of charge moment change by lightning for initiation of sprites. J. Geophys. Res. **111**(A11), A11305 (2006)

Y. Hiraki, Y. Kasai, H. Fukunishi, Chemistry of sprite discharges through ion-neutral reactions. Atmos. Chem. Phys. **8**(14), 3919–3928 (2008)

R.R. Hsu, A.B. Chen, C. Kuo, Y. Lee, H.T. Su, H. Fukunishi, Y. Takahashi, T. Adachi, K. Yamamoto, H.U. Frey, S.B. Mende, L.C. Lee, Gigantic jet observation by the ISUAL payload of FORMOSAT-2 satellite. Eos Trans. AGU **86**(52), AE23A-0992 (2005). Fall Meet. Suppl., Abstract AE23A-0992

R.R. Hsu, A.B. Chen, C.-L. Kuo, H.-T. Su, H. Frey, S. Mende, Y. Takahashi, L.-C. Lee, On the global occurrence and impacts of transient luminous events (TLEs). AIP Conf. Proc. **1118**(1), 99–107 (2009)

W.Y. Hu, S.A. Cummer, W.A. Lyons, Lightning charge moment changes for the initiation of sprites. Geophys. Res. Lett. **29**(8), 1279 (2002). doi:10.1029/2001GL014593

W.Y. Hu, S.A. Cummer, W.A. Lyons, Testing sprite initiation theory using lightning measurements and modeled electromagnetic fields. J. Geophys. Res. **112**(D13), D13115 (2007)

T.Y. Huang, C.Y. Chiang, C.L. Kuo, A.B. Chen, H.T. Su, R.R. Hsu, Investigations of lightning-induced sudden brightening in the OH airglow layer observed by ISUAL onboard FORMOSAT-II satellite. AIP Conf. Proc. **1118**(1), 21–27 (2009)

T.-Y. Huang, C.L. Kuo, C.Y. Chiang, A.B. Chen, H.T. Su, R.R. Hsu, Further investigations of lightning-induced transient emissions in the OH airglow layer. J. Geophys. Res. **115**, A10326 (2010). doi:10.1029/2010JA015558

M. Ignaccolo, T. Farges, A. Mika, T.H. Allin, O. Chanrion, E. Blanc, T. Neubert, A.C. Fraser-Smith, M. Fullekrug, The planetary rate of sprite events. Geophys. Res. Lett. **33**(11), L11808 (2006)

U.S. Inan, Lightning effects at high altitudes: sprites, elves, and terrestrial gamma ray flashes. C. R. Phys. **3**(10), 1411–1421 (2002)

U.S. Inan, T.F. Bell, J.V. Rodriguez, Heating and ionization of the lower ionosphere by lightning. Geophys. Res. Lett. **18**(4), 705–708 (1991)

U.S. Inan, C. Barrington-Leigh, S. Hansen, V.S. Glukhov, T.F. Bell, R. Rairden, Rapid lateral expansion of optical luminosity in lightning-induced ionospheric flashes referred to as 'elves'. Geophys. Res. Lett. **24**(5), 583–586 (1997)

U.S. Inan, S.A. Cummer, R.A. Marshall, A survey of elf and VLF research on lightning-ionosphere interactions and causative discharges. J. Geophys. Res. **115**, A00E36 (2010)

P. Israelevich, Y. Yair, A. Devir, J. Joseph, Z. Levin, I. Mayo, M. Moalem, C. Price, B. Ziv, A. Sternlieb, Transient airglow enhancements observed from the space shuttle Columbia during the MEIDEX sprite campaign. Geophys. Res. Lett. **31**(6), L06124 (2004). doi:10.1029/2003GL019110

T. Kanmae, H.C. Stenbaek-Nielsen, M.G. McHarg, Altitude resolved sprite spectra with 3 ms temporal resolution. Geophys. Res. Lett. **34**, L07810 (2007)

H.W. Kasemir, A contribution to the electrostatic theory of a lightning discharge. J. Geophys. Res. **65**(7), 1873–1878 (1960)

S.I. Klimov, E.A. Sharkov, L.M. Zelenyi, The tropical cyclones as the possible sources of gamma emission in the Earth's atmosphere. Eos Trans. AGU **90**, 52 (2009). Abstract AE33B-0295

P.R. Krehbiel, J.A. Riousset, V.P. Pasko, R.J. Thomas, W. Rison, M.A. Stanley, H.E. Edens, Upward electrical discharges from thunderstorms. Nat. Geosci. **1**(4), 233–237 (2008). doi:10.1038/ngeo162

C.L. Kuo, R.R. Hsu, H.T. Su, A.B. Chen, L.C. Lee, S.B. Mende, H.U. Frey, H. Fukunishi, Y. Takahashi, Electric fields and electron energies inferred from the ISUAL recorded sprites. Geophys. Res. Lett. **32**, L19103 (2005). doi:10.1029/2005GL023389

C.L. Kuo, A.B. Chen, R.R. Hsu, H.T. Su, L.C. Lee, S.B. Mende, H.U. Frey, H. Fukunishi, Y. Takahashi, Analysis of ISUAL recorded gigantic jets, in *Abstracts of Workshop on Streamers, Sprites, Leaders, Lightning: From Micro- to Macroscales*, Lorentz Center, Leiden University, 8–12 October 2007 (2007a)

C.L. Kuo, A.B. Chen, Y.J. Lee, L.Y. Tsai, R.K. Chou, R.R. Hsu, H.T. Su, L.C. Lee, S.A. Cummer, H.U. Frey, S.B. Mende, Y. Takahashi, H. Fukunishi, Modeling elves observed by FORMOSAT-2 satellite. J. Geophys. Res. **112**, A11312 (2007b)

C.L. Kuo, A.B. Chen, J.K. Chou, L.Y. Tsai, R.R. Hsu, H.T. Su, H.U. Frey, S.B. Mende, Y. Takahashi, L.C. Lee, Radiative emission and energy deposition in transient luminous events. J. Phys. D, Appl. Phys. **41**, 234014 (2008)

C.L. Kuo, J.K. Chou, L.Y. Tsai, A.B. Chen, H.T. Su, R.R. Hsu, S.A. Cummer, H.U. Frey, S.B. Mende, Y. Takahashi, L.C. Lee, Discharge processes, electric field, and electron energy in ISUAL-recorded gigantic jets. J. Geophys. Res. **114**, A04314 (2009)

A. Larsson, The effect of a large series resistance on the streamer-to-spark transition in dry air. J. Phys. D, Appl. Phys. **31**(9), 1100 (1998)

E.H. Lay, C.J. Rodger, R.H. Holzworth, M. Cho, J.N. Thomas, Temporal-spatial modeling of electron density enhancement due to successive lightning strokes. J. Geophys. Res. **115**(0), A00E59 (2010)

F. Leblanc, K.L. Aplin, Y. Yair, R.G. Harrison, J.P. Lebreton, M. Blanc, *Planetary atmospheric electricity* 532 pp. (Springer, Berlin, 2008)

L.-J. Lee, A.B. Chen, S.-C. Chang, C.-L. Kuo, H.-T. Su, R.-R. Hsu, C.-C. Wu, P.-H. Lin, H.U. Frey, S.B. Mende, Y. Takahashi, L.-C. Lee, Controlling synoptic-scale factors for the distribution of transient luminous events. J. Geophys. Res. **115**, A00E54 (2010). doi:10.1029/2009JA014823

F. Lefeuvre, E. Blanc, J.L.P.T. Team, TARANIS—a satellite project dedicated to the physics of TLEs and TGFs. AIP Conf. Proc. **1118**(1), 3–7 (2009)

C. Li, U. Ebert, W. Hundsdorfer, 3D hybrid computations for streamer discharges and production of runaway electrons. J. Phys. D, Appl. Phys. **42**(20), 202003 (2009). doi:10.1088/0022-3727/42/20/202003

J. Li, S. Cummer, Estimation of electric charge in sprites from optical and radio observations. J. Geophys. Res. **116**, A01301 (2011). doi:10.1029/2010JA015391

J. Li, S.A. Cummer, W.A. Lyons, T.E. Nelson, Coordinated analysis of delayed sprites with high-speed images and remote electromagnetic fields. J. Geophys. Res. **113**(D20), D20206 (2008)

C. Liu, E.R. Williams, E.J. Zipser, G. Burns, Diurnal variations of global thunderstorms and electrified shower clouds and their contribution to the global electrical circuit. J. Atmos. Sci. **67**(2), 309–323 (2010). doi:10.1175/2009JAS3248.1

N. Liu, Model of sprite luminous trail caused by increasing streamer current. Geophys. Res. Lett. **37**, L04102 (2010). doi:10.1029/2009GL042214

N.Y. Liu, V.P. Pasko, Effects of photoionization on propagation and branching of positive and negative streamers in sprites. J. Geophys. Res. **109**, A04301 (2004). doi:10.1029/2003JA010064

N.Y. Liu, V.P. Pasko, Molecular nitrogen LBH band system far-UV emissions of sprite streamers. Geophys. Res. Lett. **32**, L05104 (2005). doi:10.1029/2004GL022001

N.Y. Liu, V.P. Pasko, Modeling studies of NO-γ emissions of sprites. Geophys. Res. Lett. **34**, L16103 (2007)

N.Y. Liu, V.P. Pasko, No-gamma emissions from streamer discharges: direct electron impact excitation versus resonant energy transfer. J. Phys. D, Appl. Phys. **43**, 082001 (2010)

N.Y. Liu, V.P. Pasko, D.H. Burkhardt, H.U. Frey, S.B. Mende, H.-T. Su, A.B. Chen, R.-R. Hsu, L.-C. Lee, H. Fukunishi, Y. Takahashi, Comparison of results from sprite streamer modeling with spectrophotometric measurements by ISUAL instrument on FORMOSAT-2 satellite. Geophys. Res. Lett. **33**, L01101 (2006). doi:10.1029/2005GL024243

N.Y. Liu, V.P. Pasko, K. Adams, H.C. Stenbaek-Nielsen, M.G. McHarg, Comparison of acceleration, expansion, and brightness of sprite streamers obtained from modeling and high-speed video observations. J. Geophys. Res. **114**, A00E03 (2009a)

N.Y. Liu, V.P. Pasko, H.U. Frey, S.B. Mende, H.-T. Su, A.B. Chen, R.-R. Hsu, L.-C. Lee, Assessment of sprite initiating electric fields and quenching altitude of $a^1\pi_g$ state of n_2 using sprite streamer modeling and ISUAL spectrophotometric measurements. J. Geophys. Res. **114**, A00E02 (2009b)

G. Lu, S.A. Cummer, J. Li, F. Han, R.J. Blakeslee, H.J. Christian, Charge transfer and in-cloud structure of large-charge-moment positive lightning strokes in a mesoscale convective system. Geophys. Res. Lett. **36**, L15805 (2009). doi:10.1029/2009GL038880

G. Lu, R.J. Blakeslee, J. Li, D.M. Smith, X.M. Shao, E.W. McCaul, D.E. Buechler, H.J. Christian, J.M. Hall, S.A. Cummer, Lightning mapping observation of a terrestrial gamma-ray flash. Geophys. Res. Lett. **37**, L11806 (2010). doi:10.1029/2010GL043494

G. Lu, S.A. Cummer, J. Li, F. Han, D.M. Smith, B.W. Grefenstette, Characteristics of broadband lightning emissions associated with terrestrial gamma ray flashes. J. Geophys. Res. **116**, A03316 (2011a). doi:10.1029/2010JA016141

G. Lu, S.A. Cummer, W.A. Lyons, P.R. Krehbiel, J. Li, W. Rison, R.J. Thomas, H.E. Edens, M.A. Stanley, W. Beasley, D.R. MacGorman, O.A. van der Velde, M.B. Cohen, T.J. Lang, S.A. Rutledge, Lightning development associated with two negative gigantic jets. Geophys. Res. Lett. **38**, L12801 (2011b). doi:10.1029/2011GL047662

A. Luque, U. Ebert, Emergence of sprite streamers from screening-ionization waves in the lower ionosphere. Nat. Geosci. **2**(11), 757–760 (2009). doi:10.1038/NGEO662

A. Luque, U. Ebert, Sprites in varying air density: charge conservation, glowing negative trails and changing velocity. Geophys. Res. Lett. **37**, L06806 (2010). doi:10.1029/2009GL041982

A. Luque, F.J. Gordillo-Vazquez, Sprite beads originating from inhomogeneities in the mesospheric electron density. Geophys. Res. Lett. **38**, L04808 (2011). doi:10.1029/2010GL046403

W.A. Lyons, Sprite observations above the U.S. high plains in relation to their parent thunderstorm systems. J. Geophys. Res. **101**, 29641–29652 (1996)

W.A. Lyons, The meteorology of transient luminous events—an introduction and overview, in *Sprites, Elves and Intense Lightning Discharges*, ed. by M. Füllekrug, E.A. Mareev, M.J. Rycroft. NATO Science Series II: Mathematics, Physics and Chemistry, vol. 225 (Springer, Heidelberg, 2006), pp. 19–56

W.A. Lyons, T.E. Nelson, R.A. Armstrong, V.P. Pasko, M.A. Stanley, Upward electrical discharges from thunderstorm tops. Bull. Am. Meteorol. Soc. **84**(4), 445–454 (2003). doi:10.1175/BAMS-84-4-445

W.A. Lyons, M.A. Stanley, J.D. Meyer, T.E. Nelson, S.A. Rutledge, T. Lang, S.A. Cummer, The meteorological and electrical structure of TLE-producing convective storms, in *Lightning: Principles, Instruments and Applications*, ed. by H. Betz, U. Schumann, P. Laroche (Springer, Berlin, 2009), pp. 387–415

W.A. Lyons, M.A. Stanley, T.E. Nelson, S.A. Cummer, C.R. Huffines, K.C. Wiens, Supercells and sprites. Bull. Am. Meteorol. Soc. **89**(8), 1165+ (2008). doi:10.1175/2008BAMS2439.1

E.A. Mareev, Global electric circuit research: achievements and prospects. Phys. Usp. **53**(5), 504–511 (2010). doi:10.3367/UFNe.0180.201005h.0527

E.A. Mareev, S.A. Yashunin, S.S. Davydenko, T.C. Marshall, M. Stolzenburg, C.R. Maggio, On the role of transient currents in the global electric circuit. Geophys. Res. Lett. **35**(15), L15810 (2008). doi:10.1029/2008GL034554

R.A. Marshall, U.S. Inan, High-speed measurements of small-scale features in sprites: Sizes and lifetimes. Radio Sci. **41**, RS6S43 (2006). doi:10.1029/2005RS003353

R.A. Marshall, U.S. Inan, Two-dimensional frequency domain modeling of lightning EMP-induced perturbations to VLF transmitter signals. J. Geophys. Res. **115**, A00E29 (2010)

R.A. Marshall, U.S. Inan, V.S. Glukhov, Elves and associated electron density changes due to cloud-to-ground and in-cloud lightning discharges. J. Geophys. Res. **115**, A00E17 (2010)

J.D. Mathews, M.A. Stanley, V.P. Pasko, T.G. Wood, U.S. Inan, M.J. Heavner, S.A. Cummer, Electromagnetic signatures of the Puerto Rico blue jet and its parent thunderstorm. Eos Trans. AGU **83**(47) (2002). Fall Meet. Suppl., Abstract A62D-02

Y. Matsudo, I. Suzuki, K. Michimoto, K. Myokei, M. Hayakawa, Comparison of time delays of sprites induced by winter lightning flashes in the Japan Sea with those in the Pacific Ocean. J. Atmos. Sol.-Terr. Phys. **71**(1), 101–111 (2009). doi:10.1016/j.jastp.2008.09.040

V. Mazur, L.H. Ruhnke, Model of electric charges in thunderstorms and associated lightning. J. Geophys. Res. **103**(D18), 23299–23308 (1998)

M.G. McHarg, H.C. Stenbaek-Nielsen, T. Kanmae, Streamer development in sprites. Geophys. Res. Lett. **34**, L06804 (2007). doi:10.1029/2006GL027854

S.B. Mende, H.U. Frey, R.R. Hsu, H.T. Su, A.B. Chen, L.C. Lee, D.D. Sentman, Y. Takahashi, H. Fukunishi, D region ionization by lightning-induced EMP. J. Geophys. Res. **110**, A11312 (2005). doi:10.1029/2005JA011064

S.B. Mende, Y.S. Chang, A.B. Chen, H.U. Frey, H. Fukunishi, S.P. Geller, S. Harris, H. Heetderks, R.R. Hsu, L.C. Lee, H.T. Su, Y. Takahashi, Spacecraft based studies of transient luminous events, in *Sprites, Elves and Intense Lightning Discharges*, ed. by M. Füllekrug, E.A. Mareev, M.J. Rycroft. NATO Science Series II: Mathematics, Physics and Chemistry, vol. 225 (Springer, Heidelberg, 2006), pp. 123–149

A. Mika, C. Haldoupis, T. Neubert, H.T. Su, R.R. Hsu, R.J. Steiner, R.A. Marshall, Early VLF perturbations observed in association with elves. Ann. Geophys. **24**(8), 2179–2189 (2006)

G.M. Milikh, M.N. Shneider, Model of UV flashes due to gigantic blue jets. J. Phys. D, Appl. Phys. **41**(23), 234013 (2008)

E.V. Mishin, G.M. Milikh, Blue jets:Upward lightning. Space Sci. Rev. **137**(1–4), 473–488 (2008)

J. Montanya, O. van der Velde, D. Romero, V. March, G. Sola, N. Pineda, M. Arrayas, J.L. Trueba, V. Reglero, S. Soula, High-speed intensified video recordings of sprites and elves over the western Mediterranean Sea during winter thunderstorms. J. Geophys. Res. **115**, A00E18 (2010). doi:10.1029/2009JA014508

J.S. Morrill, E.J. Bucsela, V.P. Pasko, S.L. Berg, W.M. Benesch, E.M. Wescott, M.J. Heavner, Time resolved N_2 triplet state vibrational populations and emissions associated with red sprites. J. Atmos. Sol.-Terr. Phys. **60**, 811–829 (1998)

G.D. Moss, V.P. Pasko, N.Y. Liu, G. Veronis, Monte Carlo model for analysis of thermal runaway electrons in streamer tips in transient luminous events and streamer zones of lightning leaders. J. Geophys. Res. **111**, A02307 (2006). doi:10.1029/2005JA011350

G.V. Naidis, Positive and negative streamers in air: velocity-diameter relation. Phys. Rev. E **79**(5, Part 2), 057401 (2009). doi:10.1103/PhysRevE.79.057401

T. Neubert, On sprites and their exotic kin. Science **300**, 747–749 (2003)

T. Neubert, ASIM—an Instrument Suite for the International Space Station. AIP Conf. Proc. **1118**(1), 8–12 (2009)

T. Neubert, M. Rycroft, T. Farges, E. Blanc, O. Chanrion, E. Arnone, A. Odzimek, N. Arnold, C.F.E. CF, E. Turunen, T. Bosinger, A. Mika, C. Haldoupis, R.J. Steiner, O. van der Velde, O. Soula, P. Berg, F. Boberg, P. Thejll, B. Christiansen, M. Ignaccolo, M. Fullekrug, P.T. Verronen, J. Montanya, N. Crosby, Recent results from studies of electric discharges in the mesosphere. Surv. Geophys. **29**(2), 71–137 (2008)

S. Nijdam, J.S. Moerman, T.M.P. Briels, E.M. van Veldhuizen, U. Ebert, Stereo-photography of streamers in air. Appl. Phys. Lett. **92**(10), 101502 (2008). doi:10.1063/1.2894195

S. Nijdam, E.M. van Veldhuizen, U. Ebert, Comment on "NO_x production in laboratory discharges simulating blue jets and red sprites" by H. Peterson et al. J. Geophys. Res. **115**, A12305 (2010). doi:10.1029/2010JA015861

M.I. Panasyuk, V.V. Bogomolov, G.K. Garipov, O.R. Grigoryan, Y.I. Denisov, B.A. Khrenov, P.A. Klimov, L.L. Lazutin, S.I. Svertilov, N.N. Vedenkin, I.V. Yashin, S.I. Klimov, V.S. Makhmutov, Y.I. Stozkov, N.S. Svirzhevsky, V.V. Klimenko, E.A. Mareev, Y.V. Shlyugaev, V.E. Korepanov, I.H. Park, H.I. Salazar, H. Rothkaehl, Energetic particles impacting the upper atmosphere in connection with transient luminous event phenomena: Russian space experiment programs. AIP Conf. Proc. **1118**(1), 108–115 (2009)

V. Pasko, U. Inan, T. Bell, Sprites as luminous columns of ionization produced by quasi-electrostatic thundercloud fields. Geophys. Res. Lett. **23**(6), 649–652 (1996)

V.P. Pasko, Electric jets. Nature **423**, 927–929 (2003)

V.P. Pasko, Theoretical modeling of sprites and jets, in *Sprites, Elves and Intense Lightning Discharges*, ed. by M. Füllekrug, E.A. Mareev, M.J. Rycroft. NATO Science Series II: Mathematics, Physics and Chemistry, vol. 225 (Springer, Heidelberg, 2006), pp. 253–311

V.P. Pasko, Red sprite discharges in the atmosphere at high altitude: the molecular physics and the similarity with laboratory discharges. Plasma Sources Sci. Technol. **16**, S13–S29 (2007). doi:10.1088/0963-0252/16/1/S02

V.P. Pasko, Blue jets and gigantic jets: transient luminous events between thunderstorm tops and the lower ionosphere. Plasma Phys. Control. Fusion **50**, 124050 (2008)

V.P. Pasko, Recent advances in theory of transient luminous events. J. Geophys. Res. **50**, A00E35 (2010)

V.P. Pasko, A. Bourdon, Air heating associated with transient luminous events, in *Proc. 28th Int. Conf. On Phenomena in Ionized Gases (ICPIG), 5P07-12*, Institute of Plasma Physics, Prague, Czech Republic, 15–20 July 2007 (2007), pp. 1908–1911

V.P. Pasko, J.J. George, Three-dimensional modeling of blue jets and blue starters. J. Geophys. Res. **107**(A12), 1458 (2002). doi:10.1029/2002JA009473

V.P. Pasko, H.C. Stenbaek-Nielsen, Diffuse and streamer regions of sprites. Geophys. Res. Lett. **29**(10), 1440 (2002). doi:10.1029/2001GL014241

V.P. Pasko, U.S. Inan, T.F. Bell, Heating, ionization and upward discharges in the mesosphere due to intense quasi-electrostatic thundercloud fields. Geophys. Res. Lett. **22**(4), 365–368 (1995)

V.P. Pasko, U.S. Inan, T.F. Bell, Y.N. Taranenko, Sprites produced by quasi-electrostatic heating and ionization in the lower ionosphere. J. Geophys. Res. **102**(A3), 4529–4561 (1997). doi:10.1029/96JA03528

V.P. Pasko, U.S. Inan, T.F. Bell, Spatial structure of sprites. Geophys. Res. Lett. **25**, 2123–2126 (1998a)

V.P. Pasko, U.S. Inan, T.F. Bell, S.C. Reising, Mechanism of ELF radiation from sprites. Geophys. Res. Lett. **25**(18), 3493–3496 (1998b)

V.P. Pasko, U. Inan, T. Bell, Large scale modeling of sprites and blue jets. Eos Trans. AGU **80**(46), F218 (1999). Fall Meet. Suppl., Abstract A42E-11

V.P. Pasko, U.S. Inan, T.F. Bell, Fractal structure of sprites. Geophys. Res. Lett. **27**(4), 497–500 (2000). doi:10.1029/1999GL010749

V.P. Pasko, M.A. Stanley, J.D. Matthews, U.S. Inan, T.G. Wood, Electrical discharge from a thundercloud top to the lower ionosphere. Nature **416**, 152–154 (2002). doi:10.1038/416152a

H. Peterson, M. Bailey, J. Hallett, W. Beasley, NO_x production in laboratory discharges simulating blue jets and red sprites. J. Geophys. Res. **114**, A00E07 (2009). doi:10.1029/2009JA014489

H. Peterson, M. Bailey, J. Hallett, W. Beasley, Reply to comment by S. Nijdam et al. on "NO_x production in laboratory discharges simulating blue jets and red sprites". J. Geophys. Res. **115**, A12306 (2010a). doi:10.1029/2010JA015946

H. Peterson, M. Bailey, J. Hallett, W. Beasley, Reply to comment by J. de Urquijo and F. J. Gordillo-Vazquez on "NO_x production in laboratory discharges simulating blue jets and red sprites". J. Geophys. Res. **115**, A12320 (2010b). doi:10.1029/2010JA016040

N.I. Petrov, G.N. Petrova, Physical mechanisms for the development of lightning discharges between a thundercloud and the ionosphere. Tech. Phys. **44**, 472–475 (1999)

L. Pitchford, S. Oneil, J. Rumble, Extended Boltzmann analysis of electron swarm experiments. Phys. Rev. A **23**(1), 294–304 (1981)

J. Qin, S. Celestin, V.P. Pasko, On the inception of streamers from sprite halo events produced by lightning discharges with positive and negative polarity. J. Geophys. Res. **116**, A06305 (2011)

Y.P. Raizer, *Gas Discharge Physics* (Springer, New York, 1991)

Y.P. Raizer, G.M. Milikh, M.N. Shneider, S.V. Novakovski, Long streamers in the upper atmosphere above thundercloud. J. Phys. D, Appl. Phys. **31**, 3255–3264 (1998)

Y.P. Raizer, G.M. Milikh, M.N. Shneider, On the mechanism of blue jet formation and propagation. Geophys. Res. Lett. **33**(23), L23801 (2006). doi:10.1029/2006GL027697

Y.P. Raizer, G.M. Milikh, M.N. Shneider, Leader–streamers nature of blue jets. J. Atmos. Sol.-Terr. Phys. **69**(8), 925–938 (2007). doi:10.1016/j.jastp.2007.02.007

Y.P. Raizer, G.M. Milikh, M.N. Shneider, Streamer- and leader-like processes in the upper atmosphere: models of red sprites and blue jets. J. Geophys. Res. **115**, A00E42 (2010)

V.A. Rakov, M.A. Uman, *Lightning: Physics and Effects* (Cambridge University Press, Cambridge, 2003)

V.A. Rakov, D.E. Crawford, K.J. Rambo, G.H. Schnetzer, M.T.A. Uman, R. Thottappillil, M-component mode of charge transfer to ground in lightning discharge. J. Geophys. Res. **106**(D19), 22817–22831 (2001)

J.A. Riousset, V.P. Pasko, P.R. Krehbiel, R.J. Thomas, W. Rison, Three-dimensional fractal modeling of intracloud lightning discharge in a New Mexico thunderstorm and comparison with lightning mapping observations. J. Geophys. Res. **112**, D15203 (2007). doi:10.1029/2006JD007621

J.A. Riousset, V.P. Pasko, A. Bourdon, Air-density-dependent model for analysis of air heating associated with streamers, leaders, and transient luminous events. J. Geophys. Res. **115**, A12321 (2010a)

J.A. Riousset, V.P. Pasko, P.R. Krehbiel, W. Rison, M.A. Stanley, Modeling of thundercloud screening charges: Implications for blue and gigantic jets. J. Geophys. Res. **115**, A00E10 (2010b)

A. Rocco, U. Ebert, W. Hundsdorfer, Branching of negative streamers in free flight. Phys. Rev. E **66**, 035102(R) (2002). doi:10.1103/PhysRevE.66.035102

C.J. Rodger, Red sprites, upward lightning and VLF perturbations. Rev. Geophys. **37**, 317–336 (1999)

C.J. Rodger, A. Seppala, M.A. Clilverd, Significance of transient luminous events to neutral chemistry: experimental measurements. Geophys. Res. Lett. **35**(7), L07803 (2008)

R. Roussel-Dupre, J.J. Colman, E. Symbalisty, D. Sentman, V.P. Pasko, Physical processes related to discharges in planetary atmospheres. Space Sci. Rev. **137**(1–4), 51–82 (2008)

H.L. Rowland, Theories and simulations of elves, sprites and blue jets. J. Atmos. Sol.-Terr. Phys. **60**, 831–844 (1998)

M.J. Rycroft, A. Odzimek, Effects of lightning and sprites on the ionospheric potential, and threshold effects on sprite initiation, obtained using an analog model of the global atmospheric electric circuit. J. Geophys. Res. **115**, A00E37 (2010). doi:10.1029/2009JA014758

M. Sato, H. Fukunishi, Global sprite occurrence locations and rates derived from triangulation of transient Schumann resonance events. Geophys. Res. Lett. **30**(16), 1859 (2003). doi:10.1029/2003GL017291

M. Sato, T. Ushio, T. Morimoto, M. Suzuki, A. Yamazaki, K. Masayuki, R. Ishida, Y. Takahashi, U.S. Inan, Y. Hobara, Y. Sakamoto, K. Yoshida, H. Ishikawa, K. Yoshita, Science goal and mission status of JEM-GLIMS. Eos Trans. AGU **90**(52) (2009). Abstract AE23A-03

U. Schumann, H. Huntrieser, The global lightning-induced nitrogen oxides source. Atmos. Chem. Phys. **7**(14), 3823–3907 (2007)

D.D. Sentman, H.C. Stenbaek-Nielsen, Chemical effects of weak electric fields in the trailing columns of sprite streamers. Plasma Sources Sci. Technol. **18**(3), 034012 (2009)

D.D. Sentman, E.M. Wescott, D.L. Osborne, D.L. Hampton, M.J. Heavner, Preliminary results from the Sprites94 campaign: Red sprites. Geophys. Res. Lett. **22**, 1205–1208 (1995)

D.D. Sentman, E.M. Wescott, R.H. Picard, J.R. Winick, H.C. Stenbaek-Nielsen, E.M. Dewan, D.R. Moudry, F.T. Säo Sabbas, M.J. Heavner, J. Morrill, Simultaneous observations of mesospheric gravity waves and sprites generated by a Midwestern thunderstorm. J. Atmos. Sol.-Terr. Phys. **65**, 537–550 (2003)

D.D. Sentman, H.C. Stenbaek-Nielsen, M.G. McHarg, J.S. Morrill, Plasma chemistry of sprite streamers. J. Geophys. Res. **113**, D11112 (2008)

M.N. Shneider, G.M. Milikh, Analysis of UV flashes of millisecond scale detected by a low-orbit satellite. J. Geophys. Res. **115**, A00E23 (2010). doi:10.1029/2009JA014685

D. Siingh, A.K. Singh, R.P. Patel, R. Singh, R.P. Singh, B. Veenadhari, M. Mukherjee, Thunderstorms, lightning, sprites and magnetospheric whistler-mode radio waves. Surv. Geophys. **29**(6), 499–551 (2008)

D.M. Smith, L.I. Lopez, R.P. Lin, C.P. Barrington-Leigh, Terrestrial gamma-ray flashes observed up to 20 MeV. Science **307**(5712), 1085–1088 (2005). doi:10.1126/science.1107466

M. Stanley, P. Krehbiel, M. Brook, C. Moore, W. Rison, B. Abrahams, High speed video of initial sprite development. Geophys. Res. Lett. **26**, 3201–3204 (1999)

M.A. Stanley, Sprites and their parent discharges. Ph.D. thesis, New Mexico Institute of Mining and Technology, Socorro, NM (2000)

M.A. Stanley, W.A. Lyons, T.E. Nelson, P.R. Krehbiel, W. Rison, R.J. Thomas, Comparison of sprite locations with lightning channel structure. Eos Trans. AGU **88**(52) (2007). Fall Meet. Suppl., Abstract AE41A-07

H.C. Stenbaek-Nielsen, M.G. McHarg, High time-resolution sprite imaging: observations and implications. J. Phys. D, Appl. Phys. **41**, 234009 (2008)

H.C. Stenbaek-Nielsen, M.G. McHarg, T. Kanmae, D.D. Sentman, Observed emission rates in sprite streamer heads. Geophys. Res. Lett. **34**, L11105 (2007). doi:10.1029/2007GL029881

H.C. Stenbaek-Nielsen, R. Haaland, M.G. McHarg, B.A. Hensley, T. Kanmae, Sprite initiation altitude measured by triangulation. J. Geophys. Res. **115**, A00E12 (2010). doi:10.1029/2009JA014543

H.T. Su, R.R. Hsu, A.B. Chen, Y.C. Wang, W.S. Hsiao, W.C. Lai, L.C. Lee, M. Sato, H. Fukunishi, Gigantic jets between a thundercloud and the ionosphere. Nature **423**, 974–976 (2003). doi:10.1038/nature01759

A.I. Sukhorukov, P. Stubbe, Problems of blue jet theories. J. Atmos. Sol.-Terr. Phys. **23**(13), 7–9 (1998). doi:10.1016/S1364-6826(98)00021-2

L. Sutherland, H. Bass, Atmospheric absorption in the atmosphere up to 160 km. J. Acoust. Soc. Am. **115**(3), 1012–1032 (2004). doi:10.1121/1.1631937

Y. Takahashi, K. Yoshida, Y. Sakamoto, T. Sakamoi, SPRITE-SAT: A University Small Satellite for Observation of High-Altitude Luminous Events, in *Small Satellite Missions for Earth Observation*, ed. by R. Sandau, H.-P. Roeser, A. Valenzuela (Springer, Heidelberg, 2006), pp. 197–206

Y. Takahashi, A. Yoshida, M. Sato, T. Adachi, S. Kondo, R.R. Hsu, H.T. Su, A.B. Chen, S.B. Mende, H.U. Frey, L.C. Lee, Absolute optical energy of sprites and its relationship to charge moment of parent lightning discharge based on measurement by ISUAL/AP. J. Geophys. Res. **115**, A00E55 (2010). doi:10.1029/2009JA014814

P. Tardiveau, E. Marode, A. Agneray, M. Cheaib, Pressure effects on the development of an electric discharge in non-uniform fields. J. Phys. D, Appl. Phys. **34**, 1690–1696 (2001)

M. Tavani, M. Marisaldi, C. Labanti, F. Fuschino, A. Argan, A. Trois, P. Giommi, S. Colafrancesco, C. Pittori, F. Palma, M. Trifoglio, F. Gianotti, A. Bulgarelli, V. Vittorini, F. Verrecchia, L. Salotti, G. Barbiellini, P. Caraveo, P.W. Cattaneo, A. Chen, T. Contessi, E. Costa, F. D'Ammando, E. Del Monte, G. De Paris, G. Di Cocco, G. Di Persio, I. Donnarumma, Y. Evangelista, M. Feroci, A. Ferrari, M. Galli, A. Giuliani, M. Giusti, I. Lapshov, F. Lazzarotto, P. Lipari, F. Longo, S. Mereghetti, E. Morelli, E. Moretti, A. Morselli, L. Pacciani, A. Pellizzoni, F. Perotti, G. Piano, P. Picozza, M. Pilia, G. Pucella, M. Prest, M. Rapisarda, A. Rappoldi, E. Rossi, A. Rubini, S. Sabatini, E. Scalise, P. Soffitta, E. Striani, E. Vallazza, S. Vercellone, A. Zambra, D. Zanello, AGILE Team, Terrestrial Gamma-Ray Flashes as Powerful Particle Accelerators. Phys. Rev. Lett. **106**(1), 018501 (2011). doi:10.1103/PhysRevLett.106.018501

L.Z. Tong, K. Nanbu, H. Fukunishi, Simulation of gigantic jets propagating from the top of thunderclouds to the ionosphere. Earth Planets Space **57**(7), 613–617 (2005)

E. Vadislavsky, Y. Yair, C. Erlick, C. Price, E. Greenberg, R. Yaniv, B. Ziv, N. Reicher, A. Devir, Indication for circular organization of column sprite elements associated with Eastern Mediterranean winter thunderstorms. J. Atmos. Sol.-Terr. Phys. **71**(17–18), 1835–1839 (2009). doi:10.1016/j.jastp.2009.07.001

O.A. van der Velde, A. Mika, S. Soula, C. Haldoupis, T. Neubert, U.S. Inan, Observations of the relationship between sprite morphology and in-cloud lightning processes. J. Geophys. Res. **111**, D15203 (2006). doi:10.1029/2005JD006879

O.A. van der Velde, W.A. Lyons, T.E. Nelson, S.A. Cummer, J. Li, J. Bunnell, Analysis of the first gigantic jet recorded over continental North America. J. Geophys. Res. **112**, D20104 (2007). doi:10.1029/2007JD008575

O.A. van der Velde, J. Bor, J. Li, S.A. Cummer, E. Arnone, F. Zanotti, M. Fullekrug, C. Haldoupis, S. NaitAmor, T. Farges, Multi-instrumental observations of a positive gigantic jet produced by a winter thunderstorm in Europe. J. Geophys. Res. **115**, D24301 (2010a)

O.A. van der Velde, J. Montanya, S. Soula, N. Pineda, J. Bech, Spatial and temporal evolution of horizontally extensive lightning discharges associated with sprite-producing positive cloud-to-ground flashes in northeastern Spain. J. Geophys. Res. **115**, A00E56 (2010b). doi:10.1029/2009JA014773

G. Veronis, V.P. Pasko, U.S. Inan, Characteristics of mesospheric optical emissions produced by lightning discharges. J. Geophys. Res. **104**(A6), 12645–12656 (2001)

P.A. Vitello, B.M. Penetrante, J.N. Bardsley, Simulation of negative-streamer dynamics in nitrogen. Phys. Rev. E **49**, 5574–5598 (1994)

E.M. Wescott, D. Sentman, D. Osborne, D. Hampton, M. Heavner, Preliminary results from the Sprites94 aircraft campaign. 2. Blue jets. Geophys. Res. Lett. **22**(10), 1209–1212 (1995)

E.M. Wescott, D.D. Sentman, M.J. Heavner, D.L. Hampton, O.H. Vaughan Jr., Blue jets: their relationship to lightning and very large hailfall, and their physical mechanisms for their production. J. Atmos. Sol.-Terr. Phys. **60**, 713–724 (1998)

E.M. Wescott, D. Sentman, H.C. Stenbaek-Nielsen, P. Huet, M.J. Heavner, D.R. Moudry, New evidence for the brightness and ionization of blue jets and blue starters. J. Geophys. Res. **106**(A10), 21549–21554 (2001). doi:10.1029/2000JA000429

E. Williams, Y. Yair, The microphysical and electrical properties of sprite producing thunderstorms, in *Sprites, Elves and Intense Lightning Discharges*, ed. by M. Füllekrug, E.A. Mareev, M.J. Rycroft. NATO Science Series II: Mathematics, Physics and Chemistry, vol. 225 (Springer, Heidelberg, 2006), pp. 57–83

E.R. Williams, The tripolar structure of thunderstorms. J. Geophys. Res. **94**(D11), 13151–13167 (1989)

E.R. Williams, Problems in lightning physics–the role of polarity asymmetry. Plasma Sources Sci. Technol. **15**(2), S91–S108 (2006). doi:10.1088/0963-0252/15/2/S12

E.R. Williams, The global electrical circuit: a review. Atmos. Res. **91**(2–4) Sp. Iss. SI, 140–152 (2009). doi:10.1016/j.atmosres.2008.05.018

C.T.R. Wilson, The electric field of a thundercloud and some of its effects. Proc. Phys. Soc. London **37**, 32D–37D (1925)

Y. Yair, Observations of transient luminous events from earth orbit. IEEJ Trans. Fundam. Mater. **126**, 244–249 (2006)

Y. Yair, C. Price, Z. Levin, J. Joseph, P. Israelevitch, A. Devir, M. Moalem, B. Ziv, M. Asfur, Sprite observations from the space shuttle during the Mediterranean Israeli dust experiment (MEIDEX). J. Atmos. Sol.-Terr. Phys. **65**(5), 635–642 (2003). doi:10.1016/S1364-6826(02)00332-2

Y. Yair, P. Israelevich, A.D. Devir, M. Moalem, C. Price, J.H. Joseph, Z. Levin, B. Ziv, A. Sternlieb, A. Teller, New observations of sprites from the space shuttle. J. Geophys. Res. **109**, D15201 (2004). doi:10.1029/2003JD004497

Y. Yair, C. Price, M. Ganot, E. Greenberg, R. Yaniv, B. Ziv, Y. Sherez, A. Devir, J. Bor, G. Satori, Optical observations of transient luminous events associated with winter thunderstorms near the coast of Israel. Atmos. Res. **91**(2–4) Sp. Iss. SI, 529–537 (2009). doi:10.1016/j.atmosres.2008.06.018

S.A. Yashunin, E.A. Mareev, V.A. Rakov, Are lightning M components capable of initiating sprites and sprite halos?. J. Geophys. Res. **112**(D10), D10109 (2007)

N.A. Zabotin, J.W. Wright, Role of meteoric dust in sprite formation. Geophys. Res. Lett. **28**(13), 2593–2596 (2001). doi:10.1029/2000GL012699

Space Sci Rev (2012) 168:517–531
DOI 10.1007/s11214-011-9812-x

Sprite-Producing Lightning-Ionosphere Coupling and Associated Low-Frequency Phenomena

S.L. Shalimov · T. Bösinger

Received: 25 June 2011 / Accepted: 19 July 2011 / Published online: 30 August 2011

Abstract Theoretical and experimental evidence is collected for additional ionization regions in the mesosphere above lightning discharges and their connection with transient luminous events (TLE). New insights are reported into the different appearances in the Very Low Frequency (VLF) link traces as affected by the mesospheric ionization regions. Based on these findings, physical processes going on in the ionization regions and their coupling to the primary lightning discharge process are conceptualized. Thereafter, some diagnostic potential is outlined. Finally, the ionization regions are considered as a transmitter of secondary low-frequency waves. The distinction is made between the primary electromagnetic waves emitted by the lightning and the secondary waves. As a consequence the Extremely Low Frequency (ELF) and Ultra Low Frequency (ULF) transient signatures observed on the ground are understood as a composite of both wave types. In addition a novel method is introduced to extract the charge moment change of a sprite producing lightning discharge.

Keywords Lightning · Transient luminous events · Electromagnetic emission · Atmospheric-ionospheric-magnetospheric-coupling · ULF · ELF · VLF

1 Introduction

In the past two decades, thunderstorms have been found to be capable of inducing upper atmosphere flashes now carrying the names of sprites, elves, halos, blue jets and gigantic

S.L. Shalimov (✉)
Institute of Physics of the Earth, Moscow, Russia
e-mail: pmsk7@mail.ru

S.L. Shalimov
Space Research Institute, Moscow, Russia

T. Bösinger
Department of Physical Sciences, University of Oulu, Oulu, Finland
e-mail: tilmann.bosinger@oulu.fi
url: http://cc.oulu.fi/~bosinger/

jets (cf. Yair et al. (2012) in the present issue of Space Sci. Rev.). The luminous spatial structures of these transient luminous events (TLE) range from meter- to kilometer-scales and span the region from the cloud top to the lower ionosphere. TLEs are optical manifestations of the coupling between the troposphere and the upper atmosphere/lower ionosphere. It is not only a question of the optical range of electromagnetic waves giving rise to TLEs, VLF and lower frequency radio emissions are also connected in one way or another to TLE phenomena. Thus they provide additional information for studying the processes associated with TLE-producing lightning. Here we review some recent understandings of distinct phenomena in VLF/ELF/ULF ranges associated with intensive positive cloud-to-ground (+CG) lightning discharges producing sprites. We pay special attention to additional ionization that can be produced by both, the lightning electromagnetic pulse (EMP) and the induced quasi-electrostatic (QE) field in the lower ionosphere. This additional ionization is a manifestation of direct sprite-producing lightning-ionosphere coupling. Moreover, it can help in understanding some secondary low-frequency phenomena associated with the lightning discharge. We also review some diagnostics using sub-ionospherically propagating VLF signals.

2 Probing the Lower Ionosphere Above Lightning Strokes

2.1 Regions of Highly Ionized Plasma in the Upper Mesosphere

There is direct and indirect evidence of highly ionized plasma columns above sprite-associated lightning discharges in the upper mesosphere/lower ionosphere. Both numerical and experimental confirmation of electron density enhancement by one order of magnitude and well above, caused by EMP/QE field atmosphere breakdown, exists.

Inan et al. (1991) and Rodrigues et al. (1992) carried out a study of collisional heating and ionization caused by lightning using a 1D code for the electrons. Assuming a Maxwellian distribution for the heated electrons, they could estimate the number of electrons with sufficient energy to ionize the neutrals, but the change in the plasma density due to the ionization was not used to modify the propagation of the primary electromagnetic wave. This has been done in a later study (Taranenko et al. 1993) using a kinetic approach in the limit of an approximation with a weakly asymmetric distribution function. They found electrons heating up to a few eV and electron density modification in the nighttime ionosphere up to 30% at altitudes of ~80–95 km above the lightning discharge. For pulse strengths slightly higher than used by Taranenko et al. (1993), significantly higher ionization occurs, and Rowland et al. (1995, 1996) showed that lightning EMP can cause a breakdown of the neutral atmosphere. They obtained remarkably larger values of the additional ionization up to an order of magnitude between 80 and 95 km. The enhanced ionization limits the strength of the EMP that can reach higher altitudes in the ionosphere, affecting mostly the trailing edge of the pulse (see below). Additional theoretical modeling also indicates that sufficiently strong lightning EMP leads to changes in ionization (~100% or even more) with the largest increases at 80–90 km (Sukhorukov et al. 1996; Cho and Rycroft 1998; Rodger et al. 2001).

All these simulations indicated that high-altitude breakdown occurs between 70–90 km altitude and that the optical emission lasts only for a few hundred μs. Although these predictions are consistent with the enhanced airglow seen at 90 km and now known as elves (cf. Yair et al. (2012) in the present issue of Space Sci. Rev.), they are different from the longer-lived red sprites (a few ms) seen at lower altitudes. Pasko et al. (1995, 1997) were the first to compute the quasi-static breakdown caused by the transfer of charge in lightning (cf. Pasko et al. (2012) in the present issue of Space Sci. Rev.). Quasi-static breakdown can account for the longer lifetime of sprites, and it is claimed to produce a strong optical emission down to 65 km altitude and significant (several orders of magnitude) modification of the lower ionosphere conductivity (see below).

We note, however, that Adachi et al. (2004) found that the number of sprite columns in each sprite event was proportional to the peak current intensity of +CG lightning discharges while the average vertical length of columns was proportional to the charge moment change of the causative +CG discharge. This means that both the EMP and the QE field associated with lightning discharge plays a role in the generation of sprites (since the EMP is proportional to the peak current intensity, and the QE field is proportional to charge moment change, as mentioned).

Photometric data now appear to confirm that, in at least red sprite events with a tendril structure, significant ionization can be produced between 10^4–10^5 cm^{-3} (Armstrong et al. 2000; Morrill et al. 2002), over 4–6 orders of magnitude above the ambient density at sprite altitudes of ~60 km. In addition, the observed strength of the short-lived back-scattered signals proves the existence of highly ionized plasma regions at heights of 55–65 km, where electron densities were 4–6 orders of magnitude above that of the ambient nighttime ionosphere (Roussel-Dupre and Blanc 1997). Finally, VLF perturbations lasting 30–100 s indicate similar orders of additional ionization associated with plasma columns in the altitude range of 55–70 km (Dowden et al. 2001).

Recently D region ionization by lightning-induced EMP associated with elves (Mende et al. 2005) and QE field associated with sprites (Kuo et al. 2008; Adachi et al. 2008) were inferred from FORMOSAT-2 measurements (cf. Pasko et al. (2012) in the present issue of Space Sci. Rev.).

Experimental evidences of regions with highly ionized plasma in the upper mesosphere/lower ionosphere are also provided by VLF probing the regions above lightning discharges (see below).

2.2 Trimpis, Early/Fast (E/F), and Early/Slow (E/S) VLF Events

It is known that VLF transmitter signals (their amplitude and phase) propagating in the Earth-ionosphere waveguide (nighttime reflection height ~70–85 km) are affected by conductivity changes on the upper waveguide boundary/plate (ionospheric D region). Modifications in lower-ionospheric conductivity can be due to changes in collision frequency (heating) and/or electron density (ionization). A Trimpi perturbation is a change in the amplitude and/or phase of transmissions, recognized in Antarctica data by M.L. Trimpi during his time as a field scientist and finally presented in a paper by Helliwell et al. (1973). The beginning of the VLF perturbation in this case occurs about 1 s after the associated sferic (Fig. 1a). This delay being consistent with the time required for the leaking part of the sferic energy (whistler mode wave) to propagate through the ionosphere and magnetosphere to the geomagnetic equator, to interact with energetic electrons, and for these electrons to precipitate into the D region. The precipitating electrons create an ionization enhancement in the D region. This type of electron precipitation is also referred to as whistler-induced electron precipitations (WEP) or lightning-induced electron precipitations (LEP). The perturbation begins with a relatively fast (1–3 s) change in the transmission amplitude and/or phase, followed by a slower relaxation (<100 s) back to the unperturbed level (Fig. 1a). Most reported observations have amplitude changes ranging from 0.04 dB (Carpenter et al. 1984) to 6 dB (Helliwell et al. 1973) and phase changes as high as 12° (Inan and Carpenter 1987).

Armstrong (1983), however, discovered with higher time resolution data, a new class of perturbation. The new phenomenon had the same decay time as the normal Trimpi perturbation but was characterized by a very small delay (<50 ms) between the parent sferic and

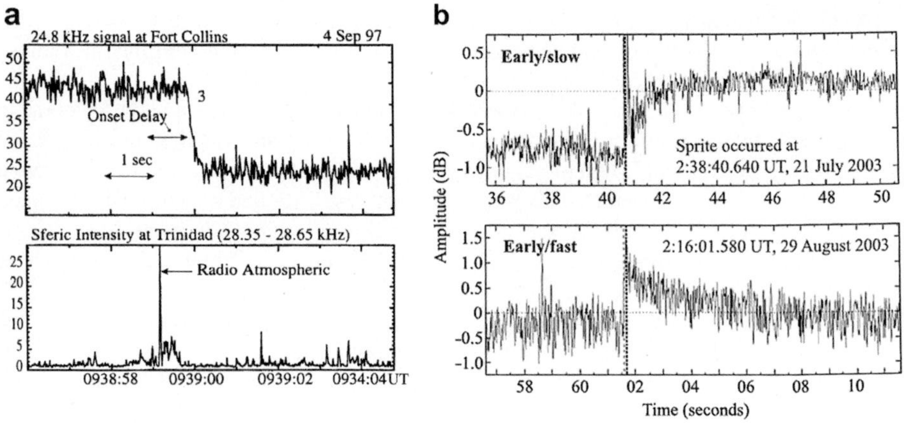

Fig. 1 An example of VLF amplitude perturbation produced by classic Trimpi (*left*) and an early fast/slow VLF events (*right*) [After Rodger (1999) and Haldoupis et al. (2006). Reproduced with permission of the American Geophysical Union]

the VLF perturbation, implying a direct effect of the lightning onto the ionosphere. One of the characteristics of the new class of perturbation was its association with whistlers, having a strong ELF tail down to 300 Hz (Armstrong 1983). These events were labeled "early" or "early/fast" (E/F) VLF events by Inan et al. (1988) who noted that the events have both early (<20 ms) onset after the causative lightning discharge and fast (<20 ms) onset duration (Fig. 1b).

After the discovery of sprites (Franz et al. 1990), Inan et al. (1995) made the first correlation between sprites and "early" events, and suggested that the physical processes that lead to sprites (i.e. QE heating; cf. Pasko et al. 1995) also alter the conductivity of the lower ionosphere. Interestingly enough, the 10–100 s recovery time of E/F events was not consistent with heating, which recovers in less than a second, but consistent with ionization recovery at 75–85 km altitude.

Meanwhile detailed studies of VLF perturbations have appeared as a result of a series of "EuroSprite" campaigns (Neubert et al. 2008). From these campaigns, Haldoupis et al. (2004, 2010) verified a one-to-one correlation between sprites and E/F events. In addition, Haldoupis et al. (2006) discovered a new category of VLF events labeled "early/slow" (E/S): like the E/F event it has an early onset time after the causative lightning discharge, but its onset duration was relatively large (0.5–2.5 s), like Trimpi perturbation (Fig. 1b).

These experimental facts provoke the question: why is there a gradual onset growth for part of the early VLF events that occur in connection with sprites? Haldoupis et al. (2006) showed experimentally that the growth phase of E/S events coincide with the occurrence of clusters (bursts) of spherics which are attributable to in-cloud (IC) lightning while onsets of E/F events were not accompanied by clusters of spherics. In fact, some E/S events appeared to have their onset growth starting somewhat earlier than the sprite, apparently because IC lightning generated EMP may start earlier, at times, by enhanced sferic clustering prior to sprite onset. This led Haldoupis et al. (2006) to suggest that IC lightning EMP, comprising of a burst of lightning pulses, can cause a slow accumulation of ionization in the D region. Consequently, both the sprite QE field and IC lightning EMP contribute to E/S events.

Recently, Marshall et al. (2010) and Marshall and Inan (2010) provided some modeling of the interaction of the in-cloud emitted EMP with the lower ionosphere. The authors showed that the resulting disturbances could be measured by VLF receivers with actually observed amplitude changes, therefore bursts of IC lightning could be made responsible for E/S and some E/F events.

Finally, the whole class of direct lightning-induced ionospheric disturbances was reclassified as "early" VLF events.

Summarizing, it may be emphasized: the significant role of bursts of VLF spherics in the course of IC lightning associated with sprite-producing discharges (Johnson and Inan 2000; Ohkubo et al. 2005), has only been recently recognized. Rowland et al. (1995) were the first who showed by numerical simulation the possibility of direct ionization by an IC EMP (see also Marshall et al. 2006) and Haldoupis et al. (2006) were the first who provided experimental evidence that the IC lightning activity is one of the agents for E/S event production. Li et al. (2008) found a nearly one-to-one relationship between sferic bursts and the so called M-components, which are intensifications in the continuing current on time scales of a fraction of an ms to many ms (Rakov and Uman 2003, p. 176). Theoretically the M-components were shown to play a role in the initiation of sprite and sprite halos (Yashunin et al. 2007) as they produce anomalously large mesospheric electric field enhancements which can be large enough to push the field near to breakdown levels. The same process was also studied by Asano et al. (2009) using numerical simulations.

All these recently developed ideas and verifications contribute to the understanding of the E/S generation mechanism. As outlined above, the +CG discharges, with charge moments smaller than some threshold value, produce a relatively weak QE field above thunderstorms causing no or low ionization levels. This is the case of the so called delayed sprites. In this situation, bursts of IC lightning, themselves associated with a weak QE field, can lead to (1) temporal raising of the QE field above the breakdown threshold and thereby triggering sprites which otherwise would not have occurred (in this scenario the IC activity serves to joint additional regions of charge and enhance the charge moment through a continuing current modulation or M-component (Li et al. 2008; Asano et al. 2009)), and (2) energizing further the electrons in the upper D region causing additional ionization (the times of interest (0.5 to 2.5 s) are much shorter than the lifetimes of the electrons at these heights), so that a gradual electron density accumulation may occur, manifesting in the long onset duration of the E/S events (Haldoupis et al. 2006).

2.3 Modeling the Relaxation of E/F VLF Events

Studies show that early VLF events, characterized by abrupt phase and/or amplitude perturbations and long recoveries, often occur in relation with TLE, especially sprites, sprite halos and elves (e.g., see Moore et al. 2003; Haldoupis et al. 2004, 2010; Mika et al. 2006; Marshall et al. 2006, 2010). Telescopic imaging of sprites (Gerken et al. 2000) shows that sprites are characterized by a lower (50–70 km altitude) streamer-like region, and often, but not always, by a higher (70–90 km altitude) region of laterally extended glow. Model estimates of electron density elevations produced in the upper regions of sprite breakdown and/or sprite halos led Barrington-Leigh et al. (2001) to suggest that they are likely to be responsible for most of the early VLF events.

In modeling the relaxation (Haldoupis et al. 2010), the early VLF events are assumed to be caused by forward scattering from areas of enhanced electron density in the D region. This conviction is reinforced by the fact that, other than the onset delay, both LEP and early events look fairly similar in narrowband VLF recordings, particularly with respect to their recoveries. This similarity offered the motivation to apply, for early events, the same D region model that has been used in LEP investigations by Glukhov et al. (1992). It relies

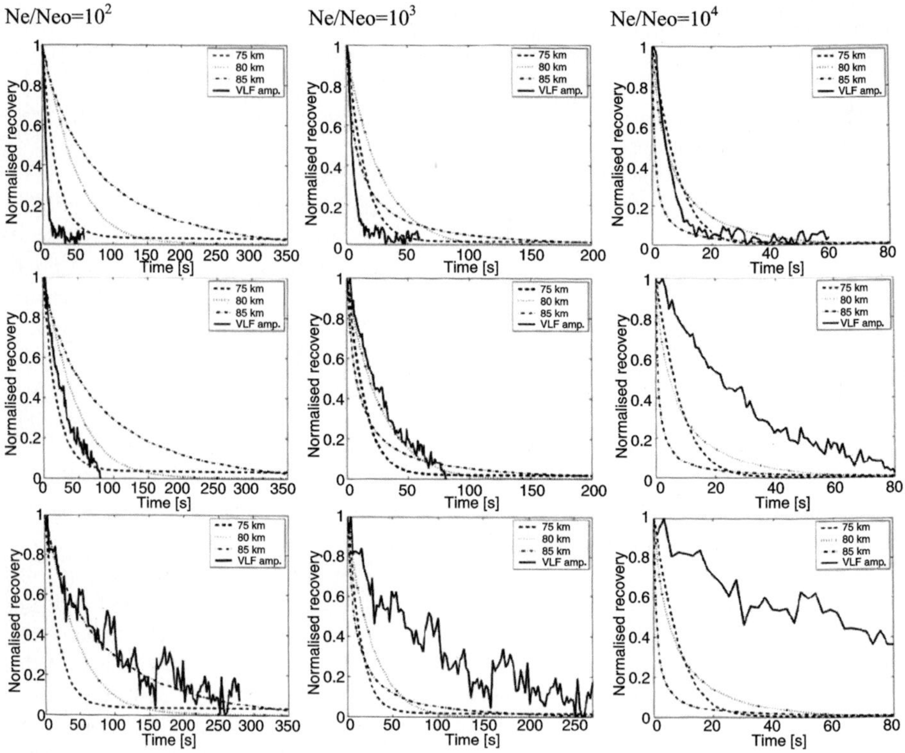

Fig. 2 Results of the Haldoupis et al. (2010) relaxation time modeling [Reproduced with permission of the American Geophysical Union]

on a simplified scheme of 4 constituents (electrons, positive ions, negative ions and positive cluster ions) as well as 6 processes (electron attachment and detachment, dissociative recombination, ion-ion recombination, recombination with cluster ions, and conversion of positive ions into positive cluster ions).

The results of the modeling showed that the relaxation time is a function of altitude and inversely proportional to the additional electron density produced after discharge; tentative estimates of electron density increase, responsible for the early VLF events, may reach values possibly higher than 10^4 electrons per cm^{-3} above the background in the upper D region ionosphere (Fig. 2). We may conclude that by knowing the background density profile one can calibrate the relaxation time dependence on altitude.

3 Using VLF Content to Extract Charge Moment Change

To move a little further, we now refer to recent and very interesting papers by Hiraki and Fukunishi (2006) and Asano et al. (2008, 2009), who suggested the importance of both the lightning discharge time and the M-component (an impulsive perturbation or transient enhancement in the continuing current Rakov and Uman 2003) in the initiation of sprites. In their numerical simulations Hiraki and Fukunishi (2006) paid special attention to the effect of the lightning discharge time. As a consequence of their simulations they suggested the following simple relation:

$$\frac{(Qz_s)^2}{\tau_s} = (Iz_s)(Qz_s) = \text{const}, \tag{1}$$

where τ_s is the discharge time, and I is the effective current that is defined approximately by $I = Q/\tau_s$. Using (1) they found that a smaller τ_s results in the initiation of sprites which is completely consistent with numerical simulations and conclusion by Asano et al. (2008, 2009), although in the latter papers they have used the rise time of the lightning current waveform rather than τ_s.

Next, we take into account the results of Adachi et al. (2004) who suggested that both EMP and QE field associated with lightning discharge plays a role in the generation of sprites. We consider a simple estimate of ionization produced by a lightning electromagnetic field taking it as a pulse with amplitude E_0 and duration τ. According to numerical simulations (Sukhorukov et al. 1996) the ionization produced by the pulse propagating upward modifies the pulse, affecting mostly the trailing edge of it. Therefore, assuming that for the pulse propagating upward its duration decreases due to loss caused (mostly) by ionization while the pulse amplitude keeps approximately constant, and using the energy conservation law we get (Borisov et al. 1986)

$$c\frac{E_0^2}{4\pi}\frac{d\tau}{dz} = -\varepsilon_i(z)N_{e0}(z)\exp(\nu_i(N_m)\tau), \quad \tau|_{z=z_0} = \tau_0, \tag{2}$$

where c is the speed of light, ε_i, ν_i is the energy and frequency of ionization, respectively, τ_0 is the initial duration of the pulse entering into the ionosphere at the altitude z_0. Here the ionization frequency $\nu_i(N_m)$ depends parametrically on the neutral density and it is computed from the following expression of Papadopoulos et al. (1993)

$$\nu_i = \nu_{am}\left(0.7 + 4.2\exp\left(-2.6\frac{E_k}{E_o}\right)\right)\left(\frac{E_o}{E_k}\right)^2 \exp\left[-4.7\left(\frac{E_k}{E_o} - 1\right)\right],$$

$$\nu_{am} = 7.6 \times 10^{-13} N_m \text{ cm}^3\text{ s}^{-1},$$

taken in the limit $\omega \ll \nu_k$, where ω is the angular frequency of the electromagnetic pulse and ν_k is an effective electron-neutral collision frequency, E_k is the air-breakdown electric field, and ν_{am} is the maximum attachment frequency. Integrating (2) one can obtain the dependence of pulse duration on altitude and estimate the resulting ionization. Specifically, for the pulse duration one obtains

$$\tau = \tau_0 - \frac{1}{\nu_i}\ln\left[1 + \frac{4\pi\varepsilon_i\nu_i}{cE_0^2}\exp(\nu_i\tau_0)\int_{z_0}^{z} N_{e0}(z)dz\right]. \tag{3}$$

For simplicity we consider a linear model of plasma density: $N_{e0}(z) = (N_{e0}/L)(z - z_0)$, where $N_{e0} = 10\text{ cm}^{-3}$, $L = 10$ km, and take $E_0 = 217\text{ V m}^{-1}$ that corresponds to the breakdown electric field at altitude $z_0 = 70$ km. The dependence (3) for different initial pulse duration 10^2 μs (black), 1 ms (red), 10 ms (blue), representing the range of characteristic time for the return stroke and charge removal time in the case of +CG lightning (Uman 1987, pp. 124, 196), is shown in Fig. 3. It is seen from this figure that the longer the pulse duration τ_0 is, the lower the altitude of the produced ionization. Taking into account both the calibrated dependence of relaxation time vs. altitude (Sect. 2.3), the dependence (3) and the relation (1) it is in principle possible to quantify the charge moment Qz_s for early VLF

Fig. 3 Impulse duration versus altitude of ionization

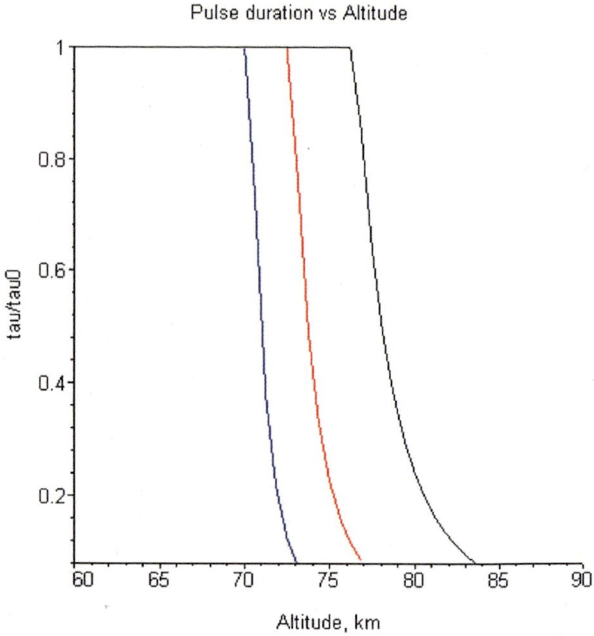

production using the following logical scheme: $\tau_r \Rightarrow h_i \Rightarrow \tau_0 \equiv \tau_s \Rightarrow Qz_s$. We note that the logic can be applied only for E/F events, because for E/S we get some additional ionization due to in-cloud discharges, and we cannot take this additional ionization into account using the given approach.

4 On Secondary ULF/ELF Waves from Mesosphere Ionization Region

The motivation for considering lightning associated low frequency (ULF/ELF) phenomena was the discovery of unexpected anomalies in this frequency range. The energy into the ULF range (0.1–10 Hz) and the lower ELF range is normally delivered by Pc1/Pi1 type magnetic pulsations, the Schumann resonance (SR), excitation of the ionospheric Alfvén resonator (IAR), and noise enhancement in the vicinity of thunderstorms. In view of the short-lived lightning discharge processes with time scales of $\sim 10^2$ μs (corresponding to frequencies in the range of 10 kHz) hardly anything is expected to be seen by a magnetometer sensitive up to 10 Hz. Ground-based observations by standard pulsation magnetometer revealed, however, a different type of physics.

Just after a lightning flash a primary unipolar pulse is detected. Its duration is dictated by the impulse response of the magnetometer ($\sim$100 ms), but thereafter the measurements typically exhibit a longer lasting decay (often referred to as a tail) following the primary pulse and lasting up to several seconds (Fukunishi et al. 1997; Füllekrug et al. 1998; Bösinger et al. 2006).

Attempts to explain the lightning discharge triggered long lasting ULF events, as well as the anomalously large amplitudes (details below), required highly ionized regions overhead the discharge (often sprite associated; see above) in the upper mesosphere/lower ionosphere. In fact the presence of such regions with extra-ionization can provide an explanation of tail phenomena in the ULF as well ELF frequency range. The extra ionization may exist for many tens of seconds and may be produced even before a sprite in the form of a halo or elve, or by sferic clusters associated with long-delayed sprites (see Sect. 2.2). From numerical simulations, two different spatial scales were extracted for regions with modified electron density (conductivity): a few km (Cho and Rycroft 2001), and a few tens of km (Rodger et al. 2001).

In tune with this conjecture, is the failure of theoretical predictions of electric field values, which were in the range of 10–100 $\mu V\,m^{-1}$ above thundercloud centers (Dejnakarinta and Park 1974; Greifinger and Greifinger 1976). The ULF field variations observed onboard rockets/satellites are in fact above these values by one to two orders (Thomas et al. 2008; Kelley et al. 1985; Burke et al. 1992).

4.1 Long Lasting (Tail-Type) ULF Events

Two types of ULF response to +CG lightning (starting <200 ms after a +CG lightning flash) were found during the EuroSprite2003 campaign: type I is a single, unipolar peak from the background level, and type II again a peak, however, with a subsequent asymptotic (tail-like) decay from the peak level to the zero line over time scales of several peak widths. As a frequent variation of these basic types, the unipolar peaks were followed by some quasi-sinusoidal oscillations. Previously, the latter were reported by Fukunishi et al. (1997) and Füllekrug et al. (1998). The observed amplitudes of the oscillation range from 0.1 to 1.0 nT at ~500 km distance (Fukunishi et al. 1997) till a few tens of pT at ~2000 km (Füllekrug et al. 1998) from the source. The oscillations last on average from less than 0.4 s up to a few seconds (the so called ultra-slow tail, UST). The two time scales suggest two different physical processes: the shorter one (0.4 s) is usually understood in terms of the magnetic signature of the continuing current (Cummer and Füllekrug 2001) whereas the longer (a few seconds) was attributed (Füllekrug et al. 1998) to the excitation of IAR by a sprite-associated +CG lightning discharge. However, IAR excitation with a subsequent propagation of the signal to the point of observation must inevitably take a few seconds and cannot explain a short delaying (<200 ms) in ULF appearance. In addition, sprites were sometimes preceded by elves and observed within the ULF wavelets lasting about 3 s (Fukunishi et al. 1997). This means that ULF-transient arrive before the sprite initiation and cannot be attributed to continuing current. We take this as an indication of the presence of an ionized layer that can radiate even before the sprite occurrence.

To interpret the early ULF signals a new approach to the UST problem was taken (Shalimov and Bösinger 2006; Bösinger and Shalimov 2008). The doughnut-like regions of enhanced ionization (with a radius R greater than its thickness h) in the nighttime lower ionosphere, imbedded into the geomagnetic field, are understood as ULF/ELF wave radiators involving the following scenario: the localized region of enhanced ionization in the altitude range of 80 km $< z <$ 95 km carries an axial (φ-component) time-varying current $j_0(r, z, t)$ induced by the EMP/QE field. The electromagnetic response of the ionized region to this axial current drive (using Maxwell equations and Ohm's law) is described by a single diffusion-like inhomogeneous differential equation. It can be interpreted as the penetration of the induced current into the ionized region along h as a diffusive type of wave, which is subject to eigen-oscillations. The observed quasi-periodic oscillations in the UST are manifestations of these eigen-oscillations, not smeared out by the strong spatial decay of the radiated signal strength over distance. In order to get the model properly working, high ionization levels in the plasma column (one order of magnitude and above) are needed, and only in cases of extremely high ionization do the quasi-periodic oscillations fall into the period range of seconds compatible with the UST phenomenon (rarely observed). The best merit of the model is, however, that it works well in the ELF range, where slow tails are more frequent. It was also here where the tail phenomenon was first observed (e.g. Rodger 1999 and references therein).

By no means are all processes understood in the ELF/ULF signature of sprite-producing lightning discharges. For example, Bösinger et al. (2006) compared measurements of ULF transients below the Schumann resonance (i.e. below 7.8 Hz) for sprite-producing and non-sprite-producing +CG lightning flashes and found no difference in waveform. Obviously the so called Q burst is not inevitably associated with the sprite producing process (as previously thought by Boccippio et al. 1995).

Another intriguing result, reported by Bösinger et al. (2006), was the encountering of two types of distant ULF responses to lightning discharge (those with and without a tail, see above). Traditionally, the tail is interpreted as a signature of the continuing current. In the case of −CG, the tail should cover only some tenths of an ms, and thus be invisible to pulsation magnetometers. However, intensive −CGs lightning can also produce a detectable ULF tail (Bösinger et al. 2006). A similarity of the ULF tail between −CGs and +CGs suggests a D region effect of the type conjectured by Shalimov and Bösinger (2006), rather than an atmospheric electricity effect in terms of a continuing current. Similarly, Greenberg et al. (2009) found that a significant number of sprite-producing lightning (33%) had no ELF transients, suggesting that long continuing current may not always be a necessary condition for sprite production. Finally, Price et al. (2004) used ELF transients to geolocate sprite-producing lightning. However they found ELF transients associated with elves, but not with sprites, contradictory to the theory of Pasko et al. (1995).

Finally the results of two balloon campaigns should be mentioned. In situ mesospheric measurements of the electromagnetic pulses that produce TLE phenomena are in fact highly wanted but difficult to achieve. Only rocket and balloon measurements could keep up with this goal, with their own obvious limitations.

The Sprites99 Campaign (Bering et al. 2002, 2004a, 2004b; Bhusal et al. 2004) was the first stratospheric balloon experiment that, along with balloon-borne electric and magnetic field measurements and optical triggering (at an altitude of about 30 km and a distance of 200–1700 km to the source), also included ground-based sprite imaging. During the campaign, stroke-delayed (by 4–6 ms) ELF pulses were measured for 90% of the CG strokes detected by the National Lightning Detection Network (NLDN) (Bering et al. 2004a, 2004b; Bhusal et al. 2004). Bering et al. (2004b) and Bhusal et al. (2004) attributed these delayed pulses to mesospheric current moments, which were substantiated by optical data of sprites or halos that occur simultaneously with only a small percentage of these pulses. Cummer et al. (1998) has measured similar delayed ELF pulses with ground-based magnetic field sensors that were directly correlated with video images of sprites. They attributed the ELF pulses to currents in the sprites.

However, the suggestion that mesospheric breakdown is a fundamental response to nearly all CG lightning (Bering et al. 2004b; Bhusal et al. 2004), is in disagreement with ground-based measurements (Cummer et al. 1998) and models (Pasko et al. 1998) which indicate that only lightning with large charge moment changes should drive mesospheric currents.

During the Brazil Sprite Balloon Campaign in 2002–2003 (Thomas et al. 2005), the vector ELF to VLF (25 Hz–8 kHz) electric and magnetic fields were measured as driven by cloud-to-ground (CG) lightning strokes. The measurements took place at horizontal distances of 75–600 km, at altitudes of 30–35 km. Unfortunately, the view of the aircraft-based imaging cameras was blocked by clouds, so sprites could not be observed during the balloon flights. In 1.4% of all 934 strokes detected by the Brazilian Integrated Lightning Network (BIN), ELF pulses occurred 4–12 ms after the retarded time of the lightning spherics.

These delayed ELF pulses were previously attributed to sprites, as mentioned. Thus, stroke-delayed ELF pulses were rare. This in turn, is in disagreement with the results from the Sprites99 Balloon Campaign. It confirms, however, the earlier ground-based measurements (Cummer et al. 1998) and the fully electromagnetic model of lightning-driven fields (Pasko et al. 1998).

In summary, in spite of the unique observational platform of balloon borne observatories, up to now, too few reports exist on such kinds of investigations to construct a systematic picture of the various findings, at least in the given context (in other contexts the reported results may be remarkable; e.g. the frequent appearance of halos, not to be confused with elves). In any case, both balloon campaigns confirm the presence of mesospheric currents driven by TLE associated lightning strokes. This is not inconsistent with our model (Shalimov and Bösinger 2006). However, more measurements are needed in order to be able to discriminate ELF pulses associated, from those not associated, with TLE.

5 Anomalously Large Amplitude of ULF E-Fields

Dejnakarinta and Park (1974) studied the upward dwelling of electric fields in the frequency range from dc to 10 Hz and showed that the electric field given at a horizontal distance of 5 km and a vertical distance of 20 km from the source (charge), suffers depression by 60 dB at 70 km, and by 140 dB at 100 km altitude. This means that an initial 100 $\mathrm{V\,m^{-1}}$ field at 20 km will be reduced to 0.1 $\mathrm{V\,m^{-1}}$ at 70 km and 10 $\mu\mathrm{V\,m^{-1}}$ at 100 km. Greifinger and Greifinger (1976) predicted a similar attenuation. Their theoretical estimates indicate ionospheric electric field values of 10–100 $\mu\mathrm{V\,m^{-1}}$ above a thundercloud. Quite contrary, Burke et al. (1992) measured, during the passage of the DE-2 satellite over Hurricane Debbie (in association with its lightning activity), ULF field variations of 0.5 $\mathrm{mV\,m^{-1}}$ at an altitude of $\sim$300 km. Moreover, electric field transients observed by a rocket in the ionosphere (at $\sim$150 km) and attributed to lightning activity had amplitudes of $\sim$10 $\mathrm{mV\,m^{-1}}$ (Kelley et al. 1985). Furthermore, Thomas et al. (2008) presented examples of lightning-driven electric field changes as obtained at 75–130 km altitude during a sounding rocket flight from Wallops Island, Virginia, in 1995. Specifically, they found that electric field changes measured at 85–100 km altitude and 255–275 km horizontal distance from the source had surprisingly large quasi-electrostatic field (QSF) components (with amplitude of 3 $\mathrm{mV\,m^{-1}}$).

There are several possible reasons for the discrepancies between theory and experiment. For example, they might be explained by the differences in the conductivity profiles in the atmosphere in cases where the lower atmospheric conductivities above the thunderstorm allow a more effective electric field penetration into the ionosphere. Indeed, scale heights of $H = 8$–11 km for the conductivity profile at altitudes below 70 km far away from the thunderstorm (100 km) were reported by Holzworth et al. (1985). These values are noticeably larger than the value of $H = 6$ km used by Greifinger and Greifinger (1976). It is, however, important to notice, that the ULF field in the ionosphere above tropospheric discharge processes is formed mainly by spatial gradients with transverse wave numbers $k_\perp$ much larger than the free wave number $k_0 = \omega/c$, and the ULF field is, therefore, roughly proportional to $1/H^3$ (Sukhorukov 1996). Thus, an increase of H would cause only a decrease of the predicted ULF fields.

We suggest, therefore, regions of highly ionized plasma in the upper mesosphere (Shalimov 2011) are responsible for the anomalously large amplitudes observed in the ionospheric ULF field. We assume that an ionization column with thickness h greater than its radius R is present in the altitude range of 80 km $< z <$ 95 km, and a strong driver current $j_0(r, z, t)$ (φ component) is imposed as a consequence of a lightning discharge. The former generates magnetosonic waves. Transformation of the magnetosonic wave in the ionospheric Hall layer, in turn, generates an outward propagating Alfvén wave. Our estimates yield Alfvén impulses compatible with amplitudes observed onboard the satellites/rockets in the presence of a highly ionized region (the density enhancement is assumed to be one order of magnitude above background values).

The relevant expressions and assumptions look like: for the current source

$$j_0 = I_0(t)\delta(r - R)\delta(z - h),$$

where $I_0(t)$ is the external current, at a distance greater than R we obtain for the E_φ and, accordingly, for the Alfvén wave amplitude E_r the following expressions (Shalimov 2011)

$$E_\phi = -\frac{\pi}{c^2}\frac{rR^2}{(r^2 + |z - h|^2)^{3/2}}\frac{\partial I_0(t)}{\partial t},$$

$$E_r = -\frac{\pi}{c^2}\frac{rR^2}{(r^2 + |z - h|^2)^{3/2}}\frac{\Sigma_H}{\Sigma_P + \Sigma_W}\frac{\partial I_0(t)}{\partial t},$$

where $\Sigma_W = c^2/4\pi c_A$ is the wave conductance. This result is valid for the time scale greater than the diffusion time (quasi-electrostatic approximation) $\tau_D \sim \pi\sigma_P R^2/c^2$, where σ_P is the Pedersen conductivity. We use the model current $I_0(t) = hR \cdot j_0 \cdot (t/T)\exp(-t/\tau)$, where T, τ is the characteristic rise time of lightning current and the time of continuing current, respectively, and some representative parameters $h = 15$ km, $R = 5$ km, $z = h$ under condition of enhanced ionization when $\Sigma_H/(\Sigma_P + \Sigma_W) \approx 1$.

Our estimates shown in Fig. 4 for Alfvén wave amplitude should be compared with the results obtained onboard rockets/satellites. With $r = 3R$, $T = 100$ μs, and for typical mesospheric currents associated with sprites $j_0 = 3\ \mu\text{A m}^{-2}$ (Cummer et al. 1998) the estimated Alfvén impulse is shown in Fig. 4 (red line). The result is in agreement with the typical rise time, duration, and amplitude of a signal detected in Kelley's rocket experiment (Kelley et al. 1985) in association with lightning (0.2 ms, 10–20 ms, and 10 mV m^{-1}, respectively). In Thomas' rocket experiment (Thomas et al. 2008) electromagnetic fields associated with elves in the lower ionosphere were investigated. Taking into account the larger horizontal distance $r = 50R$ and the elve associated currents $j_0 = 60\ \mu\text{A m}^{-2}$ (Sukhorukov et al. 1996) we are again using $T = 100$ μs. The result is shown in Fig. 4 (black line) with a typical amplitude of 3 mV m^{-1} totally in agreement with the experiment.

To our knowledge there are so far no reports which give an account of the magnetic signatures observed on the ground in the context of IC lightning. Only some intuitive common perspectives exist; that the magnetic signatures are short ranged due to the specific geometry of the lightning stroke. What is needed are means (radar, sound detection, etc.) providing not only time, coordinates and peak current, but also the length of the discharge channel and its specific alignment in the cloud and with respect to the observation point on the ground. Such instruments do in fact exist (e.g. Betz et al. 2004), but no one has so far elaborated the IC magnetic ULF signatures on the ground. It is expected that IC lightning has an effect on the waveform of the ULF signature (unipolar pulse and its tail appearance), especially in cases of intermittent, repetitive discharges or clustering, as we have seen in the case of E/S VLF events (Haldoupis et al. 2006).

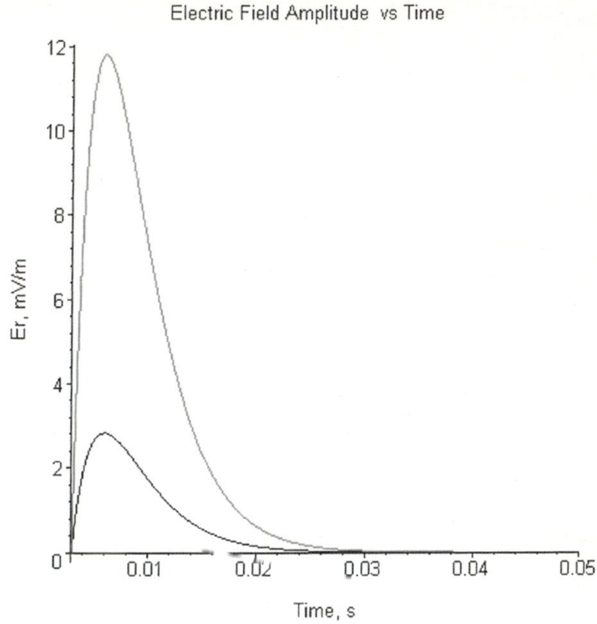

Fig. 4 Alfvén wave electric field amplitude versus time

6 Outlook

Despite almost three decades of active research, the generation mechanism of early VLF events is only rudimentarily understood and it remains a relevant topic of future research. This is even true for the whole complex of sprite-ionospheric coupling.

Only recently is the important role of IC lightning being realized, and a lot of research is still needed to specify their various interconnected mechanisms in the production of sprites and related phenomena in the VLF, ELF and ULF ranges. New instruments are needed to pinpoint the single IC lightning strokes, their temporal and spatial evolution and to evaluate the cause and effects of clustering.

Little attention has been paid, so far, to the ULF response of lightning discharges, their specific characteristics to +CG, −CG, and IC lightning activity. Clarifications are needed with respect to ULF and ELF fingerprints to sprite production, elves, and TLEs as a whole. The present uncertainties are reflected by the fact that different views and approaches can be found in the literature. In a recent paper by Schekotov et al. (2011) the authors took different theoretical and experimental approaches to the problem of IAR excitation and ULF signatures of lightning discharges. It is reviewed by Pilipenko et al. (2012) in the present issue of Space Sci. Rev. No doubt, verifications are needed of currently coexisting ULF tail models.

All this can tell that a final evaluation of the significance of sprite-producing lightning-ionosphere coupling, and how it manifests in low-frequency electromagnetic emissions, is a challenging open research field which provokes equally atmospheric and plasma physicists.

This review reported—to our knowledge for the first time—a method to extract the charge moment change of a sprite producing lightning discharge by a E/F VLF event detection.

Acknowledgements The authors thank ISSI for the invitation to the Workshop on "Coupling between the Earth's atmosphere and its plasma environment" in September/October 2010. The work of S.S. was partly supported by the grant 10-05-93107 from RFBR.

References

T. Adachi, H. Fukunishi, Y. Takahashi, M. Sato, Geophys. Res. Lett. **31**, L04107 (2004)
T. Adachi, Y. Hikaki, K. Yamamoto, Y. Takahashi et al., J. Phys. D, Appl. Phys. **41**, 234010 (2008)
W.C. Armstrong, Antarct. J. USA **18**, 281–283 (1983)
R.A. Armstrong, D.M. Suszcynsky, W.A. Lyons, T.E. Nelson, Geophys. Res. Lett. **27**(5), 653–656 (2000)
T. Asano, M. Hayakawa, M. Cho, M. Suzuki, J. Geophys. Res. **113**, A02308 (2008). doi:10.1029/2007JA012528
T. Asano, M. Suzuki, Y. Hiraki, E. Mareev et al., J. Geophys. Res. **114**, A02310 (2009). doi:10.1029/2008JA013651
C.P. Barrington-Leigh, U. Inan, M. Stanley, J. Geophys. Res. **106**(A2), 1741–1750 (2001). doi:10.1029/2000JA000073
E.A. Bering III, J.R. Benbrook, J.A. Garrett, A.M. Paredes, E.M. Wescott, D.R. Moudry, D.D. Sentman, H.C. Stenbaek-Nielsen, W.A. Lyons, Geophys. Res. Lett. **29**(5) (2002). doi:10.1029/2001GL013267
E.A. Bering III, J.R. Benbrook, L. Bhusal, J.A. Garrett, A.M. Paredes, E.M. Wescott, D.R. Moudry, D.D. Sentman, H.C. Stenbaek-Nielsen, W.A. Lyons, Geophys. Res. Lett. **31**, L05104 (2004a). doi:10.1029/2003GL018659
E.A. Bering III, L. Bhusal, J.R. Benbrook, J.A. Garrett, A.P. Jackson, E.M. Wescott, D.R. Moudry, D.D. Sentman, H.C. Stenbaek-Nielsen, W.A. Lyons, Adv. Space Res. **34**, 1782–1791 (2004b)
H.-D. Betz, K. Schmidt, P. Oettinger, M. Wirz, J. Geophys. Res. **31**, L11108 (2004). doi:10.1029/2004GL019821
L. Bhusal, E.A. Bering III, J.R. Benbrook, J.A. Garrett, A.M. Paredes, E.M. Wescott, D.R. Moudry, D.D. Sentman, H.C. Stenbaek-Nielsen, W.A. Lyons, Adv. Space Res. **34**, 1811–1814 (2004)
N. Borisov, A. Gurevich, G. Milikh, *Artificially Ionized Region in the Atmosphere* (IZMIRAN, Troitsk, 1986) (in Russian)
D.J. Boccippio, E.R. Williams, S.J. Heckman, W. Lyons et al., Science **269**, 1088–1091 (1995)
T. Bösinger, A. Mika, S. Shalimov, C. Haldoupis et al., J. Geophys. Res. **111**, A10310 (2006). doi:10.1029/2006JA011887
T. Bösinger, S. Shalimov, Space Sci. Rev. **137**, 521–532 (2008)
W.J. Burke, T.L. Aggson, N.C. Maynard, W.R. Hoegy et al., J. Geophys. Res. **97**, 6359 (1992)
D.L. Carpenter, U.S. Inan, M. Trimpi, R.A. Helliwell et al., J. Geophys. Res. **89**(11), 9857–9862 (1984). doi:10.1029/JA089iA11p09857
M. Cho, M. Rycroft, J. Atmos. Sol.-Terr. Phys. **60**, 871–888 (1998)
M. Cho, M. Rycroft, J. Atmos. Sol.-Terr. Phys. **63**, 559–580 (2001)
S.A. Cummer, U.S. Inan, T.F. Bell, C.P. Barrington-Leigh, Geophys. Res. Lett. **25**, 1281–1284 (1998)
S.A. Cummer, M. Füllekrug, Geophys. Res. Lett. **28**, 495–498 (2001)
M. Dejnakarinta, C.G. Park, J. Geophys. Res. **79**, 1903–1910 (1974)
R.L. Dowden, C. Rodger, D. Nunn, IEEE Antennas Propag. Mag. **43**(2), 12–24 (2001)
R.C. Franz, R.J. Nemzek, J.R. Winkler, Science **249**, 48–51 (1990)
H. Fukunishi, Y. Takahashi, Y. Sato, M. Shono et al., Geophys. Res. Lett. **24**, 2973–2976 (1997)
M. Füllekrug, A.C. Fraser-Smith, S. Resising, Geophys. Res. Lett. **25**, 3497–3500 (1998)
E.A. Gerken, U. Inan, C.P. Barrington-Leigh, Geophys. Res. Lett. **27**(17), 2637–2640 (2000). doi:10.1029/2000GL000035
V.S. Glukhov, V. Pasko, U. Inan, J. Geophys. Res. **97**(A11), 16,971–16,979 (1992). doi:10.1029/92JA01596
E. Greenberg, C. Price, Y. Yair, C. Haldoupis et al., J. Atmos. Sol.-Terr. Phys. **71**, 1254–1266 (2009)
C. Greifinger, P. Greifinger, J. Geophys. Res. **81**, 2237–2247 (1976)
C. Haldoupis, T. Neubert, U. Inan, A. Mika et al., J. Geophys. Res. **109**, A10303 (2004). doi:10.1029/2004JA010651
C. Haldoupis, R.J. Steiber, A. Mika, S. Shalimov et al., J. Geophys. Res. **111**, A11321 (2006). doi:10.1029/2006JA011960
C. Haldoupis, N. Amvrosiadi, B.R.T. Cotts, O.A. van der Velde et al., J. Geophys. Res. **115**, A07304 (2010). doi:10.1029/2009JA015165
R.A. Helliwell, J.P. Katsufrakis, M. Trimpi, J. Geophys. Res. **78**(22), 4679–4688 (1973)
Y. Hiraki, H. Fukunishi, J. Geophys. Res. **111**, A11305 (2006). doi:10.1029/2006JA011729
R.H. Holzworth, M.C. Kelley, C. Siefring, L. Hale et al., J. Geophys. Res. **90**, 9824–9830 (1985)
U.S. Inan, D.L. Carpenter, J. Geophys. Res. **92**(4), 3293–3303 (1987). doi:10.1029/JA092iA04p03293
U.S. Inan, D.C. Shafer, W.Y. Yip, R.E. Orvill, J. Geophys. Res. **93**(10), 11455–11472 (1988). doi:10.1029/JA093iA10p11455
U.S. Inan, T.F. Bell, J.V. Rodriguez, Geophys. Res. Lett. **18**(4), 705–708 (1991). doi:10.1029/91GL00364
U.S. Inan, T. Bell, V. Pasko, D. Sentman et al., Geophys. Res. Lett. **22**(24), 3461–3464 (1995). doi:10.1029/95GL03507

M.P. Johnson, U. Inan, Geophys. Res. Lett. **27**(9), 1391–1394 (2000). doi:10.1029/1999GL010757
M.C. Kelley, C. Siefring, R. Pfaff, P. Kintner et al., J. Geophys. Res. **90**, 9815–9823 (1985)
C.L. Kuo, A.B. Chen, J.K. Chou, L.Y. Tsai et al., J. Phys. D, Appl. Phys. **41**, 234014 (2008)
J. Li, S. Cummer, W. Lyons, T. Nelson, J. Geophys. Res. **113**, D20206 (2008). doi:10.1029/2006JD010008
R.A. Marshall, U. Inan, W.A. Lyons, J. Geophys. Res. **111**, D19108 (2006). doi:10.1029/2006JD007219
R.A. Marshall, U.S. Inan, V.S. Glukhov, J. Geophys. Res. **115**(1), A00E17 (2010). doi:10.1029/2009JA014469
R.A. Marshall, U.S. Inan, J. Geophys. Res. **115**(2), A00E29 (2010). doi:10.1029/2009JA014761
S.B. Mende, H.U. Frey, R.R. Hsu, H.T. Su et al., J. Geophys. Res. **110**, A11312 (2005). doi:10.1029/2005JA011064
A. Mika, C. Haldoupis, T. Neubert, R.R. Su et al., Ann. Geophys. **24**, 2179–2189 (2006)
R.C. Moore, C.P. Barrington-Leigh, U. Inan, T. Bell, J. Geophys. Res. **108**(A10)
J. Morrill, E. Bucsela, C. Siefring, M. Heavner et al., Geophys. Res. Lett. **29**(10), 100-1-4 (2002). doi:10.1029/2001GL014018
T. Neubert et al., Surv. Geophys. **29**, 71–137 (2008)
A. Ohkubo, H. Fukunishi, Y. Takahashi, A. Adachi, Geophys. Res. Lett. **32**, L04812 (2005). doi:10.1029/2004GL021943
C. Papadopoulos, G. Milikh, A. Gurevich et al., J. Geophys. Res. **98**, 17,591–17,596 (1993)
V.P. Pasko, U.S. Inan, Y. Taranenko, T. Bell, Geophys. Res. Lett. **22**(4), 365–368 (1995). doi:10.1029/95GL00008
V.P. Pasko, U.S. Inan, T. Bell, Y. Taranenko, J. Geophys. Res. **102**(A3), 4529–4562 (1997). doi:10.1029/96JA03528
V.P. Pasko, U.S. Inan, T.F. Bell, S.C. Reising, Geophys. Res. Lett. **18**, 3493–3496 (1998)
C. Price et al., Geophys. Res. Lett. **31**, L20107 (2004). doi:10.1029/2004GL020711
V. Rakov, M. Uman, *Lightning: Physics and Effects* (Cambridge Univ. Press, Cambridge, 2003)
C.J. Rodger, M. Cho, M.A. Clilverd, M.J. Rycroft, Geophys. Res. Lett. **28**(2), 199–202 (2001)
C. Rodger, Rev. Geophys. **37**(3), 317–336 (1999)
J.V. Rodrigues, U.S. Inan, T.F. Bell, Geophys. Res. Lett. **19**(10), 2067–2070 (1992). doi:10.1029/92GL02379
R.A. Roussel-Dupre, E. Blanc, J. Geophys. Res. **102**(A3), 4613–4622 (1997)
H.L. Rowland, R.F. Fernsler, J.D. Huba, P.A. Bernhardt, Geophys. Res. Lett. **22**(4), 361–364 (1995)
H.L. Rowland, R.F. Fernsler, P.A. Bernhardt, J. Geophys. Res. **101**(A4), 7935–7945 (1996)
A. Schekotov, V. Pilipenko, K. Shiokava, E. Fedorov, Earth Planets Space **63**, 1–10 (2011)
S. Shalimov, Cosmic Res. (2011, accepted)
S. Shalimov, T. Bösinger, J. Atmos. Sol.-Terr. Phys. **68**, 814–820 (2006)
A.I. Sukhorukov, J. Atmos. Sol.-Terr. Phys. **58**(15), 1711–1720 (1996)
A.I. Sukhorukov, E.A. Rudenchik, P. Stubbe, Geophys. Res. Lett. **23**(21), 2911–2914 (1996)
Y.N. Taranenko, U.S. Inan, T.F. Bell, Geophys. Res. Lett. **20**(15), 1539–1542 (1993). doi:10.1029/93GL01696
J.N. Thomas, R.H. Holzworth, M.P. McCarthy, O. Pinto Jr., Geophys. Res. Lett. **32**, L22807 (2005). doi:10.1029/2005GL024629
J.N. Thomas, B.H. Barnum, E. Lay, R.H. Holzworth, M. Cho, M.C. Kelley, J. Geophys. Res. **113**, A12306 (2008). doi:10.1029/2008JA013567
M.A. Uman, *The Lightning Discharge*, San Diego (1987)
S.A. Yashunin, E. Mareev, V. Rakov, J. Geophys. Res. **112**, D10109 (2007). doi:10.1029/2006JD007631

Space Sci Rev (2012) 168:533–550
DOI 10.1007/s11214-011-9859-8

Impulsive Coupling Between the Atmosphere and Ionosphere/Magnetosphere

V.A. Pilipenko

Received: 27 April 2011 / Accepted: 14 November 2011 / Published online: 9 December 2011

Abstract This review covers various aspects of the impulsive coupling in the ULF frequency range between atmospheric discharge processes and upper ionosphere. Characteristic feature of the upper ionosphere is the occurrence of the ionospheric Alfven resonator (IAR) and MHD waveguide, which can trap the electromagnetic wave energy in the range from fractions of Hz to few Hz. Induction magnetometer observations at mid-latitude stations are considered as an example of a transient ULF response to the regional and global lightning activity. For many events, besides the main impulse produced by a lightning discharge, a secondary impulse delayed about 1 sec was observed. These secondary echo-impulses are probably caused by the partial reflection of wave energy of the initial lightning pulse from the upper IAR boundary in the topside ionosphere. The multi-band spectral resonant structure (SRS) can be formed owing to the occurrence of paired pulses in analyzed time series. The statistical superposed epoch method indeed has revealed a dominance of two-pulse structure in the magnetic field background during the periods of the SRS occurrence. The numerical modeling shows that during the lightning discharge a coupled wave system comprising IAR and MHD waveguide is excited. In the lightning proximity (about few hundred km) the amplitudes of radial component is 1–2 orders less than those of the azimuthal component, and only the lowest IAR harmonics are revealed in the radial magnetic component. At distances $\sim 10^3$ km the spectral power densities of both components are comparable, and the SRS is more pronounced. The problems and further prospects of the study of the impulsive magnetosphere–ionosphere–atmosphere coupling via transient processes during thunderstorms are discussed.

Keywords ULF waves · Alfven resonator · Thunderstorm · Ionosphere · TLE · TGF

1 Introduction

Modern observations with a high temporal-spatial resolution during various geophysical processes indicate that a larger power density tends to be concentrated in smaller time and

V.A. Pilipenko (✉)
Space Research Institute, Moscow, Russia
e-mail: pilipenko_va@mail.ru

space scales. Thus, impulsive and small-scale processes, despite their relatively short duration and small scale, may involve substantial energy exchange between the geophysical media.

As an example of impulsive energy exchange in the coupled system magnetosphere–ionosphere–atmosphere one may recall the bursty process with a largest energy release in the terrestrial space plasma—the magnetospheric substorm. The mechanism of the substorm onset has not been identified yet, but probably it is related to small-scale non-MHD processes. For example, kinetic small-scale Alfvenic structures can carry a very localized, but intense, field-aligned electric field, resulting in a downward acceleration of auroral electrons (Chaston et al. 2002).

Auroral electrons can be accelerated not in downward direction only by Alfvenic small-scale structures above an auroral arc. Numerical simulation (Swift 2007) showed that intense bi-polar Alfven pulse from the magnetosphere causes a substantial acceleration in the upward direction, towards the magnetosphere, also. Thus, aurora should be also considered as a source of the seed energized electrons for the outer radiation belt, though so far there were no observations supporting this conjecture.

Atmospheric thunderstorms constitute one of the most powerful disturbance in the Earth's environment and provide the possibility of impulsive coupling of the Earth's atmosphere with the ionosphere/magnetosphere above active storm cells with a considerable energy involved. Electrical storms are known to be one of the natural sources of electromagnetic emissions in a wideband band, from ~ 0.1 Hz to ~ 30 MHz covering ULF, ELF, VLF, and HF frequency ranges. The largest spectral density is concentrated in the VLF band (few kHz), though essential spectral power is contained in the lower ELF-ULF bands. The global Earth-ionosphere cavity Schumann resonances (SR) in the ELF frequency band of several tens Hz is permanently activated by the world thunderstorm centers (Nickolaenko and Hayakawa 2002). Local lightning transients and large noise enhancement were observed in the ULF frequency range of several Hz in the thunderstorm proximity (Fraser-Smith 1993).

A peculiar feature of the geomagnetic variations in the ULF band, just below the fundamental tone of SR (~ 8 Hz), is the occurrence of multi-band Spectral Resonant Structure (SRS), observed at low (Bosinger et al. 2002), middle (Belyaev et al. 1987, 1990) and high (Belyaev et al. 1999; Yahnin et al. 2003) latitudes. The lowest frequency of the SRS is about fraction of Hz, and the difference between the spectral harmonics is ~ 0.3–0.5 Hz. The SRS is the typical nighttime phenomenon, almost all the SRS were observed between 19 and 05 LT. The SRS occurrence rate at mid-latitude demonstrates evident seasonal variation with the maximum at the September-January period and almost complete absence at the spring-early summer time (Molchanov et al. 2004). The SRS occurrence demonstrates a clear anti-correlation with the solar cycle activity (Belyaev et al. 2000).

The occurrence of SRS was attributed to the Ionospheric Alfven Resonator (IAR) in the upper ionosphere. The IAR lower boundary coincides with the ionospheric bottom-side boundary, whereas the upper boundary is located at altitude of a few thousands km due to a partial reflection of the Alfven waves from a steep gradient of the Alfven velocity vertical profile $V_A(z)$ above the maximum of the F-layer. At the upper IAR boundary a trapped Alfven wave has anti-node of the wave electric field, whereas at the low boundary the wave structure depends on the E-layer conductivity: E-field has an anti-node above the low-conductivity and a node above the high conductivity ionosphere (Hebden et al. 2005). At auroral regions, a very effective reflection boundary for the intermediate-scale Alfven waves (transverse scale $\leq 10^2$ km) can be formed by the lower edge of the auroral acceleration region located at altitudes about 1–2 R_E. In these regions a specific IAR can be formed with somewhat lower frequencies than that of a typical IAR (Pilipenko et al. 2002).

The transverse scale of IAR is limited from below by the collisional damping in the ionosphere, absorbing Alfvenic waves with wave numbers larger than $k_{\perp}^{*} \simeq \mu_o \sigma_{\parallel} / \tau_A$, where $\sigma_{\parallel}$ is the collisional field-aligned conductivity, τ_A is the Alfvenic time. According to calculations of Lessard and Knudsen (2001), Alfven waves with spatial scales less than a few km are strongly diffused and damped upon the reflection from the realistic ionosphere. Therefore, a possible occurrence in the IAR of small-scale Alfvenic structures with the transverse scales comparable to the dispersive scale λ_e (electron inertial scale), commonly invoked to interpret the mechanism of the auroral electron acceleration (e.g., Pokhotelov et al. 2000), is still disputable. From above, the transverse scale of the IAR structure is to be limited either by the ionospheric inhomogeneity or by the wavelength of the fast compressional mode in the ionospheric waveguide. However, this question was never considered because of the lack of sufficiently dense array of high-sensitive search-coil magnetometers.

The ionospheric cavity with a minimum of $V_A(z)$ works not only as a resonator for Alfven waves, but also as a waveguide for the fast magnetosonic mode. The waveguide magnetosonic modes excited by the magnetospheric (Fujita and Tamao 1988) or atmospheric (Pilipenko et al. 2005) sources can propagate to large distances (up to few thousand km) along the ionosphere. In the frequency domain the ionospheric wave guidance manifests itself as an occurrence of low-frequency cut-off corresponding to the waveguide critical frequency (lower than the IAR fundamental frequency) (Fujita and Tamao 1988).

The spectral properties of the IAR were well reproduced by numerous analytical (Polyakov and Rapoport 1981) and numerical models (Prikner et al. 2000; Hebden et al. 2005). The damping rates and Q-factors of the IAR modes strongly depend on the Alfven velocity differential in the upper ionosphere and on the lower ionosphere conductivity (Demekhov et al. 2000). In particular, more gradual vertical profile during daytime hours is one of the main reasons of the SRS suppression and weak IAR excitation during daytime hours and periods of high solar activity (Trakhtengertz et al. 2000). The SRS is very sensitive to even slight variations of the vertical profile of the ionospheric parameters, and may sometimes reveals a system of sub-resonators in the ionosphere (Demekhov 2011).

At the same time, a physical mechanism of the IAR excitation has not been established yet. The IAR can trap wave energy coming "from above"—from magnetospheric sources. At high latitudes the mechanism of resonant energy transmission from the magnetospheric convective flows to the IAR modes may be operative (Trakhtengertz and Feldstein 1991). The "feedback instability" due to the ionospheric modification by the energetic electron precipitation can also stimulate the IAR (Lysak 1991; Pokhotelov et al. 2001). Evidently, these mechanisms can operate at auroral latitudes only.

Several possible mechanisms of the IAR excitation "from below"—by atmospheric processes, have been proposed. As a possible source of the IAR excitation, the generation of the ionospheric turbulence and currents by the neutral wind fluctuations in the lower ionosphere has been suggested (Surkov et al. 2004; Molchanov et al. 2004). These authors showed theoretically that a part of the neutral gas kinetic energy can be transferred into the energy of the electric current in the conductive ionosphere, which is then converted into the energy of Alfven and fast compressional MHD modes. However, this hypothesis has not been experimentally validated yet.

The other possible energy source for the IAR excitation is related to atmospheric lightning discharges. The world thunderstorm centers in tropics were suggested to be the primary source of the IAR excitation (Belyaev et al. 1987; Bosinger et al. 2002). However, theoretical estimates have shown that the contribution of the distant world thunderstorm centers with typical charge moments of negative CG^- flashes to the mid-latitude electromagnetic field in the IAR band is about two orders of magnitude less than the observed intensity

(Surkov et al. 2005). At the same time, the theoretical estimates for the same source turned out to be in a good agreement with the observed spectral power in the SR band (Fedorov et al. 2006). To resolve this issue, Shalimov and Bosinger (2008) proposed that relatively infrequent ($\sim 10\%$), but more intense (about an order of magnitude) positive cloud-to-ground flashes CG^+ can provide a necessary ULF background at large distances. However, in this case the spectral power in SR band would be inconsistent with observations. Otherwise, SR is to be excited predominantly by CG^+ flashes as well (Surkov and Hayakawa 2010). Thus, a relative contribution of positive and negative flashes of distant thunderstorms into the IAR stimulation is still disputable.

As an alternative mechanism, it was suggested that near-by thunderstorms are able to generate the signals in the IAR range with sufficient intensities (Surkov et al. 2006). Schekotov et al. (2011) searched for a ULF magnetic field response to the regional lightning activity (few hundred km from observation site), using data from the high-sensitive induction magnetometer. Their analysis, presented in a greater detail below, has shown that the mechanism of the SRS formation is not related to the oscillatory response of the IAR, as had been commonly assumed, but is caused by the specific impulsive structure of geomagnetic disturbances during the regional lightning activity. In general, it is still uncertain how to discriminate the ULF response to distant thunderstorm centers and regional lightning activity, and what is the relative statistical contribution of those sources to the IAR stimulation at mid-latitudes.

2 Structure of the Impulsive ULF Magnetic Background

The SRS in ULF sonograms are commonly associated with the periodic response of the IAR eigenmodes to an external driving. Many researchers suppose that weak and steady harmonic responses are summed up and highlighted by the spectral analysis of sufficiently long magnetometer records, so SRS shows up in the dynamic spectra only after considerable integration.

An alternative view was recently suggested by Schekotov et al. (2011) that the multi-harmonic SRS occurrence is due not to quasi-periodic oscillations, but to a regular sequence of several impulses. They have analyzed the time intervals with evident SRS in the sonograms during October 2007 from induction magnetometer at Moshiri (MSR, $LT = UT + 9$) in Japan. The near-by lightning (≤ 100 km from the observation site) produce too strong interference on magnetic records, and the possible SRS features cannot be revealed. During regional lightning activity (from a few 10^2 km to $\sim 10^3$ km) for many events an impulse produced by a lightning discharge can be seen in the magnetometer data. A source location has been determined with the use of polarization analysis, assuming that the main axis of the polarization ellipse is to be perpendicular to the direction towards a source. The range of possible directions to a source corresponded well to the actual location of a stroke.

For many events a complicated ULF response to a lightning stroke was observed. In the example shown in Fig. 1, on Oct. 4 at 22:07:26.4 LT the main pulse was followed by the secondary impulse, seen predominantly in D-component. The delay between the main and secondary impulses was $\Delta t \sim 1$ s. This series of paired impulses results in the multi-band structure in the corresponding spectra. Both wavelet and time-integrated wavelet spectra (Fig. 1, right-hand and bottom panels) show the occurrence of spectral peaks at $f \simeq 1.5$ Hz, and 2.5 Hz. The frequency shift between the spectral lines, $\Delta f \simeq 1$ Hz, corresponds to the inverse time delay between impulses, $\Delta f \simeq 1/\Delta t$.

To visualize the average waveform of the ULF disturbance, the program, based on the superposed epoch method, selected all impulses above a certain threshold separately for H

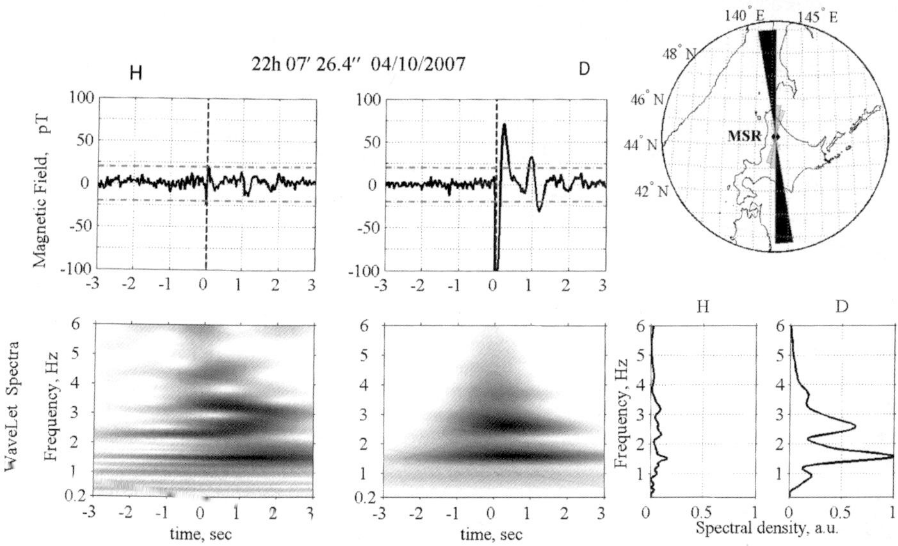

Fig. 1 The example of two-pulses pattern observed on 10/04 22:07:26.4 LT at Moshiri. The format of the plots is as follows: *Upper panels*—daily dynamic spectra for H and D components. The lightning moment is marked by a *vertical dashed line*. *Mid-panels*: wave forms of H and D components. *Bottom panels*: wavelet spectra. Time-averaged wavelet spectra are shown in the *right-hand bottom panel*. *Right-hand insert* shows the map with location of lightning, marked by *star* (the scale of star corresponds to the current intensity). The length and darkness of a lobe on the *circular diagram* corresponds to the intensity of the impulse energy flow

and D components in a running time window. Then, all the waveforms have been averaged over the time interval ± 3 s, choosing as a reference time the moment of a pulse peak value. To examine the time evolution of the impulsive background and SRS, Fig. 2 shows for six consecutive hourly intervals the quasi-3D dynamic Fourier spectra (upper panels), averaged Fourier spectra (mid-panels), and average waveforms of superposed impulses (bottom panels) for Oct. 4 from 18 LT to 24 LT. The averaged waveforms show a dominance of negative D-component pulses, which indicate the location of lightning activity southward from the station. This plot demonstrates a close correspondence between the temporal evolution of the impulsive background and SRS. Although the number of impulses is very large, $Nav = 270$, the time-averaged impulsive geomagnetic background demonstrates a regular pattern: the main bipolar impulse is followed by the secondary bipolar impulse. During the time interval 18–19 LT there are no SRS and no noticeable impulses. On 19–20 LT a weak indication on SRS occurrence has emerged, and, simultaneously, a 2-impulse structure has started to be formed. On 20–24 LT the SRS evolves and intensifies, whereas the multi-impulse structure has become evident. The average time delay and spectral gap between harmonics also vary in a consistent way, being inversely proportional to each other: on 19–20 LT $\Delta t \sim 1.2$ s and $\Delta f \sim 0.8$ Hz, whereas on 22–23 LT $\Delta t \sim 0.9$ s and $\Delta f \sim 1.1$ Hz.

The occurrence of magnetic echo-pulses after a main pulse may be related, in principle, by multiple strokes during an atmospheric electric discharge. In order to examine this possibility, Schekotov et al. (2011) have compared the time structure of the magnetic background in the SRS (0.3–3 Hz) and SR (6–15 Hz) bands using the same superposed epoch technique. Figure 3 shows the statistically averaged magnetic response in the SR (upper panel) and SRS (lower panel) bands for both components on Oct. 04, 21–22 LT. This analysis shows that an echo-impulse in the SR band is absent, while it is evident in the SRS band. The polarity

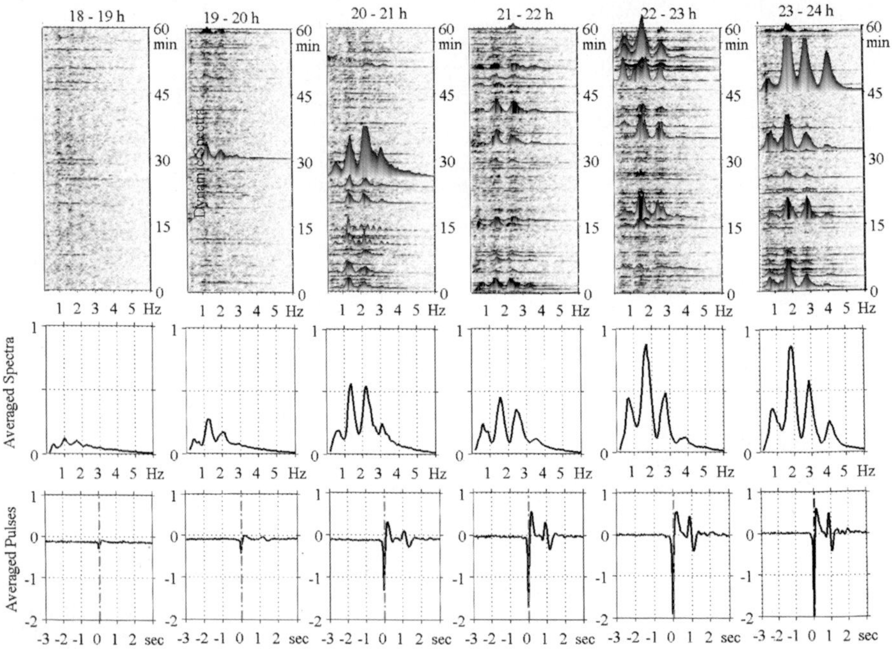

Fig. 2 The correspondence between the temporal evolution of the impulsive background and SRS for the 10/04 during the time interval 18–24 LT. The *upper panel* shows the hourly quasi-3D spectra in the range 0.1–6 Hz. The *middle panels* show the averaged spectra for the hourly intervals corresponding to the upper panel. The *bottom panel* shows the superposed waveforms of all recorded impulses

of the secondary pulse is opposite to the polarity of the main pulse. Because the SR is a sensitive indicator of the lightning discharge, the lack of its response during the occurrence of the secondary pulse in the IAR band indicates that the latter one is not caused by multiple lightning discharges.

There have been amazing observations at low latitudes showing that SRS may consist of 20 and more harmonics, even beyond SR (Bosinger et al. 2002). So many harmonics could be produced either by a very high IAR Q-factor and occurrence of sub-resonators or strongly non-linear waveforms of lightning-produced disturbances. The actual contribution of those mechanisms in the observed SRS has not been firmly established yet.

3 Modeling of the ULF Field in the System Ionosphere–Atmosphere–Ground Excited by an Electric Discharge

A typical cloud-to-ground (CG) lightning discharge starts with a downward step-wise propagating leader, carrying a negative charge. The subsequent upward-propagating return stroke produces a main breakdown of the atmosphere between the ground and cloud, carrying about few tens of Coulomb (C) negative charge to the ground. The lightning flash usually contains 3–4 return strikes with mean interval of ~ 40 ms between them, and total flash duration of about 0.2 s. The negative (CG^-) lightning with upward current are dominating, and their occurrence rate in the tropical regions is $\sim 10^2$ s^{-1}. About 5–10% of lightning is composed from relatively rare (~ 5 s^{-1}) positive (CG^+) discharges, carrying an order of magnitude

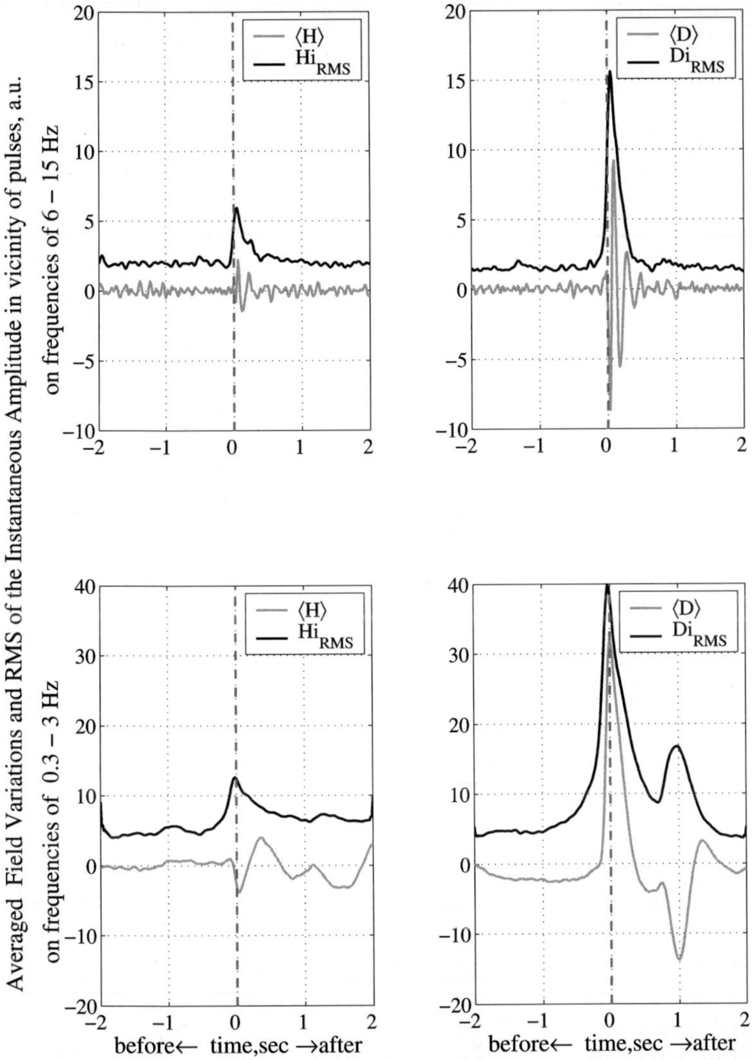

Fig. 3 Averaged field variations (*grey line*) and RMS of the instantaneous amplitude (*dark line*) for H (*left-hand panels*) and D (*right-hand panels*) components in arbitrary units. The reference moment for the superposed epoch analysis is shown by a *vertical dotted line*. *Upper panels*: magnetic response in SR 6–15 Hz (*upper panel*) and SRS 0.3–3 Hz (*lower panel*) bands on Oct. 04

larger currents I_+ (from few kA to tens kA) than CG^- discharges. Positive flashes usually consist of a single stroke followed by continuing current lasting several tens-hundreds msec. This current transfers of several tens to hundreds C positive charge.

On a qualitative basis the response of the ionosphere to the atmospheric electric discharges may be described as follows (Fig. 4). An electric discharge generates an initial TH (electric) mode pulse which propagates in the waveguide ionosphere–ground. The magnetic disturbance in this pulse is orthogonal to the direction towards a source. Upon the interaction of the initial TH pulse with the anisotropic lower ionosphere, TE mode is excited also. Therefore, the reflected from the ionosphere pulse comprises additionally the magnetic

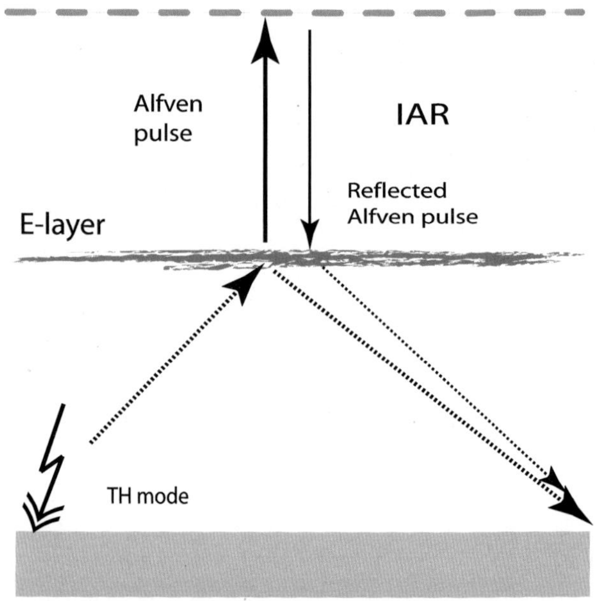

Fig. 4 A scheme of the IAR excitation by a lightning discharge and formation of the secondary echo-pulse observed on the ground

component directed towards a source. Both direct and reflected ULF pulses simultaneously reach an observation site. Meanwhile, the TH pulse partially penetrates into the ionosphere, travels up the ionosphere as an Alfven pulse, and reflects back from the upper IAR boundary. This pulse can return back to the ground as an echo-pulse in the magnetic (TE) mode owing to the mode conversion in the anisotropic lower ionosphere. A TH echo pulse cannot be detected on the Earth surface because of a low penetration of the electric mode into the highly resistive atmosphere. The delay between the echo pulse and the main pulse is about the time of the Alfven wave propagation up and down in the ionosphere. Thus, the pulse repetition period is about the IAR fundamental period.

However, this qualitative scheme is to be validated by a numerical modeling. Numerical model of the electromagnetic wave generation and propagation in the waveguide ionosphere–ground with a leaky upper boundary was developed (see references in the review by Demekhov 2011). The basics of such model is outlined here. A lightning stroke as a source of electromagnetic emission can be modeled by a current moment $M(t) = I(t)L(t)$, where $I(t)$ is the current and $L(t)$ is the lightning channel length. A vertical CG lightning (with the current moment parallel to z-axis) is the most effective source of electromagnetic fields in the Earth-ionosphere cavity. The shape of the thunderstorm-induced ULF spectrum is practically independent of the shape of single lightning discharge because the low-frequency part of the lightning spectrum is nearly constant.

We consider frequencies below the SR, and the distances not larger than few thousand km, so the Earth surface curvature can be neglected. The vertical axis z of the Cartesian coordinate system is assumed to be oriented along a homogeneous magnetic field, $\mathbf{B}_0 = B_0\mathbf{z}$, whereas x is southward, and y is eastward. The ionosphere is assumed to be horizontally stratified, so all ionospheric parameters depends on altitude z only. The atmospheric slab is bounded by the conductive ground at $z = -d$ and by the ionosphere at $z = 0$. The ground is a uniform conductor with conductivity σ_g. The vertical profiles of the ionospheric and atmospheric parameters are modeled by the IRI and MSIS models. From the local plasma parameters the vertical distributions of Alfven velocity $V_A(z)$ and anisotropic conductivity $\sigma(z)$

are deduced. The impedance boundary conditions for waves in the waveguide ionosphere–ground at the bottom side of the ionosphere are imposed.

The total field in the atmosphere is composed from fields of partial electric (E) and magnetic (H) modes, whereas in the ionosphere the wave field is composed from Alfven and fast magnetosonic waves. The primary field excited by an electric dipole in the waveguide ionosphere–ground is electric (TH) mode, which in a cylindrical coordinate system has components E_ρ, E_z, and B_ϕ. Another mode, the magnetic (TE) mode has components E_ϕ, B_ρ, and B_z. In the ionosphere Alfven mode is partially trapped into IAR, and partially penetrates into the magnetosphere. The fast magnetosonic mode can be trapped in the ionospheric waveguide and thus propagates along the ionosphere. In the anisotropically conductive ionosphere, all the modes are coupled. The Hall currents in the gyrotropic E-layer thus play a role of the secondary source exciting the magnetic (TE) mode in the atmosphere.

Taking into account the azimuthal symmetry of the problem, all quantities should depend on z and polar radius ρ. Combining Maxwell's equations and making the Fourier and Bessel transforms with respect to t and ρ, accordingly, a set of equations for time-space harmonics of E and B can be obtained. In order to get the low-frequency spectrum of the discharge one should make the inverse Bessel transform over k. The system obtained should be supplemented by the proper boundary conditions for the interfaces between media. Matching the solutions for different layers, one can find the EM field of a lightning discharge on the ground surface.

For the azimuthally-symmetric harmonic the simple estimate for the magnetic field in a static and induction zones can be obtained. In a near-field (static) zone, $\rho \ll h$, the well-known formula for the magnetic field produced by a current moment M_0, between the infinitely-conductive surfaces at $z = 0$ and $z = h$, stems

$$B_\varphi = \frac{\mu_0 M_0}{2\pi \rho h} \tag{1}$$

This simple relationship can be used for an easy evaluation of the expected ground magnetic response on lightning stroke. For estimate of the echo-pulses and related spectral features caused by the coupling of the waveguide modes with the ionospheric MHD modes a numerical modeling is necessary.

The spatial and spectral ground distributions of electromagnetic field produced by a vertical dipole in the atmosphere as provided by the numerical model for the mid-latitude nighttime ionosphere are given below. Variation of the relative magnitudes of magnetic azimuthal B_ϕ and radial B_ρ components for the frequency $f = 1$ Hz at various distances from a dipole is shown in Fig. 5.

For the frequency below the IAR range ($f \ll 1$ Hz), a weak exponential decrease of both components can be seen. Everywhere the radial component B_ρ (along the direction towards a source) is several times less than the azimuthal component B_ϕ. At $f = 1.0$ Hz, around the fundamental IAR frequency, B_ρ component demonstrates a deep interference minimum at $\rho \simeq 350$ km, and the weaker minimum at $\rho \simeq 900$ km (Fig. 5a). In this point a rapid phase variations of the radial magnetic component occurs, which should result in the change of the polarization of horizontal magnetic disturbance. The distance between these minima, ~ 550 km, corresponds to the apparent velocity of ~ 550 km/s. This value is close to the modeled V_A in the ionosphere, which indicates that the occurrence of interference minima is caused by the excitation of the magnetosonic mode. At higher $f > 2.0$ Hz, the amplitude maximum of B_ρ component at 1000–1200 km becomes comparable to the amplitude of dominant B_ϕ component (not shown).

Results of the spectra calculations in the range 0.1–3.0 Hz at various distances from a source are shown in Fig. 6. The spectral power density $B_\varphi(f)$ shows just a gradual increase

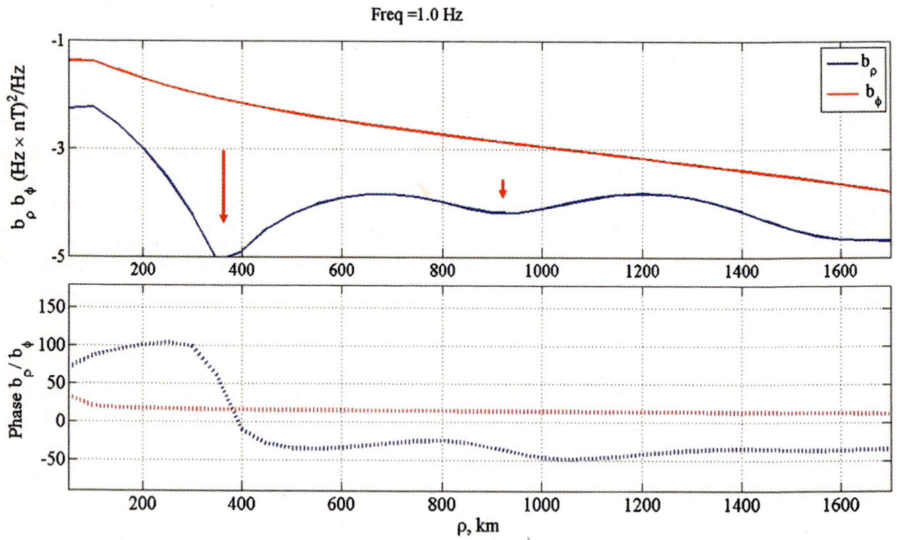

Fig. 5 The spatial ground distributions of the relative magnitudes of magnetic disturbance at frequency 1 Hz for the azimuthal B_ϕ and radial B_ρ components produced by a vertical dipole in the atmosphere as given by the numerical model for the mid-latitude nighttime ionosphere

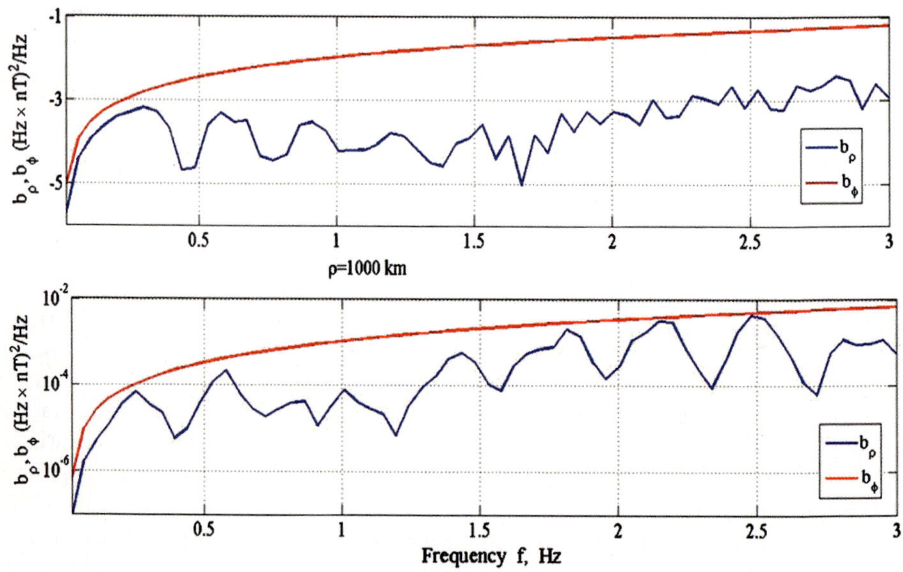

Fig. 6 Results of the spectra calculations in the range 0.1–3 Hz at distances $R = 300$ km (*upper panel*) and $R = 1000$ km (*bottom panel*) from a dipole source for the azimuthal B_ϕ and radial B_ρ components

of with frequency. The resonant structure reveals itself in the spectra $B_\rho(f)$ only. Near-by to the lightning ($R = 300$ km) the amplitudes of B_ρ is 1–2 orders less than B_ϕ component (Fig. 6, upper panel). Here only the lowest IAR harmonics are reveled, 0.3, 0.6, 0.8 Hz, whereas at higher $f > 1.5$ Hz, just a broadband enhancement occurs. At larger distances, the difference between the amplitudes of two components diminishes, and the spectral structure becomes more pronounced at higher harmonics, ≥ 1.5 Hz. At $R = 10^3$ km the spectral power densities of both components are comparable, and up to 8 SRS harmonics are well pronounced at f from 0.2 to 2.8 Hz (Fig. 6, bottom panel). At larger distances, the SRS is also evident, but it is not identical to that at 1000 km: the spectral peaks are shifted a little, though the modeled ionosphere is the same along the radio path.

Thus, these preliminary results of numerical modeling have shown that during the impact of the atmospheric electric discharge fields on the ionosphere not just a local IAR is excited, but a more global wave structure, comprising both IAR and MHD waveguide, is stimulated. Further advancement of such model should clarify whether the main contribution to the observed time pattern is produced by the ionospheric region near the stroke, or above the receiving station, or by the integral effect along the propagation path? It should provide a possibility to estimate the lateral coherency scale of the IAR signatures, a rate of the ULF energy leaking into the magnetosphere, and the occurrence of the "transparency windows".

4 Leakage of ULF/VLF Waves into the Magnetosphere and Lightning-Induced Particle Precipitation

The energy of electromagnetic wave burst produced by isolated lightning discharge, even in the optimal VLF band, seems to be too weak to cause any noticeable modification of the ionospheric plasma. Nevertheless, such effects were found indeed, and they are a good example of triggered phenomena in geophysics (Popov et al. 1989). During such phenomena a weak external disturbance releases a free energy stored in the system, and this way produces a much greater impact on the system. VLF wave burst, propagating in the magnetosphere as a whistler wave, interacts with a group of resonant energetic electrons ($\sim$ few tens of keV) and deviates their pitch-angles, without changing their energy. These small changes are sufficient to cause the electron precipitation into the ionosphere. Via this mechanism VLF burst can introduce into the ionosphere about 4 orders of magnitude larger energy that its own energy (Inan et al. 1985).

Discrete lightning-induced electron precipitation (LEP) fluxes are relatively low, 10^3–10^4 times lower than typical auroral fluxes. X-ray and optical emissions produced by such precipitating electrons are weak and difficult to detect. Secondary ionization due to expected LEP fluxes $\sim 10^4$ ergs/cm^{2}s is just a small fraction of ambient densities at E-layer, so that detection by radars, ionosonds, and riometers is not feasible. However, D-region is mostly devoid of ambient electrons, especially at night (~ 1–10 cm^{-3}), so the relatively small additional ionization by precipitation fluxes of energetic electrons (> 100 keV penetrating to < 85 km) constitutes a substantial modification. The technique suited for the detection is VLF radio remote sounding ("Trimpi-effect"): amplitude/phase of VLF signal from a radio transmitter is highly sensitive to a localized bottom ionosphere disturbance along the propagation path.

The question of possibility of ULF wave impact from atmospheric sources on the ionosphere encounters a rather controversial problem of the expected magnitudes of the thunderstorm related ULF fields in the upper ionosphere. Greifinger and Greifinger (1976) predicted only a weak strength of the ULF fields in the upper ionosphere above the thunderstorm. On the other hand, satellite observations sometimes detected very large ULF variations up to 1–10 mV/m (Burke et al. 1992). Observations of IAR-accumulated ULF wave energy in space are surprisingly rare. There is a larger discrepancy around 4–5 orders of magnitude between few reported results: at Freja at $h \sim 1700$ km in the post-noon sector of the auroral oval the spectral power density at 1 Hz (Grzesiak 2000), and at Aureol-3 (Zhao and Ni 2006) in the morning sector at $L \sim 8.7$. All these observations were made in auroral region and they are not related to mid-latitude IAR. A more detailed review of ULF-ELF electromagnetic wave phenomena related to the tropospheric-ionospheric coupling observed in space is given by Simoes et al. (2011).

Thus, lightning happened to be a powerful emitter of VLF disturbances, which effectively penetrate into the magnetosphere and cause the precipitation of trapped energetic electrons. As consideration in previous sections has shown, lightning discharge could be also a powerful emitter of ULF emission into the upper ionosphere. Therefore, should one expect the possibility of lightning-induced proton precipitation (Bosinger and Shalimov 2008)? Indeed, electromagnetic ion-cyclotron waves with frequencies about 1 Hz can resonantly interact with energetic protons (few tens of KeV) and cause their pitch-angle scattering, which eventually results in the proton aurora at sub-auroral latitudes. This effect would be hard to detect with ground photometers because of high cloudiness during thunderstorms, but low-orbiting satellite observations above powerful thunderstorms may reveal an elevated proton precipitation. On the other hand, electromagnetic ion-cyclotron waves in the Pc1 band are known to be an important factor to control the loss of relativistic electron fluxes in the magnetosphere due to the "parasitic" resonant scattering. Therefore, elevated Pc1 emissions in a flux tube conjugate with thunderstorm activity could damp the magnetospheric relativistic electrons. Advanced numerical modeling with a realistic ionospheric model should help to evaluate the feasibility of the particle precipitation due to the resonant ion-cyclotron interaction stimulated by the atmospheric electric discharges. So far, this possibility is mere a speculation.

5 Prospects of Further SRS and IAR Studies

It is generally believed that the global lightning activity concentrated in the world thunderstorm centers is the basic mechanism for excitation of the SR. However, an attempt to apply the same mechanism for the interpretation of SRS comes across with certain difficulties at middle latitudes far from the tropics. The theoretical ULF spectrum, calculated under the assumption of the distant world thunderstorms as a source, and experimental spectra agree in the SR band, but they disagree by 1–2 orders of magnitude in the IAR frequency range (Surkov et al. 2005; Fedorov et al. 2006). This fact suggests that the SRS is not entirely due to a stochastic lightning discharge process of tropical thunderstorms, but that some generation mechanisms for the IAR excitation at middle latitudes should exist. In particular, the SRS can be excited by regional thunderstorms, within thousand km from observation site. The multi-band SRS in dynamic spectra can be the result of the sequence of paired pulses in the magnetic background. The statistical superposed epoch analysis of some events has shown that magnetic disturbances during the regional lightning activity comprises regular series of paired pulses. An initial pulse caused by an atmospheric discharge is followed by 1–2 pulses with a regular time delay, reflected from the upper IAR boundary. At the same time we have to mention that though the statistical analysis undoubtedly revealed the occurrence of the secondary pulse, some individual lightning-associated pulses were not followed by any secondary pulse. The lack of a secondary pulse is hardly can be attributed to a sudden change of the IAR resonant properties, and this fact needs an additional examination. In some events, rather surprisingly, no ULF magnetic response at all to intense regional strokes was detected. The lack of ULF response is possibly related to a low intensity of the continuing current in those discharges.

The polarization of the horizontal magnetic components of the main pulse conforms the assumption that this disturbance is produced by a vertical current associated to lightning. According to the simple theoretical expectations, the secondary pulse is to be polarized along the direction rotated by $\pi/2$ from the main pulse polarization ellipse axis, because the secondary pulse is to be related to another wave mode. However, observations have shown that the difference in the orientation of the polarization ellipses of the main and secondary pulses is rather small, not more than 30°. This fact needs a more detailed theoretical modeling.

Thus, though there are evidences that regional thunderstorms do give a noticeable input to the IAR power spectrum, and may be responsible, at least partly, for the SRS occurrence at middle latitudes, this source cannot be considered as the only one. The distant thunderstorm activity, probably, contributes to the ULF impulsive background, especially at low latitudes. However, a possibility to discriminate those sources from observations at single station is unclear. Moreover, these sources do not explain all the observational results, in particular, the SRS occurrence at polar latitudes, far from any thunderstorm centers (Semenova et al. 2008). So, the rate of the relative contribution from various possible lightning sources to the IAR stimulation is still an open experimental problem. Unfortunately, the information about the lightning occurrence and their location is available from lightning monitoring systems only on a commercial basis. Therefore, any scrutiny statistical study would be possible only when those data would be freely available to the scientific community.

The relative contribution of magnetospheric and atmospheric sources into the IAR excitation still remains unknown. The intensity of magnetospheric sources should increase with the enhancement of magnetic activity. But, observations showed that any intensification of magnetic activity ceased the occurrence of SRS. The anti-correlation was observed between the SRS occurrence rate and geomagnetic Kp index at mid-latitudes (Molchanov et al. 2004) and at sub-auroral latitudes (Yahnin et al. 2003). Even a visual inspection of ULF spectrograms evidently demonstrates a suppression of the SRS by an enhancement of magnetospheric activity (Semenova and Yahnin 2008; Parent et al. 2010). At the same time, the specific auroral emission Pi1, possibly generated by fluxes of irregular precipitating auroral electrons, has a band-limited nature with a highlighted frequency about 0.3 Hz. It seems natural to interpret a band-limited nature of Pi1C pulsations by the IAR occurrence at auroral latitudes. If so, it remains unclear, why elevated magnetospheric activity stimulates IAR during Pi1 activity at auroral latitudes, but suppresses it at other latitudes. The seemingly evident explanation of this fact is that intense irregular fluctuations of the ionospheric plasma during periods of enhanced magnetic activity smear out the IAR signatures on spectrograms. Besides that, a suppression of the resonator response to an external driving by fluctuations of its parameters is a known effect (Coult et al. 2007). In line with this, even during polar nights at Svalbard (when the ionosphere is in darkness at all LT) the SRS observation probability has a deep minimum an noon hours, in the vicinity of the cusp projection, where the ionospheric plasma is highly turbulent (Semenova et al. 2008).

An intriguing possibility of the IAR stimulation by the atmospheric turbulence was indicated in theoretical models by Surkov et al. (2004), and Molchanov et al. (2004). A natural source of free energy is the neutral wind motion. These models assumed that neutral wind velocity disturbances with a spatial wave number k are dragged by the wind flow with velocity U. Neutral wind fluctuations induce electric currents in the conductive E-layer of the ionosphere. These currents effectively generate MHD wave disturbances with frequencies $\omega = kU$. A strong wind with $U = 0.5$ km/s excite MHD disturbances in the IAR frequency range $f \simeq 1$ Hz owing to the fluctuations with spatial scale $\lambda = U/f \simeq 0.5$ km. The ground magnetic response to the ionospheric disturbances with such small scale, $\lambda \ll h$, is to be substantially attenuated. The predicted enhanced level of small-scale IAR activity above the regions with a strong turbulent wind can be in principle detected by low-orbiting satellites. Therefore, this hypothesis awaits its observational validation.

A possibility of thermal processes in the bottom atmosphere to induce electromagnetic noise was indicated by Gokhberg (1998). He found a regular occurrence of 0.5–5 Hz noise with intensity ~10 pT and duration few hours during summertime only at afternoon hours at mid-latitude stations. No mechanism to interpret this phenomenon has been suggested yet. The IAR eigenmodes can be excited by the direct impact on the ionosphere in active experiments: by the chemical release experiment or rocket exhaust gases (Alpatov et al. 2005), or by modulated radiowave heating of the ionosphere (Pashin et al. 2003).

Discoveries of the transient luminous effects (TLE), such as elves, red sprites, and blue jets, possibly caused by the breakdown at altitudes ~ 90 km, revealed the possibility of a much stronger lightning coupling to the ionosphere than previously deemed possible. These processes, associated with large high-altitude nonlinear currents, may strongly enhance the weight of the ULF-ELF spectrum in the lightning transients in the ionosphere, and the IAR excitation by strong lightning discharge (Sukhorukov and Stubbe 1997). However, so far few attempts (e.g., Bosinger et al. 2006) to reveal any distinctions in ULF response to distant high-altitude TLE as compared with the response to an ordinary lightning were not successful.

6 New Opportunities to Examine the Impulsive Coupling Mechanisms: Observations from Micro-Satellites

About 2000 storms are permanently active throughout the world, producing 50–100 lightning bolts per second. Surprisingly, a whole new world of various TLEs, such as red sprites, blue jets, elves, halos, gigantic jets, associated with lightning flashes, has been discovered unexpectedly. These events are screened by clouds from a ground observer, and can be detected either from high-altitude observatories or from space. Since the discovery of these phenomena is very recent, current knowledge is mostly limited to visible light. Further studies demonstrated that visible TLEs are just a small part of a much more complex phenomenon that also involves bursts of UV emission, terrestrial gamma ray flashes (TGF), and electron acceleration up to relativistic energies (up to 30 MeV). Electron acceleration and precipitation are generally accompanied by electromagnetic emissions over a large range (0.1 Hz to several tens of MHz), which either contribute to the process or were generated by it. TGF-related lightning are the most potent natural particle accelerators on Earth, providing to a group of electrons a huge amount of energy, > 20 MeV. In contrast, colorful auroras are powered by particles with 4 orders less energy.

Despite impressive progress in the study of such phenomena, many fundamental questions still remain to be answered. So far, optical emission and TGF have not been observed simultaneously. Are TLE and TGF coupled? Probably, bursts of electromagnetic emission at the initial stage of thermal breakdown and runaway breakdown have different spectral features. How to distinguish them experimentally? Do TGFs help to trigger lightning or does lightning trigger them? What types of lightning produce TGFs, and how do they differ from non-TGF lightning? The least known questions are related to a possible coupling of high-altitude discharges with the outer space. Could TLE/TGF be responsible for some of the high-energy particles in the radiation belts?

The next few years will provide fantastic opportunities for the study of impulsive coupling between the atmospheric thunderstorms and near-Earth space. The fleet of upcoming micro-satellite missions are expected to be launched. These missions are dedicated to studies of physical effects of lightning discharges in the upper atmosphere, and they will provide inputs for the modeling of the effects of TLEs, TGFs and bursts of precipitated and accelerated electrons on the Earth's atmosphere.

FIREFLY (firefly.gsfc.nasa.gov) This micro-satellite as part of the USA NSF CubeSat program is aimed to study TGFs and their mechanism. Firefly will make simultaneous measurements of energetic electrons, γ-rays, and the radio and optical signatures of the lightning discharge. Science instrumentation comprises detector and scintillator system to measure the energy and time of arrival of X- and γ-rays. The same instrument will detect electrons in the hundreds of keV to few MeV range. The microsatellite will carry also VLF receiver to measure electromagnetic bursts from tens of Hz to tens of KHz, and photometer at high time resolution. The launch is planned in 2012.

TARANIS (smsc.cnes.fr/TARANIS) The micro-satellite TARANIS (Tool for the Analysis of RAdiations from lightNIngs and Sprites) is a project of the French national space program, dedicated to studies of impulsive transfer of energy powered by lightning between the neutral atmosphere, the ionosphere and magnetosphere. The science objectives include characterization of TLEs in the stratosphere and mesosphere above thunderstorms and TGFs. The orbit will be polar sun-synchronous at 700 km altitude. Scientific payload comprises 2 micro-cameras for visible and near-infrared (600–900 nm) emissions and 3 photometers (600 to 900 nm). X-ray, γ-ray and relativistic electron experiment will measure photon with energies 20 keV–10 MeV, and identify beams of relativistic electrons (1 MeV–10 MeV). Energetic electron detector provides high resolution electron spectra (70 keV–4 MeV) in a large dynamic range of fluxes and pitch-angle distribution. The measurement of low frequency electric field from DC to HF (35 MHz), and magnetic field in the band 0.1 Hz–1 MHz is planned. Furthermore, the study will be extended to incorporate the transient precipitations and accelerations of energy electrons, regardless of whether they are directly linked to TLEs. Launch is scheduled for 2015 and expected life time is 2–3 years.

ASIM (www.asim.dk) The Atmosphere–Space Interactions Monitor (ASIM) is an instrument suite developed within the European Space Agency and built by teams from Denmark, Spain, Norway and the US to be mounted on the external platform of the International Space Station. It will study the coupling of thunderstorms processes to the upper atmosphere, ionosphere and radiation belts, and energetic particle precipitation effects in the mesosphere and thermosphere. The scientific objectives include (1) investigations into TLEs and relativistic electron beams above thunderstorms, (2) studies of gravity waves in the thermosphere above severe thunderstorms, and (3) ozone and NO_x concentrations in the upper atmosphere. The instruments are six TV frame-rate, narrow-band optical cameras, six 100 kHz-photometers, and one X-ray sensor. The expected launch in 2011 and duration $\sim$ 2 years.

Chibis-M (chibis.cosmos.ru) This micro-satellite is designed by the Space Research Institute (Moscow) in cooperation with many other Russian and Ukrainian space centers. In this mission the multi-instrument technique will monitor lightning discharges in radio, optical and γ bands with time resolution better than msec. Data recording on circular memory is controlled by trigger signal.

The equipment comprises X-ray and γ-ray detector with energy range 50–500 keV, UV detector in emission wavelength 180 - 800 nm, digital photo camera with matrix 1000×1000 and the frame duration 0.2 s, electric field analyzer in frequency range 20–50 kHz, and ULF-VLF package with E and B sensors in the range 0.1 Hz–40 kHz. Micro-satellite is to be delivered to International Space Station by the transport vehicle "Progress", and then deployed to circular orbit at altitude 500 km. The expected launch is in January 2012 and duration > 1 year.

These upcoming micro-satellite missions could be effectively cooperated. A comprehensive understanding of the possible coupling between thunderstorms in the atmosphere and ionosphere–magnetosphere dynamics is impossible without coordinated world-wide ground-based optical and electromagnetic observations. These studies, apart fundamental scientific knowledge, will provide a valuable practical information about a possible risk for the airplane passengers from powerful γ-ray emission. These studies will synergistically unite efforts of experimentalists, data analysts, and modelers in understanding the atmospheric electricity processes, and advance physical understanding and identify the generation mechanisms for TLEs and TGFs and, in particular, the particle and wave field events which are involved in the generation processes, evaluate the effects of TLEs, TGFs, LEPs, and runaway electron beams on the Earth atmosphere and radiation belts.

Acknowledgements The contributions from Fedorov E., Schekotov A., Shiokawa K., and T. Bosinger are appreciated. This review is a result of the International Space Science Institute (Bern) workshop "Coupling between the Earth's atmosphere and its plasma environment", and is partly supported by the grant 10-05-93107 from RFBR.

References

V.V. Alpatov, M.G. Deminov, D.S. Faermark, I.A. Grebnev, M.J. Kosch, Dynamics of Alfven waves in the night-side ionospheric Alfven resonator at mid-latitudes. Ann. Geophys. **23**, 499–507 (2005)

P.P. Belyaev, S.V. Polyakov, V.O. Rapoport, V.Y. Trakhtengertz, Discovery of the resonance spectrum structure of atmospheric electromagnetic noise background in the range of short-period geomagnetic pulsations. Dokl. Akad. Nauk SSSR **297**, 840 (1987)

P.P. Belyaev, S.V. Polyakov, V.O. Rapoport, V.Y. Trakhtengertz, The ionospheric Alfven resonator. J. Atmos. Terr. Phys. **52**, 781 (1990)

P.P. Belyaev, T. Bosinger, S.V. Isaev, J. Kangas, First evidence at high latitudes for the ionospheric Alfven resonator. J. Geophys. Res. **104**, 4305 (1999)

P.P. Belyaev, S.V. Polyakov, E.N. Ermakova, S.V. Isaev, Solar cycle variations in the ionospheric Alfven resonator 1985–1995. J. Atmos. Sol.-Terr. Phys. **62**, 239–248 (2000)

T. Bosinger, C. Haldoupis, P.P. Belyaev, M.N. Yakunin, N.V. Semenova, A.G. Demekhov, V. Angelopoulos, Spectral properties of the ionospheric Alfven resonator observed at a low-latitude station ($L = 1.3$). J. Geophys. Res. **107**, 1281 (2002)

T. Bosinger, A. Mika, S.L. Shalimov, C. Haldoupis, T. Neubert, Is there a unique signature in the ULF response to sprite-associated lightning flashes? J. Geophys. Res. **111**, A10310 (2006)

T. Bosinger, S.L. Shalimov, On ULF signatures of lightning discharges. Space Sci. Rev. **137**, 521–532 (2008)

W.J. Burke, T.L. Aggson, N.C. Maynard, W.R. Hoegy, R.A. Hoffman, R.M. Candy, C. Liebrecht, E. Rodgers, Effects of a lightning discharge detected by the DE 2 satellite over hurricane Debbie. J. Geophys. Res. **97**, 6359–6367 (1992)

C.C. Chaston, J.W. Bonnell, C.W. Carlson, M. Berthomier, L.M. Peticolas, I. Roth, J.P. McFadden, R.E. Ergun, R.J. Strangeway, Electron acceleration in the ionospheric Alfven resonator. J. Geophys. Res. **107**, 1413 (2002)

N. Coult, V. Pilipenko, M. Engebretson, Suppression of resonant field line oscillations by a turbulent background. Planet. Space Sci. **55**, 694–700 (2007)

A.G. Demekhov, V.Y. Trakhtengertz, T. Bosinger, Pc 1 waves and ionospheric Alfven resonator: generation or filtration? Geophys. Res. Lett. **27**, 3805–3808 (2000)

A.G. Demekhov, Coupling at the atmosphere–ionosphere–magnetosphere interface and resonant phenomena in the ULF range. Space Sci. Rev. (2011). doi:10.1007/s11214-011-9832-6

E.N. Fedorov, A.J. Schekotov, O.A. Molchanov, M. Hayakawa, V.V. Surkov, V.A. Gladyshev, An energy source for the mid-middle latitudes IAR: World thunderstorm centers, nearby discharges or neutral wind fluctuations? Phys. Chem. Earth **31**, 462–468 (2006)

A.C. Fraser-Smith, ULF magnetic fields generated by electrical storms and their significance to geomagnetic pulsation generation. Geophys. Res. Lett. **20**, 467–470 (1993)

S. Fujita, T. Tamao, Duct propagation of hydromagnetic waves in the upper ionosphere. 1. Electromagnetic field disturbances in high latitudes associated with localized incidence of a shear Alfven wave. J. Geophys. Res. **93**, 14665–14673 (1988)
M.B. Gokhberg, New type of electromagnetic emission in the range of short-period geomagnetic pulsations. Dokl. Akad. Nauk, Ross. Akad. Nauk **359**, 543–544 (1998)
C. Greifinger, P. Greifinger, Transient ULF electric and magnetic fields following a lightning discharge. J. Geophys. Res. **81**, 2237–2247 (1976)
M. Grzesiak, Ionospheric Alfven resonator as seen by Freja satellite. Geophys. Res. Lett. **27**, 923–926 (2000)
S.R. Hebden, T.R. Robinson, D.M. Wright, T. Yeoman, T. Raita, T. Bosinger, A quantitative analysis of the diurnal evolution of Ionospheric Alfven Resonator magnetic resonance features and calculation of changing IAR parameters. Ann. Geophys. **23**, 1711–1721 (2005)
U. Inan, D. Carpenter, R. Helliwell, J. Katsufrakis, Subionospheric VLF/LF phase perturbations produced by lightning-whistler induced particle precipitation. J. Geophys. Res. **90**, 7457–7469 (1985)
M.R. Lessard, D.J. Knudsen, Ionospheric reflection of small-scale Alfven waves. Geophys. Res. Lett. **28**, 3573–3576 (2001)
R.L. Lysak, Feedback instability of the ionospheric resonator cavity. J. Geophys. Res. **96**, 1553 (1991)
O.A. Molchanov, A.Yu. Schekotov, E.N. Fedorov, M. Hayakawa, Ionospheric Alfven resonance at middle latitudes: Results of observations at Kamchatka. Phys. Chem. Earth, Parts A/B/C **29**, 649–655 (2004)
A.P. Nickolaenko, M. Hayakawa, *Resonances in the Earth-Ionosphere Cavity*, Kluwer Academic, Dordrecht (2002)
A. Parent, I.R. Mann, I.J. Rae, Effects of substorm dynamics on magnetic signatures of the ionospheric Alfven resonator. J. Geophys. Res. **115**, A02312 (2010)
A.B. Pashin, A.A. Mochalov, T. Bosinger, M.T. Rietveld, Physics of auroral phenomena, in *Proc. 26 Annual Seminar*, Apatity (2003), pp. 111–114
V. Pilipenko, E. Fedorov, M.J. Engebretson, Alfven resonator in the topside ionosphere beneath the auroral acceleration region. J. Geophys. Res. **107**, 1257 (2002)
V. Pilipenko, E. Fedorov, K. Mursula, T. Pikkarainen, Generation of magnetic noise bursts during distant rocket launches. Geophysica **41**, 57–72 (2005)
O.A. Pokhotelov, D. Pokhotelov, A. Streltsov, V. Khruschev, M. Parrot, Dispersive ionospheric Alfven resonator. J. Geophys. Res. **105**, 7737–7746 (2000)
O.A. Pokhotelov, V. Khruschev, M. Parrot, S. Senchenkov, V.P. Pavlenko, Ionospheric Alfven resonator revisited: Feedback instability. J. Geophys. Res. **106**, 813 (2001)
S.V. Polyakov, V.O. Rapoport, The ionospheric Alfven resonator. Geomagn. Aeron. **21**, 610–614 (1981)
L.N. Popov, Yu.K. Krakovezkiy, M.B. Gokhberg, V.A. Pilipenko, Terrogenic effects in the ionosphere. A review. Phys. Earth Planet. Inter. **57**, 115–128 (1989)
K. Prikner, K. Mursula, F.Z. Feygin, J. Kangas, R. Kerttula, T. Pikkarainen, O.A. Pokhotelov, V. Vagner, Non-stationary Alfven resonator: vertical profiles of wave characteristics. J. Atmos. Sol.-Terr. Phys. **62**, 311–322 (2000)
N.V. Semenova, A.G. Yahnin, Diurnal behavior of the ionospheric Alfven resonator signatures as observed at high latitude observatory Barentsburg ($L = 15$). Ann. Geophys. **26**, 2245–2251 (2008)
N.V. Semenova, A.G. Yahnin, A.N. Vasil'ev, O. Amm, Specific features of resonance structures in spectra of ULF electromagnetic noise at high latitudes (Barentsburg Observatory). Geomagn. Aeron. **48**, 36–44 (2008)
A. Schekotov, V. Pilipenko, K. Shiokawa, E. Fedorov, ULF impulsive magnetic response at mid-latitudes to lightning activity. Earth Planets Space **63**, 119–128 (2011)
F. Simoes, R. Pfaff, J.-J. Berthelier, J. Klenzing, A review of low frequency electromagnetic wave phenomena related to tropospheric–ionospheric coupling mechanisms, Space Sci. Rev. (2011), this issue
S. Shalimov, T. Bosinger, On distant excitation of the ionospheric Alfven resonator by positive cloud-to-ground lightning discharges. J. Geophys. Res. **113**, A02303 (2008)
A.I. Sukhorukov, P. Stubbe, Excitation of the ionospheric Alfven resonator by strong lightning discharges. Geophys. Res. Lett. **24**, 829–832 (1997)
V.V. Surkov, O.A. Pokhotelov, M. Parrot, E.N. Fedorov, M. Hayakawa, Excitation of the ionospheric resonance cavity by neutral winds at middle latitudes. Ann. Geophys. **22**, 2877–2889 (2004)
V.V. Surkov, O.A. Molchanov, M. Hayakawa, E.N. Fedorov, Excitation of the ionospheric resonance cavity by thunderstorms. J. Geophys. Res. **110**, A04308 (2005)
V.V. Surkov, M. Hayakawa, A.Y. Schekotov, E.N. Fedorov, O.A. Molchanov, Ionospheric Alfven resonator excitation due to nearby thunderstorms. J. Geophys. Res. **111**, A01303 (2006)
V.V. Surkov, M. Hayakawa, Schumann resonances excitation due to positive and negative cloud-to-ground lightning. J. Geophys. Res. **115**, D04101 (2010). doi:10.1029/2009JD012539
D.W. Swift, Simulation of auroral electron acceleration by inertial Alfven waves. J. Geophys. Res. **112**, A12207 (2007)

V.Yu. Trakhtengertz, A.Ya. Feldstein, Turbulent Alfven boundary layer in the polar ionosphere, 1, Excitation conditions and energetics. J. Geophys. Res. **96**, 363 (1991)

V.Y. Trakhtengertz, A.C. Demekhov, P.P. Belyaev, S.V. Polyakov, E.N. Ermakova, S.V. Isaev, A mechanism of anticorrelation in the occurrence of ULF electromagnetic noise resonance structure and Pc1 magnetic pulsations through the solar activity cycle. J. Atmos. Sol.-Terr. Phys. **62**, 253–256 (2000)

A.G. Yahnin, N.V. Semenova, A.A. Ostapenko, J. Kangas, J. Manninen, T. Turunen, Morphology of the spectral resonance structure of the electromagnetic background noise in the range of 0.1–4 Hz at $L = 5.2$. Ann. Geophys. **21**, 779 (2003)

Z.Y. Zhao, B.B. Ni, Signatures of the ionospheric Alfven resonator from Aureaol-3 ULF/ELF fluctuation measurements. J. Atmos. Sol.-Terr. Phys. **68**, 191–201 (2006)

Space Sci Rev (2012) 168:551–593
DOI 10.1007/s11214-011-9854-0

A Review of Low Frequency Electromagnetic Wave Phenomena Related to Tropospheric-Ionospheric Coupling Mechanisms

Fernando Simões · Robert Pfaff · Jean-Jacques Berthelier · Jeffrey Klenzing

Received: 18 July 2011 / Accepted: 2 November 2011 / Published online: 10 December 2011

Abstract Investigation of coupling mechanisms between the troposphere and the ionosphere requires a multidisciplinary approach involving several branches of atmospheric sciences, from meteorology, atmospheric chemistry, and fulminology to aeronomy, plasma physics, and space weather. In this work, we review low frequency electromagnetic wave observations in the Earth-ionosphere cavity from a troposphere-ionosphere coupling perspective. We discuss electromagnetic wave generation, propagation, and resonance phenomena, considering atmospheric, ionospheric and magnetospheric sources, from lightning and transient luminous events at low altitude to Alfvén waves and particle precipitation related to solar and magnetospheric processes. We review ionospheric processes as well as surface and space weather phenomena that drive the coupling between the troposphere and the ionosphere. Effects of aerosols, water vapor distribution, thermodynamic parameters, and cloud charge separation and electrification processes on atmospheric electricity and electromagnetic waves are reviewed. Regarding the role of the lower boundary of the cavity, we review transient surface phenomena, including seismic activity, earthquakes, volcanic processes and dust electrification. The role of surface perturbations and atmospheric gravity waves in ionospheric dynamics is also briefly addressed. We summarize analytical and numerical tools and techniques to model low frequency electromagnetic wave propagation and to solve inverse problems and outline in a final section a few challenging subjects that are important to advance our understanding of tropospheric-ionospheric coupling.

F. Simões (✉) · R. Pfaff · J. Klenzing
NASA/GSFC Heliophysics Science Division, Space Weather Laboratory, 8800 Greenbelt Road, Greenbelt, MD 20771, USA
e-mail: fernando.a.simoes@nasa.gov

R. Pfaff
e-mail: robert.f.pfaff@nasa.gov

J. Klenzing
e-mail: jeffrey.klenzing@nasa.gov

J.-J. Berthelier
LATMOS/IPSL, UPMC, 4 Place Jussieu, 75005 Paris, France
e-mail: jean-jacques.berthelier@latmos.ipsl.fr

Keywords Wave propagation · Lightning · Ionosphere dynamics · Aeronomy · Troposphere-ionosphere coupling

1 Introduction

Atmospheric sciences and its multiple branches have contributed to a progressive, sustained enlightenment of the physical and chemical processes that take place in the atmosphere, ionosphere, and magnetosphere and their response to space weather disturbances. Low frequency electromagnetic waves provide a very useful tool to study the global environment of the Earth from the atmosphere to the magnetosphere and the coupling of the various layers. In this work, we address wave emission and propagation in a framework that aims to bring together atmospheric and space plasma researchers in order to assess the existing connections between the troposphere and the ionosphere. To this aim, we follow an unconventional reviewing method by succinctly describing the key phenomena that play a role in this coupling and then discussing their relevance for the subject, including both recent and important earlier works.

Since this work is part of an effort to bring together current knowledge on troposphere-ionosphere coupling mechanisms, we refer to companion papers or recent reviews whenever appropriate. Barr et al. (2000) have reviewed progress in ELF and VLF wave propagation, namely waves of natural or anthropogenic origin, effects of thunderstorm electrification, lightning and sprites, transient phenomena like sferics and whistlers, Schumann resonance, and disturbances at the lower edge of the ionosphere, such as the Trimpi effect. Singh et al. (2005) have reviewed the electromagnetic coupling between the Earth atmosphere and the outer space environment, focusing on the DC electric circuit and magnetospheric phenomena. Inan et al. (2010) survey ELF and VLF research on lightning-ionosphere interactions and causative discharges, and discuss the role of optical emissions (sprites, jets, halos) and gamma ray flashes in troposphere-ionosphere coupling.

The organization of this article is as follows. Section 2 provides an overview of the most commonly observed electromagnetic waves that are used to assess the atmospheric electric environment and troposphere-ionosphere coupling. In Sect. 3, we describe the properties of the surface-ionosphere cavity/waveguide, the relevant boundary conditions, and the electromagnetic sources. Section 4 presents analytical and numerical approaches, algorithms, and techniques suitable to model wave propagation, including phenomenological atmospheric and ionospheric reference models. In Sect. 5, we discuss electromagnetic wave propagation from a global perspective, emphasizing the contributions to the understanding of the Earth's global electric environment. Finally, in Sect. 6, we summarize the significance of ULF-VLF electromagnetic wave phenomena for the investigation of troposphere-ionosphere coupling mechanisms, including not only their direct impact in energy balance and instability generation but their appropriateness for tropospheric and ionospheric in situ and remote sensing studies.

A note before proceeding: depending on the field of research, different terminology is used to define the various electromagnetic frequency ranges. In this article, we use the frequency nomenclature employed in ionosphere-magnetosphere sciences:

Ultra-Low Frequency (ULF)—Frequencies below 3 Hz,
Extremely Low Frequency (ELF)—Frequencies between 3 Hz up to 3 kHz,
Very Low Frequency (VLF)—Frequencies between 3 kHz and 30 kHz.

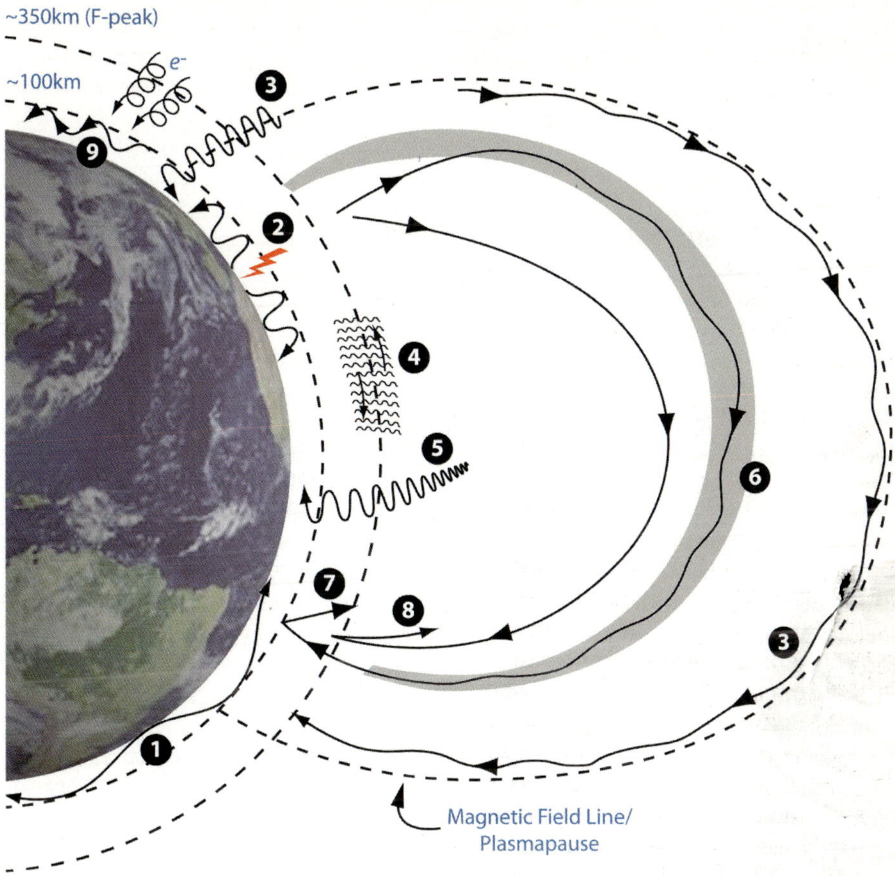

Fig. 1 Sketch of ULF-VLF electromagnetic-related waves relevant for tropospheric-ionospheric coupling mechanisms, including: (1) Schumann resonance (ELF), (2) sferics (VLF), (3) geomagnetic pulsations (ULF-ELF), (4) Alfvén waves—magnetosonic mode (ULF-ELF), (5) Alfvén waves—shear mode (ULF-ELF), (6) whistler (VLF), (7) VLF propagation back-scattered by ionospheric duct, (8) VLF wave reflected by magnetic mirror point, and (9) VLF subionospheric propagation (Trimpi effect). Figure partially inspired by Handbook of Atmospheric Electrodynamics, Edited by Hans Volland, CRC Press

2 An Overview of Waves Associated with Tropospheric-Ionospheric Coupling

A sketch of typical ULF-VLF waves that play an important role in troposphere-ionosphere coupling is presented in Fig. 1. Included are magnetohydrodynamic (MHD) waves (continuous and irregular pulsations), resonance phenomena (Schumann resonance and ionospheric Alfvén resonator), and electromagnetic wave transients (sferics, tweeks, whistlers, hiss).

2.1 Schumann Resonance

Earth can be regarded as a nearly conducting sphere, wrapped in a thin dielectric atmosphere that extends to the lower edge of the ionosphere where the conductivity is also substantial, nesting the Earth-ionosphere cavity similar to a waveguide. Two types of electromagnetic resonant modes coexist in the surface-ionosphere cavity/waveguide: (i) longitudinal modes of the cavity, corresponding to ELF wave oscillation around the planet, usually known as Schumann resonance, and (ii) transverse modes of the cavity, corresponding to nearly vertical VLF wave propagation between the surface and the ionosphere. The propagation of low frequency electromagnetic waves within a cavity bounded by two, highly conductive, concentric, spherical shells, was first studied by Schumann (1952), and the resonance signatures of the cavity later observed in ELF spectra by Balser and Wagner (1960). Several reviews have been published (Galejs 1972; Bliokh et al. 1980; Sentman 1995), including the more recent monograph by Nickolaenko and Hayakawa (2002). Simões et al. (2008a) have reviewed Schumann resonance models applied to planetary environments. A few major results are summarized owing to their specific significance for troposphere-ionosphere coupling.

Although other electromagnetic sources may play a role in excitation of Schumann resonance modes, such as transient luminous events, magnetospheric waves or volcanic eruptions (Abbas 1968; Huang et al. 1999; McNutt and Davis 2000), thunderstorm lightning are the dominant source of ELF waves in the cavity. The ELF resonant modes arise from nearly horizontal wave propagation around the globe in the transverse magnetic mode, i.e., with the wave magnetic field and the propagation vector perpendicular to each other. Lightning intensity and distribution within the cavity control the response of ELF waves and the characteristics of Schumann resonances, namely their daily and seasonal variations (Balser and Wagner 1960; Sátori 1996). Sprites produce a peculiar enhancement of the Schumann resonance spectrum (Boccippio et al. 1995; Williams et al. 2007). Nickolaenko et al. (1996) use long-term Schumann resonance data to deduce temporal variations of global lightning activity. They also estimate the lightning stroke distance to the receiver by assessing the relative intensity of the spectral lines. Boccippio et al. (1998) use lightning optical detection from orbit to calibrate single station Schumann resonance measurements. Specifically, combining conventional magnetic direction-finding techniques and Schumann resonance transient information, they compute the direction and distance of lightning strokes. Triangulation using multiple ground-based receiving stations provides of course more accurate estimates (e.g., Füllekrug and Fraser-Smith 1996) but the single-receiving point concept is quite useful for satellite measurements.

Monitoring of Schumann resonances over long periods of time contributes to inferring annual and semiannual periodicities (e.g., Nickolaenko et al. 1999; Price and Melnikov 2004). In addition to variability over daily or a few day periods, deducing long term trends is useful for climate research because of a fundamental connection between lightning and thunderstorm activity. For example, Williams (1992) claims using Schumann resonance as a global tropical thermometer to infer temperature fluctuations in the equatorial regions. Williams et al. (2000) and Williams and Sátori (2004) discuss in great detail the lightning-thunderstorm connection and implications for Schumann resonance spectral morphology, namely the 'chimneys-like' distribution of lightning as a function of local time and Schumann resonance amplitude, combining thermodynamic, hydrological, and geographic records with ELF spectral information. Anyamba et al. (2000) present evidence for the manifestation of the Madden-Julian Oscillation (MJO) in global deep convection and in Schumann resonance intensity. This result extends the connection between convective phenomena and thunderstorms to lightning and ELF wave propagation.

In addition to lightning, transient luminous events also contribute to the electric processes in the atmosphere. Huang et al. (1999) discuss criteria for sprites and elves based on Schumann resonance observations and claim that ground flashes with positive polarity associated with both sprites and elves enhance Schumann resonance response, triggering an increase of the peak amplitude by several times that of average peak structures, an effect similar to a phenomenon related to strong lightning strokes usually known as Q-burst.

Boccippio et al. (1995, 1998) report peculiar signatures produced by sprites in Schumann resonance spectra. A significant electric coupling between the troposphere and ionosphere occurs in strong cloud-to-ground lightning, and during troposphere-to-ionosphere (blue jets) and ionosphere-to-troposphere (sprites) discharges (Pasko et al. 2002).

The variability of Schumann resonances is associated with variations of not only electromagnetic sources but also atmosphere properties and cavity boundary conditions. Unlike surface changes that can frequently be neglected, variations of the lower ionosphere, which marks the upper boundary of the cavity, produce specific features in the ELF Schumann resonance spectra. Disturbances of the lower ionosphere arise from solar events and energetic solar proton precipitation (Reid 1986), solar wind-magnetosphere interaction and auroral activity (storms and auroral substorms) as well as from gravity, planetary, and tidal waves or, at a local scale, from lightning and transient luminous events (e.g., England 2011; Pasko et al. 2011; Shalimov and Bösinger 2011; Oyama and Watkins 2011), and even from anthropogenic activities such as stratospheric thermonuclear explosions (Madden and Thompson 1965). Sentman and Fraser (1991) report evidence for Schumann resonance intensity modulation by the local height of the D-region. Schlegel and Füllekrug (1999) investigate the correlation between sudden ionospheric disturbances induced by particle precipitation and Schumann resonance spectra, and conclude that ELF wave propagation is improved because the electron density profile is sharpened. Schumann resonance amplitude, frequency, and cavity Q-factor (damping factors) all often change during solar proton events. Nickolaenko and Hayakawa (2002) present an episode of classical polar cap absorption, where the frequency of the first and second eigenmodes decreases by 0.4 and 0.8 Hz, respectively. Madden and Thompson (1965) report that cavity eigenfrequencies were found to be shifted downwards by about 0.5 Hz due to high altitude nuclear explosions. The interaction between the solar wind and the ionosphere distorts and modulates (by the solar cycle) the upper boundary and influences cavity eigenfrequencies, which also respond to solar flares (Reid 1986). Finding solar cycle direct impact in Schumann resonance spectra is therefore unsurprising. Solar effects not only modulate the Schumann resonance spectrum but significantly distort the shape of the cavity. Schumann resonance responds to atmospheric and ionospheric ionization transients during solar flares and solar proton events; Schumann resonance frequency often increases with X-ray flux enhancement and decreases with solar proton rise because ionization produced by photons and protons occurs at different altitudes (Schlegel and Füllekrug 1999; Roldugin et al. 2003; De et al. 2010). Additionally, the cavity spherical symmetry is ruined by the action of solar wind, day-night asymmetry, and geomagnetic field, resulting in eigenmodes degeneracy removal (e.g., Nickolaenko and Sentman 2007). Those effects can be used for investigating the upper boundary further.

The vector electric field double probe instrument on the Communications/Navigation Outage Forecasting System (C/NOFS) satellite detected Schumann resonance signatures well beyond the upper boundary of the cavity, as shown in Fig. 2 (Simões et al. 2011a). The lowest five Schumann resonance peaks are observed at about 7.9, 14.1, 20.6, 26.8, and 32.9 Hz and match those of ground measurements. The electric field of the first peak is $\sim 0.25\ \mu\text{V}\,\text{m}^{-1}\,\text{Hz}^{-1/2}$, compared to $\sim 0.3\ \text{mV}\,\text{m}^{-1}\,\text{Hz}^{-1/2}$ measured on the ground. The Schumann resonance signatures are typically detected by C/NOFS only during nighttime conditions and are observed routinely between 400–850 km altitude, which corresponds to the satellite's perigee and apogee, respectively. Additionally, Schumann resonance signatures are not detected on the component of the electric field parallel to the geomagnetic field, B_o; for the components perpendicular to B_o, the field amplitudes of the zonal (east-west) and meridional (vertical) components are similar. The relevance of these findings for

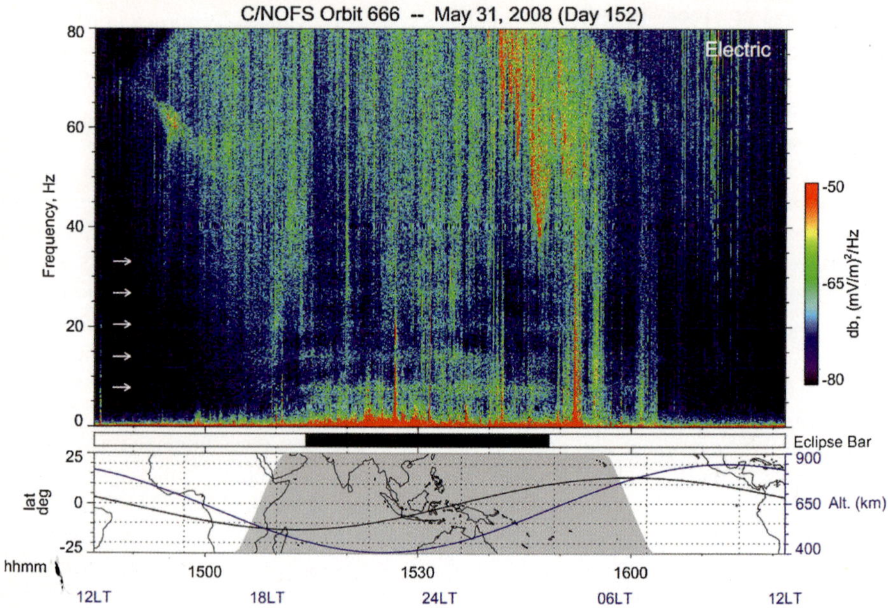

Fig. 2 Spectrogram of ELF electric field data for one complete orbit of the C/NOFS satellite. The *arrows* indicate the frequency of the Schumann resonance lowest eigenmodes. The *black bar* indicates when the satellite is in local eclipse. The *lowest panel* shows the satellite's path and altitude and the *gray-shaded* regions show where the earth's surface was in shadow (Simões et al. 2011a)

the investigation of troposphere-ionosphere coupling mechanisms is unquestionable despite the fact that the cavity leakage mechanism is poorly understood. Specifically, combination of ground-based and remote sensing ELF electric and magnetic field measurements offers new capabilities to investigate wave propagation in the Earth-ionosphere cavity.

Although large scale significant variations of surface conductivity cannot be expected, earthquakes could possibly change the local atmospheric conductivity and subsequently may produce specific signatures in the Schumann resonance spectrum. From ELF magnetic field data recorded in Japan, Hayakawa et al. (2005) reported an anomalous effect in the 4th harmonic, possibly associated to the Chi-chi earthquake in Taiwan. They observe an increase of the amplitude of this 4th eigenmode about a week before the main shock. A similar anomalous feature also appears before the second earthquake, which occurs six weeks later. According to these authors, atmospheric conductivity variations over Taiwan could enhance ELF wave scattering and modify the Schumann resonance spectrum. Since lightning distribution also affects the highest modes of the Schumann resonance spectrum, it is difficult to firmly establish the exact role of earthquake precursors and more extensive statistical studies are necessary to ascertain such an effect.

2.2 Ionospheric Alfvén Resonator

Alfvén waves are low frequency oscillations, usually below the ion cyclotron frequency, that occur in an ionized fluid permeated by a magnetic field such as the Earth ionosphere or magnetosphere. Plasma disturbances, ion displacement, and a restoring magnetic field tension to balance particle inertia lead to oscillation under the form of MHD waves. When the wave vector is parallel to the background magnetic field, Alfvén waves are transverse and termed shear Alfvén modes; if the wave vector is perpendicular to the background magnetic field, Alfvén waves are longitudinal and known as magnetosonic modes. Plasma density heterogeneities from local to global scale in the ionosphere and magnetosphere allow for the formation of waveguides and resonators for shear and magnetosonic Alfvén waves. The ionospheric magnetosonic waveguide results from the nearly total reflection of magnetosonic waves near the ionospheric F-region peak, and waves can be guided thousands of kilometers in the ionosphere due to such a ionospheric ducting (e.g., Greifinger and Greifinger 1973).

The ionospheric Alfvén resonator results from wave trapping in the vertical direction between the lower boundary of the ionosphere and the magnetosphere (Polyakov and Rapoport 1981; Belyaev et al. 1990; Lysak 1999). Although Alfvén waves play a significant role in a variety of plasma environments, namely the Sun, interstellar medium, compact astrophysical objects, Earth magnetosphere, and even nuclear fusion reactors (e.g., Allan and Poulter 1992; Narain and Ulmschneider 1996; Shukla and Stenflo 1999; Tsurutani and Ho 1999; Stasiewicz et al. 2000; Elmegreen and Scalo 2004), we merely discuss results relevant for ionosphere and troposphere-ionosphere coupling mechanisms research. Comprehensive discussions on the physics of Alfvén waves can be found in various monographs and reviews (e.g., Cross 1988; Gekelman 1999; Roberts 2000; Cramer 2001; Stasiewicz et al. 2000; Tsurutani et al. 2006). A complementary review of the ionospheric Alfvén resonator and ionospheric waveguide can be found in a companion paper by Demekhov (2011). Geomagnetic pulsations are frequently associated to Alfvén waves. In this paper we focus our attention in Pc1–Pc2 waves, associated with the ionospheric waveguide and ionospheric Alfvén resonator (IAR), and disregard waves with longer wavelengths (Pc3–Pc5), which are mainly relevant for magnetospheric research. Stasiewicz et al. (2000) present a comprehensive review of Alfvén wave models and measurements, including magnetosonic, IAR, and field line resonance results.

Since the Alfvén index of refraction shows in the F-region, magnetosonic wave propagation in the ionospheric waveguide may provide some insight into the F-region dynamics. Shvartsburg and Stenflo (1993) outline methods for assessing waveguide asymmetry, e.g., day-night and north-south dichotomy. Neudegg et al. (2000) estimate ULF wave attenuation in the waveguide at high latitudes. The latter measured geomagnetic pulsations in Antarctica, finding Pc1–Pc2 wave phase velocity and attenuation in the range 300–800 km s^{-1} and 10–95 dB Mm^{-1}, respectively, with a corresponding damping of 310 ± 220 km. Combining measurements from stations in Antarctica and satellites, Hansen et al. (1991) analyze high latitude unstructured Pc1 waves generated in the vicinity of the dayside auroral oval, obtaining different spectral morphology in the morning and afternoon sectors. They interpret Pc1 emissions as resulting from ion cyclotron waves generated in the equatorial region, entering the ionosphere within the auroral oval, and then propagating in the ionospheric waveguide to the stations. Chisham and Orr (1997) perform a statistical study of Pc5 emissions observed at mid latitudes and observed similar local time asymmetry between the morning and afternoon sectors. The asymmetry is possibly related to the fact that field line resonances driven by the solar wind interaction with the magnetosphere occurs at higher latitudes during the morning, and the most likely energy source is the Kelvin-Helmholtz instability at the magnetopause boundary. Allan and Wright (1997) develop the nonlinear Kelvin-Helmholtz instability further in the context of field line resonance and ionosphere-magnetosphere coupling. Assessment of day-night asymmetry related to ULF-VLF wave propagation is particularly useful to investigate ionosphere dynamics because empirical ionospheric models are less accurate close to the terminator (e.g., Simões et al. 2009, 2011b; Klenzing et al. 2011b).

Alfvén waves can propagate in the ionosphere not only horizontally between the northern and southern hemispheres but also in the radial direction. The magnetosonic mode is sensitive to ionospheric regional and global latitudinal variation while the Alfvén shear mode is more constrained by vertical local variations, thus both types of waves give access to different variation patterns in the ionosphere. As a matter of fact, IAR signatures contribute to constrain electron and ion density profiles between ~200 km and about one Earth radius. Belyaev et al. (1990) present convincing data of the IAR signature that confirm theoretical predictions; magnetic field data are recorded in the late afternoon, at mid latitude in Nizhny Novgorod, Russia and reveal a few peaks in the Pc1 range with $Q \sim 10$ (Q-factor is a dimensionless quantity related to wave attenuation and peak resolvability), resulting from ~15 min data averaging. Belyaev et al. (1999) show spectral resonance structures associated to IAR measured at high latitude (Kilpisjärvi, Finland). Measurements of magnetic field with linear and circular polarization show multiple (usually 3 or 4) peaks, where difference between consecutive peaks varies between ~0.5 and 1.5 Hz, preferentially with right-circular or linear polarization. Although less persuasive, IAR measurements onboard satellite at polar latitudes have been also reported (Grzesiak 2000; Chaston et al. 2002). Bösinger et al. (2004) report IAR signatures from measurements recorded in Crete, Greece. During nighttime, spectrograms show up to 20 resolved lines in the frequency range 0.1–4 Hz and ~0.2 Hz between consecutive peaks. In those spectrograms, frequency systematically increases from a fraction of 1 Hz at sunset up to 3–4 Hz about midnight, an indication of major changes of the ionospheric profile in the pre-midnight sector. Using the High Frequency Active Auroral Research Program (HAARP), Parent et al. (2010) investigate the effects of substorm dynamics on IAR. Before substorm onset, the spectral resonances display signatures in agreement with the expected variations of the ionosphere. At substorm onset, the IAR signatures disappear either buried in background noise or because the resonance ceases. The authors argue that an increase of the F-region density associated with soft electron precipitation explain the observations because, after the substorm, the signatures reappear and harmonics are shifted to lower frequencies with tighter frequency spacing. These claims corroborate results previously reported by Semenova et al. (2008), where they concluded that magnetic field measurements show a decreasing of resonance spectral structures related to an increase of ionospheric variability.

The most pertinent questions for investigating how troposphere-ionosphere coupling is related to the ionospheric Alfven resonator pertain to establishing the nature of IAR excitation sources. Surkov et al. (2006) introduce a theory of mid latitude IAR excitation showing that local lightning activity may explain the resonator excitation. Sukhorukov and Stubbe (1997) propose a mechanism to explain IAR excitation by strong lightning discharges followed by transient luminous events. Demekhov et al. (2000) discuss a different approach involving Pc1 waves. It is important to mention that the frequency range of IAR signatures and Pc1 waves is similar. Therefore, it is important to identify which Pc1 waves could be attributed to IAR phenomena and whether the resonator would act as a filter for Pc1 waves generated elsewhere. High versus low latitude systematic measurements may shed light on IAR excitation mechanisms and whether lightning-related events are relevant for the investigation of troposphere-ionosphere coupling processes.

2.3 Sferics and Tweeks

Lightning discharges produce broadband electromagnetic impulses, often termed *sferics*, that propagate in the Earth-ionosphere cavity. In addition to the ELF longitudinal modes of the cavity associated with wave propagation around the planet, transverse resonance modes can be excited locally in the VLF domain. For an ionospheric effective height of reflection of 75 km, i.e. altitude of the upper boundary, the frequency of the first transverse mode is about 2 kHz. Constructive interference resulting from multiple reflections on the surface and the ionosphere filter the broadband signal and yield specific waveforms with frequencies related to the height of the upper boundary. When escaping from the cavity and propagating in the ionosphere, possibly along ducts, the signal is slightly dispersed, producing a *tweek* that is basically a sferic that suffers small frequency dispersion when traveling through the ionosphere. Waves traveling further along geomagnetic field lines through the plasmasphere, in the right-handed mode (Storey 1953), experience larger frequency dispersion and are detected in the conjugate hemisphere as *whistlers* due to their typical whistling tones. Sferics and tweeks can be used to study the D-region of the ionosphere, tweeks provide information on the E and F-regions, and whistlers depend on the F-region and electron density profiles in the plasmasphere.

Analytical and numerical models have been developed to study VLF wave propagation in the cavity and to derive the properties of the electromagnetic source and of the lower ionosphere. Cummer et al. (1998) use a frequency domain subionospheric VLF propagation code to derive nighttime electron density profiles in the D-region and Sukhorukov (1996) develop an analytical model valid for the upper ELF and lower VLF ranges. The tweek waveforms are better resolved and suffer less attenuation at night (typically less than 0.5 dB Mm^{-1} as compared to 5 dB Mm^{-1} during day-time) due to sharper D-region electron density profiles. When detectable, higher harmonics provide useful information on propagation condition and characteristics of the lower ionosphere (Singh and Singh 1996; Kumar et al. 2008). Barr et al. (2000) have thoroughly reviewed the major results from experimental studies on sferics, including time and frequency domain methods and triangulation techniques.

Ohya et al. (2006) use tweek atmospherics observations gathered during a major geomagnetic storm in October 2000 to determine the D-region disturbances. Reeve and Rycroft (1972) use sferics and tweeks to derive the variations of the lower ionosphere during the solar eclipse of 7 March 1970 and show that the effective height of reflection moves from 69 km before eclipse to 76 km at the time of total eclipse. From transverse resonance observations during a stratospheric balloon flight in the equatorial region, Simões et al. (2009) report the occurrence of a fast depletion of the D-region between 73 and 82 km after sunset. Since the D-region is rather difficult to explore both with remote sensing and in situ techniques, systematic high resolution measurements of sferics and determination of their sources by lightning location networks provide a suitable tool for monitoring electron density profiles. Rocket-triggered lightning offers even better experimental conditions because the stroke location uncertainty would be considerably reduced.

Electromagnetic transients are useful to assess not only the ionospheric lower boundary but also the anisotropy of the medium. Hayakawa et al. (1994) use sferics and tweek characteristics, namely polarization, incidence angle, and frequency to investigate ionospheric plasma. They find that wave polarization slightly above the transverse mode cutoff frequency is always left-handed and becomes exactly circular when the cutoff is reached, confirming the role of anisotropy in wave propagation in the ionosphere. Ohya et al. (2006) use tweek atmospherics observations gathered during a major geomagnetic storm in October 2000 to determine the D-region response under high magnetic activity. Tweek reflection effective height is used to investigate ionosphere dynamics during the storm and dissimilar ionospheric dynamic conditions were identified. Shvets and Hayakawa (1998) assess tweek polarization effects in the northern and southern hemispheres employing multimode analysis, finding non-reciprocity between east-west and west-east propagation and variations in wave polarization due to the geomagnetic field. From measurements of right- and left-hand polarized ELF-VLF waves, Ostapenko et al. (2010) use tweek signatures to study the auroral region. Atmospherics can therefore contribute to assessing not only electron density profiles and D-region dynamics but also ionospheric plasma anisotropy. Injection of VLF signals from ground based transmitters nevertheless offer more reliable options for studying polarization effects and ionosphere anisotropy because electromagnetic source characteristics are well known (e.g., Inan et al. 2007).

Harrison et al. (2010) suggest using tweeks to investigate possible ionospheric disturbances related to seismic activity. Specifically, they propose a mechanism to link seismic activity and changes in the local lower ionosphere that could be tested by measuring the tweeks cutoff frequency.

2.4 Whistlers

Whistlers are circularly polarized, electromagnetic VLF waves that propagate in the "whistler mode" along the magnetic field direction with very slow phase velocities (much slower than the speed of light) and that are generated by lightning sferics that penetrate into the Earth's lower ionosphere. Since the wave velocity is a function of the electron density, the dispersion characteristics of whistlers have been used to derive the electron density in the plasmasphere. Favorable reflection conditions in the ionosphere may allow successive reflections between conjugate hemispheres with an increased dispersion observed for each travel "hop". Ion cyclotron whistlers, only detectable in space, also provide information on the ion composition (Gurnett et al. 1965). Whistlers are fundamental research tools for a number of ionospheric or magnetospheric processes such as gyroresonant wave-particle interaction (Summers and Ma 2000; Brautigam and Albert 2000), electron acceleration and precipitation from radiation belts (Meredith et al. 2003; Horne et al. 2005a, 2005b), non-linear wave-wave interaction with Alfvén waves (Sharma et al. 2010). In addition, they feature a valuable tool to study the structure of the ionized terrestrial environment (e.g., Park et al. 1978) and its dynamics in response to solar wind disturbances (e.g., Meredith et al. 2001). Monographs by Helliwell (1965), Sazhin (1993), and Ferencz et al. (2001) thoroughly cover whistler theory, measurements, and applications. In the following we briefly summarize a few results directly impacting tropospheric-ionospheric coupling.

Electron density enhancements in the lower ionosphere are produced by sources located in the cavity or outer space. In addition to transitory ionization related to cosmic ray bursts, solar particle events and flares, lightning, transient luminous events, and terrestrial gamma flashes, electron density enhancements can be produced by energetic electron precipitation driven by the interaction of whistlers with trapped particles in the radiation belt. Haldoupis et al. (2004) present a few VLF signals associated with whistler-induced electron precipitation events, which are produced by electron precipitation due to whistler wave injection into the magnetosphere by the same lightning flash that led to a sprite. Lightning also induces direct ionization enhancements in the low ionosphere. Strangeways (1999) discusses effects of D-region local ionization enhancements produced by lightning, namely ducting development associated to whistlers and amplitude perturbations on subionosphere VLF wave propagation (Trimpi effect). Lightning-generated whistlers lead to coupling between the troposphere, ionosphere, and magnetosphere. A lightning stroke can generate whistlers that interact with cyclotron resonant radiation belt electrons, producing particle acceleration and inducing particle precipitation back in the low ionosphere. Determining the depth of penetration of precipitating particles is pertinent for investigations of ozone and nitrogen oxide chemistry, for example. Rodger et al. (2007) determine the altitude range where particle precipitation plays a role and assess implications for ionization balance and neutrals chemistry in the mesosphere, by contrasting cosmic rays, solar photoionization, and whistler-induced electron precipitation contributions for the ionization budget. Whereas the ionization due to whistler-induced electron precipitation can be neglected during daytime compared to solar EUV and soft X-ray ionization, nighttime conditions offer a more complex situation. Below ~70 km the cosmic ray contribution prevails but particle precipitation usually dominates in the range 75–80 km. Despite previous research concluding that particle precipitation can lead to large-scale changes in neutral chemistry and enhance key species that drive ozone loss, studies suggest that transient electron precipitation plays an important role in some parts of the mesosphere but can be neglected in neutral chemistry models. For extensive electron energy and ionospheric conditions, Codrescu et al. (1997a, 1997b), Horne et al. (2009), and Lam et al. (2010) (and references therein) scrutinize the role of high energy electron precipitation under different geomagnetic activity conditions, observe precipitating flux increasing with geomagnetic activity, and also address possible implications for atmospheric chemistry and climate variability. Interestingly, lightning interacts with the lower ionosphere both through direct (fast) and indirect (delayed) processes. Direct heating and ionization are due to charge transfer to and from the ionosphere. Lightning-induced particle precipitation resulting from interaction with whistlers also produces ionization from above. The study by Dowden et al. (1994) details a way to distinguish between the fast and delayed effects of lightning discharges, concluding that rapid onset, rapid decay of phase and amplitude perturbations of VLF subionospheric transmissions are directly related to atmospheric discharging processes rather than whistler-induced electron precipitation.

Employing statistical analyses, Hayakawa et al. (1993) investigate the possible role of seismic activity in whistler wave generation and propagation. They claim that whistlers with dispersion that is at least twice the average are likely correlated with seismic activity. Shalimov and Gokhberg (1998) explore the topic further and argue that anomalous impulsive VLF emissions in the upper ionosphere may be related to whistler trapping in ducts, formed in the ionosphere above the seismically active region. A companion paper by Pilipenko (2011) presents a more general overview of impulsive coupling between the atmosphere, ionosphere and ionosphere.

2.5 Geomagnetic Pulsations

Geomagnetic pulsations are hydromagnetic waves that arise from resonant processes and propagate in the magnetosphere, usually observed along closed field lines or close to these regions. These waves cover the ULF-ELF range, from the longest wavelengths that the magnetospheric cavity can sustain up to ion gyrofrequencies. Geomagnetic pulsations were first observed in 1859 during aurora events (Stewart 1861) and are often split in morphological groups or frequency bands, including Pc1 (0.2–5 s), Pc2 (5–10 s), Pc3 (10–45 s), Pc4 (45–150 s), Pc5 (150–600 s), Pi1 (1–40 s), and Pi2 (40–150 s), where Pc and Pi denote continuous and irregular pulsations, respectively. These waves result from electromagnetic ion cyclotron instability, compressional fluctuations of the medium, toroidal and poloidal geometry related to propagation, field-aligned currents, magnetic field line resonances, as well as from coupling among multiple mechanisms.

Saito (1969) reviews geomagnetic pulsations and their classification, observations, and generation mechanisms. This assessment remains a consistent work in the field and was followed by a number of updated reviews on more specific topics such as the type of pulsation, frequency range of occurrence, and satellite observations, emphasizing geomagnetic pulsations relevance for a variety of Earth and space science studies (Raspopov and Lanzerotti 1976; Pilipenko 1990; Engebretson et al. 1991; Sazhin and Hayakawa 1994; Takahashi 1998; Kangas et al. 1998; Daglis et al. 1999; Olson 1999; Yahnin and Yahnina 2007; Zong et al. 2009).

Frequency characteristics (e.g., dispersion relation and harmonics structure), polarization, spatial distribution, and correlation with solar wind and geomagnetic activity make geomagnetic pulsations a valuable tool to investigate Sun-Earth connections, solar wind-magnetosphere interaction, magnetosphere dynamics, and auroral processes. Although geomagnetic pulsations play a minor role in troposphere-ionosphere coupling, a few works deserve attention owing to their interesting results.

The ionosphere in the F-region plays a significant role on geomagnetic pulsations controlling their attenuation and propagation and their coupling with ionospheric currents. Hughes and Southwood (1976) address atmospheric and ionospheric screening effects and show that only a small energy fraction of the signal, about 10%, reaches the ground showing that the strong day-night asymmetry results from the fact that the ionosphere reflects much less at night. Their work also elucidates the morphological differences between typical day Pc and night Pi signatures. Sarma and Sastry (1995) discuss the enhancement of geomagnetic pulsations near the dip equator due to the equatorial electrojet. Results indicate a sharp cutoff at about 20 s, bearing evidence for distinctive contributions in the Pc1–2 and Pc3–5 domains and the role of the equatorial electrojet in MHD wave propagation.

An interesting contribution of research on geomagnetic pulsations is related to phenomena associated with seismic activity and lightning. Iyemori et al. (2005) describe a pulsation of period 3.6 min (Pc5) observed in Phimai, Thailand, 12 min after the Sumatra earthquake on 26 December 2004. A $\sim$30 s (Pc3) period pulsation was observed in Tong Hai, China, located 10 degrees north of Phimai. They argue that the nature and period of the geomagnetic pulsation was consistent with a dynamo action in the lower ionosphere due to gravity waves triggered by the ocean floor displacement. It is also speculated that the Pc3 signal observed in Tong Hai resulted from magnetosonic waves generated by electric and magnetic perturbations of the dynamo current triggered by the earthquake. Although a cause-effect is difficult to establish, this example illustrates the complex interaction between acoustic and electromagnetic waves involving surface, atmospheric, ionospheric, and magnetospheric phenomena. Fraser-Smith (1993) presents ULF magnetic field measurements related to thunderstorm activity and discusses their significance to geomagnetic pulsation generation. A correlation between magnetic and thunderstorm activity is found, indicating that lightning could possibly generate geomagnetic pulsations.

3 The Surface-Ionosphere Cavity and Waveguide Characteristics

3.1 Boundary Conditions

The Earth's surface, where the electric conductivity changes by more than 10 orders of magnitude, is considered as the inner boundary for a variety of wave modes considered here. The upper boundary of the earth-ionosphere cavity, however, is far more complex and less well-defined. This lower edge of the ionosphere is located in the region between 70 and 110 km, where the conductivity increases by 5 to 6 orders of magnitude, but needs to be defined more precisely particularly with respect to the frequency range of interest.

Depending on the chosen formalism and on the frequency range of interest, the definition of the upper boundary of the cavity and of the investigation of its characteristics usually employs three complementary concepts: (i) propagation, reflection, and transmission coefficients, (ii) refractive index profile and (iii) skin depth. The concept of a conducting layer in the D-region/E-region playing the role of the upper boundary is valid for the DC global electric circuit (e.g., Rycroft 2006), Schumann resonance (e.g., Sentman 1995), as well as for sferics and tweeks propagation (e.g., Simões et al. 2009). In the case of the DC electric circuit, the ionosphere is considered equipotential above ~90 km to a first approximation. For ELF wave propagation, the skin depth of propagating waves is considerably smaller than the cavity thickness and a similar boundary also applies. In the ELF range, the skin depth is less than 10 km for an altitude of 100 km, where the upper boundary of the cavity is usually defined. A similar condition can be applied to the propagation of sferics and tweeks. However, a perfect electric conductor condition is no longer acceptable if anisotropy corrections have to be taken into account, because conductivity parallel to the geomagnetic field is orders of magnitude smaller than the scalar conductivity. This correction is crucial for modeling whistlers and ELF cavity leakage.

For ULF waves such as geomagnetic pulsations, the boundary is usually taken at the altitude where the Alfvén index of refraction presents a sharp variation, i.e. where strong reflection occurs. The inner and outer boundaries of the ionospheric waveguide and ionospheric Alfvén resonator are respectively located in the E-region and at several hundred/thousand kilometers above (e.g., Greifinger and Greifinger 1968; Polyakov and Rapoport 1981). Unlike the E-region inner boundary that corresponds to an altitude range with strong gradients in the index of refraction, the position of the outer boundary varies significantly with local time and latitude because the gradient of the Alfvén index of refraction above the F-peak is much smoother. Some models often set the inner boundary at the surface, thus considering wave propagation through the E-region down to the ground. This is a reasonable first-order assumption since geomagnetic pulsations related to the ionospheric Alfvén resonator are detected at the ground, implying that the sub-surface acts indeed as the lower boundary.

Unlike the static, sharp inner boundary, the ionosphere defines a fuzzy, heterogeneous, and anisotropic boundary condition. For quasi-static and electromagnetic phenomena modeling, the surface can be considered a steady, perfect electric conductor despite transient events such as earthquakes, tsunamis, and volcanic eruptions induce local atmospheric conductivity modifications or ionospheric disturbances (e.g., Smith 2011). Ionosphere dynamics, however, contributes to an intricate outer boundary because of solar wind, magnetospheric, and particle precipitation driving mechanisms from above combined with lightning, transient luminous events, and gravity waves and tides from below. Plasma irregularities occurring in the ionosphere, namely density enhancements and depletions of the medium, extend ionosphere dynamics further (e.g., de La Beaujardière et al. 2004, 2009; Klenzing et al. 2011a). As a result, the ionosphere is significantly asymmetric because of the day-night dichotomy and polar heterogeneity, and particularly dynamic due to solar activity, particle precipitation, unsettled geomagnetic field, ionospheric currents, and thunderstorm activity.

3.2 Parameterization of the Medium

The Appleton-Hartree description of plasma waves in cold magnetized plasma is the basis of the description of wave propagation for the conditions considered here. Depending on the frequency, the dispersion relation can be simplified to provide treatable, analytical solutions. In general, medium parameterization includes electron and ion density, geomagnetic field, and collision frequencies. The plasma anisotropy with respect to magnetic field must be included in Alfvén wave and whistler-mode propagation but is often neglected in ELF wave modeling. Considered simplifications can be summarized as follows:

(i) *Alfvén waves:* Heavy ions and collision frequency are typically neglected. When included, the Pedersen conductivity introduces wave attenuation and the Hall conductivity leads to coupling between shear and magnetosonic modes. In the case of IAR, analytical approximations frequently employ simplified profiles of the Alfvén velocity with propagation at constant velocity in the E-region (e.g., hundreds of $\mathrm{km\,s^{-1}}$) and the F-region (tens of km s^{-1}), and at a velocity that increases exponentially (scale height ~200 km) above the F-peak (e.g., Polyakov and Rapoport 1981). More realistic models also include propagation below the D-region, allowing propagation down to the surface.

(ii) *Schumann resonances:* Anisotropy is frequently neglected, leading to a scalar conductivity that increases with altitude. The conductivity profile is often represented by two scale heights respectively below and above a transition layer at ~50 km, the altitude where the magnitude of displacement and conduction currents is similar (e.g., Greifinger and Greifinger 1978). The conductivity variation with latitude is often neglected because the latitudinal gradient in the troposphere and stratosphere is small compared to the gradient in the vertical direction (e.g., Holzworth et al. 1985). In such a case, spherically symmetric analytical solutions may be sought (Sentman 1990). However, more subtle effects, such as eigenfrequency mode splitting, require day-night asymmetry and, possibly, polar cap specific conditions to be taken into account. In these cases, only numerical solutions are available.

(iii) *Sferics, tweeks:* Anisotropy corrections are frequently included in the investigation of sferics and tweeks. Heavy ions are sometimes neglected.

(iv) *whistlers:* Anisotropy must be taken into account for solving electron and ion whistler mode propagation. Heavy ions are included when ion cyclotron whistlers are considered. Collision frequencies are often neglected.

3.3 Electromagnetic Sources

3.3.1 Sources Within the Cavity

Lightning Lightning results from thunderstorm electrification and charge separation mechanisms. Typical discharge channel lengths are 1 to ~10 km and often involve intricate multiple branches that extend to several kilometers in the vertical and horizontal directions. A detailed description of the lightning discharges and physical processes can be found in the monograph by Rakov and Uman (2007). Lightning discharges are usually termed intracloud/intercloud (the most common), cloud-to-ground, and ground-to-cloud (the least common), and involve positive or negative charge transfer through the channel. Typically, negative cloud-to-ground discharges transfer a total of 25 C with a peak current of ~30 kA, releasing about 500 MJ. Positive discharges are less frequent though their corresponding charge and peak current are sometimes ten times larger. Lightning occurs mostly over land, in the low latitude regions, and follows a regular seasonal pattern reversing between the northern and southern hemispheres. The global lightning average rate is $44 \pm 5\ s^{-1}$, reaching a maximum of ~200 strokes per square kilometer per year in Central Africa (geographic coordinates 3°S, 28°E). Individual thunderstorms can deliver up to about 6000 strokes per hour (e.g., Rakov and Uman 2007; Schumann and Huntrieser 2007). Relevant characteristics of individual lightning involves flash duration, number of return strokes per flash, charge quantity per flash and per stroke, duration and intensity of continuing current, time to peak and return stroke current, etc.

In a seminal experiment, Kelley et al. (1985a) reported DC and wave electric fields detected by probes on two sounding rockets launched simultaneously at mid-latitudes over an active thunderstorm. A balloon-borne electric field probe (Holzworth et al. 1985) as well as ground-based measurements completed the data base. As shown in Fig. 3, these probes showed definitive evidence that a quasi-static electric field "pulse" penetrated from the troposphere into the ionosphere. Probes on the high altitude rocket showed that whistlers, associated with the lightning-generated sferics, were observed in the ionosphere but were absent

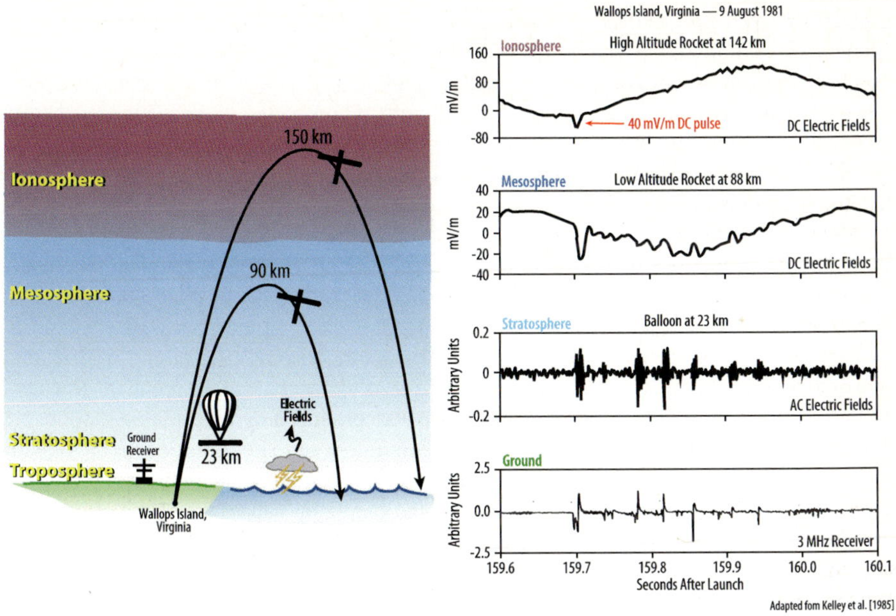

Fig. 3 Direct penetration of lightning electric fields in the ionosphere. (*Right*) Electric field measurements at various altitudes. (*Left*) Sketch of rocket trajectories and receivers position (Kelley et al. 1985a)

in the middle atmosphere, as observed in the low apogee rocket. Other rocket and balloon experiments conducted subsequent to this one have shown evidence of similar phenomena.

Lightning plays multiple roles in wave propagation. Not only does lightning generate electromagnetic waves, but it also modifies the properties of the medium through local ionization. Lightning correlation with water vapor and aerosols, namely clouds, urban pollutants, dust storms, and smoke plumes, offers complementary tools for atmospheric sciences research. In general, atmospheric conductivity is a function of aerosol content, hence providing a link to wave propagation. Fitzgerald (1991) reviews the background aerosol in the boundary layer over remote locations such as oceans to differentiate between marine and land contributions of nitrate to sea salt aerosol budget. Although the associated chemistry processes are recognized, the relative importance of the ocean, stratosphere, and lightning as a source of the nitrogen-containing precursor gases remains unclear. Over land, however, pollution plays a more obvious role in lightning and atmospheric chemistry. From an examination of the chemical characteristics of continental outflow, Talbot et al. (1996) suggest that, in addition to biomass burning, lightning or recycled reactive nitrogen may be an important source of nitric oxide (NO) to the upper troposphere. Real et al. (2010) analyze implications of biomass burning pollutants for ozone (O_3) production and conclude that, whilst the transport of pollutants to the upper troposphere is variable, pollution from biomass burning can make a supplementary contribution to photochemical production of O_3 in addition to NO_x from lightning. Bell et al. (2009) analyze extensive US lightning datasets and find evidence for a lightning weekly cycle related to storm invigoration by pollution. The weekly cycle appears to be reduced over population centers.

The lightning rate is also enhanced by the presence of smoke plumes. As a result of thunderstorms ingesting smoke from forest fires, Lyons et al. (1998) found enhanced positive cloud-to-ground lightning, triple than that of the climatological norm, and peak current amplitudes that were double those that were expected. They also claim that those thunderstorms produced an abnormally high number of sprites.

Hurricanes are important sources of lightning as well. Khain et al. (2008) discuss possible aerosol effects on lightning activity and structure of hurricanes. Although the mechanism responsible for the formation of the maximum flash rate in the periphery is poorly understood, intense and persistent lightning takes place within a 250–300 km radius ring around the hurricane center.

Nitrogen fixation is essential for the growth of ecological systems and is accomplished through both anthropogenic and natural processes. Among the natural abiogenic processes, lightning assists the conversion of molecular nitrogen to NO, hence contributing to nitrogen fixation. NO_x also plays a key role in tropospheric and stratospheric photochemistry by acting to control the concentrations of O_3 and hydroxyl (OH). Lightning contributes a few percent to the total NO_x budget but can be a leading mechanism in pollutant-free areas and the upper troposphere. NO_x radicals can be also produced by ionospheric particle precipitation in the stratosphere and mesosphere. Schumann and Huntrieser (2007) review the global lightning-induced NO_x sources and the significance for understanding and predicting O_3 distribution and trends in the troposphere, the oxidizing capability of the atmosphere, and the lifetime of trace gases destroyed by reactions with OH. They conclude that a typical thunderstorm produces 250 mol NO_x per flash. A companion review by Arnone and Hauchecorne (2011) provides complementary information on stratospheric NO_x measurements and processes.

Maximum electric fields measured in thunderclouds are 0.1–0.2 $\mathrm{MV\,m^{-1}}$, a fraction of the 1 $\mathrm{MV\,m^{-1}}$ conventional breakdown. To overcome the lack of electric field strength, two mechanisms are usually invoked for lightning initiation: (i) runaway breakdown initiated by cosmic rays (e.g., Gurevich and Zybin 2001) and (ii) positive streamers triggered by hydrometeors, i.e., products resulting from atmospheric water vapor condensation (e.g., Petersen et al. 2008). The former reflects a close relationship between cosmic rays and electrodynamic processes in the thunderstorm atmosphere. The latter is closely related to the thermodynamic properties of the convective cell, which can be used to assess the hydrometeor mass (e.g., James and Markowski 2010). Milikh and Roussel-Dupré (2010) review runaway breakdown and electrical discharges in thunderstorms and discuss observations bearing evidence for the presence of energetic particles in lightning initiation, including gamma-ray and X-ray flux intensification over thunderstorms, gamma-ray and X-ray bursts in conjunction with stepped leaders, terrestrial gamma-ray flashes, and neutrons production. Ebert et al. (2010) review the relevance of streamer discharges for lightning and sprites inception. Because large sprite discharges at the low air densities of the mesosphere are physically similar to small streamer discharges in air at standard temperature and pressure, streamers are useful for investigating the tropospheric-ionospheric electric coupling.

As far as electromagnetic waves are concerned, the lightning channel works as an effective transmitter over a wide frequency band, with a frequency spectrum peaked at a few to 10 kHz. Lightning distribution around the globe as a function of position and time is also preponderant. Lightning characterization in the time and frequency domains is fundamental to model wave propagation in the earth-ionosphere cavity. In the harmonic formalism, lightning is usually described by a Hertz dipole with a convenient frequency spectrum. Sometimes, electric tripoles are also considered to mimic lightning low frequency emission because they offer a better representation of charge distribution in the cloud. In the time domain, a time-varying current profile is considered. Several models have been used to compute the return stroke current of lightning (Nucci et al. 1990). Since the first successful lightning return stroke model which was based on a double exponential expression to facilitate analytical approximations, lightning models have been improved and classified in four major categories by Rakov and Uman (1998): (i) gas dynamics models involving coupling between gas dynamics equations (conservation of mass, momentum, and energy) and equations of state; (ii) electromagnetic models solving Maxwell equations to derive current distribution along the lightning channel, which is considered a thin wire antenna (charge channel diameter seldom exceeds 1 cm); (iii) transmission line models using resistor, capacitor, and inductor elements to compute current fields; (iv) phenomenological (engineering) models, where the spatial and temporal distribution of the return stroke current is specified from lightning characteristics. The latter model offers better approaches to compute the electromagnetic field distribution as function of space and time. Depending on relevant wave propagation modes, lightning strokes are frequently simplified by considering fast ($\sim$100 μs) and slow ($\sim$10 ms) variation components, corresponding to the return stroke and continuing current contributions.

We now briefly mention a couple of studies where lightning plays multiple roles, emphasizing multidisciplinary perspectives among atmospheric chemistry, meteorology, atmospheric electricity, and electromagnetic wave propagation. Rodriguez et al. (1992b) discuss a case-study of lightning, whistlers, and associated ionospheric effects that occurred during a particle precipitation event, and Rodriguez et al. (1992a) assess D-region disturbances caused by lightning. In these works, they claim that weak electromagnetic pulses originating from lightning subsequently heat D-region electrons; additionally, strong electromagnetic pulses would create electron density enhancements. Perturbations in artificial VLF signals originating in the proximity of storm centers are attributed to lower ionosphere heating induced by lightning. Price (2000) claims evidence for a link between global lightning activity and upper tropospheric water vapor. A positive correlation between upper troposphere water vapor variability and global lightning activity is found, suggesting that water vapor variability can be estimated from lightning activity and ELF-VLF wave monitoring.

Although marginally relevant for the present review, less familiar types of discharges, namely ball lightning, also occur in the troposphere. The interested reader can find extensive material about this topic in the monograph by Stenhoff (1999). Several theories related to waves in plasma, microwave interference and cavity modes, and transient wave propagation aiming to explain this phenomenon may be of interest.

Volcanic Eruptions Volcanic eruptions have been known to produce electrostatic and electromagnetic activity (for a review, see Johnston 1997; Mather and Harrison 2006). Impact charging plays an important role in volcanic plumes (Aizawa et al. 2010). Dust particles can get charged, leading to electrification processes similar to those acting in thunderstorm clouds. A large number of volcanic eruptions have been reported to produce lightning, with stroke rates up to 1 every 3 seconds during the eruption of Mount Spurr in Alaska (McNutt and Davis 2000). As in thunderstorms, volcano lightning occurs within the plume (equivalent to cloud-to-cloud discharges) or between the plume and ground (similar to cloud-to-ground and ground-to-cloud). Lightning-like discharges have been regularly noticed in forest fire ash plumes, too. For short periods and depending on natural conditions, namely plume speed and size, volcano lightning can be almost as intense as in thunderstorms.

At night, when observation conditions are more favorable, flashes have been observed at least 20 km far from the eruption, suggesting that significant amounts of energy are involved in the processes. Typically, discharges are hundreds of meters in length, transfer up to 0.1–0.5 C per event, produce local electric fields in the order of several kV m^{-1}, typically release about 10^6 J, and seem to begin about 20 min after the vigorous initiation of the eruptions (Anderson et al. 1965; Brook et al. 1974; Katahira 1992;

McNutt and Davis 2000). The geometric-mean peak currents of either polarity related to volcano lightning were sometimes only a factor of 2–3 lower than those associated to ordinary lightning recorded by the same network. Continuous measurements conducted at Sakurajima volcano, Japan, revealed syneruption (a period characterized by geologically instantaneous production of large volumes of volcaniclastic sediment) electric pulses, sometimes accompanied by geomagnetic pulses and lightning flashes. Transients were observed more than 10 seconds after the onset of the eruption, and tend to occur during eruptions that emit volcanic ash to high altitudes (Aizawa et al. 2010). From the perspective of atmospheric electricity investigations, namely the global electric circuit and electromagnetic wave propagation, volcanic lightning offers contrasting contributions. On the one hand, since thunderstorms generate dozens of lightning strokes per second, the average volcanic lightning contribution for the global electric circuit is negligible at the global scale. However, for short periods of time, volcanic lightning provides interesting means for assessing transient response due to its localized nature. The highest number of strikes per square kilometer per year is about 200 in central Africa though such a rate can be reached during a volcanic eruption in a matter of minutes. Although volcanic lightning is usually tenfold weaker than that of thunderstorms, larger amounts of energy can be released locally in much shorter periods of time. Investigation of strong lightning strokes generated by volcanic eruptions is therefore useful for assessing local variations of the global electric circuit and propagation of low frequency electromagnetic waves. Remarkably, the most important contribution of volcanic eruptions for the global electric circuit is indirect. Ash plumes resulting from violent volcanic eruptions are dispersed in the stratosphere at regional or global scales, where ultrafine aerosol layers are formed, leading to atmospheric conductivity decreases by one order of magnitude or more (e.g., Tinsley 2008). Effects of volcanic eruptions in electrostatic and electromagnetic noise have been investigated by Adler et al. (1999).

Corona discharges are also observed associated with turbulent lofting of dust and sand, as evidenced by $\sim$1 m long corona discharges observed during dry debris, gypsum sandstorms in New Mexico (Kamra 1972a, 1972b). Comparable scenarios involving dust devils and dust storms have been proposed for triggering corona discharging on Mars favored by the very dry Martian atmosphere. Saltation of sand particles and dust due to wind gusts induce impact charging near the surface, originating momentary sparks (Farrell et al. 1999). Massive dust storms, sometimes covering a significant part of the planet, would also be responsible for intense impact charging at much larger scales (Renno and Kok 2008). Similar effects involving ice particles and cryovolcanism in much colder environments have been suggested (James et al. 2008). However, the scenarios for how these phenomena contribute to planetary atmospheric electricity have yet to be established (Yair et al. 2008).

Anthropogenic Activity In addition to natural processes mentioned above, anthropogenic-related lightning is also possible. Lightning has been triggered, deliberately or involuntarily, by rockets and aircraft. Thermonuclear explosions have been known to produce lightning, too. Laser technology has been used to force atmospheric breakdown in thunderstorms and induce lightning strokes. Nonetheless, the most common, ubiquitous artificial source of ELF and VLF waves are power line emissions at 50 or 60 Hz and their harmonics up to more than 3 kHz. Power line radiation can be employed to investigate wave-particle interaction (for a review, see Bullough 1995). In addition to power line radiation, innumerable transmitters ranging from ELF to high frequencies and microwaves are used worldwide. Although their contribution to the global budget of electromagnetic sources is marginal, a variety of transmitters nevertheless can be used to monitor or even modify the ionosphere locally (e.g., Cohen et al. 2010).

Lightning has been triggered by rockets carrying wire bobbins through thunderstorms. The wires unwind during rocket ascent to provide preferential high conductivity paths, where discharges are more likely to occur (e.g., Biagi et al. 2009). In some cases, aircraft flying within strong convective systems can trigger lightning (Mazur 1989; Kito et al., 1985; Olsen et al. 2004; Jerauld et al. 2005). Statistically, each major aircraft is likely struck by lightning once a year and a well-known example is Apollo 12 being struck by lightning soon after takeoff (Rakov and Uman 2007). During a thermonuclear surface burst, five discrete luminous channels were observed to start from the ground or sea surface at a distance of approximately 1 km from the burst point and to subsequently penetrate into the clouds. The peak current of such lightning strokes was between 200 and 600 kA (Uman et al. 1972; Colvin et al. 1987).

Another method of inducing lightning deliberately in thunderstorms is the utilization of ultra short laser pulses for producing laser-induced breakdown in the atmosphere (Ball 1974; Khan et al. 2002). The laser pulse creates a channel of ionized gas through which the lightning stroke would be conducted to the ground. Under conditions of high electric field during two thunderstorms, Kasparian et al. 2008 observed a statistically significant number of electric events synchronized with laser pulses, at the location of the filaments. Laser-triggered lightning is intended to protect rocket launching pads, electric power facilities, airports, and other sensitive targets. Rocket- or laser-triggered events on demand would be also useful for studying ELF-VLF transient wave propagation because synchronization between the lightning stroke and receivers could be much more efficient.

3.3.2 Sources Outside the Cavity

Subsurface Phenomena The connection between surges of electric and magnetic fields and seismic activity has been advocated for a very long time though more convincing correlations are relatively recent (Gokhberg et al. 1982; Warwick et al. 1982). These observations provided new, effective means of investigating short-term precursors of earthquakes. The objective of this sub-section is to summarize possible effects associated with earthquakes and preseismic activity that may impact boundary conditions, expose atmospheric electrical properties, and represent an effective electromagnetic source generation. Reviews about seismo-electromagnetic phenomena have been published by several authors, including Johnston (1989, 1997), Parrot (1995), and Molchanov and Hayakawa (1998). The two leading hypotheses concerning electromagnetic wave generation by earthquakes are direct wave production by rock compression near the focal point (piezoelectric and piezomagnetic effects) and electric discharges induced by redistribution of electric charges in the ground (electrokinetic phenomena). However, not all earthquakes produce measurable electromagnetic emissions.

Surface and subsurface resistivity variations have been noticed during preseismic activity but conductivity changes of $\sim$0.1 mS m^{-1} (Rikitake and Yamazaki 1985; Utada et al. 1998) do not appreciably affect the cavity inner boundary. The magnetic field amplitude in the vicinity ($\sim$7 km) of the Loma Prieta earthquake (M_s6.0), for example, was $\sim$10 nT Hz$^{-1/2}$, at 0.01 Hz, several days before the initial shock (Fraser-Smith et al. 1990). A statistical analysis of several earthquakes in Japan showed that signal intensity increased by as much as 25 dBμ, at 81 kHz (Gokhberg et al. 1982). Measurements of volcano seismic activity at Izu Island, Japan, showed that ULF electric and magnetic fields were on the order of 10 μV m^{-1} Hz$^{-1/2}$ and lower than 0.1 nT before a few earthquakes with subsequent volcanic eruption (Uyeda et al. 2002).

In addition to ground-based measurements, remote sensing techniques from low Earth orbit have been dedicated to earthquake research. A comprehensive statistical analysis related to earthquake precursors onboard the satellite DEMETER shows significant decrease of VLF wave intensity shortly before some intense earthquakes (Nemec et al. 2008, 2009). While a number of other studies have claimed correlation between preseismic activity and ionospheric variability, the topic remains quite controversial due to unclear physical mechanisms linking the two. A more reasonable result, however, is the observation of ionospheric disturbances related to gravity waves triggered by earthquakes (e.g., Tanaka et al. 1984; Wolcott et al. 1984). Incoherent scatter radar observations after earthquakes have shown large amplitude vertical velocity oscillations in the thermosphere. At 300–400 km altitude, velocities as high as 100 ms^{-1} have been measured (Kelley et al. 1985a, 1985b). Fast perturbations of the Earth's crust can induce oscillatory motion in the atmosphere, generating gravity waves that propagate up to the ionosphere.

Although inducing electric and magnetic fields and modifying ground resistivity, earthquake contributions to the global electric circuit and ELF-VLF wave propagation are extremely small. For example, surface conductivity differences due to land/sea dichotomy or precipitation (dry/wet regolith) are much larger than ground resistivity variations triggered by earthquakes. Additionally, average ELF electric fields associated with earthquakes are at least 50 times lower than those attributed to Schumann resonances and IAR signatures. Nevertheless, the ELF-VLF transient response due to earthquakes deserves further investigation. Anomalous effects in Schumann resonance phenomena possibly associated with earthquakes have been claimed (Hayakawa et al. 2005, 2008). Even so, unambiguous cause-effect relations between Schumann resonance and earthquakes are difficult to establish because of the larger natural variability of lightning activity as well as the intrinsic variability of the ionosphere itself. The controversial optical emissions associated with earthquakes also deserves continued investigation, namely for assessing possible implications in the atmospheric electric circuit at a local scale.

Solar and Interplanetary Contributions Unlike the well defined, electrically stable, quasi-static surface, the outer boundary of the surface-ionosphere cavity is fuzzy, thick, irregular, and dynamic. Besides perturbations from below summarized in the previous sections, the dynamics of the outer boundary of the cavity is mostly driven by sources from above such as solar radiation, DC electric fields, neutral winds, tides, cosmic rays, energetic particle precipitation and, to a lesser extent, meteoroids. The role of high energy cosmic rays in troposphere and stratosphere ionization is addressed in the following section. Solar radiation plays a global role in atmosphere ionization with long term variations mostly associated with the solar cycle (e.g., Hunsucker and Hargreaves 2002; Kelley 2009) and short term, localized variations due to transient solar flares and other transitory phenomena. A review of ionospheric processes and dynamics can be found in a companion paper by Pfaff (2011). Here, we concentrate on those phenomena that seem to have the most impact on the variability of the lower edge of the ionosphere.

Energetic particle precipitations include mainly solar proton events and electron precipitation in the auroral zones and from the radiation belts. The mesosphere and stratosphere close to the South Atlantic anomaly are also subject to a diffuse and weak energetic particle precipitation. Solar proton events enhance atmospheric ionization below ∼150 km altitude and have a major effect in the mesosphere and stratosphere at high latitudes where they can drastically modify the conductivity profiles and the electromagnetic wave propagation with the absorption of VLF waves increasing by ∼7 dB (Verronen et al. 2005). Roble et al. (1987) found significant Joule heating of the mesosphere and lower thermosphere during solar proton events. Reagan and Watt (1976) and Reagan et al. (1981) analyze satellite and radar measurements gathered during solar particle events and investigate direct effects in stratospheric ozone production. Solar proton events induce particle ionization in the mesosphere and upper stratosphere, enhancing the concentrations of short-lived HO_x (H, OH, HO_2) and long-lived NO_x (N, NO, NO_2) constituents. These molecules are then responsible for ozone depletion, creating a polar ozone cavity. After the event, ozone concentrations decreasing in the auroral region is 46%, 16%, and 4% at altitudes of 49.5, 41, and 32 km, respectively, reducing the total columnar ozone concentration by $\sim$2%. Jackman et al. (2001) address the impact of solar flares in HO_x and NO_x concentration. Their measurements indicate short-term ($\sim$1 day) middle mesospheric ozone decreases of over 70% caused by short-lived HO_x during the event with a longer-term ($\sim$1 week) upper stratospheric ozone depletion of up to 9% caused by longer-lived NO_x. Stiller et al. (2005) discuss the effects of HNO_3 enhancements in stratosphere ion cluster chemistry due to NO_x molecules produced in the mesosphere and continuously transported downward. McConnell and Jin (2008) further investigate the impact of solar variability in not only the mesosphere and stratosphere but also down to the troposphere.

The outer radiation belt consists mostly of electrons that are injected inward by radial diffusion from the geomagnetic tail following geomagnetic storms (Shprits and Thorne 2004). Investigation of radiation belt processes is useful for lightning-related VLF wave propagation and for aeronomy studies, mainly in the mesosphere. Callis et al. (1991) discuss the effects of relativistic electron precipitation in the D-region and long-term impact on stratospheric odd nitrogen levels. Results bear evidence for a strong connection between solar variability, the state of the magnetosphere, and the chemical climatological state of the middle and lower atmosphere.

Assessment of radiation belt particle precipitation is useful not only for aeronomy but also for lightning-related wave propagation research. Imhof et al. (1983a, 1983b) report on the precipitation of energetic electrons from the radiation belts by the controlled injection of VLF radio waves from the ground. Voss et al. (1984) present satellite measurements of electron precipitation by lightning. They find a one-to-one correlation between ground-based measurements of sferics and whistlers and precipitating energetic electrons detected onboard satellites. Burgess and Inan (1993) discuss the role of ducted whistlers in the precipitation loss of radiation belt electrons from the identification of ionospheric disturbances on subionospheric low frequency signals ('Trimpi events'). The Trimpi event is a phenomenon of amplitude or phase perturbation of subionospheric VLF signal propagation produced at the time of reception of a magnetospheric whistler, which scatters particles and induces D-region electron density enhancements (e.g., Carpenter et al. 1984). For electron energies of $\sim$100 keV, ducted whistlers may contribute as significantly as hiss to radiation belt equilibrium. Results also support the idea of ducted whistlers being responsible for burst electron precipitation. Rodger et al. (2003) address the significance of lightning-generated whistlers to the inner radiation belt particle lifetimes. They estimate the contribution of whistler-induced electron precipitation for radiation belt loss rates, comparing such contributions with those associated with VLF transmitters and plasmaspheric hiss sources. Considering energetic electron precipitation fluxes driven by lightning rather than VLF wave perturbation observations, Rodger et al. (2005, 2007) provide regional and global estimates of energy deposition into the atmosphere, and study implications for ionization levels and neutral chemistry. They find that whistler-induced electron precipitation is never a significant source of ionization in the lower ionosphere for any location or altitude during the day. During nighttime conditions, however, ionization in the lower and upper mesosphere is mainly due to galactic cosmic rays and whistler-induced electron precipitation, respectively.

Horne et al. (2005a, 2005b) address the wave acceleration processes in the radiation belts. Characterization of the radiation belt particle precipitation in the low ionosphere is therefore useful for understanding ionization processes in the mesosphere.

An intricate, interesting example connecting various layers of the Earth's gaseous envelope is that associated with wave-particle interaction. Particle precipitation from the radiation belts is among the most relevant phenomena to investigate the tropospheric-ionospheric coupling. Lightning produces whistlers which subsequently interact with cyclotron resonant radiation belt electrons, leading to particle acceleration and subsequent precipitation into the atmosphere. Particle precipitation then modifies medium conductivity, which affects wave propagation conditions.

During periods of geomagnetic storms, perturbations can be observed from the magnetosphere down to the surface. For example, solar proton events cause polar cap absorption in the lower ionosphere and increase electric conductivity in the high latitude stratosphere. Lastovicka (1996) investigates effects of geomagnetic storms in the lower ionosphere, middle atmosphere, and troposphere. Frahm et al. (1997) show that diffuse aurora is a significant source of ionization in the middle atmosphere because energetic electrons can cause excitation, dissociation, ionization, and heating of neutrals. Most energy deposition, e.g., by ionization, results from direct electron impact and bremsstrahlung X-rays above and below ~50 km, respectively. In fact, in diffuse aurora and during geomagnetic storms, ionization from the bremsstrahlung component exceeds that due to the galactic cosmic ray background at altitudes as low as 30 km. These effects usually enhance electric conductivity in the mesosphere and stratosphere, and increase radio wave absorption in the medium and high frequency range.

Meteoroid and space debris ablation in the atmosphere produces important transient effects, including visible emission, radio noise, and acoustic transients known as electrophonics. Meteor showers, for example, are known to be related to all these processes including reports from long time ago (Blagdon 1784). Electrophonics, a phenomenon resembling to microphonics reversed effects, is the production of audible noise through direct conversion of very low frequency electromagnetic signals to sound through transduction phenomena (Keay 1980). In fact, sometimes, the acoustic signal on the ground precedes the corresponding optical flash observed during meteoroid ablation, which occurs in the D-region at altitudes of 80–120 km, therefore suggesting the generation of electromagnetic noise. Distinction between lightning and meteor shower transients is possible because the spectral morphology is quite different. Lightning and meteor spectra show a peak at about 5 and 0.5 kHz, respectively. Additionally, correlation between optical flashes and radio noise is significant (Price and Blum 2000). Meteor phenomena and its detection techniques are reviewed in detail by Ceplecha et al. (1998), including radar and spectroscopic observations, trajectory geometry analysis, dynamics and ablation processes, and hazard assessment. In the context of this review, we shall briefly emphasize two major concepts: (i) meteor ablation delivers high concentrations of metallic ions that then locally modify the chemistry of the D-region during brief periods of time, and thus can be investigated with radar and spectroscopy techniques; (ii) meteor ablation produces ELF electromagnetic wave signatures, which can be combined with optical flash measurements to investigate meteoroid activity, mainly during meteor outbursts.

Galactic Cosmic Rays Cosmic rays are highly-energetic charged subatomic particles arriving from outer space, from solar, galactic, and extragalactic origin. Cosmic rays include protons (~89%), alpha particles (~9%), and heavier nuclei (~1%) spanning a wide range of energies, from MeV to EeV and even higher. Electrons constitute the remaining 1% of cosmic rays. Since their discovery about 100 years ago, cosmic rays have been studied from multiple, complementary perspectives dealing with astrophysics of compact objects, nucleosynthesis and star evolution, particle physics, radioisotope science, solar dynamics, interplanetary medium, magnetospheric processes, aeronomy, atmospheric electricity, radiation hazard to humans, and radiocarbon dating (e.g., Bieber et al. 2000; Grieder 2001; Miroshnichenko 2001; Schlickeiser 2002; Stanev 2009; Dorman 2009). Here, we summarize the main implications for the surface-ionosphere cavity/waveguide, i.e., processes related with atmospheric electricity and aeronomy.

Interaction between highly-energetic cosmic rays and the Earth's gaseous envelope produces charged secondary cosmic ray components, which may involve particle multi-cascading of positron-electron pairs, protons, alpha particles, muons, photons, neutrinos, etc. Incident hadrons are subjected to strong interactions when colliding with atmospheric nuclei, mainly nitrogen and oxygen, producing nuclear spallation (Grieder 2001). Since cosmic rays are electrically charged, high-energy particles interaction with the Earth environment results from not only atmospheric density variation with altitude but also geomagnetic field and solar wind dynamics. Consequently, cosmic ray flux shows an 11 year modulation due to the solar cycle, and for certain mechanisms also a 22 year periodicity. In addition to the variation with the solar cycle, cosmic rays intensity may drop suddenly by 15–30% due to sudden changes in solar activity (timescale of few minutes to several hours) followed by a gradual recovery (several hours to a few days) to the former intensity level; this phenomenon is known as Forbush decrease (Forbush 1937, 1938, 1958) and is related to magnetic shock effects in the Earth magnetosphere due to coronal mass ejections. An opposite effect, where cosmic ray intensity increases up to 2–3 times the steady flux, is sometimes triggered by solar flares. Most galactic cosmic rays have energies too low to penetrate straight deep in the ionosphere and are channeled by the geomagnetic field toward the poles, inducing particle precipitation in the polar cap regions and contributing to aurora phenomena (Fisk et al. 1998). In addition to changes related to solar activity, a long term variation of high energy cosmic rays has been proposed to tentatively explain a 62 My periodicity in terrestrial fossil biodiversity. High energy cosmic rays also induced air showers which ionize the atmosphere leading to changes in atmospheric chemistry and microphysical processes that can lead to cloud formation, including a low altitude cloud cover (e.g., Mellot et al. 2010). Similarly, Carslaw et al. (2002) and Kirkby (2007) discuss associations between climate trends and cosmic ray variations at a much shorter timescale. Observations suggest that cloud cover may be influenced by cosmic rays and two classes of microphysical mechanisms have been proposed to relate cosmic rays to clouds: an influence of cosmic rays on (i) the production of cloud condensation nuclei and (ii) directly on the global electrical circuit.

Since cosmic ray effects in the Earth's environment are a function of atmospheric density and magnetic field, ionization rates show specific variations with altitude, magnetic latitude, longitude, azimuth, and the east-west direction. The relative intensity of cosmic rays at high northern latitudes is 50% larger compared to the equator and to southern hemisphere latitudes because of the geomagnetic cutoff imposed by the geomagnetic field to cosmic ray energies up to ~15 GeV. Although smaller, there is a 15% longitudinal modulation in cosmic ray relative intensity (+15% at 60°W and −15% at 100°E) introduced by the geomagnetic field asymmetry with respect to the equator and the South Atlantic anomaly. The geomagnetic field also drives an azimuthal response because of a cosmic ray zenith angle, θ, intensity dependence with atmospheric depth ($\propto \cos^7(\theta)$) and a dipolar shape of the field. Additionally, because of geomagnetic field polarity and positive charge predominance in cosmic rays, an east-west asymmetry also develops. As a result of cross section increasing with atmospheric depth, the particle flux due to cosmic rays increases from the magnetosphere down to the stratosphere, reaching a maximum at a height of about 20 km, which is called Pfotzer maximum (e.g., Pfotzer 1936). Unlike ionospheric ionization due to solar radiation, this cosmic ray bombardment favors the formation of atmospheric ions in the stratosphere (for a review, see Tinsley 2008).

Cosmic rays contribute to ion-pair production down to the surface, and even deeper for the most energetic particles. Another important source of ionization is provided by ground radioactivity that results from alpha, beta, and gamma radiation. One particular source is radioactive radon and daughter products emanating from the ground. This ion-pair production rate shows a 0.2 and 2–3 km scale height for ^{220}Rn and ^{222}Rn, respectively. Radioactive gas released from the ground, namely during earthquakes, increases the conductivity in the atmospheric boundary layer region by as much as a factor of ∼2. Ionization rate from radioactive gases near the surface is up to $\sim 10\ \mathrm{cm^{-3}\,s^{-1}}$ (Usoskin et al. 2009).

3.4 The AC and DC Global Electric Circuits

The concept of a global electric circuit was first proposed by Wilson (1906, 1920) to explain the existence of a vertical atmospheric electric field. The ground and ionosphere are usually regarded as equipotential surfaces and consequently define the global electric circuit boundaries. However, considering the ionosphere a perfect conductor is often too stringent. Since the ionosphere is dynamic and heterogeneous, establishing the altitude for an equipotential condition is not straightforward. The atmosphere, where the conductivity is much lower, is considered an imperfect dielectric, i.e., a medium with losses. The global electric circuit can be represented by a set of generators, resistors, and capacitors to simulate electric sources and medium behavior (e.g., Ogawa 1985; Rycroft et al. 2000). Differentiation between AC and DC electric circuits responses is determined from the relaxation time of the system, i.e., when either conduction or displacement current processes are dominant (Greifinger and Greifinger 1978). The main characteristics of the global circuit include: (i) near-surface conductivity from $\sim 10^{-6}\ \mathrm{Sm^{-1}}$ for low contaminated ice and dry, poorly conducting rocks to $\sim 10^{-2}\ \mathrm{Sm^{-1}}$ for clay or wet limestone and $\sim 1\ \mathrm{Sm^{-1}}$ for sea water; (ii) ionospheric conductivity $\sim 10^{-6}\ \mathrm{Sm^{-1}}$ in the D-region; (iii) vertical DC electric field varying between $\sim 100\ \mathrm{V\,m^{-1}}$ near the surface and a very small fraction of $1\ \mathrm{V\,m^{-1}}$ at 80 km (changing significantly with latitude and day-night dichotomy); (iv) upward currents predominantly originating from thunderstorms; (v) fair weather downward current density $\sim 1\ \mathrm{pA\,m^{-2}}$; (vi) and a 200–300 kV potential difference between the ionosphere and the surface (e.g., Bering et al. 1998; Rycroft et al. 2008). Thunderstorm activity and associated phenomena drive a DC current of ∼1 kA, and each negative and positive lightning stroke carries an average AC current of ∼30 and ∼300 kA in the circuit, respectively, in less than 0.1 ms. Singh et al. (2005) present an extensive, documented table with physical parameters of the global electric circuit.

Thunderstorms, namely those related to mesoscale convective systems, introduce significant perturbations to the global electric circuit. Strong convection produces charge separation in clouds and induces strong electric fields locally. The solar wind affects not only cloud microphysics, temperature, and dynamics in the troposphere but also the global electric circuit (e.g., Tinsley 2000). In general, aerosol formation drives ion attachment to larger particles with subsequent decreasing of ion mobility and conductivity. Whenever an atmospheric electric field threshold is reached—air electrical breakdown is $\sim 15\ \mathrm{kV\,cm^{-1}}$, depending of altitude and atmospheric conditions, but the actual threshold field is smaller due to mechanisms poorly understood—a lightning discharge occurs and broadband radiation is emitted (Rakov and Uman 2007). Under such conditions, atmospheric conductivity varies significantly within the cloud and even above the thunderstorm (e.g., Holzworth et al. 1985). Local, intense electric fields on the order of 1–5 kV m^{-1} were also observed during snow storms in Antarctica (Frank-Kamenetsky et al. 2010). Rycroft et al. (2007) argue that lightning activity contributes only 1% to maintain the high potential of the ionosphere. Specifically, moderate negative/positive lightning discharges would contribute to increase/decrease the ionospheric potential by 0.0013%/0.014%. Analyzing intense positive lightning discharges, Füllekrug (2004) obtains similar results for ~1 min time scales. The total net upward current to the ionosphere due to lightning would be only ~20 A, contrasting with the ~1000 A contribution from conduction and convection currents associated to electrified shower clouds and charge separation within thunderclouds. The time scale for ionospheric potential recovery is ~250 s, similar to the relaxation time for the global circuit (Rycroft et al. 2000, 2008). Analyzing electric field measurements corresponding to wide time spans covering most of the twentieth century, Harrison (2002) found a secular declining of the atmospheric DC electric field. The ~25% reduction of the electric field over most of that period is attributed to a decreasing of cosmic rays, which would increase the coupling resistance between the top of thunderstorms and the ionosphere. This secular variation would therefore imply modifications in global electric circuit and fair weather parameters, including ion concentrations and aerosol electrification. However, other parameters, namely surface temperature, also play a crucial role. For example, Price (2000) predicts that a 1% increase in global surface temperature may result in a ~20% increase in the ionospheric potential. The influence of cosmic rays on the global electric circuit and ramifications in climate variability in geologic timescales has been addressed from a perspective of cloud-aerosol processes (e.g., Kirkby 2007). Cosmic rays may significantly influence the production of cloud condensation nuclei and, consequently, the electric circuit at a global scale, namely by modifying the downward current density. Harrison and Märcz (2007) analyze surface atmospheric electricity data at daily resolution to investigate heliospheric and cosmic ray effects on the global circuit and find annual, semi-annual, and 1.7 year periodicities in the DC electric field magnitude. Interestingly, the latter is clearest during fair weather and reduced, often absent, when poor local weather conditions prevail.

In addition to lightning and related tropospheric phenomena, the role of transient luminous events in the global electric circuit dynamics has been addressed concurrently with theoretical and experimental developments in sprites, blue jets, and elves generation mechanisms. Füllekrug and Rycroft (2006) assess the role of sprites to the global electric circuit and estimate that sprites individual contribution is lower than 44 mV m^{-1}, and Cummer et al. (2009) report a significant charge transfer of −144 C to the lower ionosphere in a gigantic jet event. Positive blue jets contribute to the charging of the global electric circuit, while negative gigantic jets discharge the circuit (Krehbiel et al. 2008). Pasko (2010) discusses effects of sprite and jet discharges on chemical processes and the global electric circuit. Detailed assessments of global electric circuit dynamics therefore require reliable data on generators, namely thunderstorm, lightning, and transient luminous events, as well as external drivers of the properties of the medium such as cosmic rays and solar radiation.

Several reviews dedicated to the global electric circuit and connections to atmospheric electrification processes, cosmic ray and solar irradiance effects, cloud microphysics, and climate are available, offering detailed discussions of the subject (Harrison 2004; Singh et al. 2004; Tinsley et al. 2007; Rycroft et al. 2008). Williams (2010) reviews the origin and context of Wilson's pioneering ideas in atmospheric electrification and electron runaway in thunderclouds, and a companion article by Rycroft and Harrison (2011) provides a systematic analysis of the global electric circuit in troposphere-ionosphere coupling mechanisms.

4 Modeling

Accurate assessments of ULF-VLF wave propagation in the surface-ionosphere cavity/waveguide require utilization of appropriate formalism, boundary conditions, and medium parameterization. Several empirical models are available to derive medium properties, namely electron and ion density, collision frequency, and geomagnetic field distributions. While the MHD formalism is suitable for describing the ionospheric waveguide, the ionospheric Alfvén resonator, and geomagnetic pulsations, Maxwell equations are necessary for solving ELF-VLF wave propagation in the cavity. Here we review analytical and numerical approaches to solve Maxwell and MHD equations and specific phenomenological models used for deriving medium dispersion relations. Characterization of electromagnetic sources or medium properties usually benefit from inverse problem techniques.

The most familiar, straightforward approximation for solving ELF wave propagation in the surface-ionosphere cavity considers an atmospheric conductivity profile with two scale heights. The scale height transition occurs when the displacement current is commensurate with the conduction current in the medium. Greifinger and Greifinger (1978) employ the knee-model approximation and cylindrical symmetry to determine the electromagnetic field distribution and eigenfrequencies of the cavity. Sentman (1990) utilizes a similar approach employing spherical symmetry for eigenmode characterization. Mushtak and Williams (2002) extend the two-scale height approach to determine modal frequencies, wave phase velocities, and wave attenuation with improved accuracy. Nickolaenko and Hayakawa (2002) review these and asymmetric approximations of the cavity as well. Analytical approximations provide valuable results for assessing propagation and resonance mechanisms but are insufficient for obtaining accurate estimates, particularly if day-night asymmetry or polar heterogeneity is incorporated in the model. A similar situation occurs in planetary environments because thick cavities imply strong coupling between the longitudinal and transverse modes (Simões et al. 2007) and because refraction phenomena may occur in the lower atmosphere (Simões et al. 2008b), reducing model accuracy.

Modeling of the ionospheric waveguide (magnetosonic mode) and ionospheric Alfvén resonator (Shear Alfvén mode) implies solving MHD equations with specific parameterization. Although the magnetosonic and shear Alfvén modes are often modeled separately, coupling between modes may be important. Consistent analytical approaches are available for modeling the ionospheric waveguide and ionospheric Alfvén resonator. Greifinger and Greifinger (1968) address the theory of MHD wave propagation in the ionospheric waveguide, the relevance for investigating geomagnetic pulsations, namely "pearls", and implications for ionospheric ducting phenomena about the Alfvén velocity minimum near the F-peak. Polyakov and Rapoport (1981) discuss a similar theory for wave propagation in the Alfvén shear mode, as opposed to the magnetosonic mode which is connected to ionospheric ducting. Far from the poles, the ionospheric waveguide and ionospheric Alfvén resonator correspond approximately to horizontal and vertical wave propagation in the ionosphere, respectively. When losses are taken into account, the Hall conductivity couples the propagation modes and a more general formulation is employed (Lysak 1999).

Comparable formulations are used for investigating geomagnetic pulsations of lower frequency, whistler-mode propagation, and wave-particle interaction related to radiation belt electron precipitation. Carpenter and Anderson (1992) discuss an empirical model of equatorial electron density derived from satellite-based whistler signatures. The model is specially suited for investigating whistler-mode propagation, lightning, and radiation belt particles from whistler-mode dispersion measurements. Yagova et al. (1999) study diurnal variations of the magnetospheric Alfvén resonator parameters to characterize mid latitude Pc3–4 pulsations. Bortnik et al. (2006) present a wave-particle interaction model to assess temporal signatures of radiation belt particle precipitation induced by lightning-generated whistlers reflected in the magnetosphere. Although not strictly matching the scope of the present work, these models are useful for understanding the coupling between the ionosphere and the magnetosphere because the troposphere-ionosphere coupling is also driven from above, namely in the auroral region where particle precipitation happens.

A variety of time and frequency domain techniques is available to solve MHD and Maxwell equations, namely Finite Difference Time Domain (FDTD), Finite Difference Frequency Domain (FDFD), Finite Element (FE), and Transmission Line Matrix (TLM) analysis. Most techniques can be used for eigenvalue, harmonic propagation, and transient problems. While TLM provides the simplest approach, the FE technique offers robust solutions for intricate geometries. Simpson and Taflove (2007) present a review of progress in FDTD modeling of subionospheric propagation, focusing in ELF-VLF wave propagation related to lightning sources and radiation, remote sensing of localized ionospheric anomalies, and Schumann resonances. Cummer (1997) discusses FDTD methods specially suited for solving wave propagation in isotropic cold plasma and Young (1994) compares both FDTD and FDFD techniques aiming a similar purpose. For modeling field interaction and propagation in anisotropic media with generalized permittivity and permeability tensor formulation, Rappaport and McCartin (1991) discuss FDFD using unstructured mesh methods to solve ionospheric scattering problems. Simões et al. (2007, 2008b, 2008c) discuss FE methods applied to ELF-VLF wave propagation in the surface-ionosphere cavity/waveguide, namely for solving eigenvalue, harmonic propagation, and transient problems related to lightning morphology and distribution, sferics, and Schumann resonances. Christopoulos (1995) describes the TLM model in detail to solve 3D wave propagation problems, emphasizing analogies between network circuits and physical systems; distributed circuits are used to define propagation conditions and medium properties. Although FE modeling can be applied to more intricate domains than finite difference, which is more robust than transmission line methods, the TLM analysis is faster and sometimes chosen when processing time and memory constraints are crucial.

The characterization of the medium from field measurements, i.e., by solving the inverse problem, is invaluable both for troposphere and ionosphere modeling. The inverse problem can be also used for source description, namely lightning distribution in the cavity. Whereas seeking inverse problem analytical solutions is commonly impracticable, numerical modeling may provide meaningful results. He et al. (1996) show an optimization approach to solve a 1D electromagnetic inverse problem for reconstructing medium characteristics of stratified dispersive and dissipative slabs. Gustafsson and He (2000) present a similar approach to solve 2D electromagnetic inverse problems and simultaneous reconstruction of the permittivity, permeability, and conductivity of the medium by means of the iterative conjugate gradient algorithm. To investigate mode conversion in weakly magnetized plasma, Yin and Ashour-Abdalla (1999) describe the wave energy coupling and mode structure of electrostatic Langmuir and upper-hybrid branches, and electromagnetic ordinary, extraordinary, and Z modes. Sharapov et al. (2001) apply the inverse problem approach to solve MHD equations of toroidal Alfvén eigenmodes in tokamaks; similar techniques can be applied to toroidal geomagnetic pulsations (Pc5 waves) and a modified approach is useful for inferring the ion density profile from ionospheric Alfvén resonator eigenfrequency measurements. Miura (1999) proposes an MHD inverse problem to investigate phenomena related to the magnetopause Kelvin-Helmholtz instability. Ando et al. (2005) employ finite difference analyses of Schumann resonance data for lightning distribution reconstruction. Although finding reliable, accurate solutions is not straightforward, inverse problem techniques can successfully be applied to investigate source and medium characteristics from electromagnetic field measurements.

Consistent solutions of wave propagation problems in the surface-ionosphere cavity/waveguide require accurate medium parameterization. The geomagnetic field of the Earth environment can be computed from the International Geomagnetic Reference Field (IGRF). The IGRF model consists of a spherical harmonics expansion of the geomagnetic potential, whose coefficients are determined empirically (Finlay et al. 2010). The model provides the 3D geomagnetic field distribution from the core to the magnetosphere, including the secular variation rate. Another option to compute the geomagnetic field is utilizing the POtsdam Magnetic Model of the Earth (POMME), which also evaluates the magnetic field of the ionosphere and magnetosphere (Maus et al. 2006).

The collision frequencies between electrons, ions, and neutrals can be calculated from the International Standard Atmosphere (ISA), which provides the thermodynamic properties of the atmosphere, namely pressure, temperature, and density as function of altitude (Lide et al. 2010). A more elaborate approach is offered by the Naval Research Laboratory Mass Spectrometer Incoherent Scatter radar (NRLMSIS-E-00) empirical model that provides the composition, temperature, and density distribution of neutrals from ground to the thermosphere (Picone et al. 2002).

There are three main classes of models that can be used to calculate the electron density and ion composition of the ionosphere: empirical, physics-based, and assimilative. The International Reference Ionosphere (IRI) empirical model (Bilitza 2001, 2009) is the international standard for computing both electron and ion 3D distributions in the ionosphere, including electron density and temperature, and ion density, temperature, and composition. The model may also be used to provide estimates of equatorial vertical ion drift, total electron content, F-peak density, and spread-F probability. For given location, time, and date, IRI provides monthly averages of electron and ion parameterization in the altitude range from 60 to 2000 km.

There are several options for physics-based models. The US Naval Research Laboratory has developed SAMI (Sami is Another Model of the Ionosphere) under 2D and 3D formulations (SAMI2 and SAMI3, respectively). The SAMI3 code provides field line integrated conductance and is especially suited for global circulation models, namely computing DC electric fields and ion drifts at low- and mid-latitude. SAMI3 evaluates the plasma dynamics and chemistry of seven ion species in the altitude range 80–20,000 km. The temperatures of the electron population and of each component ion species are computed; the total electron content is also available. In addition, the $\mathbf{E} \times \mathbf{B}$ drift motion of the plasma is included for zonal and meridional electric fields, where $\mathbf{E}$ and $\mathbf{B}$ are the DC electric and magnetic fields (Huba et al. 2000, 2005; Krall et al. 2009). The Coupled Thermosphere-Ionosphere Plasmasphere Electrodynamics model (CTIPe) consists of four distinct component models that are fully coupled in energy, momentum, and continuity (Millward et al. 2001). The four components include a global thermosphere model, a high-latitude model, an ionosphere/plasmasphere model for the mid to low latitudes, and a model of the electrodynamic coupling between the ionosphere and thermosphere at low latitudes. CTIPe provides ion and electron densities from 140–2000 km, as well as the height and density of the F2 peak. The US Air Force Research Laboratory has been developing a Parameterized Ionospheric Model (PIM) that is capable of generating global electron number density as well as maps of total electron content. The PIM code also provides the heights of the E, F1, and F2 peaks and their respective plasma frequencies as a function of geophysical parameterization, namely solar and geomagnetic activity indices (Daniell et al. 1995; Retterer et al. 2005). The Utah State University Global Assimilation of Ionospheric Measurements (USU-GAIM) is a physics-based ionosphere-plasmasphere-polar wind model with a Kalman filter to assimilate real-time electron density measurements from a variety of ground-based and in situ platforms, including GPS, digisondes, and the DMSP satellites (Schunk et al. 2004).

Investigation of troposphere-ionosphere coupling mechanisms requires a fundamental comprehension of not only what processes take place in the Earth's gaseous envelope but how the properties of the medium are affected by such processes. Therefore, theoretical and phenomenological models, analytical and numerical approaches, direct and inverse problem solutions contribute to a better understanding of the Earth's environment because physical processes and the response of the medium are intrinsically coupled.

5 Discussion

Alfvén Waves The ionospheric magnetosonic waveguide and the ionospheric Alfvén resonator contribute to investigating connections among the troposphere, mesosphere, and thermosphere from their ULF wave signatures, namely geomagnetic pulsations. Specifically, for IAR studies, association between tropospheric and ionospheric phenomena is dual because of implications for wave propagation and electromagnetic source classification. First, we discuss relevance for medium characterization. Lysak (1999, 2004) investigates the dependence of Alfvén wave propagation on ionospheric parameters, namely the Pedersen and Hall conductivities. Ionospheric Alfvén resonator ground signatures are strongest when the Hall conductivity is greater than the Pedersen conductivity. The fundamental IAR mode is found not to couple to the magnetosonic mode, while higher harmonics of the wave are readily ducted through the ionospheric waveguide. In addition to the troposphere-ionosphere connection, investigation of IAR also contributes to an understanding of the magnetosphere-ionosphere coupling, because the resonator top boundary is sometimes located several thousand kilometers above the F-peak. Besides, this approach is useful to modeling ducted Pc1 oscillations that can propagate thousands of kilometers in the ionosphere. Surkov et al. (2005, 2006) investigate the IAR excitation mechanism due to nearby thunderstorms, and suggest lightning is the major source driving the resonator. Sharp impulses in dynamic spectrograms recorded at high magnetic latitude are in a one-to-one correspondence with IAR signatures, suggesting nearby lightning discharges are followed by transient excitation of the resonator. Additionally, they also claim that the contribution from nearby thunderstorms prevails over remote, e.g., tropical, thunderstorm activity. This result is somewhat expected due to the local nature of IAR phenomena. Lightning distribution is expected to have a more global impact in the magnetosonic mode because of ducted propagation between the northern and southern hemispheres. Demekhov et al. (2000) discuss the connection between IAR and Pc1 waves, namely the role of geomagnetic pulsations in resonator excitation. Understanding under what conditions Pc1 waves are a cause or a consequence of IAR action would present invaluable information for troposphere-ionosphere-magnetosphere coupling research. The resonator can therefore act as a filter for unstructured large band electromagnetic noise produced by lightning or as a coupled oscillator for geomagnetic pulsations. A combination of several scenarios is possible: (i) lightning may pump IAR modes directly; (ii) transient luminous events may excite the resonator; (iii) magnetospheric Pc1 waves may inject energy in the resonator; (iv) in situ ionospheric or magnetospheric processes related to the Kelvin-Helmholtz instability may produce medium turbulence and induce wave filtration in the resonator; (v) whistler-induced electron precipitation may generate electrostatic and electromagnetic noise in the lower boundary of the resonator. Either way, resonator characterization is invaluable for estimating ionospheric electron and ion density profiles.

Classification of multiple eigenfrequencies and Q-factors of the resonator provides fundamental inputs for solving the inverse problem and eventually inferring density profiles that best fit observed spectra.

ELF Waves and Schumann Resonance Global, continuous monitoring of ELF waves has been used for addressing phenomena related to not only lightning distribution but also variability of the lower ionosphere. Some of the most challenging issues are related to quantifying regional contributions to the total lightning budget. Since lightning is closely related to thunderstorm electrification processes, monitoring Schumann resonance spectral variability—amplitude, frequency, and Q-factor—contributes to investigating atmospheric electricity at a global scale, including possible connections to climate trends. Identification of correlations between ELF spectra, and lightning activity as well, and atmospheric thermodynamic parameters is valuable for understanding possible links between hydrodynamic and electrodynamic processes. Although extremely demanding, solving the ELF inverse problem to characterize electromagnetic sources from Schumann resonance spectra provides important means for unambiguous identification of trends between lightning and atmospheric parameters such as temperature and water vapor. A similar concept can be applied to investigate connections between atmospheric electricity and lower ionosphere variability related to plasma depletions, scintillations, magnetospheric phenomena, and solar activity.

Schumann Resonance and Ionospheric Variability Unambiguous detection of Schumann resonance signatures onboard the C/NOFS satellite, in the altitude range 400–850 km at equatorial latitudes, offers new means for investigating ELF wave propagation in the ionosphere (Simões et al. 2011a). Although an increase in the ambient irregularities is frequently observed at night and Schumann resonances are sometimes buried in the noise, nighttime measurements offer better propagation conditions because of significantly lower wave attenuation in the ionosphere compared to daytime. Besides, significant correlation between Schumann resonance amplitude and lightning activity is plausible. ELF-flashes and ELF-bursts (Q-bursts) have been detected from orbit. C/NOFS measurements also show that Schumann resonance electric field is approximately perpendicular to the geomagnetic field. Although propagation in the whistler and extraordinary modes are perhaps compatible with C/NOFS observations, further analysis is necessary to understand the leakage mechanism comprehensively. The leakage mechanism that allows ELF waves to escape the surface-ionosphere cavity is not understood thoroughly because analytical models currently available consider Schumann resonance modes confined to the cavity, i.e., between the surface and $\sim$100 km. It is nevertheless expected that numerical modeling and ongoing C/NOFS measurements contribute to consolidate present knowledge of ELF wave propagation in the ionosphere. Schumann resonance signatures remote detection provides invaluable information for atmospheric electricity multidisciplinary analyses: (i) ELF measurements can be combined with lightning optical detectors for nighttime operations; (ii) ELF measurements can be combined with optical cameras and spectrometers aiming thunderstorm electrification process studies; (iii) satellite measurements are complementary to ground stations because electric fields characteristics in the ionosphere are different due to anisotropy, providing additional constraints to solve wave propagation inverse problems, i.e., assessing electromagnetic sources from fields distribution; (iv) remote, concurrent ELF and DC field measurements contribute to the investigation of the global electric circuit, namely parameterization of energy budget and processes relaxation time; and (v) addressing ionospheric variability taking into account perturbations both from below (surface, troposphere, and stratosphere) and above (magnetosphere and beyond). Indeed, characterization of ELF wave propagation through the ionosphere contributes to not only an understanding of the local plasma dynamics but an assessment of the electric environment below. Moreover, Schumann resonance measurements from orbit offer supplementary means for troposphere-ionosphere coupling mechanisms research. Future Earth observation missions may explore new remote sensing capabilities employing ELF measurements for addressing atmospheric electricity and its connections to weather.

Lightning Effects on Ionospheric Irregularities and Dynamics It is well established that not only lightning produces shock waves audible as thunder but thunderstorms can trigger hydrodynamic waves. Atmospheric gravity waves can be generated by various surface and tropospheric phenomena, namely weather related—thunderstorms, hurricanes, tornadoes—impulsive auroral zone momentum injection, earthquakes, tsunamis, and volcanic eruptions, whose effects are sometimes identified in the ionosphere (e.g., see Pilipenko 2011 for a review of impulsive coupling between the atmosphere and ionosphere/magnetosphere). For example, Taylor and Hapgood (1988) identify a thunderstorm as a source of short period gravity waves in the upper atmospheric nightglow emissions, and Kelley et al. (1985b) report large-amplitude thermospheric oscillations induced by an earthquake. It is not certain, however, whether lightning or transient luminous events could trigger strong hydrodynamic effects in the mesosphere or thermosphere. On the one hand, lightning is usually associated to thunderstorms, which induce strong atmospheric convection that prevails over lightning shock wave disturbances. On the other hand, establishing causality between lightning acoustic and electromagnetic transients far from the source is not straightforward. Although connections between thermospheric gravity waves and lightning have been suggested (e.g., Chiu et al. 1979; Liao et al. 1989; Kaladze et al. 2008), a positive correlation is possibly due to deep convective activity in thunderstorms. Davis and Johnson (2005) identify a lightning-induced intensification of sporadic-E. Because no ionospheric response to low-pressure systems without lightning is detected, they conclude that sporadic-E localized intensification can be attributed to lightning. According to them, medium density enhancement could be explained by hydrodynamic or electromagnetic processes, i.e., vertically propagating gravity waves or vertical electric discharges, respectively. Various electromagnetic effects of lightning in the mesosphere, thermosphere, and beyond are well documented, namely mesospheric heating and whistler-induced particle precipitation, but hydrodynamic effects triggered by lightning are more difficult to assess. Nevertheless, a phenomenon known as explosive spread-F deserves to be discussed in detail. Spread-F is a term used to identify large ionospheric plasma depletions often observed in the equatorial regions during nighttime; spread-F is related to hydrodynamic phenomena started by the Rayleigh-Taylor instability at the bottom side of the ionospheric F-peak (Kelley 2009). Whilst conducting equatorial backscatter measurements at the Jicamarca Radio Observatory in Peru, Woodman and La Hoz (1976) and Woodman and Kudeki (1984) observed a sudden, sharp increasing in spread-F signatures sometimes accompanied by transients in a high frequency receiver. They noticed radar echo measurements attributed to spread-F occurring simultaneously with transients in a receiver. The transients in the receiver were attributed to lightning strokes. Because of morphological structure, those bursts of backscattered radar are termed explosive spread-F and are associated to plasma irregularities at altitudes near 250 km. Woodman and Kudeki (1984) and Kelley et al. (1984) suggest a causal relationship between lightning and explosive spread-F since lightning-produced transient electric fields can penetrate in the ionosphere, drive the instability mechanism, and trigger spread-F formation locally. The latter study proposes that transient electric fields cause an immediate $\mathbf{E} \times \mathbf{B}$ drift of electrons while ions remain mostly unaffected. The previous assumption is valid when the time constant of the transient is longer than the electron gyroperiod but shorter than the ion gyroperiod. The resulting differential motion generates a current that may induce medium instabilities. Because tropospheric-ionospheric coupling assessments are useful for multidisciplinary analyses, namely medium instabilities seeding and irregularities generation, measurements combining hydrodynamic and electromagnetic techniques would be valuable for validating acoustic and electromagnetic connections between lightning and spread-F, and improving our understanding of the interaction mechanisms.

ELF Waves and Tropospheric Weather Schumann resonance has been linked to various tropospheric and ionospheric phenomena related to the electromagnetic source and medium properties. Thunderstorm effects in ELF wave propagation conditions are especially important because of its connection to tropospheric water vapor, ice distribution in the tropopause, lightning, and aerosols. Williams (1992) reports an association between Schumann resonance and temperature fluctuations in the tropical regions because of a correlation between lightning flash rate and thermodynamic properties of the atmosphere. Price (2000) discusses evidence for a link between global lightning activity and upper tropospheric water vapor. Tropospheric water vapor plays an important role in climate, including a direct impact in tropospheric chemistry, namely aerosol formation, cloud albedo, and as a greenhouse gas. The latter work suggests that water vapor changes in the upper troposphere can be inferred from records of global lightning activity, because deep convective thunderstorms transport important amounts of water into the upper troposphere. In some case, water ice can be transported into the tropopause. Since lightning induces Schumann resonances, ELF wave monitoring would contribute to a better understanding of water ice transport into the tropopause and trends in climate variability. On the other hand, Anyamba et al. (2000) present evidence for a connection between tropospheric weather and ELF electromagnetic wave phenomena, specifically linking Schumann resonance amplitude with the MJO pattern. The main finding of their work is that the intraseasonal oscillations in deep convection modulate the global lightning activity inferred from Schumann resonance intensity. These principles are indeed valuable for strengthening the link between thermodynamic and electromagnetic effects in the atmosphere. In particular, they are useful for assessing the contribution of the three major tropical thunderstorm centers to global lightning budget and ELF wave propagation. Therefore, a clear identification of tropospheric weather effects in ELF spectra, namely Schumann resonance signatures, would offer an independent approach to address specific trends in climate change.

Atmospheric Electricity and Tropospheric Weather Weather fronts, hurricanes, and tornadoes are known to influence tropospheric hydrodynamics and ionospheric electrodynamics. Gravity waves, for example, have been associated to spread-F, medium-scale traveling ionospheric disturbances, and sporadic-E and are important to investigate the coupling between multiple layers, from the troposphere to the thermosphere (Pfaff 2011; Rycroft and Harrison 2011). Tropospheric phenomena, mainly thunderstorms, also drive electrodynamic processes such as charge separation, lightning, and transient luminous events. Because ELF-VLF wave signatures of the surface-ionosphere cavity are closely related to lightning activity, investigation of Schumann resonance, sferics, and whistlers contributes to address lightning variability. ELF-VLF electric and magnetic field measurements from multiple stations can be used to solve the inverse problem and estimate electromagnetic sources distribution in the cavity. Specifically, Schumann resonance Q-bursts may be valuable for assessing specific features related to hurricanes and tornadoes. In addition to the correlation between lightning and thunderstorm activity at a global scale (e.g., Rakov and

Uman 2007), investigation of local connections between lightning and hurricanes and tornadoes is useful. Cecil et al. (2002) investigate lightning characteristics of hurricane eyewalls and rainbands and find that the outer rainband region produces more lightning, a similar feature also reported by Khain et al. (2008), claiming that intense and persistent lightning takes place within a 250–300 km radius ring around the hurricane center. While investigating tornadoes and hailstorms, MacGorman and Burgess (1994) conclude that positive ground flashes can dominate in some severe storms for periods ranging from 30 min to several hours despite negative ground flashes usually dominate cloud-to-ground lightning activity. Additionally, they found a significant association between strong positive cloud-to-ground lightning and severe storms that involve hail or tornadoes. In the vicinity of severe storm, ELF Q-burst enhancement is expected. A similar conclusion might be drawn for sferics and whistlers, too. ELF-VLF spectral variability could therefore contribute to a better electrical characterization of severe thunderstorms.

Investigation of VLF wave propagation related to natural and anthropogenic sources has significantly contributed to an understanding of low ionosphere dynamics. Barr et al. (2000) review the subject thoroughly and Bösinger and Shalimov (2008) analyze lightning signatures combining the VLF and ULF ranges, discussing similarities and differences among IAR, Trimpi effect, and geomagnetic pulsations. Although subionospheric wave propagation related to VLF transmitters and whistlers present the most obvious link to study the troposphere-ionosphere coupling, other phenomena also contribute to monitoring ionospheric dynamics and the impact of tropospheric processes in the layers above. Sferics and tweeks offer additional means to investigate subionospheric wave propagation in the surface-ionosphere waveguide. Unlike VLF transmitter characteristics such as power and polarization that are adjustable and possibly synchronized with receivers, sferics can be used to investigate not only ionosphere reflection properties but also lightning characteristics. In addition to whistler-mode wave propagation and whistler-induced electron precipitation, whistler-related electrostatic waves and plasma turbulence have been studied. Baker et al. (2000) reported the generation of electrostatic emissions by lightning-induced whistlers above thunderstorms using sounding rockets that overflew active storm cells. Electric field measurements and lower-hybrid resonance frequency show similar features as function of altitude, suggesting electrostatic emissions may be responsible for ionospheric irregularities that cause explosive spread-F. Berthelier et al. (2008) discuss lightning-induced plasma turbulence and ion heating in equatorial ionospheric plasma depletions. They present observations of equatorial plasma waves that show evidence for lower-hybrid solitary structures and the simultaneous occurrence of ion heating in deep, large-scale equatorial plasma depletions that form at night during disturbed geomagnetic conditions. These phenomena are associated to lightning-induced whistlers and bear evidence for plasma turbulence and electrostatic waves playing a role in troposphere-ionosphere coupling. Along with whistler-related phenomena, a connection between subionospheric VLF signal perturbations and seismic activity has also been addressed. Molchanov and Hayakawa (1998) suggest that some subionospheric VLF signal perturbations are possibly related to earthquakes; a transient oscillation with a period of 5–10 days observed several days before a strong earthquake and lasting days or weeks after the event is reported. They claim that long-period gravity waves related to earthquakes could possibly explain the observations. Nevertheless, theories have not been able to clarify the mechanism yet, namely establishing the causality principle for those observations. Significant effort has been made to understand such kind of surface-ionosphere coupling though further comprehensive measurements are necessary to address the problem thoroughly.

Surface Phenomena Connections between hydrodynamic, electrostatic, and electromagnetic processes and earthquakes have been unequivocally established. The circumstances are nonetheless different concerning preseismic activity since establishing precursor causal nexus is rather difficult. In particular, there are several subjects where the impact of earthquakes deserves further attention, namely atmospheric conductivity, geomagnetic pulsations, whistlers, discharging phenomena, and atmospheric chemistry whenever volcanic eruptions are involved. Atmospheric conductivity enhancements within the atmospheric boundary layer have been monitored from radon content variability near the surface. However, the nature of earthquake optical phenomena remains unclear as well as its possible connection to atmospheric conductivity enhancements. Establishing whether the phenomenon is intrinsically linked to earthquake activity would be a first step. Then, identifying key processes involving conductivity, electric field, and ionization variability would be invaluable. A correlation between ELF-VLF waves and earthquakes has been addressed by the seismology community. Employing statistical analyses, Hayakawa et al. (1993, 2005, 2008) suggest anomalous effects in Schumann resonance spectra might be associated to earthquakes. They also claim that seismic activity modifies the rate and intensity of whistlers. However, it is not evident what kind of mechanisms would be implicated. Earthquakes could amplify ELF-VLF wave sources and their electrostatic and electromagnetic processes. More definitively, it has been established that earthquakes generate gravity waves that possibly modify the upper boundary of the cavity, affecting ELF wave propagation, or inject momentum in the thermosphere, disturbing whistler propagation in the ionosphere. More elaborated studies are nevertheless necessary to first corroborate previous results and then identify and characterize the mechanisms involved. For example, studying subionospheric VLF radio signals and local, medium-scale traveling ionospheric disturbances could be useful in the context of seismic activity. Investigation of major volcanic eruptions to characterize their impact in atmospheric conductivity modification, local lightning budget, and low ionosphere disturbances induced by gravity waves would be valuable, too.

6 Summary

The investigation of troposphere-ionosphere coupling mechanisms is quite challenging because it requires an understanding of a wide variety of atmospheric science branches, from meteorology to aerology, from tropospheric weather to space weather. For this reason, individual studies can often seem segmented and disjointed when compared to other fields of research. One of the most remarkable examples of a specific topic requiring integrated, coherent studies is that of lightning. The study of this seemingly simple topic testifies as to how atmospheric electricity, specifically fulminology, is interconnected to meteorology, atmospheric dynamics and chemistry, aeronomy, space plasma physics, and both hydrodynamic and electromagnetic wave propagation. Ramifications to climatology may also be implied in a broader sense. Lightning is an atmospheric discharge accompanied by thunder and usually occurring associated to thunderstorm electrification processes during deep convection. Although the mechanism is not understood thoroughly—the most widely accepted theories involve electron runaway breakdown initiated by cosmic rays or positive streamers triggered by hydrometeors—electric fields can build up until the atmospheric breakdown is reached and a discharge happens. Most discharges occur within the clouds but an electric charge can also be transferred between the cloud and the ground or the lower boundary of the ionosphere. The discharge ionizes the medium, modifying the chemistry locally—ozone increasing is a good example—and produces acoustic and electromagnetic waves. The lightning stroke radiates in a wide frequency band, showing a peak in the VLF range. Lightning is associated with several wave propagation phenomena, including electromagnetic oscillations in the ELF range (Schumann resonance), and transients in the VLF (sferics, tweeks, and whistlers) and possibly ULF (IAR signatures and geomagnetic pulsations). In the case of whistlers, lightning-induced particle precipitation in the mesosphere can occur, therefore modifying wave propagation conditions in the lower ionosphere because plasma heating and ionization are often enhanced locally. This mechanism also demonstrates the implications of atmospheric electricity for wave-particle interaction; moreover, the significance of lightning for the DC global electric field is unquestionable. In a broader sense, due to a close connection to thunderstorms, lightning is a useful tool for weather monitoring and perhaps for climate trends identification. Low frequency electromagnetic wave propagation, being closely related to lightning, can be used for identifying and characterizing weather patterns and, hypothetically, climate change. The diagram shown in Fig. 4 summarizes the most relevant sources and processes for the investigation of troposphere-ionosphere coupling processes. Two examples are given regarding diagram interpretation: (i) transient volcanic activity releases ash plumes into the atmosphere, changing ion chemistry processes and driving aerosol nucleation, which changes the electric conductivity of the medium; volcanic activity also produces lightning that directly modifies ionization processes in the troposphere and stratosphere; volcanic activity triggers lightning, which produce electromagnetic waves; (ii) meteor ablation injects heat and metallic ions in the low ionosphere, modifying ionization rates and ion chemistry.

Investigation of ULF-VLF wave propagation contributes to our understanding of the Earth atmospheric environment in a broader context. Some of the most challenging issues, possibly entangled with troposphere-ionosphere coupling mechanisms, are summarized as follows:

- Determining to what extent lightning activity contributes to pump the ionospheric Alfvén resonator;
- Establishing whether some geomagnetic pulsations, mainly Pc1, are a cause or an effect of the ionospheric Alfvén resonator;
- Solving the ionospheric Alfvén resonator inverse problem to derive electron and ion density profiles variability from ULF spectra;
- Establishing an indisputable connection between Schumann resonance spectra and thermodynamic parameters, namely tropospheric water vapor and temperature;
- Solving the Schumann resonance inverse problem to investigate lightning distribution and variability from ELF spectra;
- Identifying possible correlations between long term Schumann resonance variability and climate trends;
- Investigating ionospheric variability effects in Schumann resonance spectra, namely during the recent solar minimum;
- Understanding the surface-ionosphere cavity leakage mechanism for ELF waves;
- Accurate quantification of conductivity/ionization variability effects in ULF-VLF wave propagation, namely from galactic cosmic ray, particle precipitation, meteoroids, volcanic activity, and pollutants;
- Monitoring transient wave propagation to investigate correlations between atmospheric discharge processes, namely lightning and transient luminous events, and VLF waves;
- Validating the connection between explosive spread-F and lightning, and quantifying the contribution of electrodynamic processes to spread-F global budget;
- Confirming the role of earthquakes in Schumann resonance variability, whistler transient enhancement, and geomagnetic pulsation generation;

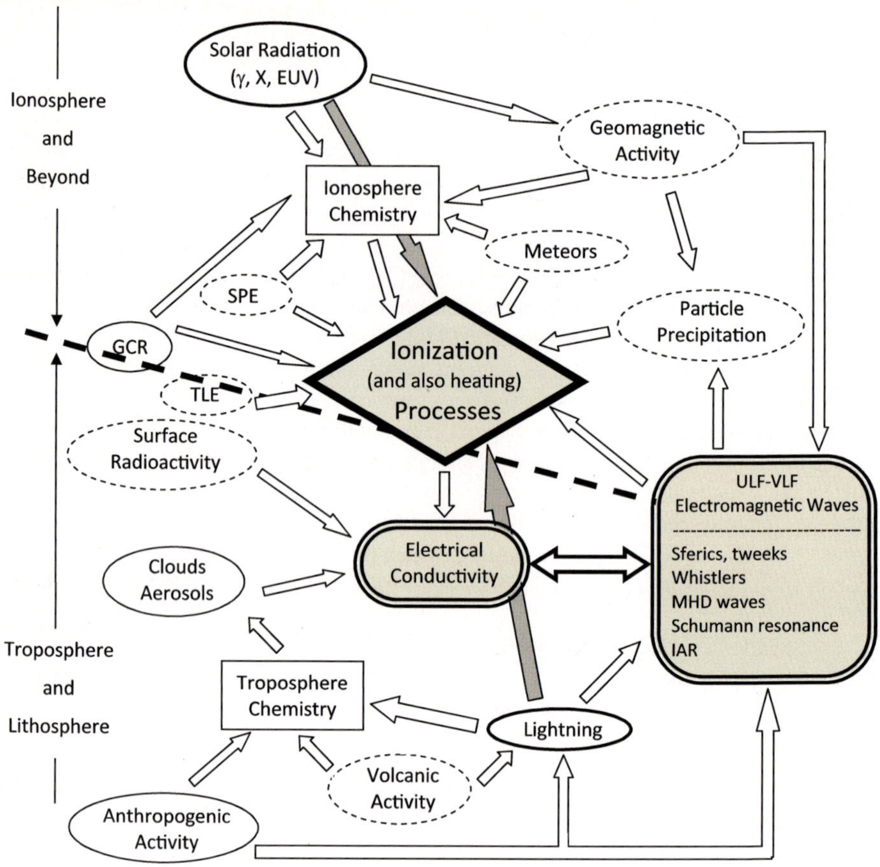

Fig. 4 Diagram of sources that play a role in tropospheric-ionospheric coupling processes. *Solid* and *dashed line* ellipses represent permanent and transient activity prevalence, respectively. SPE, TLE, and GCR mean Solar Particle Event, Transient Luminous Event, and Galactic Cosmic Ray, respectively. *The thick dashed line* discriminates between troposphere and ionosphere dominated processes despite the fact that a complete separation is too stringent for some processes, namely for TLE and GCR phenomena. The *gray-shaded* features represent core parameters related to electromagnetic wave propagation in the earth-ionosphere cavity/waveguide, i.e., electromagnetic sources, medium properties, and radiation processes

- Verifying earthquake contribution to stratospheric (and above) conductivity enhancements.

Finally, the recent, peculiar, extended minimum in solar activity also offers unique conditions to investigate the impact of solar, magnetospheric, and interplanetary phenomena in the Earth environment. Answering some of these challenges would positively contribute to expand our knowledge of the Earth atmospheric environment, namely of the coupling mechanisms between the troposphere, the ionosphere, and beyond.

Acknowledgements FS and JK are supported by an appointment to the NASA Postdoctoral Program at the Goddard Space Flight Center, administered by Oak Ridge Associated Universities through a contract with NASA.

References

M. Abbas, Planet. Space Sci. **16**, 831–844 (1968)
P.M. Adler, J.L. Le Mouel, J. Zlotnicki, Geophys. Res. Lett. **26**, 795–798 (1999)
K. Aizawa, A. Yokoo, W. Kanda, Y. Ogawa, M. Iguchi, Geophys. Res. Lett. **37**, L17301 (2010). doi:10.1029/2010GL044208
W. Allan, E.M. Poulter, Rep. Prog. Phys. **55**, 533–598 (1992)
W. Allan, A.N. Wright, J. Geophys. Res. **102**, 19927–19933 (1997)
R. Anderson, S. Björnsson, D.C. Blanchar et al., Science **148**, 1179–1189 (1965)
Y. Ando, M. Hayakawa, A.V. Shvets, A.P. Nickolaenko, Radio Sci. **40**, RS2002 (2005). doi:10.1029/2004RS003153
E. Anyamba, E. Williams, J. Susskind, A. Fraser-Smith, M. Füllekrug, J. Atmos. Sci. **57**, 1029–1044 (2000)
E. Arnone, A. Hauchecorne, Space Sci. Rev. (2011). 10.1007/s11214-011-9861-1
S.D. Baker, M.C. Kelley, C.M. Swenson, J. Bonnell, D.V. Hahn, J. Atmos. Sol.-Terr. Phys. **62**, 1393–1404 (2000)
L.M. Ball, Appl. Opt. **13**, 2292–2296 (1974)
M. Balser, C.A. Wagner, Nature **188**, 638–641 (1960)
R. Barr, D.L. Jones, C.J. Rodger, J. Atmos. Sol.-Terr. Phys. **62**, 1689–1718 (2000)
T.L. Bell, D. Rosenfel, K.M. Kim, Geophys. Res. Lett. **36**, L23805 (2009). doi:10.1029/2009GL040915
P.P. Belyaev, S.V. Polyakov, V.O. Rapoport, V.Yu. Trakhtengerts, J. Atmos. Sol.-Terr. Phys. **52**, 781–788 (1990)
P.P. Belyaev, T. Bösinger, S.V. Isaev, J. Kangas, J. Geophys. Res. **104**, 4305–4317 (1999)
E.A. Bering, J.R. Benbrook, M.J. Engebretson, R.L. Arnoldy, J. Geophys. Res. **103**, 6741–6761 (1998)
J.-J. Berthelier, M. Malingre, R. Pfaff et al., Nat. Geosci. **1**, 101–105 (2008)
C.J. Biagi, D.M. Jordan, M.A. Uman, J.D. Hill, W.H. Beasley, J. Howard, Geophys. Res. Lett. **36**, L15801 (2009). doi:10.1029/2009GL038525
J.W. Bieber, E. Eroshenko, P. Everson, E.O. Flückiger, R. Kallenbach, *Cosmic Rays and Earth*. Space Science Reviews Series, vol. 93 (Kluwer Academic, Dordrecht, 2000). ISBN 0-7923-6712-X
D. Bilitza, Radio Sci. **36**, 261–275 (2001)
D. Bilitza, Adv. Space Res. **44**, 701–706 (2009)
C. Blagdon, Philos. Trans. R. Soc. **74**, 201–232 (1784)
P.V. Bliokh, A.P. Nickolaenko, Yu.F. Filippov, in *Schumann Resonances in the Earth-Ionosphere Cavity*, ed. by D.L. Jones (Peter Peregrinus, Oxford, 1980)
D.J. Boccippio, E.R. Williams, W.A. Lyons, I. Baker, R. Boldi, Science **269**, 1088–1091 (1995)
D.J. Boccippio, C. Wong, E.R. Williams, R. Boldi, H.J. Christian, S.J. Goodman, J. Atmos. Sol.-Terr. Phys. **60**, 701–712 (1998)
J. Bortnik, U.S. Inan, T.F. Bell, J. Geophys. Res. **111**, A02204 (2006). doi:10.1029/2005JA011182
T. Bösinger, A.G. Demekhov, V.Y. Trakhtengerts, Geophys. Res. Lett. **31**, L18802 (2004). doi:10.1029/2004GL020777
D.H. Brautigam, J.M. Albert, J. Geophys. Res. **105**, 291–309 (2000)
M. Brook, C.B. Moore, T. Sigurgei, J. Geophys. Res. **79**, 472–475 (1974)
K. Bullough, in *Handbook of Atmospheric Electrodynamics II*, 1st edn., ed. by H. Volland (CRC Press, Boca Raton, 1995), pp. 291–332
W.C. Burgess, U.S. Inan, J. Geophys. Res. **98**, 15643–15665 (1993)
L.B. Callis, D.N. Baker, J.B. Blake et al., J. Geophys. Res.-Atmos. **96**, 2939–2976 (1991)
D.L. Carpenter, R.R. Anderson, J. Geophys. Res. **97**, 1097–1108 (1992)
D.L. Carpenter, U.S. Inan, M.L. Trimpi, R.A. Helliwell, J.P. Katsufrakis, J. Geophys. Res. **89**, 9857–9862 (1984)
K.S. Carslaw, R.G. Harrison, J. Kirkby, Science **298**, 1732–1737 (2002)
D.J. Cecil, E.J. Zipser, S.W. Nesbitt, Mon. Weather Rev. **130**, 769–784 (2002)
Z. Ceplecha, J. Borovicka, W.G. Elford et al., Space Sci. Rev. **84**, 327–471 (1998)
C.C. Chaston, J.W. Bonnell, C.W. Carlson, et al., J. Geophys. Res. **107**, 1413 (2002). doi:10.1029/2002JA009272
G. Chisham, D. Orr, J. Geophys. Res. **102**, 24339–24350 (1997)
Y.T. Chiu, B.C. Edgar, C.J. Rice, L.R. Sharp, Geophys. Res. Lett. **6**, 455–458 (1979)
C. Christopoulos, *The Transmission Line Modeling Method: TLM* (IEEE Press, Piscataway, 1995)
M.V. Codrescu, T.J. Fuller-Rowell, I.S. Kutiev, J. Geophys. Res. **102**, 14315–14329 (1997a). doi:10.1029/97JA00638
M.V. Codrescu, T.J. Fuller-Rowell, R.G. Roble, D.S. Evans, J. Geophys. Res. **102**, 19977–19987 (1997b). doi:10.1029/97JA01728

M.B. Cohen, U.S. Inan, M. Golkowski, N.G. Lehtinen, J. Geophys. Res. **115**, A07322 (2010). doi:10.1029/2009JA013231
J.D. Colvin, C.K. Mitchell, J.R. Greig, D.P. Murphy, R.E. Pechacek, M. Raleigh, J. Geophys. Res. **92**, 5696–5712 (1987)
N.F. Cramer, *The Physics of Alfvén Waves* (Wiley-VCH, Berlin, 2001)
R. Cross, *An Introduction to Alfvén Waves*. Adam Hilger Series on Plasma Physics (Taylor & Francis, London, 1988)
S.A. Cummer, IEEE Trans. Antennas Propag. **45**, 392–400 (1997)
S.A. Cummer, U.S. Inan, T.F. Bell, Radio Sci. **33**, 1781–1792 (1998)
S.A. Cummer, J.B. Li, F. Han et al., Nat. Geosci. **2**, 617–620 (2009)
I.A. Daglis, R.M. Thorne, W. Baumjohann, S. Orsini, Rev. Geophys. **37**, 407–438 (1999)
R.E. Daniell Jr., L.D. Brown, D.N. Anderson et al., Radio Sci. **30**, 1499–1510 (1995)
C.J. Davis, C.G. Johnson, Nature **435**, 799–801 (2005)
O. de La Beaujardière, C/NOFS Science Definition Team, J. Atmos. Sol.-Terr. Phys. **66**, 1573–1591 (2004). doi:10.1016/j.jastp.2004.07.030
O. de La Beaujardière, J.M. Retterer, R.F. Pfaff et al., Geophys. Res. Lett. **36**, L00C06 (2009). doi:10.1029/2009GL038884
S.S. De, B.K. De, B. Bandyopadhyay, S. Paul, D.K. Haldar, S. Barui, J. Atmos. Sol.-Terr. Phys. **72**, 829–836 (2010). doi:10.1016/j.jastp.2010.04.004
A.G. Demekhov, Space Sci. Rev. (2011). doi:10.1007/s11214-011-9832-6
A.G. Demekhov, V.Y. Trakhtengerts, T. Bösinger, Geophys. Res. Lett. **27**, 3805–3808 (2000)
L.I. Dorman, *Cosmic Rays in Magnetospheres of the Earth and Other Planets* (Astrophysics and Space Science Library, Heidelberg, 2009)
R.L. Dowden, C.D.D. Adams, J.B. Brundell, P.E. Dowden, J. Atmos. Sol.-Terr. Phys. **56**, 1513–1527 (1994)
U. Ebert, S. Nijdam, C. Li, A. Duque, T. Briels, E. van Veldhuizen, J. Geophys. Res. **115**, A00E43 (2010). doi:10.1029/2009JA014867
B.G. Elmegreen, J. Scalo, Annu. Rev. Astron. Astrophys. **42**, 211–273 (2004)
M.J. Engebretson, L.J. Cahill, R.L. Arnold et al., J. Geophys. Res. **96**, 1527–1542 (1991)
S. England, Space Sci. Rev. (2011). doi:10.1007/s11214-011-9842-4
W.M. Farrell, M.L. Kaiser, M.D. Desch et al., J. Geophys. Res. **104**, 3795–3801 (1999)
C. Ferencz, O.E. Ferencz, D. Hamar, J. Lichtenberger, *Whistler Phenomena—Short Impulse Propagation* (Kluwer Academic, Dordrecht, 2001)
C.C. Finlay, S. Maus, C.D. Beggan et al., Geophys. J. Int. **183**, 1216–1230 (2010)
L.A. Fisk, J.R. Jopikii, G.M. Simnett, R. von Steiger, K.-P. Wenzel, *Cosmic Rays in the Heliosphere*. Space Science Reviews Series, vol. 83 (Kluwer Academic, Dordrecht, 1998). ISBN 0-7923-5069-3
J.W. Fitzgerald, Atmos. Environ., A Gen. Top. **25**, 533–545 (1991)
S.E. Forbush, Phys. Rev. **51**, 1108–1109 (1937)
S.E. Forbush, Phys. Rev. **54**, 975–988 (1938)
S.E. Forbush, J. Geophys. Res. **63**, 651–669 (1958)
R.A. Frahm, J.D. Winningham, J.R. Sharber et al., J. Geophys. Res.-Atmos. **102**, 28203–28214 (1997)
A.V. Frank-Kamenetsky, L.N. Makarova, V.N. Morozov, A.V. Shirochkov, G.B. Burns, J. Atmos. Sol.-Terr. Phys. **72**, 419–424 (2010)
A.C. Fraser-Smith, Geophys. Res. Lett. **20**, 467–470 (1993)
A.C. Fraser-Smith, A. Bernardi, P.R. McGill, M.E. Ladd, R.A. Helliwell, O.G. Villard, Geophys. Res. Lett. **17**, 1465–1468 (1990)
M. Füllekrug, J. Atmos. Sol.-Terr. Phys. **66**, 1115–1119 (2004)
M. Füllekrug, A.C. Fraser-Smith, Geophys. Res. Lett. **23**, 2773–2776 (1996)
M. Füllekrug, M.J. Rycroft, Earth Planets Space **58**, 1193–1196 (2006)
J. Galejs, *Terrestrial Propagation of Long Electromagnetic Waves* (Pergamon, New York, 1972)
W. Gekelman, J. Geophys. Res. **104**, 14417–14435 (1999). doi:10.1029/98JA00161
M.B. Gokhberg, V.A. Morgounov, T. Yoshino, I. Tomizawa, J. Geophys. Res. **87**, 7824–7828 (1982)
C. Greifinger, P.S. Greifinger, J. Geophys. Res. **73**, 7473–7490 (1968)
C. Greifinger, P. Greifinger, J. Geophys. Res. **78**, 4611–4618 (1973)
C. Greifinger, P. Greifinger, Radio Sci. **13**, 831–837 (1978)
P.K.F. Grieder, *Cosmic Rays at Earth: Researcher's Reference Manual and Data Book* (Elsevier, Amsterdam, 2001)
M. Grzesiak, Geophys. Res. Lett. **27**, 923–926 (2000)
A.V. Gurevich, K.P. Zybin, Phys. Usp. **44**, 1119–1140 (2001)
D.A. Gurnett, S.D. Shawhan, N.M. Brice, R.L. Smith, J. Geophys. Res. **70**, 1665–1688 (1965)
M. Gustafsson, S. He, Radio Sci. **35**, 525–536 (2000)

C. Haldoupis, T. Neubert, U.S. Inan, A. Mika, T.H. Allin, R.A. Marshall, J. Geophys. Res. **109**, A10303 (2004). doi:10.1029/2004JA010651
H.J. Hansen, B.J. Fraser, F.W. Menk, Y.D. Hu, P.T. Newell, C.I. Meng, Planet. Space Sci. **39**, 709–719 (1991)
R.G. Harrison, Atmos. Environ. **36**, 159–160 (2002)
R.G. Harrison, Surv. Geophys. **25**, 441–484 (2004)
R.G. Harrison, F. Märcz, Geophys. Res. Lett. **34**, L23816 (2007). doi:10.1029/2007GL031714
R.G. Harrison, K.L. Aplin, M.J. Rycroft, J. Atmos. Sol.-Terr. Phys. **72**, 376–381 (2010)
M. Hayakawa, T. Yoshino, V.A. Morgounov, Phys. Earth Planet. Inter. **77**, 97–108 (1993)
M. Hayakawa, K. Ohta, K. Baba, J. Geophys. Res. **99**, 10733–10743 (1994)
M. Hayakawa, K. Ohta, A.P. Nickolaenko, Y. Ando, Ann. Geophys. **23**, 1335–1346 (2005)
M. Hayakawa, A.P. Nickolaenko, M. Sekiguchi, K. Yamashita, Y. Ida, M. Yano, Nat. Hazards Earth Syst. Sci. **8**, 1309–1316 (2008)
S.L. He, P. Fuks, G.W. Larson, IEEE Trans. Antennas Propag. **44**, 1277–1282 (1996)
R.A. Helliwell, *Whistlers and Related Ionospheric Phenomena* (Stanford University Press, Stanford, 1965)
R.H. Holzworth, M.C. Kelley, C.L. Siefring, L.C. Hale, J.D. Mitchell, J. Geophys. Res. **90**, 9824–9830 (1985)
R.B. Horne, R.M. Thorne, S.A. Glauert, J.M. Albert, N.P. Meredith, R.R. Anderson, J. Geophys. Res. **110**, A03225 (2005a). doi:10.1029/2005JA011516
R.B. Horne, R.M. Thorne, Y.Y. Shprits, Nature **437**, 227–230 (2005b)
R.B. Horne, M.M. Lam, J.C. Green, Geophys. Res. Lett. **36**, L19104 (2009). doi:10.1029/2009GL040236
E. Huang, E. Williams, R. Boldi et al., J. Geophys. Res.-Atmos. **104**, 16943–16964 (1999)
J.D. Huba, G. Joyce, J.A. Fedder, J. Geophys. Res. **105**, 23035–23053 (2000)
J.D. Huba, G. Joyce, S. Sazykin, R. Wolf, R. Spiro, Geophys. Res. Lett. **32**, L23101 (2005). doi:10.1029/2005GL024162
W.J. Hughes, D.J. Southwood, J. Geophys. Res. **81**, 3234–3240 (1976)
R.D. Hunsucker, J.K. Hargreaves, *The High-Latitude Ionosphere and Its Effects on Radio Propagation* (Cambridge University Press, Cambridge, 2002)
W.L. Imhof, J.B. Reagan, H.D. Voss et al., Geophys. Res. Lett. **10**, 361–364 (1983a)
W.L. Imhof, J.B. Reagan, H.D. Voss et al., Geophys. Res. Lett. **10**, 615–618 (1983b)
U.S. Inan, S.A. Cummer, R.A. Marshall, J. Geophys. Res. **115**, A00E36 (2010). doi:10.1029/2009JA014775
U.S. Inan, M. Golkowski, M.K. Casey et al., Geophys. Res. Lett. **34**, L02106 (2007). doi:10.1029/2006GL028494
T. Iyemori, M. Nose, D.S. Han et al., Geophys. Res. Lett. **32**, L20807 (2005). doi:10.1029/2005GL024083
C.H. Jackman, R.D. McPeters, G.J. Labow, E.L. Fleming, C.J. Praderas, J.M. Russell, Geophys. Res. Lett. **28**, 2883–2886 (2001)
R.P. James, P.M. Markowski, Mon. Weather Rev. **138**, 140–161 (2010)
M.R. James, L. Wilson, S.J. Lane et al., Space Sci. Rev. **137**, 399–418 (2008)
J. Jerauld, V.A. Rakov, M.A. Uman et al., J. Geophys. Res.-Atmos. **110**, D19106 (2005). doi:10.1029/2005JD005924
M.J.S. Johnston, Phys. Earth Planet. Inter. **57**, 47–63 (1989)
M.J.S. Johnston, Surv. Geophys. **18**, 441–476 (1997)
T.D. Kaladze, O.A. Pokhotelov, H.A. Shah, M.I. Khan, L. Stenflo, J. Atmos. Sol.-Terr. Phys. **70**, 1607–1616 (2008)
A.K. Kamra, Nature **240**, 143–144 (1972a)
A.K. Kamra, J. Geophys. Res. **77**, 5856–5869 (1972b)
J. Kangas, A. Guglielmi, O. Pokhotelov, Space Sci. Rev. **83**, 435–512 (1998). doi:10.1023/A:1005063911643
J. Kasparian, R. Ackermann, Y.-B. André et al., Opt. Express **16**, 5757–5763 (2008)
O. Katahira, Res. Lett. Atmos. Electr. **12**, 225–234 (1992)
C.S.L. Keay, Science **210**, 11–15 (1980)
M.C. Kelley, *The Earth's Ionosphere: Plasma Physics and Electrodynamics* (Academic Press, Oxford, 2009)
M.C. Kelley, D.T. Farley, E. Kudeki, C.L. Siefring, Geophys. Res. Lett. **11**, 1168–1171 (1984)
M.C. Kelley, R. Livingston, M. McCready, Geophys. Res. Lett. **12**, 577–580 (1985b)
M.C. Kelley, C.L. Siefring, R.F. Pfaff et al., J. Geophys. Res. **90**, 9815–9823 (1985a). doi:10.1029/JA090iA10p09815
A. Khain, N. Cohen, B. Lynn, A. Pokrovsky, J. Atmos. Sci. **65**, 3652–3677 (2008)
N. Khan, N. Mariun, I. Aris, J. Yeak, New J. Phys. **4**, 61 (2002)
J. Kirkby, Surv. Geophys. **28**, 333–375 (2007)
Y. Kito, K. Horii, Y. Higashiyama, K. Nakamura, J. Geophys. Res.-Atmos. **90**, 6147–6157 (1985)
J.H. Klenzing, D.E. Rowland, R.F. Pfaff et al., J. Geophys. Res. **116**, A09324 (2011a). doi:10.1029/2011JA016711
J. Klenzing, F. Simões, S. Ivanov et al., J. Geophys. Res.-Space (2011b). doi:10.1029/2011JA017213

J. Krall, J.D. Huba, C.R. Martinis, Geophys. Res. Lett. **36**, L10103 (2009). doi:10.1029/2009GL038441
P.R. Krehbiel, J.A. Riousset, V.P. Pasko et al., Nat. Geosci. **1**, 233–237 (2008)
S. Kumar, A. Kishore, V. Ramachandran, Ann. Geophys. **26**, 1451–1459 (2008)
M.M. Lam, R.B. Horne, N.P. Meredith, S.A. Glauert, T. Moffat-Griffin, J.C. Green, J. Geophys. Res. **115**, A00F08 (2010). doi:10.1029/2009JA014619
J. Lastovicka, J. Atmos. Sol.-Terr. Phys. **58**, 831–843 (1996)
C.P. Liao, J.P. Freidberg, M.C. Lee, J. Atmos. Sol.-Terr. Phys. **51**, 751–758 (1989)
D.R. Lide et al., *CRC Handbook of Chemistry and Physics*, 90th edn. (Taylor & Francis, Boca Raton, 2010)
W.A. Lyons, T.E. Nelson, E.R. Williams, J.A. Cramer, T.R. Turner, Science **282**, 77–80 (1998)
R.L. Lysak, J. Geophys. Res. **104**, 10017–10030 (1999)
R.L. Lysak, J. Geophys. Res. **109**, A07201 (2004). doi:10.1029/2004JA010454
D.R. MacGorman, D.W. Burgess, Mon. Weather Rev. **122**, 1671–1697 (1994)
T. Madden, W. Thompson, Rev. Geophys. **3**, 211–254 (1965)
T.A. Mather, R.G. Harrison, Surv. Geophys. **27**, 387–432 (2006)
S. Maus, M. Rother, C. Stolle et al., Geochem. Geophys. Geosyst. **7**, Q07008 (2006). doi:10.1029/2006GC001269
V. Mazur, J. Geophys. Res.-Atmos. **94**, 3326–3340 (1989)
J.C. McConnell, J.J. Jin, Atmos.-Ocean **46**, 69–92 (2008)
S.R. McNutt, C.M. Davis, J. Volcanol. Geotherm. Res. **102**, 45–65 (2000)
A.L. Mellot, D. Atri, B.C. Thomas, M.V. Medvedev, G.W. Wilson, M.J. Murray, J. Geophys. Res. **115**, E08002 (2010). doi:10.1029/2010JE003591
N.P. Meredith, R.B. Horne, R.R. Anderson, J. Geophys. Res. **106**, 13165–13178 (2001)
N.P. Meredith, M. Cain, R.B. Horne, R.M. Thorne, D. Summers, R.R. Anderson, J. Geophys. Res. **108**, A61248 (2003). doi:10.1029/2002JA009764
G. Milikh, R. Roussel-Dupré, J. Geophys. Res. **115**, A00E60 (2010). doi:10.1029/2009JA014818
G.H. Millward, I.C.F. Müller-Wodarg, A.D. Aylward et al., J. Geophys. Res. **106**, 24733–25744 (2001). doi:10.1029/2000JA000342
L.I. Miroshnichenko, *Solar Cosmic Rays* (Kluwer Academic, Dordrecht, 2001)
A. Miura, Geophys. Res. Lett. **26**, 409–412 (1999)
O.A. Molchanov, M. Hayakawa, J. Geophys. Res. **103**, 17489–17504 (1998)
V.C. Mushtak, E.R. Williams, J. Atmos. Sol.-Terr. Phys. **64**, 1989–2001 (2002)
U. Narain, P. Ulmschneider, Space Sci. Rev. **75**, 453–509 (1996)
F. Nemec, O. Santolik, M. Marrot, J.-J. Berthelier, Geophys. Res. Lett. **35**, L05109 (2008). doi:10.1029/2007GL032517
F. Nemec, O. Santolik, M. Marrot, J. Geophys. Res. **114**, A04303 (2009). doi:10.1029/2008JA013972
D.A. Neudegg, B.J. Fraser, F.W. Menk, C.L. Waters, G.B. Burns, R.J. Morris, Adv. Space Res. **25**, 1559–1565 (2000)
A.P. Nickolaenko, M. Hayakawa, *Resonances in the Earth-Ionosphere Cavity* (Kluwer Academic, Dordrecht, 2002)
A.P. Nickolaenko, D.D. Sentman, Radio Sci. **42**, RS2S13 (2007). doi:10.1029/2006RS003473
A.P. Nickolaenko, M. Hayakawa, Y. Hobara, J. Atmos. Sol.-Terr. Phys. **58**, 1699–1709 (1996)
A.P. Nickolaenko, M. Hayakawa, Y. Hobara, J. Geophys. Res.-Atmos. **104**, 27585–27591 (1999)
C.A. Nucci, G. Diendorfer, M.A. Uman, F. Rachidi, M. Ianoz, C. Mazzetti, J. Geophys. Res.-Atmos. **95**, 20395–20408 (1990)
T. Ogawa, J. Geophys. Res. **90**, 5951–5960 (1985)
H. Ohya, M. Nishino, Y. Murayama, K. Igarashi, A. Saito, J. Atmos. Sol.-Terr. Phys. **68**, 697–709 (2006)
R.C. Olsen, D.M. Jordan, V.A. Rakov, M.A. Uman, N. Grimes, Geophys. Res. Lett. **31**, L16107 (2004). doi:10.1029/2004GL020187
J.V. Olson, J. Geophys. Res. **104**, 17499–17520 (1999)
A.A. Ostapenko, E.E. Titiva, A.P. Nickolaenko, T. Turunen, J. Manninen, T. Raita, Ann. Geophys. **28**, 193–202 (2010)
S.I. Oyama, B.J. Watkins, Space Sci. Rev. (2011). doi:10.1007/s11214-011-9847-z
A. Parent, I.R. Mann, I.J. Rae, J. Geophys. Res. **115**, A02312 (2010). doi:10.1029/2009JA014673
C.G. Park, D.L. Carpenter, D.B. Wiggin, J. Geophys. Res. **83**, 3137–3144 (1978)
M. Parrot, in *Handbook of Atmospheric Electrodynamics II*, 1st edn., ed. by H. Volland (CRC Press, Boca Raton, 1995), pp. 95–116
V.P. Pasko, J. Geophys. Res. **115**, A00E35 (2010). doi:10.1029/2009JA014860
V.P. Pasko, M.A. Stanley, J.D. Mathews, U.S. Inan, T.G. Wood, Nature **416**, 152–154 (2002)
V. Pasko, Y. Yair, C.-L. Kuo, Space Sci. Rev. (2011). doi:10.1007/s11214-011-9813-9
D. Petersen, M. Bailey, W.H. Beasley, J. Hallett, J. Geophys. Res.-Atmos. **113**, D17205 (2008). doi:10.1029/2007JD009036

R. Pfaff, Space Sci. Rev. (2011), this issue
G. Pfotzer, Z. Phys. **102**, 23 (1936)
J.M. Picone, A.E. Hedin, D.P. Drob, A.C. Aikin, J. Geophys. Res. **107**, A121468 (2002). doi:10.1029/2002JA009430
V.A. Pilipenko, J. Atmos. Sol.-Terr. Phys. **52**, 1193–1209 (1990)
V.A. Pilipenko, Space Sci. Rev. (2011). 10.1007/s11214-011-9859-8
S.V. Polyakov, V.O. Rapoport, Geomagn. Aeron. **21**, 610–614 (1981)
C. Price, Nature **406**, 290–293 (2000)
C. Price, M. Blum, Earth Moon Planets **82–83**, 545–554 (2000)
C. Price, A. Melnikov, J. Atmos. Sol.-Terr. Phys. **66**, 1179–1185 (2004)
V.A. Rakov, M.A. Uman, IEEE Trans. Electromagn. Compat. **40**, 403–426 (1998)
V. Rakov, M. Uman, *Lightning* (Cambridge University Press, Cambridge, 2007)
C.M. Rappaport, B.J. McCartin, IEEE Trans. Antennas Propag. **39**, 345–349 (1991)
O.M. Raspopov, L.J. Lanzerotti, Rev. Geophys. **14**, 577–589 (1976)
J.B. Reagan, T.M. Watt, J. Geophys. Res. **81**, 4579–4596 (1976)
J.B. Reagan, R.E. Meyerott, R.W. Nightingale et al., J. Geophys. Res. **86**, 1473–1494 (1981)
E. Real, E. Orlandi, K.S. Law et al., Atmos. Chem. Phys. **10**, 3027–3046 (2010)
C.D. Reeve, M.J. Rycroft, J. Atmos. Sol.-Terr. Phys. **34**, 667–672 (1972)
G.C. Reid, in *Study in Geophysics: The Earth's Electrical Environment*, ed. by E.P. Krider, R.G. Roble (National Academy Press, Washington, 1986), pp. 183–194
N.O. Renno, J.F. Kok, Space Sci. Rev. **137**, 419–434 (2008)
J.M. Retterer, D.T. Decker, W.S. Borer, R.E. Daniell Jr., B.G. Fejer, J. Geophys. Res. **110**, A11307 (2005). doi:10.1029/2002JA009613
T. Rikitake, Y. Yamazaki, EPR, Earthqu. Predict. Res. **3**, 559–570 (1985)
B. Roberts, Sol. Phys. **193**, 139–152 (2000)
R.G. Roble, B.A. Emery, T.L. Killeen et al., J. Geophys. Res. **92**, 6083–6090 (1987)
C.J. Rodger, M.A. Clilverd, R.J. McCormick, J. Geophys. Res. **108**, A121462 (2003). doi:10.1029/2003JA009906
C.J. Rodger, M.A. Clilverd, N.R. Thomson, D. Nunn, J. Lichtenberger, Ann. Geophys. **23**, 3419–3430 (2005)
C.J. Rodger, C.-F. Enell, E. Turenen, M.A. Clilverd, N.R. Thomson, P.T. Verronen, Ann. Geophys. **25**, 1745–1757 (2007)
J.V. Rodriguez, U.S. Inan, T.F. Bell, Geophys. Res. Lett. **19**, 2067–2070 (1992a)
J.V. Rodriguez, U.S. Inan, Y.Q. Li et al., J. Geophys. Res. **97**, 65–75 (1992b)
V.C. Roldugin, Y.P. Maltsev, A.N. Vasiljev, A.V. Shvets, A.P. Nikolaenko, J. Geophys. Res. **108**, A31103 (2003). doi:10.1029/2002JA009495
M.J. Rycroft, J. Atmos. Sol.-Terr. Phys. **68**, 445–456 (2006)
M.J. Rycroft, R.G. Harrison, Space Sci. Rev. (2011). doi:10.1007/s11214-011-9830-8
M.J. Rycroft, S. Israelsson, C. Price, J. Atmos. Sol.-Terr. Phys. **62**, 1563–1576 (2000)
M.J. Rycroft, A. Odzimek, N.F. Arnold, M. Fullekrug, A. Kulak, T. Neubert, J. Atmos. Sol.-Terr. Phys. **69**, 2485–2509 (2007)
M.J. Rycroft, R.G. Harrison, K.A. Nicoll, E.A. Mareev, Space Sci. Rev. **137**, 83–105 (2008)
T. Saito, Space Sci. Rev. **10**, 319–412 (1969). doi:10.1007/BF00203620
S.V.S. Sarma, T.S. Sastry, J. Atmos. Sol.-Terr. Phys. **57**, 749–754 (1995)
G. Sátori, J. Atmos. Sol.-Terr. Phys. **58**, 1483–1488 (1996)
S.S. Sazhin, *Whistler-Mode Waves in a Hot Plasma* (Cambridge University Press, Cambridge, 1993)
S.S. Sazhin, M. Hayakawa, J. Atmos. Sol.-Terr. Phys. **56**, 735–753 (1994)
K. Schlegel, M. Füllekrug, J. Geophys. Res. **104**, 10111–10118 (1999)
R. Schlickeiser, *Cosmic Ray Astrophysics* (Springer, Berlin, 2002)
W.O. Schumann, Z. Naturforsch. A, J. Phys. Sci. **7**, 149–154 (1952) (in German)
U. Schumann, H. Huntrieser, Atmos. Chem. Phys. **7**, 3823–3907 (2007)
R.W. Schunk, L. Scherliess, J.J. Sojka et al., Radio Sci. **39**, RS1S02 (2004). doi:10.1029/2002RS002794
N.V. Semenova, A.G. Yahnin, A.N. Vasil'ev, O. Amm, Geomagn. Aeron. **48**, 36–44 (2008). doi:10.1134/S0016793208010052
D.D. Sentman, J. Atmos. Sol.-Terr. Phys. **52**, 35–46 (1990)
D.D. Sentman, in *Handbook of Atmospheric Electrodynamics I*, 1st edn., ed. by H. Volland (CRC Press, Boca Raton, 1995), pp. 267–298
D.D. Sentman, B.J. Fraser, J. Geophys. Res. **96**, 15973–15984 (1991)
S. Shalimov, T. Bösinger, Space Sci. Rev. (2011). doi:10.1007/s11214-011-9812-x
S. Shalimov, M. Gokhberg, Phys. Earth Planet. Inter. **105**, 211–218 (1998)
S.E. Sharapov, D. Testa, B. Alper et al., Phys. Lett. A **289**, 127–134 (2001)

R.P. Sharma, M.L. Goldstein, N.K. Dwivedi, P.K. Chauhan, J. Geophys. Res. **115**, A12207 (2010). doi:10.1029/2009JA015123
Y.Y. Shprits, R.M. Thorne, Geophys. Res. Lett. **31**, L08805 (2004). doi:10.1029/2004GL019591
P.K. Shukla, L. Stenflo, Lect. Notes Phys. **536**, 1–30 (1999)
A.B. Shvartsburg, L. Stenflo, J. Atmos. Sol.-Terr. Phys. **55**, 81–89 (1993)
A.V. Shvets, M. Hayakawa, J. Atmos. Sol.-Terr. Phys. **60**, 461–469 (1998)
F. Simões, R. Grard, M. Hamelin et al., Planet. Space Sci. **55**, 1978–1989 (2007)
F. Simões, M. Rycroft, N. Renno, Y. Yair, K.L. Aplin, Y. Takahashi, Space Sci. Rev. **137**, 455–471 (2008a)
F. Simões, M. Hamelin, R. Grard et al., J. Geophys. Res. **113**, E07007 (2008b). doi:10.1029/2007JE003045
F. Simões, R. Grard, M. Hamelin et al., Icarus **194**, 30–41 (2008c)
F. Simões, J.-J. Berthelier, M. Godefroy, S. Yahi, Geophys. Res. Lett. **36**, L14816 (2009). doi:10.1029/2009GL039286
F. Simões, R. Pfaff, H. Freudenreich, Geophys. Res. Lett. **38**, L22101 (2011a). doi:10.1029/2011GL049668
F. Simões, J.-J. Berthelier, R. Pfaff, D. Bilitza, J. Klenzing, in *General Assembly and Scientific Symposium, 2011 XXXth URSI*, 13–20 Aug. 2011 (2011b), pp. 1–4. doi:10.1109/URSIGASS.2011.6050887
J.J. Simpson, A. Taflove, IEEE Trans. Antennas Propag. **55**, 1582–1590 (2007)
A.K. Singh, R.P. Singh, Earth Moon Planets **73**, 277–290 (1996)
D.K. Singh, R.P. Singh, A.K. Kamra, Space Sci. Rev. **113**, 375–408 (2004)
D. Singh, R.P. Singh, A.K. Kamra et al., J. Atmos. Sol.-Terr. Phys. **67**, 637–658 (2005)
A.K. Smith, Space Sci. Rev. (2011). 10.1007/s11214-011-9791-y
T. Stanev, *High Energy Cosmic Rays* (Springer, Berlin, 2009)
K. Stasiewicz, P. Bellan, C. Chaston et al., Space Sci. Rev. **92**, 423–533 (2000)
M. Stenhoff, *Ball Lightning: an Unsolved Problem in Atmospheric Physics* (Kluwer Academic, New York, 1999)
B. Stewart, Philos. Trans. **151**, 423–430 (1861)
G.P. Stiller, G.M. Tsidu, T. von Clarmann et al., J. Geophys. Res.-Atmos. **110**, D20303 (2005). doi:10.1029/2005JD006011
L.R.O. Storey, Philos. Trans. R. Soc. Lond. A **246**, 113–141 (1953)
H.J. Strangeways, J. Geophys. Res. **61**, 1067–1080 (1999)
A.I. Sukhorukov, Ann. Geophys. **14**, 33–41 (1996)
A.I. Sukhorukov, P. Stubbe, Geophys. Res. Lett. **24**, 829–832 (1997)
D. Summers, C.Y. Ma, J. Geophys. Res. **105**, 2625–2639 (2000)
V.V. Surkov, O.A. Molchanov, M. Hayakawa, E.N. Fedorov, J. Geophys. Res. **110**, A04308 (2005). doi:10.1029/2004JA010850
V.V. Surkov, M. Hayakawa, A.Y. Schekotov, E.N. Fedorov, O.A. Molchanov, J. Geophys. Res. **111**, A01303 (2006). doi:10.1029/2005JA011320
K. Takahashi, Ann. Geophys. **16**, 787–803 (1998)
R.W. Talbot, J.D. Bradshaw, S.T. Sandholm et al., J. Geophys. Res.-Atmos. **101**, 24187–24202 (1996)
T. Tanaka, T. Ichinose, T. Okuzawa et al., J. Atmos. Sol.-Terr. Phys. **46**, 233–245 (1984)
M.J. Taylor, M.A. Hapgood, Planet. Space Sci. **36**, 975–985 (1988)
B.A. Tinsley, Space Sci. Rev. **94**, 231–258 (2000)
B.A. Tinsley, Rep. Prog. Phys. **71**, 066801 (2008). doi:10.1088/0034-4885/71/6/066801
B.A. Tinsley, G.B. Burns, L. Zhou, Adv. Space Res. **40**, 1126–1139 (2007)
B.T. Tsurutani, C.M. Ho, Rev. Geophys. **37**, 517–541 (1999)
B.T. Tsurutani, W.D. Gonzalez, A.L.C. Gonzalez et al., J. Geophys. Res. **111**, A07S01 (2006). doi:10.1029/2005JA011273
M.A. Uman, D.F. Seacord, G.H. Price, E.T. Pierce, J. Geophys. Res. **77**, 1591–1596 (1972)
I.G. Usoskin, L. Desorgher, P. Velinov et al., Acta Geophys. **57**, 88–101 (2009)
H. Utada, T. Yoshino, T. Okubo, T. Yukutake, Tectonophysics **299**, 317–331 (1998)
S. Uyeda, M. Hayakawa, T. Nagao et al., Proc. Natl. Acad. Sci. USA **99**, 7352–7355 (2002)
P.T. Verronen, A. Seppala, M.A. Clilverd et al., J. Geophys. Res. **110**, A09S32 (2005). doi:10.1029/2004JA010932
H.D. Voss, W.L. Imhof, M. Walt et al., Nature **312**, 740–742 (1984)
J.W. Warwick, C. Stoker, T.R. Meyer, J. Geophys. Res. **87**, 2851–2859 (1982)
E.R. Williams, Science **256**, 1184–1187 (1992)
E.R. Williams, J. Geophys. Res. **115**, A00E50 (2010). doi:10.1029/2009JA014581
E.R. Williams, G. Sátori, J. Atmos. Sol.-Terr. Phys. **66**, 1213–1231 (2004)
E. Williams, K. Rothkin, D. Stevenson, D. Boccippio, J. Appl. Meteorol. **39**, 2223–2230 (2000)
E.R. Williams, V.C. Mushtak, R. Boldi, R.L. Dowden, Z.-I. Kawasaki, Radio Sci. **42**, RS2S20 (2007). doi:10.1029/2006RS003498

C.T.R. Wilson, Proc. Camb. Philos. Soc. **13**, 363–382 (1906)
C.T.R. Wilson, Philos. Trans. R. Soc. Lond. A **221**, 73–115 (1920)
J.H. Wolcott, D.J. Simons, D.D. Lee, R.A. Nelson, J. Geophys. Res. **89**, 6835–6839 (1984)
R.F. Woodman, E. Kudeki, Geophys. Res. Lett. **11**, 1165–1167 (1984)
R.F. Woodman, C. La Hoz, J. Geophys. Res. **81**, 5447–5466 (1976)
N. Yagova, V. Pilipenko, E. Fedorov, M. Vellante, K. Yumoto, Earth Planets Space **51**, 129–138 (1999)
A.G. Yahnin, T.A. Yahnina, J. Atmos. Sol.-Terr. Phys. **69**, 1690–1706 (2007)
Y. Yair, G. Fischer, F. Simões, N. Renno, P. Zarka, Space Sci. Rev. **137**, 29–49 (2008)
L. Yin, M. Ashour-Abdalla, Phys. Plasmas **6**, 449–462 (1999)
J.L. Young, Radio Sci. **29**, 1513–1522 (1994)
Q.G. Zong, Y.Q. Hao, Y.F. Wang, Sci. China Ser. E **52**, 3698–3708 (2009)

Space Sci Rev (2012) 168:595–609
DOI 10.1007/s11214-011-9832-6

Coupling at the Atmosphere-Ionosphere-Magnetosphere Interface and Resonant Phenomena in the ULF Range

A.G. Demekhov

Received: 8 August 2011 / Accepted: 7 September 2011 / Published online: 15 October 2011

Abstract We discuss the electromagnetic processes in the ULF range which are important for the coupling between the atmosphere, ionosphere, and magnetosphere (AIM). The main attention is given to the Pc1–2 frequency ranges ($f \approx 0.1$–10 Hz) where some natural resonances in the AIM system are located. In particular, we consider the resonant structures in the spectra of the magnetic background noise related to the Alfvén resonances in the ionosphere as a possible diagnostic tool for studies of the ionospheric parameters. We also discuss the self-excitation of Alfvén waves in the ionosphere due to the AIM coupling and the role of such waves in the acceleration of electrons in the upper ionosphere and magnetosphere. Precipitation of magnetospheric ions due to their interaction with the ion-cyclotron waves is analyzed in relation to the ionospheric current systems, formation of partial ring current, and the influence of the ionosphere-magnetosphere feedback on the generation of such waves.

Keywords Ionosphere · Magnetosphere · Atmosphere · ULF waves · Alfvén resonator

1 Introduction

In this paper, we briefly discuss some processes in the ultra-low frequency (ULF) range which crucially influence and are influenced by the electromagnetic coupling between the Earth's atmosphere, ionosphere, and magnetosphere. The ULF range comprises several subranges, which differ greatly in their roles for the near-Earth environment, so reviewing all the relevant processes would be too much for one paper. We mainly restrict ourselves to the highest-frequency part of the ULF range which is known as Pc-1,2 band and includes the frequencies from about $f \sim 0.1$ Hz to $f \sim 10$ Hz.

The importance of this band for the coupling in the atmosphere–ionosphere–magnetosphere (AIM) system is related to the presence of some natural resonances in it. In particular, the eigenfrequencies of the ionospheric Alfvén resonator (IAR) (Polyakov and Rapoport

A.G. Demekhov (✉)
Institute of Applied Physics, Nizhny Novgorod, Russia
e-mail: andrei@appl.sci-nnov.ru

1981; Belyaev et al. 1990, see Fig. 1) and the proton gyrofrequencies in the equatorial plane fall into this frequency range. One finds here also the Schumann resonance, whose effects are covered in a separate paper in this issue (Simoes et al. 2011). The lower-frequency part of the ULF range contains one more resonance which can be of interest for the topics discussed, i.e., the magnetospheric Alfvén resonance related to the geomagnetic field-line oscillations.

The presence of the ionospheric resonances in the ULF band affects significantly the characteristics of natural and artificial signals which can be detected on the ground and on-board spacecraft. It provides an opportunity to diagnose some parameters of the ionosphere by measuring such signals. Among such signals, the most persistent is the background noise which is permanently present in the ground-based magnetic recordings. In 1980s, a resonance structure in the spectrum of this noise has been discovered and theoretically analyzed (Belyaev et al. 1987a, 1989b). Until now, studies of the properties and diagnostic potential of such spectral resonance structures (SRS) are very active. We discuss these studies in Sect. 2.

The electrodynamic coupling between the ionosphere and atmosphere at auroral latitudes, where sufficiently large convection electric field exists, can result in an instability which drives certain class of ULF waves in the ionospheric Alfvén resonator (IAR) to high amplitudes and thus gives rise to nonlinear processes involving the generation of intense field-aligned electric fields and acceleration of charged particles (mostly electrons) to suprathermal energies. This instability (Trakhtengerts and Feldstein 1981, 1984, 1987, 1991; Lysak 1991; Lysak and Song 2002; Pokhotelov et al. 2001) and its effects are considered in Sect. 3.

The Pc1 ULF waves which are generated in the magnetosphere by the ring current protons via the cyclotron resonant interactions can efficiently scatter these protons and other ring-current ions in pitch-angle and thus cause their precipitation into the ionosphere. Such ions whose energy is about 100 keV deposit it a rather narrow interval of altitudes about 110 km, i.e., in the E region. This precipitation can significantly modify the ionospheric density and hence change the ionospheric Alfvén resonator eigenfrequencies. In its turn, this change leads to a variation in the ionosphere reflection for the Pc1 waves, which can influence their generation in the magnetosphere. This feedback forms a system called Alfvén sweep maser, which was suggested as a mechanism for the formation of Pc1 pearl emissions (Polyakov et al. 1983; Belyaev et al. 1984, 1987b; Trakhtengerts et al. 2000b). It is briefly discussed in Sect. 4.1.

The ring-current proton precipitation due to the cyclotron resonant interactions with Pc1 waves depend strongly on the plasma density, since the Doppler-shifted cyclotron resonance condition has to be satisfied. Therefore, such interactions are localized in the regions with sufficiently dense cold plasma, i.e., in the plasmasphere or detached cold-plasma regions (Cornwall et al. 1970). Guided propagation of Pc1 Alfvén waves is stipulated by the plasma-density gradients (Mazur and Potapov 1983), which makes the wave-particle interaction region even more localized across the geomagnetic field, i.e., tied to the plasmapause vicinity and the edges of the detached plasma clouds. This localized precipitation can have a strong effect on the global current system in the magnetosphere and ionosphere (Trakhtengerts and Demekhov 2005), since a significant part of the ring current (about 1 MA) is driven into the ionosphere by this process. We discuss the role of Pc1 waves in the partial ring current formation and related effects in Sect. 4.2.

2 Magnetic Background Noise as a Diagnostic Tool for Ionospheric Alfvén Resonances

2.1 Ionospheric Alfvén Resonator and Spectral Resonance Structures

When the spectral resonance structures in the background noise were discovered (Belyaev et al. 1987a) and theoretically explained by the influence of the IAR (Belyaev et al. 1989b), two very important ideas were formulated. First, it was assumed that the main source of the background noise is provided by lightning discharges at the world thunderstorm centers. Second, it was suggested that their parameters are mainly determined by the local ionosphere properties above a ground-based receiver. These two points ensure the simple relationship between the SRS frequency scale and local ionosphere properties, which does not depend on the specific features of the source:

$$\Delta f \approx V_{AF}/(2l) \tag{1}$$

where l is the characteristic scale length of the upper ionosphere, and $V_{AF} = c/n_{AF}$ is the Alfvén velocity at the F-region maximum, n_{AF} is the corresponding refractive index, and c is the speed of light in vacuum. Actually, Δf is equal to the basic eigenfrequency of the local IAR in the case where the height-integrated Pedersen conductance Σ_P is much lower than the effective wave conductance $\Sigma_A = cn_{AF}/(4\pi)$.

In recent years, especially since the SRS detection in the auroral region (Belyaev et al. 1999), the observations of SRS have become almost routine and are performed at all latitudes from low (Bösinger et al. 2002) to very high (Semenova et al. 2008). These observations yielded a rich material for statistical and case studies. Extensive theoretical studies have been undertaken to improve the SRS formation model. Here, we discuss some results of those studies.

2.1.1 Sources of the ULF Background Noise

An argument for the necessity of alternative sources ULF background noise in the Pc1 frequency range was formulated by Surkov et al. (2005). They found that the measured magnetic-field amplitude is by an order of magnitude larger than the result of their calculation based on representing the primary field from distant thunderstorms as a random Poisson-like process. It allowed the authors to claim that local thunderstorms could be more important source of the ULF noise. However, Shalimov and Bösinger (2008) have demonstrated that the more accurate taking into account the temporal structure of the cloud-to-ground (CG) discharges can greatly increase the theoretical estimate of the amplitude of ULF noise generated at worldwide thunderstorm centers. Specifically, they have shown that the positive CG discharges, being much less frequent, still are most significant source of ULF fields in the Pc1 frequency range due to their high continuing (long-lasting) currents, and that the ULF noise amplitude from such discharges can be quite sufficient to explain the observed values.

Nevertheless, other sources of the ULF noise can operate. Besides the local lightnings, the currents driven by neutral wind fluctuations have been considered theoretically (Surkov et al. 2004). These calculations remain to be compared with experiment taking account the data on the actual amplitude and spatial spectrum of neutral wind fluctuations in the IAR frequency range.

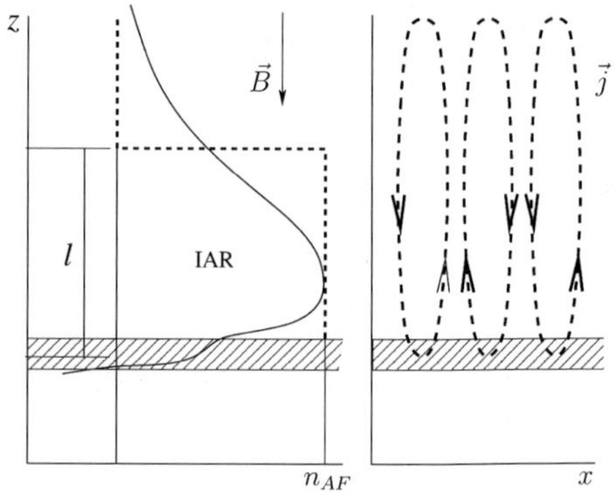

Fig. 1 Scheme of the ionospheric Alfvén resonator (*left*). *Dashed line* shows piecewise-constant profile which can be used in simplified analytical approaches. Superimposed on the right is a sketch of Alfvén vortices which can be excited in the resonator

2.1.2 Horizontal Scales of SRS Formation Region

In the first works on SRS detection it was noted (Belyaev et al. 1989a) that the spectral structure can be seen if the exciting current is located at least about one hundred km from the detection site. It was also suggested that the SRS formation is determined by a region of the horizontal size about 100 km.

Direct experimental verification of this assumption was published only recently. Such experimental tests become possible due to the high-latitude observations where the localized moving ionospheric disturbances can occur. Semenova et al. (2008) have analyzed the evolution of ULF noise spectra recorded at Barentsburg (Svalbard) during the development of two substorms. They showed that the SRS disappeared in the case where the auroral oval approached the detection site to a distance of 100–200-km. In the other case where the auroral oval expansion stopped in 300–400 km from Barentsburg, the SRS persisted during the entire substorm. This indicates that the ionosphere disturbances, such as increased ionization and, hence, absorption in the E region, destroy the resonance structure if it occurs at a distance within 100–200 km. Thus, the SRS formation region indeed occupies the region of no more than about 200 km.

On the other hand, recent observations showed that local sources with scales about 50 km do not contribute to the SRS. It was studied by comparing the signals simultaneously received at stations Novaya Zhizn' (New Life) and Fominki (near Nizhny Novgorod) spaced by 45 km and coherent subtraction of the two records (Ermakova et al. 2009). The difference signal showed no SRS whereas it was clearly seen in the separate records. Similar comparison was also made for the records obtained at the sites Novaya Zhizn' and Borok spaced by about 400 km. The SRS was found both in the separate and difference signals, which confirms in a different way that the horizontal scale of the SRS formation region is smaller than 400 km.

2.1.3 Influence of the Geomagnetic Field Inclination on the SRS Properties

The first and most widely used analytical model of SRS formation is based on the assumption of the vertical magnetic field. In this case, the shear Alfvén and compressional magnetoacoustic modes are practically uncoupled everywhere in the ionosphere above the E-layer. The coupling in the E-layer is quite significant, since it ensures the transformation from the TEM mode excited by vertical lightning discharge and the TE mode.

It is the latter component which exhibits the resonance spectral structure, while the former has flat spectrum in the simplest approximation. Such an approximation is valid if the coupling in the E-layer is weak enough and successfully be described in terms of a single-transformation approximation (Polyakov and Rapoport 1981). In this case, only the transformation of the Alfvén wave into the magnetoacoustic wave should be taken into account, and the SRS is seen only in the magnetic component along the great circle between the source and detector (Belyaev et al. 1989b).

Although the SRS indeed often dominates in only one component, in accordance with the above mentioned approach (Belyaev et al. 1989a), recent studies (Ermakova et al. 2008, 2011) showed that it can reveal much more complicated form if looked more carefully at. In particular, the maxima an minima in the spectra of the orthogonal components of the horizontal magnetic field can be shifted with respect to each other to a distance of up to one-half of the frequency scale (in this case, the maxima in one component correspond to the minima in the other). More seldom, the spectra of the orthogonal magnetic-field components can have slightly different SRS frequency scales. Such observations were mainly performed at a mid-latitude site near Nizhny Novgorod ($L \simeq 2.6$).

As we see below, these features can be explained by the effect of inclined geomagnetic field which increases significantly the effect of coupling between the Alfvén and fast magnetosonic waves. The different dispersion of these modes ensures their different phase paths in the ionosphere and, hence, their coupling can enrich the IAR spectrum by new eigenmodes which manifest themselves in the noise spectrum.

An elaborate theory of signal excitation in the Earth-ionosphere cavity allowing one to take into account the mutual transformations between the wave modes in the ionosphere has been developed (Sobchakov et al. 2003). It employs the assumptions of plane geometry and a horizontally homogeneous medium. Both these assumptions seem not too restrictive since (i) the usual distance between a thunderstorm region and an observation site does not exceed several thousands of km, and (ii) the main contribution to the spectral resonant structure is made by a fairly close area of about 100 km near the receiver (see Sect. 2.1.2). The third assumption concerns vertical direction of wave propagation in the ionosphere, and it is justified for waves incident from the Earth-ionosphere cavity by a very high refractive index in the ionosphere.

The analytical formulas obtained by Sobchakov et al. (2003) depend on the impedance of the overlying ionosphere which can be obtained numerically by employing any available model of the vertical ionosphere profile. This approach has been extensively used (Ermakova et al. 2008, 2011) for comparing the model results with recent observations. The calculations confirmed that the geomagnetic-field inclination can ensure a relative shift of the resonance spectral structures in the orthogonal magnetic-field components. This shift is maximized for the case where the source-receiver line lies in the magnetic-meridian plane and can indeed match the spectral maxima of one component to the spectral minima of the other component. Different SRS frequency scales in the orthogonal components were also obtained in the calculations.

It is noteworthy that calculations based on different ionospheric models can give significantly different results for the mentioned SRS features related to the mode coupling (Ermakova et al. 2011). For example, if the models which provide more gradual density decrease above the F-region maximum, such as IRI-2001, were used, then a notable difference in the SRS frequency scales for different components was obtained only if the magnetic-field inclination reached $\theta \approx 37^\circ$ ($L \approx 1.4$ characteristic, e.g., of Crete). At the same time, the Chapman-like profile for the upper-ionosphere density (Reinisch et al. 2007), in which the density decays steeper, yielded a fairly clear effect even for mid-latitude parameters (e.g., $\theta \approx 21^\circ$ corresponding to Novaya Zhizn' site near Nizhny Novgorod, $L \simeq 2.6$). This result can be explained by a higher IAR Q-factor in the case of a steeper density decay and, correspondingly, by more efficient mode conversion.

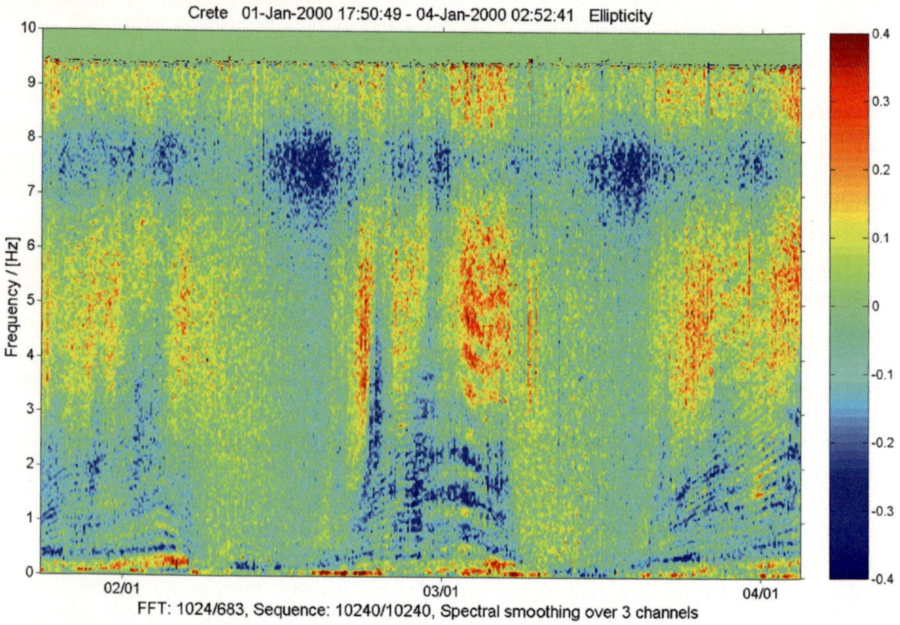

Fig. 2 Example of spectral resonance structure from Crete data (Bösinger et al. 2004). The *colour scale* corresponds to the ellipticity of the signal polarization, $\varepsilon = (H_R - H_L)/(H_R + H_L)$, where H_R and H_L are the right-hand and left-hand polarized components of the horizontal magnetic field, respectively. Both SRS and the fine spectral structure are clearly seen

As we see, careful examination of the SRS form and parameters and comparison with advanced modeling can provide additional information on not only the local ionosphere but also on the direction to the dominant source of the ULF noise. However, these opportunities require further studies. In particular, it would be very important to compare SRS observations with modeling results based on actually measured ionospheric parameters.

2.2 Fine Structure of SRS: Manifestation of the Ionosphere-Magnetosphere Alfvén Resonator?

Observations of magnetic background noise performed at Crete station ($L \approx 1.4$) revealed many interesting specific features of SRS at low latitudes (Bösinger et al. 2002, 2004). Besides the manifestations of the geomagnetic-field inclination discussed above, one of such features which has not yet been fully understood to date is the so-called fine SRS (FSRS) which is clearly seen in Fig. 2. This structure has much smaller frequency scale (about 0.12 Hz compared with about 0.8 Hz for the regular SRS which is also present in Fig. 2). Bösinger et al. (2004) proposed a mechanism of formation of such a fine spectral structure based on the influence of the entire magnetic flux tube on the noise spectrum. This system includes two IARs at conjugate ionosphere regions and the magnetospheric Alfvén resonator (MAR) between them. At low latitudes the field line is short enough, so the eigenfrequency of the field-line resonance is not so well separated from the IAR eigenfrequencies, hence, the coupling between the two IARs and the MAR is strong enough, and the entire system should be considered as one complex resonator which can be called the ionosphere-magnetosphere resonator (IMAR, Ovchinnikov 1999).

Only very simplified calculations were performed for such a low-latitude system (Bösinger et al. 2004), which indicate that such a model can indeed explain the fine spectral structure. A more detailed consideration is needed to take into account the factors neglected in the simple consideration. One of them is the wave transformation and leakage due to the field-line curvature, which affects the resonator quality. Also it would be interesting to study possible excitation of the IMAR from both ends by a ground-based source, i.e, by the field penetrating into the same flux tube at the conjugate footpoints. It can be significant because the distance between these footpoints is not so large at low latitudes, and the ULF field from, e.g., a lightning discharge can be of the same order there.

It is noteworthy that the fine spectral resonance structure has also been observed at mid-latitudes, where it has frequency scales $\Delta f \sim 0.3$ Hz. The contrast between the frequency scales of the "regular" and fine SRS is not so high here as at low latitudes. At the same time the hypothesis concerning the role of field-line oscillations is much less applicable in this case, since the field-line is longer compared with the low-latitude case but the fine frequency scale is higher. In the cases where the fine scale is only twice as narrow as the regular resonance structure, it can possibly be related to the mutual transformation of compressional and shear Alfvén waves (see Sect. 2.1.3).

2.3 Sub-Ionospheric Alfvén Resonator and Broadband Spectral Maximum

Besides the SRS, there is another spectral structure in ULF noise at frequencies below the Schumann resonance. This structure has a form of a broad spectral maximum (BSM) at about 2–4 Hz (see Fig. 3). It was first documented by Belyaev et al. (2002) who proposed that it might have been related to the ULF wave propagation in the ionospheric MHD waveguide. However, a detailed model of BSM formation (Ermakova et al. 2007) was developed on the basis of another mechanism, i.e., on the fact that the E–F1 region valley (Nusinov et al. 1996) can form one more ULF wave resonator in the ionosphere. Examples of the valley in the profiles of the Alfvén wave refractive index (n_1) and the corresponding calculations of the ULF wave spectrum on the ground are shown in Fig. 4. Comparison of these plots indicates that the position of the spectral maximum is determined by the characteristic value and spatial scale of variations in the refractive index at altitudes of about 110–250 km.

3 Self-excitation of Waves in the IAR

3.1 The Dissipative Instability of Alfvén Waves in the IAR

The self-excitation of Alfvén waves in the IAR has been discussed since 1980s (Trakhtengerts and Feldstein 1981, 1984, 1991; Lysak 1991; Pokhotelov et al. 2001). The waves can be unstable in the presence of sufficiently high convection velocity, which has a shear in the E-layer due to the friction force. The convective motion serves as a source of free energy for the waves, and the instability arises due to the modification of the wave dissipation in the E-layer caused by the relative motion of the upper and lower ionospheric layers. Thus, the instability is dissipative in its nature.

Simple estimates can be made for the spatial scales of this instability. Indeed, the characteristic vertical scale of the basic IAR eigenmodes is about the ionosphere height $l \approx 2\pi\, V_A/\omega$, and the horizontal scale is determined by the resonance condition

$$\omega = k_\perp V_c, \tag{2}$$

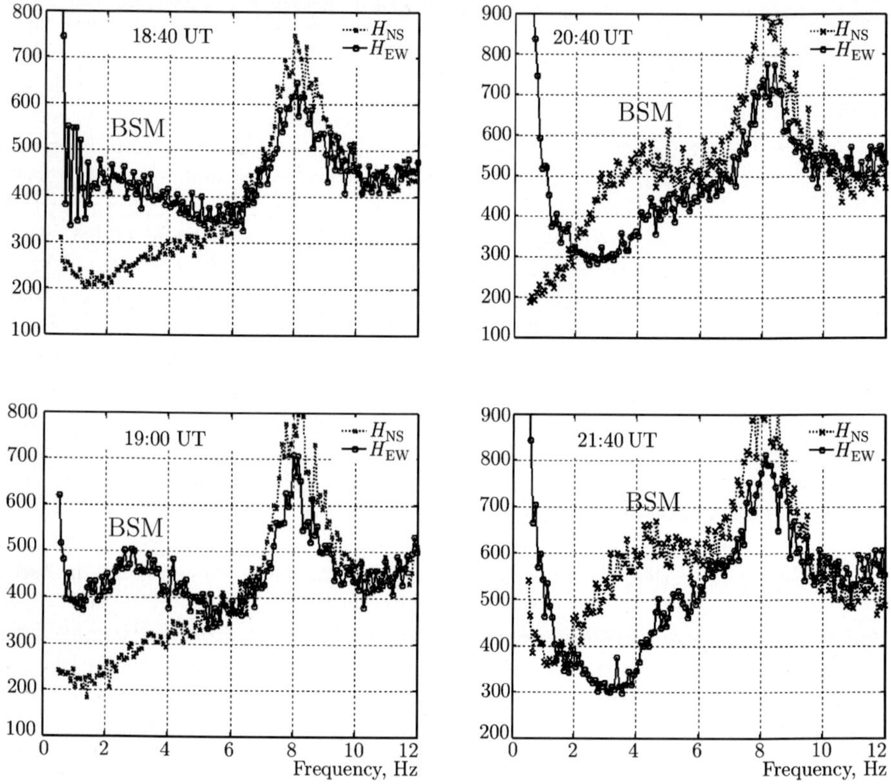

Fig. 3 Broad spectral maximum: observations on July 6, 2003 at Novaya Zhizn' near Nizhny Novgorod (adapted from Ermakova et al. 2007)

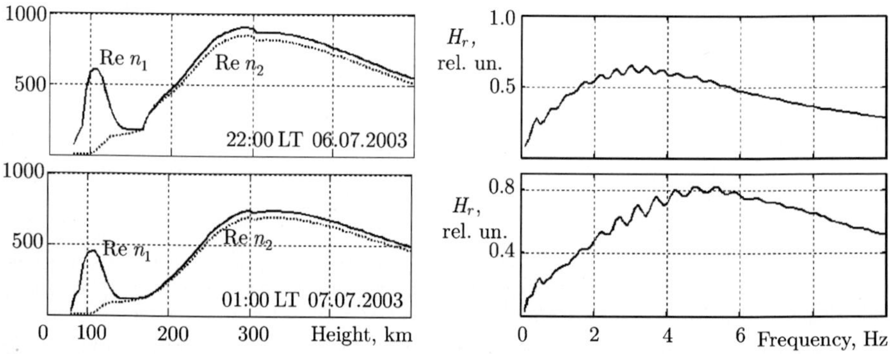

Fig. 4 Broad spectral maximum: calculations based on the IRI model for July 6, 2003 for Novaya Zhizn' near Nizhny Novgorod (adapted from Ermakova et al. 2007). Note that the Schumann resonance is not seen in the calculated spectrum since the model is based on the plane geometry

where V_c is the convection velocity at the ionospheric altitude. Hereafter, the subscripts $\perp$ and $\parallel$ denote the directions across and along the geomagnetic field. The latter can be assumed vertical at auroral latitudes where this instability can most probably occur, so the perpendicular direction almost coincides with the horizontal here. By comparing these equalities one obtains

$$k_\perp / k_\parallel \approx V_a / V_c \,. \tag{3}$$

Characteristic value of the Alfvén velocity in the ionosphere is about 500 km/s, while the typical values of the convection velocity range from about 100 m/s to about 1 km/s depending on the disturbance level. Thus, the unstable waves are strongly anisotropic, with $k_\perp \gg k_\parallel$. For the absolute values we have $l_\perp = 2\pi / k_\perp \sim$ 1–10 km. With account of the polarization properties of Alfvén waves, the unstable waves are current vortices with small transverse scales, as shown schematically in Fig. 1. Recently a detailed numerical study of the instability properties in a wide range of ionosphere parameters has been performed (Despirak et al. 2008).

3.2 Turbulent Alfvén Boundary Layer

Note that the instability discussed in Sect. 3.1 is linear, i.e., the waves do not modify the electric properties of the medium at the initial stage. Its nonlinear stage brings some qualitatively different features, in particular, the instability can develop explosively (Trakhtengerts and Feldstein 1990, 1991). Nonlinear feedback is possible because the currents in the Alfvén vortices fairly easily reach the threshold values for the ion-acoustic instability which, in its turn, produces anomalous resistivity and generates parallel electric field which is almost absent otherwise. The electric field can be strong enough to accelerate electrons in the runaway regime, and these electrons reach energies of several hundreds of eV, so they produce significant additional ionization in the E region. This additional ionization first increases the growth rate of the Alfvén vortices in the IAR (Trakhtengerts and Feldstein 1991), which closes the positive feedback loop.

In fact, we deal the small-scale stratification of the ionospheric convection whose nonlinear stage was called turbulent Alfvén boundary layer (Trakhtengerts and Feldstein 1991).

One can see that the nonlinear evolution of the IAR instability has much in common to the ionospheric feedback instability (IFI) first considered by Atkinson (1970), which is based on the field-aligned current structuring due to the electron precipitation and current closure in the magnetosphere and conjugate ionosphere, without taking into account the wave reflection from the upper ionosphere. Actually the IAR instability produces smaller-scale Alfvén-wave structures embedded in larger-scale ones considered by Atkinson (1970).

Different aspects of the interaction of small-scale Alfvén vortices with the bulk and energetic plasma components have been discussed in several publications during the last decade. We mention only some of them here, pointing the reader to the references in these papers. Kinetic theory and simulations of electron acceleration by Alfvén waves incident from the magnetosphere was developed (Lysak and Song 2003). Observational evidence for electron acceleration in the IAR based on FAST data and simulations was presented (Chaston et al. 2002). The excitation of small-scale Alfvén waves in the IAR by a large-scale electric field in the lower ionosphere was demonstrated by numerical simulations (Streltsov and Lotko 2004). The importance of wave reflection from the magnetospheric auroral acceleration region where strong parallel electric fields can exist was pointed out (Pilipenko et al. 2004). Nonlinear MHD simulations taking into account the ponderomotive forces in the Alfvén waves have indicated their important role in the formation of density cavities in the ionosphere in the vicinity of the auroral arcs (Streltsov and Lotko 2008).

Several active experiments were proposed to verify the concept of the turbulent Alfvén boundary layer (Trakhtengerts et al. 2000a). Two basic schemes suggested for stimulating the instability by the HF heating are related to (i) excitation of Alfvén vortices (e.g., by modulated ionosphere heating) and (ii) local increase of the convection electric field by the modification of the ionosphere conductivity. It was noted that the artificially excited ULF currents should have a wave-like transverse structure matching the unstable mode (see (2)) and a frequency matching the IAR eigenfrequency.

A heating experiment performed at Tromsø (Blagoveshchenskaya et al. 2001) showed auroral activation (pseudopbeakup) which was possibly affected or even stimulated by the HF pumping of the ionosphere. So far, it seems the only experiment demonstrating such an effect.

Recently an attempt to stimulate the IFI by using modulated heating with HAARP was reported (Streltsov et al. 2010). In those experiments, the modulation frequency was chosen according to IFI simulations which yielded the values of several mHz not apparently matching the IAR eigenfrequency but rather close to the magnetospheric field-line eigenfrequency (3.4 mHz). The stimulated activation was not observed in the experiment, but large-amplitude ULF waves ($\sim$100 nT) were detected near auroral arcs supporting the idea that the Alfvén waves are closely related to the substorm phenomena and associated formation of discrete auroral arcs (Streltsov et al. 2010). Perhaps the absence of the IFI stimulation in those experiments could be related to the mismatch between the spatial structure of the excited currents and the ULF eigenmode.

To summarize, the resonant response of the ionosphere to the ULF disturbances is very important for the auroral activation processes. Such processes can also produce detectable wave-like perturbations in the upper atmosphere, which is an interesting topic for future studies.

4 Ion Cyclotron Interactions in the Magnetosphere and Related Precipitation of Ring Current Ions

Very important channels of inter-layer coupling in the near-Earth space can be formed by the Pc1 waves generated in the magnetosphere. Two channels of such kind are considered in this section.

4.1 Alfvén Sweep Maser and Pc1 Pearls

The modification of the ionospheric reflection coefficient by the ring-current ions precipitated due to their interaction with ion-cyclotron waves (in the Pc1 frequency range) was long ago proposed (Polyakov et al. 1983; Belyaev et al. 1984, 1987b) as a mechanism for the formation of the so-called pearl-type Pc1 emissions having the form of regularly repeating bursts of quasi-monochromatic Pc1 oscillations with a frequency drift inside each burst (Jacobs and Watanabe 1964). The repetition period of these bursts is close to the estimated period of wave packet bouncing between conjugate ionospheres. The proposed mechanism of formation of such regular pattern is based on compensation of the wave group dispersion in the magnetosphere by the dispersive nonlinear reflection from the ionospheric Alfvén resonator. The reflection nonlinearity is related to the influence of the energetic protons generating the Pc1 waves in the magnetosphere and precipitating in the ionosphere. Actually it is a specific case of passive mode locking in the magnetospheric cyclotron maser (Bespalov 1984) which was called an Alfvén sweep maser.

This hypothesis was further elaborated by Trakhtengerts et al. (2000b) and supported by numerical modeling (Demekhov 2007). The latter paper presents a fairly detailed review

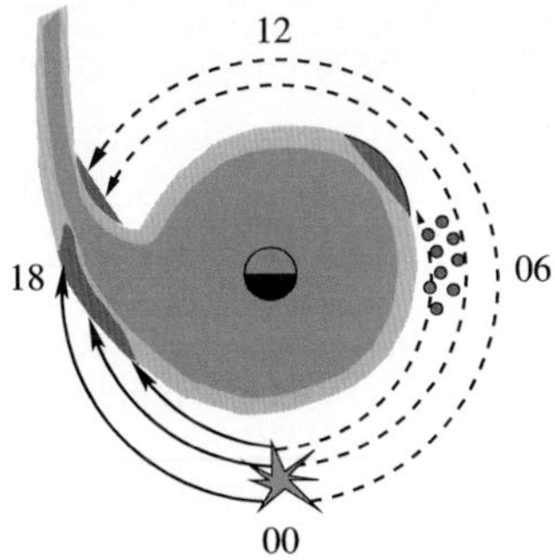

Fig. 5 Scheme of the regions of precipitation of energetic particles due to wave-particle interactions (adapted from Trakhtengerts et al. (1997)). *Gray-shaded region* represents the plasmasphere with a tail in the evening-afternoon sector formed by nonstationary convection. Trajectories of RC ions from the night-side injection region to the point of RC-plasmasphere interface are shown with *solid arrows*

of the current understanding of pearl formation, including arguments for and against the sweep maser mechanism and another popular mechanism related to the modulation of the cyclotron instability growth rate in the magnetosphere by Pc4–5 pulsations (Lyatsky and Plyasova-Bakounina 1986; Plyasova-Bakounina et al. 1996; Rasinkangas and Mursula 1998; Mursula et al. 2001). So we refer the interested reader to it for detail.

More recently, another study on the relationship between Pc1 modulated emissions and Pc5 pulsations was published (Loto'aniu et al. 2009). In that paper, Pc5 pulsations of about 2–4 nT in amplitude detected by CRRES in the magnetosphere simultaneously with Pc1 wave packets of the same repetition period as Pc5 period were reported. Those observations are interpreted in favor of Pc1 modulation by Pc5 pulsations. We note, however, that precipitated energetic protons can not only modify the ionospheric reflection of Pc1 waves but also generate field-line oscillations due to the ionosphere-conductivity variation in the presence of a background convection electric field. Model calculations show that even moderate electric field penetration at subauroral latitudes is sufficient to ensure the generation of field-line oscillations with amplitudes of up to 10 nT by modulated flux of precipitated energetic protons (Galka and Demekhov 2011). Thus, the problem of mutual relationship between short-period (Pc1) and long-period (Pc4–5) geomagnetic pulsations and the role of ionosphere-magnetosphere coupling in it deserves future studies.

4.2 Partial Ring Current and Sub-auroral Polarization Streams

Turbulent pitch-angle diffusion of the ring-current (RC) ions by electromagnetic ion-cyclotron (EMIC) waves and their related precipitation into the ionosphere has been suggested long ago as a potentially important loss mechanism for the ion ring current (Cornwall et al. 1970). The necessary condition for efficient interaction of EMIC waves with RC ions is sufficient number of resonant particles, which poses a restriction on the plasma density, magnetic field, and the characteristic particle energy (see, e.g., Bespalov and Trakhtengerts 1986; Trakhtengerts and Rycroft 2008). This condition can be written as

$$\beta_* = \frac{\Omega_p}{\Omega_B}\frac{V_0}{c} > 1, \qquad (4)$$

where Ω_p and Ω_B are the ion plasma- and gyrofrequency, and V_0 is the characteristic velocity of RC ions. Under the magnetospheric conditions, the cyclotron instability leading to the generation of EMIC waves can be efficient either in the outer plasma sheath where the magnetic field is weak but there is still sufficiently energetic ions, or in the regions of sufficiently dense cold plasma in the inner magnetosphere. Such regions exist in the plasmasphere and in the detached cold-plasma clouds. The most intense precipitation occurs near the contact of RC ion drift trajectories with the plasmaspheric bulge or tail in the evening MLT sector (see Fig. 5).

This picture has been confirmed by recent observations using IMAGE data on plasmasphere–RC interface (Spasojevic et al. 2005). Comprehensive analysis of observations clarifying the role of EMIC waves in the RC ion precipitation was recently performed by using data on precipitated energetic particles from low-orbiting NOAA satellites, ground-based ULF observations, plasmasphere and energetic-particle data from geosynchronous LANL spacecraft, and IMAGE data on proton aurora (Yahnin and Yahnina 2007; Yahnina et al. 2008; Yahnin et al. 2009). NOAA spacecraft data show that the pitch-angle diffusion often reaches the strong-diffusion limit, i.e., the distribution function in and outside the loss cone become equal.

Extensive numerical simulations have been performed to study the wave-particle interactions in the ring current (e.g., Jordanova et al. 2007). In particular, they confirmed that proton auroral arcs at the plasmapause–RC interface are related to the excitation of EMIC waves by RC ions.

The precipitated RC ions carry rather large (up to 1 MA) current into the ionosphere, and closure of this current via the ionosphere can produce significant electric fields and magnetic disturbances at the interface between the ring current and plasmasphere (Trakhtengerts et al. 1997; Trakhtengerts and Demekhov 2005). The high magnetospheric Hall conductivity in the RC region leads to the formation of the polarization current loop shown schematically in Fig. 6. The enhanced polarization electric field, which is directed poleward, prevents the ionospheric closure current from spreading quasi-uniformly. Instead, this current is concentrated in the region conjugate with partial RC which extends from the night-side injection region to the precipitation region. Analytical theory (Trakhtengerts and Demekhov 2005) shows that the polarization electric field in the ionosphere can reach about 40 mV/m if the total field-aligned current of precipitated RC ions flowing into the ionosphere amounts to 1 MA. This field causes enhanced westward drift of ionospheric ions known as subauroral polarization streams (SAPS) (Foster and Vo 2002). Earlier similar phenomenon was discovered for substorm periods and called polarization jet (Galperin et al. 1973) or subauroral ion drifts (Spiro et al. 1979). Usually the SAPS formation is explained by the enhanced polarization field due to the decreased Pedersen conductivity in the ionospheric trough region corresponding to the field-aligned currents related to the RC precipitation (e.g., Goldstein et al. 2005). This effect certainly takes place but it is obviously localized in the precipitation region, while the SAPS can also be observed in a more extended latitudinal domain. Theory involving the Hall currents in the magnetosphere-ionosphere system (Trakhtengerts and Demekhov 2005) allows one to explain both the value and spatial extent of the SAPS occurrence.

5 Conclusions

As we tried to show here, ULF waves even in a single Pc1 subrange have various impacts on the atmosphere-ionosphere-magnetosphere system. They provide very useful tools for studying several resonances in the near-Earth space. On the other hand, they can serve as important agents in the formation of large-scale structures (e.g., partial ring current) and small-scale turbulence (e.g., Alfvén vortices in the auroral ionosphere).

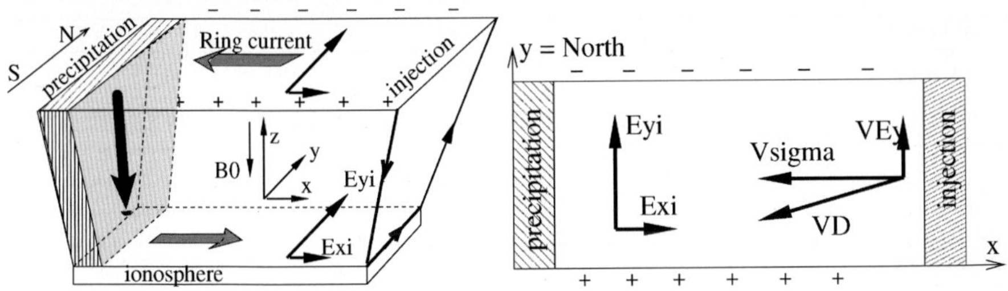

Fig. 6 Scheme of the partial ring-current loop related to the precipitation (Trakhtengerts and Demekhov 2005). Here, V_D is the magnetic-drift velocity, V_E is the electric-drift velocity, and V_Σ is the total drift velocity

It is evident from the above discussion that the considered processes of ULF coupling in the AIM system have been subject to extensive studies in recent years. Nevertheless, there are many open questions concerning such processes. We would like to mention several directions of their further studies.

Fine features in the spectral resonant structures of the background ULF noise deserve a detailed analysis, both theoretical and experimental. In particular, their detection at different sites and finding the spatial and temporal domain of their existence, as well as their relation to other geophysical factors, would be very important to understand their formation and to employ them for the diagnostic purposes.

Relation of auroral activations to the ULF resonant processes and nonlinear factors involved in M-I coupling at auroral latitudes are also not quite well understood. An interesting tool for testing the concepts in this field is provided by artificial influence on the auroral ionosphere, and we can expect interesting new results of such studies in the near future due to the increased power of heating experiments and development of complex satellite-ground observations.

Active experiments seem can also be fruitful in solving some problems of resonant wave-particle interactions in the ULF range. In particular, one can hope to influence the regime of Pc1 pearl generation by modifying the ionospheric reflection coefficient if the reflection of EMIC waves from the ionosphere is significant for their generation.

Among other problems of ULF wave-particle interactions which deserve future study we mention the interrelationship between Pc1 emissions and long-period hydromagnetic pulsations (flux-tube oscillations). Specifically, the feedback between these processes (Bespalov and Trakhtengerts 1986; Trakhtengerts and Rycroft 2008) should be studied quantitatively. Origin of discrete EMIC wave emissions like hydromagnetic chorus (Pickett et al. 2010) and their role in the energetic-particle dynamics are of basic interest. Quantifying the role of EMIC waves in formation of partial ring current and localized storm-time current systems in the ionosphere-magnetosphere system, such as subauroral polarization streams, will also attract attention of researchers.

We hope that the planned space missions like RBSP, Resonance, TARANIS, etc. will shed new light on the discussed resonant processes in the ULF range.

Acknowledgements I am very grateful to ISSI for supporting my participation in the Coupling workshop, and to Tilmann Bösinger for inspiring discussions. This work was partially supported by the General Physics Division of the Russian Academy of Sciences (the program "Plasma processes in Solar System") and by the Russian Foundation for Basic Research (project No. 11-02-00654).

References

G. Atkinson, J. Geophys. Res. **75**(25), 4746 (1970)

P.P. Belyaev, T. Bösinger, S.V. Isaev, V.Y. Trakhtengerts, J. Kangas, J. Geophys. Res. **104**(A3), 4305 (1999)

P.P. Belyaev, S.V. Polyakov, E.N. Ermakova, S.V. Isaev, M.N. Yakunin, L.A. Sobchakov, A.V. Vasilyev, N.L. Astakhova, D.N. Vladimirov, V.S. Volosevich, L.N. Protopopov, S.M. Gordyushkin, A.P. Savitsky, G.V. Red'ko, A.A. Eliseev, A.B. Fedorov, K.D. Ratnikov, Radiophys. Quantum Electron. **45**(2), 135 (2002)

P.P. Belyaev, S.V. Polyakov, V.O. Rapoport, V.Y. Trakhtengerts, Geomagn. Aeron. **24**(2), 242 (1984)

P.P. Belyaev, S.V. Polyakov, V.O. Rapoport, V.Y. Trakhtengerts, Dokl. Akad. Nauk SSSR **297**(4), 840 (1987a)

P.P. Belyaev, S.V. Polyakov, V.O. Rapoport, V.Y. Trakhtengerts, Geomagn. Aeron. **27**(4), 652 (1987b)

P.P. Belyaev, S.V. Polyakov, V.O. Rapoport, V.Y. Trakhtengerts, Radiophys. Quantum Electron. **32**(6), 491 (1989a)

P.P. Belyaev, S.V. Polyakov, V.O. Rapoport, V.Y. Trakhtengerts, Radiophys. Quantum Electron. **32**(7), 594 (1989b)

P.P. Belyaev, S.V. Polyakov, V.O. Rapoport, V.Y. Trakhtengerts, J. Atmos. Terr. Phys. **52**(9), 781 (1990)

P.A. Bespalov, Sov. Phys. JETP **60**(6), 1090 (1984)

P.A. Bespalov, V.Y. Trakhtengerts, in *Reviews of Plasma Physics*, vol. 10, ed. by M.A. Leontovich (Plenum, New York, 1986)

N.F. Blagoveshchenskaya, V.A. Kornienko, T.D. Borisova, B. Thidé, M.J. Kosch, M.T. Rietveld, E.V. Mishin, R.Y. Lukýanova, O.A. Troshichev, J. Geophys. Res. **106**(A12), 29071 (2001)

T. Bösinger, A.G. Demekhov, V.Y. Trakhtengerts, Geophys. Res. Lett. **31**(18), L18802 (2004). doi:10.1029/2004GL020777

T. Bösinger, C. Haldoupis, P.P. Belyaev, M.N. Yakunin, N.V. Semenova, A.G. Demekhov, V. Angelopoulos, J. Geophys. Res. **107**(A10), 1281 (2002). doi:1029/2001JA005076

C.C. Chaston, C.W. Carlson, M. Berthomier, L.M. Peticolas, I. Roth, J.P. McFadden, J. Geophys. Res. **107**(A11), 1413 (2002). doi:10.1029/2002JA009272

J.M. Cornwall, F.V. Coroniti, R.M. Thorne, J. Geophys. Res. **75**(25), 4699 (1970)

A.G. Demekhov, J. Atmos. Sol.-Terr. Phys. **69**(14), 1609 (2007)

I.V. Despirak, A.A. Lubchich, V.Y. Trakhtengerts, Radiophys. Quantum Electron. **51**(5), 339 (2008)

E.N. Ermakova, D.S. Kotik, S.V. Polyakov, Radiophys. Quantum Electron. **51**(7), 519 (2008)

E.N. Ermakova, D.S. Kotik, S.V. Polyakov, A.V. Shchennikov, Radiophys. Quantum Electron. **50**(7), 555 (2007)

E.N. Ermakova, S.V. Polyakov, D.S. Kotik, Radiophys. Quantum Electron. **53**(9–10), 557 (2011)

E.N. Ermakova, S.V. Polyakov, Y.V. Shluygaev, A.V. Shchennikov, in *Abstracts of 32nd Ann. Seminar "Physics of Auroral Phenomena"*, Apatity, Russia, ed. by A.G. Yahnin (2009), p. 66

J.C. Foster, H.B. Vo, J. Geophys. Res. **107**(A12), 1475 (2002). doi:10.1029/2002JA009409

A.G. Galka, A.G. Demekhov, in *Abstracts of the 34th Apatity Seminar Physics of Auroral Phenomena*, Apatity (2011), p. 30

Y.I. Galperin, V.N. Ponomarev, A.G. Zosimova, Cosm. Res. **11**, 283 (1973)

J. Goldstein, J.L. Burch, B.R. Sandel, S.B. Mende, P.C. son Brandt, M.R. Hairston, J. Geophys. Res. **110**, A03205 (2005). doi:10.1029/2004JA010712

J.A. Jacobs, T. Watanabe, J. Atmos. Terr. Phys. **26**, 825 (1964)

V.K. Jordanova, M. Spasojevic, M.F. Thomsen, J. Geophys. Res. **112**, A08209 (2007). doi:10.1029/2006JA012215

T.M. Loto'aniu, B.J. Fraser, C.L. Waters, Ann. Geophys. **27**(1), 121 (2009)

W.B. Lyatsky, T.A. Plyasova-Bakounina, Geomagn. Aeron. **26**(5), 802 (1986)

R.L. Lysak, J. Geophys. Res. **96**(A2), 1553 (1991)

R.L. Lysak, Y. Song, J. Geophys. Res. **107**(A8) (2002). doi:10.1029/2001JA000308

R.L. Lysak, Y. Song, J. Geophys. Res. **108**(A8), 1327 (2003). doi:10.1029/2003JA009859

V.A. Mazur, A.S. Potapov, Planet. Space Sci. **31**, 859 (1983)

K. Mursula, T. Bräysy, K. Niskala, C.T. Russell, J. Geophys. Res. **106**(A12), 29543 (2001)

A.A. Nusinov, L.A. Antonova, V.V. Katyushina, Radiophys. Quantum Electron. **39**(10), 838 (1996)

A.O. Ovchinnikov, Geomagn. Aeron. **39**(1), 64 (1999)

J.S. Pickett, B. Grison, Y. Omura, M.J. Engebretson, I. Dandouras, A. Masson, M.L. Adrian, O. Santolík, P.M.E. Décréau, N. Cornilleau-Wehrlin, Geophys. Res. Lett. **37**, L09104 (2010). doi:10.1029/2010GL042648

V. Pilipenko, E. Fedorov, M.J. Engebretson, K. Yumoto, J. Geophys. Res. **109**(A10), A10204 (2004). doi:10.1029/2004JA010440

T.A. Plyasova-Bakounina, J. Kangas, K. Mursula, O.A. Molchanov, J.A. Green, J. Geophys. Res. **101**, 10965 (1996)

O.A. Pokhotelov, V.V. Khruschev, M. Parrot, S.A. Senchenkov, V.P. Pavlenko, J. Geophys. Res. **106**(11), 25813 (2001)

S.V. Polyakov, V.O. Rapoport, Geomagn. Aeron. **21**(5), 610 (1981)

S.V. Polyakov, V.O. Rapoport, V.Y. Trakhtengerts, Sov. J. Plasma Phys. **9**(2), 216 (1983)

R. Rasinkangas, K. Mursula, Geophys. Res. Lett. **25**(6), 869 (1998)

B.W. Reinisch, P. Nsumei, X. Huang, D.K. Bilitza, Adv. Space Res. **39**(5), 731 (2007)

N.V. Semenova, A.G. Yahnin, A.N. Vasil'ev, O. Amm, Geomagn. Aeron. **48**(1), 36 (2008)

S.L. Shalimov, T. Bösinger, J. Geophys. Res. **113**(2), A02303 (2008). doi:10.1029/2007JA012614

F.A. Simoes, R. Pfaff, J.J. Berthelier, J. Kletzing, Space Sci. Rev. (2011, this issue)

L.A. Sobchakov, N.L. Astakhova, S.V. Polyakov, Radiophys. Quantum Electron. **46**(12), 918 (2003)

M. Spasojevic, M.F. Thomsen, P.J. Chi, B.R. Sandel, in *Inner Magnetosphere Interactions: New Perspectives from Imaging*, ed. by M. Burch, J.L. Schulz, H. Spence. AGU Geophysical Monograph, vol. 159 (AGU, Washington, 2005), p. 85

R.W. Spiro, R.A. Heelis, W.B. Hanson, Geophys. Res. Lett. **6**(8), 657 (1979)

A.V. Streltsov, W. Lotko, J. Geophys. Res. **109**(A9), A09214 (2004). doi:10.1029/2004JA010457

A.V. Streltsov, W. Lotko, J. Geophys. Res. **113**(A5), A05212 (2008). doi:10.1029/2007JA012594

A.V. Streltsov, T.R. Pedersen, E.V. Mishin, A.L. Snyder, J. Geophys. Res. **115**(A7), A07205 (2010). doi:10.1029/2009JA014961

V.V. Surkov, O.A. Molchanov, M. Hayakawa, A.Y. Schekotov, E.N. Fedorov, J. Geophys. Res. **110**(A04), A04308 (2005). doi:10.1029/2004JA010850

V.V. Surkov, O.A. Pokhotelov, M. Parrot, E.N. Fedorov, M. Hayakawa, Ann. Geophys. **22**, 2877 (2004)

V.Y. Trakhtengerts, A.G. Demekhov, Int. J. Geomagn. Aeron. **5**(3), GI3007 (2005). doi:10.1029/2004GI000091

V.Y. Trakhtengerts, A.G. Demekhov, A. Grafe, Geomagn. Aeron. **37**(4), 404 (1997)

V.Y. Trakhtengerts, P.P. Belyaev, S.V. Polyakov, A.G. Demekhov, T. Bösinger, J. Atmos. Sol.-Terr. Phys. **62**(4), 267 (2000a)

V.Y. Trakhtengerts, A.G. Demekhov, S.V. Polyakov, P.P. Belyaev, V.O. Rapoport, J. Atmos. Sol.-Terr. Phys. **62**(4), 231 (2000b)

V.Y. Trakhtengerts, A.Y. Feldstein, Geomagn. Aeron. **21**, 951 (1981)

V.Y. Trakhtengerts, A.Y. Feldstein, Planet. Space Sci. **32**(2), 127 (1984)

V.Y. Trakhtengerts, A.Y. Feldstein, Geomagn. Aeron. **27**(4), 258 (1987)

V.Y. Trakhtengerts, A.Y. Feldstein, J. Geomagn. Geoelectr. **42**(12), 1379 (1990)

V.Y. Trakhtengerts, A.Y. Feldstein, J. Geophys. Res. **96**(11), 19363 (1991)

V.Y. Trakhtengerts, M.J. Rycroft, *Whistler and Alfvén Mode Cyclotron Masers in Space*. Cambridge Atmospheric and Space Sciences Series. (Cambridge University Press, New York, 2008)

A.G. Yahnin, T.A. Yahnina, J. Atmos. Sol.-Terr. Phys. **69**(14), 1690 (2007)

A.G. Yahnin, T.A. Yahnina, H.U. Frey, T. Bösinger, J. Manninen, J. Geophys. Res. **114**, A12215 (2009). doi: 10.1029/2009JA014670

T.A. Yahnina, H.U. Frey, T. Bösinger, A.G. Yahnin, J. Geophys. Res. **113**, A07209 (2008). doi:10.1029/2008JA013099

Printed by Printforce, the Netherlands